Comprehensive Toxicology

Comprehensive Toxicology

Editors-in-Chief

I. Glenn Sipes
University of Arizona, Tucson, AZ, USA

Charlene A. McQueen
University of Arizona, Tucson, AZ, USA

A. Jay Gandolfi
University of Arizona, Tucson, AZ, USA

Volume 11

NERVOUS SYSTEM AND BEHAVIORAL TOXICOLOGY

Volume Editors

Herbert E. Lowndes
Rutgers University, Piscataway, NJ, USA

Kenneth R. Reuhl
Rutgers University, Piscataway, NJ, USA

PERGAMON

UK	Elsevier Science Ltd., The Boulevard, Langford Lane, Kidlington, Oxford, OX5 1GB, UK
USA	Elsevier Science Inc., 655 Avenue of the Americas, New York, NY 10010, USA
JAPAN	Elsevier Science Japan, 9-15 Higashi-Azabu 1-chome, Minato-ku, Tokyo 106, Japan

First edition 1997

Library of Congress Cataloging in Publication Data
Comprehensive toxicology / editors-in-chief, I. Glenn Sipes, Charlene A. McQueen, A. Jay Gandolfi.
  p. cm.
  Includes index.
  1. Toxicology. I. Sipes, I. Glenn. II. McQueen, Charlene A., 1947– .
III. Gandolfi, A. Jay.
  [DNLM: 1. Poisons. 2. Poisoning. 3. Toxicity Tests. 4. Toxicology—methods. QV 600 C737 1997]
RA1199.C648 1997
615.9—dc20
DNLM/DLC
for Library of Congress   96-33052
                                                                                    CIP

British Library Cataloguing in Publication Data
A catalogue record for this book is available from the British Library.

ISBN 0-08-042301-9 (set : alk. paper)
ISBN 0-08-042976-9 (Volume 11)

Typeset by Alden Bookset, Didcot, UK.
Printed and bound in Great Britain by Cambridge University Press, Cambridge, UK.

Contents

Preface

Toxicology is the study of the nature and actions of chemicals on biological systems. In more primitive times, it really was the study of poisons. Evidence of the use of venoms and poisonous herbs exists in both the Old World and the New World. Early recorded history documents the use of natural products as remedies for certain diseases, and as agents for suicides, execution, and murders. In those days, the "science of toxicology" certainly operated at the upper limit of the dose response curve. However, in the early 1500s, it was apparent to Paracelsus, that "the dose differentiates a poison and a remedy." Stated more directly, any chemical can be toxic if the dose is high enough. In addition, with respect to metals, he recognized that the effects they produce may differ depending upon the duration of exposure (i.e., acute vs. chronic). Clearly, the two most important tenets of toxicology were established during that time. The level of exposure (dose) and the duration of exposure (time) will determine the degree and nature of a toxicological response.

Since that time the discipline of toxicology has made major advances in identifying and characterizing toxicants. It is the scientific discipline that combines the elements of chemistry and biology. It not only addresses how chemicals affect biological systems, but how biological systems affect chemicals. Due to this broad definition, toxicology has always been an interdisciplinary science. It uses advances made in other disciplines to better describe a toxic event or to elucidate the mechanism of a toxic event. In addition, toxicology contributes to the advancement of other disciplines by providing chemicals with selective or specific mechanisms of action. These can be utilized to modify biochemical and physiological processes that advance our understanding of biology and the treatment of disease.

During the last quarter century, toxicology has made its greatest advances. Reasons for this include a demand by the public for protection of human and environmental health, an emphasis on understanding cellular and molecular mechanisms of toxicity, and use of such data in safety evaluation/risk assessment. Clearly, the explosion of information as a result of the new biology (molecular and cellular) provided the conceptual framework as well as tools to better elucidate mechanisms of toxicity and to understand interindividual differences in susceptibility to toxicants. Due to this new information, more sensitive end points of toxicity will be identified. This is critical because the real need in toxicology is to understand if/how exposure to trace levels of chemicals (i.e., low end of the dose–response curve) for long periods of time cause toxicity.

Thus, the time is appropriate for the development and publication of an extensive and authoritative work on toxicology. The concept of *Comprehensive Toxicology* was developed at Pergamon Press in 1989–1990. After market research documented a need for such a work, discussions took place to identify an Editor-in-Chief. The project was too large for one individual and the current team of co-Editors-in-Chief was identified. These individuals along with scientific and support staff from Pergamon Press (now an imprint of Elsevier Science, Ltd.), developed a tentative series of volumes to be edited by experts. These Volume Editors then solicited the Authors, who provided the chapters that appear in these several thousand pages. It is to the Authors that we are indebted, because without them, this project would still be a concept, not a product.

Although the title of this work is *Comprehensive Toxicology*, we certainly realize that it does not meet the literal definition of comprehensive. Due to space limitations, certain critical tissues are not included (skin, eye). Also, the time it takes to receive, edit, and print the various chapters limits inclusion of the latest information. Our goal was to provide a strong foundation. Similarly, our understanding of toxic mechanisms is much greater in certain tissues (liver, kidney, lung) than in others (cardiovascular, nervous system, hematopoietic). Thus, the scope of these chapters will differ. They reflect the state of the science and the specializations of the Editors/Authors. Not only will the reader find information on toxicology, but superb chapters on key physiological and biochemical processes in a variety of tissues. In all cases, a wealth of scientific material is presented that will be of interest to a wide audience.

As co-Editors-in-Chief we have appreciated the opportunity to be involved in this project. We are grateful to the very supportive staff at Elsevier and at the University of Arizona who have assisted us with the project. We trust that you will consider *Comprehensive Toxicology* a milestone in the literature of this very old, but very dynamic discipline.

I. Glenn Sipes   Charlene A. McQueen   A. Jay Gandolfi
Tucson   *Tucson*   *Tucson*

Contributors to Volume 11

Professor E. X. Albuquerque
Department of Pharmacology and Experimental Therapeutics, School of Medicine, University of Maryland at Baltimore, 655 West Baltimore Street, Baltimore, MD 21201, USA
Laboratory of Molecular Pharmacology II, Institute of Biophysics "Carlos Chagas Filho," Federal University of Rio de Janeiro, Rio de Janeiro, RJ 21944, Brazil

Dr. M. Alkondon
Department of Pharmacology and Experimental Therapeutics, School of Medicine, University of Maryland at Baltimore, 655 West Baltimore Street, Baltimore, MD 21201, USA

Dr. M. Aschner
Department of Physiology and Pharmacology, Bowman Gray School of Medicine, Medical Center Boulevard, Winston-Salem, NC 27157-1083, USA

Dr. W. D. Atchison
Department of Pharmacology and Toxicology, Neuroscience Program, B447 Life Sciences Building, Michigan State University, East Lansing, MI 48824-1317, USA

Professor G. J. Audesirk
Department of Biology, University of Colorado at Denver, PO Box 173364, Denver, CO 80217-3364, USA

Dr. M. F. Beal
Neurology Service, Warren 408, Massachusetts General Hospital, 32 Fruit Street, Boston, MA 02114, USA

Dr. S. C. Bondy
Department of Community and Environmental Medicine, University of California, Irvine, CA 92717, USA

Dr. L. W. Chang
Department of Pathology (Slot 517), University of Arkansas for Medical Sciences, 4301 West Markham Street, Little Rock, AR 72205, USA

Dr. D. A. Cory-Slechta
Department of Environmental Medicine, University of Rochester School of Medicine and Dentistry, PO Box EHSC, Rochester, NY 14642, USA

Dr. L. G. Costa
School of Public Health and Community Medicine, Department of Environmental Health, University of Washington, 4225 Roosevelt Way Northeast, #100, XD-41 PO Box 236138, Seattle, WA 98105, USA

Dr. E. F. da Cruz e Silva
Departamento de Biologia, Centro de Biologia Celular, Universidade de Aveiro, 3800 Aveiro, Portugal
Instituto de Biologia Experimental e Tecnológica, Apartado 12, 2780 Oeiras, Portugal

Dr. D. C. Dorman
Chemical Industry Institute of Toxicology, 6 Davis Drive, PO Box 12137, Research Triangle Park, NC 27709, USA

Dr. D. J. Ecobichon
Department of Pharmacology and Therapeutics, McGill University, McIntyre Medical Science Building, Montreal, Quebec H3G 1Y6, Canada

Professor A. T. Eldefrawi
Department of Pharmacology and Experimental Therapeutics, School of Medicine, University of Maryland at Baltimore, 655 West Baltimore Street, Baltimore, MD 21201, USA

Professor M. E. Eldefrawi
Department of Pharmacology and Experimental Therapeutics, School of Medicine, University of Maryland at Baltimore, 655 West Baltimore Street, Baltimore, MD 21201, USA

Dr. L. D. Fechter
College of Pharmacy, University of Oklahoma Health Sciences Center, PO Box 26901, Oklahoma City, OK 73190, USA

Dr. N. Fiedler
Environmental and Occupational Health Sciences Institute, Room 210, 681 Frelinghuysen Road, Piscataway, NJ 08855, USA

Dr. G. B. Grunwald
Departments of Anatomy, Pathology, and Cell Biology, Jefferson Medical College, 1020 Locust Street, Philadelphia, PA 19107-6731, USA

Dr. G. L. Guo
Department of Pathology (Slot 517), University of Arkansas for Medical Sciences, 4301 West Markham Street, Little Rock, AR 72205, USA

Dr. E. D. Hall
CNS Research Unit, Pharmacia & Upjohn Inc., Kalamazoo, MI 49001, USA

Dr. K. Ishihara
Department of Pharmacology and Experimental Therapeutics, School of Medicine, University of Maryland at Baltimore, 655 West Baltimore Street, Baltimore, MD 21201, USA

†Dr. R. M. Joy
Department of VM Molecular Biosciences, School of Veterinary Medicine, University of California, Davis, CA 95616, USA

Dr. Y. Liu
College of Pharmacy, University of Oklahoma Health Sciences Center, PO Box 26901, Oklahoma City, OK 73190, USA

Dr. M. Marchioro
Department of Pharmacology and Experimental Therapeutics, School of Medicine, University of Maryland at Baltimore, 655 West Baltimore Street, Baltimore, MD 21201, USA

Dr. P. Morell
Neuroscience Center, CB7250, University of North Carolina at Chapel Hill, NC 27599-7250, USA

Dr. K. T. Morgan
TOX-T1128, Glaxo Wellcome Inc., 5 Moore Drive, PO Box 13398, Research Triangle Park, NC 27709, USA

Dr. V. C. Moser
MD-74B, Neurotoxicology Division, US Environmental Protection Agency, Research Triangle Park, NC 27711, USA
1012 North Wellonsburg Place, Apex, NC 27502, USA

Dr. T. Narahashi
Department of Molecular Pharmacology and Biological Chemistry, Northwestern University Medical School, 303 East Chicago Avenue, Chicago, IL 60611-3008, USA

Dr. M. C. Newland
Department of Psychology, Auburn University, 4082 Haley Center, 110 Thach Hall, Auburn, AL 36849-5212, USA

Dr. J. P. O'Callaghan
MD-74B, Neurotoxicology Division, National Health and Environmental Effects Research Laboratory, US Environmental Protection Agency, Research Triangle Park, NC 27711, USA

Dr. J. G. Owens
Pfizer Inc., Eastern Point Road, Groton, CT 06340, USA

Dr. M. A. Pass
Department of Physiology and Pharmacology, University of Queensland, St. Lucia,
Queensland 4072, Australia

Dr. E. F. R. Pereira
Department of Pharmacology and Experimental Therapeutics, School of Medicine,
University of Maryland at Baltimore, 655 West Baltimore Street, Baltimore, MD 21201, USA

Dr. M. A. Philbert
Neurotoxicology Laboratories, University of Michigan, Room #EM6015, SPH Building
No. 2, Ann Arbor, MI 48109-2029, USA

Dr. S. J. Pyle
Department of Biology, University of North Dakota, PO Box 9019, Grand Forks,
ND 58202-9019, USA

Dr. K. R. Reuhl
Neurotoxicology Laboratories, Department of Pharmacology and Toxicology, College of
Pharmacy, Rutgers University, Livingston Campus, 41 Gordon Road, Piscataway,
NJ 08854, USA

Dr. D. C. Rice
Toxicology Research Division, Bureau of Chemical Safety, Health Protection Branch,
Health Canada, Banting Research Centre 2202D1, Tunney's Pasture, Ottawa, Ontario
K1A 0L2, Canada

Dr. T. A. Sarafian
Department of Pathology and Laboratory Medicine, University of California, Los Angeles,
Center for the Health Sciences, Box 951732, Los Angeles, CA 90095-1732, USA

Dr. W. Slikker, Jr.
Division of Neurotoxicology, National Center for Toxicological Research,
3900 NCTR Road, Jefferson, AR 72079-9502, USA

Dr. K. L. Swanson
Department of Pharmacology and Experimental Therapeutics, School of Medicine,
University of Maryland at Baltimore, 655 West Baltimore Street, Baltimore, MD 21201, USA

Dr. H. A. Tilson
MD-74B, Neurotoxicology Division, US Environmental Protection Agency, Research
Triangle Park, NC 27711, USA

Dr. A. D. Toews
Neuroscience Center, CB7250, University of North Carolina at Chapel Hill, NC
27599-7250, USA

Dr. T. J. Walsh
Department of Psychology, Rutgers University, Busch Campus, New Brunswick,
NJ 08903, USA

Dr. M. A. Verity
Department of Pathology and Laboratory Medicine, University of California, Los Angeles,
Center for the Health Sciences, Box 951732, Los Angeles, CA 90095-1732, USA

Dr. B. Weiss
Department of Environmental Medicine, University of Rochester School of Medicine,
Box EHSC, Rochester, NY 14642, USA

Dr. G. D. Zeevalk
Department of Neurology, University of Medicine and Dentistry of New Jersey-Robert
Wood Johnson Medical School, 675 Hoes Lane, Piscataway, NJ 08854, USA

Introduction

OBJECTIVES, SCOPE, AND COVERAGE

Comprehensive Toxicology is an in-depth, state-of-the-art review of toxicology with an emphasis on human systems. This series of volumes has been designed to encompass investigation from the molecular level to the intact organism. The goal is to provide a balanced presentation that integrates specific biological effects of pertinent toxicants across the various disciplines of toxicology. To accomplish this the individual chapters were written by leaders in their specific areas of toxicology or related disciplines.

These reference volumes begin with basic principles of toxicology (general aspects and biotransformation) and their utilization in toxicological testing. This is followed by a thorough systems approach to the key organs or tissues susceptible to toxic chemicals. Within the biological systems volumes the structure/function of the tissue, its response to toxic insult, approaches for evaluation of injury, and examples of specific toxicants are profiled. Particular attention will be paid to understanding the cellular and molecular mechanisms of toxicity. A portion of this treatise is dedicated to state-of-the-art *in vitro* approaches for examining the mechanisms and endpoints of toxicity. Finally, important advances in chemical carcinogenesis, including oncogenes, tumor suppressor genes, and anticarcinogens complete this major work.

In planning this series, the needs of a wide range of potential users were considered. Consequently, each area includes a thorough review as well as the latest scientific data and interpretation. Scientists and students in academic, industrial, and governmental settings will be major users of this series. University scientists will utilize this source as a basis for their research studies and as a reference text for teaching. Individuals in need of toxicological data in the chemical, pharmaceutical, agricultural, petroleum, biotechnical, mining, and semiconductor industries will find this series to be a valuable tool as will governmental scientists, regulators, and administrators. Other potential users include those with interests in environmental and medical law as well as scientists associated with research institutes and consulting firms.

The work consists of 13 volumes as described below.

Volume 1: General Principles (edited by J. Bond)

Volume 1 provides the reader with a basic overview of the field of toxicology. The volume is divided into four sections: (1) introduction, (2) toxicokinetics, (3) mechanisms of toxicity, and (4) risk assessment. There is an in-depth coverage of the field of toxicokinetics including exposure assessment, routes of exposure of toxicants, and distribution, biotransformation, and excretion of toxicants. Dosimetry modeling, including physiologically based pharmacokinetic models, and extrapolation modeling are covered as well as key mechanisms associated with the toxicity/carcinogenicity of chemicals. Concepts of risk assessment are also discussed in this volume. Volume 1 lays the groundwork for subsequent volumes of the *Comprehensive Toxicology* series where different aspects of the field of toxicology are extensively discussed.

Volume 2: Toxicological Testing and Evaluation (edited by P. D. Williams and G. H. Hottendorf)

Volume 2 provides the reader with a comprehensive overview of the routine and special toxicologic assessments performed on pharmaceutical agents, chemicals, pesticides, and consumer products. Descriptions of toxicity testing procedures emphasize experimental design considerations (dose and species selection, toxicokinetic criteria) as well as regulatory guidelines and impacts of global harmonization efforts. The risk assessment process is considered in terms of evaluating and interpreting toxicity data relative to human safety.

Volume 3: Biotransformation (edited by F. P. Guengerich)

Volume 3 deals with the enzymatic processes involved in the biotransformation of toxic and potentially toxic chemicals. The first section provides introductory material on general aspects of history, regulation, mechanism, inhibition, and stimulation of the enzymes in this group. The bulk of the volume is comprised of chapters dealing with the current status of enzymes involved in oxidation, reduction, hydrolysis, and conjugation of xenobiotic chemicals. The current status of enzyme multiplicity and nomenclature, gene and protein structure, regulation, catalytic mechanism, substrate specificity, human studies, and relevance to issues in toxicology are discussed.

Volume 4: Toxicology of the Hematopoietic System (edited by J. C. Bloom)

Volume 4, which is divided into five sections, reviews the current understanding of the hematopoietic system and toxic effects on its components. The work is arranged to provide a background on hematotoxicology, describe the components of blood and blood-forming organs, and review the complex process of hematopoiesis. Additional chapters describe toxic effects on erythrocytes, leukocytes, and hemostasis, respectively, and include discussions on putative agents and mechanisms, as well as the diagnosis and treatment of these disorders. Special attention is given to human hematopoietic stem cells, which represent an important target tissue that is amenable to *in vitro* testing. The volume concludes with risk assessment in both clinical and preclinical settings.

Volume 5: Toxicology of the Immune System (edited by D. A. Lawrence)

Volume 5 provides current knowledge regarding the intricacies of the immune system and the manner by which environmental agents can disrupt immune homeostasis and induce pathologies. The volume begins with an introduction to the field of immunotoxicology and an overview of the immune system, followed by a description of the architecture and cellular components of the immune system and how their development and reactivities can be modified. Immune functions specifically delineated include cell trafficking, processing of antigens, the manner by which cells communicate and are signaled into activation from plasma membrane events through to transcriptional events. The immunopathologies resultant from hypersensitivities are presented with emphasis on the mechanistic means by which chemicals can modify health. Animal and human examples of autoimmune diseases are included as well as the immune changes that occur with aging. Immunosuppression induced by cancers and drugs used for treatment of cancers and transplant patients are reviewed along with the immunotoxicities associated with biological response modifiers. The ability of stressors to modify immune functions help describe the importance of neuroimmunological investigations. The immunoregulatory properties of stress response proteins are also reviewed. The volume concludes with methods and applications utilized for risk assessment and the consequences of immunotoxicological analysis of humans.

Volume 6: Cardiovascular Toxicology (edited by S. P. Bishop and W. D. Kerns)

Volume 6 contains a comprehensive review of basic cardiovascular anatomy, physiology, and pharmacology as well as an analysis of the various classes of compounds affecting the heart and blood vessels. Cardiovascular biology includes embryologic development, physiology, and molecular pharmacology of the system. Methods for evaluation of the heart and blood vessels are reviewed, including both *in vivo* and *in vitro* methods of study. Separate sections deal with the response of the heart and blood vessels to injury, with emphasis on mechanisms of toxic injury. Specific compounds are classified according to their major mechanism of toxicity, and are reviewed with prime examples of those compounds for which the cellular and molecular mechanisms are best known. Emphasis is on toxicologic studies in experimental animals where mechanisms have been most thoroughly investigated, but the relationship of results from animal studies is made to toxicologic lesions in man. Finally, there is an assessment of the problems related to the use of toxicologic studies in animals for clinical application in man.

Volume 7: Renal Toxicology (edited by R. S. Goldstein)

Current concepts on mechanisms mediating chemically induced nephrotoxicity are rapidly evolving. Volume 7 is focused on capturing this rapidly growing field by providing a comprehensive, state-of-the-art review of renal pathophysiology and toxicology, written by internationally recognized experts in the field. The first section of this volume is designed to provide the reader with the required background information, including an overview of clinical nephrotoxicity, the anatomy and physiology of the kidney and urinary bladder, renal transport mechanisms, xenobiotic metabolism, and *in vivo* and *in vitro* methods used to assess renal toxicity. Current knowledge on the pathophysiology and biochemistry of acute renal failure, renal and urinary bladder carcinogenesis, and immune-mediated renal injury are covered in detail. The role of vasoactive substances, cell adhesion molecules, oxidative stress/antioxidants and membrane changes in composition/fluidity in mediating nephrotoxic and ischemic renal injury are reviewed and provide the reader with the conceptual framework for understanding key mechanisms of renal injury. In addition, the roles of gene expression and growth factors in renal injury and repair are discussed. The response of each of the major segments of the nephron to a toxic insult is discussed in detail, with chapters devoted to mechanisms mediating injury to the glomerulus, proximal tubule, collecting duct and papilla, and the tubulointerstitium. Various nephrotoxicants are discussed in detail with an emphasis on pathophysiologic and morphologic effects, and mechanisms of toxicity. Wherever possible, the relevance of experimental findings to human exposure is covered.

Volume 8: Toxicology of the Respiratory System (edited by R. A. Roth)

Volume 8 begins with a detailed description of the functional anatomy of the various regions and critical cells of the respiratory tract, from the nasal cavity to the gas exchange region. This section conveys the structural heterogeneity of the respiratory system and how it determines tissue and cellular targets for toxic agents. The next section describes functional and biochemical responses of the lungs to toxic insult and includes chapters on carcinogenic and developmental responses to chemical exposure. The final section comprises over 20 chapters describing injurious effects of selected chemicals of toxicologic interest. Human health concerns as well as mechanisms of toxicity are emphasized. Unlike other volumes that have treated this subject, attention is devoted to effects of toxic agents on both the airways and the pulmonary vasculature.

Volume 9: Hepatic and Gastrointestinal Toxicology (edited by R. S. McCuskey and D. L. Earnest)

Volume 9, which is divided into two sections, presents important new information concerning mechanisms of effect and consequences of exposure of the liver and gastrointestinal tract to a wide variety of toxicants. Sections on the liver and gastrointestinal tract provide the reader with concise introductory descriptions of relevant cellular and organ anatomy, physiology, and mechanisms of toxic injury. Specific consideration is given to methods of toxicant exposure and to processes that provide defense against injury. The discussions provide a broad focus but also contain details about molecular, cellular, and biochemical mechanisms. The section on hepatic toxicology is a comprehensive review of the toxic effects of a wide variety of specific compounds including volatile hydrocarbons, anesthetic agents, drugs of therapeutic value as well as abuse, heavy metals, natural compounds, and so on. The section on gastrointestinal tract toxicology, presenting toxicology of the esophagus, stomach, small intestine, and colon, represents the first major reference work available with a broad overview of the effect of toxicants on the mucosal, motility, and immune functions of the gastrointestinal tract as well as information about specific toxicants. The final chapter presents an overview of clinical toxicology in animals with a focus on the liver and gastrointestinal systems. This volume will be of particular interest and use to those concerned with environmental and industrial toxicology, as well as with mechanisms of toxicant injury and clinical medicine.

Volume 10: Reproductive and Endocrine Toxicology (edited by K. Boekelheide, R. Chapin, P. Hoyer, and C. Harris)

Section 1, on male reproductive toxicology, starts with a basic overview of the anatomy and physiology of the system. This is followed by reviews on the way molecules are evaluated for male reproductive toxicity in both the pharmaceutical and chemical industries, along with a strategy for evaluating the reproductive function of transgenic animals. The bulk of this section is devoted to reviewing mechanisms and manifestations of toxicants to particular targets within the male reproduction system (Sertoli cell, Leydig cell, etc.). Finally, there is an evaluation of areas of interest not often considered in the toxicologic context (fluid flow, the immune system, paracrine factors, etc.), as well as a chapter highlighting technical advances and what these mean for the field. Overall, Section 1 reviews the state of the science in a number of areas, explicitly providing investigational strategies, and identifies promising areas of future research.

Section 2 is a comprehensive overview of the field of female reproductive toxicology. Much interest has recently been focused on this area of toxicology, due to the increasing number of women in the workplace and the impact of female fertility on reproduction issues. This section begins with an overview of female reproductive physiology, with an emphasis on the complexities of its hormonal regulation. Next, the various components of the female reproductive system are described in detail, along with known and potential sites of disruption by xenobiotics. Finally, assessment of human risk is discussed from the standpoint of classical methods of evaluation, as well as recently developed, novel experimental approaches.

Section 3 provides an overview of mammalian development, from fertilization to parturition and early postnatal maturation, in terms of the stage-selective anatomical and functional characteristics of each developmental phase that may underpin the ultimate manifestations of toxicity. Several possible mechanisms of developmental toxicity are discussed in terms of the roles of biotransformation, pharmacokinetics, altered gene expression, neurobehavioral development, physiological conditions, and nutritional status. A selected list of chemical agents and environmental extremes known to produce persistent developmental abnormalities is reviewed by class and are discussed in terms of their known toxic effects and possible mechanisms of action. Strategies for the study of developmental toxicants *in vivo* and *in vitro* are described, as well as current screening and testing systems for the detection of additional potential disrupters of normal development. This section concludes with a summary and discussion of the critical need for additional understanding and the future prospects for accurate prediction, prevention, and assessment of risk in developmental toxicity.

Section 4 provides a review of the effect of toxicants on endocrine tissues. This is an emerging area of concern in toxicology. The adrenals, thyroid, parathyroid, and pancreas will be covered in this section.

Volume 11: Nervous System and Behavioral Toxicology (edited by H. E. Lowndes and K. R. Reuhl)

Neurotoxicology is among the fastest growing of the toxicological disciplines. Recent refinements of neuropathologic techniques, meticulous validation of behavioral tests, and introduction of the tools of molecular biology have led to rapid progress in understanding the mechanisms and consequences of neurotoxic injury. This volume is intended to provide the reader with an appreciation for the scope of approaches taken in the mechanistic study of neurotoxic agents. The various chapters, each contributed by recognized experts, are divided into three sections. Section 1 addresses the basic anatomical and biochemical components of the nervous system, with special attention placed on those areas where existing or emerging data have demonstrated toxic actions. This section is directed towards the nonspecialists, offering overviews of topics in neuroscience, as well as contemporary reviews of selected topics for researchers in the field. Section 2 focuses on neurobehavioral toxicology and alterations of the special issues. Specific behavioral domains and the functional consequences of their perturbation are discussed in detail. Particular emphasis has been given to research methodologies in neurobehavioral toxicology, especially the appropriate use, limitations, and interpretations. Section 3 provides a survey of selected classes of compounds demonstrated to have adverse effects on nervous system structure and function. It is intended to provide current reviews of chemical-specific neurotoxicity for practitioners in neurotoxicity.

Volume 12: Chemical Carcinogens and Anticarcinogens (edited by G. T. Bowden and S. M. Fischer)

Volume 12 provides a comprehensive overview of the field of chemical and radiation carcinogenesis with particular emphasis on molecular mechanisms. The actions of various chemical carcinogens as well as ultraviolet and ionizing radiation are discussed in the context of the multistage process of both experimental and human carcinogenesis. Background information concerning the target genes for genetic and epigenetic modulation by carcinogens are covered. These target genes include proto-oncogenes, tumor suppressor genes, and effector genes. Various classes of carcinogens including tumor initiating, promoting, and progressing agents are discussed in terms of how they bring about critical alterations in the target genes. The functional role of these gene alterations are covered in the context of the various phenotypic changes associated with each stage of carcinogenesis. Finally, strategies for intervening with the multistep process of carcinogenesis through chemoprevention are discussed.

Wherever possible the relevancy of more basic findings with experiment model systems to human carcinogenesis is emphasized.

Volume 13: Author and Subject Indexes

The Author Index contains a complete list of authors who are listed in the references at the end of each chapter and those cited in the text.

The Subject Index contains a comprehensive list of terms used throughout the other 12 volumes. The EMTREE Thesaurus has been used as a guide for the selection of preferred terms, along with the IUPAC Recommendations for the nomenclature of chemical terms.

References

References cited in text are numbered for easy identification with the complete reference (authors, title, year, volume, inclusive pages) in the Reference list at the end of each chapter. Each reference has been validated. While exhaustive referencing was not performed, sufficient references were used to support the statements and provide the reader with adequate sources for finding additional information.

SECTION I

11.01
Fundamentals of the Structure and Function of the Nervous System

GAIL D. ZEEVALK

University of Medicine and Dentistry of New Jersey-Robert Wood Johnson Medical School, Piscataway, NJ, USA

11.01.1 INTRODUCTION

The nervous system is the command center of an organism, controlling all motor and sensory activities, generating responses, integrating, learning and remembering information. Involuntary processes such as breathing, circulation, and peristalsis continue in the absence of cognitive input. Voluntary processes require the correct integration of information and wiring of neuronal circuitry. Once thought intangibles such as the neurological basis of memory and learning are beginning to be elucidated in terms of tangible physiological events.

Damage to cells within the nervous system can result in serious consequences and specific deficiencies in the organism. Loss of motor, sensory or cognitive skills will depend on the extent of damage and the particular brain region affected. Most organs of the body are subject to toxic insult from environmental sources. Despite the protective effect of a blood brain barrier, which screens out many potentially toxic substances, some neurotoxicants gain access to the nervous system and many noted examples of nerve cell damage due to toxic compounds exist. Neurotoxicants come in many forms. Some are generated by man; such is the case for pesticides, insecticides and the drug, MPTP. Some neurotoxicants are naturally occurring in our environment, for example, the heavy metals lead, manganese and mercury. Consumption of the legume *Lathyrus sativus* can result in the neurological disorder Lathyrism, a form of spastic paraparesis. 3-Nitropropionic acid, a plant and fungal toxin, has been shown to be the cause of an encephalopathy and delayed dystonia in Northern China.

As will be demonstrated in the forthcoming chapters, in some instances the toxic sequelae of events for a particular neurotoxicant are well known; for others, the search continues. Elucidation of the mechanism of toxicity induced by environmental agents may provide avenues for therapeutic intervention, as well as provide insight into mechanisms of nerve cell death in several neurodegenerative diseases. At the center of such studies is the need for a fundamental understanding of the structure and function of the nervous system, a point at which we will begin. The brevity of this chapter on structure and function, however, permits only an overview of the major components and systems in the nervous system. For more detailed descriptions of the nervous system, the reader is referred to several texts.[1–5]

11.01.2 MICROSCOPIC ANATOMY: CELLULAR COMPONENTS OF NERVOUS TISSUE

The central (CNS) and peripheral (PNS) nervous systems consist of two main classes of cells: neurons and glia. Neurons are responsible for receiving and conveying information. A neuron has four main components (Figure 1): (i) the cell body, which is similar in function and organelle content to other cells in the body, (ii) dendrites, or afferent processes, projections that receive stimuli and conduct impulses towards the cell body, (iii) a single axon, or efferent process that emerges from the nerve cell body to carry electrical impulses away from the cell to another neuron, muscle or gland, etc., and (iv) the presynaptic terminal, a specialized region at the end of the axon, which contains the transmitting elements of the neuron. Neurons communicate with one another at a synapse, through release of a chemical neurotransmitter from synaptic vesicles in the presynaptic terminal and its reception at the surface of a postsynaptic cell (Figure 1). The space between the pre- and postsynaptic elements is the synaptic cleft. Nerve cells are often organized into groups called nuclei or ganglia. Areas composed mainly of nerve cell bodies make up the gray matter, whereas areas primarily consisting of myelinated nerve processes are known as the white matter.

There are three major types of glial cells in the CNS, astrocytes, oligodendrocytes, and microglia. The Schwann cell is the major glial cell in the PNS. Oligodendrocytes form and maintain the myelin sheaths of nerve cell axons in the CNS; Schwann cells serve a similar function in the PNS (Figure 1). Myelin acts as an electrical insulator to facilitate conduction along the axon. The area between two adjacent myelin segments along an axon is known as the node of Ranvier (Figure 1). Astrocytes serve many functions: to act as a structural support element, help to remove neurotransmitter released by neurons, maintain the proper extracellular ionic composition, store and transfer metabolites from capillaries to neurons, and may aid in repair subsequent to

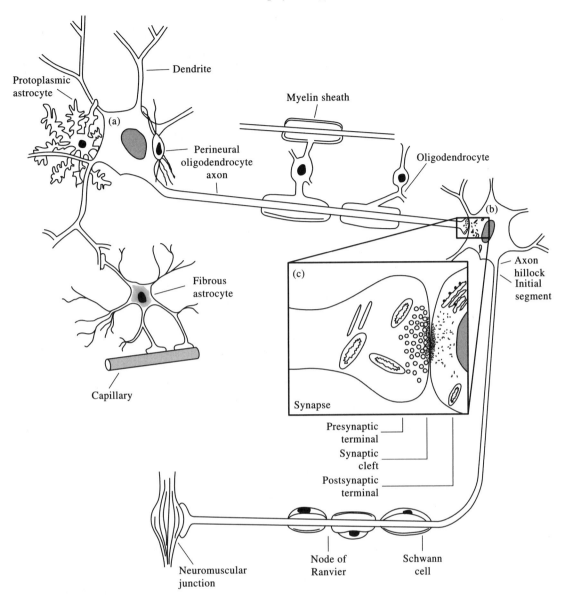

Figure 1 (a) A neuron in the cerebral cortex makes contact with (b) a motor neuron in the spinal cord at a specialized ending called a synapse (c). The axon of neuron (b) makes contact with a muscle fiber. Stimulation of the axon results in flexion or extension of the muscle. Astrocytes surround and ensheath the cell bodies of neurons (protoplasmic astrocytes, perineural oligodendrocytes) and transfer metabolites from blood capillaries to neurons (fibrous astrocytes). Oligodendrocytes in white matter form the myelin sheath of axons in the CNS; Schwann cells form the myelin covering of axons in the periphery. (c) An axon ends in a presynaptic terminal which contains vesicles filled with neurotransmitter. Upon depolarization, synaptic vesicles release their content into the synaptic cleft where the neurotransmitter makes contact with receptors in the membrane of the postsynaptic element.

trauma. Glial cells, however, may act as intermediates in the neuronal damage caused by some neurotoxicants (see Chapter 31, this volume) and may be the site of action of several toxic compounds. (see Chapter 12, this volume). Microglia are thought to be developmentally distinct from the "true glia." Derived from mesodermal rather than ectodermal sources, they are thought to invade the CNS at the time of vascularization. Microglia are considered mononuclear macrophages which act as phagocytic cells that respond to infection and injury by removing debris.

11.01.3 NEUROPHYSIOLOGY

11.01.3.1 The Resting Membrane Potential

The nerve cell, like other cells in the body, is enclosed by a plasma membrane. This

Table 1   Distribution of the major ions across the plasma membrane of the squid giant axon.

Ion	Intracellular (mM)	Extracellular (mM)
K^+	400	20
Na^+	50	440
Cl^-	52	560

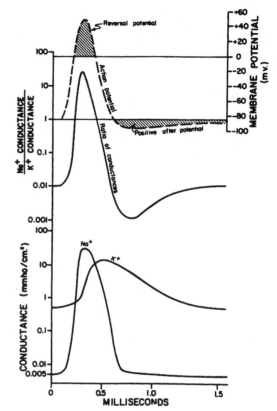

Figure 2 The upper panel shows the change in the membrane potential during an action potential. The lower panel demonstrates the changes in Na^+ and K^+ conductances during the course of an action potential (reproduced by permission of W. B. Saunders from 'Textbook of Medical Physiology,' 1991).

membrane separates ionic and other components into intra- and extracellular compartments. At rest, the outside of the cell is positively charged and the inside is negatively charged, creating a charge difference referred to as the membrane potential. The resting membrane potential varies, depending on the neuron, and may range between $-40\,mV$ and $-75\,mV$. When a neuron is hyperpolarized, its membrane potential becomes more negative. Depolarization occurs when the membrane potential becomes less negative. The membrane potential is primarily determined by the distribution of potassium (K^+) ions (high intracellularly in a resting cell), sodium (Na^+) and chloride (Cl^-) ions (high extracellularly at rest) (Table 1). The ionic gradient difference is maintained by pumps located in the membrane. The most important with regard to the membrane potential is the Na^+, K^+, ATPase. Na^+ is pumped out of the cell and K^+ into the cell at the expenditure of energy (ATP). The membrane potential of a cell can be calculated by the Nernst equation and the reader is referred to other sources for a more complete discussion of this topic.[3,6,7]

11.01.3.2   The Action Potential

The action potential is a large depolarizing signal that occurs in an all-or-none fashion (Figure 2). When action potential is generated, either by an electrical, chemical, or mechanical stimuli, Na^+ ions enter the cell through specific voltage-dependent Na^+ channels. This results in an increase in positive ions within the cell, making the charge difference across the membrane less negative. At the peak of an action potential, the membrane potential is approximately $+50\,mV$. Depolarization also results in opening of voltage-dependent K^+ channels, and the consequent efflux of K^+, but the opening of K^+ channels is slower and lags behind the opening of Na^+ channels (Figure 2, bottom). Na^+ channels eventually become inactivated; this, plus the outflow of positive charges from the cell in the form of K^+ ions helps to restore the resting membrane potential. The proper functioning of ion channels is fundamental to normal neurophysiology. Blockade of voltgage-dependent Na^+ channels by the pesticides DDT and pyrethroids, and by tetrodotoxin, the toxin in the Japanese Blowfish, Fugu, is highly neurotoxic (see Chapters 9 and 27, this volume).

Action potentials are propagated along the length of an axon. In unmyelinated axons, one segment of axon undergoing depolarization will initiate the process of depolarization in the next, adjacent segment of axon. Thus, the depolarization spreads down the length of axon with a certain velocity and with no loss of amplitude. In myelinated axons, the impulse or depolarization only flows across the axolemma at the nodes of Ranvier. The impulse is said to "jump" from node to node in a manner termed saltatory conduction. Action potentials thus travel along myelinated axons at much greater velocities than they do along unmyelinated axons.

11.01.3.3 The Receptor and Synaptic Potentials

Not all electrical signals in the nervous system are conducted without loss of amplitude, as are action potentials being conducted along an efferent, axonal process. Two types of impulses generated in the nervous system propagate with loss of amplitude along the membrane: those initiated in sensory receptor cells, which generate a receptor potential and those in postsynaptic membranes (see Figure 1), which generate a synaptic potential. In each case, the signal or charge difference across the membrane decreases progressively in amplitude with distance. This type of conduction is called electrotonic. The receptor potential can, in turn, act on nearby sensory fiber terminals, whereas the synaptic potential conveys the depolarization to the axon hillock and initial axon segment of the cell (sites along the axon important for initiation of the action potential). In most instances, the excitatory receptor or synaptic potential will attenuate sufficiently with distance so that an action potential will not be initiated. Many receptor or synaptic signals must summate in order for an action potential to be initiated.

11.01.4 NEUROCHEMISTRY

11.01.4.1 Neurotransmitters and Their Receptors

The major form of communication between neurons is mediated via release of a chemical neurotransmitter at a synapse (Figure 1) and reception of the chemical signal at the postsynaptic neurotransmitter receptor. Neurotransmitters are packaged within synaptic vesicles located in the presynaptic terminal. These vesicles can fuse with the plasma membrane and release their contents into the synaptic cleft, where neurotransmitter can then bind to the postsynaptic receptor. Neurotransmission is rapidly terminated either by uptake into surrounding cellular elements via specific transporters or by neurotransmitter degradation. There are two major categories of neurotransmitters: (i) the classical small molecule neurotransmitters such as acetylcholine, glutamate, gamma-aminobutyric acid (GABA), glycine, the catecholamines dopamine, norepinephrine and epinephrine, and the indolamines serotonin and histamine; and (ii) the neuropeptides. Most of the small molecule neurotransmitter receptors have been cloned and have revealed multiple subunits that are used to form various subtypes of receptors within a given family. The pharmacological, kinetic, and physiological responses of a particular receptor subtype can be altered, depending on the subunit composition. For a detailed discussion of the different receptor subunits for each neurotransmitter and its receptor, see Seigel et al.[8] Only the major receptor subtypes are addressed in this chapter. Details of the neuroanatomy associated with the different neurotransmitter populations are given in Sections 11.01.5 and 11.01.6.

11.01.4.1.1 Acetylcholine

Acetylcholine was the first neurotransmitter to be identified. It is synthesized from acetyl coenzyme A and choline in a reaction catalyzed by cholineacetyltransferase. Another enzyme, acetylcholinesterase, is present on the postsynaptic membrane in the synaptic cleft and is responsible for rapidly degrading acetylcholine to terminate its action. As discussed in detail in Chapter 26, this volume, several organophosphates and carbamate insecticides affect cholinergic neurons by inhibiting acetylcholinesterase.

Acetylcholine is the primary neurotransmitter in the PNS. It is released by motor neurons whose cell bodies lie in the brain stem and spinal cord and whose processes end on skeletal muscles (Figure 1); by preganglionic autonomic neurons whose processes end on autonomic ganglia and by postganglionic parasympathetic neurons whose processes end on various glands and smooth muscles (Section 11.01.6). The distribution of cholinergic neurons in the CNS is more restricted. These sites include neurons in the pons reticular formation that send projections to the thalamus, the forebrain nucleus basalis with projections to the cerebral cortex and amygdala, the limbic septal area with projections to the hippocampus and interneurons in the striatum.

Acetylcholine receptors are of two main types: (i) nicotinic receptors which form an ion channel (ionotropic), and (ii) muscarinic receptors which are linked to second messenger systems via G proteins (metabotropic). In the PNS, nicotinic receptors are present on skeletal muscles, and autonomic ganglia innervated by preganglionic sympathetic and parasympathetic fibers (Section 11.01.6.2.3). Postsynaptic autonomic effector sites innervated by all postganglionic, parasympathetic fibers (i.e., secretory glands, smooth and cardiac muscles) and postganglionic sympathetic fibers to sweat glands and the erector pili muscles have muscarinic rather than nicotinic receptors. Cholinergic innervation controls contraction of skeletal muscles. In the autonomic nervous system cholinergic innervation controls a variety of functions that involve nearly every body organ and gland. Cholinergic parasympathetic inner-vation to organs and glands often has physio-logical influence that opposes that produced by the sympathetic norepinephrine innervation (i.e., at salivary glands, parasympathetic choli-nergic innervation causes watery secretion whereas norepinephrine sympathetic innerva-tion produces thick, viscous secretion) (Section 11.01.6.2.3).

Both nicotinic and muscarinic receptors are found throughout the CNS. Cholinergic neu-rons in the striatum are involved in control of motor function. In other CNS regions, there is evidence that cholinergic neurons are involved in memory and learning, and play a critical role in the sleep-wakefulness cycle. Loss of central cholinergic neurons is a pathological feature of Alzheimer's disease.[9]

11.01.4.1.2   Monoamines

The catecholamines dopamine, norepineph-rine and epinephrine, and the indolamines serotonin and histamine comprise the mono-amine neurotransmitters. Dopamine is synthe-sized from the amino acid tyrosine. In turn, dopamine can be converted to norepinephrine and subsequently, epinephrine. Serotonin is synthesized from the amino acid tryptophan, histamine from L-histidine. Once released into the synaptic cleft, these transmitters, with the exception of histamine, are predominately inactivated by reuptake into the presynaptic terminal; histamine is degraded in the synaptic cleft by histamine methyltransferase.

Dopaminergic neurons are located in the retina and olfactory bulb where they function to reduce surrounding "noise," that is, suppress antagonistic impulses in a receptive field, which is an important feature in the "focusing" of sensory information processing. Two major dopamine projection systems are located in the midbrain. These include the substantia nigra (SN) and the ventral tegmental area (VTA). Projections from the SN terminate in the basal ganglia and are important in control of motor function. Parkinson's disease results in the loss of the dopamine neurons in the SN.[9] The dopamine neurons are also the target site for the neurotoxicant MPTP (see Chapter 31, this volume). The VTA projects to limbic and cortical systems. Disruption of dopamine pro-jections to these systems has been associated with psychotic illnesses such as schizophrenia. Dopaminergic neurons are also present in the hypothalamus and project to the pituitary gland where they function to inhibit the release of prolactin and melanocyte stimulating hor-mones. At present, five dopamine receptors

with distinct sites of distribution within the CNS are known. Dopamine receptors are of the metabotropic type, and are thought to convey inhibitory signals.

Norepinephrine (also known as noradrena-line) is the neurotransmitter in all postganglio-nic, sympathetic autonomic nerves, with the exception of sympathetic neurons that project to the sweat glands. In the CNS, noradrenergic neurons are found in only two sites in the brain stem, the locus ceruleus and the lateral teg-mental nuclei. Although small in number, these neurons send projections to all parts of the CNS (cortical forebrain, olfactory bulb, thalamus, hypothalamus, amygdala, hippocampus, cere-bellum, medulla, and spinal cord). Noradren-ergic neurons in the locus ceruleus may play a role in maintaining attention and vigilance. In other CNS areas, its action may be analogous to dopaminergic neurons in the retina, that is, to enhance signal to noise ratio by inhibiting surrounding activity. Norepinephrine in post-ganglionic sympathetic projections to various body organs and glands acts to balance the opposing cholinergic, parasympathetic input to control such autonomic activities as contrac-tion of smooth and cardiac muscle, glandular secretion and heart rate.

Epinephrine (adrenaline)-containing neurons in the CNS are found only in the medulla oblongata (lateral and dorsal parts of the tegmentum) and project to the hypothalamus and spinal cord. Norepinephrine and epineph-rine act on α and β metabotropic type receptors. The nonvascular actions of epinephrine in the CNS are at present uncertain.

Serotoninergic (or 5-hydroxytryptamine, 5HT) neurons are found in the brain stem where they are concentrated in the raphe nuclei. Projections from 5HT neurons reach the cortical forebrain, olfactory bulb, septum, hippocampus, thalamus, hypothalamus, basal ganglia (caudate, putamen and globus palli-dus), substantia nigra, cerebellum, and spinal cord. Seven 5HT receptors are known at present and can be distinguished by their pharmacology and associated second messen-ger system. The majority of 5HT receptors are metabotropic G protein-linked. The exception is the $5HT_3$ receptor which is ionotropic. Serotonin is involved in the hypothalamic control of pituitary secretion, in sleep/arousal states, in regulation of circadian rhythms and inhibition of food intake. Disturbances of the serotoninergic systems have been linked to clinical depression and obsessive-compulsive disorder.

Histaminergic neurons in the CNS are found only in the posterior hypothalamus, but they project to nearly all regions of the brain and to the spinal cord. They are also present in the PNS where they may modulate adrenergic activity. Three subtypes of histamine metabotropic receptors are known ($H_{1,2,3}$). There is evidence that either adenosine or GABA may co-exist in the same neuron with histamine and may act as a co-transmitter. The best studied actions of histamine are in the modulation of the secretion of the neuropeptides prolactin, vasopressin and corticotropin-releasing hormone. Histamine has also been implicated in control of arousal, feeding and drinking, thermoregulation, circadian rhythms, and analgesia.

been implicated in the neuropathology associated with a number of neurological conditions or neurodegenerative disorders including epilepsy, ischemia, head and spinal cord trauma, Huntington's, Parkinson's or Alzheimer's diseases, amyotrophic lateral sclerosis and AIDS dementia. Interestingly, a number of environmental neurotoxicants, as will be discussed in subsequent chapters, appear to target glutamate receptors either directly by receptor stimulation, or indirectly by producing a metabolic inhibition which in turn can trigger a secondary excitotoxicity.

11.01.4.1.3 *Glutamate*

Glutamate is the major excitatory neurotransmitter in the brain. Although present in high concentration in the blood, it does not cross the blood–brain barrier and is synthesized in the brain from glucose. Glutamate is unique among neurotransmitters in that it serves a prominent role in intermediary metabolism and protein synthesis, as well as in neurotransmission. After release into the synaptic cleft, the action of glutamate is terminated by reuptake into nerve endings and glial cells. Some glutamate receptors are ionotropic, (i.e., the NMDA, AMPA, and kainate subtypes), whereas others are metabotropic (seven identified to date; mGlu R_{1-7}). Maintaining low extracellular glutamate levels is critical since overstimulation of any of the ionotropic glutamate receptors can lead to cell death through a process known as excitotoxicity (see Chapter 31, this volume). NMDA receptors, when activated, allow a large influx of Ca^{2+} into the cell. The influx of Ca^{2+} through the NMDA receptor is thought to account for this receptor's specialized role in processes such as long term potentiation (LTP; a physiological correlate of memory and learning). Ca^{2+} influx may also contribute to the excitotoxicity produced by overstimulation of the NMDA receptor. AMPA, and perhaps kainate, receptors are responsible for fast excitatory synaptic transmission. The metabotropic glutamate receptors may modulate ionotropic glutamate receptor activity, or may have direct postsynaptic consequences of their own and have been implicated in the process of LTP. Glutamate receptor location is fairly ubiquitous throughout the brain, but with differential distribution of the various receptor subtypes. Most neurons in the brain are thought to possess one or more glutamate receptor subtypes. In the PNS, kainate receptors are present in high density in the sensory or dorsal root ganglia. Glutamate and/or glutamate receptors have

11.01.4.1.4 *Gamma-aminobutyric acid*

GABA is the major inhibitory neurotransmitter in the brain. The immediate precursor for the synthesis of GABA is glutamate. A simple one step reaction in which glutamate is decarboxylated by glutamic acid decarboxylase converts glutamate to GABA. GABAergic neurons are found throughout the brain. They are important as interneurons, that is, neurons whose processes remain confined to a particular brain structure to convey information to other neurons within that structure. GABAergic interneurons are found in the cerebral cortex, cerebellar cortex, olfactory bulb, retina, hypothalamus, and hippocampus. The striatum (caudate and putamen) contains inhibitory GABAergic interneurons as well as GABAergic projection neurons which convey inhibitory information to the dopaminergic neurons in the substantia nigra and to neurons in the globus pallidus. The striatal GABAergic projection neurons are one class of neurons that are lost in Huntington's disease.[10] Other important GABAergic projection pathways are from cerebellar Purkinje cells to the deep cerebellar nuclei and from the substantia nigra and globus pallidus to the thalamus and subthalamic nuclei.

Two distinct classes of GABA receptors have been identified, $GABA_A$ (ionotropic) and $GABA_B$ (metabotropic). Opening of the channels formed by $GABA_A$ receptors allows an influx of Cl^- into the cell, whereas activation of postsynaptic $GABA_B$ receptors is thought to indirectly activate an outwardly directed K^+ channel. Either event results in an increase in the charge differential across the membrane (inside more negative), resulting in a hyperpolarization of the membrane potential and making the postsynaptic cell less likely to initiate an action potential. Blockade of $GABA_A$ receptors with the antagonist bicuculline produces powerful convulsions. The $GABA_A$

receptor is the target site for several drugs that act as depressants or anticonvulsants. These include the barbiturates, phenobarbital and pentobarbital; the benzodiazipine agonists diazepam and chlordiazepoxide, as well as alcohol. The $GABA_A$ receptor is also the target site for cyclodiene-type insecticides and for lindane (see Chapters 9 and 27, this volume for a more complete discussion). Maintaining the proper balance between excitatory (predominately glutamatergic) and inhibitory (predominately GABAergic) information throughout the CNS is critical to normal functioning.

11.01.4.1.5 Glycine

Glycine is an important inhibitory neurotransmitter in the spinal cord and brainstem. Its immediate precursor is serine, however, like glutamate and GABA, the original source for synthesis of this amino acid is glucose. Similar to the GABA receptor, the glycine receptor forms an ion channel whose opening allows the passage of Cl^- ions into the cell, resulting in a hyperpolarization. The pharmacological profiles of the glycine and GABA receptors are distinct. For example, the plant alkaloid, strychnine, is a potent convulsant through its antagonist action at glycine, but not GABA, receptors.

11.01.4.1.6 Neuropeptides

In addition to the classical neurotransmitters, many neuropeptides are now know to serve as chemical messengers in the central and peripheral nervous systems. They range in length from as few as three to as many as 44 amino acids. A partial listing of neuropeptides is given in Table 2. Many of the peptides have been cloned and most are related to the G protein-linked superfamily of metabotropic receptors. Unlike the classical neurotransmitters, which are synthesized in the presynaptic terminal, neuropeptides are synthesized in the cell body and are transported down the axon to the nerve terminal. The various neuropeptides are differentially distributed throughout the CNS and PNS. In many instances, they co-exist with other peptide and nonpeptide neurotransmitters. The exact function(s) of the neuropeptides in the mammalian nervous system is not entirely clear. There is evidence from invertebrate studies that while neuropeptides can function as classical neurotransmitters, they may play more supportive roles as neuromodulators of the actions of the neurotransmitters.

Table 2   Biologically active neuropeptides.

Adrenocorticotropic hormone	Growth hormone-releasing hormone
Angiotensin	Leu-enkephalin
Atrial natriuretic polypeptide	Luteinizing hormone
Bradykinin	α-Melanocyte-stimulating hormone
Calcitonin	Met-enkephalin
Calcitonin gene related peptide	Neurotensin
L-Carnosine	Neuropeptide Y
Cholecystokinin octapeptide	Oxytocin
Corticotropin-releasing hormone	Somatostatin
Dynorphin A	Substance P
β-Endorphin	Thyrotropin-releasing hormone
Galanin	Vasoactive intestinal peptide
Gastrin-releasing peptide	Vasopressin
Gonadotropin-releasing hormone	

11.01.4.2 Neurotransmitter Transporters

As mentioned previously in Section 11.01.4.1, the action of many of the neurotransmitters is rapidly terminated by uptake of the transmitter either back into the presynaptic terminal or into surrounding glial cells. High affinity uptake via specific plasma membrane transporters occurs with the monoamines dopamine, norepinephrine, epinephrine and serotonin, and with the amino acid neurotransmitters glutamate, GABA and glycine. In some instances, as is the case for glutamate, distinct transporters have been identified, one found on neurons and two associated with glia. These transporters differ in their amino acid composition and affinities for glutamate. Disturbances in the transport systems can lead to excessive extracellular neurotransmitter which, for glutamate, can be toxic. One hypothesis for the loss of spinal motor neurons in amyotropic lateral sclerosis (ALS) is the decreased presence of high affinity glutamate transporters on the anterior horn cells.

11.01.5 CENTRAL NERVOUS SYSTEM: ANATOMY AND FUNCTION

11.01.5.1 Introduction

The CNS consists of the brain and spinal cord. During development of the CNS from the neural tube, five vesicles form which give rise to the major subdivisions listed in Figure 3. These vesicles are the telencephalon which forms the cerebral hemispheres; the diencephalon which forms the thalamus and hypothalamus; the mesencephalon which forms the midbrain; the metencephalon which forms the pons and cerebellum and the myelencephalon from which the medulla oblongata is derived.

11.01.5.2 Blood Supply, Blood–Brain Barrier, and Cerebrospinal Fluid

Brain metabolism is highly dependent on an adequate and continuous supply of glucose and O_2. Blood flow to the brain and spinal cord is achieved by two major pairs of arteries: the internal carotid and the vertebral arteries. Anterior portions of the brain are supplied by branches of the internal carotid. The internal carotid bifurcates to form the anterior and middle cerebral arteries. The middle cerebral artery further divides into several branches to supply blood to the lateral surface of the frontal, parietal, occipital and temporal lobes. Many small branches penetrate the brain to supply deep structures of the diencephalon (particularly the hypothalamus) and the telecephalon. The anterior cerebral artery turns medially to supply the inner (medial) surface of the frontal and parietal lobes. Thus, the internal carotid system supplies blood to the cerebral hemispheres with the exception of the medial

surface of the occipital lobe and the inferior surface of the temporal lobe.

Posterior brain structures and the spinal cord receive blood supply from branches of the vertebral arteries. The two vertebral arteries come together at the junction of the pons and medulla oblongata to form the basilar artery. Proximal to their joining to form the basilar artery, the vertebral arteries give rise to the spinal arteries which supply the spinal cord. The blood supply to the posterior brain is frequently referred to as the vertebral–basilar system. Branches from the vertebral–basilar system supply blood to the brain stem, cerebellum, parts of the diencephalon (particularly the thalamus) and areas of the cerebral hemispheres not supplied by the internal carotid system. The Circle of Willis, at the base of the brain, interconnects the internal carotid and vertebral systems. Communication between the two arterial systems and between the two sides of the brain occurs via the anterior and posterior communicating arteries.

The blood–brain barrier is a physiochemical interface that regulates the passage of many molecules from the blood into the brain. It is formed by the tight junction between adjacent endothelial cells which form the walls of the blood capillaries. The barrier is selective, allowing molecules such as glucose to readily cross while preventing the passage of other molecules. Characteristics such as size, lipid solubility and the presence of specific transport systems govern the ability of molecules to

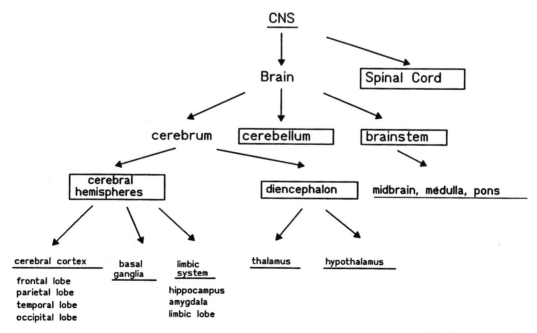

Figure 3 Major subdivisions of the CNS. Boxed areas are the main subregions described in the text. Many of these areas are further subdivided anatomically and functionally.

penetrate the barrier. Although the blood brain barrier excludes many potentially toxic substances, a number of toxic agents do gain access to the brain and exert either direct toxic effects or, as in the case of MPTP, enter the brain and are metabolized to a neurotoxic substance. Due to the lack of a completely formed blood–brain barrier in the developing fetus or young child, the immature nervous system is particularly prone to damage by neurotoxicants (see Chapters 2, 5, and 21, this volume).

The brain and spinal cord are surrounded by, and are suspended in, a clear fluid called the cerebrospinal fluid (CSF). The cavity between the brain and the skull (the subarachnoid space) contains CSF. Membranes line both sides of the cavity. The membrane covering the brain is the pia mater. Below the skull surface are two apposing membranes or meninges, the arachnoid mater and the dura mater. Four ventricles or cavities lie deep in the brain (two lateral, third and fourth ventricles) and connect with the subarachnoid space. The CSF, found in the ventricles and subarachnoid space, is a pale fluid with a low specific gravity and contains a small amount of protein, sugar, electrolytes and a few lymphocytes. CSF is formed in the ventricles by a capillary network known as the choroid plexus through a combination of filtration of the blood capillaries and active production by the epithelial cells of the choroid plexus. The CSF formed in the choroid plexi of the lateral ventricles flows into the third, then fourth ventricles. The fourth ventricle connects with the subarachnoid space. CSF flows into the subarachnoid space from the fourth ventricle and continues rostrally (towards the top of the skull) to the region of the superior sagittal venous sinus (located between the cerebral hemispheres) where the CSF will diffuse into the venous blood. The concentration of constituents in CSF differs from that in blood plasma. The CSF functions to maintain the correct extracellular milieu that surround nerve cells, act as a mechanical cushion, circulate neuroactive hormones throughout the CNS and participate in the removal of unwanted material from the brain.

11.01.5.3 Gross Anatomy and Function

11.01.5.3.1 *Cerebral hemispheres: cerebral cortex, basal ganglia, limbic system*

(i) Cerebral cortex

The cerebral cortex is divided along the midline of the brain into two hemispheres. In general, the left hemisphere is dominant for language, logic, computational and sequential analysis, whereas the right hemisphere is important for spatial patterns and musical ability. Discrete areas of the cerebral cortex are associated with specific functions (Figure 4). The lateral surface of each hemishpere of the cerebral cortex is further divided into four lobes: the frontal lobe containing the primary motor cortex, the parietal lobe containing the primary somatosensory cortex, the temporal lobe containing the auditory and speech centers and the occipital lobe containing the visual cortex (Figure 4). A fifth lobe, the limbic lobe, can only be seen by splitting the brain down the middle and viewing the medial surface (Figure 5). The limbic lobe consists of the paraterminal, cingulate, and parahippocampal gyri. The lateral lobes of the cerebral cortex primarily function to control and integrate motor and sensory information. The limbic association cortex is responsible for emotion, motivation and memory. Sensory input from the periphery (touch, vision, smell, pain etc.) reach the primary somatosensory cortex, mainly via the thalamus, which in turn projects to higher sensory and associative areas for integrative processing. From the thalamus, connections are made with the motor cortex which will ultimately govern the output response. Reaction to environmental cues, therefore, requires the cooperative participation of the sensory, associative and motor areas of the cerebral cortex. Disruption of these interconnections among the cortical areas will result in functional deficits. Apraxia, the inability to perform a desired movement; agnosia, the inability to recognize stimuli even though sensory systems are intact; or aphasia,

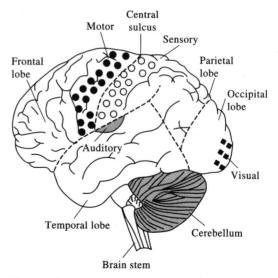

Figure 4 Lateral overview of the human brain showing the major divisions of the cerebral cortex.

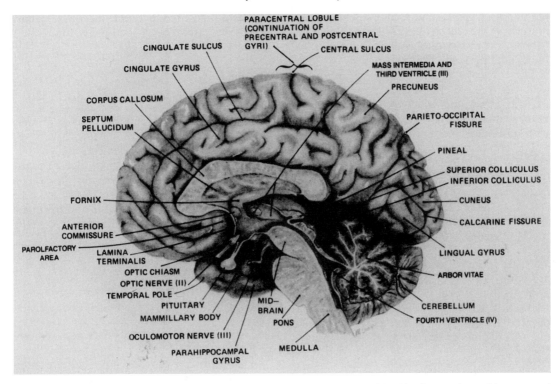

Figure 5 A midline saggital cut through the brain reveals the limbic lobe (cingulate and parahippocampal gyri) as well as the brainstem (midbrain, pons, medulla) and inner cerebellum (reproduced by permission of F. A. Davis from 'Manter and Gatz Essentials of Clinical Neuroanatomy and Neurophysiology,' 1987).

the inability to speak even though cortical areas involved in language are undamaged are some examples which demonstrate the importance of integrative cortical processing to normal functioning.

Motor neurons whose cell bodies and processes lie completely in the CNS are referred to as upper motor neurons. Lower motor neurons have cell bodies that reside in the CNS (brain and spinal cord), but have axons that extend outside the CNS. The motor neurons of the spinal cord are examples of lower motor neurons. The control of skeletal muscles throughout the body requires the coordinated effort of both upper and lower motor neurons. There are a number of descending pathways or tracts that originate from neurons in the cortex to directly or indirectly make contact with motor neurons in the spinal cord. Projections from motor neurons in the spinal cord directly contact muscle fibers to evoke muscle contraction (Figure 1).

The corticospinal (pyramidal) tract is the principal mediator of voluntary movement. Neurons in the motor cortex send direct projections to the spinal cord. At the level of the medulla, approximately 90% or the axons cross the midline to form the lateral corticospinal tract before entering the spinal cord to innervate spinal cord neurons that are contra-

lateral (opposite) to the site of origin of the neurons in the motor cortex. Some of the fibers do not cross and remain ipsilateral to form the anterior corticospinal tract. Some projections from the corticospinal tract directly contact motor neurons, whereas others terminate on interneurons, which in turn project onto the motor neurons. Excitatory input through the corticospinal tract is mediated by the neurotransmitter glutamate.

The corticorubral and rubrospinal tracts are also involved in mediation of voluntary movement. Unlike the corticospinal tract, their route to the spinal cord is indirect. Corticorubral fibers originate in the motor cortex and project onto neurons in the ipsilateral red nucleus in the midbrain. The rubrospinal tract then arises from neurons in the red nucleus. Fibers from this tract cross the midline in the upper brainstem and descend to innervate the contralateral side of the spinal cord.

The corticobulbar tract arises from neurons in the motor cortex. It initially descends with the corticospinal tract, but at the level of the midbrain takes a different pathway to ultimately influence motor nuclei of cranial nerves III (oculomotor); IV (trochlear); V (trigeminal); VI (abducens); VII (facial); IX (glossopharyngeal); X (vagus); XI (accessory); and XII (hypoglossal) (Section 11.01.6.2.2) throughout the midbrain, pons, and medulla. Thus, many muscles of the face are controlled by this tract. Most of the innervation to the motor nuclei of the cranial nerves is bilateral.

The corticotectal tract originates from cortical neurons in the occipital and inferior parietal lobes and sends projections to the brain stem. This tract is predominately involved with reflex movement of the eyes and head.

The reticulospinal tract originates from neurons in the brainstem reticular formation (found in the pons and medulla). The reticular formation receives a major input from the cortex via the corticoreticular tract. Since it also receives input from other brain regions such as the basal ganglia, red nucleus and substantia nigra, the reticulospinal tract serves as an alternate route by which spinal motor neurons are controlled. This tract also conveys autonomic information which helps influence respiration, circulation, sweating and shivering, pupil dilation, as well as sphincter muscle control of the gastrointestinal and urinary systems.

Whilst the sensory cortex exerts much influence over the output of the motor cortex, two other brain regions also greatly influence movement. The cerebellum and basal ganglia communicate with the cerebral cortex and brain stem and are intimately involved with motor function. These brain areas will be discussed in more detail below. The pathways from brain to spinal cord which control motor movement are divided into the pyramidal and the extrapyramidal systems. The pyramidal system consists of the corticospinal and corticobulbar tracts. All other systems are referred to as extrapyramidal.

(ii) Basal ganglia

The subcortical basal ganglia consists of the caudate, putamen, and globus pallidus (GP). The substantia nigra (SN) and subthalamic nucleus are closely interconnected midbrain structures and are oftentimes included as part of the basal ganglia. The SN is further divided into the substantia nigra pars compacta (SN_{pc}) region which contains dopaminergic neurons and the substantia nigra pars reticulata (SN_{pr}) which contains GABAergic neurons. The striatum refers to the caudate and putamen and is derived from the same telecephalic structure. The term corpus striatum includes the GP, a diencephalic structure, as well as the caudate and putamen. The GP is divided into an internal and external segment. The subcortical basal ganglia structures and their anatomical relationships can be seen in a cross-section through the brain as shown in Figure 6.[10]

The connections between the basal ganglia structures and cortex comprise four major circuits.

Circuit 1:

$$Cortex \rightarrow striatum \rightarrow GP \rightarrow thalamus \rightarrow cortex$$

This is the major circuit by which the basal ganglia can influence motor neurons. Conversely, the sensorimotor cortex can influence the activity of the basal ganglia. Sensorimotor input from the cortex is received by the caudate and putamen. Projections from the striatum are sent to the GP, GP to thalamus and then back to the supplementary motor and premotor cortex.

Circuit 2:

$$Cortex \rightarrow striatum \rightarrow [external\ segment\text{-}GP$$
$$\rightarrow subthalamic\ nucleus$$
$$\rightarrow internal\ segment\text{-}GP] \rightarrow thalamus \rightarrow cortex$$

This pathway provides an alternate route through the basal ganglia and interconnects the subthalamic nucleus with the GP.

Circuit 3:

$$[SN_{pr} \rightarrow thalamus \rightarrow cortex]$$
$$\uparrow$$
$$Cortex \rightarrow striatum$$
$$\uparrow$$
$$[SN_{pc}]$$

In this circuit, the SN is interconnected with the striatum. The striatum sends input to the SN_{pr}, which in turn communicates with the thalamus and cortex. The SN_{pc} send dopaminergic projections to the striatum, which influences the outflow of the striatum to the GP as indicated in circuit 1.

Circuit 4:

$$Cortex \rightarrow striatum \rightarrow GP \rightarrow [thalamus \rightarrow striatum]$$

This loop provides feedback from the thalamus to the striatum.

Lesions or nerve cell loss in any of the basal ganglia structures can lead to a variety of movement disorders. In Parkinson's disease, for example, the major finding is the loss of the dopamine neurons in the SN_{pc}. This leads to akinesia (difficulty in initiating movement), rigidity and tremor. Huntington's disease results from the loss of specific neuronal populations (mainly GABA projection neurons) in the striatum, particularly in the caudate nucleus. This produces chorea, uncontrolled flexion–extension movement of the muscles,

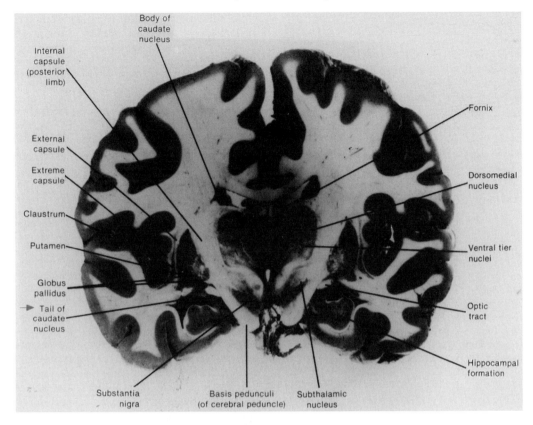

Internal capsule (posterior limb)

Body of caudate nucleus

Fornix

External capsule

Extreme capsule

Dorsomedial nucleus

Claustrum

Putamen

Ventral tier nuclei

Globus pallidus

Tail of caudate nucleus

Optic tract

Substantia nigra

Basis pedunculi (of cerebral peduncle)

Subthalamic nucleus

Hippocampal formation

Figure 6 A transverse or cross-section through the brain at approximately the level of the central sulcus. This section is at a right angle to the one shown in Figure 5. Basal ganglia structures are present and can be seen in anatomical relationship to thalamus, hippocampus and overlying cerebral cortex (reproduced by permission of Mosby Year Book from 'The Human Brain: An Introduction to its Functional Anatomy,' 3rd edn., 1993).

and dementia. Ballism or continuous wild flinging motions of the limbs is due to damage in the subthalamic nucleus. The basal ganglion structures, along with the hippocampus and certain cortical layers are extremely sensitive to metabolic poisons such as rotenone, azide, cyanide, and carbon monoxide.

In addition to the motor functions assigned to the basal ganglia, the ventral part of the striatum, the nucleus accumbens, plays a role in behaviors associated with the limbic system such as aggression and sexual behavior. This area is also thought to play a role in behavior associated with drug addiction. The nucleus accumbens receives input from the limbic and temporal lobes and from dopaminergic neurons in a midbrain region just adjacent and medial to the SN, known as the ventral tegmental area (VTA). The output from the nucleus accumbens is mainly to the ventral pallidum, an area adjacent to the GP.

(iii) Limbic system

The limbic lobe, containing the cingulate and parahippocampal gyri (Figure 5), along with

the amygdala, hippocampus and septal area (Figure 7), comprise the limbic system. This system functions in motivationally driven and emotional behaviors. The limbic lobe has numerous connections with the hypothalamus which serve to maintain homeostasis. The hypothalamus will be described in more detail in a later section. The hippocampus is an infolding of the cortex into the temporal lobe. It interconnects the association areas of the cortex via the cingulate and parahippocampal gyri with the hypothalamus and septal area and eventually, the brain stem and spinal cord (Figure 8 (upper panel)). The hypothalamus in turn, feeds back information to the cerebral cortex via the anterior thalamus. The hippocampal formation is connected to other brain regions via two pathways: the fornix and the perforant path. Each pathway contains input and output (afferent and efferent) fibers. Interconnections with the cortical areas occur through the perforant pathway, while the fornix serves as the relay between the hippocampus and hypothalamus. The hippocampus receives highly processed sensory information from the cerebral cortex and appears to play an important role in

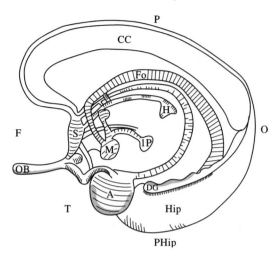

Figure 7 Limbic and associated structures: the corpus callosum (CC) is shown in both this figure and in Figure 5. When superimposed, the anatomical relationship between the limbic system and other cortical lobes can be discerned. Abbreviations: F, frontal lobe; P, parietal lobe; O, occipital lobe; PHip, parahippocampal gyrus of limbic lobe; T, temporal lobe; OB, olfactory bulb; S, septum; M mammillary body; CC, corpus callosum; Fo, fornix; H, habenula; IP, interpenduncular nucleus; DG, dentate gyrus; Hip, hippocampus, A, amygdala.

spatial and temporal awareness. Although not thoroughly understood, many studies indicate that the hippocampus is a major brain area involved in memory and learning.

At the anterior end of the hippocampus is a collection of neurons termed the amygdala (Figure 7). It is subdivided into the cortico-medial amygdala which is reciprocally connected with the olfactory system, hypothalamus and brain stem, and the basolateral amygdala which receives processed information from cortical association areas via the limbic lobe (Figure 8 (lower panel)). The exact functions of the amygdala are still under study. Experimental evidence suggests that it is important in coordinating emotional and motivational responses to external stimuli and may also serve a role in memory. The amygdala is also involved in regulating autonomic, neuroendocrine and immune functions.

11.01.5.3.2 *Diencephalon: thalamus and hypothalamus*

The diencephalon is a mass of neural tissue that lies below the cerebral hemispheres and in front of (rostral) the midbrain. The third ventricle (Figure 6) partly separates the right and left halves of the diencephalon. The major components of the diencephalon are the thalamus and the hypothalamus. In addition, the epithalamus, made up of the pineal gland (Figure 5) and habenula (Figure 7), and the subthalamus, made up of the subthalamic nucleus and zona incerta (Figure 6) are part of the diencephalon.

Hippocampus

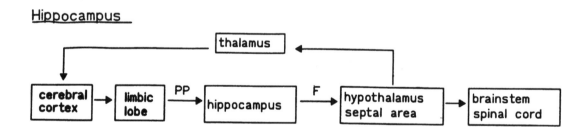

Amygdala

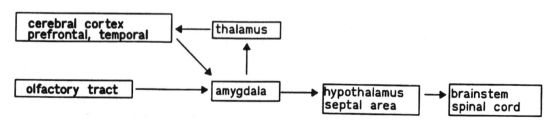

Figure 8 Schematic of the major communication pathways between the hippocampus, and amygdala, and other brain areas (PP, perforant path; F, fornix).

The thalamus is an egg-shaped structure (Figure 6). It has been termed the "gateway" to the cerebral cortex, since afferents from specific sensory, motor and limbic pathways relay onto neurons in the thalamus before going to various cortical areas. The projections from thalamus to cerebral cortex travel in the internal capsule, a fan-like arrangement of fibers coursing to and from all areas of the cerebral cortex. Decending tracts such as the corticospinal tract also travel through the internal capsule. The thalamus is divided into six major nuclear groups: lateral, medial, anterior, intralaminar, midline, and reticular. In addition, the lateral nuclear mass is further subdivided into ventral and dorsal tiers with further subdivisions of each tier based on position. Protrusions from the thalamus, known as the lateral and medial geniculate nuclei, receive information from the visual and auditory systems, respectively. The afferent and efferent connections of the various thalamic nuclei are beyond the scope of this chapter and the reader is referred to any primary neuroanatomy text for further reading. The thalamic nuclei receive input from all major sensory systems (vision, olfaction, taste, hearing, and sensation) as well as from the hypothalamus, amygdala, globus pallidus, inferior colliculus, superior colliculus, and cerebellum. The thalamus is thus an extremely important structure for information processing that influences limbic, motor and sensory modalities.

The hypothalamus rests below the thalamus and is adjacent to the floor and inferior lateral wall of the third ventricle. Regulation of autonomic function and drive-related (motivational) behavior are essential functions of this brain structure. The hypothalamus, like the thalamus, is subdivided into numerous nuclei which fall into three main regions: the supraoptic or anterior region, the tuberal or middle region and the mammillary or posterior region. The optic chiasm and several optic nuclei are found in the anterior region. The tuberal region contains the *tuber cinereum* and infundibulum or stalk of the pituitary gland. The posterior region contains the mammillary bodies. Connections with the hypothalamus fall into three functional groupings. First, interconnections with the limbic system and second, with various visceral and somatic motor and sensory nuclei in the brain stem and spinal cord are important to the role of the hypothalamus in autonomic (sympathetic and parasympathetic), emotional and somatic function. In this regard, the hypothalamus is an important center for controlling behaviors related to feeding, drinking, temperature regulation, gut motility, and sexual activity. The third functional grouping of connections include the input of the hypothalamus to the pituitary. Neurons in the supraoptic region of the hypothalamus release vasopressin (an antidiuretic hormone) and oxytocin (which causes contraction of uterine and mammary smooth muscle) into the posterior lobe of the pituitary where they then enter the circulation after passage through blood capillaries. Neurons in the *tuber cinereum* provide neuropeptides to the anterior lobe of the pituitary where they act to control the production of anterior pituitary hormones such as thyrotropin, growth hormone, luteinizing hormone, follicle stimulating hormone, and prolactin.

11.01.5.3.3 Brain stem: midbrain, medulla, and pons

The brain stem subserves the sensory and motor modalities of the head. It contains the lower motor neurons for the muscles of the head and receives sensory input from this region. It also serves as a pathway for most of the ascending and descending fiber tracts between the spinal cord and the brain. The brain stem is the location of nuclear complexes that comprise cranial nerves III through XII (Section 11.01.6.2.2). The midbrain segment of the brain stem contains nuclei important for generalized motor control. A fourth major activity of the brain stem involves integrative activity that controls respiration, cardiovascular activity and regulation of the level of consciousness.

The midbrain, pons and medulla form the brain stem (Figure 5). Each region is anatomically divided into a rostral (front) and caudal (rear) segment. The reticular formation is a diffusely organized series of nuclei and tracts that forms the central core of the three brain stem regions. The reticular formation is involved in motor, sensory and visceral control of processes such as respiration and cardiovascular activity and control of consciousness. It is also the origin of the reticulospinal tract (Section 11.01.5.3.1(i)).

The midbrain forms the segment between the diencephalon and the pons. Some of the major components of the rostral midbrain are the superior colliculus (Figure 5), important in eye movement and visual attention, the red nucleus and the substantia nigra (Figure 6), important in control of motor movement (Section 11.01.5.3.1(ii)). The cerebral peduncles (Figure 6) are two fiber bundles containing, (among others), the descending corticospinal, corticobulbar, and corticopontine fibers. The medial lemniscus ascends through all regions of the brain stem and contains fibers from the contralateral neurons in the dorsal spinal cord which receive sensory inputs from the periphery. The mass of neurons forming the third cranial nerve (occulomotor) lie in the midportion of the rostral midbrain. In the caudal midbrain, the lateral lemniscus, which contains sensory fibers involved in hearing, ends in the inferior colliculus (Figure 5). The fourth cranial nerve (trochlear) is located in the midregion of the caudal midbrain. The cerebral peduncles continue to form the anterior surface of the midbrain. The superior cerebellar peduncle is the major outflow from the cerebellum to the thalamus and red nucleus. Fibers from the superior cerebellar peduncle surround the midline, and begin to cross over or decussate at this level.

The pons is made up of a central core, the reticular formation, containing the serotonergic raphe nuclei which allow communication between the cerebellum and the cerebral cortex and the noradrenergic locus ceruleus, which projects to virtually all areas of the cerebral cortex and spinal cord. The pons is attached to the cerebellum at the superior and middle cerebellar peduncles (Section 11.01.5.3.4). The nucleus of the fifth cranial nerve (trigeminal) is located in the rostral pons. Nuclear complexes of the VI, VII, and VIII cranial nerves are in the posterior pons at the pontomedullary junction. The core region of the pons is surrounded by several peduncles.

The medulla oblongata is the location of the nuclear complexes for cranial nerves IX, X, XI, and XII. Like the midbrain and pons, the medulla contains a reticular core surrounded by fiber tracts from ascending and descending pathways, and communication pathways to and from the cerebellum. The caudal medulla is the site of the decussation (crossing) of fibers in the corticospinal tract (the decussation of the pyramids), and fibers in the medial lemniscus. The lateral reticular nucleus and inferior olivary nuclear complex in the medulla are important sources of input to the cerebellum via the inferior cerebellar peduncle. The lateral reticular nucleus receives input from the spinal cord, cranial nerves and cerebral cortex, which it relays to the cerebellum. The neurons in the inferior olive are the source of the climbing fibers in the cerebellum.

11.01.5.3.4   Cerebellum

The cerebellum is a semiattached mass of neural tissue that covers the posterior surface of the brainstem (Figure 5). It functions to coordinate motor movement, maintain posture and balance, help execute fine motor skills such as writing, and aids in refining certain properties of movement such as trajectory, velocity and acceleration. Damage to the cerebellum generally results in clumsy, slow movement.

When examined histologically, the cerebellar cortex consists of three layers. The molecular layer contains the stellate and basket cell neurons; the Purkinje layer contains the neuronal cell bodies of the Purkinje cells and, a third, innermost granule cell layer contains granule and Golgi type II neurons.

The cerebellum is attached to the brainstem via three peduncles: superior, middle, and inferior. The superior cerebellar peduncle carries most of the outflow of the cerebellum, whereas the middle and inferior peduncles carry most of the input. The outflow of the cerebellum via the superior cerebellar peduncle projects to the red nucleus (midbrain), thalamus (diencephalon), the inferior olivary nucleus and the reticular formation of the medulla. From these foci, information can be relayed to the motor cortex and spinal cord to facilitate motor adjustments that result in fine motor skills. The outflow of the cerebellum is, in turn, modified by the input to the cerebellum. There are two major sources of afferents to the cerebellum: (i) from climbing fibers, which originate from neurons in the inferior olivary nucleus in the medulla and project to the Purkinje cells, and (ii) from mossy fibers, which originate from several brain regions (spinal cord, pons, cranial nerves V and VIII, reticular formation nuclei, and deep cerebellar nuclei) and synapse on granule cells. Mossy and climbing fibers, and granule cells, are glutamatergic and, therefore, are excitatory. Golgi, basket, and Purkinje cells are GABAergic and convey inhibitory information. Information leaving the cerebellum occurs through the Purkinje cell which first synapses onto deep cerebellar nuclei. The one exception to this is the direct projection of Purkinje cells to vestibular nuclei (VIII cranial nerve).

11.01.5.3.5   Spinal cord

The terms posterior and dorsal, as well as anterior and ventral are used interchangeably when describing the spinal cord. The spinal cord begins at its junction with the medulla and extends the length of the spinal column to the first lumbar vertebra. The cord is surrounded by a continuation of the three membranes or meninges that surround the brain: pia, arachnoid, and dura mater. The subarachnoid space is found between the arachnoid and pia mater and contains CSF. When cut in crosssection, the spinal cord has a central, butterflyshaped region of gray matter (where most of the neuronal cell bodies are found), which is surrounded by white matter (predominately nerve fibers) (Figure 9). There are two dorsal and two ventral enlargements of the gray matter known as the dorsal and ventral horns. The cord is divided into a right and left half by a dorsal and ventral median fissure. The central canal of the cord is filled with CSF and is continuous with the fourth ventricle. The spinal cord is grouped into segments referred to as cervical, thoracic, lumbar, and sacral and correspond with the attached spinal nerves (Section 11.01.6.2.1).

The dorsal or posterior horn is the receptor portion of the spinal cord gray matter, its neurons receiving sensory input from the periphery via the dorsal roots and associated dorsal root or sensory ganglia (Figure 9). Projections from the neurons in the dorsal horn of the cord give rise to tracts (spinothalamic, spinocerebellar, spino-olivary, spinotectal) which project to various parts of the brain, conveying sensory information from the periphery. Each dorsal horn is capped by a specialized area of neurons call the substantia gelatinosa of Rolando. Fibers from these neurons carry pain and temperature information. The anterior or ventral horn contains the cells of origin (anterior motor neurons) of the ventral root. Axons from anterior horn cells

(motor neurons) project directly to the skeletal muscles. The lateral columns which are found in the intermediate area of gray matter contain preganglionic cells for the autonomic nervous system (Section 11.01.6.2.3) and also exit the cord via the ventral root. Disruption of the anterior motor neurons (lower motor neurons) results in a flaccid paralysis (limp, weak muscle tension and inability to move) since the muscles innervated by these neurons slowly atrophy (waste away). The cord is also the center for many monosynaptic reflex actions such as the knee jerk reflex seen in response to tapping the patellar tendon. For this reflex to occur, sensory information must reach the dorsal horn via the sensory afferents and dorsal root ganglia. Projections from the dorsal root ganglia synapse directly on motor neurons in the ventral horn which in turn project to leg muscles. Most other polysynaptic reflexes involve spinal cord interneurons which relay information to the motor neurons in the anterior horn.

In addition to the numerous sensory tracts ascending from spinal cord to the brain, the cord white matter contains the pathways descending from the brain (corticospinal, reticulospinal, etc.) (Section 11.01.5.3.1) to the cord. Fibers from these descending tracts synapse on motor neurons either directly or

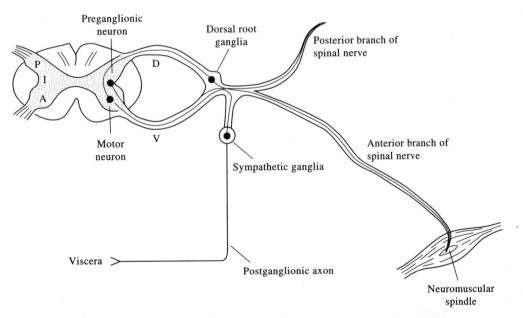

Figure 9 Cross-section through the spinal cord. Afferent sensory input is received by neurons in the dorsal root ganglia which send projections to cells in the posterior (P) or dorsal horn via the dorsal root (D). The ventral root (V) contains efferent projections from motor neurons in the anterior (A) or ventral horn and preganglionic fibers from neurons in the intermediate gray zone (I). Monosynaptic reflexes such as the knee jerk involve two neurons: a dorsal root or sensory neuron which receives stimuli from the periphery and directly projects onto a motor neuron in the anterior horn. The axonal projection from the motor neuron ends in a neuromuscular spindle on the muscle. Polysynaptic reflexes involve spinal cord interneurons which relay information from sensory neurons to the motor neurons.

via interneurons. Disruption of these upper motor neuron pathways results in a spastic paralysis (increased muscle tension and inability to move). Distinction between flaccid and spastic paralysis can aid in locating the site of nerve cell damage. Amyotrophic lateral sclerosis (ALS or Lou Gerhig's disease) usually involves loss of both upper and lower motor neurons and the corticospinal tract.

11.01.6   PERIPHERAL NERVOUS SYSTEM: ANATOMY AND FUNCTION

11.01.6.1   Introduction

The peripheral nervous system serves as a mediator between the CNS and all body tissues and organs. Peripheral nerves fall into three groupings: spinal, cranial, and autonomic. When discussing the PNS, the term somatic refers to the body wall; visceral refers to the various body organs. Afferent fibers are sensory fibers that convey information towards the CNS. Efferent fibers carry information (mostly motor) to the periphery.

11.01.6.2   Gross Anatomy and Function

11.01.6.2.1   *Spinal nerves*

The spinal nerves consist of 31 symmetrical pairs of nerves which connect the spinal cord to the periphery. There are 8 cervical, 12 thoracic, 5 lumbar, 5 sacral, and 1 coccygeal nerves. The anatomical relationship between the spinal cord, spinal nerves and autonomic sympathetic ganglia is shown in Figure 10. Each spinal nerve has dorsal and ventral roots. Dorsal roots contain the sensory fibers from neurons in the dorsal root ganglia. Fibers from neurons in the dorsal root ganglia convey information from both peripheral visceral and somatic structures to neurons in the dorsal gray horn of the spinal cord. Many ascending tracts are then formed to convey sensory information to the brain (Section 11.01.5.3.5). The ventral root of the spinal nerve contains the motor fibers of anterior motor neurons in the ventral horn and pass to the skeletal muscles. In the thoracic and lumbar regions of the cord, the ventral root also contains preganglionic sympathetic fibers from the intermediate lateral cord to the autonomic sympathetic ganglia. In the sacral region, the ventral root carries preganglionic parasympathetic fibers from neurons in the intermediate

gray of the cord to parasympathetic ganglia that innervate the pelvic region (Section 11.01.6.2.3).

11.01.6.2.2   *Cranial nerves*

There are 12 pairs of cranial nerves designated by Roman numerals I through XII. Cranial nerves III through XII have their nuclear origin in the brainstem (Section 11.01.5.3.3). Cranial nerves I and II have their cells of origin in the nose and eye, respectively. A summary of the 12 cranial nerves, their sites of origin, major destination(s) and functions are given in Table 3. Note that cranial nerves II, VII, IX, and X contain preganglionic parasympathetic fibers which contact postganglionic parasympathetic neurons of the autonomic nervous system (Section 11.01.6.2.3).

11.01.6.2.3   *Autonomic nervous system*

Many bodily functions, such as glandular secretion, cardiac muscle contraction, digestion and sweating, do not require conscious thought. They occur automatically and are controlled by the autonomic nervous system. Unlike the somatic motor neurons in the spinal cord which contact their effector skeletal muscle without synaptic interruption, the autonomic innervation to the viscera involves a two neuron relay. A preganglionic nerve cell sends out an axon which makes contact with a postganglionic neuron. The postganglionic neuron then directly contacts the muscle or gland. The autonomic nervous system is subdivided into sympathetic and parasympathetic. These divisions are distinguished by the sites of origin of their pre- and postganglionic neurons and the neurotransmitter they release.

The sympathetic nervous system consists of preganglionic neurons whose cell bodies are in the intermediate gray zone in the thoracic and first three lumbar segments of the spinal cord. The preganglionic fibers terminate on postganglionic neurons whose cell bodies are found in paravertebral chains of sympathetic ganglia. The sympathetic chain forms a bead-like structure that runs parallel to the spinal cord (Figure 10). The neurotransmitter of all preganglionic fibers is acetylcholine. Postganglionic sympathetic projections contact the viscera as shown in Figure 10. The neurotransmitter of postganglionic fibers is norepinephrine with the exception of fibers that innervate the sweat glands and erector pili (hair) muscles.

Sympathetic nervous system activation is associated with situations in which energy is

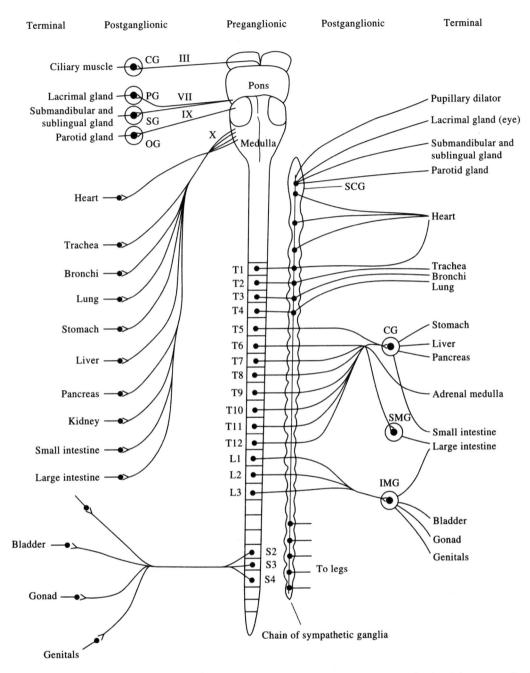

PARASYMPATHETIC

SYMPATHETIC

Figure 10 The outflow of parasympathetic (left) and sympathetic (right) preganglionic and postganglionic fibers and the viscera they innervate. Abbreviations: (Parasympathetic), CG, ciliary ganglion; OG, otic ganglion; PG, pterygopalatine ganglion; SG, submandibular ganglion; (Sympathetic),CG, celiac ganglion; IMG, inferior mesenteric ganglion; SCG, superior cervical ganglion; SMG, superior mesenteric ganglion.

needed. Activation of the sympathetic system increases heart rate, produces vasoconstriction to increase blood flow to the brain, decreases gut motility, reduces secretion from facial glands (salivary, parotid, submaxillary, submandibular); it also mediates release of epinephrine and norepinephrine from the adrenal medulla. The visceral responses elicited by activation of the sympathetic nervous system prepares an organism for "flight or fight." Its effects are widespread and long lasting.

The parasympathetic nervous system has preganglionic cell bodies in the nuclei of cranial nerves III, VII, IX, and X (located in the

Table 3   Cranial nerves.

No.	Name	Origin of cell body	Major destinations	Functions
I	Olfactory sensory	Olfactory epithelium	Olfactory bulb, primary olfactory cortex, pyriform and entorhinal cortex, amygdala, hippocampus	Smell, chemosensory control of social behaviors (aggression, sexual, maternal)
II	Optic sensory	Ganglion cells of the retina	Superior colliculus, pretectal area, lateral geniculate nucleus (thalamus), visual cortex (occipital lobe)	Vision, light reflexes, involuntary oculoskeletal reflexes
III	Occulomotor mixed motor parasympathetic	Rostral midbrain	Extraocular eye muscles ciliary ganglion	Movement of eye muscles, accommodation and constriction of the pupil
IV	Trochlear motor	Caudal midbrain	Superior oblique eye muscle	Moves eye down when turned medially (towards the nose)
V	Trigeminal mixed motor sensory	Midbrain/rostral pons	Sensory and motor modalities of the face, three major divisions: ophthalmic nerve maxillary nerve mandibular nerve	Jaw muscles for chewing pain and temperature for face, nose and mouth tactile sensation from jaw muscles and tooth sockets
VI	Abducens	Pons	Lateral rectus muscle of eye	Turns eye outward, past the midline
VII	Facial mixed motor sensory parasympathetic	Caudal pons	Facial muscles and glands	Facial muscles of expression, glands of nose, palate, eye, submaxillary and sublingual glands, taste from anterior tongue
VIII	Vestibular sensory	Pontomedullary junction	Auditory structures: organ of Corti, semicircular canals, maculas of utricle and saccule via 2 major divisions: choclear and vestibular nerves	Hearing spatial orientation, equilibrium auditory reflexes
IX	Glossopharyngeal mixed motor sensory parasympathetic	Rostral medulla	Pharynx, tongue, tonsils, carotid body, otic ganglion	Sensation to pharynx, palate, tonsils, eustachian tube, tympanic (ear) cavity; controls secretions of the parotid gland; reflex control of tongue; taste to posterior tongue
X	Vagus mixed motor sensory parasympathetic	Rostral medulla	Larynx and epiglottis; pharyngeal, cardiac, pulmonary celiac and mesenteric plexi; skin in external auditory canal	Motor innervation of pharynx, and soft palate; sensory and motor to larynx; parasympathetic control of thorax and abdominal viscera; vagal reflexes: salivation, coughing, gagging, vomiting carotid sinus
XI	Accessory motor	Caudal medulla	Inner larynx, sternomastoid and trapezius muscles	Motor to larynx, side of neck and back muscles
XII	Hypoglossal motor	Caudal medulla	Geneoglossus muscle on side of tongue	Causes tongue to deviate to opposite side when protruded

brainstem), and in the intermediate gray zone of segments 2, 3, and 4 of the sacral spinal cord (Figure 10). Preganglionic parasympathetic neurons use acetylcholine as their neurotransmitter. In contrast with the sympathetic nervous system, postganglionic parasympathetic neurons are found in ganglia that lie close to or embedded in the organs they innervate. In many instances, the parasympathetic innervation to the various organs has physiological effects opposite to those elicited by the sympathetic innervation. For instance, stimulation of the parasympathetic system results in a decreased heart rate, blood vessel dilation, watery salivary secretions, and increased gut motility. The two systems, however, do not always act in opposition. The bladder, pupil, and penile erection are predominately under parasympathetic control, whereas sweat glands and ejaculation are solely under sympathetic control.

11.01.7 SENSORY SYSTEMS

Sensory receptors evolved to allow us to sample our environment. Receptors are mostly specialized cells which convey their information via afferent neurons, modified nerve fibers or nerve fiber endings. Light, stretch, temperature, sound waves and so on, cause sensory receptors to generate a receptor potential (Section 11.01.3.3). This initiates a chain of events which evokes a depolarization impulse (action potential) in the afferent axon. The signal thus passes from receptor to neurons and transmits information from the periphery to the CNS. A general rule of sensory transduction is that information is relayed through the thalamus before reaching the cerebral cortex. One exception to this is processing of olfactory information. The sensory systems can be the recipient of the action of neurotoxicants (see Chapters 14 and 15, this volume).

11.01.7.1 Vision

The receptors for vision are the rods and cones located in the retina. Rods function mainly for night vision and do not convey any color information. Cones are used for day vision. Three types of cones: red, yellow, and blue-sensitive transmit information about the color of an object. The signal from the rods and cones passes via several orders of retinal neurons to the retinal ganglion cells. Axons of the retinal ganglion cells in turn form the optic nerve which synapses in the superior colliculus and thalamus. The visual information is then passed to the visual center in the occipital lobe of the cerebral cortex.

11.01.7.2 Olfaction

The receptors for smell are found in the mucous membranes of the nose (see also Chapter 15, this volume). Olfactory receptor neurons send fine, chemosensitive projections (dendrites) into the nasal mucosa. The axons of the olfactory receptor neurons form the olfactory nerve which terminates in the olfactory bulb. From there, olfactory input is communicated to the olfactory cortex and amygdala. The olfactory receptor neurons are continuously replaced throughout life.

11.01.7.3 Taste

Gustatory receptors are clusters of specialized epithelial cells embedded in the tongue. A cluster of approximately 50 receptors comprises a taste bud. These receptor cells are renewed approximately every 10 days. Taste buds bind the specific chemical components of food dissolved in saliva. Four basic tastes can be distinguished: bitter, salty, sour, and sweet. However, through activation of multiple receptor types in combination with olfaction, many more taste sensations can be perceived. Taste receptors are innervated by branches of cranial nerves VII, IX, and X. Chemical ligand binding to taste receptors elicits depolarization of the afferent projection of the cranial nerve. Taste information is sent to the brain stem (solitary nucleus) and thalamus before being relayed to the cerebral cortex.

11.01.7.4 Hearing

Special receptors for audition, located in the cochlea of the ear, are called hair cells (see Chapter 14, this volume). Fine hair-like projections from the receptors are sensitive to vibrations in the fluid surrounding them. Movement of the hair cells is transmitted via the eighth cranial nerve to the cochlear nucleus (origin of the cell bodies of cranial nerve VIII), at the pontomedullary junction. From there, information is transmitted to the superior olivary nucleus, inferior colliculus, thalamus, and auditory cortex. An important component of both audition and vision is converting sensory information into a conscious perception of the stimulus. The vestibular, visual, and somatosensory systems cooperate to provide information about position and movement, and help maintain orientation with respect to gravity.

11.01.7.5 Sensation

Touch, position (proprioception), pain, and temperature represent somatic sensation. Specialized receptors and nerve endings in the muscles and skin transduce sensory signals from the periphery to the spinal cord via the dorsal root ganglia. Neurons in the dorsal horn of the spinal cord send information to somatosensory areas of the parietal lobe of the cerebral cortex via the medial lemniscus, after being relayed through the thalamus. Some afferents project to the cerebellum and motor cortex, while others indirectly contact the autonomic nervous system. Unconscious responses to adjust posture, muscle tone, body temperature, and blood pressure can then be made in response to the environment.

11.01.8 CONCLUDING REMARKS

As a discipline, toxicology concerns itself with the study of the adverse effects of chemical agents on various biological systems. Its contribution to science, however, far exceeds this limited definition. Understanding the mechanism of action and relative selectivity of neurotoxic agents requires a detailed understanding of the antatomy and function of the nervous system. Knowledge in the 1990s is by no means complete. Exploration of the mechanism of action of neurotoxicants has contributed significantly to the scope of knowledge of nervous system structure and function. Agents known to cause distruction of specific cellular elements or loss of specific groups of neurons have been used to gain insight into neuroanatomy and physiology. Neurotoxicants have been employed to generate animal models with which to study disease states. Findings from such studies have contributed to the fields of neurology and medicine. Study of neurotoxicology has produced agents to selectively reduce insects and weeds resulting in increased food production. Risk assessment and the development of safety measures remain the domain of all toxicology and serve to protect the public from potentially harmful exposure to toxic agents. Many examples of agents that effect the nervous system, their mechanism and site of action, and the contribution from studies of neurotoxicants to the fields of neuroanatomy, physiology, medicine, pharmacology, and public health can be found throughout the remaining chapters of this text.

11.01.9 REFERENCES

1. J. G. Chusid, 'Correlative Neuroanatomy & Functional Neurology,' 14th edn., Lange, Los Altos, CA, 1970.
2. S. Gilman and S. W. Newman (eds.), 'Manter and Gatz Essentials of Clinical Neuroanatomy and Neurophysiology,' 8th edn., Davis, Philadelphia, PA, 1987.
3. E. R. Kandel, J. H. Schwartz and T. M. Jessell, 'Principles of Neural Science,' 3rd edn., Elsevier, Amsterdam, 1991.
4. J. Nolte, 'The Human Brain: an Introduction to its Functional Anatomy,' 3rd edn., Mosby Year Book, St. Louis, MO, 1993.
5. M. B. Carpenter, O. S. Strong and R. C. Truex, 'Human Neuroanatomy,' 7th edn., Williams & Wilkins, Baltimore, MD, 1976.
6. A. C. Guyton, 'Textbook of Medical Physiology,' 8th edn., W. B. Saunders, Harcourt Brace Jovanovich, Philadelphia, PA, 1991.
7. J. F. Stein, 'An Introduction to Neurophysiology,' Blackwell Scientific Publications, London, 1982.
8. G. J. Siegel, B. W. Agranoff, R. W. Albers *et al.* (eds.), 'Basic Neurochemistry: Molecular, Cellular, and Medical Aspects,' 5th edn., Raven Press, New York, 1994.
9. D. B. Calne (ed.), 'Neurodegenerative Diseases,' W. B. Saunders, Philadelphia, PA, 1994.
10. J. Nolte, 'A Study Guide' to accompany 'The Human Brain,' 3rd edn., Mosby Year Book, St. Louis, MO, 1993.

SECTION II

11.02

Selective Vulnerability in the Nervous System

THEODORE A. SARAFIAN and M. ANTHONY VERITY
University of California, Los Angeles, CA, USA

11.02.1 INTRODUCTION

The extraordinary complexity of the mammalian nervous system is manifested in the heterogeneity of cell types and the extensive biochemical and morphologic variability that exists within each cell type, most notably neurons. For this reason, toxic agents which affect the nervous system often do so in a highly selective manner, damaging specific cell types in specific brain regions.

The concept of selective vulnerability in the central nervous system was originally proposed by Spielmeyer in 1925[1] and by Vogt and Vogt in 1937[2] to explain the vulnerability of different nerve cell populations to hypoxia and global ischemia. The phenomenon has since been extended to include cellular variability in response to toxicants. While Spielmeyer postulated that regional cellular vulnerability could be explained on the basis of a vascular factor, the Vogts proposed that intrinsic differences in cellular metabolism underlay the selective cellular response to intoxication, a process termed pathoclisis.

Examples of pathoclisis originally presented included the selective bilateral necrosis of the globus pallidus in carbon monoxide poisoning, the primary sensitivity of retinal neurons to methyl alcohol, the unique pattern of petechial hemorrhage in phosgene poisoning affecting the corpus callosum and hemispheric white matter, and the unusual specificity of anterior horn cell injury in poliomyelitis. Since the 1930s, the concept of selective vulnerability has been broadened to embrace the identification and pathogenesis of numerous neurodegenerative disorders, characterized by well-defined anatomic and system degeneration within different hierarchal divisions of the central and peripheral nervous systems.

In this chapter the well-characterized patterns of cell destruction observed in and responsible for several of the more well-known neurodegenerative disorders are briefly reviewed. The etiologic underpinnings of these disorders have been frustratingly slow in their disclosure. However a great deal of phenomenologic description has been generated. For each disorder, a chemical or environmental agent is identified which reproduces, to some degree, the neurologic characteristics of the disorder. The subsequent section will give an overview of the various biochemical mechanisms believed to be responsible for neuronal cell death. A wide range of initiating events and target sites appear to converge onto a limited number of final common pathways which lead ultimately to neuronal death. These pathways display variable degrees of prevalence in different neuronal cell types. Finally, defense mechanisms which display heterotypic expression in brain cells will be examined and analyzed in terms of their contribution to selective vulnerability.

11.02.2 SELECTIVE VULNERABILITY IN NEURODEGENERATIVE DISEASE AND NEUROTOXIC MODELS

Important perspectives on neuronal vulnerability to toxicants are revealed by examining regional variability of neuronal involvement in various neurologic disorders. Understanding the mechanisms underlying degeneration of specific

neurons in these disorders would allow identification of biochemical targets which contribute to selective neuronal injury. The ability of a given toxicant to disrupt this target could explain or predict the selectivity of neuronal destruction. Furthermore, in many neurologic disorders the possibility of an etiologic role of unknown environmental toxicants has not been ruled out.

Human chronic neurodegenerative diseases such as Alzheimer's, Parkinson's, Huntington's, and amyotrophic lateral sclerosis (ALS) have a number of pathologic characteristics in common.[3] Each is associated with primary neuronal, as opposed to glial, degeneration and with characteristic neuronal inclusions of filamentous cytoskeletal material derived from proteins such as actin, tubulin, tau or neurofilament. Often these cytoskeletal inclusions are heavily ubiquitinated[4] and/or phosphorylated[5] suggesting abnormalities associated with post-translational modification. In Alzheimer's disease, the accumulation of β-amyloid, a 42-amino acid fragment derived from membrane-spanning protein, is also suggestive of a defect in post-translational protein processing since the normal degradation pathway for the precursor protein precludes β-amyloid formation (Figure 1).[6]

11.02.2.1 Parkinson's Disease

Parkinson's disease and the homologous syndrome caused by the chemical 1-methyl-4-phenyl-1,2,3,6-tetrahydropyridine (MPTP) offer excellent opportunities for understanding mechanisms underlying selective vulnerability. Parkinson's disease involves the loss of dopaminergic neurons of the substantia nigra. Melanin-containing pigmented neurons are particularly vulnerable.[7] Sensitive neurons uniquely possess the ability to take up dopamine and the MPTP metabolite MPP$^+$ (1-methyl-4-phyenlpyridinium). Then, through the actions of monoamine oxidase-B (MAO-B) these amines are oxidized with associated production of O_2^-.[8] MPP$^+$ also has the ability to inhibit mitochondrial complex I electron transfer, thus compromising neuronal energy production.[9,10] The compound 6-hydroxydopamine is commonly used to produce Parkinson-like destruction of striatal neurons in animals.[11] Unilateral injection of this compound in rats results in repetitive circular walking behavior. The MAO-B inhibitor, deprenyl, has been shown to prevent both MPP$^+$ and 6-hydroxydopamine toxicity and to prolong lifespan of Parkinson's disease patients.[12] Studies with these toxicants have led to the concept that an endogenous or naturally

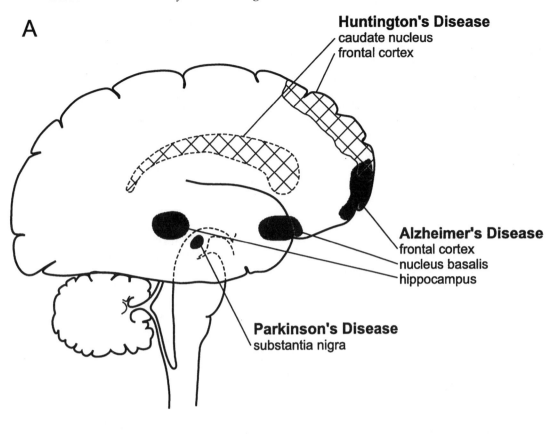

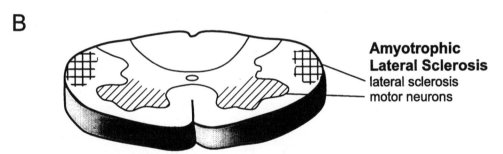

Figure 1 Diagram illustrating regional areas affected in neurodegenerative disorders of the brain (A) and spinal cord (B).

occurring neurotoxin is responsible for Parkinson's disease since no genetic factor contributes significantly to the etiology of this disorder.

11.02.2.2 Huntington's Disease

Huntington's disease is caused by mutational expansion of an unstable trinucleotide repeat (CAG) of a gene on chromosome 4 (4p16.3).[13] This mutation results in loss of neurons within the neostriatum, the caudate nucleus, and to lesser extent, the putamen and frontal cortex. Medial spiny projection neurons are particu-

larly affected, the earliest being the GABA/ enkephalin containing neurons that project to the lateral globus pallidus. The 320 kDa protein coded by this gene has been named huntingtin and its polyglutamine moiety coded by the CAG repeat becomes pathogenic when composed of more than 38 repeat units. Although the protein has been localized to neuronal synaptic vesicles, the basis for selective vulnerability of the medial spiny neurons is currently unknown. Support for an excitotoxic mechanism underlying this vulnerability has come from rat brain injection and infusion studies showing that glutamate, kainate, quinolinic acid and ibotenic acid produced patterns of neuronal destruction similar to that found in Huntington's Disease.[14,15] For example, quinolinic acid reduced levels of GABA and substance P neurotransmitter of medium spiny neurons but did not affect somatostatin (aspiny neurons). Both medium aspiny (NADPH diaphorase-positive) and large aspiny, (acetylcholinester-ase-positive) neurons were unaffected upon histochemical examination. Substantial evidence points to the involvement of the NMDA (*N*-methyl-D-aspartate) receptor as the factor primarily responsible for this selectivity.[16,17] In culture, susceptibility to toxicity correlates temporally and physiologically with expression of the NMDA-receptor but not the kainate or quisqualate-receptor. Postmortem tissue from putamen of HD patients reveals selective depletion of NMDA-type glutamate receptors.[18]

The plant toxin, 3-nitropropionic acid (3-NP), when chronically administered to rats and primates produces selective destruction of striatal neurons and abnormalities of spiny neuron dendrites very similar to those observed in Huntington's disease (see Chapters 30 and 31, this volume).[19] 3-NP acid is an irreversible inhibitor of succinate dehydrogenase (complex II) in the mitochondrial electron transport chain. Systemic administration of 3-NP selectively impairs striatal energy metabolism. Damage to striatal neurons can be prevented by blocking glutamatergic input to the striatum, suggesting again that glutamate receptors play a key role in 3-NP toxicity. It has been suggested that energy depletion of these neurons results in loss of voltage-dependent Mg^{2+} block of the NMDA receptor, with consequent Ca^{2+} entry. Thus selective vulnerability of these neurons may stem from unusually high dependence on succinate dehydrogenase for ATP generation.

11.02.2.3 Amyotrophic Lateral Sclerosis

ALS is a neurodegenerative disorder characterized by progressive, inexorable death of motor neurons and consequent wasting of muscle.[20] The age of onset is variable but usually symptoms present at about age 40. The disorder generally affects both upper and lower motor neurons which project to skeletal muscles. Some motor neurons are spared, including cells in the oculomotor nuclei and spinal neurons innervating bladder sphincter muscles. In the precentral gyrus there is distinct loss of pyramidal cells, especially large Betz cells of cortical layer V. Thus, neuronal susceptibility correlates with perikaryal size in this disorder.

While most cases of ALS occur sporadically, approximately 10% are inherited as an autosomal dominant trait. Of these approximately 50% are associated with mutations in the gene for copper/zinc-superoxide dismutase (SOD I). Many different mutations have been found in different families, associated with variable decrease in SOD1 activity.[21] Early reports suggested that failure to control accumulation of oxygen free radicals (probably O_2^-; see Chapter 3, this volume), was responsible for ALS. Large motor neurons were presumed to be particularly susceptible to O_2^--mediated injury, since all cells expressed the mutant SOD1 in affected individuals. Further study, however, revealed that the degree of loss of enzyme activity with various mutations did not correlate with severity or progression of disease.[22] The amino acid changes corresponding to the mutations occur in areas responsible for structural stability as opposed to the enzyme's active site. Furthermore expression of the mutant forms of SOD1 in cell transfection studies and transgenic mice revealed little or no loss of SOD1 activity in neuronal cells destined to die suggesting that these are gain-of-function mutations.[23,24] Identification of this newly-gained function will likely help identify the basis for the selective destruction of motor neurons.

Chronic inhibition of SOD1 in spinal cord organotypic culture using diethyldithiocarbamate (DEDTC) or antisense oligodeoxynucleotides resulted in apoptotic death of motor neurons and other neuronal types.[25] This toxicity was exacerbated by inhibitors of glutamate uptake. The latter effect is of particular interest since astrocytic glutamate transport is deficient in ALS but not in other neurologic disorders. Inhibition of the GLT-1 (astrocytic) transport system results in selective motor neurotoxicity in organotypic culture and in rats intraventricularly infused with antisense oligonucleotide to GLT-1.[26,27]

Neurofibrillary tangles are found in a number of neurologic disorders including a variant form of ALS known as Guam ALS–Parkinson's Dementia Complex (ALS–PDC). This disease was theorized to be caused by the excitotoxin β-*N*-methylamino-L-alanine (BMAA) found in the cycad seed which the Chamorro natives had used as a food source.[28] Macaques fed high doses of BMAA developed muscle tremors and severe weakness suggesting that BMAA was a slow neurotoxin requiring years of exposure to produce extensive cytotoxicity. This hypothesis has been weakened, however, by subsequent studies revealing that BMAA exposure levels in humans who ate the cycad were many orders of magnitude lower than that producing symptoms in animals.[29] Other toxins such as cycasin and methylazoxymethanol (which causes cerebellar degeneration), which damage DNA, are also present and consumed in large quantities.

Since the 1950s the incidence of Guam ALS–PDC has been declining steadily. More specifically, the pattern of expression has been changing wherein symptoms of ALS do not develop while Parkinsonism and dementia are still manifest. This shift in disease expression and, by inference, neuronal vulnerability has been ascribed to changes in social and dietary habit.

11.02.2.4 Alzheimer's Disease and Cytoskeletal Abnormalities

Alzheimer's disease is characterized by loss of several neuronal cell types, particularly cholinergic neurons from the nucleus basalis. Noradrenergic neurons from the locus ceruleus, dopaminergic neurons from the ventral tegmentum and serotonergic neurons from the raphe nuclei are all consistently and severely damaged.[30–32] Although a wide variety of neurons are lost from the cortex and hippocampus, cholinergic neurons are most prominently affected.[33]

It has been suggested that neurons with long axonal projections ending in amyloid plaque-containing regions are especially vulnerable in Alzheimer's disease.[34] Such neurons produce large amounts of neurofilament and microtubule cytoskeletal protein and the processing of these proteins may be impaired by disruptive events at the nerve ending. Inhibitory interneurons expressing Ca^{2+} binding protein may be particularly susceptible to degeneration.[35] In Alzheimer's disease and normal aging, neurofilament-rich neurons in cortical layer II of entorhinal cortex are highly vulnerable to degeneration. Transgenic mice expressing a human neurofilament gene demonstrate cell type-specific neuropathology similar to Alzheimer's disease.[36] For an unknown reason the human neurofilament protein confers vulnerability to age-related neurofilament tangle formation in specific cell types. Neurofibrillary tangles, composed primarily of hyperphosphorylated components of microtubule and microfilament cytoskeleton, are characteristic features of Alzheimer's disease and Guam ALS–PDC. Filamentous inclusions have also been reported in classic ALS (Figure 2).[37]

Aluminum has been shown to cause structural deformity and aggregation of neurofilaments in cultured neurons, cell-free systems and *in vivo* following injection of aluminum salts.[38,39] In rabbits, intraventricular injection of aluminum chloride caused perikaryal aggregation of neurofilament proteins in subsets of neurons affected in Alzheimer's disease (i.e., cortical pyramidal, basal forebrain cholinergic and upper brainstem catecholaminergic neurons).[40] Aluminum appears to interact with

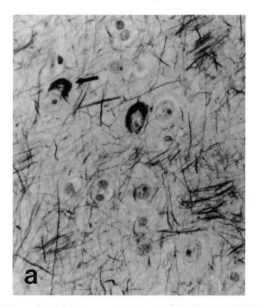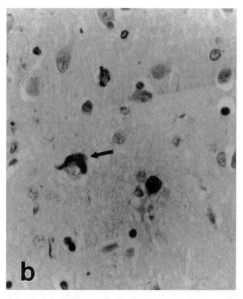

Figure 2 (a) Intraneuronal cytoplasmic neurofibrillary tangles representing pathologically altered neurofilaments, a major constituent of the neuronal cytoskeleton. Cerebral cortex in Alzheimer's disease. Bielschowsky silver impregnation reaction. (b) Immunohistochemical reaction for phosphorylated microtubule-associated tau protein. Such proteins stabilize microtubule assembly and in the hyperphosphorylated state produce abnormal aggregation leading to the paired helical filament. These changes occur with minor modification in experimental aluminum intoxication.

phosphate groups on the heavily phosphory-lated C-terminal sidearms of neurofilament proteins causing intermolecular cross-linking.[41]

Neurofibrillary tangles also contain high levels of aluminum localized in the tangle nuclear region.[42] However, an etiologic role for aluminum has not been established. Motor neurons selectively display formation of inter-woven filamentous bundles in rabbits injected chronically with aluminum maltol.[43] Alumi-num may be another candidate environmental factor in Guam ALS–PDC as high levels of this metal were found in soil and water on Guam and New Guinea. Some clinical and experi-mental studies suggest that zinc may play a role in Alzheimer's disease.[44] *In vitro* experiments demonstrated that zinc can induce amyloid-like plaques when incubated with human (but not rat) β-amyloid peptide in test tubes. If similar effects on amyloid aggregation occur *in vivo* excessive exposure to zinc may damage neurons in regions of high β-amyloid production and may be a contributory factor in Alzheimer's development.

11.02.2.5 Other Neurotoxicants

More than any other tissue, the brain is characterized by extensive regional variability with respect to gross morphology, functional activity and cellular and biochemical composi-tion. Several brain regions are also preferen-tially susceptible to attack by specific groups of toxicants (Figure 3).

11.02.2.5.1 *Cerebellum*

The selective sensitivity of the cerebellum to a variety of toxicants and xenobiotics has been established. Similar patterns of cerebellar gran-ule cell necrosis have been observed with methyl chloride,[45] 2-chloropropionic acid,[46] methyl bromide,[47] and methyl mercury.[48,49] While the mechanistic basis for primary involvement of cerebellar granule cells is unknown, changes in energy potential, glutathione depletion and/or a possible role for excitatory amino acid med-iated cell damage have all been invoked. In

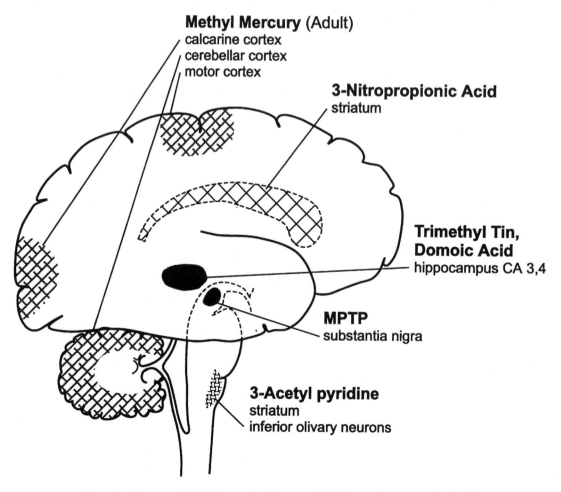

Figure 3 Diagram illustrating regional areas affected by various neurotoxicants.

contrast, Purkinje cells are relatively uninvolved but primary Purkinje cell damage has been demonstrated with 1-methyl cycloheptatriene[50] and following intragastric ethanol during postnatal development of rat.[51]

11.02.2.5.2 Hippocampus

Selective hippocampal neurotoxicity is evident following organotin and organolead administration. Trimethyl tin and triethyl lead induce aggression, hyperactivity and spontaneous seizures indicative of limbic system involvement. Variable involvement of the dentate granule cells and/or pyramidal neurons of Ammon's horn have been described.[52,53] Vulnerability appears dependent upon the functional maturity and integrity of the neuronal circuitry providing support for the concept that aberrant hyperexcitation with defective energy supply provides the mechanistic basis for the selective involvement. Similarly, the shellfish poisoning epidemic identified due to domoic acid produces a similar pattern of disorientation, confusion, neuroexcitation and persistent memory deficit. Human and experimental animals reveal hippocampal lesions limited to the pyramidal neurons of CA3–CA4 similar to those found with triethyl lead and trimethyl tin and analogous to that found with other excitotoxins.[54]

11.02.2.5.3 Myelin

In contrast to trimethyl tin, triethyl tin produces a characteristic myelin edema within the CNS. The accumulation of fluid splits the myelin sheath at the intraperiod line to form vacuoles.[55] Similar myelinotoxic agents producing intramyelinic vacuoles include cuprizone, hexachlorophene, dichloroacetate and ethidium bromide (see Chapter 11, this volume). While different brain regions may be involved, the characteristic pattern of myelinotoxicity defines the pattern of selectivity.

Thus, as with neurologic degenerative disorders and infectious disorders, neurotoxic disorders are often characterized by differences in regional susceptibility. For the vast majority of neurotoxins some form of geographical selective targeting can be demonstrated.

11.02.3 CYTOTOXIC MECHANISMS

Crucial to an understanding of selective vulnerability is the consideration of the specific cytotoxic mechanisms which underly the cellular damage caused by the toxicants. Often these mechanisms reflect a unique or prominent biochemical pathway which characterizes a specific cell type or brain region.

11.02.3.1 Hypoxia and Energy Deprivation Syndromes

Hypoxic/ischemic-induced neuronal injury provides one of the best examples of selective neuronal vulnerability and represents one of the most extensively characterized experimental systems for understanding the biochemical basis of differential[56,57] neurodegeneration. Brief periods of complete global cerebral ischemia occur in humans resuscitated after cardiac arrest. Neurons in the hippocampus, classified as CA1, CA2, CA3, and granule neurons of the dentate gyrus, show remarkably variable sensitivity to transient ischemia. After only a few minutes, CA1 neurons become committed to a delayed neuronal death program although morphologic changes are not apparent for 36–72 h after reperfusion. CA3 neurons are resistant to damage. Moreover, duration between ischemia and cell death is variable. Since immediately adjacent cells may display vastly different sensitivities to global ischemia, local variations in vascular flow cannot explain this selective vulnerability.

Other cellular mechanisms may play a role in the selectivity of hypoxic lesions. The rate, evolution, and cellular neuropathology of the hypoxic lesion ultimately depends upon energy requirement, ion (especially Ca^{2+}) homeostasis, acidosis, production of free radicals and the extracellular clearance of amino acids. Hypoxic/ischemic injury induces a fall in pH in the extracellular space.[58] The role of anaerobic tissue metabolism is revealed from studies of preischemic hyperglycemia which aggravates ischemic damage.[59] Moreover, the decrease in pH allows for lysosomal hydrolytic enzyme activation which modifies the morphological pattern of cytoplasmic vacuolization producing a "dark" cytoplasmic neuronal change. Notably, neurons are more resistant to the effects of extracellular acidosis than astrocytes. It is likely that neuronal or glial degeneration from extracellular acidosis reflects both the rapidity and degree of intracellular acidification combined with endogenous intracellular buffering capacity. Highly reactive free radicals are formed during the period of reperfusion and lead to sulfhydryl oxidation, lipid peroxidation, and membrane damage. During reperfusion, massive Ca^{2+} influx will occur, contributing to further disruption of Ca^{2+} homeostasis, phospholipase A2 and calpain activation (see Chapter 4, this volume).

Further understanding of the phenomenon of selective vulnerability can be garnered by considering the role of energy deprivation. Historically, the concept that some toxic neuropathies present as a distal axonal, "dying back" type of degeneration, invokes the role of high cellular energy metabolism needed for maintenance of long axons with a high burden of axonal cytoplasm. Experimentally, such distal axonopathies may be produced following chronic energy deprivation (arsenic intoxication, chronic thiamine deficiency), and may be reversible. In contrast, acute energy deprivation in man and experimental animals follows mitochondrial or glycolytic intoxication but has a human neuropathological counterpart in Wernicke's encephalopathy and Leigh's disease. Cavanagh[60] has cogently discussed the problem of selective vulnerability in acute energy deprivation syndromes and Table 1 summarizes the selective topography of degeneration associated with a group of metabolic poisons or deficiencies known to act on mitochondrial or glycolytic adenosine triphosphate (ATP) production. The syndromes share common features, including symmetrical brain stem involvement, predominant glio-vascular degeneration with secondary neuronal damage, variable neuronal sparing within regions of degeneration, microvascular proliferation and inhibition or reversal by administration of selective antagonists. For instance, the chlorosugars are converted to the corresponding 3-chlorolactaldehyde, an analogue of glyceraldehyde-3-phosphate, competing for the G3P dehydrogenase. Another proximal glycolytic inhibitor is 6-aminonicotinamide (6-AN) which exchanges with nicotinamide producing an ineffective product blocking NADP and NAD metabolism, especially blocking 6-phosphogluconate dehydrogenase and pentose phosphate shunt activity. Chlorosugar and 6-AN toxicity are similar (Table 1), but spare the inferior olivary nucleus in contrast to thiamine deficiency and imidazole intoxication. As mentioned, the primary lesion appears to be gliovascular: ultimate expression likely depends upon intrinsic glial sensitivity to the metabolic inhibitor with a superimposed component of regional neuronal activity producing the final common pathway of combined neuronal-glial degeneration (Figure 4).

3-Acetylpyridine (3-AP) is also an analogue of nicotinamide and inhibits NAD/NADP-dependent metabolic pathways. Systemic injection produces degeneration of inferior olivary neurons with variable involvement of hippocampus, facial and hypoglossal nuclei, and dorsal root ganglia.[61,62] The degeneration of the inferior olive and cerebellum is similar to that seen in olivopontocerebellar degeneration. Striatal injections of 3-AP produce dose and age-dependent lesions in the rat striatum.[63] The lesions spare the NADPH-diaphorase expressing neurons, (analogous to the situation in Huntington's disease), result in ATP depletion and are attenuated significantly by NMDA antagonists or glycolytic substrates capable of enhancing ATP production. These experiments suggest several important concepts: (i) *in situ* metabolic toxin injection may provide a useful experimental model for human neurodegeneration; (ii) impairments of energy metabolism may play a role in neurodegenerative disorders with regional selectivity such as Huntington's disease, Parkinson's disease and multiple system atrophy with/without cerebellar degeneration; (iii) and the evolution of 3-AP neurotoxicity, while primarily due to acute energy deprivation, is followed by secondary excitotoxicity. In this respect, numerous studies have demonstrated that energy depletion results in graded membrane depolarization and activation of excitatory amino acid receptors with subsequent excitotoxic neuronal damage (Figure 4).[64–66]

Table 1   Regional distribution of glio-vascular necrosis produced in energy deprivation syndrome.

Region	Thiamine deficiency	3-NP	6-AN	DNB	Chloro-sugars	Halo-imidazole
Spinal cord	−	−	+	−	−	+
Inferior olivary nuc	+	−	−	−	−	+
Vestibular nuc	+	−	+	+	+	+
Mamillary body	+	−	−	−	+	−
Striatum	−	+	−	−	−	−
Cortex	−	−	−	−	−	+

Source: modified from Cavanagh.[60]
3-NP, 3-nitropropionic acid; 6-AN, 6-aminonicotinamide; DNB, dinitrobenzene; Halo-imidazoles, tribromo- and trichloroimidazole.

11.02.3.2 Excitotoxicity

The discovery of neurotoxicity associated with glutamate exposure in the early 1980s led to the concept of excitotoxicity and the notion that excessive excitatory stimulation of neurons leads to cell destruction (see also Chapter 31, this volume).[67] For example, in cerebral ischemia, several observations lend support to the theory that glutamate excitotoxicity contributes directly to neuronal cell death: (i) the most vulnerable areas of the brain (i.e., hippocampus, cerebral cortex, striatum and cerebellum) are all heavily innervated[68] by glutamatergic afferents; (ii) the hippocampus contains the highest density of glutamate receptors of all brain areas;[69] (iii) experimental ischemia results in elevation of glutamate concentration in hippocampal extracellular space;[70] (iv) lesioning of innervating glutamatergic afferents prevents ischemic neurotoxicity;[71] and (v) intracerebral microinjection of NMDA antagonists protects against neuronal injury brought on by subsequent ischemia in the rat.[72] Glutamate excitotoxicity has also been implicated in a variety of other disorders including hypoglycemia and epilepsy.[73,74] Early studies of excitotoxicity focused on the NMDA-type glutamate receptor.[75] The Ca^{2+} channeling properties of this receptor were believed to be absent from the kainate and AMPA non-NMDA receptor subclasses. Subsequent studies, however, revealed that the non-NMDA class receptors can also contribute to neurotoxicity.[76] In particular, NADPH-diaphorase-containing cortical neurons may be especially vulnerable to non-NMDA receptor mediated injury. Selective sensitivity to excitotoxins is largely a function of the membrane receptor subtype expressed by a cell. For example, intracerebral injection of kainate destroys hippocampal CA3 neurons and cerebellar GABAergic neurons which are rich in high-affinity kainate binding sites.[77,78] Purkinje neurons, which express very few NMDA receptors, are resistant to NMDA toxicity but sensitive to α-amino-3-hydroxy-5-methylisoxazole-4-propionate (AMPA).[79]

An important development in this area has come through the discovery and analysis of the protein subunits of the non-NMDA receptor classes, composed of six subtypes, gluR1–gluR6.[80] Depending on the subtype composition of these receptors, channels permeable to Ca^{2+} ions may become operative. In particular the presence of a positively charged arginine residue in the intramembranous channel-forming region of the gluR2 subunit, instead of the neutral glutamine residues found in the other receptor subtypes, introduces a block to Ca^{2+} conductance. Thus, neurons expressing high levels of the gluR2 subunit of non-NMDA receptor are relatively resistant to hypoxic and other excitotoxic lesions which would normally produce elevations of intracellular Ca^{2+} concentration.

Transient forebrain ischemia in the rat causes selective decline in GluR2 subunit expression in CA1 hippocampal neurons prior to neurodegeneration.[81] Kainic acid-induced seizures produce a decline in GluR2 in the CA3/CA4 area prior to the selective degeneration of these neurons. A kainate-induced increase in GluR2 expression was observed in the dentate gyrus, a region resistant to kainate toxicity. These authors also observed a loss of GABA receptors in CA3/CA4 neurons which would also enhance excitotoxicity by decreasing inhibitory neurotransmission.

11.02.3.3 Calcium-mediated Events

Extensive investigation has been devoted to the role of Ca^{2+} in cell injury and death.[66,82–84] All cells tightly regulate their intracellular levels and transport of this ubiquitous ion. Neurons in particular devote substantial energy to the control of Ca^{2+} levels and possess unique biochemical mechanisms for regulation of Ca^{2+} metabolism. While the ability to elevate intracellular Ca^{2+} levels is mandatory for neuronal function and survival, excessive magnitude or duration of Ca^{2+} elevation is usually deleterious. A wide variety of neurotoxic agents and conditions have been linked to abnormal increases in Ca^{2+}.

Numerous pathways link elevation of Ca^{2+} to cell destruction. One such link is the activation of the neutral proteinase, calpain.[85] Activation of this nonlysosomal protease leads to degradation of several cellular proteins, including the cytoskeletal protein, spectrin, critical for maintenance of neuronal architecture. Detection of spectrin degradation fragments is often used as a measure of calpain activation and as a means of evaluating mechanism of neurotoxic action.[86] Another important mechanism of Ca^{2+}-mediated cytotoxicity involves the activation of phospholipases. These enzymes break down components of the plasma membrane which can compromise permeability functions of intracellular and plasma membranes. In addition, phospholipase A_2 activation by Ca^{2+} leads to release of arachidonic acid which results in superoxide anion production upon metabolism by cyclooxygenase and lipoxygenase.

Additional ROS are produced by high levels of Ca^{2+} via conversion of xanthine dehydrogenase to xanthine oxidase.[87] Xanthine oxidase generates both H_2O_2 and O_2^- when acting on hypoxanthine, a metabolite which accumulates under conditions of mitochondrial impairment.

Other deleterious effects of high Ca^{2+} include the suppression of mitochondrial oxidative phosphorylation[88,89] and the abnormal activation of the calmodulin-dependent nitric acid synthase (NOS), generating excessive amounts of nitric oxide.[90] The neuronal form of NOS (nNOS) is expressed in a restricted population of interneurons of the hippocampus but not in pyramidal neurons.[91] This subtype of NOS appears to enhance the cytotoxicity of the succinate dehydrogenase inhibitor, malonate, since malonate-induced injury is attenuated in genetically engineered mice lacking this form of the enzyme.[92]

11.02.3.4 Oxidative Stress

Oxidative stress injury is a prevalent form of cellular damage in the nervous system (see also Chapter 3, this volume).[93–96] Although many reports of such stress probably reflect secondary consequences of injury by other mechanisms, the number and variety of neurologic and neurotoxic disorders associated with some form of oxidative stress necessitates a thorough understanding of this disruptive process.

The nervous system is particularly vulnerable to oxidative stress for several reasons: (i) Neurons consume high and variable amounts of energy generated by oxidative metabolism. The requirement for and consumption of oxygen is especially high. (ii) Lipids in the nervous system are rich in oxidizable unsaturated fatty acids. Cells with an abundance of such lipids are susceptible to membrane damage. (iii) Neurons and oligodendrocytes contain large amounts of iron. During cell injury, this iron can catalyze hydroxyl radical formation. (iv) The brain has high levels of oxidizable small molecules such as dopamine. Cells which possess unique uptake systems for these molecules or xenobiotic homologues are more sensitive to their oxidizing effects. (v) Cells in the nervous system express different types and amounts of various antioxidant defense enzymes (discussed in Section 11.02.4). The prevalence of these factors and their cell type and regional variability of expression in the nervous system adds a great deal to the complexity and heterogeneity of damage observed in neuropathologic conditions.

11.02.3.5 Cytoskeletal Disruption

Neurons, astrocytes, and oligodendrocytes display the most unique and complex morphologies of any mammalian cell. The cytoskeleton, consisting of microtubules, microfilaments and intermediate filaments serves as the backbone underlying this morphologic complexity (see Chapter 5, this volume). Moreover, the cytoskeleton of neurons plays major functional roles such as axoplasmic transport and synaptogenesis.

Neurons of the peripheral nervous system which produce large amounts of neurofilament protein are highly susceptible to attack by hexacarbon solvents *n*-hexane, methyl *n*-butyl ketone and derivatives such as 2,5-hexanedione and β,β'-iminodipropionitrile.[97] Exposure to these agents via occupational involvement or habitual glue-sniffing can lead to peripheral neuropathy. The presence of appropriately positioned primary amine (lysine ε-amino) groups on neurofilament proteins makes these proteins highly susceptible to crosslinking by these toxicants. Acrylamide and other agents appear to act by a similar mechanism.

11.02.3.6 Trophic Factor Dependence

An emerging theory of cellular survival and functional optimization in the nervous system maintains that trophic factors and/or cell surface ligands emanating from specialized nearby cells serve to override intrinsic, primed cell suicide pathways. Teleologically, a neuron must find and interact with appropriate cellular counterparts in order to maintain maximum operational integrity of the whole organism. This survival-promoting communication is mediated by trophic factor interactions with specific cell surface receptors which, when activated, suppress the suicide pathway. Failure to maintain trophic factor–receptor interaction leads to apoptotic cell death unless the bcl-2-protein (or homologous family member) is adequately expressed.

Several examples of trophic factor specificity in the modulation of neuronal survival can be cited. Brain derived neurotrophic factor (BDNF) protects cholinergic neurons from degeneration (in rats) following lesioning of the fimbria-fornix.[98] BDNF also protects motoneurons from axotomy-induced degeneration and substantia nigra compacta dopaminergic neurons from MPTP-induced toxicity.[99] Ciliary neurotrophic factor prevents motoneuron degeneration in a mouse model for ALS.[100] Nerve growth factor (NGF) has been shown to protect against striatal neuron degeneration following lesions which mimic Huntington's disease[101] and to protect cultured hippocampal neurons against glutamate-induced excitotoxicity.[102] Studies indicate that some of the neurons which degenerate in Alzheimer's disease possess receptors for NGF and estrogen and that estrogen may help to prevent neuronal oxidative stress and degeneration in this disease.[103]

11.02.3.7 Complex Interactions

In analyzing mechanisms of toxic injury in the nervous system one must consider that biologic systems are exposed to multiple toxicants present in varying concentrations. The toxicity of a given substance can be greatly enhanced by the presence of additional agents. An example of such combinatorial effects is the heightened sensitivity to a variety of toxicants in animals exposed to glutathione-depleting agents.[104] Lead and other toxicants have been shown to greatly exacerbate glutamate-induced toxicity in neurons.[105] Environmental estrogen agonists such as PCBs display up to 100-fold greater activity when exposed in combination as opposed to individually.[106] Such synergistic effects add considerable complexity to the analysis of toxic mechanisms and require careful consideration in formulating risk assessment guidelines.

Another important consideration in analyzing toxic injury in the nervous system is the phenomenon of secondary or bystander effects. Cells of the nervous system display a high degree of interdependency and complexity of communication. Injury or death of one cell type may not be a direct effect of a given toxin, but rather the effect of an adjacent or supporting cell which has been damaged or modified by the toxin. For example, 6-aminonicotinamide produces a gliotoxicity followed by neuronal necrosis and a proposal for the neurotoxicity of methyl mercury holds that inhibition of astrocytic potassium and glutamate uptake results in neuronal excitotoxic damage due to prolonged elevation of glutamate concentration in extracellular spaces.[107]

Finally, many toxicants are capable of interacting with and disrupting more than one molecular target or biochemical pathway. Such pleiotropic effects are fairly commonly observed in the nervous system as evidenced by the multiple side effects of many neuropharmacologic agents. The existence of multiple targets greatly complicates mechanistic studies in neurobiologic systems. Observations of biochemical changes associated with neuropathologic conditions must always be interpreted with caution when attempting to make causal inferences.

11.02.4 NEURONAL DEFENSE MECHANISMS

Having described clinically relevant examples of selective neuronal injury as well as some of the biochemical mechanisms underlying this injury, we turn our attention to the various mechanisms which neurons utilize to counteract toxic insult. In many cases, these defensive strategies are found to be prevalent in select cell types, and thus may play an important role in observed patterns of selective vulnerability.

11.02.4.1 Ca^{2+} Binding Proteins

The neurotoxic effects of disturbed Ca^{2+} regulation and elevated levels of intracellular Ca^{2+} have been extensively investigated.[82] Ca^{2+} is important for cell cycle control, cell motility, cytoskeletal organization and the regulation of numerous enzymes.[108] The capacity to elevate cytoplasmic concentrations of Ca^{2+} via ligand- and voltage-gated mechanisms and by release from intracellular stores is vital for neuronal differentiation and survival. However, as discussed above, excessive or prolonged elevation of intracellular Ca^{2+} concentration is hazardous for most cells, particularly neurons.[66] Mechanisms causing disturbed Ca^{2+} homeostasis include compromise of cellular energetics and exposure to toxins. One of the major mechanisms of regulation of free cytoplasmic Ca^{2+} concentration is the expression of Ca^{2+} binding proteins.[109,110] The family of Ca^{2+} binding proteins include calmodulin, parvalbumin, calbindin-D28k, calretinin, calcineurin, calpain, and S100. Many of these proteins play a role in cytoplasmic Ca^{2+} buffering and other functions such as Ca^{2+} transport and intracellular signaling. For the majority of these proteins, specific defects have been associated with a variety of neurodegenerative disorders.

Calbindin has been found in a select population of hippocampal neurons that survive excitotoxic conditions *in vitro*.[111] Similarly calretinin-positive neurons in rat neocortical cultures are relatively resistant to excitatory amino acids and to Ca^{2+} ionophore A23187.[112]

In vivo studies, however, are somewhat contradictory. Some neurons expressing Ca^{2+} binding proteins manage to survive in animal models of neurodegeneration,[113] while others are quite sensitive.[114] Calbindin-containing neurons of the substantia nigra are spared in Parkinson's disease.[115] In Alzheimer's disease neocortical neurons containing high levels of either calbindin or calretinin are resistant to degeneration.[116]

The hippocampal distribution of chromogranin A immunoreactivity is limited to the pyramidal cells of the CA2 sector; neurons in CA1 and CA3 sectors are not immunoreactive. The predominantly immunoreactive CA2 and dentate granule cells are spared in human and experimental epilepsy, while CA1 and CA3 neurons, lacking chromogranin are destroyed.[117] An explanation for such selective vulnerability resides in an understanding of the function of chromogranin A which has been shown to be a Ca^{2+}-binding protein.[118] Because of its known neurotransmitter vesicle association, chromogranin A acts as a precursor of peptides with autocrine inhibitory function.[119] These functions are not incompatible and synergistically are likely to modify the genesis of excitotoxic injury in appropriate neuronal populations.

As alluded to above, a wide variety of chemical toxicants cause loss of neuronal Ca^{2+} homeostasis and the accompanying increase in cytoplasmic Ca^{2+} is an important contributor to cellular destruction. The plasma membrane contains numerous systems for controlling neuronal $[Ca^{2+}]_i$. These systems include the voltage-gated Ca^{2+} channels, ligand-gated ion channels, the activity of Ca^{2+}-ATPase and the efficiency of the Na^+/Ca^{2+} exchanger. The ATP-driven Ca^{2+} pump extrudes Ca^{2+} from cells, is calmodulin-dependent and is regulated by phosphorylation.[68] Variations of Ca^{2+} ATPase activity between neurons are certain to contribute to differing efficacy in maintaining homeostatic neuronal $[Ca^{2+}]$. Differing activities for the Na^+/Ca^{2+} exchanger in the membrane have been demonstrated. For instance, Mattson and Kater[120] demonstrated that hippocampal neurons in culture showed a large rise in $[Ca^{2+}]$ in response to K^+-depolarization or the Ca^{2+} ionophore, A23187, associated with marked neurotoxicity. However, neuroblastoma cells or cerebellar granular cells were markedly resistant under conditions of high external $[Na^+]$. Lowering the $[Na^+]_e$ greatly magnified the cytotoxicity revealing differing activation in the Na^+/Ca^{2+} exchanger under the influence of Ca^{2+} ionophore activation.

11.02.4.2 Antioxidant Defense

The nervous system is highly vulnerable to oxidative stress.[121–124] The unusually high rate of oxygen consumption, high levels of iron and high levels of oxidizable substrate such as low-molecular weight amines and polyunsaturated lipids contribute to a high potential for oxidative injury to neurons in contrast to glia. Thus, the nervous system must maintain strong anti-oxidant defenses in order to minimize damage to neuronal cells which must survive and function throughout the lifetime of the organism. Among these defenses are the low-molecular-weight anti-oxidants ascorbic acid, vitamin E, and glutathione. Ascorbate reacts with superoxide ion and hydroxy radical and is concentrated in the CNS by active transport

mechanisms. Vitamin E blocks membrane lipid peroxidation and deficiency of this compound in patients with abetalipoproteinemia results in progressive neuropathy and retinopathy. In both rats and humans, prolonged deficiency of vitamin E results in selective degeneration and loss of large caliber myelinated axons in peripheral nerve and spinal cord especially in posterior columns.[125]

For reasons which are unclear, glutathione levels are relatively low in the nervous system and especially low in most neurons.[126] Several reports indicate that glutathione levels are 3–10-fold lower in neurons than in glial cells. At least one report[127] indicates that granule neurons of the cerebellum and dorsal root ganglia have somewhat higher levels of glutathione than other neurons, an observation inconsistent with their recognized sensitivity to methyl mercury.[128] In patients with Parkinson's disease, levels of glutathione (GSH) are below normal in the substantia nigra while other anti-oxidant systems are normal, lending support to the notion that oxidative damage, in addition to energy deficiency, may play an important role in this disorder.

A multitude of enzyme systems are constitutively expressed in most mammalian cells to deal with the various forms of reactive oxygen. While neurons express many of these enzymes, several, including catalase and glutathione S-transferase are present in relatively low levels in neurons compared with glial cells.[129,130] The same pattern is observed with respect to several important inducible anti-oxidant enzymes including heme-oxygenase I and metallothionein MT-3. The unique sensitivity of motor neurons to hereditary mutations of the Cu^{2+}/Zn^{2+} superoxide dismutase gene was discussed above. These cells clearly appear to be deficient in the ability to cope with the likely oxidative stress caused by these mutations. Regional and neuron-specific variation in expression patterns for these enzymes have not been extensively characterized. However, subtle differences in patterns of expression or induction of these enzymes would likely produce significant differential sensitivities to the toxic free radical-generating environment of the neuron.

In this respect, enhanced NADH:quinone reductase activity prevents glutamate toxicity induced by oxidative stress.[131] In embryonic cortical neurons or neuroblastoma cell lines, quisqualate-type glutamate toxicity results from competitive inhibition of high-affinity cystine uptake, leading to depletion of cellular GSH and oxidative stress. Induction of the quinone reductase diminished glutamate toxicity while treatment with the inhibitor dicumarol potentiated glutamate toxicity. These observations reveal the involvement of oxidants in neuron degeneration, counteracted by the induction of quinone reductase. Such toxicity is not present in mature neurons. While the embryonic system is very dependent upon cystine uptake, the lack of functional glutamate receptors precludes excitotoxicity.

11.02.4.3 Bcl-2 Family

Attention has been directed to the expression of several genes involved in regulation of programmed cell death. The anti-apoptotic gene, *bcl*-2, and related genes have been found to impact viability of a variety of neuronal cell types when over-expressed *in vitro*.[132] *Bcl*-2 expression results in suppression of apoptotic death associated with growth factor withdrawal and necrotic cell death induced by a variety of toxic agents.[133] Transgenic mice over-expressing *bcl-2* in neurons under the control of the neuron-specific enolase promoter retained 30% greater neuronal cell numbers during development into adulthood. These mice displayed less motor neuron death following axotomy of the sciatic nerve. Early reports of *bcl-2* in brain indicated expression was confined to early developmental periods. Other studies, however, suggest low levels of bcl-2 mRNA expression continues throughout adulthood. Bcl-2-deficient knock-out mice apparently develop normal nervous systems.[134,135] In the brain the related protein, bcl-x, appears to be expressed far more prominently than is bcl-2. Bcl-x-deficient knockout mice die at embryonic day 13 with extensive apoptotic death of brain, spinal cord and dorsal root ganglia neurons.[136] Thus Bcl-x appears to support survival of immature neurons during development.

Some evidence suggests that *bcl-2* expression may influence survival of neurons following transient focal ischemia.[137,138] *Bcl-2* expression is induced in rat neurons which survive 60–120 min ischemia but is not induced in neurons destined to die. Studies employing 15 min ischemia have demonstrated that expression of the proapoptotic bcl-2 homolog, Bax, is up-regulated in the sensitive CA1 neurons but not CA3 prior to cell death. CA3 neurons up-regulate bcl-2 expression during this time.[139] These results suggest that bcl-2 may serve as an endogenous protective agent in cerebral ischemia.

11.02.4.4 NADPH Diaphorase/NO Synthase

A select population of neurons scattered individually throughout the mammalian central nervous system contain high levels of the enzyme NADPH diaphorase. Characterized originally by its activity in utilizing NADPH to reduce tetrazolium dyes, this enzyme was subsequently identified as nitric oxide synthase.[140] This enzyme generates the nitric oxide free radical which serves as a smooth muscle relaxing factor and possibly as a neuronal retrograde messenger.

The striatal aspiny neurons surviving in the caudate nucleus of Huntington's disease patients stain positively for this enzyme as well as for somatostatin and neuropeptide 1.[141] In cortical neuron cultures, neurons containing NO synthase were resistant to quinolinate and NMDA but are selectively vulnerable to kainate or quisqualate. These sensitivities probably reflect differences in glutamate receptor subtypes associated with the NO synthase expressing cells.

NO may be toxic when produced in excess. Inhibition of NO synthase blocks excitotoxicity in some cultured neurons.[142] The neurotoxin capsaicin (8-methyl-N-vanillyl-6-nonenamide) preferentially kills small primary sensory neurons in 2 day old rats injected subcutaneously.[143] Dorsal root ganglion neurons with unmyelinated axons are particularly sensitive. These neurons also express NO synthase mRNA and NADPH diaphorase activity. A link to a neurotoxic mechanism has been suggested through capsaicin-mediated Ca^{2+} influx coupled to Ca^{2+} activation of NO synthase. The consequent accumulation of free radical NO then induces cell death.

The paradoxical effects of NO synthase-associated protection and sensitization of select neurons may relate to differential requirements for NO in survival maintenance vs. differential expression of antiradical defense mechanisms.

11.02.5 CONCLUSION

Multiple factors play a role in the manifestation of selective vulnerability in the nervous system. These include (but are by no means restricted to): (i) differential localization/sequestration of toxicants mediated by specific transport mechanisms; (ii) the activation of excitatory neurotransmitter pathways and/or suppression of inhibitory pathways localized to specific cell types; (iii) regional and cell-variable homeostatic regulation of $[Ca^{+2}]$, including the presence of Ca-binding proteins; (iv) variable expression and inducibility of defense strategies such as anti-oxidant systems and xenobiotic- or heavy-metal-targeted proteins; (v) intrinsically different propensities to activate or suppress the

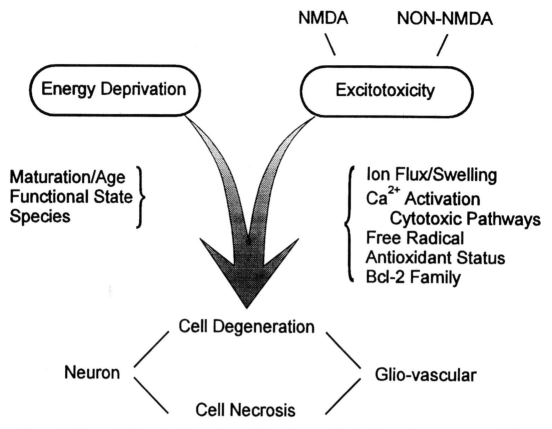

Figure 4 Schematic illustrating interactions of the major contributing factors in nervous system injury.

apoptotic cell death pathway; (vi) varying dependence on a constant supply of ATP and varying pathways for maintaining cellular energetics; and (vii) other factors such as cell morphology or cytoskeletal (Figure 4).

Any one of the above variables, or a combination of several, may ultimately underlie the specific patterns of cell damage caused by a given toxin. Vulnerability to toxic environmental agents is a property common to all living systems. However, analogous to evolutionary natural selection, when extensive variability in response to a given destructive agent is observed, significant opportunities for the advancement of understanding and the development of protective countermeasures are realized.

Cell injury in neurodegenerative disorders may hypothetically be viewed as the result of an endogenous toxicant either over-produced or inadequately controlled. Alternatively, as yet undisclosed environmental toxicants may combine with genetic factors to produce the selective destruction characteristic of neurodegenerative disorders. Understanding the biochemical and molecular bases underlying differential sensitivity to toxicants may hold the key for developing preventative or therapeutic strategies for dealing with xenobiotic environmental toxicants. Such knowledge is

also likely to be beneficial in designing strategies for treatment for numerous neurologic disorders.

11.02.6 REFERENCES

1. W. Spielmeyer, 'Zur Parthogenese ör tllch elektiver Gehirnverandevungen.' *Z. Neurol. Psychiatry*, 1925, **99**, 756–765.
2. C. Vogt and O. Vogt, 'Sitz und Wesen der Krankheiten im Lichte der topistischen Hirnforschung und des Variierens der Tiere,' Barth, Leipzig, 1937.
3. G. J. Lees, 'Contributory mechanisms in the causation of neurodegenerative disorders.' *Neuroscience*, 1993, **54**, 287–322.
4. J. M. Gallo and B. H. Anderton, 'Brain diseases. Ubiquitous variations in nerves.' *Nature*, 1989, **337**, 687–688.
5. T. Saitoh, E. Masliah, L. W. Jin *et al.*, 'Protein kinases and phosphorylation in neurologic disorders and cell death.' *Lab. Invest.*, 1991, **64**, 596–616.
6. G. L. Caporaso, S. E. Gandy, J. D. Buxbaum *et al.*, 'Chloroquine inhibits intracellular degradation but not secretion of Alzheimer β/A4 amyloid precursor protein.' *Proc. Natl. Acad. Sci. USA*, 1992, **89**, 2252–2256.
7. J. W. Langston, P. Ballard, J. W. Tetrud *et al.*, 'Chronic Parkinsonism in humans due to a product of meperidine-analog synthesis.' *Science*, 1983, **219**, 979–980.
8. J. D. Adams, Jr. and I. N. Odunze, 'Oxygen free radicals and Parkinson's disease.' *Free Radic. Biol. Med.*, 1991, **10**, 161–169.

9. K. F. Tipton and T. P. Singer, 'Advances in our understanding of the mechanisms of the neurotoxicity of MPTP and related compounds.' *J. Neurochem.*, 1993, **61**, 1191–1206.

10. C. M. Tanner, 'The role of environmental toxins in the etiology of Parkinson's disease.' *Trends Neurosci.*, 1989, **12**, 49–54.

11. S. F. Ali, K. J. Kordsmeier and B. Gough, 'Drug-induced circling preference in rats. Correlation with monoamine levels.' *Mol. Neurobiol.*, 1995, **11**, 145–154.

12. J. Knoll, 'The pharmacology of selegiline (1-deprenyl). New aspects.' *Acta Neurol. Scand. Suppl.*, 1989, **126**, 83–91.

13. Anonymous, 'A novel gene containing a trinucleotide repeat that is expanded and unstable on Huntington's disease chromosomes.' The Huntington's Disease Collaborative Research Group, *Cell*, 1993, **72**, 971–983.

14. M. F. Beal, N. W. Kowall, D. W. Ellison *et al.*, 'Replication of the neurochemical characteristics of Huntington's disease by quinolinic acid.' *Nature*, 1986, **321**, 168–171.

15. R. J. Boegman, K. Jhamandas and R. J. Beninger, 'Neurotoxicity of tryptophan metabolites.' *Ann. NY Acad. Sci.*, 1990, **585**, 261–273.

16. I. P. Lapin, 'Convulsant action of intracerebroventricularly administered L-kynurenine sulphate, quinolinic acid and other derivatives of succinic acid, and effects of amino acids: structure–activity relationships.' *Neuropharmacology*, 1982, **21**, 1227–1233.

17. R. Schwarcz, W. O. Whetsell, Jr. and R. M. Mangano, 'Quinolinic acid: an endogenous metabolite that produces axon-sparing lesions in rat brain.' *Science*, 1983, **219**, 316–318.

18. A. B. Young, J. T. Greenamyre, Z. Hollingsworth *et al.*, 'NMDA receptor losses in putamen from patients with Huntington's disease.' *Science*, 1988, **241**, 981–983.

19. M. F. Beal, E. Brouillet, B. G. Jenkins *et al.*, 'Neurochemical and histologic characterization of striatal excitotoxic lesions produced by the mitochondrial toxin 3-nitropropionic acid.' *J. Neurosci.*, 1993, **13**, 4181–4192.

20. D. B. Williams, in 'Handbook of Neurology: Diseases of the Motor System,' Elsevier Science, New York, 1990, pp. 241–251.

21. H. X. Deng, A. Hentati, J. A. Trainer *et al.*, 'Amyotrophic lateral sclerosis and structural defects in Cu, Zn superoxide dismutase.' *Science*, 1993, **261**, 1047–1051.

22. A. C. Bowling, E. E. Barkowski, D. McKenna-Yasek *et al.*, 'Superoxide dismutase concentration and activity in familial amyotrophic lateral sclerosis.' *J. Neurochem.*, 1995, **64**, 2366–2369.

23. P. C. Wong, C. A. Pardo, D. R. Borchelt *et al.*, 'An adverse property of a familial ALS-linked SOD1 mutation causes motor neuron disease characterized by vacuolar degeneration of mitochondria.' *Neuron*, 1995, **14**, 1105–1116.

24. D. R. Borchelt, M. Guarnieri, P. C. Wong *et al.*, 'Superoxide dismutase 1 subunits with mutations linked to familial amyotrophic lateral sclerosis do not affect wild-type subunit function.' *J. Biol. Chem.*, 1995, **270**, 3234–3238.

25. J. D. Rothstein, L. A. Bristol and B. Hosler, 'Chronic inhibition of superoxide dismutase produces apoptotic death of spinal neurons.' *Proc. Natl. Acad. Sci. USA*, 1994, **91**, 4155–4159.

26. J. D. Rothstein, L. J. Martin and R. W. Kuncl, 'Decreased glutamate transport by the brain and spinal cord in amyotrophic lateral sclerosis.' *N. Engl. J. Med.*, 1992, **326**, 1464–1468.

27. J. D. Rothstein, M. Van Kammen, A. I. Levey *et al.*, 'Selective loss of glial glutamate transporter GLT-1 in amyotrophic lateral sclerosis.' *Ann. Neurol.*, 1995, **38**, 73–84.

28. P. S. Spencer, P. B. Nunn, J. Hugon *et al.*, 'Guam amyotrophic lateral sclerosis-parkinsonism-dementia linked to a plant excitant neurotoxin.' *Science*, 1987, **237**, 517–522.

29. M. W. Duncan, in 'Advances in Neurology,' ed. L. P. Rowland, Raven, New York, 1991, pp. 301–310.

30. R. D. Terry, A. Peck, R. De Teresa *et al.*, 'Some morphometric aspects of the brain in senile dementia of the Alzheimer type.' *Ann. Neurol.*, 1981, **10**, 184–192.

31. D. M. A. Yates and B. Marcyniuk, 'Some morphometric observations on the cerebral cortex and hippocampus in presenile Alzheimer's disease, senile dementia of Alzheimer's type and Down's syndrome in middle age.' *J. Neurol. Sci.*, 1985, **69**, 139–159.

32. P. D. Coleman and D. G. Flood, 'Neuron numbers and dendritic extent in normal aging and Alzheimer's disease.' *Neurobiol. Aging*, 1987, **8**, 521–545.

33. R. T. Bartus, R. L. Dean, III, B. Beer *et al.*, 'The cholinergic hypothesis of geriatric memory dysfunction.' *Science*, 1982, **217**, 408–414.

34. J. Hardy, R. Adolfson, I. Alafuzoff *et al.*, 'Transmitter deficits in Alzheimer's disease.' *Neurochem. Int.*, 1985, **7**, 545–563.

35. A. C. W. Heizmann and K. Braun, 'Changes in Ca^{2+}-binding proteins in human neurodegenerative disorders.' *Trends Neurosci.*, 1992, **15**, 259–264.

36. J. C. Vickers, R. A. Lazzarini, B. M. Riederer *et al.*, 'Intraperikaryal neurofilamentous accumulations in a subset of retinal ganglion cells in aged mice that express a human neurofilament gene.' *Exp. Neurol.*, 1995, **136**, 266–269.

37. M. B. Delisle and S. Carpenter, 'Neurofibrillary axonal swellings and amyotrophic lateral sclerosis.' *J. Neurol. Sci.*, 1984, **63**, 241–250.

38. D. R. C. McLachlan, 'Aluminum and Alzheimer's disease.' *Neurobiol. Aging*, 1986, **7**, 525–532.

39. D. Langui, B. H. Anderton, J. P. Brion *et al.*, 'Effects of aluminum chloride on cultured cells from rat brain hemispheres.' *Brain Res.*, 1988, **438**, 67–76.

40. N. W. Kowall, W. W. Pendlebury, J. B. Kessler *et al.*, 'Aluminum-induced neurofibrillary degeneration affects a subset of neurons in rabbit cerebral cortex, basal forebrain and upper brainstem.' *Neuroscience*, 1989, **29**, 329–337.

41. T. B. Shea, J. F. Clarke, T. R. Wheelock *et al.*, 'Aluminum salts induce the accumulation of neurofilaments in perikarya of NB2a/dl neuroblastoma.' *Brain Res.*, 1989, **492**, 53–64.

42. S. Tokutake, H. Nagase, S. Morisaki *et al.*, 'Aluminium detected in senile plaques and neurofibrillary tangles is contained in lipofuscin granules with silicon, probably as aluminosilicate.' *Neurosci. Lett.*, 1995, **185**, 99–102.

43. C. D. Katsetos, J. Savory, M. M. Herman *et al.*, 'Neuronal cytoskeletal lesions induced in the CNS by intraventricular and intravenous aluminum maltol in rabbits.' *Neuropathol. Appl. Neurobiol.*, 1990, **16**, 511–528.

44. A. I. Bush, W. H. Pettingell, G. Multhaup *et al.*, 'Rapid induction of Alzheimer Aβ amyloid formation by zinc.' *Science*, 1994, **265**, 1464–1467.

45. T. D. Landry, J. F. Quast, T. S. Gushow *et al.*, 'Neurotoxicity of methyl chloride in continuously versus intermittently exposed female C57BL/6 mice.' *Fundam. Appl. Toxicol.*, 1985, **5**, 87–98.

46. M. G. Simpson, I. Wyatt, H. B. Jones *et al.*, 'Neuropathological changes in rat brain follow oral administration of 2-chloropropionic acid.' *Neuro-*

toxicology, 1996, **17**, 471–480.

47. S. L. Eustis, S. B. Haber, R. T. Drew *et al.*, 'Toxicology and pathology of methyl bromide in F344 rats and B6C3F1 mice following repeated inhalation exposure.' *Fundam. Appl. Toxicol.*, 1988, **11**, 594–610.

48. L. W. Chang, 'Neurotoxic effects of mercury—review.' *Environ. Res.*, 1977, **14**, 329–373.

49. K. R. Reuhl and L. W. Chang, 'Effects of methyl mercury and the development of the nervous system: a review.' *Neurotoxicology*, 1979, **1**, 21–55.

50. T. C. Marrs, I. V. Allen, H. F. Colgrave *et al.*, 'Neurotoxicity of 1-methyl cycloheptatriene-A Purkinje cell toxicant.' *Human Ex. Toxicol.*, 1991, **10**, 93–101.

51. D. R. Pierce, D. C. Serbus and K. E. Light, 'Intragastric intubation of alcohol during postnatal development of rats results in selective cell loss in the cerebellum.' *Alcoholism: Clin. Exp. Res.*, 1993, **17**, 1275–1280.

52. A. W. Brown, W. N. Aldridge, B. W. Street *et al.*, 'The behavioral and neuropathologic sequelae of intoxication by trimethyltin compounds in the rat.' *Am. J. Pathol.*, 1979, **97**, 59–82.

53. L. W. Chang, in 'Handbook of Neurotoxicology,' eds. L. W. Chang and R. S. Dyer, Dekker, New York, 1995, pp. 143–169.

54. G. R. Stewart, C. F. Zorumski, M. T. Price *et al.*, 'Domoic acid: a dementia-inducing excitotoxic food poison with kainic acid receptor specificity.' *Exp. Neurol.*, 1990, **100**, 127–138.

55. J. M. Jacobs, J. E. Cremer and J. B. Cavanagh, 'Acute effects of triethyltin on the rat myelin sheath.' *Neuropathol. Appl. Neurobiol.*, 1977, **3**, 169–181.

56. J. Cervós-Navarro and N. H. Diemer, 'Selective vulnerability in brain hypoxia.' *Crit. Rev. Neurobiol.*, 1991, **6**, 149–182.

57. N. R. Sims, 'Energy metabolism and selective neuronal vulnerability following global cerebral ischemia.' *Neurochem. Res.*, 1992, **17**, 923–931.

58. P. Cragg, L. Patterson and M. J. Purves, 'The pH of brain extracellular fluid in the cat.' *J. Physiol. (Lond.)*, 1977, **272**, 137–166.

59. R. E. Myers and S. Yamaguchi, 'Effects of serum glucose concentration on brain response to circulatory arrest.' *J. Neuropathol. Exp. Neurol.*, 1976, **35**, 301.

60. J. B. Cavanagh, 'Selective vulnerability in acute energy deprivation syndromes.' *Neuropathol. Appl. Neurobiol.*, 1993, **19**, 461–470.

61. C. D. Balaban, 'Central neurotoxic affects of intraperitoneally administered 3-acetylpyridine harmaline and niacinamide in Sprague–Dawley and Long–Evans rats: a critical review of central 3-acetylpyridine neurotoxicity.' *Brain. Res.*, 1985, **9**, 21–42.

62. C. M. Beiswanger, T. L. Roscoe-Graessle, N. Zerbe *et al.*, '3-Acetylpyridine-induced degeneration in the dorsal root ganglia involvement of small diameter neurons and influence of axotomy.' *Neuropathol. Appl. Neurobiol.*, 1993, **19**, 164–172.

63. J. B. Schulz, D. R. Henshaw, B. G. Jenkins *et al.*, '3-Acetylpyridine produces age-dependent excitotoxic lesions in rat striatum.' *J. Cereb. Blood Flow Metab.*, 1994, **14**, 1024–1029.

64. G. D. Zeevalk and W. J. Nicklas, 'Mechanisms underlying initiation of excitotoxicity associated with metabolic inhibition.' *J. Pharm. Exp. Ther.*, 1991, **257**, 870–878.

65. M. F. Beal, B. T. Hyman and W. J. Koroshetz, 'Do defects in mitochondrial energy metabolism underlie the pathology of neurodegenerative diseases?' *Trends Neurosci.*, 1993, **16**, 125–131.

66. M. A. Verity, 'Ca^{2+}-dependent processes as mediators of neurotoxicity.' *Neurotoxicology*, 1992, **13**, 139–147.

67. D. W. Choi, 'Glutamate neurotoxicity and diseases of the nervous system.' *Neuron*, 1988, **1**, 623–634.

68. G. E. Fagg and A. C. Foster, 'Amino acid neurotransmitters and their pathways in the mammalian central nervous system.' *Neuroscience*, 1983, **9**, 701–719.

69. J. T. Greenamyre, J. M. M. Olson, J. B. Penney, Jr. *et al.*, 'Autoradiographic characterization of *N*-methyl-D-aspartate; quisqualate- and kainate-sensitive glutamate binding sites.' *J. Pharmacol. Exp. Ther.*, 1985, **233**, 254–263.

70. H. Benveniste, J. Drejer, A. Schousboe *et al.*, 'Elevation of the extracellular concentrations of glutamate and aspartate in rat hippocampus during transient cerebral ischemia monitored by intracerebral microdialysis.' *J. Neurochem.*, 1984, **43**, 1369–1374.

71. H. Onodera, G. Sato and K. Kogura, 'Lesions to Schaffer collaterals prevent ischemic death of CA1 pyramidal cells.' *Neurosci. Lett.*, 1986, **68**, 169–174.

72. B. K. Siesjö, 'Cell damage in the brain: a speculative synthesis.' *J. Cereb. Blood Flow Metab.*, 1981, **1**, 155–185.

73. D. W. Choi, 'Excitotoxic cell death.' *J. Neurobiol.*, 1992, **23**, 1261–1276.

74. R. G. Fariello, G. T. Golden, G. G. Smith *et al.*, 'Potentiation of kainic acid epileptogenicity and sparing from neuronal damage by an NMDA receptor antagonist.' *Epilepsy Res.*, 1989, **3**, 206–213.

75. A. B. MacDermott, M. L. Mayer, G. L. Westbrook *et al.*, 'NMDA-receptor activation increases cytoplasmic calcium concentration in cultured spinal cord neurones.' *Nature*, 1986, **321**, 519–522.

76. J. Y. Koh, M. P. Goldberg and D. M. Hartley, 'Non-NMDA receptor-mediated neurotoxicity in cortical culture.' *J. Neurosci.*, 1990, **10**, 693–705.

77. J. W. Olney, T. Fuller and T. de Gubareff, 'Acute dendrotoxic changes in the hippocampus of kainate-treated rats.' *Brain Res.*, 1979, **176**, 91–100.

78. J. Drejer and A. Schousboe, 'Selection of a pure cerebellar granule cell culture by kainate treatment.' *Neurochem. Res.*, 1989, **14**, 751–754.

79. C. Kohler and R. Schwarcz, 'Comparison of ibotenate and kainate neurotoxicity in rat brain: a histological study.' *Neuroscience*, 1983, **8**, 819–835.

80. P. Jonas and N. Burnashev, 'Molecular mechanisms controlling calcium entry through AMPA-type glutamate receptor channels.' *Neuron*, 1995, **15**, 987–990.

81. D. E. Pellegrini-Giampietro, W. A. Pulsinelli and R. S. Zukin, 'NMDA and non-NMDA receptor gene expression following global brain ischemia in rats: effect of NMDA and non-NMDA receptor antagonists.' *J. Neurochem.*, 1994, **62**, 1067–1073.

82. F. B. Meyer, 'Calcium, neuronal hyperexcitability and ischemic injury.' *Brain Res. Rev.*, 1989, **14**, 227–243.

83. S. Orrenius, D. J. McConkey, G. Bellomo *et al.*, 'Role of Ca^{2+} in toxic cell killing.' *Trends Pharmacol. Sci.*, 1989, **10**, 281–285.

84. A. Frandsen and A. Schousboe, 'Excitatory amino acid-mediated cytotoxicity and calcium homeostasis in cultured neurons.' *J. Neurochem.*, 1993, **60**, 1202–1211.

85. E. Melloni and S. Pontremoli, 'The calpains.' *Trends Neurosci.*, 1989, **12**, 438–444.

86. B. A. Bahr, S. Tiriveedhi, G. Y. Park *et al.*, 'Induction of calpain-mediated spectrin fragments by pathogenic treatments in long-term hippocampal slices.' *J. Pharmacol. Exp. Ther.*, 1995, **273**, 902–908.

87. J. McCord, 'Oxygen-derived free radicals in postischemic tissue injury.' *N. Engl. J. Med.*, 1985, **312**, 159–163.

88. C. E. Thomas and D. J. Reed, 'Effect of extracellular Ca^{2+} omission on isolated hepatocytes II. Loss of mitochondrial membrane potential and protection by inhibitors of uniport Ca^{2+} transduction.' *J. Pharmacol. Exp. Ther.*, 1988, **245**, 501–507.

89. P. Nicotera, G. Bellomo and S. Orrenius, 'Calcium-mediated mechanisms in chemically-induced cell death.' *Annu. Rev. Pharmacol. Toxicol.*, 1992, **32**, 449–470.

90. S. A. Lipton, Y. B. Choi, Z. H. Pan *et al.*, 'A redox based mechanism for the neuroprotective and neuro-destructive effects of nitric oxide and related nitroso compounds.' *Nature*, 1993, **364**, 626–632.

91. J. L. Dinerman, T. M. Dawson, M. J. Schell *et al.*, 'Endothelial nitric oxide synthase localized to hippo-campal pyramidal cells: implications for synaptic plasticity.' *Proc. Natl. Acad. Sci.*, 1994, **91**, 4214–4218.

92. J. B. Schulz, P. L. Huang, R. T. Mathews *et al.*, 'Striatal malonate lesions are attenuated in neuronal nitric oxide synthase knockout mice.' *J. Neurochem.*, 1996, **67**, 430–433.

93. C. P. Lebel and S. C. Bondy, 'Oxygen radicals: common mediators of neurotoxicity.' *Neurotoxicol. Teratol.*, 1991, **13**, 341–346.

94. B. Halliwell, 'Reactive oxygen species and the central nervous system.' *J. Neurochem.*, 1992, **59**, 1609–1623.

95. B. H. Choi, 'Oxygen, anti-oxidants and brain dysfunction.' *Yonsei Med. J.*, 1993, **34**, 1–10.

96. P. Jenner, 'Oxidative damage in neurodegenerative disease.' *Lancet*, 1994, **344**, 796–798.

97. J. M. Jacobs and P. M. LeQuesne, in 'Greenfield's Neuropathology,' 5th edn., eds. J. H. Adams and L. W. Duchen, Oxford University Press, New York 1992, pp. 881–987.

98. B. Knusel, K. D. Beck, J. W. Winslow *et al.*, 'Brain-derived neurotrophic factor administration protects basal forebrain cholinergic but not nigral dopami-nergic neurons from degenerative changes after axotomy in the adult rat brain.' *J. Neurosci.*, 1992, **12**, 4391–4402.

99. D. M. Frim, T. A. Uhler, W. R. Galpern *et al.*, 'Implanted fibroblasts genetically engineered to produce brain-derived neurotrophic factor prevent 1-methyl-4-phenylpyridinium toxicity to dopaminer-gic neurons in the rat.' *Proc. Natl. Acad. Sci.*, 1994, **91**, 5104–5108.

100. M. Sendtner, B. Holtmann, R. Kolbeck *et al.*, 'Brain-derived neurotrophic factor prevents the death of motoneurons in newborn rats after nerve section.' *Nature*, 1992, **360**, 757–759.

101. J. M. Schumacher, M. P. Short, B. T. Hyman *et al.*, 'Intracerebral implantation of nerve-growth factor-producing fibroblasts protects striatum against neu-rotoxic levels of excitatory amino acids.' *Neuro-science*, 1991, **45**, 561–570.

102. M. P. Mattson, M. A. Lovell, K. Furukawa *et al.*, 'Neurotrophic factors attenuate glutamate-induced accumulation of peroxides, elevation of intracellular Ca^{2+} concentration, and neurotoxicity and increase anti-oxidant enzyme activities in hippocampal neurons.' *J. Neurochem.*, 1995, **65**, 1740–1751.

103. Y. Goodman, A. J. Bruce, B. Cheng *et al.*, 'Estrogens attenuate and corticosterone exacerbate excitotoxi-city, oxidative injury and amyloid B-peptide toxicity in hippocampal neurons.' *J. Neurochem.*, 1996, **66**, 1836–1844.

104. B. R. Shivakumar and V. Ravindranath, 'Selective modulation of glutathione in mouse brain regions and its effects on acrylamide-induced neurotoxicity.' *Biochem. Pharmacol.*, 1992, **43**, 263–269.

105. J. T. Naarala, J. J. Loikkanen, M. H. Ruotsalainen *et al.*, 'Lead amplifies glutamate-induced oxidative stress.' *Free Radic. Biol. Med.*, 1995, **19**, 689–693.

106. S. S. Simons, Jr., 'Environmental estrogens: can two "alrights" make a wrong?' *Science*, 1996, **272**, 1451.

107. M. Aschner, L. Rising and K. J. Mullaney, 'Differ-ential sensitivity of neonatal rat astrocyte cultures to mercuric chloride (MC) and methylmercury (MEHG): studies on K^+ and amino acid trans-port and metallothionein (MT) induction.' *Neuro-toxicology*, 1996, **17**, 107–116.

108. E. Carafoli, 'Intracellular calcium homeostasis.' *Annu. Rev. Biochem.*, 1987, **56**, 395–433.

109. K. G. Baimbridge, M. R. Celio and J. H. Rogers, 'Calcium binding proteins in the nervous system.' *Trends Neurosci.*, 1992, **15**, 303–308.

110. C. W. Heizmann and K. Braun, 'Changes in Ca (2+)-binding proteins in human neurodegenerative dis-orders.' *Trends Neurosci.*, 1992, **15**, 259–264.

111. M. P. Mattson, B. Rychlik, C. Chu *et al.*, 'Evidence for calcium-reducing and excito-protective roles for the calcium-binding protein calbindin-D28k in cul-tured hippocampal neurons.' *Neuron*, 1991, **6**, 41–51.

112. W. Lukas and K. A. Jones, 'Cortical neurons containing calretinin are selectively resistant to calcium overload and excitotoxicity *in vitro*.' *Neuro-science*, 1994, **61**, 307–316.

113. C. Leranth and C. E. Ribak, 'Calcium-binding proteins are concentrated in the CA2 field of the monkey hippocampus: a possible key to this region's resistance to epileptic damage.' *Exp. Brain Res.*, 1991, **85**, 129–136.

114. T. F. Freund and Z. Magloczky, 'Early degeneration of calretinin-containing neurons in the rat hippocam-pus after ischemia.' *Neuroscience*, 1993, **56**, 581–596.

115. T. Yamada, P. L. McGeer, K. G. Baimbridge *et al.*, 'Relative sparing in Parkinson's disease of substantia nigra dopamine neurons containing calbindin-D28K.' *Brain Res.*, 1990, **526**, 303–307.

116. P. R. Hof and J. H. Morrison, 'Neocortical neuronal subpopulations labeled by a monoclonal antibody to calbindin exhibit differential vulnerability in Alzhei-mer's disease.' *Exp. Neurol.*, 1991, **111**, 293–301.

117. D. G. Munoz, 'The distribution of chromogranin A-like immunoreactivity in the human hippocampus coincides with the pattern of resistance to epilepsy-induced neuronal damage.' *Annals Neurol.*, 1990, **27**, 266–275.

118. F. U. Reiffen and M. Gratzi, 'Chromogranins, wide-spread in endocrine and nervous tissue, bind Ca^{2+}.' *FEBS Lett.*, 1986, **195**, 327–330.

119. J. P. Simon, M. F. Bader and D. Aunis, 'Secretion from chromaffin cells is controlled by chromogranin A-derived peptides.' *Proc. Natl. Acad. Sci. USA*, 1988, **85**, 1712–1716.

120. M. P. Mattson and S. B. Kater, 'Development and selective neurodegeneration in cell cultures from different hippocampus regions.' *Brain Res.*, 1989, **490**, 110–125.

121. B. Halliwell and J. M. Gutteridge, 'Oxygen radicals and the nervous system.' *Trends Neurosci.*, 1985, **8**, 22–26.

122. L. Packer, L. Prilipko and Y. Christen (eds.), 'Free radicals in the brain, aging, neurological and mental disorders,' Springer, 1992.

123. C. P. Lebel and S. C. Bondy, 'Oxidative damage and cerebral aging.' *Progress Neurobiol.*, 1992, **38**, 601–609.

124. M. A. Verity, 'Oxidative damage and repair in the developing nervous system.' *Neurotoxicology*, 1994, **15**, 81–91.

125. M. A. Goss-Sampson and D. P. R. Muller, 'Studies on the neurobiology of vitamin E (alpha-tocopherol) and some other anti-oxidant systems in the rat.' *Neuro-pathol. Appl. Neurobiol.*, 1987, **13**, 289–296.

126. M. A. Philbert, C. M. Beiswanger, D. K. Waters *et al.*, 'Cellular and regional distribution of reduced glutathione in the nervous system of the rat: histochemical localization by mercury orange and *O*-phthaldialdialdehyde-induced histofluorescence.' *Toxicol.*

Appl. Pharmacol., 1991, **107**, 215–227.

127. M. A. Philbert, C. M. Beiswanger, M. M. Manson *et al.*, 'Glutathione *S*-transferases and gamma-glutamyl transpeptidase in the rat nervous system: a basis for differential susceptibility to neurotoxicants.' *Neurotoxicology*, 1995, **16**, 349–362.

128. T. Miyakawa and M. Deshimaru, 'Electron microscopical study of experimentally induced poisoning due to organic mercury compound. Mechanism of development of the morbid change.' *Acta Neuropathol. (Berl.)*, 1969, **14**, 126–136.

129. P. J. Carder, R. Hume, A. A. Fryer *et al.*, 'Glutathione *S*-transferase in human brain.' *Neuropathol. Appl. Neurobiol.*, 1990, **16**, 293–303.

130. S. Houdou, H. Kuruta, M. Hasegawa *et al.*, 'Developmental immunohistochemistry of catalase in the human brain.' *Brain Res.*, 1991, **556**, 267–270.

131. T. H. Murphy, M. J. De Long and J. T. Coyle, 'Enhanced NAD(P)H: quinone reductase activity prevents glutamate toxicity produced by oxidative stress.' *J. Neurochem.*, 1991, **56**, 990–995.

132. I. Garcia, I. Martinou, Y. Tsujimoto *et al.*, 'Prevention of programmed cell death of sympathetic neurons by the bcl-2 proto-oncogene.' *Science*, 1992, **258**, 302–304.

133. A. Batistatou and L. A. Greene, 'Aurintricarboxylic acid rescues PC12 cells and sympathetic neurons from cell death caused by nerve growth factor deprivation: correlation with suppression of endonuclease activity.' *J. Cell Biol.*, 1991, **115**, 461–471.

134. K. Nakayama, K. Nakayama, I. Negishi *et al.*, 'Disappearance of the lymphoid system in Bcl-2 homozygous mutant chimeric mice.' *Science*, 1993, **261**, 1584–1588.

135. D. J. Veis, C. M. Sorenson, J. R. Shutter *et al.*, 'Bcl-2-deficient mice demonstrate fulminant lymphoid apoptosis, polycystic kidneys, and hypopigmented hair.' *Cell*, 1993, **75**, 229–240.

136. N. Motoyama, F. Wang, K. A. Rothm *et al.*, 'Massive cell death of immature hematopoietic cells and neurons in Bcl-x-deficient mice.' *Science*, 1995, **267**, 1506–1510.

137. K. Shimazaki, A. Ishida and N. Kawai, 'Increase in bcl-2 oncoprotein and the tolerance to ischemia-induced neuronal death in the gerbil hippocampus.' *Neurosci. Res.*, 1994, **20**, 95–99.

138. J. Chen, S. H. Graham, P. H. Chan *et al.*, 'bcl-2 is expressed in neurons that survive focal ischemia in the rat.' *Neuroreport.*, 1995, **6**, 394–398.

139. J. Chen, R. L. Zhu, M. Nakayama *et al.*, 'Expression of the apoptosis-effector gene, Bax, is up-regulated in vulnerable CA1 neurons following global ischemia.' *J. Neurochem.*, 1996, **67**, 64–71.

140. M. N. Wallace and S. K. Bisland, 'NADPH-diaphorase activity in activated astrocytes represents inducible nitric oxide synthase.' *Neuroscience*, 1994, **59**, 905–919.

141. R. J. Ferrante, N. W. Kowall, M. F. Beal *et al.*, 'Selective sparing of a class of striatal neurons in Huntington's disease.' *Science*, 1985, **230**, 561–563.

142. V. L. Dawson, T. M. Dawson, E. D. London *et al.*, 'Nitric oxide mediates glutamate neurotoxicity in primary cortical cultures.' *Proc. Natl. Acad. Sci. USA*, 1991, **88**, 6368–6371.

143. P. S. Chard, D. Bleakman, J. R. Savidge *et al.*, 'Capsaicin-induced neurotoxicity in cultured dorsal root ganglion neurons: involvement of calcium-activated proteases.' *Neuroscience*, 1995, **65**, 1099–1108.

11.03
Degenerative and Regenerative Events in the Central and Peripheral Nervous Systems

EDWARD D. HALL

Pharmacia & Upjohn Inc., Kalamazoo, MI, USA

11.03.1 INTRODUCTION

This chapter describes the fundamental processes and molecular mechanisms involved in neuronal degeneration after neural injury, whether secondary to physical injury, ischemic insult, or chemical neurotoxicity. Regardless of the mechanism that triggers the degeneration of central or peripheral neurons, it is increasingly apparent that the degenerative biochemistry and anatomical events are, in large part, stereotypical. Thus, in the context of this chapter, the degenerative events may be described based upon extensive studies in models of experimental central nervous system (CNS) trauma or ischemia. However, these same events are equally relevant to the neural degeneration that takes place after exposure of the nervous system to a variety of neurotoxic substances. In the case of acute traumatic (i.e., mechanical) or ischemic injury, one sees the natural course of degeneration, and sometimes regeneration, in the purest pathophysiological form. In the case of the slow, chronic neurodegeneration associated with Alzheimer's, Parkinson's, or motor neuron diseases, as well as in various chemical neurotoxicities, the time course of the degenerative processes and the efficacy of reparative/regenerative responses may differ, but the molecular players are the same as those which participate in acute insults.

11.03.2 MOLECULAR MECHANISMS OF NEURONAL DEGENERATION

Following traumatic, ischemic, or neurotoxic CNS injury, neural degeneration occurs through a combination of primary and secondary mechanisms. For instance, while primary mechanical disruption of CNS parenchyma and blood vessels is obviously important in the case of mechanical injury to the brain or spinal cord, much of the neural degeneration is due to a cascade of neurochemical and pathophysiological events set in motion by the primary mechanical insult. Clinical support for this concept has been provided by the results of the second National Acute Spinal Cord Injury Study (NASCIS II) which showed that treatment of spinal cord-injured patients, beginning within the first 8 h postinjury with a 24 h dosing regimen of the glucocorticoid steroid methylprednisolone, produces a significant improvement in neurological recovery over a 3, 6, or 12 month follow-up.[1] Thus, the fact that an acute pharmacological treatment can modify the post-traumatic neurological course implies that there is indeed a modifiable secondary

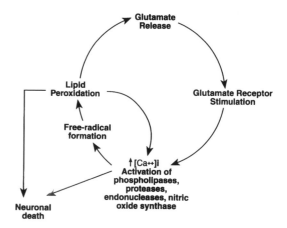

Figure 1 Interactions of excitotoxic glutamate release, glutamate receptor stimulation, intracellular Ca^{2+} overload, oxygen radical formation, and lipid peroxidation in traumatic, ischemic, or neurotoxic degeneration (modified with permission of Oxford University Press from *J. Neurosci.*, 1990, **10**, 1035–1041.)

neurodegenerative process that is triggered by the initial injury. A similar concept of secondary injury is also relevant to ischemic and hemorrhagic stroke (e.g., intracerebral hemorrhage, subarachnoid hemorrhage).

Secondary injury involves a complex interplay of multiple mechanisms. Available evidence suggests that the principal players are excessive release of the excitatory amino acid neurotransmitter glutamate, intracellular calcium overload, and the induction of iron-catalyzed, oxygen free radical-induced lipid peroxidation as represented in Figure 1. It is important to note from the outset that each of these processes are interwoven. In fact, positive feedback loops serve to amplify the preceding molecular events. The topics of excitotoxic and oxidative stress-induced degeneration will be presented in more detail in subsequent chapters. However, they need to be briefly presented here to provide a basic understanding of the fundamental mechanisms of neuronal degeneration.

11.03.2.1 Glutamate-induced Excitotoxicity

Glutamate functions as an excitatory amino acid neurotransmitter that normally elicits postsynaptic responses, mainly at dendritic sites. However, when glutamate is released in large amounts, or its reuptake is impaired, it can trigger "excitotoxic" degeneration. This topic is considered in detail in Chapter 31, this volume. Thus, it will be only briefly discussed here. Three types of glutamate receptors have been identified as being pathophysiologically important or potentially important: (i) NMDA (*N*-methyl-D-aspartate) preferring; (ii) AMPA (β-amino-3-hydroxy-5-methylisoxazole-4-propionic acid)/kainate preferring; and (iii) metabotropic.

11.03.2.1.1 NMDA receptors

The NMDA receptor, which has been the most intensely studied of the three types, appears to play an important pathophysiological role in traumatic and focal cerebral ischemic insults and, in certain chronic neurodegenerative disorders and neurotoxicities.[2] Figure 2 illustrates our present understanding of the complex regulation of the NMDA receptor, including glycine and polyamine binding modulatory sites whose function appears to be to enhance NMDA receptor activation. Pharma-

cological blockade of the glutamate recognition sites, the associated ion channel, or the glycine or polyamine regulatory sites has been reported to reduce infarct size in models of focal cerebral ischemia. Clinical trials of the competitive NMDA antagonist selfotel (CGS-19755) and the polyamine site antagonist, eliprodil, have been initiated for the treatment of head injury and thromboembolic stroke.

11.03.2.1.2 AMPA/kainate receptors

Less is known concerning the function of the AMPA/kainate receptors than the NMDA receptor, except that it is predominantly linked to a sodium-gating channel. However, overactivation of this particular glutamate receptor has been associated with neural degeneration in models of brief global and more

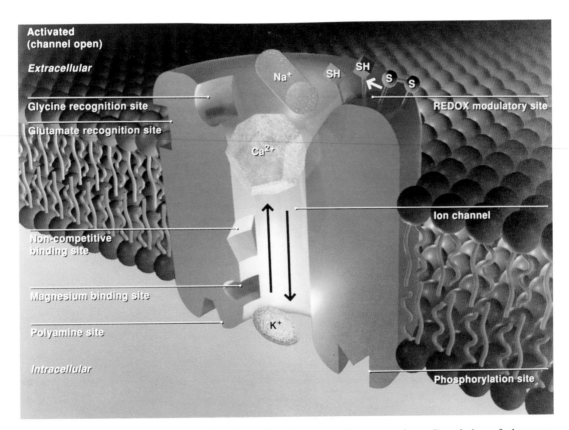

Figure 2 NMDA receptor complex showing activated receptor (i.e., open channel) and sites of pharmacological modulation. The NMDA receptor and channel possess a number of modulatory sites that can be pharmacologically manipulated to either enhance or attenuate glutamate-stimulated ionic fluxes. Interaction of glutamate or a glutamate-like agonist (e.g., NMDA) with the glutamate recognition site leads to activation of inward calcium and sodium currents that can be pathophysiologic when occurring in excess. The glycine recognition, polyamine, and phosphorylation sites, when acted upon by selective agonists, each serve to enhance the effects of glutamate receptor activation. Agents that block any of these sites will attenuate NMDA receptor-mediated ionic currents. Additionally, within the channel are noncompetitive and magnesium binding sites that attenuate calcium flux when pharmacologically blocked. There is also a REDOX modulatory site that is believed to consist of two adjacent sulfhydryl groups (cysteine residues) that must be in the reduced state for the NMDA receptor complex to be activated.

prolonged focal brain ischemia, as evidenced by the protective action of the AMPA receptor-selective antagonist 2,3-dihydroxy-6-nitro-7-sulfamoyl-benzo(*F*)quinoxaline (NBQX) in both instances.[3,4] There is some evidence that after ischemic insults, the subunit composition of the AMPA receptor is altered, allowing calcium gating and subsequent intracellular calcium overload to occur, as also seen with the NMDA receptor.

agonists (e.g., *trans*-ACPD) may enhance NMDA receptor sensitivity to glutamate and, thus, may be injury-promoting. Nevertheless, the precise consequence of metabotropic receptor activation or blockade is unclear since selective metabotropic agonists have been shown to be both neuroprotective and injury-promoting in *in vitro* and *in vivo* neuroprotection paradigms.

11.03.2.1.3 *Metabotropic receptors*

Even less is known concerning the pathophysiological role of the metabotropic glutamate receptor, except for the fact that it is not linked to ion gating, but rather to the hydrolysis of phosphatidyl inositol, which may trigger release of intracellular calcium stores.[5] Evidence suggests that it exists in close proximity to the NMDA receptor, and that metabotropic

11.03.2.2 Intracellular Calcium Overload

It has been clearly demonstrated that a major trigger for neuronal death associated with secondary injury is the intracellular accumulation of free calcium.[6] Figure 3 summarizes the aspects of cellular calcium homeostasis that are interrupted by primary and secondary neural injury leading to calcium overload. Despite the convincing body of evidence that calcium

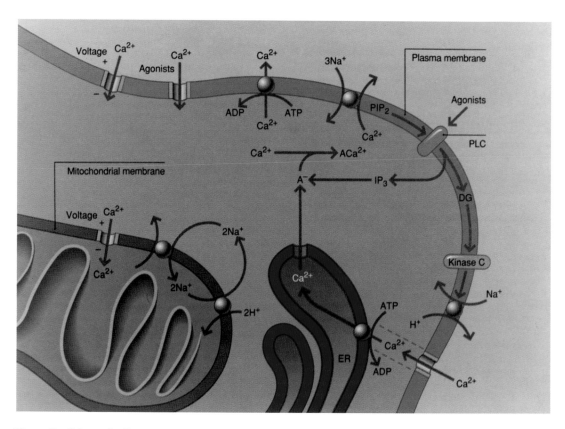

Figure 3 Schematic diagram illustrating major components of cellular calcium homeostasis. The cytosolic intracellular calcium concentration (Ca^{2+}_i) represents the balance between influx via voltage and agonist (e.g., NMDA and AMPA/kainate)-operated channels, efflux via an ATP-dependent pump (Ca^{2+}-ATPase), and a $3Na^+/Ca^{2+}$ exchanger and intracellular sequestration by endoplasmic reticulum and mitochondria. Furthermore, intracellular efflux from these storage sites can occur in response to the formation of inositol triphosphate (IP_3) in response to agonist-activated metabotropic receptors. The functional integrity of a $2Na^+/H^+$ exchange mechanism is also important for maintenance of the Na^+ gradients necessary for operation of the $3Na^+/Ca^{2+}$ exchanger.

overload is important in secondary neuronal injury, multiple trials of the dihydropyridine L-type calcium channel blocker, nimodipine, in head injury and stroke have yielded negative results. This most likely indicates that calcium entry via voltage-dependent channels only represents part of the reason for calcium overload after CNS injury, ischemia, or exposure to various neurotoxins. Glutamate receptor-operated calcium influx and/or intracellular release may be equally, or more, important in this regard.

11.03.2.3 Oxygen Free Radicals and Lipid Peroxidation

Oxygen radical-induced lipid peroxidation has been convincingly recognized as a significant factor in the pathophysiology of acute CNS traumatic and ischemic injuries[7–9] and subarachnoid hemorrhage (SAH).[10] The radical-initiated peroxidation of neuronal, glial and vascular cell membranes, and myelin is catalyzed by free iron released from hemoglobin, transferrin, and ferritin by either lowered tissue pH or by oxygen radicals. If unchecked, lipid peroxidation is a geometrically-progressing process that spreads over the surface of the cell membrane, causing impairment of phospholipid-dependent enzymes, disruption of ionic gradients and, if severe enough, membrane lysis. Much of the evidence in support of this statement is derived from studies demonstrating the early occurrence of lipid peroxidation in injured CNS tissue, and from investigations of the neuroprotective efficacy of antioxidant pharmacological compounds.[8] In addition, lipid peroxidation has been increasingly shown to be mechanistically involved in Parkinson's disease,[11,12] Alzheimer's disease,[13,14] and motor neuron diseases, most notably amyotrophic lateral sclerosis.[15,16] While proteins, nucleic acids, and carbohydrates are also susceptible to oxygen radical damage, the most avid targets of oxygen radical-induced injury are cell membrane lipids, including cholesterol and, in particular, polyunsaturated fatty acids. Central nervous tissue provides an especially vulnerable target for the occurrence of lipid peroxidative reactions due to the high content of iron found in many brain regions, which varies in parallel with the regional sensitivity to *ex vivo* lipid peroxidation.[17] Additionally, brain and spinal cord membrane phospholipids contain a higher proportion of polyunsaturated fatty acids, such as linoleic acid (18:2) and arachidonic acid (20:4), which are sensitive to peroxidation.[18]

11.03.2.3.1 *Superoxide radical and its sources*

The primary radical formed in most biological processes is superoxide anion (O_2^-). Within traumatized or ischemic nervous tissue, a number of sources of superoxide radical are operative within the first minutes and hours after the insult. These include the arachidonic acid cascade (i.e., prostaglandin synthase and 5-lipoxygenase activity), enzymatic (i.e., monoamine oxidase), or autoxidation of biogenic amine neurotransmitters (e.g., dopamine), mitochondrial leak, xanthine oxidase activity, and the oxidation of extravasated hemoglobin. Over the first few postischemic hours and days, activated microglia and infiltrating neutrophils and macrophages provide additional sources of O_2^- (Figure 4). Superoxide is also the principal radical species generated during 6-hydroxydopamine[19] and 1-methyl-4-phenyl-1,2,3,6-tetrahydropyridine (MPTP)[20] neurotoxic insult to dopaminergic neurons.

Superoxide, which is formed by the single electron reduction of oxygen, may act as either an oxidant (electron acceptor) or reductant (electron donor). While O_2^- itself is reactive, its direct reactivity toward biological substrates in aqueous environments is questionable. Moreover, once formed, O_2^- undergoes spontaneous dismutation to form H_2O_2 in a reaction that is markedly accelerated by the enzyme superoxide dismutase (SOD) ($O_2^- + O_2^- + 2H^+ \rightarrow H_2O_2 + O_2$).

However, O_2^- actually exists in equilibrium with the hydroperoxyl radical ($O_2^- + H^+ \rightarrow HO_2^\cdot$). The pK_a of this reaction is 4.8 and the relative concentrations of O_2^- and $HO_2^\cdot$ depend on the H^+ concentration. Therefore, at a pH around 6.8, the ratio of $O_2^-:HO_2^\cdot$ is 100:1, whereas at a pH of 5.8, the ratio is only 10:1. Thus, under conditions of tissue acidosis of a magnitude known to occur within the mechanically-injured or ischemic nervous system, a significant amount of the O_2^- formed exists as hydroperoxyl radical. Furthermore, compared with O_2^-, $HO_2^\cdot$ is considerably more lipid soluble and is a far more powerful oxidizing (or reducing) agent. Therefore, as the pH of a solution falls and the equilibrium between O_2^- and $HO_2^\cdot$ shifts toward greater formation of $HO_2^\cdot$, increased reactivity is observed, particularly toward lipids. In addition, while the dismutation of O_2^- to H_2O_2 is exceedingly slow at neutral pH in the absence of SOD, $HO_2^\cdot$ will dismutate to H_2O_2 far more readily at acidic pH values since the rate constant for $HO_2^\cdot$ dismutation is of the order of 10^8 times greater than for O_2^-. Thus, in an acidic environment, O_2^- is: (i) converted to the

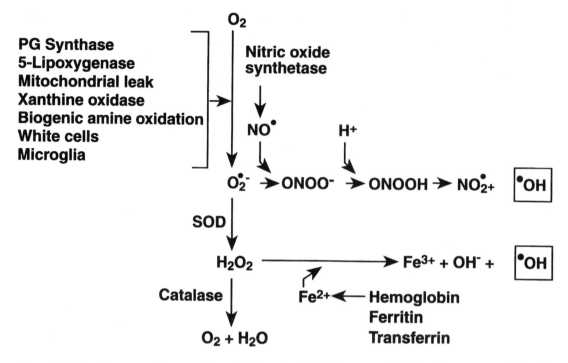

Figure 4   Potential sources of superoxide and hydroxyl radicals after traumatic, ischemic, or neurotoxic insults.

more reactive, more lipid soluble $HO_2\cdot$, and (ii) its rate of dismutation to H_2O_2 is greatly increased.

11.03.2.3.2   *Iron and the formation of hydroxyl radical and iron–oxygen complexes*

As already noted, the CNS is an extremely rich source of iron. Under normal circumstances, low molecular weight forms of redox-active iron in the brain are maintained at extremely low levels. Extracellularly, the iron transport protein, transferrin, tightly binds iron in the Fe^{3+} form. Intracellularly, Fe^{3+} is sequestered by the iron storage protein, ferritin. While both ferritin and transferrin have very high affinity for iron at neutral pH and effectively maintain iron in a noncatalytic state, both proteins readily give up their iron at pH values of 6.0 or less.[21] In the case of ferritin, its iron can also be released by reductive mobilization by O_2^- ($O_2^- + Fe^{3+} \rightarrow O_2 + Fe^{2+}$). Therefore, within the traumatized or ischemic CNS environment where pH is typically lowered and several sources of O_2^- are active, conditions are favorable for the potential release of iron from storage proteins. Similarly, neurotoxicants that generate O_2^- will trigger iron release. Once iron is released from ferritin or transferrin, it can actively catalyze oxygen radical reactions (Figure 4).

A second source of catalytically-active iron is hemoglobin. Subarachnoid or intracerebral hemorrhage places hemoglobin in contact with nervous tissue. While hemoglobin itself can stimulate oxygen radical reactions, it is more likely that iron released from hemoglobin is responsible for hemoglobin-mediated lipid peroxidation. Iron is released from hemoglobin by either H_2O_2 or lipid hydroperoxides (LOOHs), and this release is further enhanced as the pH falls to 6.5 or below. Therefore, hemoglobin may catalyze oxygen radical formation and lipid peroxidation, either directly or through the release of iron by H_2O_2, LOOH, and/or acidic pH.

Free iron or iron chelates participate in further free radical production at two levels. First, the autoxidation of Fe^{2+} provides an additional source of O_2^- ($Fe^{2+} + O_2 \rightarrow Fe^{3+} + O_2^-$). Second, Fe^{2+} is oxidized in the presence of H_2O_2 to form hydroxyl radical (Fenton's reaction; $Fe^{2+} + H_2O_2 \rightarrow Fe^{3+} + \cdot OH + OH^-$) or perhaps a ferryl ion ($Fe^{2+} + H_2O_2 \rightarrow Fe^{3+}OH + OH^-$). Both $\cdot OH$ and $Fe^{3+}OH$ are extraordinarily potent initiators of lipid peroxidation.

11.03.2.3.3   *Peroxynitrite-mediated free radical formation*

Another mechanism of hydroxyl radical formation has been identified, namely the peroxynitrite pathway.[22] This involves endothelial cells, neutrophils, macrophages and microglia which can produce two radicals, superoxide and nitric oxide (·NO), the latter from nitric oxide synthetase. The two radical species can combine to form peroxynitrite anion (ONOO⁻) which, at physiological pH, largely undergoes protonation ($pK_a = 6.8$), thus becoming peroxynitrous acid (ONOOH). However, ONOOH is an unstable acid that readily decomposes to give two potent oxidizing radicals, ·OH and nitrogen dioxide ($O_2^- + \cdot NO \rightarrow ONOO^- + H^+ \rightarrow ONOOH \rightarrow \cdot NO_2 + \cdot OH$) (Figure 4). Thus, this mechanism provides another source of the highly reactive ·OH that may be operative within the ischemic (or postischemically reperfused) CNS. Moreover, ·NO₂ may also initiate lipid peroxidation or nitrate (i.e., inactivate) cellular proteins. A particularly relevant feature

of this reaction is that ONOOH has a relatively long half-life and, thus, is potentially more diffusible compared to either superoxide or ·OH. Therefore, it may offer a mechanism by which free radical damage may occur at a site remote from the actual location of oxygen radical formation.

11.03.2.3.4 *Chemistry of lipid peroxidation*

Figure 5 provides an overview of the chemistry of the initiation and propagation phases of cell membrane lipid peroxidation. Initiation of peroxidation occurs when a radical species with significant oxidizing capability,

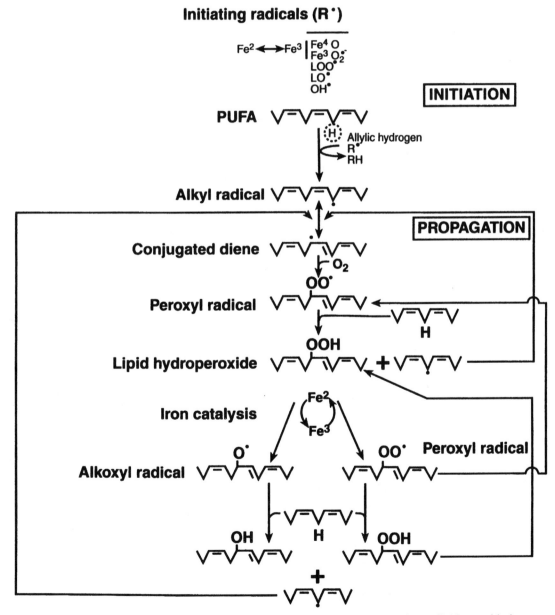

Figure 5 Chemistry of the initiation and propagation phases of cell membrane lipid peroxidation.

such as ·OH, removes an allylic hydrogen from an unsaturated fatty acid (LH), resulting in a radical chain reaction. Once the process begins, iron may participate in driving the process as lipid hydroperoxides (LOOH) formed through initiation are decomposed by reactions with either Fe^{2+}, Fe^{3+}, or their chelates.

As in the case of the formation of reactive inorganic radicals, both of the reactions of LOOH with iron have acidic pH optima, and are thus more likely to occur in the context of ischemic acidosis. Either alkoxyl (LO·) or peroxyl (LOO·) radicals arising from LOOH decomposition by iron can promote so-called LOOH-dependent lipid peroxidation, resulting in chain branching reactions (LOO· + LH → LOOH + L· or LO· + LH → LOH + L·).[23] Thus, during cerebral ischemia, LOOH-dependent lipid peroxidation may take place since sufficient LOOH and iron pre-exist in normal membranes to allow for propagatory peroxidative reactions to occur. In other words, if iron release from storage proteins occurs, propagation of lipid peroxidation may begin in the absence of an oxygen radical initiator.

11.03.2.3.5 Lipid peroxidation and compromise of ionic homeostasis

Lipid peroxidation has also been linked to dysfunctions in ionic homeostasis as schematically illustrated in Figure 6. One aspect of this dysfunction concerns the sensitivity of membrane Ca^{2+} ATPase (i.e., Ca^{2+} pump) to peroxidation-induced damage.[24] Similarly, oxidative inactivation of the membrane Na^+/K^+-ATPase can lead to intracellular Na^+ accumulation which will then reverse the direction of the Na^+/Ca^{2+} exchanger (antiporter) and exacerbate intracellular Ca^{2+} accumulation. For example, it has been shown in gerbils subjected to 3 h of unilateral carotid occlusion-induced hemispheric ischemia that there is a serious deficit in postreperfusion recovery of cortical extracellular Ca^{2+} levels (i.e., persistence of intracellular accumulation) that occurs simultaneously with postischemic brain lipid peroxidation. In contrast, pharmacological administration of the lipid antioxidant 21-aminosteroid, tirilazad, facilitates recovery of extracellular Ca^{2+} together with a reduction in ischemic neuronal damage.[25]

11.03.2.3.6 Interaction of lipid peroxidation with excitotoxic mechanisms

In addition to the association of lipid peroxidation with the loss of Ca^{2+} homeostasis, there appears to be an intimate reciprocal

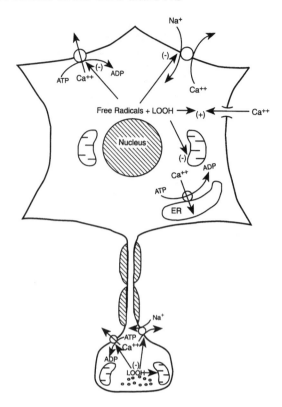

Figure 6 Schematic representation of the negative impact of oxygen free radicals and lipid peroxidation (LOOH) on Ca^{2+} influx, extrusion, and intracellular sequestration (mitochondria and endo-plasmic reticulum, ER). In this regard, the voltage-dependent and agonist-operated channels, the Ca^{2+}-ATPase pump and the $3Na^+/Ca^{2+}$ exchanger, are all sensitive to oxidative damage.

association between the excitotoxic and oxygen radical-lipid peroxidative mechanisms of neuronal degeneration. First of all, free radical mechanisms have been demonstrated to potentiate glutamate release. For example, when brain slices are exposed to the O_2^--generating system, xanthine plus xanthine oxidase, there is an enhancement in glutamate and aspartate release which is antagonized by various free radical scavengers.[26] On the other hand, glutamate infusion into rat striatum *in vivo* has been shown to stimulate the production of hydroxyl radicals as measured by the salicylate trapping method.[27] Moreover, peroxidation inhibitors have been shown to attenuate NMDA-induced damage in cortical cell cultures, implying the involvement of lipid peroxidation in glutamate excitotoxicity.[28] Thus, these and other examples have disclosed that there is an intimate association between glutamate excitotoxicity, calcium overload, and free radical formation and membrane oxidative damage. Indeed, each of these processes can enhance the others, as shown in Figure 1.

11.03.3 NECROTIC AND APOPTOTIC DEGENERATION

There are two basic forms of neuronal degeneration that have been identified, necrosis and apoptosis.[29] They are distinguishable on the basis of morphological and biochemical criteria. Necrosis typically involves the death of groups of neurons, while apoptotic degeneration is associated with single or scattered neurons within a population. In necrosis, there is a loss of cell membrane integrity and, as a result, the affected neurons swell and lyse while in apoptosis membrane blebbing occurs, but without any loss of integrity. In fact, apoptosing neurons actually shrink rather than swell.

Neurons undergoing necrosis typically show lysosomal rupture, while those manifesting apoptosis do not show lysosomal disruption. The nucleus of necrotizing neurons shows a clumpy, ill-defined, chromatin aggregation, whereas the nucleus of apoptotic neurons illustrates compaction of the chromatin into uniformly dense masses (chromatin condensation). In necrotic degeneration, there is a considerable inflammatory response, including invasion of white cells and macrophages, with significant phagocytosis of the degenerating neuronal elements. In contrast, while there is some phagocytosis by activated microglia in apoptotic degeneration, little extrinsic inflammatory invasion takes place.

Necrotic and apoptotic neuronal degeneration are also distinguishable according to biochemical criteria. Most notably, apoptosis is an active process that requires energy to drive gene transcription and protein synthesis, with tightly regulated steps and a growing list of pro- and anti-apoptotic regulatory genes having been identified. In contrast, necrosis is a passive event which lacks an energy requirement; mRNA and protein synthesis are actually severely inhibited. In necrosis, there is an early, if not immediate, loss of ionic homeostasis, whereas ionic pumping, although progressively compromised, may continue well into the apoptotic process. Finally, apoptotic and necrotic degeneration are most often distinguished by the nature of the nuclear DNA breakdown that occurs. In apoptosis, there is an enzymatic, nonrandom oligonucleosomal fragmentation of the DNA that can be nicely illustrated by "DNA laddering" in an ethidium bromide-stained agarose gel. On the other hand, the DNA digestion that occurs in necrosis is a random event with regard to the location of the DNA cleavage and the size of the resulting fragments. Table 1 summarizes and compares the characteristics of necrotic and apoptotic degeneration.

It should be noted that necrosis and apoptosis may be seen simultaneously. For instance, both processes have been shown to play a role in the context of acute traumatic and post-ischemic insults.[30–32] Likewise, there may be overlap between the two in the case of many chemical neurotoxicities.

11.03.4 DEGENERATIVE RESPONSE OF NEURONS TO AXONAL TRANSECTION

The most straightforward context in which to examine the functional and structural characteristics of neuronal degeneration is that which occurs after axonal transection (i.e., axotomy).[33] While degeneration may be observed in any neuron that undergoes axotomy, the degenerative characteristics have been most extensively studied in the setting of axotomized spinal or brainstem motor neurons.

Table 1 Comparison of the characteristics of necrotic and apoptotic neuronal degeneration.

Necrotic	*Apoptotic*
Groups of neurons	Single or scattered neurons
Loss of membrane integrity	Maintenance of membrane integrity, but blebbing occurs
Neurons swell	Neurons shrink
Lysosomal disruption	No lysosomal disruption
Clumpy, ill-defined chromatin aggregation	Chromatin condensation
Random DNA digestion	Nonrandom oligonucleosomal DNA fragmentation
DNA smear seen on agarose gels	DNA "laddering" seen on agarose gels
Inflammatory response with invasion of neutrophils and macrophages	No inflammatory response except for microglial activation
No energy requirement	Energy required
Inhibition of mRNA and protein synthesis	Active mRNA and protein synthesis
Early loss of ionic homeostasis	Later loss of ionic homeostasis

11.03.4.1 Anterograde Degeneration

Following axotomy, the distal portion of the motor axon and the nerve terminals which innervate the muscle fibers invariably undergo degeneration. Since this degeneration involves the innervating and transmitting portion of the motor neuron, it is referred to as "anterograde" degeneration (Figure 7). More commonly, it is called "Wallerian" degeneration after the nineteenth century English physician, Augustus Waller, who first described the anatomical progression of the postaxotomy anterograde axonal degenerative process. The fact that the axon and nerve terminals cannot survive separation from their cell body indicates their trophic dependence on the somatic mRNA and protein synthetic machinery. The time course of anterograde degeneration is dependent on the length of the distal axon; the longer the axon, the longer its function and morphological integrity will persist after axotomy. This no doubt reflects the fact that the longer axons possess a greater store of cell body-synthesized trophic materials which can maintain the distal axon and terminals for a longer period.

Anterograde degeneration is first manifested as a progressive loss of the function of the distal nerve terminals.[34,35] Specifically, the terminals lose their ability to maintain and recover from high frequency (i.e., tetanic) transmission. Simultaneously, their capability to show posttetanic transmitter facilitation is lost. For example, soleus nerve terminals, which are normally capable of displaying posttetanic repetitive discharge, lose that facilitatory capability as an initial degenerative event. In fact, after brief episodes of tetanic activation, the degenerative nerve terminals display a post-tetanic depression of neuromuscular transmission from which they may not recover. This is demonstrative of a compromise in their transmitter synthetic capacity (reduced acetylcholine content) and/or their membrane ionic pumping function such that their recovery of electrical excitability is inadequate. Similarly, the degenerating, but still somewhat functional nerve terminals, lose their responsiveness to pharmacological agents that enhance transmitter release and/or induce repetitive discharge (e.g., edrophonium). From an ultrastructural perspective, the nerve terminals, as they lose their transmitter function and excitability, display a decrease in the number of synaptic vesicles together with a swelling of the mitochondria, the latter change being consistent with an impairment of nerve terminal energy metabolic capacity.[36] The associated reduction in ATP synthesis contributes to the decrease in transmitter synthesis and membrane ionic homeostasis necessary for the preservation of

excitability. Consistent with these biochemical events, electrophysiological studies have documented a progressive decrease in the quantal content of the evoked neuromuscular end-plate potential and a decrease in the frequency of miniature end-plate potentials (i.e., quantal, non-evoked transmitter release) as the early degenerative events unfold.[37] A similar pattern of neuronal terminal energy metabolic compromise and consequent deficiency in neurotransmitter function are observed after exposure to mitochondrial toxicants such as 3-acetylpyridine.[38]

Following functional and structural degeneration of the nerve terminals within the first several hours or few days (depending on distal axonal length), the myelinated axons undergo progressive functional and then anatomical degeneration over the subsequent days or few weeks.[33] Since degeneration begins in the nerve terminals and then the more proximal axons, anterograde degeneration is often described as a "dying back" process. Indeed, many clinical sensory and motor neuropathies, whether caused by disease or toxin exposure, are referred to using this terminology based upon the initial loss of sensation or neuromuscular transmission followed by loss of axonal conduction. As the axonal degeneration progresses,

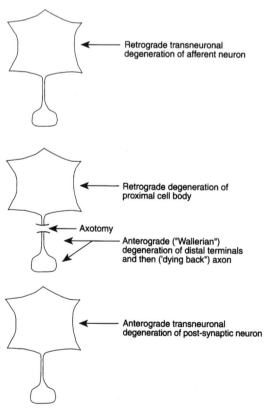

Figure 7 Illustration of retrograde and anterograde neuronal degeneration after axotomy.

the Schwann cells de-differentiate and convert to macrophages that remove the debris left from the degenerated axon and terminals. In the case of the anterograde degeneration of central axons, the astrocytes and microglia fulfill this responsibility.

The functional and anatomical aspects of anterograde "dying-back" degeneration associated with axotomy can also be produced by exposure to organophosphates, such as di-isopropylfluorophosphate (DFP).[39,40] Similarly, anterograde axonal degeneration is witnessed as a secondary event after closed (non-penetrating) head injury in the absence of actual physical axotomy.[41] Thus, the typical "Wallerian" type of degeneration can occur even when the overall neuronal morphological integrity is maintained. Indeed, chemicals that poison the cell body protein synthetic machinery (e.g., cisplatin) can elicit antero-grade degeneration by disrupting the production of materials that are necessary for the viability of the distal axons and terminals. Similarly, agents that disrupt microtubules (e.g., vincristine) can cause axonal and terminal degeneration by preventing axoplasmic transport of cell body-derived structural and trophic substances.

11.03.4.2 Retrograde Degeneration

Axonal injury or toxicity can also elicit a retrograde degenerative response (Figure 7). The anatomical aspects of the retrograde somatic response to axotomy have been historically referred to as "chromatolysis."[42] During the chromatolytic response, the cell body may swell to twice its normal size. The nucleus moves to an eccentric position and also becomes swollen. The rough endoplasmic reticulum ("Nissl substance") breaks apart and, as a consequence, the intensity of staining with basic dyes, like cresyl violet, is decreased. In actuality, the process of chromatolysis may not lead to degeneration, but rather an axonal regenerative response. The total amount of mRNA and protein synthesis increases dramatically during chromatolysis in support of the increased macromolecular requirements associated with regeneration. Most of the increase in mRNA transcription is related to the need for structural proteins. Indeed, the expression of neurotransmitter synthetic proteins may be severely downregulated during the retrograde regenerative response to axotomy.[43]

Alternatively, this regenerative chromatolytic response may give way to a degeneration response if it is insufficiently supported by external trophic factors.[33] As in the case of the anterograde degeneration of the distal nerve terminals and axon, this has been most fully studied in the context of the motor neuron. The degeneration of the motor neuronal cell body, which may take place after axotomy, demonstrates the fact that despite the advantage of possessing a nucleus and its genetic programs, as well as the full protein synthetic machinery, the cell body's well-being is, in large part, dependent on continuity with its target organ, the muscle fiber (or the innervated neuron in the case of the central nervous system). The peripheral Schwann cells also appear to exert a significant trophic influence on the motor neuronal soma. The susceptibility to retrograde degeneration is greater in immature neurons (i.e., in neonatal animals), apparently indicative of a greater dependency on, or an immaturity of, target organ or Schwann cell production of trophic factors. Similarly, if the axotomy is performed close to the cell body, retrograde degeneration is more likely.[43]

While some of the evidence is circumstantial, the characteristics of retrograde neuronal degeneration follow an apoptotic pattern. For instance, retrograde degeneration of neurons in the visual pathway has been identified as apoptotic based upon morphological and DNA fragmentation criteria.[44] Also, neonatal transgenic mice that overexpress the anti-apoptotic gene product bcl-2 show nearly complete protection from axotomy-induced death of facial motor neurons.[45] Furthermore, axotomy-induced death of neonatal rat retinal ganglion neurons is associated with apoptosis-related nuclear pyknosis and DNA fragmentation that are preventable by either inhibition of mRNA transcription with actinomycin D or protein synthesis inhibition with cycloheximide.[44]

11.03.4.3 Transneuronal Degeneration

The consequence of neuronal degeneration is often not confined to the degenerating neurons. In some instances, neuronal loss triggers degeneration of the denervated postsynaptic elements.[33] This is called anterograde transneuronal degeneration (Figure 7). While it does not occur in all instances, a notable example of the phenomenon has been documented in the visual pathway. Specifically, an optic nerve lesion which effectively denervates the lateral geniculate nucleus causes delayed atrophy of the genicular neurons which, in turn, can cause atrophy of the cortical neurons that they normally innervate. The likely reason for anterograde transneuronal degeneration is the removal of the excitatory and/or trophic influences that the presynaptic neuron exerts on the postsynaptic neuron. In addition, retrograde transneuronal degeneration can occur if the primary degenerative neuron normally exerts a trophic influence on the upstream neurons that innervate it (Figure 7).

11.03.5 ROLE OF OXYGEN RADICAL-INDUCED LIPID PEROXIDATION IN NEURONAL DEGENERATION

11.03.5.1 Lipid Peroxidation and Postaxotomy Anterograde and Retrograde Degeneration

As discussed earlier in this chapter, iron-catalyzed, oxygen radical-induced lipid peroxidation has been strongly implicated in post-traumatic and postischemic neuronal degeneration.[7-9] Similarly, evidence for a critical role of oxidative stress and peroxidative damage postaxotomy in both anterograde and retrograde degeneration has been obtained. Much of this evidence is derived from studies showing that the rate or extent of degeneration can be reduced by pharmacological administration of oxygen radical scavenging or lipid antioxidant compounds. For example, pretreatment of cats with high doses of the lipid peroxidation inhibitors vitamin E[46] or the 21-aminosteroid tirilazad[47] has been shown to slow the rate of axotomy-induced soleus motor nerve terminal functional loss. With both antioxidants, an improved maintenance of high frequency neuromuscular function has been demonstrated. While less effective, intensive pretreatment with synthetic glucocorticoid steroids, such as methylprednisolone, which possess some antioxidant efficacy, has also been shown to preserve the function of cat soleus motor nerve terminals following either axotomy[48] or exposure to neurotoxic doses of the organophosphate DFP.[49,50]

Dosing with the lipid antioxidant, tirilazad, has been shown to attenuate the extent of post-axotomy retrograde degeneration of neonatal rat facial motor neurons, pharmacologically implicating a lipid peroxidative mechanism.[51] Another compound, deprenyl, which possesses oxygen radical scavenging properties,[52] has been reported to attenuate axotomy-induced facial motor neuronal degeneration.[53] Additionally, the administration of either ciliary neurotrophic factor (CNTF),[54] brain-derived neurotrophic factor (BDNF),[55] or nerve growth factor (NGF)[56] attenuates postaxotomy degeneration in specific neuronal populations. Brain-derived neurotrophic factor has been shown to enhance the resistance of neurons to oxidative stress, implying that its trophic effects may involve regulation of cellular antioxidant defenses.[57]

Additional support for a role of oxidative mechanisms in axotomy-induced retrograde motor neuronal degeneration comes from the fact that this degeneration follows an apoptotic process. As already noted, transgenic mice that over-express the antiapoptotic bcl-2 oncoprotein, which is believed to have antioxidant properties,[58] show less motor neuron loss after axotomy than nontransgenic mice.[45] Furthermore, radical trapping agents, such *N-t*-butyl-α-phenyl nitrone (PBN), can attenuate neuronal apoptosis, further supporting a pathological role of oxygen radicals.[59]

11.03.5.2 Lipid Peroxidation in a Transgenic Model of Familial Amyotrophic Lateral Sclerosis

Accumulating evidence indicates that the motor neuronal degeneration in familial amyotrophic lateral sclerosis (ALS) may result from an increased sensitivity to the damaging effects of oxygen radicals due to a gain-of-function mutation in the antioxidant enzyme copper–zinc superoxide dismutase (Cu,Zn SOD).[15] In support of this, transgenic mice that express a mutated (Gly → Ala 93) human Cu,Zn SOD develop motor neuron disease.[60,61] Vacuolar neuropathology is observed in spinal motor neurons by 37 days of age, and clinical disease begins on the average at 91 days of age, followed by paralysis at 136 days of age. By the time of onset of clinical disease, significant loss of motor neurons is observed, and at end-stage disease, half of the motor neurons in the cervical and lumbar spinal segments are lost.

Evidence suggests that oxygen radical-induced lipid peroxidation may be involved in the degeneration of spinal motor neurons in this transgenic model. First, the mutant Cu,Zn SOD can catalyze the generation of oxygen radical species.[62] Second, there is a relative depletion of spinal cord vitamin E levels in the transgenic mice, reflecting increased antioxidant utilization. Third, chronic dietary supplementation with vitamin E delays the onset of disease symptoms.[63] However, it is apparent that excitotoxic processes are also involved since two inhibitors of glutamatergic transmission, riluzole and gabapentin, while not affecting the onset of disease in the mice, significantly prolong their survival.[63] This scenario further underscores the interrelationship of excitotoxic and oxygen radical medicated neurodegenerative mechanisms discussed earlier and illustrated in Figure 1.

11.03.6 SELECTIVE NEURONAL VULNERABILITY

It is well known that all neurons are not created equally in regard to their susceptibility to certain degenerative insults (see Chapter 2, this volume). For instance, neurotoxic chemicals, such as MPTP, 6-hydroxydopamine and methamphetamine, selectively damage dopaminergic neurons, and 5,7-dihydroxytryptamine selectively affects serotonergic neurons. This selectivity is due to the fact that each of these agents is only taken up by the neurons they affect. On the other hand, the concept of selective vulnerability is also seen in the context of other physiological and toxicological insults with the mechanistic reasons not being so readily apparent. For example, the mitochondrial neurotoxicant, 3-acetylpyridine (3-AP), causes widespread neurotoxicity, as would be expected for a nonspecific toxic mechanism. However, 3-AP causes extremely severe damage to the inferior olivary neurons, while producing lesser damage to brainstem facial and hypoglossal motor neurons, and even less injury to nigrostriatal dopamine neurons.[64] Similarly, flurothyl-induced status epilepticus selectively damages the hippocampal CA3 neuronal population, while brief episodes of near-complete forebrain ischemia in either gerbils or rats mainly affects hippocampal CA1 neurons.[65] In the latter case, increasing periods of ischemia also lead to damage in striatal interneurons, nigrostriatal dopamine neurons, and layers 3, 5, and 6 of the cerebral cortex.

Brief forebrain ischemia in the gerbil or rat has been used to create selective injury to CA1 neurons to model the neuropathological pattern of cardiac arrest/resuscitation (global ischemia/reperfusion)-induced brain damage. While the reason(s) for the selective vulnerability of CA1 neurons to ischemia is not fully known, work has documented that there is a greater basal level of oxidative stress (i.e., ·OH) in the hippocampus, and the postischemic rise in ·OH levels and lipid peroxidation products is also significantly greater compared to the less vulnerable cortex.[66] A higher level of arachidonic acid metabolism in the hippocampus, which generates oxygen radicals as a by-product, may provide an explanation. This is based on the finding that basal ·OH content in the gerbil hippocampus is suppressed by administration of the cyclo-oxygenase inhibitor, ibuprofen, which, at the same time, does not affect cortical ·OH levels.[66] Similarly, the fact that slightly more prolonged episodes of ischemia result in degeneration of nigrostriatal dopamine neurons (in addition to the CA1 population) parallels an ischemia-induced increase in dopamine turnover (monoamine oxidase activity) that is associated with oxygen radical formation.[67,68] Lipid peroxidative insult to the dopamine nerve terminals results in a dying back (retrograde) degeneration of the nigral cell bodies that is attenuated by certain lipid peroxidation inhibitors.[67] Therefore, the observation of selective neuronal vulnerability may be due, in part, to inherent differences in oxidative stress or susceptibility to oxidative damage in certain brain regions. Another aspect of the selective vulnerability of nigrostriatal dopamine neurons, in addition to the fact that dopamine metabolism provides a source of oxygen radicals, is that the substantia nigra possesses a relatively high iron content that can promote oxygen radical formation and peroxidative reactions.[11]

11.03.7 GLIAL RESPONSE TO DEGENERATION

Earlier in this chapter, a postdegenerative phagocytic role for Schwann cells, astrocytes, and microglia was briefly discussed. Glial elements are activated and may undergo proliferation in response to neurodegenerative insults, and play an exceedingly complex modulatory role. For example, microglia are massively activated after traumatic,[69] ischemic,[70] and neurotoxic[71] insults, and during retrograde degeneration of axotomized motor neurons[72] as evidenced by their expression of type I and II major histocompatibility (MHC) antigens. This activation actually precedes the onset of the neurodegeneration process and persists beyond its completion. Many of the microglial products may serve to exacerbate or hasten neurodegeneration, including oxygen radicals, proteases and cytokines.[73] Similarly, astrocytic activation, as indicated by increased expression of glial fibrillary acidic protein (GFAP), may represent a deleterious response in the sense that post-injury glial scarring may serve as an impediment to neural regeneration.

On the other hand, activated glia may play a protective role via the production of trophic factors (e.g., NGF, BDNF, CNTF), which have been shown to be capable of lessening neuronal degeneration in models of axotomy-induced retrograde degeneration.[54-56] Similarly, glial activation is a source of other potentially neuroprotective response proteins, including amyloid precursor protein (APP)[74] and apolipoprotein E (Apo E).[75] Amyloid precursor protein is a large transmembrane glycoprotein that is found in five amino acid lengths, APP677, APP695, APP714, APP751 or APP770, depending on the degree of post-transcriptional splicing. In the central nervous system, the four larger forms of APP are produced by neurons, astrocytes and microglia. An increased production of APP has been reported to occur following various experimental brain insults, including focal[76] and global[77] cerebral ischemia, excitotoxic injury with either ibotenic,[78] kainic,[79] or quinolinic[80] acids, and mechanical trauma.[81,82] This increase in APP may represent a trophic or repair response since intraventricular injection of the 695 and 751 amino acid forms of APP rescues hippocampal CA1 neurons in the rat from forebrain ischemia-induced degeneration.[83] On the other hand, post-translational cleavage of APP can give rise to β-amyloid protein (β-AP), which is found in amyloid plaques in the brains of Alzheimer's disease victims and is believed by many to play a pathophysiological role in the disease.[84] In fact, traumatic[82] and ischemia/reperfusion[77] brain injuries have also been shown to lead to increased immunostaining for β-AP, which has been repeatedly demonstrated to be neurotoxic to hippocampal and certain other types of neurons.[85] Thus, the liberation of this protein from APP may exert a negative influence on neural degeneration and repair.

Another glial-derived response protein that is increased after neural insults is apolipoprotein E (Apo E). This lipoprotein acts as a cholesterol transport protein in the periphery, a function it may also fulfill in the brain by transporting cholesterol to injured, but surviving, neurons that are undergoing repair or axonal regeneration.[86] Interestingly, a critical role of Apo E in the susceptibility to Alzheimer's disease has become apparent since the early 1990s. There are three forms of Apo E (E2, E3, and E4). Individuals who are homozygous for Apo E4 have as much as a ninefold increased risk of late-onset Alzheimer's disease in comparison to individuals who have Apo E2 or Apo E3.[87] The basis for this association of Apo E phenotype with the incidence of Alzheimer's disease is uncertain. As one possibility, Apo E4 is reported to bind to and possibly enhance the deposition of β-amyloid.[88] Indeed, an increased deposition of β-AP in cerebral cortex of Alzheimer's patients who have one or two Apo-E4 alleles has been reported.[89] Moreover, strong Apo-E immunoreactivity has been demonstrated within the amyloid cores of senile plaques,[90] and the distribution of Apo E expression correlates with that of fibrillar amyloid in cerebral cortex.[91] On the other hand, Apo E may also play a role in protection of the cytoskeleton through its purported role in preventing hyperphosphorylation of the microtubule-associated protein τ.[92] However, this action is not seen with Apo E4. Therefore, in individuals who possess Apo E4, tau proteins may be less protected from hyperphosphorylation, a prerequisite to the formation of Alzheimer's-related neurofibrillary tangle pathology, and perhaps the neurofibrillary pathologies seen with many neurotoxins. Thus, Apo E, like APP, may constitute a two-edged sword in relation to its being capable of playing a positive or a negative role in the response of the brain to injury or neurotoxicity that is largely dependent on the specific Apo E phenotype of the subject.

11.03.8   CONTRIBUTION OF STRESS AND ENDOGENOUS GLUCOCORTICOIDS TO NEURODEGENERATION

Earlier in this chapter, it was pointed out that intensive short-term treatment with synthetic glucocorticoid steroids, such as methylprednisolone, can retard the early anterograde degeneration of axotomized motor neurons, probably due to an intrinsic ability to inhibit lipid peroxidation reactions.[48] Furthermore, acute high-dose methylprednisolone treatment has been shown to inhibit posttraumatic lipid peroxidation reactions in the injured spinal cord and to facilitate long-term neurological recovery in animals and man.[93] In contrast, a considerable body of evidence has shown that glucocorticoids can exert a detrimental effect on injured neurons, or, in some instances, be directly neurotoxic.

11.03.8.1   Role of Glucocorticoid Steroids in Age-related and Neurotoxic Neuronal Degeneration

The first suggestion that glucocorticoid steroids might have deleterious effects on neuronal survival came from work in rats that showed a significant positive correlation between progressive age-related neuronal loss in the hippocampus and elevations in plasma corticosterone. The same investigators went on to show that adrenalectomy forestalled the age-related loss of hippocampal neurons.[94] These findings were replicated by others who postulated that cumulative glucocorticoid exposure over the life span may enhance age-dependent loss of central neurons (at least in the hippocampus) and that prolonged stress, by further elevating steroid levels, could accelerate brain aging.[95] Subsequently, it was shown that corticosterone can exacerbate the hippocampal neurodegenerative effects of either the mitochondrial toxin, 3-AP, or the excitotoxin, kainic acid, while adrenalectomy affords partial protection.[96]

Two potential mechanisms for the hippocampal toxicity of glucocorticoid steroids have been put forth, both of which center on a glucocorticoid receptor mechanism. The selective vulnerability of the hippocampus may be due, in part, to the high concentration of glucocorticoid receptors in that brain region.[97] The first purported mechanism relates to the finding that corticosterone can impair glucose utilization by hippocampal neurons. Administration of the sugar mannose, whose uptake by hippocampal neurons is not impaired by glucocorticoids, results in protection of hippocampal neurons against corticosterone exacerbation of kainic acid toxicity.[98] The second mechanism for glucocorticoid neurotoxicity is based upon the electrophysiological demonstration that corticosterone acts to increase intracellular calcium in hippocampal pyramidal neurons.[99] Specifically, in hippocampal slices from corticosterone-treated rats, the calcium-dependent after-hyperpolarization of pyramidal neurons is increased in comparison to that observed in normal rats. On the other hand, after-hyperpolarization amplitude and duration are less in adrenalectomized animals. Consistent with this facilitatory effect of glucocorticoids on neuronal calcium conductance, one-week treatment of cats with the potent synthetic steroid, triamcinolone, increases the amplitude of the calcium-based, after-depolarization in spinal motor neurons.[100] Similar glucocorticoid dosing acts to increase the posttetanic repetitive discharge of cat motor nerve terminals, a phenomenon also related to inward calcium fluxes.[101] Thus, there is ample evidence that glucocorticoids can augment stimulation-induced, inward calcium movement. In view of the well-known cytotoxic effects of elevated intracellular calcium, chronic glucocorticoid presence could promote neuronal damage by this mechanism, in addition to an impairment of glucose metabolism.

11.03.8.2 Glucocorticoid Exacerbation of Ischemic Neuronal Damage

In addition to a possible role in age-related neuronal damage, glucocorticoids have been shown to worsen acute postischemic neuronal necrosis. For example, in a rat, temporary global ischemia model, it has been shown that prior adrenalectomy reduces postischemic neuronal damage in cerebral cortex, striatum, and all regions of the hippocampus in comparison to intact animals, while corticosterone supplementation enhances neuronal necrosis.[102]

11.03.8.3 Glucocorticoid Impairment of Lesion-induced Axonal Sprouting and Synaptogenesis

An additional deleterious effect of glucocorticoids on the nervous system has been documented. Corticosterone treatment of rats significantly impairs axonal sprouting and synaptogenesis in the hippocampus following lesions of the entorhinal cortex.[103] Astrocytes appear to hypertrophy in lesioned and glucocorticoid-treated rats. For this reason, it has been suggested that sustained treatment (vs. acute treatment) of neural injury[99] with glucocorticoids can potentially impair recovery. The mechanism of this effect is unknown, but in view of its occurrence with modest glucocorticoid doses, a receptor-mediated action is certain.

11.03.9 SUMMARY

Neuronal degeneration secondary to traumatic, ischemic, or neurotoxic insults involves an interplay of excitotoxic, intracellular Ca^{2+} overload and free radical-induced oxidative damage. These interrelated mechanisms appear to play a role in both necrotic and apoptotic (programmed) neuronal degenerative processes. Oxidative damage, directed mainly toward cellular membrane lipids and proteins, and its impact on cellular calcium homeostasis, may be the final common pathway of the degeneration process in both instances. In particular, several lines of evidence suggest that membrane lipid peroxidation is a fundamental mechanism of degeneration of neuronal terminals, axons, and cell bodies independent of the particular primary insult. Indeed, the selective vulnerability of some neuronal populations to degeneration may involve an increased propensity to oxygen radical production and lipid peroxidation. Consistent with this view, certain antioxidant compounds have been shown to attenuate degeneration of selectively vulnerable neurons in a variety of traumatic, ischemic, and neurotoxic models. Similarly, at least some of the identified neurotrophic factors (e.g., CNTF, BDNF, NGF) may exert their neuroprotective actions via an induction of antioxidant enzymatic defenses. Additional glial-derived factors may also modulate the degenerative process, including APP and Apo E. However, in the case of both APP and Apo E, this action is phenotype-specific (i.e., APP695, APP751, Apo E2, Apo E3). Finally, stress-induced elevations in glucocorticoid hormones can enhance degeneration and impede the regeneration of selectively vulnerable neuronal populations.

11.03.10 REFERENCES

1. M. B. Bracken, 'Pharmacological treatment of acute spinal cord injury: current status and future prospects.' *Paraplegia*, 1992, **30**, 102–107.
2. J. McCulloch, 'Excitatory amino acid antagonists and their potential for the treatment of ischaemic brain damage in man [Review].' *Br. J. Clin. Pharmacol.*, 1992, **34**, 106–114.
3. M. J. Sheardown, E. O. Nielsen, A. J. Hansen *et al.*, '2,3-Dihydroxy-6-nitro-7-sulfamoyl-benzo(*F*)-quinoxaline: a neuroprotectant for cerebral ischemia.' *Science*, 1990, **247**, 571–574.
4. R. Gill, L. Nordholm and D. Lodge, 'The neuroprotective actions of 2,3-dihydroxy-6-nitro-7-sulfamoyl-benzo(*F*)quinoxaline (NBQX) in a rat focal ischaemia model.' *Brain Res.*, 1992, **580**, 35–43.
5. D. D. Schoepp and P. J. Conn, 'Metabotroic glutamate receptors in brain function and pathology.' *Trends Pharmacol. Sci.*, 1993, **14**, 13–20.
6. B. K. Siesjo and F. Bengtsson, 'Calcium fluxes, calcium antagonists, and calcium-related pathology in brain ischemia, hypoglycemia, and spreading depression: a unifying hypothesis [Review].' *J. Cereb. Blood Flow Metab.*, 1989, **9**, 127–140.
7. J. M. Braughler and E. D. Hall, 'Central nervous system trauma and stroke. I. Biochemical considerations for oxygen radical formation and lipid peroxidation [Review].' *Free Radic. Biol. Med.*, 1989, **6**, 289–301.
8. E. D. Hall and J. M. Braughler, in 'Molecular and Cellular Approaches to the Treatment of Neurological Disease,' ed. S. G. Waxman, Raven Press, New York, 1993, pp. 81–105.
9. B. K. Siesjo, C. D. Agardh and F. Bengtsson, 'Free radicals and brain damage.' *Cerebrovasc. Brain Metab. Rev.*, 1989, **1**, 165–211.
10. T. Asano, T. Matsui and Y. Takuwa, *Crit. Rev. Neurosurg.*, 1991, **1**, 361.
11. M. B. Youdim, D. Ben-Schachar and P. Riederer, 'The possible role of iron in the etiopathology of Parkinson's disease.' *Mov. Disord.*, 1993, **8**, 1–12.
12. P. Jenner, A. H. Schapira and C. D. Marsden, 'New insights into the causes of Parkinson's disease.' *Neurology*, 1992, **42**, 2241–2250.
13. M. A. Smith, P. L. Richey, S. Taneda *et al.*, 'Advanced Maillard reaction end products, free radicals, and protein oxidation in Alzheimer's disease [Review].' *Ann. NY Acad. Sci.*, 1994, **738**, 447–454.
14. K. V. Subbarao, J. S. Richardson and L. C. Ang, 'Autopsy samples of Alzheimer's cortex show increased peroxidation *in vitro*.' *J. Neurochem.*, 1990, **55**, 342–345.
15. D. R. Rosen, T. Siddique, D. Patterson *et al.*, 'Mutations in Cu/Zn superoxide dismutase gene are associated with familial amyotrophic lateral sclerosis.' *Nature*, 1993, **362**, 59–62.
16. J. D. Mitchell and M. J. Jackson, in 'Handbook of Amyotrophic Lateral Sclerosis,' ed. R. A. Smith, Marcel Dekker, New York, 1992, pp. 533–541.
17. M. M. Zaleska and R. A. Floyd, 'Regional lipid peroxidation in rat brain *in vitro*: possible role of endogenous iron.' *Neurochem. Res.*, 1985, **10**, 397–410.
18. D. A. White, in 'Function of Phospholipids,' eds. G. B. Ansell, J. N. Hawthorne and R. M. Dawson, Elsevier, Amsterdam, 1973, pp. 441–482.
19. G. Cohen, 'Monamine oxidase hydrogen peroxide and Parkinson's disease.' *Adv. Neurol.*, 1986, **45**, 119.
20. J. D. Adams, Jr. and I. N. Odunze, 'Biochemical mechanisms of l-methyl-4-phenyl-1,2,3,6-tetrahydropyridine toxicity. Could oxidative stress be involved in the brain? [Review]' *Biochem. Pharmacol.*, 1991, **41**, 1099–1105.
21. B. Halliwell and J. M. C. Gutteridge, 'Free Radicals in Biology and Medicine,' Oxford University Press, New York, 1991, pp. 1–543.
22. J. S. Beckman, 'The double-edged role of nitric oxide in brain function and superoxide-mediated injury.' *J. Dev. Physiol.*, 1991, **15**, 53–59.
23. A. Sevanian, in 'CRC Reviews, Cellular Antioxidant Defense Mechanisms,' 1992, vol. II, chap. 18, pp. 77–95.
24. T. T. Rohn, T. R. Hinds and F. F. Vincenzi, 'Ion transport ATPases as targets for free radical damage. Protection by an aminosteroid of the Ca^{2+} pump ATPase and Na^+/K^+ pump ATPase of human red blood cell membranes.' *Biochem. Pharmacol.*, 1993, **46**, 525–534.
25. E. D. Hall, K. E. Pazara and J. M. Braughler, 'Effects of tirilazad mesylate on postischemic brain lipid peroxidation and recovery of extracellular calcium in gerbils.' *Stroke*, 1991, **22**, 361–366.
26. D. E. Pelligrini-Giampietro, G. Cherici, M. Alesiani *et al.*, 'Excitatory amino acid release and free radical formation may cooperate in the genesis of ischemia-induced neuronal damage.' *J. Neurosci.*, 1990, **10**, 1035–1041.
27. D. P. C. Boisvert and C. Schreiber, in 'Pharmacology of Cerebral Ischemia,' eds. J. Krieglstein and H. Oberpichler, Wassenschaftliche Verlaggesellschaft, Stuttgart, 1992, pp. 1–10.
28. H. Monyer, D. M. Hartley and D. W. Choi, '21-Aminosteroids attenuate excitotoxic neuronal injury in cortical cell cultures.' *Neuron*, 1990, **5**, 121–126.
29. D. E. Bredesen, 'Neural apoptosis.' *Ann. Neurol.*, 1995, **38**, 839–851.
30. M. D. Linnik, R. H. Zobrist and M. D. Hatfield, 'Evidence supporting a role for programmed cell death in focal cerebral ischemia in rats.' *Stroke*, 1993, **24**, 2002–2009.
31. J. P. MacManus, A. M. Buchan, I. E. Hill *et al.*, 'Global ischemia can cause DNA fragmentation indicative of apoptosis in rat brain.' *Neurosci. Lett.*, 1993, **164**, 89–92.
32. A. Rink, K. M. Fung, J. Q. Trojanowski *et al.*, 'Evidence of apoptotic cell death after experimental traumatic brain injury in the rat.' *Am. J. Pathol.*, 1995, **147**, 1575–1583.
33. J. P. Kelly, in 'Principles of Neural Science,' eds. E. R. Kandel and J. H. Schwartz, Elsevier, New York, 1990, pp. 187–195.
34. M. Okamoto and W. F. Riker, Jr., 'Motor nerve terminals as the site of initial functional changes after denervation.' *J. Gen Physiol.*, 1969, **53**, 70–80.
35. M. Okamoto and W. F. Riker, Jr., 'Subacute denervation: a means of disclosing mammalian motor nerve terminals as critical sites of acetylcholine and facilitatory drug actions.' *J. Pharmacol. Exp. Ther.*, 1969, **166**, 217–224.
36. A. B. Drakontides and W. F. Riker, Jr., 'Functional, pharmacologic, and ultrastructural correlates in degenerating rat motor nerve terminals.' *Exp. Neurol.*, 1978, **59**, 112–123.
37. R. Miledi and C. R. Slater, 'On the degeneration of rat neuromuscular junctions after nerve section.' *J. Physiol. (Lond.)*, 1970, **207**, 507–528.
38. V. H. Sethy, H.-Y. Wu, J. A. Oostveen *et al.*, *Exp. Neurol.*, 1996, **140**, 79–83.

39. H. E. Lowndes, T. Baker and W. F. Riker, Jr., 'Motor nerve dysfunction in delayed DFP neuropathy.' *Eur. J. Pharmacol.*, 1974, **29**, 66–73.

40. H. E. Lowndes, T. Baker and W. F. Riker, Jr., 'Motor nerve terminal response to edrophonium in delayed DFP neuropathy.' *Eur. J. Pharmacol.*, 1975, **30**, 69.

41. A. D. Mendelow and G. M. Teasdale, 'Pathophysiology of head injuries.' *Br. J. Surg.*, 1983, **70**, 641–650.

42. A. R. Lieberman, 'The axon reaction: a review of the principal features of perikaryal responses to axon injury [Review].' *Int. Rev. Neurobiol.*, 1971, **14**, 49–124.

43. D. B. Hoover and J. C. Hancock, 'Effect of facial nerve transection on acetylcholinesterase, choline acetyltransferase and [³H]quinuclidinyl benzilate binding in rat facial nuclei.' *Neuroscience*, 1985, **15**, 481–487.

44. S. A. Rabacchi, L. Bonfanti, X. H. Liu *et al.*, 'Apoptotic cell death induced by optic nerve lesion in the neonatal rat.' *J. Neurosci.*, 1994, **14**, 5292–5301.

45. M. Dubois-Dauphin, H. Frankowski, Y. Tsujimoto *et al.*, 'Neonatal motoneurons overexpressing the bcl-2 protooncogene in transgenic mice are protected from axotomy-induced cell death.' *Proc. Natl. Acad. Sci. USA*, 1994, **91**, 3309–3313.

46. E. D. Hall, 'Intensive anti-oxidant pretreatment retards motor nerve degeneration.' *Brain Res.*, 1987, **413**, 175–178.

47. E. D. Hall and P. A. Yonkers, 'Preservation of motor nerve function during early degeneration by the 21-aminosteroid anti-oxidant U74006F.' *Brain Res.*, 1990, **513**, 244–247.

48. E. D. Hall and D. L. Wolf, 'Methylprednisolone preservation of motor nerve function during early degeneration.' *Exp. Neurol.*, 1984, **84**, 715–720.

49. T. Baker and A. Stanec, 'Methylprednisolone treatment of an organophosphorus-induced delayed neuropathy.' *Toxicol. Appl. Pharmacol.*, 1985, **79**, 348–352.

50. T. Baker, A. B. Drakontides and W. F. Riker, Jr., 'Prevention of the organophosphorus neuropathy by glucocorticoids.' *Exp. Neurol.*, 1982, **78**, 397–408.

51. E. D. Hall, S. L. Smith and J. A. Oostveen, *J. Neurosci. Res.*, 1996, **44**, in press.

52. C. C. Chiueh, S.-J. Huang and D. L. Murphy, 'Suppression of hydroxyl radical formation by MAO inhibitors: a novel possible neuroprotective mechanism in dopaminergic neurotoxicity.' *J. Neural Transm. Suppl.*, 1994, **41**, 189–196.

53. P. T. Salo and W. G. Tatton, 'Deprenyl reduces the death of motoneurons caused by axotomy.' *J. Neurosci. Res.*, 1992, **31**, 394–400.

54. M. Sendtner, G. W. Kreutzberg and H. Thoenen, 'Ciliary neurotrophic factor prevents the degeneration of motor neurons after axotomy.' *Nature*, 1990, **345**, 440–441.

55. M. Sendtner, B. Holtmann, R. Kolbeck *et al.*, 'Brain-derived neurotrophic factor prevents the death of motoneurons in newborn rats after nerve section.' *Nature*, 1992, **360**, 757–759.

56. L. R. Williams, S. Varon, G. M. Peterson *et al.*, 'Continuous infusion of nerve growth factor prevents basal forebrain neuronal death after fimbria fornix transection.' *Proc. Natl. Acad. Sci. USA*, 1986, **83**, 9231–9235.

57. M. B. Spina, S. P. Squinto, J. Miller *et al.*, 'Brain-derived neurotrophic factor protects dopamine neurons against 6-hydroxydopamine and *N*-methyl-4-phenylpyridinium ion toxicity: involvement of the glutathione system [see comments].' *J. Neurochem.*, 1992, **59**, 99–106.

58. H. Albrecht, J. Tschopp and C. V. Jongeneel, 'Bcl-2 protects from oxidative damage and apoptotic cell death without interfering with activation of NF-kappa B by TNF.' *FEBS Lett.*, 1994, **351**, 45–48.

59. J. L. Franklin, T. M. Miller and E. M. Johnson, 'Inhibition of programmed cell death by spin traps: evidence of a role for reactive oxygen in neuronal apoptosis.' *Soc. Neurosci. Abstr.*, 1994, **20**, 432.

60. M. E. Gurney, H. Pu, A. Y. Chiu *et al.*, 'Motor neuron degeneration in mice that express a human Cu, Zn superoxide dismutase mutation.' *Science*, 1994, **264**, 1772–1775.

61. A. Y. Chiu, P. Zhai, M. C. Dal Canto *et al.*, 'Age-dependent penetrance of disease in a transgenic mouse model of familial amyotrophic lateral sclerosis.' *Mol. Cell Neurosci.*, 1995, **6**, 349–362.

62. J. J. Wiedau-Pazos, J. J. Goto, S. Rabizadeh *et al.*, 'Altered reactivity of superoxide dismutase in familial amyotrophic lateral sclerosis.' *Science*, 1996, **271**, 515–518.

63. M. E. Gurney, F. B. Cutting, P. Zhai *et al.*, 'Antioxidants and inhibition of glutamatergic transmission have therapeutic benefit in a transgenic model of familial amyotrophic lateral sclerosis.' *Ann. Neurol.*, 1996, **39**, 147–157.

64. J. C. Desclin and J. Escubi, 'Effects of 3-acetylpyridine on the central nervous system of the rat, as demonstrated by silver methods.' *Brain Res.* 1974, **77**, 349–364.

65. R. N. Auer and B. K. Siesjo, 'Biological differences between ischemia, hypoglycemia, and epilepsy.' *Ann. Neurol.*, 1988, **24**, 699–707.

66. E. D. Hall, P. K. Andrus, J. S. Althaus *et al.*, 'Hydroxyl radical production and lipid peroxidation parallels selective post-ischemic vulnerability in gerbil brain.' *J. Neurosci. Res.*, 1993, **34**, 107–112.

67. P. K. Andrus, T. J. Fleck, J. A. Oostveen *et al.*, 'Ischemic damage to the nigrastriatal tract: relationship to increased dopamine release and oxygen radical formation and protection by the dopamine agonist pramipexole.' *Soc. Neurosci. Abstr*, 1995, **23**, 1252.

68. E. D. Hall, P. K. Andrus, J. A. Oostveen *et al.*, 'Post-ischemic degeneration of the nigrastriatal tract in gerbil brain: relationship to increased dopamine release and oxygen radical generation.' *J. Neurotrauma*, 1994, **11**, 111.

69. P. G. Popovich, W. J. Streit and B. T. Stokes, 'Differential expression of MHC Class II antigen in the contused rat spinal cord.' *J. Neurotrauma*, 1993, **10**, 37–46.

70. J. Gehrmann, P. Bonnekoh, T. Miyazawa *et al.*, 'Immunocytochemical study of an early microglial activation in ischemia.' *J. Cereb. Blood Flow Metab.*, 1992, **12**, 257–269.

71. H. Akiyama, S. Itagaki and P. L. McGeer, 'Major histocompatibility complex antigen expression on rat microglia following epidural kainic acid lesions.' *J. Neurosci. Res.*, 1988, **20**, 147–157.

72. W. J. Streit, M. B. Graeber and G. W. Kreutzberg, 'Expression of Ia antigen on perivascular and microglial cells after sublethal and lethal motor neuron injury.' *Exp. Neurol.*, 1989, **105**, 115–126.

73. J. Gehrmann, Y. Matsumoto and G. W. Kreutzberg, 'Microglia: intrinsic immuneffector cell of the brain.' *Brain Res. Rev.*, 1995, **20**, 269.

74. R. Siman, J. P. Card, R. B. Nelson *et al.*, 'Expression of beta-amyloid precursor protein in reactive astrocytes following neuronal damage.' *Neuron*, 1989, **3**, 275–285.

75. R. E. Pitas, J. K. Boyles, S. H. Lee *et al.*, 'Astrocytes synthesize apolipoprotein E and metabolize apolipoprotein E-containing lipoproteins.' *Biochim. Biophys. Acta*, 1987, **917**, 148–161.

76. K. Abe, R. E. Tanzi and K. Kogure, 'Selective induction of Kunitz-type protease inhibitor domain-containing amyloid precursor protein mRNA after persistent focal ischemia in rat cerebral cortex.' *Neurosci. Lett.*, 1991, **125**, 172–174.

77. E. D. Hall, J. A. Oostveen, E. Dunn *et al.*, 'Increased amyloid protein precursor and apolipoprotein E immunoreactivity in the selectively vulnerable hippocampus following transient forebrain ischemia in gerbils.' *Exp. Neurol.*, 1995, **135**, 17–27.

78. Y. Nakamura, M. Takeda, H. Niigawa *et al.*, 'Amyloid beta-protein precursor deposition in rat hippocampus lesioned by ibotenic acid injection.' *Neurosci. Lett.*, 1992, **136**, 95–98.

79. T. Kawarabayashi, M. Shoji, Y. Harigaya *et al.*, 'Expressing APP in the early stage of brain damage.' *Brain Res.*, 1991, **563**, 334–334.

80. R. Topper, J. Gehrmann, R. Banati *et al.* 'Rapid appearance of amyloid precursor protein immunoreactivity following excitotoxic brain injury.' *J. Neurotrauma*, 1993, **10**(Suppl. 1), S78.

81. N. Otsuka, M. Tomonaga and K. Ikeda, 'Rapid appearance of beta-amyloid precursor protein immunoreactivity in damaged axons and reactive glial cells in rat brain following needle stab injury.' *Brain Res.*, 1990, **568**, 335–338.

82. S. M. Gentleman, M. J. Nash, C. J. Sweeting *et al.*, 'Beta-amyloid precursor protein (beta APP) as a marker for axonal injury after head injury.' *Neurosci. Lett.*, 1993, **160**, 139–144.

83. V. L. Smith-Swintosky, L. C. Pettigrew, S. D. Craddock *et al.*, 'Secreted forms of beta-amyloid precursor protein protect against ischemic brain injury.' *J. Neurochem.*, 1994, **63**, 781–784.

84. B. H. Anderton, 'Expressing and processing of pathologic proteins in Alzheimer's disease.' *Hippocampus*, 1993, **3**, 227–237.

85. Y. Goodman and M. P. Mattson, 'Secreted forms of beta-amyloid precursor protein protect hippocampal neurons against amyloid beta-peptide-induced oxidative injury.' *Exp. Neurol.*, 1994, **128**, 1–12.

86. M. J. Ignatius, P. J. Gebicke-Harter, J. H. Skene *et al.*, 'Expressing apolipoprotein E during nerve degeneration and regeneration.' *Proc. Natl. Acad. Sci. USA*, 1986, **83**, 1125.

87. E. H. Corder, A. M. Saunders, W. J. Strittmatter *et al.*, 'Gene dose of apolipoprotein E type 4 allele and the risk of Alzheimer's disease in late onset families [see comments].' *Science*, 1993, **261**, 921–923.

88. W. J. Strittmatter, A. M. Saunders, D. Schmechel *et al.*, 'Apolipoprotein E: high-avidity binding to beta-amyloid and increased frequency of type 4 allele in late-onset familial Alzheimer disease.' *Proc. Natl. Acad. Sci. USA*, 1993, **90**, 1977–1981.

89. D. E. Schmechel, A. M. Saunders, W. J. Strittmatter *et al.*, 'Increased amyloid beta-peptide deposition in cerebral cortex as a consequence of apolipoprotein E genotype in late-onset Alzheimer disease.' *Proc. Natl. Acad. Sci. USA*, 1993, **90**, 9649–9653.

90. N. A. Elshourbagy, W. S. Liao, R. W. Mahley *et al.*, 'Apolipoprotein E mRNA is abundant in the brain and adrenals, as well as the liver, and is present in other peripheral tissues of rats and marmosets.' *Proc. Natl. Acad. Sci. USA*, 1985, **82**, 203–207.

91. E. Kida, A. A. Golabek, T. Wizniewski *et al.*, 'Regional differences in apolipoprotein E immunoreactivity in diffuse plaques in Alzheimer's disease brain.' *Neurosci. Lett.*, 1994, **167**, 73–76.

92. W. J. Strittmatter, A. M. Saunders, M. Goedert *et al.*, 'Isoform-specific interactions of apolipoprotein E with microtubule-associated protein tau: implications for Alzheimer disease.' *Proc. Natl. Acad. Sci. USA*, 1994, **91**, 11183–11186.

93. E. D. Hall, 'The neuroprotective pharmacology of methylprednisolone.' *J. Neurosurg.*, 1992, **76**, 13–22.

94. P. W. Landfield, R. K. Baskin and T. A. Pitler, 'Brain aging correlates: retardation by hormonal-pharmacological treatments.' *Science*, 1981, **214**, 581–584.

95. R. M. Sapolsky, L. C. Krey and B. S. McEwen, 'Prolonged glucocorticoid exposure reduces hippocampal neuron number: implications for aging.' *J. Neurosci.*, 1985, **5**, 1222–1227.

96. R. M. Sapolsky, 'A mechanism for glucocorticoid toxicity in the hippocampus: increased neuronal vulnerability to metabolic insults.' *J. Neurosci.*, 1985, **5**, 1228–1232.

97. B. S. McEwen, J. M. Weiss and L. S. Schwartz, 'Uptake of corticosterone by rat brain and its concentration by certain limbic structures.' *Brain Res.*, 1969, **16**, 227–241.

98. R. M. Sapolsky, 'Glucocorticoid toxicity in the hippocampus: reversal by supplementation with brain fuels.' *J. Neurosci.*, 1986, **6**, 2240–2244.

99. D. S. Kerr, L. W. Campbell, S. Y. Hao *et al.*, 'Corticosteroid modulation of hippocampal potentials: increased effect with aging.' *Science*, 1989, **245**, 1505–1509.

100. E. D. Hall, 'Glucocorticoid effects on central nervous excitability and synaptic transmission [Review].' *Int. Rev. Neurobiol.*, 1982, **23**, 165–195.

101. W. F. Riker, Jr., T. Baker and M. Okamoto, 'Glucocorticoids and mammalian motor nerve excitability.' *Arch. Neurol.*, 1975, **32**, 688–694.

102. R. M. Sapolsky and W. A. Pulsinelli, 'Glucocorticoids potentiate ischemic injury to neurons: therapeutic implications.' *Science*, 1985, **229**, 1397–1400.

103. S. W. Scheff and C. W. Cotman, 'Chronic glucocorticoid therapy alters axon sprouting in the hippocampal dentate gyrus.' *Exp. Neurol.*, 1982, **76**, 644–654.

11.04
Neurotoxicant-induced Oxidative Events in the Nervous System

STEPHEN C. BONDY

University of California, Irvine, CA, USA

11.04.1 INTRODUCTION

The terms free radicals or oxygen radicals have become commonly used since 1980 as a result of overwhelming data suggesting that short lived, incompletely reduced oxygen compounds are involved in a variety of disease processes. The works of Freeman and Crapo[1] and Halliwell and Gutteridge[2] especially have addressed the ubiquitous role of free radicals in the biology of disease and tissue injury. Most of the issues considered to date have dealt with the role of free radicals in the mechanisms of carcinogenesis, ischemia, and aging. In the field of toxicology, free radical research has focused predominantly on the area of pulmonary, cardiac, and hepatotoxicity.

The liver and the lung have long been known to be organs vulnerable to oxidative stress. However, the brain, with its high lipid content, high rate of oxidative metabolism, and somewhat low levels of free-radical eliminating enzymes, also may be a prime target of free-radical-mediated damage. The localization of antioxidant systems primarily in glial cells rather than in neurons,[3,4] while providing a first line of defense, may render the neurons especially susceptible to toxicants successfully traversing this barrier.

It has been known for several decades that a mammalian brain contains large amounts of substrates that are susceptible to free radical attack, such as unsaturated lipids and catecholamines. Halliwell and Gutteridge[5] were the first to discuss the potential role of oxygen radicals in the nervous system. The involvement of these reactive species in hyperoxia, ischemia, trauma, stroke, and transition metal-dependent reactions in the brain has since been a topic of considerable interest.[6,7]

The concept has been put forward that a variety of drug and chemical pathogeneses are associated with free radical mechanisms.[8] This mediation may contribute significantly to the properties of many neurotoxic agents. Any imbalance of cellular redox status in favor of greater oxidative activity can lead to several kinds of macromolecular damage, such as disruption of genomic function by alterations to DNA, or impairment of membrane properties by attack on proteins or lipids. Lipid peroxidative events are especially hazardous, since lipoperoxy radicals can initiate oxidative chain reactions.[1] Thus, the high lipid content of myelin makes nervous tissue especially susceptible to oxidative stress.

Neurotoxic compounds have characteristic and individual properties which result in distinctive morphological and biochemical lesions. However, some relatively nonspecific features also constitute part of the overall toxicity of a given agent. Attempts to unify the diffuse discipline of neurotoxicology have led to the concept of "final common pathways" that characterize frequently occurring cellular responses to disruption of homeostasis resulting from exposure to xenobiotic agents. The present work considers the possibility that oxygen radicals may be mediators of a "final common pathway" in several mechanisms of neurotoxicity.

Free radicals are defined as any species with one or more unpaired electrons (see Chapter 3, this volume). Since oxygen is ubiquitous in aerobic organisms, oxygen-centered free radicals have been implicated in several physiological, toxicological, and pathological phenomena. However, while superoxide anion and hydroxyl radical qualify as oxygen-centered radicals, hydrogen peroxide is a potent cellular toxicant that lacks unpaired electrons. The terms "reactive oxygen species" (ROS) and "oxygen radicals" have been used to describe all pro-oxidant species whether they are oxygen-centered radicals or nonradicals.

The precise nature of oxygen radicals produced in the central nervous system (CNS) is by no means clear. The difficulty of establishing this with certainty is due to the short half life and rapid interconvertibility of many of the putative key species.[9] Relatively stable species such as superoxide and hydrogen peroxide give rise to less clearly defined, highly active, transient species that are the primary oxidants. Evaluation of the free radical-forming potential of neurotoxic agents, and determination of whether oxygen radicals are common mediators of neurotoxicity should not be delayed until the identity of critical oxygen radicals is unequivocally clarified.

11.04.2 ENDOGENOUS SOURCES OF GENERATION OF REACTIVE OXYGEN SPECIES WITHIN THE CNS

11.04.2.1 Oxidative Phosphorylation

Around 2% of the oxygen consumed by mitochondria is incompletely reduced and appears as oxygen radicals.[10] This proportion may be increased when the efficient functioning of mitochondrial electron transport systems is compromised. This could account for the increased lipid peroxidation found in the brains of mice exposed to nonlethal levels of cyanide.[11]

11.04.2.2 Cytosolic Acidity

Lowered pH resulting from excess glycolytic activity may not only accelerate the process of liberating protein-bound iron in organisms, but it may also lead to an impairment of oxidative ATP generation, and to the appearance of the pro-oxidant protonated superoxide.[12] However, there is evidence that the reduction of pH during ATP depletion may be protective and enhance cell survival.[13] Chlordecone, a neurotoxic insecticide, elevates pH within synaptosomes, and depresses oxygen radical synthesis.[14]

11.04.2.3 Presence of Metal Ions with Multivalence Potential

Liberation of protein-bound iron can occur by enhanced degradation of important iron-binding proteins such as ferritin and transferrin. A small increase in levels of free iron within cells can accelerate dramatically rates of oxygen radical production.[15] A key feature in establishing the rate of production of oxygen radicals by tissue is the cytosolic concentration of free metal ions possessing the capacity to change readily their valence state.

11.04.2.4 Production of Eicosanoids and Neuromodulators

Enhanced phospholipase activity can lead to the release of arachidonic acid. This polyunsaturated fatty acid contains four ethylenic bonds and is readily autooxidizable. In fact, impure preparations of arachidonic acid may explode spontaneously on exposure to air.[2] The enzymic conversion of this compound to many bioactive prostaglandins, leukotrienes, and thromboxanes by cyclooxygenases and lipoxygenases leads to considerable oxygen radical generation.[1,16] All major catabolic pathways of arachidonic acid involve the utilization of molecular oxygen and the formation of hydroperoxide or epoxide intermediates. Subsequent metabolism by peroxidases and hydrolases can lead to further formation of free radicals. The physiological relevance of this is illustrated by the finding that antioxidants can protect against arachidonic acid-induced cerebral edema.[17]

Nitric oxide is both an important vasodilator and neuromodulator. Under some circumstances, it can interact with superoxide anion to form peroxynitrite and subsequently a very short-lived reactive oxygen species. This is recognized increasingly as a potential source of neurological damage.[18]

Increased levels of cytosolic free calcium may result from either breakdown of the steep concentration gradient of calcium across the plasma membrane, or by liberation of the large amounts of calcium bound intracellularly within mitochondria or endoplasmic reticulum. This elevation can activate phospholipases and thus stimulate oxygen radical production. In fact, the activation of phospholipase D has been linked functionally to superoxide anion production.[19] A reciprocal relation exists since free radicals can enhance phospholipase A_2 activity within cerebral capillaries.[20] Conversely, phospholipase A_2 may selectively inhibit the GABA-regulated chloride channel, and thus increase cell excitability.[21] Such mutually reinforcing changes have the potential to reach a critical level and set in motion an ever increasing entropic cascade.

11.04.2.5 Oxidases

Chemical induction of cytochrome P450-containing mixed function monooxidases can increase the rate of Phase I detoxification reactions. The oxidative metabolism of many lipophilic compounds, while necessary for their conjugation and excretion, often involves the transient formation of highly reactive oxidative intermediates such as epoxides.[22] While mixed function oxidases predominate in the liver, they are also present in the nervous system, in both neurons and glia.[23] At the intracellular level, most of these cerebral oxidases are mitochondrial rather than microsomal, and like the corresponding hepatic enzymes they are inducible.[24] Products of such enzymes may include epoxides in which the C—O—C ring strain often results in a potent oxidizing chemical.

Xanthine oxidase is a prime generator of superoxide and is produced by oxidative or proteolytic degradation of xanthine dehydrogenase.[25] It may thus be a significant exacerbating factor in several pathological states.

11.04.2.6 Phagocytosis

Extracellular formation of superoxide anion by phagocytes has long been recognized as a bactericidal mechanism. Similar oxidative activity has been observed in cerebral microglia.[2] Astroglial activation is a common event following neural trauma, and reactive astrocytes are active in clearance of cell debris and ultimately in the formation of glial scar tissue. Although the phenomenon of ROS generation has not been documented in the injured brain during neuronophagia, the ROS-enhancing potential of such events is worthy of further study.

11.04.3 INTRACELLULAR MECHANISMS FOR REGULATION OF OXIDANT SPECIES OR MITIGATION OF THEIR EFFECTS

Protection against excess levels of ROS is effected by a wide range of intracellular mechanisms. Only when such defensive processes are overwhelmed does oxidative damage ensue.

11.04.3.1 Enzymes Reducing Oxidative Stress

Enzymes such as superoxide dismutase, catalase, and peroxidase are able to destroy the superoxide radical and hydrogen peroxide respectively. While these oxidant species are not in themselves very active, they are able to interact in the presence of trace amounts of iron, and by the Haber–Weiss reaction, give rise to the highly reactive, short-lived hydroxy radical. Levels of protective enzymes are somewhat low in the brain[3] and much of the brain's antioxidant capacity lies within cerebral capillaries and glial cells.[26] Since the cerebral microvasculature is also a major site of lipid peroxidative activity,[27] and mixed function oxidase activity,[28] the capillary walls may prevent diffusion of pro-oxidants into neurons. Treatment of neurons with nerve growth factor is able to confer resistance to oxidative damage apparently by induction of antioxidant enzymes.[29]

11.04.3.2 Chelation of Metals with Pro-oxidant Potential

Iron- and copper-sequestering proteins such as ferritin, transferrin, and ceruloplasmin are important means of ensuring extremely low levels of free cytosolic metals with a potential for valence changes. Such metals are crucial cofactors in ROS generation. Lower antioxidant activity in Parkinson's disease (PD) has been related to decreased levels of ferritin in the brain.[30] Metallothionein is induced by, and can sequester, several metals with a high affinity for —SH groups, such as mercury and cadmium. This protein also acts as an antioxidant because of its high sulfhydryl content and its ability to dismutate superoxide anion.

11.04.3.3 Soluble Vitamins and Other Cytoplasmic Components

The presence of diffusible antioxidant vitamins and provitamins provides protection against oxygen radicals. Such molecules may be predominantly lipophilic (β-carotene, α-tocopherol, retinoic acid), or water soluble (ascorbic acid). A major source of reducing power is glutathione which normally exists at high concentrations intracellularly (1–5 mM).[5] Glutathione is maintained largely in the reduced form by glutathione reductase, acting in conjunction with nicotinamide adenine dinucleotide phosphate (NADPH). Inhibition of NADPH generating processes may compromise intracellular levels of glutathione, and thus reduce potential. However, NADPH is also an essential component of many oxidases, including the mixed function oxidases.

Glutathione reserves can be depleted by oxidative stress,[31] and such depletion in the brain can cause neurological deficits.[32] Glutathione distribution in the CNS is heterogeneous both at the cellular and regional level. Its content is much higher in astroglia than neurons. The brainstem has a rather low glutathione content and a high mixed function oxidase content relative to other brain regions. It has been proposed that this combination may render the region especially vulnerable to oxidative damage, and that this is relevant to nigral damage in PD.[33,34]

Some serum proteins, notably albumin, can act as free radical scavengers.[35] However, the low protein content of cerebrospinal fluid makes both albumin and ferritin largely unavailable to the CNS.[36]

11.04.4 PROMOTION OF PRODUCTION OF REACTIVE OXYGEN SPECIES WITHIN THE BRAIN

While several neurotoxic chemicals can stimulate ROS production within the nervous system, this does not imply that such stimulation represents their primary mechanism of toxic action. Pro-oxidant properties of a chemical may: (i) be the main source of its neurotoxicity; (ii) contribute to its harmfulness; or (iii) be an epiphenomenon, only secondarily relating to tissue damage.

The events subserving xenobiotic stimulation of ROS frequently involve an excess activation of one or more of the endogenous ROS-producing processes, or inhibition of the protective systems described above.

11.04.4.1 Metals

There are several means by which metal ions can catalyze the intracellular production of ROS. Perhaps the most significant of these are plurivalent metals which are able to alter their electronic configurations under biological conditions. The promotion of redox cycling by such transitions forms the basis of neurological damage effected by iron, copper, and manganese. These metals are not only essential but often owe their effectiveness in their normal biological role to those very same transition features that, when acting in an uncontrolled manner, promote adverse oxidative events. The fact that many metals are essential cofactors for several enzyme complexes ensures their continuing presence within neural tissue, and may set the stage for their aberrant distribution or metabolic handling. Oxygen species can be stabilized and rendered innocuous by the metal-containing region of many metalloenzymes concerned with oxidative and reductive processes.[37]

No metals are exclusively neurotoxic. While a wide range of tissues may be adversely affected by the abnormal presence of a metal, some organs are clearly more vulnerable to metal-induced oxidative damage. The nervous system and the kidney are, for different reasons, among the most susceptible to such injury. Both these organs receive a disproportionate amount of arterial blood and have very high oxidative metabolic rates. In the case of the kidney, the pH changes associated with glomerular filtration and tubular production of urine can lead to concentration and precipitation of metals. The brain is a target because of its relatively low levels of enzymes protecting against oxidative stress,[3] and its high myelin-associated lipid content, and consequent susceptibility to propagation of peroxidative events.[38]

Neurotoxicity induced by, and consequent to, the presence of excessive levels of metals, may be associated with several conditions including:

(i) genetic defects involving failure of normal metabolism, and thus accumulation of a metal;

(ii) xenobiotic exposure from environmental or dietary sources;

(iii) abnormal presence of a metal within tissue due to physiological disruption e.g. extravasation of hemoglobin and thus iron, into cerebral tissue;

(iv) increased excitability of neural tissue with consequent elevation of cytosolic calcium levels can be a stimulus for generation of excess pro-oxidant events.

While the blood–brain barrier is effective in limiting access of many metal cations to the brain, this protection is incomplete and can be circumvented by several means.

(i) Metals in organic form can readily enter the brain by virtue of their lipophilic properties. Amphiphilic compounds are among the most deleterious forms of metals, perhaps because possessing both a lipophilic and a water soluble component, they will concentrate in, and align themselves within intracellular membranes. Following therapeutic intervention with chelators, metals may also enter the CNS in the form of sequestered complexes. Thus, while acute toxicity is prevented in this way, the possibility of subsequent neurological damage remains. This may constitute a significant hazard especially since loosely chelated metals can be the most potent inducers of oxidative events.

(ii) Xenobiotic ions such as thallium can also gain such access to neural cells by entry through ion channels which are not completely specific for normal biologically occurring cations. This is due to the resemblance of several toxic metals to a biologically essential counterpart.

(iii) The brain has regions, especially in the hypothalamus, that have an incompletely formed blood–brain barrier, probably to permit monitoring of the endocrine composition of circulating blood. This can enable localized entry of ionic species that would otherwise be excluded. The immature brain also has no effective means of excluding charged compounds and thus the fetus may be particularly vulnerable to penetration by undesirable metal ions.

The damage effected by most neurotoxic metals is unlikely to be solely attributable to generation of excess ROS. However, most toxic metals appear to possess a free-radical inducing component relating to their harmfulness. The toxicity of each metal is distinctive, and involves a characteristic range of morphological and biochemical abnormalities. Superimposed on this may be an increased rate of ROS formation which can enhance the more selectively damaging effects of the agent. Thus, catalytic generation of ROS may represent a final common path provoked by disparate metal-containing chemicals. This overview will describe salient features of several neurotoxic metals insofar as this may involve their capacity to induce ROS and consequently provoke oxidative stress. Metals have been grouped by classes believed to involve similar mechanisms underlying their pro-oxidant potential.

11.04.4.1.1 Metals where valence flux may subserve enhanced ROS production (Fe, Cu, Mn)

Transition metals may induce oxidative stress by cycling between two valence states, thereby alternately donating and receiving electrons. Copper, iron, and manganese are in this class. This property is related to the presence of an unfilled inner electronic shell. The first metal after this series, zinc, has completely filled inner electronic orbitals and thus has no valence ambiguity and no capacity to induce ROS. Other metals with plurivalent potential that may also induce oxidative stress, such as mercury, tin, thallium, and lead, do not readily undergo such valence changes under physiological conditions, and thus are active in ROS stimulation primarily by other means.

All of the transition metals mentioned above are essential elements and constitute part of the active sites of many enzymes and transport proteins. These enzymes include those such as superoxide dismutase and catalase, which are important in removal of free radicals or their precursors. In addition, many enzymes involved in bringing about regulated oxidations are metalloenzymes. The importance of keeping the free concentration of ionic salts of these metals within biological tissues very low is implied by the existence of proteins with high affinity for a specific metal, which are thus able to sequester that metal as an inactive high molecular weight complex. Ceruloplasmin is the main copper chelating plasma protein while iron is served by several proteins, including ferritin. It has been proposed that most biological molecules are not readily autooxidized, and that low molecular weight complexes of a transition metal, especially iron, are needed to catalyze all such oxidations, whether beneficial or harmful.[39]

(i) Iron and copper

While superoxide anion and hydrogen peroxide are not in themselves very active oxidants, they are able to interact in the presence of trace amounts of iron, and by the Haber–Weiss reaction, give rise to the highly reactive, short-lived hydroxy radical.

A wide range of enzymic and nonenzymic reactive free radical generating processes can thus be activated by Fe^{3+} (or Cu^{2+} to a lesser degree) in the presence of an electron donor. Such activation can lead to lipid peroxidative events and also to the oxidative degradation of proteins, and thence to enzyme inactivation and membrane damage. Metal ion-catalyzed modification of proteins can be site-specific and may occur near an iron-binding lysine residue.[40] The susceptibility of amino acid residues within proteins to oxidation, differs from that of free amino acids since the proximity of amino acids in the peptide chain to metal binding sites largely governs their vulnerability.[41]

The situations under which copper and iron can be neurotoxic do not generally involve excessive ingestion. The penetration of iron into the brain can follow traumatic brain injury or hemorrhagic stroke when low molecular weight iron is liberated during degradation of free hemoglobin.[42] The consequent elevation of ROS may in part account for the delayed appearance of seizure activity which can result in bleeding into brain tissue.[43] Excessive levels of unsequestered iron have also been observed in several other neurological disorders including various lipofuscinoses such as Batten's disease,[44] Alzheimer's disease,[45,46] Parkinson's disease,[47] seizure disorders,[48] stroke,[49] ischemia,[50] Hallervorden–Spatz disease,[51] and

edema.[52] Evidence exists that ethanol can lead to elevated levels of lipid peroxidation in liver and brain. This has been attributed, at least in part, to liberation of protein-bound iron by ethanol.[53–56]

Chelation therapy using deferoxamine, a potent iron chelator, is being used experimentally in the treatment of Alzheimer's disease[57] and multiple sclerosis.[58] However, deferoxamine, like all chelators, is not completely specific for a single metal. In addition to metal sequestration, its mode of action as an antioxidant may also involve direct scavenging of free radicals.[59] Chelation therapy as a means of decorporating toxic metals has many adverse side effects, including the potential for enhancement of ROS formation.[60–62] Some of these undesirable side effects are probably due to removal of essential metals.[63]

A caveat concerning iron-related neurotoxicity is that association does not always imply causation. Thus, while hyperexcitation can elevate intracellular iron levels, this is not necessarily the cause of accompanying neuronal damage.[48] The molecular state of cytosolic iron is critical. Under normal homeostatic conditions iron is bound to intracellular proteins. However, oxidative events can lead to decompartmentation of this metal and consequent appearance of its pro-oxidant properties. Liberation of iron by nitric oxide has be proposed as occurring in PD.[47,64]

Sometimes pharmacological therapy has the potential for inadvertently stimulating ROS. For example, l-DOPA, used in the therapy of Parkinsonism, is capable of reducing ferric iron, the form in which iron forms a stable complex with ferritin. This may effect the liberation of ferrous iron as a low molecular weight species with significant ROS-enhancing potential.[65] Basal iron levels are especially high in the substantia nigra and globus pallidus.[66] Ferritin levels have been reported as being both elevated and depressed in these regions in PD,[30,67] implying a deranged iron metabolism.

In the case of copper which is present in bodily tissues to a much lesser extent than iron, the only clear appearance of neurotoxicity is due to a genetic defect in copper metabolism, Wilson's disease. This disorder is characterized by the virtual absence of ceruloplasmin leading to greatly elevated levels of low molecular weight diffusible copper. The psychiatric and neurological consequences of this disease can be greatly alleviated by chelation with drugs such as penicillamine. Copper is known to enhance catecholamine autooxidation and catalysis of ROS, and this distortion of the normal role of this metal is likely to underlie this many-faceted disease.[68]

Oxidases are enzymes utilizing molecular oxygen. The scission of the O_2 molecule is an intrinsically hazardous process wherein directed oxidation is required. The control of O_2 utilization is regulated at the catalytic center of these enzymes which generally contains a metal with plurivalent potential (Fe, Cu, or Mn). However all oxidases "leak" nonspecific ROS to a certain extent. There is especially efficient control in the mitochondrial respiratory chain where the uncoupled leakage rate is less than 2%.[10] The presence of copper as well as iron within cytochrome oxidase may allow this tight regulation, and enzymes containing iron alone, such as the mixed function oxidases and monoamine oxidases, may be less effective in this respect.[69] Inhibition of monoamine oxidase B has proved effective in slowing the rate of progression of PD[70] and, since antioxidant therapy has a similar effect,[71] the mechanism of the protective effect of MAO-B blockers may be by way of inhibition of free radical generation.

(ii)   Manganese

Manganese is another member of this group of metals with the potential for several valence states. Neurological disease is found following excessive exposure to this mineral in the mining, handling or processing of manganese ore. Manganism presents in several distinct phases, including an initial hyperactive, manic state ("manganese madness"), an intermediate Parkinsonian-like phase involving tremor and incoordination, and a terminal spastic, rigid phase.[72] The resemblance of the latter two stages to PD is striking, especially because of the involvement of the basal ganglia. There is increasing evidence for the role of oxidative stress in PD, and trials involving antioxidant therapy have met with some success.[73] In view of the readiness of dopamine and norepinephrine to oxidize spontaneously, catecholaminergic pathways seem to be especially susceptible to metal-catalyzed oxidative damage. This is especially true of dopamine autooxidation and such a mechanism may underlie manganism.[74] Manganese is a transition metal that can exist in at least four oxidative states (with a valence of 2, 3, 4, or 5), and, like iron, has been shown to be a potent ROS enhancer in isolated systems. There is also evidence for a role of iron in PD (see previous section).

The situation is complicated by reports describing significant free radical scavenging properties of many manganese complexes. Hydrogen peroxide and superoxide may be quenched by such chelates.[75] In fact, manganese has been stated to "fulfill the requirements of a physiologically relevant antioxidant."[76] These apparent contradictions can be reconciled by recognition of the essential role of the valence status of metal ions, and the readiness of their interconvertibility, in enabling or retarding oxidative events.[15] Although not shown neurotoxic in man, another transition metal, vanadium, has also been found to induce lipid peroxidation in the brains of experimental animals.[77]

11.04.4.1.2   Metals with a high affinity for sulfhydryl groups and for selenium (Hg, Pb, Tl, Cd, and As)

The capacity of several metals to form covalent linkages to sulfhydryl groups of peptides can lead to the formation of non-ionic complexes. A key low molecular weight peptide is glutathione, which is present at millimolar concentrations within the cytosol. Glutathione (GSH) plays several vital roles within the cell. It is a source of reducing power and thus constitutes an important water soluble defence against excess ROS. Glutathione detoxifies xenobiotic agents by direct conjugation of xenobiotic agents using glutathione transferase. Glutathione destroys peroxides in the presence of glutathione peroxidase. Under normal circumstances, the oxidized glutathione formed (GSSG) is reconverted to GSH by glutathione reductase and once again its reducing power becomes available. However, excessive ROS levels may overwhelm such regenerative capacity. GSSG usually constitutes a very minor fraction of total intracellular glutathione, but if its level is elevated this non-ionic lipophilic molecule can readily become lost to the cell by diffusion across the limiting membrane.

The heavy metals mercury, lead (see Chapter 29, this volume), thallium, and cadmium all have electron-sharing tendencies that can lead to formation of covalent attachments. Metals of this family have an avidity for sulfhydryl groups, can deplete cellular glutathione levels,[78] increase levels of lipid peroxidation,[79,80] and can accelerate lipofuscinogenesis. This latter effect is further accelerated in the presence of hyperbaric oxygen.[81]

α-Tocopherol has frequently been found protective against neurotoxicity induced by this class of metals.[82,83] In addition, these metals may displace selenium from glutathione peroxidase.[84] The consequent inactivation of this enzyme removes a key antioxidant sequence and this may account for the ability of selenium to protect against the toxicity of methyl mercury,[85] which can induce generation of ROS within the CNS,[86] but this property can be blocked by deferoxamine. Since deferoxamine does not chelate methyl mercury, iron-catalyzed pro-oxidant reactions may play a role in methyl mercury neurotoxicity.[87]

The proportion of the total toxicity of salts of these metals that is expressed as neurotoxicity is related to their ability to cross the blood–brain barrier. Thus the charged ionic mercuric chloride form of this element is predominantly harmful to non-neural tissues such as the kidney, and inorganic lead has a large range of adverse systemic effects. On the other hand, the more amphiphilic methyl mercuric halides and triethyllead salts are primarily neurotoxic.

Cadmium salts are also largely nephrotoxic but can enter the CNS by the olfactory route where the blood–brain barrier is attenuated. Anosmia is characteristic of cadmium intoxication. Interestingly, the complex that cadmium forms with diethyldithiocarbamate increases the access of this metal to the brain but reduces its neurotoxicity.[88] However, the extended presence of a relatively inert metal chelate within the CNS may have unforseeable consequences.

The next element of the sulfur-containing family in the periodic table is selenium, and metals binding to sulfur also have an affinity for selenium. Selenium has been shown to attenuate the toxicity of cadmium, arsenic, mercury, thallium, and copper. The only known mammalian enzyme in which selenium is an essential cofactor is glutathione peroxidase. This enzyme uses GSH to reduce organic hydroperoxides, and is an important defense against oxidant damage.[89] Thus, the protective properties of selenium imply oxidative events as a basis for the toxicity of the above elements.

Cadmium and mercury, but not lead, are able to induce metallothionein in several organs.[90] This cysteine-rich protein which is capable of sequestering sulfhydryl-binding metals is present only to a minor extent in neural tissues but may be important here since it is present as a CNS-specific variant, metallothionein III.

Complex interactions with both synergistic and antagonistic potential exist between cadmium, mercury, and lead.[91] Such synergism implies action at different sites, while antagonism may be related to competition of a less toxic metal with a more toxic metal, for a common —SH target.

11.04.4.1.3 Metals with the capacity to increase neural excitability Ca, Pb, Sn (organic)

The normal physiological route of elevation of rates of neuronal firing involves influx of calcium into the presynaptic area, thereby eliciting neurotransmitter release. Cytosolic calcium levels are transiently but greatly elevated before homeostasis is restored by rapid sequestration of excess calcium into endoplasmic reticulum and mitochondria, followed by eventual extrusion of calcium into the extracellular space using ion pumping or exchange processes. There is evidence that chronic excitation may result in failure to restore resting levels of calcium. This can initiate excess ROS generation by several means including activation of phospholipases and the arachidonic acid cascade. Calcium can also activate superoxide production by polymorphonuclear lymphocytes which can be present in post-ischemic neural tissues.[92] By these means, it may act as a mediator of oxidative stress in a variety of disease states involving persistent excitation including ischemia, epilepsy, and chemically induced hyperactivity.[93]

Both exogenous antioxidants and calcium chelators can inhibit lipid peroxidation and generation of superoxide anion following post-ischemic reperfusion.[94] The degree to which cytosolic calcium elevation and pro-oxidant events occur independently or act in concert remains to be unravelled.[95]

Several metals can indirectly evoke ROS by way of disruption of normal calcium homeostasis, enabling release of excess neurotransmitter (see Chapter 13, this volume). Although peroxidative events are unlikely to constitute the primary mechanism of lead toxicity, there is evidence that both inorganic and organic lead derivatives can enhance the generation of ROS. Lead is known to interfere with calcium metabolism, and lead compounds have been reported to enhance rates of lipid peroxidation within the brain.[96] Organic lead compounds, which are unlikely to closely mimic calcium, can also induce oxidant conditions in cerebral tissues.[97] In this case, depolarization-induced excitotoxicity may underlie such elevations in ROS generation. Since 5-aminolevulinic acid, a heme precursor, accumulates during lead poisoning and is capable of inducing ROS, the mechanism by which lead can generate free radicals may also be indirect.[98]

Aliphatic organic tin compounds, such as trimethyl- and triethyltin, are potent neurotoxic agents. Trimethyl tin is an excitatory agent with a rather high degree of selectivity toward the hippocampus and other limbic structures. While this agent has no ability to induce ROS in isolated preparations, it can specifically elevate hippocampal ROS in treated rats.[86] This region-specific stimulation is likely to be due to the susceptibility of hippocampal circuitry to excitatory events, rather than to any distinctive biochemical features of this area. Calcium may once again be the primary effector of excess production of ROS.

11.04.4.1.4   Aluminum and Alzheimer's Disease

Aluminum does not fit clearly into any of the classes of metals described here. However, this element may be capable of inducing ROS within neural tissues and this property may play a role in the etiology of Alzheimer's disease (AD). While such a role is controversial, aluminum has been recognized as a significant factor in dialysis dementia for some time. However, it is uncertain whether an excessive intraneuronal accumulation of aluminum is found in AD, and whether this has a causal relationship to the disease. Studies have reported the following:

(i) Aluminosilicate complexes are capable of stimulating ROS generation in isolated glial cultures.[99]

(ii) Levels of superoxide dismutase and catalase are elevated in AD brains, implying a response to oxidative stress.[100]

(iii) Chelation therapy with deferoxamine, in order to reduce the body burden of aluminum, may retard the rate of loss of intellectual performance in AD.[57] Since iron is also sequestered by this chelator, this report cannot be considered conclusive evidence for a role of aluminum in AD. While aluminum salts in themselves have no ROS-promoting capacity, they have been reported to strongly enhance the pro-oxidant potential of iron salts in an isolated system.[101–104]

(iv) Glutamine synthetase, an enzyme that is very susceptible to oxidative degradation, is depressed in brains from AD patients compared to age-matched controls.[105] This depression is confined to the frontal cortex, the location of the neuropathologic involvement.

(v) The iron binding protein, transferrin, is depressed in AD,[46] and levels of lipid peroxidation are elevated.[106]

While this evidence is circumstantial, and great advances have been made in understanding of the role of amyloid protein in AD, there is a good possibility that pro-oxidant events and derangements of metal ion balance contribute to AD. Aluminum and iron have been reported to promote the aggregation of β-amyloid peptide.[107]

The mechanism by which aluminum can promote oxidative stress is unclear. Subtle modification of membrane structures allowing increased availability of peroxidizable fatty acids has been proposed,[101,103] and another possibility is that formation of aluminum/iron-

containing mineral particulates can promote oxidative injury by stimulation of macrophage-like microglial events.[108] The issue of whether ionic or complexed aluminum is the active agent in these events is currently controversial[103,109] and its resolution would clarify processes underlying aluminum-catalyzed oxidative events.

11.04.4.2   Organic Solvents

11.04.4.2.1   Potential mechanisms for induction of ROS by organic solvents

The ability of various organic solvents to effect excess ROS synthesis has been described for several tissues,[110,111] but the brain has received little attention in this context. The neurotoxicity of two solvents of environmental significance, ethanol and toluene, will be discussed here. A large range of enzymes and other factors may account for an association between solvent exposure and excess oxidative events. Evidence for, and potential mechanisms of, solvent-induced ROS production are listed.

(i)   Glutathione levels

Reductions and also increases in glutathione levels have been found in the brain following ethanol, toluene, or *m*-dinitrobenzene dosing.[112–119] Since compensatory processes can be rapid after induction of excessive ROS, both reductions and increases in cerebral glutathione may be regarded as indices of oxidative stress.[120]

(ii)   Lipid mobilization and peroxidation

Elevations in lipid peroxidation following ethanol or toluene treatment are found in the brain.[115,116] The attenuation of some of these ethanol-induced changes by lipid-soluble antioxidant vitamins A or E, and the depletion of antioxidant vitamins in the CNS of ethanol-treated rats,[113,121] confirms the presence of induced oxidative stress.

Solvent-stimulated lipid mobilization is another phenomenon to be considered as a source of ROS. Ethanol has been shown to activate phospholipases A1 and A2 in an isolated cardiac preparation.[122] This possibility has not been reported for corresponding cerebral tissue but is likely. The effect is blocked in the presence of α-tocopherol. Phospholipase activity is also stimulated significantly in synaptosomes isolated from rats exposed to toluene.[123] The liberation of arachidonic acid by phospholipase A2 sets in motion a range of oxidative catabolic processes, the "arachidonic acid cascade." This polyunsaturated fatty acid contains four ethylenic bonds and is readily oxidizable. The enzymic conversion of this compound to prostaglandins, leukotrienes, and thromboxanes by cyclooxygenases and lipoxygenases leads to considerable ROS generation.[16] Toluene alters synaptosomal phospholipid methyltransferase activity, an event that has also been shown to be affected by oxygen radicals.[124]

(iii)   Iron mobilization

Free ionic, or incompletely sequestered iron, may be essential for the appearance of ethanol-induced ROS. Ethanol may effect the liberation of low molecular weight iron in the brain, from bound intracellular reserves,[55,56] perhaps by way of formation of the superoxide anion which can release iron from ferritin.[125] That the presence of small amounts of low molecular weight iron may lead to the formation of ROS is illustrated by the protective effects of iron chelators, such as deferoxamine, on ethanol-related changes in cerebral oxidative events.[54,126] It has even been suggested that iron-sequestering chemicals can reduce physical dependence on ethanol.[127]

It has been postulated that acetaldehyde generated during the metabolism of ethanol can initiate oxidative stress by reaction with, and depletion of, protective thiols such as cysteine and glutathione. Microsomal mixed function oxidases may also be a direct target of acetaldehyde. Binding of this ethanol metabolite to these enzymes, forming a stable adduct, can impair their properties.[128] Thus the P450 2E1 enzyme induced by ethanol is more likely to be malfunctioning.[129] The extent to which acetaldehyde found within the brain after ethanol dosing is generated intrinsically rather than systemically transported from the liver is unclear, but its ability to promote cerebral oxidative stress is apparent.[118] The reactivity of acetaldehyde with many biological constituents suggests that it is likely to be synthesized close to the site where it is detected. However, following ethanol treatment, acetaldehyde levels of brain interstitial fluid are above those present intracellularly within the CNS.[130] This suggests that acetaldehyde can cross the blood–brain barrier.

Cysteine and ascorbic acid are protective against the acute behavioral toxicity of acetaldehyde, supporting the concept of significant role for acetaldehyde-inducible oxidative stress to the brain.[131] However, such agents may also have nonantioxidant ameliorative effects by preventing the formation of adducts of acetaldehyde with protein.[132]

The catabolic steps involved in the degradation of ethanol, toluene, and other solvents are candidates for the origin of excess ROS. Several major classes of enzyme need to be considered in this context.

11.04.4.2.2   Catabolic enzymes with ROS-enhancing potential

(i)   Mixed function oxidases

Since oxidases utilize molecular oxygen directly, their induction has the potential for enhancing ROS production. Ethanol is known to induce a specific microsomal mixed function oxidase, P450 2E1, in the liver.[133] This enzyme is inducible by ethanol, and can generate oxidizing species in the absence of substrates as long as NADPH is present.[134] Under such conditions, P450 2E1 has been reported to exhibit an unusually high rate of oxidase and H_2O_2 generating activity.[133] A parallel induction of P450 2E1 has been found in the CNS where the enzyme is present in much lower concentration.[135] The distinct susceptibility of the cerebellum and the hippocampus to ethanol-induced morphological damage may relate to their relatively high content of cytochrome P450 2E1 mono-oxygenase.

The role of mixed function oxidases in solvent-stimulated ROS generation is also indicated by the finding that the capacity of toluene to stimulate cortical ROS production can be blocked by pretreatment of rats with an inhibitor of mixed function oxidases, namely metyrapone.[136]

(ii)   Superoxide dismutase

Increased levels of superoxide dismutase (SOD) an inducible enzyme, are generally taken as indirect evidence of an increased oxidant milieu. However since this is a sulfhydryl enzyme, depressed levels in a tissue can also reflect oxidative denaturation. Both ethanol-induced increases and decreases in neuronal and glial SOD have been reported.[137,138] As with glutathione, biphasic fluxes of SOD levels are common, and a change in either direction may relate to the presence of excess ROS.

(iii)   Alcohol dehydrogenase

This relatively nonspecific soluble enzyme is considered the primary initial step in the catabolism of ethanol. There is evidence that the activity of several dehydrogenases including alcohol dehydrogenase can bring about ROS formation, despite the fact that oxygen is not directly involved.[116] The inhibition of ethanol or toluene-effected free radical production by an inhibitor of alcohol dehydrogenase, 4-methylpyrazole,[116,139] suggests that ROS are generated by this enzyme. However, 4-methylpyrazole is also capable of inhibition of mixed function oxidases,[140] so use of this agent does not allow a clear distinction to be made between the two major routes of ethanol breakdown.

(iv) Aldehyde dehydrogenase

This mitochondrial NAD^+-dependent enzyme constitutes the major means of oxidation of acetaldehyde by the liver. It is also present in the brain.[130] Electron spin resonance studies reveal that this dehydrogenase is capable of producing hydroxyl ions.[116] Benzaldehyde is a potent agent in enhancing oxygen radical formation.[116] This suggests that benzaldehyde is an active metabolite responsible for the pro-oxidant properties of toluene.

(v) Aldehyde oxidase and xanthine oxidase

Both xanthine oxidase and aldehyde oxidase are molybdenum and flavin-containing enzymes capable of forming superoxide. The presence of the latter enzyme in the brain is equivocal but xanthine oxidase, which can also oxidize acetaldehyde, is present in all tissues. This enzyme is derived by the proteolytic modification of xanthine dehydrogenase. Ethanol treatment may also bring about this enzyme conversion,[127] and acetaldehyde appears to be the agent directly responsible for this.[141] The further oxidation of ethanol is primarily by way of aldehyde dehydrogenase. However, more extended dosing is also likely to effect conversion of xanthine dehydrogenase to the oxidase, thereby increasing the contribution of xanthine oxidase to ethanol catabolism.

11.04.4.2.3 Interactions of solvents with known inducers of oxidative stress

Further suggestion of the ability of ethanol to enhance induced neural oxidative stress comes from results of studies combining ethanol with other toxicants. The toxicity of manganese, 6-hydroxydopamine and 1-methyl-4-phenylpyridinium (MPP^+), three agents suspected to owe part of their neurotoxicity to oxidative events confined to the dopaminergic system, interact with ethanol or acetaldehyde, in a synergistic manner.[138,142,143]

An unresolved issue is the extent to which overall solvent toxicity is related to oxidative stress. The hepatic events consequent to prolonged and high levels of ethanol ingestion such as lipid mobilization are very likely to involve harmful pro-oxidant events. Such dramatic changes are generally not seen in nervous tissue in the absence of thiamine deficiency. However, the brain is an organ with very limited potential for cell replacement and can be vulnerable to gradual, incremental deficits. Such slowly accumulating lesions, although difficult to quantify, can be irreversible. These subtle changes may accelerate deficits resulting from normal physiological aging, and enhance susceptibility to additional neurological stressors.

Many other solvents, such as styrene and xylenes, also stimulate CNS production of ROS,[144] and extended low level exposures to aromatic and aliphatic solvent mixtures can also accomplish this.[145,146] The free-radical inducing property of solvents may not enhance their acute toxicity, but has been proposed to effect more subtle processes such as an acceleration of normal aging processes.[110]

11.04.4.3 Agents Specifically Acting on Dopaminergic Neurons

Dopaminergic circuitry is especially vulnerable to neurotoxic damage. This is at least in part due to the readiness with which dopamine is auto-oxidized in the presence of trace amounts of metals with multivalence potential. In addition, dopamine can be enzymically oxidized by monoamine oxidases to 3,4-dihydroxyphenyl acetaldehyde and H_2O_2.

A role for oxidative stress in the processes underlying 1-methyl-4-phenyl-1,2,3,6-tetrahydropyridine (MPTP) neurotoxicity has been proposed. This compound, a contaminant of an illicitly manufactured meperidine analogue, has been the subject of much interest since the neurological damage that it can cause closely resembles PD. MPTP is a very specific dopaminergic neurotoxicant. There is considerable support for the "mitochondrial theory" of MPTP toxicity, which postulates that MPP^+, the ultimate oxidation product of MPTP, blocks the reoxidation of NADH dehydrogenase by coenzyme Q_{10} and eventually leads to ATP depletion in a rotenone-like fashion.[147] However, there are also several studies that suggest oxygen radicals may play a role in the MPTP-induced neuronal damage.[148,149] Vitamin E has also been shown to protect against neural oxidative stress induced by MPTP.[149] Ganglioside GM1, a membrane-stabilizing agent effective within the CNS, may also mitigate some components of oxidative damage to neural tissue. The conflicting concepts of the primary locus of action of MPP[+], being either by inhibition of a specific mitochondrial enzyme or as an oxidative stressor, may be reconciled by the finding that the metabolic inhibitors rotenone and antimycin can increase the generation rate of oxygen radicals in crude rat synaptosomes and mitochondria.[150,151]

Emerging information concerning MPTP has led to several new ideas concerning the very prevalent neurological disorder, PD. These concepts include both the possibility of an environmental agent being contributory to the pathogenesis of Parkinsonism[152] and of the potential for antioxidant therapy of this disorder.[73] PD has been associated with abnormally high levels of superoxide dismutase within the substantia nigra,[153] implying an induced response to oxidative stress. It has been proposed that environmentally prevalent agents, such as pyridines, contribute to the incidence of this disorder. In addition, the potential utility of several antioxidant therapies is under investigation.[71,154] There is evidence that n-hexane may specifically damage dopaminergic neurons and precipitate Parkinsonian symptoms in both man and experimental animals.[155] The environmental relevance of this is underscored by the detection of the neurotoxic metabolite of n-hexane, 2,5-hexanedione, in the urine of persons not known to have been exposed to organic solvents.[156]

The neurotoxicity of another abused drug, metamphetamine, may in part be due to oxidative stress relating to dopaminergic and serotonergic circuitry.[157] This was inferred by the attenuation of metamphetamine-induced neuropathologic changes by pretreatment with a variety of antioxidants. Neuronal destruction induced by levo-DOPA and 6-hydroxydopamine is also thought to occur via free radical mechanisms,[158] and the administration of hydroxy radical scavengers such as phenylthiazolylthiourea and methimazole is protective against such induced damage.[159]

There are several reports that neuroleptics may elevate lipid peroxidation in the CNS. It has been proposed that this is a cause of tardive dyskinesia and a result of increased catecholamine metabolism.[160] α-Tocopherol has been reported to attenuate fluphenazine-induced changes in monoamine metabolism.[161]

11.04.4.4   Metabolic Inhibitors

Cerebral lipid peroxidation can be stimulated by cyanide.[11,162] This demonstrates that interruption of the respiratory chain can lead to excess free radical production. While acute cyanide exposure is rapidly lethal, more chronic exposures as a result of excess cyanogenic glycosides (e.g., found in *Cassava*) in the diet of some less developed countries can result in ataxic neuropathy.[163] This may be related to the high levels of unsaturated fatty acids in myelin that form a clear target for ROS-induced lipid peroxidation.[59]

11.04.4.5   Excitatory Compounds

Excitatory events are often associated with neurotoxic exposures. Many agents impair effective mitochondrial oxidation of energy producing substrates. The consequent depletion of ATP due to metabolic insult can lead to dissipation of ionic gradients. Entry of calcium into the cell may increase synaptic firing rates, while lowering of the sodium/potassium gradients will increase axonal excitability. Intracellular calcium overload can also set off a cascade of other events such as phospholipase activation, potentially leading to elevation of free radical production.[164,165] Some neurotoxicants (such as β-N-oxalyl-L-alanine (ODAP)) are glutamate agonists while others (such as lindane) act as GABA antagonists. Both types of interaction with neurons can directly depolarize postsynaptic cells (see Chapter 27, this volume).

Evidence suggesting a role for free radicals in excitotoxic events has been accumulating. Excess neuronal activity is known to have the potential to effect neuronal death, especially in the hippocampus (see Chapter 31, this volume). Free radicals may play a role in seizure-related brain damage.[52,166] For example, superoxide anion is generated within the brain during seizure activity.[167] In addition to the frank neuronal hyperactivity apparent in epilepsy, several other neurological disorders are associated with hyperexcitatory events. These include transient cerebral ischemia followed by restoration of the vascular oxygen supply. Attenuation of convulsive activity by phenytoin or corticosteroids also reduces cerebral levels of lipid peroxidation.[168] Free radical generation during cerebral ischemia may underlie delayed neuronal death.[169,170]

Various mechanisms have been proposed as forming the basis of such damage induced by oxidative events. Several reports indicate that there may be a direct relation between activation of various classes of glutamate receptor and free radical formation. D- and L-glutamate as well as several glutamate agonists can directly enhance ROS in isolated cerebral systems.[171] Glutamate toxicity in a neuronal cell line has been attributed to inhibition of cysteine transport and consequent oxidative stress.[172] Activation of the *N*-methyl-*d*-aspartate (NMDA) receptor site has been implicated in the postischemic elevation of lipid peroxidation in the hippocampus.[173] There is a report of exacerbation of NMDA toxicity in glutathione-deficient cortical cultures.[174]

Calcium stimulation of phospholipase A2 and thence the arachidonic acid cascade, followed by the consequent activation of ROS by lipoxygenases, may be a means by which excitatory events promote excess generation of ROS.[175] NMDA agonists are especially potent in the stimulation of NO synthetase,[176] and NO can interact with superoxide to form the intensely oxidant nitroperoxyl radical. There is evidence suggesting that the free radical nitric oxide mediates the neurotoxicity of glutamate.[177] Kainate-induced damage to cerebellar neurons may also be mediated, at least in part, by induction of superoxide.[178] The shellfish neurotoxin, domoic acid, which predominantly acts as a glutamate agonist at the kainate site,[170] is very potent in enhancing synaptosomal ROS.[171] When administered to mice, domoic acid has been found to elevate cerebral levels of superoxide dismutase,[179] which also suggests an ability to promote oxidative stress. Excitants that have not been extensively studied for their ROS-inducing potential include food additives (monosodium glutamate and aspartame), and ODAP, a glutamergic agonist found in chick-pea and a suspected cause of neurolathyrism.

Transient ischemia elevates cerebral levels of both excitatory amino acids and rates of hydroxy radical formation.[180] Whether one of these events gives rise to the second is not well established. Excitatory events may stimulate ROS, but there is also evidence that ROS can lead to release of excitatory amino acids.[181] Peroxidative damage can also impair inhibitory processes as judged by inhibition of GABA-stimulated chloride uptake.[21] Glutamine synthetase appears to be especially sensitive to ROS-induced damage, and this may increase glutamate concentrations.[182] A bidirectional cooperation between excess neuronal activity and free radicals may therefore be relatively common.[183] The functional value of such a potentially synergistic association is unclear. While there is a correlation between excitotoxicity and free radical generation, an unequivocal causal relationship remains to be established.

11.04.5 CONCLUSION

The CNS has a very high metabolic rate and also a need for this to be maintained continuously. This requirement can only be filled by an uninterrupted supply of both glucose and oxygen. Transient deficits in such materials can rapidly compromise brain function in a profound manner. Underlying this unusual energy demand is the requirement for the maintenance of appropriate ionic gradients across the neuronal plasma membrane. While all cells have such gradients, nerve tissue is distinctive in that these gradients must transiently be allowed to be attenuated in order to enable normal neuronal activity. The temporary relaxation of the normal sodium gradient is needed for the passage of the axonal action potential, and the very steep calcium gradient must be reduced in order to allow the presynaptic exocytotic release of neurotransmitters into the synaptic cleft. It has been proposed that the initial response to acute insufficiency of energy may involve sodium influx into axons, while more delayed neurodegenerative events may be enabled by the prolonged presence of excess levels of cytosolic calcium.[184]

The influx of these cations occurs by opening of ion-selective gates within the external neuronal membrane. Since this increases entropy, these are not energy-requiring events. However, the removal of excess ions from the cytosol by pumping them out into the extracellular space or, in the case of calcium, by sequestering mechanisms, are energy-demanding processes. The ability to maintain an adequate glucose supply to nerve cells is especially critical in view of the very limited intraneuronal glycogen stores and the relative inability to use other potential sources of energy such as fatty acids. The supply of oxygen to the brain is equally vital because of the rather low capacity of the brain to carry out anaerobic glycolysis.

These features of the CNS, while they may basically be only exaggerations of processes common to many cells, result in the brain possessing a very distinctive set of susceptibilities to impaired energy supply or utilization. Many neurological conditions involve either suboptimal nutrient supply or an excessive energy demand. It is for this reason that, in addition to the specific and definitive characteristics of each neurotoxic chemical, certain recurring pathological features are found, common to a wide range of unrelated neurological disorders.

The kinds of metabolic change effected by many agents deleterious to nerve tissue often have the potential for inducing excess oxygen radical production (Figure 1). The resulting oxidative damage may constitute a varying proportion of the total toxicity of a wide range of chemicals. Any chemical that disrupts membrane structure or mitochondrial function by diverse means has the potential for induction of

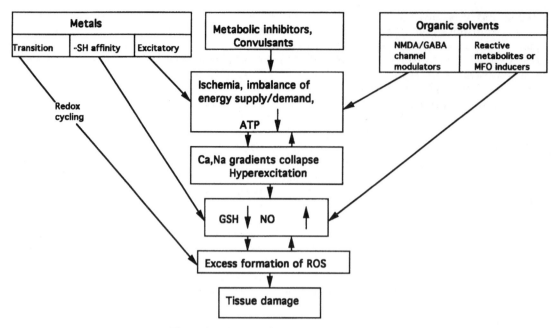

Figure 1 Neurotoxicants and free radicals.

oxygen radicals. This is especially true of the broad range of neurotoxic agents causing hyperexcitation.

Intrinsic antioxidant defensive factors must first be overwhelmed before excess ROS can damage cell functioning. This increases the likelihood that initial ROS-related damage is region specific. Such distinctive vulnerability may have either a chemical or an anatomical basis. An example of the first case involves areas rich in catecholamine containing neurons. This class of neurotransmitters is prone to oxidation and can give rise to semi-oxidized products which can result in continued and potentiated oxidative damage. An example of anatomical susceptibility is the hippocampus. The hippocampus is especially vulnerable to excitotoxic damage. This is probably because of its distinctive anatomy whereby response to repetitive stimuli is continually augmented rather than undergoing increasing suppression, an adaptive response characteristic of most brain regions.

Neurotoxic insult is not an "all-or-none" event, since subtle and insidious gradations of damage can occur. Broad, low-level exposures to hazardous agents are common, and so the magnitude of the problem across the population is very difficult to estimate. Marginal deficiencies of antioxidant vitamins over extended periods may increase vulnerability to chronic low level exposure of free-radical promoting environmental contaminants. The consequences of this may be expressed as subclinical events that are very difficult to quantify but undoubtedly relate to the well-being of an individual. Environmental factors are a suspected contributory element in many neurodegenerative disorders, and chronic excitotoxicity is the postulated mechanism of action.[185] As life expectancy increases, this represents an area of great relevance to an aging population and offers a significant challenge for future research.

ACKNOWLEDGMENT

This work was supported by grants AA8282 and ES7992 from the National Institutes of Health.

11.04.6 REFERENCES

1. B. A. Freeman and J. D. Crapo, 'Biology of disease: free radicals and tissue injury.' *Lab. Invest.*, 1982, **47**, 412–426.
2. B. Halliwell and J. M. C. Gutteridge, 'Free Radicals in Biology and Medicine,' Clarendon Press, Oxford, 1989, p. 266.
3. H. Savolainen, 'Superoxide dismutase and glutathione peroxidase activities in rat brain.' *Res. Commun. Chem. Pathol. Pharmacol.*, 1978, **21**, 173–176.
4. S. P. Raps, J. C. Lai, L. Hertz *et al.*, 'Glutathione is present in high concentrations in cultured astrocytes, but not in cultured neurons.' *Brain Res.*, 1989, **493**, 398–401.
5. B. Halliwell and J. M .C. Gutteridge, 'Oxygen radicals and the nervous system.' *Trends in Neurosci.*, 1985, **8**, 22–26.
6. J. M. Braughler and E.D. Hall, 'Central nervous system trauma and stroke. I. Biochemical considerations for oxygen radical formation and lipid peroxidation.' *Free Radic. Biol. Med.*, 1989, **6**, 289–301.

7. R. A. Floyd, 'Role of oxygen free radicals in carcinogenesis and brain ischemia.' *FASEB J.*, 1990, **4**, 2587–2597.

8. J. P. Kehrer, B. T. Mossman, A. Sevanian *et al.*, 'Free radical mechanisms in chemical pathogenesis. Summary of the symposium presented at the 1988 annual meeting of the Society of Toxicologists.' *Toxicol. Appl. Pharmacol.*, 1988, **95**, 349–362.

9. C. P. LeBel and S. C. Bondy, 'Persistent protein damage despite reduced oxygen radical formation in the aging rat brain.' *Int. J. Dev. Neurosci.*, 1991, **9**, 139–146.

10. A. Boveris and B. Chance, 'The mitochondrial generation of hydrogen peroxide. General properties and the effect of hyperbaric oxygen.' *Biochem. J.*, 1973, **134**, 707–716.

11. J. D. Johnson, W. G. Conroy, K. D. Burris *et al.*, 'Peroxidation of brain lipids following cyanide intoxication in mice.' *Toxicology*, 1987, **46**, 21–28.

12. N. J. Schisler and S. M. Singh, 'Effect of ethanol *in vivo* on enzymes which detoxify oxygen free radicals.' *Free Radic. Biol. Med.*, 1989, **7**, 117–123.

13. J. P. Kehrer, D. P. Jones, J. J. Lemasters *et al.*, 'Mechanisms of hypoxic cell injury. Summary of the symposium presented at the 1990 annual meeting of the Society of Toxicologists.' *Toxicol. Appl. Pharmacol.*, 1990, **106**, 165–178.

14. S. C. Bondy, M. McKee and C.P. LeBel, 'Changes in synaptosomal pH and rates of oxygen radical formation induced by chlordecone.' *Mol. Chem. Neuropathol.*, 1990, **13**, 95–106.

15. G. Minotti and S. D. Aust, 'The role of iron in oxygen radical mediated lipid peroxidation.' *Chem. Biol. Interact.*, 1989, **71**, 1–19.

16. R. Saunders and L. A. Horrocks, 'Eicosanoids, plasma membranes, and molecular mechanisms of spinal cord injury.' *Neurochem. Pathol.*, 1987, **7**, 1–22.

17. T. Asano, T. Koide, O. Gotoh *et al.*, 'The role of free radicals and eicosanoids in the pathogenetic mechanism underlying ischemic brain edema.' *Mol. Chem. Neuropathol.*, 1989, **10**, 101–133.

18. J. S. Beckman, 'The double-edged role of nitric oxide in brain function and superoxide-mediated injury.' *J. Dev. Physiol.*, 1991, **15**, 53–59.

19. R. W. Bonser, N. T. Thompson, R. W. Randall *et al.*, 'Phospholipase D activation is functionally linked to superoxide generation in the human neutrophil.' *Biochem. J.*, 1989, **264**, 617–620.

20. A. M. Au, P. H. Chan and R. A. Fishman, 'Stimulation of phospholipase A_2 activity by oxygen-derived free radicals in isolated brain capillaries.' *J. Cell Biochem.*, 1985, **27**, 449–453.

21. R. D. Schwartz, P. Skolnick and S. M. Paul, 'Regulation of γ-aminobutyric acid/barbiturate receptor-gated chloride ion flux in brain vesicles by phospholipase A_2: possible role of oxygen radicals.' *J. Neurochem.*, 1988, **50**, 565–571.

22. A. Sevanian, K. Nordenbrand, E. Kim *et al.*, 'Microsomal lipid peroxidation: the role of NADPH-cytochrome P450 reductase and cytochrome P450.' *Free Radic. Biol. Med.*, 1990, **8**, 145–152.

23. M. Warner, C. Kohler, T. Hansson *et al.*, 'Regional distribution of cytochrome P-450 in the rat brain: spectral quantitation and contribution of P-450b,e and P-450c,d.' *J. Neurochem.*, 1988, **50**, 1057–1065.

24. R. Perrin, A. Minn, J. F. Ghersi-Egea *et al.*, 'Distribution of cytochrome P450 activities towards alkoxyresorufin derivatives in rat brain regions, subcellular fractions and isolated cerebral microvessels.' *Biochem. Pharmacol.*, 1990, **40**, 2145–2151.

25. J. M. McCord, 'Oxygen-derived radicals: a link between reperfusion injury and inflammation.' *Fed. Proc.*, 1987, **46**, 2402–2406.

26. I. Tayarani, J. Chaudiere, J. M. Lefauconnier *et al.*, 'Enzymatic protection against peroxidative damage in isolated brain capillaries.' *J. Neurochem.*, 1987, **48**, 1399–1410.

27. E. D. Hall and J. M. Braughler, 'Central nervous system trauma and stroke. II. Physiological and pharmacological evidence for involvement of oxygen radicals and lipid peroxidation.' *Free Radic. Biol. Med.*, 1989, **6**, 303–313.

28. J. F. Ghersi-Egea, A. Minn and G. Siest, 'A new aspect of the protective functions of the blood-brain barrier: activities of four drug-metabolizing enzymes in isolated rat brain microvessels.' *Life Sci.*, 1988, **42**, 2515–2123.

29. G. R. Jackson, L. Apffel, K. Werrbach-Perez *et al.*, 'Role of nerve growth factor in oxidant-antioxidant balance and neuronal injury. I. Stimulation of hydrogen peroxide resistance.' *J. Neurosci. Res.*, 1990, **25**, 360–368.

30. D. T. Dexter, A. Carayon, M. Vidailhet *et al.*, 'Decreased ferritin levels in brain in Parkinson's disease.' *J. Neurochem.*, 1990, **55**, 16–20.

31. E. Maellaro, A. F. Casini, B. Del Bello *et al.*, 'Lipid peroxidation and antioxidant systems in liver injury produced by glutathione depleting agents.' *Biochem. Pharmacol.*, 1990, **39**, 1513–1521.

32. H. I. Calvin, C. Medvedovsky and B. V. Worgul, 'Near-total glutathione depletion and age-specific cataracts induced by buthionine sulfoximine in mice.' *Science*, 1986, **233**, 553–555.

33. T. L. Perry, D. V. Godin and S. Hansen, 'Parkinson's disease: a disorder due to nigral glutathione deficiency?' *Neurosci. Lett.*, 1982, **33**, 305–310.

34. V. Ravindranath, B. R. Shivakumar and H. K. Anandatheerthavarada, 'Low glutathione levels in brain regions of aged rats.' *Neurosci. Lett.*, 1989, **101**, 187–190.

35. B. Halliwell, 'Albumin—an important extracellular antioxidant?' *Biochem. Pharmacol.*, 1988, **37**, 569–571.

36. B. Halliwell, 'Oxidants and the central nervous system: some fundamental questions. Is oxidant damage relevant to Parkinson's disease, Alzheimer's disease, traumatic injury or stroke?' *Acta. Neurol. Scand. Suppl.*, 1989, **126**, 23–33.

37. G. Czapski and S. Goldstein, 'When do metal complexes protect the biological system from superoxide toxicity and when do they enhance it?' *Free Radic. Res. Commun.*, 1986, **1**, 157–161.

38. S. C. Bondy, 'Reactive oxygen species: relation to aging and neurotoxic damage.' *Neurotoxicology*, 1992, **13**, 87–100.

39. S. D. Aust and D. M. Miller, 'Role of iron in oxygen radical generation and reactions, "New Horizons in Molecular Toxicology,"' Lilly Research Laboratories Symposium, 1991, pp. 29–34.

40. E. R. Stadtman, 'Metal ion-catalyzed oxidation of proteins: biochemical mechanism and biological consequences.' *Free Radic. Biol. Med.*, 1990, **9**, 315–325.

41. E. R. Stadtman, 'Oxidation of free amino acids and amino acid residues in proteins by radiolysis and by metal-catalyzed reactions.' *Annu. Rev. Biochem.*, 1993, **62**, 797–821.

42. S. M. Sadrzadeh, D. K. Anderson, S. S. Panter *et al.*, 'Hemoglobin potentiates central nervous system damage.' *J. Clin. Invest.*, 1987, **79**, 662–664.

43. L. J. Willmore and W. J. Triggs, 'Iron-induced lipid peroxidation and brain injury responses.' *Int. J. Dev. Neurosci.*, 1991, **9**, 175–180.

44. J. M. Gutteridge, T. Westermarck and P. Santavuori, 'Iron and oxygen radicals in tissue damage: implications for neuronal ceroid lipofuscinoses.' *Acta Neurol. Scand.*, 1983, **68**, 365–370.

45. I. Grundke-Iqbal, J. Fleming, Y. C. Tung *et al.*, 'Ferritin is a component of the neuritic (senile) plaque in Alzheimer dementia.' *Acta Neuropath. (Berl.)*, 1990, **81**, 105–110.

46. J. R. Connor, B. S. Snyder, J. L. Beard *et al.*, 'Regional distribution of iron and iron-regulatory proteins in the brain in aging and Alzheimer's disease.' *J. Neurosci. Res.*, 1992, **31**, 327–335.

47. M. B. H. Youdim, D. Ben-Shachar and P. Reiderer, 'Is Parkinson's disease a progressive siderosis of substantia nigra resulting in iron and melanin induced neurodegeneration?' *Acta Neurol. Scand. Suppl.*, 1989, **126**, 47–54.

48. S. Shoham, E. Wertman and R. P. Ebstein, 'Iron accumulation in the rat basal ganglia after excitatory amino acid injections-dissociation from neuronal loss.' *Exp. Neurol.*, 1992, **118**, 227–241.

49. A. G. Prat and J. F. Turrens, 'Ascorbate- and hemoglobin-dependent brain chemiluminescence.' *Free Radic. Biol. Med.*, 1990, **8**, 319–325.

50. R. E. Rosenthal, R. Chanderbhan, G. Marshall *et al.*, 'Prevention of post-ischemic brain lipid conjugated diene production and neurological injury by hydroxyethyl starch-conjugated deferoxamine.' *Free Radic. Biol. Med.*, 1992, **12**, 29–33.

51. T. L. Perry, M. G. Norman, V. W. Yong *et al.*, 'Hallervorden–Spatz disease: cysteine accumulation and cysteine dioxygenase deficiency in the globus pallidus.' *Ann. Neurol.*, 1985, **18**, 482–489.

52. Y. Ikeda, K. Ikeda and D. M. Long, 'Protective effect of the iron chelator deferoxamine on cold-induced brain edema.' *J. Neurosurg.*, 1989, **71**, 233–238.

53. A. I. Cederbaum, 'Oxygen radical generation by microsomes: role of iron and implications for alcohol metabolism and toxicity.' *Free Radic. Biol. Med.*, 1989, **7**, 559–567.

54. S. Shaw, 'Lipid peroxidation, iron mobilization and free radical generation induced by alcohol.' *Free Radic. Biol. Med.*, 1989, **7**, 541–547.

55. H. Rouach, P. Houze, M. T. Orfanelli *et al.*, 'Effect of acute ethanol administration on the subcellular distribution of iron in rat liver and cerebellum.' *Biochem. Pharmacol.*, 1990, **39**, 1095–1100.

56. S. C. Bondy and K. R. Pearson, 'Ethanol-induced oxidative stress and nutritional status. *Alcohol. Clin. Exp. Res.*, 1993, **17**, 651–654.

57. D. R. A. Crapper-McLachlan, A. J. Dalton, T. P. Kruck *et al.*, 'Intramuscular desferrioxamine in pat-ients with Alzheimer's disease.' *Lancet*, 1991, **337**, 1304–1308.

58. S. M. LeVine, 'The role of reactive oxygen species in the pathogenesis of multiple sclerosis.' *Med. Hypotheses*, 1992, **39**, 271–274.

59. B. Halliwell, 'Protection against tissue damage *in vivo* by desferrioxamine: what is its mechanism of action?' *Free Radic. Biol. Med.*, 1989, **7**, 645–651.

60. S. J. Klebanoff, A. M. Waltersdorph, *et al.*, 'Oxygen-based free radical generation by ferrous ions and deferoxamine.' *J. Biol. Chem.*, 1989, **264**, 19765–19771.

61. Z. Z. Wahba, W. J. Murray and S. J. Stohs, 'Desferrioxamine-induced alterations in hepatic iron distribution, DNA damage, and lipid peroxidation in control and 2,3,7,8-tetrachlorodibenzo-*p*-dioxin-treated rats.' *J. Appl. Toxicol.*, 1990, **10**, 119–124.

62. T. P. A. Kruck, E. A. Fisher and D. R. McLachlan, 'Suppression of deferoxamine mesylate treatment-induced side effects by coadministration of isoniazid in a patient with Alzheimer's disease subject to aluminum removal by ion-specific chelation.' *Clin. Pharmacol. Therap.*, 1990, **48**, 439–446.

63. D. J. Thomas and J. Chisholm, Jr., 'Lead, zinc and copper decorporation during calcium disodium ethylenediamine tetraacetate treatment of lead-poi-soned children.' *J. Pharmacol. Exp. Ther.*, 1986, **239**, 829–835.

64. D. W. Reif and R. D. Simmons, 'Nitric oxide mediates iron release from ferritin.' *Arch. Biochem. Biophys.*, 1990, **283**, 537–541.

65. N. Ogawa, R. Edamatsu, K. Mizukawa *et al.*, 'Degeneration of dopaminergic neurons and free radicals. Possible participation of levodopa.' *Adv. Neurol.*, 1993, **60**, 242–250.

66. R. J. Wurtman and J. J. Wurtman, in 'Nutrition and the Brain,' eds. R. J. Wurtman and J. J. Wurtman, Raven Press, New York, 1990, pp. 59–74.

67. K. A. Jellinger, E. Kienzl, G. Rumpelmaier *et al.*, 'Iron and ferritin in substantia nigra in Parkinson's disease.' *Adv. Neurol.*, 1993, **60**, 267–272.

68. I. H. Scheinberg, in 'Metal Neurotoxicity,' eds. S. C. Bondy and K. N. Prasad, CRC Press, Boca Raton, FL, 1988, pp. 56–60.

69. G. T. Babcock and M. Wikstrom, 'Oxygen activation and the conservation of energy in cell respiration.' *Nature*, 1992, **356**, 301–309.

70. W. Birkmayer, J. Knoll, P. Reiderer *et al.*, 'Increased of life expectancy resulting from addition of L-deprenyl to Madopar treatment in Parkinson's disease: a longterm study.' *J. Neural. Transm.*, 1985, **64**, 113–127.

71. I. Shoulson, 'Deprenyl and tocopherol antioxidative therapy of parkinsonism (DATATOP). Parkinson Study Group.' *Acta Neurol. Scand. Suppl.*, 1989, **126**, 171–175.

72. P. K. Seth and S. V. Chandra, in 'Metal Neurotoxicity,' eds. S. C. Bondy and K. N. Prasad, CRC Press, Boca Raton, FL, 1988, pp. 19–33.

73. Anonymous, 'DATATOP: a multicenter controlled clinical trial in early Parkinson's Disease. Parkinson's Study Group.' *Arch. Neurol.*, 1989, **46**, 1052–1060.

74. J. Donaldson, F. S. Labella and D. Gessa, 'Enhanced autooxidation of dopamine as a possible basis of manganese neurotoxicity.' *Neurotoxicol.*, 1981, **2**, 53–64.

75. P. L. B. Cheton and F. S. Archibald, 'Manganese complexes and the generation and scavenging of hydroxyl free radicals.' *Free Radic. Biol. Med.*, 1988, **5**, 325–333.

76. M. Coassin, F. Ursini and A. Bindoli, 'Antioxidant effect of manganese.' *Arch. Biochem. Biophys.*, 1992, **299**, 330–333.

77. S. S. Haider and M. el-Fakhri, 'Action of alpha-tocopherol on vanadium-stimulated lipid peroxidation in rat brain.' *Neurotoxicology*, 1991, **12**, 79–85.

78. A. Naganuma, M. E. Anderson and A. Meister, 'Cellular glutathione as a determinant of sensitivity to mercuric chloride toxicity. Prevention of toxicity by giving glutathione monoester.' *Biochem. Pharmacol.*, 1990, **40**, 693–697.

79. M. Hasan and S. F. Ali, 'Effects of thallium, nickel and cobalt administration on the lipid peroxidation in different regions of the rat brain.' *Toxicol. Appl. Pharmacol.*, 1981, **57**, 8–13.

80. M. Yonaha, M. Saito and M. Sagai, 'Stimulation of lipid peroxidation by methyl mercury in rats.' *Life Sci.*, 1983, **32**, 1507–1514.

81. M. R. Marzabadi and C. B. Jones, 'Heavy metals and lipofucsinogenesis. A study on myocardial cells cultured under varying oxidative stress.' *Mech. Ageing Dev.*, 1992, **66**, 159–171.

82. L. W. Chang, M. Gilbert and J. Sprecher, 'Modification of methylmercury neurotoxicity by vitamin E.' *Environ. Res.*, 1978, **17**, 356–366.

83. G. S. Shukla, R. S. Srivastava and S. V. Chandra, 'Prevention of cadmium-induced effects on regional

glutathione status of rat brain by vitamin E.' *J. Appl. Toxicol.*, 1988, **8**, 355–359.

84. C. C. Reddy and E. J. Massaro, 'Biochemistry of selenium: a brief overview.' *Fundam. Appl. Toxicol.*, 1983, **3**, 431–436.

85. L. W. Chang and R. Suber, 'Protective effect of selenium on methylmercury toxicity: a possible mechanism.' *Bull. Environ. Contam. Toxicol.*, 1982, **29**, 285–289.

86. C. P. LeBel, S. F. Ali, M. McKee *et al.*, 'Organometal-induced increases in oxygen reactive species: the potential of 2′,7′-dichlorofluorescin diacetate as an index of neurotoxic damage.' *Toxicol. Appl. Pharmacol.*, 1990, **104**, 17–24.

87. C. P. LeBel, H. Ischiropoulos and S. C. Bondy, 'Evaluation of the probe 2′,7′-dichlorofluorescin as an indicator of reactive oxygen species formation and oxidative stress.' *Chem. Res. Toxicol.*, 1992, **5**, 227–231.

88. J. P. O'Callaghan and D. B. Miller, 'Diethyldithio-carbamate increases distribution of cadmium to brain but prevents cadmium-induced neurotoxicity.' *Brain Res.*, 1986, **370**, 354–358.

89. W. G. Hoekstra, 'Biochemical function of selenium and its relation to vitamin E.' *Fed. Proc.*, 1975, **34**, 2083–2089.

90. Y. Kojima and J. H. R. Kagi, 'Metallothionein.' *Trends Biochem. Sci.*, 1978, **3**, 90–92.

91. J. Schubert, E. J. Riley and S. A. Tyler, 'Combined effects in toxicology—a rapid systematic testing procedure: cadmium, mercury and lead.' *J. Toxicol. Environ. Health*, 1978, **4**, 763–776.

92. J. J. Zimmerman, S. M. Zuk and J. R. Millard. '*In vitro* modulation of human neutrophil superoxide anion generation by various calcium channel antagonists used in ischemia-reperfusion resuscitation.' *Biochem. Pharmacol.*, 1989, **38**, 3601–3610.

93. B. C. White, S. D. Aust, K. E. Arfors *et al.*, 'Brain injury by ischemic anoxia: hypothesis extension—a tale of two ions?' *Ann. Emerg. Med.*, 1984, **13**, 862–867.

94. A. Vanella, V. Sorrenti, C. Castorina *et al.*, 'Lipid peroxidation in rat cerebral cortex during post-ischemic reperfusion: effect of exogenous antioxidants and Ca²⁺-antagonistic drugs.' *Int. J. Dev. Neurosci.*, 1992, **10**, 75–80.

95. S. C. Bondy and C. P. LeBel, 'The relationship between excitotoxicity and oxidative stress in the central nervous system.' *Free Radic. Biol. Med.*, 1993, **14**, 633–642.

96. Shafig-Ur-Rehman, 'Lead-induced regional lipid peroxidation in brain.' *Toxicol. Lett.*, 1984, **21**, 333–337.

97. S. F. Ali and S. C. Bondy, 'Triethyl lead-induced peroxidative damage in various regions of the rat brain.' *J. Toxicol. Environ. Health*, 1989, **26**, 235–242.

98. M. Hermes-Lima, B. Pereira and E. J. H. Bechara, 'Are free radicals involved in lead poisoning?' *Xenobiotica*, 1991, **21**, 1085–1090.

99. P. H. Evans, E. Peterhans, T. Burge *et al.*, 'Alumino-silicate-induced free radical generation by murine brain glial cells *in vitro*: potential significance in the aetiopathogenesis of Alzheimer's dementia.' *Dementia*, 1992, **3**, 1–6.

100. M. A. Pappolla, R. A. Omar, K. S. Kim *et al.*, 'Immunohistochemical evidence of antioxidant stress in Alzheimer's disease.' *Am. J. Pathol.*, 1992, **140**, 621–628.

101. J. M. C. Gutteridge, G. J. Quinlan, I. Clark *et al.*, 'Aluminium salts accelerate peroxidation of membrane lipids stimulated by iron salts.' *Biochim. Biophys. Acta*, 1985, **835**, 441–447.

102. C. G. Fraga, P. I. Oteiza, M. S. Golub *et al.*, 'Effects of aluminum on brain lipid peroxidation.' *Toxicol. Lett.*, 1990, **51**, 213–219.

103. P. I. Oteiza, C. G. Fraga and C. L. Keen, 'Aluminum has both oxidant and antioxidant effects in mouse brain membranes.' *Arch. Biochem. Biophys.*, 1993, **300**, 517–521.

104. S. C. Bondy and S. Kirstein, 'The promotion of iron-induced generation of reactive oxygen species in nerve tissue by aluminum.' *Mol. Chem. Neuropathol.*, 1996, **27**, 185–194.

105. C. D. Smith, J. M. Carney, T. Tatsumo *et al.*, 'Protein oxidation in aging brain.' *Ann. NY Acad. Sci.*, 1992, **663**, 110–119.

106. K. V. Subbarao, J. S. Richardson and L. C. Ang, 'Autopsy samples of Alzheimer's cortex show increased peroxidation *in vitro*.' *J. Neurochem.*, 1990, **55**, 342–355.

107. P. W. Mantyh, J. R. Ghilardi, S. Rogers *et al.*, 'Aluminum, iron and zinc ions promote aggregation of physiological concentrations of beta-amyloid peptide.' *J. Neurochem.*, 1993, **61**, 1171–1174.

108. P. H. Evans, J. Klinowski and E. Yano, 'Cephaloco-niosis: a free radical perspective on the proposed particulate-induced etiopathogenesis of Alzheimer's dementia and related disorders.' *Med. Hypotheses*, 1991, **34**, 209–219.

109. C. Garrel, J. L. Lafond, P. Faure *et al.*, '*In vitro* study of nitric oxide production by microglial cells stimu-lated with aluminum salts.' *Proc. 2nd. Int. Symp. on Reactive Oxygen Species II*, 1993, 18.

110. F. F. Ahmad, D. L. Cowan and A. Y. Sun, 'Detection of free radical formation in various tissues after acute carbon tetrachloride administration in the gerbil.' *Life Sci.*, 1987, **41**, 2469–2475.

111. C. Cojocel, W. Beuter, W. Muller *et al.*, 'Lipid peroxidation: a possible mechanism of trichloro-ethylene-induced nephrotoxicity.' *Toxicology*, 1989, **55**, 131–141.

112. C. Guerri and S. Grisolia, 'Changes in glutathione in acute and chronic alcohol intoxication.' *Pharmacol. Biochem. Behav.*, 1980, **13**, Suppl. 1, 53–61.

113. R. Nordmann, 'Oxidative stress from alcohol in the brain.' *Alcohol Alcohol*, 1987, Suppl. 1, 75–82.

114. R. Natsuki, 'Effect of ethanol on calcium-uptake and phospholipid turnover by stimulation of adrenocep-tors and muscarinic receptors in mouse brain and heart synaptosomes.' *Biochem. Pharmacol.*, 1991, **42**, 39–44.

115. M. Uysal, G. Kutalp, G. Odzemirler *et al.*, 'Ethanol-induced changes in lipid peroxidation and glutathione content in rat brain.' *Drug Alcohol Depend.*, 1989, **23**, 227–230.

116. C. J. Mattia, J. D. Adams, Jr. and S. C. Bondy, 'Free radical induction in the brain and liver by products of toluene catabolism.' *Biochem. Pharmacol.*, 1993, **46**, 103–110.

117. D. E. Ray, N. J. Abbott, M. W. K. Chan *et al.*, 'Increased oxidative metabolism and oxidative stress in m-dinitrobenzene neurotoxicity.' *Biochem. Soc. Trans.*, 1994, **22**, 407S.

118. S. C. Bondy and S. X. Guo, 'Effect of ethanol treatment on indices of cumulative oxidative stress.' *Eur. J. Pharmacol.*, 1994, **270**, 349–355.

119. S. C. Bondy and S. X. Guo, 'Regional selectivity in ethanol-induced pro-oxidant events within the brain.' *Biochem. Pharmacol.*, 1995, **49**, 69–72.

120. J. D. Adams, L. K. Klaidman and I. N. Odunze, 'Oxidative effects of MPTP in the midbrain.' *Res. Comm. Subst. Abuse*, 1989, **10**, 169–180.

121. H. A. Nadiger, S. K. Marcus and M. V. Chavdrakala, 'Lipid peroxidation and ethanol toxicity in rat brain: effect of vitamin E deficiency and supplementation.' *Med. Sci. Res.*, 1988, **16**, 1273–1274.

122. P. C. Choy, R. Y. K. Man and A. C. Chan, 'Phosphatidylcholine metabolism in isolated rat

heart: modulation by ethanol and vitamin E.' *Biochim. Biophys. Acta*, 1989, **1005**, 225–232.

123. C. P. LeBel and R. A. Schatz, 'Altered synaptosomal phospholipid metabolism after toluene: possible relationship with membrane fluidity, Na$^+$, K$^+$-adenosine triphosphatase and phospholipid methylation.' *J. Pharmacol. Exp. Ther.*, 1990, **253**, 1189–1197.

124. M. Kaneko, V. Panagia, G. Paolillo *et al.*, 'Inhibition of cardiac phosphatidylethanolamine *N*-methylation by oxygen free radicals.' *Biochim. Biophys. Acta*, 1990, **1021**, 33–38.

125. R. Nordmann, C. Ribiere and H. Rouach, 'Involvement of iron and iron-catalyzed free radical production in ethanol metabolism and toxicity.' *Enzyme*, 1987, **37**, 57–69.

126. R. Nordmann, C. Ribiere and H. Rouach, 'Implication of free radical mechanisms in ethanol-induced cellular injury.' *Free Radic. Biol. Med.*, 1992, **12**, 219–240.

127. R.Nordmann, C. Ribiere and H. Rouach, in 'Free Radicals, Lipoproteins and Membrane Lipids,' eds. A. Crastes de Paulet, L. Douste-Blazy and R. Paoletti, Plenum Press, New York, 1990, pp. 309–319.

128. D. Lucas, Y. Lamboeuf, G. De Saint-Blanquat *et al.*, 'Ethanol-inducible cytochrome P-450 activity and increase in acetaldehyde bound to microsomes after chronic administration of acetaldehyde or ethanol.' *Alcohol Alcoholism*, 1990, **25**, 395–400.

129. U. J. Behrens, M. Hoerner, J. M. Lasker *et al.*, 'Formation of acetaldehyde adducts with ethanol-inducible P450IIEI in vivo.' *Biochem. Biophys. Res. Commun.*, 1988, **154**, 584–590.

130. J. Y. Westcott, H. Weiner, J. Shultz *et al.*, 'In vivo acetaldehyde in the brain of the rat treated with ethanol.' *Biochem. Pharmacol.*, 1980, **29**, 411–417.

131. P. J. O'Neill and R. G. Rahwan, 'Protection against acute toxicity of acetaldehyde in mice.' *Res. Commun. Chem. Pathol. Pharmacol.*, 1976, **13**, 125–128.

132. S. N. Wickramansinghe and R. Ha, '*In vitro* effects of vitamin C, thioctic acid and dihydrolipoic acid on the cytotoxicity of post-ethanol serum.' *Biochem. Pharmacol.*, 1992, **43**, 407–411.

133. G. Ekstrom and M. Ingelman-Sundberg, 'Rat liver microsomal NADPH-supported oxidase activity and lipid peroxidation dependent on ethanol-inducible cytochrome P-450 (P-450IIE1).' *Biochem. Pharmacol.*, 1989, **38**, 1313–1319.

134. H. Kuthan and V. Ullrich, 'Oxidase and oxygenase function of the microsomal cytochrome P450 monoxygenase system.' *Eur. J. Biochem.*, 1982, **126**, 583–588.

135. T. Hansson, N. Tindberg, M. Ingelman-Sundberg *et al.*, 'Regional distribution of ethanol-inducible cytochrome P450 IIEI in the rat central nervous system.' *Neuroscience.*, 1990, **34**, 451–463.

136. C. J. Mattia, C. P. LeBel and S. C. Bondy, 'Effects of toluene and its metabolites on cerebral reactive oxygen species generation.' *Biochem. Pharmacol.*, 1991, **42**, 879–882.

137. M. Ledig, J. R. M'Paria and P. Mandel, 'Superoxide dismutase activity in rat brain during acute and chronic alcohol intoxication.' *Neurochem. Res.*, 1981, **6**, 385–390.

138. M. Ledig, G. Tholey, L. Megias-Megias *et al.*, 'Combined effects of ethanol and manganese on cultured neurons and glia.' *Neurochem. Res.*, 1991, **16**, 591–596.

139. B. Gonthier, A. Jeunet and L. Barret, 'Electron spin resonance study of free radicals produced from ethanol and acetaldehyde after exposure to a Fenton system or to brain and liver microsomes.' *Alcohol*, 1991, **8**, 369–375.

140. D. E. Feierman and A. I. Cederbaum, 'Increased content of cytochrome P-450 and 4-methylpyrazole binding spectrum after 4-methylpyrazole treatment.' *Biochem. Biophys. Res. Comm.*, 1985, **126**, 1076–1081.

141. L. G. Sultatos, 'Effects of acute ethanol administration on the hepatic xanthine dehydrogenase oxidase system in the rat.' *J. Pharmacol. Exp. Ther.*, 1988, **246**, 946–949.

142. A. Zuddas, G. U. Corsini, S. Schinelli *et al.*, 'Acetaldehyde directly enhances MPP$^+$ neurotoxicity and delays its elimination from the striatum.' *Brain Res.*, 1989, **501**, 11–22

143. F. F. Oldfield, D. L. Cowan and A. Y. Sun, 'The involvement of ethanol in the free radical reaction of 6-hydroxydopamine.' *Neurochem. Res.*, 1991, **16**, 83–87.

144. C. A. Trenga, D. D. Kunkel, D. L. Eaton *et al.*, 'Effect of styrene oxide on rat brain glutathione.' *Neurotoxicology*, 1991, **12**, 165–178.

145. H. R. Lam, G. Ostergaard, S. X. Guo *et al.*, 'Three weeks' exposure of rats to dearomatized white spirit modifies indices of oxidative stress in brain, kidney, and liver.' *Biochem. Pharmacol.*, 1994, **47**, 651–657.

146. S. C. Bondy, H. R. Lam, G. Ostergaard *et al.*, 'Changes in markers of oxidative status in brain, liver and kidney of young and aged rats following exposure to aromatic white spirit.' *Arch.Toxicol.*, 1995, **69**, 410–414.

147. R. E. Heikkila, B. A. Sieber, L. Manzino *et al.*, 'Some features of the nigrostriatal dopaminergic neurotoxin 1-methyl-4-phenyl-1,2,3,6-tetrahydropyridine (MPTP) in the mouse.' *Mol. Chem. Neuropathol.*, 1989, **10**, 171–183.

148. C. Rios and R. Tapia, 'Changes in lipid peroxidation induced by 1-methyl-4-phenyl-1,2,3,6-tetrahydropyridine and 1-methyl-4-phenylpyridinium in mouse brain homogenates.' *Neurosci. Lett.*, 1987, **77**, 321–326.

149. I. N. Odunze, L. K. Klaidman and J. D. Adams, Jr., 'MPTP toxicity in the mouse brain and vitamin E.' *Neurosci. Lett.*, 1990, **108**, 346–349.

150. M. Cino and R. F. Del Maestro, 'Generation of hydrogen peroxide by brain mitochondria: the effect of reoxygenation following postdecapitative ischemia.' *Arch. Biochem. Biophys.*, 1989, **269**, 623–638.

151. S. F. Ali, C. P. LeBel and S. C. Bondy, 'Reactive oxygen species formation as a biomarker of methylmercury and trimethyltin neurotoxicity.' *Neurotoxicology*, 1992, **13**, 637–648.

152. C. M. Tanner and J. W. Langston, 'Do environmental toxins cause Parkinson's disease? A critical review.' *Neurology*, 1990, **40**, Suppl. 3, 17–30.

153. H. Saggu, J. Cooksey, D. Dexter *et al.*, 'A selective increase in particulate superoxide dismutase activity in parkinsonian substantia nigra.' *J. Neurochem.*, 1989, **53**, 692–697.

154. J. M. McCrodden, K. F. Tipton and J. P. Sullivan, 'The neurotoxicity of MPTP and the relevance to Parkinson's disease.' *Pharmacol. Toxicol.*, 1990, **67**, 8–13.

155. G. Pezzoli, S. Ricciardi, C. Masotto *et al.*, '*N*-hexane induces Parkinsonism in rodents.' *Brain Res.*, 1990, **531**, 355–357.

156. N. Fedtke and H. M. Bolt, 'Detection of 2,5-hexanedione in the urine of persons not exposed to *n*-hexane.' *Int. Arch. Occup. Environ. Health*, 1986, **57**, 143–148.

157. M. J. De Vito and G.C. Wagner, 'Metamphetamine-induced neuronal damage: a possible role for free radicals.' *Neuropharmacology*, 1989, **28**, 1145–1150.

158. J. W. Olney, C. F. Zorumski, G. R. Stewart *et al.*, 'Excitotoxicity of L-dopa and 6-OH-dopa: implications for Parkinson's and Huntington's diseases.' *Exp. Neurol.*, 1990, **108**, 269–272.

159. G. Cohen, in 'Oxygen Radicals and Tissue Injury,' ed. B. Halliwell, Clarendon Press, Oxford, 1988, pp. 130–135.

160. J. B. Lohr, R. Kuczenski, H. S. Bracha *et al.*, 'Increased indices of free radical activity in the cerebrospinal fluid of patients with tardive dyskinesia.' *Biol. Psychiatry*, 1990, **28**, 535–539.

161. V. Jackson-Lewis, S. Przedborski, V. Kostic *et al.*, 'Partial attenuation of chronic fluphenazine-induced changes in regional monoamine metabolism by D-alpha tocopherol in rat brain.' *Brain Res. Bull.*, 1991, **26**, 251–258.

162. B. K. Ardelt, J. L. Borowitz and G. E. Isom, 'Brain lipid peroxidation and antioxidant protectant mechanisms following acute cyanide intoxication.' *Toxicology*, 1989, **56**, 147–154.

163. R. D. Montgomery, in 'Handbook of Clinical Neurology, Intoxications of the Nervous System,' eds. P. J. Vinken and G.W. Bruyn, North Holland Publishing Co., Amsterdam, 1979, vol. 32, part I, pp. 515–521.

164. T. L. Pazdernik, M. Layton, S. R. Nelson *et al.*, 'The osmotic/calcium stress theory of brain damage: are free radicals involved?' *Neurochem. Res.*, 1992, **17**, 11–21.

165. J. A. Dykens, 'Isolated cerebral and cerebellar mitochondria produce free radicals when exposed to elevated Ca^{2+} and Na^+: implications for neurodegeneration.' *J. Neurochem.*, 1994, **63**, 584–591.

166. S. R. Nelson and J. P. Olson, 'Role of early edema in the development of regional seizure related-brain damage.' *Neurochem. Res.*, 1987, **12**, 561–564.

167. W. M. Armstead, R. Mirro, C. W. Leffler *et al.*, 'Cerebral superoxide anion generation during seizures in newborn pigs.' *J. Cereb. Blood Flow Metab.*, 1989, **9**, 175–179.

168. L. J. Willmore and W. J. Triggs, 'Effect of phenytoin and corticosteroids on seizures and lipid peroxidation in experimental posttraumatic epilepsy.' *J. Neurosurg.*, 1984, **60**, 467–472.

169. K. Kitagawa, M. Matsumoto, T. Oda *et al.*, 'Free radical generation during brief period of cerebral ischemia may trigger delayed neuronal death.' *Neuroscience*, 1990, **35**, 551–558.

170. R. J. Sutherland, J. M. Hoesing and I. Q. Whishaw, 'Domoic acid, an environmental toxin, produces hippocampal damage and severe memory impairment.' *Neurosci. Lett.*, 1990, **120**, 221–223.

171. S. C. Bondy and D. K. Lee, 'Oxidative stress induced by glutamate receptor agonists.' *Brain Res.*, 1993, **610**, 229–233.

172. T. H. Murphy, M. Miyamoto, A. Sastre *et al.*, 'Glutamate toxicity in a neuronal cell line involves inhibition of cystine transport leading to oxidative stress.' *Neuron*, 1989, **2**, 1547–1558.

173. K. Haba, N. Ogawa, K. Mizukawa *et al.*, 'Time course of changes in lipid peroxidation, pre- and postsynaptic cholinergic indices, NMDA receptor binding and neuronal death in the gerbil hippocampus following transient ischemia.' *Brain Res.*, 1991, **540**, 116–122.

174. R. J. Bridges, J. Y. Koh, C. G. Hatalski *et al.*, 'Increased excitotoxic vulnerability of cortical cultures with reduced levels of glutathione.' *Eur. J. Pharmacol.*, 1991, **192**, 199–200.

175. L. Pellerin and L. S. Wolfe, 'Release of arachidonic acid by NMDA-receptor activation in the rat hippocampus.' *Neurochem. Res.*, 1991, **16**, 983–989.

176. L. E. Kiedrowski, E. Costa and J. T. Wroblewski, 'Glutamate receptor agonists stimulate nitric oxide synthase in primary cultures of cerebellar granule cells.' *J. Neurochem.*, 1992, **58**, 335–341.

177. V. L. Dawson, T. M. Dawson, E. D. London, 'Nitric oxide mediates glutamate neurotoxicity in primary cortical culture.' *Proc. Nat. Acad. Sci. USA*, 1991, **88**, 6368–6371.

178. J. A. Dykens, A. Stern and E. Trenkner, 'Mechanism of kainate toxicity to cerebellar neurons *in vitro* is analogous to reperfusion tissue injury.' *J. Neurochem.*, 1987, **49**, 1222–1228.

179. R. Bose, G. R. Sutherland and C. Pinsky, 'Excitotoxins and free radicals: accomplices in post-ischemic and other neurodegeneration.' *Eur. J. Pharmacol.*, 1990, **183**, 1170–1171.

180. G. Delbarre, B. Delbarre, F. Calinon *et al.*, 'Accumulation of amino acids and hydroxyl free radicals in brain and retina of gerbil after transient ischemia.' *J. Ocul. Pharmacol.*, 1991, **7**, 147–155.

181. D. E. Pellegrini-Giampietro, G. Cherici, M. Alesiani *et al.*, 'Excitatory amino acid release and free radical formation may cooperate in the genesis of ischemia-induced neuronal damage.' *J. Neurosci.*, 1990, **10**, 1035–1041.

182. N. F. Schor, 'Inactivation of mammalian brain glutamine synthetase by oxygen radicals.' *Brain Res.*, 1988, **456**, 17–21.

183. S. M. Oh and A. L. Betz, 'Interaction between free radicals and excitatory amino acids in the formation of ischemic brain edema in rats.' *Stroke*, 1991, **22**, 915–921.

184. J. T. Coyle and P. Puttfarcken, 'Oxidative stress, glutamate and neurodegenerative disorders.' *Science*, 1993, **262**, 689–695.

185. R. Henneberry and L. Spatz, 'The role of environmental factors in neurodegenerative disorders.' *Neurobiol. Aging*, 1991, **12**, 75–79.

SECTION III

11.05
Cytoskeletal Elements in Neurotoxicity

SALLY J. PYLE and KENNETH R. REUHL
Rutgers University, Piscataway, NJ, USA

11.05.1   INTRODUCTION

The cytoskeleton is classically defined as "filaments serving to act as supportive cytoplas- mic elements."[1] This description, while essen- tially accurate, does scant justice to this remarkable network of interactive and dynamic filamentous proteins whose functions within neural cells include, but are not limited to, cell growth and division, migration, neurite extension, anterograde and retrograde axonal transport, neurotransmitter release, regulation of membrane transport proteins, and spacial organization of the cytoplasm. Virtually all neurotoxins and neurotoxicants will disturb cytoskeletal integrity or function at some stage of their pathogenetic action. This chapter describes the major cytoskeletal organelles and their associated proteins, and discusses the consequences of their disturbance by neurotoxic agents.

11.05.2 MAJOR COMPONENTS OF THE CYTOSKELETON

The filamentous proteins of the cytoskeleton are divided on the bases of size and protein structure into three families: microtubules, intermediate filaments, and actin microfilaments. The organization of these proteins within cells is fairly consistent. The actin network is juxtaposed to the cell membrane, the microtubule network is organized in spaced linear arrays extending through the axons and dendrites, and the intermediate filament network extends throughout the cell, interposed between and connected with, the other two networks (Figure 1).

11.05.2.1 Microtubules

Microtubules are the most prominent cytoskeletal element present in all cells of the mammalian nervous system. These cylindrical organelles are approximately 25 nm in diameter and are composed of 13 helically arranged protofilaments of tubulin dimers surrounding a hollow core. Tubulin, comprising 10–20% of the total soluble protein in brain and nerve, has numerous isotypes with molecular weights of approximately 55 kDa. After synthesis in the perikaryon, tubulin is transported along the axons by slow component b of axonal transport.

Microtubules are formed from dimers of α- and β-tubulin subunits. The assembly, or polymerization, of microtubules requires the coordination of nucleation (the formation of new microtubules), and elongation (the growth of existing microtubules). The rates of nucleation and elongation depend on the concentration of free tubulin dimers. Nucleation is relatively slow and begins at a specific site, usually the centrosome. In axons, which lack centrosomes, microtubule assembly occurs via elongation of existing microtubules.[2] Nucleation and elongation give the microtubules an overall polarity. At the growing or "plus-end" of the microtubule new dimers are added at a rapid rate, whereas at the minus end of the microtubule dimers are lost more rapidly. The

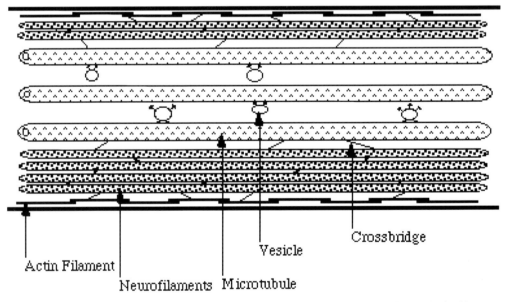

Figure 1 A conceptualized view of the normal arrangement of the axonal cytoskeleton. Actin filaments are arranged parallel to the plasma membrane, neurofilaments are longitudinally aligned throughout the axoplasm. Microtubules are longitudinally arranged in centrally located bundles. Vesicles, bound to molecular motors, move between the cell body and synaptic terminal on microtubules. Cross-bridges connect cytoskeletal proteins.

polarity of individual microtubules can be determined by adding free tubulin to tissue cross-sections. The free tubulin forms "hooks" at the ends of the cross-sectioned tubules. Clockwise hooks form at plus ends and counterclockwise hooks form at minus ends.[3,4] This method has been used to demonstrate that axonal microtubules are all oriented with their plus ends distal to the cell body, whereas the plus ends of dendritic microtubules may be oriented in either direction.[4,5] It is these differences in microtubule polarity that dictate the organelle composition of axons and dendrites (Figure 2).

Microtubules in cultured cells are constantly changing length.[6,7] At any time most microtubules are growing while a small fraction are rapidly shortening. Occasionally a microtubule will abruptly change from the elongation phase to the shortening phase; this is termed catastrophic disassembly, and its converse is termed rescue. This property of elongation and rapid collapse of microtubules has been termed dynamic instability.[8] Dynamic instability is an energy-dependent process, requiring binding of guanosine triphosphate (GTP) to β-tubulin and hydrolysis to guanosine diphosphate (GDP) after addition of a new dimer. Since GTP binding stabilizes microtubules and hydrolysis destabilizes them, microtubule stability can be regulated by altering the time between binding and hydrolysis.[9] A GTP cap, which favors polymerization, forms when hydrolysis is delayed. Loss of the cap causes catastrophic disassembly. GDP remains bound to tubulin during the elongation phase, but when a dimer is released during shortening, bound GDP can be exchanged for GTP. The new GTP-tubulin dimer can rebind to the shortening microtubule and initiate rescue.[9] Results from *in vitro* studies indicate that if GDP is not exchanged after the dimer is released, microtubules will spend less time shortening and lengthening.[10] Cells can use this principle to regulate dynamic instability. It is hypothesized that dynamic instability plays a major role in positioning microtubules, in microtubule branching during the formation of collateral axons,[11] and in axonal sprouting following degenerative injuries.[2] It is important to note that instability is an operative term based on calculated tubulin half-lives which will vary according to cell type.

11.05.2.1.1 *Post-translational modifications*

Microtubule stability can be altered through post-translational modifications of polymerized tubulin. α-Tubulin subunits can be modified by acetylation of lysyl residues or by reversible removal of the terminal tyrosine residue (detyrosination) by tubulin carboxypeptidase.[12] Detyrosination and acetylation are associated with increased microtubule stability, but the mechanism by which stability is conferred is unclear. These modifications are found on a

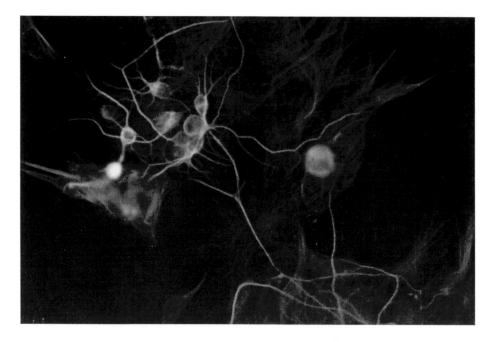

Figure 2 Immunofluorescence micrograph demonstrating the microtubule pattern in cultured neurons. Bundling of microtubules in the neurites results in intense staining, but also complicates evaluation of toxicant effects on individual microtubules.

higher proportion of subunits in stable axonal microtubules; stable dendritic microtubules are not as extensively modified.[13] These differences in post-translational modifications are important factors in the responses of axons and dendrites to microtubule toxicants.

It has been recognized that α- and β-tubulin subunits can also be modified by polyglycylation and polyglutamylation. The addition of glycyl residues may stabilize microtubules and anchor them to the axonal membrane.[14] Polyglutamylation may affect microtubule dynamics by regulating binding to the microtubule-associated protein tau.[15,16] Though phosphorylation of tubulin has long been recognized, the significance of this modification is not known. It appears to occur on the β-subunit and to be developmentally regulated.[12]

Differences in the axonal microenvironment, such as the differences between nodal and internodal regions, may be used by the cell to alter the activity or concentration of enzymes responsible for post-translational modifications, thereby tailoring microtubule dynamics to suit the localized needs of the cell (Figure 3).

11.05.2.2 Microfilaments

Actin, the protein which forms microfilaments, is reported to be the most abundant protein in eukaryotic cells, sometimes constituting greater than 5% of the total cellular protein, although this may be due to its high concentration in muscle cells. The network of actin filaments and associated proteins found lining the plasma membrane is essential for cellular strength, shape, and movement. The actin microfilaments are approximately 8 nm in diameter and are composed of actin monomers having a molecular weight of approximately 45 kDa, depending on the particular actin isotype. There are at least six types of actin found in mammalian cells, divided into three groups based on their isoelectric points.[17] The α-actin variants are a unique isoform found in muscle cells, while the β- and γ-variants are highly homologous constituents of nonmuscle cells and assemble *in vitro* into morphologically indistinguishable filaments.

Actin is a major constituent of growth cones in developing and sprouting neurons, existing

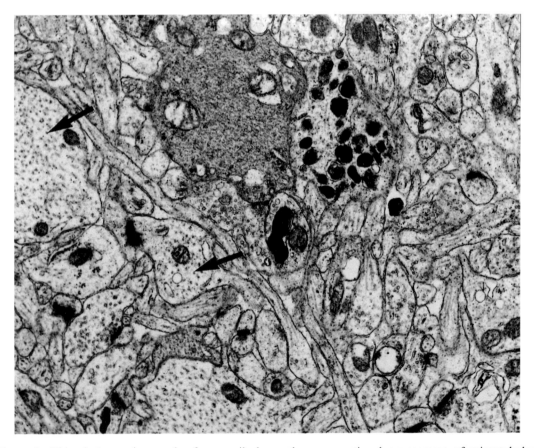

Figure 3 This electron micrograph of neuropil shows the cross-sectional arangement of microtubules (arrows) within the neuropil. Actin microfilaments are too small to be readily distinguishable, but are arranged adjacent to the plasma membrane.

in two different forms which correspond to the lamellapodia or filapodia of the actively advancing growth cone. G-actin, or globular actin, is a small polymer often found associated with other proteins, while F-actin is a fibrillar polymer composed of G-actin monomers and is usually referred to as microfilament protein. Approximately half of the actin within the cell is polymerized at any time while the other half exists as free monomers or as G-actin. In mature and developing neurons actin is synthesized in the cell body and is distributed throughout the axon by slow component b of orthograde transport.

Like microtubules, actin filaments are polar with a fast-growing or "plus" end and a slow-growing or "minus" end. This behavior is regulated by nucleotide hydrolysis and by the concentration of free monomer. Actin monomers are tightly associated with an ATP molecule which is hydrolyzed to ADP following the addition of the monomer to the filament. The hydrolysis of ATP is important in determining the critical concentration of free actin at the plus and minus ends of the filament. This leads to simultaneous assembly of new subunits at the plus end and disassembly of subunits from the minus end, a process known as "treadmilling." This phenomenon is thought to be the actin equivalent of dynamic instability in microtubules and is discussed in detail by Alberts *et al.*[17,18] The movement of organelles along actin filaments in squid axoplasm[18] is not unexpected when the similarities between actin microfilaments and microtubules are considered.

11.05.2.3 Intermediate Filaments

Intermediate filaments comprise a complex family of polypeptides. They are the most abundant proteins in large myelinated axons. Generally, intermediate filaments are cell type specific; for example, neurofilaments are expressed in neurons, vimentin is expressed in mesenchymal cells, and glial fibrillary acidic protein is expressed in glial cells. As a family, intermediate filaments are highly homologous and share the common structural features of an amino-terminal head, a central α-helical rod, and a carboxy-terminal tail. The central rod domain accounts for the fibrous shape of these proteins, imparting both strength and flexibility to the quaternary intermediate filament structure. The structures of the head and tail domains are variable and account for differences in molecular weight between intermediate filament proteins. Other intermediate filaments which are developmentally important or which are present

in small, but defined, populations of neurons, will be discussed.

Vimentin, nestin, and α-internexin are intermediate filaments expressed in some neurons during the early stages of development.[19-26] Their function in development is not fully understood, but neurites do not develop when their synthesis is inhibited.[21,22] Peripherin is an intermediate filament expressed in a subset of neurons in the peripheral nervous system.[23-25] These four intermediate filaments are of similar size (50–56 kDa), are post-translationally modified by phosphorylation of serine residues in the amino- and carboxy-terminal domains, and appear to be functionally similar to neurofilaments. The high degree of homology among intermediate filament proteins supports the argument that neurotoxicant exposure may cause similar alterations in all intermediate filaments, and indeed this has been shown with different intermediate filament types in cell culture models.[26]

11.05.3 CYTOSKELETAL ASSOCIATED PROTEINS

11.05.3.1 Microtubule Associated Proteins

Proteins specifically associated with microtubules are referred to as microtubule associated proteins (MAPs).[17] Neuronal MAPs are divided into two broad classes based on function. Fibrous proteins like MAP1 (>250 kDa), MAP2 (~200 kDa), and tau (55–62 kDa) promote microtubule assembly, stabilize assembled microtubules, and mediate their interactions with other proteins. Force-generating MAPs, such as cytoplasmic dynein (MAP1C) and kinesin provide motility for fast axoplasmic transport.[27] Most MAPs have two domains separated by a spacer region: a microtubule-binding domain and a domain that interacts with adjacent microtubules or other proteins.[28] Fibrous MAPs differ in their subcellular distribution, and may be expressed simultaneously or in a developmentally regulated sequence.[29] While tau and MAP2C are found only in the axon and MAP2 is restricted to the perikaryon and dendrites, MAP1A and 1B are ubiquitous in the neuron.[30,31] The mechanisms responsible for the sorting and localization of these MAPs are not understood; however, it has been shown that a sequence in the amino-terminal of MAP2 inhibits its entry into the axon, and that phosphorylation of a specific residue on tau interferes with binding of this MAP to microtubules in the cell body and dendrites.[32] This suggests that MAP localization is a function of protein structure and/or posttanslational modification.

Various models have been used to resolve the functions of fibrous MAPs. Transfection of nonneuronal cells with MAP2 or tau causes microtubule bundles to reorganize so that they resemble axonal microtubules.[33] Microtubule spacing in the transfected cells is 20 nm for MAP2C or tau, and 65 nm for MAP2.[28] These distances correspond to those between adjacent microtubules in axons and dendrites, respectively, suggesting that these MAPs are key determinants of neuronal microtubule spacing. Video assays of purified microtubules show that tau and MAP2 regulate dynamic instability *in vitro* by decreasing the frequency of catastrophic disassembly and increasing the frequency of rescue.[34] The addition of MAP1A to tubulin assays enhances microtubule elongation by lowering the critical concentration of tubulin necessary for assembly.[35] Post-translational phosphorylation of MAP2C or tau inhibits microtubule assembly *in vitro*.[36] These two MAPs are found exclusively in axons, and it is postulated that locally controlled cycles of MAP phosphorylation and dephosphorylation are used by the axon to regulate dynamic instability. Fibrous MAPs are important in neurite development, bundling, stability, and regulation of neuronal microtubules.

The other major neuronal function performed by MAPs is fast axonal transport. Kinesin and dynein are energy-dependent motor proteins which move vesicles and organelles in fast anterograde and retrograde transport, respectively. Kinesin moves toward the plus-end of microtubules while dynein moves toward their minus-ends. Differences in the organelle composition between axons and dendrites are regulated through binding of specific organelles to either kinesin or dynein.[27,37] For example, rough endoplasmic reticulum (RER) is found in dendrites but not axons. RER, which is assembled in the cell body, binds only to dynein, and therefore can only be transported toward the minus-end of a microtubule. This excludes RER from the axon where materials transported from the cell body can only move towards the plus-ends of microtubules.

The post-translational phosphorylation of all MAP proteins is currently an area of intense research. The phosphorylation state of MAPs appears to be developmentally regulated and may be important in neurite outgrowth.[38,39] In addition, abnormally phosphorylated tau proteins are found in the characteristic neurofibrillary tangles associated with Alzheimer's disease engendering speculation regarding the role of tau phosphorylation in this and other neurodegenerative diseases.

11.05.3.2 Actin Associated Proteins

Actin can assume a variety of macromolecular structures under different conditions even though the basic structure of the filaments does not differ. The ability to alter the basic filamentous structure is controlled by actin associated, or binding, proteins. The large number of actin binding proteins can be grouped into four distinct classes based on conserved protein structures and function: assembly regulating proteins, F-actin stabilizing and coassembly proteins, bundling and cross-linking proteins, and anchoring proteins.[40] The remainder of this section will focus on the more important isotypes from these groups present in neurons.

The proteins regulating assembly of actin are in turn loosely classified into four groups based on their functions. The monomer sequestering protein profilin is a small globular protein that binds G-actin and shifts the critical concentration required for actin assembly, thus preventing spontaneous nucleation. The association of profilin with the plasma membrane and its ability to bind phosphatidylinositol-4,5-biphosphate (PIP_2) make this protein of interest as a link in signal transduction between the plasma membrane and the actin cytoskeleton.[40,41] Cofilin and actin depolymerizing factor (ADF) can sever actin filaments in addition to sequestering G-actin, and are also sensitive to phosphoinositide signalling. ADF, a major component in growth cones, is inactive until phosphorylated in response to signal transduction messages.[40] This may be an important mechanism in axonal guidance during development. Other proteins that regulate actin polymerization include gelsolin, a calcium activated protein that severs and caps actin filaments and several proteins, such as CAP-90 and BAM-40, that bind to the barbed ends of the actin filaments and inhibit elongation.

Proteins that promote assembly and stabilize F-actin include tropomyosin and caldesmon. F-actin bound with tropomyosin does not depolymerize in the presence of ADF.[40] Caldesmon appears to enhance the binding of tropomyosin and F-actin, but when caldesmon complexes Ca^{2+}-calmodulin it dissociates from F-actin, which is then destabilized.

Numerous proteins bundle and cross-link actin filaments, including spectrin, α-actinin, synapsin I, filamin, and even MAP-2. Spectrin and filamin are closely associated with the plasma membrane and promote the formation of isotropic actin networks in contrast to α-actinin which stabilizes actin filaments into side-by-side aggregates.[42] Synapsin I is a prominent component of synaptosomes and the presynaptic terminal. There are two binding sites on synapsin, one which binds actin and another which binds synaptic vesicles. While it appears that disassembly of the actin network must occur prior to transmitter release, the role of synapsin has not yet been established.

A final major class of actin binding proteins serve to anchor F-actin to the plasma membrane. Spectrin and α-actinin may take part in this anchoring, along with ankyrin, catenin, talin, and vinculin. Spectrin appears to be important in connecting actin to transmembrane cell adhesion molecules while talin, vinculin, and α-actinin establish connections between actin and integrin.[40] Many other actin associated proteins have been reported, but their distribution in brain is not yet established.

11.05.4 THE CYTOSKELETON IN NEURITE DEVELOPMENT

Neuronal development imposes a unique set of requirements upon the cell. Growing axons and dendrites must determine their direction of growth and make the proper synaptic connections. This is accomplished by growth cones, structures found at the tips of growing neurites. Growth cones have high concentrations of soluble and filamentous actin and tubulin. Actin forms filopodia that extend and retract from the body of the growth cone. Microtubules within growth cones are unstable and, therefore, constantly experience periods of catastrophic disassembly and rescue. As this occurs the microtubules grow and retreat from the actin rich filopodia. In addition to actin and tubulin, MAP1B, MAP2, and tau are present in growth cones. Tau is thought to be an important factor in neurite development since inhibition of tau synthesis arrests neurite formation in cultured neurons.[43]

The currently favored theory regarding the mechanism of neurite guidance is based on the ability of the actin-rich filopodia to recognize and react to extrinsic signals. When a microtubule grows into a filopodia that has been activated by an extrinsic signal, the microtubule is captured and stabilized. Organelles are then transported along the newly stabilized microtubule and a new section of neurite is established. This process allows vectorial growth of the neurites in the direction of the extrinsic signal.[44] Regulation of growth depends on the degree of post-translational modification of

tubulin, the expression of MAPs within the growth cone, and ion concentrations. The stability of the cytoskeleton in both the lamellapodia and filopodia is dependent on calcium ion concentrations. The responses of actin structures to calcium concentration can be plotted on a bell-shaped curve, indicating that calcium has both stimulatory and inhibitory effects which play a direct role in axon guidance.[45] The complexity of this mechanism makes the growth cone highly sensitive to agents that disrupt microfilament and microtubule integrity.

Intermediate filaments are not present in growth cones, but are essential for neurite development and circumferential axonal growth. Vimentin, α-internexin, and nestin are expressed during early stages of development and their disappearance coincides with the appearance of neurofilament subunits.[46] Inhibition of vimentin synthesis blocks neurite outgrowth in cultured cells, implying that intermediate filaments play a crucial, early role in neurite formation.[21] The two low molecular weight neurofilament subunits, NFL and NFM, are expressed following neurite outgrowth, and the higher molecular weight neurofilament, NFH, is expressed as the processes mature. It is hypothesized that NFH may be the homeostatic subunit whose expression is an indication of neuronal maturation.[47]

11.05.5 INTERACTIONS AND REGULATION OF THE CYTOSKELETON

There is evidence that the major cytoskeletal proteins are interconnected and interactive. Freeze-fracture, freeze-etch, and quick-freeze preparations for electron microscopy reveal thin, filamentous attachments between the different cytoskeletal proteins;[48,49] these connections are transiently altered during the propagation of neuronal action potentials.[50-52] This suggests that the interconnections are dynamic and that the filamentous components of the cytoskeleton may communicate through conformational rearrangement, a contention further supported by the observation that under certain conditions disruption of microfilaments or microtubules reversibly reorganizes the intermediate filament network.[53,54]

Despite our understanding of some of the cytoskeletal regulatory proteins, there are still a great many unknown factors controlling cytoskeletal organization and interactions. Cytoskeletal proteins are exposed to cycles of phosphorylation and dephosphorylation controlled by second-messenger dependent and independent kinases or by phosphatases.[12] Phosphorylation states regulate protein–protein interactions of the cytoskeleton. Regulation can be achieved locally through differences in ion concentrations which affect the organization and bundling of microtubules and neurofilaments.[55,56] In addition, microfilaments are attached to the cell membrane through actin binding proteins and receive inputs from adhesion molecules, and from the second messenger phosphoinositide pathway.[57]

11.05.6 TOXICANT EFFECTS ON THE CYTOSKELETON

Interactions between the cytoskeleton and xenobiotics are of central importance in neurotoxicology. The preceding overview of the cytoskeleton underscores the complexity of these organelles and their interactions; this complexity gives rise to a correspondingly diverse spectrum of toxicant effects. Toxicant interactions with the major cytoskeletal proteins, associated proteins, or regulatory proteins may cause impairment of function and cell death. Moreover, differences in cytoskeletal function between developing and mature neurons contribute to the differential sensitivity to toxicant effects in the fetal and the adult nervous systems. The dynamic nature of cytoskeletal structures and their sensitivity to alterations in the cellular microenvironment make them highly vulnerable to toxins and toxicants, often with adverse effects on cellular function. Moreover, because of the structural and functional linkages between cytoskeletal elements, attack upon one component may result in secondary changes to the others. It is of interest to note that much of the work describing the normal dynamics and function of the cytoskeleton has been accomplished using neurotoxicants as tools; many unwanted side-effects of therapeutic agents and toxicities of environmental chemicals are mediated via similar mechanisms critically involving the cytoskeleton.

11.05.6.1 Pharmaceuticals

11.05.6.1.1 Antineoplastic agents

The primary goal of oncology is to destroy rapidly dividing tumor cells, thereby halting neoplastic growth. This can be accomplished by disrupting the microtubules of the mitotic spindle apparatus. Unfortunately, antineoplastic compounds react with all microtubules, including those in neurons, and produce characteristic axonopathies.[58] While axonopathies usually develop following relatively high doses, mitotic spindle formation is inhibited at much lower doses, which makes these compounds useful in cancer treatment.

The chemotherapeutic benzimidazole derivatives, nocodazole and oncodazole, bind with high specificity to the plus ends of microtubules. Nocodazole enters cells, binds to unpolymerized tubulin, and blocks microtubule assembly.[13,59] This binding results in a net loss of microtubules from the axon or dendrite due to continued disassembly of microtubules without concomitant assembly.

Vinblastine and vincristine, antineoplastic vinca alkaloids, have the same overall effect on microtubules, but accomplish this effect through tubulin aggregation and polymer stabilization. These compounds bind to free tubulin, rather than to polymerized microtubules. In treated cells free tubulin becomes bound in nontubular aggregates, the concentration of tubulin is decreased, and microtubule polymerization is inhibited. Accidental administration of large intravenous doses causes neurofilamentous accumulations in neuronal cell bodies.[60] Vinblastine stabilizes polymerized microtubules *in vitro*, and inhibits dynamic instability.[61] This may be due to steric hindrance between the bound vinblastine and the GTP/GDP cap. The result is a net loss of microtubules. Vinca alkaloids are known for their ability to cause axonal degeneration.

In contrast, taxol, an antineoplastic compound isolated from the Pacific yew, binds polymerized microtubules, stabilizes their structure, and causes the slow accumulation of microtubule bundles in the cytoplasm. There is an overall decrease in the rate of microtubule polymerization caused by a decrease in the concentration of free tubulin. Unlike the vinca alkaloids which sequester free tubulin, the decreased concentration of tubulin occurs because of a decrease in the rate of depolymerization which would normally add new dimers to the pool of free tubulin. Taxol appears to have no effect on the rate of fast axonal transport, but an inhibition of retrograde axonal transport has been reported.[62] Since the motors for anterograde and retrograde fast transport both move along microtubules it is interesting to speculate that taxol may interfere with dynein-binding. Taxol also causes an almost complete inhibition in slow anterograde tubulin transport.[63] This is not due solely to the decrease in free tubulin since the movement of actin and other proteins in slow component b is also blocked. *In vitro* studies of taxol–microtubule binding have shown that when microtubules interact with taxol their normally rigid structures become more flexible.[64] It is not known whether this contributes to the *in vivo* effects of taxol, especially since MAP-2 and tau appear to reverse this effect *in vitro*. However, this suggests that taxol induces a conformational change in microtubules which alters their cellular function.

These chemotherapeutic agents tend to decrease the rates of either polymerization or depolymerization, but they do not cause active depolymerization. Maytansine, an alkylating macrolide isolated from *Maytenus* species, and estramustine, an antineoplastic agent synthesized by the binding of nitrogen mustard to estradiol, are antimitotics that increase the rate of microtubule depolymerization.[65] Maytansine and its analogues also cause a secondary rearrangement in axonal neurofilaments leading to the formation of short neurofilament bundles.[66] Estramustine inhibits tubulin polymerization through interactions with both tubulin and MAPs.[67] These compounds are being studied as potential antineoplastic agents.

The use of *cis*-diamminedichloroplatinum (cisplatin), a highly effective chemotherapeutic agent, is limited because of its neurotoxicity. The major side effect of cisplatin is a peripheral, sensory polyneuropathy. Morphologically there is a loss of large myelinated fibers, formation of neurofilamentous axonal swellings, and Wallerian degeneration. *In vitro*, cisplatin binds irreversibly with sulfhydryl residues on tubulin[68] and inhibits the rate of microtubule disassembly.[69] It has been suggested that cisplatin induces conformational changes in microtubule structure which inhibit dynamic instability; this mechanism may be important in the development of this neuropathy.

Methotrexate, a folic acid analogue, is a widely used chemotherapeutic drug that arrests tumor growth through inhibition of dihydrofolate reductase, with a subsequent perturbation in DNA synthesis. Though it appears to be without effect on microtubules, it has been shown to cause perinuclear capping of vimentin in cultured fibroblasts.[70] While effects of methotrexate on axonal neurofilaments have not been studied, other compounds which cause perinuclear capping of vimentin are known to disrupt axonal neurofilaments.[26]

11.05.6.1.2 Colchicine

This plant alkaloid, derived from *Colchicum* species, has been used for centuries to treat gout but its tubulin binding properties have made this compound a widely utilized tool in the study of microtubule dynamics. Colchicine binds directly to free tubulin in a 1:1 ratio, inducing a conformational change in the tubulin dimer;[71] this binding is virtually irreversible. When a colchicine-bound tubulin dimer adds to the end of a growing microtubule further elongation is inhibited; the polymer is destabilized and disassembles.[12] Colchicine inhibits microtubule-based axonal transport in both axons and dendrites, possibly by disassembling the microtubule "tracks" required for the movement of the motor proteins kinesin and dynein. Not only does colchicine affect microtubule-based functions in adult cells, but it has a number of deleterious effects on developing neuronal cells. Using time course microscopy to examine cultured cells, Keith[72] showed that when colchicine acted on tubulin within growth cones, neurite extension was blocked, supporting the hypothesis that neurite extension requires microtubule polymerization and stabilization. In addition, colchicine disruption of the mitotic spindle causes cell death, depleting the number of neuronal precursor cells during the early stages of fetal development.

Colcemid acts in a manner similar to colchicine, but its effects are more rapid and more easily reversed.[12] Colchicine or colcemid induce a progressive, but reversible, reorganization of the intermediate filament network in cultured cells, leading to the formation of a juxtanuclear cap of intermediate filaments. Similar changes can be seen in intermediate filament networks following actin disruption. Colcemid-induced microtubule depolymerization has been associated with dispersion and fragmentation of the Golgi apparatus.[12] It is interesting to note that taxol causes fragmentation of the Golgi, but not dispersal, while nocodazole causes dispersal but not fragmentation.[73] This evidence provides support for the concept of microtubule-based positioning and stabilization of the Golgi apparatus.

11.05.6.1.3 Miscellaneous pharmaceuticals

Podophyllotoxins, glycosides isolated from the rhizomes of the May-apple and used as topical antiviral agents, are potent inhibitors of microtubule assembly.[74] The side groups of microtubule-bound podophyllotoxins interfere with colchicine binding through steric hindrance, indicating that the two binding sites are closely apposed. The benzophenanthridine alkaloids, such as the sanguinarine constituents of bloodroot, are related to the podophyllotoxins and are also potent antimicrotubule-assembly agents. It is thought that these alkaloids bind to the growing ends of microtubules and prevent the further addition of tubulin dimers to the polymer.[75]

Ethacrynic acid, a diuretic agent, reacts with protein sulfhydryl residues. Ethacrynic acid alters the microtubule organization of cultured cells, and inhibits assembly of purified tubulin.[76] These alterations are due to the binding of ethacrynic acid with sulfhydryl groups on β-tubulin.[77] There are no reports of neurotoxicity following the use of this drug as a diuretic.

11.05.6.2 Pesticides

A large number of organophosphate compounds (OPs) are manufactured and used as insecticides, fungicides, defoliants, and nerve gases. The anticholinesterase activity of these compounds is well documented,[78] as is the ability of select OPs to cause a delayed neuropathy, but their effects on the cytoskeleton are less well understood. The compound di-*n*-butyl-dichlorvos (DBDCVP), an insecticide and anthelminthic, is associated with a progressive deficit in retrograde axonal transport.[79] This is an early effect occurring prior to the onset of clinical signs of neuropathy or axonopathy. The mechanism has not yet been determined, but it is known that tri-*ortho*-cresyl phosphate (TOCP) increases the phosphorylation state of microtubules and neurofilament proteins *in vivo*.[80] This increase in phosphate content may result from a direct reaction with TOCP, but is more probably through TOCP modification of kinases or phosphatases involved in post-translational modification.[78]

11.05.6.3 Industrial Chemicals

Acrylamide is a neurotoxic chemical used routinely in research laboratories and industries. Prolonged exposure to high levels of acrylamide causes a distal neuropathy characterized by Wallerian-like degeneration. Acrylamide has been shown to inhibit slow anterograde and fast retrograde axonal transport,[81] deplete MAP1 and MAP2 from specific brain regions,[82,83] and alter growth cone morphology.[84] Data indicate that acrylamide binds directly to microtubules and neurofilaments *in vitro*, leading to speculation that this may also occur *in vivo*.[85] Though no alterations are observed in mitotic spindles of acrylamide-exposed fibrosarcoma cells, the exposed cells undergo mitotic arrest.[86] This suggests possible molecular mechanisms of acrylamide toxicity, including microtubule stabilization and/or inhibition of microtubule disassembly. Rats treated with acrylamide have lower levels of MAP1 and MAP2 in extrapyramidal regions of the brain,[82,83] but the significance of this finding is

not clear. Exposure to acrylamide during neural development induces gross alterations in growth cone morphology of dorsal root ganglion explants, including the profound loss of filopodial elements and inhibition of growth cone advance, suggesting that acrylamide may disrupt actin filaments.[84] Due to its effects on growth cones, acrylamide has significant potential as a developmental neurotoxicant.

Hexane, an industrial solvent, is a known neurotoxicant that causes neurofilamentous accumulations in peripheral nervous system axons and testicular atrophy. These effects are mediated by the ultimate toxic metabolite, 2,5-hexanedione (HD). HD appears to specifically target the nervous system and the testes. Studies of the testicular effects of HD show that microtubule assembly and actin distribution are altered in this organ.[87–89] It is interesting to speculate why axonal microtubules do not exhibit the same effects. It is possible that slight differences in the sequence and structure of testicular and neuronal tubulin make the axon resistant (or the testes vulnerable) to these effects. This same argument could be applied to the MAPs or regulatory proteins of these two cell types. However, it is also possible that the alterations which affect Sertoli cell tubulin and actin also affect neuronal tubulin and actin, but that these alterations do not affect protein function in the neuron (Figure 4).

N-Ethylmaleimide (NEM) is a reagent which alkylates free sulfhydryl residues on proteins and is used routinely in research to study the significance of sulfhydryls to protein function. This highly reactive compound is used only for *in vitro* studies, and is not a demonstrated neurotoxicant. However, studies show that NEM-treated kinesin is not transport-competent, implying that the sulfhydryl moiety of kinesin is important for its functionality in fast axonal transport.[90]

11.05.6.4 Biologicals Agents

Cytochalasins comprise a group of about 20 known compounds derived from molds which affect actin stress fibers, causing rapid actin depolymerization. Cytochalasins bind to the growing end of F actin, inhibit polymerization without affecting depolymerization, and cause disassembly of actin fibers. Cytochalasin B and D are highly specific and cause rapid inhibition of cytoplasmic ruffling and loss of filopodial extension in growth cones. These changes occur within minutes, but are quickly reversed when the toxins are removed.[91] There is a concomitant disruption in the microtubule organization

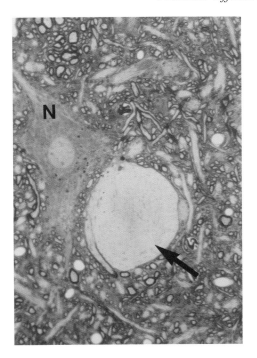

Figure 4 This plastic section from rat spinal cord, stained with toluidine blue O, shows a giant axonal swelling (arrow). A motor neuron (N) can be seen adjacent to the swollen axon. These swellings are characteristic of hexacarbon and β,β′-iminodipropionitrite intoxication.

within these growth cones, but this appears to be secondary to the effects on actin. Studies with cytochalasin B reveal that there is a disorganization of the extracellular fibronectin filaments following the alteration of the microfilament network.[12] There may be a connection between these changes and cytochalasin B-induced inhibition of the phosphoinositide cascade.[92]

Phallotoxins, another group of fungal toxins, are derived from fungi of *Amanita* species. Phalloidin and phallacidin stabilize microfilaments by binding actin polymers, reducing the amount of actin monomer recycled through depolymerization, and reducing the critical concentration of actin at the growing end of the filament. These compounds are to actin what taxol is to microtubules.

Another potent biological toxin, Toxin B from *Clostridium difficile*, has been shown to disrupt cytoskeletal organization in cultured cells. This toxin appears to affect actin organization; however, it is not known if this is due to its direct interaction with actin or to an indirect effect mediated through the action of the agent on regulatory proteins.[93]

Lectins are a diverse family of toxic plant proteins whose use is being investigated. Concanavalin A obstructs microfilament capping

and causes the dissociation of actin fibers from cell membrane receptors *in vitro*.[12]

Okadaic acid is a potent neurotoxin and phosphatase inhibitor from dinoflagellate black sponges, and is associated with seafood poisonings. In BHK-21 cells treated with okadaic acid there is a dramatic increase in the phosphorylation state of vimentin and a rapid reorganization of the intermediate filament network, followed by microtubule loss.[94] Similar findings have been reported in okadaic acid-treated dorsal root ganglion cells,[95] and an increase in MAP2 and tau phosphorylation, followed by neurite retraction, occurs in treated rat neurons and human neuroblastoma cells.[96] It remains to be determined whether these cytoskeletal changes are direct or indirect effects, but reports of alterations in intracellular calcium favor an indirect mechanism.[95] Hyperphosphorylated tau proteins are recognized as components of the characteristic neurofibrillary tangles in Alzheimer's brains. Okadaic acid and kainic acid are being used to study the relationship of these alterations in phosphorylation and intracellular calcium levels to the development of tangles.

Kainic acid is an excitotoxin isolated from the red algae *Digenea simplex*. Kainic acid causes increases in the phosphorylation state of certain cytoskeletal proteins. Hyperphosphorylated neurofilaments have been reported in the cell body and proximal neurites of rats given intrathecal kainic acid.[97] Changes in tau phosphorylation, antigenically similar to tau from neurofibrillary tangles, have been noted within 3 h of *in vivo* kainic acid injection, and become apparent before neuronal damage or degeneration.[98] Elliott *et al.*[98] have used the data from kainic acid intoxication to speculate on the development of Alzheimer's disease. They hypothesize that the formation of neurofibrillary tangles may be related to alterations in tau resulting from the increase in intracellular calcium induced by excitatory amino acids like glutamate.

11.05.6.5 Metals

Metals affect the cytoskeleton either by binding directly to cytoskeletal proteins or secondarily by inducing changes in the intracellular ionic environment necessary for cytoskeletal homeostasis. Since most metals are multifunctional toxicants capable of binding to many cellular macromolecules, it is frequently difficult to distinguish between primary and secondary events. The cytoskeletal toxicity of heavy metals has been the subject of review (see also Chapters 4, 6, 13, and 29, this volume).[99]

Microtubules, the most dynamic and labile of the major cytoskeletal proteins, are also the most frequently perturbed by metals. Tubulin possesses two nucleotide binding sites which require Mg^{2+} for optimal polymerization.[100] The substitution of other polyvalent metal cations for Mg^{2+} may alter microtubule assembly. Further, the tubulin dimer has 15 free sulfhydryl groups; binding of metals to only two of these is sufficient to inhibit microtubule assembly,[101] and to induce disassembly of the microtubule polymer. The affinity of metals such as mercury, cadmium, and chromium for sulfhydryl groups is well documented.

Changes in intermediate filaments are frequently observed following exposure to metals. In many instances, this phenomenon reflects a relatively nonspecific response of the cell to injury rather than a direct action of the metal on the filament itself. Exceptions to this generalization exist. Intermediate filament proteins have been demonstrated to bind to several metal cations *in vitro* to produce filamentous aggregates, albeit usually only at high metal concentrations.[102] Only aluminum and lead have been extensively studied for *in vivo* interactions on neurofilaments, and the effects of metals on microfilaments are essentially unexplored.

11.05.6.5.1 Mercury

Of the neurotoxic metals, mercurial compounds have been best characterized for their effects on the cytoskeleton. Both inorganic and organic mercurials have potent effects on microtubules. Inorganic mercurial salts promote disassembly of intact microtubules and inhibit assembly of microtubules from tubulin at low micromolar concentrations,[103,104] presumably by binding to tubulin sulfhydryls. Chelation of inorganic mercury by EDTA or EGTA may actually enhance the antimicrotubule action of the metal, perhaps by increasing its ability to bind with the reactive GTP-binding site E on β-tubulin.[105] Inorganic mercury also elevates intracellular calcium concentrations[106] to levels capable of inducing microtubule disassembly. The relative lack of *in vivo* neurotoxicity exhibited by inorganic mercurial salts is primarily a consequence of their failure to cross the blood–brain barrier. Mercury vapor, which readily enters the brain, may have deleterious effects on microtubule integrity.

Highly neurotoxic organomercurial compounds, particularly methylmercury, have repeatedly been shown to cause disassembly of microtubules in neurons and glia both *in vivo*

and *in vitro*.[107–109] It is thought that binding of methylmercury to —SH groups underlies this effect. Highly labile microtubules of the mitotic spindle (composed primarily of tyrosinated tubulin and possessing an extremely short half-life) are exquisitely sensitive to methylmercury. Indeed, methylmercury is one of the most potent antimitotic chemicals known. Exposure of mitotic cells to methylmercury results in arrest of cells in prometaphase–metaphase and eventual cell death. In extreme cases, the number of neurons is drastically reduced and the brain is hypoplastic. Damage to microtubules in postmitotic neurons can alter highly ordered migration. As a result, neurons miss their migratory "window" or migrate to inappropriate cortical regions, disturbing cerebral cytoarchitecture. Disturbed brain morphogenesis and associated psychomotor retardation have been reported in both human,[110,111] and experimentally induced[112] cases of methylmercury poisoning.

As the postmigratory neuron differentiates, microtubules become progressively less sensitive to methylmercury; this decreasing sensitivity correlates positively with the degree of post-translational modification of the microtubule and its decoration with MAPs.[113–115] Neuritic microtubules are more resistent to the acute effects of methylmercury than those of the perikaryon. However, even these microtubules eventually depolymerize in the continued presence of the toxicant.[115]

Specific effects of mercurials on intermediate- or microfilaments have not been reported. Secondary changes of these organelles have been observed to follow loss of microtubules.[107,116]

11.05.6.5.2 Aluminum

Brain aluminum concentrations are elevated in several iatrogenic (e.g., dialysis encephalopathy) and idiopathic neurological conditions.[117,118] These observations have led to speculation regarding the possible role of the metal in the pathogenesis of Alzheimer's disease (AD) and other neurodegenerative diseases. While clear linkage between aluminum and AD has not been established, effects of aluminum on neurofilament structure and regulation have been demonstrated repeatedly. Administration of aluminum to rabbits and nonhuman primates results in the widespread accumulation of neuofilaments, notably within pyramidal neurons of the cerebral cortex, the basal forebrain and lower motor neurons.[119] These filaments are ultrastructurally distinct from those seen in the human disease,[120] and have been reported to be abnormally phosphorylated,[121] suggesting that aluminum may induce early neurofilament phosphorylation within the neuronal perikaryon.[122] However, others have not observed enhanced phosphorylation.[123] Reports that aluminum inhibits Ca^{2+}-dependent and -independent proteolysis of neurofilaments[124] offer yet another mechanism for their accumulation (Figure 5).

Molecular studies reveal that aluminum inhibits the expression of neuron-specific NFL mRNA by as much as 75%;[125] this process appears cell specific, as mRNA levels remained normal in dorsal root ganglion neurons, which are unaffected by aluminum administration.[126]

Aluminum has been shown to support microtubule polymerization, possibly by competing with Mg^{2+}-binding site on tubulin.[127] The interaction of aluminum with tau protein and the accumulation of microtubules in brains of aluminum-intoxicated animals and in dialysis encephalopathy fueled speculation regarding the role of aluminum–microtubule interactions in the pathogenesis of several neurological conditions. Accumulation of microtubules and elevated metal concentra-

tions have been documented in the neurofibrillary tangles and senile plaques of human Alzheimer's disease patients,[128,129] but it is not clear whether the metals cause neuronal injury or merely accumulate subsequent to another, primary neurotoxic event.

11.05.6.5.3 Lead

Both inorganic and organic lead compounds inhibit the assembly of microtubules[130,131] at levels associated with *in vivo* neurotoxicity. However, it is not clear whether these changes represent primary effects of lead on the cytoskeleton or alterations consequent to the metal's documented ability to alter intracellular calcium homeostasis.[132]

Disruption of neurofilaments has been noted following *in vitro* or *in vivo* exposure to lead compounds. The effect is more pronounced following exposure to organolead compounds (e.g., triethyllead and tetraethyllead) than inorganic lead salts, possibly because of differences in bioavailability. Triethyllead was reported to cause collapse of the neurofilament network and disassembly of intact filaments into protofilaments.[133] Accumulation of filaments has been reported in rabbits following treatment with organolead compounds,[134] but the significance of this finding is unclear.

11.05.6.5.4 Other metals

Numerous metals have been shown to affect the cytoskeleton *in vitro* or in cultured cells. Cadmium,[103,135,136] chromium,[103,136] zinc, arsenite, and cobalt,[135,137] cause alterations in microtubule networks. Cadmium and zinc also cause loss of actin bundles from nonneuronal cells in culture.[138] However, few of these metals have been systematically studied for their potential effects on the neuronal cytoskeleton. Moreover, concentrations of these metals required to cause changes in nonneuronal cells are frequently high and the relevance to *in vivo* neurotoxicity remains unclear.

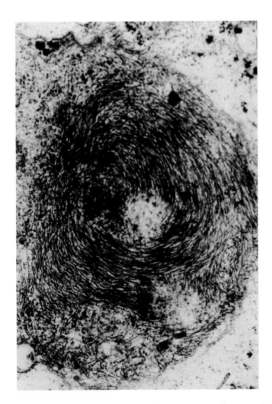

Figure 5 This micrograph demonstrates disorganization of axonal neurofilaments following aluminum exposure. The neurofilaments are no longer longitudinally arranged, but form disorganized whorls. (Micrograph courtesy of Dr. Thomas Shea, University of Massachusetts at Lowell.)

11.05.6.5.5 Metal effects on glial fibrillary acidic protein

Glial fibrillary acidic protein (GFAP) is an intermediate filament protein present in several neural cell types, most notably astrocytes. GFAP is increased in rodent brains following treatment with numerous metals, including lead, methylmercury, and trimethyltin.[139,140] The induction of this protein is rapid, frequently massive, and may result in changes to the gross morphology of the astrocyte. Both primary astrocytic injury and secondary glial responses to neuronal injury can induce increases in GFAP, supporting the use of this protein as a marker of neurotoxic injury.[139] The responses of astrocytes to toxic insult are detailed elsewhere in this volume (see Chapter 10, this volume) (Figure 6).

11.05.6.6 Others

Patients with hyperammonemia, a condition appearing secondary to renal insufficiency and high blood urea levels, often show evidence of a peripheral neuropathy.[60] Ammonia also induces an increase in the expression of tubulin in the cerebrum[141] and the proteolysis of MAP2.[142] The signal which regulates tubulin expression in this condition is unknown, but the effects on MAP2 are mediated through the activation of the *N*-methyl-D-aspartate (NMDA) receptor.[142] Activation of the NMDA receptor leads to the dephosphorylation of MAP2 and the activation of calpain I,

which is responsible for the breakdown of MAP2. Over-stimulation of NMDA receptors would lead to alterations in the phosphorylation state of the cytoskeleton and in the amounts of tubulin and MAP2 in the axon, with consequent alterations in neurological functioning.[142]

A number of vitamin deficiencies also lead to the development of axonal neuropathies, including the B vitamins: thiamine (B_1), cobalamin (B_{12}), and pyridoxine (B_6), and α-tocopherol (vitamin E). The mechanism causing these neuropathies has not been elucidated, but it appears that there are associated cytoskeletal alterations. In vitamin E deficiency there is an increase in the rate of fast anterograde transport, whereas thiamine deficiency leads to an inhibition in vesicle loading and a decrease in the amount of protein being moved in the axon by fast anterograde transport. Chronic alcoholics develop a progressive distal sensorimotor neuropathy which may be due to the thiamine deficiency that accompanies long-term alcohol abuse.[60] An excess of vitamin B_6 produces a neuronopathy characterized by necrosis of sensory neurons and axonal atrophy.[143]

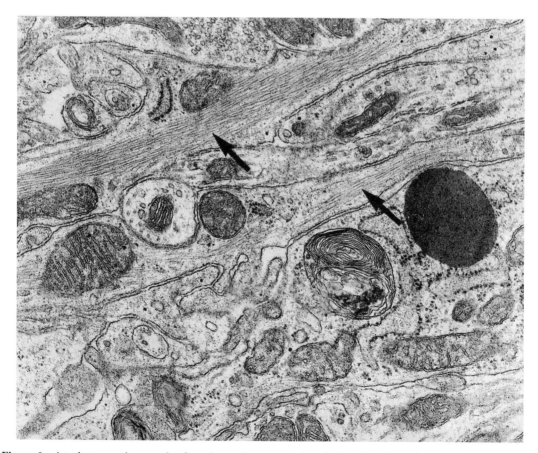

Figure 6 An electron micrograph of cerebrum from a cat chronically administered methylmercury. Neuronal damage has stimulated the production of glial filament bundles (arrows) in astrocytes.

Atrophic axons lacking neurofilament and microtubule profiles and axons swollen with neurofilament aggregates are observed in animals experimentally depleted of vitamin B_6.

Adult-onset diabetes is often associated with peripheral nerve abnormalities including distal symmetric sensory-motor or autonomic neuropathy.[60] Animals with experimentally induced diabetes develop swollen terminal axons containing large neurofilamentous aggregates, and show impairment in the rate of slow axonal transport.[144,145] The most dramatic change is seen in the rate of neurofilament transport with less change in the rates of actin and tubulin transport.[146,147] Until more detailed studies are performed or until better animal models are developed it is not possible to determine whether these cytoskeletal alterations are due to changes in the major cytoskeletal proteins, the associated and/or regulatory proteins, or whether these changes are only secondary modifications of the neuron in response to an as yet unknown primary injury.

11.05.7 CONCLUSIONS

It is clear that the cytoskeleton is an interactive complex of proteins with multiple points of contact and varied responses to toxicant-induced injury. Due to this complexity, alterations of the cytoskeleton must be carefully studied to determine whether they are primary or secondary responses to toxicant exposure. As our understanding of the interactions of the cytoskeleton continues to expand so will our understanding of the mechanisms of toxicant injury.

ACKNOWLEDGMENTS

Preparation of this chapter was supported by the Johnson & Johnson Research Fellowship in Developmental Neurotoxicology (SJP) and NIH ES04976 (KRR). The authors thank Dr. Herbert Lowndes and Ms. Kathleen Roberts for their assistance.

11.05.8 REFERENCES

1. T. L. Stedman, 'Stedman's Medical Dictionary,' illustrated 24th edn., Williams & Wilkins, Baltimore, MD, 1982, p. 1678.
2. P. W. Baas and F. J. Ahmad, 'The plus ends of stable microtubules are the exclusive nucleating structures for microtubules in the axon.' *J. Cell Biol.*, 1992, **116**, 1231–1241.
3. P. W. Baas, M. M. Black and G. A. Banker, 'Changes in microtubule polarity orientation during the development of hippocampal neurons in culture.' *J. Cell Biol.*, 1989, **109**, 3085–3094.
4. M. M. Black and P. W. Baas, 'The basis of polarity in neurons.' *Trends Neurosci.*, 1989, **12**, 211–214.
5. P. B. Sargent, 'What distinguishes axons from dendrites? Neurons know more than we do.' *Trends Neurosci.*, 1989, **12**, 203–205.
6. T. Horio and H. Hotani, 'Visualization of the dynamic instability of individual microtubules by dark-field microscopy.' *Nature*, 1986, **321**, 605–607.
7. P. J. Sammak and G. G. Borisy, 'Direct observation of microtubule dynamics in living cells.' *Nature*, 1988, **332**, 724–726.
8. T. Mitchison and M. Kirschner, 'Dynamic instability of microtubule growth.' *Nature*, 1984, **312**, 237–242.
9. H. P. Erickson and E. T. O'Brien, 'Microtubule dynamic instability and GTP hydrolysis.' *Annu. Rev. Biophys. Biomol. Struct.*, 1992, **21**, 145–166.
10. A. Vandecandelaere, S. R. Martin and P. M. Bayley, 'Regulation of microtubule dynamic instability by tubulin-GDP.' *Biochemistry*, 1995, **34**, 1332–1343.
11. W. Yu, F. J. Ahmad and P. W. Baas, 'Microtubule fragmentation and partitioning in the axon during collateral branch formation.' *J. Neurosci.*, 1994, **14**, 5872–5884.
12. M. Schliwa, 'The Cytoskeleton. An Introductory Survey,' Springer-Verlag, New York, 1986, p. 326.
13. P. W. Baas and M. M. Black, 'Individual microtubules in the axon consist of domains that differ in both composition and stability.' *J. Cell Biol.*, 1990, **111**, 495–509.
14. V. Redeker, N. Levilliers, J. M. Schmitter *et al.*, 'Polyglycylation of tubulin: a post-translational modification in axonemal microtubules.' *Science*, 1994, **266**, 1688–1691.
15. B. Edde, J. Rossier, J. P. Le Caer *et al.*, 'Posttranslational glutamylation of α-tubulin.' *Science*, 1990, **247**, 83–85.
16. D. Boucher, J.-C. Larcher, F. Gros *et al.*, 'Polyglutamylation of tubulin as a progressive regulator of *in vitro* interactions between the microtubule-associated protein tau and tubulin.' *Biochemistry*, 1993, **33**, 12471–12477.
17. B. Alberts, D. Bray, J. Lewis *et al.*, 'Molecular Biology of the Cell,' 3rd edn., Garland, New York, 1994, p. 1294.
18. S. A. Kuznetsov, G. M. Langford and D. G. Weiss, 'Actin-dependent organelle movement in squid axoplasm.' *Nature*, 1992, **356**, 722–725.
19. M. P. Kaplan, S. S. M. Chin, K. H. Fliegner *et al.*, 'α-Internexin, a novel neuronal intermediate filament protein, precedes the low molecular weight neurofilament protein (NF-L) in the developing rat brain.' *J. Neurosci.*, 1990, **10**, 2735–2748.
20. J. Dahlstrand, L. B. Zimmerman, R. D. G. McKay *et al.*, 'Characterization of the human nestin gene reveals a close evolutionary relationship to neurofilaments.' *J. Cell Sci.*, 1992, **103**, 589–597.
21. T. B. Shea, M. L. Beermann and I. Fischer, 'Transient requirement for vimentin in neuritogenesis: intracellular delivery of anti-vimentin antibodies and antisense oligonucleotides inhibit neurite initiation but not elongation of existing neurites in neuroblastoma.' *J. Neurosci. Res.*, 1993, **36**, 66–76.
22. C. M. Troy, L. A. Greene and M. L. Shelanski, 'Neurite outgrowth in peripherin-depleted PC12 cells.' *J. Cell Biol.*, 1992, **117**, 1085–1092.
23. M. Escurat, K. Djabali, M. Gumpel *et al.*, 'Differential expression of two neuronal intermediate-filament proteins, peripherin and the low-molecular-mass neurofilament protein (NF-L), during the development of the rat.' *J. Neurosci.*, 1990, **10**, 764–784.

24. C. M. Troy, N. A. Muma, L. A. Greene *et al.*, 'Regulation of peripherin and neurofilament expression in regenerating rat motor neurons.' *Brain Res.*, 1990, **529**, 232–238.

25. W. A. Pedersen, L. E. Becker and H. Yeger, 'Expression and distribution of peripherin protein in human neuroblastoma cell lines.' *Int. J. Cancer*, 1993, **53**, 463–470.

26. H. D. Durham, 'Aggregation of intermediate filaments by 2,5-hexanedione: comparison of effects on neurofilaments, GFAP-filaments and vimentin-filaments in dissociated cultures of mouse spinal cord-dorsal root ganglia.' *J. Neuropathol. Exp. Neurol.*, 1988, **47**, 432–442.

27. M. P. Sheetz and C. H. Martenson, 'Axonal transport: beyond kinesin and cytoplasmic dynein.' *Curr. Opin. Neurobiol.*, 1991, **1**, 393–398.

28. J. Chen, Y. Kanai, N. J. Cowan *et al.*, 'Projection domains of MAP2 and tau determine spacings between microtubules in dendrites and axons.' *Nature*, 1992, **360**, 674–677.

29. N. Hirokawa, 'Neuronal Cytoskeleton Morphogenesis, Transport and Synaptic Transmission,' CRC Press, Boca Raton, FL, 1993, p. 335.

30. D. W. Cleveland, 'Microtubule MAPping.' *Cell*, 1990, **60**, 701–702.

31. N. Hirokawa, 'Microtubule organization and dynamics dependent on microtubule-associated proteins.' *Curr. Opin. Cell Biol.*, 1994, **6**, 74–81.

32. Y. Kanai and N. Hirokawa, 'Sorting mechanisms of tau and MAP2 in neurons: suppressed axonal transit of MAP2 and locally regulated microtubule binding.' *Neuron*, 1995, **14**, 421–432.

33. N. Hirokawa, 'The Neuronal Cytoskeleton: Its Role in Neuronal Morphogenesis and Organelle Transport,' ed. N. Hirokawa, CRC Press, Boca Raton, FL, 1993, p. 3.

34. N. K. Pryer, R. A. Walker, V. P. Skeen *et al.*, 'Brain microtubule-associated proteins modulate microtubule dynamic instability *in vitro*. Real-time observations using video microscopy.' *J. Cell Sci.*, 1992, **103**, 965–976.

35. B. Pedrotti and K. Islam, 'Purified native microtubule associated protein MAP1A: kinetics of microtubule assembly and MAP1A/tubulin stoichiometry.' *Biochemistry*, 1994, **33**, 12463–12470.

36. H. Yamamoto, Y. Saitoh, K. Fukunaga *et al.*, 'Dephosphorylation of microtubule proteins by brain protein phosphatases 1 and 2A, and its effect on microtubule assembly.' *J. Neurochem.*, 1988, **50**, 1614–1623.

37. M. P. Sheetz, E. R. Steuer and T. A. Schroer, 'The mechanism and regulation of fast axonal transport.' *Trends Neurosci.*, 1989, **12**, 474–478.

38. J. Avila, J. Dominguez and J. Diaz-Nido, 'Regulation of microtubule dynamics by microtubule-associated protein expression and phosphorylation during neuronal development.' *Int. J. Dev. Biol.*, 1994, **38**, 13–25.

39. K. D. Lee and P. J. Hollenbeck, 'Phosphorylation of kinesin *in vivo* correlates with organelle association and neurite outgrowth.' *J. Biol. Chem.*, 1995, **270**, 5600–5605.

40. J. R. Bamburg and B. W. Bernstein, 'Actin and Actin-binding Proteins in Neurons,' ed. R. D. Burgoyne, Wiley-Liss, New York, 1991, pp. 121.

41. T. Kreis and R. Vale, 'Guidebook to the Cytoskeletal and Motor Proteins,' Oxford University Press, 1993, p. 276.

42. T. P. Stossel, C. Chaponnier, R. M. Ezzell *et al.*, 'Nonmuscle actin-binding proteins.' *Annu. Rev. Cell Biol.*, 1985, **1**, 353–402.

43. A. Caceres, J. Mautino and K. S. Kosik, 'Suppression of MAP2 in cultured cerebellar macroneurons inhibits minor neurite formation.' *Neuron*, 1992, **9**, 607–618.

44. P. R. Gordon-Weeks, 'Organization of microtubules in axonal growth cones: a role for microtubule-associated protein MAP 1B.' *J. Neurocytol.*, 1993, **22**, 717–725.

45. K. L. Lankford and P. C. Letourneau, 'Evidence that calcium may control neurite outgrowth by regulating the stability of actin filaments.' *J. Cell Biol.*, 1989, **109**, 1229–1243.

46. T. B. Shea and R. A. Nixon, 'Differential distribution of vimentin and neurofilament protein immunoreactivity in NB2a/d1 neuroblastoma cells following neurite retraction distinguishes two separate intermediate filament systems.' *Brain Res.*, 1988, **469**, 298–302.

47. R. A. Nixon and T. B. Shea, 'Dynamics of neuronal intermediate filaments: a developmental perspective.' *Cell Motil. Cytoskeleton*, 1992, **22**, 81–91.

48. N. Hirokawa, 'Cross-linker system between neurofilaments, microtubules, and membranous organelles in frog axons revealed by the quick-freeze, deep-etching method.' *J. Cell Biol.*, 1982, **94**, 129–142.

49. N. Hirokawa, S. Hisanaga and Y. Shiomura, 'MAP2 is a component of crossbridges between microtubules and neurofilaments in the neuronal cytoskeleton: quick-freeze, deep-etch immunoelectron microscopy and reconstitution studies.' *J. Neurosci.*, 1988, **8**, 2769–2779.

50. C. C. Wurtz and M. H. Ellisman, 'Alterations in the ultrastructure of peripheral nodes of Ranvier associated with repetitive action potential propagation.' *J. Neurosci.*, 1986, **6**, 3133–3143.

51. C. C. Wurtz and M. H. Ellisman, 'Activity associated with ultrastructural changes in peripheral nodes of Ranvier are independent of fixation.' *Exp. Neurol.*, 1988, **101**, 87–106.

52. T. Ichimura and M. H. Ellisman, 'Three-dimensional fine structure of cytoskeletal-membrane interactions at nodes of Ranvier.' *J. Neurocytol.*, 1991, **20**, 667–681.

53. A. D. Bershadsky and I. M. Vasil'ev, 'Cytoskeleton,' Plenum, New York, 1988, p. 298.

54. I. S. Tint, P. J. Hollenbeck, A. B. Verkhovsky *et al.*, 'Evidence that intermediate filament reorganization is induced by ATP-dependent contraction of the actomyosin cortex in permeabilized fibroblasts.' *J. Cell Sci.*, 1991, **98**, 375–384.

55. S. Hisanaga, A. Ikai and N. Hirokawa, 'Molecular architecture of the neurofilament. I. Subunit arrangement of neurofilament L protein in the intermediate-sized filament.' *J. Mol. Biol.*, 1990, **211**, 857–869.

56. B. M. Riederer, F. Monnet-Tschudi and P. Honegger, 'Development and maintenance of the neuronal cytoskeleton in aggregated cell cultures of fetal rat telencephalon and influence of elevated K^+ concentrations.' *J. Neurochem.*, 1992, **58**, 649–658.

57. P. Forscher, 'Calcium and polyphosphoinositide control of cytoskeletal dynamics.' *Trends Neurosci.*, 1989, **12**, 468–474.

58. R. P. Gupta and M. B. Abou-Donia, 'Axonopathy,' ed. L. W. Chang, Dekker, New York, 1994, pp. 135–151.

59. P. W. Baas, T. Slaughter, A. Brown *et al.*, 'Microtubule dynamics in axons and dendrites.' *J. Neurosci. Res.*, 1991, **30**, 134–153.

60. U. De Girolami, D. C. Anthony and M. P. Frosch, 'Peripheral Nerve and Skeletal Muscle,' eds. R. S. Cotran, V. Kumar and S. L. Robbins, Saunders, Philadelphia, PA, 1994, pp. 1273–1293.

61. R. J. Toso, M. A. Jordan, K. W. Farrell *et al.*, 'Kinetic stabilization of microtubule dynamic instability *in vitro* by vinblastine.' *Biochemistry*, 1993, **32**, 1285–1293.

62. I. Nennesmo and F. P. Reinholt, 'Effects of intraneuronal injection of taxol on retrograde axonal transport and morphology of corresponding nerve cell bodies.' *Virchows Arch. B: Cell Pathol. Incl. Mol. Pathol.*, 1988, **55**, 241–246.

63. Y. Komiya and T. Tashiro, 'Effects of taxol on slow and fast axonal transport.' *Cell Motil. Cytoskeleton*, 1988, **11**, 151–156.

64. R. B. Dye, S. P. Find and R. C. Williams, Jr., 'Taxol-induced flexibility of microtubules and its reversal by MAP-2 and tau.' *J. Biol. Chem.*, 1993, **268**, 6847–6850.

65. B. Dahllof, A. Billstrom, F. Cabral *et al.*, 'Estramustine depolymerizes microtubules by binding to tubulin.' *Cancer Res.*, 1993, **53**, 4573–4581.

66. Y. Sato, S. U. Kim and B. Ghetti, 'Induction of neurofibrillary tangles in cultured mouse neurons by maytanprine.' *J. Neurol. Sci.*, 1985, **68**, 191–203.

67. M. E. Stearns, M. Wang, K. D. Tew *et al.*, 'Estramustine binds a MAP-1-like protein to inhibit microtubule assembly *in vitro* and disrupt microtubule organization in DU 145 cells.' *J. Cell Biol.*, 1988, **107**, 2647–2656.

68. V. Peyrot, C. Braind, R. Momburg *et al.*, '*In vitro* mechanism study of microtubule assembly inhibition by cis-dichlorodiammine-platinum (II).' *Biochem. Pharmacol.*, 1986, **35**, 371–375.

69. K. Boekelheide, M. E. Arcila and J. Eveleth, '*cis*-Diamminedichloroplatinum (II) (cisplatin) alters microtubule assembly dynamics.' *Toxicol. Appl. Pharmacol.*, 1992, **116**, 146–151.

70. D. A. Jackson, C. K. Pearson, D. C. Fraser *et al.*, 'Methotrexate-induced morphological changes mimic those seen after heat shock.' *J. Cell Sci.*, 1989, **92**, 37–49.

71. M. C. Roach, S. Bane and R. F. Luduena, 'The effects of colchicine analogues on the reaction of tubulin with iodo[^{14}C]acetamide and *N,N'*-ethylenebis(iodoacetamide).' *J. Biol. Chem.*, 1985, **260**, 3015–3023.

72. C. H. Keith, 'Neurite elongation is blocked if microtubule polymerization is inhibited in PC12 cells.' *Cell Motil. Cytoskeleton*, 1990, **17**, 95–105.

73. A. A. Rogalski and S. J. Singer, 'Associations of elements of the Golgi apparatus with microtubules.' *J. Cell Biol.*, 1984, **99**, 1092–1100.

74. J.-Y. Chang, F.-S. Han, S.-Y. Liu *et al.*, 'Effect of 4b-arylamino derivatives of 4'-*O*-demethylepipodophyllotoxin on human DNA topoisomerase II, tubulin polymerization, KB cells, and their resistant variants.' *Cancer Res.*, 1991, **51**, 1755–1759.

75. J. Wolff and L. Knipling, 'Antimicrotubule properties of benzophenanthridine alkaloids.' *Biochemistry*, 1993, **32**, 13334–13339.

76. S. Xu, S. Roychowdhury, F. Gaskin *et al.*, 'Ethacrynic acid inhibition of microtubule assembly *in vitro*.' *Arch. Biochem. Biophys.*, 1992, **296**, 462–467.

77. R. F. Luduena, M. C. Roach and D. L. Epstein, 'Interaction of ethacrynic acid with bovine brain tubulin.' *Biochem. Pharmacol.*, 1994, **47**, 1677–1681.

78. M. B. Abou-Donia and R. P. Gupta, 'Involvement of Cytoskeletal Proteins in Chemically Induced Neuropathies,' ed. L. W. Chang, Dekker, New York, 1994, pp. 153–210.

79. A. Moretto, M. Lotti, M. I. Sabri *et al.*, 'Progressive deficit of retrograde axonal transport is associated with the pathogenesis of di-*n*-butyl-dichlorvos axonopathy.' *J. Neurochem.*, 1987, **49**, 1515–1522.

80. E. Suwita, D. M. Lapadula and M. B. Abou-Donia, 'Calcium and calmodulin-enhanced *in vitro* phosphorylation of hen brain cold-stable microtubules and spinal cord neurofilament triplet proteins after a single oral dose of tri-*o*-cresyl phosphate.' *Proc. Natl. Acad. Sci. USA*, 1986, **83**, 6174–6178.

81. A. Moretto and M. I. Sabri, 'Progressive deficits in retrograde axon transport precede degeneration of motor axons in acrylamide neuropathy.' *Brain Res.*, 1988, **440**, 18–24.

82. N. B. Chauhan, P. S. Spencer and M. I. Sabri, 'Acrylamide-induced depletion of microtubule-associated proteins (MAP1 and MAP2) in the rat extrapyramidal system.' *Brain Res.*, 1993, **602**, 111–118.

83. N. B. Chauhan, P. S. Spencer and M. I. Sabri, 'Effect of acrylamide on the distribution of microtubule-associated proteins (MAP1 and MAP2) in selected regions of rat brain.' *Mol. Chem. Neuropathol.*, 1993, **18**, 225–245.

84. C. H. Martenson, M. P. Sheetz and D. G. Graham, '*In vitro* acrylamide exposure alters growth cone morphology.' *Toxicol. Appl. Pharmacol.*, 1995, **131**, 119–129.

85. D. M. Lapadula, M. Bowe, C. D. Carrington *et al.*, '*In vitro* binding of [^{14}C]acrylamide to neurofilament and microtubule proteins of rats.' *Brain Res.*, 1989, **481**, 157–161.

86. D. W. Sickles, D. A. Welter and M. A. Friedman, 'Acrylamide arrests mitosis and prevents chromosome migration in the absence of changes in spindle microtubules.' *J. Toxicol. Environ. Health*, 1995, **44**, 73–86.

87. K. Boekelheide, '2,5-Hexanedione alters microtubule assembly. I. Testicular atrophy, not nervous system toxicity, correlates with enhanced tubulin polymerization.' *Toxicol. Appl. Pharmacol.*, 1987, **88**, 370–382.

88. K. Boekelheide, '2,5-Hexanedione alters microtubule assembly. II. Enhanced polymerization of cross-linked tubulin.' *Toxicol. Appl. Pharmacol.*, 1987, **88**, 383–396.

89. E. S. Hall, S. J. Hall and K. Boekelheide, 'Sertoli cells isolated from adult 2,5-hexanedione-exposed rats exhibit atypical morphology and actin distribution.' *Toxicol. Appl. Pharmacol.*, 1992, **117**, 9–18.

90. K. K. Pfister, M. C. Wagern, G. S. Bloom *et al.*, 'Modification of the microtubule-binding and ATPase activities of kinesin by *N*-ethylmaleimide (NEM) suggests a role for sulfhydryls in fast axonal transport.' *Biochemistry*, 1989, **28**, 9006–9012.

91. P. Forscher and S. J. Smith, 'Actions of cytochalasins on the organization of actin filaments and microtubules in a neuronal growth cone.' *J. Cell Biol.*, 1988, **107**, 1505–1516.

92. J. Kitanaka, S. Maeda and A. Baba, 'Cytochalasin B inhibits phosphoinositied hydrolysis in rat hippocampal slices.' *Neurochem. Res.*, 1993, **18**, 225–229.

93. W. Malorni, C. Fiorentini, S. Paradisi *et al.*, 'Surface blebbing and cytoskeletal changes induced *in vitro* by toxin B from *Clostridium difficile*: an immunochemical and ultrastructural study.' *Exp. Mol. Pathol.*, 1990, **52**, 340–356.

94. J. E. Eriksson, D. L. Brautigan, R. Vallee *et al.*, 'Cytoskeletal integrity in interphase cells requires protein phosphatase activity.' *Proc. Natl. Acad. Sci. USA*, 1992, **89**, 11093–11097.

95. M. G. Sacher, E. S. Athlan and W. E. Mushynski, 'Okadaic acid induces the rapid and reversible disruption of the neurofilament network in rat dorsal root ganglion neurons.' *Biochem. Biophys. Res. Commun.*, 1992, **186**, 524–530.

96. C. Arias, N. Sharma, P. Davies *et al.*, 'Okadaic acid induces early changes in microtubule-associated protein 2 and tau phosphorylation prior to neurodegeneration in cultured cortical neurons.' *J. Neurochem.*, 1993, **61**, 673–682.

97. J. Hugon and J. M. Vallat, 'Abnormal distribution of phosphorylated neurofilaments in neuronal degeneration induced by kainic acid.' *Neurosci. Lett.*, 1990, **119**, 45–48.

98. E. M. Elliott, M. P. Mattson, P. Vanderklish *et al.*, 'Corticosterone exacerbates kainate-induced alterations in hippocampal tau immunoreactivity and spectrin proteolysis *in vivo*.' *J. Neurochem.*, 1993, **61**, 57–67.

99. R. D. Graff and K. R. Reuhl, in 'Neurotoxicology of Heavy Metals,' ed. L. W. Chang, CRC Press, Boca Raton, FL, 1996, pp. 639–658.

100. J. B. Olmsted and G. G. Borisy, 'Characterization of microtubule assembly in porcine brain extracts by viscometry.' *Biochemistry*, 1973, **12**, 4282–4289.

101. R. Kuriyama and H. Sakai, 'Role of tubulin —SH groups in polymerization to microtubules. Functional —SH groups in tubulin for polymerization.' *J. Biochem. (Tokyo)*, 1974, **76**, 651–654.

102. J. C. Troncoso, J. L. March, M. Haner *et al.*, 'Effect of aluminum and other multivalent cations on neurofilaments *in vitro*: an electron microscopic study.' *J. Struct. Biol.*, 1990, **103**, 2–12.

103. K. Miura, M. Inokawa and N. Imura, 'Effects of methylmercury and some metal ions on microtubule networks in mouse glioma cells and *in vitro* tubulin polymerization.' *Toxicol. Appl. Pharmacol.*, 1984, **73**, 218–231.

104. K. Miura and N. Imura, 'Mechanisms of cytotoxicity of methylmercury. With special reference to microtubule disruption.' *Biol. Trace Elem. Res.*, 1989, **21**, 313–316.

105. E. F. Duhr, J. C. Pendergrass, J. T. Slevin *et al.*, 'HgEDTA complex inhibits GTP interactions with the E-site of brain beta-tubulin.' *Toxicol. Appl. Pharmacol.*, 1993, **122**, 273–280.

106. M. F. Hare, K. M. McGinnis and W. D. Atchison, 'Methylmercury increases intracellular concentrations of Ca^{2+} and heavy metals in NG108-15 cells.' *J. Pharmacol. Exp. Ther.*, 1993, **266**, 1626–1635.

107. P. R. Sager, R. A. Doherty and J. B. Olmsted, 'Interaction of methylmercury with microtubules in cultured cells and *in vitro*.' *Exp. Cell Res.*, 1983, **146**, 127–137.

108. P. R. Sager and T. L. Syversen, 'Differential responses to methylmercury exposure and recovery in neuroblastoma and glioma cells and fibroblasts.' *Exp. Neurol.*, 1984, **85**, 371–382.

109. M. Cadrin, G. O. Wasteneys, E. Jones-Villeneuve *et al.*, 'Effects of methylmercury on retinoic acid-induced neuroectodermal derivatives of embryonal carcinoma cells.' *Cell Biol. Toxicol.*, 1988, **4**, 61–80.

110. H. Matsumoto, G. Koya and T. Takeuchi, 'Fetal Minamata disease. A neuropathological study of two cases of intrauterine intoxication by methylmercury compound.' *J. Neuropathol. Exp. Neurol.*, 1965, **24**, 563–575.

111. B. H. Choi, L. W. Lapham, L. Amin-Zaki *et al.*, 'Abnormal neuronal migration, deranges cerebral cortical organization, and diffuse white matter astrocytosis of human fetal brain: a major effect of methylmercury poisoning *in utero*.' *J. Neuropathol. Exp. Neurol.*, 1978, **37**, 719–733.

112. K. R. Reuhl and L. W. Chang, 'Effects of methylmercury on the development of the nervous system: a review.' *Neurotoxicology*, 1979, **1**, 21–55.

113. M. M. Falconer, A. Vaillant, K. R. Reuhl *et al.*, 'The molecular basis of microtubule stability in neurons.' *Neurotoxicology*, 1994, **15**, 109–122.

114. K. R. Reuhl, L. A. Lagunowich and D. L. Brown, 'Cytoskeleton and cell adhesion molecules: critical targets of toxic agents.' *Neurotoxicology*, 1994, **15**, 133–145.

115. T. Abe, T. Haga and M. Kurokawa, 'Blockage of axoplasmic transport and depolymerisation of reassembled microtubules by methyl mercury.' *Brain Res.*, 1975, **86**, 504–508.

116. G. O. Wasteneys, M. Cadrin, E. M. Jones-Villeneuve *et al.*, 'The effects of methylmercury on the cytoskeleton of embryonal carcinoma cells.' *Cell Biol. Toxicol.*, 1988, **4**, 41–60.

117. A. C. Alfrey, G. R. LeGendre and W. D. Kheany, 'The dialysis encephalopathy syndrome. Possible aluminum intoxication.' *New Engl. J. Med.*, 1976, **294**, 184–188.

118. D. R. McLachlin and U. De Boni, 'Aluminum in human brain disease—an overview.' *Alum. Neurotoxic. Pap. Symp.*, 1980, 3–16.

119. W. W. Pendlebury, M. F. Beal, N. W. Kowall *et al.*, 'Neuropathologic, neurochemical, and immunocytochemical characteristics of aluminum-induced neurofilamentous degeneration.' *Neurotoxicology*, 1988, **9**, 503–510.

120. C. D. Katsetos, J. Savory, M. M. Herman *et al.*, 'Neuronal cytoskeletal lesions induced in the CNS by intraventricular and intravenous aluminum maltol in rabbits.' *Neuropathol. Appl. Neurobiol.*, 1990, **16**, 511–528.

121. J. C. Troncoso, P. N. Hoffman, J. W. Griffin *et al.*, 'Aluminum intoxication: a disorder of neurofilament transport in motor neurons.' *Brain Res.*, 1985, **342**, 172–175.

122. A. Bizzi and P. Gambetti, 'Phosphorylation of neurofilaments is altered in aluminum intoxication.' *Acta Neuropathol. (Berl.)*, 1986, **71**, 154–158.

123. J. Savory, M. M. Herman, J. C. Hundley *et al.*, 'Quantitative studies on aluminum deposition and its effects on neurofilament protein expression and phosphorylation, following the intraventricular administration of aluminum maltolate to adult rabbits.' *Neurotoxicology*, 1993, **14**, 9–12.

124. T. B. Shea, P. Balikian and M. L. Beermann, 'Aluminum inhibits neurofilament protein degradation by multiple cytoskeleton-associated proteases.' *FEBS Lett.*, 1992, **307**, 195–198.

125. J. C. Troncoso, P. N. Hoffman, J. W. Griffin *et al.*, 'Aluminum intoxication: a disorder of neurofilament transport in motor neurons.' *Brain Res.*, 1985, **342**, 172–175.

126. N. A. Muma, J. C. Troncoso, P. N. Hoffman, 'Aluminum neurotoxicity: altered expression of cytoskeletal genes.' *Brain Res.*, 1988, **427**, 115–121.

127. T. L. MacDonald, W. G. Humphreys and R. B. Martin, 'Promotion of tubulin assembly by aluminum ion *in vitro*.' *Science*, 1987, **236**, 183–186.

128. D. L. Price, R. J. Altschuler, R. G. Struble *et al.*, 'Sequestration of tubulin in neurons in Alzheimer's Disease.' *Brain Res.*, 1986, **385**, 305–310.

129. D. Wenstrup, W. D. Ehmann and W. R. Markesbery, 'Trace element imbalances in isolated subcellular fractions of Alzheimer's disease brains.' *Brain Res.*, 1990, **533**, 125–131.

130. G. Roderer and K. H. Doenger, 'Influence of trimethyl lead and inorganic lead on the *in vitro* assembly of microtubules from mammalian brain.' *Neurotoxicology*, 1983, **4**, 171–180.

131. H. P. Zimmerman, K. H. Doenges and G. Roderer, 'Interaction of triethyl lead chloride with microtubules *in vitro* and in mammalian cells.' *Exp. Cell Res.*, 1985, **156**, 140–152.

132. J. G. Pounds, 'Effect of lead intoxication on calcium homeostasis and calcium-mediated cell function: a review.' *Neurotoxicology*, 1984, **5**, 295–331.

133. H. P. Zimmermann, U. Plagens and P. Traub, 'Influence of triethyllead on neurofilaments *in vivo* and *in vitro*.' *Neurotoxicology*, 1987, **8**, 569–578.

134. W. J. Niklowitz, 'Ultrastructural effect of acute tetraethyllead poisoning on nerve cells of the rabbit brain.' *Environ. Res.*, 1974, **8**, 17–36.

135. I. N. Chou, 'Distinct cytoskeletal injuries induced by As, Cd, Co, Cr, and Ni compounds.' *Biomed. Environ. Sci.*, 1989, **2**, 358–365.

136. W. Li, H. M. Kagan and I. N. Chou, 'Alterations in cytoskeletal organization and homeostasis of cellular thiols in cadmium-resistant cells.' *Toxicol. Appl. Pharmacol.*, 1994, **126**, 114–123.

137. W. Li and I. N. Chou, 'Effects of sodium arsenite on the cytoskeleton and cellular glutathione levels in cultured cells.' *Toxicol. Appl. Pharmacol.*, 1992, **114**, 132–139.

138. J. W. Mills, J. H. Zhou, L. Cardoza *et al.*, 'Zinc alters actin filaments in Madin–Darby canine kidney cells.' *Toxicol. Appl. Pharmacol.*, 1992, **116**, 92–100.

139. J. P. O'Callahan, in 'Neurotoxicology,' ed. M. B. Abou-Donia, CRC Press, Boca Raton, FL, 1996, pp. 61–78.

140. J. P. O'Callahan, K. F. Jensen, and D. B. Miller, 'Quantitative aspects of drug and toxicant-induced astrogliosis.' *Neurochem. Int.*, 1995, **26**, 115–124.

141. M. D. Minana, V. Felipo, R. Wallace *et al.*, 'High ammonia levels in brain induce tubulin in cerebrum but not cerebellum.' *J. Neurochem.*, 1988, **51**, 1839–1842.

142. V. Felipo, E. Grau, M. D. Minana *et al.*, 'Ammonium injection induces an *N*-methyl-D-aspartate receptor-mediated proteolysis of the microtubule-associated protein MAP-2.' *J. Neurochem.*, 1993, **60**, 1626–1630.

143. Y. Xu, J. T. Sladky and M. J. Brown, 'Dose-dependent expression of neuronopathy after experimental pyridoxine intoxication.' *Neurology*, 1989, **39**, 1077–1083.

144. R. Medori, L. Autilio-Gambetti, S. Monaco *et al.*, 'Experimental diabetic neuropathy: impairment of slow transport with changes in axon cross-sectional area.' *Proc. Natl. Acad. Sci.*, 1985, **82**, 7716–7720.

145. R. E. Schmidt, S. B. Plurad, C. A. Parvin *et al.*, 'Effect of diabetes and aging on human sympathetic autonomic ganglia.' *Am. J. Pathol.*, 1993, **143**, 143–153.

146. J. R. Larsen and P. Sidenius, 'Slow axonal transport of structural polypeptides in rat, early changes in streptozocin diabetes, and effect of insulin treatment.' *J. Neurochem.*, 1989, **52**, 390–401.

147. P. Macioce, G. Filliatreau, B. Figliomeni *et al.*, 'Slow axonal transport impairment of cytoskeletal proteins in streptozocin-induced diabetic neuropathy.' *J. Neurochem.*, 1989, **53**, 1261–1267.

11.06
Role of Cell Signaling in Neurotoxicity

LUCIO G. COSTA

University of Washington, Seattle, WA, USA

11.06.1 INTRODUCTION

The mechanisms by which extracellular signals are transferred to the cell cytosol and nucleus, commonly known as "cell signaling," is receiving much attention in all areas of biology and medical sciences. Two Nobel prizes for Medicine have been awarded to scientists whose investigations have contributed greatly to our understanding of key cellular processes in cell signaling, that is, protein phosphorylation and regulation of G proteins. Intracellular elements of cell signaling are targeted for developing new therapeutics for a number of diseases, as their selectivity is higher than initially thought.[1] Increasing attention is also being given to the possibility that alterations of cell signaling may be involved in the mechanisms of toxicity of numerous chemicals. This chapter will focus on the potential role of cell signalling in neurotoxicity. Some intracellular pathways linked to the activation of receptors for neurotransmitters, cytokines, and growth factors will be reviewed, and examples of interactions of neurotoxic chemicals with second- and third-messenger systems will be presented. Additional discussions of these areas of investigations can also be found in other reviews.[2-5]

11.06.2 METABOLISM OF PHOSPHOLIPIDS

11.06.2.1 The Phosphoinositide/Calcium/ PKC Pathway

The first observation that agonists for several receptors can stimulate the metabolism of inositol phospholipids was made in the 1950s. However, this field of research languished until 1975, when it was suggested that the stimulation of inositol phospholipid breakdown was associated with a rise in cytosolic Ca^{2+} ion concentration in the stimulated cell.[6] Since then there has been a rapid advancement of knowledge in this area, and these findings have been

summarized in several reviews.[7-9] Phosphoinositide metabolism in the nervous system has received particular attention because of the presence of several receptor systems coupled to this response in both nerve and glial cells, and because of the high concentrations of protein kinase C (PKC) present in the brain.[2,10-12]

When an agonist interacts with the appropriate receptor (e.g., cholinergic muscarinic M_1 or M_3 receptors, or metabotropic excitatory amino acid receptors), it activates a phosphoinositidase (phospholipase C) that hydrolyzes phosphoinositide 4,5-bisphosphate (PtdIns 4,5-P_2) to form inositol 1,4,5-trisphosphate (Ins 1,4,5-P_3) and 1,2-diacylglycerol (DG) (Figure 1). A guanosine triphosphate (GTP)-binding protein couples the receptor to phospholipase C. Evidence suggests that more than one protein exists, depending on the receptor subtype and the cell type.[13,14] Ins 1,4,5-P_3 binds to specific and saturable receptor sites located in the endoplasmic reticulum and causes the mobilization of calcium ions in the cytosol. By the action of phosphatases, Ins 1,4,5-P_3 is then dephosphorylated to Ins 1,4,-P_2, Ins 1-P, and inositol.[15] Inositol 1,4,5-trisphosphate can also be phosphorylated by a 3-kinase to form inositol 1,3,4,5-tetrakisphosphate (Ins 1,3,4,5-P_4), which may play a role in modulating the mobilization of intracellular Ca^{2+} by Ins 1,4,5-P_3 and in regulating Ca^{2+} entry from outside the cell.[16] Other inositol phosphates that have been

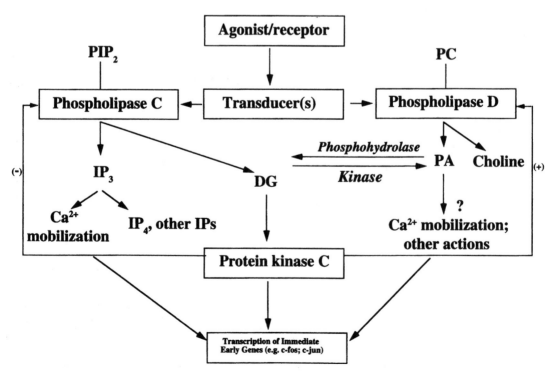

Figure 1 Agonist-induced activation of phospholipase C and phospholipase D leads to calcium mobilization, activation of protein kinase C and induction of immediate early genes (after Shukla and Halenda[17]).

identified include Ins 1,3,4-P_3, Ins 1,3,4,5,6-P_5, Ins-P_6, and cyclic inositol 1,2,4,5-trisphosphate. Although all have been shown to exert some cellular effect, their precise roles in the phosphoinositide pathway are still far from clear.[2,12]

A main consequence of the activation of the phosphoinositide pathway is a change in the intracellular concentration of calcium deriving from Ins 1,4,5-P_3-sensitive stores or from Ca^{2+}- and/or Ins 1,3,4,5-P_4-sensitive stores, or entering the cell from receptor-operated calcium channels.[16,18] Intracellular calcium can bind to two types of broadly defined receptor protein. One type (e.g., calmodulin) does not possess enzymatic activity, but after binding calcium is capable of activating various enzymes, such as the Ca^{2+}-calmodulin-activated protein kinases or the calcium adenosinetriphosphatase (ATPase).[19] The second type of protein, represented for example by PKC, possesses intrinsic enzymatic activity that can be activated by calcium.[19] While a transient increase in intracellular calcium is necessary for the normal physiological functions of the cell, a sustained increase can produce toxicity. In particular, calcium can activate phospholipases, proteases, and endonucleases, leading to fragmentation of phospholipids, proteins, and DNA,[20,21] and sustained elevated calcium levels may play relevant roles in both cell necrosis and apoptosis.[21,22]

DG, the other product of phosphoinositide hydrolysis, activates a protein kinase, PKC. Activation of PKC requires the presence of calcium and phospholipids, particularly phosphatidylserine, and involves its translocation from the inactive form in the cytosol to the membrane.[23] As with other enzymes of the phosphoinositide cycle, several isozymes of PKC exist. At least seven subspecies of PKC have been found in nervous tissue, and their structures, deduced by analysis of their DNA sequences, have been elucidated.[24] PKC is the receptor for, and can be activated by, phorbol esters, a known class of tumor promoters.[25] The gamma form of PKC, which is expressed only in the brain and spinal cord, can also be activated by arachidonic acid and by other eicosanoids. A large number of proteins can be phosphorylated by PKC, including receptors, ion channels, cytoskeletal proteins, and membrane enzymes.[26] PKC has been involved in the release of a variety of neurotransmitters, an effect strongly correlated with phosphorylation of a synaptic protein known as B-50 or GAP43.[27] This action may be linked to the role of PKC in the maintenance of long-term potentiation, suggesting a role for this enzyme in the process of memory.[28] Another important role of PKC, and of the phosphoinositide

pathway in general, is in cell proliferation.[29,30] In the nervous system, proliferation of astrocytes and Schwann cells has been shown to be associated with activation of the phosphoinositide/PKC system.[31,32] Direct activation of PKC by phorbol esters has also been shown to cause neurodegeneration of human cortical neurones in culture.[33] Whether this effect is due to excessive phosphorylation of regulatory proteins by PKC or, conversely, associated with a downregulation of PKC, leading to overactivation of other kinases, which in turn would overphosphorylate cytoskeletal proteins, remains to be determined.

11.06.2.2 Metabolism of Phosphatidylcholine

Several agents that activate the metabolism of phosphoinositides also induce hydrolysis of phosphatidylcholine.[34] This membrane lipid, which accounts for up to 40% of the total cellular phospholipid content, can be hydrolyzed by both phospholipase C and phospholipase D, generating DG and phosphocholine or phosphatidic acid (PA) and choline, respectively[35] (Figure 1). PA, which itself could be a potential intracellular mediator, as it causes calcium influx across the plasma membrane and can stimulate DNA synthesis,[36] can be further degraded by a propanolol-sensitive phosphohydrolase to DG. Metabolism of phosphatidylcholine may play a role in long-term cellular effects, such as cell proliferation, mediated by DG activation of PKC and by PA. In addition to hydrolysis, phospholipase D also catalyzes a transphosphatidylation reaction whereby the phosphatidyl moiety of the phospholipid substrate is transferred to primary alcohols to produce phosphatidylalcohols.[37] In the presence of ethanol, this reaction leads to the formation of phosphatidylethanol, which can substitute for phosphatidylserine in activating a brain-specific PKCγ.[38]

11.06.2.3 Activation of Phospholipase A_2

Another intracellular pathway whose importance is increasingly recognized, but which has received little attention in a neurotoxicology context, is the metabolism of arachidonic acid (AA). AA is liberated from membrane phospholipids by two pathways: a Ca^{2+}-activated phospholipase A_2 generates AA and lysophospholipid, while phospholipase C generates DG, which is then converted to AA and glycerol by lipases.[39] Free AA can diffuse out of the cells, can be reincorporated into phospholipids, or can undergo intracellular metabolism. The three metabolic pathways of AA are initially catalyzed by cyclo-oxygenase, lipoxygenase, and cytochrome P450.[40] The cyclo-oxygenase pathway leads to the formation of prostaglandins, thromboxanes, and prostacyclin, and is inhibited by anti-inflammatory drugs such as aspirin or indomethacin. Cytochrome P450 catalyzes the conversion of AA into epoxyeicosatrienoic acids. Lipoxygenase metabolizes AA to hydroxyperoxyeicosatetraenoic acids (HPETE), which are then converted to leukotrienes.

Various neurotransmitters (e.g., glutamate via NMDA (*N*-methyl-D-asparatate) receptors, acetylcholine via muscarinic receptors) cause the formation of lipoxygenase metabolites in nervous tissue.[41] Increases in lipoxygenase products occur as a consequence of ischemic insult and epileptic seizures.[41] In certain systems (e.g., N1E-115 murine neuroblastoma cells), lipoxygenase products may also mediate the stimulation of cyclic GMP (guanosine monophosphate) synthesis.[42] Inhibition of neurotransmitter release by 12-HPETE (formed from AA by the action of a 12-lipoxygenase) has been documented. A cytochrome P450-generated metabolite of 12-HPETE can activate K^+ channels, leading to decreased calcium entry and decreased neurotransmitter release. 12-HPETE can also inhibit calmodulin-dependent protein kinase II and thus reduce phosphorylation of synapsin I and neurotransmitter release.[43] Arachidonic acid itself may also have a role in long-term potentiation.[43] Eicosanoids can also inhibit Na^+, K^+-ATPase, thus affecting ion gradients across cell membranes and may play a role in cell proliferation.[44] As knowledge of the functional role(s) of AA and its several metabolites in nerve tissue increases, their potential involvement in neurotoxic events will become more apparent.

11.06.3 REGULATION OF CYCLIC NUCLEOTIDE METABOLISM

11.06.3.1 The Cyclic AMP/Protein Kinase A Pathway

The role of cyclic adenosine monophosphate (AMP) as a mediator of several physiological processes has been established since the early 1970s.[45] Activation of a number of receptors (e.g., β-adrenoceptors), coupled through the GTP-binding protein G_s to the enzyme adenylate cyclase, causes the conversion of ATP to cyclic AMP. Before being inactivated by phosphodiesterases, cyclic AMP can activate a cyclic AMP-dependent protein kinase (PKA) that in turn can phosphorylate a number of

cellular substrates, including proteins involved in neurotransmitter release, receptors, ion channels, cytoskeletal proteins, and other enzymes.[46] One of the substrates of this kinase is a protein known as CREB (cAMP response element-binding protein).[47] Interestingly, CREB can also be phosphorylated by calcium–calmodulin protein kinase, making it a common link between the cyclic AMP and the calcium-signaling pathways. Among the genes regulated by CREB are those coding for the neuropeptide somatostatin and the immediate early genes, including the proto-oncogenes *c-fos* and *c-myc*, which are believed to play a fundamental role in directing the growth and differentiation of nerve cells. Activation of certain receptors (e.g., muscarinic M_2, A_1-adenosine) can also cause inhibition of adenylate cyclase, leading to a decrease in the intracellular levels of cyclic AMP.[48] This "negative coupling" to adenylate cyclase is mediated by a G protein known as G_i.

11.06.3.2 Nitric Oxide and Cyclic GMP Metabolism

Activation of other receptors (e.g., glutamate, muscarinic M_1 or M_3) causes elevation in intracellular cyclic GMP levels. The effect has been considered to be indirect, involving an additional second messenger, since receptor agonists, differing from compounds such as sodium nitroprusside, cannot activate guanylate cyclase (the soluble enzyme that converts GTP to cyclic GMP) in broken cell preparations. Two likely candidates considered for this role were calcium ions and a metabolite of arachidonic acid. A role for calcium was suggested by experiments showing that extracellular calcium ions are necessary for receptor stimulation of cyclic GMP and that calcium channel antagonists block this effect.[49] An involvement of a metabolite of arachidonic acid was suggested by the finding that inhibitors of lipoxygenase (but not cyclo-oxygenase) block the cyclic GMP response.[42]

The cyclic GMP pathway has received strong new attention in recent years, since it has been discovered that nitric oxide (NO) synthesized from L-arginine by NO synthase binds to a heme moiety attached to guanylate cyclase and stimulates the accumulation of cyclic GMP.[50–55] NO is a free radical gas, formed in neurons and in glial cells by the action of NO synthase (NOS). Several NOS have been cloned, and they are usually classified as constitutive (cNOS) and inducible (iNOS). NO has been shown to be involved in a number of central nervous system (CNS) effects including neurotransmitter release, long-term potentiation, synaptic plasticity, nociception, and cerebrovascular functions. Of interest to this discussion is that NO may possess both neuroprotective and neurotoxic properties, depending on its oxidative state (Figure 2). NO^+ is thought to be neuroprotective through the nitrosylation of free sulfhydryl groups in the redox modulatory site of the NMDA receptor, thus inactivating the NMDA receptors. On the other hand, $NO^{\cdot}$ can react with superoxide anions ($O_2^{-\cdot}$ to form peroxynitrite ($ONOO^-$). Peroxynitrite is a potent oxidant, and can also decompose to the hydroxyl and nitrogen dioxide free radicals. The importance of peroxynitrite in damage during ischemia is supported by the observation that superoxide dismutase markedly attenuates the infarct volume after focal ischemia.[56] NO has been shown to mediate the "delayed neurotoxicity" of NMDA in cortical neurons,[57] as indicated by the ability of NOS inhibitors and hemoglobin, which binds extracellular NO, to antagonize NMDA neurotoxicity. The free radicals generated from NO can damage DNA by causing DNA breaks, with subsequent activation of the nuclear enzyme poly (ADP-ribose) synthetase (PARS). As activation of PARS consumes four moles of ATP (which are necessary to

regenerate one mole of NAD consumed during the transfer of ADP-ribose), the cell's energy stores are rapidly depleted. PARS inhibitors have been found to offer protection against NMDA neurotoxicity.[58] NO has also been implicated as a mediator of oxygen CNS toxicity,[59] and may play a role in the pathogenesis of acquired immunodeficiency syndrome (AIDS) dementia.[60] It is thus apparent that NO may play a role in various neurotoxic effects. Judging from the recent developments and discoveries, it is also apparent that the study of the role of NO in neurotoxicity of a variety of chemicals is fertile terrain for further investigations.

11.06.4 PROTEIN PHOSPHORYLATION

Protein phosphorylation is a major additional step in the second-messenger cascade, and its importance in neuronal function is increasingly recognized[46] (see also Chapter 10, this volume). In addition to the aforementioned calcium/phospholipid-dependent protein kinase (PKC) and the cyclic AMP-dependent protein kinase, other well-characterized kinases are the cyclic GMP-dependent protein kinase and the Ca^{2+}-calmodulin-dependent protein

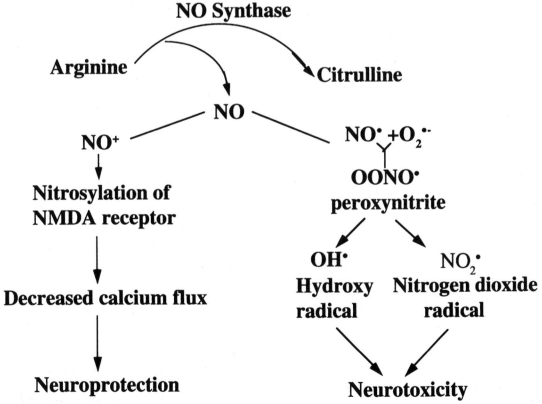

Figure 2 Synthesis of nitric oxide and its dual action of (neuroprotective/neurotoxic) depending on its oxidative state (adapted from T. M. Dawson and V. L. Dawson, *The Neuroscientist*, 1995, **1**, 7–18).

kinases (I, II, and III).[46] A protein kinase phosphorylates a substrate protein by transferring a terminal phosphate from ATP. The so-formed phosphoprotein changes its function, presumably by changing its structure, and is then reversed through removal of the phosphate groups by phosphatases. A large number of proteins are regulated by phosphorylation, including enzymes involved in neurotransmitter synthesis, ion channels, receptors, cytoskeletal proteins, and proteins involved in the regulation of transcription and translation. A number of protein kinases also exist in brain tissue that can phosphorylate endogenous proteins on serine and threonine residues and do not appear to be regulated by any known second-messenger system, but may be of neurotoxicological significance. For example, casein kinase II can phosphorylate DARPP-32, a region- and cell-specific phosphoprotein; glycogen synthase kinase 3 can catalyze the *in vitro* phosphorylation of the neuronal cell adhesion molecule *N*-CAM; and several neurofilament protein kinases can phosphorylate the three neurofilament proteins.[46] An additional class of protein kinases is represented by the tyrosine-specific protein kinases, which include proto-oncogene products (e.g., pp60[c-src], the gene product of *c-src*) as well as growth factor receptors (e.g., the insulin-like growth factor I).[46,61]

Surprisingly, while phosphorylation of proteins by protein kinases has been long recognized as a primary dynamic regulatory process for post-translational modifications in the nervous system, its possible role in neurotoxicity has received only limited attention. A role for endogenous protein phosphorylation in the delayed polyneuropathy induced by certain organophosphorus compounds was suggested in 1984.[62] Triethyltin increases the phosphorylation of synapsin I, possibly by activation of cyclic AMP-dependent protein kinases.[63] Alterations of phosphorylation have been found in various clinical disorders (e.g., diabetes, cystic fibrosis) and may also be relevant in Alzheimer's disease.[46] Administration of the antimitotic agent methylazoxymethanol to rats at day 15 of gestation has been shown to reduce phosphorylation of the neuronal phosphoprotein B50 (GAP43) without altering the relative amount of the protein or the levels of mRNA.[64] Several neuron- and glia-specific phosphoproteins are also being used as markers for neurotoxic damage.[65]

11.06.5 LIPIDATION OF PROTEINS

Since the mid-1980s, numerous studies have demonstrated that proteins are modified post-translationally by lipids.[66–68] These "modifiers" include fatty acids (myristate and palmitate), glycophospholipids (glycophosphatidylinositol and GPI), and isoprenoids (farnesol and geranylgeranol). The alpha subunits of G proteins have been shown to be palmitoylated and myristoylated. Addition of GPI (consisting of glycophosphatidylinositol linked to a tetrasaccharide) to many different proteins increases their half-life on the cell surface. Prenylation is a very common form of post-translational protein modification and involves the attachment of isoprenoids via a thioether linkage to cysteine residues at or near the carboxy terminus. Two types of modifications have been identified: the addition of farnesyl or geranylgeranyl moieties. Farnesol exists within the cell as farnesyl pyrophosphate, an intermediate of cholesterol biosynthesis. Addition of geranylgeranol, which is also present in the cells as a pyrophosphate, but is not an intermediate in cholesterol biosynthesis, represents the most abundant form of protein prenylation. The molecular details of protein prenylation have been elucidated. Proteins containing a cystein residue fourth from the COOH-terminus (the Cys-A-A-X motif) can be farnesylated or geranylgeranylated by specific transferases, depending on the nature of the COOH terminal residue (the X).[68] Following prenylation, proteins are directed to a membrane compartment for further processing, which includes removal of the three COOH-terminal residues and methylation of the now-exposed COOH of prenylcysteine. A number of proteins have been shown to be modified by prenylation, among them laminin, G-proteins, and Ras.[69] In the case of Ras, it has been shown that mutant Ras proteins which do not undergo isoprenoid modification no longer associate with the plasma membrane and are not transforming.[66] Though protein prenylation has not been investigated as a possible target for neurotoxic compounds, it is increasingly considered with regard to its role in brain tumors, particularly gliomas.

11.06.6 INDUCTION OF IMMEDIATE–EARLY GENES

Several receptors that directly (e.g., via the phosphoinositide pathway and/or direct opening of calcium channels) or indirectly (e.g., via phosphorylations that follow an initial activation of the adenylate cyclase pathway) cause an increase in intracellular calcium levels lead to the activation of immediate–early genes[70] (Figure 1). Expression of immediate–early genes (e.g., *c-fos*, *c-jun*) is very low in quiescent cells but is rapidly induced at the transcriptional level. The mRNAs accumulate for 1–2 h in the cytoplasm, and the respective proteins (e.g., Fos and Jun) are translated and then translocated to the nucleus, where they form a heterodimeric protein complex that binds to the DNA regulatory element known as the AP-1 binding site.[71] The binding of the Fos–Jun complex to DNA causes the activation of target genes that may code for a variety of structural or other proteins.[70] The transient nature of the mRNAs and their proteins suggests that these may act as a signaling system, i.e., as a third-messenger system within an intracellular cascade linking extracellular stimuli to long-term adaptive processes.[71] Seizure activity induced by the convulsant metrazole or by electroconvulsive shock has been shown to cause a dramatic induction of *c-fos*.[72,73] The organochlorine insecticide lindane has been reported to cause an elevation in the expression of *c-fos* at doses that do not induce seizure activity, suggesting that *c-fos* induction may, in some cases, represent an early marker of neurotoxicity.[74]

11.06.7 EFFECT OF NEUROTOXICANTS ON SECOND-MESSENGER SYSTEMS

11.06.7.1 Metals

Several metals have been shown to interact with PKC or other elements of the phosphoinositide pathway[2,75] (Table 1). Inorganic mercury and various forms of organic mercury have been shown to inhibit PKC activity and the binding of ^{3}H-phorbol 12,13-dibutyrate at low micromolar concentrations.[76] This is not surprising, since PKC differs from other kinases because of its rich cysteine content, and mercurials are known to have a high affinity for sulfhydryl groups, which are present in the

catalytic and regulatory domains of PKC. Inhibition of phorbol ester binding in brain regions was not observed following acute or repeated exposure of mice to methylmercury,[77] though PKC activity was significantly decreased.[78] Inorganic mercury has also been shown to potentiate nerve growth factor (NGF)-induced DG production in neuroadrenergic PC12 cells and to have an effect on its own at submicromolar concentrations.[79] It appears that, in these cells, the effect of NGF on cellular differentiation is mediated by stimulation of glycosylphosphatidylinositol hydrolysis, which generates inositol glycan and DG. In cultured cerebellar granule neurons methylmercury, but not mercury chloride, has been shown to increase inositol phosphates and free intracellular calcium levels, and it has been suggested that these actions are associated with the mechanism of neurotoxic injury of this compound.[80] Increases in free cytosolic calcium by methylmercury have also been observed in synaptosomal preparations,[81] and have been ascribed to either membrane or intracellular actions. In NG108-15 cells, methylmercury increases intracellular calcium by causing its release from Ins 1,4,5-P_3-sensitive pools, as well as by causing calcium influx.[82] Methylmercury has also been found to stimulate protein phosphorylation in cerebellar granule cells, but to inhibit it in cerebellar glial cells[83] and to stimulate phospholipase A_2 activity.[84] Although several of these effects may be involved in mercury neurotoxicity, the known ability of mercury compounds to affect a large number of biochemical and physiological processes[85] calls for caution in evaluating the relative contribution of each effect to its overall neurotoxicity.[82]

Lead is known to alter calcium-mediated processes and may also mimic calcium's effects[95] (see Chapter 13 and Chapter 28, this volume). For example, lead has a higher affinity than calcium for calmodulin and can activate some

Table 1 Neurotoxic metals affecting signal transduction.

Metal	Effect	Ref.
Mercury	Inhibition of PKC	76
Methylmercury	Increase in $[Ca^{2+}]_i$	81,82
	Modulation of protein phosphorylation	83
	Stimulation of PLA_2	84
Lead	Activation of PKC	86–88
	Inhibition of cNOS	89
Aluminum	Activation of basal phosphoinositide metabolism	90
	Inhibition of agonist-stimulated phosphoinositide metabolism	91–94

calmodulin-dependent processes.[96] Lead can also substitute for calcium in stimulating PKC activity.[86–88] In isolated immature rat microvessels, lead caused activation of PKC in a concentration range of $0.1–10 \mu M$ and also caused translocation of PKC from the cytosol to the particulate fraction. More surprisingly, lead has been shown to activate PKC, partially purified from rat brain, at concentrations as low as $10^{-15} M$, being several orders of magnitude more potent than calcium itself.[87] The reason for the apparent discrepancy between the potency of lead in the brain and in isolated microvessels is not clear, but it may be linked to the relative distribution of subspecies of PKC. It should be noted, however, that an inhibitory effect of lead on PKC activity has also been reported.[97] As long-term potentiation, a possible functional equivalent of memory storage, is blocked by inhibitors of PKC, a lead-induced decrease in PKC activity (via inhibition or downregulation due to sustained stimulation), may relate to learning and memory deficits caused by this compound. Activation of PKC by lead has also been suggested to be involved in its inhibition of astroglia-induced microvessel formation *in vitro*.[98] Developmental exposure to lead has been shown to alter the binding of phorbol esters,[99] a finding that deserves further study because of the known developmental neurotoxicity of this metal. Potential interactions of lead with the calcium-releasing ability of Ins $1,4,5-P_3$ could also be of interest. At micromolar concentrations, lead has been shown to enhance the binding of ^{3}H-Ins $1,4,5-P_3$ and ^{3}H-Ins $1,3,4,5-P_4$ to cerebellar membranes, and the effect is antagonized by adenosine triphosphate (ATP).[100] Similarly, exposure to lead during development has also been shown to decrease cortical Ins $1,4,5-P_3$ receptors and to reduce the ability of Ins $1,4,5-P_3$ to increase calcium in permeabilized cortical neurons.[101] Lead, at concentrations as low as $10 nM$, *in vitro*, has been found to inhibit cNOS in brain microvascular endothelial cells, without affecting iNOS induced by interferon or tumor necrosis factor α.[89] Inhibition of the calcium-dependent cNOS was prevented by the addition of calcium. Several of these effects of lead on cell signaling are of much interest, and are worth further investigation as they may also explain other neurochemical changes associated with lead exposure, such as changes in cell surface receptors.

Another neurotoxic metal, aluminum, can activate phosphoinositide metabolism where present alone or complexed with fluoride to form AlF_4^-.[90] On the other hand, aluminum chloride has been shown to inhibit agonist-stimulated phosphoinositide metabolism in brain slices and neuroblastoma cells.[91,92,102] The exact mechanism(s) involved in this inhibition have not been fully elucidated, but may relate to an interaction of aluminum with G proteins and/or with phospholipase C.[93,94] Though aluminum can inhibit brain PKC activity *in vitro* at micromolar concentrations,[103,104] this effect may not be responsible for the observed inhibition of agonist-stimulated phosphoinositide metabolism.[94] Aluminum has also been reported to affect phosphorylation of cytoskeletal proteins[105] and the metabolism of cAMP and cGMP.[106] Again, however, the possible physiological or pathological implications of these actions of aluminum are far from clear.

Other metals (e.g., cadmium, manganese, organotins) have also been shown to activate or inhibit PKC *in vitro*.[2,107] In the case of cadmium, it has been shown that this metal can potentiate the effect of phorbol ester on nuclear binding and activation of PKC at nanomolar concentrations.[108] This action of cadmium may play a role in its carcinogenicity by activating epigenetic mechanisms; however, the possible involvement of PKC in cadmium neurotoxicity is unclear. Vanadate has been shown to prevent the hydrolysis of Ins $1,4,5-P_3$ to Ins $1,4-P_2$, and of Ins $1,3,4,5-P_4$ to Ins $1,3,4-P_3$, thereby potentiating the basal and stimulated accumulation of inositol phosphates.[109] These effects of vanadium may be involved in the alleged role played by this metal in the etiology of some forms of depression and bipolar disorder.[109]

11.06.7.2 Pesticides

A number of studies have investigated the interactions of pesticides, particularly organophosphates, with the metabolism of phosphoinositides. Repeated exposures to organophosphates cause a decrease in carbachol-stimulated phosphoinositide metabolism in the rat cerebral cortex, which parallels a decrease in the density of muscarinic receptors.[110,111] A series of studies by Savolainen *et al.*[112,113] have investigated the involvement of phosphoinositides and calcium in the convulsant effect of organophosphates. An increase in cerebral Ins 1-P was observed in convulsing rats but was not present (or occurred only transiently) in nonconvulsing animals. Increased levels of calcium and morphological changes were also observed, and all effects, including convulsions, were potentiated by lithium, suggesting that the phosphoinositide signaling pathway may be involved in convulsions and, through elevation of intracellular calcium, contribute to neuronal injury. Potentiation of other direct (pilocarpine) or indirect (physostigmine) cholinergic agonists by lithium has also been reported,[114–116] confirming a role for the phosphoinositide pathway in cholinergic-induced seizures and brain damage.

Pyrethroids, of both type I and type II, can stimulate phosphoinositide breakdown in guinea pig synaptosomes; for type II compounds, this effect is due to activation of sodium channels, while type I compounds may act through a different, still unknown, mechanism.[117] The type II pyrethroid deltamethrin has also been shown to stimulate protein phosphorylation in rat brain synaptosomes, an effect believed to be secondary to its increase in intracellular calcium.[118] Several chlorinated hydrocarbons (e.g., chlordane, lindane) stimulate PKC from rat brain, an effect that may be related to the hepatic tumor-promoting action of some of these neurotoxic organochlorines.[119] Lindane was also reported to increase inositol phosphate formation in brain slices[120] and to increase calcium levels in synaptosomes.[121]

11.06.7.3 Alcohol and Drugs of Abuse

Several studies have investigated the interaction of ethanol with second-messenger systems, particularly cAMP metabolism and, more recently, phosphoinositide metabolism.[122,123] Interpretation of the results of these studies is complicated by the different experimental systems utilized (e.g., *in vitro* cell culture vs. *ex vivo* brain homogenates or slices) and the type of exposure to ethanol (e.g., acute vs. chronic). With regard to adenylate cyclase, the interaction of ethanol appears to involve in particular G_s.[123] Acute exposure to ethanol increases the amount of cAMP produced on stimulation of receptors coupled to adenylate cyclase, while chronic exposure leads to an opposite effect. The latter has been associated with a decrease of the α submit of G_s, both at the protein and mRNA levels.[124]

Many of the studies on the interaction of ethanol with phosphoinositide metabolism, calcium homeostasis, and PKC were aimed at defining their possible role in the processes of tolerance and dependence or specific organ effects (e.g., hepatotoxicity).[122] In certain systems, for example, intact hepatocytes and human platelets, ethanol appears to activate phospholipase C and to increase intracellular calcium.[125] However, this effect has not been consistently found in brain tissue. On the other hand, in most tissue, ethanol has been found to inhibit agonist-induced phosphoinositide metabolism and elevation of intracellular calcium.[2]

Another aspect of the interaction of ethanol with phosphoinositides is the formation of phosphatidylethanol by the action of phospholipase D[126] (Table 2). Phospatidylethanol can activate PKC, particularly the gamma subtypes,[38] and may contribute to the downregulation of PKC observed following chronic *in vivo* exposure to ethanol.[127] Ethanol may also directly interact with PKC. However, the exact nature of this interaction is not clear, since both stimulatory[128] and inhibitory[129] effects have been reported.

A series of studies have suggested the hypothesis that phosphoinositide metabolism stimulated by activation of cholinergic muscarinic receptors in the developing rat brain may represent a relevant target for the developmental neurotoxicity of ethanol.[130,131] The hypothesis stemmed from the observation that the developmental profile of muscarinic receptor-stimulated phosphoinositide metabolism in the rat has a striking resemblance to that of the brain growth spurt, suggesting a role for this system in the regulation of neuromorphogenesis, synaptogenesis, and glial cell proliferation.[132] The age-, brain region-, and receptor-specific inhibitory effects of ethanol on this system are intriguing and may relate to some of the effects found in the fetal alcohol syndrome. Of particular interest is the finding that acetylcholine is a mitogen for developing astrocytes,[133] and that ethanol is a potent ($IC_{50} = 10$ mM) inhibitor of muscarinic receptor-induced astrocyte proliferation.[134] This action of ethanol may be relevant for ethanol-induced microencephaly and/or for abnormal neuronal development involving impaired astrocyte functions.

Cannabinoids, such as Δ^8-tetrahydrocannabinol, bind to specific receptors whose activation leads to inhibition of adenylate cyclase.

Table 2 Formation of phosphatidylethanol and phosphatidic acid in rat cortical astrocytes.

Agonist	EtOH	PEtOH	PA
		% of control	
None	−	100	100
Carbachol	−	−	191
TPA	−	−	250
Carbachol	+	243	100
TPA	+	443	97

Source: Costa *et al.*.[135]
Both the cholinergic muscarinic agonist carbachol (1 mM) and the phorbol ester TPA (12-*O*-tetradecanoyl-phorbol-13-acetate; 10 μM), an activator of PKC, stimulate the formation of PA (phosphatidic acid). In the presence of ethanol (EtOH; 200 mM), the formation of PA is inhibited and phosphatidylethanol (PEtOH) is formed.

This effect occurs at micromolar concentrations, which are pharmacologically relevant, and involves activation of the GTP binding protein G_i.[136] At the same concentrations, cannabinoids also inhibit phosphoinositide metabolism induced by carbachol, glutamate, and norepinephrine in rat hippocampal cultures.[137] However, the psychoinactive cannabidiol had a similar effect, raising questions about the pharmacological significance of this observation. Cocaine, its metabolite norcocaine, and cocaethylene, a product of the reaction between cocaine and ethanol, have been found to inhibit muscarinic receptor-stimulated phosphoinositide in rat cortical slices[138] and it has been suggested that this effect may be involved, at least in part, in the developmental neurotoxicity of cocaine.[139]

11.06.7.4 Other Compounds

11.06.7.4.1 Colchicine

A series of studies by Tilson and co-workers[140,141] have shown that direct intradentate administration of colchicine, a neurotoxicant known to block mitosis and axoplasmic transport, results in an alteration of carbachol-mediated phosphoinositide hydrolysis in the hippocampus. This change occurred between 3 and 12 weeks after infusion of colchicine and persisted for up to one year. This increased sensitivity is believed to be a compensatory change associated with a loss of muscarinic receptors due to the destruction of granule cells and mossy fibers. Similar changes in ibotenate- and glutamate-induced phosphoinositide hydrolysis have also been reported following colchicine administration.[142]

11.06.7.4.2 Polychlorinated biphenyls

A series of studies by Kodavanti *et al.*[143-145] have investigated the effects of polychlorinated biphenyls on calcium and PKC in rat cerebellar granule cells. 2,2'-Dichlorobiphenyl caused a sustained increase in intracellular calcium, but did not alter basal phosphoinositide metabolism. This same compound also increased phorbol ester binding, which reflects translocation of PKC. On the other hand, 3,3',4,4',5-pentachlorobiphenyl, a non-*ortho*-substituted congener, had no effect on either calcium homeostasis or PKC. Though these observations are of interest, their interpretation in light of the evidence of neurotoxicity of polychlorinated biphenyls is not yet clear. Nevertheless, these cellular and molecular effects are worth considering also with regard to other toxic effects of these agents.

11.06.7.4.3 Excitatory amino acids

Interactions with phosphoinositide metabolism may also be involved in the neurotoxicity of excitatory amino acids[2] (see Chapter 31, this volume). In hippocampal slices, quisqualate- and glutamate-stimulated phosphoinositide breakdown is more pronounced in the neonatal than in the adult rat.[146] This is of interest, since quisqualic acid is particularly toxic to the immature brain.[147] Excessive and prolonged translocation of PKC from the cytosol in the neuronal membrane and destabilization of calcium homeostasis have been suggested as important elements in the neurotoxicity of glutamate in cerebellar granule cells.[148,149] A link between excitatory amino acid receptors, phosphoinositide metabolism, and neurotoxicity has been proposed. For example, hypoxia-ischemia causes a specific increase in quisqualate-stimulated phosphoinositide metabolism in hippocampal and striatal slices of the neonatal rat.[150] This enhanced phosphoinositide hydrolysis may contribute to several potentially detrimental steps, including an increased availability of arachidonic acid, an important substrate for free radical formation, and a metabolic derangement due to the increased metabolic cost of phosphoinositide breakdown and resynthesis.[150] On the other hand, it has also been reported that activation of the metabotropic glutamate receptor attenuates the neurotoxicity caused by NMDA in murine cortical cultures.[151] Furthermore, studies with ibotenic acid have shown that stimulation of phosphoinositide metabolism was neither necessary nor sufficient for neurotoxicity in cortical neurons.[152] Subcutaneous treatment of neonatal or adult rats with 2-amino-3-phosphonopropionic acid (AP3), a metabotropic receptor antagonist, caused neuronal degeneration. On the other hand, the metabotropic agonist *trans*-ACPD (1-aminocyclopentene-1,3-dicarboxylic acid) was also found to cause degeneration of neurons in the lateral septal nucleus when injected into the lateral ventricle, and its effect was blocked by AP3.[153] It has been suggested that the glutamate metabotropic receptor harbors neurotoxic potential that can be unleashed either by its suppression or its hyperstimulation.[153] Some other apparently contradictory results may be explained by the activation of different subtypes of excitatory amino acid receptors by different compounds. This has been shown by Nicoletti *et al.*,[154] who reported that activation of the metabotropic glutamate receptors 1 and 5 leads to activation of phosphoinositide hydrolysis as well as neurotoxicity, while other metabotropic glutamate receptors, negatively linked to adenylate cyclase, may exert neuroprotective effects. The involvement of calcium and nitric oxide in the neurotoxicity of glutamate mediated by activation of NMDA receptors has been discussed earlier.

11.06.7.4.4 *Others*

Cyanide induces PKC translocation in rat cerebellum and hippocampus.[155] This action is blocked by NMDA antagonists and may mediate the neurotoxicity of cyanide, since the latter is also antagonized by NMDA receptor antagonists.[156] It was also noted that cyanide-induced convulsions are inhibited by N^G-nitro-L-arginine,[157] which would imply an NMDA-mediated activation of the nitric oxide pathway. An increase in cytosolic PKC activity in rat sciatic nerve following repeated administrations of acrylamide has also been reported;[158] however, the significance of such findings remains obscure.

11.06.8 OTHER FACTORS AFFECTING CELL SIGNALING

In a discussion on the effects of neurotoxicants on cell signaling, it should also be mentioned that alterations of cell signaling systems may occur secondarily to other cellular effects. For example, it is well established that certain neurotoxicants, particularly metals, induce oxidative stress in nervous system cells.[159] The primary products of lipid peroxidation are fatty acid hydroperoxides such as derivatives of oleic, linoleic, and arachidonic acid that stimulate the activity of PKC.[160] Oxidation of critical intracellular thiols has been shown to inhibit carbachol-stimulated phosphoinositide metabolism.[161] Thus, an additional consequence of oxidative stress is an impairment of second-messenger systems.

11.06.9 CONCLUSIONS

Cell signaling is the process by which specific information is transferred from the cell surface to the cytosol and ultimately to the nucleus, leading to changes in gene expression.[162] Since these chains of biochemical and molecular steps control the normal function of each cell, it is apparent that any disruption of these processes will have a significant impact on normal cell physiology. In the last few years, cell signaling has been studied in search of possible targets for neurotoxic chemicals, and a few examples have been reviewed in this chapter. As the exact mechanisms involved in the neurotoxicity and developmental neurotoxicity of most chemicals are still unknown, the examples may provide initial suggestions and leads for further investigations.

The study of signal transduction mechanisms is among the most actively investigated fields of neurobiology in the mid-1990s, and in addition to the pathways reviewed in this chapter, several others (e.g., the tyrosine kinase pathway, the MAP (mitogen-activatd protein) kinases, etc.) are subject to ever-increasing investigations. As progress in basic neurosciences unravels the molecular steps involved in cell signaling and their significance in normal cell function, neurotoxicologists will have novel potential targets for mechanistic studies. Alterations of cell signaling by neurotoxicants may in some cases be causally linked to cell damage, cell death, or functional impairment. In other instances, effects on signal transduction may serve as early indicators of toxicity that could find a role as useful biomarkers.[163] For such investigations to be successful, however, there is a need for cogent, unifying hypotheses that consider the various aspects of cell signaling in the context of their physiological relevance and of the known signs and symptoms of neurotoxicity of the studied compound.

ACKNOWLEDGMENTS

Research by the author was supported by grants from NIEHS, NIAAA, the Alcohol and Drug Abuse Institute, University of Washington, and the Fondazione Clinica del Lavoro, Pavia. Ms. Chris Sievanen provided valuable secretarial assistance.

11.06.10 REFERENCES

1. D. Corda, A. Luini and S. Garattini, 'Selectivity of action can be achieved with compounds acting at second messenger targets.' *Trends Pharmacol. Sci.*, 1990, **11**, 471–473.
2. L. G. Costa, 'The phosphoinositide/protein kinase C system as a potential target for neurotoxicity.' *Pharmacol. Res.*, 1990, **22**, 393–408.
3. L. G. Costa, in 'Neurotoxicology,' eds. H. A. Tilson and C. L. Mitchell, Raven Press, New York, 1992, pp. 101–123.
4. L. G. Costa, in 'Principles of Neurotoxicology,' ed. L. W. Chang, Marcel Dekker, New York, 1994, pp. 475–493.
5. L. G. Costa, 'Signal transduction mechanisms in developmental neurotoxicity: the phosphoinositide pathway.' *Neurotoxicology*, 1994, **15**, 19–27.
6. R. H. Michell, 'Inositol phospholipids and cell surface receptor function.' *Biochim. Biophys. Acta*, 1975, **415**, 81–147.
7. M. J. Berridge and R. J. Irvine, 'Inositol phosphates and cell signalling.' *Nature*, 1989, **341**, 197–205.

8. R. S. Rana and L. E. Hokin, 'Role of phosphoinositides in transmembrane signalling.' *Physiol. Rev.*, 1990, **70**, 115–164.

9. T. F. Martin, 'Receptor regulation of phosphoinositidase C.' *Pharmacol. Ther.*, 1991, **49**, 329–345.

10. D. M. Chuang, 'Neurotransmitter receptors and phosphoinositide turnover.' *Annu. Rev. Pharmacol. Toxicol.*, 1989, **29**, 71–110.

11. C. J. Fowler and G. Tiger, 'Modulation of receptor-mediated inositol phospholipid breakdown in the brain.' *Neurochem. Int.*, 1991, **19**, 171–206.

12. S. K. Fisher, A. M. Heacock and B. W. Agranoff, 'Inositol lipids and signal transduction in the nervous system: an update.' *J. Neurochem.*, 1992, **58**, 18–38.

13. W. W. Y. Lo and J. Hughes, 'Receptor-phosphoinositidase C coupling. Multiple G proteins?' *FEBS Lett.*, 1987, **224**, 1–3.

14. A. Ashkenazi, E. G. Peralta, J. W. Winslow *et al.*, 'Functionally distinct G proteins selectively couple different receptors to PI hydrolysis in the same cell.' *Cell*, 1989, **56**, 487–493.

15. P. W. Majerus, T. M. Connolly, V. S. Bansal *et al.*, 'Inositol phosphates: synthesis and degradation.' *J. Biol. Chem.*, 1988, **263**, 3051–3054.

16. F. S. Menniti, G. St. J. Bird, M. C. Glennon *et al.*, 'Inositol polyphosphates and calcium signalling.' *Mol. Cell Neurosci.*, 1992, **3**, 1–10.

17. S. D. Shukla and S. P. Halenda, 'Phospholipase D in cell signalling and its relationship to phospholipase C.' *Life Sci.*, 1991, **48**, 851–866.

18. V. Henzi and A. B. MacDermott, 'Characteristics and functioning of Ca^{2+} and inositol 1,4,5-triphosphate-releasable stores of Ca^{2+} in neurons.' *Neuroscience*, 1992, **46**, 251–273.

19. H. Rasmussen, P. Barrett, J. Smallwood *et al.*, 'Calcium ion as intracellular messenger and cellular toxin.' *Environ. Health Perspect.*, 1990, **84**, 17–25.

20. S. Orrenius, M. J. Burkitt, G. E. N. Kass *et al.*, 'Calcium ions and oxidative cell injury.' *Ann. Neurol.*, 1992, **32**, Suppl., S33–S42.

21. P. Nicotera, G. Bellomo and S. Orrenius, 'Calcium-mediated mechanisms in chemically induced cell death.' *Annu. Rev. Pharmacol. Toxicol.*, 1992, **32**, 449–470.

22. G. B. Corcoran, L. Fix, D. P. Jones *et al.*, 'Apoptosis: molecular control point in toxicity.' *Toxicol. Appl. Pharmacol.*, 1994, **128**, 169–181.

23. Y. Nishizuka, 'Intracellular signalling by hydrolysis of phospholipids and activation of protein kinase C.' *Science*, 1992, **258**, 607–614.

24. S. Stabel and P. J. Parker, 'Protein kinase C.' *Pharmacol. Ther.*, 1991, **51**, 71–95.

25. U. Kikkawa, Y. Takai, Y. Tanaka *et al.*, 'Protein kinase C as a possible receptor protein of tumor-promoting phorbol esters.' *J. Biol. Chem.*, 1983, **258**, 11442–11445.

26. C. Tanaka and Y. Nishizuka, 'The protein kinase C family for neuronal signalling.' *Annu. Rev. Neurosci.*, 1994, **17**, 551–567.

27. L. V. Dekker, P. N. E. De Graan, D. H. G. Versteeg *et al.*, 'Phosphorylation of B-50 (GAP43) is correlated with neurotransmitter release in rat hippocampal slices.' *J. Neurochem.*, 1989, **52**, 24–30.

28. V. P. Chiarugi, M. Ruggiero and R. Corradetti, 'Oncogenes, protein kinase C, neuronal differentiation and memory.' *Neurochem. Int.*, 1989, **14**, 1–9.

29. M. Whitman and L. Cantley, 'Phosphoinositide metabolism and the control of cell proliferation.' *Biochim. Biophys. Acta*, 1989, **948**, 327–344.

30. N. R. Bhat, 'Role of protein kinase C in glial cell proliferation.' *J. Neurosci. Res.*, 1989, **22**, 20–27.

31. S. Murphy, N. McCabe, C. Morrow *et al.*, 'Phorbol ester stimulates proliferation of astrocytes in primary culture.' *Dev. Brain Res.*, 1987, **428**, 133–135.

32. R. D. Saunders and G. H. DeVries, 'Schwann cell proliferation is accompanied by enhanced inositol phospholipid metabolism.' *J. Neurochem.*, 1988, **50**, 876–882.

33. M. P. Mattson, 'Evidence for the involvement of protein kinase C in neurodegenerative changes in cultured human cortical neurons.' *Exp. Neurol.*, 1991, **112**, 95–103.

34. M. M. Billah and J. C. Anthes, 'The regulation and cellular functions of phosphatidylcholine hyrolysis.' *Biochem. J.*, 1990, **269**, 281–291.

35. S. D. Shukla and S. P. Halenda, Phospholipase D in cell signalling and its relationship to phospholipase C. *Life Sci.*, 1991, **48**, 851–866.

36. W. H. Moolenaar, W. Kruijer, B. C. Tilly *et al.*, 'Growth factor-like action of phosphatidic acid.' *Nature*, 1986, **323**, 171–173.

37. N. T. Thompson, R. W. Bonser and L. G. Garland, 'Receptor-coupled phospholipase D and its inhibition.' *Trends Pharmacol. Sci.*, 1991, **12**, 404–408.

38. Y. Asaoka, U. Kikkawa, K. Sekiguchi *et al.*, 'Activation of brain-specific protein kinase C subspecies in the presence of phosphatidylethanol.' *FEBS Lett.*, 1988, **231**, 221–224.

39. A. B. Mukherjee, L. Miele and N. Pattabiraman, 'Phospholipase A_2 enzymes: regulation and physiological role.' *Biochem. Pharmacol.*, 1994, **48**, 1–10.

40. T. Shimizu and L. S. Wolfe, 'Arachidonic acid cascade and signal transduction.' *J. Neurochem.*, 1990, **55**, 1–15.

41. T. Simmet and B. A. Peskar, 'Lipoxygenase products of polyunsaturated fatty acid metabolism in the central nervous system: biosynthesis and putative functions.' *Pharmacol. Res.*, 1990, **22**, 667–682.

42. M. McKinney and E. Richelson, in 'The Muscarinic Receptors,' ed. J. H. Brown, Humana Press, Clifton, NJ, 1989, pp. 309–340.

43. D. Piomelli and P. Greengard, 'Lipoxygenase metabolites of arachidonic acid in neuronal transmembrane signalling.' *Trends Pharmacol. Sci.*, 1990, **11**, 367–373.

44. D. Piomelli, 'Arachidonic acid in cell signalling.' *Curr. Opin. Cell Biol.*, 1993, **5**, 274–280.

45. G. A. Robison, R. W. Butcher and E. W. Sutherland, 'Cyclic AMP,' Academic Press, New York, 1971, p. 316.

46. S. I. Walaas and P. Greengard, 'Protein phosphorylation and neuronal function.' *Pharmacol. Rev.*, 1991, **43**, 299–349.

47. M. Sheng, G. McFadden and M. E. Greenberg, 'Membrane depolarization and calcium induce c-fos transcription via phosphorylation of transcription factor CREB.' *Neuron*, 1990, **4**, 571–582.

48. T. Kendall-Harden, in 'The Muscarinic Receptors,' ed. J. H. Brown, Humana Press, Clifton, NJ, 1989, pp. 221–258.

49. E. El-Fakahany and E. Richelson, 'Effect of some calcium antagonists on muscarinic receptor-mediated cyclic GMP formation.' *J. Neurochem.*, 1983, **40**, 705–710.

50. S. Moncada, R. M. J. Palmer and E. A. Higgs, 'Nitric oxide: physiology, pathophysiology and pharmacology.' *Pharmacol. Rev.*, 1991, **43**, 109–142.

51. J. Bruhwyler, E. Chleide, J. F. Liegeois *et al.*, 'Nitric oxide: a new messenger in the brain.' *Neurosci. Biobehav. Rev.*, 1993, **17**, 373–384.

52. E. Southam and J. Garthwaite, 'The nitric oxide-cyclic GMP signalling pathway in rat brain.' *Neuropharmacology*, 1993, **32**, 1267–1277.

53. H. H. H. W. Schmidt, S. M. Lohmann and U. Walter, 'The nitric oxide and cGMP signal

transduction system: regulation and mechanism of action.' *Biochim. Biophys. Acta*, 1993, **1178**, 153–175.

54. E. M. Schuman and D. V. Madison, 'Nitric oxide and synaptic function.' *Annu. Rev. Neurosci.*, 1994, **17**, 153–183.

55. D. S. Bredt and S. H. Snyder, 'Nitric oxide: a physiologic messenger molecule.' *Annu. Rev. Biochem.*, 1994, **63**, 175–195.

56. H. Kinouchi, C. J. Epstein, T. Mizui *et al.*, 'Attenuation of focal cerebral ischemic injury in transgenic mice overexpressing Cu Zn superoxide dismutase.' *Proc. Natl. Acad. Sci. USA*, 1991, **88**, 11158–11162.

57. V. L. Dawson, T. M. Dawson, E. D. London *et al.*, 'Nitric oxide mediates glutamate neurotoxicity in primary cortical cultures.' *Proc. Natl. Acad. Sci. USA*, 1991, **88**, 6368–6371.

58. J. Zhang, V. L. Dawson, T. M. Dawson *et al.*, 'Nitric oxide activation of poly (ADP-ribose) synthetase in neurotoxicity.' *Science*, 1994, **263**, 687–689.

59. T. D. Oury, Y. S. Ho, C. A. Piantadosi *et al.*, 'Extracellular superoxide dismutase, nitric oxide, and central nervous system O_2 toxicity.' *Proc. Natl. Acad. Sci. USA*, 1992, **89**, 9715–9719.

60. V. L. Dawson, T. M. Dawson, G. R. Uhl *et al.*, 'Human immunodeficiency virus type I coat protein neurotoxicity mediated by nitric oxide in primary cortical cultures.' *Proc. Natl. Acad. Sci. USA*, 1993, **90**, 3256–3259.

61. B. J. Druker, H. J. Mamon and T. M. Roberts, 'Oncogenes, growth factors and signal transduction.' *N. Engl. J. Med.*, 1989, **321**, 1383–1391.

62. M. B. Abou-Donia, S. E. Patton and D. M. Lapadula, in 'Cellular and Molecular Neurotoxicology,' ed. T. Narahashi, Raven Press, New York, 1984, pp. 265–283.

63. P. E. Neuman and F. Taketa, 'Effects of triethyltin bromide on protein phosphorylation in subcellular fractions from rat and rabbit brain.' *Brain Res.*, 1987, **388**, 83–87.

64. M. Di Luca, M. Cimino, P. N. E. De Graan *et al.*, 'Microencephaly reduces the phosphorylation of the PKC substrate B-50/GAP43 in rat cortex and hippocampus.' *Brain Res.*, 1991, **538**, 95–101.

65. J. P. O'Callaghan, in 'Neurotoxicology,' eds. H. A. Tilson and C. L. Mitchell, Raven Press, New York, 1992, pp. 88–100.

66. M. Chow, C. J. Der and J. E. Buss, 'Structure and biological effects of lipid modifications on proteins.' *Curr. Opin. Cell Biol.*, 1992, **4**, 629–636.

67. P. J. Casey, 'Lipid modifications of G proteins.' *Curr. Opin. Cell Biol.*, 1994, **6**, 219–225.

68. P. J. Casey, 'Protein lipidation in cell signalling.' *Science*, 1995, **268**, 221–225.

69. W. R. Schafer and J. Rine, 'Protein prenylation: genes, enzymes, targets and functions.' *Annu. Rev. Genet.*, 1992, **26**, 209–237.

70. M. Sheng and M. E. Greenberg, 'The regulation and function of c-fos and other immediate early genes in the nervous system.' *Neuron*, 1990, **4**, 477–485.

71. J. I. Morgan and T. Curran, 'Stimulus-transcription coupling in neurons: role of cellular immediate early genes.' *Trends Neurosci.*, 1989, **12**, 459.

72. J. I. Morgan, D. R. Cohen, J. L. Hempstead *et al.*, 'Mapping patterns of c-fos expression in the central nervous system after seizure.' *Science*, 1987, **237**, 192–197.

73. N. H. Zawia and S. C. Bondy, 'Electrically stimulated rapid gene expression in the brain: ornithine decarboxylase and c-fos.' *Brain Res. Mol. Brain Res.*, 1990, **7**, 243–247.

74. M. Vendrell, N. H. Zawia, J. Serratosa *et al.*, 'c-fos and ornithine decarboxylase gene expression in brain as early markers of neurotoxicity.' *Brain Res.*, 1991, **544**, 291–296.

75. D. Minnema, in 'The Vulnerable Brain and Environmental Risks,' eds. R. L. Isaacson and K. F. Jensen, Plenum Press, New York, 1992, vol. 2, pp. 83–109.

76. Y. Inoue, K. Saijoh and K. Sumino, 'Action of mercurials on activity of partially purified soluble protein kinase C from mice brain.' *Pharmacol. Toxicol.*, 1988, **62**, 278–281.

77. H. Katsuyama, K. Saijoh, Y. Inoue *et al.*, '³H-PDBu binding after administration of methylmercury to mice.' *Bull. Environ. Contam. Toxicol.*, 1989, **43**, 886–892.

78. K. Saijoh, T. Fukunaga, H. Katsuyama *et al.*, 'Effects of methylmercury on protein kinase A and protein kinase C in the mouse brain.' *Environ. Res.*, 1993, **63**, 264–273.

79. A. Rossi, L. Manzo, S. Orrenius *et al.*, 'Modifications of cell signalling in the cytotoxicity of metals.' *Pharmacol. Toxicol.*, 1991, **68**, 424–429.

80. T. A. Sarafian, 'Methyl mercury specifically increases intracellular Ca^{2+} and inositol phosphate levels in cultured cerebellar granule neurons.' *J. Neurochem.*, 1993, **61**, 648–657.

81. H. Komulainen and S. C. Bondy, 'Increased free intrasynaptosomal Ca^{2+} by neurotoxic organometals: distinctive mechanisms.' *Toxicol. Appl. Pharmacol.*, 1987, **88**, 77–86.

82. W. D. Atchison and M. F. Hare, 'Mechanisms of methylmercury-induced neurotoxicity.' *FASEB J.*, 1994, **8**, 622–629.

83. T. Sarafian and M. A. Verity, 'Methyl mercury stimulates protein ³²P phospholabeling in cerebellar granule cell culture.' *J. Neurochem.*, 1990, **55**, 913–921.

84. M. A. Verity, T. Sarafian, E. H. K. Pacifici *et al.*, 'Phospholipase A₂ stimulation by methyl mercury in neuron culture.' *J. Neurochem.*, 1994, **62**, 705–714.

85. L. G. Costa, 'Interactions of neurotoxicants with neurotransmitter systems.' *Toxicology*, 1988, **49**, 359–366.

86. J. Markovac and G. W. Goldstein, 'Lead activates protein kinase C in immature rat brain microvessels.' *Toxicol. Appl. Pharmacol.*, 1988, **96**, 14–23.

87. J. Markovac and G. W. Goldstein, 'Picomolar concentrations of lead stimulate brain protein kinase C.' *Nature*, 1988, **334**, 71–73.

88. G. J. Long, J. F. Rosen and F. A. X. Schanne, 'Lead activation of protein kinase C from rat brain. Determination of free calcium, lead and zinc by ¹⁹F NMR.' *J. Biol. Chem.*, 1994, **269**, 834–837.

89. M. E. Blazka, G. J. Harry and M. I. Luster, 'Effect of lead acetate on nitrite production by murine brain endothelial cell cultures.' *Toxicol Appl. Pharmacol.*, 1994, **126**, 191–194.

90. S. M. Candura, A. F. Castoldi, L. Manzo *et al.*, 'Interaction of aluminum ions with phosphoinositide metabolism in rat cerebral cortical membranes.' *Life Sci.*, 1991, **49**, 1245–1252.

91. T. J. Shafer, W. R. Mundy and H. A. Tilson, 'Aluminum decreases muscarinic, adrenergic, and metabotropic receptor-stimulated phosphoinositide hydrolysis in hippocampal and cortical slices from rat brain.' *Brain Res.*, 1993, **629**, 133–140.

92. P. C. Wood, R. J. H. Wojcikiewicz, J. Burgess *et al.*, 'Aluminum inhibits muscarinic agonist-induced inositol 1,4,5-trisphosphate production and calcium mobilization in permeabilized SH-SY5Y human neuroblastoma cells.' *J. Neurochem.*, 1994, **62**, 2219–2223.

93. A. Haug, B. Shi and V. Vitorello, 'Aluminum interaction with phosphoinositide-associated signal transduction.' *Arch. Toxicol.*, 1994, **68**, 1–7.

94. T. J. Shafer, A. C. Nostrandt, H. A. Tilson *et al.*, 'Mechanisms underlying AlCl$_3$ inhibition of agonist-stimulated inositol phosphate accumulation. Role of calcium, G-proteins, phospholipase C and protein kinase C.' *Biochem. Pharmacol.*, 1994, **47**, 1417–1425.

95. J. P. Bressler and G. W. Goldstein, 'Mechanisms of lead neurotoxicity.' *Biochem. Pharmacol.*, 1991, **41**, 479–484.

96. E. Habermann, K. Crowell and P. Janicki, 'Lead and other metals can substitute for Ca^{2+} in calmodulin.' *Arch. Toxicol.*, 1983, **54**, 61–70.

97. K. Murakami, G. Feng and S. G. Chen, 'Inhibition of brain protein kinase C subtypes by lead.' *J. Pharmacol. Exp. Ther.*, 1993, **264**, 757–761.

98. J. Laterra, J. P. Bressler, R. R. Indurti *et al.*, 'Inhibition of astroglia-induced endothelial differentiation by inorganic lead: a role for protein kinase C.' *Proc. Natl. Acad. Sci. USA*, 1992, **89**, 10748–10752.

99. S. J. Farmer and T. R. Guilarte, 'Quantitative autoradiography of ^{3}H-PDBu binding to hippocampal membrane-bound PKC in lead exposed rats.' *Toxicologist*, 1994, **14**, 143.

100. P. J. S. Vig, S. N. Pentyala, C. S. Chetty *et al.*, 'Lead alters inositol polyphosphate receptor activities: protection by ATP.' *Pharmacol. Toxicol.*, 1994, **75**, 17–22.

101. A. K. Singh, 'Age-dependent neurotoxicity in rats chronically exposed to low levels of lead: calcium homeostasis in central neurons.' *Neurotoxicology*, 1993, **14**, 417–427.

102. B. Shi and A. Haug, 'Aluminum interferes with signal transduction in neuroblastoma cells.' *Pharmacol. Toxicol.*, 1992, **71**, 308–313.

103. H. Katsuyama, K. Saijoh, Y. Inoue *et al.*, 'The interaction of aluminum with soluble protein kinase C from mouse brain.' *Arch. Toxicol.*, 1989, **63**, 474–478.

104. M. Cochran, D. C. Elliott, P. Brennan *et al.*, 'Inhibition of protein kinase C activation by low concentrations of aluminum.' *Clin. Chim. Acta*, 1990, **194**, 167–172.

105. G. V. W. Johnson and R. S. Jope, 'Phosphorylation of rat brain cytoskeletal proteins is increased after orally administered aluminum.' *Brain Res.*, 1988, **456**, 95–103.

106. G. V. W. Johnson, 'The effects of aluminum on agonist-induced alterations in cyclic AMP and cyclic GMP concentrations in rat brain regions *in vivo*.' *Toxicology*, 1988, **51**, 299–308.

107. L. A. Speizer, M. J. Watson, J. R. Kanter *et al.*, 'Inhibition of phorbol ester binding and protein kinase C activity by heavy metals.' *J. Biol. Chem.*, 1989, **264**, 5581–5585.

108. D. Beyersmann, C. Block and A. N. Malviya, 'Effects of cadmium on nuclear protein kinase C.' *Environ. Health Perspect.*, 1994, **102**, Suppl. 3, 177–180.

109. M. Bencherif and R. J. Lukas, 'Vanadate amplifies receptor-mediated accumulation of inositol triphosphates and inhibits inositol tris-and tetrakis-phosphatase activities.' *Neurosci. Lett.*, 1992, **134**, 157–160.

110. L. G. Costa, G. Kaylor and S. D. Murphy, 'Carbachol- and norepinephrine-stimulated phosphoinositide metabolism in rat brain: effect of chronic cholinesterase inhibition.' *J. Pharmacol. Exp. Ther.*, 1986, **239**, 32–37.

111. W. R. Mundy, T. R. Ward, V. F. Dulchinos *et al.*, 'Effect of repeated organophosphate administration on carbachol-stimulated phosphoinositide hydrolysis in the rat brain.' *Pharmacol. Biochem. Behav.*, 1993, **45**, 309–314.

112. M. R. Hirvonen, L. Paljarvi, A. Naukkarinin *et al.*, 'Potentiation of malaoxon-induced convulsions by lithium: early neuronal injury, phosphoinositide signalling and calcium.' *Toxicol. Appl. Pharmacol.*, 1990, **104**, 276–289.

113. K. M. Savolainen, O. Muona, S. R. Nelson *et al.*, 'Lithium modifies convulsions and brain phosphoinositide turnover induced by organophosphates.' *Pharmacol. Toxicol.*, 1991, **68**, 346–354.

114. M. P. Honchar, J. W. Olney and W. R. Sherman, 'Systemic cholinergic agents induce seizures and brain damage in lithium-treated rats.' *Science*, 1983, **220**, 323–325.

115. R. S. Jope, K. Kolasa, L. Song *et al.*, 'Seizures selectively impair agonist-stimulated phosphoinositide hydrolysis without affecting protein kinase C activity in rat brain.' *Neurotoxicology*, 1992, **13**, 389–400.

116. M. R. Hirvonen, L. Paljarvi and K. M. Savolainen, 'Sustained effects of pilocarpine-induced convulsions on brain inositol and inositol monophosphate levels and brain morphology in young and old male rats.' *Toxicol. Appl. Pharmacol.*, 1993, **122**, 290–299.

117. F. Gusovsky, S. F. Secunda and J. W. Daly, 'Pyrethroids: involvement of sodium channels in effects on inositol phosphate formation in guinea pig synaptoneurosomes.' *Brain Res.*, 1989, **492**, 72–78.

118. E. Enan and F. Matsumura, 'Stimulation of protein phosphorylation in intact rat brain synaptosomes by a pyrethroid insecticide, deltamethrin.' *Pest. Biochem. Physiol.*, 1991, **39**, 182–195.

119. G. I. Moser and R. C. Smart, 'Hepatic tumor promoting chlorinated hydrocarbons stimulate protein kinase C activity.' *Toxicologist*, 1989, **9**, 125.

120. R. M. Cristofol, E. Rodriguez - Farre and C. Sanfeliu, 'Effects of γ and δ hexachlorocyclohexane isomers on inositol phosphate formation in cerebral cortex and hippocampus slices from developing and adult rats.' *Neurotoxicology*, 1993, **14**, 451–458.

121. S. C. Bondy and L. Halsall, 'Lindane-induced modulation of calcium levels within synaptosomes.' *Neurotoxicology*, 1988, **9**, 645–652.

122. J. B. Hoek and E. Rubin, 'Alcohol and membrane-associated signal transduction.' *Alcohol. Alcohol*, 1990, **25**, 143–156.

123. P. L. Hoffman and B. Tabakoff, 'Ethanol and guanine nucleotide binding proteins: a selective interaction.' *FASEB J.*, 1990, **4**, 2612–2622.

124. D. Mochly-Rosen, F. H. Chang, L. Cheever *et al.*, 'Chronic ethanol causes heterologous desensitization of receptors by reducing α$_s$ messenger RNA.' *Nature*, 1988, **333**, 848–850.

125. J. B. Hoek, A. P. Thomas, T. A. Rooney *et al.*, 'Ethanol and signal transduction in the liver.' *FASEB J.*, 1992, **6**, 2386–2396.

126. L. Gustavsson and C. Alling, 'Formation of phosphatidylethanol in rat brain by phospholipase D.' *Biochem. Biophys. Res. Commun.*, 1987, **142**, 958–963.

127. F. Battaini, R. Del Vesco, S. Govoni *et al.*, 'Chronic alcohol intake modifies phorbol ester binding in selected rat brain areas.' *Alcohol*, 1989, **6**, 169–172.

128. M. Virmani and B. Ahluwalia, 'Biphasic protein kinase C translocation in PC12 cells in response to short-term and long-term ethanol exposure.' *Alcohol, Alcohol*, 1992, **27**, 393–401.

129. S. J. Slater, K. J. A. Cox, J. V. Lombardi *et al.*, 'Inhibition of protein kinase C by alcohols and anesthetics.' *Nature*, 1993, **364**, 82–84.

130. W. Balduini and L. G. Costa, 'Effects of ethanol on muscarinic receptor-stimulated phosphoinositide metabolism during brain development.' *J. Pharmacol. Exp. Ther.*, 1989, **250**, 541–547.

131. W. Balduini, S. M. Candura, L. Manzo *et al.*, 'Time-, concentration-, and age-dependent inhibition of mus-

carinic receptor-stimulated phosphoinositide metabolism by ethanol in the developing rat brain.' *Neurochem. Res.*, 1991, **16**, 1235–1240.

132. W. Balduini, S. D. Murphy and L. G. Costa, 'Developmental changes in muscarinic receptor-stimulated phosphoinositide metabolism in rat brain.' *J. Pharmacol. Exp. Ther.*, 1987, **241**, 421–427.

133. M. Guizzetti, P. Costa, J. Peters *et al.*, 'Acetylcholine as a mitogen: muscarinic receptor-mediated proliferation of rat astrocytes and human astrocytoma cells.' *Eur. J. Pharmacol. Mol. Pharmacol.*, 1996, **297**, 265–273.

134. M. Guizzetti and L. G. Costa, 'Inhibition of muscarinic receptor-stimulated glial cell proliferation by ethanol.' *J. Neurochem.*, 1996, **67**, 2236–2245.

135. L. G. Costa, W. Baldini and F. Renò, 'Muscarinic receptor stimulation of phospholipase D activity in the developing brain.' *Neurosci. Res. Commun.*, 1995, **17**, 169–176.

136. A. C. Howlett, J. M. Qualy and L. L. Khachatrian, 'Involvement of Gi in the inhibition of adenylate cyclase by cannabimimetic drugs.' *Mol. Pharmacol.*, 1986, **29**, 307–313.

137. S. Y. Nah, D. Saya and Z. Vogel, 'Cannabinoids inhibit agonist-stimulated formation of inositol phosphates in rat hippocampal cultures.' *Eur. J. Pharmacol.*, 1993, **246**, 19–24.

138. X. X. Tan and L. G. Costa, 'Inhibition of muscarinic receptor - stimulated phosphoinositide metabolism by cocaine, norcocaine an cocaethylene in rat brain.' *Brain Res. Dev. Brain Res.*, 1994, **79**, 132–135.

139. X. X. Tan and L. G. Costa, 'Long-lasting microencephaly following exposure to cocaine during brain growth spurt in the rat.' *Brain Res. Dev. Brain Res.*, 1995, **84**, 179–184.

140. P. Tandon, G. J. Harry and H. A. Tilson, 'Colchicine-induced alterations in receptor-stimulated phosphoinositide hydrolysis in the rat hippocampus.' *Brain Res.*, 1989, **477**, 308–313.

141. P. Tandon, S. Barone, Jr., E. G. Drust *et al.*, 'Long-term behavioral and neurochemical effects of intradentate administration of colchicine in rats.' *Neurotoxicology*, 1991, **12**, 67–77.

142. F. Nicoletti, J. T. Wroblewski, H. Alho *et al.*, 'Lesions of putative glutamatergic pathways potentiate the increase of inositol phospholipid hydrolysis elicited by excitatory amino acids.' *Brain Res.*, 1987, **436**, 103–112.

143. P. R. S. Kodavanti, D. S. Shin, H. A. Tilson *et al.*, 'Comparative effects of two polychlorinated biphenyl congeners on Ca^{2+} homeostasis in rat cerebellar granule cells.' *Toxicol. Appl. Pharmacol.*, 1993, **123**, 97–106.

144. P. R. S. Kodavanti, T. J. Shafer, T. R. Ward *et al.*, 'Differential effects of polychlorinated biphenyl congeners on phosphoinositide hydrolysis and protein kinase C translocation in rat cerebellar granule cells.' *Brain Res.*, 1994, **662**, 75–82.

145. P. R. S. Kodavanti, T. R. Ward, J. D. McKinney *et al.*, 'Increased ^{3}H-phorbol ester binding in rat cerebellar granule cells by polychlorinated biphenyl mixtures and congeners: structure-activity relationships.' *Toxicol. Appl. Pharmacol.*, 1995, **130**, 140–148.

146. W. Balduini, S. M. Candura and L. G. Costa, 'Regional development of carbachol-, glutamate-, norepinephrine-, and serotonin-stimulated phosphoinositide metabolism in rat brain.' *Brain Res. Dev. Brain Res.*, 1991, **62**, 115–120.

147. F. S. Silverstein, R. Chen and M. V. Johnston, 'The glutamate analogue quisqualic acid is neurotoxic in striatum and hippocampus of immature rat brain.' *Neurosci. Lett.*, 1986, **71**, 13–18.

148. H. Manev, M. Favaron, M. Bertolino *et al.*, in 'Neurotoxicity of Excitatory Amino Acids,' ed. A. Guidotti, Raven Press, New York, 1990, pp. 63–78.

149. D. W. Choi, in 'Neurotoxicity of Excitatory Amino Acids,' ed. A. Guidotti, Raven Press, New York, 1990, pp. 235–242.

150. C. K. Chen, F. S. Silverstein, S. K. Fisher *et al.*, 'Perinatal hypoxic-ischemic brain injury enhances quisqualic acid-stimulated phosphoinositide turnover.' *J. Neurochem.*, 1988, **51**, 353–359.

151. J. Y. Koh, E. Palmer and C. W. Cotman, 'Activation of the metabotropic glutamate receptor attenuates *N*-methyl-D-aspartate neurotoxicity in cortical cultures.' *Proc. Natl. Acad. Sci. USA*, 1991, **88**, 9431–9435.

152. W. C. Zinkand, W. C. Moore, C. Thompson *et al.*, 'Ibotenic acid mediates neurotoxicity and phosphoinositide hydrolysis by independent receptor mechanisms.' *Mol. Chem. Neuropathol.*, 1992, **16**, 1–10.

153. M. T. Price, C. Ikonomidou, J. Labruyere *et al.*, 'Neurotoxicity linked to the glutamate metabotropic receptor.' *Soc. Neurosci. Abst.*, 1992, **18**, 93.

154. F. Nicoletti, G. Aleppo, V. Bruno *et al.*, 'Metabotropic glutamate receptors: neurotoxic or neuroprotective?' *Toxicol. Lett.*, 1992, Suppl. 1, 79–80.

155. A. Rathinavelu, P. Sun, G. Pavlakovic *et al.*, 'Cyanide induces protein kinase C translocation: blockade by NMDA antagonists.' *J. Biochem. Toxicol.*, 1994, **9**, 235–240.

156. M. N. Patel, G. K. Yim and G. E. Isom, 'Blockade of *N*-methyl-D-aspartate receptors prevents cyanide-induced neuronal injury in primary hippocampal cultures.' *Toxicol. Appl. Pharmacol.*, 1992, **115**, 124–129.

157. H. Yamamoto, 'Protective effect of N^G-nitro-L-arginine (N^5[imino(nitroamino) methyl]-L-ornithine) against cyanide-induced convulsions in mice.' *Toxicology*, 1992, **71**, 277–283.

158. E. J. Lehning, R. M. LoPachin, J. Mathew *et al.*, 'Changes in Na-K-ATPase and protein kinase C activities in peripheral nerve of acrylamide-treated rats.' *J. Toxicol. Environ. Health*, 1994, **42**, 331–342.

159. C. P. LeBel, S. F. Ali, M. McKee *et al.*, 'Organometal-induced increases in oxygen reactive species: the potential of $2',7'$-dichlorofluorescin diacetate as an index of neurotoxic damage.' *Toxicol. Appl. Pharmacol.*, 1990, **104**, 17–24.

160. C. A. O'Brian, N. E. Ward, I. B. Weinstein *et al.*, 'Activation of rat brain protein kinase C by lipid oxidation products.' *Biochem. Biophys. Res. Commun.*, 1988, **155**, 1374–1380.

161. W. Balduini, F. Reno, F. Cattabeni *et al.*, 'Oxidative transition of intracellular protein thiols modulates muscarinic receptor-stimulated phosphoinositide breakdown in rat cortical slices.' *Neurosci. Res. Commun.*, 1994, **15**, 125–132.

162. D. R. Edwards, 'Cell signalling and the control of gene transcription.' *Trends Pharmacol. Sci.*, 1994, **15**, 239–244.

163. A. F. Castoldi, T. Coccini, A. Rossi *et al.*, 'Biomarkers in environmental medicine: alterations of cell signaling as early indicators of neurotoxicity.' *Funct. Neurol.*, 1994, **9**, 101–109.

11.07
Cell Adhesion Molecules as Targets of Neurotoxicants

KENNETH R. REUHL

Rutgers University, Piscataway, NJ, USA

and

GERALD B. GRUNWALD

Jefferson Medical College, Philadelphia, PA, USA

11.07.1 INTRODUCTION

11.07.1.1 Identification and Classification of Cell Adhesion Molecule Families

A central objective of neurotoxicological research is the elucidation of the cellular and molecular mechanisms by which neurotoxicants affect the embryonic development and mature function of the nervous system. One approach towards this objective is the identification of key molecules which regulate neuronal development and whose structural and functional characteristics suggest that they may be targets of neurotoxicologic insult. Among such candidates are members of the cadherin and immunoglobulin (IgCAM) cell adhesion molecule families, which play critical roles in the development and maintenance of central and peripheral nervous system organization. The characterization of these cell adhesion molecules culminated from more than a century of investigation into the nature of cell interactions.[1] This chapter reviews our current understanding of the structure and function of these major cell adhesion molecule families and examines particularly the direct and indirect interactions of these molecules with various classes of neurotoxicants and the possible consequences of such interaction.

11.07.1.2 Functions of Cell Adhesion Molecules in Development and Disease

Since the discovery of the cadherin and IgCAM cell adhesion molecules, a variety of immunological and molecular genetic probes have been generated for analysis of their expression and function. Descriptive studies of the spatiotemporal patterns of cell adhesion molecule expression, in combination with functional perturbation experiments, have elucidated the critical roles these molecules play in the development of a wide variety of tissues in both human and experimental animal models. Changes in the expression and function of members of both adhesion molecule families regulate the intricate cellular rearrangements associated with the earliest stages of embryogenesis, continuing through gastrulation and neurulation, and drive the more specialized histogenetic processes of highly differentiated tissues as well. Within the developing nervous system, these two families of cell adhesion molecules mediate processes as diverse as neuronal, glial, and growth cone migration, neural–glial cell adhesion, axon elongation, neurite fasciculation, and cell–cell signaling and synaptogenesis. In addition to their roles in controlling the intricate morphogenetic movements associated with embryonic development, cell adhesion molecules continue to be expressed in adult tissues. Indeed, a growing number of diseases are known to be associated with altered expression and function of cell adhesion molecules, and these molecules are targets of investigation for both assessment of diagnostic utility and as therapeutic interventions. The critical roles played by cell adhesion molecules at key stages in the developing nervous system predicts that their perturbation by neurotoxicants would have profound effects on histogenesis of neural tissues.

11.07.1.3 Cell Adhesion Molecule– Cytoskeletal Interactions

In addition to functioning as cell surface ligands binding cells to other cells and to extracellular matrices, cell adhesion molecules are transmembrane glycoproteins that serve as structural and functional links between the extracellular and intracellular environments. This linkage, often direct, is mediated through the interaction of the cytoplasmic domains of cell adhesion molecules with proteins of the cytoskeleton. While cytoskeletal linkage is a fundamental property of many members of the cadherin family, and is critical to many of their functions, this is less so for the IgCAMs. Certain proteins of the cytoskeleton interact with members of multiple adhesion molecule families; others preferentially associate with members of a particular cell adhesion molecule type. The functional linkages between cell adhesion molecules and their cytoskeletal partners are bi-directional, insofar as extracellular ligation of cell adhesion molecules can effect changes intracellularly, and perturbation of the cytoskeletal interactions of adhesion molecules likewise affects cell adhesion molecule binding activity. Thus, in considering potential mechanisms of neurotoxicant action via perturbations of neural cell adhesion molecule function, both intracellular and extracellular sites of action need to be considered. The cytoskeleton as a locus of neurotoxicant action, and the effects of such perturbations on cell adhesion molecule function and nerve cell biology, are subjects considered below.

11.07.1.4 Adhesion Receptors and Signal Transduction

In addition to their direct integration of extracellular and intracellular activities by means of cytoskeletal linkages, cell adhesion molecules of the cadherin and IgCAM families provide indirect linkage through second messenger-mediated signal transduction pathways. This pathway too has been shown to be bidirectional, insofar as changes in cell adhesion molecule activity have been demonstrated to cause changes in cytoplasmic second messenger levels, and activation or inhibition of signal transduction pathways in turn can affect cell adhesion molecule expression and function. This raises the possibility that the actions of certain neurotoxicants may result from their effects on cell adhesion molecules as transmitted, or amplified, through second messenger pathways, or conversely, such effects may result from perturbation of signaling pathway control of cell adhesion molecule activity.

11.07.2 CALCIUM-DEPENDENT CELL ADHESION MOLECULES: THE CADHERIN GENE SUPERFAMILY

11.07.2.1 Diversity of Structure and Function Among Cadherins

Cadherins are cell surface integral membrane glycoproteins which mediate calcium-dependent intercellular adhesion. A number of reviews are available which provide considerable detail and different perspectives on the biology of cadherins.[1–15] The cadherins comprise a large and diverse gene superfamily whose ranks continue to grow with the identification of new members (see Table 1). Cadherins have been identified in all vertebrate classes, as well as among invertebrates, and a significant number of different cadherins are expressed in the nervous system.

The term cadherin was coined in 1984 by Takeichi and colleagues to refer to the emerging group of related cell adhesion molecules which were distinguished by their calcium-dependence

Table 1 Cadherins expressed in neural tissues.

Cadherin subtype	Ref.
N-Cadherin	16,17
R-Cadherin	18
B-Cadherin	19
T-Cadherin	20
Cadherins 5–11	21
E-Cadherin	22
Protocadherins 42, 43	23
C-Ret	24

for proper function.[25] It had been recognized in the late nineteenth century that the integrity of solid tissues was compromised if they became depleted of calcium, and the early conceptual and experimental history of cell adhesion research connecting these early observations to the ultimate identification of specific cell adhesion molecules has been reviewed.[1] The demonstration that the morphogenetic movements of early embryonic development could be attributed to cell-autonomous properties of mutual recognition and affinity arose from the experiments of Holtfreter, which clearly showed that dissociated cells were capable of recombining and sorting into tissue-like aggregates.[26,27] Adhesive properties were subsequently demonstrated among cells derived from more highly differentiated tissues, including those of the nervous system.[28] These latter experiments utilized the proteolytic enzyme trypsin to dissociate tissues, a process found to be more efficient when combined with the removal of divalent cations. When embryonic neural or other tissues were digested with trypsin without removing calcium, the cells were found to retain a significant degree of mutual adhesiveness.[29] However, at that time, the molecules mediating these calcium-dependent adhesive interactions remained to be identified.

The identification of the first cadherins was accomplished through a combination of biochemical and immunological methods. It was demonstrated that digestion of tissues with trypsin in the presence of calcium resulted in the retention of residual adhesiveness because such digestion failed to efficiently remove certain proteins from the cell surface.[30] It was these proteins which were later identified as the cadherins. Interestingly, digestion of cells or tissues with low levels of trypsin, even without calcium present, results in production of cells that retain weak adhesive interactions that occur in a calcium-independent fashion. These calcium-independent adhesions are now known to be mediated in part by members of the immunoglobulin gene superfamily. Cadherins bind calcium, but assume an altered configuration in its absence which renders them more susceptible to proteolytic digestion and incapable of full adhesive function. When tissues are digested with trypsin in the presence of calcium, any cadherins which are present are retained at the cell surface. The intimate interaction of calcium with the cadherins provides a mechanism for perturbation of cadherin function by agents which affect calcium metabolism.

While Takeichi's original studies were done with fibroblastic cells, these studies were soon extended to embryonic neural cells. Identification of neural cell-surface proteins exhibiting

calcium-dependent resistance to tryptic digestion, as well as their further identification
with function blocking antibodies, led to the
identification of the first neural cadherin now
known as N-cadherin.[16,17,31,32] While similar
approaches originally led to the identification of
the three original E- (epithelial), P- (placental),
and N- (neural) cadherins, molecular genetic
analyses made possible by the cloning of these
three original cadherins has since led to the
recognition of a large number of structurally
diverse members of the cadherin family. Many
of the distinct cadherins are expressed in the
developing and mature nervous system, from
which N-cadherin was first cloned.[34] Other
cadherins isolated by similar molecular genetic
methods include R- (retinal), B- (brain), and
T- (truncated) cadherins, as well as more
distantly related proteins such as the c-ret proto-
oncogene, and a number of additional cadherins and so-called protocadherins.[18–21,23,35,36]

The cloning and structural analysis of the
various members of the cadherin gene family
has led to the distinction of two major groups,
the classical and nonclassical cadherins. Included among the classical cadherins expressed
in nervous tissues are the N-, R-, and B-
cadherins. These molecules have considerable
structural homology and are all glycoproteins
of about 130 kDa in size, each possessing a
large extracellular domain, a single transmembrane domain, and a short intracellular domain
(Figure 1). The extracellular domain consists of
five homologous subdomains referred to as
EC1–5. EC1 contains critical amino acid sequences for recognition and adhesion, near the
amino terminus of the protein. The extracellular domain also contains conserved regions
which are the binding sites for calcium. Towards the carboxyl terminus within the cytoplasmic domain, regions of high homology
among the cadherins have been identified which

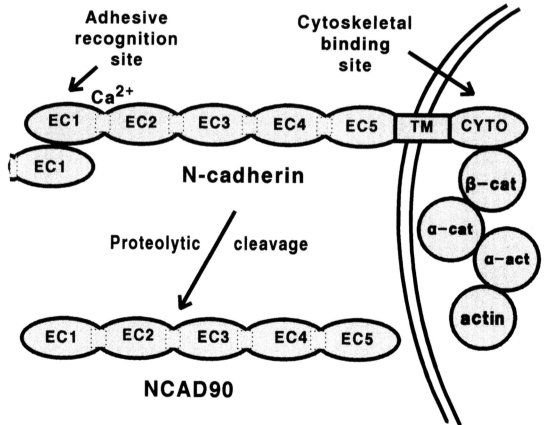

Figure 1   Structure of N-cadherin. The N-cadherin protein is 130 kDa in size and is composed of extracellular, transmembrane (TM) and cytoplasmic (CYTO) domains. The extracellular domain is itself composed of
five repeating subdomains referred to as EC1–EC5. EC1 contains the adhesive recognition and binding
region, and along with EC2 contains critical sites for calcium binding. Adhesive interactions likely occur
through binding of EC1 domains between two cadherins on opposite cells. N-cadherin is processed by
metalloproteolytic cleavage at the cell surface to yield a soluble 90 kDa N-terminal fragment called
NCAD90. The C-terminal domain contains binding sites for cytoskeletal proteins, and binds directly to β-
catenin, γ-catenin, (plakoglobin), or p120[cas]. This in turn can complex with α-catenin, α-actinin, and actin.
The detailed stoichiometry and order of interactions is still under investigation.

are critical for interaction with the cytoskeleton. Cadherin structural elements are conserved across a wide variety of species, and cadherins are subject to a variety of posttranslational modifications, including glycosylation, phosphorylation, and sulfation which has been described for N-cadherin.[37,38]

X-ray crystallographic data have provided additional detailed insights into cadherin structure and evidence for a zipper-like mode of intermolecular interactions by which the cadherins combine to form their characteristic homophilic adhesions.[39] While basic structural features are shared by all the classical cadherins, interesting variations on the cadherin theme are exhibited by nonclassical cadherins. For example, T-cadherin is truncated at the extracellular juxtamembrane region, thus lacking transmembrane and cytoplasmic domains, and is linked to the cytoskeleton by a glycosylphosphatidylinositol linkage. C-ret possesses a cadherin-like extracellular domain, but has an intracellular domain, encoding a tyrosine kinase. Many of the additional cadherins that have been identified remain to be characterized in detail and are still only recognized by fragmentary cloning sequences. These structural variations indicate cadherins play a variety of roles in the developing and mature nervous system, with the common features of mediating calcium-dependent adhesive and signaling between cells.

11.07.2.2 Cadherin Expression and its Regulation

The availability of specific antibody probes for various members of the cadherin family has permitted the determination of patterns of cadherin expression in a wide variety of tissues and species, both during development and in adult organisms. These studies have led to the realization that while many cadherins bear the names of the tissues in which they were first identified, these names are misnomers in the sense that a given cadherin may be expressed in multiple tissues, and a given tissue can express multiple cadherins. The diversity and dynamics of cadherin expression is particularly striking in the nervous system. The analysis of cadherin expression by polymerase chain reaction (PCR) amplification of brain tissue resulted in the identification of eight novel cadherin-like sequences in addition to three previously identified cadherins.[21] Of these, N-cadherin remains the best characterized neural cadherin in terms of its detailed expression patterns. In general, three periods of N-cadherin expression in the developing central nervous system (CNS) can

be described.[17,33,40–42] The first period is that of neural induction and neurulation, when E-cadherin, originally expressed in the embryonic ectoderm, is replaced by N-cadherin in the neural plate cells. This pattern of expression suggests a role for cadherin switching and differential expression in tissue segregation and sorting, such as occurs during neurulation, as later born out by functional studies. The second period of expression is characterized by a high level of N-cadherin expression throughout the developing nervous system, suggesting a central role in its histogenesis, also born out by experimental analysis. The third period of expression of N-cadherin in the developing CNS is characterized by a general downregulation of expression, but with retention at specific regions of the nervous system characterized by high concentrations of cell junctions, such as the ependymal lining of the ventricles, the outer limiting region of the retina, and the choroid plexus. Examples of cadherin expression in the nervous system are illustrated in Figure 2.

Other cadherins are also found in the developing CNS and their patterns of expression provide an interesting contrast with N-cadherin. Interestingly, E-cadherin, which disappears from the ectoderm during neurulation, reappears in a restricted fashion in discrete regions of the developing CNS.[22] Comparative studies of N- and R-cadherin have indicated a complementary pattern of expression, both temporally and spatially.[43] These patterns of expression suggest distinct roles for different cadherins in formation of functionally distinct neural pathways, such as motor vs. sensory, and in the formation of discrete CNS nuclear groups. The dynamics of cadherin expression are also particularly striking during development of the neural crest, where N- and T-cadherin also show differential patterns of expression, suggesting a role in modeling of neural crest-derived tissues.[40,44] The expression of multiple cadherins in the CNS is not simply a reflection of the variety of cell types found there, since even a single neural-tube derived cell type, such as the retinal pigment epithelium, can express multiple cadherins simultaneously.[45–47] This suggests that each cadherin likely has unique functional roles to play in the CNS which will require further comparative studies for their elucidation.

Expression of cadherins in the developing nervous system is regulated by both genetic and epigenetic mechanisms.[48] This was demonstrated in a study of retinal development in which N-cadherin expression was shown to be correlated with the level of mRNA expression. In addition, expression in the retina was affected by insulin which downregulated the

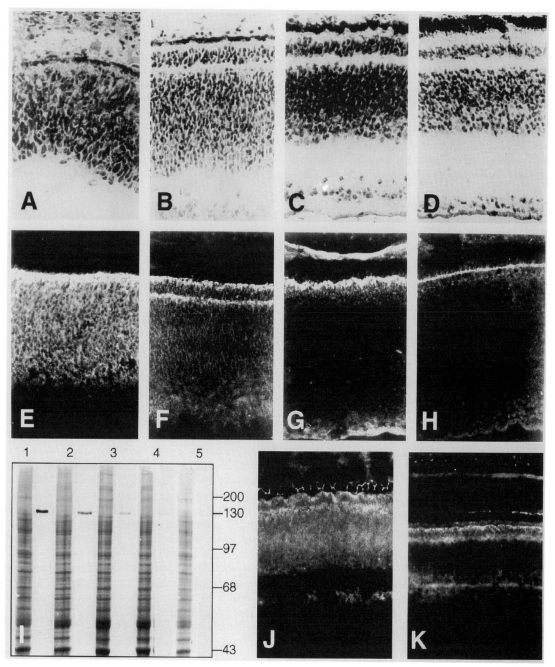

Figure 2 Cadherin expression during neural development. The developing chick neural retina provides an example of cadherin dynamics during neural development. Shown here are stages of chick retina histogenesis at 7 d (A, E), 10 d (B, F, J), 14 d (C, G), and 21 d (D, H, K) *in ovo*. Hematoxylin and eosin stained sections are shown in panels (A–D), while immunohistochemically stained tissue sections are shown in panels (E–H) for N-cadherin and panels (J, K) for R-cadherin. N- and R-Cadherin have complementary expression patterns. In the neural retina, N-cadherin is expressed at high levels across the retina early in development, but in the mature retina is restricted to the outer limiting membrane (compare panels E and H). The biochemical expression pattern of N-cadherin is illustrated in panel I, which shows paired total protein and anti-N-cadherin immunoblot patterns for developing retinas of (i) 7 d, (ii) 10 d, (iii) 14 d, (iv) 17 d, and (v) 21 d development. N-Cadherin is expressed at only very low levels in the retinal pigment epithelium. In contrast, R-cadherin is expressed in the immature retina at very high levels in the retinal pigment epithelium, but at low levels in the neural retina. Later in development, R-cadherin is expressed strongly in a laminar fashion in the neural retina (compare panels J and K). Antibodies to N- and R-cadherin were generously provided by Drs. M. Takeichi and C Redies (panels A–I are reproduced by permission of Academic Press from *Dev. Biol.*, 1989, **135**, 158–171).

level of N-cadherin. Interestingly, cytokines affect expression of other cadherins as well, as was shown for Wnt-1 and hepatocyte growth factor for E-cadherin and P-cadherin.[49-50] The role of cytokines is further complicated by the observations that cadherins interact with cytokine receptors such as those for EGF and FGF.[51,52] Downregulation of cadherins is associated with migration of cells during normal development, as with the neural crest, but has also been associated with cancer progression.[53] While this association has been amply documented for E-cadherin and carcinomas, preliminary evidence suggests this may also be the case for neural tumors and N-cadherin.[54]

11.07.2.3 Cadherin Function in Neuronal Cell Adhesion and Signaling

The fundamental, but not exclusive, function of cadherins expressed among neurons is to mediate cell–cell adhesive interactions. This has been amply demonstrated utilizing function-blocking polyclonal and monoclonal antibodies such as those directed against N-cadherin which interfere with the adhesive interactions of neurons *in vitro*.[16,17] In addition to mediating adhesive interactions between neurons, cadherins also mediate adhesion between neurons and glial cells.[55] Given these observations, perturbation of cadherin expression would

thus be expected to lead to a disruption of neural histogenesis. This was demonstrated for N-cadherin in the developing retina where perturbation of N-cadherin led to a disruption of the normal, layered histological appearance of this tissue[41] (see Figure 3). Inhibition of E-cadherin has a similar effect of disrupting organization of neural tissues, although on a more anatomically local level, given the restricted expression of E-cadherin in the brain.[22] Similar adhesion-blocking antibodies have also indicated roles in cell–cell adhesion for the R-, B-, and T-cadherins.

A critical phase of neural development is the elaboration, directed outgrowth, and synapse formation among axons and dendrites. A number of cell adhesion molecules, including the cadherins, have been demonstrated to play roles in guidance of neurite growth during neural development, including the cadherins.[56,57] Again, because of the availability of reagents for analysis, this has been most clearly demonstrated for N-cadherin. Cells transfected to express N-cadherin on their surface provide a potent substrate for neurite growth, while their control normal counterparts do not.[58] That this effect of N-cadherin on stimulation of neurite outgrowth is a direct effect was demonstrated using purified, intact N-cadherin protein or N-cadherin protein fragments as a substrate, which also proved a potent stimulus for neurite growth[59,60] (see Figure 4). Cells respond uniquely to N-cadherin compared to

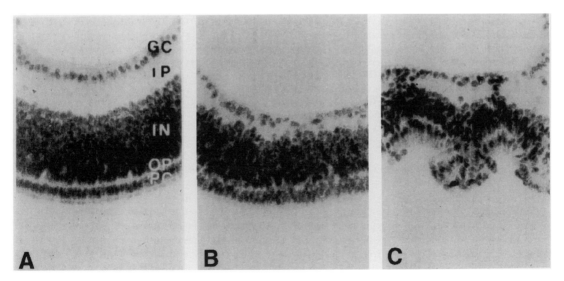

Figure 3 Anti-N-cadherin antibodies perturb neural histogenesis. Intact fragments of chick embryonic retinal tissues were explanted at day 6.5 of development and incubated for 4 d (A) without antibodies, (B) with anti-N-cadherin monoclonal antibody NCAD-2 at $100\,\mu g\,mL^{-1}$, and (C) with anti-N-cadherin polyclonal antibody Fab fragments at $2\,mg\,mL^{-1}$. While both types of antibodies perturb retinal histogenesis, the monoclonal antibodies have a slight effect on layering of cells, while the polyclonal Fab fragments cause a profound disturbance of retinal cytoarchitecture. GC, ganglion cells; IP, inner plexiform layer; IN, inner nuclear layer; OP, outer plexiform layer; PC, photoreceptor cell layer (after Matsunaga *et al.*[41]).

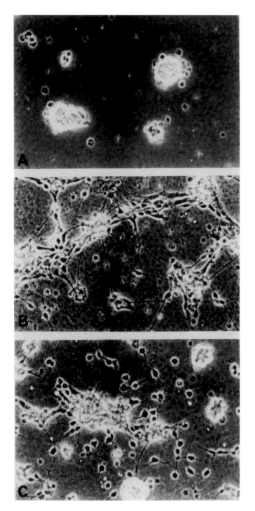

Figure 4 N-Cadherin promotes neurite growth among neurons cultured *in vitro*. Chick 9-day embryo retinal cells were cultured on substrates of various purified proteins. (A) On albumin, the retinal cells bind to each other rather than the dish, and form small floating clusters. (B) On laminin, the retinal cells adhere to the dish, and tend to form clusters from which long, straight neurites extend between clusters. (C) On NCAD90, many individual retinal cells adhere to the dish and extended numerous, branched and curving neurites (reproduced by permission of Wiley-Liss from *J. Neurosci. Res.*, 1993, **36**, 33–45).

other substrates, forming curving, highly branched neurites on N-cadherin, whereas on laminin they form more frequent straight, unbranched, fasciculated neurites. Interestingly, there may be selectivity in the manner in which different types of neurons respond to N-cadherin, since different responses in the extent of neurite outgrowth have been observed in different classes of retinal cells.[61]

The cytoplasmic domains of cadherins are highly conserved and provide a linkage site for interaction with the actin cytoskeleton, with

which the cadherins colocalize in cells. Indeed, interaction of cadherins with the cytoskeleton is critically important in their maximum function for formation of adhesive bonds.[10,62] Cadherins were initially shown to interact with a group of three cytoskeletal proteins termed α-, β-, and γ-catenins.[63] These proteins are related to vinculin, armadillo, and plakoglobin, respectively. In addition, the cytoskeletal protein α-actinin has been identified as the possible linker protein between the cadherin–catenin complex, and the more distal elements of the actin cytoskeleton.[64] Another protein has been added to the catenin family, called p120^{cas}, which is an src kinase substrate.[65] While the stoichiometry and ordering of interactions of all these proteins is still being established, dynamic changes in the components of the cadherin–catenin complex are likely to be important in overall expression and function of the cadherins.[66] Interestingly, a further function of the catenins has been suggested by the demonstration of signaling functions for β-catenin involved in embryonic pattern formation.[67] The cytoskeletal proteins which interact with N-cadherin are illustrated in Figure 1.

The requirement for interaction with cytoskeleton for full function of cadherins has been exploited in the analysis of their function utilizing dominant-negative mutations in the form of truncated proteins lacking either the extracellular adhesive recognition domain or the intracellular cytoplasmic cytoskeletal-binding domain.[68–71] The former type of mutant lacks an adhesive domain, and seems to inhibit binding by providing nonfunctional sites for catenin binding and could potentially inhibit all cadherins. The second type of mutation is more specific in that it expresses the cadherin-specific binding domain, but lacks the cytoskeletal linking segment, thus preventing full function of the engaged cadherin. These studies have been carried out in cell lines as well as in early *Xenopus* embryos, the latter studies clearly demonstrating disruption of normal epithelial organization and subsequent misregulation of morphogenetic changes during gastrulation and neurulation.

Gene knockout experiments have been reported for N-cadherin, but the resulting null mutant mice generated have heart defects which result in early embryonic lethality.[72] Although fairly normal through 8 d of development, these mice are thus not suited to a detailed analysis of N-cadherin function in the later steps of nervous system development. To circumvent this problem, preliminary studies using the dominant-negative approach coupled with liposome-mediated transfection of limited numbers of cells have been carried out in more highly differentiated nervous tissues using the retina as a model system. In these experiments perturbation of cadherin function was found to disturb patterns of neurite growth.[73,74] An interesting study, pointing to another approach for future experiments avoiding problems inherent in null knockout mutations, analyzed E-cadherin function in intestinal development, taking advantage of a tissue-specific promoter to drive localized disruption of E-cadherin function.[75] Such studies using neural tissue-specific promoters are currently underway.

Cadherins are the major transmembrane structural component of adherens junctions, which in addition to their obvious structural function, have also been implicated in more dynamic aspects of cell signaling.[76] Indeed, the response of cells to cadherin stimulation, such as neurite outgrowth, has been suggested to result not so much from direct adhesive changes *per se*, but rather from second messenger stimulation activated by cadherin and other cell adhesion molecules. Evidence for this has been provided in experiments where some of these actions were shown to occur via direct activation of the second messenger pathways or alternatively were inhibited by blockade of calcium-channel or G-protein action.[77,78] A direct role for cadherins in mediating calcium signaling in neurons has been revealed by experiments where purified NCAD90 was used as a stimulus for cultured neurons, which responded using elevated intracellular calcium levels in growth cones and cell bodies[79] (see Figure 5). These observations provide insights into functions of cadherin beyond directing cell–cell adhesion in neural development. They also suggest that disruption of cadherin function can occur through several mechanisms besides direct actions on adhesion; this provides an additional avenue for perturbation by neurotoxic agents, such as, lead, which may work via calcium antagonism.

Metalloproteases have been identified as a factor in the invasiveness of cancer cells, with a clear role in modification of extracellular matrices upon and through which metastatic cells migrate.[80] Evidence indicates that cadherin-mediated cell–cell interactions in the developing nervous system may also be regulated by proteolysis.[48,60,81] These studies showed that in the developing retina, the downregulation of N-cadherin expression associated with neural histogenesis was due in part to the turnover of N-cadherin at the neural cell surface to yield a soluble 90 kDa N-terminal fragment of N-cadherin, called NCAD90. Production of NCAD90 appears to be due to cleavage of intact 130 kDa N-cadherin by a cell-surface associated metalloprotease. In addition to providing a mechanisms

for localized regulation of N-cadherin expression, these studies also showed that NCAD90 retains biological activity and mediates adhesion and neurite growth when presented as a bound substrate. The function of a zinc-dependent metalloprotease in regulating cadherin expression is a potential mechanism of developmental neurotoxicity.

The regulation of N-cadherin by proteolysis has been linked to earlier observations that N-cadherin is a phosphoprotein.[38] Further studies have shown that N-cadherin is the target of several classes of kinases and phosphatases, and that the tyrosine phosphorylation state of N-cadherin is related to its rate of turnover into NCAD90.[71,82,83] These studies also suggested a mechanism for the earlier observation that insulin downregulates N-cadherin, since the insulin/IGF-1 receptors are themselves tyrosine kinases. Interestingly, it has been demonstrated that tyrosine phosphatase (rPTPl) has a direct linkage with the cadherins,[84] and that expression of rPTPl is itself regulated by cell–cell interactions.[85] Kinase signaling pathways are another potential locus of neurotoxicant action, and both positive and negative kinase-mediated effects on cadherin function can be anticipated, given the bifunctional and complex signaling pathways controlled by phosphorylation.[86]

11.07.2.4   Cadherins as Potential Molecular Targets of Neurotoxicants

The structure of cadherins, the mechanisms of regulation of their expression and function, and their interactions with other proteins provide a variety of potential direct and indirect targets for neurotoxicant action (see Figure 6). Since the conformational state of the cadherins is an important determinant of their functional state, direct toxicant effects on cadherin function are most likely to be effected through modification of cadherin structure. Given the critical role which has been established for bound calcium ions in determination of cadherin structure, displacement of calcium ions from their binding sites is potentially a major mechanism by which neurotoxicants could affect neural cadherins. Among the known neurotoxicants which have the potential to affect cadherins in this fashion are the heavy metals, such as inorganic lead. Lead is an established neurotoxicant with the dual ability to affect the nervous system both during its development as well as in its mature functional state.[87–89] Many reviews have considered both the neurotoxicologic consequences of lead

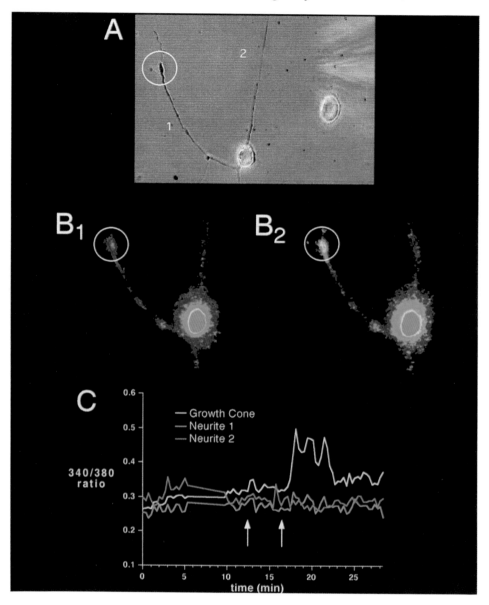

Figure 5 N-cadherin stimulates calcium signaling in cultured neurons and their growth cones. Ciliary ganglion neurons have been loaded with the calcium-sensitive dye fura2 prior to exposure to NCAD90 protein. In panel A, a single neuron with several neurites (1 and 2) and a prominent growth cone (circle) is evident. To the right of the neuron, out of the plane of focus, is the tip of the micropipette containing NCAD90. In panel B, the same neuron is shown before (B$_1$) and after (B$_2$) application of NCAD90. The pseudocolor images indicate the increase of intracellular calcium in the growth cone (circles). In panel C, fluorescence measurements in the growth cone and two neurites of the neuron in panels A and B is indicated, demonstrating a response in the growth cone, but not the neurites, following NCAD90 application (arrows). The circles in panels A and B are 25 µm in diameter (reproduced by permission of Rockefeller University Press from *J. Cell Biol.*, 1994, **27**, 1461–1475).

intoxication and the potential mechanisms by which neurotoxicity is evoked.[90–92] However, the specific mechanism(s) by which this occurs remains to be elucidated. Common to theories regarding mechanisms of lead toxicity is interference with calcium metabolism, especially competition for calcium among calcium-binding metalloproteins.[93,94]

Besides direct displacement of calcium from cadherin binding sites, several less direct pathways for perturbation of cadherin function by heavy metals are also possible. Lead has been shown to affect calcium channel function.[95,96] Cadherin interactions stimulate calcium flux changes in cells, and inhibition of calcium channel activity affects downstream responses

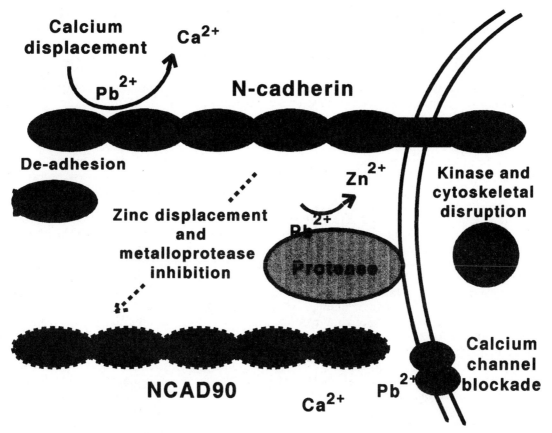

Figure 6 Hypothetical routes of lead perturbation of N-cadherin function. The ability of lead to displace other divalent cations could result in the displacement of calcium from N-cadherin itself, resulting in conformational changes and lack of adhesion between opposing cadherin proteins. Lead could also interfere with the processing of N-cadherin into NCAD90 via inhibition of the zinc-dependent metalloprotease. Signaling activities of the cadherin complex could be perturbed by lead blockade of calcium channels or modulation of kinase activity which regulates cadherin–catenin interaction or downstream signaling events triggered by cadherin binding.

to cadherin stimulation, providing another avenue for lead effects on cadherin function. Further, since phosphorylation state of cadherins is an important determinant of their expression and function, perturbation of the kinase and phosphatase enzymes which control these post-translational modifications could also provide a mechanism for neurotoxicant-mediated cadherin perturbation.[97] Lead at very low levels is known to stimulate protein kinase C activity[98,99] (see Chapters 6, 13, and 28, this volume). Cadherins are known to be subject to additional post-translational modifications such as glycosylation. In common with NCAM (neural cell adhesion molecule), a member of the immunoglobulin gene superfamily, lead has been demonstrated to affect glycosylation of NCAM, and could potentially affect N-cadherin as well, with subsequent effects on its structural organization.[100]

In addition to its capacity for mimicking or displacing calcium, lead also has the capacity to similarly affect metalloproteins which complex with zinc. A further regulatory control which has been identified for N-cadherin in the developing nervous system is cleavage of 130 kDa N-cadherin at the cell surface leading to the release of NCAD90. While identification of the enzyme mediating this cleavage is under active investigation, it has been established that it has the properties of a zinc-containing metalloprotease. Production of NCAD90 is inhibited by zinc chelators, and it is possible that lead or other heavy metals could disturb this enzymatic activity by virtue of displacement of zinc from the responsible enzyme.

11.07.2.5 Cadherins and the Developmental Neurotoxicology of Lead

The problem of lead intoxication remains a worldwide issue of major significance.[88,89] Indeed, lead exposure is purported to be the leading environmental health hazard for children.[101] Behavioral studies of both experimental animal models and human populations exposed to neurotoxicants can provide insights into large-scale, long-term effects of neurotoxicants whose effects may be traced back to perturbations of cadherin or other cell adhesion molecule function.[102] However, causal connections between perturbations of cell adhesion molecules early in development and large-scale delayed effects, such as altered adult behavior and intelligence, remain speculative in the case of the cadherins.

Several studies have examined the morphological consequences of lead intoxication in a variety of experimental animal model systems.[103–106] Studies of early embryonic development have identified malformations consistent with disruption of normal epithelial cell interactions with resultant tissue swelling and degeneration. Interestingly, a number of these studies reported damage to cerebral blood vessels resulting in hemorrhage that could reflect, in part, a selective effect of lead on cadherins of the vascular endothelium.[66] Some of these studies were carried out at relatively high doses of lead and certain studies report little overt morphological effect of low-level lead on selected regions of the developing nervous system, such as the cerebellum.[107] However, subtle differences in CNS function can occur in the absence of overt morphological changes in the CNS on a gross level. The continuous fine-tuning of cell patterning and synaptology during development, as well as the continuing role for cadherins in the mature nervous system, suggest that there may be a middle ground where perturbations of cadherin or other cell adhesion molecule function may lead to altered CNS function in the absence of catastrophic alterations of tissue structure.

Given the central roles of cadherins in organization and stabilization of the CNS, it will be of critical importance to determine if there are indeed effects of lead on cadherin function which could account for its neurotoxicologic effects. One revealing and relevant series of studies has been carried out[108] (see Figure 7). Utilizing the chick embryo neural retina as a model system, the effect of lead on neural cell adhesion was examined under conditions where N-cadherin is known to mediate the reaggregation of dissociated embryonic neural retinal cells, and to be required for the stabilization of the resulting histotypic aggregates. This study reported that 10 μM lead acetate (a physiologically attainable concentration) inhibited the ability of dissociated cells to form such aggregates. In addition, if such aggregates were allowed to preform from dissociated cells under normal conditions and then exposed to lead, they became destabilized with a resulting decompaction of cells reminiscent of that which occurs following disruption of cadherin-mediated adhesions. In another study, cadmium was shown to perturb epithelial cell–cell adhesion and the distribution of E-cadherin on the cell surface.[109] While these studies suggest that lead may interfere with cadherin function among embryonic neural cells, further investigations directed towards unraveling interactions of heavy metals with the cadherin adhesion system are required to ascertain whether such effects result directly or indirectly via one or more of the cadherin regulatory mechanisms.

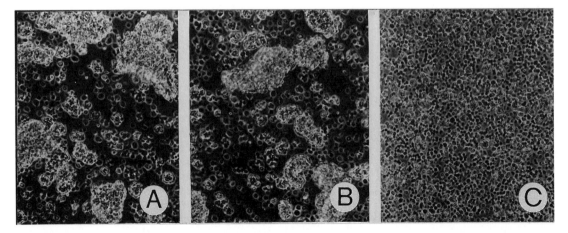

Figure 7 Effect of lead (10 μM) on aggregation of embryonic neural retina. Untreated cells (A) form large aggregates. Lead causes a loss of tight cell aggregation (B), leading eventually to disaggregation into small clusters or single cells (C).

11.07.3 CALCIUM-INDEPENDENT CELL ADHESION MOLECULES: THE IMMUNOGLOBULIN SUPERFAMILY

The immunoglobulin gene superfamily (IgCAMs) represents a large number of proteins which share structural homology with classical immunoglobulins. Members of this family possess one or more immunoglobulin domains, termed Ig loops, consisting of approximately 100 amino acids bridged by a disulfide linkage. The nervous system expresses a number of these IgCAMs, including Thy-1, P0, myelin associated glycoprotein (MAG), the neural cell adhesion molecule, and L1. Representative structures of the IgCAMs are shown in Figure 8. The present discussion will focus primarily on NCAM and to a lesser extent L1, as representative members of this superfamily

for which toxicological data exist. It is likely that toxic interactions with other members will emerge with further study.

11.07.3.1 Neural Cell Adhesion Molecules

Neural cell adhesion molecules (NCAM) are the best characterized of a growing family of calcium-dependent adhesion molecules. More than 20 isoforms of NCAM have been identified, all of which are generated by alternative splicing of a single gene possessing 26 exons.[110] The isoforms have been classified according to their molecular weights into three groups of approximately 180, 140, and 120 kDa, respectively. The isoforms share identical extracellular amino terminal domains, but differ in the size of the intracellular domain. NCAM 180, the largest, has a long intracellular domain

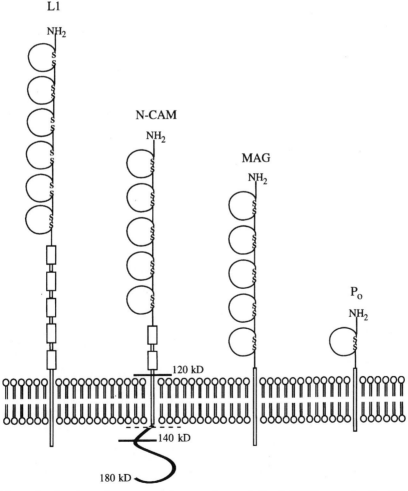

Figure 8 Schematic drawing of representative members of the IgCAM superfamily. Members possess immunoglobulin loops. Some members also have fibronectin repeats (boxes) and membrane-spanning domains (reproduced by permission of CRC Press from 'Toxicology of Metals,' 1996).

which may link with spectrin, a component of the cytoskeleton.[111] NCAM 140 also possesses an intracellular domain, but this is shorter than that of NCAM 180 by 267 amino acids. NCAM 120 is the smallest of the isoforms and completely lacks a membrane-spanning domain; rather, the molecule is linked to the plasma membrane by a phosphatidylinositol linkage.[112]

The conserved extracellular domain of NCAM contains five immunoglobulin loops and two fibronectin Type III domains. The extracellular domain of NCAM may be post-translationally phosphorylated or sulfated.[113] However, the dominant post-translational modification is the developmentally regulated expression of polymers of polysialic acids (PSA) on the extracellular domain near the fifth immunoglobulin loop. The PSA chains are linked to NCAM by an α-2,8-linkage. During development NCAM is heavily decorated with PSA, which may constitute as much as 30% of the weight of the molecule. This heavily sialylated form of NCAM is referred to as embryonal or eNCAM. The PSA surrounds the extracellular domain with a large, negatively charged volume which inhibits binding with adjacent cells expressing complimentary eNCAM.[114] By preventing cell adhesion, the sialic acids serve a permissive regulatory function, allowing relative movement of migrating neurons or growing neurites during morphogenesis and preventing inappropriate interactions with cells encountered *en passant*.[115,116]

Upon completion of development the level of PSA decreases markedly on most neurons; the resulting poorly sialylated NCAM is termed adult or aNCAM and signals the establishment of stable adult cytoarchitecture. Several brain regions, however, retain eNCAM during adult life. These areas, which include hippocampus and other limbic areas,[117–119] the thalamo-neurohypophysial system,[120] and the olfactory system,[121,122] are noted for their ability to alter synaptic architecture ("remodel") in response to physiological or neurotoxic stimuli.

NCAM binds with the complementary molecule expressed on adjacent cells, a form of binding termed homophilic. The specific details of NCAM binding remain obscure. *In vitro* aggregation assays in which microspheres were coated with combinations of soluble recombinant NCAM proteins corresponding to each immunoglobulin domain suggested that all five Ig domains are involved in binding.[123] The binding rate of NCAM is fifth order; relatively small alterations of expression of NCAM on a cell surface can have very significant impact on the cell's adhesive character.[124] The inhibition of cell–cell contact by PSA thus represents an important means of regulating NCAM-mediated adhesion.

11.07.3.2 NCAM Expression and Function

While originally described in the brain, NCAM is expressed at least transiently by a variety of non-neuroectodermal tissues, including kidney, heart, and skeletal muscle. Within neural tissues, NCAM displays a complex and occasionally cell-specific pattern of expression, a full discussion of which is beyond the scope of this chapter and which has been reviewed elsewhere.[125]

During very early development, NCAM is expressed in a poorly sialylated state which favors adhesion and helps maintain the structural integrity of the developing embryonic cell mass. With neurulation, NCAM shows a developmental stage-dependent wave of NCAM expression on proliferating neuroblasts. Administration of blocking antibodies to NCAM or certain toxicants (Figure 9) results in defective formation of neural tube and defects in neural crest migration.[126] eNCAM is expressed on postmitotic neurons and on neurites as they migrate through the neuropil. NCAM is also expressed on astrocytes which form pathways for migration or advancing axonal fibers. The importance of eNCAM in the migration process is confirmed by studies of mice possessing deletion mutations of NCAM180, which display defective neuronal migration. Injection of endoneuraminidase N, which cleaves PSA from NCAM, produces a phenocopy of the NCAM180 mutant.[127]

In the adult, NCAM contributes to the maintenance of mature cytoarchitecture by

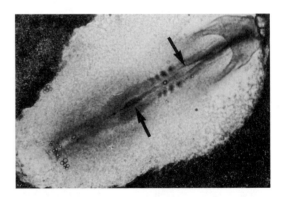

Figure 9 Chick embryo following *in ovo* administration of inorganic lead. Neural tube fusion is impaired, leading to dysraphism. Both N-cadherin and NCAM are involved in neural tube fusion, and their disturbance by toxicants may result in neural malformations (reproduced by permission of CRC Press from 'Toxicology of Metals,' 1996).

serving as a homophilic ligand between neural elements. It also serves prominently in the stabilization of brain structure and particularly in the formation and maintenance of synaptic structure.

11.07.3.3 NCAM in Learning and Memory

Current theories of learning place central importance on the ability of the brain to form new synaptic contacts and to modify existing ones as a means of making information acquisition "permanent."[128] NCAM expression appears to play an essential role in this process. During specific stages of memory acquisition and consolidation, increased levels of heavily sialylated NCAM are observed in those regions thought to be involved in memory. This upregulation of sialylation coincides with a dramatic increase in the number of new synapses. The level of sialylation subsequently returns to control levels as selected synapses are stabilized and the remainder are lost.[129]

Several lines of evidence implicate NCAM in synaptic regulation. NCAM180, but not the 140 and 120 isoforms, is concentrated in the postsynaptic density of the synapse of most neuronal populations, where it is believed to stabilize the synaptic surfaces by linkage with brain spectrin.[111,130,131] Mutant mice unable to express NCAM180 manifest distinct behavioral deficits, particularly on learning tasks.[127,132,133]

Perhaps the strongest evidence implicating NCAM in learning and memory comes from antibody blockage studies in which NCAM antibodies were injected into the brain of rats at different times prior to, or following learning a specific behavioral task (often a passive avoidance paradigm). Anti-NCAM antibody administered before or during training had no significant effect on recall of the task when tested 48 h later. However, antibody administered 6–8 h post-training resulted in amnesia of the behavioral task at 48 h.[129,134,135] These results have been confirmed in the chick.[136,137] It was hypothesized that administration of NCAM antibodies interfered with a transient overproduction and subsequent deletion of heavily sialylated synapses during the period of memory consolidation.[129,138] This hypothesis is supported by experiments utilizing pharmacological agents. Administration of the muscarinic antagonist scopolamine during training of a behavioral task disrupts memory consolidation; this effect is associated with inhibition of the anticipated increase in NCAM sialylation following training. Administration of the piracetam-related cognition-enhancing drug, nefiracetam, blocks the amnesic action and sialylation-suppressing effects of scopolamine.[135]

11.07.4 NCAM AS A TARGET FOR NEUROTOXICANTS

Based solely on structural and functional considerations, NCAM and other members of the IgCAM superfamily represent attractive targets for neurotoxic compounds. These adhesion molecules undergo complex, age-dependent regulation requiring transcriptional, translational, and post-translational fidelity for normal physiological function. They also have extensive extracellular domains responsible for intercellular binding behavior; these may be altered by direct action of toxicants or secondarily altered by extracellular enzymes activated by the toxicant. The transmembrane and intracellular domains of the CAMs, which mediate signal transduction and physical anchoring within the membrane, are similarly vulnerable to protease activation or other secondary events. Many neurotoxicants are highly reactive molecules with multiple sites of action. Any or all of these toxic actions may alter the ability of IgCAMs to effect the level of membrane–membrane intimacy required for appropriate cell behavior.

Among the postulated roles for NCAM is interaction with elements of the cytoskeleton analogous to that identified for the cadherins. NCAM180 has been shown to link with spectrin/fodrin.[111,130] This linkage serves to inhibit lateral diffusion of NCAM by anchoring it in the membrane, thus enhancing the strength of the synaptic contact. NCAM (and perhaps L1) may also influence other elements of the cytoskeleton. Atashi *et al.*[139] found that antibodies against the extracellular domain of NCAM inhibited tyrosine phosphorylation of tubulin *in vitro*. Alteration of tubulin phosphorylation state in developing (or regenerating) neurites may influence the kinetics of microtubule assembly and consequently axonal growth.

Also similar to the cadherins, interactions between NCAM and second messenger systems may be crucial cellular events, particularly during development. Doherty and co-workers identified a homology domain within fibroblast growth factor (FGF) receptor which bound NCAM, L1 and N-cadherin.[52,140] Activation of the FGF receptor stimulated neurite outgrowth. NCAM and L1 are also involved in regulation of inositol phosphates and G-protein dependent activation of calcium receptors.[78] Involvement of these pathways provide numerous avenues by which IgCAMs can have direct and indirect regulatory actions on cell function.[141]

Evidence is rapidly accumulating linking IgCAM perturbation to the expression of neurotoxic states. While the most conclusive data concern the toxic heavy metals, there is an emerging consensus that both cadherin and IgCAMs are vulnerable to a broad spectrum of agents.

11.07.4.1 Lead

The neurotoxic effects of lead on both developing and adult nervous systems have been recognized since the late nineteenth century. The metal causes a wide range of neurological injury, the character of which depends primarily upon the age during which exposure occurs. Developmental lead exposure is by far the most serious and results in damage ranging in severity from subtle psychomotor deficit to acute, life-threatening encephalopathy or congenital malformation. Adults are significantly more resistant to lead, with neurotoxicity usually expressed as decrements in intelligence and cognitive ability (see Chapters 6, 13, and 28, this volume).

Lead was the first environmental neurotoxicant for which an interaction with NCAM was identified. In 1987, Regan and co-workers[142,143] demonstrated that exposure of nursing pups to lead via dam's milk delayed the developmentally regulated conversion of NCAM from the heavily sialylated embryonal form to the less sialylated adult form, a conversion which normally occurs coincident with postnatal synaptic stabilization. This inhibition occurred at blood-lead levels in the pup of $20\,\mu g\,dL^{-1}$, concentrations relevant to human exposures. Subsequent experiments demonstrated that lead stimulated Golgi sialyltransferase, the enzyme regulating the sialylation state of NCAM.[144] Continued activity of this enzyme would be expected to give rise to heavily-sialylated eNCAM and to delay the normal appearance of aNCAM. Persistence of sialylated NCAM throughout the period of synaptogenesis would impair the ability of the lead-intoxicated animal to acquire and consolidate information by preventing the formation of permanent synapses.[145] Continued expression of eNCAM may also prolong periods of neurogenesis and/or migration by preventing appropriate contact-mediated "stop" signals, resulting in increased numbers of cells in specific regions of the brain[146] and further complicating synaptic ordering.

This hypothesis of NCAM-mediated synaptic pathology is supported by morphological studies demonstrating alterations in synaptic number and structure following lead exposure.[147,148] Since learning difficulties and lowered IQs are dominant manifestations of human lead exposure, particularly in children, lead–CAM interactions are of significant clinical interest.

11.07.4.2 Methylmercury

Methylmercury is a highly toxic environmental contaminant. Since the 1950s, several major outbreaks of poisoning have been reported, involving over 7000 individuals.[150] The major target of methylmercury is the CNS; both the immature and the mature nervous systems are highly susceptible. The mechanisms of methylmercury neurotoxicity have been extensively reviewed.[151] This reactive toxicant targets numerous cellular processes, particularly those dependent on sulfhydryl-containing molecules.

Adverse effects of methylmercury on NCAM expression and function have been described both *in vivo* and *in vitro*. Mouse pups administered methylmercury on postnatal days 1–6 exhibit delayed conversion of the embryonal to the adult isoforms of CAM.[152,153] The period of delay correlates well with a transient delay of neuronal migration and increased neuronal death in the cerebrum and cerebellum. In adult rodents, acute intoxication with methylmercury causes a transient decrease in total NCAM180 in hippocampus and to a lesser extent cerebellum. These alterations may correspond to marked synaptic dysgenesis noted in rats following prenatal methylmercury treatment.[154] Immunohistochemical studies and Western blot analyses using antibodies specific for sialic acids reveal increased amounts of PSA in hippocampus. These data suggest that the brain may display two distinct responses to methylmercury. Reduction of total NCAM180 may represent the toxicant's effect on intracellular homeostasis or direct binding with the antigen recognition site, while stimulation of sialyation may represent a protective or reparative response of axons and dendrites. Both effects appear reversible following cessation of methylmercury exposure.

Neurons in cell culture have proven useful tools with which to study methylmercury interactions with NCAM. Using neurons derived from P19 embryonal carcinoma cells, Graff *et al.*[155] demonstrated a rapid (within 1 h) and dose-dependent decrease in NCAM staining, beginning on neurites and proceeding centripetally to the neuronal perikaryon. Biochemical studies confirmed that a decrease of

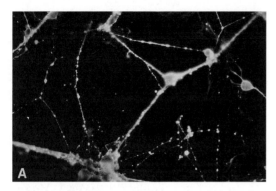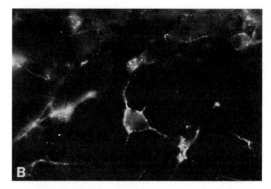

Figure 10 Effect of methylmercury on NCAM staining in cultured neurons. Control (A) are well-labeled with antibody to NCAM, while exposure to 2 μM (B) methylmercury for 2 h caused loss of immunofluorescence. The effect was slowly reversible following removal of methylmercury from the medium.

NCAM protein accompanied the loss of staining. When methylmercury was removed from the culture medium, levels of NCAM protein in neurons recovered after 2 h but staining of NCAM on neurites was not evident until 4 h (see Figure 10). This suggests that the methylmercury might remove or irreversibly mask NCAM on the membrane, requiring replacement by newly synthesized protein.

Methylmercury is known to alter cellular calcium homeostasis and a range of second messenger systems.[151] While it is premature to generalize from these studies, it is tempting to speculate that some of methylmercury's more prominent neurological effects, such as aberrant cell migration and pronounced psychomotor retardation, may originate from the toxicant's interaction with NCAM-dependent functions during developmentally critical periods of morphogenesis.

11.07.4.3 Trimethyltin

Trimethyltin is a highly toxic organometal that selectivity injures regions of the limbic system, particularly the hippocampus and pyriform cortex.[156–159] Long-term behavioral alterations may occur as a result of low-level trimethyltin exposure;[160,161] these are thought to result from disruption of neural processes underlying learning and memory[162] and can occur even in the absence of microscopic evidence of neuronal injury.

Trimethyltin induces a selective loss of NCAM180, but not the 140 or 120 isoforms in the hippocampus and cerebellum of adult mice.[163] In the hippocampus this loss occurs rapidly, within 4 h of administration. By 24 h, 75–100% of NCAM180 may disappear. Reappearance, likely representing *de novo* synthesis of the molecule, is observed beginning 48–72 h later and control levels are regained within 1–2

weeks (see Figure 11). A similar but less robust sequence of disappearance and recovery is observed in cerebellum. NCAM loss is not merely a consequence of neuronal death, as it occurs prior to morphologically evident cell injury and recovery of NCAM levels occurs even with significant concurrent neuronal death.

The mechanism by which trimethyltin induces loss of NCAM180 is unclear. Disruption of the Golgi apparatus by trimethyltin has been reported,[158] and this may contribute to alterations in post-translational processing of NCAM. However, loss of NCAM induced by trimethyltin is too rapid to be solely the consequence of Golgi-dependent processes. Rather, Dey *et al.*[163] postulated the activation of intracellular proteases at the synapse as a possible mechanism for the effect. NCAM180 is susceptible to proteolysis at micromolar calcium concentrations, suggesting that it might be a substrate for calpain I.[164,165] NCAM180 and NCAM140 contains PEST sequences which may serve as targets for neutral proteases; why NCAM180 appears preferentially vulnerable is not known. Trimethyltin-induced elevations of synaptosomal calcium may activate calcium-

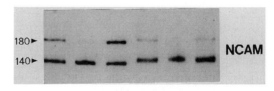

Figure 11 Effect of trimethyltin on expression of NCAM in hippocampus (lanes 1–3) and cerebellum (lanes 4–6). Controls (lanes 1 and 4) show strong labeling for both NCAM 180 and 140. Within 36 h of trimethyltin treatment, NCAM 180 (but not NCAM140) has disappeared from both brain structures (lanes 2 and 5), and returns to normal levels (lanes 3 and 6) within several weeks (reproduced by permission of Academic Press from *Toxicol. Appl. Pharmacol.*, 1994, **126**, 69–74).

dependent proteases, cleaving the anchoring linkage between NCAM and the spectrin. This effect would likely be most pronounced in those brain areas with the greatest propensity for synaptic plasticity.

11.07.4.4 Other Neurotoxicants

Neurotoxic injury of the brain elicits a spectrum of reparative and adaptive responses by neural tissue. Modulation of NCAM expression appears to be one of the most universal of these responses, particularly in tissues capable of substantial plasticity. Alterations of NCAM have been reported following surgical lesions,[166] in response to experimentally induced cortical ischemia,[167] kainic acid-induced seizures,[119,168] following treatment with cytotoxic and cytostatic drugs such as cytosine arabinoside,[169] and during the progression of certain idiopathic neurologic disorders such as schizophrenia[170–171] or Huntington's disease.[172] While involvement of NCAM in these conditions is likely to be related more to the brain response to injury than to the mechanism of the injury itself, the ability of damaged cells to modulate NCAM functions is likely to be a critical factor in the ultimate clinical expression of injury response and the fidelity of repair.

In addition to metals, solvents may mediate neurotoxicity via effects on IgCAMs. Ethanol has been shown to disturb function of both NCAM and L1. Administration of ethanol to chicks *in ovo* increased the amount of heavily sialated NCAM in cerebrum and disturbance of neurite fasciculation.[173]

Using NG108–15 cells (a neuroblastoma X glioma hybrid), Charness and co-workers[174] observed a strong induction of NCAM and L1 expression by recombinant human osteogenic protein-1 (hOP-1), resulting in enhanced cell–cell adhesion and clustering. Ethanol at levels achieved during social drinking inhibited the cell–cell adhesion without affecting induction or membrane expression of either L1 or NCAM. Ethanol also inhibited adhesion of cerebellar granule cells to L1-transfected 3T3 fibroblasts.[175] Molecular analysis suggests that ethanol can block L1-mediated adhesion by binding at a small hydrophobic site within L1, inhibiting function without resulting in removal of the molecule from the membrane. The authors speculate that perturbation of L1 by ethanol may contribute to the spectrum of defects seen in fetal alcohol syndrome. While *in vivo* confirmation of this hypothesis has not yet appeared, several cross-linked defects of L1, such as MASA (mental retardation, adducted thumbs, spasticity and aphasia) and HSAS[176]

(cross-linked hydrocephalus), share some phenotypic features in common with fetal alcohol syndrome, implying similar molecular disturbances.

11.07.5 CONCLUSIONS

Since the mid-1980s there has been extraordinary growth in our understanding of the role of cell adhesion molecules in neurobiology. More than simple ligands of adhesion between cells or between cells and matrix, the CAMs are now recognized as integral to the development and regulation of neural function under normal and pathological conditions.[153,177,178] The roles of CAMs in neurotoxic injury are largely unexplored, and the crucial interactions between the different superfamilies of CAMs, and between members of the same superfamily, are just beginning to be identified. Moreover, the number of CAMs is expanding. Whole families of molecules, such as the integrins, remain to be examined for their susceptibility to neurotoxic insult. Perturbations of these essential and multifaceted molecules by xenobiotics present attractive mechanistic explanations for neurotoxicity.

ACKNOWLEDGMENTS

The authors would like to thank Dr. H. E. Lowndes, Kathleen Roberts, and P. M. Dey for their careful review of this chapter, and Ms. Patricia Pailing for her help in preparing the manuscript. This work was partially supported by NIH grants ES04976 and ES05022 to KRR and EY06658 and EY10965 to GBG.

11.07.6 REFERENCES

1. G. B. Grunwald, in 'A Conceptual History of Modern Embryology,' ed S. F. Gilbert, Plenum, New York, 1991, vol. 7, pp. 129–158.
2. B. Geiger and O. Ayalon, 'Cadherins.' *Annu. Rev. Cell Biol.*, 1992, **8**, 307–332.
3. G. B. Grunwald, in 'Fundamentals of Medical Cell Biology,' ed. E. E. Bittar, JAI Press, Greenwich, CT, 1992, vol. 7, pp. 103–132.
4. G. B. Grunwald, 'The structural and functional analysis of cadherin calcium-dependent cell adhesion molecules.' *Curr. Opin. Cell Biol.*, 1993, **5**, 797–805.
5. G. B. Grunwald, in 'Principles of Medical Cell Biology,' ed. E. E. Bittar, JAI Press, Greenwich, CT, 1996, in press.
6. G. B. Grunwald, 'Cadherin cell adhesion molecules in retinal development and pathology.' *Prog. Retinal Eye Res.*, 1996, **15**, 363–392.
7. G. B. Grunwald, in 'Adhesion Molecules,' ed. D. R. Colman, JAI Press, Greenwich, CT, 1996, in press.
8. B. M. Gumbiner, 'Cell adhesion: the molecular basis

of tissue architecture and morphogenesis.' *Cell*, 1996, **84**, 345–357.

9. R. Kemler, 'Classical cadherins.' *Semin. Cell Biol.*, 1992, **3**, 149–155.

10. R. Kemler, 'From cadherins to catenins: cytoplasmic protein interactions and regulation of cell adhesion.' *Trends Genet.*, 1993, **9**, 317–321.

11. B. Ranscht, 'Cadherin cell adhesion molecules in vertebrate neural development.' *Semin. Neurosci.*, 1991, **3**, 285–296.

12. M. Takeichi, 'Cadherins: a molecular family important in selective cell–cell adhesion.' *Annu. Rev. Biochem.*, 1990, **59**, 237–252.

13. M. Takeichi, 'Cadherin cell adhesion receptors as a morphogenetic regulator.' *Science*, 1991, **251**, 1451–1455.

14. M. Takeichi, 'The cadherins: cell–cell adhesion molecules controlling animal morphogenesis.' *Development*, 1992, **102**, 639–655.

15. M. Takeichi, 'Cadherins in cancer: implications for invasion and metastasis.' *Curr. Opin. Cell Biol.*, 1993, **5**, 806–811.

16. G. B. Grunwald, R. S. Pratt and J. Lilien, 'Enzymatic dissection of embryonic cell adhesive mechanisms. III. Immunological identification of a component of the calcium-dependent adhesive system of embryonic chick neural retina cells.' *J. Cell Sci.*, 1982, **55**, 69–83.

17. K. Hatta and M. Takeichi, 'Expression of N-cadherin adhesion molecules associated with early morphogenetic events in chick development.' *Nature*, 1986, **320**, 447–449.

18. H. Inuzuka, S. Miyatani and M. Takeichi, 'R-cadherin: a novel Ca^{2+}-dependent cell–cell adhesion molecule expressed in the retina.' *Neuron*, 1991, **7**, 69–79.

19. E. W. Napolitano, K. Venstrom, E. F. Wheeler *et al.*, 'Molecular cloning and characterization of B-cadherin, a novel chick cadherin.' *J. Cell Biol.*, 1991, **113**, 893–905.

20. B. Ranscht and M. T. Dours-Zimmermann, 'T-cadherin, a novel cadherin cell adhesion molecule in the nervous system, lacks the conserved cytoplasmic region.' *Neuron*, 1991, **7**, 391–402.

21. S. Suzuki, K., Sano and H. Tanihara, 'Diversity of the cadherin family: evidence for eight new cadherins in nervous tissue.' *Cell Regul.*, 1991, **2**, 261–270.

22. K. Shimamura and M. Takeichi, 'Local and transient expression of E-cadherin involved in mouse embryonic brain morphogenesis.' *Development*, 1992, **116**, 1011–1019.

23. K. Sano, H. Tanihara, R. L. Heimark *et al.*, 'Protocadherins: a large family of cadherin-related molecules in central nervous system.' *EMBO J.*, 1993, **12**, 2249–2256.

24. V. Pachnis, B. Mankoo and F. Costantini, 'Expression of the c-ret protooncogene during mouse embryogenesis.' *Development*, 1993, **119**, 1005–1017.

25. C. Yoshida-Noro, N. Suzuki and M. Takeichi, 'Molecular nature of the calcium-dependent cell–cell adhesion system in mouse teratocarcinoma and embryonic cells studied with a monoclonal antibody.' *Dev. Biol.*, 1984, **101**, 19–27.

26. M. Holtfreter, 'Tissue affinity, a means of embryonic morphogenesis.' *Arch. Exp. Zellforsch.*, 1939, **23**, 169–209. (English translation in 'Foundations of Experimental Embryology,' ed. B. H. Willier and J. M. Oppenheimer, Prentice-Hall, NJ, 1964, pp. 186–255.)

27. P. L. Townes and J. Holtfreter, 'Directed movements and selective adhesion of embryonic amphibian cells.' *J. Exp. Zool.*, 1955, **128**, 53–120.

28. R. Moscona, 'Analysis of all recombinations in experimental synthesis of tissues *in vitro*.' *J. Cell Comp. Physiol.*, 1962, **60**, 65–80.

29. M. S. Steinberg, P. B. Armstrong and R. E. Granger, 'On the recovery of adhesiveness by trypsin-dissociated cells.' *J. Membr. Biol.*, 1973, **13**, 97–128.

30. M. Takeichi, 'Functional correlation between cell adhesive properties and some cell surface proteins.' *J. Cell Biol.*, 1977, **75**, 464–474.

31. G. B. Grunwald, R. L. Geller and J. Lilien, 'Enzymatic dissection of embryonic cell adhesive mechanisms.' *J. Cell Biol.*, 1980, **85**, 766–776.

32. G. B. Grunwald, R. E. Bromberg, N. J. Crowley *et al.*, 'Enzymatic dissection of embryonic cell adhesive mechanisms. II. Developmental regulation of an endogenous adhesive system in the chick neural retina.' *Dev. Biol.*, 1981, **86**, 327–338.

33. L. A. Lagunowich and G. B. Grunwald, 'Expression of calcium-dependent cell adhesion during ocular development: a biochemical, histochemical and functional analysis.' *Dev. Biol.*, 1989, **135**, 158–171.

34. K. Hatta, A. Nose, A. Nagafuchi *et al.*, 'Cloning and expression of cDNA encoding a neural calcium-dependent cell adhesion molecule: its identity in the cadherin gene family.' *J. Cell Biol.*, 1988, **106**, 873–881.

35. T. Iwamoto, M. Taniguchi, N. Asai *et al.*, 'cDNA cloning of mouse ret protooncogene and its sequence similarity to the cadherin superfamily.' *Oncogene*, 1993, **8**, 1087–1091.

36. H. Tanihara, M. Kido, S. Obata *et al.*, 'Characterization of cadherin-4 and cadherin-5 reveals new aspects of cadherins.' *J. Cell Sci.*, 1994, **107**, 1697–1704.

37. L. A. Lagunowich, L. A. Donoso and G. B. Grunwald, 'Identification of mammalian and invertebrate analogues of the avian calcium-dependent cell adhesion protein N-cadherin with synthetic-peptide directed antibodies against a conserved cytoplasmic domain.' *Biochem. Biophys. Res. Commun.*, 1990, **172**, 313–320.

38. L. A. Lagunowich and G. B. Grunwald, 'Tissue and age-specificity of post-translational modifications of N-cadherin during chick embryo development.' *Differentiation*, 1991, **47**, 19–27.

39. L. Shapiro, A. M. Fannon, P. D. Kwong *et al.*, 'Structural basis of cell–cell adhesion by cadherins.' *Nature*, 1995, **374**, 327–337.

40. M. Takeichi, 'The cadherins: cell–cell adhesion molecules controlling animal morphogenesis.' *Development*, 1988, **102**, 639–655.

41. M. Matsunaga, K. Hatta and M. Takeichi, 'Role of N-cadherin cell adhesion molecules in the histogenesis of neural retina.' *Neuron*, 1988, **1**, 289–295.

42. L. A. Lagunowich, J. C. Schneider, S. Chasen *et al.*, 'Immunohistochemical and biochemical analysis of N-cadherin expression during CNS development.' *J. Neurosci. Res.*, 1992, **32**, 202–208.

43. C. Redies, H. Inuzuka and M. Takeichi, 'Restricted expression of N- and R-cadherin on neurites of the developing chicken CNS.' *J. Neurosci.*, 1992, **12**, 3525–3534.

44. B. Ranscht and M. Bronner-Fraser, 'T-cadherin expression alternates with migrating neural crest cells in the trunk of the avian embryo.' *Development*, 1991, **111**, 15–22.

45. C. Murphy-Erdosh, E. W. Napolitano and L. F. Reichardt, 'The expression of B-cadherin during embryonic chick development.' *Dev. Biol.*, 1994, **161**, 107–125.

46. J. S. Schiffman and G. B. Grunwald, 'Molecular genetic analysis of cadherin diversity in the developing avian retinal pigment epithelium.' *Soc. Neurosci.*, 1993, **19**, 459.

47. E. A. Capper, J. S. Schiffman and G. B. Grunwald, 'Multiple cadherins are differentially expressed during *in vitro* development of the retinal pigment epithelium.' *Mol. Cell Biol.*, 1995, **6**, 171a.

48. E. F. Roark, N. E. Paradies, L. A. Lagunowich *et al.*, 'Evidence for endogenous proteases, mRNA level and insulin as multiple mechanisms of N-cadherin downregulation during retinal development.' *Development*, 1992, **114**, 973–984.

49. K. Shimamura, S. Hirano, A. P. McMahon *et al.*, 'Wnt-1-dependent regulation of local E-cadherin and αN-catenin expression in the embryonic mouse brain.' *Development*, 1994, **120**, 2225–2234.

50. A. Tannapfel, W. Yasui, H. Yokozaki *et al.*, 'Effect of hepatocyte growth factor on the expression of E- and P-cadherin in gastric carcinoma cell lines.' *Virchows Arch.*, 1994, **425**, 139–144.

51. H. Hoschuetsky, H. Aberle and R. Kemler, 'β-catenin mediates the interaction of the cadherin-catenin complex with epidermal growth factor receptor.' *J. Cell Biol.*, 1994, **127**, 1375–1380.

52. E. J. Williams, J. Furness, F. S. Walsh *et al.*, 'Activation of the FGF receptor underlies neurite outgrowth stimulated by L1, NCAM and N-cadherin.' *Neuron*, 1994, **13**, 583–594.

53. W. Birchmeier and J. Behrens, 'Cadherin expression in carcinomas: role in the formation of cell junctions and the prevention of invasiveness.' *Biochim. Biophys. Acta*, 1994, **1198**, 11–26.

54. J. S. Schiffman and G. B. Grunwald, 'Differential cell adhesion and expression of N-cadherin among retinoblastoma cell lines.' *Invest. Ophthalmol. Vis. Sci.*, 1992, **33**, 1568–1574.

55. J. Drazba and V. Lemmon, 'The role of cell adhesion molecules in neurite outgrowth on Muller cells.' *Dev. Biol.*, 1990, **138**, 82–93.

56. R. O. Hynes and A. D. Lander, 'Contact and adhesive specificities in the associations, migrations, and targeting of cells and axons.' *Cell*, 1992, **68**, 303–322.

57. C. E. Holt and W. A. Harris, 'Position, guidance and mapping in the developing visual system.' *J. Neurobiol.*, 1993, **24**, 1400–1422.

58. M. Matsunaga, K. Hatta, A. Nagafuchi *et al.*, 'Guidance of optic nerve fibres by N-cadherin adhesion molecules.' *Nature*, 1988, **334**, 62–64.

59. J. L. Bixby and R. Zhang, 'Purified N-cadherin is a potent substrate for the rapid induction of neurite outgrowth.' *J. Cell Biol.*, 1990, **110**, 1253–1260.

60. N. E. Paradies and G. B. Grunwald, 'Purification and characterization of NCAD90, a soluble endogenous form of N-cadherin, which is generated by proteolysis during retinal development and retains adhesive and neurite-promoting function.' *J. Neurosci. Res.*, 1993, **36**, 33–45.

61. I. J. Kljavin, C. Lagenaur, J. L. Bixby *et al.*, 'Cell adhesion molecules regulating neurite growth from amacrine and rod photoreceptor cells.' *J. Neurosci.*, 1994, **14**, 5035–5049.

62. F. M. Pavalko and C. A. Otey, 'Role of adhesion molecule cytoplasmic domains in mediating interactions with the cytoskeleton.' *Proc. Soc. Exp. Biol. Med.*, 1994, **205**, 282–293.

63. M. Ozawa and R. Kemler, 'Molecular organization of the uvomorulin–catenin complex.' *J. Cell Biol.*, 1992, **116**, 989–996.

64. K. A. Knudsen, A. P. Soler, K. R. Johnson *et al.*, 'Interaction of α-actinin with the cadherin–catenin cell–cell adhesion complex via α-catenin.' *J. Cell Biol.*, 1995, **130**, 65–77.

65. A. B. Reynolds, J. Daniel, P. D. McCrea *et al.*, 'Identification of a new catenin: the tyrosine kinase substrate p120^cas associates with E-cadherin complexes.' *Mol. Cell Biol.*, 1994, **14**, 8333–8342.

66. M. G. Lampugnani, M. Corada, L. Caveda *et al.*, 'The molecular organization of endothelial cell to cell junctions: differential association of plakoglobin, β-catenin, and α-catenin with vascular endothelial cadherin (VE-cadherin).' *J. Cell Biol.*, 1995, **129**, 203–217.

67. K. A. Guger and B. M. Gumbiner, 'β-Catenin has Wnt-like activity and mimics the Nieuwkoop signaling center in *Xenopus* dorsal-ventral patterning.' *Dev. Biol.*, 1995, **172**, 115–125.

68. T. Fujimori, S. Miyatani and M. Takeichi, 'Ectopic expression of N-cadherin perturbs histogenesis in *Xenopus* embryos.' *Development*, 1990, **110**, 97–104.

69. C. Kintner, 'Regulation of embryonic cell adhesion by the cadherin cytoplasmic domain.' *Cell*, 1992, **69**, 225–236.

70. S. Dufour, J. P. Saint-Jeannet, F. Broders *et al.*, 'Differential perturbations in the morphogenesis of anterior structures induced by overexpression on truncated XB- and N-cadherins in *Xenopus* embryos.' *J. Cell Biol.*, 1994, **127**, 521–535.

71. M. M. Lee and G. B. Grunwald, 'Biochemical analysis of N-cadherin phosphorylation in developing chick embryo tissues.' *Soc. Neurosci.*, 1993, **19**, 460.

72. G. Radice, 'Genetic analysis of cadherin function in mice.' *Mol. Biol. Cell* 1995, **6** (Suppl.), 171a.

73. A. Lilienbaum, R. H. Riehl, A. A. Rezka *et al.*, 'Expression of mutant β1 integrin and N-cadherin in the developing retina.' *Soc. Neurosci.*, 1993, **19**, 1089.

74. M. C. Ferreira, R. J. Buono and G. B. Grunwald, 'Neurite morphology is altered by molecular genetic perturbation of N-cadherin expression and function during neural retinal development.' *Mol. Cell Biol.*, 1994, **5** (Suppl.), 233a.

75. M. L. Hermiston and J. I. Gordon, '*In vivo* analysis of cadherin function in the mouse intestinal epithelium: essential roles in adhesion, maintenance of differentiation, and regulation of programmed cell death.' *J. Cell Biol.*, 1995, **129**, 489–506.

76. D. F. Woods and P. J. Bryant, 'Apical junctions and cell signalling in epithelia.' *J. Cell Sci.*, 1993, **17** (Suppl.), 171–181.

77. J. L. Saffell, F. S. Walsh and P. Doherty, 'Direct activation of second messenger pathways mimics cell adhesion molecule-dependent neurite outgrowth.' *J. Cell Biol.*, 1992, **118**, 663–670.

78. P. Doherty, S. V. Ashton, S. E. Moore *et al.*, 'Morphoregulatory activities of NCAM and N-cadherin can be accounted for by G protein-dependent activation of L- and N-type neuronal Ca²⁺ channels.' *Cell*, 1991, **67**, 21–33.

79. J. L. Bixby, G. B. Grunwald and R. J. Bookman, 'Ca²⁺ influx and neurite growth in response to purified N-cadherin and laminin.' *J. Cell Biol.*, 1994, **127**, 1461–1475.

80. G. J. Rucklidge, K. Edvardsen and E. Bock, 'Cell-adhesion molecules and metalloproteinases: a linked role in tumour cell invasiveness.' *Biochem. Soc. Trans.*, 1993, **22**, 63–68.

81. M. C. Ferreira, N. E. Paradies and G. B. Grunwald, 'Characterization of the endogenous retinal protease which catalyzes N-cadherin turnover.' *Soc. Neurosci.*, 1993, **19**, 459.

82. M. M. Lee and G. B. Grunwald, 'Generation of NCAD90 from N-cadherin in the developing retina is modulated by phosphorylation state and IGF-1.' *Mol. Cell. Biol.*, 1995, **6** (Suppl.), 170a.

83. M. M. Lee, B. D. Fink and G. B. Grunwald, 'Evidence that tyrosine phosphorylation regulates N-cadherin turnover during retinal development.' Submitted for publication.

84. S. M. Brady-Kalnay, D. L. Rimm and N. K. Tonks, 'Receptor protein tyrosine phosphatase PTPμ associates with cadherins and catenins *in vivo.*' *J. Cell Biol.*, 1995, **130**, 977–986.

85. M. F. Gebbink, G. C. Zondag, G. M. Koningstein *et al.*, 'Cell surface expression of receptor protein tyrosine phosphatase RPTPμ is regulated by cell–cell contact.' *J. Cell Biol.*, 1995, **131**, 251–260.

86. T. Hunter, 'Protein kinases and phosphatases: the yin and yang of protein phosphorylation and signaling.' *Cell*, 1995, **80**, 225–236.

87. G. Audesirk, 'Effects of lead exposure on the physiology of neurons.' *Prog. Neurobiol.*, 1985, **24**, 199–231.

88. W. Yule, 'Review: neurotoxicity of lead.' *ChildCare, Health Dev.*, 1992, **18**, 321–337.

89. I. A. S. Al-Saleh, 'The biochemical and clinical consequences of lead poisoning.' *Med. Res. Rev.*, 1994, **14**, 415–486.

90. J. P. Bressler and G. W. Goldstein, 'Mechanisms of lead neurotoxicity.' *Biochem. Pharmacol.*, 1991, **41**, 479–484.

91. E. K. Silbergeld, 'Mechanisms of lead neurotoxicity, or looking beyond the lamppost.' *FASEB J.*, 1992, **6**, 3201–3206.

92. J. S. Cranmer, 'New dimensions of lead neurotoxicity: redefining mechanisms and effects. Proceedings of the 9th International Neurotoxicology Conference, Little Rock, Arkansas, October 28–31, 1991.' *Neurotoxicology*, 1993, **14** (2–3), 1–364.

93. P. L. Goering, 'Lead–protein interactions as a basis for lead toxicity.' *Neurotoxicology*, 1993, **14**, 45–60.

94. T. J. Simons, 'Lead–calcium interactions in cellular lead toxicity.' *Neurotoxicology*, 1993, **14**, 77–85.

95. G. Audesirk, 'Electrophysiology of lead intoxication: effects on voltage-sensitive ion channels.' *Neurotoxicology*, 1993, **14**, 137–147.

96. D. Busselberg, M. L. Evans, H. L. Haas *et al.*, 'Blockade of mammalian and invertebrate calcium channels by lead.' *Neurotoxicology*, 1993, **14**, 249–258.

97. J. P. O'Callaghan, 'A potential role for altered protein phosphorylation in the mediation of developmental neurotoxicity.' *Neurotoxicology*, 1994, **15**, 29–40.

98. J. Markovac and G. W. Goldstein, 'Picomolar concentrations of lead stimulate brain protein kinase C.' *Nature*, 1988, **334**, 71–73.

99. G. W. Goldstein, 'Evidence that lead acts as a calcium substitute in second messenger metabolism.' *Neurotoxicology*, 1993, **14**, 97–101.

100. C. M. Regan, 'Neural cell adhesion molecules, neuronal development and lead toxicity.' *Neurotoxicology*, 1993, **14**, 69–74.

101. J. A. Riess and H. L. Needleman, in 'The Vulnerable Brain and Environmental Risks. Vol. 2: Toxins in Food,' eds. R. L. Isaacson and K. F. Jensen, Plenum, New York, 1992, pp. 111–126.

102. K. R. Reuhl, 'Delayed expression of neurotoxicity: the problem of silent damage.' *Neurotoxicology*, 1991, **12**, 341–346.

103. A. Hirano and J. A. Kochen, 'Neurotoxic effects of lead in the chick embryo: morphologic studies.' *Lab. Invest.*, 1973, **29**, 659–668.

104. S. J. Carpenter and V. H. Ferm, 'Embryopathic effects of lead in the hamster: morphologic analysis.' *Lab. Invest.*, 1977, **37**, 369–385.

105. L. D. De Gennaro, 'The effects of lead nitrate on the central nervous system of the chick embryo. I. Observations of light and electron microscopy.' *Growth*, 1978, **42**, 141–155.

106. P. Jacquet, 'Early embryonic development in lead-intoxicated mice.' *Arch. Pathol. Lab. Med.*, 1977, **101**, 641–643.

107. F. Hasan, G. R. Cookman, G. J. Keane *et al.*, 'The effect of low-level lead exposure on the postnatal structuring of the rat cerebellum.' *Neurotoxicol. Teratol.*, 1989, **11**, 433–440.

108. L. A. Lagunowich, A. P. Stein and K. R. Reuhl, 'N-cadherin in normal and abnormal brain development.' *Neurotoxicology*, 1994, **15**, 123–132.

109. W. C. Prozialeck and R. J. Niewenhuis, 'Cadmium (Cd^{2+}) disrupts Ca^{2+}-dependent cell–cell junctions and alters the pattern of E-cadherin immunofluorescence in LLC-PK1 cells.' *Biochem. Biophys. Res. Commun.*, 1991, **181**, 1118–1124.

110. F. S. Walsh, 'The NCAM gene is a complex transcriptional unit.' *Neurochem. Int.*, 1988, **12**, 263–267.

111. G. E. Pollerberg, K. Burridge, K. E. Krebs *et al.*, 'The 180 kDa component of the neural cell adhesion molecule NCAM is involved in cell–cell contacts and cytoskeleton-membrane interactions.' *Cell Tissue Res.*, 1987, **250**, 227–236.

112. H. T. He, J. Finne and C. Goridis, 'Biosynthesis, membrane association, and release of NCAM-120, a phosphatidylinositol-linked form of the neural cell adhesion molecule.' *J. Cell Biol.*, 1987, **105**, 2489–2500.

113. S. Hoffman and G. M. Edelman, 'Kinetics of homophilic binding of embryonic and adult forms of the neural cell adhesion molecule.' *Proc. Natl. Acad. Sci. USA*, 1983, **80**, 5762–5766.

114. U. Rutishauser, A. Acheson, A. K. Hall *et al.*, in 'Current Issues in Neural Regeneration Research,' eds. P. J. Reier, R. P. Bungeand and F. T. Seil, Liss, New York, 1988, pp. 229–236.

115. U. Rutishauser, A. Acheson, A. K. Hall *et al.*, 'The neural cell adhesion molecule (NCAM) as a regulator of cell–cell interactions.' *Science*, 1988, **240**, 53–57.

116. L. Landmesser, L. Dahm, J. C. Tang *et al.*, 'Polysialic acid as a regulator of intramuscular nerve branching during embryonic development.' *Neuron*, 1990, **4**, 655–667.

117. T. Seki and Y. Arai, 'The persistent expression of a highly polysialylated NCAM in the dentate gyrus of the adult rat.' *Neurosci. Res.*, 1991, **12**, 503–513.

118. T. Seki and Y. Arai, 'Expression of highly polysialylated NCAM in the neocortex and piriform cortex of the developing and the adult rat.' *Anat. Embryol. (Berlin)*, 1991, **184**, 395–401.

119. G. Le Gal La Salle, G. Rougon and A. Valin, 'The embryonic form of neural cell surface molecule (E-NCAM) in the rat hippocampus and its reexpression on glial cells following kainic acid-induced status epilepticus.' *J. Neurosci.*, 1992, **12**, 872–882.

120. D. T. Theodosis, G. Rougon and D. A. Poulain, 'Retention of embryonic features by an adult neuronal system capable of plasticity: polysialylated neural cell adhesion molecule in the hypothalamo–neurohypophysial system.' *Proc. Natl. Acad. Sci. USA*, 1991, **88**, 5494–5498.

121. F. Miragall, G. Kadmon, M. Husmann *et al.*, 'Expression of cell adhesion molecules in the olfactory system of the adult mouse: presence of the embryonic form of NCAM.' *Dev. Biol.*, 1988, **129**, 516–531.

122. G. Rougon, S. Olive and D. Figarella-Branger, in 'Polysialic Acid,' eds. J. Roth, U. Rutishauer and F. A. Troy, II, Birkhauser Verlag, Basel, 1993, pp. 323–333.

123. T. S. Ranheim, G. M. Edelman and B. A. Cunningham, 'Homophilic adhesion mediated by the neural cell adhesion molecule involves multiple immunoglobulin domains.' *Proc. Natl. Acad. Sci. USA*, 1996, **93**, 4071–4075.

124. G. M. Edelman and K. L. Crossin, 'Cell adhesion molecules: implications for a molecular histology.' *Annu. Rev. Biochem.*, 1991, **60**, 155–190.

125. R. Brackenbury, B. C. Sorkin and B. A. Cunningham,

in 'Molecular Neurobiology in Neurology and Psychiatry,' ed. E. Kandel, Raven Press, New York, 1987.

126. M. Bronner-Fraser, J. J. Wolf and B. A. Murray, 'Effects of antibodies against N-cadherin and NCAM on the cranial neural crest and neural tube.' *Dev. Biol.*, 1992, **153**, 291–301.

127. K. Ono, H. Tomasiewicz, T. Magnuson *et al.*, 'N-CAM mutation inhibits tangential neuronal migration and is phenocopied by enzymatic removal of polysialic acid.' *Neuron*, 1994, **13**, 595–609.

128. S. P. Rose, in 'Synaptic Plasticity: Molecular, Cellular, and Functional Aspects,' eds. M. Baudry, R. F. Thompson and J. L. Davis, MIT Press, Cambridge, MA, 1993, pp. 209–229.

129. C. M. Regan, 'in 'Polysialic Acid,' eds. J. Roth, U. Rutishauser and F. A. Troy, II, Birkhauser Verlag, Basel, 1993, pp. 313–321.

130. G. E. Pollerberg, M. Schachner and J. Davoust, 'Differentiation state-dependent surface mobilities of two forms of the neural cell adhesion molecule.' *Nature*, 1986, **324**, 462–465.

131. E. Persohn, G. E. Pollerberg and M. Schachner, 'Immunoelectron-microscopic localization of the 180 kDa component of the neural cell adhesion molecule NCAM in postsynaptic membranes.' *J. Comp. Neurol.*, 1989, **288**, 92–100.

132. H. Tomasiewicz, K. Ono, D. Yee *et al.*, 'Genetic deletion of a neural cell adhesion molecule variant (NCAM-180) produces distinct defects in the central nervous system.' *Neuron*, 1993, **11**, 1163–1174.

133. H. Cremer, R. Lange, A. Christoph *et al.*, 'Inactivation of the NCAM gene in mice results in size reduction of the olfactory bulb and deficits in spatial learning.' *Nature*, 1994, **367**, 455–459.

134. E. Doyle, P. M. Nolan, R. Bell *et al.*, 'Intraventricular inclusions of antineural cell adhesion molecules in a discrete posttraining period impair consolidation of a passive avoidance response in the rat.' *J. Neurochem.*, 1992, **59**, 1570–1573.

135. E. Doyle, C. M. Regan and T. Shiotani, 'Nefiracetam (DM-9384) preserves hippocampal neural cell adhesion molecule-mediated memory consolidation processes during scopolamine disruption of passive avoidance training in the rat.' *J. Neurochem.*, 1993, **61**, 266–272.

136. A. B. Scholey, S. P. R. Rose, M. R. Zamani *et al.*, 'A role for the neural cell adhesion molecule in a late, consolidating phase of glycoprotein synthesis 6 h following passive avoidance training of the young chick.' *Neuroscience*, 1993, **55**, 499–509.

137. R. Mileusnic, S. P. R. Rose, C. Lancashire *et al.*, 'Characterisation of antibodies specific for chick brain neural cell adhesion molecules which cause amnesia for a passive avoidance task.' *J. Neurochem.*, 1995, **64**, 2598–2606.

138. C. M. Regan and G. B. Fox, 'Polysialylation as a regulator of neural plasticity in rodent learning and aging.' *Neurochem. Res.*, 1995, **20**, 593–598.

139. J. R. Atashi, S. G. Klinz, C. A. Ingraham *et al.*, 'Neural cell adhesion molecules modulate tyrosine phosphorylation of tubulin in nerve growth cone membranes.' *Neuron*, 1992, **8**, 831–842.

140. P. Doherty, E. Williams and F. S. Walsh, 'A soluble chimeric form of the L1 glycoprotein stimulates neurite outgrowth.' *Neuron*, 1995, **14**, 57–66.

141. U. Schuch, M. J. Lohse and M. Schachner, 'Neural cell adhesion molecules influence second messenger systems.' *Neuron*, 1989, **3**, 13–20.

142. G. R. Cookman, W. King and C. M. Regan, 'Chronic low-level lead exposure impairs embryonic to adult conversion of the neural cell adhesion molecule.' *J. Neurochem.*, 1987, **49**, 399–403.

143. C. M. Regan, 'Lead-impaired neurodevelopment. Mechanisms and threshold values in the rodent.' *Neurotoxicol. Teratol.*, 1989, **11**, 533–537.

144. K. C. Breen and C. M. Regan, 'Lead stimulates Golgi sialyltransferase at times coincident with the embryonic to adult conversion of the neural cell adhesion molecule (N-CAM).' *Toxicology*, 1988, **49**, 71–76.

145. C. M. Regan and K. Keegan, 'Neuroteratological consequences of chronic low-level lead exposure.' *Dev. Pharmacol. Ther.*, 1990, **15**, 189–195.

146. K. J. Murphy, G. B. Fox, J. Kelly *et al.*, 'Influence of toxicants on neural cell adhesion molecule-mediated neuroplasticity in the developing and adult animal: persistent effects of chronic perinatal low-level lead exposure.' *Toxicol. Lett.*, 1995, **82/83**, 271–276.

147. P. T. McCauley, R. J. Bull, A. P. Tonti *et al.*, 'The effect of prenatal and postnatal lead exposure on neonatal synaptogenesis in rat cerebral cortex.' *J. Toxicol. Environ. Health*, 1982, **10**, 639–651.

148. K. R. Reuhl, D. C. Rice, S. G. Gilbert *et al.*, 'Effects of chronic developmental lead exposure on monkey neuroanatomy: visual system.' *Toxicol. Appl. Pharmacol.*, 1989, **99**, 501–509.

149. C. A. Drew, I. Spence and G. A. Johnston, 'Effect of chronic exposure to lead on GABA binding in developing rat brain.' *Neurochem. Int.*, 1990, **17**, 43–51.

150. F. Bakir, S. F. Damluji, L. Amin-Zaki *et al.*, 'Methylmercury poisoning in Iraq.' *Science*, 1973, **181**, 230–241.

151. L. W. Chang and M. A. Verity, in 'Handbook of Neurotoxicology,' eds. L. W. Chang and R. S. Dyer, Dekker, New York, 1995, pp. 31–59.

152. L. A. Lagunowich, S. Bhambhani, R. D. Graff *et al.*, 'Cell adhesion molecules in the cerebellum: targets of methylmercury toxicity?' *Soc. Neurosci.*, 1991, **17**, 515.

153. K. R. Reuhl, L. A. Lagunowich and D. L. Brown, 'Cytoskeleton and cell adhesion molecules: critical targets of toxic agents.' *Neurotoxicology*, 1994, **15**, 133–145.

154. G. Stoltenburg-Didinger and S. Markwort, 'Prenatal methylmercury exposure results in dendritic spine dysgenesis in rats.' *Neurotoxicol. Teratol.*, 1990, **12**, 573–576.

155. R. D. Graff, J. A. Elzer, L. A. Lagunowich *et al.*, 'Methylmercury alters expression of the neural cell adhesion molecule (NCAM) in cultured neurons.' *Toxicologist*, 1993, **13**, 168.

156. L. W. Chang, T. W. Tiemeyer, G. R. Wenger *et al.*, 'Neuropathology of trimethyltin intoxication. I. Light microscopy study.' *Environ. Res.*, 1982, **29**, 435–444.

157. W. N. Aldridge, A. W. Brown, J. B. Brierley *et al.*, 'Brain damage due to trimethyltin compounds.' *Lancet*, 1981, **2**, 692–693.

158. T. W. Bouldin, N. D. Goines, R. C. Bagnell *et al.*, 'Pathogenesis of trimethyltin neuronal toxicity. Ultrastructural and cytochemical observations.' *Am. J. Pathol.*, 1981, **104**, 237–249.

159. K. R. Reuhl, E. A. Smallridge, L. W. Chang *et al.*, 'Developmental effects of trimethyltin intoxication in the neonatal mouse. I. Light microscopic studies.' *Neurotoxicology*, 1983, **4**, 19–28.

160. T. J. Walsh, D. B. Miller and R. S. Dyer, 'Trimethyltin, a selective limbic system neurotoxicant, impairs radial-arm maze performance.' *Neurobehav. Toxicol. Teratol.*, 1982, **4**, 177–183.

161. D. E. McMillan and G. R. Wenger, 'Neurobehavioral toxicology of trialkyltins.' *Pharmacol. Rev.*, 1985, **37**, 365–379.

162. B. Earley, M. Burke and B. E. Leonard, 'Effects of captopril on locomotor activity, passive avoidance behaviour and spatial memory tasks in the trimethyltin-treated rat.' *Neuropsychobiology*, 1989, **22**, 49–56.

163. P. M. Dey, R. D. Graff, L. A. Lagunowich *et al.*, 'Selective loss of the 180 kDa form of the neural cell adhesion molecule in hippocampus and cerebellum of the adult mouse following trimethyltin administration.' *Toxicol. Appl. Pharmacol.*, 1994, **126**, 69–74.

164. J. Covault, Q. Y. Liu and S. El-Deeb, 'Calcium-activated proteolysis of intracellular domains in the cell adhesion molecules NCAM and N-cadherin.' *Brain Res. Mol. Brain Res.*, 1991, **11**, 11–16.

165. A. Sheppard, J. Wu, U. Rutishauser *et al.*, 'Proteolytic modification of neural cell adhesion molecule (NCAM) by the intracellular proteinase calpain.' *Biochim. Biophys. Acta*, 1991, **1076**, 156–160.

166. S. Lehmann, S. Kuchler, S. Gobaille *et al.*, 'Lesion-induced reexpression of neonatal recognition molecules in adult rat cerebellum.' *Brain Res. Bull.*, 1993, **30**, 515–521.

167. M. Jucker, C. Mondadori, H. Mohajeri *et al.*, 'Transient upregulation of NCAM mRNA in astrocytes in response to entorhinal cortex lesions and ischemia.' *Brain Res. Mol. Brain Res.*, 1995, **28**, 149–156.

168. A. Represa, J. Niquet, H. Pollard *et al.*, 'Cell death, gliosis, and synaptic remodeling in the hippocampus of epileptic rats.' *J. Neurobiol.*, 1995, **26**, 413–425.

169. K. Ono, A. Tokunaga, K. Mizukawa *et al.*, 'Abnormal expression of embryonic neural cell adhesion molecule (NCAM) in the developing mouse cerebellum after neonatal administration of cytosine arabinoside.' *Brain Res. Dev. Brain Res.*, 1992, **65**, 119–122.

170. D. Barbeau, J. J. Liang, Y. Robitalille *et al.*, 'Decreased expression of the embryonic form of the neural cell adhesion molecule in schizophrenic brains.' *Proc. Natl. Acad. Sci. USA*, 1995, **92**, 2785–2789.

171. M. Poltorak, I. Khoja, J. J. Hemperly *et al.*, 'Disturbances in cell recognition molecules (NCAM and L1 antigen) in the CSF of patients with schizophrenia.' *Exp. Neurol.*, 1995, **131**, 266–272.

172. K. Nihei and N. W. Kowall, 'Neurofilament and neural cell adhesion molecule immunocytochemistry of Huntington's disease striatum.' *Ann. Neurol.*, 1992, **31**, 59–63.

173. S. Kentroti, H. Rahman, J. Grove *et al.*, 'Ethanol neuronotoxicity in the embryonic chick brain *in ovo* and in culture: interaction of the neural cell adhesion molecule (NCAM).' *Int. J. Dev. Neurosci.*, 1995, **13**, 859–870.

174. M. E. Charness, R. M. Safran and G. Perides, 'Ethanol inhibits neural cell–cell adhesion.' *J. Biol. Chem.*, 1994, **269**, 9304–9309.

175. R. Ramanathan, M. F. Wilkemeyer, B. Mittal *et al.*, 'Alcohol inhibits cell–cell adhesion mediated by human L1.' *J. Cell Biol.*, 1996, **133**, 381–390.

176. A. Rosenthal, M. Jouet and S. Kenwrick, 'Aberrant splicing of neural cell adhesion molecule L1 mRNA in a family with X-linked hydrocephalus.' *Nat. Genet.*, 1992, **2**, 107–112.

177. K. R. Reuhl and P. M. Dey, in 'Toxicology of Metals,' ed. L. W. Chang, CRC Press, Baton Roca, FL, 1996, pp. 1097–1110.

178. O. S. Jorgensen, 'Neural cell adhesion molecule (NCAM) as a quantitative marker in synaptic remodeling.' *Neurochem. Res.*, 1995, **20**, 533–547.

11.08
Neurotransmitter Receptors

AMIRA T. ELDEFRAWI and MOHYEE E. ELDEFRAWI
University of Maryland at Baltimore, MD, USA

11.08.1   INTRODUCTION

Nerve cells communicate with each other and with effector cells (e.g., muscles, glands) via neurotransmitters, which are the small chemicals that nerve cells synthesize and release to activate or inhibit other cells. Neurotransmitters are recognized by highly specialized cell membrane proteins; for example, the neurotransmitter receptors. These receptors act as transducers that turn on or off the cell in which they reside upon reversible binding of their high affinity neurotransmitters. This produces changes in the receptors' three-dimensional structure (i.e., conformation), that triggers cellular response. Most neurotransmitter receptors are located in the synapse, mainly in the postsynaptic membranes of nerve, muscle, and gland cells. Few receptors are located in the presynaptic nerve terminal that adjust and control the amount of transmitter released. Neurotransmitter receptors are molecular targets for many drugs, toxins, and toxicants.

There are two major classes of neurotransmitter receptors that differ in the transduction mechanism by which the stoichiometric binding of the transmitter to the receptor initiates the cellular response, as well as structural and functional features. The ionotropic receptors are ligand-gated ion channels that have multiple subunit structures of membrane-spanning subunits. The subunits form a central aqueous pore that spans the width of the cell membrane. It is the critical amino acid residues that line the receptor's ionic channel that determine whether the channel is cationic or anionic, while the minimum diameter of the channel determines the maximum size of the hydrated ions that can pass through. These receptors are fairly large glycoproteins (M_r 250 kDa but 500 kDa for glutamate receptors) (Figure 1). At rest, the ion channel is closed and does not allow exchange of ions between the cytoplasm and the external medium in the synapse. Na^+, Ca^{2+}, and Cl^- concentrations are higher outside the cell membrane than inside, while the reverse is true for K^+. Upon binding of the neurotransmitter to its recognition site in the receptor's extracellular domain, the receptor undergoes a nanosecond conformational change from a resting to an active state that results in opening of its ion channel that allows thousands of ions to flow along their electrochemical gradients in a few milliseconds, thereby amplifying the binding signal. The ionic currents initiated by binding of excitatory neurotransmitters (e.g., acetylcholine, glutamate, ATP, and serotonin) activate the receptor's cationic channels, resulting in net cation influx. The net flux of both monovalent (Na^+ inwards and K^+ outwards) and divalent (Ca^{2+} inwards) cations contributes

to the excitatory postsynaptic potentials that cause membrane depolarization. On the other hand, ionic currents initiated by binding of the inhibitory neurotransmitters, γ-aminobutyric acid (GABA) or glycine, to their respective receptors produce Cl^- influx that leads to an inhibitory postsynaptic potential, causing membrane hyperpolarization (Chapter 9, this volume). All ionotropic receptors carry multiple binding sites for agonists, competitive antagonists, noncompetitive antagonists, and allosteric modulators, including naturally occurring toxins, drugs, and synthetic neurotoxicants.

The second class is metabotropic (m) receptors, which respond to a variety of hormones and neurotransmitters, ranging from biogenic amines (e.g., epinephrine, histamine, retinol) to peptides (e.g., substance P) and large glycoprotein hormones (e.g., parathyroid hormone). These receptors are small glycoproteins, each made of a single subunit (M_r from 50 kDa to 100 kDa, but larger for the metabotropic glutamate receptor (mGluR)), where transduction requires that a G-protein binds to the activated receptor and activates or inhibits

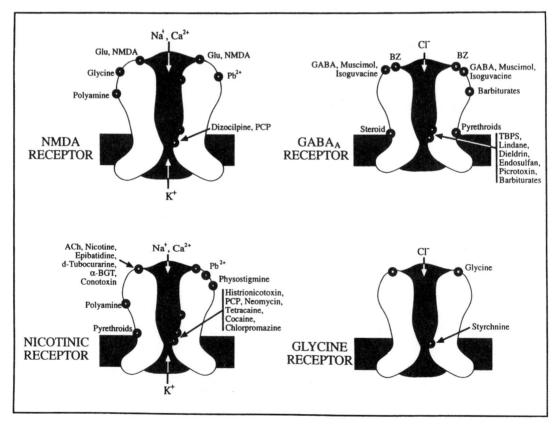

Figure 1 Models of ionotropic receptors, shown as vertical cross-sections through the cell membrane with binding sites (circles) for agonists, antagonists and modulators, including toxicants and toxins. Note that the size of NMDAR is about double that of the other receptors. Glu, glutamate; TBPS, *t*-butylbicyclophosphorothioate, PCP, phencyclidine; BZ, benzodiazepine.

second messengers that regulate cellular processes (i.e., synthesis of cAMP or inositol triphosphate. These receptors (Figure 2) share a common membrane topography consisting of seven hydrophobic transmembrane segments (TM1 → TM7) connected by three extracellular (e1, e2, e3) and three intracellular loops (i1, i2, i3), an extracellular amino terminal and an intracellular carboxy terminal. This structure transmits (via conformational changes) the extracellular signal to the cytoplasmic surface where the interaction with the G protein occurs. A major feature is the large i3 that carries most of the binding domain of G proteins in the regions proximal to the membrane. The metabotropic receptor carries a single agonist and competitive antagonist binding site which is in a hydrophobic pocket buried in the transmembrane core of the receptor, while peptide ligands bind to both the extracellular and transmembrane domains. Tryptophan and proline residues, within their hydrophobic transmembrane domains, are highly conserved across the entire superfamily of G-protein-coupled receptors and play key roles in receptor activation,

expression, and ligand binding.[1] Almost all metabotropic receptors have an Asp residue in TM2, and several cyst residues in the extracellular loops that stabilize the receptor in a functional conformation. There is an Asp in mid TM3 that is conserved in all biogenic amine receptors, and is implicated in binding of protonated amines. In a few cases there is also a noncompetitive binding site that allosterically inhibits or modulates the agonist site.

Binding of the agonist to its receptor causes a conformational change that allows the receptor to interact with the G-protein in the cell membrane. This protein is composed of an α subunit and tightly connected β–γ subunits. The high affinity agonist-receptor-G-protein catalyzes guanine nucleotide exchange on the α subunit of the G protein leading to dissociation into its α and βγ subunits. The activated GTP-bound form dissociates from the receptor and activates an effector protein, modulating levels of intracellular second messengers, thereby amplifying and transducing the signal. Pertussis toxin causes ADP ribosylation of the α subunit, thereby preventing the G-protein

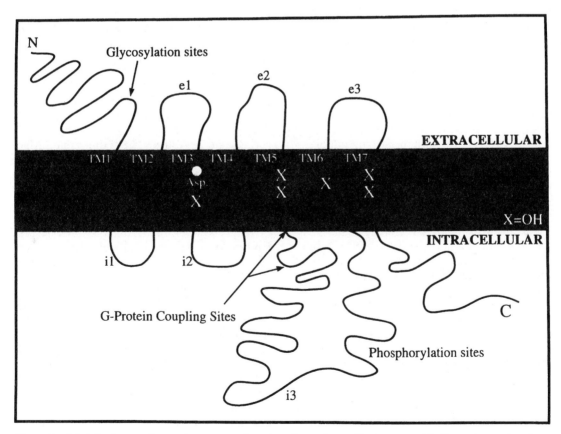

Figure 2 Model of a metabotropic receptor shown as the amino acid sequence with the amino and carboxy termini outside and inside the cell membrane, respectively, the 7 transmembrane (TM) regions, 3 extracellular (e) and 3 intracellular (i) regions. The agonist binding site is shown as an aspartate (COOH) and several hydroxyl groups (X) in different TM regions.

from being activated by the receptor. At least 25 different G protein subunits (15 α, five β, and five γ subunits) have been identified that mediate the stimulation or inhibition of a variety of enzymes (adenylyl cyclase, guanylyl cyclase, phospholipases C and A_2) and ion channels (Ca^{2+} and K^+). A single receptor can activate several of these pathways in a cell, but the predominant one may vary between cells by coupling to different G proteins.[2] Because of the involvement of several proteins and reactions, metabotropic receptor responses are much slower than those of ionotropic receptors, which take <10 ms.

Both classes of receptors have sites of post-translational modification (e.g., for glycosylation on the extracellular portions of the protein, palmitoylation and disulfide bonding). These receptors also have internal sites for phosphorylation by several protein kinases, thereby imparting stabilization of the protein conformation in the membrane. After activation by an agonist, the receptors are rapidly "desensitized" to nonresponsive conformations, which results from this phosphorylation. The rates of desensitization of ionotropic receptors are much faster than those of metabotropic receptors. If agonist exposure persists, the cell reduces receptor numbers by internalizing them; they may be either returned to the surface or catabolized by the cell if exposure to the agonist is for hours or days. This process results in "downregulation" of receptor numbers and may be a result of, or accompanied by, a reduction in receptor gene expression and receptor synthesis. The opposite (i.e., "up regulation") occurs with persistent inhibition by an antagonist. Thus, cells with upregulated receptors are highly sensitized when they are exposed to an agonist. This receptor plasticity allows cells to function under adverse conditions of exposure to drugs or toxicants.

Advances in molecular biology made possible site-directed mutagenesis, deletion mutations and formation of chimeras, expression of the genes in cell lines and measurement of microchanges in ionic conductances produced by drugs or toxicants. This in turn has led to better understanding of the molecular basis of receptor structure and function as well as pharmacological and toxicological properties. Such information is necessary to understand neurotoxicity and represents a major portion of this chapter. Considering the tremendous complexity of the nervous system, its control of body function and behavior, and the major roles played by neurotransmitter receptors, the activation, inhibition or modulation of these receptors by toxicants, toxins, or drugs have major impact on body function. However, the nervous system also has plasticity and adaptive mechanisms which modulate the adverse effects of toxicants. The actions of only a few toxicants and toxins on neurotransmitter receptors have been studied; these affect receptors directly or indirectly via changes in transmitter concentration in the synapse.

11.08.2　IONOTROPIC RECEPTORS

11.08.2.1　Nicotinic Acetylcholine Receptors

Acetylcholine is the transmitter for two subclasses of receptors: The nicotinic acetylcholine receptor (nAChR), which is selectively activated by nicotine and the muscarinic acetylcholine receptor (mAChR), which is selectively activated by muscarine and inhibited by atropine. There are more mAChRs than nAChRs in mammalian brain, but the nAChR also controls skeletal muscle function, and its major role in ganglia magnifies its impact on various body functions. Significant actions of brain nAChRs include memory and learning and mediation of nicotine addiction.

The nAChR was the first neurotransmitter receptor to be identified in membranes, purified,[3] cloned, and its subunit composition, amino acid sequences, binding sites and structure determined. It led the way in receptor discoveries and is frequently used as a model for understanding homologous and other ligand-gated ion channels. Two dimensional crystalline arrays of this receptor revealed a 120 Å by 80 Å protein, 60 Å outside the extracellular surface and 20 Å inside the cytoplasmic surface surrounding a 25 Å channel.[4] Each of its five subunits has four transmembrane regions, with the second regions of all subunits lining the channel (Figure 1).

α-Bungarotoxin (αBGT), the 8 kDa peptide in the venom of the banded krate *Bungarus multicintus,* is an irreversible inhibitor of skeletal muscle and *Torpedo* nAChRs. Its introduction as a selective probe,[5] has resulted in classification of nAChRs into: (i) skeletal muscle and *Torpedo* nAChRs that are inhibited irreversibly by α-BGT, (ii) neuronal nAChRs that bind α-BGT, and (iii) neuronal nAChRs that do not bind α-BGT. These subtypes differ in subunit composition, properties of their ion channels, function, pharmacology and toxicology.[6,7] Four genes encode the *Torpedo* and adult skeletal muscle nAChRs, which are composed of two α1, one β1, one γ, and one δ subunits. Only the α subunit carries the ACh-binding site externally and binding of ACh to the two α subunits opens the channel fully. Sixteen genes have been cloned that encode nAChRs subunits (α, β, γ, δ, and ϵ). The nAChR may be formed by several identical subunits (i.e., homomeric). Example are the α7, α8, or α9 homomeric neuronal nAChRs, which have ionic channels with identical properties and a unique feature, that is, a high ion channel selectivity for Ca^{2+}, 11-fold that of *Torpedo* $(\alpha 1)_2\beta 1\gamma\delta$ receptor channel. The affinities that the receptor has for ACh and nicotine and its desensitization rates vary between the different nAChRs based on their subunit composition. α-BGT binds only to a few α subunits, namely α1 of skeletal muscle and *Torpedo* receptor and neuronal α7, α8, and α9.

The ACh-binding site is extracellular, stereospecific and fully conserved in the α subunit across species. It includes three tyrosine and one tryptophan, as well as two cysteine residues that form a disulfide bond. When the nAChR is activated its channel opens as a result in part of movement of residues forming the ion selectivity gate in the ion channel (close to the cytoplasmic end).[8] The neuronal homomeric α7 nAChR changes from cationic to anionic by only three mutations in each of its five α subunits.[9] Partial agonists do not open the channel to the extent produced by ACh. There are full agonists such as the insecticide ($-$)nicotine, and ($+$)anatoxin (produced by blue-green algae *Anabaena flos aquae* that are toxic to water fowl),[10] cytosine and epibatidine, that has highest selectivity on two hippocampal nAChR subtypes: the α7-based nAChR (inhibited by αBGT) and an α4β2-based receptor (inhibited by dihdryo-β-erythroidine). A novel agonist site was discovered on the α subunit that binds the carbamate anticholinesterase physostigmine and its action is inhibited only by the open channel blockers.[7] However unlike ACh, physostigmine elicits a single-channel current without leading to endplate potentials.

In general, there are two groups of antagonists (Figure 1): (i) Competitive antagonists, for example, tubocurarine and α-conotoxin (from the *Conus* marine snail) and the 8 kDa α-BGT, that bind to the ACh-binding site and inhibit ACh binding and opening of the ionic channel in a voltage-independent manner (ii) noncompetitive antagonists that bind to other sites on the nAChR and allosterically modulate its function. Many have voltage-dependent effects, inhibiting ion channel conductance (maximal current and/or duration). These bind to high affinity sites within the ionic channel, for example, the frog skin neurotoxin perhydrohistrionicotoxin,[11] phencyclidine, opiates, antipsychotics, antibiotics, and antiviral drugs.[12] Others bind to different low-affinity sites, for example, inhalation anesthetics and ethanol.[13] Organophosphate anticholinesterases have been shown not only to inhibit acetylcholinesterase, but also to bind directly to nAChRs, acting as agonists and/or blockers of their ionic channels and enhancers of receptor desensitization.[14] Diisopropylfluorophosphate was found to bind to a unique site on the receptor that is different from the sites that bind ACh, phencyclidine, or histrionicotoxin. Binding increases the receptor's affinity for these other ligands. Low concentrations of lead ($3\,\mu M$) potently inhibit the channel subunit of only one subtype, the fast-desensitizing, methylcaconitine-sensitive nAChR in the hippocampus of the developing rat brain, that has high permeability to Ca^{2+}.[15]

11.08.2.2 Glutamate Receptors

Glutamate is the major excitatory neurotransmitter in mammalian brains, and its receptors play important roles in development, learning and memory as well as in mediating neurodegenerative consequences of hypoxemia, epilepsy and several diseases. Many excitotoxins (e.g., quisqualic, domoic, and kainic acids) are agonists of glutamate receptors and cause death and degeneration of neurons. They have been used to develop models for brain degenerative diseases. Glutamate action is mediated by at least three families of ionotropic glutamate receptors (iGluRs) and a family of G-protein coupled metabotropic glutamate receptors (mGluRs). All these GluRs have larger subunits than those of other neurotransmitter receptors in their classes. At least 22 different functional genes for iGluR subunits have been cloned.[16] There are three iGluR families which are activated by glutamate but differ in sensitivites to other agonists: (i) AMPA/kainate receptors that are activated by either AMPA or kainate. They are composed of either subunits Glu 1–4 and have higher affinity for AMPA than kainate or subunits Glu 5–7 and have low affinity for AMPA, but high affinity for kainic and domoic acids. (ii) Kainate receptors made of subunits KA_1, KA_2, and GluR, that have high affinity only for kainate. (iii) *N*-methyl-D-asparate (NMDA) receptors. This receptor and the AMPA/kainate receptors, that do not have the GluR2 subunit, have significant permeability to Ca^{2+} in addition to Na^+. This is attributable to a single amino acid in M2, that is, arginine in GluR2 vs. glutamate in the others.[17] The excessive rise in cytoplasmic Ca^{2+} that results from NMDAR activation, which can lead to neuronal death, occurs from Ca^{2+} influx as well as Ca^{2+} release from an intracellular pool, that can be inhibited by the skeletal muscle relaxant dantrolene.[18]

The NMDARS have pharmacological and physiological properties that distinguish them from the AMPA/kainate receptors. They require glycine as a co-agonist that potentiates glutamate action and are activated by glutamate and NMDA but not AMPA or kainate. This receptor's channel is permeable to Na^+, K^+ and Ca^{2+} and allows two to three times the current of AMPA/kainate receptors. The NMDAR-mediated responses are selectively antagonized competitively by 2-amino-5-phosphonopentanoate. The endogenous tryptophan metabolites quinolinic and kynurenic acids, acitvate and inhibit NMDA responses, respectively. The anesthetic drugs phencyclidine and ketamine and the anticonvulsant dizocilpine block NMDA responses noncompetitively, acting as channel blockers. NMDA responses have low sensitivity to quinoxalinediones, which are competitive antagonists of AMPA/kainate receptors.

Six NMDAR subunit genes have been cloned: NR1 and NR2A, 2B, 2C, $2D_1$, and $2D_2$.[19] This receptor is likely to have a heteromeric pentameric structure and its subunits are like those of the AMPA/kainate receptors ($M_r \approx 100$ kDa), which are twice as large as nAChR subunits. The NR-1 subunit carries the glutamate and glycine binding sites (at least seven isoforms identified). The NR-2 subunit is larger with four isoforms (NR-2A-D). By analogy with the nAChR, the receptor may carry on its extracellular portion two glutamate-binding sites on two α-subunits and two glycine binding sites on the two NR-1 subunits, that increase the receptor's affinity for glutamate. At the glycine site, agonists are (R)serine and (R)alanine. It is increasingly evident that TM2 in GluRs enters but does not cross the membrane.[20]

The NMDAR also has one or more polyamine binding sites, one in the pore, which explains their dual actions. Binding of spermine increases glycine's binding affinity, potentiates the maximal effect of NMDA and glycine and also produces voltage dependent block. In the resting state, the receptor channel is blocked by Mg^{2+}, which dissociates if the membrane is partially depolarized by activation of other excitatory receptors as may occur with excessive release of glutamate. There are also polyamine binding sites on the quisqualate-sensitive GluR, that activates locust muscle: it is inhibited in a voltage-dependent manner by polyamine-containing venom neurotoxins such as those produced by orb web spider *Argiope* sp.[21] and the predatory wasp *Philanthus triangulum*.[22] The latter toxins were also found to modulate and noncompetitively inhibit *Torpedo* and muscle nAChRs[23] and rat brain kainate, NMDA, and quisqualate GluR.[24]

Polyamine containing chemicals cannot be used as insecticides, because they do not penetrate into the nervous system as a result of their polyionic nature, but are potent if injected into insects.

Both nitric oxide (NO) and carbon monoxide (CO) are gases that are not stored in vesicles or released, but are produced in neurons and permeate cell membranes easily. Both dilate blood vessels by stimulating guanylyl cyclases. Yet these gases are suggested to function as "retrograde messengers" from the postsynaptic to the presynaptic neuron. Stimulation of the Ca-permeable NMDAR results in activation of a Ca^{2+}-dependent NO synthase.[25] Similar observations were made with CO,[26] which is produced by heme oxygenase and detected in the hippocampus. Both gases have been shown to be possibly involved in long-term potentiation (LTP), and may mediate glutamate neurotoxicity. LTP is a form of synaptic plasticity that is accepted as a cellular model for stabilization of synapses during development, learning and memory and can spread to neighboring neurons by a retrograde signal. Enhanced release of neurotransmitter possibly occurs in presynaptic boutons that are associated with the potentiated synapses, thereby affecting all synapses that are located on these boutons.

The strong voltage dependence and high Ca^{2+} permeability of NMDARs confer integrative properties utilized in generation of motor activity, regulation of neuronal development in embryonic nervous system and stabilization of synapses that are important for memory. Excessive activation of glutamate receptors produce convulsions and may trigger neuronal cell death, such as observed in the degenerative consequences of stroke, head or spinal cord injury, cardiac arrest, pathogenesis of epilepsy and possibly Parkinson's, Huntington's and Alzheimer's diseases. An example of excessive activation of iGluRs was observed in an incidence of poisoning after eating blue mussels in Canada (which contained diatoms that produce domoic acid). Other amino acid excitotoxins are kainate and quisqualate as well as beta-oxaloamino-alanine in grass peas, often eaten during famine in Asia and Africa and may cause lathyrism and beta-methylamine-alanine in Guam, which is linked to ALS and Parkinson's diseases (see Chapter 31, this volume). Lead is a potent inhibitor of the NMDAR as it is of the α7 homomeric nAChR, that also has high permeability for Ca^{2+} (see Chapter 28, this volume). The cellular mechanism leading to cell death is believed to be due to Ca^{2+} influx and possibly NO that increases presynaptic release of glutamate. It is likely that both major classes of iGluRs and mGluRs contribute to acute and chronic neurotoxicities. Several drugs that inhibit NMDAR noncompetitively (e.g., phencyclidine and dizocilpine) have been shown to impair memory and learning and so has Pb^{2+}.[27,28]

11.08.2.3 GABA Receptors

GABA is the major inhibitory transmitter in the central nervous system, with 20–50% of all brain synapses using it. Most GABARs are ionotropic (i.e., $GABA_AR$), but one is metabotropic (i.e., $GABA_BR$). The latter is inhibited by baclofen and not bicuculline, is located presynaptically and postsynaptically and its subtypes differ in drug sensitivities. The $GABA_AR$ is composed of five different subunits, responds to the agonists muscimol and isoguvacine and is inhibited by the convulsants bicuculline competitively and picrotoxinin noncompetitively.[29] Up to 1996, 15 structurally related $GABA_A$ subunits have been identified: 6α, 3β, 3γ, 1δ, and 2ρ subunits, which can theoretically form up to 150 000 different subunit arrangements. Although the subunit composition of any $GABA_AR$ subtype is still unknown, it is suggested that the receptor with the composition α_1, β_2 (or β_3), γ_2 is the most abundant subtype. The α subunit, is most critical since it binds GABA and agonists (muscimol and isoguvacine), competitive antagonists (e.g., bicuculline) and benzodiazepine. The γ subunit sets the overall conformation. Benzodiazepines increase the receptor's affinity for GABA. "Inverse benzodiazepine receptor agonists" bind to these sites and reduce GABA-induced Cl^- influx, while "benzodiazepine receptor antagonists" do not, but inhibit the actions of both benzodiazepine agonists and inverse agonists. There are sites close to or within the receptor's Cl^- channel that bind picrotoxinin, some bicyclic cage convulsants (e.g., butylbicyclophosphorothionate (TBPS)), barbiturates and insecticides. Binding to these sites inhibits the receptor's function (Figure 1). These toxicants may inhibit GABA binding allosterically and only partially modulate the benzodiazepine site. There are also sites that bind sedative hypnotic barbiturates (e.g., pentobarbital) and enhance GABA's action by increasing the average channel open duration. Steroids (e.g., anesthetic alphaxalone) bind to different sites at the lipid–receptor interface.

Insecticides are designed to rapidly kill insects and thus the majority are neurotoxicants. Cyclodienes were produced in the 1950s but most were banned from use in the 1970s (e.g., aldrin, dieldrin, endrin) because of persistence in the environment and biomagnifica-

tion. This was long before the discovery that their acute toxicity is due to stereospecific inhibition of $GABA_ARs$ in mammalian brain (Figure 3)[30–32] and in insect skeletal muscle, by binding to sites in the receptor's Cl^- channel. Another cyclodiene, α-endosulfan is still in use. In addition, lindane, bicycloorthobenzoates, and dithianes are potent insecticides that also bind to the receptor's Cl^- channel.[32] By inhibiting this major inhibitory receptor, these insecticides produce convulsions.

Other insecticides that target $GABA_ARs$ includes moxidectin and avermectins, which are a family of macrocyclic lactones used in control of nematodes and arthropods. Ivermectin is widely used as an anthelmintic in veterinary medicine and to treat onchocerciasis (river blindness) in humans. Ivermectin is a miticide and insecticide used in crop protection, that acts on $GABA_ARs$ by binding to a unique site that opens the GABA receptor channel. It noncompetitively inhibits binding of the channel inhibitor ethynylbicycloorthobenzoate. This site is unaffected in a *Drosophila melanogaster* strain, with a point mutation in its $GABA_AR$ that makes it resistant to cyclodiene insecticides. Such mutations have reduced the potencies for other channel blockers, for example, lindane, chlordane, and picrotoxinin.[33]

Fungi capable of producing tremorgenic mycotoxins, including metabolites produced by *Aspergillus flavus*, contaminate forages, corn and silage. In cattle, these mycotoxins cause "staggers" syndrome (muscle tremor, uncoordinated movements, weakness in hind legs and stiff forelegs). Several tremorgenic mycotoxins (aflatrem, paspalinine, paxilline and verruculogen are potent inhibitors of GABA-induced $^{36}Cl^-$ flux and binding of $[^{35}S]t$-butylbicyclophosphorothioate to the $GABA_AR$'s channel in rat brain. The nontremorgenic verruculotoxin have no effect.[34]

11.08.2.4 Glycine Receptors

Glycine is the major inhibitory neurotransmitter in the spinal cord. The glycine receptor (GlyR) is a glycine-gated Cl^- channel, that is selectively blocked by strychnine (Figure 1). There are four known GlyR subunits: three α and one β. The $\alpha1$ is abundant in spinal cord but less so in brain. The $\alpha2$ transcripts are found in several brain regions, but much less $\alpha3$ are expressed. By contrast, the β subunit transcripts are abundant in all brain regions, even in areas devoid of $[^3H]$strychnine binding,[35] which is a selective noncompetitive inhibitor of GlyRs. $[^3H]$Strychnine binds to sites in the receptor's ionic channel, similar to the mechanism of

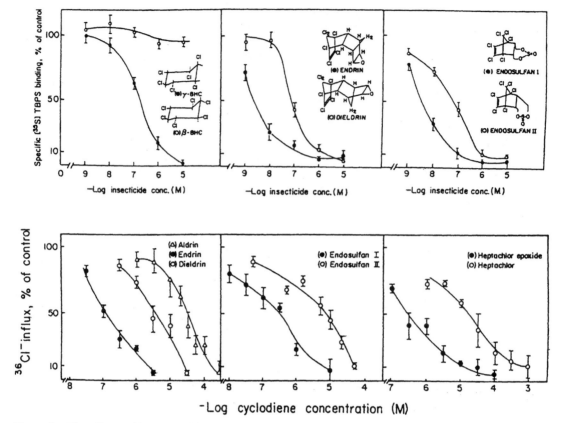

Figure 3 The effects of isomers of hexachlorocyclohexane (BHC) and cyclodiene insecticides on GABA$_A$ receptor. (Top) Inhibition of the specific binding of [^{35}S]TBPS (2 nM) to GABA$_A$ receptors of rat brain (reproduced by permission of Academic Press from *Pesticide Biochem. Physiol.*, 1985, **24**, 95–102.). (Bottom) Inhibition of ^{36}Cl$^-$ influx into resealed rat brain membranes (i.e., microsacs) is a biochemical correlate of the physiological function of the GABA$_A$ receptor (reproduced by permission of Academic Press from *Toxicol. Appl. Pharmacol.*, 1987, **88**, 313–321).

action of picrotoxinin on the GABA$_A$R, histrionicotoxins on the nAChR and dizocilpine on the NMDAR. All subunits have the four TM segments typical of pentameric receptor structure, except that both amino and carboxy terminals are outside the cell. The GlyR protein, isolated from mammalian spinal cords, has two peptides: α (48 kDa) and β (58 kDa). It is possible that the functional receptor has the α$_2$, β$_3$ structure. Although little is known about the pharmacology or toxicology of GlyRs, based on its structural similarity to other ionotropic receptors, it is expected that many other drugs, toxicants and toxins would also be found to affect GlyR function.

11.08.3 METABOTROPIC RECEPTORS

11.08.3.1 Muscarinic Acetylcholine Receptors

The five cloned mAChR genes (m$_1$–m$_5$) share the highest degree of identity within the TM domains and short connecting loops.[36] The amino acid sequence at the N-terminus of i3 (mainly the first 16–21 amino acids) differs clearly between the PI- (M$_1$, M$_3$, and M$_5$) and cAMP- (M$_2$ and M$_4$) coupled mAChRs, but is well conserved within the two functional receptor classes. All five subtypes share ~145 amino acid residues, but virtually no sequence similarity within the third cytoplasmic loop excepting areas proximal to the membrane. The mAChR subtypes differ in their distribution, drug specificities, transduction mechanisms and functions. They have similar affinities for ACh, the antagonists quinuclidinyl benzilate and atropine, but different affinities for selective antagonists. The M$_1$ receptor (found in ganglia, brain) has a high affinity for pirenzepine but low affinity for himbacine, while the reverse is true for the M$_2$ receptor (found in cardiac and smooth muscle and brain). It is suggested that the positively charged ammonium group of muscarinic ligands binds to the carboxylate side chain of Asp147 residue, while a series of hydroxyl-containing Thr and Tyr residues in TM 3, 5, 6, and 7 (conserved among all mAChRs) are critical for high-affinity agonist binding, possibly via hydrogen binding. While several drugs are more selective on one of the subtypes than others, none has more than twofold selectivity for a single subtype over all others. The subtype selectivity depends on several regions located on different TM domains, which differ from tricyclic antagonists (e.g., pirenzepine, himbacine) to the 4-diphenylacetoxy-*N*-methylpiperidine type (with high affinity for M_3). Certain mAChRs have in addition, an allosteric antagonist binding site for gallamine on the heart M_2 receptor. There is also the positive allosteric effect of alcuronium, another neuromuscular relaxant, whose binding to the mAChR increases the ACh-binding site's affinity for agonists.

The various subtypes can also be differentiated based on their selectivities for G-proteins and functional responses.[37] The M_1, M_3, and M_5 receptors selectively couple with a PTX-insensitive G-protein to stimulate phospholipase C or to open Ca^{2+}-dependent K^+ channels, while M_2 and M_4 (postsynaptic and presynaptic in brain) couple with a PTX-sensitive G-protein that inhibits adenylyl cyclase. The M_2 and M_4 receptors also weakly stimulate phospholipase C, open nonspecific cation conductances, inwardly rectify K^+ conductances and inhibit Ca^{2+} conductances. Conversely, choleratoxin activates Gs proteins, which couple to M_1, M_3 and M_5 receptors. A single mAChR can activate multiple signaling effectors simultaneously, including phospholipases A2, C and D, tyrosine kinases and voltage-insensitive Ca^{2+} channels. Activation of presynaptic M_2 receptors on cholinergic neurons in the heart or on noradrenergic neurons inhibits release of these transmitters.

Although some neurotoxicants affect the subtypes similarly, others are highly selective. Muscarine, found in *Amanita muscaria* and *Inocybe* mushrooms, activates all mAChRs and produces excessive stimulation. The resulting lowering of heart rate and blood pressure can be lethal. Inhibition of acetylcholinesterases by organophosphates and carbamates produces similar toxicity symptoms as does muscarine activation of mAChRs, with atropine acting as an effective antidote in both cases. Repeated exposure to organophosphate anticholinesterase increases ACh concentration in the synapse and produces downregulation of mAChRs[38,39] (as a result of reduced gene transcription)[40] and nAChRs.[41] Not only do organophosphates indirectly activate mAChRs via the increased ACh concentration in the synapse, but they also do so by binding directly to mAChRs. Examples are the paraoxon activation of presynaptic M_4[42] and M_2 subtypes,[43] which would reduce ACh release thereby ameliorating the excitatory paraoxon effect on postsynaptic mAChRs and associated toxicity. Conversely, paraoxon would exaggerate the effect of excessive stimulation by ACh if it activates directly the inhibitory postsynaptic M_2 receptor in cardiac muscle, for which it has a high affinity[44] (see also Chapter 26, this volume).

11.08.3.2 Adrenergic Receptors

The sympathetic system, as opposed to the parasympathetic, is normally active continuously with degrees that vary between organs and with time. This allows rapid adjustments to changes in the environment that affect heart rate, blood pressure and flow, temperature, and so on, which prepares the organism for "fight or flight." Adrenergic receptors (ARs) mediate the central and peripheral actions of the neurotransmitter norepinephrine (noradrenaline) and epinephrine (adrenaline). There are three major types of ARs for 10 subtypes that exhibit receptor diversity:[45,46] α_{1A-D}, α_{2A-C}, β_{1-3}. The α_1ARs active phospholipase C or Ca^{2+} channels, while α_2ARs inhibit adenyl cyclase. βARs activate adenylyl cyclase. The subtypes have different pharmacological profiles, whish is critical for the development of therapies and understanding neurotoxicities. No agonist shows selectivity for only one subtype. While αARs have higher affinity for norepinephrine than epinephrine, the reverse is true for βARs. Phenylephrine has high affinity for α_1, but not α_2, and clonidine has high selectivity for α_2. Albuterol has higher selectivity for β_2 than β_1, while dobutamine has the reverse selectivity for β_1. Antagonists are highly selective between α and β, but most do not differentiate much between the various subtypes within each. Thus, αARs have high affinities for phentolamine, phenoxybenzamine and ergotamine, but very low for propranolol or pindolol. The reverse is true for βARs. Oxymetazoline is 100-fold more selective for the α_{2A} than α_{2B}, while prazosin is 60-fold more selective for α_{2A} than α_{2B}.

The toxicity of methylxanthines is partially mediated by ARs. The ARs enhance adrenergic responses via their inhibition of phosphodiesterase which raises cAMP levels initially triggered by ARs. The highly toxic ergot alkaloids act as partial agonists (ergotamine) and antagonists (dihydroxyergotamine) of αARs. These alkaloids are produced by *Claviceps purpurea* fungus that infects cereals and are responsible for epidemics of poisoning. The symptoms include mental disturbances and intense painful peripheral vasoconstriction that leads to gangrene.

11.08.3.3 Glutamate Receptors

The metabotropic glutamate receptors (mGluRs) have no discernible homology to any subfamily of G-protein linked receptors. They are much larger in molecular weight due to an extended N-terminal region with many conserved cysteine residues that possibly function as the glutamate binding region. In this respect, they are more homologous to the iGluRs than they are to other metabotropic receptors. There appears to be three subgroups of mGluRs:[47] (i) $mGluR_1$ and $mGluR_5$, which stimulate IP3 synthesis and their agonist potency is quisqualate > glutamate > ibotenate; (ii) $mGluR_2$, and $mGluR_3$, which inhibit cAMP synthesis, but their agonist potency is glutamate > ibotenate > quisqualate; and (iii) $mGluR_4$, $mGluR_6$, and $mGluR_7$ which also inhibit cAMP synthesis and their agonist potency is L-AP4 > glutamate. It is suggested that $mGluR_4$, $mGluR_6$, and $mGluR_7$ act as inhibitory autoreceptors on glutamatergic neurons, while $mGluR_2$ inhibits GABA release from GABAergic dendrites of granule cells. These mGluRs have variable distribution in the brain and are found on glia as well as neurons. Activation of mGluRs produces numerous responses including increased K^+ and Ca^{2+} channel conductances. They vary in their transduction mechanisms, such that $mGluR_1$ stimulates IP synthesis, while $mGluR_2$ inhibits cAMP formation. They contribute to synaptic efficacy and mediate excitotoxicities of glutamic, quisqualic, kainic, and domoic acids.

11.08.3.4 Serotonin Receptors

Serotonin (or 5-hydroxytryptamine (5-HT)) is an autacoid that acts both at microdistances as neurotransmitters and at long distances as a hormone. Most of the serotonin in the body is synthesized and stored in enterochromaffin-tissue associated with the gastrointestinal tract, and is released in the blood as a potent vasoconstricting agent, with >90% of it sequestered in platelets. It is also synthesized and released by neurons, serving as a neurotransmitter in both the central and peripheral nervous system. Serotonin produces numerous pharmacological effects mainly because of the diversity of its receptors, that are either ionotropic or metabotropic.[48] There is also an ionotropic $5-HT_3R$. This ligand-gated monvalent cation channel is present in high density in

the brain region that contains the emetic centres and its antagonists (e.g., ondansetron) are potent antiemetics. The metabotropic 5-HTRs are important targets in the brain for action of numerous therapeutics including antidepressants, anxiolytic, and antimigraine drugs. By analogy with nAChRs, these drugs are likely to act as channel blockers. The metabotropic 5HTRs are linked to either Gp protein and their activation decreases cAMP synthesis ($5-HT_{1A-F}$) or G_s protein, that activates phospholipase C and increases synthesis of IP_3 and diacylglycerol ($5-HT_{2A-C}$). Although there are many high-affinity agonists and antagonists for all subtypes, there are none that are totally selective for one subtype. Ketanserin for $5-HT_{2A}$ is an example of an antagonist.

11.08.3.5 Purine Receptors

The purine nucleotides, adenosine, adenosine diphosphate (ADP), and adenosine triphosphate (ATP), are autacoids that serve as neurotransmitters in addition to their well known cellular functions as intermediates in energy pathways and mediators of enzymatic reactions or cofactors. Adenosine-induced bradycardia and decrease in blood pressure has led to its common use in treatment of supraventricular arrhythmias. The P_1-purinoceptors are the adenosine receptors that are more responsive to adenosine and AMP than to ATP and ADP and are antagonized by methylxanthines (e.g., theophylline and caffeine). The P_2-purinoceptors are more responsive to ATP and ADP than to AMP and adenosine and are not recognized by methylxanthines. Four adenosine receptors have been cloned (A_1, A_{2A}, A_{2B}, and A_3).[49] Activatiaon of A_1 and A_3 receptors is associated with inhibition of cAMP synthesis, whereas that of A_{2A} and A_{2B} stimulates cAMP synthesis; the latter is located on presynaptic membranes.

While ATP depolarizes neuronal membranes, ADP produces hyperpolarization; thus its antagonists are stimulants. ATP can mediate fast neurotransmission via ligand-gated ion channels that cause membrane depolarization as well as pre- and postsynaptic modulatory actions. ATP is also a co-transmitter for ACh, norepinephrine and substance P. It is stored in the vesicles and released from neurons and from activated target cells. Five P_2 receptor subtypes have been cloned P_{2X-u},[50] which include ionotropic and metabotropic receptors. The P_{2x} gates a cation channel for Na^+, K^+, and Ca^{2+}, and its high Ca^{2+} permeability raises the possibility that ATP may be involved in neurotoxicity similar to glutamate. The metabotropic receptors P_{2y}, P_{2T}, and P_{2u} activate phospholipase C, while the first two in addition inhibit adenylyl cyclase. The pharmacology of these receptors is not yet well developed and selectivity is based on agonist responses. The most potent agonist on P_{2x} is α,β-methylene ATP, on P_{2y} it is 2-methylene SATP, on P_{2u}, it is UTP or ATP and on P_{2T} 2-methylene SADP. No natural toxins of ATP receptors are known.

11.08.3.6 Histamine Receptors

Histamine is an autacoid that is closely associated with mast cells and functions as a mediator of inflammation. Like serotonin, it is also a neurotransmitter in the central and peripheral nervous systems. Its effects are mediated by three receptor subtypes with differential selectivities for both agonists and antagonists (e.g., mepyramine for H_1, ranitidine for H_2, and thiperamide for H_3).[51] They share structural and membrane topography features with other metabotropic receptors. Histamine receptors H_1 and H_2 are postsynaptic whereas H_3 is presynaptic. There are no known natural toxins or toxicants of HRs but H_1 and H_2 antagonists are potent therapeutics. The potent therapeutic H_1 agonists are effective for the treatment of allergies, but their side effects and toxicities include sedation and anticholinergic actions. Antagonists for H_2 receptor are excellent therapeutics for peptic and gastric ulcers, because of their ability to block histamine-induced gastric acid production.

11.08.3.7 Dopamine Receptors

Dopamine has diverse central and peripheral effects that include motor control, cognition, emotion, and neuroendocrine and cardiovascular regulation. Drugs affecting dopamine receptors (DARs) are used in psychiatric (schizophrenia), neurologic (Parkinson's and Huntington's diseases) and endocrine disorders. Dopamine indirectly mediates the reinforcing and addictive effects of cocaine and amphetamine, whose inhibitory actions on presynaptic amine transporters result in increased synaptic dopamine. Knowledge of their molecular and pharmacological properties should lead to more selective agents to produce desirable modulation of DA function and to better understand their potential role in mediating neurotoxicities. Dopamine may play a role in mediating CNS tissue damage since denervation of dopaminergic neurons in the striatum by 6-hydroxydopamine reduces the toxicity produced by injected ibotenate.[52] No natural toxicants or toxins are known to target DARs.

Five DARs (D_1–D_5) are known to exist in the brain, each having several variants.[53] The DARs are divided into two major classes: The D_1 and D_5 are linked to G_S protein and their activation results in increased levels of cAMP. The D_2 is coupled to a Gi protein and its activation results in decreased levels of cAMP, activation of K^+ channels and inhibition of Ca^{2+} channels. Both the D_1 and D_2 receptors appear to exist mostly as well as postsynaptically. The D_2, D_3 and D_4 receptors are unlike all other G protein-coupled receptors in that they have introns within their coding regions like the rhodopsin receptor gene. The D_3 and D_4 receptors may be presynaptic autoreceptors and are like D_2 receptors although their effector systems are not clearly defined. Amongst agonists, apomorphine has highest selectivity for D_1 and D_2 receptors, dopamine and SKF 38393 are most potent on D_1 and D_5 receptors, while bromocriptine is most potent on D_2 and D_3[54] receptors. Amongst antagonists, raclopride is most potent on D_2 and D_3 receptors. Antipsychotic drugs, that are used to treat schizophrenia, are D_2 antagonists and their long term use results in elevated D_2 receptor.

11.08.3.8 Opioid Receptors

There are three major endogenous opioid peptides that are widely distributed as neurotransmitters in the brain: the enkephalins (Met and Leu), β-endorphin and the dynorphins (A and B). Their discovery resulted from studies of the mechanism of action of the powerful analgesic morphine, which acts as an agonist to produce analgesia, but it is also a powerful reinforcer. This is a form of behavioral plasticity where changes occur in response to acute exposure to a drug which elicits a drug seeking response. Drug addiction results from activating reward pathways where opioid receptors play major roles. The actions of opioid peptides are mediated by μ, δ, and κ opioid receptors, which are PTX-sensitive G protein-linked receptors, which inhibit adenylyl cyclase, enhance K^+, and inhibit Ca^{2+} channels.[55] They have different pharmacological profiles and regional distribution in the brain. The met- and leu-enkephalins have highest potencies on the δ receptor, β-endorphin on the μ and the dynorphines on the κ receptors The antagonist naloxone is most potent on κ and μ receptors, while naltrindole is best on the δ receptor. The toxicological significance of opioid receptors is their association with the abuse of natural (morphine and codeine) and synthetic (e.g., heroin) opioids. The addictive opioids and methadone (used to treat heroin addiction) bind specifically to the μ receptor. Amino acid sequences in TM 2, 3, and 7 and the i2 loop of the subtypes are highly conserved, while the amino and carboxyl termini differ considerably. It is suggested that the amino terminal and the second and third extracellular domains constitute the peptide agonist-binding domain.

11.08.4 CONCLUDING REMARKS

Neurotransmitter receptors are vital for nervous system function. Toxicants often have multiple targets. Thus, in order to establish that a receptor is the primary target for a toxicant, several factors should be considered: (i) toxicity symptoms (behavioral or organic) should correlate with the toxicant's effect on the receptor (e.g., agonist or antagonist); (ii) excellent QSAR data; (iii) correlation of receptor affinity and reversibility of the effect with the toxicant concentration *in vitro* and the reversibility of its action; and (iv) occupancy level of the target receptor required to produce an effect. Although each receptor subtype differs in sensitivities to drugs and toxicants, there are pharmacological similarities in channel sites of some ionotropic receptors. For example, certain subtypes of both nAChR and NMDAR channels are inhibited by phencyclidine, amantadine, dizocilpine, antipsychotics, antidepressants, naltrexone and Pb^{2+}. Also, alcohol and philanthotoxins affect both receptors. It is likely that several of the noncompetitive inhibitors that block the $GABA_A R$, may also inhibit GlyR function. It may also be possible to make an intelligent guess as to which receptor a certain chemical may affect, based on the toxicity symptoms it produces.

It is important to note that the complexity of neuronal circuit interactions raises the prospects that some receptors may modulate neurotoxicities mediated by other receptors. The diversity of receptor subtypes usually reflects different drug selectivities, locations and transduction mechanisms. Accordingly, $\alpha_2 AR$ agonists were successful in preventing neurotoxicity resulting from inhibition of NMDAR by dizocilpine.[56] Another target for toxicants is the presynaptic autoreceptors (e.g., M_2 and M_4 mAChRs) on neurons that inhibit release of neurotransmitters. Presynaptic receptors are also often located on neuronal endings where other neurotransmitters are released, for example, mGluR2 inhibiting GABAergic release. Thus, when an organophosphate insecticide inhibits anticholinesterases it increases synaptic ACh concentration thereby activating postsynaptic mAChRs and nAChRs as well as presynaptic mAChRs. Since they may also directly activate mAChRs, their toxicities will depend on the toxicant's relative actions on different subtypes.

Neurotoxicants and toxins can be grouped on the basis of their mechanisms of action, mole-cular targets, symptoms produced or effects on a certain function. It is strikingly clear that many more toxicants and toxins target ionotropic receptors (primarily or secondarily) than metabotropic receptors. This is possibly because the ionotropic receptor carries many different binding sites, is a much larger protein, and is made of different subunits. On the other hand, the functions of metabotropic receptors may also be targeted via action on second messenger systems, for example, pertussis and cholera toxins. There are toxins (e.g., αBGT, α-conotoxin, anatoxin, histrionicotoxin) which act primarily on a certain neurotransmitter receptor (e.g., nAChR) and there are toxicants that affect multiple receptors, whether metals (e.g., Pb^{2+}, Zn^{2+}, Mg^{2+}) or insecticides.

Most insecticides in use today are neurotoxic such as anticholinesterases (e.g., organophosphates and carbamates), and inhibitors of Na^+ channels (e.g., pyrethroids) and neurotransmitter receptors, as primary (nicotinoids on nAChR; α-endosulfan, lindane and ivermectin on $GABA_A R$) or secondary targets (e.g., pyrethroids on $GABA_A R$ and nAChR and organophosphates on mAChR and $GABA_A R$) (see Chapters 26 and 27, this volume). They are designed to rapidly kill insects with minimal acute and chronic toxicities to mammals, birds. and fish. Since their molecular targets are generally the same in insects and other species, selective toxicity to insects is achieved by increased penetration and bioactivation, reduced catabolism, increased sensitivity or uniqueness of the insect receptor subtype (e.g., inhibitory glutamate-gated Cl^- channel) or receptor location. An example is the local control of insect skeletal muscles by *in situ* stimulatory GluR and inhibitory $GABA_A R$, rather than an *in situ* nAChR in mammalian skeletal muscles and inhibitory $GABA_A Rs$ in ganglia.

Memory is a very complex brain function[57] (see Chapters 20 and 21, this volume). It progresses from short-lived labile to long-term stable form by consolidation. Multiple components emerge at different times after training and different brain parts contribute to different memories. Various receptors serve as feedback transducers to help formulate and strengthen important connections during development, and eliminate superfluous or incorrect ones. Effects on the developing brain may be irreversible through interference with receptor expression at critical periods, and may underlie certain psychiatric disorders and cognitive deficits. Receptors that are pivotal in memory, learning, and synaptogenesis include the NMDAR, $GABA_A R$, (important for memory consolidation processes), and mAChRs for controlling interference with learning of new memories.[58] Chronic exposure to

Pb^{2+} produces cognitive deficits in humans and animals.[28] Because of the potent inhibition by Pb^{2+} of NMDAR,[27] a nAChR subtype[15] (both of which have high permeability to Ca^{2+}), a mAChR,[59] and its effect on the dopamine system it is likely that multiple molecular mechanisms are involved in Pb^{2+}-induced cognitive deficits. It has been observed that the density of dendritic spines in neurons changes dynamically by neuronal activity and they act as an independent cellular Ca^{2+} compartment, leading to the suggestion that they are not only a storage device for long-term memory but possibly for neuroprotection.[60] Such spines are rich in the Ca^{2+}-fluxing Pb^{2+}-sensitive nAChRs.[15]

Occupational exposure to organophosphate insecticides has been shown to produce cognitive deficits and psychiatric episodes.[61] Rats dosed repeatedly with low DFP, that did not produce visible cholinergic signs, had progressively impaired working memory.[62] Exposure to an organophosphate downregulates mAChRs due to reduced receptor gene transcription,[40] with decreases in inositol phosphate accumulation in the case of M_1 and M_3 receptors, and increased *myo*inositol incorporation into phospholipids.[63] Degeneration in the hippocampus was also observed in adult rats subchronically exposed to fenthion and more severely in exposed aged animals.[64] Although receptor regulation occurs specifically in the receptor that is activated or inhibited, there may also be cross-regulation, such as the activation of ARs that results in down-regulation of M_2 and M_4 mAChRs in embryonic chicken hearts.[65] Two observations made on reduced vulnerability of hippocampal mAChRs compared to mAChRs in other brain regions are intriguing. These were made in postnatal rats exposed to low-level Pb^{2+} [59] and in adult rats exposed to low-level parathion, where the difference was not due to a lower *in situ* concentration.[40] Exposure to a toxicant that adversely affects cognitive functions is most critical for the developing brain and also for the already compromised brains of the aged, with their reduced receptor numbers (see Chapter 21, this volume).

Neurotransmitter receptors are also targets for other heavy metals, because of their content of sulfhydryl and hydroxyl groups at critical sites. Examples include mAChR inhibition with mercuric chloride ($0.1\,\mu M$).[66] Zinc, which is essential for gene expression, growth, and fetal development, inhibits NMDAR,[67] thus it can act as a neuroprotective agent against excessive excitation. Zn^{2+} and Ca^{2+} also inhibit $GABA_A R$ at a site likely to be located near the channel's external orifice.[68]

The relative paucity of information on toxicants and toxins that affect neurotransmit-ter receptors is due in part to the relatively few studies that address this area or routinely test suspect chemicals on receptors. The multitude of receptor subtypes being discovered, each with a different drug and toxicant selectivity adds to the problem. Most testing in the past utilized a radioligand which labeled all the subtypes to identify a change in a receptor that is induced by the toxicant. This can easily mask a potent effect on only one subtype. The availability of subtype-selective ligands in the 1990s and the ability to measure the function of a single receptor subtype in a patch clamp of a brain neuron, should facilitate the discovery of important mechanisms of toxicant actions on neurotransmitter receptors.

11.08.5 REFERENCES

1. J. Wess, S. Nanavati, Z. Vogel *et al.*, 'Functional role of proline and tryptophan residues highly conserved among G protein-coupled receptors studied by mutational analysis of the m3 muscarinic receptor.' *EMBO J.*, 1993, **12**, 331–338.
2. C. D. Strader, T. M. Fong, M. P. Graziano *et al.*, 'The family of G-protein-coupled receptors.' *FASEB J.*, 1995, **9**, 745–754.
3. M. E. Eldefrawi and A. T. Eldefrawi, 'Purification and molecular properties of the acetylcholine receptor from *Torpedo* electroplax.' *Arch. Biochem. Biophys.*, 1973, **159**, 362–372.
4. N. Unwin, 'Neurotransmitter action: opening of ligand-gated ion channels.' *Cell*, 1993, **72** (Suppl.), 31–41.
5. R. Miledi, P. Molinoff and L. T. Potter, 'Isolation of the cholinergic receptor protein of *Torpedo* electric tissue.' *Nature*, 1971, **229**, 554–557.
6. J. Lindstrom, R. Anand, X. Peng *et al.*, 'Neuronal nicotinic receptor subtypes.' *Ann. NY Acad. Sci.*, 1995, **757**, 100–116.
7. E. X. Albuquerque, E. F. R. Pereira, N. G. Castro *et al.*, 'Neuronal nicotinic receptors: function, modulation and structure.' *Semin. Neurosci.*, 1995, **7**, 91–101.
8. D. Bertrand and J. P. Changeux, 'Nicotinic receptor: an allosteric protein specialized for intercellular communication.' *Semin. Neurosci.*, 1995, **7**, 75–90.
9. A. Villarroel, N. Burnashev and B. Sakmann, 'Dimensions of the narrow portion of a recombinant NMDA receptor channel.' *Biophys. J.*, 1995, **68**, 866–875.
10. K. L. Swanson, Y. Aracava, F. J. Sardina *et al.*, '*N*-methylanatoxinol isomers: derivatives of the agonist (+)-anatoxin-a block the nicotinic acetylcholine receptor ion channel.' *Mol. Pharmacol.*, 1989, **35**, 223–231.
11. R. S. Aronstam, A. T. Eldefrawi, I. N. Pessah *et al.*, 'Regulation of [^{3}H]-perhydrohistrionicotoxin binding to *Torpedo orellata* electroplax by effectors of the acetylcholine receptor.' *J. Biol. Chem.*, 1981, **256**, 2843–2850.
12. A. T. Eldefrawi, E. R. Miller, D. L. Murphy *et al.*, '[^{3}H] Phencyclidine interactions with the nicotinic acetylcholine receptor channel and its inhibition by psychotropic, antipsychotic, opiate, antidepressant, antibiotic, antiviral and antiarrhythmic drugs.' *Mol. Pharmacol.*, 1982, **22**, 72–81.
13. E. F. El-Fakahany, E. R. Miller, M. A. Abbassy *et al.*, 'Alcohol modulation of drug binding to the channel sites of the nicotinic acetylcholine receptor.' *J. Pharm. Exp. Ther.*, 1983, **224**, 289–296.

14. M. E. Eldefrawi, G. Schweizer, N. M. Bakry et al., 'Desensitization of the nicotinic acetylcholine receptor by diisopropylfluorophosphate.' *J. Biochem. Toxicol.*, 1988, **3**, 21–32.

15. K. Ishihara, M. Alkondon, J. G. Montes et al., 'Nicotinic responses in acutely dissociated rat hippocampal neurons and the selective blockade of fast-desensitizing nicotinic currents by lead.' *J. Pharmacol. Exp. Ther.*, 1995, **273**, 1471–1482.

16. M. Hollmann and S. Heinemann, 'Cloned glutamate receptors.' *Annu. Rev. Neurosci.*, 1994, **17**, 31–108.

17. P. H. Seeburg, 'The TIPS/TINS Lecture: the molecular biology of mammalian glutamate receptor channels.' *Trends Pharmacol. Sci.*, 1993, **14**, 297–303.

18. I. Mody and J. F. MacDonald, 'NMDA receptor-dependent excitotoxicity: the role of intracellular Ca^{2+} release.' *Trends Pharmacol. Sci.*, 1995, **16**, 356–359.

19. C. J. McBain and M. L. Mayer, 'N-Methyl-D-aspartic acid receptor structure and function.' *Physiol. Rev.*, 1994, **74**, 723–760.

20. T. E. Hughes, 'Transmembrane topology of the glutamate receptors. A tale of novel twists and turns.' *J. Mol. Neurosci.*, 1994–1995, **5**, 211–217.

21. H. Jackson and P. N. Usherwood, 'Spider toxins as tools for dissecting elements of excitatory amino acid transmission.' *Trends Neurosci.*, 1988, **11**, 278–283.

22. A. T. Eldefrawi, M. E. Eldefrawi, K. Konno et al., 'Structure and synthesis of a potent glutamate receptor antagonist in wasp venom.' *Proc. Natl. Acad. Sci. USA*, 1988, **85**, 4910–4913.

23. R. Rozental, G. T. Scoble, E. X. Albuquerque et al., 'Allosteric inhibition of nicotinic acetylcholine receptors of vertebrates and insects by philanthotoxin.' *J. Pharm. Exp. Ther.*, 1989, **249**, 123–130.

24. D. Ragsdale, D. B. Gant, N. A. Anis et al., 'Inhibition of rat brain glutamate receptors by philanthotoxin.' *J. Pharm. Exp. Ther.*, 1989, **251**, 156–163.

25. D. S. Bredt, and S. H. Snyder, 'Nitric oxide: a physiologic messenger molecule.' *Annu. Rev. Biochem.*, 1994, **63**, 175–195.

26. A. Verma, D. J. Hirsch, C. E. Glatt et al., 'Carbon monoxide: a putative neural messenger.' *Science*, 1993, **259**, 381–384.

27. M. Alkondon, A. C. Costa, V. Radhakrishnan et al., 'Selective blockade of NMDA-activated channel currents may be implicated in learning deficits caused by lead.' *FEBS Lett.*, 1990, **261**, 124–130.

28. D. A. Cory-Slechta, 'Relationships between lead-induced learning impairments and changes in dopaminergic, cholinergic and glutamatergic neurotransmitter system function.' *Annu. Rev. Pharmacol. Toxicol.*, 1995, **35**, 391–415.

29. W. Sieghart, 'Structure and pharmacology of γ-aminobutyric acid$_A$ receptor subtypes.' *Pharmacol. Rev.*, 1995, **47**, 181–234.

30. I. M. Abalis, M. E. Eldefrawi and A. T. Eldefrawi, 'High affinity stereospecific binding of cyclodiene insecticides and γ-hexachlorocyclohexane to γ-aminobutyric acid receptors of rat brain.' *Pesticide Biochem. Physiol.*, 1985, **24**, 95–102.

31. D. B. Gant, M. E. Eldefrawi and A. T. Eldefrawi, 'Cyclodiene insecticides inhibit GABA$_A$ receptor-regulated chloride transport.' *Toxicol. Appl. Pharmacol.*, 1987, **88**, 313–321.

32. J. E. Hawkinson and J. E. Casida, 'Binding kinetics of gamma-aminobutyric acidA receptor noncompetitive antagonists: trioxabicyclooctane, dithiane, and cyclodiene insecticide-induced slow transition to blocked chloride channel conformation.' *Mol. Pharmacol.*, 1992, **42**, 1069–1070.

33. L. M. Cole, R. T. Roush and J. E. Casida, 'Drosophila GABA-gated chloride channel: modified [^{3}H]EBOB binding site associated with Ala→Ser or Gly mutants of *Rdl* subunit.' *Life Sci.*, 1995, **56**, 757–765.

34. D. B. Gant, R. J. Cole, J. J. Valdes et al., 'Action of tremorgenic mycotoxins on GABA$_A$ receptor.' *Life Sci.*, 1987, **41**, 2207–2214.

35. H. Betz, 'Glycine receptors: heterogenous and widespread in the mammalian brain.' *Trends Neurosci.*, 1991, **14**, 458–461.

36. J. Wess, 'Molecular basis of muscarinic acetylcholine receptor function.' *Trends Pharmacol. Sci.*, 1993, **14**, 308–313.

37. C. C. Felder, 'Muscarinic acetylcholine receptors: signal transduction through multiple effectors.' *FASEB J.*, 1995, **9**, 619–625.

38. L. G. Costa, B. W. Schwab, H. Hand et al., Reduced [^{3}H]quinuclidinyl benzilate binding to muscarinic receptors in disulfoton-tolerant mice.' *Toxicol. Appl. Pharmacol.*, 1981, **60**, 441–450.

39. D. A. Jett, E. F. Hill, J. C. Fernando et al., 'Down-regulation of muscarinic receptors and the m3 subtype in white-footed mice by dietary exposure to parathion.' *J. Toxicol. Environ. Health*, 1993, **39**, 395–415.

40. D. A. Jett, J. C. Fernando, M. E. Eldefrawi et al., 'Differential regulation of muscarinic receptor subtypes in rat brain regions by repeated injections of parathion.' *Toxicol. Lett.*, 1994, **73**, 33–41.

41. L. G. Costa and S. D. Murphy, '[^{3}H]-Nicotine binding in rat brain: alteration after chronic acetylcholinesterase inhibition.' *J. Pharmacol. Exp. Ther.*, 1983, **226**, 392–397.

42. D. A. Jett, E. A. Abdallah, E. E. El-Fakahany et al., 'High affinity activation by paraoxon of a muscarinic receptor subtype in rat brain striatum.' *Pestic. Biochem. Physiol.*, 1991, **39**, 149–157.

43. R. A. Huff, J. J. Corcoran, J. K. Anderson et al., 'Chlorpyrifos oxon binds directly to muscarinic receptors and inhibits cAMP accumulation in rat striatum.' *J. Pharmacol. Exp. Ther.*, 1994, **269**, 329–335.

44. C. L. Silveira, A. T. Eldefrawi and M. E. Eldefrawi, 'Putative M_2 muscarinic receptors of rat heart have high affinity for organophosphorus anticholinesterases.' *Toxicol. Appl. Pharmacol.*, 1990, **103**, 474–481.

45. D. B. Bylund. 'Subtypes of α_1- and α_2-adrenergic receptors.' *FASEB J.*, 1992, **6**, 832–839.

46. A. P. Ford, T. J. Williams, D. R. Blue et al., 'α_1-Adrenoreceptor classification: sharpening Occam's razor.' *Trends Pharmacol. Sci.*, 1994, **15**, 167–170.

47. R. J. Miller, 'G-Protein linked glutamate receptors.' *Seminars in Neurosci.*, 1994, **6**, 105–115.

48. S. J. Peroutka, 'Serotonin receptor subtypes and neuropsychiatric diseases: focus on 5-HT$_{1D}$ and 5-HT$_3$ receptor agents.' *Pharmacol. Rev.*, 1991, **43**, 579–586.

49. M. E. Olah and G. L. Stiles, 'Adenosine receptor subtypes: characterization and therapeutic regulation.' *Annu. Rev. Pharmacol. Toxicol.*, 1995, **35**, 581–606.

50. T. K. Harden, J. L. Boyer and R. A. Nicholas, 'P$_2$-purinergic receptors: subtype-associated signaling responses and structure.' *Annu Rev. Pharmacol. Toxicol.*, 1995, **35**, 541–579.

51. S. J. Hill, 'Distribution, properties and functional characteristics of three classes of histamine receptor.' *Pharmacol. Rev.*, 1990, **42**, 45–83.

52. F. Filloux and J. K. Wamsley, 'Dopaminergic modulation of excitotoxicity in rat striatum: evidence from nigrostriatal lesions.' *Synapse*, 1991, **8**, 281–288.

53. D. R. Sibley and F. J. Monsma, Jr., 'Molecular biology of dopamine receptors.' *Trends Pharmacol. Sci.*, 1992, **31**, 61–69.

54. P. Seeman and H. H. Van Tol, 'Dopamine receptor pharmacology.' *Trends Pharmacol. Sci.*, 1994, **15**, 264–270.

55. T. Reisine and G. I. Bell, 'Molecular biology of opioid receptors.' *Trends Neurosci.*, 1993, **16**, 506–520.

56. N. B. Farber, J. Foster, N. L. Duhan *et al.*, 'α_2-Adrenergic agonists prevent MK-801 neurotoxicity.' *Neuropsychopharmacology*, 1995, **12**, 347–349.

57. J. DeZazzo and T. Tully, 'Dissection of memory formation: from behavioral pharmacology to molecular genetics.' *Trends Neurosci.*, 1995, **18**, 212–218.

58. M. E. Hasselmo and J. M. Bower, 'Acetylcholine and memory.' *Trends Neurosci.*, 1993, **16**, 218–222.

59. H. Bielarczyk, J. L. Tonsig and J. B. Suszkiw, 'Perinatal low-level lead exposure and the septo-hippocampal cholinergic system: selective reduction of muscarinic receptors and cholineacetyltransferase in the rat septum.' *Brain Res.*, 1994, **643**, 211–217.

60. M. Segal, 'Dendritic spines for neuroprotection: a hypothesis.' *Trends Neurosci.*, 1995, **18**, 468.

61. U. K. Misra, D. Nag, V. Bhushan *et al.*, 'Clinical and biochemical changes in chronically exposed organophosphate workers.' *Toxicol. Lett.*, 1985, **24**, 187–193.

62. P. J. Bushnell, S. S. Padilla, T. Ward *et al.*, 'Behavioral and neurochemical changes in rats dosed repeatedly with diisopropylfluorophosphate.' *J. Pharmacol. Exp. Ther.*, 1991, **256**, 741–750.

63. W. R. Mundy, T. R. Ward, V. F. Dulchinos *et al.*, 'Effect of repeated organophosphate administration on carbachol-stimulated phosphoinositide hydrolysis in the rat brain.' *Pharmacol. Biochem. Beh.*, 1993, **45**, 309–314.

64. B. Veronesi, K. Jones and C. Pope, 'The neurotoxicity of subchronic acetylcholinesterase (AChE) inhibition in rat hippocampus.' *Toxicol. Appl. Pharmacol.*, 1990, **104**, 440–456.

65. B. A. Habecker and N. M. Nathanson, 'Regulation of muscarinic acetylcholine receptor mRNA expression by activation of homologous and heterologous receptors.' *Proc. Natl. Acad. Sci. USA*, 1992, **89**, 5035–5038.

66. R. S. Aronstam and M. E. Eldefrawi, 'Transition and heavy metal inhibition of ligand binding to muscarinic acetylcholine receptors from rat brain.' *Toxicol. Appl. Pharmacol.*, 1979, **48**, 489–496.

67. T. G. Smart, X. Xie and B. J. Krishek, 'Modulation of inhibitory and excitatory amino acid receptor ion channels by zinc.' *Prog. Neurobiol.*, 1994, **42**, 393–341.

68. J. Y. Ma and T. Narahashi, 'Differential modulation of GABA$_A$ receptor-channel complex by polyvalent cations in rat dorsal root ganglion neurons.' *Brain Res.*, 1993, **607**, 222–232.

11.09
Ion Channels

TOSHIO NARAHASHI

Northwestern University Medical School, Chicago, IL, USA

11.09.1 INTRODUCTION

Ion channels play vital roles in a variety of physiological functions of cells. In excitable cells, ion channels are the critical sites for generation of action potentials, neurotransmitter release, and muscle contraction. Thus, it is not surprising to see the increasing evidence that ion channels are the target sites of a variety of chemicals, including therapeutic drugs, toxicants, and natural toxins.

Whereas the study of ion channels has a long history of almost half a century, it was not until around 1980 that toxicology and pharmacology of ion channels became one of the most important fields in biomedical sciences. This is due largely to developments of two major techniques, patch clamp and molecular biology, as applied to ion channel toxicology and pharmacology. Patch clamp techniques which were originally developed by Neher and Sakmann[1] and later improved by Hamill *et al.*[2] have made it possible to measure ionic currents through any channels and in any types of cells.

Although the patch clamp is in principle a form of the voltage clamp technique invented by Cole,[3] the applicability of the original voltage clamp methods is limited because of the clamp space which is required for recording of ionic currents. Molecular biology has made it possible to clone a variety of receptors and ion channels providing us with insight into the molecular structural aspects. Thus, it is now possible to combine patch clamp and molecular biology techniques to elucidate the mechanism of action of chemicals on ion channels from both structural and functional points of view.

Ion channels may be classified into two large groups: one includes voltage-activated ion channels in which the gating mechanism is controlled by the membrane potential; and the other comprises transmitter-activated ion channels in which the gating mechanism is operated by binding of the neurotransmitter to its receptor (see Chapter 8, this volume). The voltage-activated channels include sodium channels, potassium channels, and calcium channels, and the transmitter-activated channels include those activated by acetylcholine (ACh), l-glutamate, γ-aminobutyric acid (GABA), glycine, serotonin, dopamine, norepinephrine, and histamine. The chemicals that cause neurotoxicity comprise a wide variety of agents. Practically all therapeutic drugs are toxic if given at excessive doses, and many of them cause toxicity to the nervous system.

This chapter will begin with a brief historical account of the development of ion channel toxicology. It will be followed by methodological developments that have led us to the present status. The major part of the chapter will be divided into two large sections, voltage-gated ion channels and ligand-gated ion channels. Several important channels in each category will be covered as related to neurotoxicity of various chemicals. Emphasis will be placed on environmentally important neurotoxicants such as insecticides and heavy

metals, but some therapeutic drugs and natural toxins that act on ion channels will also be included. Some of these chemicals will be described in detail to illustrate the importance of ion channels in neurotoxicity.

11.09.2 DEVELOPMENT OF ION CHANNEL TOXICOLOGY

The development of ion channel toxicology began with the discovery of tetrodotoxin (TTX) block of voltage-gated sodium channels.[4,5] TTX blocked the sodium channels selectively without any effect on other types of channels (Figure 1). Although TTX failed to be developed into a useful therapeutic drug due to its side effects and a lack of antidotes, the discovery aroused interest by neurophysiologists and neuropharmacologists to use it as a chemical tool in the laboratory and to search for other chemicals as potential tools. The initial TTX studies also set a prototype of channel pharmacology and

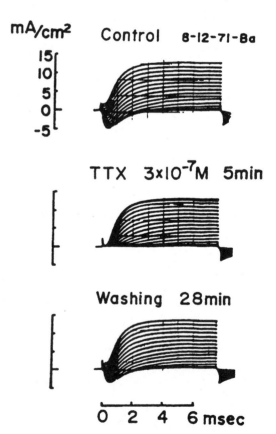

Figure 1 Selective block of sodium currents of a squid giant axon by tetrodotoxin (TTX). In control, the membrane currents associated with step depolarizations to various levels are composed of transient inward or outward sodium currents and steady-state outward potassium currents. TTX blocks sodium currents without any effect on potassium currents.

toxicology which in turn triggered voltage clamp studies of various chemicals that acted on ion channels.

As far as the toxicology of ion channels is concerned, significant developments following TTX studies occurred in the study of DDT and pyrethroid insecticides.[6–9] It was found that the sodium channels were a major target site of these insecticides, and these studies laid ground work for further progress in this field. The interactions of natural neurotoxins with the sodium channels soon became a hot subject of investigation, including those of the sea anemone *Condylactis* toxin,[10] batrachotoxin,[11,12] and scorpion toxin.[13,14] Environmental heavy metals such as mercury and lead had been a subject of electrophysiological investigations for some time, yet the study at the ion channel level commenced only in mid-1970s (see Chapters 8, 13, 28, and 29, this volume).[15]

The early studies of ion channel toxicology quoted above and many others that followed were performed by voltage clamp techniques using invertebrate nerve fibers notably giant axons isolated from squid, crayfish and lobsters. The development of patch clamp techniques[1,2] has indeed made a quantum leap in ion channel toxicology and pharmacology, because it became possible to utilize mammalian neurons or any other types of cells. Effective applications of molecular biology techniques to receptors/channels[16,17] set a precedent for the study of molecular structure of channels. A variety of information about the toxicology and pharmacology of ion channels has since been gained in an explosive manner. Whereas we have not witnessed any further quantum leap since patch clamp and molecular biology, recent accumulation of knowledge using these techniques has been quite substantial.

11.09.3 VOLTAGE-GATED ION CHANNELS

Three major classes of channels have been studied most extensively among voltage-gated ion channels: sodium, potassium, and calcium channels. The sodium and calcium channels have been shown to be important target sites of certain environmental toxicants.

11.09.3.1 Sodium Channels

Studies of pharmacology and toxicology of voltage-gated sodium channels have a long history for at least three reasons. First, a variety of natural toxins and therapeutic drugs modulate the function of sodium channels. Second, TTX and saxitoxin (STX), potent and selective sodium channel blockers, set a prototype of studies of chemical–sodium channel interactions. Third, classical voltage clamp techniques were applicable to nerve fibers and muscle endplates, and the former is endowed primarily with sodium and potassium channels. Several review articles have appeared.[18–22]

11.09.3.1.1 Effects of natural toxins on sodium channels

(i) Tetrodotoxin and saxitoxin

As briefly described earlier, TTX was demonstrated to block the sodium channel selectively and potently.[4,5] STX contained in the dinoflagellate *Gonyaulax catenella* also exerts the same sodium channel blocking action.[23] The mechanism of action of TTX and STX was reviewed.[18,20,24] They have been used extensively in the laboratory as chemical tools for the study of sodium and other channels. For example, pure potassium, calcium or any other channel activity can be analyzed by blocking sodium channel currents. The mechanism of neurotransmitter release can be studied in the presence of TTX or STX without complications caused by action potentials, because they do not interfere with the release process and postsynaptic receptors. The density of sodium channels has been estimated in various preparations by measuring the binding of TTX or STX.

TTX and STX bind to the same sodium channel site as the binding of STX to the sodium channel was inhibited by TTX.[25] Furthermore, both TTX and STX act on the sodium channel in the cationic form.[26] Thus, it is expected that the binding of TTX/STX is influenced by various cations present in the external solution. This was actually the case as will be described later.

The sodium channels of certain organs or cells are known to be relatively insensitive to TTX and STX. Classical examples include the sodium channels of cardiac muscle and denervated skeletal muscle.[20,24] Sodium channels of the brain, skeletal muscle and heart exhibit different pharmacological properties. The brain and skeletal muscle sodium channels are more sensitive to TTX and STX than the cardiac sodium channels, whereas only the skeletal muscle sodium channels are sensitive to the blocking action of μ-conotoxin.[27] The skeletal muscle and cardiac sodium channels have the same affinity for the block caused by divalent cations, Mg^{2+}, Ca^{2+}, Sr^{2+}, Ba^{2+}, Mn^{2+}, Co^{2+}, and Ni^{2+}, whereas the cardiac sodium channels are more sensitive to Cd^{2+} and Zn^{2+} than the skeletal muscle sodium channels.[28,29] Competitive interactions for binding between TTX/STX and cations have been shown for the alkali cations Li^+, Na^+, K^+, Rb^+ and Cs^+,[30] and for Zn^{2+} and Ca^{2+}.[31,32] The concentration of Zn^{2+} needed for the interaction with STX is too low to produce significant screening effect.[31]

The amino acids that are responsible for the differential actions of TTX/STX and Zn^{2+}/Cd^{2+} have been studied. Iodoacetamide, a cysteine-specific alkylating agent, abolished the blocking action of Zn^{2+} and modified the kinetics of STX binding in the cardiac sodium channels. It was found that a cysteine group is responsible for the low sensitivity to TTX/STX and the high sensitivity to Zn^{2+},[31] and that a tyrosine (adult skeletal muscle specific) or phenylalanine (brain specific) group confers on the channel high TTX and low metal sensitivity.[33-36] In the brain TTX/STX and divalent cations block the sodium channel at overlapping sites.[37] Although TTX and STX have been assumed to bind to the same sodium channel site, a study of cardiac sodium channels using membrane-impermeant, cysteine-specific, methanethiosulfonate analogues does not support the notion that STX and TTX are interchangeable.[38] A model has also been proposed to predict that TTX interacts directly with the SS1–SS2 segments of repeats I and II and that STX additionally interacts with the segments of repeats III and IV.[39]

The voltage dependence is an important parameter in understanding the mechanism of action as it provides us with a clue to the site and mode of interaction of chemical with the channel. For instance, it is possible to calculate the site of drug binding within the membrane potential field from the data of voltage dependence.[40] TTX/STX block of sodium channels was previously found to be independent of the membrane potential.[41,42] However, some reported voltage-dependent block of sodium channels by TTX.[43] Rando and Strichartz[44] have shown that although the STX block of normal sodium channels is not voltage dependent, the STX block of the sodium channels modified by batrachotoxin (BTX) is voltage dependent. The results were interpreted as indicating that STX binding to open channels was voltage dependent whereas STX affinity for closed and inactivated channels was voltage independent. A study with rat brain sodium channels has shown an increase in the affinity for STX/TTX with hyperpolarization.[37] Our study with crayfish giant axons has also shown that STX block is voltage dependent.[45] STX block was use dependent, and was enhanced with an increasing negativity of holding membrane potential from $-100\,mV$ to $-160\,mV$. The resting STX block was dependent on both the membrane potential and external calcium concentration in a complex manner. With $10\,mM$

Ca^{2+} outside, no voltage dependence of STX block was observed. However, with $50\,mM\ Ca^{2+}$, the apparent dissociation constant for STX block was markedly increased and the block exhibited a dual voltage dependence increasing with either depolarization or hyperpolarization from $-120\,mV$. The results are interpreted as indicating that the interaction between STX and calcium ions is modulated by membrane potential.

Tetrodotoxin-resistant (TTX-R) sodium channels have also been found in neuronal tissues,[46] including group C sensory neurons,[47] rat nodose ganglion neurons,[48] and human, rat, and mouse dorsal root ganglion (DRG) neurons.[49-51] Tetrodotoxin-sensitive (TTX-S) and TTX-R sodium channels of rat DRG neurons have been analyzed in detail.[52-54] They were different from each other in several aspects. First, the sensitivity to TTX and STX was different between TTX-R and TTX-S sodium channels by five orders of magnitude. Second, TTX-R sodium channels activated and inactivated more slowly than TTX-S sodium channels (Figure 2). Third, TTX-R sodium channels

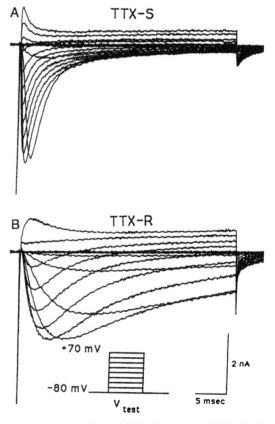

Figure 2 Families of TTX-S (A) and TTX-R (B) sodium currents associated with step depolarizations to various levels from a holding potential of $-80\,mV$. Rat dorsal root ganglion neurons (reproduced by permission of the Society for Neuroscience from *J. Neurosci.*, 1992, **12**, 2104–2111).

activated and inactivated at more positive membrane potential then TTX-S sodium channels. There also were large differences in the sensitivity to drug actions between TTX-R and TTX-S sodium channels. Lidocaine blocked TTX-S sodium channels four times more potently that TTX-R sodium channels.[52] By contrast, TTX-R sodium channels were blocked by Pb^{2+} and Cd^{2+} more potently than TTX-S sodium channels.[52] More drastic differences between TTX-S and TTX-R sodium channels were found in the action of the pyrethroid insecticides in modulating the channel kinetics.[55,56] This will be described later in detail.

Striatal and hippocampal neurons of 21–40 day old rats generated a slow sodium current which was insensitive to $12\,\mu M$ TTX.[57] This TTX-R sodium channel current activated and inactivated very slowly. Neurons isolated from the superficial layers of the medial entorhinal cortex of the rat generated sodium currents which were relatively insensitive to TTX (IC_{50} $\sim 146\,nM$).[58] This TTX-insensitive sodium channel had a higher sensitivity to the blocking action of Cd^{2+}, Zn^{2+}, and La^{3+} than the TTX-S sodium channel.

(ii) Conotoxins

Marine snails that belong to the genus *Conus* contain various toxins. Some of them block acetylcholine receptors, skeletal muscle sodium channels or calcium channels, and some others open sodium channels. A review of these toxins and their actions is given by Wu and Narahashi.[59] μ-Conotoxins are isolated from *Conus geopraphus*, and include GIIIA (geographutoxin I), GIIIB (geographutoxin II), and GIIIC. This group of toxins preferentially blocks the sodium channels of skeletal muscle and electroplax. Whereas the action potential recorded from guinea pig extensor digitorum longus muscle is completely blocked by $0.5\,\mu M$ conotoxin GIIIB, the action potentials from the crayfish giant axon and the papillary muscle of guinea pig heart are not affected at concentrations of $5\,\mu M$ and $1\,\mu M$, respectively.[60] Voltage clamp experiments with bullfrog sartorius muscle fibers clearly showed block of sodium channel currents.[60]

(iii) Sea anemone toxins

Venoms from various species of sea anemones contain mixtures of polypeptide toxins. These polypeptides are classified into four groups based on their molecular weight. Two of them with lower molecular weights ($<3000\,Da$ and $4000–6000\,Da$) are potent neurotoxins modulating the sodium channel gating kinetics, whereas the other two groups have no effect on ion channels.[59] The former two groups bind to site 3 of the sodium channel.[18] The selective block of sodium channel inactivation was demonstrated first with a toxin isolated from the sea anemone *Condylactis gigantea*,[10] and later with *Anemonia sulcata* toxin II,[61] and anthopleurin A isolated from *Anthopleura xanthogrammica*.[62] Interestingly, sea anemone toxins are not effective in changing the sodium channel gating kinetics in squid giant axons, although they are potent on crustacean axons, vertebrate neurons and cardiac muscles.[10,61] Binding experiments indicated that sea anemone toxins and α-scorpion toxins share a common site in the sodium channel.[63]

(iv) Scorpion toxins

A number of toxins have been isolated from various species of scorpions, and their structures and physiological effects on sodium channels were studied.[18,64,65] α-Scorpion toxins that inhibit the sodium channel inactivation include: toxin v, toxin var. 1–3 and toxin IIα from *Centruroides sculpturatus*; toxins M_7 and 2001 from *Buthus eupeus*; toxins V and XII from *Buthus tamulus*; and toxins I and II from *Androctonus australis*. At the node of Ranvier, these scorpion toxins also shift the steady-state sodium inactivation curve in the hyperpolarizing direction and elevate the curve at positive potentials so that inactivation is incomplete with depolarization. α-Scorpion toxins bind to site 3 of the sodium channel, the same site as that of sea anemone toxins.[18]

β-Scorpion toxins cause a shift of the sodium channel activation–voltage relationship in the hyperpolarizing direction without slowing the inactivation kinetics. Whole venom from *Centruroides sculpturatus* was used for the first time to demonstrate this effect.[66] β-Scorpion toxins include toxins I, II, IV, VI isolated from *C. sculpturaturs*; toxins 1α–IIIα and IIIβ from the same species; toxin II from *C. suffusus suffusus*; and toxin γ from *Titus serrulatus*. However, toxin γ is somewhat different altering both activation and inactivation kinetics in neuroblastoma cells.[67] Both activation and inactivation curves are shifted by this toxin in the hyperpolarizing direction. These toxins bind to site 4 of sodium channels.[18]

(v) Batrachotoxin

Batrachotoxin (BTX) is one of the toxic components contained in the skin secretion of the Colombian poison arrow frog *Phyllobates aurotaenia*.[11,68] BTX has a potent and irreversible depolarizing action on nerve, skeletal muscle and cardiac muscle membranes through opening of sodium channels.[12] The sodium channel inactivation is removed by BTX,[69,70] and the sodium channel activation curve is greatly shifted in the hyperpolarizing direction.[70] These two effects account for membrane depolarization. BTX modifies the sodium channel in its open configuration. Single-channel analyses with neuroblastoma cells have revealed that the mean open time of sodium channels is greatly prolonged by BTX, and that in the presence of BTX the channels open at large negative potentials where no channels are open in normal, BTX-untreated conditions.[71] Batrachotoxin decreases the apparent single-sodium channel conductance of squid giant axons without much changing ion selectivity.[72] In planar lipid bilayers into which sodium channel polypetides from the electric organ were incorporated, BTX, grayanotoxin-I (GTX-I), and veratridine modulated the sodium channel activity in a similar manner in removing the sodium inactivation and shifting the activating voltage in the hyperpolarizing direction.[73] BTX binds to site 2 of sodium channels.[18]

(vi) Grayanotoxins

Grayanotoxins (GTXs) are the toxic principles isolated from the leaves of plants belonging to the family Ericaceae (*Leucothoe, Rhododendron, Andromeda, Kalmia*). In squid giant axons, GTX I and α-dihydrograyanotoxin II (α-H$_2$-GTX II) cause a large membrane depolarization due to a specific increase in resting sodium permeability.[74,75] Voltage clamp measurements have shown that the resting sodium permeability is increased by GTX by a factor of 10–100 depending on the concentration. Upon step depolarization of the squid axon membrane exposed to GTX, there appears a peak sodium current, which is decreased in amplitude, and a secondary slow steady-state sodium current.[76] The latter current represents the current flowing through the open GTX-modified sodium channels. The modified sodium channels open at large negative membrane potentials where the normal unmodified sodium channels do not open. This explains a large membrane depolarization caused by GTX. The ionic selectivity of the modified sodium channels is decreased; the P_x/P_{Na} ratio where P_x and P_{Na} refer to the permeabilities to X cations and Na ions, respectively, is greatly increased, the P_x/P_{Na} ratios for hydroxylamine, formamidine, and ammonium being estimated to be 1.58, 0.71, and 1.10, respectively.[76] GTX binds to site 2 of sodium channels.[18]

(vii) Aconitine

Aconitine is a toxic component contained in the plant *Aconitum napellus*. It causes an inhibition of the sodium channel inactivation and a shift of the sodium channel activation voltage in the hyperpolarizing direction.[77,78] Aconitine decreases the ionic selectivity of sodium channels.[78] It binds to site 2 of sodium channels.[18]

(viii) Brevetoxins

Brevetoxins are produced by the dinoflagellate *Ptychodiscus brevis* which causes red tides killing fish.[59,79] Whereas the nomenclature of brevetoxins is confusing, eight components have been identified called PbTX-1 through PbTX-8.[59] Although the potency and efficacy of these brevetoxins are different, their effects on sodium channels are characterized by the inhibition of inactivation and the shift of activation voltage in the hyperpolarizing direction, which are responsible for membrane deplorization observed in the presence of brevetoxins.[80] Binding experiments have shown that brevetoxins bind to site 5 of sodium channels.[18]

(ix) Versutoxin

Versutoxin (VTX) is a toxin purified from the venom of the Australian Blue Mountains funnel-web spider *Hadronyche versutus*. VTX had no effect on TTX-R sodium currents or potassium currents of rat DRG neurons. In contrast VTX caused a dose-dependent slowing or removal of TTX-S sodium current inactivation, and a reduction in peak TTX-S sodium current, but did not markedly slow tail current kinetics of TTX-S sodium currents.[81] In the presence of VTX, the voltage dependence of steady-state sodium channel inactivation also showed a 7 mV shift in the hyperpolarizing direction with no change in the slope factor. These selective actions of VTX on sodium channel gating kinetics are similar to, but distinct from, those of α-scorpion and sea anemone toxins.

11.09.3.1.2 *Effects of insecticides on sodium channels*

(i) Pyrethroids

Pyrethroids are synthetic derivatives of the natural toxins pyrethrins contained in the flowers of *Chrysanthemum cinerariaefolium* which were used as natural insecticides before and during World War II. A number of potent and long-lasting synthetic insecticides were developed and used extensively after the war, yet serious concern over environmental effects and human toxicity that emerged in the mid-1960s led to the development of synthetic pyrethroids which were potent on insects, much less toxic to humans, and biodegradable. Many pyrethroids are widely being used as near-ideal insecticides (see Chapter 27, this volume).

As in the case of most toxic substances, pyrethroids have multiple actions on various receptors and ion channels. However, over many years of studies, it has become abundantly clear that the major target site of pyrethroids is the voltage-gated sodium channels, modification of which causes severe symptoms of poisoning in animals characterized by hyperexcitation.[82-90] Thus, the mechanisms underlying pyrethroid–sodium channel interactions are emphasized with some descriptions of other effects of pyrethroids.

(a) Pyrethroid modulation of sodium channels. Pyrethroids may be divided into two groups: type I pyrethroids do not have a cyano group in the α position, and their symptoms of poisoning are characterized by hyperexcitation, ataxia, convulsions and paralysis; type II pyrethroids have an α cyano group and cause hypersensitivity, choreoathetosis, tremors and paralysis. However, the primary target for both types of pyrethroids is the sodium channel. Single-channel and gating current analyses have clearly demonstrated that pyrethroids slow the kinetics of both opening and closing of individual sodium channels resulting in delayed and prolonged openings (Figure 3).[91] This causes prolongation of whole-cell sodium current during a depolarizing pulse and marked slowing of the tail sodium current upon repolarization (Figure 4).[92] Pyrethroids also cause a shift of the activation voltage in the direction of hyperpolarization. These changes in sodium channel kinetics lead to a membrane depolarization and an increase in depolarizing after-potential. The latter reaches the threshold for excitation causing repetitive after-discharges. The membrane depolarization of sensory neurons increases discharge frequency, and that of nerve terminals increases the release of transmitter and the frequency of spontaneous miniature postsynaptic potentials. All of

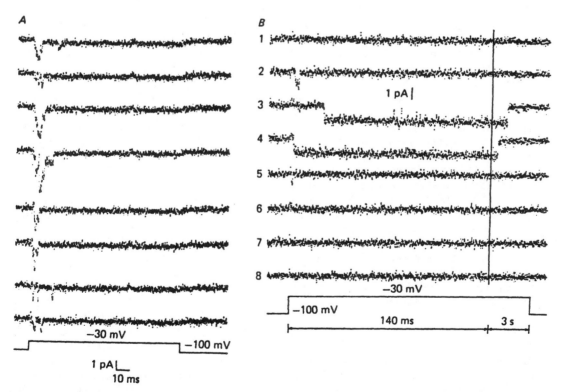

Figure 3 Effects of deltamethrin on single sodium channel currents of a neuroblastoma N1E-115 cell. (A) Currents from a cell before drug treatment in response to 140 ms depolarizing pulses from a holding potential of −100 mV to −30 mV. (B) Currents from the same cell after exposure to 10 μM deltamethrin in response to 3140 ms depolarizing pulses from a holding potential of −100 mV to −30 mV. Note sizable prolongation of channel opening and delay in onset of current after exposure to deltamethrin (reproduced by permission of the Physiological Society from *J. Physiol. (Lond.)*, 1986, **380**, 191–207).

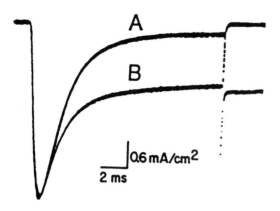

Figure 4 Effects of allethrin on squid axon sodium current. (A) Control current in response to step depolarization from a holding potential of $-100\,mV$ to $-20\,mV$. (B) After exposure to $1\,\mu M$ allethrin. Note a large increase in steady-state sodium current during step depolarization and an increase and prolongation of tail sodium current after repolarization (reproduced by permission of Little Brown & Company from *Ann. Neurol.*, 1984, **16**, S39–S51).

these changes can account for the various symptoms of poisoning including hyperexcitation, hypersensitivity, convulsions and tremors. DDT resembles type I pyrethroids in its action on sodium channels. Details of these effects of pyrethroids on the sodium channel are given in review articles.[84–87,90] Experiments by measuring [22]Na$^+$ uptake and batrachotoxin binding provide additional support to the sodium channel target theory.[93–96]

The question of closed vs. open sodium channel modification by pyrethroids and DDT has largely been settled. Hille[9] originally proposed open channel modification by DDT, and the concept has been elaborated to include allethrin.[97] Although tetramethrin was proposed initially to modify both open and closed channels,[98,99] more extensive analyses using different approaches have clearly indicated that pyrethroids modify the sodium channel largely in its closed state before opening.[100–103] However, since the sodium channels have a higher affinity for pyrethroids in the open state, more channels are modified by pyrethroids upon opening, thus explaining use-dependent modification.[104,105]

Pyrethroids have been shown to bind to a sodium channel site not shared by TTX, local anesthetics, n-octylguanidine, BTX, and GTXs.[102,106,107] It appears that pyrethroids are dissolved in the lipid phase of the nerve membrane and get access to the channel gating machinery. The use of stereoisomers of tetramethrin has revealed at least three binding sites in the sodium channel, that is, a *trans* site, a *cis* site, and a negative allosteric site.[106] The

insecticidally active (+)-*trans* and (+)-*cis* tetramethrin bind to the *trans* site and *cis* site, respectively, to modify the sodium channel. The insecticidally inactive (−)-*trans* and (−)-*cis* isomers bind to the *trans* site and the *cis* site, respectively, as well as to the allosteric *trans/cis* site, yet cause no physiological effect. Pyrethroids do not appear to enter the lumen of the sodium channel, because the cation selectivity of the open channel is not altered by the pyrethroids.[108]

Certain local anesthetics such as lidocaine, tetracaine and QX314 have been found to exert differential blocking actions; deltamethrin-modified sodium channels are blocked more than normal sodium channels.[109] This is worth pursuing for potential antidotes. Although the underlying mechanism is unknown, the prolonged opening in the pyrethroid-modified channel may facilitate open channel block by local anesthetics. Paresthesia caused by pyrethroids is one of the practical problems associated with their use as insecticides, and vitamin E has been used for the treatment. Our recent patch clamp experiments have clearly demonstrated that vitamin E blocks the pyrethroid-modified sodium channel without affecting the normal sodium channel as will be described later.

(b)   Differential modulation of tetrodotoxin-sensitive and tetrodotoxin-resistant sodium channels by pyrethroids. Tetramethrin and allethrin modify both TTX-S and TTX-R sodium channels, but in somewhat different manners.[55,56] In TTX-S sodium channels, the slowly inactivating component of sodium current during step depolarization is increased somewhat by tetramethrin, and a tail sodium current with a slowly rising and falling phase appears upon repolarization (Figure 5(A)). In TTX-R sodium channels, the slow sodium current during step depolarization is markedly increased by tetramethrin, and upon repolarization a large instantaneous tail sodium current is generated and decays slowly (Figure 5(B)). The steady-state sodium inactivation curve is shifted by tetramethrin in the hyperpolarizing direction in both TTX-S and TTX-R channels. The sodium conductance–voltage curve is also shifted by tetramethrin in the hyperpolarizing direction in both TTX-S and TTX-R channels and the latter is affected more strongly than the former.

A method to measure the percentage of sodium channels modified by pyrethroids was developed.[56] The slow tail current associated with step repolarization is indicative of the activity of the pyrethroid-modified sodium channels, whereas the peak current during step depolarization is a result of the activity of the normal unmodified sodium channels. Thus

A

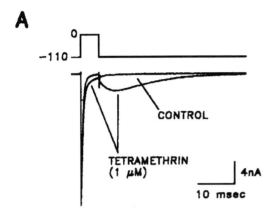

B

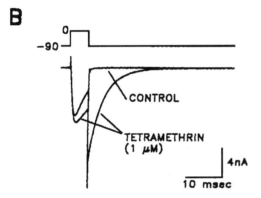

Figure 5 Effects of 1 μM tetramethrin on TTX-S sodium current (A) and TTX-R sodium current (B) recorded from rat dorsal root ganglion neurons. See text for further explanation (reproduced by permission of the American Society for Pharmacology and Experimental Therapeutics from *J. Pharmacol. Exp. Ther.*, 1994, **270**, 595–603).

the sodium conductance upon repolarization can be divided by the peak conductance during step depolarization to calculate the percentage of the modified sodium channels. The percentage of the sodium channels modified by tetramethrin (*M*) is calculated by the following equation:

$$M = [\{I_{tail}/(E_h - E_{Na})\}/\{I_{Na}/(E_t - E_{Na})\}] \times 100,$$

where I_{tail} is the tail current amplitude obtained by extrapolation of the slowly decaying phase of the tail current to the moment of membrane repolarization assuming a single exponential decay, E_h is the potential to which the membrane is repolarized, E_{Na} is the equilibrium potential for sodium ions obtained as the reversal potential for sodium current, and E_t is the potential of step deplorization. The percentage of sodium channel modification by various concentrations of tetramethrin are given in Table 1. The concentrations of tetramethrin required to modify 1.3% of sodium channels are 10 nM for TTX-R channels and 300 nM

Table 1 Percentages of the fraction of the TTX-S and TTX-R sodium channels modified by various concentrations of TM.

| TM (μM) | Modification of sodium channels | |
	TTX-S (%)	TTX-R (%)
0.01		1.31 ± 0.28
0.03	0	5.15 ± 0.30
0.1	0.24 ± 0.10	15.35 ± 0.79
0.3	1.25 ± 0.13	35.48 ± 2.70
1	3.53 ± 0.66	57.82 ± 2.29
3	7.70 ± 1.20	74.85 ± 1.23
10	12.03 ± 1.89	81.20 ± 1.57

Source: Tatebayashi and Narahashi.[56]
Each value indicates the mean ± SEM (*n* = 4).

for TTX-S channels. To modify 12–15% of channels, 100 nM tetramethrin is required for TTX-R channels, whereas as much as 10 μM tetramethrin is required for TTX-S channels. Therefore, the difference in tetramethrin sensitivity between TTX-R and TTX-S sodium channels is as much as 30–100-fold. Invertebrate nerve sodium channels, which are sensitive to TTX, are almost equally sensitive to pyrethroids to DRG TTX-R channels.

(c) Pyrethroid toxicity amplification in rat cerebellar Purkinje neurons. Whereas the data with rat DRG neurons are highly informative, a question is raised as to how the sodium channels in the brain behave in response to pyrethroids. Cerebellar Purkinje neurons of the rat were found to be endowed with TTX-S sodium channels which were only moderately sensitive to tetramethrin like TTX-S sodium channels of DRG neurons.[110] The percentages of the sodium channels modified by tetramethrin range from 0.6% at 0.1 μM tetramethrin to 25% at 30 μM. The threshold concentration of tetramethrin to induce repetitive after-discharges is 0.1 μM. Therefore, modification of 0.6% of the sodium channels is enough to cause repetitive after-discharges leading to hyperexcitatory symptoms of poisoning animals. The "toxicity ampliation" theory, which was originally proposed with squid giant axons on the basis of several assumptions,[106] is now firmly established in mammalian neurons based on solid experimental data. This concept has important implications and applicability to evaluate effective concentrations of toxicants or therapeutic drugs that act via a threshold phenomenon. Take an example in which a slow membrane depolarization caused by channel openings reaches the threshold and induces repetitive discharges. Discharges may be

stopped once the slow depolarization is brought down by a drug just below the threshold for excitation. The concentration of the drug for this action will be much less than the EC_{50}. Thus, as far as drugs that act through a threshold phenomenon are concerned, we need to reconsider the traditional pharmacological concept which dictates comparison of *in vitro* EC_{50} value with the drug concentration in the patient's serum.

(d) Temperature dependence of pyrethroid action. It is well known that temperature has a profound effect on the insecticidal activity of pyrethroids with their potencies increasing with lowering the temperature.[82,83,88] The negative temperature dependence of the insecticidal activity of pyrethroids is due largely to an increase in nerve sensitivity to the insecticide with lowering the temperature.[82,83,86] However, no data are available for mammalian brain neurons and no mechanism underlying the temperature dependence of pyrethroid sensitivity of sodium channels has been elucidated by the mid-1990s.

Both current clamp and voltage clamp experiments were performed with rat cerebellar Purkinje neurons.[110] Tetramethrin at 100 nM and 300 nM, which modifies 0.62% and 2.19% of sodium channels, respectively, induces repetitive after-discharges at low temperatures (15–20 °C) but repetitive responsiveness subsides at higher temperatures (30–35 °C). Voltage clamp experiments disclosed profound influence of temperature on tetramethrin-induced slow tail currents. While the peak amplitude of tail current is not drastically changed by temperature change, both the rising and falling phases of slow tail current are greatly slowed by lowering the temperature. The Q_{10} values for the percentage of the sodium channels modified by 3 µM tetramethrin are about the same in the two temperature ranges, 0.77 and 0.79 for 20–30 °C and 25–35 °C, respectively. The time constant for the decay of slow tail current has a large negative temperature coefficient, the Q_{10} showing 0.07 and 0.11 for 20–30 °C and 25–30 °C, respectively. As expected from the Q_{10} values for the time constant of tail current decay, the charge movement during tail current also shows a large negative temperature coefficient, with Q_{10} values of <0.22 and 0.18 for 20–30 °C and 25–35 °C, respectively. Thus the increased and prolonged flow of sodium ions through tetramethrin-modified sodium channels at low temperatures augments the depolarizing after-potential which in turn reaches the threshold for repetitive after-discharges. Although the effects of pyrethroids on the nerve as a function of temperature have been studied using various materials by observing either

action potentials or sodium currents, the present study is the first to successfully correlate changes in action potentials and sodium currents in the presence of pyrethroids and as a function of temperature in the same preparation.

(e) Mechanism of vitamin E alleviation of pyrethroid paresthesia. One annoying symptom often noted by humans exposed to pyrethroids, usually in an occupational setting, is abnormal skin sensation or paresthesia, which is due to repetitive firing of sensory nerve endings.[90] In a human assay system using the topical application of pyrethroids to the lobule of the ear, vitamin E oil was found to be an effective prophylactic and therapeutic agent for cutaneous exposure to pyrethroids.[111,112] Explanations were sought for the mechanism of inhibitory action of vitamin E oil in its antioxidant and biomembrane stabilizing effect. Furthermore, Oortgiesen *et al.*[109] reported that vitamin E had no effect on deltamethrin (a type II pyrethroid)-modified sodium current in neuroblastoma cells. It was found that vitamin E selectively blocks the tetramethrin (a type I pyrethroid)-modified sodium current in rat cerebellar Purkinje cells and DRG cells without any effect on normal sodium channel currents.[113] ($\pm$)-α-Tocopherol at 10 µM and 30 µM blocks 10 µM tetramethrin-modified TTX-S sodium channels in rat cerebellar Purkinje cells by 31% and 77%, respectively, and in rat DRG cells by 34% and 76%, respectively (Figure 6). The concentration–response curves for tetramethrin modification of the sodium channels are shifted in the direction of higher concentrations by ($\pm$)-α-tocopherol in a competitive manner. Elevated depolarizing after-potential or repetitive after-discharges caused by tetramethrin in both cerebellar Purkinje cells and DRG cells are effectively blocked by ($\pm$)-α-tocopherol (Figure 7). The selective block of tetramethrin-modified sodium channels by ($\pm$)-α-tocopherol is one of the important mechanisms underlying ($\pm$)-α-tocopherol alleviation of paresthesia.

(f) Selective toxicity of pyrethroids. The results of experiments with mammalian brain neurons are compared with those with various invertebrate neurons obtained previously.[110] First, since there is a 10 °C difference in temperature between mammals and invertebrates and also since the Q_{10} value for pyrethroid action on the nerve membrane is 0.2 in both cases, the difference in toxicity ascribed to temperature difference is estimated to be fivefold. Second, the intrinsic sensitivity of sodium channels to tetramethrin is at least 10 times lower in mammals than in invertebrates. Third, the recovery of sodium channels from tetramethrin intoxication after washing *in vitro* is at

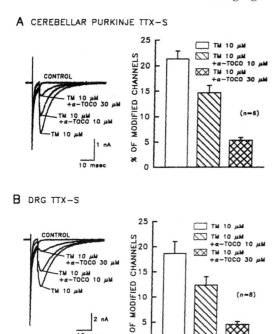

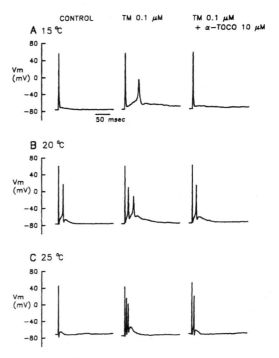

Figure 6 Suppression of 10 μM tetramethrin-induced tail currents by 10 μM and 30 μM (±)-α-tocopherol in TTX-S sodium channels of rat cerebellar Purkinje cells (A) and DRG cells (B). Currents were evoked by depolarizing the membrane to 0 mV for 5 ms from a holding potential of −110 mV. Cells were first treated with 10 μM tetramethrin, and then 10 μM or 30 μM (±)-α-tocopherol was added to the perfusion solution containing 10 μM tetramethrin. Records were taken 5 min after the addition of each chemical. Mean ± SEM with $n = 6$ (a) and $n = 8$ (b) (reproduced by permission of the American Society for Pharmacology and Experimental Therapeutics from *J. Pharmacol. Exp. Ther.*, 1995, **275**, 1402–1411).

Figure 7 (±)-α-Tocopherol at 10 μM selectively blocks tetramethrin-induced depolarizing after-potential and repetitive after-discharges in a cerebellar Purkinje cell without affecting the action potential at 15 °C, 20 °C and 25 °C. All action potentials shown here were recorded from a single cell. Action potentials were evoked by injecting a current pulse (200 pA, 2 ms). Vm represents the membrane potential. TM, tetramethrin; TOCO, tocopherol (reproduced by permission of the American Society for Pharmacology and Experimental Therapeutics from *J. Pharmacol. Exp. Ther.*, 1995, **275**, 1402–1411).

least five times faster in mammals than in invertebrates. A fourth contributing factor is enzymatic detoxication of pyrethroids which is estimated to be three times faster in mammals than in invertebrates due to temperature difference. A fifth factor is body size; pyrethroids have more chance to be detoxified before reaching the target site in mammals than in invertebrates with an estimated difference of at least threefold. Thus, the overall difference in tetramethrin toxicity is estimated to be 2250-fold between mammals and invertebrates which is in the same order of magnitude as the differences in LD_{50} values of 500–4500-fold for tetramethrin.[114–117]

11.09.3.2 Potassium Channels

Whereas potassium channels, most notably delayed rectifying potassium channels, have been studied for a long time, not much progress

has been made as related to toxicology. This is partly because not many toxicants have been shown to affect potassium channels as a major target site. Potassium channels were indeed "pharmacological orphans."[118] However, some natural toxins and chemical agents have been developed into useful chemical tools in the laboratory and therapeutic drugs. Many types of potassium channels are now known to exist. Some are activated by membrane potential and some others by ligands including Ca^{2+}, ACh, 5-hydroxytryptamin, ATP, and Na^+. Several reviews have been published.[19,22,119–129] Only a brief account will be given regarding the interactions of natural toxins with some of the potassium channels.

11.09.3.2.1 Delayed rectifying potassium channels

Several chemical agents are known to block the delayed rectifying potassium channel, including tetraethylammonium and its derivatives, 4-aminopyridine and its derivative, and cesium ions. Several toxins also block the delayed rectifying potassium channel, but none of them specifically. Some of them modulate the sodium channel as well, and some others act also on other types of potassium channels. Block of the delayed rectifying potassium channel prolongs the action potential and causes hyperexcitation.

Toxins from scorpions *Leiurus quinquestriatus*, *Buthus tamulus*, and *Centruroides noxius* block the delayed rectifying potassium channels. Charybdotoxin isolated from the scorpion *Leiurus quinquestriatus* var. *hebraeus* is generally recognized as a specific blocker of Ca^{2+}-activated potassium channels,[130] yet it also blocks Ca^{2+}-insensitive potassium channels, the A-current potassium channels, and the component f1 of the delayed rectifying potassium current. Charybdotoxin is not a single entity of toxin and derived from various species of scorpions, and is divided into several subfamilies: charybdotoxin-type (subfamily 1), noxiustoxin-type (subfamily 2), kaliotoxin-type (subfamily 3), and tytiustoxin-type (subfamily 4).[118]

Snake venoms contain various toxins, and some of them block the delayed rectifying potassium channels, albeit not selectively. Dendrotoxins isolated from the mambas *Dendroaspis angusticeps* and *D. polylepis* block the delayed rectifying potassium current more potently than the transient A-current and voltage-activated transient Ca^{2+}-dependent potassium currents.

11.09.3.2.2 *Transient potassium channels (A-currents)*

The transient potassium channel generates an A-current upon membrane depolarization. A-current attains a peak and decays slowly, and is deemed responsible for modulating the firing rate of neurons. Some of the toxins known to block other types of potassium channels also block the transient potassium channel, for example, aminopyridines which also block the delayed rectifying potassium channels, charybdotoxin which also blocks the Ca^{2+}-activated potassium channels, dendrotoxin which also blocks the delayed rectifying potassium channels.[19]

11.09.3.3 Calcium Channels

Voltage-gated calcium channels are classified into at least five types based on their physiological and pharmacological properties, that is,

T-, L-, N-, P-, and Q-types.[131] Whereas the physiological role of each type of calcium channel is not completely understood, some information is available: T-type calcium channels generate slow membrane depolarization in neurons; L-type channels play a role in muscle contraction as well as transmitter release from nerve terminals; N-type channels participate in transmitter release from nerve terminals; and P- and Q-types play a role in neuronal activity including transmitter release in the brain. Calcium channels are important for the action of some toxicants including lead and mercury as well as natural toxins and therapeutic drugs.[132–134]

11.09.3.3.1 *T-type calcium channels*

T-type (or type I) calcium channels are widely distributed in various cells including neuroblastoma cells, dorsal root ganglion neurons, endocrine cells, cardiac muscle, and vascular smooth muscle, and play a role in generating a burst of action potentials, in pacemaker activity and in hormone secretion.[19] Dihydropyridines which are regarded as specific blockers of L-type calcium channels can block T-type calcium channels as well, albeit with a low potency.[19] Phenytoin blocks T-type calcium channels without much effect on L-type calcium channels in neuroblastoma cells.[135]

Environmental toxicants that block T-type calcium channels are the pyrethroid tetramethrin[136,137] and polyvalent cations including La^{3+}, Pb^{2+}, Cd^{2+}, Ni^{2+}, and Co^{2+}.[138] However, these effects are not specific for T-type calcium channels, and the toxicological significance remains largely to be seen.

11.09.3.3.2 *L-type calcium channels*

L-type (type II) calcium channels have been found in many types of cells including neuroblastoma cells, skeletal muscle, cardiac cells vascular smooth muscle, DRG neurons, and endocrine GH_3 cells.[19] Several types of therapeutic drugs are known to block L-type calcium channels: the calcium antagonists verapamil, D600, dihydropyridines such as nifedipine, nimodipine, and nitrendipine; benzodiazepines such as diazepam and nitrazepam but not clonazepam; barbiturates such as pentobarbital and phenobarbital; ethanol; chlorpromazine; and opioids such as leucine-enkephalin.[139] The polyvalent cations La^{3+}, Pb^{2+}, Cd^{2+}, Ni^{2+}, and Co^{2+} block L-type calcium channels (Figure 8).[138]

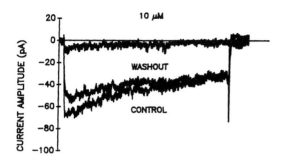

Figure 8 Effect of 10 μM lead on L/N-type calcium channel current recorded from a human neuroblastoma SH-SY5Y cell. Lead blocks the current reversibly (reproduced by permission of Elsevier Science from *Brain Res.*, 1991, **545**, 312–314).

11.09.3.3.3 N-type calcium channels

N-type calcium channels are also widespread among various types of cells, and are blocked by ω-conotoxin GVIA and ω-conotoxin MVIIA isolated from the venom of the marine snails *Conus geographus* and *C. magus*, respectively.[139,140] Lead also blocks N-type calcium channels (Figure 8).[139] The μ-opioids such as [D-Ala,[2] N-Me-Phe,[4] Gly,[5] ol]-enkephalin (DAMGO) blocks N-type calcium channels through activation of the opioid receptor.[141]

11.09.3.3.4 P-, Q-, and R-type calcium channels

P-type calcium channels were originally discovered in Purkinje cells,[142,143] and are insensitive to dihydropyridines or ω-conotoxin GVIA.[139] We now know that both P-type and Q-type calcium channels are widespread. However, the situation is somewhat confusing as to how to distinguish these two types. ω-Conotoxin MVIIC blocks N-, P-, and Q-type calcium channels, and Q-type is most sensitive. ω-Agatoxin IVA blocks P- and Q-type calcium channels, the former being more sensitive than the latter. ω-Grammotoxin SIA blocks N-, P-, and Q-type clacium channels amost equally,[144–151] A calcium current component resistant to all of these toxins has also been found and is designated R-type.[147]

11.09.4 LIGAND-GATED ION CHANNELS

A variety of ligand-gated ion channels have been discovered and studied. However, the progress in this field as related to toxicology has so far been limited to several types, that is, the glutamate, GABA, and ACh systems (see Chapters 8 and 31, this volume).

11.09.4.1 Glutamate Receptor Channels

The excitatory glutamate receptor-channel system is widely distributed in the CNS of mammals and invertebrates and in invertebrate neuromuscular junctions. Studies of this system were greatly accelerated by the introduction of patch clamp techniques in the early 1980s, which allowed the measurement of the activity with any type of cell. The glutamate system has since been studied extensively for several reasons: (i) it is related to neurotoxicity, especially cell death caused by ischemic brain damage; (ii) it is related to epileptic discharges; (iii) it is also related to long-term potentiation and memory; and (iv) there are drugs that act primarily on this system. Many review articles have been published.[139,152–168]

11.09.4.1.1 Receptor classification and characterization

The glutamate receptor channels are classified into four groups: *N*-methyl-D-aspartate (NMDA) receptor, kainate receptor, quisqualate or α-amino-3-hydroxy-5-methyl-4-isoxazole pronpionic acid (AMPA) receptor, and 2-amino-4-phosphonobutanoate (AP4) receptor. The former three receptors have been studied most thoroughly as related to toxicology.

The NMDA receptor channels are activated by NMDA, l-glutamate, l-aspartate, and ibotenate, modulated by glycine, and blocked by Mg^{2+}, Zn^{2+}, 2-amino-5-phosphonovalerate (AP5), phencyclidine (PCP), ketamine, and alcohols.[139] The channel opened by NMDA is highly permeable to Ca^{2+}, with the Ca^{2+}/Na^+ permeability ratio of 10.[169] Thus channel opening causes a massive influx of Ca^{2+} ions leading to cell death as a result of a cascade of events ranging from the activation of phospholipase C to the release of Ca^{2+} from intracellular stores. The NMDA receptor channels mediate slow processes of synaptic transmission including the formation of memory.

The kainate receptor channels are activated by domoate, kainate, quisqualate and l-glutamate, in the order of decreasing potencies, and blocked by 6-cyano-7-nitro-quinoxaline-2, 3-dione (CNQX) and 6,7-dinitro-quinoxaline-2,3-dione (DNQX).[139] Although the open channels are permeable to Ca^{2+}, the Ca^{2+}/Na^+ permeability ratio is only 0.15. The AMPA receptor channels are activated by quisqualate,

AMPA, l-glutamate and kainate, in the order of decreasing potencies, and blocked by CNQX and DNQX.[139] The Ca^{2+}/Na^{2+} permeability ratio is 0.15. The kainate and AMPA receptors mediate fast synaptic transmission.

11.09.4.1.2 Drugs that act on glutamate receptor channels

Several divalent cations are known to block the NMDA receptor channels by diverse mechanisms.[139] Mg^{2+} block has physiological significance as it is dependent on membrane potential with a hyperpolarization releasing the block. Zn^{2+} also blocks the channels but in a voltage-independent manner, indicating that it binds to a site located on the external surface of the receptor-channel complex. Some divalent cations block the channels as they give up their water slowly (e.g., Ni^{2+}, Co^{2+}, Mg^{2+}). Other divalent cations that dehydrate more quickly can permeate (e.g., Ca^{2+}, Ba^{2+}, Cd^{2+}, Sr^{2+}). Some of them can permeate and block (e.g., Ca^{2+}, Mg^{2+}, Cd^{2+}). It should be noted that any ions permeant to a channel could also block while passing through the open channel.[170,171]

Ketamine and PCP block the NMDA receptor channels in a voltage- and use-dependent manner without affecting the kainate and AMPA receptor channels.[139] Ketamine appears to enter the open channel and to be trapped within the channel until it is reopened by NMDA.[172] Dizocilpine (MK-801) is an anticonvulsant, and causes an open-channel block in a voltage-dependent manner.[173] Some of the antidepressants, neuroleptic drugs, and phencyclidine-like drugs also block the NMDA receptor channels, including imipramine, chlorimipramine, desmethylimipramine, nortrptyline, protriptyline, chlorpromazine, dextromethorphan, cyclazocine, ($\pm$)-SKF 10 047, promazine, pentobarbital, and infenprodil, whereas haloperidol augments NMDA-induced currents.[139] The NMDA receptor channels represent one of the sites of action of ethanol and are blocked by ethanol.[174] Spider and wasp toxins block glutamate receptor channels.[175]

Pyrethroids have been shown to inhibit the binding of [^{3}H]kainate to mouse brain homogenates[176] and to suppress the glutamate sensitivity of the muscle of housefly larvae.[177] Toxicological significance of these actions remains to be seen.

Endogenous polyamines have important implication for ischemic neuronal damage.[139] The postischemic overshoot in putrescine formation may be related to an increase in mitochondrial calcium uptake and NMDA-

dependent calcium influx. The activation of the NMDA receptor increases the ornithine decarboxylase activity which in turn increases putrescine. Purtrescine augments L-type calcium channel currents.[178]

11.09.4.2 GABA Receptor Channels

The GABA receptor system may be classified into two large groups, $GABA_A$ and $GABA_B$ receptors. The $GABA_A$ receptor is associated with a chloride channel, and is present primarily on the postsynaptic side. The $GABA_B$ receptor is associated with a potassium channel, and is present on both the presynaptic and the postsynaptic sides. Several reviews have been published (see Chapter 8, this volume).[179–183]

11.09.4.2.1 Drugs that act on GABA_A receptor channels

The $GABA_A$ receptor channels are an important target site for a variety of drugs and toxicants, and unique in that the activity is easily augmented or inhibited depending on the chemicals.[139] At least five binding sites have been identified, they are, the GABA site, the benzodiazepine site, the barbiturate site, the picrotoxin site, and the chloride channel.[184–187] The GABA site is blocked by bicuculline. Benzodiazepines augment GABA-induced currents through binding to the benzodiazepine site which is inhibited by Ro15-1788. β-Carboline 3-carboxylate esters behave like inverse agonists. Barbiturates also augment GABA-induced currents through binding to the barbiturate site. Picrotoxin, pentylenetetrazol, and *t*-butyl-bicyclo-phosphorothionate (TBPS) block the chloride channel. Other drugs known to potentiate GABA-induced currents include alcohols, the inhalational general anesthetics halothane, enflurane, and isoflurane, and the general anesthetic propofol.

Environmental toxicants that act on the $GABA_A$ receptor channels include cyclodienes, hexachlorocyclohexane (HCH), pyrethroids, Hg^{2+}, lanthanides, and avermectin B_{1a}.[139]

11.09.4.2.2 Cyclodienes and lindane

It was not until the early 1980s that the $GABA_A$ system was suggested to be the target site of dieldrin and lindane following the demonstration of synaptic facilitation by these insecticides in 1950s.[188,189] Lindane and cyclodiences were shown to antagonize GABA-induced $^{36}Cl^-$ uptake and to inhibit [^{35}S]

TBPS binding, and these actions run parallel with each other indicating that they act at the TBPS binding site.[139]

The first patch clamp experiments for these insecticides have clearly demonstrated that the GABA-induced chloride current is suppressed by lindane.[190] Noise analyses have shown that lindane and dieldrin decrease the frequency of GABA-chloride channel openings without changing the mean open time.[191] Poly (A)$^+$RNA from rat cerebral cortex expressed GABA-induced currents ($I_{G\text{-}Actx}$) in *Xenopus* oocytes, and RNA from bovine retina expressed two types of GABA-induced currents, $I_{G\text{-}Aret}$ and $I_{G\text{-}BR}$, the latter of which was atypical in that it was resistant to bicuculline.[192] γ-HCH or lindane potently inhibited all three types of GABA-induced currents, whereas α-, δ-, and β-HCH exhibited differential effects: these three isomers potentiated $I_{G\text{-}Actx}$ and $I_{G\text{-}Aret}$, but had little or no effect on $I_{G\text{-}BR}$.

Dieldrin exerts a dual action on the GABA receptor-channel complex of rat DRG neurons (Figure 9). GABA-induced currents were first augmented and then suppressed slowly. In human embryonic kidney (HEK-293) cells expressing various combinations of GABA receptor subunits, dieldrin augments GABA-induced currents in a combination containing the γ$_2$ subunit, whereas only suppression is observed in the absence of the γ$_2$ subunit (Figure 10).[193,194] Comparison of the α, β, γ, and δ isomers of HCH has demonstrated that their stimulating and/or inhibitory effect on the GABA system can explain behavioral changes caused by each isomer.[195]

Lindane is known to inhibit gap junctional communication in rodent hepatocytes[196,197] and fibroblasts.[198] It has been shown that lindane's action to inhibit gap junctional communication is partially mediated through activation of PKC.[199] This raises a strong possibility that lindane and cyclodienes modulate the GABA system through second messengers including PKC and PKA.

11.09.4.2.3 Pyrethroids

Type II pyrethroids have been shown to block the GABA$_A$ receptor-channel complex.[139] The GABA hypothesis has been a matter of extensive debate, partly because the potency and efficacy of the action are low. Initial patch clamp experiments with rat dorsal root ganglion neurons have unequivocally shown that while 10 μM deltamethrin greatly prolongs the sodium current as expected, the GABA-induced chloride current recorded from the very same cell remains totally unchanged.[190] However, patch

A

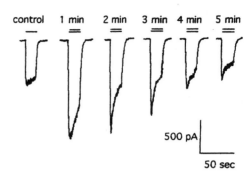

B

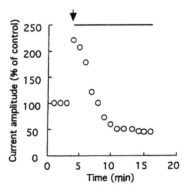

Figure 9 Dual effect of dieldrin on GABA-induced chloride currents recorded from a rat dorsal root ganglion neuron. (A) Currents in response to 20 s application of 10 μM GABA (solid bar) and to co-application of 10 μM GABA and 1 μM dieldrin (dotted bar) at the time indicated after taking control record. The peak current is greatly enhanced but gradually decreases during repeated co-applications. Desensitization of current is accelerated. (B) Time course of the change in peak amplitude before and during repeated co-applications (dotted line) (reproduced by permission of the American Society for Pharmacology and Experimental Therapeutics from *J. Pharmacol. Exp. Ther.*, 1994, **269**, 164–171).

clamp experiments with rat hippocampal neurons have indicated that there is a transient increase in GABA-induced chloride current following application of pyrethroids.[139]

11.09.4.2.4 Mercury

Mercury affects synaptic transmission.[139] The frequency of spontaneous miniature end-plate potentials (MEPPs) increases, while the amplitude of the nerve-evoked end-plate potential (EPP) decreases. Although binding of ACh and quinuclidinyl benzilate (QNB) to the nicotinic and muscarinic receptors, respectively, is inhibited by mercuric chloride and methylmercury, the end-plate sensitivity to ACh is not affected by them. It appears that

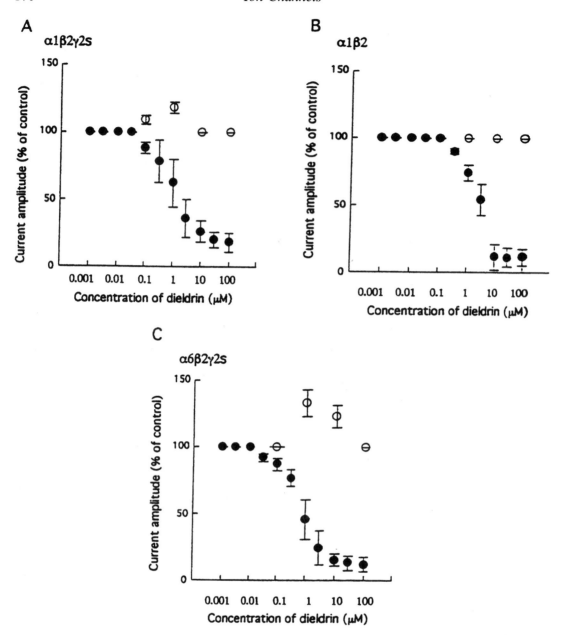

Figure 10 Dose–response relationships for dieldrin effects on GABA-induced chloride currents in the α1β2γ2s (A), α1β2 (B) and α6β2γ2s (C) combinations of GABA$_A$ receptor transfected in human embryonic kidney (HEK 293) cells. Current enhancement (open circles) is observed in the α1β2γ2s and α6β2γ2s combinations, but not in the α1β2 combination. Current suppression (closed circles) is observed in all three combinations (reproduced by permission of Elsevier Science from *Brain Res.*, 1994, **645**, 19–26).

the decrease in EPP amplitude is due to block of calcium entry to the nerve terminals, and that intracellular release of Ca^{2+} from nerve terminal mitochondria is responsible for the increase in the MEPP frequency by mercury.

(i) Mercury modulation of ion channels

In contrast to the extensive studies performed with mercury-intoxicated neuromuscular prep-

arations, only a limited amount of data is available for the action of mercury on ion channels. Sodium and potassium currents were suppressed by methylmercury in squid axons and neuroblastoma cells.[15,200] ^{45}Ca^{2+} influx into synaptosome and PC12 cells was inhibited by methylmercury.[201] N-type and L-type calcium channels of PC12 cells were blocked by low concentrations of methylmercury, 10 μM methylmercury causing 77% and 70% block of N- and L-type, respectively.[202] Kaninateactivated currents were suppressed by low concentrations of Hg^{2+} ($K_i = 70\,nM$).[203,204] In contrast to the insensitivity of the end-plate nicotinic ACh receptors to mercury, the neuronal nicotinic ACh receptor and two kinds of muscarinic ACh receptors were suppressed by methylmercury.[200]

(ii) Mercury modulation of $GABA_A$ receptor channels

The $GABA_A$ receptor channels have been found to be strongly stimulated by mercuric chloride at low concentrations (Figure 11).[205] At the minimum effective concentration of $0.1\,\mu M$, GABA-induced currents were increased to 115% of control, and at $10\,\mu M$ the current was 200% of control. In contrast, methylmercury at $100\,\mu M$ suppressed the current. Both types of mercury also generated an inward current which was carried through nonspecific channels. Competition experiments have led to the conclusion that mercury binds to a site different from the GABA site, the picrotoxin site, and the La^{3+} site on the GABA receptor.

(iii) Active form(s) of mercury

Mercury exists in a variety of forms. Studies with humans (the Minamata case)[206] and monkeys have clearly indicated that the percentages of inorganic mercury of the total mercury in the brain are very high, ranging from 72% to 88%.[207] Inorganic mercury is formed in the brain as a result of biotransformation from methylmercury and has a very long half-life extending several years. This is important, as mercuric chloride is more potent than methylmercury on various receptors and channels.[208,209] It is reasonable to assume that methylmercury, after crossing the blood–brain barrier, is partially converted into inorganic mercury, which in turn exerts a potent stimulatory effect on the GABA system.

A question still remains as to which form(s) of mercuric chloride is(are) responsible for the stimulating action on the GABA system. Mercuric chloride exists in an aqueous solution in various forms such as Hg^{2+}, $HgCl_2$, $HgCl_3^-$, $HgCl_4^{2-}$, and $Hg(OH)_2$, depending on the chloride concentration and pH.[210] This problem has been largely neglected, yet it is of critical importance as it is related to the molecular mechanism of action on the GABA system.

11.09.4.2.5 Copper and zinc

Copper and zinc are known to exert multiple actions on various receptors and ion channels, including the $GABA_A$ receptor channels. Both Cu^{2+} and Zn^{2+} suppress GABA-induced currents with an IC_{50} value of $19\,\mu M$.[139,211] They exert a very similar action, and competition experiments have shown that they share a common binding site on or near the external surface of the GABA receptor–channel complex.

11.09.4.2.6 Lanthanides

Lanthanides, or rare-earth metals, comprise a series of 15 metals starting with lanthanum (atomic number 57) and ending with lutetium (atomic number 71). Lanthanides have been used extensively in industry, and their therapeutic use is being explored. All of the seven lanthanides tested out of 15 exert similar actions on the GABA system.[139,212] Most patch clamp experiments were performed with lanthanum and terbium.

Lanthanum augments GABA-induced currents of rat DRG neurons with an EC_{50} of $230\,\mu M$ by shifting the GABA dose–response curve in the direction of lower concentrations (Figure 12). Competition experiments have shown that lanthanum binds to a site that is distinct from the GABA site, the picrotoxin site, the benzodiazepine site, the barbiturate site, and the Cu^{2+}/Zn^{2+} site, and that it is

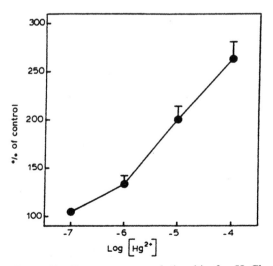

Figure 11 Dose–response relationship for $HgCl_2$-induced enhancement of the peak current evoked by $30\,\mu M$ GABA in rat dorsal root ganglion neurons. Mean $\pm$ SD ($n = 4$–5). At $0.1\,\mu M$, SD is smaller than the size of the symbol (reproduced by permission of Elsevier Science from *Brain Res.*, 1991, **551**, 58–63).

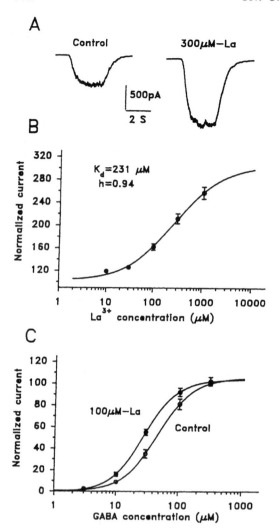

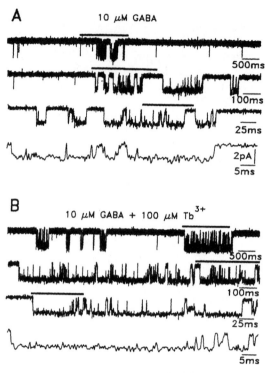

Figure 13 GABA (10 μM)-induced single-channel currents before (A) and during co-application of 100 μM terbium (Tb^{3+}) (B) in a membrane patch isolated from a rat dorsal root ganglion neuron. The currents are shown on various time scales. The portion of current record under the horizontal line in each current trace is shown on an expanded time scale in the trace below (reproduced by permission of the Society for Neuroscience from *J. Neurosci.*, 1994, **14**, 3835–3841).

Figure 12 La^{3+} potentiation of GABA-induced currents in rat dorsal root ganglion neurons. (A) Enhancement by 300 μM La^{3+} of the current induced by 10 μM GABA. (B) The dose–response relationship for La^{3+} potentiation of the current induced by 30 μM GABA ($n=4$). The current amplitude is normalized to the control current in the absence of La^{3+}. (C) Shift of the dose–response curve for GABA-induced currents by 100 μM La^{3+} ($n=4$) (reproduced by permission of Elsevier Science from *Brain Res.*, 1993, **607**, 222–232).

located on or near the external surface of the GABA receptor–channel complex.

Single-channel analyses of the GABA receptor channels have led to the conclusion that the augmentation of the whole-cell GABA-induced current by Tb^{3+} is ascribed to three changes in single-channel activity (Figure 13), that is, the increase in the overall mean open time due to an increase in the relative fraction of the longest time constant, the decrease in the overall mean closed time due to a decrease in the relative fraction of the longest time constant, and the increase in the overall mean burst time due to

an increase in the relative fraction of the longest time constant.[213]

11.09.4.3 Acetylcholine Receptor Channels

11.09.4.3.1 *Role of acetylcholine receptor channels in intoxication with anticholinesterases*

It has long been known that following intoxication with organophosphorus (OP) compounds the cholinesterase (ChE) level remains inhibited to a very low level even after the symptoms of poisoning in animals have disappeared.[214] Significant loss in blood AChE alone produced no detectable neurobehavioral deficits in rats injected with paraoxon.[215] It is also widely recognized that animals develop tolerance after repeated exposure to various OP insecticides; the initial symptoms of poisoning disappear following repeated applications of the OP compounds. It was proposed that the mechanism of tolerance was due to a decreased sensitivity of the cholinergic receptors to OP compounds.[216,217] OP compounds were indeed demonstrated to reduce the density of muscarinic receptors in both central and peripheral nerve tissues after repeated applications.[218]

Another line of evidence in support of the nonacetylcholinesterase (nonAChE), direct action of carbamate has been obtained by electrophysiological techniques. Patch clamp experiments with isolated skeletal muscle fibers have demonstrated that physostigmine, neostigmine, pyridostigmine, and erdophonium exhibit potent noncompetitive antagonistic activities on muscle nicotinic actylcholine receptor (AChR).[219,220] At the single channel level, many brief flickers appeared suggesting channel block at its open configuration. The EC_{50} values for channel block were roughly in the same order of magnitude as those for AChE inhibition. It was proposed that carbamate block of AChR channels plays an important role in counteracting cholinergic hyperexcitation caused by OP compounds.

Reactivation of AChE has been regarded as the major mechanism by which oximes protect against OP intoxication. However, oximes themselves exert both excitatory and inhibitory actions on AChR.[219] Pralidoxime (2-PAM) has been shown to cause flickerings in muscle AChR single channels, indicative of open channel block.[219] HI-6, another ChE reactivator, has the same effect. It was suggested that these direct actions of oximes on AChR channels contribute significantly to their protective action against OP and carbamate poisoning.

11.09.4.3.2 Neuronal acetylcholine receptor channels

The aforementioned patch clamp experiments were conducted with skeletal muscles. However, it is well known that nicotinic muscle AChR is different from neuronal nicotinic AChR in various respects. Most remarkable are drastic differences in pharmacological sensitivity. For instance, muscle AChR is highly sensitive to benzoquinonium, decamethonium, succinylcholine and pancuronium, whereas neuronal AChR is very sensitive to hexamethonium and mecamylamine. Although neuronal nicotinic AChR exhibits low sensitivity to α-bungarotoxin which is a potent blocker of muscle nicotinic AChR,[221] κ-bungarotoxin blocks neuronal AChR potently.[222] Thus, in order to elucidate the mechanism of action of anti-ChE OP and carbamate insecticides, it is of utmost importance to study their actions on neuronal AChR channels.

Neuronal nicotinic AChR is not only different from muscle nicotinic AChR in terms of physiology and pharmacology, but is also diverse in nature.[223] For example, rat hippocampal neurons express four different types, IA, IB, II, and III, which show different current kinetics and different sensitivities to the blocking actions of α-bungarotoxin, κ-bungarotoxin, methyllycaconitine, and dihydro-β-erythroidine.[224]

Nitromethylene heterocyles are a novel and distinct class of insecticides.[225] They exhibit selective toxicity towards insects, with relatively low toxicity in mammals and fish.[226] 1-(Pyridine-3-yl-methyl)-2-nitromethylene-imidazolidine (PMNI) exerted agonistic effects followed by antagonistic effects in locust neurons, but was lacking agonistic effects in N1E-115 neuroblastoma cells with only weak blocking effects.[227] Comparison of six nitromethylene heterocycle compounds also revealed selective actions on locust neurons as compared with mammalian muscle nicotinic AChR and N1E-115 AChR.

ACh-activated currents (nicotinic) of neuroblastoma cells have been found very sensitive to lead with an IC_{50} of 19 nM.[228] However, the block was relieved with increasing concentration of lead with an EC_{50} of 21 μM resulting in a bell-shaped dose–response curve.[228,229] Contrary to the neuronal nicotinic AChR, frog endplate AChR was not affected by 10 μM lead, with some decrease only at 100 μM.[230] Thus, one cannot necessarily extrapolate the data on the muscle AChR to the CNS.

Binding of QNB to the muscarinic AChR of rat brain was inhibited by lead at ~50 μM.[231] In keeping with the electrophysiological data, K-stimulated $^{45}Ca^{2+}$ influx to rat brain synaptosomes was inhibited by lead with a K_i of 1.1 μM.[232] As might be expected from these calcium data, the release of ACh was suppressed by lead with a K_i at 16 μM.[232] Mercury also binds to AChR. QNB or pilocarpine binding to the muscarinic AChR was inhibited by mercuric chloride, but methylmercury was much less potent.[233] Nicotine or ACh binding to the nicotinic ACh receptors was also inhibited by methylmercury.[234] In keeping with these binding data, the responses of nicotinic ACh receptor and two kinds of muscarinic ACh receptors were suppressed by methylmercury.[200]

11.09.4.3.3 Effects of pyrethroids on acetylcholine receptor channels

The ACh receptor–channel complex has been claimed to be a target site of pyrethroids. The binding of [³H]perhydrohistrionicotoxin

([^{3}H]H$_{12}$-HTX) to the *Torpedo* electric organ
was inhibited to various extents by type I and
type II pyrethroids.[235] These binding data are
at variance with the electrophysiological data
on frog skeletal muscle end-plates.[236] The
amplitude of the end-plate potential was not
affected by 10 µM allethrin, although repetitive
end-plate potentials were evoked by a single
nerve stimulus as a result of repetitive discharges at the nerve terminals. One possible
explanation for the controversy is that binding
does not necessarily lead to response. This
problem remains to be solved. More important
experiments would be to study the effects of
pyrethroids on neuronal nicotinic and muscarinic ACh receptor channels. In fact, there is
some evidence to indicate pyrethroid interactions with muscarinic ACh receptors.[237] In
Helix neurons, deltamethrin depressed the
ACh-induced current.[238]

11.09.4.4 Ligand-gated Potassium Channels

11.09.4.4.1 Ca^{2+}-activated potassium channels

Three types of Ca^{2+}-activated potassium
channels are known: high conductance Ca^{2+}
activated potassium channel, sometimes called
big or maxi potassium channel; intermediate
conductance Ca^{2+}-activated potassium channel; and small conductance Ca^{2+}-activated
potassium channel sometimes called small
potassium channel. Big potassium channels
exhibit a high single-channel conductance of
100–250 pS, intermediate potassium channels
have an 18–50 pS conductance, and small
potassium channels have a 10–14 pS conductance.[19]
Charybdotoxin isolated from the scorpion
Leiurus quinquestriatus var. *hebraeus* blocks big
and intermediate Ca^{2+}-activated potassium
channels, and leiurotoxin I, another toxic
component from the same species of scorpion,
blocks small Ca^{2+}-activated potassium channels. Apamin isolated from the venom of the
honeybee *Apis mellifera* blocks small Ca^{2+}-
activated potassium channels.[19]

11.09.4.4.2 ATP-sensitive potassium channels

ATP-sensitive potassium channels were originally discovered in cardiac membranes but
were later found in other cells including skeletal
muscle, pancreatic β-cells, arterial smooth
muscle, and cortical neurons. This type of
potassium channels is inhibited by an increase
in intracellular ATP concentration, and plays
an important role in various functions of cells.

Hypoglycemic sulfonylureas such as tolbuta-
mide and glibenclamide, which have been used
in the treatment of diabetes, block the ATP-
sensitive potassium channel.[19]

11.09.4.4.3 Potassium channel openers

Drugs that open potassium channels have
received much attention, because they could be
used in the treatment of cardiovascular disorders. Cromakalim, pinacidil, nicorandil, min-
oxidil, diazoxide, and RP52891 are among
them, and cause vasodilatation through opening of potassium channels. It appears that their
target is the ATP-sensitive potassium channel.
However, some of them may block voltage-
gated calcium channels and Ca^{2+}-activated
potassium channels.[19]

11.09.4.4.4 Muscarinic-inactivated potassium channels (M-current)

M-currents have been found in a variety of
neurons. Although it does not generate an
action potential, it affects excitability by modulating the frequency and pattern of discharges.
It is inhibited via activation of G proteins by
muscarinic ACh receptor agonists such as ACh,
muscarine, methacholine, and oxotremorine,
and by peptides such as *t*-luteinizing hormone-
releasing hormone, substance P, substance K,
eledoisin, kassinin, and physalaemin.[19]

ACKNOWLEDGMENTS

The author wishes to thank Julia Irizarry for
secretarial assistance. The author's studies cited
in this article were supported by grants from the
National Institutes of Health NS14143 and
NS14144.

11.09.5 REFERENCES

1. E. Neher and B. Sakmann, 'Single-channel currents
   recorded from membrane of denervated frog muscle
   fibres.' *Nature*, 1976, **260**, 799–802.
2. O. P. Hamill, A. Marty, E. Neher *et al.*, 'Improved
   patch-clamp techniques for high-resolution current
   recording from cells and cell-free membrane patches.'
   Pflügers Arch., 1981, **391**, 85–100.
3. K. S. Cole, 'Dynamic electrical characteristics of the
   squid axon membrane.' *Arch. Sci. Physiol.*, 1949, **3**,
   253–258.
4. T. Narahashi, T. Deguchi, N. Urakawa *et al.*,
   'Stabilization and rectification of muscle fiber membrane by tetrodotoxin.' *Am. J. Physiol.*, 1960, **198**,
   934–938.
5. T. Narahashi, J. W. Moore and W. R. Scott,

'Tetrodotoxin blockage of sodium conductance increase in lobster giant axons.' *J. Gen. Physiol.*, 1964, **47**, 965–974.

6. T. Narahashi and N. C. Anderson, 'Mechanism of excitation block by the insecticide allethrin applied externally and internally to squid giant axons.' *Toxicol. Appl. Pharmacol.*, 1967, 10, 529–547.

7. T. Narahashi and H. G. Haas, 'DDT: interaction with nerve membrane conductance changes.' *Science*, 1967, **157**, 1438–1440.

8. T. Narahashi and H. G. Haas, 'Interaction of DDT with the components of lobster nerve membrane conductance.' *J. Gen. Physiol.*, 1968, **51**, 177–198.

9. B. Hille, 'Pharmacological modifications of the sodium channels of frog nerve.' *J. Gen. Physiol.*, 1968, **51**, 199–219.

10. T. Narahashi, J. W. Moore and B. I. Shapiro, '*Condylactis* toxin: interaction with nerve membrane ionic conductances.' *Science*, 1969, **163**, 680–681.

11. E. X. Albuquerque, J. W. Daly and B. Witkop, 'Batrachotoxin: chemistry and pharmacology.' *Science*, 1971, **172**, 995–1002.

12. T. Narahashi, E. X. Albuquerque and T. Deguchi, 'Effects of batrachotoxin on membrane potential and conductance of squid giant axons.' *J. Gen. Physiol.*, 1971, **58**, 54–70.

13. E. Koppenhöfer and H. Schmidt, 'Die Wirkung von Skorpiongift auf die Ionenströme des Ranvierschen Schnürrings. I. Die Permeabilitäten P_{Na} und P_K. [Effect of scorpion venom on ionic currents of the node of Ranvier. I. The permeabilities P_{Na} and P_K.]' *Pflügers Arch.*, 1968, **303**, 133–149.

14. E. Koppenhöfer and H. Schmidt, 'Die Wirkung von Skorpiongift auf die Ionenströme des Ranvierschen Schnürrings. II. Unvollständige Natrium-Inactivierung. [Effects of scorpion venom on ionic currents of the node of Ranvier. II. Incomplete sodium inactivations.]' *Pflügers Arch.*, 1968, **303**, 150–161.

15. B. B. Shrivastav, M. S. Brodowick and T. Narahashi, 'Methylmercury: effects on electrical properties of squid axon membranes.' *Life Sci.*, 1976, **18**, 1077–1081.

16. M. Noda, H. Takahashi, T. Tanabe *et al.*, 'Primary structures of β- and δ-subunit precursors of *Torpedo californica* acetylcholine receptor deduced from cDNA sequences.' *Nature*, 1983, **301**, 251–255.

17. M. Noda, H. Takahashi, T. Tanabe *et al.*, 'Primary structure of the α-subunit precursor of *Torpedo californica* acetylcholine receptor deduced from cDNA sequence.' *Nature*, 1982, **299**, 793–797.

18. W. A. Catterall, 'Cellular and molecular biology of voltage-gated sodium channels.' *Physiol. Rev.*, 1992, **72** Suppl., S15–S48.

19. T. Narahashi and M. D. Herman, in 'Methods in Enzymology, Ion Channels.' eds. B. Rudy and L. E. Iverson. Academic Press, San Diego, CA., 1992, vol. 207, pp. 620–643.

20. T. Narahashi, 'Chemicals as tools in the study of excitable membranes.' *Physiol. Rev.,* 1974, 54, 813–889.

21. S. A. Cohen and R. L. Barchi, 'Voltage-dependent sodium channels.' *Int. Rev. Cytol.*, 1993, **137C**, 55–103.

22. K. T. Wann, Neuronal sodium and potassium channels: structure and function.' *Br. J. Anaesth.*, 1993, **71**, 2–14.

23. T. Narahashi, H. G. Haas and E. F. Therrien, 'Saxitoxin and tetrodotoxin: comparison of nerve blocking mechanism.' *Science*, 1967, **157**, 1441–1442.

24. T. Narahashi, in 'Handbook of Natural Toxins, Marine Toxins and Venoms,' ed. A. T. Tu, Dekker, New York, 1988, pp.185–210

25. J. M. Ritchie and R. B. Rogart, 'The binding of labelled saxitoxin to the sodium channels in normal and denervated mammalian muscle, and in amphibian muscle.' *J. Physiol. (Lond.)*, 1977, **269**, 341–354.

26. T. Narahashi, J. W. Moore and D. T. Frazier, 'Dependence of tetrodotoxin blockage of nerve membrane conductance on external pH.' *J. Pharmacol. Exp. Ther.*, 1969, **169**, 224–228.

27. J. S. Trimmer and W. S. Agnew, 'Molecular diversity of voltage-sensitive Na channels.' *Annu. Rev. Physiol.*, 1989, **51**, 401–418.

28. D. D. Doyle, Y. Guo, S. L. Lustig *et al.*, 'Divalent cation competition with [^{3}H]saxitoxin binding to tetrodotoxin-resistant and -sensitive sodium channels. A live-site structural model of ion/toxin interaction.' *J. Gen. Physiol.*, 1993, **101**, 153–182.

29. A. Ravindran, L. Schild and E. Moczydlowski, 'Divalent cation selectivity for external block of voltage-dependent Na$^+$ channels prolonged by batrachotoxin. Zn^{2+} induces discrete substates in cardiac Na$^+$ channels.' *J. Gen Physiol.*, 1991, **97**, 89–115.

30. R. L. Barchi and J. B. Weigele, 'Characteristics of saxitoxin binding to the sodium channel of sarcolemma isolated from rat skeletal muscle.' *J. Physiol. (Lond.)*, 1979, **295**, 383–396.

31. L. Schild and E. Moczydlowski, Competitive binding interaction between Zn^{2+} and saxitoxin in cardiac Na$^+$ channels. Evidence for a sulfhydryl group in the Zn^{2+} saxitoxin binding site.' *Biophys. J.*, 1991, **59**, 523–537.

32. J. F. Worley, III, R. J. French and B. K. Krueger, 'Trimethyloxonium modification of single batrachotoxin-activated sodium channels in planar bilayers. Changes in unit conductance and in block by saxitoxin and calcium.' *J. Gen. Physiol.*, 1986, **87**, 327–349.

33. P. H. Backx, D. T. Yue, J. H. Lawrence *et al.*, 'Molecular localization of an ion-binding site within the pore of mammalian sodium channels.' *Science*, 1992, **257**, 248–251.

34. J. Satin, J. W. Kyle, M. Chen *et al.*, 'A mutant of TTX-resistant cardiac sodium channels with TTX-sensitive properties.' *Science*, 1992, **256**, 1202–1205.

35. J. Satin, J. W. Kyle, M. Chen *et al.*, 'The cloned cardiac Na channel α-subunit expressed in *Xenopus* oocytes show gating and blocking properties of native channels.' *J. Membr. Biol.*, 1992, **130**, 11–22.

36. J. Satin, J. W. Kyle, Z. Fan *et al.*, 'Post- repolarization block of cloned sodium channels by saxitoxin: the contribution of pore-region amino acids.' *Biophys. J.*, 1994, **66**, 1353–1363.

37. J. Satin, J. T. Limberies, J. W. Kyle *et al.*, 'The saxitoxin/tetrodotoxin binding site on cloned rat brain IIa Na channels is in the transmembrane electric fluid.' *Biophys. J.*, 1994, **66**, 1007–1014.

38. G. E. Kirsch, M. Alam and H. A. Hartmann, 'Differential effects of sulfhydryl reagents on saxitoxin and tetrodotoxin block of voltage-dependent Na channels.' *Biophys. J.*, 1994, 67, 2305–2315.

39. G. M. Lipkind and H. A. Fozzard, 'A structural model of the tetrodotoxin and saxitoxin binding site of the Na$^+$ channel.' *Biophys. J.*, 1994, 66, 1–13.

40. A. M. Woodhull, 'Ionic blockage of sodium channels in nerve.' *J. Gen. Physiol.*, 1973, **61**, 687–708.

41. R. J. French, J. F. Worley, III and B. K. Krueger, 'Voltage-dependent block by saxitoxin of sodium channels incorporated into planar lipid bilayers. *Biophys. J.*, 1984, **45**, 301–310.

42. E. Moczydlowski, S. Hall, S. S. Garber *et al.*, 'Voltage-dependent blockage of muscle Na$^+$ channels by guanidinium toxins.' *J. Gen. Physiol.*, 1984, **84**, 687–704.

43. M. Baer, P. M. Best and H. Reuter, 'Voltage-dependent action of tetrodotoxin in mammalian cardiac muscle.' *Nature*, 1976, **263**, 344–345.

44. T. A. Rando and G. R. Strichartz, 'Saxitoxin blocks batrachotoxin-modified sodium channels in the node of Ranvier in a voltage-dependent manner.' *Biophys. J.*, 1986, 49, 785–794.

45. V. L. Salgado, J. Z. Yeh and T. Narahashi, 'Use- and voltage-dependent block of sodium channel by saxitoxin.' *Ann. NY Acad. Sci.*, 1986, 479, 84–95.

46. S. Yoshida, 'Tetrodotoxin-resistant sodium channels.' *Cell. Mol. Neurobiol.*, 1994, 14, 227–244.

47. J. L. Bossu and A. Feltz, 'Patch-clamp study of the tetrodotoxin-resistant sodium current in group C sensory neurones.' *Neurosci. Lett.*, 1984, 51, 241–246.

48. S. R. Ikeda and G. G. Schofield, 'Tetrodotoxin-resistant sodium current of rat nodose neurones: monovalent cation selectivity and divalent cation block.' *J. Physiol. (Lond.)*, 1987, 389, 255–270.

49. P. G. Kostyuk, N. S. Veselovsky and A. Y. Tsyndrenko, 'Ionic currents in the somatic membrane of rat dorsal root ganglion neurons—I. Sodium currents.' *Neuroscience*, 1981, 6, 2423–2430.

50. M. J. McLean, P. B. Bennett and R. M. Thomas, 'Subtypes of dorsal root ganglion neurons based on different inward currents as measured by whole-cell voltage clamp.' *Mol. Cell. Biochem.*, 1988, 80, 95–107.

51. A. Schwartz, Y. Palti and H. Meiri, 'Structural and developmental differences between three types of Na channels in dorsal root ganglion cells of new born rats.' *J. Memb. Biol.*, 1990, 116, 117–128.

52. M. L. Roy and T. Narahashi, 'Differential properties of tetrodotoxin-sensitive and tetrodotoxin-resistant sodium channels in rat dorsal root ganglion neurons.' *J. Neurosci.*, 1992, 12, 2104–2111.

53. N. Ogata and H. Tatebayashi, 'Kinetic analysis of two types of Na⁺ channels in rat dorsal root ganglia.' *J. Physiol. (Lond.)*, 1993, 466, 9–37.

54. A. A. Elliott and J. R. Elliott, 'Characterization of TTX-sensitive and TTX-resistant sodium currents in small cells from adult rat dorsal root ganglia.' *J. Physiol. (Lond.)*, 1993, 463, 39–56.

55. K. S. Ginsburg and T. Narahashi, 'Differential sensitivity of tetrodotoxin-sensitive and tetrodotoxin-resistant sodium channels to the insecticide allethrin in rat dorsal root ganglion neurons.' *Brain Res.*, 1993, 627, 239–248.

56. H. Tatebayashi and T. Narahashi, 'Differential mechanism of action of the pyrethroid tetramethrin on tetrodotoxin-sensitive and tetrodotoxin-resistant sodium channels.' *J. Pharmacol. Exp. Ther.*, 1994, 270, 595–603.

57. K. Hoehn, T. W. Watson and B. A. MacVicar, 'A novel tetrodotoxin-insensitive, slow sodium current in striatal and hippocampal neurons.' *Neuron*, 1993, 10, 543–552.

58. J. A. White, A. Alonso and A. R. Kay, 'A heart-like Na⁺ current in the medial entorhinal cortex.' *Neuron*, 1993, 11, 1037–1047.

59. C. H. Wu and T. Narahashi, 'Mechanism of action of novel marine neurotoxins on ion channels.' *Annu. Rev. Pharmacol. Toxicol.*, 1988, 28, 141–161.

60. M. Kobayashi, C. H. Wu, M. Yoshii *et al.*, 'Preferential block of skeletal muscle sodium channels by geographutoxin II, a new peptide toxin from *Conus geographus.' Pflügers Arch.*, 1986, 407, 241–243.

61. G. Romey, J. P. Abita, H. Schweitz *et al.*, 'Sea anemone toxin: a tool to study molecular mechanisms of nerve conduction and excitation–secretion coupling.' *Proc. Natl. Acad. Sci. USA*, 1976, 73, 4055–4063.

62. P. A. Low, C. H. Wu and T. Narahashi, 'The effect of anthopleurin-A on crayfish giant axon.' *J. Pharmacol. Exp. Ther.*, 1979, 210, 417–421.

63. W. A. Catterall and L. Beress, 'Sea anemone toxin and scorpion toxin share a common receptor site associated with the action potential sodium ionophore.' *J. Biol. Chem.*, 1978, 253, 7393–7396.

64. H. Meves, J. M. Simard and D. D. Watt, 'Interactions of scorpion toxins with the sodium channel.' *Ann. NY Acad. Sci.*, 1986, 479, 113–132.

65. T. Narahashi, B. I. Shapiro, T. Deguchi *et al.*, 'Effects of scorpion venom on squid axon membranes.' *Am. J. Physiol.*, 1972, 222, 850–857.

66. M. D. Cahalan, 'Modification of sodium channel gating in frog myelinated nerve fibres by *Centruroides sculpturatus* scorpion venom.' *J. Physiol. (Lond.)*, 1975, 244, 511–534.

67. H. P. M. Vijverberg, D. Pauron and M. Lazdunski, 'The effect of *Tityus serrulatus* scorpion toxin γ on Na channels in neuroblastoma cells.' *Pflügers Arch.*, 1984, 401, 297–303.

68. B. I. Khodorov, 'Batrachotoxin as a tool to study voltage-sensitive sodium channels of excitable membranes.' *Prog. Biophys. Mol. Biol.*, 1985, 45, 57–148.

69. B. I. Khodorov and S. V. Revenko, 'Further analysis of the mechanisms of action of batrachotoxin on the membrane of myelinated nerve.' *Neuroscience*, 1979, 4, 1315–1330.

70. J. Tanguy and J. Z. Yeh, 'BTX modification of Na Channels in squid axons. I. State dependence of BTX action.' *J. Gen. Physiol.*, 1991, 97, 499–519.

71. F. N. Quandt and T. Narhashi, 'Modification of single Na⁺ channels by batrachotoxin.' *Proc. Nat. Acad. Sci. USA*, 1982, 79, 6732–6736.

72. A. M. Correa, R. Latorre and F. Bezanilla, 'Ion permeation in normal and batrachotoxin-modified Na⁺ channels in the squid giant axon.' *J. Gen Physiol.*, 1991, 97, 605–625.

73. D. S. Duch, A. Hernandez, S. R. Levinson *et al.*, 'Grayanotoxin-I-modified eel electroplax sodium channels. Correlation with batrachotoxin and veratridine modifications.' *J. Gen. Physiol.*, 1992, 100, 623–645.

74. T. Hironaka and T. Narahashi, 'Cation permeability ratios of sodium channels in normal and grayanotoxin-treated squid axon membranes.' *J. Membr. Biol.*, 1977, 31, 359–381.

75. T. Narahashi and I. Seyama, 'Mechanism of nerve membrane depolarization caused by grayanotoxin I.' *J. Physiol. (Lond.)*, 1974, 242, 471–487.

76. I. Seyama and T. Narahashi, 'Modulation of sodium channels of squid nerve membranes by grayanotoxin I.' *J. Pharmacol. Exp. Ther.*, 1981, 219, 614–624.

77. H. Schmidt and O. Schmitt, 'Effect of aconitine on the sodium permeability at the node of Ranvier.' *Pflügers Arch.*, 1974, 349, 133–148.

78. D. T. Campbell, 'Modified kinetics and selectivity of sodium channels in frog skeletal muscle fibers treated with aconitine.' *J. Gen. Physiol.*, 1982, 80, 713–731.

79. G. Strichartz, T. Rando and G. K. Wang, 'An integrated view of the molecular toxicology of sodium channel gating in excitable cells.' *Annu Rev. Neurosci.*, 1987, 10, 237–267

80. W. D. Atchison, V. S. Luke, T. Narahashi *et al.*, 'Nerve membrane sodium channels as the target site of brevetoxins at neuromuscular junctions.' *Br. J. Pharmacol.*, 1986, 89, 731–738.

81. G. M. Nicholson, M. Willow, M. E. H. Howden *et al.*, 'Modification of sodium channel gating and kinetics by versutoxin from the Australian funnel-web spider *Hadronyche versuta.' Pflügers Arch.*, 1994, 428, 400–409.

82. T. Narahashi, in 'Advances in Insect Physiology,' eds. J. W. L. Beament, J. E. Treherne and V. B. Wigglesworth, Academic Press, London, 1971, vol. 8, pp. 1–93.

83. T. Narahashi, in 'Insecticide Biochemistry and Physiology,' ed. C. F. Wilkinson, Plenum Press, New York, 1976, pp. 327–352.

84. T. Narahashi, 'Nerve membrane ionic channels as the primary target of pyrethroids.' *Neurotoxicology*, 1985, **6**, 3–22.

85. T. Narahashi, in 'Sites of Action for Neurotoxic Pesticides,' eds. R. M. Hollingworth and M. B. Green, American Chemical Society symposium series, no. 356. American Chemical Society, Washtington, DC, 1987, pp. 226–250.

86. T. Narahashi, 'Nerve membrane Na$^+$ channels as targets of insecticides.' *Trends Pharmacol. Sci.*, 1992, **13**, 236–241.

87. T. Narahashi, J. M. Frey, K. S. Ginsburg *et al.*, in 'Sodium channels and GABA-activated channels as the target sites of insecticides. ACS Symposium Series 591, Molecular Action of Insecticides on Ion Channels,' ed. J. M. Clark, American Chemical Society, Washington, DC, 1995, pp. 26–43.

88. G. S. F. Ruigt, in 'Comprehensive Insect Physiology, Biochemistry and Pharmacology,' eds. G. A. Kerkut and L. I. Gilbert, Pergamon Press, Oxford, 1984, vol. 12, chap. 7, pp. 183–263.

89. D. M. Soderlund and J. R. Bloomquist, 'Neurotoxic actions of pyrethroid insecticides.' *Annu. Rev. Entomol.*, 1989, **34**, 77–96.

90. H. P. M. Vijverberg and J. van den Bercken, 'Neurotoxicological effects and the mode of action of pyrethroid insecticides.' *Crit. Rev. Toxicol.*, 1990, **21**, 105–126.

91. K. Chinn and T. Narahashi, 'Stabilization of sodium channel states by deltamethrin in mouse neuroblastoma cells.' *J. Physiol. (Lond.)*, 1986, **380**, 191–207.

92. T. Narahashi, in 'Basic Mechanisms of the Epilepsies. Annals of Neurology,' eds. A. V. Delgado-Escueta, A. A. Ward, Jr. and D. Woodbury, 1984, vol. 16, Suppl., pp. S39–S51.

93. J. R. Bloomquist and D. M. Soderlund, 'Pyrethroid insecticides and DDT modify alkaloid-dependent sodium channel activation and its enhancement by sea anemone toxin.' *Mol. Pharmacol.*, 1988, **33**, 543–550.

94. G. B. Brown, J. E. Gaupp and R. W. Olsen, 'Pyrethroid insecticides: stereospecific allosteric interaction with the batrachotoxin-A benzoate binding site of mammalian voltage-sensitive sodium channels.' *Mol. Pharmacol.*, 1988, **34**, 54–59.

95. S. M. Ghiasuddin and D. M. Soderlund, 'Pyrethroid insecticides: potent, stereospecific enhancers of mouse brain sodium channel activation.' *Pesticide Biochem. Physiol.*, 1985, **24**, 200–206.

96. A. Lombet, C. Mourre and M. Lazdunski, 'Interaction of insecticides of the pyrethroid family with specific binding sites on the voltage-dependent sodium channel from mammalian brain.' *Brain Res.*, 1988, **459**, 44–53.

97. M. D. Leibowitz, J. B. Sutro and B. Hille, 'Voltage-dependent gating of veratridine-modified Na channels.' *J. Gen. Physiol.*, 1986, **87**, 25–46.

98. A. E. Lund and T. Narahashi, 'Modification of sodium channel kinetics by the insecticide tetramethrin in crayfish giant axons.' *Neurotoxicology*, 1981, **2**, 213–229.

99. A. E. Lund and T. Narahashi, 'Kinetics of sodium channel modification by the insecticide tetramethrin in squid axon membranes.' *J. Pharmacol. Exp. Ther.*, 1981, **219**, 464–473.

100. L. D. Brown and T. Narahashi, 'Modulation of nerve membrane sodium channel activation by deltamethrin.' *Brain Res.*, 1992, **584**, 71–76.

101. S. F. Holloway, V. L. Salgado, C. H. Wu *et al.*, 'Kinetic properties of single sodium channels modified by fenvalerate in mouse neuroblastoma cells.' *Pflügers Arch.*, 1989, **414**, 613–621.

102. J. R. de Weille, H. P. M. Vijverberg and T. Narahashi, 'Interactions of pyrethroids and octylguanidine with sodium channels of squid giant axons.' *Brain Res.*, 1988, **445**, 1–11.

103. J. R. de Weille and T. Leinders, 'The action of pyrethroids on sodium channels in myelinated nerve fibres and spinal ganglion cells of the frog.' *Brain Res.*, 1989, **482**, 324–332.

104. A. E. Lund and T. Narahashi, 'Kinetics of sodium channel modification as the basis for the variation in the nerve membrane effects of pyrethroids and DDT analogs.' *Pesticide Biochem. Physiol.*, 1983, **20**, 203–216.

105. V. L. Salgado and T. Narahashi, 'Immobilization of sodium channel gating charge in crayfish giant axons by the insecticide fenvalerate.' *Mol. Pharmacol.*, 1993, **43**, 626–634.

106. A. E. Lund and T. Narahashi, 'Dose-dependent interaction of the pyrethroid isomers with sodium channels of squid axon membranes.' *Neurotoxicology*, 1982, **3**, 11–24.

107. K. Takeda and T. Narahashi, 'Chemical modification of sodium channel inactivation: separate sites for the action of grayanotoxin and tetramethrin.' *Brain Res.*, 1988, **448**, 308–312.

108. D. Yamamoto, J. Z. Yeh and T. Narahashi, 'Ion permeation and selectivity of squid axon sodium channels modified by tetramethrin.' *Brain Res.*, 1986, **372**, 193–197.

109. M. Oortgiesen, R. G. D. M. van Kleef and H. P. M. Vijverberg, 'Block of deltamethrin-modified sodium current in cultured mouse neuroblastoma cells: local anesthetics as potential antidotes.' *Brain Res.*, 1990, **518**, 11–18.

110. J. H. Song and T. Narahashi, 'Modulation of sodium channels of rat cerebellar Purkinje neurons by the pyrethroid tetramethrin.' *J. Pharmacol. Exp. Ther.*, 1996, **277**, 445–451.

111. S. B. Tucker, S. A. Flannigan and C. E. Ross, 'Inhibition of cutaneous paresthesia resulting from synthetic pyrethroid exposure.' *Int. J. Dermatol.*, 1984, **23**, 686–689.

112. S. A. Flannigan and S. B. Tucker, 'Variation in cutaneous sensation between synthetic pyrethroid insecticides.' *Contact Dermatitis.*, 1985, **13**, 140–147.

113. J. H. Song and T. Narahashi, 'Selective block of tetramethrin-modified sodium channels by ($\pm$)-α-tocopherol (vitamin E).' *J. Pharmacol. Exp. Ther.*, 1995, **275**, 1402–1411.

114. M. Elliott, in 'Synthetic Pyrethroids, ACS Symposium Series 42,' ed. M. Elliott, American Chemical Society, Washington, DC, 1977, pp. 1–28.

115. H. Hirai, 'ETOC, a new pyrethroid,' *Sumitomo Pyrethroid World*, 1987, **9**, 2–4.

116. J. Miyamoto, 'A risk assessment of household insecticides,' *Sumitomo Pyrethroid World*, 1993, **20**, 14–19.

117. W. J. Wiswesser (ed.), 'Pesticide Index,' 5th edn., Entomological Society of America, College Park, MD, 1976, 328pp.

118. C. Miller, 'The charybdotoxin family of K$^+$ channel-blocking peptides.' *Neuron*, 1995, **15**, 5–10.

119. G. Edwards and A. H. Weston, in 'Handbook of Experimental Pharmacology,' eds. L. Szerkeres and J. G. Papp, Springer, Berlin, 1994, vol. 111, pp. 469–531.

120. C. G. Nichols, 'The inner core of inwardly rectifying K$^+$ channel.' *Trends Pharmacol. Sci.*, 1993, **14**, 320–323.

121. O. Pongs, 'Structural basis of voltage-gated K$^+$ channel pharmacology.' *Trends Pharmacol. Sci.*, 1992, **13**, 359–365.

122. B. Rudy, 'Diversity and ubiquity of K$^+$ channels.' *Neuroscience*, 1988, **25**, 729–749.

123. A. Terzic, R. T. Tung and Y. Kurachi, 'Nucleotide regulation of ATP sensitive potassium channels.' *Cardiovasc. Res.*, 1994, **28**, 746–753.

124. M. Taglialatela and A. M. Brown, 'Structural correlates of K+ channel function.' *News Physiol. Sci.*, 1994, **9**, 169–173.

125. E. Honoré, F. Lesage and G. Romey, 'Molecular biology of voltage-gated K+ channels in heart.' *Fundam. Chin. Pharmacol.*, 1994, **8**, 108–116.

126. S. E. Dryer, 'Na(+)-activated K+ channels: a new family of large-conductance ion channels.' *Trends Neurosci.*, 1994, **17**, 155–160.

127. A. M. Brown, 'Funtional bases for interpreting amino acid sequences of voltage-dependent K+ channels.' *Annu. Rev. Biophys. Biomol. Struct.*, 1993, **22**, 173–198.

128. U. Quast, 'Do the K+ channel openers relax smooth muscle by opening K+ channels? *Trends Pharmacol. Sci.*, 1993, **14**, 332–337.

129. K. Ho, C. G. Nichols, W. J. Lederer *et al.*, 'Cloning and expression of an inwardly rectifying ATP-regulated potassium channel.' *Nature*, 1993, **362**, 31–38.

130. C. Miller, E. Moczydlowski, R. Latorre *et al.*, 'Charybdotoxin a protein inhibitor of single Ca2+-activated K+ channels from mammalian skeletal muscle.' *Nature*, 1985, **313**, 316–318.

131. B. M. Olivera, G. P. Miljanich, J. Ramachandran *et al.*, 'Calcium channel diversity and neurotransmitter release: the ω-conotoxins and ω-agatoxins.' *Annu. Rev. Biochem.*, 1994, **63**, 823–867.

132. D. J. Triggle, M. Hawthorn, M. Gopalakrishnan *et al.*, 'Synthetic organic ligands active at voltage-gated calcium channels.' *Ann. NY Acad. Sci.*, 1991, **635**, 123–138.

133. G. Varadi, Y. Mori, G. Mikala *et al.* 'Molecular determinants of Ca2+ channel function and drug action.' *Trends Pharmacol. Sci.*, 1995, **16**, 43–49.

134. E. Reuveny and T. Narahashi, 'Potent blocking action of lead on voltage-activated calcium channels in human neuroblastoma cells SH-SY5Y.' *Brain Res.*, 1991, **545**, 312–314.

135. D. A. Twombly, M. Yoshii and T. Narahashi, 'Mechanisms of calcium channel block by phenytoin.' *J. Pharmacol., Exp. Ther.*, 1988, **246**, 189–195.

136. M. Yoshii, A. Tsunoo and T. Narahashi, 'Effects of pyrethroids and veratridine on two types of Ca channels in neuroblastoma cells.' *Soc. Neurosci. Abstr.*, 1985, **11**, 518.

137. N. Hagiwara, H. Irisawa and M. Kameyama, 'Contribution of two types of calcium currents to the pacemaker potentials of rabbit sino-atrial node cells.' *J. Physiol. (Lond.)*, 1988, **395**, 233–253.

138. T. Narahashi, A. Tsunoo and M. Yoshii, 'Characterization of two types of calcium channels in mouse neuroblastoma cells.' *J. Physiol. (Lond.)*, 1987, **383**, 231–249.

139. T. Narahashi, in 'Principles of Neurotoxicology,' ed. Louis W. Chang, Dekker, New York, 1994, pp. 609–655.

140. B. M. Olivera, L. J. Cruz, V. de Santos *et al.*, 'Neuronal calcium channel antagonists. Discrimination between calcium channel subtypes using ω-conotoxin from *Coinus magus* venom.' *Biochemistry*, 1987, **26**, 2086–2090.

141. K. Nomura, E. Reuveny and T. Narahashi, 'Opioid inhibition and desensitization of calcium channel currents in rat dorsal root ganglion neurons.' *J. Pharmacol. Exp. Ther.*, 1994, **270**, 466–474.

142. R. Llinás, M. Sugimori, J. W. Lin *et al.*, 'Blocking and isolation of a calcium channel from neurons in mammals and cephalopods utilizing a toxin fraction (FTX) from funnel-web spider poison.' *Proc. Natl. Acad Sci. USA*, 1989, **86**, 1689–1693.

143. R. Llinás, M. Sugimori, D. E. Hillman *et al.*, 'Distribution and functional significance of the P-type, voltage-dependent Ca2+ channels in the mammalian central nervous system.' *Trends Neurosci.*, 1992, **15**, 351–355.

144. T. J. Turner, R. A. Lampe and K. Dunlap, 'Characterization of presynaptic calcium channels with ω-conotoxin MVIIC and ω-grammotoxin SIA: role for a resistant calcium channel type in neurosecretion.' *Mol. Pharmacol.*, 1995, **47**, 348–353.

145. T. M. Piser, R. A. Lampe, R. A. Keith *et al.*, 'ω-Grammotoxin blocks action-potential-induced Ca2+ influx and whole-cell Ca2+ current in rat dorsal-root ganglion neurons.' *Pflügers Arch.*, 1994, **426**, 214–220.

146. R. A. Lampe, P. A. Defeo, M. D. Davison *et al.*, 'Isolation and pharmacological characterization of ω-grammotoxin SIA, a novel peptide inhibitor of neuronal voltage-sensitive calcium channel responses.' *Mol. Pharmacol.*, 1993, **44**, 451–460.

147. A. Randall and R. W. Tsien, 'Pharmacological dissection of multiple types of Ca2+ channel currents in rat cerebellar granule neurons.' *J. Neurosci.*, 1995, **15**, 2995–3012.

148. A. B. Malmberg and T. L. Yaksh, 'Voltage-sensitive calcium channels in spinal nociceptive processing: blockade of N- and P-type channels inhibits formalin-induced nociception.' *J. Neurosci.*, 1994, **14**, 4882–4890.

149. M. M. Usowicz, M. Sugimori, B. Cherskey *et al.*, 'P-type calcium channels in the somata and dendrites of adult cerebellar Purkinje cells.' *Neuron*, 1992, **9**, 1185–1199.

150. I. M. Mintz, M. E. Adams and B. P. Bean, 'P-type calcium channels in rat central and peripheral neurons.' *Neuron*, 1992, **9**, 85–95.

151. I. M. Mintz, V. J. Venema, K. M. Swiderek *et al.*, 'P-type calcium channels blocked by the spider toxin ω-Aga-IVA.' *Nature*, 1992, **355**, 827–829.

152. G. L. Collingridge and R. A. J. Lester, 'Excitatory amino acid receptors in the vertebrate central nervous system.' *Pharmacol. Rev.*, 1989, **41**, 143–210.

153. K. M. Johnson and S. M. Jones, 'Neuropharmacology of phencyclidine: basic mechanism and therapeutic potential.' *Annu. Rev. Pharmacol. Toxicol.*, 1990, **30**, 707–750.

154. J. F. MacDonald and L. M. Nowak, 'Mechanisms of blockade of excitatory amino acid receptor channels.' *Trends Pharmacol. Sci.*, 1990, **11**, 167–172.

155. M. L. Mayer and R. J. Miller, 'Excitatory amino acid receptors, second messengers and regulation of intracellular Ca2+ in mammalian neurons.' *Trends Pharmacol. Sci.*, 1990, **11**, 254–60.

156. D. T. Monaghan, R. J. Bridges and C. W. Cotman, 'The excitatory amino acid receptors: their classes, pharmacology, and distinct properties in the function of the central nervous system.' *Annu. Rev. Pharmacol. Toxicol.*, 1989, **29**, 365–402.

157. T. Narahashi, in 'Methods in Enzymology: Ion Channels,' eds. B. Rudy and L. E. Iverson, Academic Press, Orlando, FL, 1992, pp. 643–658.

158. J. W. Olney, 'Excitotoxic amino acids and neuropsychiatric disorders.' *Annu. Rev. Pharmacol. Toxicol.*, 1990, **30**, 47–71.

159. E. H. F. Wong and J. A. Kemp, 'Sites for antagonism on the N-methyl-D-asparate receptor channel complex.' *Annu. Rev. Pharmacol. Toxicol.*, 1991, **31**, 401–425.

160. J. T. Wroblewski and W. Danysz, 'Modulation of glutamate receptors: molecular mechanisms and functional implications.' *Annu. Rev. Pharmacol. Toxicol.*, 1989, **29**, 441–474.

161. A. Baskys, 'Metabotopic receptors and "slow" excitatory actions of glutamate agonists in the hippocampus.' *Trends Neurosci.*, 1992, **15**, 92–96.

162. J. A. Kemp and P. D. Leeson, 'The glycine site of the NMDA receptor—five years on.' *Trends Pharmacol. Sci.*, 1993, **14**, 20–25.

163. S. Nakanishi and M. Masu, 'Molecular diversity and functions of glutamate receptors.' *Annu. Rev. Biophys. Biomol. Struct.*, 1994, **23**, 319–348.

164. D. D. Schoepp and P. J. Conn, 'Metabotropic glutamate receptors in brain function and pathology.' *Trends Pharmacol. Sci.*, 1993, **14**, 13–20.

165. J. Watkins and G. Collingridge, 'Phenylglycine derivatives as antagonists of metabotropic glutamate receptors.' *Trends Pharmacol. Sci.*, 1994, **15**, 333–342.

166. S. Nakanishi, 'Metabotropic glutamate receptors: synaptic transmission, modulation, and plasticity.' *Neuron*, 1994, **13**, 1031–1037.

167. M. Hollmann and S. Heinemann, 'Cloned glutamate receptors.' *Annu. Rev. Neurosci.*, 1994, **17**, 31–108.

168. S. Ozawa, 'Glutamate receptor channels in hippocampal neurons.' *Jpn. J. Physiol.*, 1993, **43**, 141–159.

169. M. L. Mayer and G. L. Westbrook, 'Permeation and block of N-methyl-D-aspartic acid receptor channels by divalent cations in mouse cultured central neurones.' *J. Physiol. (Lond.)*, 1987, **394**, 501–527.

170. D. Yamamoto, J. Z. Yeh and T. Narahashi, 'Voltage-dependent calcium block of normal and tetramethrin-modified single sodium channels.' *Biophys J.*, 1984, **45**, 337–344.

171. D. Yamamoto, J. Z. Yeh and T. Narahashi, 'Interactions of permeant cations with sodium channels of squid axon membranes.' *Biophys. J.*, 1985, **48**, 361–368.

172. M. L. Mayer, G. L. Westbrook and L. Vyklicky, Jr., 'Sites of antagonist action on N-methyl-D-aspartic acid receptors studied using fluctuation analysis and a rapid perfusion technique.' *J. Neurophysiol.*, 1988, **60**, 645–663.

173. J. E. Huettner and B. P. Bean, 'Block of N-methyl-D-aspartate-activated current by the anticonvulsant MK-801: selective binding to open channels.' *Proc. Natl. Acad. Sci. USA*, 1988, **85**, 1307–1311.

174. S. M. Lovinger, G. White and F. F. Weight, 'Ethanol inhibits NMDA-activated ion current in hippocampal neurons.' *Science*, 1989, **243**, 1721–1724.

175. I. S. Blagbrough, P. T. H. Brackley, M. Bruce *et al.*, 'Arthropod toxins as leads for novel insecticides: an assessment of polyamine amides as glutamate antagonist.' *Toxicon*, 1992, **30**, 303–322.

176. C. G. Staatz, A. S. Bloom and J. J. Lech, Effect of pyrethroids on [³H]kainic acid binding to mouse forebrain membranes.' *Toxicol. Appl. Pharmacol.*, 1982, **64**, 566–569.

177. G. R. Seabrook, I. R. Duce and S. N. Irving, 'Effects of the pyrethroid cypermethrin on L-glutamate-induced changes in the input conductance of the ventrolateral muscles of the larval house fly, *Musca domestica*.' *Pesticide Biochem. Physiol.*, 1988, **32**, 232–239.

178. M. D. Herman, E. Reuveny and T. Narahashi, 'The effect of polyamines on voltage-activated calcium channels in mouse neuroblastoma cells.' *J. Physiol. (Lond.)*, 1993, **462**, 645–660.

179. H. Bittiger, W. Froestl, S. Mickel *et al.*, 'GABA$_B$ receptor antagonists: from synthesis to therapeutic applications.' *Trends Pharmacol. Sci.*, 1993, **14**, 391–394.

180. N. G. Bowery, 'GABA$_B$ receptor pharmacology.' *Annu. Rev. Pharmacol. Toxicol.*, 1993, **33**, 109–147.

181. K. Kaila, 'Ionic basis of GABA$_A$ receptor channel function in the nervous system.' *Prog. Neurobiol.*, 1994, **42**, 489–537.

182. R. L. Macdonald and R. W. Olsen, 'GABA$_A$ receptor channels.' *Annu. Rev. Neurosci.*, 1994, **17**, 569–602.

183. G. Bonanno and M. Raiteri, 'Multiple GABA$_B$ receptors.' *Trends Pharmacol. Sci.*, 1993, **14**, 259–261.

184. W. N. Green and N. S. Millar, 'Ion-channel assembly.' *Trends Neurosci.*, 1995, **18**, 280–287.

185. R. W. Olsen, M. Bureau, R. W. Ransom *et al.*, in 'Neuroreceptors and Signal Transduction,' eds., S. Kito, T. Segawa, K. Kuriyama *et al.*, Plenum Press, New York, 1988, 1–14.

186. R. W. Olsen, D. M. Sapp, M. H. Bureau *et al.*, 'Allosteric actions of central nervous system depressants including anesthetics on subtypes of the inhibitory γ-aminobutyric acid$_A$ receptor–chloride channel complex.' *Ann. N.Y. Acad. Sci.*, 1991, **625**, 145–154.

187. G. B. Smith and R. W. Olsen, 'Functional domains of GABA$_A$ receptors.' *Trends Pharmacol. Sci.*, 1995, **16**, 162–168.

188. T. Yamasaki and T. (Narahashi) Ishii, 'Studies on the mechanism of action of insecticides. X. Nervous activity as a factor of development of γ-BHC symptoms in the cockroach.' *Botyu-Kagaku (Scientific Insect Control)*, 1954, **19**, 106–112.

189. T. Yamasaki and T. Narahashi, 'Nervous activity as a factor of development of dieldrin symptoms in the cockroach. Studies on the mechanism of action of insecticides XVI.' *Botyu-Kagaku (Scientific Insect Control)*, 1958, **23**, 47–54.

190. N. Ogata, S. M. Vogel and T. Narahashi, 'Lindane but not deltamethrin blocks a component of GABA-activated chloride channels.' *FASEB J.*, 1988, **2**, 2895–2900.

191. I. Bermudez, C. A. Hawkins, A. M. Taylor *et al.*, 'Actions of insecticides on the insect GABA receptor complex.' *J. Recept. Res.*, 1991, **11**, 221–232.

192. R. M. Woodward, L. Polenzani and R. Miledi, 'Effects of hexachlorocyclohexanes on γ-aminobutyric acid receptors expressed in *Xenopus* oocytes by RNA from mammalian brain and retina.' *Mol. Pharmacol.*, 1992, **41**, 1107–1115.

193. K. Nagata and T. Narahashi, 'Dual action of the cyclodiene insecticide dieldrin on the γ-aminobutyric acid receptor–chloride channel complex of rat dorsal root ganglion neurons.' *J. Pharmacol. Exp. Ther.*, 1994, **269**, 164–171.

194. K. Nagata, B. J. Hamilton, D. B. Carter *et al.*, 'Selective effects of dieldrin on the GABA$_A$ receptor-channel subunits expressed in human embryonic kidney cells.' *Brain Res.*, 1994, **645**, 19–26.

195. K. Nagata and T. Narahashi, 'Differential effects of hexachlorocyclohexane isomers on the GABA receptor–chloride channel complex in rat dorsal root ganglion neurons.' *Brain Res.*, 1995, **704**, 85–91.

196. R. J. Ruch, J. E. Klaunig and M. A. 'Pereira, Inhibition of intercellular communication between mouse hepatocytes by tumor promoters.' *Toxicol. Appl. Pharmacol.*, 1987, **87**, 111–120.

197. J. E. Klaunig, R. J. Ruch and C. M. Weghorst, 'Comparative effects of phenobarbital, DDT, and lindane on mouse hepatocyte gap junctional intercellular communication.' *Toxicol. Appl. Pharmacol.*, 1990, **102**, 553–563.

198. G. Tsushimoto, C. C. Chang, J. E. Trosko *et al.*, 'Cytotoxic, mutagenic, and cell–cell communication inhibitory properties of DDT, lindane and chlordane on Chinese hamster cells *in vitro*.' *Arch. Environ. Contam. Toxicol.*, 1983, **12**, 721–729.

199. K. A. Criswell, R. Loch-Caruso and E. L. 'Stuenkel, Lindane inhibition of gap junctional communication in myometrial myocytes is partially dependent on phosphoinositide-generated second messengers.' *Toxicol. Appl. Pharmacol.*, 1995, **130**, 280–293.

200. F. N. Quandt, E. Kato and T. Narahashi, 'Effects of methylmercury on electrical responses of neuroblastoma cells.' *Neurotoxicology*, 1982, **3**, 205–220.

201. T. J. Shafer and W. D. Atchison, 'Methylmercury blocks currents mediated by voltage-dependent Ca channels in nerve growth factor-differentiated

pheochromocytoma cells.' *Soc. Neurosci. Abstr.*, 1990, **16**, 512.

202. T. J. Shafer and W. D. Atchison, 'Methylmercury blocks N- and L-type Ca^{++} channels in nerve growth factor-differentiated pheochromocytoma (PC12) cells.' *J. Pharmacol. Exp. Ther.*, 1991, **258**, 149–157.

203. N. I. Kiskin, O. A. Krishtal, A. Y. Tsyndrenko *et al.*, 'Are sulphydryl groups essential for the function of the glutamate-operated receptor–ionophore complex? *Neurosci. Lett.*, 1986, **66**, 305–310.

204. J. A. Umbach and C. B. Gundersen, 'Mercuric ions are potent noncompetitive antagonists of human brain kainate receptors expressed in *Xenopus* oocytes.' *Mol. Pharmacol.*, 1989, **36**, 582–588.

205. O. Arakawa, M. Nakahiro and T. Narahashi, 'Mercury modulation of GABA-activated chloride channels and non-specific cation channels in rat dorsal root ganglion neurons.' *Brain Res.*, 1991, **551**, 58–63.

206. T. Takeuchi, N. Morikawa, H. Matsumoto, *et al.*, 'A pathological study of Minamata disease in Japan.' *Acta Neuropathol.*, 1962, **2**, 40–57.

207. L. Friberg and N. K. Mottet, 'Accumulation of methylmercury and inorganic mercury in the brain.' *Biol. Trace Elem. Res.*, 1989, **21**, 201–206.

208. A. S. Abd-Elfattah and A. E. Shamoo, 'Regeneration of a functionally active rat brain muscarinic receptor by D-penicillamine after inhibition with methylmercury and mercuric chloride.' *Mol. Pharmacol.*, 1981, **20**, 492–497.

209. R. Von Burg, F. K. Northington and A. Shamoo, 'Methylmercury inhibition of rat brain muscarinic receptors.' *Toxicol. Appl. Pharmacol.*, 1980, **53**, 285–292.

210. H. C. H. Hahne and W. Kroontje, 'Significance of pH and chloride concentration on behavior of heavy metal pollutants: Mercury (II), cadmium (II), zinc (II), and lead (II).' *J. Environ. Qual.*, 1973, **2**, 444–450.

211. J. Y. Ma and T. Narahashi, 'Differential modulation of GABA$_A$ receptor-channel complex by polyvalent cations in rat dorsal root ganglion neurons.' *Brain Res.*, 1993, **607**, 222–232.

212. J. Y. Ma and T. Narahashi, 'Enhancement of γ-aminobutyric acid-activated chloride channel currents by lanthanides in rat dorsal root ganglion neurons.' *J. Neurosci.*, 1993, **13**, 4872–4879.

213. J. Y. Ma, E. Reuveny and T. Narahashi, 'Terbium modulation of single γ-aminobutyric acid-activated chloride channels in rat dorsal root ganglion neurons.' *J. Neurosci.*, 1994, **14**, 3835–3841.

214. S. D. Murphy, L. G. Costa and C. Wang, in 'Cellular and Molecular Neurotoxicology,' ed. T. Narahashi, Raven Press, New York, 1984, pp. 165–176.

215. S. Padilla, V. C. Moser, C. N. Pope *et al.*, 'Paraoxon toxicity is not potentiated by prior reduction in blood acetylcholinesterase.' *Toxicol. Appl. Pharmacol.*, 1992, **117**, 110–115.

216. J. Brodeur and K. P. DuBois, 'Studies on the mechanism of acquired tolerance by rats to *o,o*-diethyl *S*-2-(ethiolthio) ethyl phosphorodithioate (di-syston).' *Arch. Internat. Pharmacol.*, 1964, **149**, 560–570.

217. J. J. McPhillips, 'Altered sensitivity to drugs following repeated injections of a cholinesterase inhibitor to rats.' *Toxicol. Appl. Pharmacol.*, 1969, **14**, 67–73.

218. L. G. Costa, B. W. Schwab, H. Hand *et al.*, 'Reduced [³H]quinuclidinyl benzilate binding to muscarinic receptors in disulfoton-tolerant mice.' *Toxicol. Appl. Pharmacol.*, 1981, **60**, 441–450.

219. E. X. Albuquerque, M. Alkondon, S. S. Deshpande *et al.*, in 'Neurotox '88: Molecular Basis of Drug and Pesticide Action,' ed. G. G. Lunt, Elsevier, Amsterdam, 1988, pp. 349–373.

220. E. X. Albuquerque, J. W. Daly and J. E. Warwick, in 'Ion Channels,' ed. T. Narahashi, Plenum, New York, 1988, vol. 1, pp. 95–162.

221. P. M. Ravdin and D. K. Berg, 'Inhibition of neuronal acetylcholine sensitivity by α-toxins from *Bungarus multicinctus* venom.' *Proc. Natl. Acad. Sci. USA*, 1979, **76**, 2072–2076.

222. V. A. Chiappinelli, 'Kappa-bungarotoxin: a probe for the neuronal nicotinic receptor in avian ciliary ganglion.' *Brain Res.*, 1983, **277**, 9–22.

223. P. B. Sargent, 'The diversity of neuronal nicotinic acetylcholine receptors.' *Annu. Rev. Neurosci.*, 1993, 16, 403–443.

224. M. Alkondon and E. X. Albuquerque, Diversity of nicotinic acetylcholine receptors in rat hippocampal neurons. I. Pharmacological and functional evidence for distinct structural subtypes.' *J. Pharmacol. Exp. Ther.*, 1993, **265**, 1455–1473.

225. A. S. Moffat, 'New chemicals to outwit insect pests.' *Science*, 1993, **261**, 550–551.

226. S. B. Soloway, A. C. Henrey, W. D. Kollmeyer *et al.*, in 'Pesticide and Venom Neurotoxicity,' eds. D. L. Shankland, R. M. Hollingworth and T. Smyth, Plenum Press, New York, 1978, pp. 153–158.

227. R. Zwart, M. Oortgiesen and H. P. M. Vijverberg, 'The nitromethylene heterocycle 1-(pyridin-3-yl-methyl)-2-nitromethylene-imidazolidine distinguishes mammalian from insect nicotinic receptor subtypes.' *Eur. J. Pharmacol.*, 1992, **228**, 165–169.

228. M. Oortgiesen, R. G. D. M. van Kleef, R. B. Bajnath *et al.*, 'Nanomolar concentrations of lead selectively block neuronal nicotinic acetylcholine responses in mouse neuroblastoma cells.' *Toxicol. Appl. Pharmacol.*, 1990, **103**, 165–174.

229. M. Oortgiesen, T. Leinders, R. G. D. M. van Kleef *et al.*, 'Differential neurotoxicological effects of lead on voltage-dependent and receptor-operated ion channels.' *Neurotoxicology*, 1993, **14**, 87–96.

230. R. S. Manalis and G. P. Cooper, 'Presynaptic and postsynaptic effects of lead at the frog neuromuscular junction.' *Nature*, 1973, **243**, 354–356.

231. R. S. Aronstam and M. E. Eldefrawi, 'Transition and heavy metal inhibition of ligand binding to muscarinic acetylcholine receptors from rat brain.' *Toxicol. Appl. Pharmacol.*, 1979, **48**, 489–496.

232. J. Suszkiw, G. Toth, M. Murawsky *et al.*, 'Effects of Pb^{2+} and Cd^{2+} on acetylcholine release and Ca^{2+} movements in synaptosomes and subcellular fractions from rat brain and *Torpedo* electric organ.' *Brain Res.*, 1984, **323**, 31–46.

233. E. A. M. Abdallah and A. E. Shamoo, 'Protective effect of dimercaptosuccinic acid on methylmercury and mercuric chloride inhibition of rat brain muscarinic acetylcholine receptors.' *Pesticide Biochem. Physiol.*, 1984, **21**, 385–393.

234. M. E. Eldefrawi, N. A. Mansour and A. T. Eldefrawi, in 'Membrane Toxicology,' eds. M. W. Miller and A. E. Shamoo, Plenum Press, New York, 1977, pp. 449–463.

235. M. A. Abbassy, M. E. Eldefrawi and A. T. Eldefrawi, 'Phyrethroid action on the nicotinic acetylcholine receptor/channel.' *Pesticide Biochem. Physiol.*, 1983, **19**, 229–308.

236. W. Wouters, J. Van den Bercken and A. Van Ginneken, 'Presynaptic action of the pyrethroid insecticide allethrin in the frog motor end plate.' *Eur. J. Pharmacol.*, 1977, **43**, 163–171.

237. P. Eriksson and A. Nordberg, 'Effects of two pyrethroids, bioallethrin and deltamethrin, on subpopulations of muscarinic and nicotinic receptors in the neonatal mouse brain.' *Toxicol. Appl. Pharmacol.*, 1990, **102**, 456–463.

238. T. Kiss and O. N. Osipenko, 'Effect of deltamethrin on acetylcholine-operated ionic channels in identified *Helix pomatia* L. neurons.' *Pesticide Biochem. Physiol.*, 1991, **39**, 196–204.

11.10
Phosphoprotein Phosphatases as Potential Mediators of Neurotoxicity

EDGAR F. DA CRUZ E SILVA
Universidade de Aveiro, Portugal and Instituto de Biologia Experimental e Tecnológica, Oeiras, Portugal

and

JAMES P. O'CALLAGHAN
US Environmental Protection Agency, Research Triangle Park, NC, USA

11.10.1 INTRODUCTION

Reversible protein phosphorylation is a dynamic process whereby phosphate groups are incorporated into proteins by protein kinases and removed by protein phosphatases. Protein phosphorylation is, therefore, a reversible process that can alter the function of target proteins and in this fashion control biological function. Indeed, protein phosphorylation is now recognized as the major post-translational modification through which numerous physiological processes are regulated. The protein kinases and protein phosphatases, the key controlling elements, are regulated by a myriad of extracellular and intracellular signals. Interestingly, of all mammalian tissues, it is probably the brain that expresses the greatest diversity of protein kinases and protein phosphatases. Many extracellular messengers exert their effects in the CNS by regulating the intracellular concentration of specific second messengers, which in turn activate specific kinases and/or phosphatases. Many of the complex functions of the mammalian CNS are now known to be regulated by protein phosphorylation.[1,2] Furthermore, defects in CNS phosphorylation mechanisms may be a common feature of a variety of neurological and chemical-induced disease states.[3-5]

The role of protein kinases as mediators of neurotoxicity has been discussed previously,[6] and will not be considered here. Unlike the protein kinases, that all belong to a single gene family, the protein phosphatases are divided into several distinct and unrelated protein/gene families. Ser/Thr-specific protein phosphatases comprise two distinct families, the PP1/PP2A/PP2B gene family (that also includes the bacteriophage λ orf221 phosphatase) and the PP2C gene family. The Tyr-specific phosphatase family, as well as including the Tyr-specific phosphatases, also comprises the so-called dual specificity phosphatases (capable of dephosphorylating Ser, Thr, and Tyr residues). Besides these intracellular phosphatases involved in signal transduction, there are also other nonspecific alkaline and acid phosphatases that are unrelated, and usually found either in specialized intracellular compartments or in the extracellular milieu. Given the identification, from remarkably diverse sources, of a large variety of potent inhibitors of the catalytic subunits of the three major types of Ser/Thr protein phosphatases (namely PP1, PP2A, and PP2B), this chapter will focus specifically on these phosphatases as possible mediators of neurotoxicity.

11.10.2 PROTEIN PHOSPHATASES AND PROTEIN PHOSPHORYLATION SYSTEMS

Reversible protein phosphorylation systems are tripartite, consisting of a substrate, a phosphorylating kinase, and a dephosphorylating phosphatase. Historically, the protein kinases were almost always regarded as the key regulatory component of this tripartite phosphorylation system.[7] However, one cannot explain the exquisite complexity observed in the control of diverse metabolic processes via protein phosphorylation by considering kinases alone. Thus, our appreciation of protein phosphatases has evolved from regarding them as molecules of secondary importance, merely responsible for indiscriminately removing phosphate groups that had been previously incorporated by the highly regulated kinases, to our view of phosphatases as equal players in partnership with the kinases. That is, ultimately it is the balance between kinase and phosphatase activities that determines the level of phosphorylation of a particular substrate. The regulation of any phosphorylation system in response to a particular stimulus can occur by regulating the level and/or activity of the kinase and/or phosphatase involved, and by regulating the level and/or availability of the substrate. With respect to the protein phosphatases, there are many examples that illustrate the nature of the regulation of their activity in response to extracellular stimuli (for a review, see Ref. 8). In particular, given their broad and overlapping substrate specificity *in vitro*, it appears that mechanisms must exist to restrict and modulate their action *in vivo*.

Extensive evidence indicates that it is the interaction of the phosphatase catalytic subunits with targeting and/or regulatory subunits that serves to determine their cellular location and to control their catalytic activity and substrate specificity.[9,10]

The available data demonstrate the highly regulatable nature of the Ser/Thr-specific protein phosphatases and their involvement in the control of many neuronal functions. Many of the immediate early events following depolarization or receptor stimulation involve the dephosphorylation of specific effector proteins, indicating the central importance of the control of protein phosphatase activity in the control of normal neuronal function. The focus on protein phosphatases has not only raised our awareness of their importance, but has also served to identify them as possible targets for neurotoxins. Several natural toxins and toxicants of environmental, agricultural, and medical importance have been identified as potent and specific inhibitors of the Ser/Thr-specific protein phosphatases. Significantly, some of these toxins (e.g., okadaic acid, microcystin, etc.) have also been identified as powerful tumor promoters.[11-13]

11.10.3 CLASSIFICATION OF THE SER/THR-SPECIFIC PROTEIN PHOSPHATASES

The description in 1943 of an activity responsible for the conversion of active glycogen phosphorylase *a* to its inactive *b* form probably constitutes the first phosphatase to be detected.[14] Thus, the detection of protein phosphatases probably predates the discovery of the first kinase. Until the advent of molecular cloning techniques, only four different types of Ser/Thr-specific protein phosphatase (PP) were identified, as classified originally by Ingebritsen and Cohen[15] on the basis of substrate specificity, sensitivity to specific inhibitors and cation requirement. The four Ser/Thr-specific protein phosphatase types (termed PP1, PP2A, PP2B, and PP2C) were all identified in a variety of tissues and cell types, including the nervous system. However, the application of recombinant DNA techniques to the field has yielded not only the primary structure of all four phosphatase types, but also documented the existence of isoforms for each type and revealed the existence of previously undetected phosphatases in a variety of eukaryotic cells. Although a number of excellent and comprehensive reviews are available,[8,10,16-18] a brief description of this aspect is presented below in order to make this chapter self-explanatory.

11.10.3.1 Isoforms

Each individual activity type (PP1, PP2A, PP2B, PP2C) comprises several molecular isoforms. After the initial discovery of PP2A isoforms,[19,20] isoforms were also described for every other phosphatase type (e.g., see Ref. 21). However, an early report[22] describing the existence of two isoforms of PP1 produced from the same gene and then termed PP1α and PP1β is incorrect, since it resulted from deficient interpretation of experimental data. Three genes are now known to encode type 1 phosphatase catalytic subunits, termed PP1α, PP1β, and PP1γ. At least PP1γ is known to undergo tissue-specific processing to yield an ubiquitously expressed PP1γ$_1$ isoform and a testis-specific PP1γ$_2$ isoform.[23] Two genes are known for the catalytic subunit of PP2A, termed PP2Aα and PP2Aβ. The three known PP2B catalytic subunit genes (Aα, Aβ, and Aγ) are also subject to complex regulation to yield several alternatively spliced isoforms.

11.10.3.2 Gene Families

Primary sequence analysis identified the PP1, PP2A, and PP2B phosphatases as members of the same family. However, PP2C was found to be unrelated to the others, and thus belongs to a distinct and unrelated family. A curious finding was the discovery that the genome of bacteriophage λ encoded a phosphatase (orf221) related to the PP1/PP2A/PP2B family.[24] A homologous phosphatase also occurs in the genome of the bacteriophage φ80.

11.10.3.3 Novel Phosphatases

Perhaps more surprising was the discovery (from a variety of tissues and species) of other phosphatase catalytic subunit isoenzymes, with some being more closely related to the type 1 isoforms and others to the type 2A isoforms: these were termed "novel" phosphatases. The first such phosphatase to be identified (from rabbit liver) was termed protein phosphatase X or PPX and was more closely related to PP2A.[25] Subsequently, several others were cloned and isolated from a variety of mammalian and other sources. Thus, PPY is a *Drosophila* phosphatase[26] and PPZ is a yeast phosphatase,[27] both being more closely related to PP1. Many others are now known.

While the functional classification of Ingebritsen and Cohen[15] is not informative in terms of the molecular data now available, it continues to be widely used due to the lack of a credible alternative classification and/or nomenclature. For the remainder of this chapter we will refer almost exclusively only to PP1, PP2A, and PP2B, since none of the compounds described affect the activity of PP2C. While many of the phosphatase inhibitory compounds referred to below have been tested against the different phosphatase types, it should be emphasized that for the most part they have not been tested against the different isoforms of each type. However, these are unlikely to differ appreciably in their sensitivities due to their highly conserved nature. Indeed, this has been confirmed, for example, with the inhibition of the different PP1 isoforms by okadaic acid, microcystin, calyculin A, tautomycin, and cantharidin (E. F. da Cruz e Silva, unpublished observations).

11.10.4 LOCALIZATION OF NEURONAL PROTEIN PHOSPHATASES

11.10.4.1 PP1

PP1 is one of the most highly conserved eukaryotic proteins, with yeast PP1 being 80–90% identical to its mammalian counterparts, and it has been detected in virtually all eukaryotic cells examined to date (including neurons and glia). This exceptionally high degree of conservation may be due in part to its fundamental role in a variety of cellular processes. Thus, PP1 has been implicated in the control of many and diverse cellular processes, including neuronal long-term depression and synaptic plasticity.[28–30] Most of the initial studies on PP1 concentrated on the protein from skeletal muscle and liver, and therefore relatively less is known about brain PP1. In general, neuronal PP1 activity, like its non-neuronal counterpart, is mostly found in particulate fractions. However, increasing attention has been paid to neuronal phosphatases in general. Thus, the isoforms of PP1 expressed in neurons have been identified and the distribution of their mRNAs studied by *in situ* hybridization.[21,23] In this way, PP1α, PP1β, and PP1γ$_1$ were all cloned from neuronal sources, and their respective mRNAs were found to be particularly abundant in the hippocampus and cerebellum. The development of isoform-specific antibodies against PP1α and PP1γ$_1$ also allowed a detailed study of their distribution at the protein level. Both were found to be more highly expressed in brain than in peripheral tissues. The highest levels were measured in the striatum, where they were shown to be relatively enriched in the medium-sized spiny neurons,[23] also known

to be highly enriched in DARPP-32 (dopamine and cAMP-regulated phosphoprotein, M_r 32 kDa). DARPP-32 is a potent inhibitor of PP1 when phosphorylated at a specific site (Thr34) by protein kinase A (PKA). Interestingly, several first messenger pathways (dopaminergic, glutamatergic, GABAergic, etc.) have been shown to achieve part of their striatal effects through the regulation of the phosphorylation state of DARPP-32. The convergence of major neurotransmitter pathways on the PP1/DARPP-32 system is consistent with PP1 having a critical role in the mediation of their actions.

Consistent with previous biochemical studies,[31,32] Ouimet *et al.*[33] reported the highest levels of immunoreactivity for PP1α and PP1γ$_1$ in the neostriatum and the hippocampal formation. At the electron microscopic level, PP1 immunoreactivity was demonstrated in dendritic spine heads and spine necks, and possibly also in postsynaptic density (PSDs). The robust immunoreactivity observed in spines accounts for the small immunoreactive puncta (1 μm or less) seen at the light microscopic level. Other processes, such as dendritic shafts, axons, and axon terminals, were found to be mostly weakly immunolabeled.[33] PP1 immunoreactivity has also been reported in human hippocampal neuronal cytoplasm.[34] In addition, most neuronal nuclei were not immunoreactive for PP1γ$_1$ but were usually strongly immunoreactive for PP1α.[33] These findings are in agreement with previous biochemical and genetic findings in nonneuronal tissues, indicating nuclear roles for PP1.[35–38] Thus, the PP1α and PP1γ$_1$ isoforms were shown to be highly and specifically enriched in dendritic spines of the neostriatum. The average concentrations of DARPP-32 and inhibitor-1 in the striatum were estimated to be around 20–50 μM[39] and 1–2 μM,[40] respectively. In contrast to PP1, its inhibitory proteins appear to be more uniformly distributed through the neurons, and the extremely high concentrations of inhibitors present in the striatum are probably explained by the incredible enrichment of PP1 in dendritic spines. Overall, the average combined concentration of PP1α and PP1γ$_1$ in the striatum was estimated to be around 5–10 μM,[41] which implies a much higher concentration in dendritic spines. These results provided strong evidence that dendritic spines probably evolved as discrete biochemical compartments (specialized organelles) to facilitate interactions among various signal transduction pathways and, therefore, enabling the integration of the multiplicity of signals reaching the neuron from a variety of afferent nerve terminals and leading to an appropriate physiological response. Likewise, the D$_1$ dopamine receptor and the gluR1 glutamate receptor were also shown to be enriched in spines present on the DARPP-32/PP1-enriched medium-sized spiny neurons of the neostriatum,[42–44] consistent with PP1 being centrally involved in the mediation of the actions of these neurotransmitters. Additionally, PP1 immunoreactive spines were found throughout the brain,[33] suggesting that PP1 activity can be also regulated via these and other neurotransmitter pathways in other brain regions.

11.10.4.2 PP2A

PP2A, unlike PP1, is thought to be mainly cytoplasmic[45] and, of all tissues tested, its activity was highest in brain extracts.[46] Similar results were also observed at the mRNA level[19] and by immunoblotting.[47] The catalytic subunit of PP2A was found to have a wide regional distribution in brain, with the highest immunoreactivity being present in neurons, and particular enrichment in the cytosolic and synaptosolic subcellular fractions.[48] However, the work of Shields *et al.*[31] indicated that brain PP2A is likely to be predominantly a cytosolic enzyme since fivefold less PP2A was detected in synaptic junction and synaptic plasma membrane fractions compared with PP1. Unlike for PP1, the mechanisms responsible for the *in vivo* regulation of PP2A activity remain largely unknown. However, the catalytic subunit of PP2A has been purified in association with several regulatory subunits, constituting several distinct holoenzyme forms. There are regulatory subunits which appear to have a restricted tissue distribution (some may be brain-specific), and molecular isoforms also have been described.[49] To some extent, the various holoenzyme forms of PP2A have different substrate specificities.

11.10.4.3 PP2B

That PP2B (calcineurin) is not only expressed in neuronal tissue was demonstrated by the detection of high levels of Ca^{2+}-dependent/calmodulin-stimulated phosphatase activity in a variety of mammalian tissues, including brain, muscle, heart, and liver.[46] However, experiments using antibodies raised against the brain enzyme often failed to detect the larger catalytic A subunit of calcineurin, suggesting the possible existence of tissue-specific isozymes. Indeed, the application of molecular cloning techniques to the field confirmed the existence of a series of calcineurin isoforms. At least three genes are now known to encode isoforms of the A subunit (termed Aα, Aβ, and Aγ), and two genes are known for the B subunit. Whereas Aα is by far the most abundantly expressed in neurons, the Aγ isoform appears to be largely testis-specific. In addition, at least three alternatively spliced forms are known for each A subunit isoform. Thus, although not every spliced form has been demonstrated for each of the three A subunit isoforms, likely there is a total of nine distinct catalytic subunits that can each complex with either of the two different regulatory B subunits. This incredible level of complexity is unique among the Ser/Thr phosphatase family. Calcineurin is also known to occur in a variety of lower eukaryotes including squid, *Drosophila*, *Paramecium*, and yeast.[27]

Calcineurin phosphatase activity has been detected in both the soluble and particulate fractions of brain homogenates.[50] A dual cytoplasmic and membrane distribution was revealed by a series of immunohistochemical studies. Calcineurin has been detected in association with postsynaptic densities,[51,52] enriched at postsynaptic loci, plasma membranes, and dendritic microtubules, with a more diffuse cytoplasmic distribution,[53] and has also been reported in axons.[54] In chicken brain, low levels of calcineurin were reported in synaptic membranes and synaptic junction preparations, with high concentrations having been reported in synaptosomes.[55,56] Thus, at the cellular level calcineurin is found throughout the CNS, but with a highly heterogeneous distribution in the various regions. It is present in neurons but not all neurons contain similar levels.[53,57] The highest level of immunoreactivity was observed in the caudate-putamen, hippocampus, and substantia nigra of rat,[53,58,59] with high concentrations being detected in the major neuronal population of the caudate nucleus and the pyramidal cells of the CA1–CA2 subregions of the hippocampus. Calcineurin is either absent or present at very low levels in glia.[60] In rat hippocampus, the catalytic A subunit was localized in the stratum lucidum, the mossy fiber terminals forming giant synaptic boutons.[61] Calcineurin is found throughout the cytoplasm of each individual cell and, therefore, the overall pattern of staining indicates immunoreactive perikarya and dendrites in areas containing calcineurin-rich cells or immunoreactive nerve terminals in areas that are innervated by such cells. Loss of calcineurin immunoreactivity has been reported in the brains of both Huntington's disease[62] and Parkinson's disease[63] patients. Thus far, it appears that the Aα and Aβ isoforms of the catalytic subunit of calcineurin may have different regional distributions in mammalian brain.[64,65] The future development of antibodies recognizing specific isoforms of calcineurin should allow significant progress in this field.

11.10.5   PROTEIN PHOSPHATASES AND NEURONAL FUNCTION

Pharmacological, viral, and genetic investigations, as well as several *in vitro* studies using purified phosphatases, have implicated the Ser/Thr-specific phosphatases in numerous signal transduction pathways. Indeed, many aspects of neuronal function have been shown to be regulated by protein dephosphorylation, including the control of neurotransmitter release, neurotransmitter receptors, and ion channels. Furthermore, a number of neuronal-specific inhibitors of phosphatases have been identified, including the previously mentioned striatally-enriched DARPP-32 and its cerebellar Purkinje cell homologue, G-substrate. The latter is a substrate for cGMP-dependent protein kinase and an inhibitor of PP2A. Given the obvious importance of protein phosphatases in the nervous system, abnormalities associated with these enzymes might be expected to be responsible, or at least contribute, to the onset or the severity of specific neuropathological conditions. It also raises the possibility that they may be specifically targeted by several toxicants known to alter neuronal protein phosphorylation levels. The text below describes only a few selected and representative examples of the roles played by phosphatases in the nervous system.

11.10.5.1   Regulation of Neurotransmitter Release

Although the exact molecular mechanisms involved in neurotransmitter release remain to be fully elucidated, several lines of evidence implicate protein phosphorylation in general and protein phosphatases specifically in the control of this process.[66,67] The rapid influx of calcium through voltage-gated calcium channels that results from depolarization of synaptosomes leads to the phosphorylation of several proteins and the rapid dephosphorylation of others. Among the latter, one of 96 kDa known as dephosphin has received considerable attention. The depolarization-induced dephosphorylation of dephosphin was shown to be only partly inhibited by 1 μM okadaic acid (a selective phosphatase inhibitor), with full inhibition requiring 5 μM okadaic acid.[68] Abdul-Ghani *et al.*[69] also showed that okadaic acid produced an enhancement of neurotransmitter release at frog (cholinergic) and lobster (glutamatergic and GABAergic) neuromuscular junc-

tions. Quantal analysis indicated that the effect of okadaic acid was presynaptic, and therefore implicated its targets (PP1 and/or PP2A) in the control of neurotransmitter release from nerve terminals.

11.10.5.2   Regulation of K⁺ Channels

While PKA phosphorylation was shown to up-regulate the type 1 "fast" large conductance Ca^{2+}-activated K^+ channels and to down-regulate the type 2 channels, this effect was specifically reversed by the catalytic subunit of PP2A but not by PP1.[70] In contrast, phosphorylation of a slower gating, large-conductance Ca^{2+}-activated K^+ channel by a tightly associated kinase increased the probability of channel opening, an effect that was specifically reversed by PP1 but not by PP2A.[71] There is also some evidence that neurotransmitters may be able to modulate K^+ channels by activation of a phosphatase. The activation of K^+ channels, previously inactivated by PKA phosphorylation, by the neuropeptide somatostatin can be blocked by the highly specific phosphatase inhibitor okadaic acid,[72] suggesting that somatostatin activates an okadaic acid-sensitive phosphatase. In addition, several studies in *Aplysia* have suggested that phosphatases may be the dominant regulators of the serotonin-sensitive K^+ channels under basal conditions.[73,74]

11.10.5.3   Regulation of Ligand-gated Receptor Channels

The phosphorylation of non-NMDA (*N*-methyl-D-aspartate) glutamate receptors responsible for mediating fast excitatory synaptic transmission may represent a mechanism for controlling the efficacy of synaptic transmission and neuron excitability. Recent studies showed that PKA could stimulate the response of non-NMDA receptor-gated channels,[75,76] and okadaic acid also produced an identical stimulation.[75]

11.10.6   ALTERED PROTEIN PHOSPHORYLATION AND NEUROTOXICITY

Given the central role played by protein phosphorylation in many aspects of neural function,[2] it might be expected that alterations in protein phosphorylation pathways will result in subtle to overt deficits in brain function. Such defects could occur at the level of a particular kinase, substrate, phosphatase, or any combination thereof. Linkage of specific neurotoxic conditions to changes in neural protein phosphorylation, however, has been rare. Most efforts have been devoted to neurologic disorders such as Alzheimer's disease and ischemia; relatively little attention has been directed toward the role of protein phosphorylation in toxin- or toxicant-induced changes in neural systems.[6]

A number of protein kinases have been implicated in the etiology of Alzheimer's disease, including PKA, protein kinase C (PKC), casein kinase, calcium/calmodulin-dependent kinases and tyrosine kinases.[5,77,78] The substrates phosphorylated by these kinases, especially those associated with the paired helical filaments (PHF) characteristic of this disease, such as the isoforms of tau, have come under particularly close scrutiny. A relatively consistent finding in these studies has been a high incidence of hyperphosphorylation of tau and other PHF phosphoproteins, effects which often are attributed to excessive kinase-generated phosphorylation. The recent demonstration that okadaic acid-sensitive phosphatases are involved in hyperphosphorylation of human tau,[79] suggests that phosphatases may be an overlooked component in aberrant phosphorylation associated with Alzheimer's disease.

Phosphatases also may be a neglected variable in the altered phosphorylation observed in cerebral ischemia and neurotrauma. The pathophysiology of traumatic and ischemic brain injury often is associated with increased extracellular levels of excitatory amino acids, most notably glutamate, which in turn is linked to excessive stimulation of NMDA and AMPA (alpha-amino-3-hydroxy-5-methyl-4-isoxazole proprionic acid) receptors and an increase in intracellular calcium concentration. Enhanced intracellular calcium levels can result in calcium-regulated proteolysis and phospholipid degradation, processes that may underlie neuronal injury. The latter process is associated with translocation and activation of specific isoforms of PKC, suggesting that membrane PKC may play a role in neuronal damage following traumatic and ischemic brain injury.[80,81] However, enhanced calcium influx linked to excessive activation of NMDA receptors may also be associated with activation of calcineurin. Indeed, Halpain *et al.*[82] showed that calcineurin was activated in intact neurons in response to NMDA receptor stimulation or depolarization. Thus, although an emphasis has been placed on protein kinases, especially PKC, calcineurin and other phosphatases may play a role in traumatic and

ischemic brain injury that has yet to be recognized.

The known and putative sites of action of neurotoxins and neurotoxicants are as diverse and varied as the cell types and processes that compose the developing and adult CNS.[83] The central and fundamental role played by protein phosphorylation systems in neural function, however, makes it likely that any action of an environmental toxin or toxicant likely will be reflected, at some level, by a change in protein phosphorylation. As with neurological disease states and traumatic and ischemic brain injury, protein kinases and selected phosphoprotein substrates have been the focus of limited investigations on the role of protein phosphorylation and neurotoxicity.[6] With the exception of the recent discovery of toxins that selectively inhibit specific phosphatases (see below), the role of phosphatases in neurotoxicity has received the least attention of the three components of the protein phosphorylation equation. Yet, the sites of action of neurotoxins and neurotoxicants are known to be regulated by phosphoprotein phosphatases. These include, for example, ion channels, neurotransmitter receptors, neurotransmitter transporters, neurotransmitter biosynthesis, neurotransmitter release, axonal cytoskeleton, mitochondrial energy metabolism, and various targets in the neurotransmitter signal transduction cascade.[8,18] Below are detailed the actions of selected toxins and toxicants on neural phosphoprotein phosphatases.

11.10.7 NEUROTOXINS AND NEUROTOXICANTS ACTING ON PHOSPHATASES

11.10.7.1 Toxins

One of the most significant advances in the study of the Ser/Thr protein phosphatases, and the elucidation of the cellular events they control, was the identification of several naturally occurring toxins as powerful and specific phosphatase inhibitors.[84,85] However, some of the earliest observations on the production of such toxins by cyanobacteria were unwittingly made over 100 years ago.[86] Below we discuss some of the more commonly occurring and more frequently used phosphatase inhibitor toxins.

11.10.7.1.1 Okadaic acid

Okadaic acid (Figure 1) is a complex polyether fatty acid synthesized by marine dinoflagellates, for example, *Prorocentrum lima*,[87]

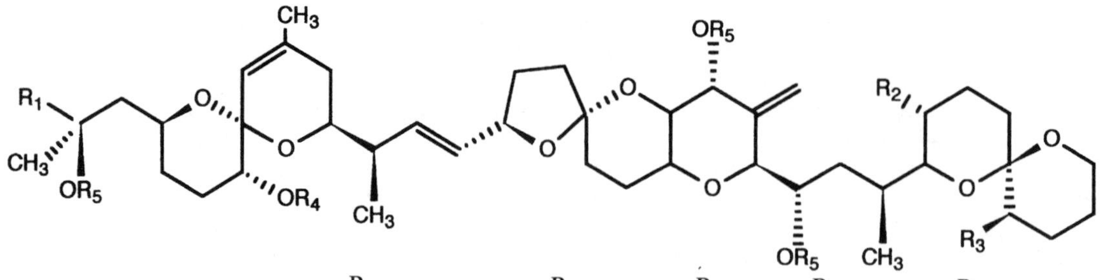

	R_1	R_2	R_3	R_4	R_5
Okadaic acid	HOOC	CH_3	H	H	H
DTX-1	HOOC	CH_3	CH_3	H	H
Methyl okadaate	CH_3OOC	CH_3	H	H	H
Okadaic acid tetraacetate	HOOC	CH_3	H	CH_3CO	CH_3CO

Figure 1   Structure of okadaic acid and some of its derivatives (adapted from Shestowsky et al.[94]).

that accumulate in several filter feeding organisms such as the black sponge *Halichondria okadaii* from which it was first isolated and from which it derives its name.[88] It is thought to be the causative agent of diarrhetic shellfish poisoning,[89] resulting from the ingestion of contaminated shellfish, specially mussels and scallops where it accumulates in the digestive glands. Okadaic acid was shown to be a potent inhibitor of PP2A, being 10–100-fold less effective at inhibiting PP1,[84,85] and is also a powerful tumor promoter.[11,13] In this respect, a study aimed at determining the mutagenicity of okadaic acid found that, whereas it did not induce mutations in *Salmonella typhimurium* (with or without a microsomal activation system), it was strongly mutagenic to Chinese hamster lung cells even in the absence of a microsomal activation system.[90] Indeed, the values obtained indicated that okadaic acid mutagenicity is comparable to that of 2-amino-N^6-hydroxyadenine, one of the strongest mutagens known. Being cell permeable, it has been used extensively to study protein phosphatases, and the cellular processes they control, both in cell extracts and intact cells.[91,92] Dinophysistoxin-1 (DTX-1) is an analogue (35-*S*-methylokadaic acid) of okadaic acid, isolated from members of the *Dinophysis* species, that shares the same biochemical properties.[89,93]

Several inactive analogues of okadaic acid are commercially available and have proved very useful as negative controls. The best characterized are methyl okadaate (the methyl ester of okadaic acid), 1-norokadaone (with similar physical and chemical properties to okadaic acid but lacking phosphatase inhibitory activity), and okadaic acid, 7,10,24,28-tetraacetate.[95,96]

Holmes and Boland[97] reviewed a large number of studies in which okadaic acid was used to implicate protein phosphorylation (and more specifically PP1 and/or PP2A) in the control of specific cellular processes. It should be noted that the free acid form of okadaic acid, as well as being highly unstable other than in its crystallized form, also requires preparation in solvents such as DMSO whose usage may sometimes be undesirable. Thus, the availability of water soluble, salt forms (sodium, potassium, or ammonium) of okadaic acid is now recommended for most applications.

11.10.7.1.2   Tautomycin

Tautomycin (Figure 2) was first isolated from *Streptomyces spiroverticillatus* and, like okadaic acid, it can also enter intact cells and inhibit PP1 and PP2A.[98,99] Of all known phosphatase inhibitory toxins, it appears to display the highest specificity against PP1 (PP1 $IC_{50} \approx 1\,nM$; PP2A $IC_{50} \approx 10\,nM$).

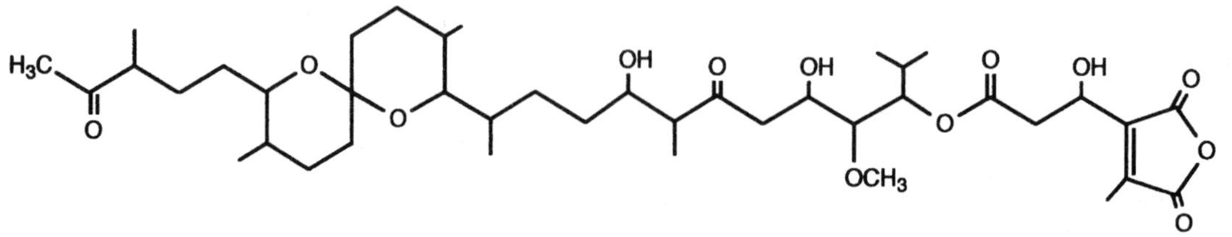

Figure 2   Structure of tautomycin.

11.10.7.1.3 Calyculin A

Calyculin A (Figure 3) is a multifunctional complex molecule isolated from the marine sponge *Discodermia calyx*,[100] whose structure is therefore different from that of okadaic acid. However, like okadaic acid, it is a potent protein phosphatase inhibitor and tumor promoter.[101,102] Since it also readily enters intact cells and shows approximately equal inhibitory potency against PP1 and PP2A, it is often used in parallel with okadaic acid in attempting to discern which phosphatase type might be involved in a particular cellular process (e.g., see Ref. 103). Several other related compounds have also been isolated—calyculins E, F, G, and H.

11.10.7.1.4 Microcystins

Microcystins are cyclic heptapeptide hepatotoxins isolated from fresh water cyanobacteria such as *Microcystis aeruginosa*.[104,105] The microcystins are found worldwide and can be responsible for extensive wildlife fatalities, as well as adverse effects to human health in countries where drinking water supplies are contaminated with the producing cyanobacteria. A large number of different microcystins

have been identified and characterized, differing mostly in the nature of the two variable amino acids indicated by the suffix letters (Figure 4). Thus, the best known form is microcystin-LR (i.e., L = leucine and R = arginine), which shows similar potencies against PP1 and PP2A.[92,106,107] Microcystin-RR is a less toxic Arg–Arg analogue of microcystin-LR, which has been used for comparative structure and function studies. Microcystin-YR is the Tyr–Arg analogue which shows similar toxicity to microcystin-LR. The presence of Tyr makes it potentially useful for radiolabeling with iodine for subsequent studies.

Unlike the okadaic acid family of polyether-like protein phosphatase inhibitors, the microcystins are not generally cell permeable. Thus, they are not usually toxic to neuronal cells, except in experimental situations where the cells are permeabilized or the toxin is microinjected into the cell. The potent hepatotoxicity of the microcystins is apparently due to the presence of a specific uptake system in hepatocytes. Hence, in a perfused rat liver model Runnegar *et al.*[108] described differential toxicity effects for microcystin and calyculin A. This probably resulted from the restricted action of the former to hepatocytes, whereas calyculin A likely affected all the different cell types due to its

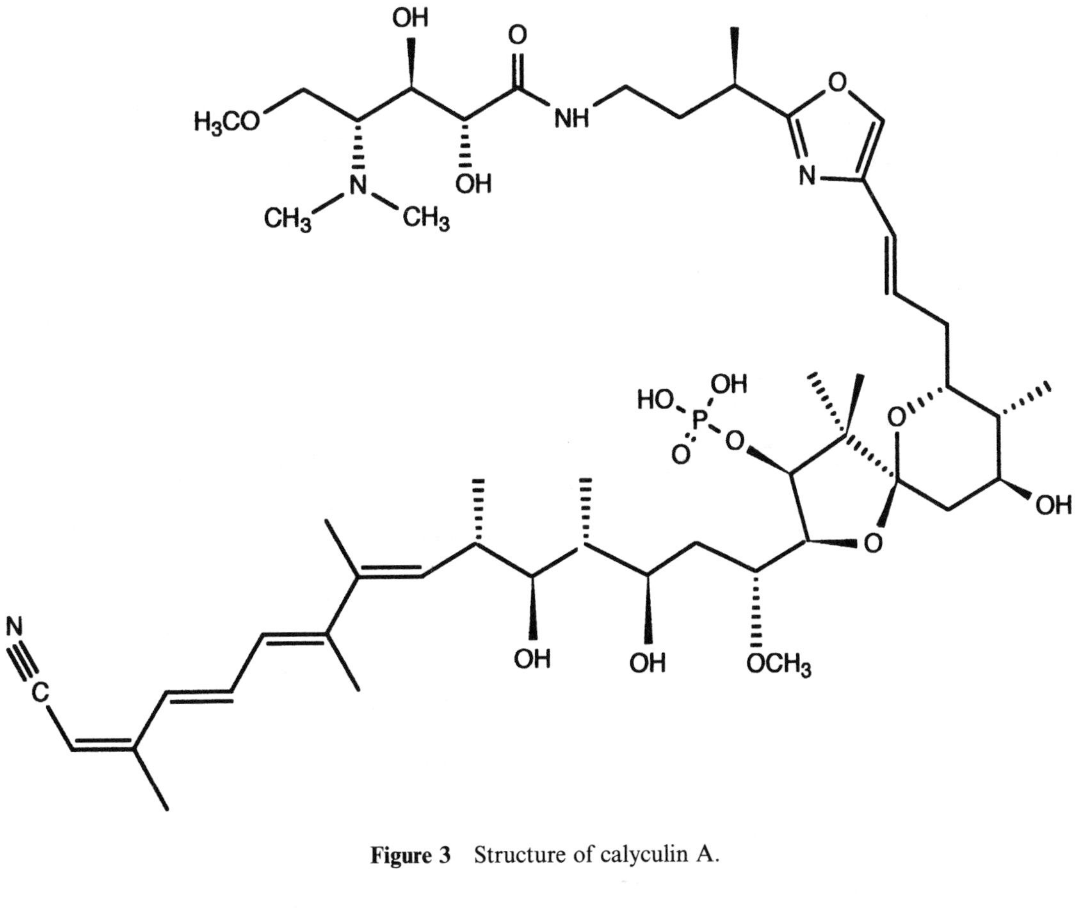

Figure 3 Structure of calyculin A.

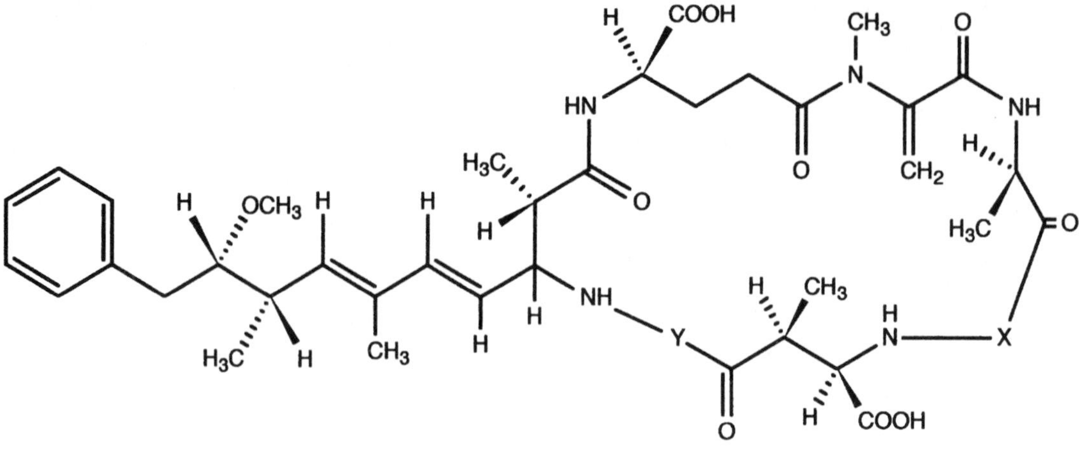

Figure 4   Structure of microcystins (adapted from Carmichael[105]).

	X	Y
Microcystin-LR	Leucine	Arginine
Microcystin-RR	Arginine	Arginine
Microcystin-YR	Tyrosine	Arginine

high cell permeability. However, some of the less common microcystins contain a variety of hydrophobic amino acids in the variable residues, and may therefore enter intact cells much more readily and generally than the more commonly occurring microcystin-LR.

The recent determination of the crystal structure of recombinant PP1α complexed with microcystin yielded fresh insights on the interaction of the toxin with phosphatases.[109] A covalent link between Cys273 of PP1α and the terminal carbon of the MDHA (*N*-methyl-

dehydroalanine) side chain of the toxin was confirmed, although it appears not to be necessary for inhibition to occur. The interaction between microcystin and recombinant PP1α appears to involve three distinct surface regions: the metal-binding site, the hydrophobic groove, and the edge of the C-terminal groove near the active site.[109] Since the essential chemical and conformational features of microcystin are preserved in its active nodularin homologue, the latter probably interacts in a similar fashion with protein phosphatases.

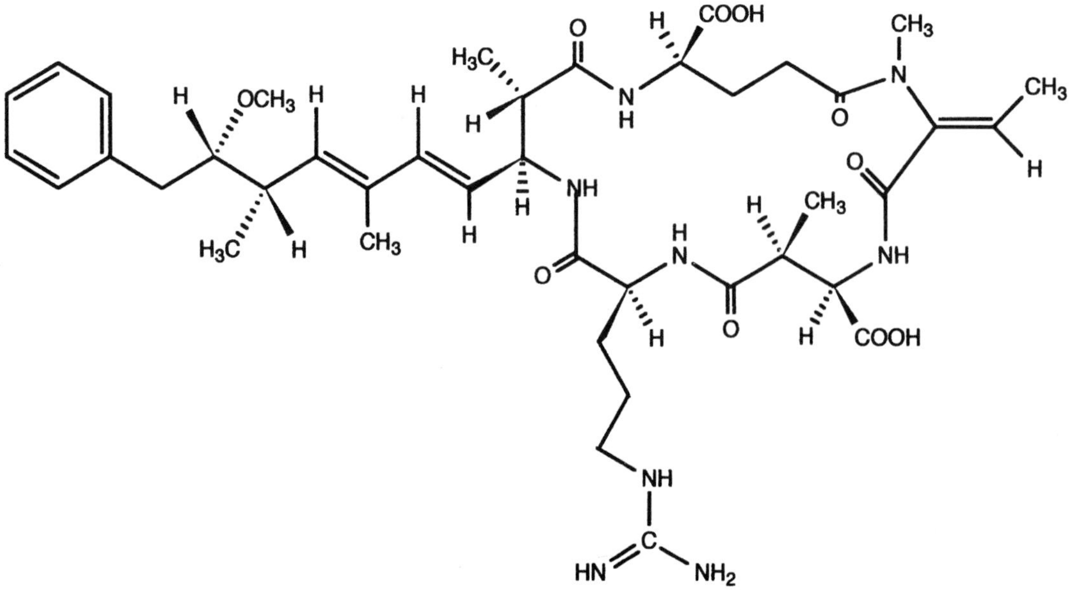

Figure 5   Structure of nodularin.

11.10.7.1.5 Nodularin

Nodularin (Figure 5) is a cyclic pentapeptide toxin isolated from the cyanobacterium *Nodularia spumigena*. It is also an inhibitor of PP1 and PP2A.[110,111] Nodularin-V and nodularin-I (also termed motuporins) have been isolated from the marine sponge *Thoenella swinhoei*,[112] the first identification of a protein phosphatase inhibitor of the cyclic peptide type in the marine environment.

11.10.7.2 Pesticides and Herbicides

Some commonly used insecticides and herbicides have been identified as potent and specific inhibitors of the Ser/Thr-specific protein phosphatases. Thus, our concerns in this respect surpass naturally occurring environmental toxins and toxicants, and include also some of the compounds described below that are of agricultural, environmental, and even of medical interest.

11.10.7.2.1 Pyrethroids

Pyrethroids are widely used as potent insecticides (see Chapter 27, this volume). The older compounds (termed Type I pyrethroids) can be distinguished from the more recently developed compounds (termed Type II pyrethroids) by their differing effects on animals, and by the unusual *in vivo* symptoms produced by the Type II compounds.[113–115] Type II pyrethroids are characteristically able to stimulate neurotransmitter release,[116,117] an effect that is only partly explained by their known interaction with sodium channels,[118] since tetrodotoxin fails to abolish the pyrethroid-induced stimulation.[119] The demonstration that a typical Type II pyrethroid (deltamethrin) was capable of producing a persistent increase in protein phosphorylation in intact brain synaptosomes,[120] led to an investigation of the ability of these compounds to inhibit calcineurin, a protein phosphatase that is particularly abundant in nervous tissue.[121] The latter study provided convincing evidence that calcineurin could be specifically inhibited by the Type II pyrethroids (cypermethrin, deltamethrin, and fenvalerate) at subnanomolar concentrations, whereas their noninsecticidal chiral isomers and the Type I pyrethroids were essentially unable to inhibit this phosphatase. Furthermore, there was excellent correlation between the insecticidal potencies of the various compounds tested and their ability to inhibit calcineurin.

Calcineurin can constitute up to 1% of total mammalian brain protein, being the major calmodulin-binding protein,[57] and its inhibition probably explains the prolonged increase in neurotransmitter release in the CNS observed with Type II pyrethroids. In contrast, calcineurin appears to be much less abundant in peripheral tissues, constituting only 0.03% of the soluble skeletal muscle protein.[57] Calcineurin is a heterodimer composed of the catalytic A subunit and a regulatory B subunit that is related to calmodulin.[122] This phosphatase can be activated synergistically by Ca^{2+} binding to the B subunit, or by calmodulin (in the presence of calcium) binding to the A subunit. Interestingly, one of the proteins whose phosphorylation is stimulated by Type II pyrethroids is synapsin I.[120] Synapsin I is one of the main players in the control of neurotransmitter release at the synaptic terminal, a function that is controlled via its phosphorylation state.[1,2] Thus, the observed effects of Type II pyrethroids on intact and semiintact neuronal preparations, coupled with the known localization of calcineurin at the nerve terminal are all consistent with the observed neurotoxicological effects of these compounds and with calcineurin being the mediator *in vivo*.

Since Type II pyrethroids can also affect sodium channels, care must be exercised when using these compounds as specific calcineurin inhibitors for *in vivo* experiments. The Type I compounds may be appropriate negative controls since they also affect sodium channels but do not inhibit calcineurin. Some of the commonly available Type II pyrethroids include cypermethrin, deltamethrin, and fenvalerate, whose IC_{50} values against calcineurin are approximately 40 pM, 100 pM, and 2 nM, respectively. Useful Type I pyrethroids showing little or no activity against calcineurin include allethrin, permethrin, and resmethrin. Bioallethrin is also a Type I pyrethroid but, unlike other members of this class, it appears to inhibit calcineurin with a potency approaching that of the weaker Type II pyrethroids.

11.10.7.2.2 Cantharidin

Cantharidin is an oxabicycloheptanedicarboxylic acid and the natural toxicant isolated from blister beetles.[123] Interestingly, cantharidin is thought to be the active ingredient of the purported aphrodisiac known as "Spanish fly" and, as a consequence, numerous cases of human cantharidin poisoning have occurred. Work by Casida and colleagues led to the identification of the cantharidin-binding protein as PP2A and subsequently, cantharidin and a series of its derivatives were shown to be potent inhibitors of both PP2A and PP1.[124–126] Intraperitoneal treatment of mice with cantharidin was shown to result in a dose-dependent decrease in phosphorylase *a* phosphatase activity and to increase the phosphorylation state of several phosphoproteins.[127] Related compounds, such as endothal, are widely used as herbicides.[128] The phosphatase specificity of these compounds is very similar to okadaic acid (inhibition of PP2A at concentrations 5–10-fold lower than for PP1 and essentially inactive against calcineurin and PP2C), but unlike okadaic acid they have a considerably lower cost.

Sulfur mustard (2,2'-dichlorodiethyl sulfide) is an alkylating agent and potent vesicant, in use since World War I for chemical warfare, whose mechanism of vesicant action was not fully understood until the mid-1990s. However, the observations of Casida and co-workers[124,126,129] on cantharidin, also a vesicant, led Brimfield[130] to examine the ability of sulfur mustard and its hydrolysis product (TDG, bishydroxyethyl sulfide) to inhibit the Ser/Thr-specific phosphatases. Thus, the nonalkylating hydrolysis product of sulfur mustard, TDG, was shown to inhibit the Ser/Thr phosphatase activity present in mouse liver cytosol in a dose-dependent manner. This led Brimfield[130] to suggest that phosphatase inhibition may therefore represent a general mechanism of vesicant action.

11.10.7.3   Metals

Most studies on metal toxicity and protein phosphorylation have failed to examine their effect at the level of Ser/Thr protein phosphatases. Furthermore, the interaction between the Ser/Thr protein phosphatases and metal ions is controversial. King and Huang[131] indicated the presence of iron and zinc in calcineurin, but the presence of these or other metal ions in other Ser/Thr phosphatase types lacks convincing evidence. It is even uncertain whether phosphatases require metal ions in order to function. However, two recent studies seem to lend credence to the idea of the requirement of metal ions for phosphatase activity. The x-ray structure of calcineurin inhibited by the FKBP12–FK506 complex indicates the occurrence of two metal ions at the active site, which were modeled as iron and zinc.[132] The recent crystallization of recombinant PP1α also indicated the presence of manganese at the active site.[109] Thus, the role of metal ions in phosphatase catalysis appears to be confirmed but, since manganese was included in the PP1α crystallization buffer, the identity of the metal

ions in the native enzyme remains uncertain. It may be that, as Goldberg *et al.*[109] suggested, the interaction is relatively insensitive to the nature of the metal ions. Below, and in alphabetical order, we discuss some specific aspects of metal ion toxicity as they relate to phosphatase activity.

11.10.7.3.1   *Aluminum*

Aluminum intoxication has been shown to result in altered protein phosphorylation and is thought to be associated with many of the pathological features associated with Alzheimer's disease.[133,134] The neurofilaments of aluminum-treated animals appear hyperphosphorylated, like the cytoskeletal elements of plaques and tangles from Alzheimer's disease subjects (reviewed by Saitoh *et al.*[5]). Interestingly, the secretion of the Alzheimer's amyloid precursor protein can be stimulated both by phorbol esters, that stimulate PKC-mediated phosphorylation, and by okadaic acid, calyculin A, and cantharidin, that are known inhibitors of Ser/Thr protein phosphatases.[103] Although many studies have demonstrated the ability of aluminum to induce neuronal cytoskeletal abnormalities consistent with altered neurofilament phosphorylation,[135,136] the molecular mechanism responsible for the observed aluminum-induced hyperphosphorylation remains to be fully elucidated. However, several studies have suggested that decreased phosphatase activity (i.e., aluminum-induced inhibition of phosphatase activity) might explain the observed neuropathological effects of aluminum.[79,137]

11.10.7.3.2   *Lead*

The effects of lead exposure on brain cell function have been reviewed in detail by Goldstein (see also Chapters 5, 8, 9, 13, 19, and 28, this volume).[138] Although it is not clear whether lead can directly affect protein phosphatase activity, several lines of evidence suggest that lead toxicity may indeed be mediated by inhibition of protein phosphatase activity. For example, Kern and Audesirk[139] showed that inorganic lead could inhibit neurite initiation in cultured rat hippocampal neurons. Significantly, this effect could also be mimicked with okadaic acid, a relatively specific inhibitor of PP1 and PP2A. These results are in agreement with the previous observation that okadaic acid also inhibited nerve growth factor-stimulated neurite development from PC12 cells.[140] Indeed, okadaic acid has been shown to produce hyperphosphorylation of several cytoskeletal proteins, including MAP2, tau, and neurofilaments,[141,142] known to play a role in neurite extension and growth.

The hippocampus, an important structure for certain types of learning, is also a target for chronic developmental lead neurotoxicity, according to both morphological[143–145] and biochemical criteria.[146–148] The hippocampus also has one of the highest concentrations of protein phosphatases in the mammalian brain, including the various isoforms of PP1.[23] Not surprisingly, experimental lead intoxication results in altered learning and complex behaviors in primates and rodents,[149,150] including long-term potentiation.[146] Consistent with the animal studies, low-level lead exposure in children produces a variety of neurotoxic symptoms, including low IQ and poor fine motor control.[151–156] It is interesting to note that, based on theoretical grounds, Lisman[28] postulated a key role for PP1 in the control of long-term potentiation and long-term depression. Experimental evidence supports Lisman's hypothesis,[29,30] and PP1 is now known to be highly and specifically enriched in dendritic spines.[33] Thus, the observed neurotoxicological effects of lead are consistent with its inhibition of phosphatase activity (probably PP1), whose mRNA isoforms are known to be relatively abundant in the hippocampal formation[23] and whose corresponding protein isoforms are dramatically enriched in dendritic spines.[33]

11.10.7.3.3 Manganese

Manganese is an essential element, usually acquired via dietary means, whose deficiency in animals and humans can result in severe nervous system dysfunction. Furthermore, epileptic patients have been reported to show significantly lower blood levels of manganese than control subjects.[157] Similarly, high level exposure to manganese also results in neurologic disorders, affecting mainly the extrapyramidal system. Indeed, "manganism," which has been described over the years in miners and manganese alloy production workers, shows many similarities to Parkinson's Disease. Current data suggest that there is a dose-dependent relationship between the level of manganese exposure and the type and severity of the resulting clinical symptoms. The localization of manganese in intoxicated humans and primates reveals the globus pallidus, the striatum, and the substantia nigra as the preferential sites for manganese deposition, whereas most,

but not all, neuropathological lesions appear to be concentrated in the striatum, the pallidum, and the subthalamic nucleus.

Several hypothesis have been formulated to explain the effects of manganese on dopaminergic neurotransmission, but the available data may also be consistent with manganese neurotoxicity being mediated via its effect on PP1. Over the years, researchers have observed that while Mn^{2+} is usually slightly inhibitory to the activity of PP1,[46,158] during storage or after prolonged exposure to manganese, PP1 gradually converts to a manganese-dependent form whose specific activity is considerably higher than that of the manganese-independent form.[146] Although only low levels of manganese were found in PP1[159] and the enzyme apparently can not bind stoichiometric amounts of manganese,[160,161] the recent crystallization and elucidation of the three-dimensional structure of recombinant PP1α indicates that this phosphatase is a metalloenzyme possessing two metal ions at the active site.[109] The phosphatase activity of the recombinant form of the enzyme used in this study is dependent on manganese for full activity and manganese was also included in the crystallization buffer. Thus, while the physiological interaction of PP1 with its active site metal ions remains to be more fully elucidated, there is no doubt that prolonged exposure of PP1 to manganese results in an abnormally "activated" form of the enzyme. In this respect it is noteworthy that the striatum is one of the main target areas for both manganese contamination and location of the derived neuropathology. As already discussed, the main isoforms of PP1 are relatively more abundant in the striatum,[23] where they appear to be particularly enriched in dendritic spines.[33] The observed distribution of PP1 is also consistent with a study of manganese-intoxicated monkeys that revealed alterations in dopaminergic postsynaptic structures in the striatum.[162]

Incidents of water and food supply contamination with high levels of natural manganese have occurred, but the main sources of manganese contamination in the environment are usually mining, industrial emissions, and manganese-containing pesticides. However, the recent approval of MMT (an organomanganese compound) as an antiknock agent for unleaded gasoline in the USA, coupled with its use in Canada since 1977, may provide a more widespread source of manganese exposure and bring manganese contamination to the fore in forthcoming years.[163] Indeed, one of the most valid public health concerns is that manganese-related disorders appear to be progressive, even after exposure to the metal has ceased.[164,165]

11.10.7.4 Organophosphates

Exposure to organophosphates is known to result in hyperphosphorylation of cytoskeletal proteins such as the neurofilaments, β-tubulin and MAP-2.[3] Jensen *et al.*[166] showed anomalous aggregation of neurofilaments within the CNS and PNS of animals exposed to organophosphates by using antibodies directed against the phosphorylated form of neurofilament proteins. It will be of some interest to determine whether organophosphates are indeed capable of inhibiting phosphatase activity and therefore explain the observed organophosphate-induced hyperphosphorylation.

11.10.7.5 Others

Cyclosporin A and FK506 (also known as tacrolimus) are two macrocyclic natural products widely used as immunosuppressants for the prevention of graft rejection, that have been shown to be inhibitors of calcineurin.[167,168] These compounds are only active inhibitors of calcineurin in association with cyclophilin and FK506-binding protein (FKBP), respectively. It is the complex of immunosuppressant/ immunophilin that binds and inhibits calcineurin. Another related immunosuppressant, rapamycin, can also bind FKBP but does not inhibit calcineurin. A compound related to FK506 and equally capable of inhibiting calcineurin is FK520. Since calcineurin is uniquely modulated by Ca^{2+} and calmodulin, its activity can also be inhibited by phenothiazines and other "anticalmodulin" drugs. However, this inhibition is much less specific than that achieved by the highly specific immunosuppressants.

Sodium orthovanadate is generally used as a relatively specific inhibitor of the Tyr-specific family of phosphatases in both homogenates and intact cells.[169] Care must be taken to ensure that its proper oxidation state is preserved and its use is further complicated by its effects on other enzymes systems, such as the Na^+/K^+-ATPase. Fluoride is also often used as a nonspecific inhibitor of the Ser/Thr-specific protein phosphatases. Again its use is complicated by possible effects on other systems.

11.10.8 CONCLUSIONS

A large variety of compounds is now available capable of specifically inhibiting Ser/Thr-specific protein phosphatases. Most can readily

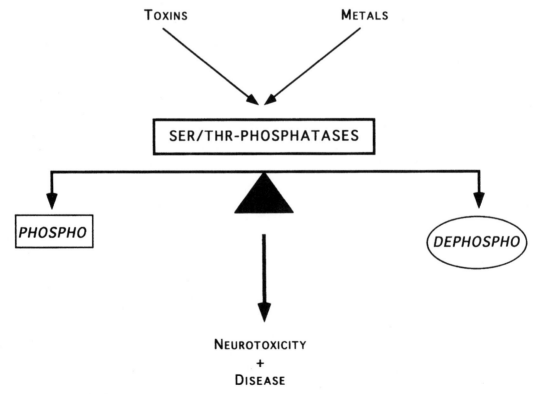

Figure 6 Hypothesized central role of the Ser/Thr-specific protein phosphatases as key mediators of neurotoxicity. Neurotoxic agents acting on phosphatases upset the delicate balance of intracellular neuronal protein phosphorylation systems, leading to the development of a cascade of events that results in neurotoxicity and/or disease.

enter intact cells and their use has produced new and important insights into the physiological importance of this class of proteins. While new and more specific phosphatase inhibitors continue to be actively sought from natural sources, sufficient knowledge is already available for the design of synthetic phosphatase inhibitors soon to become a reality. The development or discovery of highly type/isoform-specific phosphatase inhibitors should prove useful in the research laboratory and find widespread therapeutic applications. Many of the known phosphatase inhibitory compounds are readily found in nature, a factor that is of some concern since they may function as a trigger for human neurotoxicity and disease.

It is becoming increasingly clear that Ser/Thr-specific protein phosphatases represent potential targets of broad classes of known and suspect neurotoxicants. Figure 6 highlights and summarizes the role played by the Ser/Thr-specific protein phosphatases in mediating some of the action of neurotoxic agents. Central to this scenario is the dominant role played by protein phosphorylation. Inherent to this scheme is the now widely accepted view that the specificity of signal transduction cascades (see Chapter 6, this volume) is integrated through the overlapping regulatory mechanisms controlling protein kinases and phosphatases. Thus, any factors that affect protein kinases or protein phosphatases have the potential to disturb the delicate intracellular protein phosphorylation balance. The emergence of phosphoprotein phosphatases as equal partners with protein kinases in maintaining this balance dictates the need to consider their role in neurotoxicant-induced alterations in protein phosphorylation systems.

While significant progress has been made in the study of the Ser/Thr-specific protein phosphatases expressed in the nervous system, the identification of their physiological substrates and of their regulatory and targeting subunits represents a major challenge for the future. The possible role of Ser/Thr-specific phosphatases as key players in the cascade of events leading to neurotoxicity opens new avenues for research on the molecular mechanisms involved and will hopefully allow the future development of rational therapeutic approaches.

11.10.9 REFERENCES

1. E. J. Nestler and P. Greengard, 'Protein Phosphorylation in the Nervous System,' Wiley, New York, 1984.
2. S. I. Walaas and P. Greengard, 'Protein phosphorylation and neuronal function.' *Pharmacol. Rev.*, 1991, **43**, 299–349.
3. M. B. Abou-Donia, D. M. Lapadula and E. Suwita, 'Cytoskeletal proteins as targets for organophosphorus compound and aliphatic hexacarbon-induced neurotoxicity.' *Toxicology*, 1988, **49**, 469–477.
4. L. G. Costa, 'The phosphoinositide/protein kinase C system as a potential target for neurotoxicity.' *Pharmacol. Res.*, 1990, **22**, 393–408.
5. T. Saitoh, E. Masliah, L. W. Jin *et al.*, 'Biology of disease. Protein kinases and phosphorylation in neurologic disorders and cell death.' *Lab. Invest.*, 1991, **64**, 596–616.
6. J. P. O'Callaghan, 'A potential role for altered protein phosphorylation in the mediation of developmental neurotoxicity.' *Neurotoxicology*, 1994, **15**, 29–40.
7. E. G. Krebs, in 'The Enzymes,' 3rd edn., eds. P. D. Boyer and E. G. Krebs, Academic Press, Orlando, FL, 1986, XVIIa, pp. 3–19.
8. S. Shenolikar and A. C. Nairn, 'Protein phosphatases: recent progress.' *Adv. Second Messenger Phosphoprotein Res.*, 1991, **23**, 1–121.
9. M. J. Hubbard and P. Cohen, 'On target with a new mechanism for the regulation of protein phosphorylation.' *Trends Biochem. Sci.*, 1993, **18**, 172–177.
10. M. Bollen and W. Stalmans, 'The structure, role, and regulation of type 1 protein phosphatases.' *Crit. Rev. Biochem. Mol. Biol.*, 1992, **27**, 227–281.
11. M. Suganuma, H. Fujiki, H. Suguri *et al.*, 'Okadaic acid: an additional non-phorbol-12-tetradecanoate-13-acetate-type tumor promoter.' *Proc. Natl. Acad. Sci. USA*, 1988, **85**, 1768–1771.
12. R. Nishiwaki-Matsushima, S. Nishiwaki, T. Ohta *et al.*, 'Structure–function relationships of microcystins, liver tumor promoters, in interaction with protein phosphatase.' *Jpn. J. Cancer Res.*, 1992, **82**, 993–996.
13. H. Fujiki and M. Suganuma, 'Tumor promotion by inhibitors of protein phosphatases 1 and 2A: the okadaic acid class of compounds.' *Adv. Cancer Res.*, 1993, **91**, 143–194.
14. G. T. Cori and A. A. Green, 'Crystalline muscle phosphorylase. II. Prosthetic group.' *J. Biol. Chem.*, 1943, **151**, 31–38.
15. T. S. Ingebritsen and P. Cohen, 'Protein phosphatases: properties and role in cellular regulation.' *Science*, 1983, **221**, 331–338.
16. P. Cohen, 'The structure and regulation of protein phosphatases.' *Annu. Rev. Biochem.*, 1989, **58**, 453–508.
17. P. Cohen, 'Classification of protein-serine/threonine phosphatases: identification and quantitation in cell extracts.' *Meth. Enzymol.*, 1991, **201**, 389–398.
18. A. T. R. Sim, 'The regulation and function of protein phosphatases in the brain.' *Mol. Neurobiol.*, 1992, **5**, 229–246.
19. O. B. da Cruz e Silva, S. Alemany, D. G. Campbell *et al.*, 'Isolation and sequence analysis of a cDNA clone encoding the entire catalytic subunit of a type-2A protein phosphatase.' *FEBS Lett.*, 1987, **221**, 415–422.
20. O. B. da Cruz e Silva and P. T. W. Cohen, 'A second catalytic subunit of type-2A protein phosphatase from rabbit skeletal muscle.' *FEBS Lett.*, 1987, **226**, 176–178.
21. E. F. da Cruz e Silva and P. Greengard, 'Cloning of neuronal isoforms of protein phosphatase 1 by low-stringency screening of cDNA libraries.' *Neuroprotocols*, 1995, **6**, 2–10.
22. P. T. W. Cohen, 'Two isoforms of protein phosphatase 1 may be produced from the same gene.' *FEBS Lett.*, 1988, **232**, 17–23.
23. E. F. da Cruz e Silva, C. A. Fox, C. C. Ouimet *et al.*, 'Differential expression of protein phosphatase 1 isoforms in mammalian brain.' *J. Neurosci.*, 1995, **15**, 3375–3389.
24. P. T. W. Cohen, J. F. Collins, A. F. Coulson *et al.*, 'Segments of bacteriophage lambda (orf 221)

and phi 80 are homologous to genes coding for mammalian protein phosphatases.' *Gene*, 1988, **69**, 131–134.

25. O. B. da Cruz e Silva, E. F. da Cruz e Silva and P. T. W. Cohen, 'Identification of a novel protein phosphatase catalytic subunit by cDNA cloning.' *FEBS Lett.*, 1988, **242**, 106–110.

26. V. Dombradi, J. M. Axton, D. M. Glover *et al.*, 'Molecular cloning and chromosomal localization of a novel *Drosophila* protein phosphatase.' *FEBS Lett.*, 1989, **247**, 391–395.

27. E. F. da Cruz e Silva, V. Hughes, P. McDonald *et al.*, 'Protein phosphatase 2B_W and protein phosphatase Z are *Saccharomyces cerevisiae* enzymes.' *Biochim. Biophys. Acta*, 1991, **1089**, 269–272.

28. J. Lisman, 'A mechanism for the Hebb and the anti-Hebb processes underlying learning and memory.' *Proc. Natl. Acad. Sci. USA*, 1989, **86**, 9574–9578.

29. R. M. Mulkey, S. Endo, S. Shenolikar *et al.*, 'Involvement of a calcineurin/inhibitor-1 phosphatase cascade in hippocampal long-term depression.' *Nature*, 1994, **369**, 486–488.

30. R. C. Malenka, 'Synaptic plasticity in the hippocampus: LTP and LTD.' *Cell*, 1994, **78**, 535–538.

31. S. M. Shields, T. S. Ingebritsen and P. T. Kelly, 'Identification of protein phosphatase 1 in synaptic junctions: dephosphorylation of endogenous calmodulin-dependent kinase II and synapse-enriched phosphoproteins.' *J. Neurosci.*, 1985, **5**, 3414–3422.

32. L. A. Dokas, M. R. Pisano, L. H. Schrama *et al.*, 'Dephosphorylation of B-50 in synaptic plasma membranes.' *Brain Res. Bull.*, 1990, **24**, 321–329.

33. C. C. Ouimet, E. F. da Cruz e Silva and P. Greengard, 'The α and γ_1 isoforms of protein phosphatase 1 are highly and specifically concentrated in dendritic spines.' *Proc. Natl. Acad. Sci. USA*, 1995, **92**, 3396–3400.

34. J. J. Pei, E. Sersen, K. Iqbal *et al.*, 'Expression of protein phosphatases (PP-1, PP-2A, PP-2B and PTP-1B) and protein kinases (MAP kinase and P34^{cdc2}) in the hippocampus of patients with Alzheimer disease and normal aged individuals.' *Brain Res.*, 1994, **655**, 70–76.

35. J. H. Doonan and N. R. Morris, 'The bimG gene of *Aspergillus nidulans*, required for completion of anaphase, encodes a homolog of mammalian phosphoprotein phosphatase 1.' *Cell*, 1989, **57**, 987–996.

36. H. Ohkura, N. Kinoshita, S. Miyatani *et al.*, 'The fission yeast dis2+ gene required for chromosome disjoining encodes one of two putative type 1 protein phosphatases.' *Cell*, 1989, **57**, 997–1007.

37. J. M. Axton, V. Dombradi, P. T. W. Cohen *et al.*, 'One of the protein phosphatase 1 isoenzymes in *Drosophila* is essential for mitosis.' *Cell*, 1990, **63**, 33–46.

38. V. Dombradi, J. M. Axton, N. D. Brewis *et al.*, '*Drosophila* contains three genes that encode distinct isoforms of protein phosphatase 1.' *Eur. J. Biochem.*, 1990, **194**, 739–745.

39. H. C. Hemmings, Jr. and P. Greengard, 'DARPP-32, a dopamine- and adenosine 3',5'-monophosphate-regulated phosphoprotein: regional, tissue, and phylogenetic distribution.' *J. Neurosci.*, 1986, **6**, 1469–1481.

40. A. C. Nairn, H. C. Hemmings Jr., S. I. Walaas *et al.*, 'DARPP-32 and phosphatase inhibitor-1, two structurally related inhibitors of protein phosphatase-1, are both present in striatonigral neurons.' *J. Neurochem.*, 1988, **50**, 257–262.

41. F. Desdouits, J. J. Cheetham, H.-B. Huang *et al.*, 'Mechanism of inhibition of protein phosphatase 1 by DARPP-32: studies with recombinant DARPP-32 and synthetic peptides.' *Biochem. Biophys. Res. Commun.*, 1995, **206**, 652–658.

42. A. L. Levey, S. M. Hersch, D. B. Rye *et al.*, 'Localization of D_1 and D_2 dopamine receptors in brain with subtype-specific antibodies.' *Proc. Natl. Acad. Sci. USA*, 1993, **90**, 8861–8865.

43. J. F. Smiley, A. I. Levey, B. J. Ciliax *et al.*, 'D_1 dopamine receptor immunoreactivity in human and monkey cerebral cortex: predominant and extrasynaptic localization in dendritic spines.' *Proc. Natl. Acad. Sci. USA*, 1994, **91**, 5720–5724.

44. L. J. Martin, C. D. Blackstone, A. I. Levey *et al.*, 'AMPA glutamate receptor subunits are differentially distributed in rat brain.' *Neuroscience*, 1993, **53**, 327–358.

45. E. Waelkens, P. Agostinis, J. Goris *et al.*, 'The polycation-stimulated protein phosphatases: regulation and specificity.' *Adv. Enzyme Reg.*, 1987, **26**, 241–270.

46. T. S. Ingebritsen, A. A. Stewart and P. Cohen, 'The protein phosphatases involved in cellular regulation. 6. Measurement of type-1 and type-2 protein phosphatases in extracts of mammalian tissues: an assessment of their physiological roles.' *Eur. J. Biochem.*, 1983, **132**, 297–307.

47. M. C. Mumby, D. D. Green and K. L. Russell, 'Structural characterization of cardiac protein phosphatase with a monoclonal antibody.' *J. Biol. Chem.*, 1985, **260**, 13763–13770.

48. Y. Saitoh, H. Yamamoto, Y. Ushio *et al.*, 'Characterization of polyclonal antibodies to brain protein phosphatase 2A and immunohistochemical localization of the enzyme in rat brain.' *Brain Res.*, 1989, **489**, 291–301.

49. R. E. Mayer, P. Hendrix, P. Cron *et al.*, 'Structure of the 55 kDa regulatory subunit of protein phosphatase 2A: evidence for a neuronal-specific isoform.' *Biochemistry*, 1991, **30**, 3589–3597.

50. E. A. Tallant and W. Y. Cheung, 'Calmodulin-dependent phosphatase: a developmental study.' *Biochemistry*, 1983, **22**, 3630–3635.

51. R. K. Carlin, D. J. Grab and P. Siekevitz, 'Function of calmodulin in postsynaptic densities. III. Calmodulin-binding proteins of the postsynaptic density.' *J. Cell Biol.*, 1981, **89**, 449–455.

52. D. J. Grab, R. K. Carlin and P. Siekevitz, 'Function of calmodulin in postsynaptic densities. I. Presence of a calmodulin-activatable cyclic nucleotide phosphodiesterase activity.' *J. Cell Biol.*, 1981, **89**, 433–439.

53. J. G. Wood, R. W. Wallace, J. N. Whitaker *et al.*, 'Immunocytochemical localization of calmodulin and a heat-labile calmodulin-binding protein (CaM-BP_{80}) in basal ganglia of mouse brain.' *J. Cell Biol.*, 1980, **84**, 66–76.

54. R. L. Kincaid, M. L. Billingsley and C. D. Balaban, 'Immunocytochemical localization of calmodulin-dependent phosphodiesterase and calcineurin in rat brain.' *Fed. Proc.*, 1985, **44**, 6979.

55. N. G. Cooper, B. J. McLaughlin, E. A. Tallant *et al.*, 'Calmodulin-dependent protein phosphatase: immunocytochemical localization in chick retina.' *J. Cell Biol.*, 1985, **101**, 1212–1218.

56. F. A. Anthony, M. A. Winkler, H. H. Edwards *et al.*, 'Quantitative subcellular localization of calmodulin-dependent phosphatase in chick forebrain.' *J. Neurosci.*, 1988, **8**, 1245–1253.

57. C. B. Klee and P. Cohen, 'The calmodulin-regulated protein phosphatase.' *Mol. Asp. Cell Reg.*, 1988, **5**, 225–248.

58. R. W. Wallace, E. A. Tallant and W. Y. Cheung, 'High levels of a heat-labile calmodulin-binding protein in bovine neostriatum.' *Biochemistry*, 1980, **19**, 1831–1837.

59. S. Goto, Y. Matsukado, S. Uemura *et al.*, 'A comparative immunohistochemical study of calcineurin and S-100 protein in mammalian and avian brains.' *Exp. Brain Res.*, 1988, **69**, 645–650.

60. S. Goto, Y. Matsukado, Y. Mihara *et al.*, 'The distribution of calcineurin in rat brain by light and electron microscopic immunohistochemistry and enzyme-immunoassay.' *Brain Res.*, 1986, **397**, 161–172.

61. H. Matsui, A. Doi, T. Itano *et al.*, 'Immunohistochemical localization of calcineurin, calmodulin-stimulated phosphatase, in the rat hippocampus using a monoclonal antibody.' *Brain Res.*, 1987, **402**, 193–196.

62. S. Goto, A. Hirano and R. R. Rojas-Corona, 'An immunohistochemical investigation of the human neostriatum in Huntington's disease.' *Ann. Neurol.*, 1989, **25**, 298–304.

63. S. Goto, A. Hirano and R. R. Rojas-Corona, 'Calcineurin immunoreactivity in striatonigral degeneration.' *Acta Neuropathol. (Berl.)*, 1989, **78**, 65–71.

64. T. Takaishi, N. Saito, T. Kuno *et al.*, 'Differential distribution of the mRNA encoding two isoforms of the catalytic subunit of calcineurin in the rat brain.' *Biochem. Biophys. Res. Commun.*, 1991, **174**, 393–398.

65. R. L. Kincaid, P. R. Giri, S. Higuchi *et al.*, 'Cloning and characterization of molecular isoforms of the catalytic subunit of calcineurin using non-isotopic methods.' *J. Biol. Chem.*, 1990, **265**, 11312–11319.

66. P. J. Robinson and P. R. Dunkley, 'Depolarisation-dependent protein phosphorylation and dephosphorylation in rat cortical synaptosomes is modulated by calcium.' *J. Neurochem.*, 1985, **44**, 338–348.

67. P. J. Robinson, R. Hauptschein, W. Lovenberg *et al.*, 'Dephosphorylation of synaptosomal proteins P96 and P139 is regulated by both depolarization and calcium, but not by a rise in cytosolic calcium alone.' *J. Neurochem.*, 1987, **48**, 187–195.

68. A. T. R. Sim, P. R. Dunkley, P. E. Jarvie *et al.*, 'Modulation of synaptosomal protein phosphorylation/dephosphorylation by calcium is antagonised by inhibition of protein phosphatases with okadaic acid.' *Neurosci. Lett.*, 1991, **126**, 203–206.

69. M. Abdul-Ghani, E. A. Kravitz, H. Meiri *et al.*, 'Protein phosphatase inhibitor okadaic acid enhances transmitter release at neuromuscular junctions.' *Proc. Natl. Acad. Sci. USA*, 1991, **88**, 1803–1807.

70. P. H. Reinhart, S. Chung, B. L. Martin *et al.*, 'Modulation of calcium-activated potassium channels from rat brain by protein kinase A and phosphatase 2A.' *J. Neurosci.*, 1991, **11**, 1627–1635.

71. S. K. Chung, P. H. Reinhart, B. L. Martin *et al.*, 'Protein kinase activity closely associated with a reconstituted calcium-activated potassium channel.' *Science*, 1991, **253**, 560–562.

72. R. E. White, A. Schonbrunn and D. L. Armstrong, 'Somatostatin stimulates Ca^{2+}-activated K^+ channels through protein dephosphorylation.' *Nature*, 1991, **351**, 570–573.

73. S. Endo, M. Ichinose, S. D. Critz *et al.*, 'Protein phosphatases and their role in control of membrane currents in *Aplysia* neurons.' *Adv. Protein Phosphatases*, 1991, **6**, 411–432.

74. M. Ichinose and J. H. Byrne, 'Role of protein phosphatases in the modulation of neuronal membrane currents.' *Brain Res.*, 1991, **549**, 146–150.

75. L.-Y. Wang, M. W. Salter and J. F. Macdonald, 'Regulation of kainate receptors by cAMP-dependent protein kinase and phosphatases.' *Science*, 1991, **253**, 1132–1135.

76. P. Greengard, J. Jen, A. C. Nairn *et al.*, 'Enhancement of the glutamate response by cAMP-dependent protein kinase in hippocampal neurons.' *Science*, 1991, **253**, 1135–1138.

77. S. Gandy, A. J. Czernik and P. Greengard, 'Phosphorylation of Alzheimer disease amyloid precursor peptide by protein kinase C and Ca^{++}/calmodulin-dependent protein kinase II.' *Proc. Natl. Acad. Sci. USA*, 1988, **85**, 6218–6221.

78. S. E. Gandy, G. L. Caporaso, J. D. Buxbaum *et al.*, 'Protein phosphorylation regulates utilization of processing pathways for Alzheimer β/A4 amyloid precursor protein.' *Ann. NY Acad. Sci.*, 1993, **695**, 117–121.

79. K. A. Harris, G. A. Oyler, G. M. Doolittle *et al.*, 'Okadaic acid induces hyperphosphorylated forms of tau protein in human brain slices.' *Ann. Neurol.*, 1993, **33**, 77–87.

80. J. P. Durkin, R. Tremblay, A. Buchan *et al.*, 'An early loss in membrane protein kinase C activity precedes the excitatory amino acid-induced death of primary cortical neurons.' *J. Neurochem.*, 1996, **66**, 951–962.

81. B. Padmaperuma, R. Mark, H. S. Dhillon *et al.*, 'Alterations in brain protein kinase C after experimental brain injury.' *Brain Res.*, 1996, **714**, 19–26.

82. S. Halpain, J. A. Girault and P. Greengard, 'Activation of NMDA receptors induces dephosphorylation of DARPP-32 in rat striatal slices.' *Nature*, 1990, **343**, 369–372.

83. J. Luthman, H. Andersen and L. Olson, 'Environmental neurotoxicology: a review on identification and action of neurotoxicants.' *Technol. J. Franklin Inst.*, 1995, **332A**, 151–182.

84. C. Bialojan and A. Takai, 'Inhibitory effect of a marine-sponge toxin, okadaic acid, on protein phosphatases. Specificity and kinetics.' *Biochem. J.*, 1988, **256**, 283–290.

85. J. Hescheler, G. Mieskes, J. C. Ruegg *et al.*, 'Effects of a protein phosphatase inhibitor, okadaic acid, on membrane currents of isolated guinea-pig cardiac myocytes.' *Pflugers Arch.*, 1988, **412**, 248–252.

86. G. Francis, 'Poisonous Australian lake.' *Nature*, 1878, **18**, 11–12.

87. Y. Murakami, Y. Oshima and T. Yasumoto, 'Identification of okadaic acid as a toxic component of a marine dinoflagellate.' *Bull. Jpn. Soc. Sci. Fish.*, 1982, **48**, 69–72.

88. K. Tachibana, P. J. Scheuer, Y. Tsukitani *et al.*, 'Okadaic acid, a cytotoxic polyether from two marine sponges of the genus *Halichondria*.' *J. Am. Chem. Soc.*, 1981, **103**, 2469–2471.

89. M. Murata, M. Shimatani, H. Sugitani *et al.*, *Bull. Jpn. Soc. Sci. Fish.*, 1982, **48**, 549–552.

90. S. Aonuma, T. Ushijima, M. Nakayasu *et al.*, 'Mutation induction by okadaic acid, a protein phosphatase inhibitor, in CHL cells, but not in *S. typhimurium*.' *Mutat. Res.*, 1991, **250**, 375–381.

91. T. A. J. Haystead, A. T. R. Sim, D. Carling *et al.*, 'Effects of the tumour promoter okadaic acid on intracellular protein phosphorylation and metabolism.' *Nature*, 1989, **337**, 78–81.

92. P. Cohen, C. F. B. Holmes and Y. Tsukitani, 'Okadaic acid: a new probe for the study of cellular regulation.' *Trends Biochem. Sci.*, 1990, **15**, 98–102.

93. J. E. Eriksson, D. L. Brautigan, R. Vallee *et al.*, 'Cytoskeletal integrity in interphase cells requires protein phosphatase activity.' *Proc. Natl. Acad. Sci. USA*, 1992, **89**, 11093–11097.

94. W. S. Shestowsky, C. F. B. Holmes, T. Hu *et al.*, 'An anti-okadaic acid-anti-idiotypic antibody bearing an internal image of okadaic acid inhibits protein phosphatase PP1 and PP2A catalytic activity.' *Biochem. Biophys. Res. Commun.*, 1993, **192**, 302–310.

95. S. Nishiwaki, H. Fujiki, M. Suganuma *et al.*, 'Structure–activity relationship within a series of okadaic acid derivatives.' *Carcinogenesis*, 1990, **11**, 1837–1841.

96. J. E. Swain, R. Robitaille, G. R. Dass *et al.*,

'Phosphatases modulate transmission and serotonin facilitation at synapses: studies with the inhibitor okadaic acid.' *J. Neurobiol.*, 1991, **22**, 855–864.

97. C. F. B. Holmes and M. P. Boland, 'Inhibitors of protein phosphatase-1 and -2A; two of the major serine/threonine protein phosphatases involved in cellular regulation.' *Curr. Opin. Struct. Biol.*, 1993, **3**, 934–943.

98. M. C. Gong, P. Cohen, T. Kitazawa *et al.*, 'Myosin light chain phosphatase activities and the effects of phosphatase inhibitors in tonic and phasic smooth muscle.' *J. Biol. Chem.*, 1992, **267**, 14662–14668.

99. C. MacKintosh and S. Klumpp, 'Tautomycin from the bacterium *Streptomyces verticillatus*. Another potent and specific inhibitor of protein phosphatases 1 and 2A.' *FEBS Lett.*, 1990, **277**, 137–140.

100. Y. Kato, N. Fusetani, S. Matsunaga *et al.*, 'Calyculins, potent antitumor metabolites from the marine sponge *Discodermia calyx*: biological activities.' *Drugs Exp. Clin. Res.*, 1988, **14**, 723–728.

101. H. Ishihara, B. L. Martin, D. L. Brautigan *et al.*, 'Calyculin A and okadaic acid: inhibitors of protein phosphatase activity.' *Biochem. Biophys. Res. Commun.*, 1989, **159**, 871–877.

102. M. Suganuma, H. Fujiki, H. Furuya-Suguri *et al.*, 'Calyculin A, an inhibitor of protein phosphatases, a potent tumor promoter on CD-1 mouse skin.' *Cancer Res.*, 1990, **50**, 3521–3525.

103. E. F. da Cruz e Silva, O. A. B. da Cruz e Silva, C. T. B. V. Zaia *et al.*, 'Inhibition of protein phosphatase 1 stimulates secretion of Alzheimer amyloid precursor protein.' *Mol. Med.*, 1995, **1**, 535–541.

104. W. W. Carmichael, in 'Natural Toxins: Characterization, Pharmacology and Therapeutics,' eds. C. L. Ownby and G. V. O'Dell, Pergamon, Oxford, 1989, pp. 3–16.

105. W. W. Carmichael, 'Cyanobacteria secondary metabolites—the cyanotoxins.' *J. Appl. Bacteriol.*, 1992, **72**, 445–459.

106. C. MacKintosh, K. A. Beattie, S. Klumpp *et al.*, 'Cyanobacterial microcystin-LR is a potent and specific inhibitor of protein phosphatases 1 and 2A from both mammals and higher plants.' *FEBS Lett.*, 1990, **264**, 187–192.

107. P. Cohen, S. Klumpp and D. L. Schelling, 'An improved procedure for identifying and quantitating protein phosphatases in mammalian tissues.' *FEBS Lett.*, 1989, **250**, 596–600.

108. M. T. Runnegar, T. Maddatu, L. D. Deleve *et al.*, 'Differential toxicity of the protein phosphatase inhibitors microcystin and calyculin A.' *J. Pharmacol. Exp. Ther.*, 1995, **273**, 545–553.

109. J. Goldberg, H. B. Huang, Y. G. Kwon *et al.*, 'Three-dimensional structure of the catalytic subunit of protein serine/threonine phosphatase-1.' *Nature*, 1995, **376**, 745–753.

110. R. Matsushima, S. Yoshizawa, M. F. Watanabe *et al.*, '*In vitro* and *in vivo* effects of protein phosphatase inhibitors microcystins and nodularin on mouse skin and fibroblasts.' *Biochem. Biophys. Res. Commun.*, 1990, **171**, 867–874.

111. S. Yoshizawa, R. Matsushima, M. F. Watanabe *et al.*, 'Inhibition of protein phosphatases by microcystins and nodularin associated with hepatotoxicity.' *J. Cancer Res. Clin. Oncol.*, 1990, **116**, 609–614.

112. S. D. DeSilva, D. E. Williams, R. J. Andersen *et al.*, 'Motuporin, a potent new protein phosphatase inhibitor from the Papua New Guinea sponge *Theonella swinhoei* Gray.' *Tetrahedron Lett.*, 1992, **33**, 1561–1564.

113. M. Elliott, 'Synthetic Pyrethroids,' American Chemical Society Symposium Services No. 42, American Chemical Society, Washington, DC, 1977, pp. 1–229.

114. D. W. Gammon, M. A. Brown and J. E. Casida, 'Two classes of pyrethroid action in the cockroach.' *Pestic. Biochem. Physiol.*, 1981, **15**, 181–191.

115. J. G. Scott and F. Matsumura, 'Evidence of two types of toxic actions of pyrethroids on susceptible and DDT-resistant German cockroaches.' *Pestic. Biochem. Physiol.*, 1983, **19**, 141–150.

116. V. L. Salgado, S. N. Irving and T. A. Miller, 'The importance of nerve terminal depolarization in pyrethroid poisoning of insects.' *Pestic. Biochem. Physiol.*, 1983, **20**, 169–175.

117. M. W. Brooks and J. M. Clark, 'Enhancement of norepinephrine release from rat brain synaptosomes by alpha-cyano pyrethroids.' *Pestic. Biochem. Physiol.*, 1987, **28**, 127–139.

118. R. A. Nicholson, R. G. Wilson, C. Potter *et al.*, in 'Pesticide Chemistry,' eds. J. Miyamoto and P. C. Kearney, Pergamon, Elmsford, NY, 1983, vol. 3, pp. 75–78.

119. J. M. Clark and M. W. Brooks, 'Neurotoxicology of pyrethroids: single or multiple mechanisms of action?' *Environ. Toxicol. Chem.*, 1989, **8**, 361–372.

120. E. Enan and F. Matsumura, 'Stimulation of protein phosphorylation in intact rat brain synaptosomes by a pyrethroid insecticide, deltamethrin.' *Pestic. Biochem. Physiol.*, 1991, **39**, 182–195.

121. E. Enan and F. Matsumura, 'Specific inhibition of calcineurin by Type II synthetic pyrethroid insecticides.' *Biochem. Pharmacol.*, 1992, **43**, 1777–1784.

122. C. B. Klee, T. H. Crouch and M. H. Krinks, 'Calcineurin: a calcium- and calmodulin-binding protein of the nervous system.' *Proc. Natl. Acad. Sci. USA*, 1979, **76**, 6270–6273.

123. J. E. Carrel and T. Eisner, 'Cantharidin: potent feeding deterrent to insects.' *Science*, 1974, **183**, 755–757.

124. Y. M. Li and J. E. Casida, 'Cantharidin-binding protein: identification as protein phosphatase 2A.' *Proc. Natl. Acad. Sci. USA*, 1992, **89**, 11867–11870.

125. R. E. Honkanen, 'Cantharidin, another natural toxin that inhibits the activity of serine/threonine protein phosphatases types 1 and 2A.' *FEBS Lett.*, 1993, **330**, 283–286.

126. Y. M. Li, C. MacKintosh and J. E. Casida, 'Protein phosphatase 2A and its [³H] cantharidin/[³H] endothall thioanhydride binding site. Inhibitor specificity of cantharidin and ATP analogues.' *Biochem. Pharmacol.*, 1993, **46**, 1435–1443.

127. R. Eldridge and J. E. Casida, 'Cantharidin effects on protein phosphatases and the phosphorylation state of phosphoproteins in mice.' *Toxicol. Appl. Pharmacol.*, 1995, **130**, 95–100.

128. M. Matsuzawa, M. J. Graziano and J. E. Casida, 'Endothal and cantharidin analogues: Relation of structure to herbicidal activity and mammalian toxicity.' *J. Agric. Food Chem.*, 1987, **35**, 823–829.

129. M. J. Graziano, I. N. Pessah, M. Matsuzawa *et al.*, 'Partial characterization of specific cantharidin binding sites in mouse tissues.' *Mol. Pharmacol.*, 1988, **33**, 706–712.

130. A. A. Brimfield, 'Possible protein phosphatase inhibition by bis(hydroxyethyl)sulfide, a hydrolysis product of mustard gas.' *Toxicol. Lett.*, 1995, **78**, 43–48.

131. M. M. King and C. Y. Huang, 'The calmodulin-dependent activation and deactivation of the phosphoprotein phosphatase, calcineurin, and the effect of nucleotides, pyrophosphate, and divalent metal ions. Identification of calcineurin as a Zn and Fe metalloenzyme.' *J. Biol. Chem.*, 1984, **259**, 8847–8856.

132. J. P. Griffith, J. L. Kim, E. E. Kim *et al.*, 'X-ray structure of calcineurin inhibited by the immunophilin-immunosuppressant FKBP12–FK506 complex.' *Cell*, 1995, **82**, 507–522.

133. H. M. Wisniewski, J. A. Sturman and J. W. Shek,

'Aluminum chloride induced neurofibrillary changes in the developing rabbit: a chronic animal model.' *Ann. Neurol.*, 1980, **8**, 479–490.

134. J. F. Leterrier, D. Langui, A. Probst *et al.*, 'Molecular mechanism for the induction of neurofilament bundling by aluminum ions.' *J. Neurochem.*, 1992, **58**, 2060–2070.

135. T. B. Shea and I. Fischer, 'Aluminum-induced cytoskeletal abnormalities in PC12 cells.' *Neurosci. Res. Commun.*, 1991, **9**, 21–26.

136. T. B. Shea, M. L. Beermann and R. A. Nixon, 'Aluminum alters the electrophoretic properties of neurofilament proteins; role of phosphorylation state.' *J. Neurochem.*, 1992, **58**, 542–547.

137. K. T. Shetty, Veeranna and S. C. Guru, 'Phosphatase activity against neurofilament proteins from bovine spinal cord: effect of aluminium and neuropsychoactive drugs.' *Neurosci. Lett.*, 1992, **137**, 83–86.

138. G. W. Goldstein, 'Lead poisoning and brain cell function.' *Environ. Health Perspect.*, 1990, **89**, 91–94.

139. M. Kern and G. Audesirk, 'Inorganic lead may inhibit neurite development in cultured rat hippocampal neurons through hyperphosphorylation.' *Toxicol. Appl. Pharmacol.*, 1995, **134**, 111–123.

140. J. Y. Chiou and E. W. Westhead, 'Okadaic acid, a protein phosphatase inhibitor, inhibits nerve growth factor-directed neurite outgrowth in PC12 cells.' *J. Neurochem.*, 1992, **59**, 1963–1966.

141. C. Arias, N. Sharma, P. Davies *et al.*, 'Okadaic acid induces early changes in microtubule-associated protein 2 and phosphorylation prior to neurodegeneration in cultured cortical neurons.' *J. Neurochem.*, 1993, **61**, 673–682.

142. M. G. Sacher, E. S. Athlan and W. E. Mushynski, 'Okadaic acid induces the rapid and reversible disruption of the neurofilament network in rat dorsal root ganglion neurons.' *Biochem. Biophys. Res. Commun.*, 1992, **186**, 524–530.

143. D. P. Alfano, J. C. LeBoutillier and T. L. Petit, 'Hippocampal mossy fiber pathway development in normal and postnatally lead-exposed rats.' *Exp. Neurol.*, 1982, **75**, 308–319.

144. D. P. Alfano and T. L. Petit, 'Neonatal lead exposure alters the dendritic development of hippocampal dentate granule cells.' *Exp. Neurol.*, 1982, **75**, 275–288.

145. J. B. Campbell, D. E. Woolley, V. K. Vijayan *et al.*, 'Morphometric effects of postnatal lead exposure on hippocampal development of the 15-day-old rat.' *Brain Res.*, 1982, **255**, 595–612.

146. L. Altmann, F. Weinsberg, K. Sveinsson *et al.*, 'Impairment of long-term potentiation and learning following chronic lead exposure.' *Toxicol. Lett.*, 1993, **66**, 105–112.

147. W. J. Brooks, T. L. Petit, J. C. Leboutillier *et al.*, 'Differential effects of early chronic lead exposure on postnatal rat brain NMDA, PCP, and adenosine A_1 receptors: an autoradiographic study.' *Drug Dev. Res.*, 1993, **29**, 40–47.

148. S. M. Lasley, J. Polan-Curtain and D. L. Armstrong, 'Chronic exposure to environmental levels of lead impairs *in vivo* induction of long-term potentiation in rat hippocampal dentate.' *Brain Res.*, 1993, **614**, 347–351.

149. J. M. Davis, D. A. Otto, D. E. Weil *et al.*, 'The comparative developmental neurotoxicity of lead in humans and animals.' *Neurotoxicol. Teratol.*, 1990, **12**, 215–229.

150. J. Cohn and D. A. Cory-Slechta, 'Subsensitivity of lead-exposed rats to accuracy-impairing and rate-altering effects of MK-801 on a multiple schedule of repeated learning and performance.' *Brain Res.*, 1993, **600**, 208–218.

151. J. M. Davis and D. J. Svendsgaard, 'Lead and child development.' *Nature*, 1987, **329**, 297–300.

152. S. K. Cummins and L. R. Goldman, 'Even advantaged children show cognitive deficits from low-level lead toxicity.' *Pediatrics*, 1992, **90**, 995–997.

153. D. C. Bellinger and K. M. Stiles, 'Epidemiologic approaches to assessing the developmental toxicity of lead.' *Neurotoxicology*, 1993, **14**, 151–160.

154. R. A. Goyer, 'Lead toxicity: current concerns.' *Environ. Health Perspect.*, 1993, **100**, 177–187.

155. H. L. Needleman, 'The current status of childhood low-level lead toxicity.' *Neurotoxicology*, 1993, **14**, 161–166.

156. K. M. Stiles and D. C. Bellinger, 'Neuropsychological correlates of low-level lead exposure in school-age children: a prospective study.' *Neurotoxicol. Teratol.*, 1993, **15**, 27–35.

157. G. F. Carl, C. L. Keen, B. B. Gallagher *et al.*, 'Association of low blood manganese concentrations with epilepsy.' *Neurology*, 1986, **36**, 1584–1587.

158. T. J. Resink, B. A. Hemmings, H. Y. L. Tung *et al.*, 'Characterisation of a reconstituted Mg-ATP-dependent protein phosphatase.' *Eur. J. Biochem.*, 1983, **133**, 455–461.

159. S. C. B. Yan and D. J. Graves, 'Inactivation and reactivation of phosphoprotein phosphatase.' *Mol. Cell Biochem.*, 1982, **42**, 21–29.

160. D. L. Brautigan, C. Picton and E. H. Fischer, 'Phosphorylase phosphatase complex from skeletal muscle. Activation of one of two catalytic subunits by manganese ions.' *Biochemistry*, 1980, **19**, 5787–5794.

161. E. Villa-Moruzzi, L. M. Ballou and E. H. Fischer, 'Phosphorylase phosphatase. Interconversion of active and inactive forms.' *J. Biol. Chem.*, 1984, **259**, 5857–5863.

162. H. Eriksson, P. G. Gillberg, S. M. Aquilonius *et al.*, 'Receptor alterations in manganese intoxicated monkeys.' *Arch. Toxicol.*, 1992, **66**, 359–364.

163. D. Mergler, 'Manganese: the controversial metal— at what levels can deleterious effects occur?' *Can. J. Neurol. Sci.*, 1996, **23**, 93–94.

164. C. C. Huang, C. S. Lu, N. S. Chu *et al.*, 'Progression after chronic manganese exposure.' *Neurology*, 1993, **43**, 1479–1483.

165. K. Nelson, J. Golnick, T. Korn *et al.*, 'Manganese encephalopathy: utility of early magnetic resonance imaging.' *Br. J. Indust. Med.*, 1993, **50**, 510–513.

166. K. F. Jensen, D. M. Lapadula, J. K. Anderson *et al.*, 'Anomalous phosphorylated neurofilament aggregations in central and peripheral axons of hens treated with tri-ortho-cresyl phosphate (TOCP).' *J. Neurosci. Res.*, 1992, **33**, 455–460.

167. J. Liu, J. D. Farmer Jr., W. S. Lane *et al.*, 'Calcineurin is a common target of cyclophilin–cyclosporin A and FKBP–FK506 complexes.' *Cell*, 1991, **66**, 807–815.

168. N. A. Clipstone and G. R. Crabtree, 'Identification of calcineurin as a key signaling enzyme in T-lymphocyte activation.' *Nature*, 1992, **357**, 695–697.

169. J. A. Gordon, 'Use of vanadate as protein-phosphotyrosine phosphatase inhibitor.' *Meth. Enzymol.*, 1991, **201**, 477–482.

11.11
Myelin and Myelination as Affected by Toxicants

PIERRE MORELL and ARREL D. TOEWS

University of North Carolina at Chapel Hill, NC, USA

11.11.1 MYELIN

A compound may be neurotoxic by virtue of being targeted to perturb some metabolic, structural, or functional property characteristic of myelin or myelinating cells. Myelin and myelination are described in this section, with emphasis on points relevant to understanding neurotoxic mechanisms. More information is readily available in a general review,[1] or in collections of more specialized reviews of subtopics relevant to myelin.[2,3]

11.11.1.1 Morphology

The gross distribution of myelin in the nervous system can be detected by the unaided eye. Most dramatically, a cross-section of brain or spinal cord reveals areas of "white matter." The glistening white appearance giving rise to this name is because of the high content of myelinated axons. This term is in contrast to "gray matter," areas of brain and spinal cord enriched in nerve cell bodies and their extensive dendritic arborizations. The concentration of myelin in brain increases as one ascends the evolutionary tree; in humans myelin accounts for half of the dry weight of white matter and, therefore, about a quarter of the dry weight of brain. The high concentration of myelin in nerves of the peripheral nervous system (e.g., sciatic nerve) is also obvious from their gross appearance. Peripheral nerves consist largely of

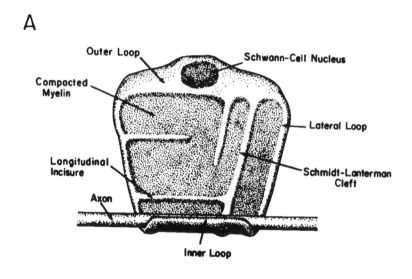

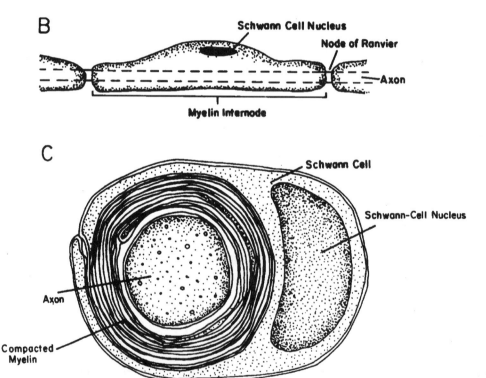

Figure 1   The myelin sheath of a peripheral nerve axon is formed by a Schwann cell. (A) Schematic drawing depicting the unrolled myelin sheath, showing its relationship to the cell body. Cytoplasmic channels course though myelin (Schmidt–Lanterman clefts) and are depicted as extending from the cytoplasm of the peri-karyon, with longitudinal incisures continuous with the thin loop of cytoplasm at the edge of the sheath (lateral loop). (B) Intact myelin sheath, (C) schematic cross-section (adapted from Morell and Norton[4]).

myelinated axons and bundles of smaller, unmyelinated axons.

Myelin is arranged as an interrupted electrical insulator along the axon. Each segment of myelin is an internode, with the periodic gaps between adjacent internodes being known as nodes of Ranvier. This periodically interrupted arrangement is critical to the function of myelin, which is to facilitate the movement of waves of depolarization along the axon. Sodium channels are concentrated at these nodes and, when activated, sodium ions flow through and depolarize the paranodal area of the axonal membrane. The local circuit generated cannot flow through the high-resistance myelin sheath and, therefore, depolarizes the membrane at the next node—which is of the order of 1 mm away. This discontinuous conduction (saltatory conduction) is much more rapid, and much more efficient, than the continuous movement of the wave of depolarization characteristic of unmyelinated axons. For example, in the 500 μm diameter unmyelinated giant axons of the squid (the model system initially used for deciphering the mechanisms of electrical excitability), waves of depolarization move at $25\,m\,s^{-1}$ at room temperature. In contrast, in frogs, information can be propagated at the same rate by a 12 μm diameter myelinated axon occupying 0.07% of the volume and requiring 0.02% as much energy. Clearly, it is difficult to conceive of a nervous system of any complexity which is not myelinated.

11.11.1.2 Cellular Origin and Development

Myelin of the peripheral nervous system (PNS) originates from the Schwann cells. During development, precursors of Schwann cells are associated with the developing nerve as it grows forth to make connections. Schwann cells fated to produce myelin (not all Schwann cells make myelin) line up along the axon. They occupy a domain along the axon and synthesize vast amounts of plasma membrane which is differentiated to form mature myelin as it wraps around the axon many times. Many of the morphological features of myelin, such as a suggestion of multilayer structure and presence of nodes of Ranvier, can be differentiated at the light microscopic level. A more detailed understanding of the structure, and observations concerning its formation during development, have been derived using electron microscopy and other physical techniques. The relationship between the Schwann cells and development of peripheral myelin is summarized in a highly schematic manner in Figure 1.

The cell involved in making myelin of the central nervous system (CNS) is the oligodendroglial cell. Superficially, a cross-section of a tract in the spinal cord or brain resembles that of the peripheral nervous system in that there are axons which are surrounded by myelin. There are, however, many points of difference relative to the situation in the PNS. Unlike the one-to-one relationship between the Schwann cell and a myelin internode, the oligodendroglial cell extends many processes and may myelinate segments of as many as 50 axons (although generally not more than one segment on any given axon). The possibility of deposition of myelin by rotation of the oligodendroglial cell body around an axon is therefore excluded. The relationship between the oligodendroglial cell and the many axons it may myelinate is shown in Figure 2.

It is evident that development of both CNS and PNS myelin is morphologically a complex and highly specialized process. The myelinating cell must make contact with the axon and some feature of the axon must indicate that it is appropriate for myelination. The myelinating cell must then synthesize an enormous amount of plasma membrane and, therefore, of the specialized components needed to form mature

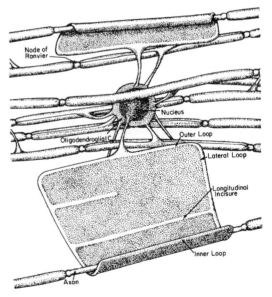

Figure 2 This is a schematic illustration of an oligodendroglial cell, the myelin-forming cell of the CNS. Oligodendroglial cells extend many processes to form multiple segments of myelin. The actual number of myelin segments formed on average is considerably greater than the 10 illustrated. One of the segments has been unwrapped to illustrate the continuity of myelin with the oligodendroglial cell plasma membrane. Longitudinal incisures and lateral loops are present. Adapted from Morell and Norton.[4]

myelin (see Section 11.11.1.3). The nature of the molecular motor which drags or pushes the growing sheath underneath the proceeding layer remains a mystery. There has long been evidence that maintenance of myelin–axon contact is an active process depending on availability of some signals from the axon. If the axon is injured or destroyed, the myelinating cell withdraws its support of the myelin sheath which is then rapidly degraded.

A distinction between PNS and CNS relates to capacity for remyelination. Myelinating cells in the periphery remain viable after a demyelinating insult and retain the ability to proliferate and differentiate into new myelinating cells. The oligodendroglial cells of the CNS are restricted with respect to capacity to remyelinate. This view has, however, undergone a challenge, and the extent to which regeneration of oligodendroglial cells and remyelination in the CNS can occur is the subject of research.[5,6] There is some evidence of remyelination in the CNS and a population of oligodendroglial cell precursors may persist in adulthood, even in humans.[7]

11.11.1.3 Composition

Myelin of the CNS is about 70% lipid and the composition of myelin of the PNS is even more biased, with almost 80% of it as lipid. This is in contrast to most plasma membranes which have a 50% or greater content of protein. The high proportion of lipid makes myelin preferentially vulnerable to toxicants which are hydrophobic in nature and can accumulate in the myelin sheath. The lipid composition of myelin of the CNS and PNS is similar qualitatively, although there are some quantitative differences. In each case, major lipids include cholesterol, cerebroside (galactosylceramide, a sphingolipid found in high concentrations only in the myelin sheath), and various phospholipids. The phospholipid composition is unusual in that most of the ethanolamine phosphatides are present as plasmalogens (glycerophospholipids in which the hydrocarbon moiety at the first carbon has a vinyl-ether linkage rather than the more generally found ester bond).

Although there are minor differences in lipid composition between CNS and PNS myelin, they are not as dramatic as the differences in structural protein composition. In the CNS an unusual and highly hydrophobic protein called proteolipid protein predominates, with myelin basic protein next in quantitative significance.

This latter protein exists in several forms as a result of alternative splicing of a common mRNA precursor. Myelin basic protein has long been of interest because it is antigenic and can be made the target of an experimentally-induced autoimmune disorder known as experimental allergic encephalomyelitis. This highly antigenic protein is partially hidden from the immune surveillance system and, if immune cells become sensitized to it by injection of the antigen, in the presence of nonspecific adjuvant, immune cells may attack myelin. Thus, it is possible that this antigen plays a role in natural autoimmune disorders; immune reaction to myelin basic protein may even exist as a complication of traumatic or toxicant-induced disorders in which the blood–brain barrier is breached and myelin is exposed to the immune system. Research[8] suggests myelin proteolipid protein may also be used to induce experimental allergic encephalomyelitis.

Peripheral myelin contains some of the same proteins as does myelin of the CNS. Proteolipid protein, however, is absent. Instead, the major protein present is P_0 protein which, surprisingly, has little in common with proteolipid protein in terms of its physical properties. These, and other myelin proteins, have been cloned and cDNA sequences are available. There is also some gene structure information for a number of them. A number of human and animal diseases are the result of mutations in some of these myelin protein genes. There is a suggestion that neurotoxicity of methylmercury is related to properties of a myelin-specific protein.[9]

Although the structural proteins of myelin account for the majority of the protein, there are a large number of proteins which, even though intrinsic to myelin, do not appear to have primarily structural roles. Some of these, such as myelin-associated glycoprotein (usually referred to as MAG), may have a role in recognition between Schwann cells and axons. Another important component is $2',3'$-cyclic nucleotide phosphodiesterase, so called because of its activity against an artificial substrate (a nonphysiological activity since naturally occurring cyclic nucleotides are of $3',5'$ structure). Studies suggest this protein may play a role in the cytoskeletal network supporting myelin[10] and, indeed, some cytoskeletal proteins (e.g., tubulin[11]) are present in myelin. Importantly, myelin also contains enzymes that function in lipid metabolism.[12] Of special interest with respect to mechanisms of neurotoxicity is the presence of carbonic anhydrase, which is postulated to have a role in ion transport, at the periaxolemmal domain.[13]

11.11.1.4 Metabolism

A noteworthy feature of myelin metabolism is the slow rate of turnover of certain of its components. Myelinating cells are similar to neurons in that, once differentiated, most of them survive for the life of the organism and are not easily replaced as are, for example, hepatocytes. The perikaryon of a myelinating cell is usually quite small relative to the mass of myelin it supports. Furthermore, the specialized structure of myelin, with cytoplasm excluded from the tightly compacted apposing cytoplasmic faces, suggests that degradation and replacement of membrane components is limited by restraints on the physical accessibility of membrane to degradative enzymes, and of access to the systems needed for replacement of lipids and proteins. Metabolic data from animal models indicates that in the CNS, proteolipid protein and myelin basic protein, as well as cholesterol and cerebroside, have half-lives of the order of months or longer.[14] Thus, hydrophobic toxicants can accumulate in this lipid-rich membrane over a long period of time.

The comments regarding stability of some lipids and proteins do not imply that myelin is metabolically inert. To the contrary, most components of myelin are involved in metabolic turnover. In the CNS, the glycerophospholipids turn over with half-lives ranging from days to many weeks. The turnover of any particular myelin glycerophospholipid is a function both of its head-group structure and its fatty acid composition. Phospholipids differing only in that one fatty acid is a few carbons longer than the other may have markedly different metabolic half-lives.[15] Also noteworthy is that there is a brisk metabolism, with a half-life of the order of minutes, of the monoesterified phosphates of glycerophosphoinositides.[16] Similarly, although the peptide backbone of major proteins may be relatively stable, some moieties involved in their post-translational modification may be rapidly metabolized. For example, many of the phosphate groups on myelin basic protein turn over with a half-life of the order of minutes or

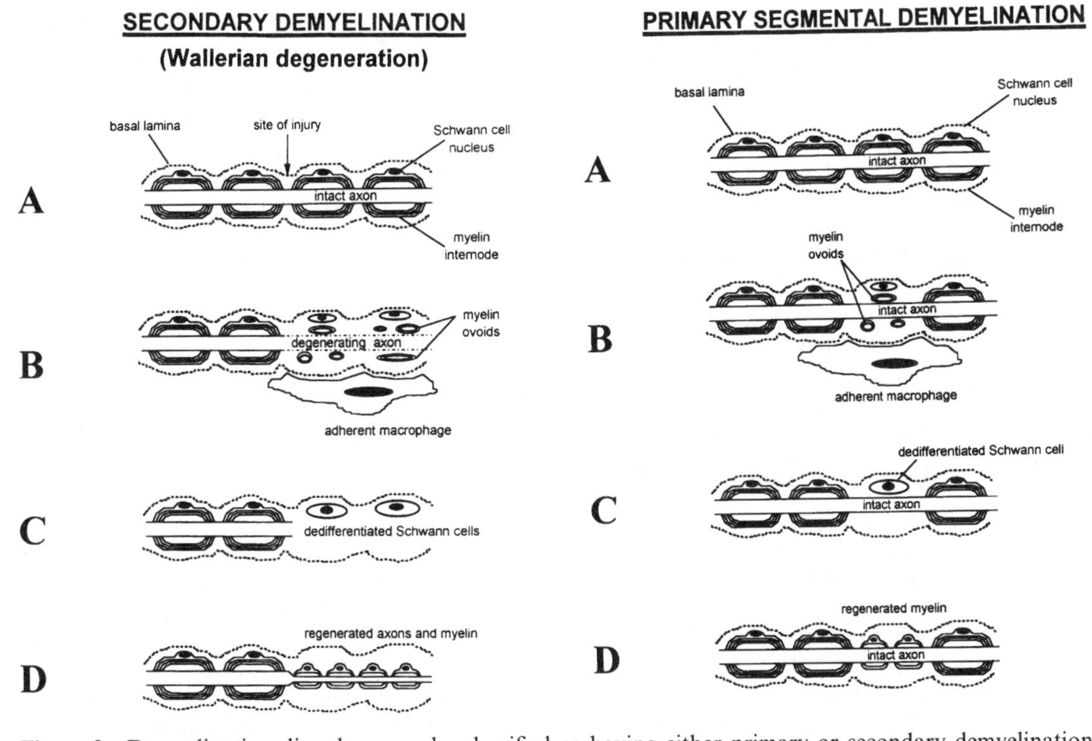

Figure 3 Demyelinating disorders may be classified as having either primary or secondary demyelination. Secondary demyelination (left panel) occurs subsequent to initial damage to the underlying axon. Axonal degeneration subsequent to a nerve cut or crush, with its associated secondary demyelination, is sometimes termed Wallerian degeneration. Axonal regeneration is a prerequisite for remyelination in this case. Primary demyelination (right panel) involves initial damage to the myelin itself or to the myelinating cell supporting it. Note that in this case, the underlying axon remains intact. If only occasional myelin internodes are lost, the condition is termed a primary segmental demyelination. In both types of demyelination, macrophages invade the damaged tissue and clear the debris. This is followed by a period of Schwann cell proliferation, differentiation of these cells to a myelinating phenotype, and remyelination of the axons.

less.[17] What biological processes might be tied in to these rapid metabolic events? One possible role of this rapid metabolism may relate to active extrusion of water from between the faces of the myelin sheath, thus, maintaining compaction. The mechanism relating this phosphate turnover to ion transport, assuming such exists, has not been elucidated.

Another aspect of rapid metabolism in myelin may relate to the observations that myelinating cells, originally conceived of as existing primarily to offer metabolic support for myelin, probably interact with neurons at the level of "neurotransmitter" recognition. Myelinating Schwann cells have receptors for ATP; their occupation (presumably by release of ATP from neurons) initiates a message cascade involving mobilization of calcium into the cytoplasm.[18,19] Oligodendroglial cells have been shown to contain cholinergic and purinergic receptors, among others, and also have the ability to mobilize calcium.[20] There is a body of work suggesting that cholinergic receptors exist on myelin[21,22] and G-proteins are also present.[23] Glial cells contain ion channels[24] and these may be present in myelin.[25]

11.11.2 CLASSIFICATIONS OF TOXIC DISORDERS OF MYELIN

Disorders of myelin are traditionally classified in morphological terms. The initial decision to be made is whether a demyelination is primary or secondary (Figure 3). As mentioned previously, damage to the neuron, or its axon, is rapidly followed by demyelination. Operationally, classification as a secondary demyelination is made on the basis of damage to the neuron or axon being observed prior to signs of demyelination. This designation is in contrast to primary demyelination (direct attack on myelin or myelinating cells), in which case damage to myelinating glia or directly to myelin is observed prior to evidence of damage in neurons or axons. Especially with respect to the PNS, in which myelinated axons may be "teased" apart and examined longitudinally, information is available from the distribution of demyelinating internodes. If, following a systemic insult, only some internodes are demyelinated, this is segmental demyelination and there is presumed to be a primary effect on myelin or Schwann cells. It is also sometimes the case that demyelination is seen as proceeding from the distal end of the axon towards the proximal end. This is generally taken as evidence for a secondary demyelination in which the axon is dying back from the point of its furthest reach, with demyelination as an obligatory consequence. Another distinction

made is between damage directly to myelin and damage to the myelinating cell; in the peripheral nervous system, the specific name Schwannopathy is given to this classification.

In many cases however, the above noted designation as primary or secondary demyelination involves a rather subjective evaluation. The "first" signs of damage to be visible are subtle, their quantitative evaluation is difficult, and the location of the underlying insult is usually not obvious. Furthermore, it seems likely that many toxicant-induced events may have elements of both primary and secondary demyelination.

A potentially important category of myelin disorders not covered by the above classification is hypomyelination, a term signifying failure of myelin to accumulate to normal levels during development. Hypomyelination is difficult to define quantitatively by morphological techniques unless the deficit is very severe. Work with animal models, in which decreased yield of myelin can be determined by chemical assays, suggests there can be a number of causes of such disorders, but it is difficult to gauge the extent of hypomyelination in the human population. The term dysmyelination is also often encountered in the literature. Strictly speaking, a dysmyelinating disorder is an inborn error of metabolism in which a catabolic defect causes abnormal accumulation of some myelin component. The consequent perturbation in myelin composition causes a collapse of the ordered myelin structure and a resulting myelin deficit. The term dysmyelination is, however, sometimes used less specifically to suggest any metabolic disorder of myelin.

11.11.3 ASPECTS OF VULNERABILITY OF MYELIN TO TOXICANTS

In order for a compound to be myelinotoxic, it presumably targets some specialized feature of myelin biology or chemistry. Considerations of the pathophysiology of lesions induced by such a toxicant include:

(i) Accessibility to nervous system—to access the CNS the toxicant must cross the blood–brain barrier. In order to do so it must either be lipid-soluble, undergo transport across the barrier by some existing transporter system, or have a mechanism of action which involves lesioning of the barrier. Similar considerations pertain with respect to the PNS and the blood–nerve barrier.

(ii) Vulnerable period of development—a relatively nonspecific toxicant, once it enters the nervous system, may preferentially perturb whatever developmental processes are most active at the time of exposure. There are two stages during development when myelination is preferentially vulnerable. The first is a period of rapid proliferation of precursors of myelinating cells. In the CNS this occurs in different regions at different times—in rats most of this occurs during the first 2 postnatal weeks. There is also a later period when these cells are particularly stressed because of the need to synthesize large amounts of myelin in a limited period of time; in rats, for example, although there is some accumulation of myelin throughout life, the peak rate of accumulation of myelin in the brain is at about postnatal day 20. Myelination is particularly vulnerable at these times. However, another reason for lower than normal levels of myelin is that metabolic insult earlier in development may preferentially impair the size and number of axons. Toxicant-induced insults with this result may occur even earlier in development. A secondary result of such damage is decreased myelin accumulation.

(iii) Solubility of toxicants in lipid—a lipid-soluble toxicant both gains access to the nervous system and may become concentrated in the lipid-rich myelin sheath. Several examples of lipid solubility being a contributing factor in myelin toxicity are discussed.

(iv) Metabolic pathways related to the specialized chemical composition of myelin may be a target. Myelinating cells must synthesize large quantities of specialized lipids and at least one toxicant (tellurium) is known to cause demyelination by interfering with such synthesis.

(v) Other metabolic functions specialized in myelinating cells may be targeted. This includes, for example, the active pumping mechanism which is presumed to keep fluids from accumulating at the intraperiod line of the compact myelin sheath; toxic mechanisms may interfere with this process.

(vi) Other points of vulnerability—we do not yet understand enough about the process of myelin formation to define likely sites of toxicant action. It is well known that there are highly specific molecular interactions involved in recognition between the myelinating cell and axon; understanding of the mechanisms involved is just beginning to accumulate. As indicated earlier, even in the mature axon-myelin unit there must be some active communication between the partners since damage to the axon rapidly results in damage to myelin. This communication may involve neuroligands; as noted earlier, there are specific neuroligand receptors on the surface of Schwann cells and oligodendroglial cells. The possibility that a toxicant might preferentially affect such communication is noted with respect to Vigabatrin

(see Section 11.11.4.9). There is also evidence that astrocytes may play a role in regulation of myelination by oligodendrocytes,[25] so toxic insults to astrocytes might ultimately lead to alterations in myelin.

As will become evident, there are few examples where a single neurotoxic mechanism can unambiguously be isolated as the primary contributor to a toxicant-induced myelinopathy. In many cases it is evident that several of these factors are involved. Nevertheless, it is useful to consider these parameters when considering the pathophysiology of a particular neurotoxicant.

11.11.4 SPECIFIC EXAMPLES OF TOXICANTS WHICH AFFECT MYELIN

The following toxicants were selected for comment because there are reasonable hypotheses suggestive of their modes of action. Many of these points were raised in an earlier review by one of the authors[27] and in another survey of the literature.[28]

11.11.4.1 Hexachlorophene

Hexachlorophene (2,2′-methylenebis(3,4,6-trichlorophenol)) was for many years used at high concentrations as an antimicrobial agent in certain soaps and detergents. Hexachlorophene is very hydrophobic, a property enabling it to be absorbed through the skin and also promoting its accumulation in the nervous system. This property, coupled with the practice (common in the 1950s) of bathing newborn babies in solutions containing hexachlorophene to prevent bacterial infection, led to untoward consequences.[29–31] A model neurotoxicity in adult rats was produced by chronic exposure through a diet containing 500 ppm hexachlorophene; progressive paralysis developed over several weeks.[32] The myelin edema due to splitting of the intraperiod line which is observed in humans[33] is duplicated in the animal models. Both CNS and PNS are affected.[34] Insight into the mechanism of action obtained by Cammer and co-workers[28] still seem appropriate. Hexachlorophene is concentrated in myelin and even low concentrations of this substance are sufficient to uncouple oxidative phosphorylation.[35] This presumably diminishes availability of energy for exclusion of water from the intraperiod line. Cammer et al.[36] also demonstrated that hexachlorophene brings about direct inhibition of carbonic anhydrase activity, again potentially inhibiting the presumptive mechanism for exclusion of water from myelin. Hexachlorophene is also toxic in young rats even before there is sufficient myelin to cause localized accumulation (about 15 d of age). This toxicity may be due to many of the same factors as target it towards myelin—accumulation in membranes, interference with ion transport, and so on—but distributed generally over other brain membranes.

11.11.4.2 Triethyltin

The toxicity of triethyltin in humans was dramatically thrust into the spotlight with the 100 deaths reported as a result of contamination of a medication by triethyltin.[37,38] Administering rats drinking water containing triethyltin has provided a model system for study of this disorder. Swelling of myelin due to splitting of the intraperiod line[39] is a readily differentiable morphological feature, and is clearly relevant to the pathophysiology (Figure 4). In this model about a quarter of the isolatable myelin may be lost over a period of several weeks[40,41] and there is a corresponding loss in myelin-specific components.[40] Myelin of the PNS is affected only at dosage regimens more aggressive than those sufficient to induce CNS pathology.[42] Exposure of developing rats to triethyltin results in decreased brain weights and even more marked deficits in myelin, suggesting a preferential effect on myelination.[43] Myelin deficits resulting from such developmental exposure persist into adulthood.[44] In metabolic studies it has been shown that synthesis of myelin lipids is depressed initially, but there may be a increase (perhaps compensatory in nature) after several weeks.[41] Myelin protein synthesis is also perturbed[41,43] as is the level of mRNA for myelin basic protein.[45]

Considerable information is available on which to base a mechanism for triethyltin neurotoxicity. This molecule would be expected to concentrate in myelin due to its hydrophobicity. The actual specificity is greater than suggested just by lipid solubility; Lock and Aldridge[47] have identified a high-affinity binding site in myelin for binding of triethyltin. What is the relationship between this localized reservoir of triethyltin and the observed swelling of myelin? Triethyltin may interfere in two ways to cause this failure in osmotic regulation. A postulated failure in oxidative phosphorylation[48] could be mediated by a specific inhibition of mitochondrial Mg^{2+}-ATPase.[49] Triethyltin also acts as an ionophore, demonstrated to be able to catalyze Cl^-/OH^- exchange.[50] The relationship of this to inhibition of ATP production is discussed by Lock[51]

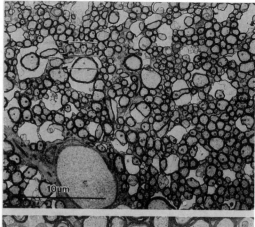

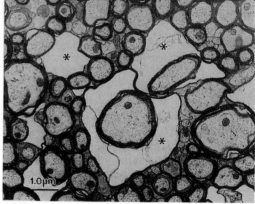

Figure 4 (Upper panel) Electron micrograph of spinal cord white matter from a 58-day-old mouse, 12 h after intraperitoneal injection of triethyltin at a dosage level of $10\,mg\,kg^{-1}$.[46] Intramyelinic edema is evident as electron-lucent spaces between myelin lamellae of many myelinated nerve fibers. The underlying axons appear normal. (Lower panel) Higher magnification of the area indicated by arrows in the upper panel. * indicates intramyelinic vacuoles formed by splitting of the myelin lamellae along the intraperiod line (photographs courtesy of Dr. Kinuko Suzuki).

and Cammer.[28] Another mechanism of action of triethyltin is on a mitochondrial inner-membrane anion channel.[52] Finally, triethyltin may act directly to allow Cl^- entry into the intraperiod line region, thereby, bringing in water to cause the observed swelling.

The general vulnerability of developing brain to triethyltin is, therefore, presumably due to targeting of mitochondrial function which is essential to support the high metabolic activity of brain. Although this toxicity is observed during earlier development, the disorder becomes more focused when enough myelin has accumulated so that triethyltin preferentially accumulates at this location and causes a more discrete pathophysiology. By both indirect means (inhibition of ATP formation) and direct action on myelin, triethyltin interferes with the energy-dependent exclusion of fluid from the intraperiod line. The fluid accumulation can be demonstrated by x-ray diffraction techniques.[53] The suggestion that it is the failure of osmotic regulation which is centrally involved in the pathophysiology receives support from the observation that recovery from triethyltin-induced edema is blocked by acetazolamide, a blocker of carbonic anhydrase activity. Carbonic anhydrase is presumed to be involved in osmotic control both within the myelin sheath[36,52] as well as in glial cells.[13,55]

11.11.4.3 Triethyllead

Although there are similarities in chemical properties, triethyllead given to adult rats does not target myelin as specifically as does triethyltin. Triethyllead given to young rats does block accumulation of myelin.[56] Animals treated in this manner have a deficit in synthesis of myelin proteins, relative to total protein synthesis.[57] Although the amount of myelin synthesized is less than normal, it is of normal composition.[58] The apparent preference for inhibiting myelination may be a combination of some preferential localization of the toxicant due to hydrophobicity, and the fact that myelination is such an active ongoing process and therefore more susceptible to perturbation. It is also the case that hypomyelination may not indicate much specificity for myelin, but might be secondary to interference with later differentiation of neurons. A decrease in the number of axons, or their average diameter, would bring about hypomyelination as a secondary result.

11.11.4.4 Lead

In humans, the symptoms of lead toxicity are not initially specific to myelin. Chronic lead intoxication results in myelin degeneration that is clearly secondary to axonal degeneration. It is only in certain animal models, when segmental demyelination of the PNS can be observed, that specificity for myelinating cells can be addressed. Most commonly used are models in which lead salts are administered in the drinking water or by intraperitoneal injection.[59–63] Initial changes observed include Schwann cell hypertrophy.[64] Splitting of myelin occurs at both the intraperiod and the major dense lines, in contrast to some of the models mentioned above in which the intraperiod lines are preferentially affected. Chronic dosing with lead acetate ($400 \, mg \, kg^{-1} \, d^{-1}$ by gavage) during development decreases accumulation of brain myelin relative to nourishment-matched controls[65] and this decrement in myelin persists into adulthood.[44]

There is no clear-cut hypothesis concerning specificity of the toxic effect for Schwann cells or PNS myelin. The competition of lead with calcium ions with respect to energy requiring transport has been postulated to play a role in a number of toxic affects of lead,[66] and this may be the case with respect to damage to myelin.

11.11.4.5 Methylmercury

Methylmercury inhibits many neurobiological processes[67] but as with lead, there is an indication that under some conditions there may be some specificity for myelination. There are data[68] suggesting that in chronically-treated, developing animals there may be some preferential depression of activity of enzymes associated with myelin biosynthesis. Acute dosing with methylmercury ($10 \, mg \, kg^{-1} \, d^{-1}$ of mercury chloride by injection) may inhibit the normal phosphorylation of proteins of peripheral myelin. It has been demonstrated[9] that the sciatic nerve of several species contains a glycoprotein of molecular weight 21 kDa which has a high affinity for methylmercury. There are two species of this glycoprotein. The N-terminal sequences are related to those of P_0 protein and PMP-22 (a characterized minor protein component of PNS myelin), respectively.

11.11.4.6 Cuprizone

The copper chelator Cuprizone (biscyclohexanoneoxalyhydrazone) causes CNS demyelination in weanling mice. The loss of myelin in the cerebrum approaches 70%, as determined by loss of galactosylceramide.[69] Since Cuprizone renders unavailable the copper needed for cytochrome oxidase activity, it presumably reduces ATP levels, therefore, depressing energy-driven ion transport. The reason for its specificity for myelin is not clear. It is not markedly hydrophobic. Perhaps specificity is at the level of inhibition of carbonic anhydrase activity, which takes place well before demyelination is observed.[70]

The remarkable specificity for loss of myelin, coupled with the fact that this is one of the few models in which there is extensive remyelination after a CNS demyelinating event, make this system popular as a model for study of demyelination. A study has been conducted[71] of the effects of Cuprizone treatment on steady-state levels of mRNA for myelin-associated glycoprotein—a molecule, presumed to be involved in the axon–glial cell recognition process. Both spliced variants of this message are downregulated during demyelination and approach or exceed normal levels during the remyelinating phase. A study by Komoly et al.[72] focused on insulin-like growth factor-1 gene expression. Upregulation of the production of mRNA for this protein, and of the peptide produced from this message, took place in astrocytes in areas correlating with myelin breakdown. During recovery, levels of the peptide decreased. In contrast, receptor for insulin-like growth factor was increased in immature oligodendroglia during the recovery period. Although alternative interpretations of these data are possible, this suggests the possibility that myelin metabolism is, in part, regulated by astrocytes and that it was the astrocyte which was the original target for neurotoxicity, with myelin damage being a secondary consequence.

11.11.4.7   2′,3′-Dideoxycytidine

Dideoxycytidine (ddC) has undergone clinical trials as a possible chemotherapeutic agent for use in the treatment of AIDS. A complication of its use is a dose-limiting peripheral neuropathy. Although in humans there does not seem to be a targeting towards myelin, a rabbit model developed by Anderson et al.[73] exhibits specificity for myelin. At a dose-level of above $50\,\mathrm{mg\,kg^{-1}\,d^{-1}}$ by oral intubation, peripheral nerves exhibit myelin splitting and edema. These authors followed the development of the disease and suggested that the observed axonopathy is actually secondary to the myelinopathy. The mechanisms for targeting myelin or myelinating cells is obscure. It is its activity as an inhibitor of reverse transcriptase which suggests its utility in AIDS therapy. The rabbit neuropathy is, however, clearly related to damage to mitochondria of Schwann cells.[74] There may be targeting of the mitochondrial DNA polymerase. Changes in metabolism of the major myelin protein, P_0, have also been observed.[74]

11.11.4.8   Tellurium

Tellurium, element 52, is a metalloid used at low concentrations in the formulation of various steel, rubber, and ceramic products, among others. Tellurium toxicity is rare,[75] but it is of interest as a model system for study of the type of metabolic insult which can lead to specific demyelination. Administration to weanling rats of a diet containing about 1% by weight of elemental tellurium induces a neuropathy within several days. The axons are not affected; the neurological manifestations are clearly the result of a segmental demyelination of peripheral nerves involving up to 20% of the internodes.[76–79] Since a single demyelinated internode is sufficient to alter or block a wave of depolarization moving down the axon, longer nerves are affected preferentially, and hindlimb neuropathy is the primary initial symptom. The largest myelin internodes are the most vulnerable.[80] Tellurium-induced demyelination is a developmental phenomena in that only young animals undergo demyelination; adults given tellurium are relatively resistant. The CNS is not generally affected although, in some species, high levels of dosing over a prolonged period of time can induce pathological alterations.

Ex vivo studies[81,82] suggest that the demyelination is secondary to a block in cholesterol biosynthesis. *In vitro* studies demonstrate that tellurite (tellurium at the +4 oxidation state), at micromolar concentrations, inhibits squalene epoxidase, an enzyme in the cholesterol biosynthesis pathway. This is a mixed-function oxidase which receives its necessary reducing equivalents from NADPH via NADPH cytochrome P450 reductase. Other mixed function oxidases, such as benzo(a)pyrene and aniline hydroxylase are not inhibited at concentrations of tellurite which severely inhibit squalene epoxidase activity.[83]

Since the metabolic lesion is systemic (squalene epoxidase activity is initially inhibited in all organs), why is there preference for the PNS demyelination? The period of greatest vulnerability is at weaning—rats go on to solid food and initiate a period of very rapid growth. The sciatic nerve stretches greatly to accommodate for this growth. As the internodes lengthen there is a need for very rapid deposition of myelin and consequent need for cholesterol. In fact, the demand for cholesterol in sciatic nerve of postweaning animals is an order of magnitude greater than that of the CNS,[84] thus, accounting for the preferential vulnerability of the PNS.

Another factor in the targeting of the toxic effects of tellurium to the nervous system is that, in response to the tellurium challenge, the liver, but not sciatic nerve, greatly upregulates the cholesterol biosynthesis pathway and maintains normal levels of cholesterol production and of circulating cholesterol. This is because the primary rate-limiting enzyme in the pathway of cholesterol biosynthesis, hydroxymethylglutaryl-CoA reductase, is greatly up-regulated—presumably a feedback loop related to the need for liver to maintain levels of circulating cholesterol.[85] Although this cholesterol can be utilized to varying extents by different organs, access to the nervous system is denied by the blood–brain and blood–nerve barriers.[86]

This model has given insight into processes controlling the assembly of myelin. Inhibition of cholesterol synthesis rapidly results in a decrease in levels of mRNA for "myelin genes," such as those for protein P_0, myelin basic protein, and UDP-galactose : ceramide-galactosyl transferase.[87] This downregulation takes place in all myelinating Schwann cells of the sciatic nerve, even though only about 20% of the internodes are involved in frank demyelination.[88] Thus, downregulation of the message for myelin components is not just a secondary response to injury, but rather reflects the normal coordinated control of gene expression. When the tellurium challenge is lifted, there is a rapid remyelination and an associated upregulation of the "myelin genes." In summary, tellurium feeding results in peripheral neuropathies because: (i) it inhibits biosynthesis of cholesterol, a component enriched in the myelin sheath; (ii) control of hydroxymethylglutaryl-CoA reductase is organ-specific, so the nervous system does not upregulate cholesterol synthesis in response to this challenge; (iii) the nervous system is isolated by barriers from access to circulating cholesterol; and (iv) there is a particular developmental stage of vulnerability at the time when the rate of synthesis of nerve myelin is high.

11.11.4.9 Vigabatrin

Vigabatrin (γ-vinyl-γ-aminobutyric acid) is an enzyme-activated, irreversible inhibitor of γ-aminobutyric acid (GABA) transaminase. An antiepileptic agent, this drug acts by increasing GABA levels and is without serious toxicity at therapeutically useful levels.[89] Similar dosage levels in rats and dogs result in intramyelin edema.[90] The edema is extensive enough so that it can be assayed *ex vivo* by magnetic resonance imaging.[91] The reason for susceptibility of oligodendroglial cells to this drug has not been definitively determined. It does, however, seem reasonable to suspect that the excess GABA levels induced by treatment with vigabatrin act via GABA receptors, since a subclass of these receptors appears to be specific to oligodendroglial cells[92] and it is known that cultured oligodendroglial cells are depolarized by GABA.[93]

11.11.5 INTERACTIONS BETWEEN MYELINATING GLIAL CELLS AND AXONS—BIOLOGICAL MARKERS?

During development, glial cells or their precursors must find axons and recognize signals directing myelination. Furthermore, the relationship between axons and glia is an active one since, as noted above, destruction of the axon causes collapse of the overlying myelin sheath and rapid removal of myelin debris by phagocytic cells. This implies that the axon must directly or indirectly communicate with the myelinating cell, a process which may involve cell–cell recognition through surface receptors and/or diffusion of soluble substances, such as neuroligands or growth factors. The processes of recognition and communication might themselves be targets for neurotoxicants. Close physical proximity of myelin and axon is essential for support of the biological functioning of this unit (a review on the relationship between Schwann cell and axon, focussing on aspects relevant to neurotoxicity is available[94]). Monitoring this relationship might provide a sensitive indicator of onset of neuropathy. This approach is not meant to challenge the primacy of morphological and, ultimately, behavioral studies in defining neurotoxicity. Biochemical screening methods, however, have the potential of being relatively quick and inexpensive as well as providing insight into mechanisms.

One biological marker which reflects the integrity of the myelin–axon relationship is nerve growth factor receptor (NGF-R), which is expressed in precursor Schwann cells and non-myelinating Schwann cells, but not in myelinating Schwann cells. In connection with this, the mRNA levels for NGF-R are normally very low in sciatic nerve, but are markedly upregulated in distal stumps of transected nerves as a secondary response to axonal degeneration. NGF-R message levels also increase during primary demyelination (no axonal degeneration) induced by exposure to tellurium.[88] In this model, partial paranodal demyelination is sufficient for upregulation, and newly synthesized NGF-R protein appears to be targeted to local areas of myelin loss. This suggests that increased NGF-R expression in Schwann cells occurs in response to alterations at the axon–Schwann cell interface.

The sensitivity of the assay suggested it might be of utility in monitoring subtle alterations in the axon–myelinating Schwann cell relationship, irrespective of whether the physical separation of the two units is a function of primary demyelination or axonopathy. Young rats were dosed with either isoniazid or acrylamide at doses known to eventually cause both behaviorally detectable neuropathy and a clear-cut pattern of morphological changes. Upregulation of NGF-R, however, was present prior to either morphological or behavioral disturbances.[95] In a later study it was also shown that chronic exposure to low doses of carbon disulfide upregulate NGF-R mRNA expression in sciatic nerve prior to morphological alterations.[96] Thus, it appears that assay of NGF-R mRNA is a rapid (relative to quantitative morphological assays) and sensitive screen for compounds which are toxic to the PNS.

An advantage of using mRNA as a marker, rather than immunoassay for the protein itself, is that once the RNA is prepared and separated on a gel, several probes can be assayed sequentially on the same sample. Upregulation of steady-state levels of NGF-R indicates that there is some ongoing neuropathic process. A parallel marked downregulation of some gene specific for myelination (e.g., P_0 protein or myelin basic protein) would suggest this is a primary demyelination. If the downregulation of genes for myelin components is not significant until the neuropathy is advanced, then the demyelination is probably secondary. A potentially more sensitive way of distinguishing primary from secondary demyelination is to determine whether there is a proximal-distal gradient of upregulation of NGF-R. In the dying back axonopathy induced by isoniazid, the initial marked elevation of NGF-R mRNA is, as expected, preferential to the distal region of the nerve.[95]

The use of NGF-R as a marker for a disturbed axonal–glial relationship may be valid only for the PNS—nerve regions can be dissected and mRNA for Schwann cell functions assayed. In brain or spinal cord sections the presence of message in neurons would complicate such analysis. Damage resulting in significant loss of cells or cell processes is detectable in the CNS by upregulation of glial fibrillary acidic protein, a marker of reactive gliosis.[97] Whether a more specific marker is available to determine deterioration of the oligodendroglial–axon relationship is worthy of further study.

cell types. Such considerations are, however, of central importance in the study of neurotoxicants. The intimate physical and functional relationships between various classes of neural cells means that toxic insult specific for one cell type may cause metabolic and other perturbations in other cell types on a very rapid time scale. Another consideration dictated by the complexity of the nervous system is that the tightly programmed development of structure and function extends the period of developmental vulnerability later than is the case for other organs. Furthermore, plasticity—in this context the ability to regenerate functionality after damage—is more restricted in nervous system than in other organs.

It is increasingly clear that an understanding of mechanisms of myelinotoxicity depends not only on knowledge of metabolic processes intrinsic to myelin formation; understanding of processes involved in cell–cell interactions is also crucial. In particular, the interactions between axons and the cells that synthesize and maintain myelin need to be better understood at the molecular level. Furthermore, it should be noted that, in the CNS, perinodal astrocytes are also in close physical proximity with the neurons and oligodendroglia. These cells contact axons at the node of Ranvier and have a metabolic relationship with the axon.[98] They may also interact with oligodendroglia. Microglia in the CNS and resident macrophages in the PNS may also play a role in modulating some aspects of the glia–axon relationship; they certainly participate in metabolic events following demyelination. Interaction between microglia and oligodendroglia is an emerging field of research;[99] see also a publication on microglia and prostaglandins.[100] Knowledge of mechanisms of neurotoxicity is limited by the boundaries of knowledge of how the nervous system is developed and how it operates. Fortunately, investigators are meeting this challenge with mechanistically based investigations of the mode of action of neurotoxicants, so that these increasingly provide information both about how the nervous system works in the normal state, as well as the consequences of perturbing the system with a toxicant.

11.11.6 CONCLUDING REMARKS AND FUTURE DIRECTIONS

Studies of mechanisms of toxicity almost routinely include consideration of interactions between various organ systems and different

ACKNOWLEDGMENTS

The authors thank R. Northcott for secretarial support. The authors have received funding for their studies of neurotoxicity from USPHS grants ES-01104 and NS-11615 in a center receiving core support from HD03110.

11.11.7 REFERENCES

1. P. Morell, R. H. Quarles and W. T Norton, in 'Basic Neurochemistry: Molecular, Cellular, and Medical Aspects,' 5th edn., eds. R. W. Albers, G. W. Siegel, P. Molinoff *et al.*, Raven Press, New York, 1994, pp. 117–143.

2. P. Morell (ed.) 'Myelin,' 2nd edn., Plenum, New York, 1984.

3. R. E. Martenson (ed.) 'Myelin: Biology and Chemistry,' CRC Press, Boca Raton, FL, 1992.

4. P. Morell and W. T. Norton, 'Myelin.' *Sci. Am.*, 1980, **242**, 88–90.

5. M. Dubois-Dalcq and R. Armstrong, 'The cellular and molecular events of central nervous system remyelination.' *Bioessays*, 1990, **12**, 569–576.

6. R. S. Vick, T. J. Neuberger and G. H. DeVries, 'Role of adult oligodendrocytes in remyelination after neural injury.' *J. Neurotrauma*, 1992, **9**, S93–S104.

7. R. C. Armstrong, H. H. Dorn, C. V. Kufta *et al.*, 'Pre-oligodendrocytes from adult human CNS.' *J. Neurosci.*, 1992, **12**, 1538–1547.

8. R. C. van der Veen, R. A. Sobel and M. B. Lees, 'Chronic experimental allergic encephalomyelitis and antibody responses in rabbits immunized with bovine proteolipid apoprotein.' *J. Neuroimmunol.*, 1986, **11**, 321–333.

9. S. Ozaki, T. Ichimura, T. Isobe *et al.*, 'Identification and partial characterization of a glycoprotein species with high affinity for methylmercury in peripheral nervous tissues of man and experimental animals.' *Arch Toxicol.*, 1993, **67**, 268–274.

10. D. A. De Angelis and P. E. Braun, 'Isoprenylation of brain 2′,3′-cyclic nucleotide 3′-phosphodiesterase modulates cell morphology.' *J. Neurosci. Res.*, 1994, **39**, 386–397.

11. I. Gozes and C. Richter-Landsberg, 'Identification of tubulin associated with rat brain myelin.' *FEBS Lett.*, 1978, **95**, 169–172.

12. R. W. Ledeen, 'Enzymes and receptors of myelin,' in 'Myelin: Biology and Chemistry,' ed. R. E. Martenson, CRC Press, Boca Raton, FL, 1992, pp. 531–570.

13. V. S. Sapirstein, R. Durrie, C. E. Nolan *et al.*, 'Identification of membrane-bound carbonic anhydrase in white matter coated vesicles: the fate of carbonic anhydrase and other white matter coated vesicle protein in triethyl tin-induced leukoencephalopathy.' *J. Neurosci. Res.*, 1993, **35**, 83–91.

14. J. A. Benjamins and M.E Smith, in 'Myelin,' 2nd edn., ed. P. Morell, Plenum, New York, 1984, pp. 225–258.

15. A. H. Ousley and P. Morell, 'Individual molecular species of phosphatidylcholine and phosphatidylethanolamine in myelin turn over at different rates.' *J. Biol. Chem.*, 1992, **267**, 10362–10369.

16. D. S. Deshmukh, S. Kuizon, W. D. Bear *et al.*, 'Rapid incorporation *in vivo* of intracerebrally injected $^{32}P_i$ into polyphosphoinositides of three subfractions of rat brain myelin.' *J. Neurochem.*, 1981, **36**, 594–601.

17. K. C. DesJardins and P. Morell, 'Phosphate groups modifying myelin basic proteins are metabolically labile; methyl groups are stable.' *J Cell Biol.*, 1983, **97**, 438–446.

18. S. A. Lyons, P. Morell, and K. D. McCarthy, 'Schwann cells exhibit P_{2Y} purinergic receptors that regulate intracellular calcium and are upregulated by cyclic AMP analogs.' *J. Neurochem.*, 1994, **63**, 552–560.

19. S. A. Lyons, P. Morell and K. D. McCarthy, 'Schwann cell ATP-mediated calcium increases *in vitro* and *in situ* are dependent on contact with neurons.' *Glia*, 1995, **13**, 27–38.

20. M. He and K. D. McCarthy, 'Oligodendroglial signal transduction systems are developmentally regulated.' *J. Neurochem.*, 1994, **63**, 501–508.

21. J. N. Larocca, R. W. Ledeen, B. Dvorkin *et al.*, 'Muscarinic receptor binding and muscarinic receptor-mediated inhibition of adenylate cyclase in rat brain myelin.' *J. Neurosci.*, 1987, **7**, 3869–3876.

22. D. W. Kahn and P. Morell, 'Phosphatidic acid and phosphoinositide turnover in myelin and its stimulation by acetylcholine.' *J. Neurochem.*, 1988, **50**, 1542–1550.

23. P. E. Braun, E. Horvath, V. W. Yong *et al.*, 'Identification of GTP-binding proteins in myelin and oligodendrocyte membranes.' *J. Neurosci. Res.*, 1990, **26**, 16–23.

24. B. A. Barres, 'New roles for glia.' *J. Neurosci.*, 1991, **11**, 3685–3894.

25. B. Cherksey, R. Durrie, P. E. Braun *et al.*, 'In vitro analysis of ion channels in periaxolemmal-myelin and white matter clathrin coated vesicles: modulation by calcium and GTP gamma S.' *Neurochem. Res.*, 1994, **19**, 1101–1106.

26. R. P. Skoff and P. E. Knapp, 'Myelin protein gene mutations: nurture overcomes nature.' *J. Neurochem.*, 1995, **64** (Suppl.), S82A.

27. P. Morell, in 'Principles of Neurotoxicology,' ed. L. W. Chang, Marcel Dekker, 1994, pp. 583–608.

28. W. Cammer, in 'Experimental and Clinical Neurotoxicology,' eds. P. S. Spencer and H. H. Schaumburg, Williams & Williams, Baltimore, MD, 1980, 239–256.

29. W. B. Herter, 'Hexachlorophene poisoning.' *Kaiser Found. Med. Bull.*, 1959, **7**, 228.

30. W. S. Gump, 'Toxicological properties of hexachlorophene.' *J. Soc. Cosmet. Chem.*, 1969, **20**, 173.

31. J. Towfighi, in 'Experimental and Clinical Neurotoxicology,' eds. P. S. Spencer and H. H. Schaumburg, Williams & Williams, Baltimore, MD, 1980, pp. 440–455.

32. R. D. Kimbrough, in 'Essays in Toxicology,' ed. W. J. Hayes, Jr., Academic Press, New York, 1976, vol. VII, pp. 99–120.

33. H. Powell, O. Swarner, L. Gluck *et al.*, 'Hexacholrophene myelinopathy in premature infants.' *J. Pediatr.*, 1973, **82**, 976–981.

34. J. Towfighi, N. K. Gonatas and L. McCree, 'Hexachlorophene-induced changes in central and peripheral myelinated axons of developing and adult rats.' *Lab. Invest.*, 1974, **31**, 712–721.

35. W. Cammer and C. L. Moore, 'The effect of hexachlorophene on the respiration of brain and liver mitochondria.' *Biochem. Biophys. Res. Commun.*, 1972, **46**, 1887–1894.

36. W. Cammer, T. Fredman, A. L. Rose *et al.*, 'Brain carbonic anhydrase: acitivity in isolated myelin and the effect of hexachlorophene.' *J. Neurochem.*, 1976, **27**, 165–171.

37. M. Rouzaud and J. Lutier, 'Oédeme subaigu cérébroméningé du a une intoxication d'actualité.' *Presse Médicale*, 1954, **62**, 1075.

38. I. Watanabe, in 'Experimental and Clinical Neurotoxicology,' eds. P. S. Spencer and H. H. Schaumburg, Williams & Williams, Baltimore, MD, 1980, pp. 545–557.

39. F. P. Aleu, R. Katzman and R. D. Terry, 'Fine structure and electrolyte analysis of cerebral edema induced by alkyl tin intoxication.' *J. Neuropathol. Exp. Neurol.*, 1963, **22**, 403.

40. Y. Eto, K. Suzuki and K. Suzuki, 'Lipid composition of rat brain myelin in triethyl tin-induced edema.' *J. Lipid. Res.*, 1971, **12**, 570–579.

41. M. E. Smith, 'Studies on the mechanism of demyelination: triethyl tin-induced demyelination.' *J. Neurochem.*, 1973, **21**, 357–372.

42. D. I. Graham and N. K. Gonatas, 'Triethyltin sulfate-induced splitting of peripheral myelin in rats.' *Lab. Invest.*, 1973, **29**, 628–632.

43. W. D. Blaker, M. R. Krigman, D. J. Thomas *et al.*, 'Effect of triethyl tin on myelination in the developing rat.' *J. Neurochem.*, 1981, **36**, 44–52.

44. A. D. Toews, W. D. Blaker, D. J. Thomas *et al.*, 'Myelin deficits produced by early postnatal exposure to inorganic lead or triethyltin are persistent.' *J. Neurochem.*, 1983, **41**, 816–822.

45. B. Veronesi, K. Jones, S. Gupta *et al.*, 'Myelin basic protein-messenger RNA (MBP-mRNA) expression during triethyltin-induced myelin edema.' *Neurotoxicology*, 1991, **12**, 265–276.

46. H. Nagara, K. Suzuki, C. W. Tiffany *et al.*, 'Triethyl tin does not induce intramyelinic vacuoles in the CNS of the quaking mouse.' *Brain Res.*, 1981, **225**, 413–420.

47. E. A. Lock, and W. N. Aldridge, 'The binding of triethyltin to rat brain myelin.' *J. Neurochem.*, 1975, **25**, 871–876.

48. W. N. Aldridge and J. E. Cremer, 'The biochemistry of organo-tin compounds: diethyltin dichloride and triethyltin sulfate.' *Biochem. J.*, 1955, **61**, 406.

49. J. S. Wassenaar, and A. M. Kroon, 'Effects of triethyltin on different ATPases, 5'-nucleotidase and phosphodiesterases in grey and white matter of rabbit brain and their relation with brain edema.' *Eur. Neurol.*, 1973, **10**, 349–370.

50. M. J. Selwyn, A. P. Dawson, M. Stockdale *et al.*, 'Chloride–hydroxide exchange across mitochondrial, erythrocyte and artificial lipid membranes mediated by trialkyl- and triphenyltin compounds.' *Eur. J. Biochem.*, 1970, **14**, 120–126.

51. E. A. Lock, 'The action of triethyltin on the respiration of rat brain cortex slices.' *J. Neurochem.*, 1976, **26**, 887–892.

52. M. F. Powers and A. D. Beavis, 'Triorganotins inhibit the mitochondrial inner membrane anion channel.' *J. Biol. Chem.*, 1991, **266**, 17250–17256.

53. D. A. Kirschner and V. S. Sapirstein, 'Triethyl tin-induced myelin oedema: an intermediate swelling state detected by x-ray diffraction.' *J. Neurocytol.*, 1982, **11**, 559–569.

54. D. S. Reiss, M. B. Lees and V. S. Sapirstein, 'Is Na$^+$-K$^+$-ATPase a myelin-associated enzyme?' *J. Neurochem.*, 1981, **36**, 1418–1426.

55. R. S. Bourke and H. G. Kimelberg, 'The effect of HCO$_3^-$ on the swelling and ion uptake of monkey cerebral cortex under conditions of raised extracellular potassium.' *J. Neurochem.*, 1975, **25**, 323–328.

56. G. Konat and J. Clausen, 'Triethyllead-induced hypomyelination in the developing rat forebrain.' *Exp. Neurol.*, 1976, **50**, 124–133.

57. G. Konat, H. Offner and J. Clausen, 'Triethyllead-restrained myelin deposition and protein synthesis in the developing rat forebrain.' *Exp. Neurol.*, 1976, **52**, 58–65.

58. G. Konat and J. Clausen, 'Protein composition of forebrain myelin isolated from triethyllead-intoxicated young rats.' *J. Neurochem.*, 1978, **30**, 907–909.

59. P. J. Dyck, P. C. O'Brien and A. Ohnishi, 'Lead neuropathy: 2. random distribution of segmental demyelination among "old internodes" of myelinated fibers.' *J. Neuropathol. Exp. Neurol.*, 1977, **36**, 570–575.

60. P. W. Lampert and S. S. Schochet, Jr., 'Demyelination and remyelination in lead neuropathy. Electron microscopic studies.' *J. Neuropathol. Exp. Neurol.*, 1968, **27**, 527–545.

61. A. Ohnishi, K. Schilling, W. S. Brimijoin *et al.*, 'Lead neuropathy. 1. Morphometry, nerve conduction, and

choline acetyltransferase transport: new finding of endoneurial edema associated with segmental demyelination.' *J. Neuropathol. Exp. Neurol.*, 1977, **36**, 499–518.

62. W. W. Schlaepfer, 'Experimental lead neuropathy: a disease of the supporting cells in the peripheral nervous system.' *J. Neuropathol. Exp. Neurol.*, 1969, **28**, 401–418.

63. M. R. Krigman, T. W. Bouldin and P. Mushak, in 'Experimental and Clinical Neurotoxicology,' eds. P. S. Spencer and H. H. Schaumburg, Williams & Williams, Baltimore, MD, 1980, pp. 490–507.

64. F. Monton and F. Coria, 'Reversible Schwann cell hypertrophy in lead neuropathy.' *Neuropathol. Appl. Neurobiol.*, 1991, **17**, 231–236.

65. A. D. Toews, M. R. Krigman, D. J. Thomas *et al.*, 'Effect of inorganic lead exposure on myelination in the rat.' *Neurochem. Res.*, 1980, **5**, 605–616.

66. E. K. Silbergeld, in 'Cellular and Molecular Neurotoxicology,' ed. T. Narahashi, Raven Press, New York, 1984, pp. 153–164.

67. L. W. Chang, 'The neurotoxicology and pathology of organomercury, organolead, and organotin.' *J. Toxicol. Sci.*, 1990, **15** (Suppl. 4), 125–151.

68. I. K. Grundt and N. M. Neskovic, 'Comparison of the inhibition by methylmercury and triethyllead of galactolipid accumulation in rat brain.' *Environ. Res.*, 1980, **23**, 282–291.

69. E. M. Carey, and N. M. Freeman, 'Biochemical changes in Cuprizone-induced spongiform encephalopathy. I. Changes in the activities of 2',3'-cyclic nucleotide 3'-phosphohydrolase, oligodendroglial ceramide galactosyl transferase, and the hydrolysis of the alkenyl group of alkenyl, acyl-glycerophospholipids by plasmalogenase in different regions of the brain.' *Neurochem. Res.*, 1983, **8**, 1029–1044.

70. S. Komoly, M. D. Jeyasingham, O. E. Pratt *et al.*, 'Decrease in oligodendrocyte carbonic anhydrase activity preceding myelin degeneration in cuprizone induced demyelination.' *J. Neurol. Sci.*, 1987, **79**, 141–148.

71. N. Fujita, H. Ishiguro, S. Sato *et al.*, 'Induction of myelin-associated glycoprotein mRNA in experimental remyelination.' *Brain Res.*, 1990, **513**, 152–155.

72. S. Komoly, L. D. Hudson, H. D. Webster *et al.*, 'Insulin-like growth factor I gene expression is induced in astrocytes during experimental demyelination.' *Proc. Natl. Acad. Sci. USA*, 1992, **89**, 1894–1898.

73. T. D. Anderson, A. Davidovich, R. Arceo *et al.*, 'Peripheral neuropathy induced by 2',3'-dideoxycytidine. A rabbit model of 2',3'-dideoxycytidine neurotoxicity.' *Lab. Invest.*, 1992, **66**, 63–74.

74. T. D. Anderson, A. Davidovich, D. Feldman *et al.*, 'Mitochondrial Schwannopathy and peripheral myelinopathy in a rabbit model of dideoxycytidine neurotoxicity.' *Lab. Invest.*, 1994, **70**, 724–739.

75. A. J. Larner, 'The biological effects of tellurium: a review.' *Trace Elements Electrolytes*, 1995, **12**, 26.

76. P. Lampert, F. Garro and A. Pentschew, 'Tellurium neuropathy.' *Acta Neuropathol. (Berl.)*, 1970, **15**, 308–317.

77. S. Duckett, G. Said, L. G. Streletz *et al.*, 'Tellurium-induced neuropathy: correlative physiological, morphological and electron microprobe studies.' *Neuropathol. Appl. Neurobiol.*, 1979, **5**, 265–278.

78. T. Takahashi, 'Experimental study on segmental demyelination in tellurium neuropathy.' *Hokkaido Igaku Zasshi*, 1981, **56**, 105–131.

79. G. Said and S. Duckett, 'Tellurium-induced myelinopathy in adult rats.' *Muscle Nerve*, 1981, **4**, 319–325.

80. T. W. Bouldin, G. Samsa, T. S. Earnhardt *et al.*, 'Schwann cell vulnerability to demyelination is associated with internodal length in tellurium neuropathy.' *J. Neuropathol. Exp. Neurol.*, 1988, **47**, 41–47.

81. G. J. Harry, J. F. Goodrum, T. W. Bouldin *et al.*, 'Tellurium-induced neuropathy: metabolic alterations associated with demyelination and remyelination in rat sciatic nerve.' *J Neurochem.*, 1989, **52**, 938–945.

82. M. Wagner-Recio, A. D. Toews and P. Morell, 'Tellurium blocks cholesterol synthesis by inhibiting squalene metabolism: preferential vulnerability to this metabolic block leads to peripheral nervous system demyelination.' *J. Neurochem.*, 1991, **57**, 1891–1901.

83. M. Wagner, A. D. Toews and P. Morell, 'Tellurite specifically affects squalene epoxidase: investigations examining the mechanism of tellurium-induced neuropathy.' *J. Neurochem.*, 1995, **64**, 2169–2176.

84. F. A. Rawlins and M. E. Smith, 'Myelin synthesis *in vitro*: a comparative study of central and peripheral nervous tissue.' *J. Neurochem.*, 1971, **18**, 1861–1870.

85. J. L. Goldstein and M. S. Brown, 'Regulation of the mevalonate pathway.' *Nature*, 1990, **343**, 425–430.

86. H. Jurevics and P. Morell, 'Cholesterol for synthesis of myelin is made locally, not imported into brain.' *J. Neurochem.*, 1995, **64**, 895–901.

87. A. D. Toews, S. Y. Lee, B. Popko *et al.*, 'Tellurium-induced neuropathy: a model for reversible reductions in myelin protein gene expression.' *J. Neurosci Res.*, 1990, **26**, 501–507.

88. A. D. Toews, I. R. Griffiths, E. Kyriakides *et al.*, 'Primary demyelination induced by exposure to tellurium alters Schwann cell gene expression: a model for intracellular targeting of NGF receptor.' *J. Neurosci.*, 1992, **12**, 3676–3687.

89. S. M. Grant and R. C. Heel, 'Vigabatrin. A review of its pharmacodynamic and pharmacokinetic properties, and therapeutic potential in epilepsy and disorders of motor control.' *Drugs*, 1991, **41**, 889–926.

90. J. P. Gibson, J. T. Yarrington, D. E. Loudy *et al.*, 'Chronic toxicity studies with vigabatrin, a GABA-transaminase inhibitor.' *Toxicol. Pathol.*, 1990, **18**, 225–238.

91. R. G. Peyster, N. M. Sussman, B. L. Hershey *et al.*, 'Use of *ex vivo* magnetic resonance imaging to detect onset of vigabatrin-induced intramyelinic edema in canine brain.' *Epilepsia*, 1995, **36**, 93–100.

92. G. Von Blankenfeld, J. Trotter and H. Kettenmann, 'Expression and developmental regulation of a GABAA receptor in clutured murine cells of the olidodendrocyte lineage.' *Eur. J. Neurosci.*, 1990, **3**, 310.

93. P. Gilbert, H. Kettenmann and M. Schachner, 'γ-Aminobutyric acid directly depolarizes cultured oligodendrocytes.' *J. Neurosci.*, 1984, **4**, 561–569.

94. R. M. LoPachin and E. J. Lehning, in 'Principles of Neurotoxicology,' ed. L. W. Chang, Marcel Dekker, 1994, pp. 237–264.

95. M. D. Roberson, A. D. Toews, T. W. Bouldin *et al.*, 'NGFR-mRNA expression in sciatic nerve: a sensitive indicator of early stages of axonopathy.' *Brain Res. Mol. Brain Res.*, 1995, **28**, 231–238.

96. A. D. Toews, G. J. Harry, K. B. Lowrey *et al.*, 'Low doses of carbon disulfide upregulate NGF-R mRNA expression in sciatic nerve.' *Trans. Am. Soc. Neurochem.*, 1995, **26**, 64, S93.

97. J. P. O'Callaghan and D. B. Miller, 'Quantification of reactive gliosis as an approach to neurotoxicity assessment.' *NIDA Res. Monogr.*, 1993, **136**, 188–212.

98. S. G. Waxman, J. A. Black, H. Sontheimer *et al.*, 'Glial cells and axo–glial interactions: implications for demyelinating disorders.' *Clin. Neurosci.*, 1994, **2**, 202–210.

99. J. E. Merrill, L. J. Ignarro, M. P. Sherman *et al.*, 'Microglial cell cytotoxicity of oligodendrocytes is mediated through nitric oxide.' *J. Immunol.*, 1993, **151**, 2132–2142.

100. M. Matsuo, Y. Hamasaki, F. Fujiyama *et al.*, 'Eicosanoids are produced by microglia, not by astrocytes, in rat glial cell cultures.' *Brain Res.*, 1995, **685**, 201–204.

11.12
Glial Cells

MARTIN A. PHILBERT
University of Michigan, Ann Arbor, MI, USA

and

MICHAEL ASCHNER
Bowman Gray School of Medicine, Winston-Salem, NC, USA

11.12.1 INTRODUCTION

A concept unique to the nervous system is that its functions are overwhelmingly due to the properties of its electrically excitable cells, the neurons. However, there is an even more numerous class of nonexcitable cells in the nervous system, collectively referred to as the glia or neuroglia. These cells comprise the astroglia or astrocytes, oligodendroglia or oligodendrocytes, as well as the microglia. The term neuroglia comes from an essentially erroneous concept of Virchow which he put forward in 1850, namely that neurons were embedded in a connective tissue to which he gave the name neuroglia or nerve glue. Elucidation of the true nature of the neuroglia depended on advances in histological staining by Golgi and Ramon y Cajal around 1870 and 1890, respectively. Neuronocentric views of the nervous system further cemented notions that glial cells provided little else than mechanical support and a nutritive gel for the neurons. However, involvement of glial cells in a number of idiopathic and toxicant-induced injuries prompted investigation of the participation of glia in the etiology and progression of altered neural states. Recent developments in neurobiology have revealed active specialized roles for glial cells in the management of regional ionic and neurotransmitter homeostasis in addition to participation in intermediary and xenobiotic metabolism. Furthermore, glial cells may serve to activate endogenous or xenobiotic neurotoxicants to injurious intermediates which are subsequently transported to neighboring neurons. It is now generally held that glia play a dynamic role in the maintenance of normal neural tissues following exposure to xenobiotics and are intimately involved in degenerative disease processes. Several extensive reviews have been published on the anatomy, function, and physiology of glial cells.[1–6] More recent attention has been directed to the role of glia in neurotoxicity.[7] This chapter will, therefore, provide an overview of glia as targets and mediators of neurotoxicity in the nervous system.

11.12.1.1 Astrocytes

The general term "nervenkitt" or nerve-glue was coined by Virchow in 1856 to describe any non-neuronal cell in the central nervous system (CNS) and includes astrocytes, oligodendrocytes, and microglia. Astrocytes are further subdivided by morphology into protoplasmic and fibrous phenotypes with the former predominating in the gray and the latter in the white matter of the CNS.

Astrocytes play a pivotal role in protecting the brain from attack by a wide variety of hydrophilic chemicals in the systemic circulation. The intimate physical association between astrocytes and endothelial cells lining blood vessels in the brain gives rise to the morphological, physical, and biochemical properties of the blood–brain barrier (BBB). In contrast to vasculature elsewhere in the body, brain endothelia display large numbers of mitochondria, limited pinocytosis, and absence of cell membrane fenestrations.[8–12] Pericellular movement of water-soluble substances into the brain is further restricted by the presence of *zonulae occludentes* or tight junctions between the cell membranes of adjacent endothelial cells.[13] The net result of this morphological arrangement is that facilitated diffusion of ions, sugars, amino acids, macromolecules, and xenobiotics is restricted to a variety of receptors, transporters, and channels. However, nonpolar, lipid-soluble substances generally gain free access to the brain by simple diffusion across the membranes of the vascular endothelial cells.[10–12,14,15]

Although astrocytes do not possess intercellular junctions of the *occludentes* type[16] they are required for the formation and maintenance of the BBB. The use of chimeric models provided confirmation of the role of astrocytes in the establishment of the BBB.[17] Transplantation of fragments of quail brain into the chick celomic cavity results in the loss of tight junctions and other characteristics of the BBB. Conversely, transplantation of quail somites into chick ventricles results in the investment of transplanted vessels with concomitant increases in endothelial cholinesterase and alkaline phosphatase activity, and the formation of *zonulae occludentes*. Endothelial differentiation to a BBB type occurs with investment of the transplanted vasculature with astrocytic foot processes. The pioneering studies of Stewart and Wiley[17] have since been confirmed by a number of investigators.[18,19] Studies *in vitro* have shown a requirement for physical contact between astrocyte foot processes and endothelial cells for differentiation to microcapillaries and development of tight junctions.[20]

Although endothelia are not strictly considered neural cells, there is mounting evidence that the vasculature in neural tissues is both a physical and metabolic defense against xenobiotics. The BBB is a target of many neurotoxicants which either circumvent the barrier by using a compatible transporter/receptor to gain access to the CNS, or directly open extracellular spaces between endothelial cells. It is often unclear whether opening of the BBB is a primary mechanism of toxicity or secondary to parenchymal events. However, once opened, the CNS becomes freely accessible to a variety of potentially deleterious endogenous or exogenous compounds. Any event which results in separation of the vascular endothelial cell from its ensheathing astrocytic foot process will likely result in weakening of the BBB and further complicate the neuropathological picture.

Investigations suggest that the vasculature contains enzyme systems which are capable of metabolizing endogenous compounds, drugs, and xenobiotics.[21–24] The presence of drug-metabolizing enzymes in the vasculature may either (i) serve to detoxify potentially hazardous moieties entering the CNS, or (ii) provide an enzymatic means of activation to reactive intermediates.[25,26]

In addition to their role in the establishment and maintenance of the BBB, astrocytes perform a variety of functions in the central nervous system of the developing and adult organisms (Table 1). These functions include:

(i) Regulation of glycogen metabolism following a variety of pharmacologic or physical manipulations.[27–30]

(ii) Regulation of the microenvironment by uptake of released amino acids such as glutamate, gamma-aminobutyric acid (GABA), taurine, to name a few;[31] transmitters such as catecholamines and serotonin;[32,33] and choline[34] (Table 1).

(iii) Regulation of brain volume by the removal of ions such as K^+ from the extracellular space, see later.[35-38]

(iv) Guidance of migratory neurons from germinal layers during development of the brain and spinal cord.[39] This function is performed by radial glia which are present during migration and, with the exception of cerebellar Bermann glia and glia in the vicinity of the third ventricle and median raphe of the brainstem, disappear once the neurons have come to rest.[40-42]

(v) Reaction to and repair of neural damage induced by physical trauma or chemical insult.

11.12.1.2 Oligodendrocytes

First described in 1921 by del Rio Hortega,[43] oligodendrocytes are the myelin-producing cells of the CNS (see Chapter 11, this volume). They are most frequently found in the white matter and in association with axons in the gray matter. Three types of oligodendrocyte have been identified by morphological criteria. The least electron-dense oligodendrocytes are immature and most actively divide in response to a variety of stimuli.[44-46] The other two are more mature phenotypes, the darkest being the most fully developed. The primary functions of oligodendrocytes are nutritive support for neurons and the formation of the myelin sheath.

Theories on symbiotic relationships between oligodendroglia and neurons were first expounded by Holmgren[47] and later elaborated by Hydén and Pigon.[48] The idea of neuron–oligodendrocyte interdependence was based on the measurement of enzymatic activities in neuroglia surrounding the giant neurons of Dieter's nucleus. Hydén and co-workers determined that although neuroglia isolated from Dieter's nucleus had less RNA than their neuronal counterparts, cytochrome oxidase and succinyl oxidase activities were approximately twice that of neighboring neurons.[48-50] These studies were among the first to demonstrate cellular compartmentalization of enzymatic processes which are an intrinsic part of ubiquitous metabolic pathways. Later studies also demonstrated oligodendrocytic distribution of ion channels which were previously thought to be located in neuronal membranes. The membranes of adult oligodendrocytes exhibit a wide variety of molecules which serve to regulate the periaxonal and, in particular, the perinodal spaces. These include voltage-sensitive sodium channels, fast and slow potassium channels, "inward rectifying" channels, Na^+/K^+- and Ca^{2+}-ATPases, and a specialized Na^+/Ca^{2+} antiporter in central white matter.[51]

The advent of electron microscopy permitted resolution of the cytoplasmic composition of oligodendrocytes and the intramyelinic substance. These morphological observations, in combination with studies performed in various

Table 1 Present views on some of the properties and roles of astrocytes in CNS physiology and pathology.

Development
　Neuronal and axonal guidance and migration in development
　Influence on synaptogenesis and neuronal development and survival
　Induction of tight junctions between endothelial cells (blood–brain barrier)
Ion and pH homeostasis involving both voltage- and ligand-gated ion channels
Receptors for transmitters
Transmitter uptake and homeostasis
Release of transmitters and synthesis of neuropeptides
Immune responses—release of immune system signals, response to interleukins, and possible presentation of antigens and recruitment of peripheral immune cells
Glycogen storage and metabolic interactions, including supply of substrates to neurons and perhaps oligodendrocytes and microglia
Pathology
　Swelling of astrocytes associated with trauma, ischemia, and hepatic encephalopathy
　Structural cellular injury leads to astrogliosis or glial scars formed by reactive astrocytes
　Environmentally induced neuropathies, e.g., Parkinsonism via an MPTP-like induced mechanism
　Role in growth or transplantation of neural or non-neural tissue to the brain and regeneration of damaged neurons
　Epilepsy and psychiatric disorders due to involvement with transmitter functions and neuronal excitability
　Protective or promotive roles in the aging brain and degenerative diseases such as Huntington's disease and multiple sclerosis
Neuroprotection by sequestration of neurotoxic substances and their metabolism
Neuroprotection may also be afforded by astrocyte induction of stress-related proteins such as heat shock protein and/or metalloproteins, specifically metallothioneins

mammalian embryos, have led to the conclusion by several investigators that myelin is elaborated by oligodendrocytes. Oligodendrocytes appear just prior to the myelination of axons in the CNS, and the rate of myelination appears to be proportional to the rate of proliferation and differentiation of these glia.[52,53] During myelination of axons, oligodendrocyte processes are in intimate contact with axis cylinders and are involved with assembly and insertion of proteins into nodal membranes.[54] Following brain injury involving edematous changes or demyelination in the white matter, molecular events associated with remyelination are reminiscent of those observed during organogenesis in the embryo.[55,56]

Since oligodendrocytes are intimately involved in the electrical insulation of axons and in the maintenance of nodal membrane proteins, alteration in their function results in impaired conduction velocity and/or cross-talk between neighboring axons. This concept has been clearly demonstrated in a number of experimental and idiopathic demyelinating diseases.[57-60]

11.12.1.3 Microglia

The origins and even the presence of resting microglia in the CNS have been a topic of much debate. However, immunocytochemical techniques have permitted the identification of Fc and complement receptors on the surface of resting cells in the brain.[61] As such, these cells are immunological and undergo rapid mitotic proliferation in degenerative and inflammatory conditions. The presence of resting microglia in the brains of germ-free animals strongly suggest that microglia are a normal component of the neuropil.[62] Microglia comprise 4-5% of the total population of glial cells in the white matter and 6-18% of the gray matter depending upon region and species.[46,63,64] The primary function of microglial cells is as a macrophage precursor. Activated microglial cells are frequently seen engulfing myelin fragments or with lipid inclusion bodies in Wallerian degeneration. Later in the disease process, activated cells are frequently seen with a wide variety of cell debris within their cytoplasm.[65] Microglia have also been observed to penetrate deep into degenerating motor neurons in the spinal cords of experimental animals.[66] Microglia have received relatively little attention in studies of neurotoxicology and have not been considered an active part of the neurotoxic process. Additional details on their potential role in neurotoxic injury are provided below.

11.12.1.4 Ependymal Cells

Ependymal cells form an epithelial layer which lines the ventricles and central canal of the brain and spinal cord, respectively. The mammalian ependyma are derived from the ventricular zone which produce ependymal cells, neuroblasts, and glial cells during active development of the brain.[67] Ependymal cells with processes which extend to the pial surface are frequently observed in the developing nervous system. These cells persist in the walls of the third and fourth ventricles, cerebral aqueduct, and central spinal canal of adult mammals and are referred to as tanycytes.[68-70] The subependymal region contains many actively dividing neuroglial cells during development and provides an intermediate zone between migrating neuroblasts and the ventricular layers. The subependymal layers may persist in areas such as the lateral angle of the lateral ventricles.[71] Regions subtending a persistent subventricular zone are particularly susceptible to chemical or viral agents which induce glial tumors in experimental animals.[72-75]

In adults, the ependyma consist of a single layer of ciliated cuboidal, columnar, or squamous cells, depending on their anatomical location. The cilia are arranged in a central configuration on the free/ventricular surface of the cell and are attached to basilar bodies located in the apical cytoplasm of the ependymal cell.[71] The cilia move with a coordinated metachronal rhythm, thereby facilitating movement of cerebrospinal fluid (CFS) through the ventricles, aqueducts, and central canal.[76,77]

The lateral surfaces of ependymal cells interdigitate and possess two types of junction, namely: gap junctions and *zonulae adherentia*.[78,79] Gap junctions, which were initially mistaken for *zonulae adherentes*, are usually found between pairs of adherent junctions and provide intracellular communication between 2-70 adjacent ependymal cells.[80] However, since no tight junctions exist between apposed lateral membranes of ependymal cells, transit of macromolecules from ventricles to the parenchyma is possible.[13] Xenobiotics in the CSF might also gain access to specific populations of neurons which have transependymal projections into the ventricular spaces.[81]

Consistent with their function of physically moving large volumes of cerebrospinal fluid, ependymal cells possess a large number of mitochondria which are concentrated in the apical portions of the cytoplasm.[82] Also, a large network of intermediate filaments (8-10 nm) which do not react with antibodies directed against glial fibrillary acidic protein (GFAP) are arranged in swirled fascicles throughout the cytoplasm.[83] These fascicles are part of an extensive cytoskeleton which coordinate the considerable mechanical work carried out by the ependyma. As such, the ependyma are likely an important component of any neurotoxicological condition which arises from exposure to agents which interfere with either energy metabolism or components of the cytoskeleton. While probably not of primary consequence in mechanisms of neurotoxicity, perturbations of ependymal function may contribute to the overall picture of neurological damage.

11.12.2 ROLE OF GLIA IN DEVELOPMENT

11.12.2.1 Role of Glia in Neuronal Differentiation

The concept that trophic molecules serve as determinants of survival, morphology, function, and circuitry has been eloquently articulated by Black.[84,85] Glia participate in the differentiation of neurons during development by physical contact and the transient, sequential elaboration and secretion of soluble factors in both the central and peripheral nervous systems.[86–88] Induction of choline acetyl transferase in cholinergic neurons by physical contact is mediated by a membrane-derived factor and other soluble factors.[89,90] Concomitant with expression and release of growth and other factors in the brain is the transient expression of neuron-specific receptors in developing cerebellar Purkinje cells, suggesting local delivery of neurotrophins.[91] If the terminal projections of neurons express the appropriate receptor, cell bodies distal from the primary site of glial secretion may benefit from binding to neurotrophins.[92] There appears to be a reciprocal glial advantage in binding neurotrophins such as nerve growth factor (NGF). Glial cells express low-affinity receptors for NGF and their growth is stimulated by the addition of exogenous NGF in culture. Furthermore, NGF induces the expression of GFAP.[93] Studies *in vitro* provide compelling evidence for a complex collaborative interaction between NGF, ciliary neurotrophic factor (CNTF), and fibroblastic growth factor (FGF) in the terminal differentiation of neuronal cell progenitors.[94] However, little information is available on the nondevelopmentally-linked expression of either neurotrophins and/or receptors following either physical or chemical injury. It is reasonable to suggest that glial mechanisms of neurotrophic-mediated repair might be critical to re-establishment of neuronal function following neurotoxic insult. Thus, investigation of interactions of neurotrophins and their receptors with neurotoxicants is crucial to a complete understanding of recovery processes invoked following neural damage.

11.12.2.2 Role of Glia in Neuronal Migration

The structure and organization of the brain requires sequential and well-organized movement of neural progenitor cells within the germinal layers. Fibrous astrocytes have been identified as the primary glial cell-type involved in the guidance of immature neuroblasts and neurites to their ultimate destination in the CNS. This guidance is achieved both by the formation of physical astrocytic boundaries and the secretion of a variety of glycoproteins, glycolipids, and glycosaminoglycans by fibrous astrocytes into the extracellular space.[95–97] Components of the extracellular matrix which inhibit neuronal migration include J1/tenascin and 473 proteoglycan; both are transiently expressed at various times thoughout development.[98] J1/tenascin and the associated chondroitin sulfate-containing 473 proteoglycan are developmentally regulated and appear surrounding whisker barrels, neostratal striosomes, and nuclear boundaries prior to and during neuronal migration.[99] However, these inhibitory boundaries disappear only after regional synaptic contacts are established.[98]

The association of neuroblasts with radial arms of fibrous astrocytes in the developing nervous system were postulated by Golgi.[100] Later studies by Rakic[101,102] largely confirmed astrocytic roles in the guidance of migrating neurons to their ultimate destination (see also Refs. 103 and 104). Experiments *in vitro* provide strong evidence for the participation of astrocytes and other neural cell types in determining the kind and rate of neurite extension.[105–108] The most rapid rates of cerebellar granule cell neurite extension have been observed either on astrocyte monolayers or extracellular matrix derived from astrocytes.[109–111] The processes of migration and neurite extension are likely mediated by adhesion molecules on apposed astrocyte surfaces (see Chapter 7, this volume). While the neural cell adhesion molecule (*N*-CAM), the calcium-dependent adhesion molecule (*N*-cadherin), and L1 facilitate homotypic fasciculation of neurites, binding to astrocytes appears to be mediated by astrotactin, a 100 kDa glycoprotein expressed on the neuronal cell surface.[112–115]

Following physical or chemical neural injury resulting in the loss of synaptic contacts, many of the transient guidance/adhesion events observed during development are repeated during the process of re-establishment of neuron–neuron contact. However, the specific cellular locus of secretion of many chemotaxic or inhibitory extracellular molecules remains to be elucidated.[114]

11.12.3 ROLE OF GLIA IN THE MAINTENANCE AND REGENERATION OF NEURONS

Developments in molecular biology have permitted the identification and characterization of glia-derived molecules which are secreted during development and in response to injury. Coculture of mesencephalic neurons with glia from the same region results in a fourfold increase in neuron survival.[117,118] Soluble factors released from nigral astrocytes *in vitro* promote the survival of mesencephalic neurons by direct binding to neuronal receptors, a mechanism distinct from that of basic fibroblastic growth factor which simply causes mitotic division of astrocytes.[119] Elliot and co-workers[120] demonstrated rapid induction of brain-derived neurotrophic factor and a somewhat more sluggish induction of NGF following depolarization by K^+. In addition to NGF, mesencephalic glial cells have been shown to elaborate glial-derived neurotrophic factors (GDNF) which selectively increase the survival rates of nigral dopaminergic neurons.[121] Expression of GDNF transcripts is region specific and appears to be confined to protoplasmic astrocytes.[121,122]

CNTF is a cytosolic protein expressed postnatally by myelinating Schwann cells and a subset of astrocytes (see Ref. 123 for review). Although distinct from brain-derived neurotrophic factor (BDNF), CNTF similarly ameliorates motor nerve cell degeneration induced by physical, genetic, or chemical damage.[124–127] In the brain, CNTF released by astrocytes has also been implicated in maturation and maintenance of oligodendrocytes and inhibition of apoptotic death.[128] Increased expression of mRNA for CNTF and its receptor during development and following axotomy of peripheral nerves appears to be region- and cell-specific.[129–131] Evidence points to a CNTF-like protein in the nuclei of neurons in the CNS of adult rats and mice, suggesting astrocyte–neuron interplay in the expression of neurotrophic factors.[132]

There is much interest in the exogenous application of glial-derived neurotrophins for the clinical management of a variety of degenerative central and peripheral nervous

system diseases. However, with the possible exception of lead, relatively little is known about the role of neurotoxicants in the inhibition or stimulation of expression of these maintenance and regenerative factors.

11.12.4 ROLE OF GLIA IN REGULATION OF THE MICROENVIRONMENT

Both astrocytes and oligodendrocytes participate in regulating the composition of the extracellular milieu. This is achieved by transporter/receptor-mediated movement of ions, amino acids, and neurotransmitters.[133] Maintenance of pH is one of the best characterized regulatory mechanisms. Protein kinase C-linked proton pumps in astrocyte membranes aid in the regulation of extracellular pH $[pH]_o$. Astrocytes secrete H^+ ions into the extracellular space[134] and are thought to participate in post-excitotoxic compensation by mildly acidifying the perineuronal space, and thereby ameliorating the neurotoxic effects of glutamate receptor agonists.[135,136] Lowering $[pH]_o$ inhibits the opening of N-methyl-D-aspartate (NMDA)-coupled ion channels, attenuating the effects of excessive excitation following ischemia/reperfusion injury or direct excitotoxicity.[137] Other buffering systems described in the membranes of astrocytes in culture include Ca^{2+};[138] Na^+,[139–141] K^+,[142,143] and Cl^-.[140,144] Many of these ion-regulatory proteins are expressed in oligodendrocytes, Schwann cells, and their precursor cells.[145–147] Glial cells are responsive to a variety of neurotransmitters,[148–152] and release taurine in response to ionic stimulation.[153,154] It is also important to note that the distribution of ion channels is heterogeneous with respect to region, cell type, and cellular contact with neuronal subtype.[155–158]

11.12.5 GLIA AS MEDIATORS OF NEUROTOXICITY

Section 11.12.4 highlighted the role of glial cells in neurophysiology, extending it to well beyond passive structural support and sensitivity to axon commands. From an early developmental period, glia and neurons establish a highly dynamic and reciprocal relationship, which is maintained throughout life, influencing nervous system growth, morphology, behavior, repair, and aging. Considering the diversity and complexity of this relationship, it is reasonable to postulate that the glial–neuronal interdependency represents at least one of the functional units of nervous tissue (*vis-à-vis* neuron–neuron interactions, glial–glial interactions, endothelial–glial interactions, and so on). Perturbation of homeostatic functions in glial or neuronal cell types is likely, therefore, to render the unit dysfunctional. A neurotoxic event might disrupt adhesion molecule interactions and ultimately glial–neuronal contact; intercellular communication or exchange of macromolecules might be altered, leading to both structural and functional consequences for each cell type. Glial uptake of a pro-neurotoxicant could result in its metabolic conversion to an active toxicant, its release and ensuing neuronal toxicity, while initially sparing glial function.[114,115]

11.12.5.1 Microglia as Mediators of Neurotoxicity

The microglia appear to be involved mainly in the response of the brain to injury, exhibiting important immunological properties. Microglia constitute approximately 10% of the total glial cell population and are considered to be the macrophages of the brain. Their major function is as scavenger cells, ingesting cellular debris, a process which may be important for tissue modeling in both the developing and damaged CNS. Microglia are also involved in inflammation and repair processes in the adult CNS due to their phagocytic ability, release of neutral proteinases, and production of oxidative radicals. Microglia have also been shown to express multihistocompatability (MHC) antigens upon activation,[159–161] and while a matter for some debate, they are also likely to constitute the brain's antigen presenting cells (APCs), as they are known to secrete a number of immuno-regulatory cytokines, and respond to cytokine stimulation.[162,163] Those who favor microglia (over the astrocytes) as the brain's APCs argue that they are the more likely source of inter-leukin-1 (IL-1) during acute-phase brain injury, because microglia are the first brain cells to appear in increased numbers at sites of trauma or infection. In addition, IL-1 appears to be produced more efficiently in microglia than in astrocytes. Given their primary functions in phagocytosis and remodeling, it is believed that if microglia were readily activated then juxtaposed neurons would be at a constant risk of microglial-mediated damage. It is almost certain, therefore, that *in vivo* microglial responses are held under strict regulatory control.

Stimulated by *in vitro* observations on the ability of microglial cultures to release a number of neurotoxic substances, renewed interest has focused on their potential to mediate secondary neurodegenerative processes via the elaboration and release of neurotoxic substances during ischemic or traumatic brain injury. A significant increase in spinal cord concentrations of quinolinic acid, an excitatory amino acid, was noted upon a contusion injury.[164] Resident microglia (rather than blood-circulating) were implicated in this response and as the source of the quinolinic acid. Other potentially neurotoxic substances elaborated and released by microglia include glutamate, nitric oxide, reactive oxygen species, and agents of unknown composition with a molecular weight $<500\,\mathrm{Da}$.[165–168] Production of these neurotoxins in culture is constitutive and can be further enhanced by microglial treatment with lipopolysaccharide (LPS), interferon-γ (IFN-γ), and zymosan A particles (for review see Ref. 169). However, since it is also well established that microglial cells in culture release a number of neuron-survival promoting compounds,[170,171] it would appear that the precise role of microglia in neurotoxic injury will have to await further *in vivo* experimentation. The balance between neurotoxicity and neurotrophism is likely to depend on the nature of the injury, and it may progress by divergent pathways which by themselves may be dependent upon the nature of the released substances and the ability of the "substrate" cells (i.e., neurons) to respond to them.

11.12.5.2 Astrocytes as Mediators of Neurotoxicity

The modern experimental approach to the role of astrocytes in brain function is generally considered to have started with the pioneering electrophysiological studies of Kuffler and his colleagues on astrocyte-like glial cells in primitive animals such as the leech.[142] Disposing of perhaps what was the last remnant of the old dogma which had led electron microscopists to propose that astrocytes formed the extracellular space of the brain ("enlarged watery structures"), Kuffler's work clearly refuted this concept, at least for the glia in the leech's nervous system and the optic nerve of amphibians.[142] The amphibian glial cells were found to have a normal high intracellular K^+ and, in fact, to be characterized by a membrane potential which was essentially the same as the Nernst potential for K^+ (-80 to $-90\,\mathrm{mV}$). Later work on "electrically silent" cells in the mammalian CNS were identified as putative glial cells. These had the same characteristic, namely a nonexcitable cell with large negative membrane potential, apparently sensitive only to changes in $[K^+]_o$. This led to one of the earliest functions proposed for glial cells, and specifically astrocytes, namely, the control of extracellular K^+.[142] Throughout this early period, perhaps stemming from a desire to justify the prevailing "neuronocentric" approach, there was a tendency to incorrectly assert that the astrocyte contributed little to the overall function of the nervous system, other than a general support role for neurons. Work since the 1970s has shown this narrow interpretation to be totally incorrect. Table 1 represents some of the roles and functions that are attributed to astrocytes, both in normal brain physiology and development, as well as in the etiology of pervasive brain dysfunction.

Major progress in the modern experimental period of astrocyte research commenced in the early 1970s with the availability of primary cultures prepared from neonatal or late-stage fetal rodent brains. These cultures were found to contain predominately, and with appropriate manipulations, almost exclusively, monolayers of astrocytes.[172] Confirmation of the astrocytic nature of these cultures came with the discovery

that the intermediate filaments of astrocytes were comprised of a unique protein, glial fibrillary acidic protein (GFAP), and that almost all the cells in these primary cultures stained positively for GFAP.[173] This protein found clinical and experimental use in the differential identification of primary neural neoplasms (Figure 1). The development of methods for routinely preparing >95% GFAP-positive cell cultures from healthy brains by methods such as that of Frangakis and Kimelberg[174] enabled experimental techniques employed in neuroscience and cell biology (i.e., electrophysiology, molecular biology, transport studies using radiolabeled compounds, immunocytochemistry) to be applied to the study of these cultures.

Classically conceived neurotoxic mechanisms have generally proposed a neuron-specific action for chemicals or neuropathic diseases. These mechanisms define glial involvement as reactive (e.g., proliferation, removal of axon debris) and/or sustaining (e.g., neurite support).

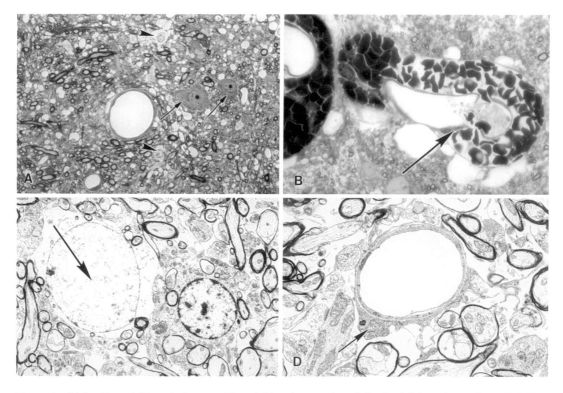

Figure 1 Light (A and B) and electron (C and D) micrographs of the fastigial nucleus of a ♂ F-344 rat given 5 doses of 25 mg kg^{-1} 1,3-dinitrobenzene per os. Astrocytic edema in the deep cerebellar roof nuclei of the rat is an early hallmark of 1,3-dinitrobenzene intoxication (A). Watery astrocytes (arrow-heads) can be observed in the vacuolated neuropil. Apparently normal neuronal somata (arrows) are partially denuded by the retraction of astrocytic end-feet. Later stages of intoxication are characterized by opening of the Virchow–Robin space and perivascular hemorrhage (B). Rarely, perforation of the capillary wall with transient erythrocytes can be observed (arrow). Electron microscopic examination reveals significantly swollen astrocytic cell bodies and nuclei (C; large arrow) and mildly edematous oligodendrocytes (small arrow). Condensation and retraction of perivasular astrocyte foot-processes is a common feature of severely affected brainstem nuclei (D). A and B, 1 μm plastic section, cresyl violet.

However, it has been shown that a direct insult of neurotoxic compounds on astrocytes leading to their compromised function may represent the primary biochemical trigger for the ensuing neurotoxicity. Investigations of the effects of neurotoxic compounds on astrocytes in the CNS *in situ* are limited and are focused on the neuropathological effects of these compounds (for review see Ref. 175). Functional consequences have only been studied *in vitro*, and several of those will be summarized below. There appears to be a sizable body of work to suggest that these cells, when adversely affected by neurotoxins or physical trauma (i.e., swelling, energy depletion, etc.), lose many of their important protective or homeostatic functions, such as uptake of extracellular K^+ and excitatory amino acids (EAAs). Other data suggest that astrocytes actively metabolize proneurotoxicants, leading to the formation and release of toxic intermediates, which can directly compromise neuronal activity while sparing the astrocytes themselves.

Excessive ammonia accumulation within the CNS, such as occurs in hepatic encephalopathy (HE) or congenital and acquired hyperammonemia, will serve as the first example for the role of astrocytes in mediating neurotoxic injury. Mammalian astrocytes undergo morphological changes upon chronic exposure to ammonia, yielding the so-called Alzheimer type II astrocytes common to most hyperammonemic conditions, and which precedes any other morphological change in the CNS.[176–178] Astrocytes serve as the exclusive site for the detoxification of ammonia and glutamate (Figure 3), a process requiring ATP-dependent amidation of glutamate to glutamine, and mediated by the astrocyte-specific enzyme, glutamine synthetase (reviewed Refs. 179 and 180). Chronic exposure of astrocytes to ammonia *in vivo* leads to diminished glutamine metabolism and impairment of astrocytic energy metabolism (reviewed Ref. 181). Furthermore, the reduced astrocytic capacity to metabolize ammonia leads to ammonia-induced cytotoxicity in juxtaposed neurons, and promotes accumulation of glutamine which, in turn, leads to decreased cerebral glucose consumption and amino acid imbalances.[182,183] Additional processes which appear to be sensitive to increased intracellular ammonia concentrations include inhibition of the synthesis of neuronal glutamate precursors, potentially resulting in diminished glutamatergic neurotransmission, changes in neurotransmitter uptake (glutamate), stimulation of taurine release which is likely to promote inhibitory neurotransmission by way of GABA, as well as changes in receptor-mediated metabolic responses of astrocytes to neuronal signals.[180]

A second example of the role of astrocytes in mediating neurotoxicity is illustrated by the intracellular conversion of a proneurotoxicant to an active metabolite capable of inflicting selective neuronal damage. Perhaps best documented is the mechanism associated with the metabolic activation of the designer drug, 1-methyl-4-phenyl-1,2,3,6-tetrahydropyridine (MPTP). MPTP intoxication is associated with selective destruction of nigrostriatal dopaminergic neurons.[184–186] The molecular mechanism of MPTP toxicity is only partially understood, although several features have emerged (for reviews see Refs. 187 and, 188). (i) Conversion of MPTP to 1-methyl-4-phenylpyridinium ion (MPP^+) is required for neurotoxicity, (ii) MPP^+ accumulates intraneuronally by the high-affinity dopamine uptake mechanism, (iii) a dopamine uptake blocker such as mazindol protects against the neurotoxicity of MPTP, and (iv) MPTP is not neurotoxic when conversion to MPP^+ is blocked by inhibitors of monoamine oxidase type B (MAO-B).

Astrocytes *in situ* are immunocytochemically positive for MAO-B[189] and actively metabolize MPTP to MPP^+.[188,190] That astrocytes mediate an essential step in MPTP toxicity is also supported by *in vivo* studies with the astroglial-selective toxicant, α-aminoadipic acid (α-AA), a six carbon chemical analogue of glutamate.[191] When MPTP is microinjected into the substantia nigra, a loss of nigral neurons is observed as revealed by fluorescent retrograde axonal tracing. In contrast, coinjection of MPTP plus α-AA into the substantia nigra is associated with reduced neuronal degeneration, presumably due to the initial destruction of resident astrocytes and reduced conversion of MPTP to MPP^+. The protective effect of α-AA is curtailed, however, if reactive astrocytosis is allowed to proceed and MPTP is injected into the substantia nigra 1 week after the initial injection of α-AA. Thus, once repopulated with astrocytes (astrocytic scar), MPTP is again oxidized to MPP^+, in turn, leading to nigral damage and loss of dopaminergic neurons. Furthermore, because of the increased number of astrocytes upon the formation of the gliotic scar, MPTP damage is increased compared to the experimental paradigm where it is coinjected with α-AA, reflecting increased metabolic conversion of MPTP to MPP^+. Accordingly, a model of MPTP toxicity has been proposed invoking the diffusion of MPTP into the brain where it is taken up by MAO-B positive astrocytes and neurons and subsequently metabolized to the reactive neurotoxic metabolite, MPP^+. MPP^+ is subsequently released from the astrocytes and concentrated in nigrostriatal neurons via dopamine uptake.

Huntington's disease has been associated with elevated CNS levels of quinolinic acid,[192–195] a normal by-product of the metabolism of the amino acid tryptophan (see Ref. 196 for review). Quinolinic acid excites neurons by interacting with the NMDA receptor.[197] At high doses, particularly after intrahippocampal injections, quinolinic acid causes convulsions and neurodegenerative changes that are remarkably similar to those observed in human temporal lobe epilepsy.[193,198] Because of its potential role in excitatory amino acid neurotransmission and excitotoxic-induced brain damage, the neurobiology of quinolinic acid and its "astrocytic connection" has been investigated in detail. The enzyme that synthesizes quinolinic acid, 3-hydroxyanthranilate oxygenase (3-HAO) is present predominantly, or perhaps even exclusively, in astrocytes.[199] Evidence also suggests that activated microglia/macrophages may also produce quinolinic acid under immunocompromised conditions (reviewed in Ref. 196). While quinolinic acid is readily released from astrocytes, it is not subject to regulation by EAAs. Astrocytes are also implicated in the enzymatic degradation of quinolinic acid, carried out by quinolinic acid phosphoribosyltransferase (QPRT).[200] Hence, abnormalities in either the synthesizing or degrading enzymes of the kynurenine pathway can shift the balance of quinolinic acid concentrations in the extracellular fluid, triggering NMDA receptor activation and neuronal excitotoxicity. Astrocytes (and perhaps microglia as well) harbor the machinery responsible for quinolinic acid synthesis and release, and therefore may serve to modulate excitatory amino acid receptor function and, in pathological conditions, as determinants of excitotoxic neuronal vulnerability.

Neurotoxic mechanisms may also involve an effect on astrocyte physiology that either exacerbates coincidental neuron injury or initiates secondary neurotoxicity. For example, astrocytic swelling and the ensuing release of glutamate may have important implications for neuronal homeostasis.[201] Astrocytic swelling is observed in a number of pathological conditions such as hypoxia, CNS trauma, hepatic encephalopathy, ischemia,[175,202–206] and following exposure to certain neurotoxicants such as lead, methylmercury, 1,3-dinitrobenzene, and triethyltin (Figures 2 and 3).[178,207–213]

Astrocytic swelling occurs rapidly, within 1 h of injury (*in vivo*), but may reverse slowly with time. It is believed that in its exaggerated form astrocytic swelling is deleterious, directly contributing to CNS dysfunction and pathology. Persistent astrocytic swelling can be viewed as a pathological extension of more limited and controlled volume changes which are part of the normal homeostatic function of astrocytes. A number of mechanisms for the physiologic/pathologic swelling have been proposed (Figure 3).[204,214] These include swelling associated with acid–base changes, glutamate-induced swelling, and K^+-induced swelling. In the first model, the uptake of Na^+ and Cl^- occur by simultaneous operation of Cl^-/HCO_3^- and Na^+/H^+ transport, with H^+ and HCO_3^- cycling from the intra- to extracellular spaces via membrane permeant CO_2.[204,214] Thus, the reduced clearance of CO_2 due to ischemia and/or lactic acid-induced acidification of HCO_3^- to H_2CO_3 and then CO_2 and H_2O, coupled with the reduced ability of the astrocytes to export intracellular Na^+ along with Cl^- (due to falling intracellular energy stores), would lead to astrocytic swelling. Glutamate-induced astrocytic swelling may operate via the same mechanism, namely increased production of CO_2 due to increased astrocytic metabolism. Alternatively, glutamate-induced astrocytic swelling may be associated with its efficient clearance from the extracellular space by a Na^+-dependent transport system.[215] This uptake could lead to swelling driven by Na^+-loading of the cells. Finally, increased extracellular K^+ concentrations, if coupled with significant Cl^- conductance, will also lead to astrocytic swelling because of depolarization-induced net uptake of KCl. Such uptake is commonly caused by Donnan forces, driving K^+ and Cl^- through separate K^+ and Cl^- channels.[206,216]

Since astrocytic swelling is so prevalent it is important to understand its consequences to juxtaposed neurons. When swollen, astrocytes show regulatory volume decrease (RVD)[204,217–220] as do most cells, re-establishing their pre-swelling volume by losing ions such as K^+ and Cl^-, as well as amino acids. This leads to the hypothesis that swollen astrocytes might also promote the unregulated release of endogenous neuroactive and excitatory substances which mediate nerve injury (Figure 3). Astrocytic swelling and compensatory volume regulation are associated with the efflux of taurine and EAAs such L-glutamate and D-aspartate.[175,201,219] The origin of glutamate (as well as other EAAs such as L-aspartate) has been tacitly assumed to be presynaptic nerve endings.[221,222] However, it is known that astrocytes take up glutamate by a Na^+-dependent mech-anism.[215] In the presence of ammonia, glutamate is metabolized to glutamine by the astrocyte-specific enzyme glutamine synthetase (GS),[223–225] maintaining [glutamate]$_o$ at $0.3\,\mu M$.[226,227] This represents a $10\,000$-fold gradient vs. [glutamate]$_i$ ($3\,mM$). This glutamate–glutamine pathway constitutes the

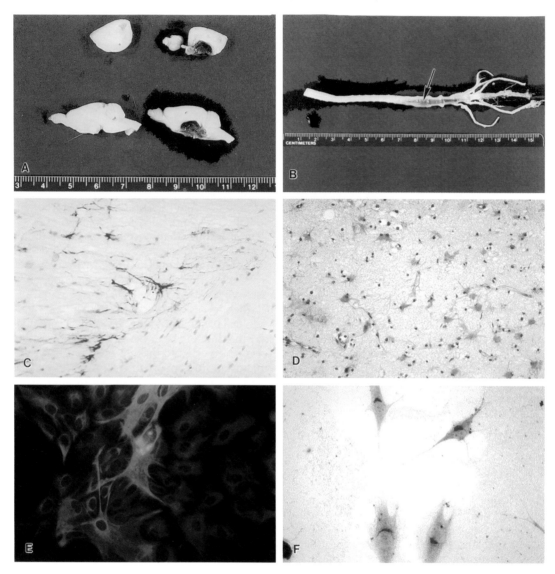

Figure 2 Exposure to N,N'-ethylnitrosourea (ENU) *in utero* results in the development of central nervous system tumors in rats (A and B). Sagittal and parasagittal sections (A) through a control (left side) and treated rat (right side) reveal well-defined hemorrhagic astrocytic tumors in treated rats. Oligodendrogliomas are frequently formed in the lumber region of the spinal cord (B, arrow). Light microscopic examination reveals aberrantly-shaped, glial fibrillary acidic protein-positive astrocytes within the tumor (C). Although normally confined to neuronal nuclei, α-glutathione *S*-transferase can be seen in astrocytes within the astrocytoma (D). Primary cultures of an astrocytoma isolated from the occipital lobes of a treated rat display strong immunoreactivity for glial fibrillary acidic protein (E). Primary cultures of a malignant astrocytoma isolated from an ENU-treated rat show highly localized immunoreactivity for the estrogen receptor (F). Note the localization of estrogen receptor to the nuclear poles.

locus of a glutamate pool in brain described in the early 1960s.[228] Release from this pool occurs as a result of astrocytic swelling, a feature common to a number of neuropathological states (see above).

As a simple model for studying the consequences of astrocytic swelling in the brain, hypotonic buffer may be applied to the cells to initiate swelling.[174,204,205] *In vivo*, significant hypo-osmolarity is almost always due to low-

ered plasma Na^+ concentration, that is, hyponatremia. This condition also leads to loss of taurine, aspartic acid, and glutamic acids from the brain. Although it is not proposed that astrocytes *in vivo*, other than in hyponatremia or in some form of neonatal renal disease, swell because of decreased extracellular osmolarity, a working hypothesis is that the swelling of the astrocytes *in situ*, by whatever mechanism, activates the same ion or amino acid transport

Models of astrocytic swelling

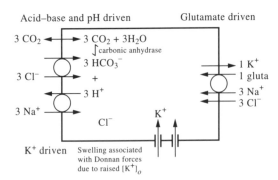

Figure 3 Mechanisms of astrocyte swelling. For further details, refer to the text.

processes activated by hypotonic media-induced swelling *in vitro*. Experimental evidence on the effect of a number of neurotoxic compounds, such as methylmercury,[212,214] suggests that astrocytic swelling and the ensuing release of amino acids (glutamate, aspartate) are likely to play an important role in neurotoxicity.

While it is reasonable to accept that acute brain insult, causing massive release of glutamate could damage neurons expressing NMDA receptors, it is more difficult to link the neurotoxic action of glutamate to the long latent period associated with low-level xenobiotic exposure and the slow progression of neurodegenerative disorders. There is a mismatch between the millisecond dynamics of glutamate-gated ion kinetics and the years, perhaps decades, involved in the progressive neuronal injury associated with these disorders. Environmental levels insufficient to cause acute, overt CNS toxicity, but associated with chronic low-level elevation of glutamate, could nevertheless accelerate the processes of excitotoxic neurodegeneration associated with disease or aging. Evidence is emerging that activation of glutamate-gated cation channels may be an important source of oxidative stress (reviewed in Ref. 229), referring to the cytotoxic consequences of oxygen radicals (superoxide anion, hydroxy radical, and hydrogen peroxide) which are generated as by-products of normal and aberrant metabolic processes that use molecular O_2. Thus, the convergence of sequential and interactive mechanisms (glutamate-gated ion channel activation and oxidative stress) may provide a link between astrocytic injury and glutamate (or other EAA)-induced neuronal toxicity. From this perspective, glutamate is acting as a long-term "pulsing" stimulus, progressively and cumulatively leading to

neuronal damage. The latter would then be measurable as the sum total of the life-time glutamate-linked events.

Direct physical contact between astrocytes and neurons is mediated by adhesion molecules, facilitating intercellular communication, in turn, determining morphological, biochemical, and functional differentiation of each cell type. Thus, "uncoupling" glial–neuronal contact would be expected to impact on the function of both cell types.[230,231] Neurotoxicants may interfere with intermolecular binding processes or with biochemical expression of adhesion (recognition) molecule. Evidence for the latter was provided by Cookman *et al.*,[232] demonstrating that chronic exposure of rat pups *in utero* to lead impairs *N*-CAM desialylation. Transformation of *N*-CAM from a sialic acid-rich form to an oligosialylate congener is associated with increased adhesive potential and is, therefore, a primary determinant of stable cell–cell contact.[233] Neurotoxicant-induced impairment of adhesion molecule function would also be expected to prevent or retard nerve cell migration. In support of this, the heavy metal methylmercury has been shown to alter membrane *N*-CAM distributions and to reduce neuronal aggregation and neurite fasciculation.[234]

11.12.5.3 Oligodendrocytes as Mediators of Neurotoxicity

The oligodendroglia, or oligodendrocytes as they are alternatively termed, are known to be responsible for the formation of CNS myelin, which functions predominantly to speed up nerve conduction for a given cross-sectional axonal area, thus, greatly conserving space within the brain and allowing for complex interneuronal connections. Perturbations in functions (reviewed Ref. 235) or direct damage to myelin that cannot be compensated for by normal myelin production and maintenance will, therefore, result in loss of the myelin sheath or demyelination. The latter can result from mechanisms that directly impair oligodendrocyte or myelin sheath functions (reviewed Ref. 236). Viral infection of oligodendrocytes,[237] autoimmune responses directed against myelin components,[238,239] and toxic metabolic (triethyltin and alcohol) or hereditary diseases (reviewed Ref. 240) exemplify primary demyelination disorders that directly impair oligodendrocytic function.

Acute exposure of adult rats to triethyltin (TET) produces severe edema of CNS myelin.[241–245] Specifically, TET induces a unique picture of edema fluid splitting myelin lamellae at intraperiod lines. Primary mechanisms associated with this effect include impairment of glial energy-related metabolism, specifically glucose oxidation,[246,247] uncoupling of mitochondrial oxidative phosphorylation[248,249] and inhibition of Na^+/K^+-ATPase activity.[250,251] Loss of oligodendrocytes has also been reported.[252] Complications associated with alcohol exposure are also correlated with myelin disorders, presenting as vitamin B_{12} deficiency, central pontine myelinolysis, or Marchiafava–Bignami disease.[235,236,240]

Hexachlorophene and acetylethyltetramethyltraline (AETT) can also induce myelin damage (reviewed in Ref. 253). The former inhibits various enzymes including carbonic anhydrase and cytochrome oxidase. Another demyelinating agent is diphtheria toxin, which causes a demyelinating lesion when directly injected into the CNS. Phenothiazines which are commonly used as sedatives, anesthetics, and analgesics, as well as blockers of dopamine receptor blockers, are also capable of damaging CNS myelin due to alteration in biochemical processes such as ATP generation and peroxisomal fatty acid oxidation. For an excellent review on CNS myelin disorders during development and adulthood in the CNS the reader is referred de los Monteros and de Vellis.[253]

Cues originating within axons can also lead to demyelination (termed indirect demyelination). For example, axons undergoing Wallerian degeneration trigger simultaneous degeneration in their associated myelin sheaths (see Refs. 253 and 254 for a review). A second form of indirect demyelination referred to as "bystander demyelination" occurs in the process of cell-mediated immune responses, as invading macrophages release lytic enzymes, damaging neighboring oligodendrocytes and myelin.[240] It has also been suggested that nitric oxide and tumor necrosis factor released from activated microglia in the course of CNS inflammation may lead to oligodendrocyte lysis.[255–260]

11.12.6 CONCLUSIONS

In contrast to the previously emphasized support roles of neuroglial cells for neurons, a wide body of literature now provides clear evidence that neuroglia and, in particular, astrocytes and oligodendrocytes directly contact neurons, significantly affecting the morphological and functional differentiation of the latter. This reciprocity between neurons and the neuroglia suggests that the morphological and physiological attributes of neurons are a product of this cell–cell interaction. The wide diversity of neuroglial functions in maintaining

homeostasis and the number of functions assumed by these cells is also quite large. Their potential in modulating damage and repair is also reflected in this chapter. The small amount of information available on the effects of neurotoxic compounds makes the expansion of such studies timely and worthwhile in view of the large number of properties which we attribute to neuroglia and which can be experimentally examined, providing a rich field for investigations of the action of toxic compounds on neuroglial function. Expanded investigations on neuroglial involvement in neurotoxicity is clearly warranted, and as new experimental tools are developed it is likely that further strides will be made in understanding neuroglial-mediated mechanisms of neurotoxicity and neurodegeneration.

ACKNOWLEDGMENTS

Preparation of this chapter was supported in part by NIEHS grant 07331 and USEPA 819210 (MA), and NIEHS grants 06103 and 05022 (MAP). The authors would like to acknowledge the help of Ms. Catherine Porter in the final stages of the preparation of this manuscript.

11.12.7 REFERENCES

1. S. Fedoroff and A. Vernadakis (eds.), 'Astrocytes.' Academic Press, Orlando, FL, 1986, vols. I and II.
2. N. J. Abbott (ed.), 'Glial–Neuronal Interaction,' New York Academy of Sciences, New York, 1991.
3. A. Peters, S. L. Palay and H. de F. Webster, 'The Fine Structure of the Nervous System; Neurons and their Supporting Cells,' 3rd edn., Oxford University Press, New York, 1991.
4. R. Martenson (ed.), 'Myelin: Biology and Chemistry,' CRC Press, Boca Raton, FL, 1992.
5. S. Murphy (ed.), 'Astrocytes: Pharmacology and Function,' Academic Press, New York, 1993.
6. S. Fedoroff, B. H. J. Juurlink and R. Doucette (eds.), 'Biology and Pathology of Astrocyte–Neuron Interactions,' Plenum Press, New York, 1993.
7. M. Aschner and H. K. Kimelberg (eds.), 'The Role of Glia in Neurotoxicity,' CRC Press, Boca Raton, FL, 1996.
8. U. Friedemann, 'Blood–brain barrier.' *Physiol. Rev.*, 1942, **223**, 125–145.
9. T. S. Reese and M. J. Karnovsky, 'Fine structural localization of a blood–brain barrier to exogenous peroxidase.' *J. Cell Biol.*, 1967, **34**, 207–217.
10. G. W. Goldstein and A. L. Betz, 'Recent advances in understanding brain capillary function.' *Ann. Neurol.*, 1983, **14**, 389–395.
11. W. M. Pardridge, W. H. Oldendorf, P. A. Cancilla *et al.*, 'Blood–brain barrier: interface between internal medicine and the brain.' *Ann. Intern. Med.*, 1986, **105**, 82–95.
12. W. M. Pardridge, D. Triguero, J. Yang *et al.*, 'Comparison of *in vitro* and *in vivo* models of drug

transcytosis through the blood–brain barrier.'
J. Pharmacol. Exp. Ther., 1990, **253**, 884–891.

13. M. W. Brightman and T. S. Reese, 'Junctions between intimately apposed cell membranes in the vertebrate brain.' *J. Cell Biol.*, 1969, **40**, 648–677.

14. H. Davson and W. H. Oldendorf, 'Symposium on membrane transport. Transport in the nervous system.' *Proc. R. Soc. Med.*, 1967, **60**, 326–329.

15. W. H. Oldendorf, 'Brain uptake of radiolabeled amino acids, amines and hexoses after arterial injection.' *Am. J. Physiol.*, 1971, **221**, 1629–1639.

16. S. W. Kuffler and J. G. Nicholls, 'The physiology of glial cells.' *Ergeb. Physiol.*, 1966, **57**, 1–90.

17. P. A. Stewart and M. J. Wiley, 'Developing nervous tissue induces formation of blood–brain barrier characteristics in invading endothelial cells: a study using quail–chick transplantation chimeras.' *Dev. Biol.*, 1981, **84**, 183–192.

18. R. C. Janzer and M. C. Raff, 'Astrocytes induce blood-brain barrier properties in endothelial cells.' *Nature*, 1987, **325**, 253–257.

19. J. H. Tao-Cheng, Z. Nagy and M. W. Brightman, 'Tight junctions of brain endothelium *in vitro* are enhanced by astroglia.' *J. Neurosci.*, 1987, **7**, 3293–3299.

20. J. Laterra, C. Guerin and G. W. Goldstein, 'Astrocytes induce neural microvascular endothelial cells to form capillary-like structures *in vitro*.' *J. Cell Physiol.*, 1990, **144**, 204–215.

21. A. Meister, 'On the enzymology of amino acid transport.' *Science*, 1973, **180**, 33–39.

22. J. A. Johnson, A. el Barbary, S. E. Kornguth *et al.*, 'Glutathione *S*-transferase isoenzymes in rat brain neurons and glia.' *J. Neurosci.*, 1993, **13**, 2013–2023.

23. M. A. Philbert, C. M. Beiswanger, M. M. Manson *et al.*, 'Glutathione *S*-transferases and gamma-glutamyl transpeptidase in the rat nervous system: a basis for differential susceptibility to neurotoxicants.' *Neurotoxicology*, 1995, **16**, 349–362.

24. H. E. Lowndes, M. A. Philbert, C. M. Beiswanger *et al.*, in 'Handbook of Toxicology,' eds. L. W. Chang and R. S. Dyer, Marcel Dekker, New York, 1995, pp. 1–27.

25. R. D. White, R. Norton and J. S. Bus, 'Evidence for *S*-methyl glutathione in mediating the acute toxicity of methyl chloride (MeCl).' *Pharmacologist*, 1982, **24**, 172.

26. G. J. Chellman, R. D. White, R. M. Norton *et al.*, 'Inhibition of the acute toxicity of methyl chloride in male B6C3F1 mice by glutathione depletion.' *Toxicol. Appl. Pharmacol.*, 1986, **86**, 93–104.

27. C. H. Phelps, 'Barbiturate-induced glycogen accumulation in brain. An electron microscope study.' *Brain Res.*, 1972, **39**, 225–234.

28. C. H. Phelps, 'An ultrastructural study of methionine sulfoximine-induced glycogen accumulation in astrocytes of the mouse cerebral cortex.' *J. Neurocytol.*, 1975, **4**, 479–490.

29. H. Watanabe and S. Ishii. 'The effect of brain ischemia on the levels of cyclic AMP and glycogen metabolism in gerbil brain *in vivo*.' *Brain Res.*, 1976, **102**, 385–389

30. P. R. Lundgren and J. Miquel, 'The incorporation of isotopic carbon (^{14}C) into the cerebral glycogen of normal and x-irradiated rats.' *J. Neurochem.*, 1970, **17**, 1383–1386.

31. E. Hösli, A. Schousboe and L. Hösli, in 'Astrocytes,' eds. S. Fedoroff and A Vernadakis, Academic Press, Orlando, FL, 1986, vol. 2, pp. 133–143.

32. H. K. Kimelberg, in 'Astrocytes,' eds. S. Fedoroff and A. Vernadakis, Academic Press, Orlando, FL, 1986, vol. 2, pp. 107–127.

33. H. K. Kimelberg, in 'Neuronal Serotonin,' eds. N. N. Osborne and M. Hamon, Wiley, New York, 1988, pp. 347–366

34. R. Massarelli, S. Mykita and G. Sorentino, in 'Astrocytes,' eds. S. Ferdoroff and A. Vernadakis, Academic Press, Orlando, FL, 1986, vol. 2, pp. 155–178.

35. W. Walz, 'Role of glial cells in the regulation of the brain ion microenvironment.' *Prog. Neurobiol.*, 1989, **33**, 309–333.

36. H. Brew, P. T. A. Gray, P. Mobbs *et al.*, 'Endfeet of retinal glial cells have higher densities of ion channels that mediate K^+ buffering.' *Nature*, 1986, **324**, 466–468.

37. E. Newman, 'Inward rectifying potassium channels in retinal Muller (glial) cells.' *Abstr. Soc. Neurosci.*, 1989, 353.

38. B. Nilius and A. Reichenbach, 'Efficient K^+ buffering by mammalian retinal glial cells is due to cooperation of specialized ion channels.' *Pflugers Arch.*, 1988, **411**, 654–660.

39. P. Rakic, 'Neuron–glia relationship during granule cell migration in developing cerebellar cortex: a Golgi and electron microscopic study in Macacus Rhesus.' *J. Comp. Neurol.*, 1971, **141**, 283–312.

40. M. E. Hatten, 'Riding the glial monorail: a common mechanism for glial-guided neuronal migration in different regions of the developing mammalian brain.' *Trends Neurosci.*, 1990, **13**, 179–184.

41. K. Mori, J. Ikeda and O. Hayaishi, 'Monoclonal antibody R2D5 reveals midsagittal radial glial system in postnatally developing and adult brainstem.' *Proc. Natl. Acad. Sci. USA*, 1990, **87**, 5489–5493.

42. L. Seress, 'Development and structure of the radial glia in the postnatal rat brain.' *Anat. Embryol. (Berl.)*, 1980, **160**, 213–226.

43. del Rio Hortega, 'Estudos sobre la neuroglia. La glía de escasas radiaciones.' *Bol. Real Soc. Espan. Hist. Nat.*, 1921, **21**, 63–92.

44. J. E. Vaughn, 'An electron microscopic analysis of gliogenesis in rat optic nerves.' *Z. Zellforsch. Mirosk. Anat.*, 1969, **94**, 293–324.

45. S. Mori, and C. P. Leblond, 'Electron microscopic identification of three classes of oligodendrocytes and a preliminary study of their proliferative activity in the corpus callosum of young rats.' *J. Comp. Neurol.*, 1970, **139**, 1–28.

46. D. W. Vaughn and A. Peters, 'Neuroglial cells in the cerebral cortex of rats from young adulthood to old age: an electron microscope study.' *J. Neurocytol.*, 1974, **3**, 405–429.

47. E. Holmgren, 'Weitere Mitteilungen über die "Saftkanälchen" der Nervenzellen.' *Anat. Anz.*, 1900, **18**, 290–296

48. H. Hydén and A. Pigon, 'A cytophysiological study of the functional relationship between oligodendroglial cells and nerve cells of Deiter's nucleus.' *J. Neurochem.*, 1960, **6**, 57–72.

49. J. Cummins and H. Hydén, 'Adenosine triphosphate levels and adenosine triphosphatases in neurons, glia and neuronal membranes of the vestibular nucleus.' *Biochim. Biophys. Acta*, 1962, **60**, 271–283.

50. H. Hydén and E. Egyházi, 'Glial RNA changes during a learning experiment in rats.' *Proc. Natl. Acad. Sci. USA*, 1963, **49**, 618–624

51. S. G. Waxman and J. M. Ritchie, 'Molecular dissection of the myelinated axon.' *Ann. Neurol.*, 1993, **33**, 121–136.

52. G. Hirose and N. H. Bass, 'Maturation of oligodendroglia and myelinogenesis in rat optic nerve: a quantitative histochemical study.' *J. Comp Neurol.*, 1973, **152**, 201–209.

53. R. P. Skoff, D. Toland, and E. Nast, 'Pattern of myelination and distribution of neuroglial cells along developing optic system of the rat and rabbit.' *J. Comp. Neurol.*, 1980, **191**, 237–253.

54. S. G. Waxman, T. J. Sims, and S. A. Gilmore, 'Cytoplasmic membrane elaborations in oligodendrocytes during myelination of spinal motor neuron axons.' *Glia*, 1988, **1**, 286–291.

55. A. Hirano, 'A confirmation of the oligodendroglial origin of myelin in the adult rat.' *J. Cell Biol.*, 1968, **38**, 637–640.

56. S. K. Ludwin, and D. A. Bakker, 'Can oligodendrocytes attached to myelin proliferate?' *J. Neurosci.*, 1988, **8**, 1239–1244.

57. S. G. Waxman, 'Prerequisites for conduction in demyelinated fibers.' *Neurology*, 1978, **28**, 27–33.

58. C. M. Bowe, J. D. Kocsis, E. F. Targ *et al.*, 'Physiological effects of 4-aminopyridine on demyelinated mammalian motor and sensory fibers.' *Ann. Neurol.*, 1987, **22**, 264–268.

59. P. D. Clouston, L. Kiers, G. Zuniga *et al.*, 'Quantitative analysis of the compound muscle action potential in early acute inflammatory demyelinating polyneuropathy.' *Electroencephalogr. Clin. Neurophysiol.*, 1994, **93**, 245–254.

60. E. A. Grana and G. H. Kraft, 'Electrodiagnostic abnormalities in patients with multiple sclerosis.' *Arch. Phys. Med. Rehabil.*, 1994, **75**, 778–782.

61. V. H. Perry, D. A. Hume and S. Gordon, 'Immunohistochemical localization of macrophages and microglia in the adult and developing mouse brain.' *Neuroscience*, 1985, **15**, 313–326.

62. J. Cammermeyer, in 'Neurosciences Research,' eds. S. Ehrenpreis and O. C. Solnitzky, Academic Press, New York, 1970, vol. 3, pp. 44–130.

63. J. M. Kerns and E. J. Hinsman, 'Neuroglial response to sciatic neurectomy. I. Light microscopy and autoradiography.' *J. Comp. Neurol.*, 1973, **151**, 237–254.

64. J. O'Kusky and M. Colonnier, 'Postnatal changes in the number of astrocytes, oligodendrocytes and microglia in the visual cortex (area 17) of the Macaque monkey: a stereological analysis in normal and monocularly deprived animals.' *J. Comp. Neurol.*, 1982, **210**, 307–315.

65. J. E. Vaughn, P. L. Hinds and R. P. Skoff, 'Electron microscopic studies of Wallerian degeneration in rat optic nerves. I. The multipotential glia.' *J. Comp. Neurol.*, 1970, **140**, 175–206.

66. W. J. Streit and G. W. Kreutzberg, 'Lectin binding by resting and reactive microglia.' *J. Neurocytol.*, 1987, **16**, 249–260.

67. Anonymous, 'Embryonic vertebrate central nervous system: revised terminology. The Boulder Committee.' *Anat. Rec.*, 1970, **166**, 257–261.

68. E. Horstmann, 'Die Faserglia des Selachiergehirns.' *Z. Zellforsch. Mikosk. Anat.*, 1954, **39**, 588–617.

69. O. E. Millhouse, 'A Golgi study of third ventricle tanycytes in the adult rodent brain.' *Z. Zellforsch. Mikrosk. Anat.*, 1971, **121**, 1–13.

70. O. E. Millhouse, 'Light and electron microscope studies of the ventricular wall.' *Z. Zellforsch. Mikrosk. Anat.*, 1972, **127**, 149–174.

71. E. Westergaard, Doctoral dissertation, University of Aarhus, 1970.

72. H. M. Zimmerman, 'The natural history of intracranial neoplasms, with special reference to the gliomas.' *Am. J. Surg.*, 1957, **93**, 913–924.

73. K. C. Snell, H. L. Stewart, H. P. Morris *et al.*, 'Intracranial neurilemmoma and medulloblastoma induced in rats by the dietary adminstration of *N,N'*-2,7-fluorenylenebis(acetamide).' *Natl. Cancer Inst. Monogr.*, 1961, **5**, 85.

74. F. Ikuta and H. M. Zimmerman, 'Virus particles in reactive cells induced by intracerebral implantation of dibenzanthrecene.' *J. Neuropath. Exp. Neurol.*, 1965, **24**, 225–243.

75. A. S. Grove, Jr., G. Di Chiro and G. F. Rabotti, 'Experimental brain tumors, with a report of those induced in dogs by Rouse sarcoma virus.' *J. Neurosurg.*, 1967, **26**, 465–477.

76. W. C. Worthington and R. S. Cathcart, 'Ependymal cilia: distribution and activity in the human brain.' *Science*, 1963, **139**, 221–222.

77. R. S. Cathcart and W. C. Worthington, 'Ciliary movement in the rat cerebral ventricles: clearing action and direction of current.' *J. Neuropathol. Exp. Neurol.*, 1964, **23**, 609–618.

78. M. W. Brightman and S. L. Palay, 'The fine structure of ependyma in the rat brain.' *J. Cell Biol.*, 1963, **19**, 415–439.

79. G. N. Klinkerfuss, 'An electron microscopic study of the ependymal glia of the lateral ventricle of the cat.' *Am. J. Anat.*, 1964, **115**, 71–100.

80. C. R. Jarvis and R. D. Andrew, 'Correlated electrophysiology and morphology of the ependyma in rat hypothalamus.' *J. Neurosci.*, 1988, **8**, 3691–3702.

81. E. Westergaard, 'The fine structure of nerve fibers and endings in the lateral cerebral ventricles of the rat.' *J. Comp. Neurol.*, 1972, **144**, 345–354.

82. E. Thomas and A. G. E. Pearse, 'The fine localization of dehydrogenases in the nervous system.' *Histochimie*, 1961, **2**, 266–282.

83. F. Hajós and E. Bascó, 'The surface-contact glia.' *Adv. Anat. Embryol. Cell Biol.*, 1984, **84**, 1–79.

84. I. B. Black, 'Trophic molecules and evolution of the nervous system.' *Proc. Natl. Acad. Sci. USA*, 1986, **83**, 8249–8252.

85. G. M. Jonakait and I. B. Black, 'Neurotransmitter phenotypic plasticity in the mammalian embryo.' *Curr. Topics Dev. Biol.*, 1986, **20**, 165–175.

86. J. A. Kessler and I. B. Black, 'Similarities in development of substance P and somatostatin in periperal sensory neurons: effects of capsaicin and nerve growth factor.' *Proc. Natl. Acad. Sci. USA*, 1981, **78**, 4644–4647.

87. C. R. Buck, H. J. Martinez. M. V. Chao *et al.*, 'Differential expression of the nerve growth factor receptor gene in multiple brain areas.' *Brain Res. Dev. Brain Res.*, 1988, **44**, 259–268.

88. B. Lu, J. M. Lee, R. Elliott *et al.*, 'Regulation of NGF gene expression in CNS glial by cell–cell contact.' *Brain Res. Mol. Brain Res.*, 1991, **11**, 359–362.

89. H. J. Martinez, C. F. Dreyfus, G. M. Jonakait *et al.*, 'Nerve growth factor selectively increases cholinergic markers but not neuropeptides in rat basal forebrain in culture.' *Brain Res.*, 1987, **412**, 295–301.

90. J. M. Lee, J. E. Adler and I. B. Black, 'Regulation of neurotransmitter expression by a membrane-derived factor.' *Exp. Neurol.*, 1990, **108**, 109–113.

91. S. Cohen-Cory, C. F. Dreyfus and I. B. Black, 'Expression of high- and low-affinity nerve growth factor receptors by Purkinje cells in the developing rat cerebelum.' *Exp. Neurol.*, 1989, **105**, 104–109.

92. B. Lu, C. R. Buck, C. F. Dreyfus *et al.*, 'Expression of NGF and NGF receptor mRNAs in the developing brain: evidence for local delivery and action of NGF.' *Exp. Neurol.*, 1989, **104**, 191–199.

93. M. Yokoyama, I. B. Black and C. F. Dreyfus, 'NGF increases brain astrocyte number in culture.' *Exp. Neurol.*, 1993, **124**, 377–380.

94. N. Y. Ip, T. G. Boulton, Y. Li *et al.*, 'CNTF, FGF and NGF collaborate to drive the terminal differentiation of MAH cells into postmitotic neurons.' *Neuron*, 1994, **13**, 443–455.

95. N. G. Cooper and D. A. Steindler, 'Monoclonal antibody to glial fibrillary acidic protein reveals a parcellation of individual barrels in the early postnatal mouse somatosensory cortex.' *Brain Res.*, 1986, **380**, 341–348.

96. H. Imai, M. R. Park, D. A. Steindler *et al.*, 'The morphology and divergent axonal organization of midbrain raphe projection neurons in the rat.' *Brain Dev.*, 1986, **8**, 343–354.

97. H. Imai, D. A. Steindler and S. T. Kitai, 'The organization of divergent axonal projections from the midbrain raphe nuclei in the rat.' *J. Comp. Neurol.*, 1986, **243**, 363–380.

98. D. A. Steindler and N. G. Cooper, 'Glial and glycoconjugate boundaries during postnatal development of the central nervous system.' *Brain Res.*, 1987, **433**, 27–38.

99. D. A. Steindler, N. G. Cooper, A. Faissner *et al.*, 'Boundaries defined by adhesion molecules during development of the cerebral cortex: the J/1tenascin glycoproten in the mouse somatosensory cortical barrel field.' *Dev. Biol.*, 1989, **131**, 243–260.

100. C. Golgi, 'Opera Omnina,' Hoepli, Milan, 1903.

101. P. Rakic, 'Neuronal–glial interaction during brain development.' *Trends Neurosci.*, 1981, **4**, 184–187.

102. P. Rakic, in 'Cellular and Molecular Biology of Neuronal Development,' ed. I. B. Black, Plenum, New York, 1984, pp. 29–50.

103. M. E. Hatten and R. K. Liem, 'Astroglia cells provide a template for the positioning of developing cerebellar neurons *in vitro*.' *J. Cell Biol.*, 1981, **90**, 622–630.

104. U. E. Gasser and M. E. Hatten, 'Neuron-glia interactions of rat hippocampal cells *in vitro*: glial-guided neuronal migration and neuronal regulation of glial differentiation.' *J. Neurosci.*, 1990, **10**, 1276–1285.

105. M. E. Hatten, M. B. Furie and D. B. Rifkin, 'Binding of developing mouse cerebellar cells to fibronectin: a possible mechanism for the formation of the external granular layer.' *J. Neurosci.*, 1982, **2**, 1195–1206.

106. S. Huck and M. E. Hatten, 'Developmental stage-specific changes in lectin binding to mouse cerebellar cells *in vitro*.' *J. Neurosci.*, 1981, **1**, 1075–1084.

107. W. A. Gregory, J. C. Edmondson, M. E. Hatten *et al.*, 'Cytology and neuron–glial apposition of migrating cerebellar granule cells *in vitro*.' *J. Neurosci.*, 1988, **8**, 1728–1738.

108. D. H. Baird, M. E. Hatten and C. A. Mason, 'Cerebellar target neurons provide a stop signal for afferent neurite extension *in vitro*.' *J. Neurosci.*, 1992, **12**, 619–634.

109. C. A. Mason, J. C. Edmonson and M. E. Hatten, 'The extending astroglial process: development of glial cell shape, the growing tip and interactions with neurons.' *J. Neurosci.*, 1988, **8**, 3124–3134.

110. C. A. Baptista, M. E. Hatten, R. Blazeski *et al.*, Cell–cell interactions influence survival and differentiation of purified Purkinje cells *in vitro*.' *Neuron*, 1994, **12**, 243–260.

111. F. G. Rathjen and M. Schachner, 'Immunocytochemical and biochemical characterization of a new neuronal cell surface component (L1 antigen) which is involved in cell adhesion.' *EMBO J.*, 1984, **3**, 1–10.

112. T. N. Stitt and M. E. Hatten, 'Antibodies that recognize astrotactin block granule neuron binding to astroglia.' *Neuron*, 1990, **5**, 639–649.

113. G. Fishell and M. E. Hatten, 'Astrotactin provides a receptor system for CNS neuronal migration.' *Development*, 1991, **113**, 755–765.

114. J. C. Edmondson, R. K. H. Liem, J. E. Kuster *et al.*, 'Astrotactin: a novel neuronal cell surface antigen that mediates neuron–astroglial interactions in cerebellar microcultures.' *J. Cell Biol.*, 1988, **106**, 505–517.

115. W. O. Gao, N. Heintz and M. E. Hatten, 'Cerebellar granule cell neurogenesis is regulated by cell–cell interactions *in vitro*.' *Neuron*, 1991, **6**, 705–715.

116. E. D. Laywell and D. A. Steindler, 'Boundaries and wounds, glia and glyconconjugates: cellular and molecular analyses of developmental partitions and adult brain lesions.' *Ann. NY Acad. Sci.*, 1991, **633**, 122–141.

117. E. K. O'Malley, I. B. Black and C. F. Dreyfus, 'Local support cells promote survival of substantia nigra dopaminergic neurons in culture.' *Exp. Neurol.*, 1991, **112**, 40–48.

118. E. K. O'Malley, B. A. Sieber, I. B. Black *et al.*, 'Mesencephalic type I astrocytes mediate the survival of substantia nigra dopaminergic neurons in culture.' *Brain Res.*, 1992, **582**, 65–70.

119. E. K. O'Malley, B. A. Sieber, R. S. Morrison *et al.*, 'Nigral type I astrocytes release a soluble factor that increases dopaminergic neuron survival through mechanisms distinct from basic fibroblastic growth factor.' *Brain Res.*, 1994, **647**, 83–90.

120. R. C. Elliott, C. E. Inturrisi, I. B. Black *et al.*, 'An improved method detects differential NGF and BDNF gene expression in response to depolarization in cultured hippocampal neurons.' *Brain Res. Mol. Brain Res.*, 1994, **26**, 81–88.

121. D. G. Schaar, B. A. Sieber, A. C. Sherwood *et al.*, 'Multiple astrocyte transcripts encode nigral trophic factors in rat and human.' *Exp. Neurol.*, 1994, **130**, 387–393.

122. D. G. Schaar, B. A. Sieber, C. F. Dreyfus *et al.* Regional and cell-specific expression of GDNF in rat brain.' *Exp. Neurol.*, 1993, **124**, 368–371.

123. S. Davis and G. D. Yancopoulos, 'The molecular biology of the CNTF receptor.' *Curr. Opin. Cell Biol.*, 1993, **5**, 281–285.

124. Y. Arakawa, M. Sendtner and H. Thoenen, 'Survival effect of ciliary neurotrophic factor (CNTF) on chick embryonic motor neurons in culture: comparison with other neurotrophic factors and cytokines.' *J. Neurosci.*, 1990, **10**, 3507–3515.

125. Y. Masu, E. Wolf, B. Holtmann *et al.*, 'Disruption of the CNTF gene results in motor neuron degeneration.' *Nature*, 1993, **365**, 27–32.

126. H. Mitsumoto, K. Ikeda, B. Klinkosz *et al.*, 'Arrest of motor neuron disease in wobbler mice cotreated with CNTF and BDNF.' *Science*, 1994, **265**, 1052–1053.

127. Z. Sahenk, J. Seharaseyon and J. R. Mendell, 'CNTF potentiates peripheral nerve regeneration.' *Brain Res.*, 1994, **655**, 246–250.

128. J. C. Louis, E. Magal, S. Takayama *et al.*, 'CNTF protection of oligodendrocytes against natural and tumor necrosis factor-induced death.' *Science*, 1993, **259**, 689–692.

129. G. M. Dobrea, J. R. Unnerstall and M. S. Rao, 'The expression of CNTF message and immunoreactivity in the central and peripheral nervous system of the rat.' *Brain Res. Dev. Brain. Res.*, 1992, **66**, 209–219.

130. M. Mata, C. F. Jin and D. J. Fink, 'Axotomy increases CNTF receptor mRNA in rat spinal cord.' *Brain Res.*, 1993, **610**, 162–165.

131. A. J. MacLennan, A. J. Gaskin and D. C. Lado, 'CNTF receptor alpha mRNA expression in rodent cell lines and developing rat.' *Brain Res. Mol. Brain Res.*, 1994, **25**, 251–256.

132. J. T. Henderson, N. A. Seniuk and J. C. Roder, 'Localization of CNTF immunoreactivity to neurons and astroglia in the CNS.' *Brain Res. Mol. Brain Res.*, 1994, **22**, 151–165.

133. H. K. Kimelberg (ed.), 'Glial Cell Receptors,' Raven Press, New York, 1988.

134. M. Chesler and R. P. Kraig, 'Intracellular pH

transients of mammalian astrocytes.' *J. Neurosci.,* 1989, **9**, 2011–2019.

135. R. G. Giffard, H. Monyer, C. W. Christine *et al.,* 'Acidosis reduces NMDA receptor activation, glutamate neurotoxicity and oxygen–glucose deprivation neuronal injury in cortical cultures.' *Brain Res.,* 1990, **506**, 339–342.

136. G. C. Tombaugh and R. M. Sapolsky, 'Mild acidosis protects hippocampal neurons from injury induced by oxygen and glucose deprivation.' *Brain Res.,* 1990, **506**, 343–345.

137. S. F. Traynellis and S. G. Cull-Candy, 'Proton inhibition of *N*-methyl-D-aspartate receptors in cerebellar neurons.' *Nature,* 1990, **345**, 347–350.

138. R. W. Tsien, P. T. Ellinor and W. A. Horne, 'Molecular diversity of voltage-dependent Ca^{2+} channels.' *Trends Pharmacol. Sci.,* 1991, **12**, 349–354.

139. C. L. Bowman, H. K. Kimelberg, M. V. Frangakis *et al.,* 'Astroctyes in primary culture have chemically activated sodium channels.' *J. Neurosci.,* 1984, **4**, 1527–1534.

140. S. Bevan, S. Y. Chui, P. T. A. Grey *et al.,* 'The presence of voltage-gated sodium, potassium and chloride channels in rat cultured astrocytes.' *Proc. R. Soc. Lond. B Biol. Sci.,* 1985, **225**, 299–313.

141. H. Sontheimer, B. R. Ransom, A. H. Cornell-Bell *et al.,* 'Na^+-current expression in rat hippocampal astrocytes *in vitro*: alterations during development.' *J. Neurophysiol.,* 1991, **65**, 3–19.

142. S. W. Kuffler, J. G. Nicholls and R. K. Orkand, 'Physiological properties of glial cells in the central nervous system of amphibia.' *J. Neurophysiol.,* 1966, **29**, 768–787.

143. R. K. Orkand, J. G. Nicholls and S. W. Kuffler, 'Effects of nerve impulses on the membrane potential of glial cells in the central nervous system of amphibia.' *J. Neurophysiol.,* 1966, **29**, 788–806.

144. P. T. A. Grey and J. M. Ritchie, 'Voltage-gated ion channels in Schwann and glial cells.' *Trends Neurosci.,* 1985, **15**, 345–351.

145. H. Sontheimer, J. Trotter, M. Schachner *et al.,* 'Channel expression correlates with differentiation stage during the development of oligodendrocytes from their precursor cells in culture.' *Neuron,* 1989, **2**, 1135–1145.

146. H. Hortnagl, M. L. Berger, G. Sperk *et al.,* 'Regional heterogeneity in the distribution of neurotransmitter markers in the rat hippocampus.' *Neuroscience,* 1991, **45**, 261–272.

147. J. G. McLarnon and S. U. Kim, 'Ion channels in cultured adult human Schwann cells.' *Glia,* 1991, **4**, 534–539.

148. H. K. Kimelberg, T. Jalonen and W. Walz, in 'Astrocytes: Pharmacology and Function,' ed. S. Murphy, Academic Press, New York, 1993.

149. W. Walz and W. R. Schlue, 'Ionic mechanisms of a hyperpolarizing 5-hydroxytryptamine effect on leech neuropile glial cells.' *Brain Res.,* 1982, **250**, 111–121.

150. V. W. Pentreath, T. Radojcic, L. H. Seal *et al.,* 'The glial cells and glia–neuron relations in the buccal ganglia of *Planorbis corneus* L: cytological, qualitative and quantitative changes during growth and ageing.' *Philos. Trans. R. Soc. Lond. B Biol. Sci.,* 1985, **307**, 399–455.

151. M. Cambray-Deakin, B. Pearce, C Morrow *et al.,* 'Effects of neurotransmitters on astrocyte glycogen stores *in vitro*.' *J. Neurochem.,* 1988, **51**, 1852–1857.

152. L. Arbones, F. Picatoste and A. Garcia, 'Histamine stimulates glycogen breakdown and increases $^{45}Ca^{2+}$ permeability in rat astrocytes in primary culture.' *Mol. Pharmacol.,* 1990, **37**, 921–927.

153. R. A. Philibert, K. L. Rogers, A. J. Allen *et al.,* 'Dose-dependent K^+-stimulated efflux of endogenous

154. R. A. Philibert, K. L. Rogers and G. R. Dutten, 'K^+-evoked taurine efflux from cerebellar astrocytes: on the roles of Ca^{2+} and Na^+.' *Neurochem. Res.,* 1989, **14**, 43–48.

155. J. A. Black, B. Friedman, S. G. Waxman *et al.* 'Immuno-ultrastructural localization of sodium channels at nodes of Ranvier and perinodal astrocytes in rat optic nerve.' *Proc. R. Soc. Lond. B. Biol. Sci.,* 1989, **238**, 39–51.

156. J. A. Black, S. G. Waxman, B. Friedman *et al.,* 'Sodium channels in astrocytes of rat optic nerve *in situ*: immuno-electron microscopic studies.' *Glia,* 1989, **2**, 353–369.

157. B. A. Barres, L. L. Y. Chun and D. P. Cory, 'Ion channel expression by white matter glia. I. Type 2 astrocytes and oligodendrocytes.' *Glia,* 1988, **1**, 10–30.

158. B. A. Barres, W. J. Koroshetz, L. L. Y. Chun *et al.,* 'Ion channel expression by white matter glia: the type-1 astrocyte.' *Neuron,* 1990, **5**, 527–544.

159. W. J. Streit, M. B. Graeber and G. W. Kreutzberg, 'Peripheral nerve lesion produces increased levels of major histocompatability complex antigens in the central nervous system.' *J. Neuroimmunol.,* 1989, **21**, 117–123.

160. W. J. Streit, M. B. Graeber and G. W. Kreutzberg, 'Expression of Ia antigens on perivascular and microglial cells after sublethal and lethal neuron injury.' *Exp. Neurol.,* 1989, **105**, 115–126.

161. W. J. Streit and M. B. Graeber, 'Heterogeneity of microglial and perivascular cell populations: insights gained from the facial nucleus paradigm.' *Glia,* 1993, **7**, 68–74.

162. W. F. Hickey and H. Kimura, 'Perivascular microglial cells of the CNS are bone marrow-derived and present antigen *in vivo*.' *Science,* 1988, **239**, 290–292.

163. W. J. Streit, M. B. Graeber and G. W. Kreutzberg, 'Functional plasticity of microglia: a review.' *Glia,* 1988, **1**, 301–307.

164. P. G. Popovich, J. F. Reinhard, Jr., E. M. Flanagan *et al.,* 'Elevation of the neurotoxin quinolinic acid occurs following spinal cord trauma.' *Brain Res.,* 1994, **633**, 348–352.

165. D. Piani, K. Frei, K. Q. Do *et al.,* 'Murine brain macrophages induced NMDA receptor mediated neurotoxicity *in vitro* by secreting glutamate.' *Neurosci. Lett.,* 1991, **133**, 159–162.

166. K. M. Boje and P. K. Arora, 'Microglial-produced nitric oxide and reactive nitrogen oxides mediate neuronal cell death.' *Brain Res.,* 1992, **587**, 250–256.

167. C. C. Chao, S. Hu, T. W. Molitor *et al.,* 'Activated microglia mediate neuronal cell injury via a nitric oxide mechanism.' *J. Immunol.,* 1992, **149**, 2736–2741.

168. D. Giulian, M. Corpuz, S. Chapman *et al.,* Reactive mononuclear phagocytes release neurotoxins after ischemic and traumatic injury to the central nervous system.' *J. Neurosci. Res.,* 1993, **36**, 681–693.

169. W. J. Streit, in 'The Role of Glia in Neurotoxicity,' eds. M. Aschner and H. K. Kimelberg, CRC Press, Boca Raton, FL, 1996, pp. 3–13.

170. K. Nagata, N. Takei, K. Nakajima *et al.,* 'Microglial conditioned medium promotes survival and development of cultured mesencephalic neurons from embryonic rat brain.' *J. Neurosci. Res.,* 1993, **34**, 357–363.

171. B. Chamak, V. Morandi and M. Mallat, 'Brain macrophages stimulate neurite growth and regeneration by secreting thrombospondin.' *J. Neurosci. Res.,* 1994, **38**, 221–233.

172. M. Sensenbrenner, J. Booher and P. Mandel, 'Cultivation and growth of dissociated neurons from chick

embryo cerebral cortex in the presence of different substrates.' *Z. Zellforsch. Mikrosk. Anat.*, 1971, **117**, 559–569

173. A. Bignami, L. Forno and D. Dahl, 'The neuroglial response to injury following spinal cord transection in the goldfish.' *Exp. Neurol.*, 1974, **44**, 60–70

174. M. V. Frangakis and H. K. Kimelberg, 'Dissociation of neonatal rat brain by dispase for preparation of primary astrocyte cultures.' *Neurochem. Res.*, 1984, **9**, 1689–1698.

175. H. K. Kimelberg, D. Vitarella and M. Aschner, in 'The Role of Glia in Neurotoxicity,' eds. M. Aschner and H. K. Kimelberg, CRC Press, Boca Raton, FL, 1996, pp. 311–334.

176. H. Martin, K. Voss, P. Hufnagl *et al.*, 'Morphometric and densitometric investigations of protoplasmic astrocytes and neurons in human hepatic encephalopathy.' *Exp. Pathol.*, 1987, **32**, 241–250.

177. M. J. Mossakowski, K. Renkawek, Z. Krasnicka *et al.*, 'Morphology and histochemistry of Wilsonian and hepatogenic gliopathy in tissue culture.' *Acta Neuropathol. (Berl.)*, 1970, **16**, 1–16.

178. M. D. Norenberg, in 'Advances in Cellular Neurology,' eds. S. Fedoroff and L. Hertz, Academic Press, New York, 1981, vol. 2, pp. 304–338.

179. M. D. Norenberg, in 'Astrocytes: Cell Biology and Pathology of Astrocytes,' eds. S. Fedoroff and A. Vernadakis, Academic Press, New York, 1986, vol. 3, pp. 425–460.

180. J. Albrecht, in 'The Role of Glia in Neurotoxicity,' eds. M. Aschner and H. K. Kimelberg, CRC Press, Boca Raton, FL, 1996, pp. 137–153.

181. J. Albrecht, W. Hilgier, J. W. Lazarewicz *et al.*, in 'Biochemical Pathology of Astrocytes,' eds. M. D. Norenberg, L. Hertz and A. Schousboe, Alan R. Liss, New York, 1988, pp. 465–476.

182. R. A. Hawkins, J. Jessy, A. M. Mans *et al.*, 'Effect of reducing brain glutamine synthesis on metabolic symptoms of hepatic encephalopathy.' *J. Neurochem.*, 1993, **60**, 1000–1006.

183. R. A. Hawkins and J. Jessy, 'Hyperammonemia does not impair brain function in the absence of net glutamine synthesis.' *Biochem. J.*, 1991, **277**, 697–703.

184. R. S. Burns, C. C. Chiueh, S. P. Markey *et al.* 'A primate model of parkinsonism: selective destruction of dopaminergic neurons in the pars compacta of the substantia nigra by *N*-methyl-4-phenyl-1,2,3,6-tetrahydropyridine.' *Proc. Natl. Acad. Sci. USA*, 1983, **80**, 4546–4550.

185. R. E. Heikkila, A. Hess and R. C. Duvoisin, 'Dopaminergic neurotoxicity of 1-methyl-4-phenyl-1,2,5,6-tetrahydropyrine in mice.' *Science*, 1984, **224**, 1451–1453.

186. J. W. Langston, L. S. Forno, C. S. Rebert *et al.*, 'Selective nigral toxicity after systemic administration of 1-methyl-4-phenyl-1,2,5,6-tetrahydropyrine (MPTP) in the squirrel monkey.' *Brain Res.*, 1984, **292**, 390–394.

187. A. M. Marini, J. P. Schwartz and I. J. Kopin, 'The neurotoxicity of 1-methyl-4-phenylpyridinium in cultured cerebellar granule cells.' *J. Neurosci.*, 1989, **9**, 3665–3672.

188. D. De Monte, J. E. Reyland, I. Irwin *et al.*, 'Astrocytes as the site for bioactivation of neurotoxins.' *Neurotoxicology*, 1996, **17**, 697–704.

189. P. Levitt, J. E. Pintar and X. O. Breakefield, 'Immunocytochemical demonstration of monoamine oxidase B in brain astrocytes and serotonergic neurons.' *Proc. Natl. Acad. Sci. USA*, 1982, **79**, 6385–6389.

190. B. R. Ransom, D. M. Kunis, I. Irwin *et al.*, 'Astrocytes convert the parkinsonism inducing neu-rotoxin, MPTP, to its active metabolite, MPP$^+$.' *Neurosci. Lett.*, 1987, **75**, 323–328.

191. M. Takada, Z. K. Li and T. Hattori, 'Astroglial ablation prevents MPTP-induced nigrostriatal neuronal death.' *Brain Res.*, 1990, **509**, 55–61.

192. R. Schwarcz, W. O. Whetsell, Jr. and R. M. Mangano, 'Quinolinic acid: an endogenous metabolite that produces axon-sparing lesions in rat brain.' *Science*, 1983, **219**, 316–318.

193. R. Schwarcz, A. C. Foster, E. D. French *et al.*, 'Excitotoxic models for neurodegenerative disorders.' *Life Sci.*, 1984, **35**, 19–32.

194. R. Schwarcz, C. A. Tamminga, R. Kurlan *et al.*, 'Cerebrospinal fluid levels of quinolinic acid in Huntington's disease and schizophrenia.' *Ann. Neurol.*, 1988, **24**, 580–582.

195. M. F. Beal, W. R. Matson, K. J. Swartz *et al.*, 'Kynurenine pathway measurements in Huntington's disease striatum: evidence for reduced formation of kynurenic acid.' *J. Neurochem.*, 1990, **55**, 1327–1339.

196. R. Schwarcz, P. Guidetti and R. C. Roberts, in 'The Role of Glia in Neurotoxicity,' eds. M. Aschner and H. K. Kimelberg, CRC Press, Boca Raton, FL, 1996, pp. 245–262.

197. A. B. Young, J. T. Greenamyre, Z. Hollingsworth *et al.*, 'NMDA receptor losses in putamen from patients with Huntington's Disease.' *Science*, 1988, **241**, 981–983.

198. A. M. Dam, 'Epilepsy and neuron loss in the hippocampus.' *Epilepsia*, 1980, **21**, 617–629.

199. R. Schwarcz, E. Okuno, R. J. White *et al.*, '3-Hydroxyanthranilate oxygenase activity is increased in the brains of Huntington's disease victims.' *Proc. Natl. Acad. Sci. USA*, 1988, **85**, 4079–4081.

200. F. Du, E. Okuno, W. O. Whetsell Jr. *et al.*, 'Distribution of quinolinic acid phosphoribosyltransferase in the human hippocampal formation and parahippocampal gyrus.' *J. Comp. Neurol.*, 1990, **295**, 71–82.

201. H. K. Kimelberg, S. K. Goderie, S. Higman *et al.*, 'Swelling-induced release of glutamate, aspartate, and taurine from astrocyte cultures.' *J. Neurosci.*, 1990, **10**, 1583–1591.

202. D. W. Choi, 'Glutamate neurotoxicity and diseases of the nervous system.' *Neuron*, 1988, **1**, 623–634.

203. D. W. Choi, 'Bench to bedside: the glutamate connection.' *Science*, 1992, **258**, 241–243.

204. H. K. Kimelberg, in 'Advances in Comparative and Environmental Physiology, Volume and Osmolality Control in Animal Cells,' eds. R. Gilles, E. K. Hoffmann and L. Bolis, Springer-Verlag, Berlin, 1991, vol. 9, pp. 81–117.

205. H. K. Kimelberg, 'Astrocytic edema in CNS trauma.' *J. Neurotrauma*, 1992, **9**, S71–S81.

206. H. K. Kimelberg and B. R. Ransom, in 'Astrocytes: Cell Biology and Pathology of Astrocytes,' eds. S. Fedoroff and A. Vernadakis, Academic Press, New York, 1986, vol. 3, pp. 129–166.

207. H. M. Gerschenfeld, F. Wald, A. Zadunaisky *et al.*, 'Function of astroglia in water-ion metabolism of the central nervous system.' *Neurology*, 1959, **9**, 412.

208. D. Holtzman, C. DeVries, H. Nguyen *et al.*, Maturation of resistance to lead encephalopathy: cellular and subcellular mechanisms.' *Neurotoxicology*, 1984, **5**, 97–124.

209. M. A. Philbert, C. C. Nolan, J. E. Cremer *et al.* '1,3-Dinitrobenzene-induced encephalopathy in rats.' *Neuropathol. Appl. Neurobiol.*, 1987, **13**, 371–389.

210. I. Watanabe, in 'Experimental and Clinical Neurotoxicology,' eds. P. S. Spencer and H. H. Schaumburg, Williams and Wilkins, Baltimore, MD, 1980, pp. 545–557.

211. M. C. Yu, L. Bakay and J. C. Lee, 'Ultrastructure of the central nervous system after prolonged hypoxia. II. Neuroglia and blood vessels.' *Acta Neuropathol. (Berl.)*, 1972, **22**, 235–244.

212. M. Aschner, N. B. Eberle, K. Miller *et al.*, 'Interactions of methylmercury with rat primary astrocyte cultures: effects on rubidium and glutamate uptake and efflux and induction of swelling.' *Brain Res.*, 1990, **530**, 245–250.

213. R. M. LoPachin, Jr. and M. Aschner, 'Glial–neuronal interactions: relevance to neurotoxic mechanisms.' *Toxicol. Appl. Pharmacol.*, 1993, **118**, 141–158.

214. D. Vitarella, D. J. DiRisio, H. K. Kimelberg *et al.*, 'Regulatory volume decrease in primary astrocyte cultures: relevance to methylmercury neurotoxicity.' *Neurotoxicology*, 1996, **17**, 117–123.

215. L. Hertz, 'Functional interactions between neurons and astrocytes I. Turnover and metabolism of putative amino acid transmitters.' *Prog. Neurobiol.*, 1979, **13**, 277–323.

216. H. K. Kimelberg, P. Sankar, E. R. O'Connor *et al.*, in 'Progress in Brain Research. Neuron–astrocyte interactions,' eds. A. C. H. Yu, L. Hertz, M. D. Norenberg *et al.*, Elsevier, Amsterdam, 1992, vol. 94, pp. 57–68.

217. E. K. Hoffmann and H.-A. Kolb, in 'Comparative and Environmental Physiology,' eds. R. Gilles, E. K. Hoffmann and L. Bolis, Springer-Verlag, New York, 1991, vol. 9, pp. 140–185.

218. J. E. Olson, R. Sankar, D. Holtzman *et al.* 'Energy-dependent volume regulation in primary cultured cerebral astrocytes.' *J. Cell Physiol.*, 1986, **128**, 209–215.

219. H. Pasantes-Morales and A. Schousboe, 'Volume regulation in astrocytes: a role for taurine as an osmoeffector.' *J. Neurosci. Res.*, 1988, **20**, 503–509.

220. R. Gilles, E. K. Hoffman and L. Bolis (eds.), in 'Advances in Comparative and Environmental Physiology: Volume and Osmolality Control in Animal Cells,' Springer Verlag, Berlin, 1991.

221. S. M. Rothman and J. W. Olney, 'Excitotoxicity and the NMDA receptor—still lethal after eight years.' *Trends Neurosci.*, 1995, **18**, 57–58.

222. B. Meldrum and J. Garthwaite, 'Excitatory amino acid neurotoxicity and neurodegenerative disease.' *Trends Pharmacol. Sci.*, 1990, **11**, 379–387.

223. A. Martinez-Hernandez, K. P. Bell and M. D. Norenberg, 'Glutamine synthetase: glial localization in brain.' *Science*, 1977, **195**, 1356–1358.

224. M. D. Norenberg, 'Distribution of glutamine synthetase in the central nervous system.' *J. Histochem. Cytochem.*, 1979, **27**, 756–762.

225. M. D. Norenberg and A. Martinez-Hernandez, 'Fine structural localization of glutamine synthetase in astrocytes of rat brain.' *Brain Res.*, 1979, **161**, 303–313.

226. A. Schousboe and I. Divac, 'Difference in glutamate uptake in astrocytes cultured from different brain regions.' *Brain Res.*, 1979, **177**, 407–409.

227. R. A. Waniewski and D. L. Martin, 'Exogenous glutamate is metabolized to glutamine and exported by rat primary astrocyte cultures.' *J. Neurochem.*, 1986, **47**, 304–313.

228. S. Berl, A. Lajtha and A. Waelsch, 'Amino acid and protein metabolism. VI. Cerebral compartments of glutamic acid metabolism.' *J. Neurochem.*, 1961, **7**, 186.

229. J. T. Coyle and P. Puttfarcken, 'Oxidative stress, glutamate, and neurodegenerative disorders.' *Science*, 1993, **262**, 689–695.

230. K. R. Reuhl and H. E. Lowndes, in 'Neurotoxicology,' eds. H. A. Tilson and C. L. Mitchell, Raven Press, New York, 1992, pp. 67–81.

231. M. A. Verity, in 'Neurotoxicology,' eds. H. A. Tilson and C. L. Mitchell, Raven Press, New York, 1992, pp. 1–20.

232. G. R. Cookman, W. King and C. M. Regan, 'Chronic low-level lead exposure impairs embryonic to adult conversion of the neural cell adhesion molecule.' *J. Neurochem.*, 1987, **49**, 399–403.

233. S. Hoffman and G. M. Edelman, 'Kinetics of homophilic binding by embryonic and adult forms of the neural cell adhesion molecule.' *Proc. Natl. Acad. Sci. USA*, 1983, **80**, 5762–5766.

234. K. R. Reuhl and R. Borgeson, 'Methylmercury, N-CAM expression and dysmorphogenesis.' *Toxicologist*, 1990, **10**, 136.

235. R. C. Armstrong, in 'National Institute on Alcohol Abuse and Alcoholism Research Monograph, Alcohol and Glial Cells,' NIH Publication No. 94-3742, Bethesda, MD, Monograph 27, 1994, pp. 41–53.

236. C. S. Raine, in 'Myelin,' 2nd edn., ed. P. Morell, Plenum Press, New York, 1984, pp. 259–310.

237. J. K. Fazakerly and M. J. Buchmeier, in 'Myelin: Biology and Chemistry,' ed. R. E. Martenson, CRC Press, Boca Raton, FL, 1992, pp. 893–934.

238. M. L. Shin and C. L. Koski, in 'Myelin: Biology and Chemistry,' ed R. E. Martenson, CRC Press, Boca Raton, FL, 1992, pp. 801–832.

239. E. C. Alvord, L. M. Rose and T. L. Richards, in 'Myelin: Biology and Chemistry,' ed. R. E. Martenson, CRC Press, Boca Raton, FL, 1992, pp. 849–892.

240. W. T. Norton and W. Cammer, in 'Myelin,' 2nd edn., ed. P. Morell, Plenum Press, New York, 1984, pp. 369–404.

241. F. P. Aleu, R. Katman and R. D. Terry, 'Fine structure and electrolyte analyses of cerebral edema induced by alkyl tin intoxication.' *J. Neuropathol. Exp. Neurol.*, 1963, **22**, 403.

242. I. Klatzo, 'Presidential address. Neuropathological aspects of brain edema.' *J. Neuropathol. Exp. Neurol.*, 1967, **26**, 1–14.

243. R. N. Magee, H. B. Stoner and J. M. Barnes, 'The experimental production of edema in the central nervous system of the rat by triethyltin compounds.' *Pathol. Bacteriol.*, 1957, **73**, 107.

244. B. Veronesi and S. Bondy, 'Triethyltin-induced neuronal damage in neonatally exposed rats.' *Neurotoxicology*, 1986, **7**, 69–76.

245. W. D. Blaker, M. R. Krigman, D. J. Thomas *et al.*, 'Effect of triethyltin on myelination of the developing rat.' *J. Neurochem.*, 1981, **36**, 44–52.

246. J. E. Cremer, 'Selective inhibition of glucose oxidation by triethyltin in rat brain *in vivo*.' *Biochem. J.*, 1970, **119**, 95–102.

247. A. E. Lock, 'The action of triethyltin in the respiration of rat brain cortex slices.' *J. Neurochem.*, 1976, **26**, 887–892.

248. W. N. Aldridge, B. W. Street and D. W. Skicleter, 'Oxidative phosphorylation. Halide-dependent and halide-independent effects of triorganotin and triorganolead compounds on mitochondrial functions.' *Biochem. J.*, 1977, **168**, 353–364.

249. M. Stockdale, A. P. Dawson and M. J. Selwyn, 'Effects of trialkyltin and triphenyltin compounds on mitochondrial respiration.' *Eur. J. Biochem.*, 1970, **15**, 342–351.

250. K. S. Jacobs, J. J. Lemasters and L. W. Reiter, in 'Developments in the Science and Practice of Toxicology,' eds. A. Hayes, R. C. Schnell and T. S. Miya, Elsevier, Amsterdam, 1983.

251. R. M. Torak, 'The relationship between adenosine trisphosphate activity and triethyltin toxicity in the production of cerebral edema of the rat.' *Am. J. Pathol.*, 1965, **46**, 245.

252. US, DHHS, PHS, NIH, Monograph, 'Alcohol-induced brain damage.' *Research Monograph*, 22.

253. A. de los Monteros and J. de Vellis, in 'The Role of Glia in Neurotoxicity,' eds. M. Aschner and H. K. Kimelberg, CRC Press, Boca Raton, FL, 1996, in press.

254. M. E. Smith and J. A. Benjamins, in 'Myelin,' 2nd edn., ed. P. Morell, Plenum Press, New York, 1984, pp. 441–487.

255. A. Compston, 'Limiting and repairing the damage in multiple sclerosis.' *Schweitz. Med. Wochenschr.*, 1993, **123**, 1145–1152.

256. J. E. Merrill, L. J. Ignarro, M. P. Sherman *et al.*, Microglial cell cytotoxicity of oligodendrocytes is mediated through nitric oxide.' *J. Immunol.*, 1993, **151**, 2132–2141.

257. H. K. Kimelberg and M. D. Norenberg, 'Astrocytes.' *Sci. Am.*, 1989, **260**, 66–72; 74; 76.

258. H. K. Kimelberg and M. Aschner, in 'National Institute on Alcohol Abuse and Alcoholism Research Monograph, Alcohol and Glial Cells,' NIH Publication No. 94-3742, Bethesda, MD, Monograph 27, 1994, pp. 1–40.

259. D. Vitarella, D. J. DiRisio, H. K. Kimelberg *et al.*, 'Potassium and taurine release are highly correlated with regulatory volume decrease in neonatal primary rat astrocyte cultures.' *J. Neurochem.*, 1994, **63**, 1143–1149.

260. W. O. Whetsell, Jr., C. Köhler and R. Schwarcz, in 'The Biochemical Pathology of Astrocytes,' eds. M. D. Norenberg, L. Hertz and A. Schousboe, Alan R. Liss, New York, 1988, pp. 191–202.

SECTION IV

11.13
Toxicology of the
Neuromuscular Junction

WILLIAM D. ATCHISON

Michigan State University, East Lansing, MI, USA

11.13.1 INTRODUCTION

Because of the importance of skeletal muscle function to respiration, as well as voluntary movement, it is not surprising that a number of naturally occurring toxins such as botulinum toxin or anthropogenic toxicants such as cholinesterase inhibitors have actions directed at the synapse made by motor axons on skeletal muscle or neuromuscular junctions. Additionally, due to its well-characterized physiology, biochemistry, and ultrastructure, the neuromuscular junction has been commonly used as a model synapse to examine the actions of putative neurotoxicants or neurotoxins on synaptic function.

Toxicants acting at the neuromuscular junction typically alter the presynaptic processes of synthesis, storage, or release of the transmitter acetylcholine (ACh), and/or disrupt the postsynaptic processes of recognition of ACh by its target receptor, activation of conductances through the receptor-associated ion channel, or impair breakdown of ACh.

11.13.1.1 Neuromuscular Junction Morphology

As shown in the electron micrograph of Figure 1, the morphological features of the neuromuscular junction are typical of chemical synapses. The presynaptic motor nerve terminal contains characteristic synaptic vesicles. The terminal is separated from the postsynaptic muscle endplate membrane by a synaptic cleft. The postsynaptic membrane is highly invaginated and contains the nicotinic cholinergic receptors for ACh.

Numerous excellent reviews and chapters have examined neuromuscular transmission.[1-4] Moreover, the cell biology of transmitter release is evolving rapidly[5-15] and will not be dealt with in this chapter. As such, only a brief synopsis of normal transmission will be given. Chemical neurotransmission can be subdivided into: (i) processes associated directly with the synthesis and release of neurotransmitter, and (ii) processes associated with the binding of transmitter to its receptor sites with subsequent opening of ionic conductance channels leading to the postsynaptic polarization. The former occur in the presynaptic nerve terminal (presynaptic processes) while the latter occur in the receiving cells (postsynaptic processes). Effects at either site alter neuromuscular transmission and hence depress muscle activity. Consequently, to obtain further information as to the nature of potential effects on neurotransmission, both processes must be studied independently.

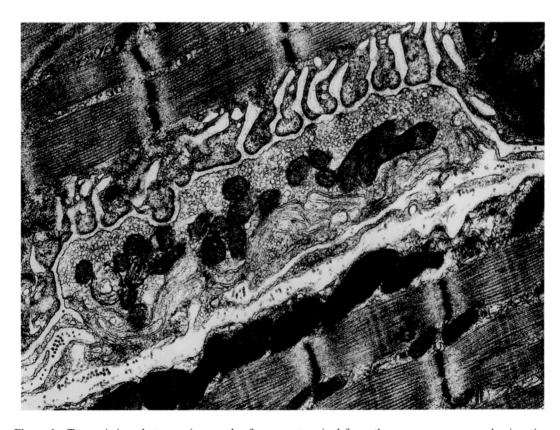

Figure 1 Transmission electron micrograph of a nerve terminal from the mouse neuromuscular junction. Magnification, 14 000× (after Atchison *et al.*, 1994).

11.13.1.2 Neuromuscular Junction Physiology

As shown in Figure 2, the ionic changes in the nerve terminal membrane associated with propagation of an action potential into the terminal cause depolarization of the terminal. This induces entry of Ca^{2+} through voltage-dependent ion channels.[16,17] It is believed that the Ca^{2+} channels are closely apposed to transmitter release sites.[18–20] so Ca^{2+} influx increases the probability of transmitter release by interacting with Ca^{2+}-binding molecules—the exact details of which are only now being elucidated.[5,6] The pharmacological identity of Ca^{2+} channels at mammalian motor nerve terminals is still unclear, although the predominant form(s) appear to be sensitive to the spider venom toxin ω-agatoxin IVA[21] and the cone snail toxin ω-conotoxin MVIIC. Conversely, while the "L" and "N" channel subtypes can contribute to mammalian neuromuscular transmission under certain circumstances,[22–24] they do not appear to be primarily responsible for regulating ACh release at mammalian neuromuscular junctions.[22,23,25–27] At amphibian neuromuscular junctions, "N" type channels do play a prominent role in transmitter release.[18,28] Irrespective of the type of Ca^{2+} channel involved, influx of Ca^{2+} likely raises the local $[Ca^{2+}]$ around the release sites to very high levels ($\sim 100\ \mu M$),[20] inducing the release of numerous packets or quanta each containing approximately 10 000–20 000 ACh molecules, into the synaptic cleft. Following diffusion of the ACh across the synaptic cleft, it interacts with receptor macromolecules located in the postjunctional muscle membrane (acetylcholine receptors; AChR). This results in the increased permeability of the endplate membrane to Na^+, K^+, and to a lesser extent, Ca^{2+}. Because permeability to Na^+ exceeds that of K^+ and ionic movement is passive, the net charge movement is inward, and the muscle endplate membrane depolarizes. Action of ACh is terminated due to rapid hydrolysis by acetylcholinesterase. If ACh-induced depolarization of the endplate reaches a threshold level, a muscle action potential propagates from the endplate region leading to muscle contraction. Choline generated by the breakdown of ACh is taken up into the nerve terminal by high-affinity transport systems. Ca^{2+} which entered the terminal during the action potential is buffered by intracellular binding proteins or organelles and slowly dissipated back to the extracellular fluid. Synaptic vesicles are thought to move from a reserve pool to release sites at the terminal membrane.

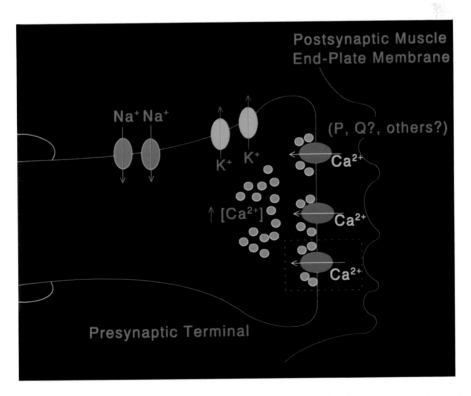

Figure 2 The ionic changes in the nerve terminal membrane associated with propagation of an action potential

11.13.1.2.1 *Quantal release of ACh*

There are two forms of quantal release of ACh at the neuromuscular junction.[29] The first is known as the endplate potential or EPP, and occurs in response to nerve stimulation. The EPP is presumed to result from the synchronous discharge of multiple vesicles in responce to Ca^{2+} entry induced by the action potential. It gives rise to the graded depolarization of the endplate. The second form, known as the miniature endplate potential (MEPP), occurs spontaneously and is thought to result from random discharge of a single quantum of transmitter, although some still debate this. Unlike the EPP, for which the function is obvious, the function of MEPPs in normal synaptic physiology is less clear. They have been suggested to play a trophic role in maintaining normal synaptic structure.[30–32] Factors which reduce the sensitivity of the postsynaptic ACh receptors decrease the amplitude of MEPPs, whereas agents which either facilitate flow of ions through the receptor-activated channel, or which prolong the duration of action of ACh, such as acetylcholinesterase inhibitors, increase MEPP amplitude. Similarly, chemicals which impair the presynaptic synthesis or storage of ACh in synaptic vesicles alter MEPP amplitude. For example, blocking of high-affinity transport of choline into the axon terminal by hemicholinium-3,[33] or blocking of active uptake of ACh into cholinergic vesicles by vesamicol both reduce MEPP amplitude.

MEPPs occur randomly with a normal frequency of occurrence of approximately $0.5-2$ MEPPs s^{-1} (Hz). MEPP frequency is increased by treatments which increase the $[Ca^{2+}]$ within the nerve terminal cytosol.[2,34] Thus, agents which increase the entry of Ca^{2+} [35,36] or those which release Ca^{2+} from internal bound stores[37,38] increase MEPP frequency. Following a train of high-frequency stimuli delivered to the nerve (tetanus), a transient increase in MEPP frequency occurs,[39] an effect referred to as post-tetanic potentiation.[40–43] Study of the effects of chemicals on these processes has provided clues to the ways in which chemicals can disrupt presynaptic processes associated with mobilization of transmitter[44] or maintenance of Ca^{2+} homeostasis within the axon terminal.[45]

Two abnormal forms of quantal release also occur. One of these is characterized by abnormally small MEPPs or "sub-MEPPs."[46] This class of small MEPPs gives rise to a separate distribution of amplitudes which has been proposed to represent subunits of the typical MEPP. Other hypotheses to explain the presence of "subMEPPs" include incompletely filled vesicles or quanta released from release sites nonadjacent to active zones.

The other form of MEPP observed is the so-called "giant MEPP," (gMEPP),[47] which is observed at a variable rate at normal neuromuscular junctions. These responses occur in a modified form more frequently at junctions during poisoning with agents such as botulinum toxin and dithibiuret (DTB).[44,48,49] Giant MEPPs differ in several important respects from normal MEPPs.[50]

Motor nerve terminals can release ACh in a nonquantal form,[51] the source of which is presumed to be the cytoplasm. In neurochemical experiments, over 95% of the ACh release measured from the nerve terminal at rest is nonquantal and not associated with MEPPs.

11.13.1.2.2 *Postsynaptic nicotinic receptor*

The postsynaptic nicotinic AChR to which ACh binds following its release is an integral membrane protein consisting of five subunits: two α subunits and one each of β, γ or ϵ, and δ. The γ and ϵ subunits replace one another during different stages of development. The γ subunit is expressed in developing or denervated muscle, and is replaced in adult muscle by the ϵ subunit.[52] The five subunits combine to form a transmembrane pore, through which Na^+, K^+, and Ca^{2+} ions flow in response to ACh. The α subunits also contain the binding sites for ACh; thus each receptor–channel complex contains two binding sites for ACh.

Binding of ACh to the receptor leads to the opening of receptor-associated ion channels. These channels open for several milliseconds and allow Na^+ entry and K^+ efflux from the muscle cell. Other positively charged molecules can pass through the channel but do not contribute appreciably to the current under physiological conditions.[53,54] At a normal resting membrane potential of approximately -90 mV, the predominantly inward cationic current leads to depolarization of the endplate region known as the EPP.

Using the patch voltage clamp method, the behavior of single AChR channels can be studied. After two ACh molecules bind to the AChR, the channel undergoes a burst of activity, during which time it opens and closes several times. Openings are brief and triggered by binding of two ACh molecules to the receptor. Dissociation of one or both agonist molecules from the receptor leads to channel closure and the cycle can repeat. Examination of single ACh channel currents reveals that most events appear as a burst of channel openings separated by brief closures. Each burst of openings apparently represents an oscillation between the closed and open state, in which the channel opens, closes, and reopens before the agonist molecule can dissociate. The AChR has a greater affinity (up to 100×) for binding of the second agonist molecule once one molecule is bound.[55] Thus, openings associated with the singly liganded AChR channel are normally observed at a low frequency.

In early studies in which the effects of D-tubocurarine (curare) on transmission were examined, changes in transmission were observed which appeared to represent presynaptic effects of curare on transmitter release. These early studies indicated that ACh appeared to act presynaptically to mobilize transmitter stores (feed-forward mechanism) and that D-tubocurarine blocked this effect causing a decline in transmitter release during repetitive stimulation.[3,56] Evidence indicates that both nicotinic and muscarinic receptors are present presynaptically at the neuromuscular junction and that these receptors play a role in the modulation of transmitter release.[57,58]

11.13.2 TOXIC AGENTS AFFECTING NEUROMUSCULAR FUNCTION BY PREDOMINANTLY PREJUNCTIONAL EFFECTS

11.13.2.1 Biological Toxins

11.13.2.1.1 Botulinum toxin

Botulinum toxin is a generic term applied to a group of toxins produced by the bacterium *Clostridium botulinum*. Currently, eight different botulinum toxins have been identified: types A, B, C_1, C_2, D, E, F, and G. The molecular pharmacology and biochemistry of botulinum toxins have been the subject of a number of excellent reviews[59–62] and readers are referred to these for more detailed information.

Botulinum toxins are extremely potent blockers of neuromuscular transmission. Block is relatively rapid in onset. If death does not occur, paralysis is long-lived. The re-establishment of normal transmission occurs because of reinnervation due to nerve sprouting.[63,64]

The neuromuscular blocking effects have been studied most for botulinum toxin A.[65] Slight differences, especially in potency, have been reported for other serotypes of botulinum toxin, most notably type B.[66,67] Specifically, botulinum toxin B was reported to have a shorter duration of action than A, and to be less effectively antagonized by potassium channel blockers. Botulinum toxins D and E were similarly reported to be less potent than type A.[66–69]

The actions of botulinum toxin are directed at ACh secretion. The characteristic effect of the toxins is to reduce evoked release of ACh to virtually a complete block. Thus EPPs recorded at botulinum toxin-poisoned junctions are subthreshold and usually very small. Botulinum toxins do not block Ca^{2+} influx.[70]

Several investigators have reported that during the early stages of poisoning with botulinum toxins that small-amplitude sub-MEPPs occur, resulting in a shift of amplitude histograms to the left.[68,71,72] In addition, after several days of poisoning, gMEPPs appear, with amplitudes occasionally approaching 10 mV.[48,73,74] These MEPPs are also characterized by their slow rise time compared with MEPPs at unpoisoned junctions. Slow MEPPs or gMEPPs increased in incidence 6–10 days after intoxication with botulinum toxin A, but their occurrence differs for soleus compared to *extensor digitorum longus* muscles.[74]

A three-step process involving binding of the toxin to a membrane receptor, internalization of the toxin, and interaction with the component in the release process has been proposed to underlie the actions of botulinum toxins on ACh release.[59]

The unique identity of binding of botulinum toxin from its paralytic action was first hypothesized based on the observation that for a given concentration of botulinum toxin onset of paralysis was more rapid when the nerve was stimulated frequently.[75] Conversely, paralysis occurred after much longer latencies during inactivity. It was subsequently shown that botulinum antitoxin could not antagonize paralysis induced by the toxin, but could prevent paralysis when present during the initial binding stages.[76] Presumably toxin bound to the membrane surface was exposed to the antitoxin, while that internalized would be refractory to action of antitoxin.

The heavy chain of the toxin molecule is evidently responsible for binding. Addition of the heavy chain itself to a nerve-muscle preparation protects the preparation from paralysis due to the intact toxin molecule,[60] but does not paralyze the preparation. Thus, the heavy chain binds apparently to the receptor, occluding it from the native toxin. Although the specific binding site for botulinum toxin has not yet been identified, the toxin binds with high-affinity to gangliosids,[77] specifically, ganglioside G_{T1b}.[78] However, this effect is somewhat dependent on the toxin serotype.[79]

Internalization of the botulinum toxin molecule is the second step in the process of neuromuscular dysfunction. The notion that internalization preceded the induction of block was based on the observation that there was a marked difference in sensitivity of neuromuscular transmission to botulinum toxin depending upon the temperature. Incubation of diaphragm preparations with toxin at 4 °C resulted in the toxin remaining accessible to a neutralizing antibody, whereas incubation at more physiological temperatures resulted in rapid disappearance of toxin from accessibility to the antibody. Studies on isolated lipid membranes suggest that the heavy chain of the toxin forms a pore through which a peptide such as the light chain passes.

The final step in the poisoning process is intracellular. The extraordinary sensitivity of neuromuscular transmission to block by botulinum toxin implies a unique and long-lasting action. An enzymatic action was proposed by Simpson,[59] although the specific identity of this proposed action was unclear. More recently, three substrates have been identified for different botulinum toxin serotypes (Figure 3). The light chains of botulinum toxins A–E contain a Zn^{2+} binding site.[80–84] Moreover, blocking protease activity slows the onset of activity of the toxin, thus a sensitive Zn^{2+} protease site on a component of the release apparatus has been proposed. Synaptobrevin I/II has been implicated as one potential substrate, since botulinum toxin B and tetanus toxin have both been shown to cleave synaptobrevin,[86–88] and this cleavage was prevented by the chelation of zinc. Synaptobrevin has been implicated in the formation of a fusion pore between the synaptic vesicle and plasma membrane at the active zone. The specific protein molecules identified as substrates for the remaining botulinal toxins are synaptosomal protein of 25 kDa (SNAP 25) and syntaxin. The exact roles of these proteins in exocytosis are not yet known with certainty, but it is believed that they make up part of a prefusion complex whose association is needed for docking of the synaptic vesicle (see Figure 4(A)–(D)).

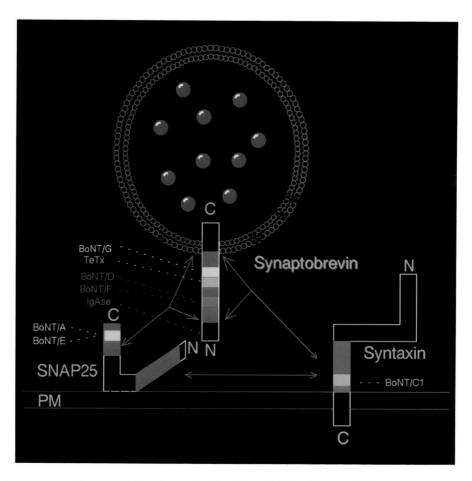

Figure 3 Diagram of proposed target proteins for *Clostridial botulinum* and *Clostridial tetani* toxins. Syntaxin and SNAP 25 (snaptosomal-associated protein, 25 kDa) are localized to the nerve terminal membrane, whereas synaptobrevin is localized to the synaptic vesicles. Concensus target sequences for the binding and proteolytic action of the various toxins are depicted (after Hayashi *et al.*[85]).

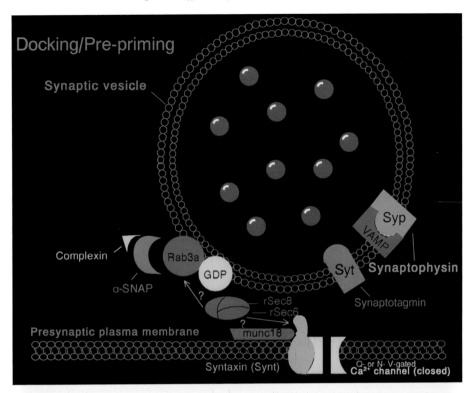

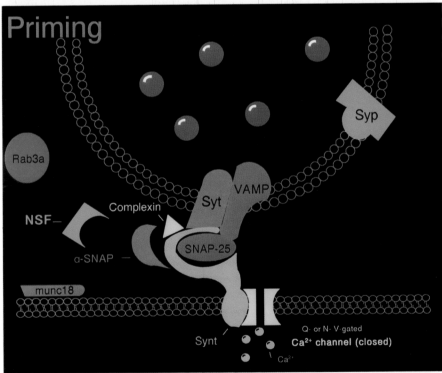

Figure 4 A proposed mechanism of the proteins SNAP 25 and syntaxin in exocytosis.

11.13.2.1.2 Tetanus toxin

Tetanus toxin is another highly potent toxin produced by the *Clostridial* family of bacterium (*Clostridium tetani*). Like botulinum toxin, tetanus toxin binds to the presynaptic motor nerve terminal and disrupts the release of ACh.[89,90] MEPP frequency is reduced and nerve-evoked release of ACh is blocked.[91,92] However, tetanus toxin does not affect non-

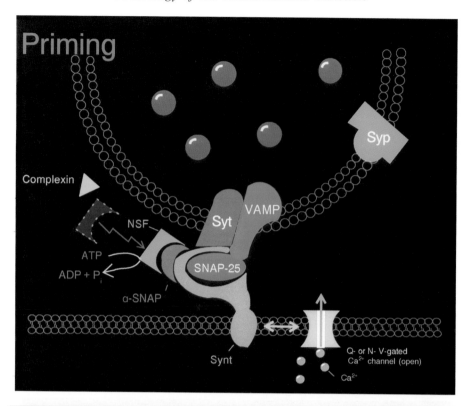

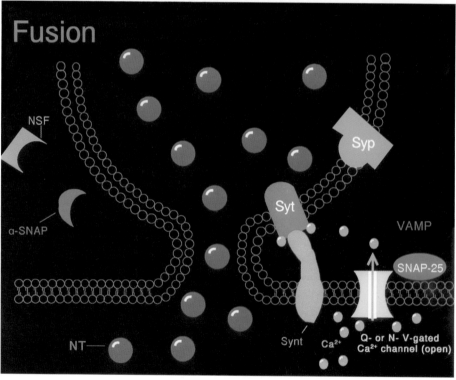

Figure 4   (continued)

quantal release of ACh, nor does it induce appearance of the large-amplitude, slow rise time gMEPPs. Tetanus toxin is apparently transported retrogradely to the central nervous system where it exerts its primary action, namely to block release of the inhibitory amino acid transmitters GABA and glycine, leading ultimately to a spastic form of paralysis.

The general pattern of effects of tetanus toxin on the neuromuscular junction resembles that of botulinum toxin.[92] For both toxins, the paralytic process involves the three-step sequence of binding, internalization, and block. Like botulinum toxin, tetanus toxin is believed to bind to gangliosides of the 1b group.[94–96] Again like botulinum toxin, tetanus toxin does not block nerve terminal Ca^{2+} channels.[70] Despite the fact that the two toxins are reported to block neuromuscular transmission by different paths,[97] tetanus toxin has been shown to interact with synaptobrevin in what appears to be a zinc-dependent manner, just as is the case for certain of the botulinum toxins[86,88] (see Figure 3).

11.13.2.1.3 *Snake venom toxins*

A number of snake venoms affect the neuromuscular junction. The actions of these venom toxins can be broadly divided into those with presynaptic and postsynaptic loci. The postsynaptically acting toxins characterized by the snake alpha toxins (α-bungarotoxin) will be considered later in this chapter. These toxins have proven to be vital tools in characterizing, identifying, and isolating the nicotinic AChR.

Neurotoxins which block neuromuscular transmission by presynaptic actions are found in venom of many of the same species of snakes that contain postsynaptic toxins.[97,98] The best known of these, and the first to be described, was β-bungarotoxin, from the snake *Bungarus multicinctus*. This toxin blocks neuromuscular transmission irreversibly by reducing ACh release.[99,100] Other neurotoxins with similar presynaptic actions are notexin from the Australian tiger snake, *Notechis scutatus*,[101,102] taipoxin from the Australian taipan, *Oxyuranus scutellatus*,[103,104] and a myotoxin from the Asian sea snake *Enhyodrina schistosa*.[105,106] β-Bungarotoxin has a triphasic pattern of effects on ACh release. Initially, release is decreased, this is followed by a transient increase in release, with a subsequent progressive inhibition to block.[107,108] Actions of β-bungarotoxin have also been demonstrated in the central nervous system (CNS).[109] However, much higher concentrations of the toxin are needed to alter transmitter release in the brain. Its actions in the peripheral nervous system are proposed to be specific for cholinergic neurons;[110] peripheral nonadrenergic terminals are unaffected by β-bungarotoxin.[111,112]

The precise mechanisms underlying the various actions of β-bungarotoxin are not yet completely understood. Several snake venoms contain toxins with phospholipase A_2 (PLA_2) activity. These include notexin, β-bungarotoxin, and taipoxin. There is considerable sequence homology among amino acids for β-bungaro-

toxin and other snake venom PLA_2 enzymes.[113] Snake venom PLA_2 neurotoxins are thought to hydrolyze membrane phospholipids.[107,114] However, several studies suggest that the PLA_2 activity in and of itself does not explain the toxin's effects.[115,116] For example, a potassium channel blocking action has also been reported for β-bungarotoxin.[117] This effect occurs independently of the phospholipase activity, and has been proposed to be responsible for the transient increase in ACh release. However, it is unclear how this action could be specific for peripheral cholinergic terminals since potassium channels would be found on other terminals as well. The cause of the ultimate block of release of ACh by β-bungarotoxin and other PLA_2-type neurotoxins is unknown. Suggested mechanisms include inhibition of oxidative phosphorylation,[118] inhibition of cytoskeletal phosphorylation,[119] and cytoskeleton disruption.[120] β-bungarotoxin also inhibits phosphorylation of synapsin I in rat brain synaptosomes.[121] Synapsin I is a neuronal phosphoprotein associated with synaptic vesicles[122] that has been implicated in moving vesicles from a reserve to a releasable pool near the active zones.[123]

11.13.2.1.4 α-*Latrotoxin*

The venom of the black widow spider *Latrodectus mactans* also causes profound effects on vertebrate neuromuscular transmission.[124–126] The syndrome is characterized by excessive cholinergic activity and is now known to be due to presynaptic effects of the venom, especially on the peripheral motor and autonomic nerve terminals. Inasmuch as these effects differ markedly from those caused by any other known group to toxins, α-latrotoxin (α-LTx), and its constituents have been major tools in studying the dynamics of ACh release, as well as its origin.[127]

α-LTx is the main protein component of black widow spider venom.[128] It has a molecular weight of approximately 130 kDa. α-LTx is thought to be responsible for all the major effects caused by the venom on the neuromuscular junction. The potency of the peptide is very high; effects are observed at isolated synapses in the 10^{-10} M range. The actions are not limited to the neuromuscular junction, and are generalizable to all synapses thus far tested. When the purified toxin or crude venom is applied to the neuromuscular junction, it induces a dramatic increase in MEPP frequency which eventually progresses to complete block of junctional transmission and depletion of synaptic vesicles.[124,125] Thus α-LTx acts by greatly increasing the probability of vesicle fusion with the presynaptic membrane, yet simultaneously appears to inhibit synaptic vesicle recycling. The actions of the toxin are focused at the active zones, those sites at which transmitter release is thought to occur specifically; vesicle fusions do not occur randomly throughout the presynaptic terminal.

The action of α-LTx is not blocked by botulinum toxin.[129,130] Thus, the clostridial neurotoxins and α-LTx have unique receptors, and the α-LTx receptor appears to be "downstream" from the site of action of botulinum and tetanus toxin.

The actions of the venom are not reversed by prolonged washing of the preparation with toxin-free solution,[131] but are prevented by preincubation of neuromuscular preparations with antivenin. Moreover, application of antivenin at the time of peak increase in MEPP frequency returns the frequency to control levels even one hour after treatment.[132] Thus during activity, toxin still remains accessible to antitoxin.

α-LTx is thought to act in a dual manner to increase vesicular release of ACh at the neuromuscular junction, increasing cation permeability as well as directly by-passing the Ca^{2+}-dependent step.[133]

Early suggestions were made that α-LTx increased MEPP frequency by increasing entry of Na^+ or Ca^{2+} into the nerve terminal. However, the venom stimulates MEPP frequency, blocks neuromuscular transmission, and depletes synaptic vesicles even in solutions deficient in Na^+ and Ca^{2+}.[131,133] An observation which is consistent with an increase in cation permeability within the nerve terminal is the swelling of some nerve terminals. Increased surface area of the terminal membrane was explained as being due to incorporation of vesicle membrane into the plasma membrane; however, increased terminal volume could not be explained as such. In the absence of Na^+ in the extracellular medium, nerve terminal swelling in response to α-LTx was not pronounced, suggesting an increase in permeability to Na^+ as being responsible for the nerve terminal swelling. The possibility that α-LTx alters membrane cation permeability has been explored in more detail. α-LTx activates inward currents that cannot be blocked with tetrodotoxin, a potent blocker of voltage-gated Na^+ channels, or verapamil, which blocks a variety of different types of Ca^{2+} channels.[134] The channels appear not to discriminate between Na^+ and K^+. Isolated nerve terminals from the brain (synaptosomes) contain the α-LTx receptor.[135] The affinity of the toxin for this receptor is in the same range as the potency of toxin for blocking neuromuscular transmission. Furthermore, reconstitution of α-LTx receptors into liposomes

causes Ca^{2+} transport in the presence of toxin, whereas toxin alone does not induce transport in receptor-deficient liposomes.[136] Taken together, these results are consistent with the hypothesis that the α-LTx receptor is not only a binding site for toxin, but actually participates in part of the toxin-mediated response. In this respect the action of α-LTx differs from that of the clostridial neurotoxins.

The α-LTx binding protein is thought to be the putative receptor[137,138] and appears to belong to a family of neuronal cell surface proteins known as the neurexins.[139] One of these proteins, neurexin Iα, has been suggested to be localized specifically to synapses, and in particular to the presynaptic membrane,[139] the same localization as the putative α-LTx receptor.[140] This receptor has been shown to interact with synaptotagmin, an integral membrane protein of synaptic vesicles thought to play a role in exocytotic fusion of the vesicle with the membrane.[141] This has led to the proposal that perhaps α-LTx causes transmitter release by a direct interaction with synaptotagmin, by-passing the normal Ca^{2+}-dependence.[138,142]

In summary, it is believed that α-LTx blocks neuromuscular transmission by two effects. One effect involves an ionophoric-like action to increase cationic conductances, while the other effect is independent of $[Ca^{2+}]_e$ and remains as yet undefined, but may involve interactions with synaptic vesicle-specific proteins involved in the secretory process.

11.13.2.1.5 Miscellaneous toxins

Because of the dependence of the EPP on functional conduction of normal nerve action potentials, any chemical which impairs nerve terminal excitation will impair neuromuscular function. Thus, a number of toxins which alter invasion of the action potential into the nerve terminal by actions on Na^+ or K^+ channels, or agents which prevent Ca^{2+} entry by action on Ca^{2+} channels, will disrupt neuromuscular transmission.

Several toxins which interact specifically with Ca^{2+} channels have been isolated from venom preparations. A family of polypeptide toxins with Ca^{2+} channel activity has been isolated from the venom of marine snails of the *Conus* family.[143] ω-Conotoxin GVIA (ω-CTx-GVIA), isolated from the venom of *Conus geographus*[144] is a potent inhibitor of nonmammalian Ca^{2+} channels, binding irreversibly in subpicomolar concentrations[145] and inhibiting neuromuscular transmission in the frog.[26,28] In mammalian preparations, the actions of ω-conotoxin GVIA are less clear. This peptide does not block normal neurotransmitter release at the neuromuscular junction,[25–27] although under certain conditions it can modulate release. This suggests that there may be differences in the Ca^{2+} channels at different terminal types or that the peptide may not gain access to certain nerve endings. In conotoxin-sensitive preparations, ω-conotoxin GVIA has different effects on "N" and "L" types of Ca^{2+} channels:[146] "L" type currents are insensitive to ω-CTx-GVIA, and "N" currents can be separated into a component blocked irreversibly by and one blocked reversibly by the toxin.[26,147–149] Other components of *Conus* toxins have also been isolated. A fraction from the venom of *Conus magnus* known as ω-conotoxin MVIIC (ω-CTx MVII-C) inhibits the ω-conotoxin GVI-A-insensitive fraction of Ca^{2+} channels present in mammalian synaptosomes and Purkinje cells.[150] This peptide also blocks mammalian neuromuscular transmission and blocks perineurial currents recorded at mouse motor axon terminals.[151–154] Maitotoxin, produced by the marine dinoflagelate *Gambierdiscus toxicus*, activates neuronal Ca^{2+} channels in neuronal cells in culture;[155,156] as such it would be expected to increase evoked release of ACh at the neuromuscular junction. Spider venoms also contain peptides and polyamines which block Ca^{2+} channels. ω-Agatoxin IVA has been isolated from the venom of funnel web spider (*Agelenopsis aperta*) and is a potent blocker of Ca^{2+} channels[157,158] and of mammalian neuromuscular transmission. Moreover, a polyamine funnel web toxin (FTX), which blocks Ca^{2+} channels and release of ACh at mammalian endplates, has also been purified from the venom of funnel web spiders.[159] Several synthetic compounds with similar activities but markedly reduced potency have been synthesized. FTX and ω-agatoxin have been shown to bind to the "P"-class of the Ca^{2+} channel which is insensitive to both dihydropyridine-type blockers *and* ω-CgTx-GVIA.[158,159]

Toxins which interact with voltage-gated Na^+ or K^+ channels also affect neuromuscular transmission. The classic sodium channel blockers, tetrodotoxin (TTx) and saxitoxin (STx), will both block nerve evoked release of ACh and hence junctional transmission by blocking depolarization of the terminal subsequent to entry of sodium. However, toxins such as TTx or STx have no prominent effect on resting MEPP amplitude or frequency. Conversely, activation of sodium channels by toxins such as batrachotoxin,[160,161] veratrum alkaloids such as veratridine, certain gymnodinium (red tide)[162] or scorpion venoms have prominent effects on both resting and evoked release of ACh. These compounds prolong the open state of the Na^+ channel, thus they increase the duration of depolarization at the nerve terminal, subsequently maintaining the voltage-gated Ca^{2+} channels in the terminal open for longer time. This enhances Ca^{2+} entry and increased MEPP frequency dramatically.

Several snake, insect, and arthropod venoms contain toxins which can block voltage-gated K^+ channels. Included in this category is dendrotoxin, a component of the venom of *Dendroaspis angusticeps*[163–165] and charybdotoxin from *Leiurus quinquestriatus*.[166] As mentioned previously, at least one component of action of β-bungarotoxin also consists of K^+ channel block. By blocking K^+ efflux during the repolarization phase of the action potential, these toxins prolong the duration of depolarization and thus increase Ca^{2+} entry. This results in an increased EPP amplitude due to increased release of transmitter.

11.13.2.2 Antibiotics

Neuromuscular block is a side effect of the use of antibiotics of the aminoglycoside, tetracycline, lincosamide, and polymyxin classes.[167,168] Block occurs both clinically, and experimentally.[169–184] The precise mechanisms are still poorly understood,[167,185] but include: (i) diminished release of ACh from the motor nerve terminal in response to motor nerve stimulation, (ii) decreased sensitivity of the postjunctional ACh receptor due to competitive block, and (iii) decreased conductance through the ACh receptor-activated ionic channel due to block of the channel. Each of these factors alone would theoretically be sufficient to reduce the amplitude of the EPP, below the threshold necessary to evoke a muscle action potential, and thus, block the contraction of the skeletal muscle. For a particular antibiotic, neuromuscular block may involve any or all of these mechanisms. For aminogylcoside antibiotics (neomycin, streptomycin) and clindamycin the prejunctional effects predominate at low doses and postjunctional block occurs at high doses.[174–176] For lincomycin and perhaps polymyxin B, postjunctional block is thought to be more prevalent.[173,176] Several thorough studies have examined the relative contribution of presynaptic and postsynaptic effects of neomycin, streptomycin,[172,174,175] and the lincosamides as a function of concentration,[176] but for the most part, detailed studies of mechanisms responsible for the presynaptic effects of these and other antibiotics are lacking.

Aminoglycosides act presynaptically to decrease release of ACh, evidently due to competitive block of Ca^{2+} entry into the presynaptic nerve terminal during nerve stimulation.[186]

In studies comparing the binding of $[^{125}I]$-ω-conotoxin GVIA, a putative blocker of "N"-type Ca^{2+} channels in some species,[143,144,187,188] aminoglycosides and polymyxin B were able to inhibit binding of the toxin to brain membrane fragments with IC_{50}s in the low micromolar range ($<25\,\mu M$).[189] These antibiotics also block the depolarization-dependent uptake of $^{45}Ca^{2+}$ into isolated nerve terminals from the brain (synaptosomes) via pathways associated with voltage-dependent Ca^{2+} channels.[190] This effect occurs at concentrations which block transmitter release at the neuromuscular junction.

11.13.2.3   Effects of Heavy Metals on Neuromuscular Transmission

Early mechanistic studies of neurotoxic inorganic heavy metals such as Pb^{2+} and Hg^{2+} as well as other heavy metals such as Co^{2+}, Cd^{2+}, Mn^{2+}, and La^{3+} focused on the effect of these metals on neuromuscular transmission. These studies arose in part from the well-characterized physiology, microscopic anatomy, and biochemistry of the neuromuscular junction, and in part due to the reported incidence of neuromuscular weakness as a side effect of poisoning with Pb^{2+} and methylmercury (MeHg). The extent to which these effects observed acutely occur during intoxication remains unknown. However, these studies have contributed greatly to our understanding of the basic processes of neuromuscular transmission, as well as providing clues to cellular mechanisms by which these neurotoxic metals might act.

11.13.2.3.1   Mercury and methylmercury

In isolated nerve-skeletal muscle preparations, acute-bath application of mercuric chloride and MeHg chloride blocked twitches evoked by stimulation of the motor nerve.[191,192] Responses evoked by direct stimulation of the muscle were less affected, or unaffected, suggesting that mercurials either impaired conduction of the nerve action potential, disrupted synaptic transmission directly, or both. Muscle contractility measured *in situ* as tension of individual twitches and tetanic tension from the gastrocnemius of rats treated acutely ($10\,mg\,kg^{-1}\,d^{-1}$, Sub Q, 7 days) or subchronically ($2\,mg\,kg^{-1}\,d^{-1}$, Sub Q, 5 days week^{-1} for 3–4 week) with

MeHg was diminished compared to pair-fed controls.[193] As responses to direct electrical stimulation of the muscle were not measured, it is impossible to determine whether the reductions were due to direct effects of MeHg on skeletal muscle or effects on the motor nerve impulse or neuromuscular junction. However, these results signify that effects of MeHg on skeletal muscle contractility occur not only with direct bath application but also following systemic application. These studies have provided the basis for further experimentation on effects of mercurials on synaptic transmission.

Numerous studies have been done using conventional intracellular microelectrode recording techniques to examine the effects of mercurials on neuromuscular transmission. In the frog sartorius[194–196] and rat[197,198] and mouse diaphragm,[198] the primary effect of both mercurials is to decrease nerve-evoked release of ACh and increase, then decrease spontaneous quantal release.

At the time of EPP block, MEPPs of normal amplitude and duration still occur.[197] Block of nerve-evoked transmitter release by inorganic Hg^{2+} or MeHg is not reversible by washing the preparation with mercury-free solutions.[197,199]

Effects of inorganic and MeHg on nerve-evoked transmitter release are time-dependent, but exhibit an unusual concentration dependence. Specifically, higher concentrations reduce the time required to produce an effect, suggesting a time–concentration interaction and implying that a critical concentration of mercurial must be attained before toxicity occurs. At low concentrations (0.1–$10\,\mu M$), mercurials cause a transient increase in EPP amplitude which precedes block of the EPP; complete block of the EPP was not seen for 60 min or more.[194,195] Concentrations of 20–$100\,\mu M$ MeHg virtually blocked completely nerve-evoked postsynaptic responses between 5 and 30 min.[197,199,200] EPP amplitude declines progressively with time, to complete block, and does not appear to attain a steady state short of complete block. Consequently, a strict concentration dependence does not occur. A lower threshold concentration below which block of evoked release of ACh does not occur may exist,[197] but studies of synaptic transmission at the CA1 region of hippocampal slice suggest that if the microelectrode impalement can be sustained long enough, even low bath concentrations of MeHg ($4\,\mu M$) will block the excitatory postsynaptic potential (EPSP).[201] This may simply reflect the fact that an apparent lack of effect was due to the latent period to produce block at these concentrations being longer than the period over which the measurements were made. Thus, a potential effect might have been missed. As such, one cannot say whether or not that there is a threshold concentration of mercury needed to block evoked release of transmitter.

Mercurials also reportedly increase transiently EPP amplitude prior to reducing it.[194,195,197] Similar stimulatory effects of MeHg have been observed on transmission at autonomic ganglia[202] and in hippocampal slices.[201,203] Effects of MeHg and $HgCl_2$ on mean EPP amplitude parallel those on mean quantal content (m), indicating that a predominant component of mercurial-induced block of synaptic transmission is due to presynaptic effects.[195,197,204] The depression of m produced by MeHg is due primarily to depression of the release statistic n—the immediately available store of neurotransmitter,[197] an index of an intracellular action of MeHg on the release process. The release statistic p—the probability of transmitter release—was actually increased by MeHg.[197] This is consistent with the observation that MeHg increases intracellular $[Ca^{2+}]$.[205–209] Release parameters have not been examined in the presence of divalent inorganic mercury.

EPP block by mercurials is initially noncompetitive with external $[Ca^{2+}]$; increasing the bath $[Ca^{2+}]$ from 2 mM to 4 or 8 mM did not prolong the time to block of the EPP, or decrease the degree of block produced by MeHg.[200] However, this observation is somewhat complicated by the fact that MeHg impairs axonal conduction, albeit at high concentrations.[199,210,211] Nonetheless, once membrane excitability is restored, raising the $[Ca^{2+}]_e$ partially reverses the effects of MeHg on EPP amplitude,[199] suggesting that there are complex effects of MeHg on both membrane excitability and Ca^{2+}-dependent ACh release. A Ca^{2+} channel blocking action of MeHg was demonstrated by block of Ca^{2+} entry during K^+-induced depolarization of isolated nerve terminals (synaptosomes) derived from rat forebrain[200,211] and of voltage-dependent Ca^{2+} currents in PC12 cells[212] and cerebellar granule cells.[213]

The effects of mercurials on spontaneous transmitter release are biphasic with continued exposure. The time course for effects on MEPP frequency differs from that observed for effects on the EPP amplitude. Initially, MEPP frequency is increased from control values of $0.3–3.0\,s^{-1}$ to frequencies of $10–100\,s^{-1}$ in the frog, rat, and mouse.[195,197,198,214,215] This effect usually occurs following a latency of 2–40 min, depending on the concentration of mercury employed. Binah *et al.*[204] and Cooper and Manalis[196] reported that $HgCl_2$ increased MEPP frequency immediately upon exposure of frog neuromuscular preparations, but this

has not been reported by others. Given that MeHg clearly increases $[Ca^{2+}]_i$, and does so very rapidly in preparations of cells in culture or isolated nerve terminals, this difference may reflect a more rapid absorption of the mercurial into the amphibian as opposed to mammalian nerve/muscle preparations. The latent period can be cut dramatically by depolarizing the preparation using elevated extracellular K^+ (15–20 mM).[215,216] The increase in spontaneous release does not appear to be concentration dependent. That is, similar frequencies are produced by concentrations of 10–100 μM. Reports vary as to whether $HgCl_2$ is more or less potent than MeHg. Miyamoto[215] reported that $HgCl_2$ increased MEPP frequency at a threshold concentration of 3 μM compared to 100 μM for MeHg, while Binah *et al.*[204] reported that $HgCl_2$ was more potent than the organomercurial mersalyl. On the other hand, Juang[195] showed that $HgCl_2$ and MeHg at 10 μM were approximately equieffective in increasing MEPP frequency at the frog endplate but $HgCl_2$ was less effective than MeHg in stimulating transmitter release at the isolated guinea pig ganglia.[202] In this latter case, increased spontaneous release of ACh was also seen when mercury was given systemically as opposed to by bath application.[202]

With continued bath application of mercurials, spontaneous transmitter release declines until eventually no further MEPPs can be recorded. This effect is not due to depletion of vesicular neurotransmitter stores, because treatment of the preparation at the time that MEPPs disappeared with $LaCl_3$, which is a profound stimulator of spontaneous release,[217] was able to induce high frequencies of MEPPs.[216] Moreover, since normal appearing MEPPs were evoked by La^{3+} and since ACh depolarizations were normal at the time that spontaneous release ceased,[197] diminished postsynaptic sensitivity to ACh does not seem to account for the block of spontaneous release.

Mercurial-induced increase in MEPP frequency is not prevented by blocking either of the axon membrane Na^+ channels[197,215] or Ca^{2+} channels.[197,215] Simultaneous block of both Na^+ and Ca^{2+} channels prevents the increase in MEPP frequency associated with $HgCl_2$ but not that associated with MeHg,[215] implying that either inorganic mercuric ions enter the nerve terminal through Na^+ and Ca^{2+} channels or that depolarization of the terminal[208] opens Na^+ and Ca^{2+} channels, enhancing the intra-terminal divalent cation concentration. MeHg, on the other hand, may enter the cell not only through the channels but also directly through the membrane due to its lipophilicity. The latent period preceding increased MEPP frequency can be shortened dramatically by facilitation of Ca^{2+} entry into the axon terminal[216,218] or in the presence of increased $[Ca^{2+}]_e$[216] as well as prolonged when the preparation is deficient of Ca^{2+}.[216] However, the increase in spontaneous release is not dependent upon extracellular Ca^{2+} (Ca^{2+}_e), because increases in MEPP frequency still occur with MeHg in the absence of Ca^{2+} or in Ca^{2+}-deficient solutions.[216] The shortened latency obtained with a Ca^{2+} ionophore, a Ca^{2+} channel agonist, and depolarization are consistent with the notion that the latency is related to the time required for MeHg to enter the nerve terminal and elevate $[Ca^{2+}]_i$.[218] Results in Ca^{2+}-deficient and Ca-free solutions suggest that Ca^{2+}_i stores may be the source of increased Ca^{2+} responsible for the increased MEPP frequency. The ability of mitochondria to release Ca^{2+} in response to MeHg has been implicated in the increase in MEPP frequency.[219–221] However, whether or not this is the only contributor, or even a major contributor remains to be seen.

Effects of mercurials on stimulus-evoked release of ACh could be due either to presynaptic or postsynaptic effects; the results of measurements of MEPP amplitudes taken at the time of depressed EPP amplitude have indicated in general no significant effect of mercurials.[197,198,215] Since the postsynaptic processes responsible for both the MEPP and EPP are identical, a lack of effect of mercury on MEPP amplitude suggests that there is no significant postsynaptic blocking effect of mercurials at the time the EPP is blocked. This, coupled with the measurements of statistical release parameters implies that the block of synaptic transmission is primarily presynaptic.

Iontophoretic application of ACh onto the motor endplate of mercurial poisoned preparations was used to test unequivocally for direct effects of mercurials on the AChR.[194,197] Endplate depolarizations due to ACh are not decreased in amplitude by concentrations of mercurials as high as $100\,\mu M$ or for periods of exposure for up to 60 min, by which time MEPPs are no longer observed. Thus, at the motor endplate, mercurials appear to produce few postsynaptic effects. In contrast to this, other experiments have indicated that mercurials do affect the postsynaptic membrane. For example, iontophoretic application of the organomercurial *p*-chloromercuribenzoate depolarized the postsynaptic membrane of the electroplax in a manner similar to ACh and carbachol.[222] These findings contrast with several biochemical studies which indicate that MeHg decreases the binding of cholinergic agonists to nicotinic receptors.[223] The apparent lack of effect of mercurials on the postsynaptic membrane at the neuromuscular junction is puzzling given the well-known affinity of inorganic and organic mercurials for sulfhydryl groups. The ACh receptor is known to contain sulfhydryl groups, whose modification leads to decreased affinity of the receptor to cholinergic agonists.[224] Presumably, MeHg should interact with these sulfhydryl groups to modify the ACh receptor and decrease the postsynaptic response to ACh.

11.13.2.3.2 Lead

Because of the interest in lead as an environmental neurotoxicant, it was one of the first agents whose effects on synaptic function were examined, and it is one of the most widely studied of the heavy metals.

Manalis and Cooper[225] first used intracellular microelectrode recording techniques to examine the effects of Pb^{2+} at the frog neuromuscular junction. Pb^{2+} caused a pattern of effects, which is now recognized to be similar to that of a number of toxic heavy metals. Pb^{2+} first decreased the amplitude of the EPP and increased the frequency of occurrence of MEPPs. Both effects occurred in a concentration-dependent manner, both were observed in the micromolar range, and both could be reversed by washing out the Pb^{2+} from the bath. Differences in the time course of effects of Pb^{2+} on EPPs and MEPPs led to the conclusion that the effects were mediated by distinct actions. Iontophoretic application of ACh to the endplate in the presence of $100\,\mu M$ Pb^{2+} induced normal depolarizing responses of amplitude similar to those elicited in Pb^{2+}-free solutions, implying a lack of postjunctional effect on the muscle nicotinic receptors. Subsequent studies revealed that Pb^{2+} had a similar spectrum of effects on the mammalian neuromuscular junction.[226,227]

The effects of Pb^{2+} on the EPP are Ca-dependent.[196,228–230] Varying the Ca concentration caused the amplitude of the EPP to increase. Addition of $1\,\mu M$ Pb^{2+} caused a parallel shift of the $[Ca^{2+}]$ response curve to the right, indicating that Pb^{2+} was a competitive antagonist to Ca^{2+} in the release process. A similar competitive relationship was observed at the rat neuromuscular junction.[227] One major difference between the mammal and amphibian preparations was that much higher concentrations of lead were needed to block transmission at rat neuromuscular junctions. The rapid, reversible nature of the interaction between Pb^{2+} and Ca^{2+} led to the proposal that block of the EPP occurs because of a Ca^{2+} channel-blocking effect of Pb^{2+}. This hypothesis was strengthened by the observation that Pb^{2+} blocked uptake of ^{45}Ca into nerve terminals isolated from the CNS (synaptosomes)[231] and more recently by the demonstration of block of Ca^{2+} channels in isolated neural cell somas by Pb^{2+}.[232,233]

In contrast to the inhibitory effect of Pb^{2+} on the EPP, MEPP frequency was increased dramatically by Pb^{2+}. As was the case with MeHg, the effects on the two forms of release differed dramatically in their time courses. Increased MEPP frequency by Pb^{2+} occurred after a latent period that presumably reflected the time necessary for Pb^{2+} to enter the axon terminal.

Extracellular Ca^{2+} is not necessary for Pb^{2+} to increase MEPP frequency.[227,234] This led to the speculation that Pb^{2+} releases Ca^{2+} from bound intracellular stores thereby increasing the free $[Ca^{2+}]_i$. This hypothesis has remained as a major tenet until the startling observation that Pb^{2+} can substitute for Ca^{2+} as a secretagogue in isolated chromaffin cells.[235] Whether or not this also occurs at the neuromuscular junction is unknown, but there is no reason *a priori* to expect that it will not. This would have profound implications for the basic process of transmitter release, especially with regard to the Ca^{2+}-dependent substrates.

11.13.2.3.3 Effects of other heavy metals on neuromuscular transmission

Other polyvalent cations besides lead and mercury also affect neuromuscular transmission. The principal effect is to block evoked transmitter release. This effect has been described for magnesium,[236,237] manganese,[238] cobalt,[239] lanthanum,[240] cadmium,[228,241] nickel, praseodymium,[242] zinc,[243] thallium,[244] gadolinium,[245] triethyl tin,[246] and erbium.[247] Strontium is an exception in that it supports evoked release, although it is much less effective than equimolar concentrations of Ca^{2+} in generating an EPP.[248–250] Barium has an unusual effect on evoked release; it cannot support normal phasic evoked release but instead causes an asynchronous discharge of quanta, detectable as MEPPs, following nerve stimulation.[251–254] Despite inhibition of evoked release, MEPP frequency is increased by virtually all of these polyvalent cations.[228,244,255–261] Thus, these other cations affect the neuromuscular junction in a manner similar to that of Pb^{2+} and MeHg. The mechanisms responsible for the effects of

these divalent and trivalent cations are assumed to be on two distinct sites: block of entry of Ca^{2+} through membrane Ca^{2+} channels and displacement of intracellular Ca^{2+} stores. The ability of several of these polyvalent cations to block influx of Ca^{2+} through voltage-dependent membrane Ca^{2+} channels has been demonstrated.[262–266] It is this effect which is thought to be responsible for block of nerve-evoked transmitter release. This block is thought to be competitive and can be antagonized by higher concentrations of Ca^{2+}.[227,238,267] However, for at least thallium, Ca^{2+} channel block is not thought to be the mechanism responsible for block of the EPP.[268] The metal is then postulated to gain entry to the nerve terminal, presumably through the same Ca^{2+} channels, and act intraterminally to increase spontaneous release. This notion is supported by studies in which, following substitution of Ni^{2+}, Mn^{2+}, or Mg^{2+} for Ca^{2+}, increased spontaneous release is produced by these cations under conditions which facilitate entry of the cation into the nerve terminal such as depolarization, use of Ca^{2+} ionophores,[35] or repetitive stimulation.[269]

11.13.2.4 Other Agents Affecting Neuromuscular Transmission by Presynaptic Effects

11.13.2.4.1 Dithiobiuret

2,4-Dithiobiuret (DTB) is a substituted thiourea derivative which produces a delayed onset, flaccid neuromuscular weakness in rats.[270–272] DTB is not widely used, so exposure to it does not appear to be a common toxicological problem. However, DTB causes what appears to be a purely motor-directed neurodysfunction; sensory systems do not appear to be significantly affected.[271,273] Thus, DTB is unique from other chemicals inducing delayed-onset neurotoxicity, with the sole exception of botulinum toxin. The effect of DTB typically requires accumulation of a threshold toxic dose as well as a minimum time of exposure to cause manifestation of the syndrome.[272] When treatment is stopped, rats can recover apparently normal motor function in 4–10 days. The time course and severity of DTB-induced paralysis can be altered chemically by daily treatment of rats with certain sulfur-containing chelators.[274] Moreover, rats paralyzed by DTB recover in approximately 2–7 days if DTB treatment is supplemented by treatment with these chelators. The mechanism responsible for protection against and reversal of DTB-induced paralysis is not known, but does not appear to entail altered distribution or bioavailability of DTB.[274]

A neuromuscular deficit is thought to be responsible for this weakness, as twitches of skeletal muscles isolated from rats affected by DTB are reduced in amplitude or blocked when elicited by electrical stimulation of the motor nerve, but not when elicited by direct stimulation of the muscle itself.[275]

Results of both contractility and quantal release measurements indicate that DTB acts prejunctionally.[44,276,277] Both nerve-evoked and spontaneous release of ACh from motor nerve terminals are depressed in muscles taken from rats affected by DTB.[44,277] DTB-induced paresis was associated with a diminished quantal content (m).[44] an effect attributable to a reduction of the immediately available store (n), and not the probability of release (p). No frequency-dependent fatigue was evident in evoked quantal release in the DTB-treated group. Increasing $[Ca^{2+}]_e$ or facilitating Ca^{2+} influx into the axon terminal with 4-aminopyridine improved transmission in the DTB-poisoned group, but did not restore quantal release to control levels.[44] $[K^+]_e$-evoked MEPP frequency was lower for DTB-treated rats, as was mean MEPP frequency in the absence of stimulation. DTB-paralyzed preparations were characterized by the presence of very large MEPPs (giant MEPPs) with prolonged rise and decay.[44] Thus, paralysis induced by chronic DTB treatment is associated with impairment of presynaptic processes resulting in a diminished number of quanta liberated spontaneously, and in response to motor nerve stimulation.

Evidence suggests that DTB may have additional postjunctional effects. Rise and decay rates for MEPPs and EPPs are slowed at the time of muscle weakness,[44] while the amplitude of endplate currents (EPCs) and miniature endplate currents (MEPCs) recorded from rats paralyzed with DTB and from rats following a single, nonparalytic dose of DTB[278] were both decreased. The decrease in EPC amplitude was associated with a decrease in quantal content while mean MEPC amplitude was unaffected. The frequency of gMEPPs or MEPCs was increased in muscles from chronically treated rats at both 1 h and 24 h following a single large dose of DTB. Thus, exposure to DTB affects current flow at the neuromuscular junction.[278] The neuromuscular toxicology of this interesting chemical is reviewed in greater detail in Atchison and Spitsbergen.[279] Interested readers are referred to this review.

11.13.2.4.2 Hexanedione

Several solvents including methyl *n*-butyl-ketone and *n*-hexane induce a peripheral neuropathy through the common metabolite 2,5-hexanedione.[280] Potential neuromuscular toxicity during the early stages of intoxication with 2,5-hexanedione was examined following administration of the toxicant to rats for up to 34 days.[281] A number of fibers from the treated rats lacked EPPs due to denervation. In the denervated myofibers, no change was observed in the resting membrane potential until the end of the 34 day treatment regimen. At earlier periods (14 days), MEPP frequency and amplitude of both soleus (a "slow"-twitch muscle) and *extensor digitorum longus* (EDL, a "fast"-twitch muscle) were increased. Mean quantal content (m) was reduced significantly in EDL, but not soleus muscles. Cangiano *et al.*[281] suggested that inhibition of glycolytic enzymes was responsible for the differences in sensitivity of fast and slow twitch fibers, and that this inhibition ultimately led to depolarization of the myofibers and increased MEPP frequency.

11.13.3 EFFECTS OF TOXINS AND NEUROTOXICANTS WHICH ACT THROUGH PREDOMINANTLY POSTJUNCTIONAL MECHANISMS

By studying how chemicals, toxins, and other agents alter EPPs, EPCs, or single-channel currents, information can be obtained concerning the mechanism of action by which these chemicals affect the neuromuscular junction.

Agents which act postjunctionally can be classified into two broad categories: (i) those which compete with ACh for binding sites on the receptor complex (competitive antagonists) and (ii) those which bind irreversibly to ACh binding sites or act at different sites from ACh to alter its effectiveness as agonist (noncompetitive antagonists). There are additional ways in which the AChR–channel complex behavior can be altered. Compounds can bind to the same sites on the receptor as ACh and prevent ACh from exerting its effects; they can bind to regions other than those to which ACh binds and alter the receptor's affinity for ACh; or they can chemically modify the ACh receptor region to alter the affinity of the receptor for ACh. Finally, toxicants can bind to the channel portion of the receptor–channel complex and alter the ionic conductance, without altering the binding of ACh to the receptor.

11.13.3.1 Competitive Inhibition

11.13.3.1.1 Curare

Curare (*d*-tubocurarine), is probably the most well known of the neuromuscular blocking agents. It is a competitive antagonist at nicotinic ACh, that is, it prevents ACh from binding and activating the receptor–channel complex.[282] Curare decreases the amplitude of EPPs or EPCs, while decreasing the frequency of single-channel currents. However, curare also affects channel conductance, and may actually activate ACh channels in some preparations,[283,284] Additionally, there are numerous reports of prejunctional effects of curare, presumably on presynaptic nicotinic AChRs.[3,58]

11.13.3.1.2 α-Bungarotoxin

α-Bungarotoxin (α-BGT), a polypeptide toxin purified from venom of the snake *Bungarus multicinctus*, binds to postjunctional nicotinic AChR causing a nondepolarizing block that is essentially irreversible. Neuromuscular block induced by α-BGT occurs on a slower time course than that of curare; however, block by curare is readily reversible while that of α-BGT is not. This property has made α-BGT an important tool for the characterization and isolation of neuromuscular type nicotinic AChR.

11.13.3.2 Noncompetitive Inhibition

Several agents affect AChR function noncompetitively by interacting with the AChR at sites different from those at which ACh binds. Histrionicotoxin, phencyclidine, and the local anesthetics bind to the channel portion of the AChR.[285,286] Block by several of these agents is increased in the presence of agonist, indicating that they bind preferentially to the open channel. In addition, the agents increase the affinity of the AChR for ACh, and thus increase the rate of desensitization.

11.13.3.3 Chemical Modification of the Nicotinic Receptor

The receptor region of the AChR–channel complex contains disulfide bonds which are critical to normal function.[287,288] Reduction of these disulfides by dithiothreitol (DTT) decreases the responsiveness to ACh, while reoxidation with 5,5′-dithio-bis(2-nitrobenzoate) (DTNB) reverses these effects.[287,288] The reduction of the AChRs from *Torpedo electroplax* with DTT decreased the binding affinity of AChRs for carbamylcholine (CCh) and shifted the dose–response curve for CCh-induced increases in $^{22}Na^+$ permeability to higher CCh concentrations.[289] In further studies on the effects of thiol-group modification on ion flux activation and inactivation, Walker *et al.*[290] examined CCh binding and ^{88}Rb influx into vesicles with reconstituted AChR-channels purified from *Torpedo californica* before and after reduction with DTT. The main effect of DTT was to shift the EC_{50} values for activation and slow inactivation to higher agonist concentrations.

Following exposure of muscles to DTT, the amplitude and decay times for EPPs and MEPPs were decreased,[291,292] the time which single-channels remain open and the conductance for these channels are decreased.[293] Thus reduction of critical disulfide groups located on the AChR leads to a decrease in affinity of the receptor for agonist. Associated with reduction is a decrease in single-channel conductance and open time. Oxidation or sulfonation[294,295] of the postjunctional nicotinic receptor also cause alterations of decay kinetics of the EPP or EPC.

Lophotoxin, an uncharged molecule found in coral, reacts covalently with the AChR to block the ACh recognition site. Prior reduction of cysteine groups in the binding region followed by alkylation prevents the activity of lophotoxin.[296] Lophotoxin reduces the amplitude of nerve-evoked EPCs as well as spontaneously occurring MEPCs.[297]

11.13.3.4 Chemicals Which Induce Channel Block

Many compounds block the AChR-channel following its opening. This includes all of the agonist molecules tested thus far.[298,299] Moreover, the classic receptor blocker *d*-tubocurarine, also blocks the ion channel,[284,300] although not at the same concentrations that block the AChR. A variety of charged and uncharged molecules[301–303] also appear to block the channel physically. During open-channel block, characteristic effects on both macroscopic (i.e., EPCs and MEPCs) and single-channel currents can be observed. Single-channel currents exhibit characteristic bursts of activity, which represent a single channel opening and becoming blocked and unblocked in rapid succession, followed by channel closure. The effects of open-channel block on EPCs and MEPCs appear as a biphasic decay of the current. In the presence of open-channel blockers, the normal number of ion channels open following release of ACh. Thus, the initial amplitude of the current is normal. However, a portion of the channels rapidly become blocked, leading to a rapid initial decay of the current. As channels become unblocked over time, this allows current to flow through them prior to their final closure, leading to the prolonged phase.[304,305]

11.13.3.5 Cholinesterase Inhibition

The action of ACh is terminated by acetylcholinesterase (AChE). This enzyme has an extremely high turnover rate for ACh, and is present in very high concentrations in the synaptic cleft.[306] Thus the released ACh is removed from the synaptic cleft within a few hundred microseconds. Normally, when ACh is released from the nerve terminal, it binds only a single time to receptors on the postsynaptic membrane before it is hydrolyzed by AChE upon unbinding from the receptor. Thus, inhibition of AChE leads to an increase in the amplitude and a prolongation of rise and decay times of synaptic potentials.[51,307,308] Cholinesterase inhibition also leads to pathological alterations in skeletal muscles which can be observed as early as 30 min following AChE exposure and which progress to encompass up to 7% of muscles in the body with chronic drug administration.[309] Following 30 min of exposure to paraoxon (an irreversible cholinesterase inhibitor), mitochondria located within end-plate regions of muscle become dilated, sarcoplasmic reticulum is increased, subsynaptic folds become widened and fused, and there is an increase in coated cleft vesicles. Twenty-four hours following exposure to paraoxon, muscle fiber architecture displays a generalized disruption and there is an accompanying infiltration of phagocytes.[310] Addition prejunctional effects have been described for some inhibitors of AChE. Within 30 min following paraoxon exposure, evoked release of transmitter is blocked, while MEPP frequency is increased and rise and decay times for MEPPs are prolonged.[308,311] AChE from *Torpedo electroplax* has been shown to contain a single free sulfhydryl group on its catalytic subunit which can react with a number of sulfhydryl reagents. Modification of this group by thiol reagents has been found to inactivate the enzyme.[312]

11.13.4 CONCLUSION

Numerous chemicals affect transmssion at the neuromuscular junction, and do so by a bewildering multiplicity of mechanisms. However, functionally, the result of these various actions is typically the same: impaired muscle contractility. The range of chemicals and toxins which affect this crucial synapse is broad, encompassing therapeutic agents, biological toxins, important environmental pollutants, and frequently used agricultural pesticides.

Functionally, toxicity at the neuromuscular junction is often masked by other toxic manifestations, or does not occur at the lowest toxic concentrations, so for some agents such as the heavy metals, neuromuscular toxicity may be less relevant to the organism than other toxicities. On the other hand, for most venoms, or toxins, as well as for acetylcholinesterase inhibitors, neuromuscular toxicity is among the hallmark clinical signs, and leads directly to death of the organism.

As noted in Section 11.13.1, the neuromuscular junction has been commonly used as a model synapse to study the actions of toxic chemicals on information transfer (synaptic transmission) within the nervous system. This is still a valuable use of this preparation. However, with the advent of use of equally well-defined preparations from the CNS such as brain slices, or the ability to make synaptic connections in co-culture, this use has and will probably continue to decline. Nonetheless, the neuromuscular junction remains a preparation which has provided an untold wealth of information regarding the actions of chemicals on ion channels, intracellular ion homeostasis, signal transduction, release mechansims, and receptor channels.

ACKNOWLEDGMENTS

The author gratefully acknowledges the secretarial assistance of Emily Wisler, Dana Goryl, and Sarah Bloomfield. This work was supported by NIH grants NS20683, ES03299, and ES05822.

11.13.5 REFERENCES

1. A. R. Martin, 'Quantal nature of synaptic transmission.' *Physiol. Rev.*, 1966, **46**, 51–66.
2. J. I. Hubbard, 'Mechanisms of transmitter release.' *Prog. Biophys. Mol. Biol.*, 1970, **21**, 33–124.
3. J. I. Hubbard and D. F. Wilson, 'Neuromuscular transmission in a mammalian preparation in the

absence of blocking drugs and the effect of D-tubo-curarine.' *J. Physiol. (Lond.)*, 1973, **228**, 307–325.

4. E. M. Silinsky, 'The biophysical pharmacology of calcium-dependent acetylcholine secretion.' *Pharmacol. Rev.*, 1985, **37**, 81–132.

5. T. C. Südhof and R. Jahn, 'Proteins of synaptic vesicles involved in exocytosis and membrane recycling.' *Neuron*, 1991, **6**, 665–677.

6. W. S. Trimble, M. Linial and R. H. Scheller, 'Cellular and molecular biology of the presynaptic nerve terminal.' *Annu. Rev. Neurosci.*, 1991, **14**, 93–122.

7. S. Ferro-Novick and R. Jahn, 'Vesicle fusion from yeast to man.' *Nature*, 1994, **370**, 191–193.

8. R. Jahn and T. C. Südhof, 'Synaptic vesicles and exocytosis.' *Annu. Rev. Neurosci.*, 1994, **17**, 219–246.

9. M. K. Bennett and R. H. Scheller, 'The molecular machinery for secretion is conserved from yeast to neurons.' *Proc. Natl. Acad. Sci. USA*, 1993, **90**, 2559–2563.

10. M. K. Bennett and R. A. Scheller, 'A molecular description of synaptic vesicle membrane trafficking.' *Annu. Rev. Biochem.*, 1994, **63**, 63–100.

11. H. Niemann, J. Blasi and R. Jahn, '*Clostridial* neurotoxins: new tools for dissecting exocytosis.' *Trends Cell Biol.*, 1994, **4**, 179–185.

12. T. Söllner and J. E. Rothman, 'Neurotransmission: harnessing fusion machinery at the synapse.' *Trends Neurosci.*, 1994, **17**, 344–348.

13. T. Söllner, S. W. Whiteheart, M. Brunner *et al.*, 'SNAP receptors implicated in vesicle targeting and fusion.' *Nature*, 1993, **362**, 318–324.

14. R. H. Scheller, 'Membrane trafficking in the presynaptic nerve terminal.' *Neuron*, 1995, **14**, 893–897.

15. P. I. Hanson, H. Otto, N. Barton *et al.*, 'The *N*-ethylmaaleimide-sensitive fusion protein and α-SNAP induce a conformational change in syntaxin.' *J. Biol. Chem.*, 1995, **270**, 16955–16961.

16. B. Katz and R. Miledi, 'The timing of calcium action during neuromuscular transmission.' *J. Physiol. (Lond.)*, 1967, **189**, 535–544.

17. B. Katz and R. Miledi, 'Tetrodotoxin-resistant electric activity in presynaptic terminals.' *J. Physiol. (Lond.)*, 1969, **203**, 459–487.

18. R. Robitaille, E. M. Adler and M. P. Charlton, 'Strategic location of calcium channels at transmitter release sites of frog neuromuscular synapses.' *Neuron*, 1990, **5**, 773–779.

19. M. W. Cohen, O. T. Jones and K. J. Angelides, 'Distribution of Ca^{2+} channels on frog motor nerve terminals revealed by fluorescent ω-conotoxin.' *J. Neurosci.*, 1991, **11**, 1032–1039.

20. R. Llinás, M. Sugimori and R. B. Silver, 'Microdomains of high calcium concentration in a presynaptic terminal.' *Science*, 1992, **256**, 677–679.

21. O. D. Uchitel, D. A. Protti, V. Sanchez *et al.*, 'P-type voltage-dependent calcium channel mediates presynaptic calcium influx and transmitter release in mammalian synapses.' *Proc. Natl. Acad. Sci. USA*, 1992, **89**, 3330–3333.

22. W. D. Atchison and S. M. O'Leary, 'BayK 8644 increases release of acetylcholine at the murine neuromuscular junction.' *Brain Res.*, 1987, **419**, 315–319.

23. W. D. Atchison, 'Diyhdropyridine-sensitive and -insensitive components of acetylcholine release from rat motor nerve terminals.' *J. Pharmacol. Exp. Ther.*, 1989, **251**, 672–678.

24. I. Wessler, D. J. Dooley, H. Osswald *et al.*, 'Differential blockade by nifedipine and ω-conotoxin GVIA of alpha₁- and beta₁-adrenoceptor-controlled calcium channels on motor nerve terminals of the rat.' *Neurosci. Lett.*, 1990, **108**, 173–178.

25. A. J. Anderson and J. M. Harvey, 'ω-Conotoxin does not block the verapamil-sensitive calcium channels at mouse motor nerve terminals.' *Neurosci. Lett.*, 1987, **82**, 177–180.

26. K. Sano, K. Enomoto and T. Maeno, 'Effects of synthetic ω-conotoxin, a new type Ca^{2+} antagonist, on frog and mouse neuromuscular transmission.' *Eur. J. Pharmacol.*, 1987, **141**, 235–241.

27. D. A. Protti, L. Szczupak, F. S. Scornik *et al.*, 'Effect of ω-conotoxin GVIA on neuro-transmitter release at the mouse neuromuscular junction.' *Brain Res.*, 1991, **557**, 336–339.

28. L. M. Kerr and D. Yoshikami, 'A venom peptide with a novel presynaptic blocking action.' *Nature*, 1984, **308**, 282–284.

29. B. Katz and R. Miledi, 'Spontaneous and evoked activity of motor nerve endings in calcium Ringer.' *J. Physiol. (Lond.)*, 1969, **203**, 689–706.

30. R. Miledi and C. R. Slater, 'Electrophysiology and electron microscopy of rat neuromuscular junction after nerve degeneration.' *Proc. Roy. Soc. Ser. B.*, 1968, **169**, 289–306.

31. C. Colmeus, S. Gomez, J. Molgó *et al.*, 'Discrepancies between spontaneous and evoked synaptic potentials at normal, regenerating and botulinum toxin poisoned mammalian neuromuscular junctions.' *Proc. R. Soc. Lond. B. Biol. Sci.*, 1982, **215**, 63–74.

32. C. G. Muniak, M. E. Kriebel and C. G. Carlson, 'Changes in MEPP and EPP amplitude distributions in the mouse diaphragm during synapse formation and degeneration.' *Brain Res.*, 1982, **5**, 123–138.

33. I. G. Marshall, 'Studies on the blocking action of 2-(4-phenyl piperidino) cyclohexanol (AH5183).' *Br. J. Pharmacol.*, 1970, **38**, 503–516.

34. J. Del Castillo and B. Katz, 'The effect of magnesium on the activity of motor nerve endings.' *J. Physiol. (Lond.)*, 1954, **124**, 553–559.

35. H. Kita and W. Van Der Kloot, 'Effects of the ionophore X-537A on acetylcholine release at the frog neuromuscular junction.' *J. Physiol. (Lond.)*, 1976, **259**, 177–198.

36. H. Kita, K. Narita and W. Van Der Kloot, 'Tetanic stimulation increases the frequency of miniature end-plate potentials at the frog neuromuscular junction in Mn^{2+}-, Co^{2+}-, and Ni^{2+}-saline solutions.' *Brain Res.*, 1981, **205**, 111–121.

37. P. F. Baker and A. C. Crawford, 'A note on the mechanism by which inhibitors of the sodium pump accelerate spontaneous release of transmitter from motor nerve terminals.' *J. Physiol. (Lond.)*, 1975, **247**, 209–226.

38. E. Alnaes and R. Rahamimoff, 'On the role of mitochondria in transmitter release from motor nerve terminals.' *J. Physiol. (Lond.)*, 1975, **248**, 285–306.

39. A. W. Liley, 'The effects of presynaptic polarization on the spontaneous activity at the mammalian neuromuscular junction.' *J. Physiol. (Lond.)*, 1956, **134**, 427–443.

40. K. L. Magleby and J. E. Zengel, 'A quantitative description of tetanic and post-tetanic potentiation of transmitter release at the frog neuromuscular junction.' *J. Physiol. (Lond.)*, 1975, **245**, 183–208.

41. J. E. Zengel and K. L. Magleby, 'Transmitter release during repetitive stimulation: selective changes produced by Sr^{2+} and Ba^{2+}.' *Science*, 1977, **197**, 67–69.

42. J. E. Zengel and K. L. Magleby, 'Changes in miniature endplate potential frequency during repetitive nerve stimulation in the presence of Ca^{2+}, Ba^{2+} and Sr^{2+} at the frog neuromuscular junction.' *J. Gen. Physiol.*, 1981, **77**, 503–529.

43. J. E. Zengel and K. L. Magleby, 'Augmentation and facilitation of transmitter release. A quantitative

description at the frog neuromuscular junction.' *J. Gen. Physiol.*, 1982, **80**, 583–611.

44. W. D. Atchison, 'Alterations of spontaneous and evoked release of acetylcholine during dithiobiuret-induced neuromuscular weakness.' *J. Pharmacol. Exp. Ther.*, 1989, **29**, 735–743.

45. D. L. Traxinger and W. D. Atchison, 'Reversal of methylmercury-induced block of nerve-evoked release of acetylcholine at the neuromuscular junction.' *Toxicol. Appl. Pharmacol.*, 1987, **90**, 23–33.

46. M. E. Kriebel, F. Llados and D. R. Matteson, 'Histograms of the unitary evoked potential of the mouse diaphragm show multiple peaks.' *J. Physiol. (Lond.)*, 1982, **322**, 211–222.

47. A. W. Liley, 'An investigation of spontaneous activity at the neuromuscular junction of the rat.' *J. Physiol. (Lond.)*, 1956, **132**, 650–666.

48. S. Thesleff, J. Molgó and H. Lundh, 'Botulinum toxin and 4-aminoquinoline induce a similar abnormal type of spontaneous quantal transmitter release at the rat neuromuscular junction.' *Brain Res.*, 1983, **264**, 89–97.

49. K. A. Alkadhi, 'Giant miniature end-plate potentials at the untreated and emetine-treated frog neuromuscular junction.' *J. Physiol. (Lond.)*, 1989, **412**, 475–491.

50. P. C. Molenaar, B. S. Oen, R. L. Polak *et al.*, 'Surplus acetylcholine and acetylcholine release in the rat diaphragm.' *J. Physiol. (Lond.)*, 1987, **385**, 147–167.

51. B. Katz and R. Miledi, 'The nature of the prolonged endplate depolarization in anti-esterase treated muscle.' *Proc. R. Soc. Lond. B Biol. Sci.*, 1975, **192**, 27–38.

52. B. Mishina, T. Takai, K. Imoto *et al.*, 'Molecular distinction between fetal and adult forms of muscle acetylcholine receptor.' *Nature (Lond.).*, 1986, **321**, 406–411.

53. D. J. Adams, T. M. Dwyer and B. Hille, 'The permeability of endplate channels to monovalent and divalent metal cations.' *J. Gen. Physiol.*, 1980, **75**, 493–510.

54. T. M. Dwyer, D. J. Adams and B. Hille, 'The permeability of the endplate channel to organic cations in frog muscle.' *J. Gen. Physiol.*, 1980, **75**, 469–492.

55. S. M. Sine, T. Claudio and F. J. Sigworth, 'Activation of Torpedo acetylcholine receptors expressed in mouse fibroblasts: single channel current kinetics reveal distinct agonist binding affinities.' *J. Gen. Physiol.*, 1990, **96**, 395–437.

56. M. I. Glavinovic, 'Presynaptic action of curare.' *J. Physiol. (Lond.)*, 1979, **290**, 499–506.

57. I. Wessler, 'Control of transmitter release from the motor nerve by presynaptic nicotinic and muscarinic autoreceptors.' *Trends Pharmacol. Sci.*, 1989, **10**, 110–114.

58. W. C. Bowman, C. Prior and I. G. Marshall, 'Presynaptic receptors in the neuromuscular junction.' *Ann. NY Acad. Sci.*, 1990, **604**, 69–81.

59. L. L. Simpson, 'The origin, structure, and pharmacological activity of botulinum toxin.' *Pharmacol. Rev.*, 1981, **33**, 155–188.

60. L. L. Simpson, 'Molecular pharmacology of botulinum toxin and tetanus toxin.' *Annu. Rev. Pharmacol. Toxicol.*, 1986, **26**, 427–453.

61. B. D. Howard and C. B. Gundersen, Jr., 'Effects and mechanisms of polypeptide neurotoxins that act presynaptically.' *Annu. Rev. Pharmacol. Toxicol.*, 1980, **20**, 307–336.

62. G. Sakaguchi, '*Clostridium botulinum* toxins.' *Pharmacol. Ther.*, 1982, **19**, 165–194.

63. L. W. Duchen and S. J. Strich, 'The effects of botulinum toxin on the pattern of innervation of skeletal muscles in the mouse.' *J. Exp. Physiol. Cogn. Med. Sci.*, 1968, **53**, 84–89.

64. L. W. Duchen, 'An electron microscopic study of the changes induced by botulinum toxin in the motor end-plates of slow and fast skeletal muscle fibers of the mouse.' *J. Neurol. Sci.*, 1971, **14**, 47–60.

65. L. L. Simpson 'Pharmacological studies on the subcellular site of action of botulinum toxin type A.' *J. Pharmacol. Exp. Ther.*, 1978, **206**, 661–669.

66. L. C. Sellin, S. Thesleff and B. R. Dasgupta, 'Different effects of types A and B botulinum toxin on transmitter release at the rat neuromuscular junction.' *Acta. Physiol. Scand.*, 1983, **119**, 127–133.

67. L. C. Sellin J. A. Kauffman and B. R. DasGupta, 'Comparison of the effects of botulinum neurotoxin types A and E at the neuromuscular junction.' *Med. Biol.*, 1983, **61**, 120–125.

68. A. J. Harris and R. Miledi, 'The effect of type D botulinum toxin on frog neuromuscular junctions.' *J. Physiol. (Lond.)*, 1971, **217**, 497–515.

69. L. L. Simpson and B. R. Dasgupta, 'Botulinum neurotoxin type E: studies on mechanism of action and on structure-activity relationships.' *J. Pharmacol. Exp. Ther.*, 1983, **224**, 135–140.

70. F. Dreyer, A. Mallart and J. L. Brigant, 'Botulinum A toxin and tetanus toxin do not affect presynaptic membrane currents in mammalian motor nerve endings.' *Brain Res.*, 1983, **270**, 373–375.

71. D. A. Boroff, J. Del Castillo, J. H. Evoy *et al.*, 'Observations on the action of type A botulinum toxin on frog neuromuscular junctions.' *J. Physiol. (Lond.)*, 1974, **240**, 227–253.

72. M. E. Kriebel and C. E. Gross, 'Multimodal distribution of frog miniature endplate potentials in adult, denervated, and tadpole leg muscle.' *J. Gen. Physiol.*, 1974, **64**, 85–103.

73. L. C. Sellin and S. Thesleff, 'Pre- and post-synaptic actions of botulinum toxin at the rat neuromuscular junctions.' *J. Physiol. (Lond.)*, 1981, **317**, 487–495.

74. Y. I. Kim, T. Lomo, M. T. Luga *et al.*, 'Miniature end-plate potentials in rat skeletal muscle poisoned with botulinum toxin.' *J. Physiol. (Lond.)*, 1984, **356**, 587–599.

75. R. Hughes and B. C. Whaler, 'Influence of nerve-ending activity and of drugs on the rate of paralysis of rat diaphragm preparations by *Cl. botulinum* type A toxin.' *J. Physiol. (Lond.)*, 1962, **160**, 221–233.

76. L. L. Simpson, 'Studies on the binding of botulinum toxin type A to the rat phrenic nerve-hemidiaphragm preparation.' *Neuropharmacology*, 1974, **13**, 683–691.

77. L. L. Simpson and M. M. Rapport, 'The binding of botulinum toxin to membrane lipids: phospholipids and proteolipid.' *J. Neurochem.*, 1971, **18**, 1671–1677.

78. M. Kitamura, M. Iwamori and Y. Nagai, 'Interaction between *Clostridium botulinum* neurotoxin and gangliosides.' *Biochim. Biophys. Acta.*, 1980, **628**, 328–335.

79. S. Kozaki, G. Sakaguchi, M. Nishimura *et al.*, 'Inhibitory effect of ganglioside G_{TIB} on the activities of *Clostridium botulinum* toxins.' *FEMS Microbiol. Lett.*, 1984, **21**, 219–223.

80. T. Binz, H. Kurazono, M. R. Popoff *et al.*, 'Nucleotide sequence of the gene encoding *Clostridium botulinum* neurotoxin type D.' *Nucleic Acids Res.*, 1990, **18**, 5556.

81. T. Binz, H. Kurazono, M. Wille *et al.*, 'The complete sequence of botulinum neurotoxin type D.' *Nucleic. Acids Res.*, 1990, **265**, 9153–9158.

82. S. M. Whelan, M. J. Elmore, N. J. Bodsworth *et al.*, 'Molecular cloning of the *Clostridium botulinum* structural gene encoding the type B neurotoxin and determination of its entire nucleotide sequence.' *Appl. Environ. Microbiol.*, 1992, **58**, 2345–2354.

83. D. Hauser, M. W. Eklund, H. Kurazono *et al.*, 'Nucleotide sequence of *Clostridium botulinum* Cl neurotoxin.' *Nucleic Acids Res.*, 1990, **18**, 4924.

84. S. Poulet, D. Hauser, M. Quanz, *et al.*, 'Sequences of the botulinal neurotoxin E derived from *Clostridium botulinum* type E (strain Beluga) and *Clostridium botulinum* (strains ATCC 43181 and ATCC 43755).' *Biochem. Biophys. Res. Commun.*, 1992, **183**, 107–113.

85. T. Hayashi, H. McMahon, S. Yamasaki *et al.*, 'Synaptic vesicle membrane-fusion complex. Action of clostridial neurotoxins on assembly.' *EMBO J.*, 1994, **13**, 5051–5061.

86. G. Schiavo, F. Benfenati, B. Poulain *et al.*, 'Tetanus and botulinum-B neurotoxins block neurotransmitter release by proteolytic cleavage of synaptobrevin.' *Nature*, 1992, **359**, 832–835.

87. G. Schiavo, B. Poulain, O. Rossetto *et al.*, 'Tetanus toxin is a zinc protein and its inhibition of neurotransmitter release and protease activity depend on zinc.' *EMBO J.*, 1992, **11**, 3577–3583.

88. E. Link, L. Edelmann, J. H. Chou *et al.*, 'Tetanus toxin inhibition of neurotransmitter release linked to synaptobrevin proteolysis.' *Biochem. Biophys. Res. Commun.*, 1992, **189**, 1017–1023.

89. L. W. Duchen and D. A. Tonge, 'The effect of tetanus toxin on neuromuscular transmission and on the morphology of motor end-plates in slow and fast skeletal muscle of the mouse.' *J. Physiol. (Lond.)*, 1973, **228**, 157–172.

90. E. Habermann, F. Dreyer and H. Bigalke, 'Tetanus toxin blocks the neuromuscular transmission *in vitro* like botulinum A toxin.' *Naunyn Schmiedebergs Arch. Pharmacol.*, 1980, **311**, 33–40.

91. F. Dreyer and A. Schmitt, 'Transmitter release in tetanus and botulinum A toxin-poisoned mammalian motor endplates and its dependence on nerve stimulation and temperature.' *Pflügers Arch.*, 1983, **399**, 228–234.

92. F. Dreyer, C. Becker, H. Bigalke *et al.*, 'Action of botulinum A toxin and tetanus toxin on synaptic transmission.' *J. Physiol. (Paris)*, 1984, **79**, 252–258.

93. W. E. Van Heyningen, 'Gangliosides as membrane receptors for tetanus toxin, cholera toxin and serotinin.' *Nature (Lond.)*, 1974, **249**, 415–417.

94. L. L. Simpson, 'Botulinum toxin and tetanus toxin recognize similar membrane determinants.' *Brain Res.*, 1984, **305**, 177–180.

95. L. L. Simpson, 'Fragment C of tetanus toxin antagonizes the neuromuscular blocking properties of native tetanus toxin.' *J. Pharmacol. Exp. Ther.*, 1984, **228**, 600–604.

96. M. Gansel, R. Penner and F. Dreyer, 'Distinct sites of action of *clostridial* neurotoxins revealed by double-poisoning of mouse motor nerve terminals.' *Pflügers Arch.*, 1987, **409**, 533–539.

97. M. E. Datyner and P. W. Gage, 'Presynaptic and postsynaptic effects of the venom of the Australian tiger snake at the neuromuscular junction.' *Br. J. Pharmacol.*, 1973, **49**, 340–354.

98. C. C. Chang and J. D. Lee, 'Crotoxin, the neurotoxin of South American rattlesnake venom, is a presynaptic toxin acting like β-bungarotoxin.' *Naunyn Schmiedebergs Arch. Pharmacol.*, 1977, **296**, 159–168.

99. C. C. Chang and C. Y. Lee, 'Isolation of neurotoxins from the venom of *Bungarus multicinctus* and their modes of neuromuscular blocking agents.' *Arch. Int. Pharmacodyn.*, 1963, **144**, 241–257.

100. C. C. Chang, T. F. Chen and C. Y. Lee, 'Studies of the presynaptic effect of α-bungarotoxin on neuromuscular transmission.' *J. Pharmacol. Exp. Ther.*, 1973, **184**, 339–345.

101. J. Halpert and D. Eaker, 'Amino acid sequence of a presynaptic neurotoxin from the venom of *Notechis scutatus scutatus* (Australian tiger snake).' *J. Biol. Chem.*, 1975, **250**, 6990–6997.

102. J. Halpert, D. Eaker and E. Karlsson, 'The role of phospholipase activity in the action of a presynaptic neurotoxin from the venom of *Notechis scutatus scutatus* (Australian tiger snake).' *FEBS Lett.*, 1976, **61**, 72–76.

103. S. G. Oberg and R. B. Kelly, 'The mechanism of β-bungarotoxin action. I. Modification of transmitter release at the neuromuscular junction.' *J. Neurobiol.*, 1976, **7**, 129–141.

104. J. Fohlman, D. Eaker, E. Karlsson *et al.*, 'Taipoxin, an extremely potent presynaptic neurotoxin from the venom of the Australian snake taipan (*Oxyuranus s. scutellatus*). Isolation, characterization, quaternary structure and pharmacological properties.' *Eur. J. Biochem.*, 1976, **68**, 457–469.

105. J. Fohlman and D. Eaker, 'Isolation and characterization of a lethal myotoxic phospholipase A from the venom of the common sea snake *Enhydrina schistosa* causing myoglobiniuria in mice.' *Toxicon*, 1977, **15**, 385–393.

106. J. Fohlman, D. Eaker, E. Karlsson *et al.*, 'Taipoxin, an extremely potent presynaptic neurotoxin from the venom of the Australian snake taipan (*Oxyuranus s. scutellatus*). Isolation, characterization, guaternary structure and pharmacological properties.' *Eur. J. Biochem*, 1976, **68**, 457–469.

107. T. Abe, A. R. Limbrick and R. Miledi, 'Acute muxcle denervation by bungarotoxin.' *Proc. R. Soc. Lond. Ser. B.*, 1976, **194**, 545–553.

108. T. Abe, S. Alema and R. Miledi, 'Isolation and characterization of presynaptically acting neurotoxins from the venom of *Bungarus* snakes.' *Eur. J. Biochem.*, 1977, **80**, 1–12.

109. J. V. Halliwell and J. O. Dolly, 'Preferential action of â-BuTX at nerve terminal regions in the hippocampus.' *Neurosci. Lett.*, 1982, **30**, 321–327.

110. R. Chapell and P. Rosenberg, 'Specificity of action of beta-bungarotoxin on acetylcholine release from synaptosomes.' *Toxicon*, 1992, **30**, 621–633.

111. A. C. Kato, J. E. B. Pinto, M. Glavinovic *et al.*, 'Action of a bungarotoxin on autopnomic ganglia and adrenergic neurotransmission.' *Can. J. Physiol. Pharmacol.*, 1977, **55**, 574–584.

112. A. Miura, I. Muramatsu, M. Fuliwara *et al.*, 'Species and regional differences in cholinergic blocking actions of bungarotoxin.' *J. Pharmacol. Exp. Ther.*, 1981, **217**, 505–509.

113. H. M. Verheij, A. J. Slotboom and G. H. DeHaas, 'Structure and function of phospholipase A2.' *Rev. Physiol. Biochem. Pharmacol.*, 1981, **91**, 91–203.

114. P. N. Strong and R. B. Kelly, 'Membranes undergoing phase transitions are preferentially hydrolyzed by bungarotoxin.' *Biochim. Biophys. Acta.*, 1977, **469**(2), 231–235.

115. C. G. Caratsch, B. Maranda, R. Miledi *et al.*, 'A further study of the phospholipase-independent action of beta-bungarotoxin at frog end-plates.' *J. Physiol. (Lond.)*, 1981, **319**, 179–191.

116. C. G. Caratsch, R. Miledi and P. N. Strong, 'Influence of divalent cations on the phospholipase-independent action of beta-bungarotoxin at frog neuromuscular junctions.' *J. Physiol. (Lond.)*, 1985, **363**, 169–179.

117. F. Dreyer and R. Penner, 'The actions of presynaptic snake toxins of membrane currents of mouse motor nerve terminals.' *J. Physiol. (Lond.)*, 1987, **386**, 455–463.

118. J. F. Wernicke, A. D. Vanker and B. D. Howard, 'The mechanism of action of bungarotoxin.' *J. Neurochem.*, 1975, **25**, 483–496.

119. E. Ueno and P. Rosenberg, 'Inhibition of phosphorylation of rat synaptosomal proteins by snake venom phospholipase A2 neurotoxins (bungarotoxin, notexin) and enzymes (*Naja naja atra, Naja nigricollis*).' *Toxicon*, 1990, **28**, 1423–1437.

120. I. Sen, P. A. Grantham and J. R. Cooper, 'Mechanism of action of β-bungarotoxin on synaptosomal preparations.' *Proc. Natl. Acad. Sci. USA*, 1976, **73**, 2664–2668.

121. E. Ueno and P. Rosenberg, 'Inhibition of phosphorylation of synapsin I and other synaptosomal proteins by beta-bungarotoxin, a phospholipase A2 neurotoxin.' *J. Neurochem.*, 1992, **59**, 2030–2039.

122. W. B. Huttner, W. Schiebler, P. Greengard *et al.*, 'Synapsin I (protein I), a nerve terminal-specific phosphoprotein. III. Its association with synaptic vesicles studied in a highly purified synaptic vesicle preparation.' *J. Cell Biol.*, 1983, **96**, 1374–1388.

123. R. Llinás, T. L. McGuinness, C. S. Leonard *et al.*, 'Intraterminal injection of synapsin I or calcium/calmodulin-dependent protein kinase II alters neurotransmitter release at the squid giant synapse.' *Proc. Natl. Acad. Sci. USA*, 1985, **82**, 3035–3039.

124. A. W. Clark, W. P. Hurlbut and A. Mauro, 'Changes in the fine structure of the neuromuscular junction of the frog caused by black widow spider venom.' *J. Cell Biol.*, 1972, **52**, 1–14.

125. A. W. Clark, A. Mauro, H. E. Longenecker *et al.*, 'Effects of black widow spider venom on the frog neuromuscular junction.' *Nature (Lond.)*, 1970, **225**, 703–705.

126. B. Ceccarelli, F. Grohovaz and W. P. Hurlbut, 'Freeze-fracture studies of the frog neuromuscular junctions during intense release of neurotransmitter. Effects of black widow spider venom and Ca²⁺-free solutions on the structure of the active zone.' *J. Cell. Biol.*, 1979, **81**, 163–177.

127. A. Gorio, W. P. Hurlbut and B. Ceccarelli, 'Acetylcholine compartments in mouse diaphragm; comparison of the effects of black widow spider venom, electrical stimulation, and high concentration of potassium.' *J. Cell Biol.*, 1978, **78**, 716–733.

128. N. Frontali, B. Ceccarelli, A. Gorio *et al.*, 'Purification from black widow spider venom of a protein factor causing the depletion of synaptic vesicles at neuromuscular junctions.' *J. Cell Biol.*, 1976, **68**, 462–479.

129. S. G. Cull-Candy, H. Lundh and S. Thesleff, 'Effects of botulinum toxin on neuromuscular transmission in the rat.' *J. Physiol. (Lond.)*, 1976, **260**, 177–203.

130. B. Figliomeni and A. Grasso, 'Tetanus toxin affects the K⁺-stimulated release of catecholamines from nerve growth factor-treated PC12 cells.' *Biochem. Biophys. Res. Commun.*, 1985, **128**, 249–256.

131. A. Gorio and A. Mauro, 'Reversibility and mode of action of Black Widow spider venom on the vertebrate neuromuscular junction.' *J. Gen. Physiol.*, 1979, **73**, 245–263.

132. H. B. Longenecker, W. P. Hurlbut, A. Mauro *et al.*, 'Effects of black widow spider venom on the frog neuromuscular junction.' *Nature (Lond.)*, 1970, **225**, 701–703.

133. A. Gorio, L. L. Rubin and A. Mauro, 'Double action of black widow spider venom on frog neuromuscular junction.' *J. Neurocytol.*, 1978, **7**, 193–205.

134. E. Wanke, A. Ferroni, P. Gattanini *et al.*, 'α-Latrotoxin of the black widow spider venom opens a small, non-closing cation channel.' *Biochem. Biophys. Res. Commun.*, 1986, **134**, 320–325.

135. H. Scheer and J. Meldolesi, 'Purification of the putative α-latrotoxin receptor from bovine synaptosomal membranes in an active binding form.' *EMBO J.*, 1985, **4**, 323–327.

136. H. Scheer, G. Prestipino and J. Meldolesi, 'Reconstitution of the purified α-latrotoxin receptor in liposomes and planar lipid membranes. Clues to the mechanism of toxin action.' *EMBO J.*, 1986, **5**, 2643–2648.

137. A. G. Petrenko, M. S. Perrin, B. A. Davletov *et al.*, 'Binding of synaptotagmin to the α-latrotoxin receptor implicates both in synaptic vesicle exocytosis.' *Nature (Lond.)*, 1990, **353**, 65–68.

138. A. G. Petrenko, M. S. Perin, B. A. Davletov *et al.*, 'Binding of synaptotagmin to the alpha-latrotoxin receptor implicates both in synaptic vesicle exocytosis.' *Nature*, 1991, **353**, 65–68.

139. Y. A. Ushkaryov, A. G. Petrenko, M. Geppert *et al.*, 'Neurexins: synaptic cell surface proteins related to the alpha-latrotoxin receptor and laminin.' *Science*, 1992, **257**, 50–56.

140. F. Valtorta, L. Madeddu, J. Meldolesi *et al.*, 'Specific localization of the α-latrotoxin receptor in the nerve terminal plasma membrane.' *J. Biol. Chem.*, 1984, **99**, 124–132.

141. N. Brose, A. G. Petrenko, T. C. Südhof *et al.*, 'Synaptotagmin: a calcium sensor on the synaptic vesicle surface.' *Science*, 1992, **256**, 1021–1025.

142. Y. Hata, B. Davletov, A. G. Petrenko *et al.*, 'Interaction of synaptotagmin with the cytoplasmic domains of neurexins.' *Neuron*, 1993, **10**, 307–315.

143. B. M. Olivera, L. J. Cruz, V. de Santos *et al.*, 'Neuronal calcium channel antagonists. Discrimination between calcium channel subtypes using ω-conotoxin from *Conus magnus* venom.' *Biochemistry*, 1987, **26**, 2086–2090.

144. B. M. Olivera, W. R. Gray, R. Zeikus *et al.*, 'Peptide neurotoxins from fish-hunting cone snails.' *Science*, 1985, **230**, 1338–1343.

145. L. J. Cruz and B. M. Olivera, 'Calcium channel antagonists. Omega-conotoxin defines a new high-affinity site.' *J. Biol. Chem.*, 1986, **261**, 6230–6233.

146. M. C. Nowycky, A. P. Fox and R. W. Tsien, 'Three types of neuronal calcium channel with different calcium agonist sensitivity.' *Nature*, 1985, **316**, 440–443.

147. T. Abe, K. Koyano, H. Saisu *et al.*, 'Binding of ω-conotoxin to receptor sites associated with the voltage-sensitive calcium channel.' *Neurosci. Lett.*, 1986, **71**, 203–208.

148. E. W. McCleskey, A. P. Fox, D. H. Feldman *et al.*, 'ω-Conotoxin: direct and persistant blockade of specific types of calcium channels in neurons but not muscle.' *Proc. Natl. Acad. Sci. USA*, 1987, **84**, 4327–4331.

149. A. P. Fox, M. C. Nowycky and R. W. Tsien, 'Kinetic and pharmacological properties distinguishing three types of calcium currents in chick sensory neurones.' *J. Physiol. (Lond.)*, 1987, **394**, 149–172.

150. D. R. Hillyard, V. D. Monje, I. M. Mintz *et al.*, 'A new *Conus* peptide ligand for mammalian presynaptic Ca²⁺ channels.' *Neuron*, 1992, **9**, 69–77.

151. S. J. Hong and C. C. Chang, 'Inhibition of acetylcholine release from mouse motor nerve by a P-type calcium channel blocker, ω-agatoxin IVA.' *J. Physiol. (Lond.)*, 1995, **482**, 283–290.

152. C. H. Yan, Y. F. Xu and W. D. Atchison, 'Comparative blocking actions of ω-agatoxin IVA and ω-conotoxin MVIIC on rat forebrain synaptosomes and neuromuscular junctions.' *Soc. Neurosci. Abstr.*, 1994, **20**, 71.

153. C. H. Yan and W. D. Atchison, 'Block of ⁴⁵Ca uptake into rat brain synaptosomes by ω-conotoxin MVIIC: K⁺, Ca²⁺ and Na⁺ dependence.' *Soc. Neurosci. Abstr.*, 1995, **21**, 339.

154. Y. F. Xu and W. D. Atchison, 'Characterization of giant MEPPs induced at neuromuscular junctions by 2,4-dithiobiuret.' *Soc. Neurosci. Abstr.*, 1995, **21**, 830.

155. M. Takahashi, Y. Ohizumi and T. Yasumoto, Maitotoxin, a Ca^{2+} channel activator candidate. *J. Biol. Chem.*, 1982, **257**, 7287–7289.

156. M. Takahashi, M. Tatsumi, Y. Ohizumi *et al.*, 'Ca^{2+} channel activating function of maitotoxin, the most potent marine toxin known, in clonal rat pheochromocytoma cells.' *J. Biol. Chem.*, 1983, **258**, 10944–10949.

157. M. E. Adams, V. P. Bindokas, L. Hasegawa *et al.*, 'ω-Agatoxins: novel calcium channel antagonists of two subtypes from funnel web spider (*Agelenopsis aperta*) venom.' *J. Biol. Chem.*, 1990, **265**, 861–867.

158. I. M. Mintz, M. E. Adams and B. P. Bean, 'P-type calcium channels in rat central and peripheral neurons.' *Neuron*, 1992, **9**, 85–95.

159. R. Llinás, M. Sugimori, J. W. Lin *et al.*, 'Blocking and isolation of a calcium channel from neurons in mammals and cephalopods utilizing a toxin fraction (FTX) from funnel-web spider poison.' *Proc. Natl. Acad. Sci. USA*, 1989, **86**, 1689–1693.

160. E. X. Albuquerque, J. Warnick and F. N. Sansone, 'The pharmacology of batrachotoxin. II. Effect on electrical properties of the mammalian nerve and skeletal muscle membranes.' *J. Pharmacol. Exp. Ther.*, 1971, **176**, 511–528.

161. E. X. Albuquerque and J. Warnick, 'The pharmacology of batrachotoxin. IV. Interaction with tetrodotoxin on innervated and chronically deverrated rat skeletal muscle.' *J. Pharmacol. Exp. Ther.*, 1972, **180**, 683–697.

162. W. D. Atchison, V. S. Luke, T. Narahashi *et al.*, 'Nerve membrane sodium channels as the target site of brevetoxins at neuromuscular junctions.' *Br. J. Pharmacol.*, 1986, **89**, 731–738.

163. A. L. Harvey and P. W. Gage, 'Increase of evoked release of acetylcholine at the neuromuscular junction by a fraction from the venom of the Eastern green mamba snake (*Dendroaspis angusticeps*).' *Toxicon*, 1981, **19**, 373–381.

164. A. L. Harvey and E. Karlsson, 'Protease inhibitor homologues from mamba venoms: facilitation of acetylcholine release and interactions with pre-juctional blocking toxins.' *Br. J. Pharmacol.*, 1982, **77**, 153–161.

165. J. V. Halliwell, I. B. Othman, A. Pelchen-Matthews *et al.*, 'Central action of dendrotoxin: selective reduction of a transient K conductance in hippocampus and binding to localized acceptors.' *Proc. Natl. Acad. Sci. USA*, 1986, **83**, 493–497.

166. C. Miller, E. Moczydlowski, R. Latorre *et al.*, 'Charybdotoxin, a protein inhibitor of single Ca^{2+}-activated K$^+$ channels from mammalian skeletal muscle.' *Nature*, 1985, **313**, 316–318.

167. C. Pittinger and R. Adamson, 'Antibiotic blockade of neuromuscular function.' *Annu. Rev. Pharmacol.*, 1972, **12**, 169–184.

168. M. D. Sokoll and S. D. Gergis, 'Antibiotics and neuromuscular function.' *Anesthesiology*, 1981, **55**, 148–159.

169. A. J. Caputy, Y. I. Kim and D. B. Sanders, 'The neuromuscular blocking effects of therapeutic concentrations of various antibiotics on normal rat skeletal muscle: a quantitative comparison.' *J. Pharmacol. Exp. Ther.*, 1981, **217**, 369–378.

170. B. Dunkley, I. Sanghvi and G. Goldstein, 'Characterization of neuromuscular block produced by streptomycin.' *Arch. Int. Pharmacodyn. Ther.*, 1973, **201**, 213–223.

171. N. N. Durant and J. J. Lambert, 'The action of polymyxin B at the frog neuromuscular junction.' *Br. J. Pharmacol.*, 1981, **72**, 41–47.

172. J. M. Farley, C. H. Wu and T. Narahashi, 'Mechanism of neuromuscular block by streptomycin: a voltage clamp analysis.' *J. Pharmacol. Exp. Ther.*, 1982, **222**, 488–493.

173. J. F. Fiekers, 'Neuromuscular block produced by polymyxin B: interactions with end-plate channels.' *Eur. J. Pharmacol.*, 1981, **70**, 77–81.

174. J. F. Fiekers, 'Effects of the aminoglycoside antibiotics, streptomycin and neomycin, on neuromuscular transmission. I. Presynaptic considerations.' *J. Pharmacol. Exp. Ther.*, 1983, **225**, 487–495.

175. J. F. Fiekers, 'Effects of the aminoglycoside antibiotics, streptomycin and neomycin, on neuromuscular transmission. II. Postsynaptic considerations.' *J. Pharmacol. Exp. Ther.*, 1983, **225**, 496–502.

176. J. F. Fiekers, F. Henderson, I. G. Marshall *et al.*, 'Comparative effects of clindamycin and lincomycin on end-plate currents and quantal content at the neuromuscular junction.' *J. Pharmacol. Exp. Ther.*, 1983, **227**, 308–315.

177. C. Lee, D. Chen and R. L. Katz, 'Characteristics of nondepolarizing neuromuscular block: (I) Post-junctional block by alpha-bungarotoxin.' *Can. Anaesth. Soc. J.*, 1977, **24**, 212–219.

178. Y. N. Singh, I. G. Marshall and A. L. Harvey, 'Some effects of the aminoglycoside antibiotic amikacin on neuromuscular and autonomic transmission.' *Br. J. Anaesth.*, 1978, **50**, 109–117.

179. Y. N. Singh, I. G. Marshall and A. L. Harvey, 'Depression of transmitter release and postjunctional sensitivity during neuromuscular block produced by antibiotics.' *Br. J. Anaesth.*, 1979, **51**, 1027–1033.

180. Y. N. Singh, I. G. Marshall and A. L. Harvey, 'Pre- and postjunctional blocking effects of aminoglycoside, polymyxin, tetracycline and lincosamide antibiotics.' *Br. J. Anaesth.*, 1982, **54**, 1295–1306.

181. T. Uchiyama, J. Molgó and M. Lemeignan, 'Presynaptic effects of bekanamycin at the frog neuromuscular junction. Reversibility by calcium and aminopyridines.' *Eur. J. Pharmacol.*, 1981, **72**, 271–280.

182. J. M. Wright and B. Collier, 'Characterization of the neuromuscular block produced by clindamycin and lincomycin.' *Can. J. Physiol. Pharmacol.*, 1976, **54**, 937–944.

183. J. M. Wright and B. Collier, 'The site of the neuromuscular block produced by polymyxin B and roli-tetracycline.' *Can. J. Physiol. Pharmacol.*, 1976, **54**, 926–936.

184. J. M. Wright and B. Collier, 'The effects of neomycin upon transmitter release and action.' *J. Pharmacol. Exp. Ther.*, 1977, **200**, 576–587.

185. W. E. Sanders and C. C. Sanders, 'Toxicity of antibacterial agents: mechanism of action on mammalian cells.' *Annu. Rev. Pharmacol. Toxicol.*, 1979, **19**, 53–83.

186. R. Redman and E. M. Silinsky, 'Decrease in calcium currents induced by aminoglycoside antibiotics in frog motor nerve endings.' *Br. J. Pharmacol.*, 1994, **113**, 375–378.

187. D. J. Dooley, A. Lupp and G. Hertting, 'Inhibition of central neurotransmitter release by ω-conotoxin GVIA, a peptide modulator of the N-type voltage-sensitive calcium channel.' *Naunyn Schmiedebergs Arch. Pharmacol.*, 1987, **336**, 467–470.

188. I. J. Reynolds, J. A. Wagner, S. H. Snyder *et al.*, 'Brain voltage-sensitive calcium channel subtypes differentiated by ω-conotoxin fraction GVIA.' *Proc. Natl. Acad. Sci. USA*, 1986, **83**, 8804–8807.

189. H. G. Knaus, J. Striessnig, A. Koza *et al.*, 'Neurotoxic aminoglycoside antibiotics are potent inhibitors of [^{125}I]-ω-conotoxin GVIA binding to guinea-pig cerebral cortex membranes.' *Naunyn Schmiedebergs Arch. Pharmacol.*, 1987, **336**, 583–586.

190. W. D. Atchison, L. Adgate and C. M. Beaman, 'Effects of antibiotics on uptake of calcium into

isolated nerve terminals.' *J. Pharmacol. Exp. Ther.*, 1988, **245**, 394–401.

191. R. Von Burg and T. Landry, 'Methylmercury and the skeletal muscle receptor.' *J. Pharm. Pharmacol.*, 1976, **28**, 548–551.

192. M. S. Juang, 'Depression of frog muscle contraction by methylmercuric chloride and mercuric chloride.' *Toxicol. Appl. Pharmacol.*, 1976, **35**, 183–185.

193. G. G. Somjen, S. P. Herman and R. Klein, 'Electrophysiology of methyl mercury poisoning.' *J. Pharmacol. Exp. Ther.*, 1973, **186**, 579–592.

194. R. S. Manalis and G. P. Cooper, 'Evoked transmitter release increased by inorganic mercury at frog neuromuscular junction (Letter).' *Nature*, 1975, **257**, 690–691.

195. M. S. Juang, 'An electrophysiological study of the action of methylmercuric chloride and mercuric chloride on the sciatic nerve-sartorius muscle preparation of the frog.' *Toxicol. Appl. Pharmacol.*, 1976, **37**, 339–348.

196. G. P. Cooper and R. S. Manalis, 'Influence of heavy metals on synaptic transmission: a review.' *Neurotoxicology*, 1983, **4**, 69–83.

197. W. D. Atchison and T. Narahashi, 'Methylmercuryinduced depression of neuromuscular transmission in the rat.' *Neurotoxicology*, 1982, **3**, 35–50.

198. W. D. Atchison, A. W. Clark and T. Narahashi, in 'Cellular and Molecular Neurotoxicology,' ed. T. Narahashi, Raven Press, New York, 1984, pp. 23–43.

199. D. L. Traxinger and W. D. Atchison, 'Comparative effects of divalent cations on the methylmercuryinduced alterations of acetylcholine release.' *J. Pharmacol. Exp. Ther.*, 1987, **240**, 451–459.

200. W. D. Atchison, U. Joshi and J. E. Thornburg, 'Irreversible suppression of calcium entry into nerve terminals by methylmercury.' *J. Pharmacol. Exp. Ther.*, 1986, **238**, 618–624.

201. Y. Yuan and W. D. Atchison, 'Methylmercury acts at multiple sites to block hippocampal synaptic transmission.' *J. Pharmacol. Exp. Ther.*, 1995, **275**, 1308–1316.

202. M. S. Juang and K. Yonemura, 'Increased spontaneous transmitter release from presynaptic nerve terminal by methylmercuric chloride.' *Nature*, 1975, **256**, 211–213.

203. Y. Yuan and W. D. Atchison, 'Disruption by methylmercury of membrane excitability and synaptic transmission in hippocampal slices of the rat.' *Toxicol. Appl. Pharmacol.*, 1993, **120**, 203–215.

204. O. Binah, U. Meiri and H. Rahamimoff, 'The effects of $HgCl_2$ and mersalyl on mechanisms regulating intracellular calcium and transmitter release.' *Eur. J. Pharmacol.*, 1978, **51**, 453–457.

205. H. Komulainen and S. C. Bondy, 'Increased free intrasynaptosomal Ca^{2+} by neurotoxic organometals: distinctive mechanisms.' *Toxicol. Appl. Pharmacol.*, 1987, **88**, 77–86.

206. M. F. Hare, K. M. McGinnis and W. D. Atchison, 'Methylmercury increases intracellular concentrations of Ca^{++} and heavy metals in NG108-15 cells.' *J. Pharmacol. Exp. Ther.*, 1993, **266**, 1626–1635.

207. M. F. Denny, M. F. Hare and W. D. Atchison, 'Methylmercury alters intrasynaptosomal endogenous concentrations of polyvalent cations.' *Toxicol. Appl. Pharmacol.*, 1993, **122**, 222–232.

208. M. F. Hare and W. D. Atchison, 'Comparative action of methylmercury and divalent inorganic mercury on nerve terminal and intraterminal mitochondrial membrane potentials.' *J. Pharmacol. Exp. Ther.*, 1992, **261**, 166–172.

209. M. F. Hare and W. D. Atchison, 'Methylmercury mobilizes Ca^{++} from intracellular stores sensitive to inositol 1,4,5-trisphosphate in NG108-15 cells.' *J. Pharmacol. Exp. Ther.*, 1995, **272**, 1016–1023.

210. B. B. Shrivastav, M. S. Brodwick and T. Narahashi, 'Methylmercury: effects on electrical properties of squid axon membranes.' *Life Sci.*, 1976, **18**, 1077–1081.

211. T. J. Shafer and W. D. Atchison, 'Effects of methylmercury on perineurial Na^+- and Ca^{2+}-dependent potentials at neuromuscular junctions of the mouse.' *Brain Res.*, 1992, **595**, 215–219.

212. T. J. Shafer and W. D. Atchison, 'Methylmercury blocks N- and L-type calcium channels in nervegrowth factor-differentiated pheochromocytoma (PC12) cells.' *J. Pharmacol. Exp. Ther.*, 1991, **258**, 149–157.

213. J. E. Sirois and W. D. Atchison, 'Methylmercury decreases whole cell barium currents in cerebellar granule neurons.' *Mol. Pharmacol.*, submitted, 1996.

214. J. Barrett, D. Botz and D. B. Chang, in 'Behavioral Toxicology: Early Detection of Occupational Hazards,' eds. C. Xintaras, B. L. Johnson and I. de Groot, US Department of Health, Education, and Welfare, 1974, pp. 277–287.

215. M. D. Miyamoto, 'Hg^{2+} causes neurotoxicity at an intracellular site following entry through Na and Ca channels.' *Brain Res.*, 1983, **267**, 375–379.

216. W. D. Atchison, 'Extracellular calcium-dependent and -independent effects of methylmercury on spontaneous and potassium-evoked release of acetylcholine at the neuromuscular junction.' *J. Pharmacol. Exp. Ther.*, 1986, **237**, 672–680.

217. J. Heuser and F. R. S. Miledi, 'Effect of lanthanum ions on function and structure of frog neuromuscular junctions.' *Proc. R. Soc. Lond. Ser. B.*, 1971, **179**, 247–260.

218. W. D. Atchison, 'Effects of activation of sodium and calcium entry on spontaneous release of acetylcholine induced by methylmercury.' *J. Pharmacol. Exp. Ther.*, 1987, **241**, 131–139.

219. P. C. Levesque and W. D. Atchison, 'Interactions of mitochondrial inhibitors with methylmercury on spontaneous quantal release of acetylcholine.' *Toxicol. App. Pharmacol.*, 1987, **87**, 315–324.

220. P. C. Levesque and W. D. Atchison, 'Effect of alteration of nerve terminal Ca^{2+} regulation on increased spontaneous quantal release of acetylcholine by methyl mercury.' *Toxicol. Appl. Pharmacol.*, 1988, **94**, 55–65.

221. P. C. Levesque and W. D. Atchison, 'Disruption of brain mitochondrial calcium sequestration by methylmercury.' *J. Pharmacol. Exp. Ther.*, 1990, **256**, 236–242.

222. J. D. Del Castillo, E. Bartels and J. A. Sobrino, 'Microelectrophoretic application of cholinergic compounds, protein oxidizing agents, and mercurials to the chemically excitable membrane of the electroplax.' *Proc. Natl. Acad. Sci. USA*, 1972, **8**, 2081–2085.

223. M. E. Eldefrawi, N. A. Mansour and A. T. Eldefrawi, in 'Membrane Toxicity,' eds. M. Miller and A. E. Shamoo, Plenum Press, New York, 1977, pp. 499–462.

224. A. Karlin and E. Bartels, 'Effects of blocking sulfhydryl groups and of reducing disulfide bonds on the acetylcholine-activated permeability system of the electroplax.' *Biochem. Biophys. Acta*, 1966, **126**, 525–535.

225. R. S. Manalis and G. P. Cooper, 'Presynaptic and postsynaptic effects of lead at the frog neuromuscular junction.' *Nature*, 1973, **243**, 354–356.

226. J. B. Pickett and J. C. Bornstein, 'Some effects of lead at mammalian neuromuscular junction.' *Am. J. Physiol.*, 1984, **246**, C271–C276.

227. W. D. Atchison and T. Narahashi, 'Mechanism of action of lead on neuromuscular junctions.' *Neurotoxicology*, 1984, **5**, 267–282.

228. G. P. Cooper and R. S. Manalis, 'Interactions of lead and cadmium on acetylcholine release at the frog neuromuscular junction.' *Toxicol. Appl. Pharmacol.*, 1984, **74**, 411–416.

229. G. P. Cooper and R. S. Manalis, 'Cadmium: effects on transmitter release at the frog neuromuscular junction.' *Eur. J. Pharmacol.*, 1984, **99**, 251–256.

230. R. S. Manalis, G. P. Cooper and S. L. Pomeroy, 'Effect of lead on neuromuscular transmission in the frog.' *Brain Res.*, 1984, **294**, 95–109.

231. J. Suszkiw, G. Toth, M. Murawsky *et al.*, 'Effects of Pb^{2+} and Cd^{2+} on acetylcholine release and Ca^{2+} movements in synaptosomes and subcellular fractions from rat brain and *Torpedo* electric organ.' *Brain Res.*, 1984, **323**, 31–46.

232. E. Reuveny and T. Narahashi, 'Potent blocking action of lead on voltage-activated calcium channels in human neuroblastoma cells SH-SY5Y.' *Brain Res.*, 1991, **545**, 312–314.

233. D. Büsselberg, M. L. Evans, H. L. Haas *et al.*, 'Blockade of mammalian and invertebrate calcium channels by lead.' *Neurotoxicology*, 1993, **14**, 249–258.

234. L. Kolton and Yaari, 'Sites of action of lead on spontaneous transmitter release from motor nerve terminals.' *Isr. J. Med. Sci.*, 1982, **18**, 165–170.

235. Z. Shao and J. B. Suszkiw, 'Ca^{2+}-surrogate action of Pb^{2+} on acetylcholine release from rat brain synaptosomes.' *J. Neurochem.*, 1991, **56**, 568–574.

236. J. Del Castillo and L. Engbaek, 'The nature of the neuromuscular block produced by magnesium.' *J. Physiol. (Lond.)*, 1954, **124**, 370–384.

237. D. H. Jenkinson, 'The nature of the antagonism between calcium and magnesium ions at the neuromuscular junction.' *J. Physiol.*, 1957, **138**, 434–444.

238. R. J. Balnave and P. W. Gage, 'The inhibitory effect of manganese on transmitter release at the neuromuscular junction of the toad.' *Br. J. Pharmacol.*, 1973, **47**, 339–352.

239. J. N. Weakly, 'The action of cobalt ions on neuromuscular transmission in the frog.' *J. Physiol. (Lond.)*, 1973, **234**, 597–612.

240. N. Kajimoto and S. M. Kirpekar, 'Effect of manganese and lanthanum on spontaneous release of acetylcholine at frog motor nerve terminals.' *Nat. New Biol.*, 1972, **235**, 29–30.

241. P. J. Forshaw, 'The inhibitory effect of cadmium on neuromuscular transmission in the rat.' *Eur. J. Pharmacol.*, 1977, **42**, 371–377.

242. E. Alnaes and R. Rahamimoff, 'Dual action of praseodymium (Pr^{3+}) on transmitter release at the frog neuromuscular synapse.' *Nature*, 1974, **247**, 478–479.

243. P. R. Benoit and J. Mambrini, 'Modification of transmitter release by ions which prolong the presynaptic action potential.' *J. Physiol. (Lond.)*, 1970, **210**, 681–695.

244. H. Wiegand, R. Papadopoulos, M. Csicsaky *et al.*, 'The action of thallium acetate of spontaneous transmitter release in the rat neuromuscular junction.' *Toxicol. Arch.*, 1984, **55**, 253–257.

245. J. Molgó, E. del Pozo, J. E. Banos *et al.*, 'Changes of quantal transmitter release caused by gadolinium ions at the frog neuromuscular junction.' *Br. J. Pharmacol.*, 1991, **104**, 133–138.

246. J. E. Allen, P. W. Gage, D. D. Leaver *et al.*, 'Triethyltin depresses evoked transmitter release at the mouse neuromuscular junction.' *Chem. Biol. Interact.*, 1980, **31**, 227–231.

247. S. Metral, C. Bonneton, C. Hort-Legrand *et al.*, 'Dual action of erbium on transmitter release at the frog neuromuscular synapse.' *Nature*, 1978, **271**, 773–775.

248. R. Miledi, 'Strontium as a substitute for calcium in the process of transmitter release at the neuro-muscular junction.' *Nature*, 1966, **212**, 1233–1234.

249. F. A. Dodge, Jr., R. Miledi and R. Rahamimoff, 'Strontium and quantal release of transmitter at the neuromuscular junction.' *J. Physiol. (Lond.)*, 1969, **200**, 267–283.

250. U. Meiri and R. Rahamimoff, 'Neuromuscular transmission: inhibition by manganese ions.' *Science*, 1972, **176**, 308–309.

251. E. M. Silinsky, 'Can barium support the release of acetylcholine by nerve impulses?' *Br. J. Pharmacol.*, 1977, **59**, 215–217.

252. E. M. Silinsky, 'An estimate of the equilibrium dissociation constant for calcium as an antagonist of evoked acetylcholine release: implications for excitation-secretion coupling.' *Br. J. Pharmacol.*, 1977, **61**, 691–693.

253. E. M. Silinsky, 'On the role of barium in supporting the asynchronous release of acetylcholine quanta by motor nerve impulses.' *J. Physiol. (Lond.)*, 1978, **274**, 157–171.

254. E. M. Silinsky, 'Enhancement by an antagonist of transmitter release from frog motor nerve terminals.' *Br. J. Pharmacol.*, 1978, **63**, 485–493.

255. R. Anwyl, T. Kelly and F. Sweeney, 'Alterations of spontaneous quantal transmitter release at the mammalian neuromuscular junction induced by divalent and trivalent ions.' *Brain Res.*, 1982, **246**, 127–132.

256. G. P. Cooper, J. B. Suszkiw and R. S. Manalis, in 'Cellular and Molecular Basis of Neurotoxicity of Environmental Agents,' ed. T. Narahashi, Raven Press, New York, 1984, pp. 1–21.

257. M. Nishimura, I. Tsutsui, O. Yagasaki *et al.*, 'Transmitter release at the mouse neuromuscular junction stimulated by cadmium ions.' *Arch. Int. Pharmacodyn. Ther.*, 1984, **271**, 106–121.

258. H. Wiegand, M. Csicsaky and U. Kramer, 'The action of thallium acetate on neuromuscular transmission in the rat phrenic nerve-diaphragm preparation.' *Arch. Toxicol.*, 1984, **55**, 55–58.

259. H. Wiegand, R. Papadopoulos, M. Csicsaky *et al.*, 'The action of thallium acetate on spontaneous transmitter release in the rat neuromuscular junction.' *Arch. Toxicol.*, 1984, **55**, 253–257.

260. M. Nishimura, 'Zn^{2+} stimulates spontaneous transmitter release at mouse neuromuscular junctions.' *Br. J. Pharmacol.*, 1988, **93**, 430–436.

261. Y. X. Wang and D. M. Quastel, 'Multiple actions of zinc on transmitter release at mouse end-plates.' *Pflügers Arch.*, 1990, **415**, 582–587.

262. P. F. Baker, A. L. Hodgkin and E. B. Ridgway, 'Depolarization and calcium entry in squid giant axons.' *J. Physiol. (Lond.)*, 1971, **218**, 709–755.

263. P. G. Kostyuk, O. A. Krishtal, Y. A. Shakhovalor *et al.*, 'Separation of sodium and calcium currents in the somatic membrane of mollusc neurones.' *J. Physiol. (Lond.)*, 1977, **270**, 545–568.

264. D. A. Nachshen, 'Selectivity of the Ca binding site in synaptosome Ca channels. Inhibition of Ca influx by multivalent metal cations.' *J. Gen. Physiol.*, 1984, **83**, 941–967.

265. A. P. Fox, M. C. Nowycky and R. Tsien, 'Single-channel recordings on three types of calcium channels in chick sensory neurones.' *J. Physiol. (Lond.)*, 1987, **394**, 173–200.

266. M. L. Evans, D. Büsselberg and D. O. Carpenter, 'Pb^{2+} blocks calcium currents of cultured dorsal root ganglion cells.' *Neurosci. Lett.*, 1991, **129**, 103–106.

267. T. E. Kober and G. P. Cooper, 'Lead competitively inhibits calcium-dependent synaptic transmission in the bullfrog sympathetic ganglion.' *Nature*, 1976, **262**, 704–705.

268. H. Wiegand, S. Uhlig, U. Gotzsch, 'The action of cobalt, cadmium and thallium on presynaptic currents in mouse motor nerve endings.' *Neurotoxicol. Teratol.*, 1990, **12**, 313–318.

269. W. P. Hurlbut, H. E. Longenecker, Jr. and A. Mauro, 'Effects of calcium and magnesium on the frequency of miniature end-plate potentials during prolonged tetanization.' *J. Physiol. (Lond.)*, 1971, **219**, 17–38.

270. E. B. Astwood, A. M. Hughes, M. Lubin et al., 'Reversible paralysis of motor function in rats from the chronic administration of dithiobiuret.' *Science*, 1945, **102**, 196–197.

271. W. D. Atchison and R. E. Peterson, 'Potential neuromuscular toxicity of 2,4-dithiobiuret in the rat.' *Toxicol. Appl. Pharmacol.*, 1981, **57**, 63–68.

272. W. D. Atchison, K. H. Yang and R. E. Peterson, 'Dithiobiuret toxicity in the rat: evidence for latency and cumulative dose thresholds.' *Toxicol. Appl. Pharmacol.*, 1981, **61**, 166–171.

273. K. M. Crofton, K. F. Dean, R. C. Hamrick et al., 'The effects of 2,4-dithiobiuret on sensory and motor function.' *Fundam. Appl. Toxicol.*, 1991, **16**, 469–481.

274. K. D. Williams, R. M. LoPachin, W. D. Atchison et al., 'Antagonism of dithiobiuret toxicity in rats.' *Neurotoxicology*, 1986, **7**, 33–49.

275. W. D. Atchison, P. M. Lalley, R. G. Cassens et al., 'Depression of neuromuscular function in the rat by chronic 2,4-dithiobiuret treatment.' *Neurotoxicology*, 1981, **2**, 329–346.

276. W. D. Atchison, W. S. Mellon, P. M. Lalley et al., 'Dithiobiuret-induced muscle weakness in rats: evidence for a prejunctional effect.' *Neurotoxicology*, 1982, **3**, 44–54.

277. H. M. Weiler, K. D. Williams and R. E. Peterson, 'Effects of 2,4-dithiobiuret treatment in rats on cholinergic function and metabolism of the *extensor digitorum longus* muscle.' *Toxicol. Appl. Pharmacol.*, 1986, **84**, 220–231.

278. J. M. Spitsbergen and W. D. Atchison, 'Voltage clamp analysis reveals multiple populations of quanta released at neuromuscular junctions of rats treated with 2,4-dithiobiuret.' *J. Pharmacol. Exp. Ther.*, 1991, **256**, 159–163.

279. W. D. Atchison and J. M. Spitsbergen, in 'Principles of Neurotoxicology,' ed. L. W. Chang, Marcel Dekker, New York, 1994, pp. 265–307.

280. P. S. Spencer and H. H. Schaumburg, 'Experimental neuropathy produced by 2,5-hexanedione—a major metabolite of the neurotoxic industrial solvent methyl *n*-butyl ketone.' *J. Neurol. Neurosurg. Psych.*, 1975, **38**, 771–775.

281. A. Cangiano, L. Lutzemberger, N. Rizzuto et al., 'Neurotoxic effects of 2,5-hexanedione in rats: early morphological and functional changes in nerve fibres and neuromuscular junctions.' *Neurotoxicology*, 1980, **2**, 25–32.

282. P. Taylor, in 'The Pharmacological Basis of Therapeutics,' eds. A. Gilman, L. S. Goodman, T. W. Rall et al., Macmillan, New York, 1985, pp. 222–235.

283. M. B. Jackson, H. Lecar, V. Askanas et al., 'Single cholinergic receptor channel currents in cultured human muscle.' *J. Neurosci.*, 1982, **2**, 1465–1473.

284. G. J. Strecker and M. B. Jackson, 'Curare binding and the curare-induced subconductance state of the acetylcholine receptor channel.' *Biophys. J.* 1989, **56**, 795–806.

285. S. M. Sine and P. Taylor, 'Local anesthetics and histrionicotoxin are allosteric inhibitors of the acetylcholine receptor. Studies of clonal muscle cells.' *J. Biol. Chem.*, 1982, **257**, 8106–8114.

286. R. L. Papke and R. E. Oswald, 'Mechanisms of noncompetitive inhibition of acetylcholine-induced single-channel currents.' *J. Gen. Physiol.*, 1989, **93**, 785–811.

287. A. Karlin and M. Winnik, 'Reduction and specific alkylation of the receptor for acetylcholine.' *Proc. Natl. Acad. Sci. USA*, 1968, **60**, 668–674.

288. H. P. Rang and J. M. Ritter, 'The effect of disulfide bond reduction on the properties of cholinergic receptors in chick muscle.' *Mol. Pharmacol.*, 1971, **7**, 620–631.

289. J. W. Walker, R. J. Lukas and M. G. McNamee, 'Effects of thio-group modifications on the ion permeability control and ligand binding properties of *Torpedo californica* acetylcholine receptor.' *Biochemistry.*, 1981, **23**, 2191–2199.

290. J. W. Walker, C. A. Richardson and M. G. McNamee, 'Effects of thio-group modifications of *Torpedo californica* acetylcholine receptor on ion flux activation and inactivation kinetics.' *Biochemistry.*, 1984, **23**, 2329–2338.

291. D. Ben-Haim, E. M. Landau and I. Silman, 'The role of a reactive disulphide bond in the function of the acetylcholine receptor at the frog neuromuscular junction.' *J. Physiol. (Lond.)*, 1973, **234**, 305–325.

292. D. A. Terrar, 'Effects of dithiothreitol on end-plate currents.' *J. Physiol. (Lond.)*, 1978, **276**, 403–417.

293. D. Ben-Haim, F. Dreyer and K. Peper, 'Acetylcholine receptor: modification of synaptic gating mechanism after treatment with a disulfide bond reducing agent.' *Pflügers Arch.*, 1975, **355**, 19–26.

294. A. Steinacker and C. Zuazaga, 'Changes in neuromuscular junction endplate current time constants produced by sulfhydryl reagents.' *Proc. Natl. Acad. Sci. USA*, 1981, **78**, 7806–7809.

295. A. Steinacker and C. Zuazaga, 'Further kinetic analysis of the chemically modified acetylcholine receptor.' *Pflügers Arch.*, 1987, **409**, 555–560.

296. S. N. Abramson, P. Culver, T. Kline et al., 'Lophotoxin and related coral toxins covalently label the alpha-subunit of the nicotinic acetylcholine receptor.' *J. Biol. Chem.*, 1988, **263**, 18568–18573.

297. W. D. Atchison, T. Narahashi and S. M. Vogel, 'Endplate blocking actions of lophotoxin.' *Br. J. Pharmacol.*, 1984, **82**, 667–672.

298. D. C. Ogden and D. Colquhoun, 'Ion channel block by acetylcholine, carbachol and suberyldicholine at the frog neuromuscular junction.' *Proc. R. Soc. Lond. (Biol.).*, 1985, **225**, 329–355.

299. S. M. Sine and J. H. Steinbach, 'Activation of a nicotinic acetylcholine receptor.' *Biophys. J.*, 1984, **45**, 175–185.

300. D. Colquhoun, F. Dreyer and R. E. Sheridan, 'The actions of tubocurarine at the frog neuromuscular junction.' *J. Physiol. (Lond.)*, 1979, **293**, 247–284.

301. E. Neher and J. H. Steinbach, 'Local anaesthetics transiently block currents through single acetylcholine-receptor channels.' *J. Physiol. (Lond.)*, 1978, **277**, 153–176.

302. D. C. Ogden, S. A. Siegelbaum and D. Colquhoun, 'Block of acetylcholine-activated ion channels by an unchared local anaesthetic.' *Nature*, 1981, **289**, 596–598.

303. E. Neher, 'The charge carried by single-channel currents of rat cultured muscle cells in the presence of local anaesthetics.' *J. Physiol. (Lond.)*, 1983, **339**, 663–678.

304. P. R. Adams and B. Sakmann, 'Decamethonium both

opens and blocks endplate channels.' *Proc. Natl. Acad. Sci. USA*, 1978, **75**, 2994–2998.

305. J. J. Lambert, N. N. Durant, L. S. Reynolds *et al.*, 'Characterization of end-plate conductance in transected frog muscle: modification by drugs.' *J. Pharmacol. Exp. Ther.*, 1981, **216**, 62–69.

306. J. R. Cooper, F. E. Bloom and R. H. Roth, in 'The Biochemical Basis of Neuropharmacology,' Oxford University Press, New York, 1982, chap. 5, pp 77–108.

307. J. C. Eccles and W. V. MacFarlane, 'Actions of anticholinesterases on endplate potential of frog muscle.' *J. Neurophysiol.*, 1948, 59–80.

308. M. B. Laskowski and W. D. Dettbarn, 'An electrophysiological analysis of the effects of paraoxon at the neuromuscular junction.' *J. Pharmacol. Exp. Ther.*, 1979, **210**, 269–274.

309. M. Gwilt and D. Wray, 'The effect of chronic neostigmine treatment on channel properties at the rat skeletal neuromuscular junction.' *Br. J. Pharmacol.*, 1986, **88**, 25–31.

310. L. Wecker, B. Laskowski and W. D. Dettbarn, 'Neuromuscular dysfunction induced by acetylcholinesterase inhibition.' *Fed. Proc.*, 1978, **37**, 2818–2822.

311. M. B. Laskowski and W. D. Dettbarn, 'Presynaptic effects of neuromuscular cholinesterase inhibition.' *J. Pharmacol. Exp. Ther.*, 1975, **194**, 351–361.

312. N. Steinberg, E. Roth and I. Silman, '*Torpedo* acetylcholinesterase is inactivated by thiol reagents.' *Biochem. Int.*, 1990, **21**, 1043–1050.

11.14
Auditory Toxicology

LAURENCE D. FECHTER and YE LIU

University of Oklahoma Health Sciences Center, Oklahoma City, OK, USA

11.14.1 BACKGROUND

Auditory impairments include the loss of auditory sensitivity, marked by an elevation in the detection threshold for sound; tinnitus, a perceived sound in the ears in the absence of external sound stimulation; and loudness recruitment or a shrinking of the dynamic range between sound detection and the pain threshold for sound. Such impairments may result from damage along any portion of the auditory pathway, but specific ototoxic impairments have been identified most closely with the peripheral auditory system and, specifically, with the cochlea or inner ear. This chapter discusses mechanisms that can account for cochlear impairment. While some chemicals, such as lead[1] and hexanes,[2,3] do disrupt central auditory pathways, it is assumed here that such effects relate to general neurotoxic processes rather than to ototoxic processes. Therefore, agents having such central effects are identified in this chapter, but the mechanisms that underlie central auditory system injury are not discussed here.

11.14.2 ANATOMICAL BASIS OF AUDITORY IMPAIRMENTS

Space does not permit more than a cursory description of the structure and function of the cochlea, but the interested reader is referred to two excellent books for a fuller understanding.[4,5] The cochlea consists of the sensory epithelium for audition (the organ of Corti), the spiral ganglion cells or neuronal cell bodies with which these receptors synapse, the supporting tissue and fluid spaces that maintain structural integrity and biochemical homeostasis, and the bony otic capsule that contains this soft tissue (see Figure 1). Sound frequency is encoded by a "place code" such that high frequency tones stimulate cells near the basal turn of the cochlea, while low frequency sound is encoded at more apical locations along the cochlear partition. Sound intensity is encoded in terms of unit entrainment with increased synchronous firing of spiral ganglion cells yielding a larger compound action potential (CAP), the summed propagated electrical output of the cochlea. The sensory cells consist of the inner and outer hair cells. The inner hair cells, the primary sensory receptors for sound, make synaptic contact with the Type 1 spiral ganglion cells and these, here in turn, contribute approximately 95% of the afferent information to the auditory branch of the eighth cranial

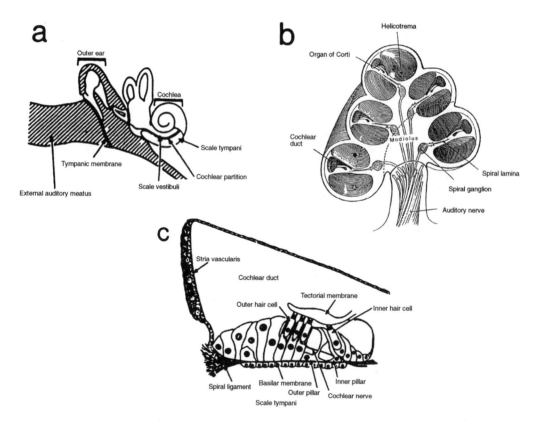

Figure 1 Diagrams of mammalian cochlea showing its gross structure and location relative to the outer ear (a), the cochlea in cross-section demonstrating the spiral nature of the structure, the position of the organ of Corti, spiral ganglion cells, stria vascularis, and auditory nerve (b), and section through one turn of the cochlea showing the detailed structure of the organ of Corti (c). Figure 1(a) adapted from von Bekesy[6] (reproduced with permission of McGraw-Hill from 'Experiments in Hearing,' 1960). Figure 1(b) adapted from Langley *et al.*[7] (reproduced with permission of McGraw-Hill from 'Dynamic Anatomy and Physiology,' 1969). Figure 1(c) adapted from Borysenko and Beringer[8] (reproduced by permission of Little & Brown from 'Functional Histology,' 1984).

nerve. The inner hair cells are embedded within supporting cells and extend stereocilia toward the tectorial membrane. The bending of inner hair cell stereocilia when the cochlea is stimulated by a pressure wave is coupled to calcium channels in the inner hair cell, such that release of neurotransmitter substances is enhanced or inhibited depending upon whether the cell is depolarized or hyperpolarized (see Figure 2).

The outer hair cells are far more numerous than the inner hair cells, and are viewed as important ancillary cells that influence the sensitivity of the true sense receptor, the inner hair cell. Damage to the outer hair cells may be a prominent consequence of some ototoxicant exposures and can account for an approximate loss of 40 dB in auditory sensitivity.[10] The outer hair cells are cylindrically shaped cells that are fixed at the base by supporting cells and at the apex by the stereocilia that they project into the gelatinous tectorial membrane (see Figure 1). The lateral surfaces of these cells are exposed to perilymphatic fluid rather than being embedded in a cellular matrix. Thus, a considerable surface area of the outer hair cell would be exposed to ototoxicants present in perilymphatic fluid. A unique system of subsurface cysternae, specialized structures subserving the active contractile response that these cells exhibit in response to stimulation, and mitochondria are localized along the longitudinal axis of the cell.[11]

A prominent supportive tissue bed localized along the lateral wall of the cochlea, known as the stria vascularis, is responsible both for maintaining the ionic milieu of the cochlea and

for delivery of key nutrients (see Figure 1). This tissue has an intensely high metabolic activity and has far higher levels of carbonic anhydrase, Na^+-K^+-ATPase, and adenylate cyclase, for example, than does the organ of Corti.[12,13] The stria vascularis is responsible for cycling K^+ into the scala media containing the hair cells, such that the endolymph fluid contained in scala media has a K^+ level of 140 mM. The K^+ level in perilymph is approximately 5 mM.[14] In addition, the stria vascularis is responsible for maintaining a positive resting d.c. potential of +80 mV called the endocochlear potential. The intermediate cells of the stria vascularis have tight junctions providing a blood–labyrinth barrier analogous to the blood–brain barrier.

11.14.3 PATTERNS OF TOXIC IMPAIRMENT IDENTIFIED IN THE AUDITORY SYSTEM

The study of traditional ototoxic agents (loop diuretics, aminoglycoside antibiotics, chemotherapeutics, quinine, and salicylate) has provided three primary patterns of cochlear impairment and injury that have served, for many years as the only models of ototoxicity. Aminoglycoside antibiotic and cisplatin chemotherapy are associated with permanent, preferentially high frequency, hearing impairment and with the loss or injury of outer hair cells, especially in the basal (high-frequency) turn of the cochlea (cf. Ref. 11). Loop diuretics, for example, target the stria vascularis disrupting Na^+-K^+-

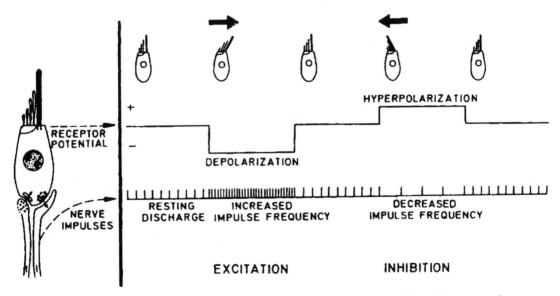

Figure 2 Diagrammatic representation of the electrophysiological response of the auditory nerve in response to movement of the stereocilia in a manner producing hyperpolarization and depolarization of a hair cell. Drawing adapted from Flock[9] (reproduced by permission of Cold Spring Harbor Laboratory from Cold Spring Harbor Symposium on Quantitative Biology, 1965, **30**, 135–146).

ATPase, leading to impairment of K^+ currents in the cochlea and a loss of the endocochlear potential.[15-17] Impairments observed here tend to be transitory because of the reversible nature of enzyme inhibition and the repair capacity of vascular tissue. Salicylates and quinine toxicity are most closely associated with the development of tinnitus—a phenomenon that includes significant outer hair cell dysfunction.[18-20]

Since the mid-1980s, ototoxic environmental contaminants have been identified that demonstrate additional cellular targets and patterns of ototoxicity than the well-studied ototoxicants noted above. Among these, many organic solvents preferentially disrupt mid-frequency rather than high frequency auditory function,[21] and some organic metals and chemical asphyxiants preferentially disrupt inner hair cells and spiral ganglion cells rather than outer hair cells.[22,23]

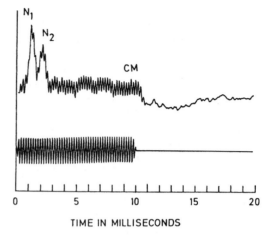

Figure 3 Representation of the normal electrophysiological signal recorded at the round window (top trace) elicited by a pure tone stimulus (bottom trace). The N_1 response triggered at the Type 1 spiral ganglion cell represents the initial feature of the propagated compound action potential (CAP) and the primary output of the cochlea to the auditory cortex. The nonpropagated cochlear microphonic potential (CM) generated predominantly from outer hair cells is also depicted. Note that the CM is phase locked to the stimulus (reproduced with permission of Academic Press from 'Auditory Physiology,' 1983).

11.14.4 OUTER HAIR CELL IMPAIRMENT: FUNCTIONAL AND STRUCTURAL MARKERS

11.14.4.1 *In Vivo* Markers

Loss of outer hair cells is easily recognized and quantified (albeit through a laborious process) when the organ of Corti is prepared as a surface preparation (see Figure 3). Outer hair cell function can be evaluated *in vivo* by measurement of the cochlear microphonic (CM) potential, and by measurement of distortion product otoacoustic emissions (DPOAE). The CM is an evoked a.c. wave that is produced in response to sound stimulation, and can be recorded in the anesthetized surgically prepared animal either at the round window or differentially along the length of the cochlear partition (see Figure 4). The CM is phase-locked to the stimulating sound frequency, such that it mirrors closely the sound frequency delivered to the ear. The outer hair cell is the primary generator for the nonpropagated CM potential[24] and when disrupted, the amplitude of the CM is diminished, such that a louder tone must be delivered in order to obtain some criterion voltage. This potential is dependent upon the presence of a normal resting endocochlear potential of $+80\,mV$ generated by the stria vascularis. Thus, depression of the CM potential can be attributed to outer hair cells only after it is assured that the endocochlear potential is normal. The CM measured at the round window will be heavily weighted toward measuring the function of outer hair cells near

that structure—namely the outer hair cells at the base of the cochlea. Disruption in the CM in response to low frequency sound cannot, therefore, be attributed to a loss of outer hair cells in apical cochlear turns.

Measurement of DPOAE is an acoustical method which relies on the fact that the active cochlea generates sound energy in response to external sound stimulation. The frequency of the sound which the cochlea generates, however, is shifted away from the stimulating sound in a predictable fashion. For example, when stimulated with tones of frequencies f_1 and f_2, a prominent distortion product is generated at a frequency of $2(f_1 - f_2)$. This distortion product can be detected by a microphone placed into the external ear canal of a human or laboratory animal subject. The primary source of the DPOAE is believed to be the outer hair cells when relatively low intensity stimuli are employed.[25,26] Disruption of outer hair cell integrity is associated with reduction in DPOAE amplitude (see Figure 5).

11.14.4.2 *In Vitro* Methods

Outer hair cells are capable of at least two sorts of motility when maintained in primary

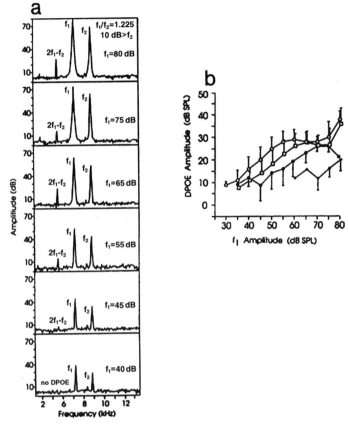

Figure 4 Representation of the normal distortion product at $2(f_1 - f_2)$ elicited by stimulus pairs of increasing intensity between 40–80 dB (a) showing an increase in DPOAE amplitude as a function of stimulus intensity. Impairment of the DPOAE amplitude following toluene exposure (1400 ppm 16 h d^{-1}) is presented in panel b, prior to exposure (open circles), after 3 d of exposure (open squares), after 5 d of exposure (closed circles) and 4 d following an 8 d exposure. Note the decrement in DPOAE amplitude across f_1 amplitude following toluene exposure. Data from Johnson[27] (reproduced by permission of Academic Press from *Scand. Audiol.*, 1994, **39** (Suppl.), 1–40).

culture (and presumably *in situ*, as well). These are the rapid electromotile responses elicited by electrical stimulation,[28,29] mechanical stimulation,[28,30] and by the uncaging of calcium ions resulting in increased concentration of free intracellular calcium $[Ca^{2+}]_i$.[31] There is also a slower contraction that is seen following ototoxic challenge and perturbation of extracellular osmolality.[32,33] These slow contractile shifts are accompanied by an increase in cell volume and are believed to serve as a mechanism for maintaining a positive pressure within the cell relative to the surrounding medium.[33] Ototoxicants, such as salicylates,[20] trimethyltin and triethyltin,[34] and toluene[35] can produce a shortening of isolated outer hair cells at concentrations in the micromolar range (see Figure 6). Unidentified metabolites of aminoglycoside antibiotics generated by a hepatic S9 fraction reduce isolated outer hair cell length, although the parent compound may not alter this measure.[36]

11.14.5 INNER HAIR CELL-TYPE 1 SPIRAL GANGLION CELL TOXICITY: MARKERS OF IMPAIRMENT

Inner hair cells and Type 1 spiral ganglion cells constitute the pre- and postsynaptic elements in the cochlea. The loss in sensitivity of the compound action potential (CAP) generated at this synapse when the CM is unchanged is consistent with dysfunction at the inner hair cell and/or Type 1 spiral ganglion cell. This outcome has been reported following such ototoxic (and neurotoxic) compounds as trimethyltin,[37] hypoxia,[38,39] and trichloroethylene.[40] *In vitro* study of the isolated inner hair cell has been largely ignored because of the relative difficulty in culturing inner hair cells, and a preoccupation with understanding processes of outer hair cell motility. The inner hair cell is not motile.

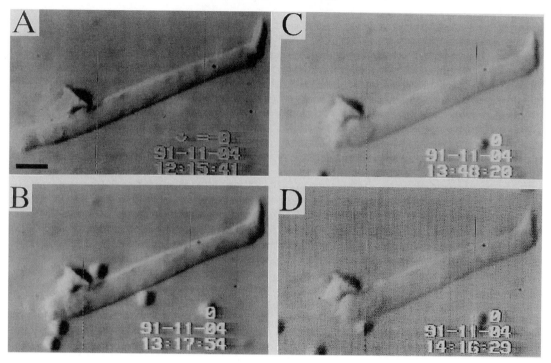

Figure 5   Reduction in outer hair cell length of an isolated cell exposed to 30 M trimethyltin for 0 min (a), 30 min (b), 60 min (c), and 90 min (d). Data from Clerici *et al.*[34] (reproduced by permission of Academic Press from *Toxicol. Appl. Pharmacol.*, 1993, **120**, 193–202).

The Type 1 spiral ganglion cell serves as the generator of the action potential within the auditory nerve. As a postsynaptic cell body with a glutamatergic input, the Type 1 spiral ganglion cell is vulnerable to excitotoxic influences mediated via glutamate receptors, as well as perturbations in ion flux.[41,42]

11.14.6   MECHANISMS OF OTOTOXICITY

11.14.6.1   Excitotoxicity

Excessive release and/or impaired inactivation of excitatory neurotransmitters have been viewed as basic mechanisms of postsynaptic injury to neurons having glutamate receptor sites in the central nervous system (see Chapter 4, this volume). Evidence supports the role of glutamate as the afferent neurotransmitter at the inner hair cell–Type 1 spiral ganglion cell synaptic junction.[43,44] Glutamate has been localized within inner hair cells.[45] It is released in a Ca^{2+}-dependent fashion[46] and is excitatory to the type 1 spiral ganglion cell.[47] Uptake of the precursor, glutamine, occurs at the inner hair cell.[48] Glutamate receptor blockers infused into the cochlea disrupt neurotransmission and impair the CAP.[49–52] Littman *et al.*[50] found evidence for kainate and quisqualate receptors based on the ability of 6,7-dinitro-quinoxaline-

2,3-dione (DNQX) and 6-cyano-7 nitroquinoxaline-2,3-dione (CNQX) to impair auditory nerve conduction when infused directly into the cochlea. Evidence for *N*-methyl-D-aspartate (NMDA) receptors has also been published.[51,53] Immunocytochemistry and *in situ* hybridization histochemistry studies for cellular localization of glutamatergic receptor expression have detected AMPA (GluR2–4), Kainate (GluR5&6 and KA1&2) and NMDA (NMDAR1 and NMDAR2A-D, NMDAR-K) receptor subunit expression.[54,55]

Janssen *et al.*[56] showed that administration of exogenous glutamate to neonatal subjects results in a loss of high frequency auditory function as measured by the auditory brainstem response. While this measure cannot pinpoint the site of dysfunction, histopathology was restricted to the Type 1 spiral ganglion cell in the corresponding basal turn of the cochlea. No evidence of presynaptic inner hair cell or outer hair cell loss was reported.

Juiz *et al.*[57] demonstrated that kainic acid (5 nM) perfused through the cochlea and allowed to remain there for 10 d or more selectively destroyed Type 1 spiral ganglion cells, but not Type 2 spiral ganglion cells, inner hair cells, outer hair cells, or supporting cells in the rat. The selectivity of this effect to the Type 1 spiral ganglion cell population strongly supports postsynaptic excitotoxicity.

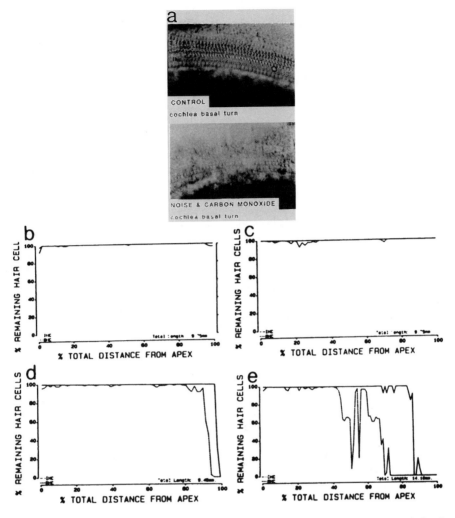

Figure 6 Light micrographs (a) showing damage to organ of Corti from a rat exposed simultaneously to noise (105 dBA) and carbon monoxide (1200 ppm) for 3.5 h compared to a control subject. Subjects were allowed an 8 week recovery period before subjects were sacrificed for histopathology. Note the absence of hair cells in the basal turn compared to the control subject. Quantification of hair cell loss (b–e) shows a far more profound loss of hair cells in subjects receiving combined exposure than those exposed to noise only. Subjects exposed to carbon monoxide only and unexposed controls did not show a significant loss of hair cells at any point along the cochlea.

11.14.6.2 Enhanced Intracellular Calcium

Cells expend a significant amount of energy maintaining $[Ca^{2+}]_i$ levels far lower than extracellular levels. Shifts in $[Ca^{2+}]_i$ can occur through a variety of mechanisms including dynamic changes in the normal function of active calcium channels, effectiveness of $[Ca^{2+}]_i$ storage mechanisms, and intactness of the cellular membrane. Because elevated $[Ca^{2+}]_i$ may produce cell injury and cell death,[58] measurement of $[Ca^{2+}]_i$ becomes not only a marker for alterations in cell physiology, but also a predictor of cell injury and death when potential xenobiotics are delivered to the cells. Increased $[Ca^{2+}]_i$ in the outer hair cell (at doses

below those producing cell injury) is associated with rapid outer hair cell contraction,[31] a condition with functional consequences.

The study of several known ototoxicants disrupt $[Ca^{2+}]_i$ homeostasis *in vitro*. Dulon *et al.*[32] demonstrated that the ototoxic aminoglycoside antibiotics, gentamicin and neomycin, block the K^+-induced increase in $[Ca^{2+}]_i$ in isolated outer hair cells using the calcium indicator Fluo-3. Yamamoto *et al.*[59] showed that the ototoxicant, cisplatin, was able to suppress calcium currents in patch clamped isolated outer hair cells in a concentration-dependent manner. Free radical generation via hypoxanthine plus xanthine oxidase appears to increase the uptake of extracellular Ca^{2+} into isolated outer hair cells.[60] A variety of ototoxicants including trimethyltin[61] increase $[Ca^{2+}]_i$ levels in outer hair cells by enhancing release from intracellular stores. Trimethyltin can also increase $[Ca^{2+}]_i$ levels in spiral ganglion cells by increasing both uptake via Ca^{2+} channels and release from intracellular stores.

11.14.6.3 Free Radical Injury

Free radical injury may be a prominent mechanism of ototoxic injury as it is in other organ systems (see also Chapter 4, this volume). However, the relative sensitivity of the cochlea to free radicals as compared to other tissue beds has not been well studied. Superoxide dismutase, glutathione reductase, glutathione *S*-transferase, and catalase have been identified in the guinea pig cochlea (including the organ of Corti) at levels at or above those found in brain.[62,63] It has been assumed that their presence is attributable to a protective role with respect to reactive oxygen species.

Two approaches have been used to assess the role of free radical injury in the cochlea. First, the susceptibility of cochlear function to known ototoxic drugs has been assessed following presumed depletion of glutathione activity in the cochlea using systemic manipulations.[64,65] However, it is difficult to determine that the protective effects observed result from elimination of free radicals in the cochlea rather than from alteration in peripheral chemical metabolism and, thereby, the dose of toxicant reaching the cochlea. Data on the administration of free radical scavengers and blockers of free radical formation have been presented. Seidman and Quirk,[66] for example, determined that both allopurinol, a free radical blocker, and superoxide dismutase-polyethylene glycol, a free radical scavenger, could reduce the disruptive effects of ischemia and reperfusion on pure tone auditory thresholds. The lipid peroxidation inhibitor (U74006F) was able to protect the cochlea against ischemic injury.[67] Garetz *et al.*[68] reported that the survival of outer hair cells exposed to a toxic metabolite of gentamycin *in vitro* was enhanced if glutathione, dithiothreitol, or the antioxidants ascorbic acid, phenylene diamine, or trolox were included in the incubation medium. Clerici *et al.*[69] have demonstrated bleb formation and shortening of outer hair cells in primary culture exposed to hydroxyl radical and hydrogen peroxide, and the blockade of such effects by free radical scavengers. Ikeda *et al.*[60] demonstrated that hypoxanthine plus xanthine oxidase increased $[Ca^{2+}]_i$ in isolated outer hair cells.

11.14.7 OTOTOXIC AGENTS

11.14.7.1 Chemical Asphyxiants

The impairment of normal oxidative metabolism by hypoxia or direct impairment of metabolism at the cellular level degrades the sensitivity of the cochlea to sound stimuli.[38,70] Acute, high level carbon monoxide exposure also reduces sensitivity of the cochlea to sound, as measured either by measurement of the CAP or the nonpropagated CM.[71] However, recovery of function is observed as carboxyhemoglobin levels return toward normal. Sensitivity to high frequency tones is impaired preferentially by carbon monoxide exposure. Studies demonstrating a protective effect of MK 801 against carbon monoxide-induced threshold shifts, suggest that carbon monoxide may disrupt cochlear function by enhancing the release of glutamate, the putative neurotransmitter at the inner hair cell–spiral ganglion cell synapse.[72] The production of hypoxia by creation of methemoglobin by butyl nitrite exposure[73] also impairs auditory function and, finally, cyanide administration can interfere with cochlear function.[74]

Carbon monoxide can potentiate noise-induced hearing loss in laboratory animals when it is presented simultaneously with a persistent noise stimulus.[71,75,76] While carbon monoxide exposure produces a transient loss of high-frequency auditory function, combined exposure to noise and carbon monoxide yields a profound high-frequency hearing loss that is far greater than the permanent auditory impairments produced by noise alone. Histopathological data show that this dysfunction is accompanied by a significant increase in hair cell loss in subjects receiving combined exposure as compared to those receiving noise only. Carbon monoxide by itself does not produce a measurable elevation in hair cell loss above control subjects (see Figure 3).

11.14.7.2 Organic Solvents

A sizable list of ototoxic organic solvents has emerged from research in several laboratories since the mid-1980s. These agents are heterogeneous with respect to their target within the auditory system, with some agents producing central auditory injury, while others selectively target the cochlea. Among the latter group, ototoxicity appears to occur preferentially at middle rather than high auditory frequencies (see Figure 7). This pattern represents an interesting puzzle in that it is difficult to identify a feature of cells in the middle turns of the cochlea that would render them vulnerable to

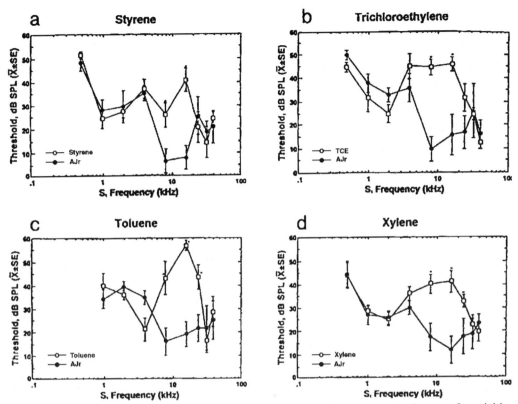

Figure 7 Ototoxic pattern produced by exposure to selected organic solvents showing a preferential impairment of auditory function at mid-frequencies. Data from Crofton *et al.*[21] (reproduced by permission of Elsevier from *Hear. Res.*, 1994, **80**, 25–30).

any toxicant. Hair cells vary in a variety of ways, but a consistent gradient of any given characteristic is expressed typically with the extreme values observed at the most apical and most basal portions of the cochlea.

11.14.7.2.1 Solvents having a presumed cochlear target

(i) Toluene

Of all solvents, toluene has received the most attention as an ototoxicant both in laboratory animals exposed for generally short durations at high doses, and also in humans who are commonly exposed to low doses, but over extended periods of time. Workers exposed to toluene show loss of pure tone threshold sensitivity and at least an additive (if not synergistic) impairment of hearing when noise exposure occurs simultaneously.[77,78] Hearing loss has also been reported in cases of glue sniffers who are exposed to high doses of toluene.[79] Data from several laboratories confirm that toluene exposure in rats in the range of 1000–2000 ppm over 3–5 d exposure results in permanent auditory impairments, as reflected by disruption of the

early (cochlear) components of the auditory brainstem response (ABR) or by behavioral testing.[21,27,80–82] Sullivan *et al.*[82] demonstrated midfrequency impairments of the ABR in rats exposed to toluene by gavage ($1\,\text{mL}\,\text{kg}^{-1}\,\text{d}^{-1}$ for 8 wk) and presented limited histopathological evidence of outer hair cell damage. Johnson and Canlon[83] published data showing impairment of the DPOAE in laboratory rats exposed to toluene (see Figure 5), bolstering the claim for a cochlear site of toluene ototoxicity and also suggesting that the outer hair cell is a prominent target of toxicity.

(ii) Styrene

Styrene appears to be ototoxic under some exposure regimens in laboratory rats,[21,84,85] and there are also preliminary findings in humans which suggest that high frequency hearing impairment may also occur among exposed workers.[86] It has been suggested that the site of styrene ototoxicity is the outer hair cell,[85] but other possible targets have not been investigated. The dosing regimen necessary to observe styrene ototoxicity has not been adequately described. For example, there has been at least one failure to find an ototoxic effect of styrene (1200 ppm for 7 h) at the level of the cochlea, and a study using combined exposure to styrene with noise failed to find either a styrene effect or an increment in the noise-induced auditory threshold shift by styrene.[87]

(iii) Trichloroethylene

Trichloroethylene does produce a midfrequency hearing loss in laboratory animals.[21,88] Crofton and Zhao[88] detected midfrequency impairments following trichloroethylene exposure (5 d exposures 8 h d^{-1} at 4000 ppm) using a behavioral measure of hearing, and these findings were corroborated using direct electrophysiological measurements from the cochlea (the round window CAP and CM potential).[40] Long-term (12 week) exposures to 1600 or 3200 ppm trichloroethylene depressed the amplitude of the ABR measured at 1–2 week following exposure.[89]

11.14.7.2.2 Solvents having a presumed central auditory target

(i) Carbon disulfide

Carbon disulfide has an illustrious history as a neurotoxicant producing impairments both to the central and peripheral nerves, and in producing psychiatric disturbances. Carbon disulfide may also impair auditory function among human workers.[77] While this compound has been suggested to be ototoxic, the relevant studies actually show a disruption restricted to the late potentials within the ABR,[90,91] suggesting impairment of central sensory pathways. It is not certain that the auditory impairment is especially sizable, however, as disruption of motor function has been reported to occur prior to sensory impairments.[88]

(ii) Hexanes and xylene

Hearing loss has also been reported in laboratory animals exposed to *n*-hexane[3] and xylenes.[84] However, the endpoints studied, late auditory brainstem response potentials, are more indicative of central auditory dysfunction than of cochlear impairment.

(iii) Polychlorinated biphenyl and antithyroid agents

The normal development of the cochlea is dependent, in part, upon the presence of an adequate level of thyroid hormone at the time of ontogeny, and structural and functional anomalies have been reported both following administration of the antithyroid drug, propylthiouracil to rats[93] as well as idiopathic hypothyroidism.[94,95] Data suggest that Aroclor, a polychlorinated biphenyl mixture, given early in development may also function as an antithyroid treatment that can disrupt auditory thresholds in laboratory animals.[96]

11.14.7.3 Metals

11.14.7.3.1 Trialkyltins

Trimethyltin chloride (TMT) is a neurotoxicant which has prominent toxic consequences in the peripheral auditory system. The initial target of TMT ototoxicity is the inner hair cell and/or Type 1 spiral ganglion cell[22,23] as TMT elevates the CAP threshold and attenuates the growth of the N_1 amplitude as sound levels are increased above threshold. These effects are observed within 30 min of either intraperitoneal injection or topical administration of TMT to the round window membrane.[22,37,97] The CM is not affected by TMT until several hours following *in vivo* exposure.[23] Although the effect of TMT on the auditory system has been fully described,[22,23,98–100] the mechanism of ototoxicity still remains uncertain. TMT induces a marked and sustained elevation in $[Ca^{2+}]_i$ level in the isolated spiral ganglion cell that appears to have a rapid initial phase and a slower saturating phase.[101] Studies performed using calcium free medium and with the L-type calcium channel blocker, nifedipine, indicate that TMT elevates $[Ca^{2+}]_i$ in isolated spiral ganglion cells both by increasing extracellular uptake via Ca^{2+} channels, and also by releasing Ca^{2+} from intracellular stores. Thus, TMT ototoxicity appears to include a direct postsynaptic toxic event.

11.14.7.3.2 Methylmercury

The neurotoxicant, methylmercury, can produce auditory impairments in man[102,103] and in various laboratory species. Rice and Gilbert,[104] for example, demonstrated a preferential impairment of high frequency auditory function in cynomolgus monkeys dosed from birth to 7 years of age with 50 µg kg^{-1} d^{-1} methylmercuric chloride and tested at 14 years of age. Auditory impairment has also been seen in rats[105] and guinea pigs.[106,107] While methylmercury produces profound central nervous system injury, it is uncertain that damage occurs in the peripheral auditory system as well. Falk *et al.*[107] suggested a cochlear lesion restricted to outer hair cells at a fairly apical (low-frequency) location. However, the extent of the lesion was extremely small (approximately 6%) even in rats exposed to daily injections of $2\,mg\,kg^{-1}$ methyl mercury hydroxide for 6 weeks. No relationship was found between duration of treatment and extent of lesion. Konishi and Hamrick[106] showed a loss in the CM output for high frequency tones at very high levels of stimulation, which might be consistent with outer hair cell injury. However, the absence of effect at low tone frequencies is inconsistent with the histopathological data reported by Falk *et al.*[107] The CAP amplitude was markedly suppressed at low sound levels in the treated subjects indicating an impairment either within the cochlea or in synaptic transmission at the inner hair cell–spiral ganglion cell synapse.

11.14.7.3.3 Lead

Lead exposure has been associated with auditory impairment in children and with a preclinical marker of altered auditory function in occupationally exposed adults. Schwartz and Otto[108,109] reported a significant elevation in pure tone auditory thresholds that was correlated with blood lead levels. Even blood lead levels in the range of $10\,\mu g\,dL^{-1}$ were associated with a small loss of auditory acuity. Because traditional audiometric methods were used, it is not clear whether the effect reflected central or peripheral auditory system effects. Dietrich *et al.*[110] showed evidence of impaired central auditory processing associated with lead exposure in 5 year old children.

Discalzi *et al.*[111] reported a modest increase in absolute brainstem auditory evoked potential latency and an increase in interpeak latency in lead exposed workers with blood lead levels in the range of $50\,\mu g\,dL^{-1}$. While the small delay in wave I might reflect a cochlear effect, such as a shift in threshold detection, the alterations in later peak latencies and interpeak latencies is consistent with effects at more central areas of the auditory system.

11.14.7.4 Aminoglycoside Antibiotics

Millions of doses of aminoglycoside antibiotics are given annually and published reports indicate that these agents produce auditory impairments in approximately 1–10% of the subjects.[112–114] Several excellent reviews[15,115,116] deal extensively with the broad issues related to aminoglycoside antibiotic ototoxicity and this review will provide only some highlights that have not been widely disseminated. Aminoglycoside antibiotic ototoxicity is marked by a preferential high frequency hearing loss and preferential loss of outer hair cells from the basal turn of the cochlea. While the precise mechanism(s) of ototoxic impairment by these agents is not known, it has been proposed that a key feature of toxicity includes disruption of phospholipid metabolism.[117] Further, viability of isolated outer hair cells under exposure to a gentamicin metabolite could be improved if glutathione, dithioerythritol, or the antioxidants ascorbic acid, phenylene diamine, or trolox were included in the incubation mixture.[118] Based upon data showing that aminoglycosides themselves do not affect isolated hair cells,[36] but that cell toxicity can be seen in extracts incubated with a liver postmitochondrial fraction,[119] it has been proposed that some metabolite rather than the parent compound is ototoxic.

Data have also suggested that ototoxicity found in a small percentage of highly sensitive individuals might reflect a mutation in the mitochondrial ribosomal RNA. Studies of familial aminoglycoside ototoxicity[120–122] lend credence to a maternally inheritable susceptibility to aminoglycoside ototoxicity. Polymerase chain reaction (PCR) amplification identified specific anomalous mitochondrial DNA sequences characteristic of sensitive individuals. A mutation at the 1555 position on the 12S mitochondrial rRNA has been identified as a likely candidate whereby enhanced binding of aminoglycosides might occur.[123]

11.14.7.5 Loop Diuretics

The ototoxicity of loop diuretics, such as ethacrynic acid and furosemide, has been reviewed in detail.[124] These agents preferentially disrupt stria vascularis function producing a decrement in the standing d.c. endocochlear potential that leads, consequently, to loss of CAP sensitivity. An additional outcome of loop diuretic treatment is impairment of the K^+ current resulting from the active transport of this ion into endolymph. Fortunately, the effects of loop diuretics on hearing are temporary. However, the loop diuretics are able to potentiate permanent auditory impairments produced by aminoglycoside antibiotics[125] and cisplatin,[126] possibly by increasing the dose of these latter agents reaching the cochlea.

11.14.7.6 Cisplatin and Related Chemotherapeutics

The antineoplastic agent, cisplatin, and some analogues, such as carboplatin, are potent ototoxicants at therapeutic doses, producing permanent preferentially high-frequency hearing loss. The incidence of such effects in clinical trials is high, often reaching 50% of treated patients.[127,128] Histological study in laboratory animals[129-131] as well as cochlear electrophysiological studies[132] suggest that cisplatin preferentially disrupts outer hair cells in the basal turn of the cochlea and can block transduction channels in outer hair cells.[132] Attempts to block cisplatin ototoxicity in laboratory animals suggest that sulfur-rich compounds, such as dithiocarbamate,[133] that can bind platinum may be effective. However, it has also been proposed that the protective effects of dithiocarbamate might be mediated via protection of antioxidant enzyme systems.[134] Fosfomycin has also been used preclinically to block cisplatin ototoxicity with some success reported.[135]

11.14.7.7 Salicylates

Salicylates given at high doses for treatment of arthritis frequently produce tinnitus.[136] The effects are typically reversible when dose levels are adjusted downward. In experimental animals, the effects of salicylates can be observed as a loss in auditory threshold sensitivity and of the CM response[137,138] as well as a loss in spontaneous otoacoustic emissions.[18] In isolated outer hair cells, perfusion with salicylate in artificial perilymph impairs electromotility and produces ultrastructural changes in the subsurface cysternae[20] along with a slow shortening of outer hair cell length.[19]

11.14.8 CONCLUSION

The rapid development of the ototoxicity literature since the mid-1980s has enhanced greatly our understanding of the process by which the auditory system may be impaired. It is apparent that the consequences of ototoxic exposure are not due immutably to the loss of outer hair cells from the base of the cochlea resulting in high frequency hearing loss. The identification of specific agents that preferentially disrupt more apical turns of the cochlea, such as several organic solvents, as well as agents that target the inner hair cell and Type 1 spiral ganglion cell promise to facilitate the identification of mechanisms responsible for hearing impairment. Studies also cast doubt on

the significance of earlier dogma that correlated ototoxicity with renal toxicity. While many classic ototoxicants, including loop diuretics, aminoglycoside antibiotics, and cisplatin, can produce significant toxicity to the renal tubule and in the cochlea, there are now a significant number of compounds that have limited or no marked effect on the kidney while they are potent ototoxicants. Finally, studies underscore similarities in mechanisms responsible for ototoxicity and for toxicity in other organs including the central nervous system. Thus, free radical injury and excitotoxicity are being proposed as significant bases of ototoxicity. Determination of relative sensitivity of the inner ear to these mechanisms with other organs and the elucidation of how such mechanisms result in specific forms of cochlear injury are a task for coming research.

11.14.9 REFERENCES

1. H. Lillienthal, G. Winneke and T. Ewert, 'Effects of lead on neurophysiological and performance measures: animal and human data.' *Environ. Health Perspect.*, 1990, **89**, 21–25.
2. G. T. Pryor, J. Dickinson, R. A. Howd *et al.*, 'Neurobehavioral effect of subchronic exposure of weanling rats to toluene or hexane.' *Neurobehav. Toxicol. Teratol.*, 1983, **5**, 47–52.
3. C. S. Rebert, P. W. Houghton, R. A. Howd *et al.*, 'Effects of hexane on the brainstem auditory response and caudal nerve action potential.' *Neurobehav. Toxicol. Teratol.*, 1982, **4**, 79.
4. A. R. Møller, 'Auditory Physiology,' Academic Press, New York, 1983.
5. R. A. Altschuler, R. P. Bobbin and D. W. Hoffman (eds.), 'Neurobiology of Hearing: The Cochlea,' Raven Press, New York, 1986.
6. G. von Beksey, in 'Experiments in Hearing,' ed. E. G. Wever, McGraw-Hill, New York, 1960, p. 11.
7. L. L. Langley, I. R. Telford and J. B. Christensen, in 'Dynamic Anatomy and Physiology,' 3rd edn., McGraw-Hill, New York, NY, 1969, p. 354.
8. M. Borysenko and T. Beringer (eds.), in 'Functional Histology,' 2nd edn., Little & Brown, Boston, MA, 1988, p. 489.
9. A. Flock *et al.*, Cold Spring Harbor Symposium on Quantitative Biology, 1965, **30**, 135–146.
10. A. Ryan and P. Dallos, 'Effect of absence of cochlear outer hair cells: effects on behavioural auditory threshold.' *Nature*, 1975, **253**, 44–46.
11. D. J. Lim and A. Flock, 'Ultrastructural morphology of enzyme-dissociated cochlear sensory cells.' *Acta Otolaryngol. (Stockh.)*, 1985, **99**, 478–492.
12. J. Schacht, 'Hormonal regulation of adenylate cyclase in the stria vascularis of the mouse.' *Hear. Res.*, 1985, **20**, 9–13.
13. D. Bagger-Sjoback, C. S. Filipek and J. Schacht, 'Characteristics and drug responses of cochlear and vestibular adenylate cyclase.' *Arch. Otorhinolaryngol.*, 1980, **228**, 217–222.
14. B. M. Johnstone and P. M. Sellick, 'The peripheral auditory apparatus.' *Q. Rev. Biophy.*, 1972, **5**, 1–57.
15. L. P. Rybak, 'Drug ototoxicity.' *Annu. Rev. Pharmacol. Toxicol.*, 1986, **26**, 79–99.
16. L. P. Rybak and C. Whitworth, 'Comparative oto-

toxicity of furosemide and piretanide.' *Acta Otolaryngol. (Stockh.)*, 1986, **101**, 59–65.

17. L. P. Rybak, C. Whitworth and V. Scott, 'Comparative acute ototoxicity of loop diuretic compounds.' *Eur. Arch. Otorhinolaryngol.*, 1991, **248**, 353–357.

18. D. McFadden and H. S. Plattsmier, 'Aspirin abolishes spontaneous otoacoustic emissions.' *J. Acoust. Soc. Am.*, 1984, **76**, 443–448.

19. W. E. Shehata, W. E. Brownell and R. Dieler, 'Effects of salicylate on shape, electromotility and membrane characteristics of isolated outer hair cells from guinea pig cochlea.' *Acta Otolaryngol. (Stockh.)*, 1991, **111**, 707–718.

20 R. Dieler, W. E. Shehata-Dieler and W. E. Brownell, 'Concomitant salicylate-induced alterations of outer hair cell subsurface cisternae and electromotility.' *J. Neurocytol.*, 1991, **20**, 637–653.

21. K. M. Crofton, T. L. Lassiter and C. S. Rebert, 'Solvent-induced ototoxicity in rats: an atypical selective midfrequency hearing deficit.' *Hear. Res.*, 1994, **80**, 25–30.

22. W. J. Clerici, B. Ross, Jr. and L. D. Fechter, 'Acute ototoxicity of trialkyltins in the guinea pig.' *Toxicol. Appl. Pharmacol.*, 1991, **109**, 547–556.

23. L. D. Fechter, W. J. Clerici, L. Yao *et al.*, 'Rapid disrupition of cochlear function and structure by trimethyltin in the guinea pig.' *Hear. Res.*, 1992, **58**, 166–174.

24. P. Dallos and M. A. Cheatham, 'Production of cochlear potentials by inner and outer hair cells.' *J. Acoust. Soc. Am.*, 1976, **60**, 510–512.

25. D. O. Kim, in 'Hearing Science: Recent Advances,' ed. C. I. Berlin, College Hill Press, San Diego, CA, 1984, p. 241.

26. D. T. Kemp and M. Souter, 'A new rapid component in the cochlear response to brief electrical efferent stimulation: CM and otoacoustic observations.' *Hear. Res.*, 1988, **34**, 49–62.

27. A. C. Johnson, 'The ototoxic effect of toluene and the influence of noise, acetyl salicylic acid, or genotype. A study in rats and mice.' *Scand. Audiol.*, 1994, **39** (Suppl.), 1–40.

28. W. E. Brownell, C. R. Bader, D. Bertrand *et al.*, 'Evoked mechanical responses of isolated cochlear outer hair cells.' *Science*, 1985, **227**, 194–196.

29. J. Santos-Sacchi and J. P. Dilger, 'Whole cell currents and mechanical responses of isolated outer hair cells.' *Hear. Res.*, 1988, **35**, 143–150.

30. L. Brundin and I. Russell, 'Tuned phasic and tonic motile responses of isolated outer hair cells to direct mechanical stimulation of the cell body.' *Hear. Res.*, 1994, **73**, 35–45.

31. D. Dulon, C. Blanchet and E. Laffon, 'Photo-released intracellular Ca^{2+} evokes reversible mechanical responses in supporting cells of the guinea-pig organ of Corti.' *Biochem. Biophys. Res. Commun.*, 1994, **201**, 1263–1269.

32. D. Dulon, G. Zajic, J. M. Aran *et al.*, 'Aminoglycoside antibiotics impair calcium entry but not viability and motility in isolated cochlear outer hair cells.' *J. Neurosci. Res.*, 1989, **24**, 338–346.

33. M. E. Chertoff and W. E. Brownell, 'Characterization of cochlear outer hair cell turgor.' *Am. J. Physiol.*, 1994, **266**, C467–C479.

34. W. J. Clerici, M. E. Chertoff, W. E. Brownell *et al.*, '*In vitro* organotin administration alters guinea pig cochlear outer hair cell shape and viability.' *Toxicol. Appl. Pharmacol.*, 1993, **120**, 193–202.

35. L. D. Fechter and Y. Liu, in 'Proceedings of the VII International Congress of Toxicology, Seattle, Washington,' 1995, p. 64–PD7.

36. M. Y. Huang and J. Schacht, 'Formation of a cyto-

toxic metabolite from gentamicin by liver.' *Biochem. Pharmacol.*, 1990, **40**, R11–R14.

37. L. D. Fechter and Y. Liu, 'Trimethyltin disrupts N_1 sensitivity, but has limited effects on the summating potential and cochlear microphonic.' *Hear. Res.*, 1994, **78**, 189–196.

38. M. C. Brown, A. L. Nuttall, R. I. Masta *et al.*, 'Cochlear inner hair cells: effects of transient asphyxia on intracellular potentials.' *Hear. Res.*, 1983, **9**, 131–144.

39. L. D. Fechter, P. R. Thorne and A. L. Nuttall, 'Effects of carbon monoxide on cochlear electrophysiology and blood flow.' *Hear. Res.*, 1987, **27**, 37–45.

40. K. M. Crofton, L. Kehn and L. D. Fechter, in 'Proceedings of the 16th Midwinter Research meeting on Association for Research in Otolaryngology, St. Petersburg Beach,' ed. D. J. Lim, St. Petersburg Beach, 1993, p. 143.

41. T. Nakagawa, S. Kakehata, N. Akaike *et al.*, 'Calcium channel in isolated outer hair cells of guinea pig cochlea.' *Neurosci. Lett.*, 1991, **125**, 81–84.

42. D. Y. Han, T.Yamashita, N. Harada *et al.*, 'Calcium mobilization in isolated cochlear spiral ganglion cells of the guinea pig.' *Acta Otolaryng. (Stockh.)*, 1993, **506** (Suppl.), 26–29.

43. M. Eybalin and R. Pujol, 'Cochlear neuroactive substances.' *Arch. Otorhinolaryngol.*, 1989, **246**, 228–234.

44. R. P. Bobbin and M. H. Thompson, 'Effects of putative transmitters on afferent cochlear transmission.' *Ann. Otol. Rhinol. Laryngol.*, 1978, **87**, 185–190.

45. R. A. Altschuler, C. E. Sheridan, J. W. Horn *et al.*, 'Immunocytochemical localization of glutamate immunoreactivity in the guinea pig cochlea.' *Hear. Res.*, 1989, **42**, 167–173.

46. G. L. Jenison, R. P. Bobbin and R. Thalmann, 'Potassium-induced release of endogenous amino acids in the guinea pig cochlea.' *J. Neurochem.*, 1985, **44**, 1845–1853.

47. R. P. Bobbin, 'Glutamate and aspartate mimic the afferent transmitter in the cochlear.' *Exp. Brain Res.*, 1979, **34**, 389–393.

48. I. R. Schwartz and A. F. Ryan, 'Amino acid labeling patterns in the efferent innervation of the cochlea: an electron microscopic autoradiographic study.' *J. Comp. Neurol.*, 1986, **246**, 500–512.

49. R. P. Bobbin and G. Ceasar, 'Kynurenic acid and gamma-D-glutamylaminomethylsulfonic acid suppress the compound action potential of the auditory nerve.' *Hear. Res.*, 1987, **25**, 77–81.

50. T. Littman, R. P. Bobbin, M. Fallon *et al.*, 'The quinoxalinediones DNOX, CNOX and two related congeners suppress hair cell-to-auditory nerve transmission.' *Hear. Res.*, 1989, **40**, 45–53.

51. J. L. Puel, S. Ladrech, R. Chabert *et al.*, 'Electrophysiological evidence for the presence of NMDA receptors in the guinea pig cochlea.' *Hear. Res.*, 1991, **51**, 255–264.

52. J. L. Puel, R. Pujol, S. Ladrech *et al.*, 'Alpha-amino-3-hydroxy-5-methyl-4-isoxazole propionic acid electrophysiological and neurotoxic effects in the guinea-pig cochlea.' *Neuroscience*, 1991, **45**, 63–72.

53. P. P. Lefebvre, T. Weber, P. Leprince *et al.*, 'Kainate and NMDA toxicity for cultured developing and adult rat spiral ganglion neurons: further evidence for a glutamatergic excitatory neurotransmission at the inner hair cell synapse.' *Brain Res.*, 1991, **555**, 75–83.

54. H. Kuriyama, R. L. Albin and R. A. Altschuler, 'Expression of NMDA-receptor mRNA in the rat cochlea.' *Hear. Res.*, 1993, **69**(1–2), 215–220.

55. A. S. Niedzielski and R. J. Wenthold, 'Expression of AMPA, kainate, and NMDA receptor subunits in

cochlear and vestibular ganglia.' *J. Neurosci.*, 1995, **15**, 2338–2353.

56. R. Janssen, L. Schweitzer and K. F. Jensen, 'Glutamate neurotoxicity in the developing rat cochlea: physiological and morphological approaches.' *Brain Res.*, 1991, **552**, 255–264.

57. J. M. Juiz, J. Rueda, J. A. Merchan *et al.*, 'The effects of kainic acid on the cochlear ganglion of the rat.' *Hear. Res.*, 1989, **40**, 65–74.

58. S. Orrenius, D. J. McConkey, G. Bellomo *et al.*, 'Role of Ca^{2+} in toxic cell killing.' *Trends Pharmacol. Sci.*, 1989, **10**, 281–285.

59. T. Yamamoto, S. Kakehata, T. Saito *et al.*, 'Cisplatin blocks voltage-dependent calcium current in dissociated outer hair cells of guinea-pig cochlea.' *Brain Res.*, 1994, **648**, 296–298.

60. K. Ikeda, H. Sunose and T. Takasaka, 'Effects of free radicals on the intracellular calcium concentration in the isolated outer hair cell of the guinea pig cochlea.' *Acta Otolaryngol. (Stockh.)*, 1993, **113**, 137–141.

61. Y. Liu and L. D. Fechter, 'Elevation of calcium levels in outer hair cells by trimethyltin.' *Toxicology In Vitro*, 1996, **10**, 567–576.

62. M. G. Pierson and B. H. Gray, 'Superoxide dismutase activity in the cochlea.' *Hear. Res.*, 1982, **6**, 141–151.

63. A. el Barbary, R. A. Altschuler and J. Schacht, 'Glutathione *S*-transferases in the organ of Corti of the rat: enzymatic activity, subunit composition and immunohistochemical localization.' *Hear. Res.*, 1993, **71**, 80–90.

64. D. W. Hoffman, C. A. Whitworth, K. L. Jones *et al.*, 'Nutritional status, glutathione levels, and ototoxicity of loop diuretics and aminoglycoside antibiotics.' *Hear. Res.*, 1987, **31**, 217–222.

65. D. W. Hoffman, C. A. Whitworth, K. L. Jones-King *et al.*, 'Potentiation of ototoxicity by glutathione depletion.' *Ann. Otol. Rhinol. Laryngol.*, 1988, **97**, 36–41.

66. M. D. Seidman and W. S. Quirk, 'The protective effects of tirilated mesylate (U74006F) on ischemic and reperfusion-induced cochlear damage.' *Otolaryngol. Head Neck Surg.*, 1991, **105**, 511–516.

67. M. D. Seidman, W. S. Quirk, A. L. Nuttall *et al.*, 'The protective effects of allopurinol and superoxide dismutase-polyethyleneglycol on ischemic and reperfusion-induced cochlear damage.' *Otolaryngol. Head Neck Surg.*, 1991, **105**, 457–463.

68. S. L. Garetz, R. A. Altschuler and J. Schacht, 'Attenuation of gentamicin ototoxicity by glutathione in the guinea pig *in vivo*.' *Hear. Res.*, 1994, **77**, 81–87.

69. W. J. Clerici, D. L. DiMartino and M. R. Prasad, 'Direct effects of reactive oxygen species on cochlear outer hair cell shape *in vitro*.' *Hear. Res.*, 1995, **84**, 30–40.

70. M. Lawrence, A. L. Nuttall and P. A. Burgio, 'Cochlear potentials and oxygen associated with hypoxia.' *Ann. Otol. Rhinol. Laryngol.*, 1975, **84**, 499–512.

71. L. D. Fechter, J. S. Young and L. Carlisle, 'Potentiation of noise induced threshold shifts and hair cell loss by carbon monoxide.' *Hear Res.*, 1988, **34**, 39–47.

72. Y. Liu and L. D. Fechter, 'MK-801 protects against carbon monoxide-induced hearing loss.' *Toxicol. Appl. Pharmacol.*, 1995, **132**, 196–202.

73. L. D. Fechter, C. L. Richard, M. Mungekar *et al.*, 'Disruption of auditory function by acute administration of a "room odorizer" containing butyl nitrite in rats.' *Fundam. Appl. Toxicol.*, 1989, **12**, 56–61.

74. T. Konishi and E. Kelsey, 'Effect of cyanide on cochlear potentials.' *Acta Otolaryngol. (Stockh.)*, 1968, **65**, 381–390.

75. L. D. Fechter, 'A mechanistic basis for interactions between noise and chemical exposure.' *Arch. Complex Environ. Studies*, 1989, **1**, 23.

76. J. S. Young, M. B. Upchurch, M. J. Kaufman *et al.*, 'Carbon monoxide exposure potentiates high-frequency auditory threshold shifts induced by noise.' *Hear Res.*, 1987, **26**, 37–43.

77. T. C. Morata, 'Study of the effects of simultaneous exposure to noise and carbon disulfide on workers' hearing.' *Scand. Audiol.*, 1989, **18**, 53–58.

78. T. C. Morata, D. E. Dunn, L. W. Kretschmer *et al.*, 'Effects of occupational exposure to organic solvents and noise on hearing.' *Scand. J. Work Environ. Health.*, 1993, **19**, 245–254.

79. A. Ehyai, and F. R. Freemon, 'Progressive optic neuropathy and sensorineural hearing loss due to chronic glue sniffing.' *J. Neurol. Neurosurg. Psychiatry*, 1983, **46**, 349–351.

80. C. S. Rebert, S. S. Sorenson, R. A. Howd *et al.*, 'Toluene-induced hearing loss in rats evidenced by the brainstem auditory-evoked response.' *Neurobehav. Toxicol. Teratol.*, 1983, **5**, 59–62.

81. G. T. Pryor, C. S. Rebert, J. Dickinson *et al.*, 'Factors affecting toluene-induced ototoxicity in rats.' *Neurobehav. Toxicol. Teratol.*, 1984, **6**, 223–238.

82. M. J. Sullivan, K. E. Rarey and R. B. Conolly, 'Ototoxicity of toluene in rats.' *Neurotoxicol. Teratol.*, 1988, **10**, 525–530.

83. A. C. Johnson and B. Canlon, 'Toluene exposure affects the functional activity of the outer hair cells.' *Hear. Res.*, 1994, **72**, 189–196.

84. G. T. Pryor, C. S. Rebert and R. A. Howd, 'Hearing loss in rats caused by inhalation of mixed xylenes and styrene.' *J. Appl. Toxicol.*, 1987, **7**, 55–61.

85. B. L. Yano, D. A. Dittenber, R. R. Albee *et al.*, 'Abnormal auditory brainstem responses and cochlear pathology in rats induced by an exaggerated styrene exposure regimen.' *Toxicol. Pathol.*, 1992, **20**, 1–6.

86. H. Muijser, E. M. Hoogendijk and J. Hooisma, 'The effects of occupational exposure to styrene on high-frequency hearing thresholds.' *Toxicology*, 1988, **49**, 331–340.

87. L. D. Fechter, 'Effects of acute styrene and simultaneous noise exposure on auditory function in the guinea pig.' *Neurotoxicol. Teratol.*, 1993, **15**, 151–155.

88. K. M. Crofton and X. Zhao, 'Midfrequency hearing loss in rats following inhalation exposure to trichloroethylene: evidence from reflex modification audiometry.' *Neurotoxicol. Teratol.*, 1993, **15**, 413–423.

89. C. S. Rebert, V. L. Day, M. J. Matteucci *et al.*, 'Sensory-evoked potentials in rats chronically exposed to trichloroethylene: predominant auditory dysfunction.' *Neurotoxicol. Teratol.*, 1991, **13**, 83–90.

90. C. S. Rebert, S. S. Sorenson and G. T. Pryor, 'Effects of intraperitoneal carbon disulfide on sensory-evoked potentials of Fischer-344 rats.' *Neurobehav. Toxicol. Teratol.*, 1986, **8**, 543–549.

91. C. S. Rebert and E. Becker, 'Effects of inhaled carbon disulfide on sensory-evoked potentials of Long–Evans rats.' *Neurobehav. Toxicol. Teratol.*, 1986, **8**, 533–541.

92. W. J. Clerici and L. D. Fechter, 'Effects of chronic carbon disulfide inhalation on sensory and motor function in the rat.' *Neurotoxicol. Teratol.*, 1991, **13**, 249–255.

93. L. Van Middlesworth and C. H. Norris, 'Audiogenic seizures and cochlear damage in rats after perinatal antithyroid treatment.' *Endocrinology*, 1980, **106**, 1686–1690.

94. A. Uziel, C. Legrand and A. Rabie, 'Corrective effects of thyroxine on cochlear abnormalities induced by congenital hypothyroidism in the rat. II. Electrophysiological study.' *Brain Res.*, 1985, **351**, 123–127.

95. A. Uziel, C. Legrand and A. Rabie, 'Corrective effects of thyroxine on cochlear abnormalities induced by congenital hypothyroidism in the rat. I. Morphological study.' *Brain Res.*, 1985, **351**, 111–112.

96. E. S. Golden, L. S. Kehn, C. Lau *et al.*, 'Developmental exposure to polychlorinated biphenyls (Aroclor 1254) reduces circulating thyroid hormone concentrations and causes hearing deficites in rats.' *Toxicol. Appl. Pharmacol.*, 1995, **135**, 77–88.

97. Y. Liu and L. D. Fechter, 'Trimethyltin disrupts loudness recruitment and auditory threshold sensitivity in guinea pigs.' *Neurotoxicol. Teratol.*, 1995, **17**, 281–287.

98. L. W. Chang and R. S. Dyer, 'A time-course study of trimethyltin induced neuropathology in rats.' *Neurobehav. Toxicol. Teratol.*, 1983, **5**, 443–459.

99. K. M. Crofton, K. F. Dean, M. G. Menache *et al.*, 'Trimethyltin effects on auditory function and cochlear morphology.' *Toxicol. Appl. Pharmacol.*, 1990, **105**, 123–132.

100. J. S. Young and L. D. Fechter, 'Reflex inhibition procedures for animal audiometry: a technique for assessing ototoxicity.' *J. Acoust. Soc. Am.*, 1983, **73**, 1686–1693.

101. L. D. Fechter and Y. Liu, 'Elevation of intracellular calcium levels in spiral ganglion cells by trimethyltin.' *Hear. Res.*, 1995, **91**, 101–109.

102. L. Amin-zaki, M. A. Majeed, T. W. Clarkson *et al.*, 'Methylmercury poisoning in Iraqi children: clinical observations over two years.' *Br. Med. J.*, 1978, **1**, 613–616.

103. K. Mizukoshi, Y. Watanabe, H. Kobayashi *et al.*, 'Neurological follow-up studies upon Minamata disease.' *Acta. Otolaryngol (Stockh.)*, 1989, **468** (Suppl.), 353–357.

104. D. C. Rice and S. G. Gilbert, 'Exposure to methylmercury from birth to adulthood impairs high-frequency hearing in monkeys.' *Toxicol. Appl. Pharmacol.*, 1992, **115**, 6–10.

105. M. F. Wu, J. R. Ison, J. R. Wecker *et al.*, 'Cutaneous and auditory function in rats following methylmercury poisoning.' *Toxicol. Appl. Pharmacol.*, 1985, **79**, 377–388.

106. T. Konishi and P. E. Hamrick, 'The uptake of methylmercury in guinea pig cochlea in relation to its ototoxic effect.' *Acta Otolaryngol. (Stockh.)*, 1979, **88**, 203–210.

107. S. A. Falk, R. Klein, J. K. Haseman *et al.*, 'Acute methyl-mercury intoxication and ototoxicity in guinea pigs.' *Arch. Pathol.*, 1974, **97**, 297–305.

108. J. Schwartz and D. Otto, 'Blood lead, hearing thresholds, and neurobehavioral development in children and youth.' *Arch. Environ. Health*, 1987, **42**, 153–160.

109. J. Schwartz and D. Otto, 'Lead and minor hearing impairment.' *Arch. Environ. Health*, 1991, **46**, 300–305.

110. K. N. Dietrich, P. A. Succop, O. G. Berger *et al.*, 'Lead exposure and the central auditory processing abilities and cognitive development of urban children: the Cincinnati Lead Study cohort at age 5 years.' *Neurotoxicol. Teratol.*, 1992, **14**, 51–56.

111. G. Discalzi, D. Fabbro, F. Meliga *et al.*, 'Effects of occupational exposure to mercury and lead on brainstem auditory evoked potentials.' *Int. J. Psychophysiol.*, 1993, **14**, 21–25.

112. M. Barza and R. T. Scheife, 'Drug therapy reviews: Antimicrobial spectrum, pharmacology and therapeutic use of antibiotics—part 4: aminoglycosides.' *Am. J. Hosp. Pharm.*, 1977, **34**, 723–737.

113. G. G. Jackson and G. Arcieri, 'Ototoxicity of gentamicin in man: a survey and controlled analysis of clinical experience in the United States.' *J. Infect. Dis.*, 1971, **124** (Suppl.), 130.

114. C. R. Smith, K. L. Baughman, C. Q. Edwards *et al.*, 'Controlled comparison of amikacin and gentamicin.' *N. Engl. J. Med.*, 1977, **296**, 349–353.

115. R. D. Brown and A. M. Feldman, 'Pharmacology of hearing and ototoxicity.' *Annu. Rev. Pharmacol. Toxicol.*, 1978, **18**, 233–252.

116. E. S. Harpur, 'The pathophysiology of hearing.' *Br. Med. Bull.*, 1987, **43**, 871–886.

117. J. Schacht, 'Biochemical basis of aminoglycoside ototoxicity.' *Otolaryngol. Clin. North Am.*, 1993, **26**, 845–856.

118. S. L. Garetz, D. J. Rhee and J. Schacht, 'Sulfhydryl compounds and antioxidants inhibit cytotoxicity to outer hair cells of a gentamicin metabolite *in vitro.*' *Hear. Res.*, 1994, **77**, 75–80.

119. S. A. Crann, M. Y. Huang, J. D. McLaren *et al.*, 'Formation of a toxic metabolite from gentamicin by a hepatic cytosolic fraction.' *Biochem. Pharmacol.*, 1992, **43**, 1835–1839.

120. D. N. Hu, W. Q. Qui, B. T. Wu *et al.*, 'Genetic aspects of antibiotic induced deafness: mitochondrial inheritance.' *J. Med. Genet.*, 1991, **28**, 79–83.

121. T. R. Prezant, J. V. Agapian, M. C. Bohlman *et al.*, 'Mitochondrial ribosomal RNA mutation associated with both antibiotic-induced and nonsyndromic deafness.' *Nat. Genet.*, 1993, **4**, 289–294.

122. G. Cortopassi and T. Hutchin, 'A molecular and cellular hypothesis for aminoglycoside-induced deafness.' *Hear. Res.*, 1994, **78**, 27–30.

123. T. Hutchin, I. Haworth, K. Higashi *et al.*, 'A molecular basis for human hypersensitivity to aminoglycoside antibiotics.' *Nucleic Acids Res.*, 1993, **21**, 4174–4179.

124. L. P. Rybak, 'Ototoxicity of loop diuretics.' *Otolaryngol. Clin. North Am.*, 1993, **26**, 829–844.

125. S. J. Lee and E. S. Harpur, 'Abolition of the negative endocochlear potential a consequence of the gentamicin–furosemide interaction.' *Hear. Res.*, 1985, **20**, 37–43.

126. G. Laurell and B. Engstrm, 'The combined effect of cisplatin and furosemide on hearing function in guinea pigs.' *Hear. Res.*, 1989, **38**, 19–26.

127. R. L. Brown, R. C. Nuss, R. Patterson *et al.*, 'Audiometric monitoring of *cis*-platinum ototoxicity.' *Gynecol. Oncol.*, 1983, **16**, 254–262.

128. S. D. Schaefer, J. D. Post, L. G. Close *et al.*, 'Ototoxicity of low- and moderate-dose cisplatin.' *Cancer*, 1985, **56**, 1934–1939.

129. R. W. Fleischman, S. W. Stadnicki, M. F. Ethier *et al.*, 'Ototoxicity of *cis*-dichlorodiammine platinum (II) in the guinea pig.' *Toxicol. Appl. Pharmacol.*, 1975, **33**, 320–332.

130. S. Komune, S. Asakuma and J. B. Snow, Jr., 'Pathophysiology of the ototoxicity of *cis*-diamminedichloroplatinum.' *Otolaryngol. Head Neck Surg.*, 1981, **89**, 275–282.

131. S. E. Barron and E. A. Daigneault, 'Effect of cisplatin on hair cell morphology and lateral wall Na,K-ATPase activity.' *Hear. Res.*, 1987, **26**, 131–137.

132. D. McAlpine and B. M. Johnstone, 'The ototoxic mechanism of cisplatin.' *Hear. Res.*, 1990, **47**, 191–203.

133. S. Yee, M. Fazekas-May, E. M. Walker *et al.*, 'Inhibition of cisplatin toxicity without decreasing antitumor efficacy. Use of a dithiocarbamate.' *Arch. Otolaryngol. Head Neck Surg.*, 1994, **120**, 1248–1252.

134. L. P. Rybak, R. Ravi and S. M. Somani, 'Mechanism of protection by diethyldithiocarbamate against cisplatin ototoxicity: antioxidant system.' *Fundam. Appl. Toxicol.*, 1995, **26**, 293–300.

135. V. G. Schweitzer, D. F. Dolan, G. E. Abrams *et al.*, 'Amelioration of cisplatin-induced ototoxicity by fosfomycin.' *Laryngoscope*, 1986, **96**, 948–958.

136. F. A. Boettcher and R. J. Salvi, 'Salicylate ototoxicity: review and synthesis.' *Am. J. Otolaryngol.*, 1991, **12**, 33–47.

137. C. Mitchell, R. Brummett, D. Himes *et al.*, 'Electro-physiological study of the effect of sodium salicylate upon the cochlea.' *Arch. Otolaryngol.*, 1973, **98**, 297–301.

138. J. L. Puel, R. P. Bobbin and M. Fallon, 'Salicylate, mefenamate, meclofenamate, and quinine on cochlear potentials.' *Otolaryngol. Head Neck Surg.*, 1990, **102**, 66–73.

11.15
Olfactory System

DAVID C. DORMAN
Chemical Industry Institute of Toxicology, Research Triangle Park, NC, USA

JANE G. OWENS
Pfizer Inc., Groton, CT, USA

and

KEVIN T. MORGAN
Glaxo Wellcome, Research Triangle Park, NC, USA

11.15.1 INTRODUCTION

Neurotoxicological studies of sensory function are relatively rare compared with the number of studies that use learning and memory or motor activity as behavioral end points. Most studies of sensory function have dealt with the visual or auditory systems or, less commonly, the somatosensory system.[1] Neurotoxicological tests for evaluating olfactory function rarely are considered, and standardized testing procedures are almost nonexistent. One reason why so little attention has been paid to the effects of toxicants or chemicals on the olfactory system is the common perception that the sense of olfaction is less important since humans rely chiefly on visual and auditory input. In spite of this perception, over 200 000 people annually in the United States consult physicians about impairment of the sense of smell.[2] In contrast to other sensory deficits, many individuals with olfactory dysfunction often are unaware of their impairment. Olfactory impairment, however, can lead to diminished quality of life.[3] Deems *et al.*,[4] in a study including 750 patients with chemosensory dysfunction, found that 68% of the patients reported that the disorder affected their quality of life, resulting in impaired daily living or psychological well-being. Almost half of these people noted changes in either appetite or body weight. Loss of olfactory capabilities by workers within the chemical industries may also adversely affect job safety since a functional olfactory system often provides them with an early detector or warning system of the presence of volatile chemicals. Although the potential for sensory neurotoxicity is high, only a limited number of experimental studies have examined the potential olfactory neurotoxicity of chemicals. This chapter will address current understanding of chemically-induced olfactory toxicity with a special emphasis on research of interest to neurotoxicologists. No attempt has been made to provide an exhaustive review of each of the issues addressed, but wherever possible references to review articles are provided.

11.15.2 NORMAL STRUCTURE AND FUNCTION

11.15.2.1 Anatomy and Physiology

The anatomy and physiology of the mammalian olfactory system have been reviewed in detail[5–7] and will be considered briefly, with special reference to the olfactory mucosa and issues of toxicologic significance. Both the anatomy and physiology of the nose are related in many ways to the functions of this complex region of the upper respiratory tract. The nose functions to provide a passageway to the remainder of the respiratory tract, which conditions both inhaled (e.g., warming, humidifying, and cleaning) and exhaled air (e.g., dehumidification). The nose also contains receptors, including those essential for the sense of smell, pheromone detection (located in the vomeronasal organ), and trigeminal nerve function. The nose also plays a role in immunosurveillance and provides a drainage route for lachrymal secretions and cerebrospinal fluid.[8]

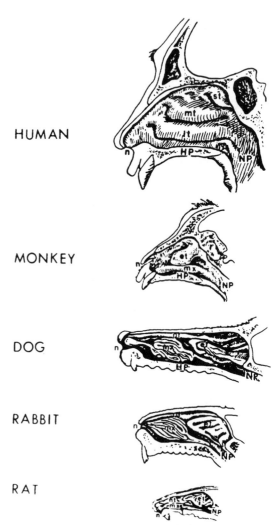

Figure 1 Diagrammatic representations of the mucosal surface of the nasal passages of five species to show variation in internal anatomy and the location of selected structures. HP, hard palate; n, naris; NP, nasopharynx; et, ethmoid turbinate; nt, naso-turbinate; mx, maxilloturbinate; mt, middle turbinate; it, inferior turbinate; st, superior turbinate (reproduced by permission of Toxicologic Pathology from *Toxicol. Pathol.*, 1991, **19**, 321–336).

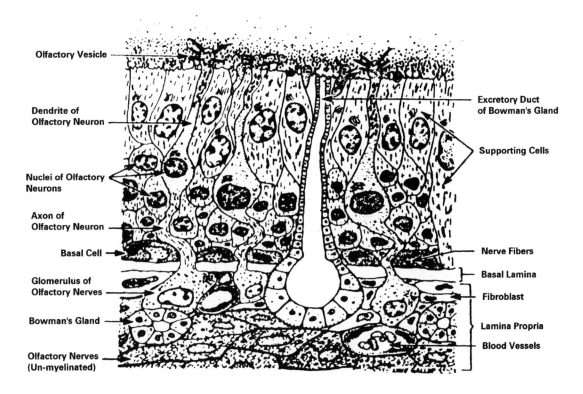

Figure 2 Schematic drawing of the olfactory mucosa of the rat depicting the major structures (reproduced by permission of the National Institute of Environmental Health Sciences from *Environ. Health Perspect.*, 1990, **85**, 187–208).

Humans and certain nonhuman primates such as rhesus monkeys have relatively simple nasal anatomy in regions of the nose lined by olfactory mucosa.[9] In contrast, the rat, which is the most frequently used animal species employed in toxicology studies, has a highly complex set of ethmoid turbinates (Figure 1),[10,125] as do the mouse, dog, and rabbit. In spite of considerable differences in nasal anatomy,[11] the basic structure of the olfactory mucosa in humans is similar to that of the majority of mammalian laboratory animals studied to date (Figure 2).

11.15.2.2 Olfactory System

The olfactory system begins with the olfactory epithelium, a patch of cells in the dorsal regions of the nasal cavity. The olfactory epithelium is pseudostratified and composed of three principal cell types: (i) receptor cells, (ii) supporting or sustentacular cells, and (iii) basal cells.[6] The olfactory epithelium is distinguished from other nervous tissues by several features. Unlike the majority of postembryonic neurons, receptor cells have a lifetime of 2–4 weeks and are

replaced continually by replication from underlying basal cells.[12] Olfactory sensory cells located in the nasal airway lining detect odorants at or near the airway surface and transmit this information as modifications of patterns of action potentials passing along their axons to synapses in the olfactory bulb of the brain. The bipolar odorant receptors are located in the membranes of sensory cilia, which extend into the airway lining mucus from the free extremity of sensory cell dendrites (olfactory vesicles). The olfactory mucosa is covered by a continuous sheet of mucus in which the olfactory cilia are embedded. The role of olfactory mucus in the sense of smell has been reviewed.[13] The cell bodies of the olfactory sensory cells lie in the nasal olfactory epithelium and are supported by sustentacular cells. Olfactory sensory cells are bipolar neurons, with a short apical dendrite and a long, thin (~0.2 μm diameter) unmyelinated axon that passes into the underlying lamina propria to form prominent olfactory nerve bundles. The bundles of sensory axons, supported by Schwann cells, pass through the lamina propria towards the cribriform plate, via which they penetrate the cranial cavity to synapse with the dendrites of mitral cells in the olfactory bulb. The olfactory bulb is the terminal nucleus of the olfactory nerve. In the olfactory bulb, the receptor cell axons synapse with mitral cells, the most prominent cell type of the olfactory bulb. Receptor cell axons, mitral cell dendrites, and smaller, tufted cells form brush-like terminals (olfactory glomeruli). About 25 000 axons enter each glomerulus and synapse with approximately 25 mitral cells that send signals to the brain.[6] The axons of the tufted and mitral cells largely make up the olfactory tract to the cerebrum. Unlike the other sensory systems, olfactory impulses reach the cerebral cortex without relay through the thalamus. Similar to other sensory systems, the olfactory system has an area of associated neocortex.

11.15.2.3 Airflow

The majority of odorants and many olfactory toxicants reach the olfactory mucosa via the inspired air. Airflow patterns in the nose are determined largely by internal nasal anatomy combined with inspiratory airflow rate, which is a function of the inspiratory cycle.[14–16] Many animals, including rodents, are obligate nose breathers, and patterns of breathing are different among species. In models of the nasal passages of humans,[17] rats, and rhesus monkeys,[18] only a very small proportion of the inspired air reaches the olfactory mucosa. The olfactory streams are derived from a specific and small intake source at the nostril of these species. Movements of the alar and other cartilages of the nostril are thought to have a significant influence on the amount of air reaching the olfactory mucosa during sniffing maneuvers.

11.15.2.4 Blood Flow

Delivery of chemicals or toxic metabolites via the blood supply will depend upon the kinetics of concentrations in the blood and the local blood perfusion rate. The olfactory mucosa receives arterial blood from the ethmoidal arterial branches of the ophthalmic artery,[9,19] which, in turn, are derived from the internal carotid artery. This region of the nose thus receives a blood supply that is independent of that to other epithelial regions. For rats and mice, less than 1% of the cardiac output is delivered to the nasal passages.[20] The proportion of this blood that reaches the olfactory mucosa is unknown, but such information is critical for understanding the role of regional dosimetry of systemically delivered olfactory toxicants.

11.15.2.5 Olfactory Signal Transduction

Odorants may first be recognized by olfactory binding proteins which transport them to olfactory neuronal receptors. In contrast to odorant receptors, olfactory binding protein is an abundant carrier of small lipophilic molecules and constitutes more than 1% of soluble nasal protein.[21] The olfactory receptor neuron is highly specialized to accommodate its function. Odor detection thresholds in animals and humans are reported to average in the picomolar range.[22] The initial site of signal transduction occurs in the olfactory receptor neuronal cilia.[23] As with other sensory neurons, the olfactory receptor neuron responds to odorant stimulation with the generation of a graded, receptor-mediated action potential. After odorants bind to receptors, the phosphoinositide and adenylate cyclase messenger systems are triggered. Odorant receptors are coupled to phospholipase C via a G_s-like G protein, resulting in the cleavage of phosphatidylinositol 4,5-biphosphate (PIP_2), releasing inositol 1,4,5-triphosphate (IP_3), which may act at the ciliary plasma membrane via an inositol 1,4,5-triphosphate receptor (IP_3R) to mediate influx of intracellular Ca^{2+}. Calcium may then augment the activities of phosphodiesterase or adenylyl cyclase. Alternatively, odorant receptors may be directly coupled to adenylyl cyclase via G_{olf}, a G_s-like protein.[21] The adenylyl cyclase activated by this interaction with odorants produces cAMP, which can open nonspecific cation channels, leading to the generation of an action potential. In addition to the action of these second-messenger systems, the action potential firing rate can also be influenced by the odorant concentration.

11.15.2.6 Metabolism

The olfactory mucosa has considerable capacity to undertake enzymatic biotransformation of xenobiotics, which in some respects rivals that of the liver. The activities of biotransforming enzymes in the olfactory mucosa have been reviewed in detail with respect to both biochemical[24,25] and histochemical parameters.[26] Detoxification of xenobiotics via phase one (hydroxylation and other reactions increasing hydrophilicity) and phase two (conjugation) reactions occurs extensively in the nose. Saturation of local metabolic capacity may lead to toxic responses, as has been shown for formaldehyde following glutathione depletion.[27] Local metabolism probably accounts for much of the cellular specificity of toxic responses to xenobiotics in the nose through bioactivation.[25]

11.15.3 OLFACTORY SYSTEM TOXICOLOGY

Loss or impairment of olfactory function in humans has been attributed to exposure to environmental pollutants, but this appears to account for only a minority of cases.[7] Other factors include genetic defects, damage to olfactory nerves at the level of cribriform plate during traumatic head injury, upper respiratory tract infections, or lesions of the central olfactory pathways. In contrast, an increasing number of chemicals are being found to induce olfactory damage in laboratory animals. These lesions exhibit distinct morphologic characteristics, specific cellular targets, and chemical-specific distribution patterns in the olfactory mucosa.[28] The location of olfactory mucosal lesions is attributable to local dose of the chemical, tissue susceptibility, or, more frequently, a combination of these factors.[29] Nasal neoplasia, including changes in the olfactory mucosa, have been reviewed extensively.[30–32]

A wide range of olfactory mucosal lesions have been reported, with only selected examples cited here.[33–47] Olfactory mucosal damage exhibits clear concentration or dose–response relationships and generally presents chronologically as progressive degeneration, with or without inflammation, followed by regeneration with or without residual damage or adaptive responses such as respiratory metaplasia. The nature of degenerative changes depends upon the primary target site. For instance, damage to sustentacular cells frequently results in epithelial vacuolation.[44] Severe sustentacular cell damage leads to characteristic folding of the epithelium before detachment occurs along with the sensory cell layer.[39] Loss of sensory cells may occur while sustentacular cells remain intact, and frequently these cells become engorged with eosinophilic proteinaceous deposits in chronic studies.[48,49] The nature and significance of these deposits is unknown. Numerous examples of non-neoplastic olfactory lesions are provided by a number of reviews.[10,36,50] Determination of the distribution of non-neoplastic and neoplastic (see below) responses can play a critical role in assessing their relevance to human risk.[10,28]

11.15.4 ASSESSMENT OF OLFACTORY SYSTEM FUNCTION

11.15.4.1 Imaging Techniques

The application of imaging techniques to the study of chemical-induced olfactory toxicity has been limited. Two promising technologies that may receive increased use in the evaluation of smell disorders are magnetic resonance imaging (MRI) and positron emission tomography (PET). The cerebral structures associated with olfactory stimulation have been identified in human volunteers using PET imaging following $H_2^{15}O$ administration.[51] Functional evaluation of the human olfactory cortex may also be achieved using MRI techniques coupled to odor stimulation.[52]

11.15.4.2 Behavioral Testing Methods

Standardized testing to evaluate and quantitate olfactory function in humans is primarily subjective and often includes tests of odor discrimination and odor detection threshold measurement.[53,54] The University of Pennsylvania Smell Identification Test (UPSIT, commercially marketed as the Smell Identification Test) is the most widely used test to assess clinically the ability of humans to identify odors. This self-administered scratch-and-sniff test consists of 40 chemically microencapsulated odor patches that release an odor when scratched.[55] Olfactory discrimination can also be tested by the use of items such as cinnamon, peanut butter, and mothballs placed in opaque plastic jars.[56] Human olfactory thresholds can be measured by using glass flasks or plastic bottles containing serial dilutions of an odorant (e.g., pyridine) in mineral oil or water.[57] The lowest concentration reliably selected corresponds to the detection threshold for that odorant. Similar methods using odor-impregnated paper strips have been used to compare performance in the olfactory discrimination and threshold tests in humans and squirrel monkeys.[58] Apparently, humans can resolve smaller odor concentration changes than the rat, as people can resolve differences in odor concentration in the range of approximately 10–50%.[59] Rats[60] and dogs[61] have a 2.5 log unit lower absolute odor detection threshold than humans. Devices to generate olfactory stimuli (olfactometers) are usually custom-fabricated. The most precise are dynamic flow systems using the principle of flow dilution.[59,62] Odor sources may be odorant-impregnated permeation tubes maintained at constant temperatures for odor dispersion,[63] glass tubes that function as vapor saturators as air passes through them,[64,65] or Tedlar bags containing an odorant with a low boiling point.[66]

Of the large number of olfactory mucosal toxicants discovered, very few have been examined for evidence of functional impairment. Methods to assess olfactory function in animals have ranged from simple tests involving the presentation of a cotton swab containing an odorant to the animal, followed by evaluation for orientation toward and sniffing of the swab, to more elaborate methods that generate and deliver diluted odorants at a constant concentration while allowing for the measurement of more complex odorant-cued behaviors. Simple tests of olfactory discrimination may be assessed in neonatal rats after postnatal day eight.[67] Rat pups placed in the middle of a container with clean shavings on one side and soiled home cage bedding on the other will orient and move toward the side containing the home cage bedding.[68] This technique has been used to measure concurrently olfactory discrimination and motor activity.[69] Other investigators have trained rats to

retrieve food and have used these methods to report decrements in performance associated with toxicants such as methyl bromide.[39,70] Functional deficits have also been identified using olfactory discrimination tasks in rats following 3-methylindole administration.[46] An active avoidance-based behavioral olfactometer that delivers diluted odorants to the subject has been developed in our laboratory.[66] High-dose 3-methylindole administration results in a marked reduction in the number of odorant-cued correct responses in treated rats (Figure 3).

11.15.4.3 Neurophysiological Techniques

One alternative to the use of behavioral tests to study olfactory dysfunction is the use of

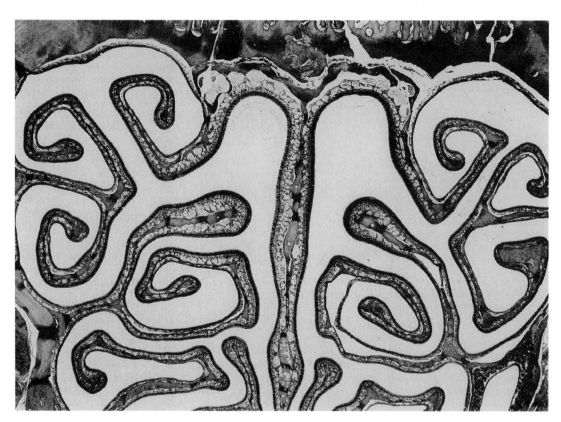

Figure 3 Nasal histopathology and loss of olfactory function following exposure to 400 mg kg^{-1} 3-methylindole are presented. Light micrograph of the ethmoid region of the nose of (this page) a young mature male F344 rat to show normal structure, and (next page, top) a rat 7 days following i.p. treatment. See Mery *et al.*[10] for classification of the turbinates in this region. Note the even dark staining of the olfactory epithelium lining much of this region of the nose in the control and extensive loss of this epithelium in the 3-MI treated animal, especially in the dorsal medial airways. The treated rat also shows formation of intralumenal proliferative changes with fusion of the dorsal scroll of the third ethmoturbinate to the nasal septum. Hematoxylin and eosin × 24. Odor-cued avoidance behavior data (*n* = 6; ± SEM) (next page, bottom) are shown from 4 weeks of training (predosing) and week 5 following treatment. Acetaldehyde was presented as the odorant at 1.86 ppm for 120 trials each day. Animals were trained to avoid a mild footshock that followed presentation of the odorant. Avoidance responses to acetaldehyde were reduced by 3-MI administration at 7 days postdose.

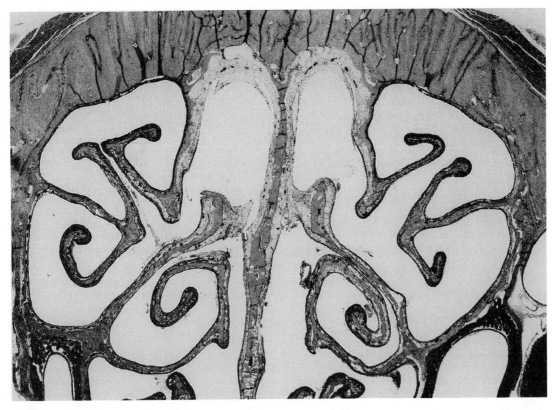

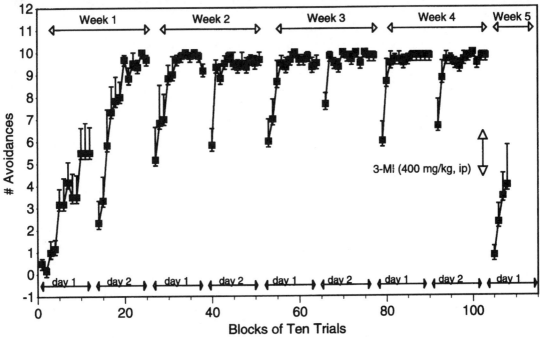

Figure 3 (continued)

neurophysiologic modalities. Two approaches have been suggested for the evaluation of humans with anosmia.[71] First, direct excitations of the receptor cells could be recorded from the nasal mucosal surface (electro-olfactogram). Second, electroencephalographic (EEG) recordings of the olfactory activity of cortical neurons in response to odorant presentation could be obtained from the skull (olfactory EEG). These neurophysiological techniques integrate the neural activity of multiple neural circuits whose magnitude is in part dependent upon the odorant concentration. This odorant-induced dose-response relationship may be applicable to toxicology since a quantifiable decrement vs. all-or-none response in olfactory function may be observable following toxicant exposure in either the electro-olfactogram[72] or olfactory EEG. Clinical application of electro-olfactograms and olfactory EEGs have been reported in anosmic dogs[73–75] and humans,[71,76] and similar use in toxicologic studies may improve human risk assessment. Olfactory EEG was reportedly more sensitive than behavioral olfactometry in determining odor thresholds in dog.[77] In humans (and possibly rodents), olfactory EEGs may also allow distinction between olfactory-evoked potentials (i.e., those mediated by the olfactory nerve such as vanillin odor) and chemosomatosensory-evoked potentials (i.e., those mediated by the trigeminal nerve such as carbon dioxide).

11.15.4.4 Histopathology

When assessing olfactory mucosal damage using histopathology, good fixation is essential;[78] awareness of histological artifacts, to which the olfactory mucosa is particularly prone, is of prime importance. The majority of olfactory mucosal lesions found in laboratory animals are detected initially histopathologically. However, in the case of β,β′-iminodiproprionitrile (IDPN), evidence of olfactory damage was first detected during olfactory function studies,[79] and olfactory mucosal lesions were found subsequently.[38] In many of these toxicological studies that assess olfactory function in rodents, extensive histologic damage (~90%) was required before a functional deficit was noted in an animal's ability to detect an odorant since only a very small percentage of the olfactory epithelium is required to permit nearly normal olfactory function.[70,80] Thus, the tremendous sensitivity of the olfactory system makes correlation of histopathologic lesions to olfactory function as assessed by simple tests problematic.

11.15.5 OLFACTORY TOXICANTS

Most reports of the effects of chemicals on olfaction are clinical or anecdotal in nature, and most of the compounds implicated have not been subjected to any rigorous scientific study. Olfactory damage may occur following systemic or inhalation exposure to many compounds. The olfactory system has been reported to be adversely affected by approxi-

mately 200 compounds of toxicological interest (Amoore, 1986). What follows is a brief review of several common chemical olfactory toxicants with potential significance to man (see Chapters 26, 27, and 29, this volume).

11.15.5.1 Pesticides

Methyl bromide is a broad-spectrum fumigant effective at controlling insects, nematodes, and fungi and as such is one of the most widely used pesticides in the world. Methyl bromide is a pre-emergent pesticide and is applied primarily to soil before crop planting. Humans are most commonly exposed to methyl bromide through inhalation or dermal absorption. The primary clinical signs observed in humans following high-dose methyl bromide exposure include ataxia, paralysis, central nervous system depression, apathy, and retardation of speech and movement.[70] Exposure of rodents to high concentrations of methyl bromide results in extensive damage to the olfactory epithelium without changes in the respiratory epithelium.[39,70] Studies of olfactory function in animals following chemical exposure are limited. Hurtt et al.[39] as well as Hastings et al.[70] noted decreased food retrieval behaviors in rats exposed to 200 ppm methyl bromide for 5–8 days at 4–6 h per day. Hastings et al.[70] reported that olfactory function in rats exposed to methyl bromide was severely impaired after only one 4 h exposure period. Neuronal degeneration of the cerebral cortex[81] and granule cells of the cerebellar cortex[82] have also been reported in rodents following methyl bromide exposure.

2,6-Dichlorobenzonitrile (dichlobenil) is an herbicide with numerous noncrop terrestrial and aquatic uses. Intraperitoneal or dermal administration of dichlobenil results in Bowman's gland necrosis followed by secondary degeneration of the olfactory neuroepithelium in mice.[30,35] The olfactory toxicity of dichlobenil is dependent upon cytochrome P450 activity.[30,83] A structural analogue, chlorthiamide (2,6-dichlorothiobenzamide), is a major environmental metabolite of dichlobenil that also induces olfactory toxicity similar to dichlobenil following administration to mice.[33] To our knowledge, human toxicity has not been reported with dichlobenil. However, experimental studies in rodents indicate a potential hazard.

Benomyl (methyl-1-(butylcarbamoyl)-2-benzimidazole-carbamate) is a widely used agricultural fungicide that has been shown to induce olfactory degeneration following a subchronic (90 day) inhalation exposure of up to 200 mg m^{-3} in rats.[84] The nasal mucosal damage was related to airflow patterns, particle size, and resultant deposition in the nasal passages. Olfactory toxicity could not be replicated following a high-dose dietary exposure (5000–15 000 ppm).[85]

11.15.5.2 Solvents and Related Chemicals

Inhalation exposures of rodents to a wide variety of volatile organic compounds has been associated with olfactory damage. Solvents that induce olfactory damage in rodents include 3-trifluoromethyl pyridine,[86,87] methyl methacrylate,[88] 2-methyl-1,3-butadiene,[89] *n*-hexane,[90] 1,3-dichloropropene,[91] and pyridine.[44] In addition to direct olfactory toxicity, several solvents, including toluene and xylene, may undergo axonal flow-mediated transport from the nasal mucosa to the olfactory lobe of the brain.[92] Evaluation of the toxicological significance of this finding awaits further study.

The olfactory mucosal response to certain esters provides an example of metabolically mediated toxicity for which a considerable body of information is available on mechanism of action using a wide range of techniques. Dibasic esters are a solvent mixture of dimethyl adipate, dimethyl glutarate, and dimethyl succinate used in the paint and coating industry. A subchronic inhalation toxicity study has demonstrated that a dibasic ester mixture of 16.5% dimethyl adipate, 16.9% dimethyl succinate, and 66.6% dimethyl glutarate induces a mild degeneration of the olfactory epithelium of the rat nasal cavity at exposure concentrations ranging from 20 to 390 mg m^{-3}.[93] Histochemical studies of carboxylesterase revealed considerable activity of this enzyme in the target olfactory sustentacular cells.[94] Subsequent ultrastructural studies demonstrated that these cells are the primary cellular target for such esters,[95] which supports the proposal that enzymatic generation of carboxylic acids in sustentacular cells is responsible for the destructive effects of these esters. The presence of these esterases has been demonstrated in human nasal mucosa.[96]

Schwartz *et al.*[97] reported that chemical manufacturing workers exposed to acrylic acid, methacrylic acid, acrylates, or methacrylates had decreased odor identification capabilities when tested using the UPSIT. Olfactory dysfunction increased with prolonged exposure and these effects may be reversible. In a similar study using the UPSIT, Schwartz *et al.*[98] identified cumulative exposure-related decrements in odor identification ability among never-smoking workers exposed to toluene, xylene, and methyl ethyl ketone at two paint formulation plants. In contrast, Sandmark *et al.*[99] did not find any significant differences in odor identification ability in solvent-exposed painters. Experimental studies using human volunteers have demonstrated temporary loss of smell following a 7 h inhalation exposure to 20–40 ppm methyl isobutyl ketone.[100] Similarly, Mergler and Beauvais[101] have also demonstrated a shift in the olfactory perception threshold in human volunteers following a 7 h inhalation exposure to toluene (50 ppm) or xylene (40 ppm), either singly or in combination.

11.15.5.3 Metals

A variety of metals have been implicated in the development of chemically-induced olfactory toxicity. Zinc sulfate ablation of the olfactory mucosa has been used extensively as an experimental model of olfactory toxicity.[34] Nickel sulfate is a soluble compound used for nickel electroplating. Inhalation of nickel sulfate, nickel subsulfide, or nickel sulfate hexahydrate results in atrophy of the olfactory epithelium in rodents.[63,90,102] Although atrophy of the olfactory epithelium has not been noted in nickel refinery workers,[103] nickel electrolysis workers have reported increased incidences of anosmia.[104] Olfactory degeneration has also been reported in rodents following inhalation exposure to cobalt sulfate[105] and ferrocene (dicyclopentadienyl iron).[106]

Cadmium, a highly toxic trace mineral found widely in industry and the environment, is frequently cited as an olfactory toxicant in humans.[107,108] The National Institute for Occupational Safety and Health (NIOSH) estimates that 100 000 American workers are exposed occupationally to cadmium fumes or dusts in such industries as electroplating and battery manufacturing as well as during the manufacture of pigments, plastics, and metal alloys.[108] Rose *et al.* have reported an increased incidence of hyposmia in workers engaged in brazing operations when assessed using tests of olfactory identification and butanol detection thresholds.[108] Interestingly, olfaction as measured using a behavioral olfactometer was not affected in rats after inhalation exposure to cadmium oxide, even though significant elevations in the amount of cadmium in olfactory bulbs and olfactory epithelium occurred.[109] Unfortunately, this research did not examine the olfactory epithelium histopathologically. Transneural transport of cadmium has been reported in a variety of animal species.[110,111]

11.15.5.4 Hydrogen Sulfide

Hydrogen sulfide is a common and potent toxic agent that is the primary chemical hazard in the production of sour gas (i.e., sulfur-containing) production. Hydrogen sulfide toxicity is also encountered in swine containment facilities and industries involved in manure and sewage-handling.[112] Transient loss of consciousness following exposure to high concentrations of hydrogen sulfide occur in the petrochemical industries, with fatalities occurring occasionally each year. Hydrogen sulfide interacts with a variety of enzymes and macromolecules, with cytochrome oxidase being the most critical target enzyme.[113] Hydrogen sulfide is a potent neurotoxicant,[112,114] and at ambient air concentrations above 200 ppm induces olfactory nerve paralysis. Prolonged exposure may also result in lethal respiratory paralysis.

11.15.5.5 3-Methylindole

3-Methylindole (3-MI, skatole) is a ruminal metabolite of L-tryptophan and is one cause of acute bovine pulmonary emphysema and interstitial edema. Histopathologic lesions in the respiratory tract include pulmonary congestion and edema, type II pneumocyte hyperplasia, and destruction of Clara cells.[115] In addition to its pulmonary effects, 3-MI also selectively damages the olfactory mucosa in rodents and other animals. The compound is structurally related to 3-methylfuran, another known olfactotoxicant, that exists as a component of atmospheric smog.[116] In addition to its veterinary toxicologic significance, 3-MI has become a commonly used chemical model for the induction of olfactory mucosal damage. Advantages of 3-MI include the fact that nasal toxicity is induced following systemic exposure and, in contrast to methyl bromide, is not a central nervous system toxicant. 3-Methylindole is thought to be metabolized by mixed-function oxidases.[115] High-dose exposure (400 mg kg^{-1}, i.p.) of rats to 3-MI results in severe degeneration of sensory olfactory epithelium as well as functional deficits identified using an olfactory discrimination task.[46]

11.15.6 EXPERIMENTAL MODELS FOR THE STUDY OF OLFACTORY TOXICITY

11.15.6.1 Animal Models

As stated by Ache,[117] studies of olfactory function are rendered challenging as a consequence of the fact that (i) chemical senses research has focused largely upon a limited number of species, (ii) gustatory sensations are not always easily distinguishable from olfactory sensations, (iii) animals that are taxonomically closely related may have markedly different chemosensitivity as a consequence of evolutionary divergence, and (iv) the search for similarities across phylogenetically diverse animal species may be confounded by evolutionary convergence of function with very different underlying cellular or molecular mechanisms. Thus, when selecting an animal model for assessment of risk of olfactory toxicity in humans, care should be taken to consider the potential impact of anatomical, physiological, biochemical, and other differences between animals and humans.

11.15.6.2 Cell Cultures and Other *In Vitro* Techniques

The olfactory epithelium is used as a model system for studying neuronal development and synaptogenesis, electrophysiology of sensory transduction, and the physiological and morphological properties of olfactory neurons. In short-term monolayer culture systems of olfactory tissues, olfactory receptor neurons (ORNs) have been identified based on morphology and immunostaining with neuron-specific enolase. These ORNs were determined by patch clamp analysis to be electrically active and responded to application of an odorant mixture.[118] ORNs have been difficult to grow and maintain in long-term cultures free of supporting cells and the nasal mucosa. Staining for the olfactory marker protein (OMP), which is found almost exclusively in mature ORNs, has not been found consistently in monolayer cultures, and most ORNs disappeared after six days.[119] However, disaggregated nasal cells from newborn rats grown in coculture with astrocytes have been shown to differentiate into OMP-expressing ORNs, by 10 days after plating. These ORNs were also shown to be immunopositive for the adhesion molecules N-CAM and L1[120] as well as for neuron-specific tubulin[119] and more closely resemble the olfactory epithelium *in situ* than other methods using only nasal cells.[121] The low number of viable olfactory neuroepithelial cells that can be obtained from neonatal rats may be a significant limiting factor in the use of olfactory cell cultures for neurotoxicological assessment.

11.15.6.3 Dosimetry Modeling

A major issue for toxicology studies is the extrapolation of results obtained following high-dose exposures in rodents to human exposures, generally at concentrations several orders of magnitude lower than those used experimentally. Dosimetry, the measurement of this dose, is a complex but critical issue in toxicology that has direct relevance to inter-species extrapolation and human risk assessment for potential olfactory toxicants. The role of dosimetry in nasal toxicology, including effects on the olfactory system, has been reviewed extensively.[122] Mathematical modeling can be used to provide a rational basis for both animal-to-human and low-dose to high-dose extrapolations. Information on the relationship between local dosimetry and metabolism has resulted in the development of mathematical models of these processes that are needed for appropriate human risk assessment.[123] These models are used to predict regions in which increased nasal deposition and uptake (i.e., hot spots) of a chemical may occur. The effect of dose on the extent of olfactory mucosal lesions is well demonstrated by studies of methimazole.[37] In addition to toxicant-induced damage to the olfactory system, environmental agents such as cadmium,[110,111] aluminum,[124] and xylene[92] may enter the brain via the olfactory nerve.

11.15.7 REFERENCES

1. L. Hastings, 'Sensory neurotoxicology: use of the olfactory system in the assessment of toxicity.' *Neurotoxicol. Teratol.*, 1990, **12**, 455–459.
2. National Advisory Neurological and Communicative Disorders and Stroke Council Panel on Communicative Disorders, 'Report of the Panel on Communicative Disorders to the National Advisory Neurological and Communicative Disorders and Stroke Council. National Institute of Neurological and Communicative Disorders and Stroke Council,' US Department of Health, Education, and Welfare, Bethesda, MD, NIH Publication 79-1914, 319.
3. T. Engen, 'The Perception of Odors,' Academic Press, New York, 1982, pp. 1–16.
4. D. A. Deems, R. L. Doty, R. G. Settle *et al.*, 'Smell and taste disorders, a study of 750 patients from the University of Pennsylvania Smell and Taste Center.' *Arch. Otolaryngol. Head Neck Surg.*, 1991, **117**, 519–528.
5. D. Ackerman, 'A Natural History of the Senses,' 1st edn., Random House, New York, 1990.
6. C. A. Greer, in 'Smell and Taste in Health and Disease,' eds. T. V. Getchell, R. L. Doty, L. M. Bartoshuk *et al.*, Raven Press, New York, 1991, pp. 65–81.
7. M. J. Serby and K. L. Chobar (eds.), 'Science of Olfaction,' Springer-Verlag, New York, 1992.
8. H. F. Cserr and P. M. Knopf, 'Cervical lymphatics, the blood–brain barrier and the immunoreactivity of the brain: a new view.' *Immunol. Today*, 1992, **13**, 507–512.
9. R. Warwick and P. L. Williams (eds.), 'Gray's Anatomy,' W. B. Saunders, Philadelphia, PA, 1973, pp. 1086–1091.
10. S. Mery, E. A. Gross, D. R. Joyner *et al.*, 'Nasal diagrams: a tool for recording the distribution of nasal lesions in rats and mice.' *Toxicol. Pathol.*, 1994, **22**, 353–372.
11. V. Negus, 'The Comparative Anatomy and Physiology of the Nose and Paranasal Sinuses,' Livingstone, Edinburgh, 1958, pp. 34–81.
12. A. Mackay-Sim and P. Kittel, 'Cell dynamics in the adult mouse olfactory epithelium: a quantitative autoradiographic study.' *J. Neurosci.*, 1991, **11**, 979–984.
13. M. L. Getchell and T. K. Mellert, in 'Smell and Taste in Health and Disease,' eds. T. V. Getchell, R. L. Doty, L. M. Bartoshuk *et al.*, Raven Press, New York, 1991, pp. 83–95.
14. P. Cole, in 'The Nose, Upper Airway Physiology and the Atmospheric Environment,' eds. D. F. Proctor and I. Anderson, Elsevier, New York, 1982, pp. 163–189.
15. M. M. Mozell, P. F. Kent, P. W. Scherer *et al.*, in 'Smell and Taste in Health and Disease,' eds. T. V. Getchell, R. L. Doty, L. M. Bartoshuk *et al.*, Raven Press, New York, 1991, pp. 481–492.
16. A. W. Proetz, 'Air currents in the upper respiratory tract and their clinical importance.' *Ann. Otol. Rhinol. Laryngol.*, 1951, **60**, 439–441.
17. H. Masing, 'Experimentelle Untersuchungen uber den Stromungsverlauf im Nosenmodell' ['Experimental studies on the flow in a nose model'] *Arch. Klin. Exp. Ohren. Nasen. Kehlkopfheilk.*, 1967, **189**, 371–381.
18. K. T. Morgan, J. S. Kimbell, T. M. Monticello *et al.*, 'Studies of inspiratory airflow patterns in the nasal passages of the F344 rat and rhesus monkey using nasal molds: relevance to formaldehyde toxicity.' *Toxicol. Appl. Pharmacol.*, 1991, **110**, 223–240.
19. R. Hebel and M. W. Stromberg, in 'Anatomy and Embryology of the Laboratory Rat,' BioMed, Worthsee, 1986, pp. 58–59.
20. W. T. Stott, M. D. Dryzga and J. C. Ramsey, 'Blood-flow distribution in the mouse.' *J. Appl. Toxicol.*, 1983, **3**, 310–312.
21. G. V. Ronnett and S. H. Snyder, 'Molecular messengers of olfaction.' *Trends Neurosci.*, 1992, **15**, 508–513.
22. V. E. Dionne and A. E. Dubin, 'Transduction diversity in olfaction.' *J. Exp. Biol.*, 1994, **194**, 1–21.
23. D. Lancet, 'Vertebrate olfactory reception.' *Annu. Rev. Neurosci.*, 1986, **9**, 329–355.
24. A. R. Dahl and W. M. Hadley, 'Nasal cavity enzymes involved in xenobiotic metabolism: effects on the toxicity of inhalants.' *Crit. Rev. Toxicol.*, 1991, **21**, 345–372.
25. C. J. Reed, 'Drug metabolism in the nasal cavity: relevance to toxicology.' *Drug Metab. Rev.*, 1993, **25**, 173–205.
26. M. S. Bogdanffy, 'Biotransformation enzymes in the rodent nasal mucosa: the value of a histochemical approach.' *Environ. Health Perspect.*, 1990, **85**, 177–186.
27. M. Casanova and H. d'A. Heck, 'Further studies of the metabolic incorporation and covalent binding of inhaled [^{3}H]- and [^{14}C]formaldehyde in Fischer-344 rats: effects of glutathione depletion.' *Toxicol. Appl. Pharmacol.*, 1987, **89**, 105–121.
28. K. T. Morgan, 'Nasal dosimetry, lesion distribution, and the toxicologic pathologist: a brief review.' *Inhal. Toxicol.*, 1994, **6**, 41–57.
29. K. T. Morgan and T. M. Monticello, 'Airflow, gas deposition, and lesion distribution in the nasal passages.' *Environ. Health Perspect.*, 1990, **85**, 209–218.
30. I. Brandt, E. B. Brittebo, V. J. Feil *et al.*, 'Irreversible binding and toxicity of the herbicide dichlobenil (2,6-dichlorobenzonitrile) in the olfactory mucosa of mice.' *Toxicol. Appl. Pharmacol.*, 1990, **103**, 491–501.
31. H. R. Brown, T. M. Monticello, R. R. Maronpot

et al., 'Proliferative and neoplastic lesions in the rodent nasal cavity.' *Toxicol. Pathol.*, 1991, **19**, 358–372.

32. V. J. Feron, R. A. Woutersen and B. J. Spit, in 'Toxicology of the Nasal Passages,' ed. C. S. Barrow, Hemisphere Publishing Corporation, Washington, DC, 1986, pp. 67–89.

33. E. B. Brittebo, C. Eriksson, V. Feil *et al.*, 'Toxicity of 2,6-dichlorothiobenzamide (chlorthiamid) and 2,6-dichlorobenzamide in the olfactory nasal mucosa of mice.' *Fundam. Appl. Toxicol.*, 1991, **17**, 92–102.

34. P. Cancalon, 'Degeneration and regeneration of olfactory cells induced by ZnSO₄ and other chemicals.' *Tissue Cell.*, 1982, **14**, 717–733.

35. N. J. Deamer, J. P. O'Callaghan and M. B. Genter, 'Olfactory toxicity resulting from dermal application of 2,6-dichlorobenzonitrile (dichlobenil) in the C57BL mouse.' *Neurotoxicology*, 1994, **15**, 287–293.

36. B. A. Gaskell, 'Nonneoplastic changes in the olfactory epithelium—experimental studies.' *Environ. Health Perspect.*, 1990, **85**, 275–289.

37. M. B. Genter, N. J. Deamer, B. L. Blake *et al.*, 'Olfactory toxicity of methimazole: dose-response and structure-activity studies and characterization of flavin-containing monooxygenase activity in the Long-Evans rat olfactory mucosa.' *Toxicol. Pathol.*, 1995, **23**, 477–486.

38. M. B. Genter, J. Llorens, J. P. O'Callaghan *et al.*, 'Olfactory toxicity of β,β'-iminodiproprionitrile (IDPN) in the rat.' *J. Pharmacol. Exp. Ther.*, 1992, **263**, 1432–1439.

39. M. E. Hurtt, D. A. Thomas, P. K. Working *et al.*, 'Degeneration and regeneration of the olfactory epithelium following inhalation exposure to methyl bromide: pathology, cell kinetics, and olfactory function.' *Toxicol. Appl. Pharmacol.*, 1988, **94**, 311–328.

40. X. Z. Jiang, L. A. Buckley and K. T. Morgan, 'Pathology of toxic responses to the RD50 concentration of chlorine gas in the nasal passages of rats and mice.' *Toxicol. Appl. Pharmacol.*, 1983, **71**, 225–236.

41. K. P. Lee, R. Valentine and M. S. Bogdanffy, 'Nasal lesion development and reversibility in rats exposed to aerosols of dibasic esters.' *Toxicol. Pathol.*, 1992, **20**, 376–393.

42. D. M. Mann and M. M. Esiri, 'The site of the earliest lesions of Alzheimer's disease.' *N. Engl. J. Med.*, 1988, **318**, 789–790.

43. R. R. Miller, J. T. Young, R. J. Kociba *et al.*, 'Chronic toxicity and oncogenicity bioassay of inhaled ethyl acrylate in Fischer 344 rats and B6C3F1 mice.' *Drug Chem. Toxicol.*, 1985, **8**, 1–42.

44. K. J. Nikula and J. L. Lewis, 'Olfactory mucosal lesions in F344 rats following inhalation exposure to pyridine at threshold limit value concentrations.' *Fundam. Appl. Toxicol.*, 1994, **23**, 510–517.

45. Y. Ohashi, Y. Nakai, S. Horiguchi *et al.*, 'Studies on industrial styrene poisoning (part XII). Electron-microscopic observations on the mucosal membrane of respiratory tracts of rats exposed to styrene'. *Sanyo Igaku*, 1981, **23**, 513–529.

46. D. B. Peele, S. D. Allison, B. Bolon *et al.*, 'Functional deficits produced by 3-methylindole-induced olfactory mucosal damage revealed by a simple olfactory learning task.' *Toxicol. Appl. Pharmacol.*, 1991, **107**, 191–202.

47. S. S. Schiffman and H. T. Nagle, 'Effect of environmental pollutants on taste and smell.' *Otolaryngol. Head Neck Surg.*, 1992, **106**, 693–700.

48. E. A. Gross, D. L. Patterson and K. T. Morgan, 'Effects of acute and chronic dimethylamine exposure on the nasal mucociliary apparatus of F-344 rats.' *Toxicol. Appl. Pharmacol.*, 1987, **90**, 359–376.

49. D. C. Wolf, K. T. Morgan, E. A. Gross *et al.*, 'Two-

year inhalation exposure of female and male B6C3F1 mice and F344 rats to chlorine gas induces lesions confined to the nose.' *Fundam. Appl. Toxicol.*, 1995, **24**, 111–131.

50. T. M. Monticello, K. T. Morgan and L. Uraih, 'Nonneoplastic nasal lesions in rats and mice.' *Environ. Health Perspect.*, 1990, **85**, 249–274.

51. R. J. Zatorre, M. Jones-Gotman, A. C. Evans *et al.*, 'Functional localization and lateralization of human olfactory cortex.' *Nature*, 1992, **360**, 339–340.

52. I. Koizuka, H. Yano, M. Nagahara *et al.*, 'Functional imaging of the human olfactory cortex by magnetic resonance imaging.' *ORL J. Otorhinolaryngol. Relat. Spec.*, 1994, **56**, 273–275.

53. R. L. Doty, 'Diagnostic tests and assessment.' *J. Head Trauma Rehabil.*, 1992, **7**, 47–52.

54. R. L. Doty, D. P. Perl, J. C. Steele *et al.*, 'Olfactory dysfunction in three neurodegenerative diseases.' *Geriatrics*, 1991, **46**(Suppl. 1), 47–51.

55. R. L. Doty, P. Shaman, S. L. Applebaum *et al.*, 'Smell identification ability: changes with age.' *Science*, 1984, **226**, 1441–1443.

56. W. S. Cain, J. F. Gent, R. B. Goodspeed *et al.*, 'Evaluation of olfactory dysfunction in the Connecticut Chemosensory Clinical Research Center.' *Laryngoscope*, 1988, **98**, 83–88.

57. J. E. Amoore, in 'Toxicology of the Nasal Passages,' ed. C. S. Barrow, Hemisphere Publishing Corporation, Washington, DC, 1986, pp. 155–190.

58. M. Laska and R. Hudson, 'Assessing olfactory performance in a New World primate, *Saimirre sciureus*.' *Physiol. Behav.*, 1993, **53**, 89–95.

59. B. Berglund, U. Berglund and T. Lindvall, 'Theory and methods for odor evaluation.' *Experientia*, 1986, **42**, 280–287.

60. R. G. Davis, 'Olfactory psychophysical parameters in man, rat, dog, and pigeon.' *J. Comp. Physiol. Psychol.*, 1973, **85**, 221–232.

61. D. Krestel, D. Passe, J. C. Smith *et al.*, 'Behavioral determination of olfactory thresholds to amyl acetate in dogs'. *Neurosci. Biobehav. Rev.*, 1984, **8**, 169–174.

62. A. Dravnieks, in 'Methods in Olfactory Research,' eds. J. W. Johnston, Jr., D. G. Moulton and A. Turk Academic Press, New York, 1975, pp. 1–61.

63. J. E. Evans, M. L. Miller, A. Andringa *et al.*, 'Behavioral, histological, and neurochemical effects on nickel(II) on the rat olfactory system.' *Toxicol. Appl. Pharmacol.*, 1995, **130**, 209–220.

64. B. M. Slotnick and F. W. Schoonover, 'Olfactory thresholds in unilaterally bulbectomized rats.' *Chem. Senses*, 1994, **9**, 325–340.

65. J. W. Fraser, 'Measurement of olfactory signal delectability using an air-dilution olfactometer.' *Percept. Mot. Skills*, 1988, **67**, 827–830.

66. J. G. Owens, R. A. James, O. R. Moss *et al.*, 'Design and evaluation of an olfactometer for the assessment of 3-methylindole-induced hyposmia.' *Fundam. Appl. Toxicol.*, 1996, **33**, 60–70.

67. E. H. Gregory and D. W. Pfaff, 'Development of olfactory-guided behavior in infant rats'. *Physiol. Behav.*, 1971, **6**, 573–576.

68. J. Adams, J. Buelke-Sam, C. A. Kimmel *et al.*, 'Collaborative behavioral teratology study: protocol design and testing procedures.' *Neurobehav. Toxicol. Teratol.*, 1985, **7**, 579–586.

69. J. Altman, R. L. Brunner, G. G. Bulert *et al.*, in 'Drugs and the Developing Brain,' eds. A. Vernadakis and N. Weiner, Plenum, New York, 1974, pp. 321–348.

70. L. Hastings, M. L. Miller, D. J. Minnerma *et al.*, 'Effects of methyl bromide on the rat olfactory system.' *Chem. Senses*, 1991, **16**, 43–56.

71. G. Kobal and T. Hummel, in 'Smell and Taste in

Health and Disease,' eds. T. V. Getchell, R. L. Doty, L. M. Bartoshuk *et al.*, Raven Press, New York, 1991, pp. 255–275.

72. S. G. Shirley, E. H. Polak, D. A. Edwards *et al.*, 'The effect of concanavalin A on the rat electro-olfactogram at various odorant concentrations.' *Biochem. J.*, 1987, **245**, 185–189.

73. L. J. Myers, R. Nash and H. S. Elledge, 'Electro-olfactography: a technique with potential for diagnosis of anosmia in the dog.' *Am. J. Vet. Res.*, 1984, **45**, 2296–2298.

74. L. J. Myers, L. A. Hanrahan, L. J. Swango *et al.*, 'Anosmia associated with canine distemper.' *Am. J. Vet. Res.*, 1988, **49**, 1295–1297.

75. L. J. Myers, K. E. Nusbaum, L. J. Swango *et al.*, 'Dysfunction of sense of smell caused by canine parainfluenza virus infection in dogs.' *Am. J. Vet. Res.*, 1988, **49**, 188–190.

76. M. Furukawa, M. Kamide, T. Ohkado *et al.*, 'Electro-olfactogram (EOG) in olfactometry.' *Auris Nasus Larynx*, 1989, **16**, 33–38.

77. L. J. Myers and R. Pugh, 'Thresholds of the dog for detection of inhaled eugenol and benzaldehyde determined by electroencephalographic and behavioral olfactometry.' *Am. J. Vet. Res.*, 1985, **46**, 2409–2412.

78. L. C. Uraih and R. R. Maronpot, 'Normal histology of the nasal cavity and application of special techniques.' *Environ. Health Perspect.*, 1990, **85**, 187–208.

79. D. B. Peele, S. D. Allison and K. M. Crofton, 'Learning and memory deficits in rats following exposure to 3,3'-iminodipropionitrile.' *Toxicol. Appl. Pharmacol.*, 1990, **105**, 321–332.

80. J. W. Harding, T. V. Getchell and F. L. Margolis, 'Denervation of the primary olfactory pathway in mice. V. Long-term effect of intranasal $ZnSO_4$ irrigation on behavior, biochemistry and morphology.' *Brain Res.*, 1978, **140**, 271–285.

81. S. L. Eustis, S. B. Haber, R. T. Drew *et al.*, 'Toxicology and pathology of methyl bromide in F344 rats and B6C3F1 mice following repeated inhalation exposure.' *Fundam. Appl. Toxicol.*, 1988, **11**, 594–610.

82. M. E. Hurtt, K. T. Morgan and P. K. Working, 'Histopathology of acute toxic responses in selected tissues from rats exposed by inhalation to methyl bromide.' *Fundam. Appl. Toxicol.*, 1987, **9**, 352–365.

83. C. Eriksson and E. B. Brittebo, 'Metabolic activation of the olfactory toxicant, dichlobenil, in rat olfactory microsomes: comparative studies with *p*-nitrophenol.' *Chem. Biol. Interact.*, 1995, **94**, 183–196.

84. D. B. Warheit, D. P. Kelly, M. C. Carakostas *et al.*, 'A 90-day inhalation toxicity study with benomyl in rats.' *Fundam. Appl. Toxicol.*, 1989, **12**, 333–345.

85. M. E. Hurtt, C. A. Mebus and M. S. Bogdanffy, 'Investigation of the effects of benomyl on rat nasal mucosa.' *Fundam. Appl. Toxicol.*, 1993, **21**, 253–255.

86. B. A. Gaskell, P. M. Hext, G. H. Pigott *et al.*, 'Olfactory and hepatic changes following inhalation of 3-trifluoromethyl pyridine in rats.' *Toxicology*, 1988, **50**, 57–68.

87. B. A. Gaskell, P. M. Hext, G. H. Pigott *et al.*, 'Olfactory and hepatic changes following a single inhalation exposure of 3-trifluoromethyl pyridine in rats: concentration and temporal aspects.' *Toxicology*, 1990, **62**, 35–51.

88. P. C. Chan, S. L. Eustis, J. E. Huff *et al.*, 'Two-year inhalation carcinogenesis studies of methyl methacrylate in rats and mice: inflammation and degeneration of nasal epithelium.' *Toxicology*, 1988, **52**, 237–252.

89. R. L. Melnick, J. H. Roycroft, B. J. Chou *et al.*, 'Inhalation toxicology of isoprene in F344 rats and B6C3F1 mice following two-week exposures.' *Environ. Health Perspect.*, 1990, **86**, 93–98.

90. J. K. Dunnick, D. G. Graham, R. S. H. Yang *et al.*, 'Thirteen-week toxicity study of n-hexane in B6C3F1 mice after inhalation exposure.' *Toxicology*, 1989, **57**, 163–172.

91. W. T. Stott, J. T. Young, L. L. Calhoun *et al.*, 'Subchronic toxicity of inhaled technical grade 1,3-dichloropropene in rats and mice.' *Fundam. Appl. Toxicol.*, 1988, **11**, 207–220.

92. H. Ghantous, L. Dencker, J. Gabrielsson *et al.*, 'Accumulation and turnover of metabolites of toluene and xylene in nasal mucosa and olfactory bulb in the mouse.' *Pharmacol. Toxicol.*, 1990, **66**, 87–92.

93. C. M. Keenan, D. P. Kelly and M. S. Bogdanffy, 'Degeneration and recovery of olfactory epithelium following inhalation of dibasic esters.' *Fundam. Appl. Toxicol.*, 1990, **15**, 381–393.

94. M. S. Bogdanffy, H. W. Randall and K. T. Morgan, 'Biochemical quantitation and histochemical localization of carboxylesterase in the nasal passages of the Fischer-344 rat and B6C3F1 mouse.' *Toxicol. Appl. Pharmacol.*, 1987, **88**, 183–194.

95. B. A. Trela, S. R. Frame and M. S. Bogdanffy, 'A microscopic and ultrastructural evaluation of dibasic esters (DBE) toxicity in rat nasal explants.' *Exp. Mol. Pathol.*, 1992, **56**, 208–218.

96. P. M. Mattes and W. B. Mattes, 'α-Naphthyl butyrate carboxylesterase activity in human and rat nasal tissue'. *Toxicol. Appl. Pharmacol.*, 1992, **114**, 71–76.

97. B. S. Schwartz, R. L. Doty, C. Monroe *et al.*, 'Olfactory function in chemical workers exposed to acrylate and methacrylate vapors.' *Am. J. Public Health*, 1989, **79**, 613–618.

98. B. S. Schwartz, D. P. Ford, K. I. Bolla *et al.*, 'Solvent-associated decrements in olfactory function in paint manufacturing workers.' *Am. J. Ind. Med.*, 1990, **18**, 697–706.

99. B. Sandmark, I. Broms, L. Lofgren *et al.*, 'Olfactory function in painters exposed to organic solvents.' *Scand. J. Work Environ. Health*, 1989, **15**, 60–63.

100. P. Gagnon, D. Mergler and S. Lapare, 'Olfactory adaptation, threshold shift and recovery at low levels of exposure to methyl isobutyl ketone (MIBK).' *Neurotoxicology*, 1994, **15**, 637–642.

101. D. Mergler and B. Beauvais, 'Olfactory threshold shift following controlled 7-hour exposure to toluene and/or xylene.' *Neurotoxicology*, 1992, **13**, 211–215.

102. J. M. Benson, D. G. Burt, R. L. Carpenter *et al.*, 'Comparative inhalation toxicity of nickel sulfate to F344/N rats and B6C3F1 mice exposed for twelve days.' *Fundam. Appl. Toxicol.*, 1988, **10**, 164–178.

103. W. Torjussen, L. A. Solberg and A. C. Hogeveit, 'Histopathological changes of the nasal mucosa in active and retired nickel workers.' *Br. J. Cancer*, 1979, **40**, 568–580.

104. F. W. Sunderman, Jr., in 'Clinical Chemistry and Chemical Toxicology of Metals,' Proceedings of the first international symposium organized by the Commission in Toxicology, IUPAC Section on Clinical Chemistry held at Monte Carlo, ed. S. S. Brown, Elsevier, Amsterdam, 1977, pp. 231–260.

105. J. R. Bucher, M. R. Elwell, M. B. Thompson *et al.*, 'Inhalation toxicity studies of cobalt sulfate in F344/N rats and B6C3F1 mice.' *Fundam. Appl. Toxicol.*, 1990, **15**, 357–372.

106. J. D. Sun, A. R. Dahl, N. A. Gillett *et al.*, 'Two-week, repeated inhalation exposure of F344/N rats and B6C3F1 mice to ferrocene.' *Fundam. Appl. Toxicol.*, 1991, **17**, 150–158.

107. R. J. Wood, in 'Nervous System Toxicology,' ed. C. L. Mitchell, Raven Press, New York, 1982, pp. 199–212.

108. C. S. Rose, P. G. Heywood and R. M. Costanzo, 'Olfactory impairment after chronic occupational cadmium exposure.' *J. Occup. Med.*, 1992, **34**, 600–605.

109. L. Hastings and T. J. Sun, 'Effects of cadmium on the rat olfactory system.' *Ann. NY Acad. Sci.*, 1987, **510**, 355.

110. J. Gottofrey and H. Tjalve, 'Axonal transport of cadmium in the olfactory nerve of the pike'. *Pharmacol. Toxicol.*, 1991, **69**, 242–252.

111. L. Hastings and J. Evans, 'Transaxonal transport of cadmium in the olfactory system.' *Chem. Senses*, 1988, **13**, 696.

112. R. O. Beauchamp, Jr., J. S. Bus, J. A. Popp *et al.*, 'A critical review of the literature on hydrogen sulfide toxicity.' *Crit. Rev. Toxicol.*, 1984, **13**, 25–97.

113. A. A. Khan, M. M. Schuler, M. G. Prior *et al.*, 'Effects of hydrogen sulfide exposure on lung mitochondrial respiratory chain enzymes in rats.' *Toxicol. Appl. Pharmacol.*, 1990, **103**, 482–490.

114. T. L. Guidotti, 'Occupational exposure to hydrogen sulfide in the sour gas industry: some unresolved issues.' *Int. Arch. Occup. Environ. Health*, 1994, **66**, 153–160.

115. M. A. M. Turk, W. Flory and W. G. Henk, 'Chemical modulation of 3-methylindole toxicosis in mice: effect on bronchiolar and olfactory mucosal injury.' *Vet. Pathol.*, 1986, **23**, 563–570.

116. W. M. Haschek, C. C. Morse, M. R. Boyd *et al.*, 'Pathology of acute inhalation exposure to 3-methyl-furan in the rat and hamster.' *Exp. Mol. Pathol.*, 1983, **39**, 342–354.

117. B. W. Ache, in 'Smell and Taste in Health and Disease,' eds. T. V. Getchell, R. L. Doty, L. M. Bartoshuk *et al.*, Raven Press, New York, 1991, pp. 3–18.

118. S. K. Pixley and R. Y. K. Pun, 'Cultured rat olfactory neurons are excitable and respond to odors.' *Brain Res. Dev. Brain Res.*, 1990, **53**, 125–130.

119. S. K. Pixley, 'CNS glial cells support *in vitro* survival, division, and differentiation of dissociated olfactory neuronal progenitor cells.' *Neuron*, 1992, **8**, 1191–1204.

120. M. I. Chuah, S. David and O. Blaschuk, 'Differentiation and survival of rat olfactory epithelial neurons in dissociated cell culture.' *Brain Res. Dev. Brain Res.*, 1991, **60**, 123–132.

121. S. K. Pixley, M. Bage, D. Miller *et al.*, 'Olfactory neurons *in vitro* show phenotypic orientation in epithelial spheres.' *Neuroreport*, 1994, **5**, 543–548.

122. F. Miller, 'Nasal toxicity and dosimetry of inhaled xenobiotics: implications for human health.' *Inhal. Toxicol.*, 1994, **6**, 1.

123. C. B. Frederick, J. B. Morris, J. S. Kimbell *et al.*, 'Comparison of four biologically-based dosimetry models for the rodent nasal cavity to describe the deposition and distribution of rapidly metabolized vapors.' *Inhal. Toxicol.*, 1994, **6**(Suppl.), 135–157.

124. D. P. Perl and P. F. Good, 'Uptake of aluminum into central nervous system along nasal-olfactory pathways.' *Lancet*, 1987, **1**, 1028.

125. J. R. Harkema, 'Comparative aspects of nasal airway anatomy: relevance to inhalation toxicology.' *Toxicol. Pathol.*, 1991, **19**, 321–336.

11.16
The Developing Nervous System

WILLIAM SLIKKER, JR.

National Center for Toxicological Research, Jefferson, AR, USA

11.16.1 INTRODUCTION

The cytoarchiterture of the nervous system exhibits an intricate and interconnective pattern and is more complex than most organ systems. Not only does the nervous system consists of a number of specialized cell types of various structures and functions, but the majority of these cells are contained within a specialized, semipermeable barrier that separates them from other cells of the body. In addition, the developing nervous system cells and their interactions undergo a well orchestrated maturation.

The developing nervous system possesses certain characteristics that may alter its susceptibility to toxicants as compared to the adult. In the developing nervous system, where diverse cell types are differentiating from various primordial cells, the opportunity for differential toxicant effects is enhanced compared to mature, generally nondividing cells of the adult nervous system. In addition to the blood–brain barrier, during prenatal mammalian development the placenta provides another semipermeable barrier to toxicant exposure. Due to placental growth and maturation, the influence of the placenta changes with gestational age. Therefore, the developing nervous system poses a unique challenge to toxicologists in that it can be more or less sensitive to toxicants responsible for damage to other organ systems in the adult or developing animal.

11.16.2   DEVELOPMENTAL PERIODS OF SUSCEPTIBILITY

For developmental toxicity in general, the nature and extent of neurotoxic effects are often dependent on the timing of exposure. As stages of nervous system development vary significantly between species in relation to the time of birth, variations in neurotoxic outcome across species are to be expected. One of the critical periods of susceptibility when many organ systems are formed, organogenesis, varies in time from conception and overall duration from species to species. Organogenesis in the mouse is considered to occur from days 7–16 of gestation, in the monkey from days 20–45, and in the human from days 20–55. Species differences also exist in the various stages of brain growth and development. It is imperative in development neurotoxicology studies, therefore, that the time and duration of exposure in the animal model be selected to match the window of exposure in the human situation.[1]

11.16.3   ANATOMICAL AND PHYSIOLOGICAL CHARACTERISTICS

Between the site of maternal administration and the nervous system of the conceptus are two specialized sets of membranes that chemicals must cross. Both the placenta and the fetal blood–brain barrier have their origins in the conceptus and have anatomical and functional features that influence chemical transfer to the developing conceptus. These two sets of membranes share the additional features that both undergo considerable change with development and also exhibit substantial species differences.

11.16.3.1   Placenta

Morphologically, the placenta may be defined as the fusion or apposition of fetal membranes to the uterine mucous membrane.[2] As in many species, the human actually exhibits two placentas, the early yolk sac placenta (day 13–80 of gestation) which overlaps developmentally with the later chorioallantoic placenta (day 21 to term).[3] The human yolk sac placenta consists of endoderm (inner lining) and mesothelium (outer lining), both possessing microvilli. The interposed mesenchyme contains vessels and blood-forming tissues from day 16 onward. This temporary but physiologically important structure is believed to have several absorptive and secretory functions. Differences between the yolk sac placenta of the human and that of the rat include an opposite layering of the endoderm and mesothelium in the rat inverted visceral yolk sac, and absence of direct contact between the human yolk sac and the maternal circulation as is observed in the rat.[3]

The human chorioallantoic placenta consists of two morphologically and physiologically distinguished cell layers; the mononucleated, inner layer of cytotrophoblasts which produces the multinucleated, outer layer of syncytiotrophoblasts.[3] During the first trimester, the placenta develops fetal cotyledons which consist of a special maternal artery that discharges blood into loose centers of fetal lobules with numerous villi. These villi arise from the cytotrophoblast cell proliferation into the maternal decidual tissue. As pregnancy progresses, the syncytiotrophoblast layer (in direct contact with maternal blood and relatively thick in early pregnancy) becomes progressively thinner.[3] The cytotrophoblast layer becomes more discontinuous and the endothelium of the embryo/fetal vessels within the villi becomes thinner. Thus, as pregnancy progresses, there is closer contact between the fetal blood and the syncytiotrophoblast, that placental cell layer most important to placental function and maternal-embryo/fetal exchange. On the macroscopic level, the effects of gestational age can be exemplified (Figure 1) by comparing the tremendous change in the ratio of placental/fetal weight (ratio equals 4 at 10 weeks and 0.2 at 40 weeks).[4]

Based merely on the number of layers of cells between fetal and maternal circulations, it is tempting to define similarities and differences in placental transfer of chemicals by anatomy alone. Neither the function nor the anatomical thickness of the placenta, however, is consistently related to the number of layers separating the conceptus from the mother. Importantly, any substance in the maternal circulation can, to some extent, be transferred across the placenta unless it is metabolized or detoxified before or during its placental passage.[3,5]

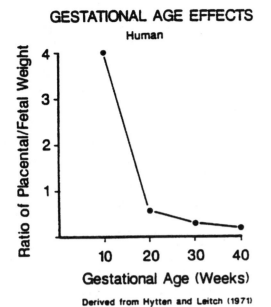

GESTATIONAL AGE EFFECTS

Human

Derived from Hytten and Leitch (1971)

Figure 1 The influence of gestational age on the ratio of placental/fetal weight in the human.

One may define species differences from a functional perspective by focusing on the placental transfer of a series of nonmetabolized, model compounds of increasing molecular weight by the isolated, dually-perfused placenta. Placental transfer, as defined by a clearance ratio, decreases as the molecular weight of the model compounds increases.[6] Under these very controlled conditions without maternal or fetal involvement, differences in placental transfer between species are evident, with sheep being more different from human than is the guinea pig. Thus, while anatomical differences alone may not fully predict species differences in rates of placental transfer of chemicals, functional assessments have been shown to be useful in defining the exposure of the conceptus after maternal dose administration.

11.16.3.2 Blood–Brain Barrier

With the exception of the circumventricular organs, the central nervous system (CNS) and cerebrospinal fluid (CSF) are isolated from the blood by the blood–brain barrier. The choroid plexus, the blood vessels of the brain and subarachnoid membrane are the interfaces between the blood and the CSF and brain.[7] Peripheral nerves also possess a barrier composed of the blood vessels that surround nerve bundles. Intercellular diffusion is restricted at all these barrier sites by cells with tight junctions. The cells of the blood vessels are endothelial, whereas those of the choroid plexus and arachnoid and perineurium in peripheral nerves are epithelial.[7] These barrier cells are connected

together by tight junctions that block paracellular diffusion and contain few transfer vesicles.[8] Due to these features, blood-borne solutes must course through both endothelial and epithelial cell membranes as well as the cytoplasm of the endothelial cells before reaching the extracellular space of the brain. Molecules may pass through membranes by passive diffusion as a function of the lipid solubility of their unionized form. Ionized forms of these molecules do not transverse membranes readily, but are in equilibrium with their unionized forms, based on the pH of the environment and their dissociation constants. Additional regulators of chemical blood–brain transfer include astrocytic processes that ensheath capillaries within the CNS, specific carriers located on cell membranes (e.g., neurotransmitter uptake sites), and cytoplasmic metabolic enzymes.[7–9]

11.16.4 CHEMICAL EXPOSURE

A chemical must pass through and may interact with several anatomical compartments on its journey from the site of maternal administration to the nervous system of the developing conceptus. Therefore, these primary anatomical sites will be used as focal points for discussion of the principles of pharmacokinetics as they relate to developmental neurotoxicants.

11.16.4.1 Maternal Considerations

One of the primary reasons for applying pharmacokinetic principles to the study of developmental neurotoxicology is to determine the concentration of chemicals at the suspected site of action as a function of time. Maternal factors act to either enhance or diminish the concentration of an active chemical in the nervous system of the conceptus. For example, maternal detoxification would tend to decrease the concentration of the active agent, whereas maternal bioactivation would enhance its concentration. Maternal factors including absorption, distribution, serum binding, and elimination also influence the concentration of active agent at the target site. Due to the physiological changes that occur during pregnancy, the influence of these maternal factors on chemical delivery may also change during gestation.[10–14] These issues have been reviewed.[13,15–18]

11.16.4.2 Placental Transfer

Factors which influence the placental transfer of chemicals include uterine–placental blood flow, placental permeability, and placental metabolism.[19–23] These factors are not static during pregnancy but may change as gestation progresses. In addition, placental morphology varies among species so that interspecies extrapolation is not always straightforward. Several reviews have discussed the factors affecting placental transfer of chemicals.[21–24] The following discussion will focus on how these factors may influence the delivery of chemicals to the nervous system of the conceptus.

11.16.4.2.1 Blood flow

Chemical delivery to the developing conceptus relies primarily on blood flow to the placenta. Although chemicals may transfer from mother to fetus via the amniotic fluid after crossing the amnion, the majority of agents gain access to the conceptus via placental passage.[25] In addition to the changes in placental blood flow that occur during gestation, changes in blood flow as a result of chemical exposure must also be considered. Experimentally induced changes in blood flow have been shown to alter normal development of the conceptus.[26–28]

11.16.4.2.2 Permeability

Placental permeability to a chemical is influenced both by placental characteristics such as thickness, surface area, carrier systems, and lipid/protein content of the membranes, and by characteristics of the chemical agent such as degree of ionization, lipid solubility, protein binding, and molecular weight.[19,22,24] It is now recognized that almost any maternally administered compound has the potential to cross the placenta. The rate of placental transfer is rapid for nonionized, lipid-soluble chemicals of low molecular weight (less than 1000) and is largely controlled by placental blood flow.[19,22] However, charged molecules, such as tubocurarine, have also been shown to enter the fetus.[29,30] Likewise, chemicals that are highly ionized at normal blood pH, such as the salicylates, readily cross the placenta.[31] The question is, thus, not whether a compound crosses the placenta, but rather at what rate.

The degree of plasma protein binding of a chemical also influences its rate of placental transfer because it is generally only the free drug that crosses the membranes of the placenta.[17,24] Protein binding is usually reversible and there is a finite number of binding sites; thus, binding is saturable and equilibrium may be described by the law of mass action.[17,20] As long as binding is reversible, it does not prevent the chemical from crossing membranes, but only slows the rate at which the transfer occurs.[32]

11.16.4.2.3 Biotransformation

Chemical biotransformation has been subdivided into two major phases. Phase I reactions include oxidation, reduction, hydrolysis, or other reactions which modify chemicals in such a way that phase II reactions may occur. Phase II reactions involve the conjugation of chemicals with an endogenous moiety, such as glucuronic acid or sulfate.[33] The overall effect of these biotransformations is to increase chemical polarity (i.e., water solubility), resulting in more rapid excretion, biological inactivation, or both. In some cases, however, biotransformation results in the formation of a more potent or toxic agent capable of producing deleterious effects (e.g., cholestasis produced by estradiol-17β-glucuronide).

Placental metabolism may be the most critical of several factors influencing the delivery of chemicals to the developing conceptus. Placental biotransformation of a chemical occurring prior to fetal delivery may dramatically alter the chemical profile observed in the conceptus from that in the maternal organism.[34] Equilibrium factors, which may clearly influence the rate of placental transfer, can result in quantitative differences of exposure; placental metabolism, however, can qualitatively alter the exposure of the conceptus to potentially toxic chemicals.

Placental metabolism is less well characterized than hepatic metabolism, but existing data suggest that placental metabolic capacity is severely limited compared to adult hepatic capacity.[19,35] A listing of some of the human placental xenobiotic and hormone metabolizing enzymes or isoenzymes is presented in Table 1.

11.16.4.3 Embryo/Fetal Considerations

The biotransformation of chemicals affects the concentration of biologically active or inactive compounds at the cellular locus of action. Knowledge of the activities of chemical-metabolizing enzymes in various tissues aids in the understanding of tissue susceptibilities and in the ability to extrapolate between species. As with most organ systems during development, however, the various chemical-metabolizing systems undergo quantitative if not qualitative changes. The ontogeny and regulatory factors

Table 1 Partial list of human placental xenobiotic and hormone metabolizing enzymes or isoenzymes.

Phase	Type	Reaction (gene)	Substrate	Constitutive	Inducer
I	MFO	*O*-de-ethylase (CYP1A1)	7-Ethoxycoumarin	(+)	Cigarette smoke
I	MFO	Aryl-hydrocarbon hydroxylase (CYP1A1)	PAH	(?)	Cigarette smoke
I	MFO	Hydroxylase (CYP3A7)	Cortisol	+	
I	MFO	Aromatase (CYP19)	Androgens	(+)	
I	MFO	Cholesterol side chain cleavage (CYP11A)	Cholesterol	(+)	
I	MFO	Estrogen catechol formation, 2-hydroxylation and/or 4-hydroxylation	Estrogens	(+)	Cigarette smoke
I	MFO	25-Hydroxycholecalciferol hydroxylase	25-Hydroxy-cholecalciferol		
I	Oxidoreductase	17β-Hydroxydehydrogenase	Estradiol/estrone	(+)	
I	Oxidoreductase	11β-Hydroxydehydrogenase	Cortisol/cortisone	(+)	
I	Oxidation	Dehydrogenase	Alcohol/acetaldehyde	(+)	
I	Oxidation	Monoamine	Norepinephrine	(+)	
II	Sulfatase	Sulfate cleavage	Steroid sulfates	(+)	
II	Conjugation	Glutathione-*S*-transferase	Epoxides	(+)	
II	Conjugation	Catechol-*O*-methyl-transferase	Catecholamines, catechol estrogens	(+)	

Source: Slikker and Miller.[14,18]
MFO, mixed function oxidase; PAH, polycyclic aromatic hydrocarbon.

of chemical metabolism have been reviewed and the expression of fetal cytochrome P450 isozymes is believed to be under both developmental and tissue specific regulatory factors.[18,36–38]

11.16.4.3.1 Delivery

The majority of chemicals entering the fetal circulation do so via the umbilical vein after passage through the placenta. A portion of the blood flow entering the liver of the conceptus is shunted via the ductus venosus directly to the inferior vena cava and to the heart for total body distribution.[39] As in the adult, approximately 16% of the fetal cardiac output is directed toward the fetal brain.[40] The remaining umbilical flow enters hepatic tissue and exits to the vena cava via the portal vein.[41] Therefore, a fraction of the blood-borne chemical first passes through the developing liver, while another fraction passes directly to the remaining tissues, including the nervous system of the developing conceptus.[42] Even with this large blood flow, as demonstrated in the fetal mon-

key, brain exposure to certain chemicals, such as dideoxynucleosides, may be limited compared to other fetal organs.[43]

11.16.4.3.2 Biotransformation

Chemical biotransformation by the developing conceptus has been extensively reviewed.[5,13,36,37,44–51] Despite the fact that data have been collected using a variety of techniques and some "data gaps" exist because of technical or ethical reasons, several general conclusions may be drawn from the literature: (i) during prenatal development, the activities of most, but not all, systems which catalyze both phase I and II reactions are lower than in the respective adult of the species; (ii) as in the adult, the conceptus exhibits substrate specificity in its ability to metabolize chemicals, suggesting the existence of several sets of enzymes or isozymes which may or may not be the same as in the adult; (iii) these enzyme systems may be inhibited or induced by pretreatment with a variety of chemicals; (iv) enzyme activity generally increases with gestational age; (v) the

Table 2   Some cytochrome P450 (CYP) isozymes and their occurrence in human tissues.

CYP	Adult liver	Adult brain	Fetal liver	Fetal brain
1A1	+	+	+	
1A2	+	+		
2A6/2A7	+			
2B1/2B2	+	+		
2B6/2B7	+			
2C8–19	+		+	+
2D6	+	+	+	
2E1	+	+	+	
2F1				
3A3/3A4	+	+	+	
3A7	+		+	
4B1				

Source: Murray *et al.*,[52] Farin and Omiecinski,[53] Maenpaa *et al.*,[54] Hakkola *et al.*,[55] Ravindranath and Boyd.[56]

ontogeny of each individual enzyme may be different and the controlling mechanisms of maturation of enzyme activity are incompletely understood; (vi) prenatal human and nonhuman primates exhibit higher levels of many metabolizing enzymes (especially P450s) than do other commonly used laboratory species; (vii) as in the adult, the liver of the conceptus appears to have the greatest capacity for chemical metabolism, although the fetal adrenal, kidney, lung and brain also exhibit metabolic capabilities. Table 2 summarizes data concerning some cytochrome P450 (CYP) isozymes and their occurrence in human tissues.

11.16.4.3.3   Elimination

Just as placental transfer of chemicals is the predominate pathway from the maternal system to the conceptus, placental transfer is also the predominate route for embryo/fetal chemical elimination, and the same pharmacokinetic rules apply; those agents which are nonionized and lipid soluble will leave the conceptus, determined by the concentration gradient from conceptus to mother. If, however, a chemical has been conjugated by the fetus (e.g., glucuronidation, sulfation, etc.) or otherwise metabolized to a more polar form, the rate of return to the maternal circulation will be slower than that for the parent compound.[57–60] Physiologically based pharmacokinetic models to predict the fate of neurotoxicants in fetal brain regions as a function of development are emerging.[61,62]

11.16.5   NERVOUS SYSTEM CONSIDERATIONS

The blood–brain barrier, much like the placenta, is not an impermeable barrier but rather a regulatory interface between blood and the nervous system. The regulatory function of this interface is superimposed on baseline permeability restrictions and controls the immediate environment of the glial and neuronal cells.[7] The description of the role of the CSF, choroid plexus and cerebrovascular endothelium in the regulatory capacity of the blood–brain barrier is beyond the scope of this chapter. With a focus on developmental neurotoxicity, there are two areas of special concern: the circumventricular organs (CVOs) of the brain that lie outside the normal protection of the blood–brain barrier, and the developmental changes that alter susceptibility to neurotoxicant exposure.

The CVOs are highly vascularized structures that are in contact with the brain median ventricular system and, therefore, CSF, but which lack the usual protection of the blood–brain barrier. These structures include the subfornical organ, pineal gland, median eminence, choroid plexus, suprapineal recess, neurohypophysis, area postrema, subcommisural organ, and supraoptic crest.[63] Agents that generally do not have access to most brain areas because of the blood–brain barrier pass readily into the CVOs due, at least in part, to the fenestrated endothelial walls and widened perivascular spaces of the CVO blood vessels.[64] Several agents, such as glutamate and 6-hydroxydopamine, that normally require direct brain injection into brain to produce neurotoxicity have been shown to exhibit CVO damage when administered systemically.[65,66]

The ontogeny of the blood–brain barrier is complicated in that several of its features mature at different rates. These include the tight junctions between endothelial cells of the cerebral blood vessels, astrocytic processes or end-feet that may completely ensheath cerebral capillaries, several transport mechanisms (e.g., small ion, glucose, amino acids, certain monocarboxylic acids, deoxynucleosides, vitamins, etc.) and metabolism by endothelial and choroid plexus enzymes.[67] Determination of the impact of each of these features on the functionality of the blood–brain barrier during development is confounded by the concomitant, dramatic changes in brain mass, and in CSF volume and its turnover.[7]

One of the major components of the blood–brain barrier, the tight junctions between endothelial cells, appears early in fetal development and does not subsequently undergo

Disposition of Chemicals during Pregnancy

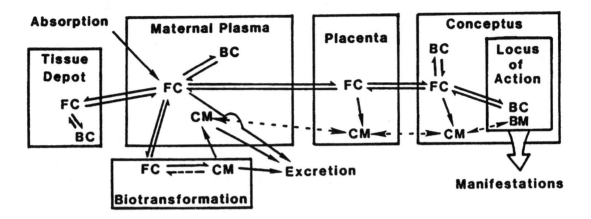

C = Chemical; F = Free; B = Bound; M = Metabolite.

Figure 2 A compartmental model describing the disposition of chemicals during pregnancy.

detectable changes.[67] In contrast, the development of astrocytic processes appears to correlate with the physiological data reflecting maturation of barrier function.[68] Certain transport functions also exhibit a maturational component. The selective, carrier-mediated transport of 19 amino acids was studied in the developing rat. The transport systems were present in the early neonatal rat, but entry rates for most of the amino acids decreased with age.[69] The capacity to move lactate across the blood–brain barrier in the neonatal rat is several fold greater than in the adult.[70]

In contrast, the carrier-mediated influx of glucose into rat brain is low at birth and increases to maximal levels in the young adult.[71] Although little is known about developmental changes, there is evidence for two nucleoside transport systems: a facilitated diffusion system for certain nucleosides excluding thymidine, deoxyuridine, and deoxycytidine at the blood–brain barrier; and an active transport system at the blood–CSF barrier (the choroid plexus) for thymidine and deoxycytidine.[43,72,73] Enzymes such as NADH tetrazolium reductase and ATPase increase and alkaline phosphatases decrease in cerebral capillaries with fetal development. Both dihydroxyphenylalanine (DOPA)-decarboxylase and monoamine oxidase are present in the vascular endothelium of the immature brain and become more effective with age.[7] In general, little information is available concerning the ontogeny of many features of the blood–brain barrier and choroid plexus because of the technical problems associated with studying these phenomena. The available information does suggest that, with the exception of the tight junctions, several features of the blood–brain barrier responsible for controlling brain exposure to chemicals change with maturation. Whether these maturational changes alter CNS susceptibility to neurotoxicants, however, are issues that will continue to be determined on a case-by-case basis until a more comprehensive database is available.

11.16.6 RESEARCH STRATEGIES FOR DEVELOPMENTAL NEUROTOXICANTS

The overall strategy to understand developmental neurotoxicity is based on several assumptions: (i) the developing CNS may be more or less susceptible to neurotoxic insult than the adult CNS depending on the stage of development, and (ii) neuropathological, neurochemical, neurophysiological, and behavioral evaluations or endpoints are necessary and complementary approaches to determining the type(s) and degree(s) of CNS toxicity. The research approach consists of the following four steps: (i) gather information from all available endpoints and to use it to generate a developmental neurotoxicity profile; (ii) correlate structural and chemical lesions with overt behavioral manifestations of neurotoxicity; (iii) compare with adult neurotoxicity profiles to determine relative susceptibility of the developing organism, and (iv) develop pharmacokinetic/metabolic bases for interspecies extrapolation (Figure 2).

There are numerous examples in the literature that indicate that exposure of the pregnant animal to some chemicals, such as certain pesticides, drugs, or even vitamins, will result in neurotoxicity in the offspring, but not necessarily the mother.[74] In other cases, the adult of the species has proved more susceptible to agents, such as trimethyltin and methylenedioxymethamphetamine (MDMA), than the developing animal.[75,76] It has also been proposed that perinatal exposure to certain agents may result in silent toxicity that manifests itself with aging or environmental challenge.[77]

11.16.7 DIFFERENTIAL NEUROTOXICITY DURING DEVELOPMENT: SELECTED EXAMPLES

11.16.7.1 Methylenedioxymethamphetamine

In order to exemplify the principles of developmental neurotoxicology outlined above, several examples will be presented. MDMA is an amphetamine derivative and a popular drug of abuse.[78,79] Administration of this amphetamine derivative to adult laboratory animals produces a set of highly reproducible long-term effects. The long-term effects of MDMA exposure include persistent reductions in concentrations of 5-hydroxytryptamine (serotonin, 5-HT) and its metabolite, 5-hydroxyindoleacetic (5-HIAA), persistent losses of tryptophan hydroxylase activity, the rate limiting enzyme for 5-HT synthesis, and loss of the 5-HT reuptake transporter.[80-88] Neurohistological examination of laboratory animals exposed to MDMA reveals a loss of serotonergic immunoreactivity in the forebrain.[89-91] Thus, it is hypothesized that MDMA exposure results in a selective ablation of serotonergic reuptake sites on nerve terminals.

Specific facets of serotonergic and dopaminergic function play a role in the expression of the persistent serotonergic alterations induced by MDMA exposure. If MDMA-induced 5-HT release is blocked by the administration of the serotonin reuptake inhibitor fluoxetine, the serotonergic neurotransmitter system is spared from the neurotoxic effects of MDMA.[82,83,92,93] In general, treatments that enhance dopaminergic release tend to increase the neurotoxic effect of MDMA on the serotonergic neurotransmitter system and treatments that reduce dopaminergic outflow tend to decrease neurotoxicity.[94]

Both the serotonergic and dopaminergic neurotransmitter systems begin synaptogenesis during late prenatal development in the rat. This developmental period, which continues through weaning, exhibits increasing concentrations of 5-HT and dopamine (DA) as well as their respective receptor and reuptake systems.[95,96] Therefore, at birth, concentrations of these neurotransmitters are low, with whole brain concentrations being less than 10% of the adult values. Concentrations of 5-HT and DA increase rapidly during the first four postnatal weeks, being preceded by the development of the rate-limiting enzymes responsible for their synthesis, tryptophan hydroxylase and tyrosine hydroxylase, respectively.[97-99] As concentrations of 5-HT and DA increase, so do populations of their respective pre- and postsynaptic receptors and high-affinity reuptake sites.[100-102]

As 5-HT$_2$ receptors, 5-HT reuptake sites, and both 5-HT and DA concentrations are implicated in MDMA-induced serotonergic neurotoxicity it, therefore, seems likely that the effects of MDMA on the serotonergic neurotransmitter system would vary with developmental status.[103] In support of this hypothesis, a potential role for developmental age as a modulator of MDMA-induced serotonergic neurotoxicity was recognized when multiple doses of MDMA administered during early and midgestation in the rat had no effect on the postnatal neurochemical development of the serotonergic neurotransmitter system.[75] Similarly, p-chloroamphetamine and fenfluramine, chemical relatives of MDMA, do not produce long-term deficits in 5-HT concentrations when administered to the rat during early postnatal development.[104-106]

Dose–response data reported by Broening et al.[107] demonstrates that the serotonergic neurotransmitter system is not sensitive to the MDMA-induced loss of 5-HT content until approximately postnatal day 30 in the rat. Furthermore, 5-HT reuptake sites are not sensitive to reduction after MDMA administration until approximately postnatal day 40. This data supports the hypothesis that the development of the sensitivity of the serotonergic neurotransmitter system to MDMA-induced neurotoxicity is a postnatal phenomenon. It is surprising to note, however, how late in the development of the rat that the serotonergic neurotransmitter system became susceptible to the neurotoxic effects of MDMA. It is generally thought that the neurochemical and morphological development of the serotonergic neurotransmitter system in the rat is almost complete by 25–30 d of age.[95,98,100-102,108,109] Thus, it can be concluded that the development of sensitivity to MDMA-induced serotonergic neurotoxicity does not incrementally follow the ontogeny of the serotonergic neurotransmitter system. Instead, the sensitivity of these serotonergic neurotransmitter systems to MDMA-induced neurotoxicity becomes apparent only after the majority of its development has occurred.

11.16.7.2 Retinoids

The isomer of vitamin A, *cis*-retinoic acid (isotretinoin) produces a variety of congenital defects involving craniofacial and CNS structures in human infants exposed to this agent *in utero*.[110] No toxicity of this nature is observed in the adult. The effects in infants and children of prenatal exposure to *cis*-retinoic acid have been described and include mental retardation, and learning and attention deficits.[111] The results of this clinical study and others using animal models suggest that developmental CNS toxicants can produce neuropsychological deficits in the absence of anatomically detectable pathology in the CNS.[112] However, more sophisticated neurohistological techniques, when appropriately applied, may uncover hitherto undetected lesions.

Studies in a variety of animal models have corroborated the human findings. Early studies with pregnant rats and monkeys demonstrated vitamin A (all-*trans*-retinoic acid) to be a developmental neurotoxicant over a decade before clinical cases were reported.[113,114] Human and monkey studies have described embryo/fetal exposure to vitamin A and metabolites after maternal administration of either *cis*-retinoic or all-*trans*-retinoic acid.[115,116] Although fetal brain concentrations have not been systematically evaluated, preliminary data indicate that the brain of the early monkey fetus is a target for maternally administered all-*trans*-retinoic acid.[115] Although not completely understood, the effect of retinoids on developing brain are probably mediated via complex hormone receptor-mediated gene regulation pathways to alter developmental events in the CNS.[117]

11.16.8 CONCLUSION

Other agents besides MDMA produce dramatic neurotoxicity in the adult, but have either no or lessor effects on the developing or young nervous system. A few examples would include methylphenyltetrahydropyridine (MPTP),[118] trimethyltin (TMT),[119,120] domoic acid,[121] and amphetamine.[122,123] However, a variety of controlling features combine to alter the sensitivity of the developing nervous system compared to the adult. In some cases, pharmacokinetic differences may play a role. For example, adolescent monkeys have a several fold higher clearance of domoic acid than adults, and are considerably less sensitive to its neurotoxic effects.[121,124] In other cases, the developing nervous system is exposed to measurable or adult equivalent concentrations of the toxicant (e.g., TMT) and yet little neurotoxicity is exhibited.[119] The answer may reside not with what the developing nervous system contains, but what it does not contain. As with the MDMA example, critical receptors, neurotransmitter synthesis or metabolizing enzymes and fully functioning neurotransmitter systems are frequently not present during early development and, therefore, can not be acted upon or used by the chemical to manifest its neurotoxicity. In addition, the bioactivation of a chemical may not occur in the developing brain.[43,75]

On the other hand, there are many agents such as the retinoids that produce developmental neurotoxicity, but little or no adult neurotoxicity. A partial list includes alcohol, polychlorinated biphenyls and selected metals (e.g., lead and methylmercury) at low doses.[70] As most of the above mentioned agents produce deleterious effects when administered at critical times (susceptible periods) and the toxicity is generally not reversible, they may be acting on endogenous cell signaling mechanisms. The gradient of critical signaling molecules, their receptors or second messenger systems in developing tissues may well be the target for some developmental neurotoxicants. This case has been made for retinoids and alcohol by Vorhees.[117] Another mechanism of action for developmental neurotoxicants is via apoptosis, a mode of cell death typical of physiological neuronal elimination during development.[125] Whether activated by oxidative stress,[126] excitotoxicity,[125] or alkylation of DNA,[127] apoptosis may extend beyond its normally well control role of shaper of the nervous system and remove more cells than appropriate for normal nervous system development. However, whatever the mechanism, the changing morphology and function of the developing nervous system and the genetic control on which this ontogeny is based, remain as critical characteristics that must be considered in understanding and predicting toxicity to the developing nervous system.

11.16.9 REFERENCES

1. J. G. Wilson, in 'Principles and Techniques,' eds. J. G. Wilson and J. Warkany, University of Chicago Press, Chicago, IL, 1964, pp. 251–261.

2. P. Kaufmann and B. F. King, 'Structural and Functional Organization of the Placenta,' Karger, Basel, 1982.

3. J. M. Garbis-Berkvens and P. W. Peters, in 'Pharmacokinetics in Teratogenesis,' eds. H. Nau and W. J. Scott, 1987, vol. 1, pp. 14–44

4. F. E. Hytten and I. Leitch (eds.), in 'The Physiology of Human Pregnancy,' Blackwell Scientific Publications, Oxford, 1971, p. 322.

5. W. Slikker, Jr., 'The role of metabolism in the testing of developmental toxicants.' *Regul. Toxicol. Pharmacol.*, 1987, **7**, 390–413.

6. N. P. Illsley, S. Hall, P. Penfold *et al.*, 'Diffusional permeability of the human placenta.' *Contrib. Gynecol. Obstet.*, 1985, **13**, 92–97.

7. M. D. Rapoport, 'Blood–Brain Barrier in Physiology and Medicine,' Raven Press, New York, 1976, pp. 1–86.

8. J. M. Lefauconnier, Y. Tayarani and G. Bernard, 'Blood–brain barrier permeability to excitatory amino acids.' *Adv. Exp. Med. Biol.*, 1986, **203**, 191–198.

9. D. D. Heistad, 'The blood–brain barrier.' *Fed. Proc.*, 1984, **43**, 185.

10. H. Noschel, G. Peiker, R. Voigt *et al.*, 'Research on pharmacokinetics during pregnancy.' *Arch. Toxicol.*, 1980, **4**, Suppl., 380–384.

11. A. J. Cummings, 'A survey of pharmacokinetic data from pregnant women.' *Clin. Pharmacokinet.*, 1983, **8**, 344–354.

12. M. C. Bogaert and M. Thiery, 'Pharmacokinetics and pregnancy.' *Eur. J. Obstet. Gynecol. Reprod. Biol.*, 1983, **16**, 229–235.

13. M. R. Juchau and E. Faustman-Watts, 'Pharmacokinetic considerations in the maternal–placental–fetal unit.' *Clin. Obstet. Gynecol.*, 1983, **26**, 379–390.

14. M. R. Juchau, in 'Placental Toxicology,' ed. B. V. Rama Sastry, CRC Press, Boca Raton, FL, 1995, pp. 197–212.

15. G. Levy, 'Pharmacokinetics of fetal and neonatal exposure to drugs.' *Obstet. Gynecol.*, 1981, **58**, Suppl. 5, 9S–16S.

16. J. G. Brock-Utne, J. W. Downing and E. Mankowitz, 'Drugs and the fetomaternal unit.' *S. Afr. Med. J.*, 1980, **58**(9), 366–369.

17. B. Krauer, F. Krauer and F. E. Hytten, 'Drug disposition and pharmacokinetics in the maternal–placental–fetal unit.' *Pharmacol. Ther.*, 1980, **10**, 301–328.

18. W. Slikker, Jr. and R. K. Miller, in 'Developmental Toxicology,' eds. C. A. Kimmel and J. Buelke-Sam, Raven Press, New York, 1994, pp. 245–283.

19. B. L. Mirkin and S. Singh, in 'Perinatal Pharmacology and Therapeutics,' ed. B. L. Mirkin, Academic Press, New York, 1976, pp. 1–69.

20. R. K. Miller, T. R. Koszalka and R. L. Brent, in 'The Cell Surface in Animal Embryogenesis and Development,' eds. G. Post and G. L. Nicolson, Elsevier/North-Holland Biomedical Press, New York, 1976, pp. 145–223.

21. W. J. Waddell and C. Marlowe, 'Transfer of drugs across the placenta.' *Pharmacol. Ther.*, 1981, **14**, 375–390.

22. G. W. Mihaly and D. J. Morgan, 'Placental drug transfer: effects of gestational age and species.' *Pharmacol. Ther.*, 1984, **23**, 253–266.

23. M. R. Juchau, in 'Extrahepatic Metabolism of Drugs and other Foreign Compounds,' ed. T. E. Gram, S. P. Medical and Scientific Books, New York, 1980, pp. 211–238.

24. F. Welsch, 'Placental and fetal uptake of drugs.' *J. Vet. Pharmacol. Ther.*, 1982, **5**, 91–104.

25. H. Nau, and C. Liddiard, 'in 'Role of Pharmaco-kinetics in Prenatal and Perinatal Toxicology,' eds. D. Neubert, H.-J. Merker, H. Nau *et al.*, Theime, Stuttgart, 1978, pp. 465–481.

26. F. C. Greiss, Jr. and F. L. Gobble, Jr., 'Effect of sympathetic nerve stimulation on the uterine vascular bed.' *Am. J. Obstet. Gynecol.*, 1967, **92**, 962–967.

27. M. Barr and R. L. Brent, in 'Handbook of Teratology,' eds. J. G. Wilson and F. C. Fraser, Plenum, New York, 1978, vol. 4, pp. 275–304.

28. G. Millicovsky and J. M. DeSesso, 'Differential embryonic cardiovascular responses to acute maternal uterine ischemia: an *in vivo* microscopic study of rabbit embryos with either intact or clamped umbilical cords.' *Teratology*, 1980, **22**, 335–343.

29. I. Kivalo and S. Saarikoski, 'Placental transmission and foetal uptake of ^{14}C-dimethyltubocurarine.' *Br. J. Anaesth.*, 1972, **44**, 557–561.

30. I. Kivalo and S. Saarikoski, 'Placental transfer of ^{14}C-dimethyltubocurarine during caesarean section.' *Br. J. Anaesth.*, 1976, **48**, 239–242.

31. J. G. Wilson, E. J. Ritter, W. J. Scott *et al.*, 'Comparative distribution and embryotoxicity of acetylsalicylic acid in pregnant rats and Rhesus monkeys.' *Toxicol. Appl. Pharmacol.*, 1977, **41**, 67–78.

32. R. R. Levine, 'Pharmacology Drug Actions and Reactions,' Little, Brown, Boston, MA, 1973, p. 103.

33. R. T. Williams, 'Detoxication Mechanisms: the Metabolism and Detoxication of Drugs, Toxic Substances and Other Organic Compounds,' 2nd edn. Chapman & Hall, London, 1959.

34. W. Slikker, Jr., D. E. Hill and J. F. Young, 'Comparison of the transplacental pharmacokinetics of 17β-estradiol and diethylstilbestrol in the subhuman primate.' *J. Pharmacol. Exp. Ther.*, 1982, **221**, 173–182.

35. M. R. Juchau, 'Drug biotransformation in the placenta.' *Pharmacol. Ther.*, 1980, **8**, 501–524.

36. J. E. A. Leakey, in 'Biological Basis of Detoxication,' ed. J. Caldwell, Academic Press, New York, 1983, pp. 77–103.

37. G. J. Dutton and J. E. A. Leakey, 'The perinatal development of drug-metabolizing enzymes: what factors trigger their onset?' *Prog. Drug Res.*, 1981, **25**, 189–273.

38. S. E. Eltom, J. G. Babish and W. S. Schwark, 'The postnatal development of drug-metabolizing enzymes in hepatic, pulmonary and renal tissues of the goat.' *J. Vet. Pharmacol. Ther.*, 1992, **16**(2), 152–163.

39. A. M. Rudolph and M. A. Heymann, 'The circulation of the fetus *in utero*. Methods for studying distribution of blood flow, cardiac output and organ blood flow.' *Circ. Res.*, 1967, **21**, 163–184.

40. R. E. Behrman, M. H. Lees, E. N. Peterson *et al.*, 'Distribution of the circulation in the normal and asphyxiated fetal primate.' *Am. J. Obstet. Gynecol.*, 1970, **108**, 956–969.

41. G. S. Dawes, in 'Foetal and Neonatal Physiology,' ed. G. S. Dawes, Year Book Medical Publishers, Chicago, IL, 1968, pp. 69.

42. G. G. Power and L. D. Longo, 'Fetal circulation times and their implications for tissue oxygenation.' *Gynecol. Invest.*, 1975, **6**, 342–355.

43. J. A. Sandberg, Z. Binienda, G. Lipe *et al.*, 'Placental transfer and fetal disposition of 2′,3′-dideoxycytidine and 2′,3′-dideoxyinosine in the Rhesus monkey.' *Drug Metab. Dispos.*, 1995, **23**, 881–884.

44. A. Rane and G. Tomson, 'Prenatal and neonatal drug metabolism in man.' *Eur. J. Clin. Pharmacol.*, 1980, **18**, 9–15.

45. A. H. Neims, M. Warner, P. M. Loughnan *et al.*, 'Developmental aspects of the hepatic cytochrome P450 monooxygenase system.' *Annu. Rev. Pharmacol. Toxicol.*, 1976, **16**, 427–445.

46. O. Pelkonen, 'Transplacental transfer of foreign compounds and their metabolism by the foetus.' *Prog. Drug. Metab.*, 1977, **2**, 119–161.

47. G. J. Dutton, 'Developmental aspects of drug conjugation, with special reference to glucuronidation.' *Annu. Rev. Pharmacol. Toxicol.*, 1978, **18**, 17–35.

48. H. Nau and D. Neubert, in 'Role of Pharmacokinetics in Prenatal and Perinatal Toxicology,' eds. D. Neubert, H.-J. Merker, H. Nau *et al.*, Theime, Stuttgart, 1978, pp. 13–44.

49. O. Pelkonen, 'Biotransformation of xenobiotics in the fetus.' *Pharmacol. Ther.*, 1980, **10**, 261–281.

50. B. H. Dvorchik, in 'Drug Metabolism in the Immature Human,' eds L. F. Soyka and G. P. Redmond, Raven Press, New York, 1981, pp. 146–162.

51. W. Slikker, Jr., in 'Principles of Neurotoxicology,' ed. L. W. Chang, Dekker, New York, 1994, pp. 659–680.

52. G. I. Murray, C. O. Foster, T. S. Barnes *et al.*, 'Cytochrome P4501A expression in adult and fetal human liver.' *Carcinogenesis*, 1992, **13**(2), 165–169.

53. F. M. Farin and C. J. Omiecinski, 'Regiospecific expression of cytochrome P450s and microsomal epoxide hydrolase in human brain tissue.' *J. Toxicol. Environ. Health*, 1993, **40**, 317–335.

54. J. Maenpaa, A. Rane, H. Raunio *et al.*, 'Cytochrome P450 isoforms in human fetal tissues related to phenobarbital-inducible forms in the mouse.' *Biochem. Pharmacol.*, 1993, **45**, 899–907.

55. J. Hakkola, M. Pasanen, R. Purkunen *et al.*, 'Expression of xenobiotic-metabolizing cytochrome P450 forms in human adult and fetal liver.' *Biochem. Pharmacol.*, 1994, **48**, 59–64.

56. V. Ravindranath and M. R. Boyd, 'Xenobiotic metabolism in brain.' *Drug Metab. Rev.*, 1995, **27**(3), 419–448.

57. J. Dancis, W. L. Money, G. P. Condon *et al.*, 'The relative transfer of estrogens and their glucuronides across the placenta in the guinea pig.' *J. Clin. Invest.*, 1958, **37**, 1373–1378.

58. M. Levitz, G. P. Condon, W. L. Money *et al.*, 'The relative transfer of estrogens and their sulfates across the guinea pig placenta: sulfurylation of estrogens by the placenta.' *J. Biol. Chem.*, 1960, **235**, 973–977.

59. U. Goebelsmann, N. Wiqvist and E. Diczfalusy, 'Placental transfer of estriol glucosiduronates.' *Acta Endocrinol. (Copenh.)*, 1968, **59**, 426–432.

60. U. Goebelsmann, J. M. Roberts and R. B. Jaffe, 'Placental transfer of ³H-oestriol-3-sulphate-16-glucosiduronate and ³H-oestriol-16-glucosiduronate-¹⁴C.' *Acta Endocrinol. (Copenh.)*, 1972, **70**, 132–142.

61. J. A. Sandberg, H. M. Duhart, G. Lipe *et al.*, 'Disposition of 2,4-dichlorophenoxyacetic acid (2,4-D) in maternal and fetal rabbits.' *J. Toxicol. Environ. Health*, 1996, **49**, 497–509.

62. C. S. Kim, Z. Binienda and J. A. Sandberg, 'Construction of a physiologically based pharmacokinetic model for 2,4-dichlorophenoxyacetic acid dosimetry in the developing rabbit brain.' *Toxicol. Appl. Pharmacol.*, 1996, **136**, 250–259.

63. M. H. Mark and P. M. Farmer, 'The human subfornical organ: an anatomic and ultrastructural study.' *Ann. Clin. Lab. Sci.*, 1984, **14**, 427–442.

64. M. T. Price, J. W. Olney, O. H. Lowry *et al.*, 'Uptake of exogenous glutamate and aspartate by circumventricular organs but not other regions of brain.' *J. Neurochem.*, 1981, **36**, 1774–1780.

65. J. W. Olney, V. Rhee and T. D. Gubareff, 'Neurotoxic effects of glutamate on mouse area postrema.' *Brain Res.*, 1977, **120**, 151–157.

66. M. C. Palazzo, K. R. Brizzee, H. Hofer *et al.*, 'Effects of systemic administration of 6-hydroxydopamine on the circumventricular organs in nonhuman primates.

II. Subfornical organ.' *Cell Tissue Res.*, 1978, **191**, 141–150.

67. N. R. Saunders, 'Ontogeny of the blood–brain barrier.' *Exp. Eye Res.*, 1977, **25** , Suppl., 523–550.

68. C. E. Johanson, 'Permeability and vascularity of the developing brain: cerebellum vs. cerebral cortex.' *Brain Res.*, 1980, **190**, 3–16.

69. G. Banos, P. M. Daniel, S. R. Moorhouse *et al.*, 'The entry of amino acids into the brain of the rat during the postnatal period.' *J. Physiol. (London)*, 1971, **213**, 45–46.

70. J. E. Cremer, V. J. Cunningham, W. M. Pardridge *et al.*, 'Kinetics of blood–brain barrier transport of pyruvate, lactate and glucose in suckling, weanling and adult rats.' *J. Neurochem.*, 1979, **33**, 439–445.

71. P. M. Daniel, E. R. Lover and O. E. Pratt, 'The effect of age upon the influx of glucose into the brain.' *J. Physiol. (London)*, 1978, **274**, 141–148.

72. R. Spector and J. Eells, 'Deoxynucleoside and vitamin transport into the central nervous system.' *Fed. Proc.*, 1984, **43**, 196–200.

73. J. A. Sandberg and W. Slikker, Jr., 'Developmental pharmacology and toxicology of anti-HIV therapeutic agents: dideoxynucleosides.' *FASEB J.*, 1995, **9**, 1157–1163.

74. W. Slikker, Jr., 'Principles of developmental neurotoxicology.' *Neurotoxicology*, 1994, **15**(1), 11–16.

75. V. E. V. St. Omer, S. F. Ali, R. R. Holson *et al.*, 'Behavioral and neurochemical effects of prenatal methylenedioxymethamphetamine (MDMA) exposure in rats.' *Neurotoxicol. Teratol.*, 1991, **13**, 13–20.

76. M. G. Paule, K. Reuhl, J. J. Chen *et al.*, 'Developmental toxicology of trimethyltin in the rat.' *Toxicol. Appl. Pharmacol.*, 1986, **84**, 412–417.

77. B. Weiss, 'Risk assessment: the insidious nature of neurotoxicity and the aging brain.' *Neurotoxicology*, 1990, **11**, 305–313.

78. D. M. Barnes, 'New data intensify the agony over ecstasy.' *Science*, 1988, **239**, 864–866.

79. S. J. Peroutka, 'Incidence of recreational use of 3,4-methylenedioxymethamphetamine (MDMA, "Ecstasy") on an undergraduate campus.' *N. Engl. J. Med.*, 1987, **317**, 1542–1543.

80. S. F. Ali, A. C. Scallet, G. D. Newport *et al.*, 'Persistent neurochemical and structural changes in rat brain after oral administration of MDMA.' *Res. Commun. Subst. Abuse*, 1989, **10**, 225–235.

81. G. Battaglia, S. Y. Yeh, E. O'Hearn *et al.*, '3,4-Methylenedioxymethamphetamine and 3,4-methylenedioxyamphetamine destroy serotonin terminals in rat brain: quantification of neurodegeneration by measurement of (³H)-paroxetine-labeled serotonin uptakes sites.' *J. Pharmacol. Exp. Ther.*, 1987, **242**, 911–916.

82. C. J. Schmidt, 'Neurotoxicity of the psychedelic amphetamine, methylenedioxymethamphetamine.' *J. Pharmacol. Exp. Ther.*, 1987, **240**, 1–7.

83. C. J. Schmidt, J. A. Levin and W. Lovenberg, '*In vitro* and *in vivo* neurochemical effects of methylenedioxymethamphetamine on striatal monoaminergic systems in the rat brain.' *Biochem. Pharmacol.*, 1987, **36**, 747–755.

84. C. J. Schmidt and V. L. Taylor, 'Depression of rat brain tryptophan hydroxylase activity following the acute administration of methylenedioxymethamphetamine.' *Biochem. Pharmacol.*, 1987, **36**, 4095–4102.

85. W. Slikker, Jr., R. R. Holson, S. F. Ali *et al.*, 'Behavioral and neurochemical effects of orally administered MDMA in the rodent and nonhuman primate.' *Neurotoxicology*, 1989, **10**, 529–542.

86. W. Slikker, Jr., S. F. Ali, A. C. Scallet *et al.*, 'Neuro-

chemical and neurohistological alterations in the rat and monkey produced by orally administered methylenedioxymethamphetamine (MDMA).' *Toxicol. Appl. Pharmacol.*, 1988, **94**, 448–457.

87. W. Slikker, Jr., M. G. Paule and H. W. Broening, in 'Neurotoxicology: Approaches and Methods,' eds. L. W. Chang and W. Slikker Jr., Academic Press, San Diego, CA, 1995, pp. 371–380.

88. W. Slikker, Jr., in 'Experimental and Clinical Neurotoxicology,' 2nd edn., eds. J. S. Spencer, H. H. Schaumburg and Ludolph, Williams & Wilkins, Baltimore, MD, 1996.

89. D. L. Commins, G. Vosmer, R. M. Virus *et al.*, 'Biochemical and histological evidence that methylenedioxymethamphetamine (MDMA) is toxic to neurons in the rat brain.' *J. Pharmacol. Exp. Ther.*, 1987, **241**, 338–345.

90. E. O'Hearn, G. Battaglia, E. B. De Souza *et al.*, 'Methylenedioxyamphetamine (MDA) and methylenedioxymethamphetamine (MDMA) cause selective ablation of serotonergic axon terminals in forebrain: immunocytochemical evidence for neurotoxicity.' *J. Neurosci.*, 1988, **8**, 2788–2803.

91. A. C. Scallet, G. W. Lipe, S. F. Ali *et al.*, 'Neuropathological evaluation by combined immunohistochemistry and degeneration-specific methods: application to methylenedioxymethamphetamine.' *Neurotoxicology*, 1988, **9**, 529–537.

92. U. V. Berger, X. F. Gu and E. C. Azmitia, 'The substituted amphetamines 3,4-methylenedioxymethamphetamine, methamphetamine, *p*-chloroamphetamine and fenfluramine induce 5-hydroxytryptamine release via a commonmechanism blocked by fluoxetine and cocaine.' *Eur. J. Pharmacol.*, 1992, **215**, 153–160.

93. C. R. Hekmatpanah and S. J. Peoutka, '5-hydroxytryptamine uptake blockers attenuate the 5-hydroxytryptamine releasing effect of 3,4-methylenedioxymethamphetamine and related agents.' *Eur. J. Pharmacol.*, 1990, **177**, 95–98.

94. H. W. Broening and W. Slikker, Jr., 'Age-related sensitivity to the neurotoxic effects of methylenedioxymethamphetamine.' *Neurotoxicol. Teratol.*, 1996, submitted.

95. P. Herregodts, B. Velkeniers, G. Ebinger *et al.*, 'Development of monoaminergic neurotransmitters in fetal and postnatal rat brain: analysis by HPLC with electrochemical detection.' *J. Neurochem.*, 1990, **55**, 774–779.

96. Y. Nomura, F. Naitoh and T. Segawa, 'Regional changes in monoamine content and uptake of the rat brain during postnatal development.' *Brain Res.*, 1976, **101**, 305–315.

97. G. R. Breese and T. D. Traylor, 'Developmental characteristics of brain catecholamines and tyrosine hydroxylase in the rat: effects of 6-hydroxydopamine.' *Br. J. Pharmacol.*, 1972, **44**, 210–222.

98. T. Deguchi and J. Barchas, 'Regional distribution and developmental change of tryptophan hydroxylase activity in rat brain.' *J. Neurochem.*, 1972, **19**, 927–929.

99. W. Porcher and A. Heller, 'Regional development of catecholamine biosynthesis in rat brain.' *J. Neurochem.*, 1972, **19**, 1917–1930.

100. A. Bruinink, W. Lichtensteiger and M. Schlumpf, 'Pre- and postnatal ontogeny and characterization of dopaminergic D_2 serotonergic S_2 and spirodecanone binding sites in rat forebrain.' *J. Neurochem.*, 1983, **40**, 1227–1236.

101. D. F. Kirksey and T. A. Slotkin, 'Concomitant development of (^{3}H)-dopamine and (^{3}H)-5-hydroxytryptamine uptake systems in rat brain regions.' *Br. J. Pharmacol.*, 1979, **67**, 387–391.

102. P. M. Whitaker-Azmitia, A. V. Shemer, J. Caruso *et al.*, 'Role of high affinity serotonin receptors in neuronal growth.' *Ann. NY Acad. Sci.*, 1990, **600**, 315–330.

103. H. W. Broening, L. Bacon and W. Slikker Jr., 'Age modulates the long-term but not the acute effects of the serotonergic neurotoxicant 3,4-methylenedioxymethamphetamine.' *J. Pharmacol. Exp. Ther.*, 1994, **271**(1), 285–293.

104. J. A. Clemens, R. W. Fuller, K. W. Perry *et al.*, 'Effects of *p*-chloroamphetamine on brain serotonin in immature rats.' *Commun. Psychopharmacol.*, 1978, **2**, 11–16.

105. B. V. Clineschmidt, A. G. Zacchei, J. A. Totaro *et al.*, 'Fenfluramine and brain serotonin.' *Ann. NY Acad. Sci.*, 1978, **305**, 222–241.

106. J. B. Lucot, J. Horwitz and L. S. Seiden, 'The effects of *p*-chloroamphetamineadministration on locomotor activity and serotonin in neonatal and adult rats.' *J. Pharmacol. Exp. Ther.*, 1981, **217**, 738–744.

107. H. W. Broening, J. F. Bowyer and W. Slikker, Jr., 'Age-dependent sensitivity of rats to the long-term effects of the serotonergic neurotoxicant ($\pm$) 3,4-methylenedioxymethamphetamine (MDMA) correlates with the magnitude of MDMA-induced thermal response.' *J. Pharmacol. Exp. Ther.*, 1995, **275**, 325–333.

108. D. S. Bennett and N. J. Giaman, 'Schedule of appearance of 5-hydroxy-tryptamine (serotonin) and associated enzymes in the developing rat brain.' *J. Neurochem.*, 1965, **12**, 911–918.

109. H. G. W. Lidov and M. E. Molliver, 'An immunohistochemical study of serotonin neuron development in the rat: ascending pathways and terminal fields.' *Brain Res. Bull.*, 1982, **8**, 389–430.

110. E. J. Lammer, D. T. Chen, R. M. Hoar *et al.*, 'Retinoic acid embryopathy.' *N. Engl. J. Med.*, 1985, **313**, 837–841.

111. J. Adams and E. J. Lammer, in 'Functional Neuroteratology of Short-Term Exposure to Drugs,' eds. E. Fujii and C. J. Barr, Tokyo University Press, Tokyo, 1991, pp. 159–170.

112. D. Ecobichon, J. E. Davies, J. Doull *et al.*, in 'The Effects of Pesticides on Human Health,' eds. S. R. Baker and C. F. Wilkinson, Princeton Scientific Publishing, Princeton, NJ, 1990, pp. 136–181.

113. D. E. Hutchings, J. Gibbon and M. A. Kaufman, 'Maternal vitamin A excess during the early fetal period: effects on learning and development in the offspring.' *Dev. Psycholbiol.*, 1973, **6**, 445–457.

114. C. V. Vorhees, R. L. Brunner and R. E. Butcher, 'Psychotropic drugs as behavioral tertogens.' *Science*, 1979, **205**, 1220–1225.

115. C. Eckhoff, J. R. Bailey, W. Slikker Jr. *et al.*, 'Placental transfer of retinoids following high-dose vitamin during early pregnancy in the monkey.' *Teratology*, 43, 371–408.

116. J. Creech Kraft, H. Nau, E. Lammer *et al.*, 'Human embryo retinoid concentrations after maternal intake of isotretinoin.' *N. Engl. J. Med.*, 1989, **321**, 262.

117. C. V. Vorhees, in 'Principles of Neurotoxicology,' ed. L. W. Chang, Dekker, New York, 1994, pp. 733–763.

118. S. F. Ali, S. N. David, G. D. Newport *et al.*, 'MPTP-induced oxidative stress and neurotoxicity are age-dependent: evidence from measures of reactive oxygen species and striatal dopamine levels.' *Synapse*, 1994, **18**, 27–34.

119. M. G. Paule, K. Reuhl, J. J. Chen *et al.*, 'Developmental toxicology of trimethyltin in the rat.' *Toxicol. Appl. Pharmacol.*, 1986, **84**(2), 412–417.

120. J. C. Matthews and A. C. Scallet, 'Nutrition, neurotoxicants and age-related neurodegeneration.' *Neurotoxicology*, 1991, **12**, 547–558.

121. A. C. Scallet, Z. Binienda, F. A. Caputo *et al.*, 'Domoic acid-treated cynomolgus monkeys (*M. fascicularis*): effects of dose on hippocampal neuronal and terminal degeneration.' *Brain Res.*, 1993, **627**, 307–313.

122. M. G. Kolta, F. M. Scalzo, S. F. Ali *et al.*, 'Ontogeny of the enhanced behavioral response to amphetamine in amphetamine-pretreated rats.' *Psychopharmacology (Berlin)*, 1990, **100**, 377–382.

123. J. F. Bowyer, D. L. Davies, L. Schmued *et al.*, 'Further studies of the role of hyperthermia in methamphetamine neurotoxicity.' *J. Pharmacol. Exp. Ther.*, 1994, **268**(3), 1571–1580.

124. W. Slikker, Jr., J. A. Sandberg, L. Holder *et al.*, *Toxicologist*, 1994, **14**, 117.

125. F. Finiels, J. J. Robert, M. L. Samolyk *et al.*, 'Induction of neuronal apoptosis by excitotoxins associated with long-lasting increase of 12-*O*-tetradecanoylphorbol 13-acetate-responsive element-binding activity.' *J. Neurochem.*, 1995, **65**(3), 1027–1034.

126. R. R. Ratan, T. H. Murphy and J. M. Baraban, 'Oxidative stress induces apoptosis in embryonic cortical neurons.' *J. Neurochem.*, 1994, **62**(1), 376–379.

127. J. M. Thayer and P. E. Mirkes, 'Programmed cell death and *N*-acetoxy-2-acethylaminofluorene-induced apoptosis in the rat embryo.' *Teratology*, 1995, **51**(6), 418–429.

11.17
Neural, Behavioral, and Measurement Considerations in the Detection of Motor Impairment

M. CHRISTOPHER NEWLAND
Auburn University, AL, USA

11.17.1 INTRODUCTORY COMMENTS

In 1984 Anger[1] summarized the criteria by which the American Conference of Governmental Industrial Hygienists (ACGIH) set threshold limit values (TLVs). Of those cited for neurotoxic effects, about one half (89 substances) had some form of motor effect included as a contributor to TLV considerations. A wide range of motor effects were cited but four dominated the profiles: tremor, weakness or fatigue, convulsions or spasms, and incoordination. This list alone suggests that an ability to measure and characterize chemical-induced motor deficits would be of value in any neurotoxicity evaluation.

The reasons for considering motor endpoints come from other sources, not least of which is that behavior is the point of contact between the nervous system and the environment. The execution of any behavioral act involves one or more component of the motor systems and this means that motor endpoints are always present in behavioral measures. Accordingly, a full understanding of a neurotoxic effect entails a separation of motor from other possible effects. This is a complication that careful experimental design can circumvent, and it is also an advantage. The intimate relationship between motor acts and behavior means that our understanding of the principles governing behavior can be applied to an understanding of motor function. Indeed, a full characterization of how neurotoxicants impair motor function will draw from behavioral, measurement, neural, and toxicological principles.

In this chapter the contributions of these different domains to the development of tactics for measuring and comprehending motor dysfunction as it appears in behavior will be considered. First the basic components of the motor system and some screening tests that have been used to identify damage to these systems will be presented. Then, considerations that bear on the measurement of motor, or any other, behavioral act will be examined. Certain indirect effects of reinforcement schedules that enhance the sensitivity to motor dysfunction will be examined in general, and as they have revealed the consequences of low-level exposure to manganese. Finally, the quantitative assessment of an important and perhaps apical motor endpoint, tremor, will be described.

11.17.2 THE NEURAL COMPONENTS OF MOTOR SYSTEMS

Control over motor function is governed by three regions in the central nervous system and the peripheral outflow to muscles. The three central regions are the primary motor cortex, the basal ganglia, and the cerebellum. Of these, the basal ganglia, cerebellum, and descending motor pathways have received the most attention in neurotoxicological investigations, perhaps because of their importance in organizing and regulating motor acts or perhaps because they are especially vulnerable to toxic insult. The result of centrally-mediated decisions about what motor acts to conduct, and how, are communicated to muscles through the descending spinal tracts and ultimately through motor axons to the muscles.

Damage to these regions has been associated with different types of functional deficits, some of which are listed in Table 1. Most of the tests listed in Table 1 have been conducted with rodent species, but some deficits seem easier to identify in primates. The tests in Table 1 are among those that should be considered when selecting screening procedures for characterizing neurotoxicants that are poorly understood, since it is desirable that a variety of neurotoxic effects be caught in the screen. Care should be taken in interpretation, however, since motor function is integrative. Abnormal performance on a particular test might suggest motor involvement but single tests are not specific enough to permit an investigator to relate poor performance to specific impairment.

11.17.2.1 Motor Cortex

A map, sometimes called the motor homunculus, of the motor neurons can be found on the surface of the primary motor cortex.[19] Perpendicular to the surface at a particular location is a column of neurons responsible for the contraction of a particular muscle. The column contains afferents (inputs) from other regions of the brain and some of its efferents (output) eventually reach motor neurons in the spinal cord. When a motor neuron fires, the muscle contracts after a short delay.[19] Because of the close linkage between a cortical neuron and a motor unit (the motor fiber and the muscle it innervates), there is a relationship between the amount of surface area of the precentral gyrus devoted to a region and the precision of movement in the muscles controlled by that region. Much cortical area is devoted to the lips and fingers in the human, for example, while very little is devoted to the trunk and back.

Cortical damage in the adult is most commonly associated with stroke or physical trauma. Damage to a specific region of the primary motor cortex impairs function of the individual neurons controlled by that region, so functional impairment is specific to the locus of damage. Cortical damage in the young organism, which might be diagnosed as cerebral palsy, could result from a variety of causes including birth trauma or intrauterine exposure to chemicals. Toxicant-induced damage can result from exposure during development and the consequences can be subtle. For example, methylmercury impairs the mechanism by which the cortical neurons develop and this damage is manifested as cerebral palsy if exposure levels during gestation are sufficiently high.[20,21]

Table 1   Motor effects of lesions in certain parts of the nervous system.

Motor system	Clinical signs	Tests	Intervention	Ref.

Cortical	Palsy
	Specific losses of
	   movement
	Loss of integrated
	   movement
		Licking (tongue reach,
		   but not grooming)
		Forelimb reach
		Subtle impairment of
		   grooming
		Postural changes
		Flexed limbs when
		   held by tail
		Trapped by corner
			Surgical lesions	2–5

Basal ganglia	Low-frequency resting
	   or action tremor
	Bradykinesia
	Akinesia
	Loss of postural
	   reflexes
	Oral–facial movements
	Dystonia
	Athetosis
	Altered blinking
	Stereotypies
		Turning
		Catalepsy
		Stereotypy
		Tremor
		Blink rate
		Licking (tongue
		   control)
		Forelimb reach
		Dystonic postures
		Vacuous chewing
			Haloperidol
			Manganese
			Carbon disulfide
			MPTP
			Hypoxia
				6–12

Cerebellum	Intention tremor at
	   physiological
	   frequency
	Lowered tremor
	   amplitude?
	Nystagmus
	Loss of postural
	   reflexes
	Uncoordination
	Disordered gait
		Gait analysis
		Rotorod
		Action tremor
		Nystagmus
		Running wheel
		Rotorod
		Landing foot spread
			Ethanol
			3-Acetyl pyridine
			Organic mercury
				13–16

Peripheral
Nerve
Neuromuscular
   junction
	Tremor, often low
	   frequency or
	   broad-band
	   high frequency
	Weakness
	Loss of use of limbs
	Peripheral neuropathy
		Grip strength
		Rotorod
		Gait analysis
		Landing foot spread
		Low-frequency tremor
		Leg splay
			2,5-Hexanedione-
			   acrylamide
			Organophosphates
			Carbamate
				12–14,17,18

11.17.2.2 Basal Ganglia

The basal ganglia are a collection of four or five subcortical bodies: caudate and putamen, collectively called the striatum, subthalamic nucleus (in some sources not included as part of the basal ganglia), globus pallidus, and substantia nigra. Despite their intimate role in mediating motor function, the basal ganglia send no axons directly to motor neurons, receive no afferents from proprioceptors monitoring muscle activity and movement, and the activity of single neurons in basal ganglia seem to be unrelated to ongoing movement.[22,23] Instead, afferents to the basal ganglia derive from diverse regions of the cortex and efferents return to the cortex by way of the thalamus.

The basal ganglia seem to funnel and organize information from many cortical regions about what motor acts are required, and return the result to premotor or motor cortices via thalamic pathways. The basal ganglia are involved in coordinated movements, behavioral sequences, and locomotion.[24–26]

The complexities of basal ganglia interconnections cannot be described here, but a general picture can be described of a circuit originating from cortex, passing through striatum, leaving through the globus pallidus or substantia nigra, and returning to the cortex by way of the thalamus.[23,27,28] Inputs to the basal ganglia comprise excitatory nerve fibers that terminate in the striatum. The striatum sends inhibitory fibers to the globus pallidus and substantia nigra, which send inhibitory fibers to the thalamus which returns excitatory pathways to the cortex. As a first approximation to function, if the fibers to the cortex are excited then bradykinesia (slow movement) and increased muscle tone are the result. If they are excited uncontrollably then akinesia (absence of movement) and rigidity is the result.[27] Such motor patterns is characteristic of Parkinson's disease, which damages substantia nigra, disinhibits the thalamus, and results in bradykinesia, akinesia, rigidity, and a reliance on external stimuli to guide otherwise common movements. Excessive exposure to manganese, which has an affinity for pallidum, the other output region of the basal ganglia, results in dystonic postures and disturbances in gait.[29,30] Carbon monoxide and carbon disulfide poisoning also damage the globus pallidus.[31]

The basal ganglia have been hypothesized to be involved in movement that is not directly under the control of external stimuli.[28] Parkinsonian patients, who experience the results of an impaired substantia nigra, show an enhanced reliance upon external stimuli to guide movements and even gait.[32–35] Experimental evidence suggests that this deficit may be related to dopaminergic pathways which originate in the substantia nigra and terminate in the striatum. A model of this in nonhuman species can be seen in the acute actions of major tranquilizers, which block the action of dopamine on the striatum. These compounds produce a characteristic reduction in control over behavior by exteroceptive stimuli.[36,37]

Bradykinesia (slowed movement) and excessive muscle tone results from damage to the output regions to the basal ganglia but removing the inhibition over these output regions results in roughly opposite effects.[27] Huntington's disease damages the striatum, which inhibits the globus pallidus and substantia nigra, and produces uncontrolled movements such as chorea (abrupt movements of limbs and facial muscles) and athetosis (uncontrolled writhing movements) as well as dementia.

Of the circuits interconnecting regions of the basal ganglia, the nigral–striatal loop has been among the most investigated. This loop includes a major dopaminergic pathway beginning with dopamine-containing cell bodies in the substantia nigra located in the midbrain and whose terminals synapse onto cholinergic cell bodies in the striatum. Major tranquilizers such as chlorpromazine and haloperidol block the dopaminergic synapses in the striatum, an action that produces a neurochemical lesion similar to removing the substantia nigra. The result is a pseudoparkinsonism, a common side effect of treatment with these drugs. With chronic use of major tranquilizers, even when interposed by "drug holidays," receptor upregulation occurs and supersensitivity is the result. This increases the dopaminergic influence over the striatum and the result is tardive dyskinesia: uncontrolled movements of face, tongue, and sometimes limbs.[38,39]

The transition from pseudoparkinsonism to tardive dyskinesia illustrates the dynamism in the motor effects of a compound: the early stages of a syndrome may be quite different, and perhaps opposite from later stages. A full characterization of a toxicant should allow for such dynamism to appear if it is present, and when the basal ganglia are involved there is a fair chance that toxicity will change over time.

Basal ganglia dysfunction has been difficult to identify in nonprimate species, even though the rodent has all the structural components of the basal ganglia observed in primates. The effects of methylphenyltetrahydropyridine (MPTP), for example, which produces profound Parkinsonism in human and nonhuman primates, has not produced such effects in rodents.[40] One problem is that rodents do not exhibit basal ganglia dysfunction in the same manner as primates but the difficulty is sometimes with other considerations, such as difficulties in the passage of some basal ganglia neurotoxicants such as MPTP through the blood–brain barrier[41] or that the important pathologies present in the portions of the primate basal ganglia have evolved separately from those of the rodents.[23] It is also possible that in some cases, experimentalists have been using the wrong tests to identify dysfunction in rodents. For example, the measurement of tongue-reach as an indicator of motor dysfunction in rodents has been extended and expanded to characterize neuroleptics.[42] Parkinsonian effects of neuroleptics are also reflected in response durations, tremor, reflexive and operant licking, and within-session declines in a wide variety of behavior observed in rodents.[43–46]

11.17.2.3 Cerebellum

The cerebellum is phylogenetically older but shares with the basal ganglia the function of structuring and organizing movements. Its position on the dorsal (back) side of the brainstem places it in a convenient location to monitor traffic of the spinal cord, one of its major functions. The cerebellum adjusts ongoing movements by monitoring motor commands traveling down the spinal cord and sensory feedbacks traveling up the spinal cord. The cerebellum also helps to maintain posture and participates in motor learning.

The cerebellum is subdivided into three functional components: vestibulocerebellum, spinocerebellum, and cerebrocerebellum.[27] The vestibulocerebellum receives afferents from the vestibular nuclei in the brainstem and returns efferents back to that region. This region receives information about acceleration of the head due to body movements or the actions of gravity. It maintains posture and coordinates head and eye movements, the latter in order to maintain a consistent position of visual stimuli on the retina.

The spinocerebellum receives sensory information about muscle contraction and joint position which is contained in sensory fibers originating in proprioceptor and traveling up the spinal cord. The spinocerebellum also monitors motor information from descending motor fibers originating in the cortex and terminating on motor units in the spinal cord. Commands to muscles contained in the motor fibers are compared with the results contained in the sensory fibers and corrections are issued according to discrepancies between these two. The corticocerebellum receives information from the cortex and projects back to motor areas of the cortex. This region participates in the planning of motor acts.

Cerebellar control over movement is plastic, despite this structure's rather rigid functional organization. Respondent conditioning has been observed in the cerebellum and the learning of motor skills requires cerebellar involvement.[47,48] Testing for cerebellar dysfunction should therefore accommodate the possibility of adjustment to impairments or learning of the task, and advanced testing should be used to examine the degree to which such adjustments are impaired.

Some of the signs of cerebellar impairment can be seen in dose-related changes in motor function produced by ethyl alcohol intoxication, since the cerebellum seems especially sensitive to this drug. Early stages appear as unstable posture, postural sway, and deficits in dexterous, well-learned actions like typing. In later stages, gait disorders and even nystagmus can be seen. Table 1 summarizes some procedures used to detect behavioral deficits when the cerebellum has been lesioned, and indicates some specific characteristics of cerebellar tremor.

11.17.2.4 Descending Pathways and Neuromuscular Junction

In order for a motor command to effect action it must travel from the central nervous system to peripherally located muscles through the descending motor pathways. In order for the cerebellum or other motor regions to adjust motor actions, it is necessary to receive sensory information from proprioceptors in joints and muscles. This information travels from the periphery to the central nervous system through ascending fibers. The ascending and descending pathways themselves, spinal reflexes influenced by the descending pathways, and the neuromuscular junction are all targets for neurotoxic action. If sensory information about motor acts is impaired, small adjustments in movement cannot take place. Examples of neurotoxicants that impair the sensory and motor fibers are carbon disulfide, acrylamide, 2,5-hexanedione, and organophosphorus compounds. These compounds can damage sensory or motor fibers through various mechanisms including dying-back axonopathies or destruction of the myelin sheath that surrounds axons and facilitates the conduction of an action potential.[49] The consequences of such damage could appear as alterations in forelimb or hindlimb grip strength.[18] Elsner *et al.*[50] described a procedure whereby rats were trained to press on a lever with some force, and the force required for reinforcer delivery was titrated upward according to performance. This general approach permits the separation of motivational and motor variables in the description of strength.

The neuromuscular junction contains nicotinic receptors for the neurotransmitter acetylcholine, so compounds that disrupt some stage of cholinergic neurotransmission or that act on nicotinic receptors can influence this portion of the motor system. Botulinum toxin prevents the release of acetylcholine,[51] functionally denervating the muscle and preventing the motor neuron from contracting it. Organophosphate and carbamate pesticides inhibit the activity of target cholinergic junctions by preventing the termination of acetylcholine activity since these pesticides inhibit the activity of acetylcholine ester-ase, which initially overstimulates the neuromuscular junction producing muscle twitching due to overactivity, but eventually weakness and paralysis appear.

11.17.3 MEASUREMENT

Measurement issues are always present, whether acknowledged or not, when assessing motor function. These considerations arise both with screening tests and advanced applications designed to identify mechanisms and they appear in the generality of the physical measures selected for study and in their scaling properties.[52-54]

11.17.3.1 Physical Measures

To take the assessment of motor dysfunction beyond qualitative description it is necessary to provide physical measures such as movement, positioning, or strength. The physical dimensions involved are mass, displacement, and time or combinations of these into velocity (displacement/time), acceleration (rate of change of velocity), force (mass times acceleration), and frequency (events/time). Transducers are used to convert a physical property into an electrical signal that can be monitored by a computer or other electronic devices. Some transducers associated with different measurements are listed in Table 2. Such transducers measure physical phenomena so that dimensions of motor function can and should be described quantitatively in appropriate dimensions. Only then can measurements taken across different settings be compared meaningfully. If "arbitrary" dimensions such as voltage read from the transducer are reported, then comparisons across laboratories can only be nominal (a change was noted) or ordinal (a higher dose produced a greater increase). More precise comparisons such as those reported below comparing the properties of tremor are impossible unless measurement considerations are carefully applied.

11.17.3.2 Behavioral Measures

Behavioral measures of motor dysfunction are a necessary component of characterization. Measures of rate, response duration, or topography are coupled with physical measures, whether acknowledged or not. As described below, the behavioral component of physical measures is influenced by the conditions maintaining behavior. Behavioral measures like response rate can serve as an adjunct to detailed physical measures of motor function. If the topography of the response is engineered to constrain its form and reveal specific nervous system action, then rate itself can be a measure of motor dysfunction and should be designed in order to target the sorts of dysfunction that are likely to occur. The sensitivity of advanced applications aimed at identifying mechanisms or examining effects of low levels of exposure can be increased by designing a response according to known effects at high levels of exposure or according to the neural system thought to underlie behavioral effects.[53,54] Kallman and Fowler[55] summarize several different approaches to designing manipulanda and applying behavioral technology to examine in detail the effects of drugs or toxic substances.

11.17.3.3 Quality of Measurement and Consistency Across Applications

The exposure that many scientists have to measurement considerations can be exemplified by the sorts of numbers that come from rulers, balances, and stopwatches. These are all ideal forms of measurement, in part because they correspond to our everyday understanding of how numbers are used. Differences in ways of

Table 2 Physical measurements used to describe motor function.

Physical measure	Representative units	Transducer	Representative application
Mass	Grams (g)	Load cell	Mass not often applied in isolation
Displacement	Centimeters (cm)	Potentiometer Rotary transformer Magnet in electric field Platform on speaker coil	Limb movement Precision of positioning
Time	Seconds (s)	Clock Timer	Lever-press duration Reaction time
Velocity	Centimeters per second (cm s^{-1})	Speedometer Tachometer	Running speed Speed of limb movements
Acceleration	Centimeters per second squared (cm s^{-2})	Accelerometer	Change in speed of limb movement Tremor
Force	Dynes, newtons (dyn, N)	Load cell Strain gauge	Press with limb to measure strength, precision Tremor

measuring phenomena quickly become apparent in the behavioral domain, so it is no accident that a nomenclature for measurement has arisen out of experimental psychology.

Johnston and Pennypacker[56] offer a helpful distinction between two measurement approaches that have arisen in the natural and behavioral sciences. Measurement entails both the thing measured and the numbers used to describe it. In the common examples described above these things are separate: standard and independent scales for length are used to measure such things as height. Such scales, termed idemnotic from a word derived from the Latin idem meaning the same, are anchored to a constant and support comparisons across settings and applications. Well-known examples of idemnotic scales are the physical measurements of length, time, or mass but other examples include reaction time, and combinations of measures such as force (combinations of mass, distance, and time), response rate (count or number of events per unit time), or response duration.

These measures stand in contrast to an approach called vaganotic, a term drawing from the a Latin root word vagare, meaning to wander. Vaganotic measurement is the creation of a measurement scale on the basis of variability (or error) in the underlying observations. For example, a score may be registered as the number of standard deviation units from a mean, and so the score can change when different populations are used as the norm. With this form of measurement, the units of measure and the thing measured are inseparable. Scores on personality or IQ scales are examples. Such scales are sometimes dimensionless and standardized, if at all, to a local norm. If the heights of individuals were measured by finding the mean in some population and describing individuals according to standard deviation units, then height would be vaganotic, and much more limited as a measurement. Another example is the attempt to

rank, on a scale of 1 to 5, the degree of motor impairment. The numbers are not anchored to a separate, independent physical constant. The transition from a 2 to a 3 is driven by uncertainties and variability within a rank and cannot be tied to a physical dimension. As a result of such vague measurement units, the rankings may vary from setting to setting.

A second set of issues related to measurement is the establishment of different types of measurement scales. Stine[57] following Stevens[58] has described different scales along which measurements are taken and the constraints that apply when using them. Table 3 summarizes the common measurement scales and provides an example from the measurement of tremor. Nominal scales are categorical and do not contain measurement information *per se*. Descriptions of tremor based upon the conditions under which it is seen, action tremor as opposed to resting tremor, for example, is an example of nominal classification.

Most clinical examinations and many observational batteries produce ordinal numbers. The Webster scale assigns numbers of 0 to 3 according to the degree of tremor. On this scale a 0 means no detectable tremor (there is always some tremor); a 1 is assigned if tremor is 2 cm in displacement; a 2 if tremor is less than 10 cm; and a 3 if tremor is more than 10 cm. This illustrates a limitation of the ordinal scale, which is that equal spacings between numbers on the scale do not reflect equal spacing between the events being measured. A value of "2" is not twice the value of "1" and the difference between 1 and 2 is not the same as the difference between 2 and 3. In addition, the scale cannot accommodate a decrease in tremor, and decreases are indicative of several types of dysfunction. Summarizing ordinal numbers with a mean and a standard deviation is incorrect and misleading because doing so requires addition,[57] which is a meaningless operation when the spacing between numbers is inconsistent.

Table 3 Types of measurement scales.

Name of scale	Information conveyed	Example
Nominal	Category	Presence or type of tremor (resting, action)
Ordinal	Order	Degree of tremor (1–5 scale)
Interval or affine	Intervals (order, addition, and subtraction)	Amount of water spilled by tremorous patient holding a glass
Ratio	Proportions (order, addition, and subtraction, multiplication and division)	RMS acceleration in 7.5–8 Hz band of power spectrum

Numbers on interval scales are spaced consistently so differences are meaningful and addition and subtraction denote meaningful operations, but zero is arbitrary so statements about proportions are meaningless. Ratio scaled data contain the full potential that quantification provides and greatly facilitate descriptions of the phenomena under investigation. With tremor, root-mean-squared acceleration is an example of ratio-scaled data so it is meaningful to talk about multiplicative relationships such as one type of tremor being twice another. The numbers are anchored to physical constants and have physical dimensions and a zero denotes something natural. The superb measurement properties of such approaches to measuring tremor facilitates the comparison of tremor measures taken across laboratories, as described in Table 6.

The mechanics behind determining the type of measurement scale with which one is working are beyond the scope of this chapter (see Ref. 57), but some simple rules of thumb are possible. If events can be arranged alphabetically (A = no tremor, B = some tremor, etc.) without a resultant loss of information, then the numbers are ordinally scaled. If the zero is arbitrary, such as the lack of spilled water in Table 3, then the scale is probably interval. It is worth noting that such a determination depends fundamentally upon what one does with the numbers. Indeed, numbers may be ordinal with respect to one set of phenomena and ratio with respect to another (e.g., amount of spilled water in milliliters vs. spilled water as a measure of tremor).

The distinctions described in this section are useful guides to evaluating different forms of measurement of motor dysfunction. Measurement considerations appear in every stage of an investigation, including the decisions of how to detect and describe a phenomenon, how to compare these descriptions with those deriving from other settings, and what inferential techniques to apply at the end of an investigation.

11.17.4 BEHAVIOR

The measurement of motor function at the level of the whole organism requires the use of behavior. Motor function might be evaluated as a component of behavioral screens, and a strategy to selecting screening procedures has been described above. Motor function can also be evaluated as a component of other behavioral testing if measures of the physical dimensions of behavior are added. In addition, behavioral technology can be applied to design advanced applications for assessing motor

function. The last two tactics will be described in this section.

Operant behavior can be described using the three-term contingency of reinforcement. Stated simply: in the presence of a stimulus, a response occurs that has a consequence, and that consequence influences the future likelihood of the response re-occurring in the presence of that stimulus. When evaluating motor function, attention is especially applied to the physical characteristics of the response. These physical characteristics are jointly influenced by the capabilities of the organism and by the conditions supporting operant behavior.[54] An understanding of direct and indirect effects of reinforcement contingencies can be applied to enhance the likelihood of gaining information about motor capabilities. The schedule of reinforcement exerts enormous direct and indirect influences over the physical characteristics of responding. Direct effects are contained in the schedule definition and can be used to shape a response to a particular specification. Indirect effects are not specified directly, but consistently appear nonetheless.

11.17.4.1 Indirect Effects of Reinforcement Contingencies on the Operant

Reinforcement schedules specify the conditions required for a reinforcing stimulus to be delivered. An appreciation of linkages between the reinforcement contingencies as specified in the schedule and the type of response engendered is related to motor function because the schedule determines the physical characteristics of behavior and thereby the ability to detect motor effects.[54]

In one class of reinforcement schedule, response-based schedules, reinforcer delivery is directly linked to the number of responses that have occurred. For example, under a fixed ratio 10 schedule, 10 responses result in reinforcer delivery. Under such response based schedules there is a "molar" relationship between the overall rate of responding and the overall rate of reinforcement and a "molecular" relationship between responding and reinforcer delivery such that reinforcers tend to follow bursts of high-rate responding, or short inter-response times, and therefore increase the likelihood of such bursts.[59] Both relationships contribute to a high rate and vigorous pattern of responding under ratio schedules.

Under time-based schedules, time is an integral component of the reinforcement contingencies. For example, under a fixed interval 60 s schedule, a single response after 60 s have passed results in reinforcer delivery. In such a case, the link between response rate and reinforcer rate has been weakened considerably and, accordingly, response rates are lower, more variable, and durations are longer.[46,60,61] Other common schedules are variable ratio and variable interval schedules which share the properties that the response or time requirements are variable, unpredictable, and maintain relatively constant rates of responding.[62]

Responding maintained by ratio schedules is characterized by its high overall response rates, short inter-response times, and short response durations. This characteristic of ratio schedules is true regardless of how the response is defined and shows generality across species.[45,60,61] Such behavior can be especially sensitive to neurotoxicants known to produce motor dysfunction. Newland and Weiss[63] examined manganese's effect on behavior maintained under both a fixed ratio and a fixed-interval schedule. The response entailed pulling through a displacement of about 10 cm against a spring that resisted with a force approximating the subject's body weight. Manganese at levels well below those that produced overt clinical signs greatly enhanced the number of responses that failed to meet the physical requirement under the ratio schedule, which maintained a highly vigorous form of responding (Figure 1). Such an effect was not seen under the interval schedule, which maintained a slower and

more variable pattern of responding. Effects that appeared on the fixed-interval schedule were transient and did not occur until a cumulative dose of $30\,\mathrm{mg\,kg^{-1}}$ was administered for the monkey illustrated. Physical exertion and rapid flexion and extension are indirect effects of the fixed-ratio schedule, but they are not directly required by it, and this response characteristic was probably the reason for the sensitivity of the ratio schedule.

The appearance of pronounced impairment in this effortful, back-and-forth response was correlated with shortened spin–lattice relaxation time (T_1) as detected by magnetic resonance imaging (MRI). Shortened T_1 values reflect increased concentrations of paramagnetic metals such as manganese. The effect can be seen in Figure 2, which shows a coronal image of the brain of one of the monkeys in the study. In this image, there is a clear highlighting of the globus pallidus, but not other visible regions of the basal ganglia, shortly after exposure to $5\,\mathrm{mg\,kg^{-1}}$ of manganese. The caudate and putamen were highlighted only after increasing the cumulative dose of manganese. Shortened spin–lattice relaxation times in these different regions revealed the presence of manganese in the globus pallidus and pituitary at the lowest levels of exposure. Manganese remained in these two regions for the longest time after

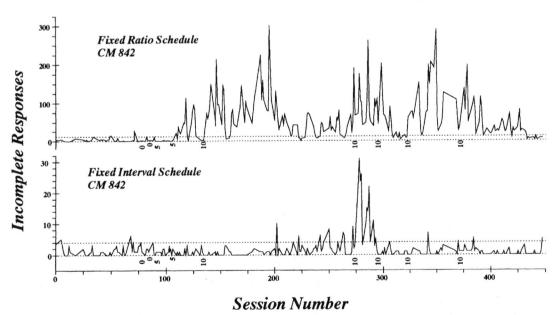

Session Number

Figure 1 Number of incomplete responses during the fixed ratio (top) and fixed interval (bottom) components for monkey CM 842. The scales on the ordinates are different for each figure. Dotted lines show the 5th and 95th percentiles taken from baseline sessions. There were usually five sessions per week. The numbers immediately above the horizontal axis show the dose of manganese (mg Mn kg⁻¹ body weight, i.v.). Note the 100-fold increase in incomplete responses under the fixed-ratio component, that declined and then increased again at about session 275. As this was occurring there was little detectable effect under the fixed interval schedule (reproduced by permission of Academic Press from *Toxicol. Appl. Pharmacol.*, 1992, **113**, 87–97).

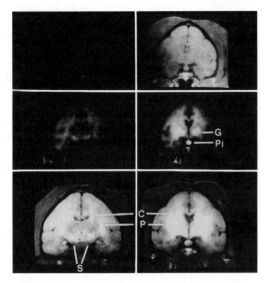

Figure 2 T_1-weighted magnetic resonance images from a monkey exposed to manganese. The top row shows an image taken before manganese, the middle row after a cumulative dose of $5\,mg\,kg^{-1}$ and the bottom row after a cumulative dose of $50\,mg\,kg^{-1}$. Note the darkening of globus pallidus (G), pituitary (Pi), and substantia nigra (S) after only $5\,mg\,kg^{-1}$ and of the putamen and caudate after $50\,mg\,kg^{-1}$ (reproduced by permission of Academic Press from *Toxicol. Appl. Pharmacol.*, 1992, **113**, 87–97).

exposure terminated. The early appearance of manganese in the images were associated with subclinical signs of manganese toxicity.[64]

The relative sensitivity of the fixed-ratio and fixed-interval schedules illustrated in Figure 1 can be compared with the effect of manganese as reported in other published reports (Figure 3). Increased incomplete responses under the fixed-ratio schedule and enhanced signals on magnetic resonance images appeared at manganese levels approximately 10 times lower than those that altered MRI signals in caudate and produced the clinical sign, action tremor. This comparison within a single study indicates both the sensitivity of some behavioral baselines and the importance of carefully consideration of the reinforcement contingencies. Comparisons across studies are also shown in Figure 3 and these comparisons indicate that other studies identified overt neurological signs at much higher doses, sometimes more than two orders of magnitude higher than those reported by Newland and Weiss.[63] It is not known whether the alterations in dopamine binding or levels represents a lowest effect adverse effect level because those were determined after the monkeys already showed neurological signs. Important differences across studies that could have

influenced the results include the behavior observed species and dosing rate.

Operant behavior under the control of reinforcement schedules, most likely some variant of ratio and interval schedules, may be a component of neurotoxicity assessment.[67,68] The characteristic patterns and rates of responding seen under different schedules of reinforcement have been the endpoints of most concern. As noted in this section, however, these schedules also produce different physical response characteristics and these, too, might be included in neurotoxicity assessment. It is plausible that motor effects will be reflected as alterations in response durations and that these effects might be separable from other behavioral effects, but a broader database than is currently available is required to secure such claims.

11.17.4.2   Direct Schedule Effects and Response Synthesis

The three-term contingency may well represent a fundamental mechanism by which behavior is molded by contact with the environment. This plasticity can be exploited to synthesize, or piece together, behavior that reflects specific nervous system function, a general approach that has been successful in the analysis of sensory function.[69,70] The literature on motor function includes impressive examples of training specific tasks, but our understanding of the determinants of behavior has yet to be applied to its full potential.

The tactic of training a specific response for examination always includes shaping by successive approximation. A variety of movements have been trained, depending upon the specific question of interest. For example, monkeys have been trained to exert continuous movements using a joystick or of some other tracking device in order to examine accuracy, to distinguish ballistic movements from slower, more guided movements, to examine the role of visual or other sensory control in movement, and to examine control over movement by velocity, trajectory, or past movements, control subsumed under the term "prediction;"[71–74] and to lift small weights to examine proprioceptive control over the estimation of weight.[75] Rats have been trained to reach, release levers in a reaction time task, lick, and press a load cell with a specific force. With a good understanding of behavior, a specific task can be designed for a research question.[42,55]

Rather than review different ways in which specific applications might be trained, information that can be found in other sources,[45,55,76]

Manganese Neurotoxicity in Primates

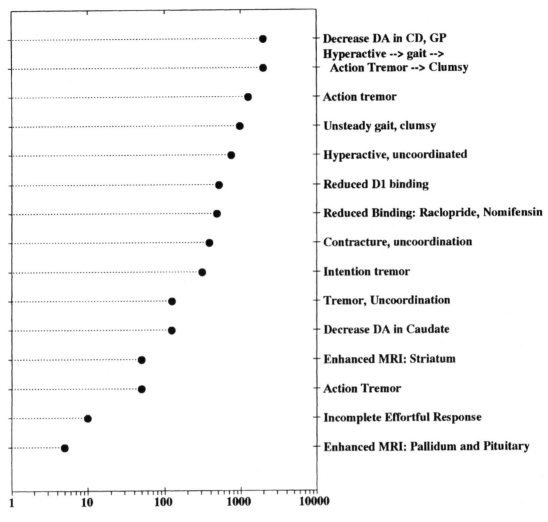

Figure 3 Lowest cumulative dose at which certain effects of manganese appeared. For example, enhanced MRI appeared at 5 mg kg[-1] (after Newland and Weiss,[63] Eriksson *et al.*,[65] and Suzuki *et al.*[66]).

this section will summarize some ways in which behavior and motor components of the nervous system can interact. To do so, studies of operant behavior will be drawn upon. Respondent behavior (produced by respondent or Pavlovian conditioning), habituation, and reflexive modifications all provide excellent opportunities for analysis,[77] but these will not be covered here.

11.17.4.3   Acquisition

Acquisition has been described as the selection of behavior by its consequences.[78] Because the number of activities that an organism can

do simultaneously is limited, the selection (training) of one activity necessarily involves the elimination of others, but, just as with natural selection, both variability and a mechanism of selection is required. The interaction between variation and selection by reinforcement contingencies can be seen in a report by Notterman and Mintz[61] (reproduced in Ref. 54) who describe the acquisition and extinction of a force-based response by a rat. Presses on a load cell with a force exceeding 3 g, an easy response for a rat, were reinforced. Prior to this training the force with which the rat pressed the load cell ranged from about 2 to 50 g. After training, however, the range was greatly reduced by the imposition of a simple contingency of reinforcing any force greater than 3 g; after about 15 reinforcers the range was reduced to about 8 g. Extinction (cessation of the reinforcement contingency) restored variability to 50 g. The transition from highly variable responding to tight control over the physical execution of a response was abrupt and spanned only a few responses.

A similar acquisition phenomenon was reported by Brooks et al.[74] who trained a monkey to move its forearm in a horizontal arc from a center position to the left or the right. Throughout the course of training, erratic, incomplete, and discontinuous movements decreased while smooth, rapid movements increased. The change took place so rapidly (although more slowly than the rat experiment described above) that it occasioned the researchers to describe it as reflecting insight and understanding. The impressive process seen in the shaping of a new response reflects a point of contact between the contingencies of reinforcement and the neural organization of a motor act and is a potential target for neurotoxic investigation. Acquisition processes are necessarily difficult to study, but when they have been studied they have proven to be both quantifiable and sensitive.[79,80]

11.17.4.4 Stimulus Control

The three-term contingency describing operant behavior includes discriminative and reinforcing roles for stimuli, and these influence motor acts. A stimulus can function discriminatively, occasioning the initiation of a movement or guiding the next movement in a sequence. A stimulus can be a consequence of movement, a situation sometimes described as feedback in a closed-loop system. Finally, a stimulus can function in both capacities. Discussions of the influence of stimuli over behavior usually refers to exteroceptive events but it can be expected that control by more private stimuli will follow similar behavioral principles, even if they do so along different neural substrates.

Behavior under the control of exteroceptive stimuli is so resistant to acute and chronic disruption by exposure to drugs or to toxic substances[81,82] that such stimuli might be called "behavioral prosthesis."[37] The ability of exteroceptive control to provide resistance to perturbation is not restricted to motor endpoints but instead shows generality across several behavioral measures, even as it shows some pharmacological specificity. For example, neuroleptics like chlorpromazine and promazine, which block dopamine receptors and produce both antipsychotic and Parkinsonian-like effects when administered acutely, interact with exteroceptive stimulus control in ways different from that seen with drugs from other classes.[36,37] This interaction could appear with interoceptive stimuli as well. In rats trained to press a load cell with a force of either 8 or 32 g, the high-force task was more resistant to disruption by haloperidol than the low-force task, an effect attributed to the greater stimulus control proprioceptively produced by pressing with a high force.[44]

Reliance on exteroceptive stimuli to guide, or function as discriminative stimuli for, behavior has been described in association with Parkinson's disease. Such patients become increasingly reliant upon exteroceptive cues to carry out tasks that previously were conducted unconsciously, such as walking. The simple placement of strips of adhesive on the ground at about the distance of a stride can greatly improve the gait of a Parkinsonian patient.[35] Disorders that damage the substantia nigra or globus pallidus result in enormous and relatively selective impairment in tracking-like tasks in which the movement is regular and predictable but for which there is no stimulus functioning in a discriminative or reinforcing way.[32,83,84] Adding a stimulus in these situations results in a pronounced improvement in performance.

11.17.4.5 Adjustment to Impairment

Sometimes the behavioral disruption produced by effects of chemical exposure subsides with continued opportunities to behave while under the influence of the chemical. This effect appears as a diminished sensitivity to the dose of a drug or toxicant and accordingly meets the definition of tolerance, although the mechanism may be found in behavior rather than in the liver or at the synapse. Behavioral tolerance is most likely to appear when reinforcer loss is a consequence of the behavioral disruption.[85] For example, Schuster et al.[86] showed that amphetamine's rate increasing effects decreased the reinforcement rate under a differential reinforcement of the low-rate schedule but not under a fixed-interval schedule. Tolerance to these rate-increasing effects appeared only under differential reinforcement of the low-rate schedule, the schedule for which the drug-induced increases in response rate reduced the reinforcement rate.

Adjustment to motor effects has also been reported under conditions in which it appears that the consequence of chemical exposure is more closely related to increased effort in producing a response, even while reinforcer loss is minimal. Tang *et al.*[87] trained rats to execute a response with a force of 0.147–0.265 N (equivalent to 14.7–26.5 g). Behavioral tolerance to midazolam appeared in the accuracy of the response but not to measures of rate. LeBlanc *et al.*[88] administered ethanol to rats either before or after being placed on a moving belt. The rats who behaved while under the influence of ethanol acquired tolerance considerably faster than those who received ethanol after the moving belt test. Both groups received the same doses of ethanol at the same frequency; the only difference was whether they had the opportunity to behave while the drug was active. In this case, there is no change in the delivery of a reinforcer, but an argument can certainly be offered that the aversiveness of falling on the moving belt could operate in a similar manner (see also Ref. 89).

In the experiment on manganese neurotoxicity described above, Newland and Weiss[63] observed dynamics that were interpreted as adjustment to impairment. A large increase in incomplete responses showed a decline (although not to baseline) over the course of several months after exposure. This decline was reversed at the same time that action tremor appeared. A second decline in incomplete responses followed, even though action tremor continued. The pattern of response durations through the course of the 20 responses comprising the fixed-ratio schedule also showed some adjustment to severe perturbations. Before exposure the durations showed a U-shaped distribution such that they were longer earlier and later in the ratio than in the middle. Manganese made all durations long and abolished the pattern, but the pattern was partially restored with continued exposure to the schedule. Manganese also disrupted inter-response time distribution during the ratio component but this disruption did not change, so the adjustment to impairment may have been specific to the response duration.

Fowler *et al.*[90] examined the effects of triazolam, a benzodiazepine minor tranquilizer, on the pressing of a conventional lever maintained under a fixed-ratio schedule. Some rats received triazolam chronically before a session and others chronically after the session, a design similar to that used by Leblanc *et al.*,[88] and a design advocated as a requirement for concluding that behavioral mechanisms contribute to tolerance.[85] No tolerance occurred to the rate-reducing effects of the drug, even though the effect reduced the rate of reinforcement and behavioral tolerance would have counteracted this effect. Acute triazolam nearly doubled the duration of responses in the group administered triazolam after the session but the rats that behaved while the drug was behaviorally active, that is, when the drug was administered before a session, showed no such increase in response duration. This study provides an example of behavioral adjustment to impairment in the absence of a sizable change in reinforcer rate. Response durations were fairly short so even their doubling did not appreciably change the reinforcement rate, while the decline in overall response rate, which showed no tolerance, certainly changed it.

The control conditions required to confirm the involvement of behavioral tolerance should be applied where possible to demonstrate compellingly that behaviorally mediated adjustment to impairment has occurred. That is, there should be one condition in which the drug is administered before behavioral sessions, so the subject can behave while the drug is active. A second group should be administered the drug at the same frequency, but after behavioral sessions. This permits a comparison of tolerance or adjustment through behavior or physiological (e.g., alteration in liver enzymes or receptor sensitivity) mechanisms.

With toxicants that produce chronic impairment, such control conditions cannot be accomplished directly so adjustment to impairment must be shown by other means. An irreversible lesion of the dorsal root ganglia in nonhuman primates prevents sensory control over the use of one arm. When a second arm was still available, use of the lesioned arm declines considerably, but when the second arm can no longer be used, the lesioned arm must be used and adjustment to the impairment develops.[91] This observation led to a successful clinical intervention with stroke victims.[92] In the lesion studies, the lesion was clearly in place and, in one condition unilateral so that some control over the duration of the lesion was observable. In related studies, reorganization of the cerebral cortex was also noted in monkeys long after deafferentation.

There is much to be learned about adjustment to impairment to determine if it follows the same principles as behavioral tolerance. If it does, then it can be expected that an opportunity to behave in the presence of impairment is necessary but not sufficient for adjustment to impairment to develop. Consequences must follow behavior although the consequences could be in the form of reduced effort rather than direct increases in reinforcement rate. It might also be expected that the environmental conditions under which adjustment occurs are an important component. The degree to which induction from the originally trained response follows such adjustment to impairment needs to be assessed. More suitable and efficient control conditions also need to be developed. The extent to which such adjustment alters dose–effect relationships when there is ample opportunity to behave is also unknown, although by definitions, such adjustment would shift dose–effect curves to the right.

11.17.5 THE USES AND MEASUREMENT OF TREMOR

Alterations in normal physiological tremor is a consequence of exposure to a wide variety of neurological disorders[93,94] as well as exposure neurotoxicants including manganese,[29,65,95] ethanol,[96] pesticides,[97,98] neuroleptics,[99] and mercury vapor.[100] As shown in Table 4, chemicals can increase tremor, change the pattern of tremor, and even decrease its amplitude. Limb oscillations that appear as tremor occur up to about 25–30 Hz in frequency. In humans there is a dominant mode, typically around 8–12 Hz but often more than one mode is observed.

Tremor represents the action of a limb, and as such represents the action of all components of the motor system. Tremor has several causes, including the pulsing of the blood, the duration of the stretch reflex, central oscillators that dominate frequencies up to about 12 Hz, and the activity of motor units associated with muscle contraction appearing in the higher frequencies.[91,102,103] It has been suggested that the frequency of tremor represents the maximum speed at which the nervous system can move voluntarily.[102–104]

In general, dysfunction has been associated with increases in tremor, but this may be due to the fact that increases are easier to see than decreases. Broad-band decreases in tremor have been reported after acute ethanol exposure;[96] decreases in high frequencies have been reported in muscular dystrophy;[102,103] disruptions in the high-frequency band, including decreases in the relative power in high frequencies may be associated with developmental disability;[101] and chronic neuroleptic use by humans[101] and acute neuroleptic exposure in rats have been associated with such high-frequency alterations.[55] Tremor approaches being an apical test and that alone makes the effort behind quantifying it worthwhile, even though its anatomical bases, function, and even the utility of various classification schemes are still poorly understood. Perhaps neurotoxicology can contribute to an understanding of tremor. This section describes considerations to be used in the measurement of tremor and the value of providing measures of tremor with good measurement properties.

11.17.5.1 Quantification of Tremor

When an outstretched limb is observed carefully it can be seen to oscillate at some frequency (in cycles per second, or Hertz) and amplitude (e.g., millimeters). What is being observed is the dominant frequency of normal, physiological tremor and it is seen as changes in position. The observation takes place for a brief period of time, called a temporal window. Such

Table 4   Effects of some representative interventions on tremor.

Intervention	Type of tremor	Frequencies affected
Cerebellar lesions	Increased kinetic ("intention") Postural	3–5 Hz 3–5 Hz, may be slower
Basal ganglia, Parkinson's disease	Increased resting, especially "pill-rolling"	3–5 Hz
Peripheral neuropathy	Increased physiological Lack of exacerbation with anxiogenics	8–12 Hz
Anxiogenics, ethanol withdrawal	Increased broad-band Influenced by anxiolytics	8–12 and others
Acute ethanol	Decreased broad-band	2–25
Tardive dyskinesia	Altered profile in higher frequencies	>7 Hz
Acute haloperidol	Elevated high frequency (rat)	>10 Hz
Cholinergic agonist (e.g., chlordecone, oxotremorine)	Increased high frequency	7–20 Hz

Source: Refs. 53, 55, 93, 94, and 101.

an observation captures the most obvious element of tremor, the dominant frequency, but it misses subtleties. Large increases would be detectable easily, but chemical-induced reductions would be impossible to see this way, making precise measurement critical.

Spectral analysis converts the temporal signal into a combination of frequencies that describe the composition of the signal as a linear sum of sine waves.[105] The spectral form has the great advantage of quantifying the dynamic nature of tremor to permit comparisons across exposure conditions or linkages to nervous system function. The analytic tool most commonly used to describe the frequency spectrum of tremor is the fast Fourier transform, which converts a time-domain signal (changes in position/unit time) into a sum of sine waves in which the sine waves oscillate at different frequencies. Each sine term is characterized by three elements: magnitude, frequency, and phase (which is rarely used when

measuring tremor). The graphical representation of a tremor spectrum shows a measure of amplitude on the vertical axis and frequency on the horizontal axis. Often amplitude is described in units of variance whose dimensions are amplitude squared; a technique, called power spectral analysis, resembles the analysis of variance. The analysis of variance breaks down total variance into orthogonal contributions from different experimental conditions, while power spectral analysis breaks down total variance into orthogonal contributions from different frequencies.[105,106]

11.17.5.2 Considerations in the Measurement of Tremor

The first step in the measurement of tremor is the selection of a transducer. The most common classes of transducers that have been used are those that measure position, accelera-

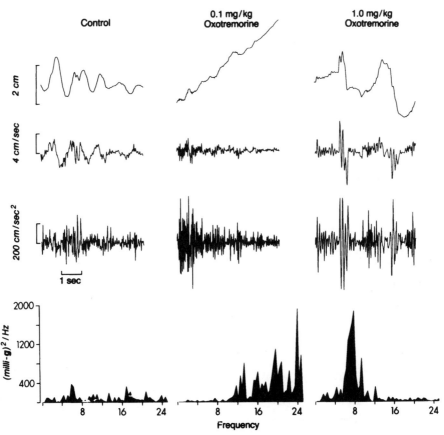

Figure 4 Representative signals taken from a squirrel monkey using a displacement transducer. Each column contains data from a different dosing condition. The first row shows displacement of the bar through a 5.12 s sampling interval. The second row shows the first derivative (velocity) and the third row shows the second derivative in units of acceleration. The bottom row is the power spectrum of the acceleration signal in row 3. A dose of 0.1 mg kg^{-1} of oxotremorine elevated power in the higher frequencies while a dose of 1 mg kg^{-1} resulted in a single, pronounced mode at about 6 Hz (reproduced by permission of Raven Press from 'Neurobehavioral Toxicity: Analysis and Interpretation,' 1994).

tion, or force (see Ref. 95, Table 1). Accelerometers are most commonly used when measuring human tremor because they are light and easily attached to a limb. Accelerometers can also measure tremor in moving or stationary limbs, an advantage if the category of tremor (moving, resting, etc.) is also of interest and the acceleration signal emphasizes frequencies in the range of tremor (greater than about 6 Hz) while de-emphasizing slower movements (less than about 2 Hz). Load cells, which are sometimes used, measure force, which is the product of mass and acceleration. Therefore, the output from load cells emphasizes the same frequencies as an accelerometer.[99,100] Load cells offer the further advantage of permitting the assessment of tremor under conditions of exertion, but since they are isometric, movement cannot be included in the measurement, and enough force must be applied to avoid damping by soft fleshy tissue-like finger pads.

When using transducers to detect tremor in nonhuman species, the ease of operating the transducer and the likelihood that the subject will destroy the transducer are additional considerations. Load cells[99,107] or position transducers[96,105] have been selected on the basis of these considerations and also have advantages if tracking is of interest. Figure 4 illustrates tremor as detected in a monkey gripping a bar that was attached to a rotary variable differential transformer, a transducer that measures angular position (described and illustrated by Newland[96,105]). The monkey reached with its arm to grip a handle positioned directly in front of it. Radial movement in the vertical dimension was permitted by the transducer. Holding the bar within an angular range for 10 s was reinforced with fruit juice.

The top row in Figure 4 illustrates the signal in units of position. A 5.12 s sample was taken from a control session and after a low or high dose of oxotremorine, a muscarinic cholinergic agonist. The position signal in the first row emphasizes slow changes in position but more rapid oscillations caused by tremor are difficult to see. The second row shows a velocity signal, which is the first derivative of position or change in position per unit time. The third row shows an acceleration signal such as would be obtained from an accelerometer or load cell and is simply the derivative of velocity. Slow changes in position have been greatly de-

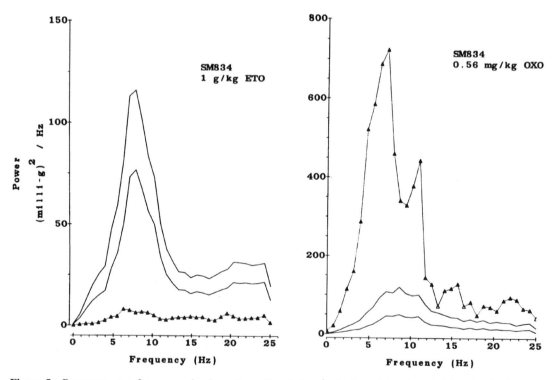

Figure 5 Power spectra from a squirrel monkey after 1 g kg^{-1} of ethanol (p.o.) or 0.56 mg kg^{-1} of oxotremorine (i.m.). In each panel the mean spectrum from the drug conditions (triangles) is compared with the upper and lower 95% confidence intervals (lines) taken from days preceding drug administration. Note the change of scale on the ordinates. The control sessions for oxotremorine and ethanol were taken from sessions separated by about a year but the profile and magnitude almost match (reproduced by permission of Alcohol Research Documentation Inc., Center for Alcohol Studies at Rutgers University from *J. Stud. Alcohol*, 1991, **52**, 492–499).

emphasized in the acceleration form of the signal (although it could be recovered by integration), while higher frequency changes that are of interest in the assessment of tremor are visible. Differences across drug conditions can be seen most clearly in the bottom row, which shows the spectral content of the acceleration signals on the third row. The low dose of oxotremorine greatly increased high frequencies across a broad band, a sort of nonspecific jitter, which is nearly invisible in the position signal on the top row. The higher dose produced regular undulations which appear as a pronounced mode at 6 Hz in the frequency spectrum at the bottom, and which are visible, on close inspection, in the top row.

Differentiating the position signal to achieve data in units of acceleration, as was accomplished in Figure 4, amplifies the effects of noise, because the magnitude of the influence of higher frequencies increases with the value of the frequency. Filtering and averaging across spectra, called ensemble averaging, can ameliorate some of these effects since true noise will average to zero. A comparison of human tremor taken from an accelerometer and the

position transducer used to produce the spectra in Figure 4 can be seen in Ref. 105.

Figure 5 compares tremor in a squirrel monkey after ethanol with tremor seen after oxotremorine. Dose–effect information on oxotremorine can be seen in Newland[54] and for ethanol in Newland and Weiss.[96] The right panel of Figure 5 shows a control profile remarkably similar to the bottom right panel in Figure 4. The magnitudes are smaller, probably reflecting a lower dose, but the spectra share the presence of a pronounced dominant frequency, whose magnitudes are similar to one another. By contrast, note the spectrum taken after ethanol in the left panel of Figure 5. This reflects a broad-band reduction in tremor, changes that would be impossible to detect visually by noticing displacement but that are striking in the spectrum. This effect, which has been noted in patients with essential tremor, begins to appear at doses of ethanol well below those that produce other behavioral effects (a comparison of this effect with others of ethanol can be seen in Ref. 53). The neural mechanism for this reduction is unknown but it might be related to the observations that low levels of

Table 5 Considerations when applying spectral analysis to a digital record using fast Fourier techniques.

Resolution
  The lowest frequency that can be estimated is 1/(record length)
  Analyzed frequencies are integer multiples of the lowest frequency

Aliasing
  The highest frequency that can be estimated is 1/2 the sampling frequency
  Higher frequencies must be removed before digitizing

Smoothing
  Frequency smoothing is required to overcome the effect of sampling in a temporal window
    In the frequency domain this involves estimating power in a frequency by averaging over that and
      adjacent windows (e.g., Hanning)
    In the time domain this appears as tapering the onset and offset of the sample

  Ensemble averaging
    The more frequency-domain records that enter an estimate, the smaller the standard error of the
      estimate
    The standard error can be estimated empirically from the samples or theoretically from a formula
    Record length is unrelated to the standard error of a frequency estimate

Number of samples
  Equals the product of sampling rate and record length
  Should be a power of 2 for most efficient use of the data and best estimate

Common measures
  Acceleration (the standard)
  Force (product of acceleration and mass)
  Position (convert to acceleration by differentiating twice and smoothing)

Remove the contribution of the measuring device
  Accelerometers are light enough that their contribution is insignificant
  If an animal operates a manipulandum, or is placed in a cage on a load cell, the dynamics of that system
    must be accounted for

ethanol interfere with the activities of components in the stretch reflex[108] or to ethanol's inhibition of the neuron by enhancing chloride conductance. Functionally, this reduction in tremor is likely to be linked to such impairments in motor function as increased reaction time, postural sway, and impaired adjustment to perturbations in posture or gait.[53,109] Insofar as tremor represents the preparation of the nervous system to react quickly, ethanol-induced reductions could act to make such reactions more sluggish.

Table 5 lays out some of the measurement considerations that must be made in the assessment of tremor. Some elaboration on these issues can be found in Ref. 53 or an advanced text[106] on signal processing can be consulted. These considerations apply when tremor is assessed by (i) creating an analogue measure of tremor using a transducer, (ii) digitizing that signal and storing the digital representation as a record of some temporal length, and (iii) analyzing the digital record using a fast Fourier transform.

Consider the use of a 5.12 s, 100 Hz sample of the output of an accelerometer. The output from a position transducer would work well, too, after the position signal is differentiated twice to produce acceleration. The lowest frequency that can be estimated from this signal is $1/5.12\,s = 0.1953\,Hz$. At least one complete cycle is required to form an unambiguous estimate. All other estimates will be integer multiples of this base frequency.

The highest frequency that can be estimated in this example is 50 Hz, and it is called the Nyquist folding frequency. At least two digitized samples per cycle must be provided (100 Hz will sample a 50 Hz signal twice for each cycle). Since there is little information in tremor above 25–30 Hz, the sampling rate of 100 Hz, which produces a Nyquist frequency of 50 Hz, is adequate for this purpose. If any higher frequencies are present in the signal, then they will contribute to the estimate in the 0.1953–50 Hz range, creating an error called aliasing. For example, a 60 Hz signal (10 Hz higher than the 50 Hz maximum) will appear in the frequency estimate at 40 Hz (10 Hz lower). Aliasing is an outcome of the digitizing process and its cause can be seen visually in Ref 53. Frequencies above the upper cutoff must be filtered out of the signal before digitizing takes place. This can be achieved by analogue filtering or by sampling at a rate higher than any that will be present, and that value must be known. One might sample at say 10 000 Hz, apply a digital filter to the 10 000 Hz sample, and then down-sample to the 100 Hz by taking every 100th point.

A signal, or the spectrum, must be smoothed in two ways. One, called ensemble averaging, is the familiar average except that it is taken at each frequency component. In the example, suppose that 100 samples of 5.12 s are taken and Fourier transformed. This produces 100 frequency spectra. The 100 estimates at 0.1953 Hz can be averaged together, as can the 100 estimates at each other frequency up to 50 Hz. The standard error can also be determined in this averaging process, or by a theoretical consideration. Bendat and Piersol[106] have shown that if the data are log-transformed before averaging then the normalized standard error of the estimate is the square root of $(2/N)$, where N is the number of estimates. In this example it would be the square root of $2/100 = 0.14$. If the log of the magnitude of the estimate is 10, then the standard error is 1.4.

The second form of smoothing applies to the temporal window. If sampling is initiated abruptly then the analysis routine will treat the signal as instantaneously moving from 0 to some value. Similarly, if sampling is terminated abruptly then the signal will move instantaneously from some value to 0. In the frequency domain such rapid changes are represented by high frequencies. These high frequencies contaminate the spectral estimate since they are an artifact of the abrupt initiation of sampling, and not present in the signal being sampled. The solution is to taper the onset and offset of the signal. In the frequency domain this appears as a moving average. One common approach is called Hanning, in which the estimate at some frequency is half the value at that frequency plus half the mean of the value of the frequencies on either side of it. This process removes some of the power, but the amount is known and the actual value can be restored by multiplying each estimate by 1.56.[110]

For optimal performance of the fast-Fourier algorithm, the number of samples should be a power of 2. That is why the sample in the example was 5.12 s: 100 samples s^{-1} times $5.12\,s = 512$ samples, a power of 2.

The transducer measures all contributions to the signal. This includes the tremoring limb, the mass of the transducer, and anything in between. When attaching an accelerometer to a human limb, the contribution of the dynamic properties of the accelerometer are negligible. If tremor is assessed by placing a rat in a cage and mounting the cage on a load cell, then temporal changes in the measures from the load cell will be a function of the mass of the cage, vibrations in the cage, as well as the tremoring animal. If the rat suddenly falls down, the cage may continue oscillating even while the rat is lying still. As a quick test to determine the "transfer function" of this system, that is, its spectral profile, place the whole system, including a static mass the same as the animal to be used, on the transducer. Then tap it with a hammer and measure the spectral content. Any frequencies present in the band to be measured must be accounted for, but the techniques for doing this are beyond the scope of this discussion (see Ref. 106).

The assessment of tremor sounds very involved but the presence of digital computers and software packages makes it much easier to perform this sort of analysis. Caution must be exerted as software contains errors (always a painful lesson) and a careless choice of options can render the outcome uninterpretable. The above discussion can provide a guide to the choices. Two approaches to testing a software package are mentioned in Newland.[53] First, generate a signal containing a known spectrum by summing a number of sine waves together and analyzing the output. Second, take advantage of the fact that the integral of the power spectrum equals the variance in the time-domain signal. These numbers are often produced by separate analytical paths so the variance should be compared with the area under the power curve, and both of these numbers are usually readily available. These provide a check on the software and on the particular application.

11.17.5.3 Human Tremor

As described in the discussion on measurement scales, if dependent measures have reliable measurement properties then comparisons across applications can be made more precisely. Spectral analyses of tremor contain both sets of measurement properties. The numbers are ratio scaled and if the investigator takes the trouble to describe the outcome in units of acceleration, rather than arbitrary units like volts, then the numbers are anchored to a physical constant and can be compared across applications.

Table 6 summarizes some measures of human tremor under different conditions drawn from the frequency of the dominant peak, when one can be estimated,[53] the magnitude of the dominant peak in mgs, and the power in (mg)^2. One g is the acceleration due to gravity, 981 cm s^{-1} s^{-1}. A mg is 1/1000 of that and is a more convenient unit for describing tremor. The typeface is made distinctive to distinguish the abbreviation from that for grams. A footnote in Table 6 mentions a difficulty in estimating the power because of an ambiguous notation in describing power. A more comprehensive discussion of the units of tremor can be found in Ref. 53.

The variation in the total magnitude of normal tremor spans approximately a 30-fold

Table 6   Descriptions of human postural tremor from selected experiments that use accelerometers.[a]

| Condition and limb | Total power (mg)^2 | Characteristics of spectral peak | | | Source |
		Frequency (Hz)	Power (mg)^2 Hz^{-1}	RMS power (mg)	
Normal hand	1369 (445–2400)	6 (4.9–7.1)	543 (180–992)	23 (3–31)	115
Normal hand	NA	7.0 (5.8–8.3)	611[b] (2–1225)	17[b] (1–35)	117
Normal hand	3370 (1170–11 160)	7.2[b] (6–8.3)	570[c]	24[c]	111,113
Parkinson's, hand	93 000 (27 000–279 000)	5.7[b] (4.5–6.9)	28 000[c]	167[c]	113
Essential tremor, hand	195 000 (45 000–604 000)	7.2[b] (5.7–8.6)	108 000[c]	329[c]	113
Normal, finger	73 (16–162)	6.4 (3.2–9.3)	11 (2–20)	3 (1–5)	114
Normal, ankle	12 (4–49)	7.5[b] (6–9)	23[c]	4	116

[a]Selected reports from which sufficient information could be extracted. The value 1$g^2 \times 10^{-3}$ was interpreted to by 0.001g^2 and is entered as 1000 (mg)^2. When a range is available, it is shown in parentheses beneath entries.   [b]The mid range was used since no estimate of central tendency was available.   [c] Populations statistics were unavailable so values from figures describing representative subjects are used.

range, although most values are of the order of 1000–2000 mg^2, which would translate to a root-mean-square (RMS, and equivalent to the standard deviation) of about 32–95 mg. This number describes total variance in the signal before it has been isolated into different frequencies. The RMS power in the peak frequency, when one could be identified, was of the order of about 15–25 mg, although the values also spanned about a 30-fold range. Despite the wide range, distinguishing pathological states such as Parkinsonian or essential tremor could be readily accomplished as these values were 10–100 times greater.

In contrast, the estimates of the peak frequency, when one could be detected, were more stable across subjects and generally fell in the 6–8 Hz range for finger or hand tremor. About 5% of adults and 60% of children have no clearly definable peak and many have more than one.[112] The content of other elements of the frequency spectrum may derive from other sources. For example, power above about 12 Hz seems to reflect the firing of motor neurons.

An appreciation of how such tremor might appear on visual inspection can be gained by converting acceleration to displacement using the relationship:

$$\text{cm of displacement} = \frac{0.981(a)}{(2\pi f)^2} = \frac{0.025a}{f^2} \quad (1)$$

In this relationship, acceleration (a) is in units of cm s^{-2} and frequency (f) is in units of cycles s^{-1}. The magnitude of displacement is directly related to the magnitude of acceleration and inversely related to the square of the frequency. Applying this relation to the normative data in Table 6 indicates that 1–5 mg of tremor in the finger results in about 0.0005–0.0025 cm displacement at 7 Hz, while the displacement in essential tremor in the hand ranges from about 0.02 cm to 0.2 cm at 7 Hz. It has been said that at about 7 mg, tremor becomes noticeable and at about 30 mg, it becomes socially embarrassing.[113] The small displacements involved indicate why a reduction in tremor would be so difficult to visualize directly, unless the baseline is large as in essential tremor. The dependence on 1/(frequency^2) reveals why alterations in high-frequency tremor are so difficult to see; the displacements are tiny. The low-frequency tremor in Parkinson's disease is striking because the displacements are so large. The importance of each form of tremor, however, is not necessarily reflected in how visible it is.

Despite the superior measurement properties of tremor and the potential for providing norm-

ative data on it, we are still a long way from using tremor to make precise diagnoses. Two difficulties present themselves immediately. The first is that there is still considerable uncertainty about the mechanisms of the different components of tremor.[111,112] The second reason is that there is a lack of agreement about how to provide summary measures of tremor. The total power or variance is a crude measure but does not capture important information contained in the shape of the spectrum. The frequency and magnitude of the peak captures a portion of the spectrum but misses potentially important information available from the higher or lower frequencies, and sometimes a clearly-defined peak is not available. This is not to say that tremor should not be used. On the contrary, spectral analyses of tremor should be used more and greater care should be taken to describe tremor quantitatively in units that permit comparisons across experiments. As an adequate base arises in the published literature, methods for summarizing descriptions of tremor are more likely to suggest themselves.

11.17.6 CONCLUDING REMARKS

This chapter has been an attempt to summarize some of the various considerations that enter an examination of motor dysfunction at the level of the intact, behaving subject. It has drawn from the study of behavior, of measurement, and of the neural sciences to do so. The general subject of assessing motor dysfunction offers many opportunities to bring together two areas of research that previously have shown little awareness of each other: environmental determinants of behavior and neural control over the execution of a motor act.

There are many ways in which practical concerns will arise when measures of motor dysfunction enter the regulatory arena. Although these concerns have not been addressed directly, they certainly lie behind the discussion offered. Before risk assessment and management can be conducted, it is necessary to have reliable and meaningful measures to work with. The goal of this chapter has been to bring neural, behavioral, and measurement sciences into the development of motor endpoints that are reliable, valid, comparable across settings, and quantitatively useful.

11.17.7 REFERENCES

1. W. K. Anger, 'Neurobehavioral testing of chemicals: impact on recommended standards.' *Neurobehav. Toxicol. Teratol.*, 1984, **6**, 147–153.
2. C. H. Vanderwolf, B. Kolb and R. K. Cooley,

'Behavior of the rat after removal of the neocortex and hippocampal formation.' *J. Comp. Physiol. Psychol.*, 1978, **92** 156–175.

3. I. Q. Whishaw in 'The Cerebral Cortex of the Rat,' eds. B. Kolb and R. C. Tees, MIT Press, Cambridge, MA, 1990, pp. 239–267.

4. I. Q. Whishaw, W. T. O'Connor and S. B. Dunnett, 'The contributions of motor cortex, nigrostriatal dopamine, and caudate-putamen to skilled forelimb use in the rat.' *Brain*, 1986, **109**, 805–843.

5. I. Q. Whishaw, T. Schallert and B. Kolb, 'An analysis of feeding and sensorimotor abilities of rats after decortication.' *J. Comp. Physiol. Psychol.*, 1981, **95**, 85–103.

6. R. Richter, 'Degeneration of the basal ganglia in monkeys from chronic carbon disulfide poisoning.' *J. Neuropath. Exp. Neurol.*, 1945, **4**, 324–353.

7. I. Creese and S. D. Iversen, 'Blockage of amphetamine induced motor stimulation and stereotypy in the adult rat following neonatal treatment with 6-hydroxy-dopamine.' *Brain Res.*, 1973, **55**, 369–382.

8. I. Creese and S. D. Iversen, 'The role of forebrain dopamine systems in amphetamine induced stereotyped behavior in the rat.' *Psychopharmacologia*, 1974, **39**, 345–357.

9. C. N. Karson, 'Spontaneous eye-blink rates and dopaminergic systems.' *Brain*, 1983, **106**, 643–653.

10. I. Q. Whishaw and B. Kolb, ' "Stick out your tongue": tongue protrusion in neocortex and hypothalamic damaged rats.' *Physiol. Behav.*, 1983, **30**, 471–480.

11. M. Pisa, 'Motor functions of the striatum in the rat: critical role of the lateral region in tongue and forelimb reaching.' *Neuroscience*, 1988, **24**, 453–463.

12. R. T. O'Keeffe and K. Lifshitz, 'Nonhuman primates in neurotoxicity: screening and neurobehavioral toxicity studies.' *J. Am. Coll. Toxicol.*, 1989, **8**, 127–140.

13. F. B. Jolicoeur, D. B. Rondeau, E. Hamel *et al.*, 'Measurement of ataxia and related neurological signs in the laboratory rat.' *Can. J. Neurol. Sci.*, 1979, **6**, 209–215.

14. S. G. Gilbert and J. P. J. Maurissen, 'Assessment of the effects of acrylamide, methylmercury, and 2,5-hexanedione on motor functions in mice.' *J. Toxicol. Environ. Health*, 1982, **10**, 31–41.

15. B. M. Kulig, R. A. P. Vanwersch and O. L. Wolthuis, 'The automated analysis of coordinated hindlimb movement in rats during acute and prolonged exposure to toxic agents.' *Toxicol. Appl. Pharmacol.*, 1985, **80**, 1–10.

16. A. F. Youssef, B. Weiss and C. Cox, 'Neurobehavioral toxicity of methanol reflected by operant running.' *Neurotoxicol. Teratol.*, 1993, **15**, 223–227.

17. P. S. Spencer and H. H. Schaumburg, 'Experimental and clinical neurotoxicology,' Williams & Wilkins, Baltimore, MD, 1980.

18. G. T. Pryor, E. T. Uyeno, H. A. Tilson *et al.*, 'Assessment of chemicals using a battery of neurobehavioral tests: a comparative study.' *Neurobehav. Toxicol.*, 1983, **5**, 91–117.

19. C. Ghez, in 'Principles of Neural Science,' eds. E. R. Kandel, J. H. Schwartz and T. M. Jessell, Appleton & Lange, Norwalk, CT, 1991, pp. 609–625.

20. K. R. Reuhl and L. W. Chang, 'Effects of methylmercury on the development of the nervous system: a review.' *Neurotoxicology*, 1979, **1**, 21–55.

21. D. O. Marsh, in 'The Toxicity of Methyl Mercury,' eds. C. U. Eccles, Z. Annau and Z. Annau, Johns Hopkins University Press, Baltimore, MD, 1987, pp. 45–53.

22. P. Brotchie, R. Iansek and M. K. Horne, 'Motor function of the monkey globus pallidus.' *Brain*, 1991, **114**, 1667–1683.

23. A. Parent, 'Extrinsic connections of the basal ganglia.' *Trends Neurosci.*, 1990, **13**, 254–258.

24. G. E. Alexander, M. R. DeLong and P. L. Strick, 'Parallel organization of functionally segregated circuits linking basal ganglia and cortex.' *Ann. Rev. Neurosci.*, 1986, **9**, 357–381.

25. M. R. DeLong and A. P. Georgopoulos, in 'Handbook of Physiology, Section 1: The Nervous System, Vol. 2: Motor Control,' ed. V. B. Brooks, American Physiological Society, Bethesda, MD, 1981, pp. 1017–1061.

26. M. R. Delong, A. P. Georgopoulos, M. D. Crutcher *et al.*, in 'Functions of the Basal Ganglia,' eds. D. Evered and M. O'Connor, Pitman, London, 1984, pp. 74–77.

27. W. T. Thach and E. B. Montgomery, in 'Neurobiology of Disease,' eds. A. L. Pearlman and R. C. Collins, Oxford University, New York, 1990, pp. 168–196.

28. L. Cote and M. D. Crutcher, in 'Principles of Neural Science,' eds. E. R. Kandel, J. H. Schwartz and T. M. Jessell, Appleton & Lange, E. Norwalk, CT, 1991, pp. 647–659.

29. A. Barbeau, 'Manganese and extrapyramidal disorders.' *Neurotoxicology*, 1984, **5**, 13–35.

30. A. Barbeau, N. Inoue and T. Cloutier, 'Role of manganese in dystonia.' *Adv. Neurol.*, 1976, **14**, 339–352.

31. K. Jellinger, 'Handbook of Clinical Neurology,' Elsevier, Amsterdam, 1986, vol. 5, pp. 465–489.

32. K. A. Flowers, 'Visual "closed-loop" and "open-loop" characteristics of voluntary movement in patients with Parkinsonism and intention tremor.' *Brain*, 1976, **99**, 269–310.

33. K. A. Flowers, 'Lack of prediction in the motor behavior of Parkinsonism.' *Brain*, **101**, 1978, 35–52.

34. J. P. Martin, 'The Basal Ganglia and Posture,' Pitman, London, 1967.

35. H. Forssberg, B. Johnels and G. Steg, 'Is Parkinsonian gait caused by a regression to an immature walking pattern.' *Adv. Neurol.*, 1984, **40**, 375–379.

36. V. G. Laties and B. Weiss, 'Influence of drugs on behavior controlled by internal and external stimuli.' *J. Pharmacol. Exp. Ther.*, 1966, **152**, 388–396.

37. V. G. Laties, 'The role of discriminative stimuli in modulating drug action.' *Fed. Proc.*, 1975, 1880–1888.

38. T. R. Barnes, 'The present status of tardive dyskinesia and akathisia in the treatment of schizophrenia.' *Psychiatr. Dev.*, 1987, **5**, 301–319.

39. B. Weiss, S. Santelli and G. Lusink, 'Movement disorders induced in monkeys by chronic haloperidol treatment.' *Psychopharmacology (Berl.)*, 1977, **53**, 289–293.

40. R. S. Burns, C. C. Chiueh, S. P. Markey *et al.*, 'A primate model of parkinsonism: selective destruction of dopaminergic neurons in the pars compacta of the substantia nigra by *N*-methyl-4-phenyl-1,2,3,6-tetrahydropyridine.' *Proc. Natl. Acad. Sci. USA*, 1983, **80**, 4546–4550.

41. N. J. Raichi, S. I. Harik, R. N. Kalaria *et al.*, 'On the mechanisms underlying 1-methyl-4-phenyl-1,2,3,6-tetrahydropyridine neurotoxicity. II. Susceptibility among mammalian species correlates with the toxin's metabolic patterns in brain microvessels and liver.' *J. Pharmacol. Exp. Ther.*, 1988, **244**, 443–448.

42. S. C. Fowler and C. Mortell, 'Low doses of haloperidol interfere with rat tongue extensions during licking: a quantitative analysis.' *Behav. Neurosci.*, 1992, **106**, 386–395.

43. S. C. Fowler, 'Neuroleptics produce within-session response decrements: facts and theories.' *Drug Dev. Res.*, 1990, **20**, 1–15.

44. S. C. Fowler, P. D. Skjoldager, R. M. Liao *et al.* 'Distinguishing between haloperidol's and decamethonium's disruptive effects on operant behavior

in rats: use of measurements that complement response rate.' *J. Exp. Anal. Behav.*, 1991, **56**, 239–260.

45. S. C. Fowler, in 'Advances in Behavioral Pharmacology,' eds. T. Thompson and P. B. Dews, Erlbaum, New York, 1987.

46. S. C. Fowler, R. J. Filewich and M. R. Leberer, 'Drug effects upon force and duration of response during fixed-ratio performance in rats.' *Pharmacol. Biochem. Behav.*, 1977, **6**, 421–426.

47. P. F. Gilbert and W. T. Thach, 'Purkinje cell activity during motor learning.' *Brain Res.*, 1977, **128**, 309–328.

48. W. T. Thach, H. P. Goodkin and J. G. Keating, 'The cerebellum and the adaptive coordination of movement.' *Ann. Rev. Neurosci.*, 1992, **15**, 403–442.

49. P. H. Ayres and W. D. Taylor, in 'Principles and Methods of Toxicology,' ed. A. W. Hayes, Raven Press, New York, 1989, pp. 111–135.

50. J. Elsner, C. Fellmann and G. Zbinden, 'Response force titration for the assessment of the neuromuscular toxicity of 2,5-hexanedione in rats.' *Neurotoxicol. Teratol.*, 1988, **10**, 3–13.

51. J. R. Cooper, F. E. Bloom and R. H. Roth, 'The Biochemical Basis of Neuropharmacology,' 6th edn., Oxford University Press, New York, 1991.

52. H. A. Tilson and V. C. Moser, 'Comparison of screening approaches.' *Neurotoxicology.*, 1992, **13**, 1–14.

53. M. C. Newland, in 'Neurobehavioral Toxicity: Analysis and Interpretation,' eds. B. Weiss and J. L. O'Donohgue, Raven Press, New York, 1994, pp. 273–298.

54. M. C. Newland, in 'Neurotoxicology: Approaches and Methods,' eds. L. W. Chang and W. Slikker, Academic Press, San Diego, CA, 1995, pp. 265–300.

55. M. J. Kallman and S. C. Fowler, in 'Principles of Neurotoxicology,' ed. L. W. Chang, Marcel Dekker, New York, 1994, pp. 373–396.

56. J. M. Johnston and H. S. Pennypacker, 'Strategies and Tactics of Human Behavioral Research,' Lawrence Erlbaum Associates, Hillsdale, NJ, 1980.

57. W. W. Stine, 'Meaningful inference: the role of measurement in statistics.' *Psychol. Bull.*, 1989, **105**, 147–155.

58. S. S. Stevens, 'Handbook of Experimental Psychology,' Wiley, New York, 1951.

59. M. Zeiler, in 'Handbook of Operant Behavior,' eds. W. K. Honig and J. E. R. Staddon, Prentice-Hall, Englewood Cliffs, NJ, 1977.

60. M. C. Newland and B. Weiss, 'Drug effects on an effortful operant: pentobarbital and amphetamine.' *Pharmacol. Biochem. Behav.*, 1990, **36**, 381–387.

61. J. M. Notterman and D. E. Mintz, 'Dynamics of Response,' Wiley, New York, 1965.

62. B. Weiss and D. A. Cory-Slechta, in 'Principles and Methods of Toxicology,' 3rd edn., ed. A. W. Hayes, Raven Press, New York, 1994, pp. 1091–1155.

63. M. C. Newland and B. Weiss, 'Persistent effects of manganese on effortful responding and their relationship to manganese accumulation in the primate globus pallidus.' *Toxicol. Appl. Pharmacol.*, 1992, **113**, 87–97.

64. M. C. Newland, T. L. Ceckler, J. H. Kordower *et al.*, 'Visualizing manganese in the primate basal ganglia with magnetic resonance imaging.' *Exp. Neurol.*, 1989, **106**, 251–258.

65. H. Eriksson, K. Magiste, L. O. Plantin *et al.*, 'Effects of manganese oxide on monkeys as revealed by a combined neurochemical, histological, and neurophysiological evaluation.' *Arch. Toxicol.*, 1987, **61**, 46–52.

66. Y. Suzuki, T. Mouri, Y. Suzuki *et al.*, 'Study of subacute toxicity of manganese dioxide in monkeys.' *Tokushima J. Exp. Med.*, 1975, **22**, 5–10.

67. W. F. Sette, in 'Neurobehavioral Toxicity: Analysis

and Interpretation,' eds. B. Weiss and J. L. O'Donoghue, Raven Press, New York, 1994, pp. 231–242.

68. W. F. Sette and T. E. Levine, in 'Neurobehavioral Toxicology,' ed. Z. Annau, Johns Hopkins University Press, Baltimore, MD, 1986, pp. 391–403.

69. D. Blough and P. Blough, in 'Handbook of Operant Behavior,' eds. W. K. Honig and J. E. R. Staddon, Prentice-Hall, Englewood Cliffs, NJ, 1977, pp. 514–539.

70. D. C. Rice in 'Neurobehavioral Toxicity: Analysis and Interpretation,' eds. B. Weiss and J. L. O'Donoghue, Raven Press, New York, 1994, pp. 299–318.

71. R. C. Miall, D. J. Weir and J. F. Stein, 'Planning of movement parameters in a visuo-motor tracking task.' *Behav. Brain Res.*, 1988, **27**, 1–8.

72. T. E. Milner, 'Judgment and control of velocity in rapid voluntary movements.' *Exp. Brain Res.*, 1986, **62**, 99–110.

73. W. G. Darling, K. J. Cole and J. H. Abbs, 'Kinematic variability of grasp movements as a function of practice and movement speed.' *Exp. Brain Res.*, 1988, **73**, 225–235.

74. V. B. Brooks, P. R. Kennedy and H. G. Ross, 'Movement programming depends on understanding of behavioral requirements.' *Physiol. Behav.*, 1983, **31**, 561–563.

75. R. M. Wylie and C. F. Tyner, 'Performance of a weight-lifting task by normal and deafferented monkeys.' *Behav. Neurosci.*, 1989, **103**, 273–282.

76. V. B. Brooks, 'The Neural Basis of Motor Control,' Oxford Univerisity Press, New York, 1986.

77. L. D. Fechter and J. S. Young in 'Neurobehavioral Toxicology,' ed. Z. Annau, Johns Hopkins University Press, Baltimore, MD, 1986, pp. 23–42.

78. B. F. Skinner, 'Selection by consequences.' *Science*, 1981, **213**, 501–504.

79. M. C. Newland, Y. Sheng, B. Logdberg *et al.*, 'Prolonged behavioral effects of *in utero* exposure to lead or methylmercury: reduced sensitivity to changes in reinforcing stimuli during behavioral transitions and in steady state.' *Toxicol. Appl. Pharmacol.*, 1994, **126**, 6–15.

80. D. M. Thompson and J. M. Moerschbaecher, 'An experimental analysis of the effects of d-amphetamine and cocaine on the acquisition and performance of response chains in monkeys.' *J. Exp. Anal. Behav.*, 1979, **32**, 433–444.

81. D. C. Rees, R. W. Wood and V. G. Laties, 'The roles of stimulus control and reinforcement frequency in modulating the behavioral effects of d-amphetamine in the rat.' *J. Exp. Anal. Behav.*, 1985, **43**, 243–255.

82. V. G. Laties, 'The modification of drug effects on behavior by external discriminative stimuli.' *J. Pharmacol. Exp. Ther.*, 1972, **183**, 1–13.

83. C. A. Bloxham, T. A. Mindel and C. D. Frith, 'Initiation and execution of predictable and unpredictable movements in Parkinson's disease.' *Brain*, 1984, **107**, 371–384.

84. Y. Stern, R. Mayeux, J. Rosen *et al.*, 'Perceptual motor dysfunction in Parkinson's disease: a deficit in sequential and predictive voluntary movement.' *J. Neurol. Neurosurg. Psych.*, 1983, **46**, 145–151.

85. P. K. Corfield-Sumner and I. P. Stolerman, in 'Contemporary Research in Behavioral Pharmacology,' eds. D. E. Blackman and D. J. Sanger, Plenum, New York, 1978, pp. 391–448.

86. C. R. Schuster, W. S. Dockens and J. H. Woods, 'Behavioral variables affecting the development of amphetamine tolerance.' *Psychopharmacologia*, 1966, **9**, 170–182.

87. M. Tang, C. E. Lau and J. L. Falk, 'Midazolam and discriminative motor control: chronic administration, withdrawal, and modulation by the antago-

nist Ro 15-1788.' *J. Pharmacol. Exp. Ther.*, 1988, **246**, 1053–1060.

88. A. E. LeBlanc, H. Kalant and R. J. Gibbins, 'Acquisition and loss of behaviorally augmented tolerance to ethanol in the rat.' *Psychopharmacology (Berl.)*, 1976, **48**, 153–158.

89. J. R. Wenger, P. M. McEvoy and S. C. Woods, 'Sodium pentobarbital-induced cross-tolerance to ethanol is learned in the rat.' *Pharmacol. Biochem. Behav.*, 1986, **25**, 35–40.

90. S. C. Fowler, S. E. Bowen and M. J. Kallman, 'Practice-augmented tolerance to triazolam: evidence from an anlaysis of operant response durations and interresponse times.' *Behav. Pharmacol.*, 1993, **4**, 147–157.

91. E. Taub and A. J. Berman, in 'The Neuropsychology of Spatially Oriented Behavior,' ed. S. J. Freedman, Dorsey, Homewood, IL, 1968, pp. 173–192.

92. E. Taub, N. E. Miller, T. A. Novack *et al.*, 'Technique to improve chronic motor deficit after stroke.' *Arch. Phys. Med. Rehabil.*, 1993, **74**, 347–354.

93. R. J. Elble and W. C. Koller, 'Tremor,' Johns Hopkins University Press, Baltimore, MD, 1990.

94. L. J. Findley and R. Capildeo (eds.), 'Movement Disorders: Tremor,' Oxford University Press, New York, 1984.

95. N. H. Neff, R. E. Barrett and E. Costa, 'Selective depletion of caudate nucleus dopamine and serotonin during chronic manganese dioxide administration to squirrel monkeys.' *Experientia*, 1969, **25**, 1140–1141.

96. M. C. Newland and B. Weiss, 'Ethanol's effects on tremor and positioning in squirrel monkeys.' *J. Stud. Alcohol*, 1991, **52**, 492–499.

97. J. M. Gerhart, J. S. Hong and H. A. Tilson, 'Studies on the possible sites of chlordecone-induced tremor in rats.' *Toxicol. Appl. Pharmacol.*, 1983, **70**, 382–389.

98. D. W. Herr, J. S. Hong, P. Chen *et al.*, 'Pharmacological modification of DDT-induced tremor and hyperthermia in rats: distributional factors.' *Psychopharmacology (Berl.)*, 1986, **89**, 278–283.

99. S. C. Fowler, R. M. Liao and P. Skjoldager, 'A new rodent model for neuroleptic-induced pseudo-parkinsonism: low doses of haloperidol increase forelimb tremor in the rat.' *Behav. Neurosci.*, 1990, **104**, 449–456.

100. R. W. Wood, A. B. Weiss and B. Weiss, 'Hand tremor induced by industrial exposure to inorganic mercury.' *Arch. Environ. Health*, 1973, **26**, 249–252.

101. R. E. A. van Emmerik, R. L. Sprague and K. M. Newell, 'Finger tremor and tardive dyskinesia.' *Exp. Clin. Psychopharmacol.*, 1993, **1**, 259–268.

102. H. J. Freund, H. Hefter, V. Homberg *et al.*, in 'Movement Disorders: Tremor,' eds. L. J. Findley and R. Capildeo, Oxford University Press, New York, 1984, pp. 27–36.

103. H. J. Freund, H. Hefter, V. Homberg *et al.*, in 'Movement Disorders: Tremor,' eds. L. J. Findley and R. Capildeo, Oxford University Press, New York, 1984, pp. 195–204.

104. E. Logigian, H. Hefter, K. Reiners *et al.*, 'Does tremor pace repetitive voluntary motor behavior in Parkinson's disease?' *Ann. Neurol.*, 1991, **30**, 172–179.

105. M. C. Newland, 'Quantification of motor function in toxicology.' *Toxicol Lett.*, 1988, **43**, 295–319.

106. J. S. Bendat and A. G. Piersol, 'Random Data: Analysis and Measurement Procedures,' Wiley-Interscience, New York, 1971.

107. S. C. Fowler, C. Morgenstern and J. M. Notterman, 'Spectral analysis of variations in force during a bar-pressing time discrimination.' *Science*, 1972, 1126–1127.

108. C. M. Lathers and C. M. Smith, 'Ethanol effects on muscle spindle afferent activity and spinal reflexes.' *J. Pharmacol. Exp. Ther.*, 1976, **197**, 126–134.

109. D. B. Goldstein, 'Pharmacology of Alcohol,' Oxford University Press, New York, 1983.

110. F. J. Harris, 'On the use of windows for harmonic analysis with the discrete Fourier transform.' *Proc. IEEE*, 1978, **66**, 51–83.

111. C. D. Marsden, in 'Movement Disorders: Tremor,' eds. L. J. Findley and R. Capildeo, Oxford University Press, New York, 1984, pp. 37–84.

112. C. D. Marsden, J. C. Meadows, G. W. Lange *et al.*, 'Variations in human physiological finger tremor with particular reference to changes with age.' *Electroencephalogr. Clin. Neurophysiol.*, 1969, **27**, 169–178.

113. P. Wade, M. A. Gresty and L. J. Findley, 'A normative study of postural tremor of the hand.' *Arch. Neurol.*, 1982, **39**, 358–362.

114. R. H. Wyatt, 'A study of power spectra analysis of normal finger tremors.' *IEEE Trans. Biomed. Eng.*, 1968, **BMD-15**, 33–45.

115. H. Roels, J. Malchaire, J. P. van Wambeke *et al.*, 'Development of a quantitative test for hand tremor measurement. Study of factors influencing the response in a control population.' *J. Occup. Med.*, 1983, **25**, 481–487.

116. R. S. Pozos, P. A. Iaizzo and R. W. Petry, 'Physiological action tremor of the ankle.' *J. Appl. Physiol.*, 1982, **52**, 226–230.

117. V. Homberg, H. Hefter, K. Reiners *et al.*, 'Differential effects of changes in mechanical limb properties on physiological and pathological tremor.' *J. Neurol. Neurosurg. Psych.*, 1987, **50**, 568–579.

11.18
Somatosensory Neurotoxicity: Agents and Assessment Methodology

DEBORAH C. RICE
Bureau of Chemical Safety, Ottawa, ON, Canada

11.18.1   INTRODUCTION

Impairment in somatosensory system function is a frequent consequence of exposure to environmental, industrial, or agricultural agents, as well as some classes of therapeutic drugs. For example, Anger[1] reported tactile disorders to be a possible consequence of exposure to 77 of 750 industrial chemicals identified as neurotoxic. Despite this, batteries developed for assessment of industrial neurotoxicity, including the World Health Organization (WHO), the Neurobehavioral Evaluation System (NES), and Finland's Institute of Occupational Health (FIOH) batteries do not include direct assessment of any somatosensory function.[2] On the other hand, the US Agency for Toxic Substances and Disease Registry

(ATSDR) has developed a neurobehavioral battery for use in environmental health field studies in adults which includes assessment of vibrotactile function as a recommended part of the battery.[3–5] While determination of vibration sensitivity assesses only one modality of somatosensory function, its inclusion represents a recognition of the importance of the potential for somatosensory dysfunction as a consequence of exposure to neurotoxic agents.

The somatosensory system is composed of several modalities, including light touch, pain, temperature, and vibration sensitivities (Table 1). For many agents, the only available data consist of reports of symptoms such as numbness, a sensation of pins and needles, or pain of various types. Such reports have sometimes been pursued by physicians using relatively crude and subjective methods of clinical assessment. Objective assessment of various modalities of somatosensory function has been pursued in both animals and humans in order to directly assess the functional integrity of the system. Functional testing provides the most readily interpretable data concerning sensory deficits, yielding answers of direct relevance to the subject being tested, the clinician, and the epidemiologist. Electrophysiological endpoints, on the other hand, provide only indirect evidence of dysfunction. In addition, such techniques may be painful or uncomfortable to the subject, as well as invasive, and so are often unsuitable as tests for epidemiological research. Similarly, neuropathological techniques are both invasive and difficult to interpret, again providing indirect evidence of functional impairment.

This chapter discusses the various methodologies used in clinical and experimental settings to assess the sensation actually experienced by the subject, as well as the relevant methodological issues, strengths and weaknesses associated with the available procedures. Discussion is restricted to consideration of cutaneous and position senses, and excludes visceral sensations. A survey of the chemical agents reported to produce somatosensory dysfunction is also provided.

11.18.2 METHODOLOGICAL TECHNIQUES AND ISSUES

11.18.2.1 Anatomical Considerations

The somatosensory system includes components of the central nervous system (spinal column, somatosensory cortex) as well as the peripheral nervous system (see also Chapter 1, this volume). Therefore assessment of somatosensory function, while often performed with the objective of assessing peripheral neuropathy, in fact detects impairment at all levels of the nervous system. Different cutaneous modalities are subserved by several types of nerve fibers and end-organ receptors. Vibration, pressure, and proprioception are conveyed by large-diameter, heavily myelinated fibers, while pain and temperature are conveyed by thinly myelinated or unmyelinated fibers. Light touch is subserved by a spectrum of fiber sizes and dispersed tracts. It may often be the case that one fiber type is affected preferentially. Both taxol and cisplatin affect vibration and proprioception with relative sparing of pain and temperature,[6] while acrylamide affects vibration while sparing pain modalities.[7,8] It is important, therefore, to test more than one cutaneous sense. A number of clinical investigators have assessed vibration and temperature within a study, for example, which provides the opportunity to assess functions subserved by both large and small diameter peripheral fibers. Even within a modality, function may be subserved by different structures. Vibration above approximately 100 Hz is detected by Pacinian corpuscles or RA-II receptors, while lower frequencies are sensed by RA-I receptors; therefore assessment of vibration at a single frequency yields an incomplete picture of even this one cutaneous function. In addition, some agents produce motor or sensory impairment preferentially, while other agents impair both functions to an approximately equal degree. For example, n-hexane produces somatosensory deficits before motor signs such as muscle weakness,[9,10] while lead affects motor systems before cutaneous sensation. Acrylamide, on the other hand, affects both systems approximately equally.[11]

In most cases cutaneous dysfunction begins and may remain confined to distal structures. Since upper or lower limbs may be affected first

Table 1   Categories of somatosensory function.

Cutaneous	Light touch
	Pressure
	Vibration
	Temperature
	Superficial pain
Deep	Deep pain
	Deep pressure
	Joint, tendon, muscle proprioception
Visceral	Visceral pain
	Visceral sensation (e.g., hunger, thirst)

depending on the agent, it is advisable to test both. Occasionally, perioral numbness or paresthesia is reported: for example with synthetic pyrethroids (see Table 3) or methylmercury (see Table 2).

11.18.2.2 Clinical Methodology

The first line of assessment of somatosensory function in the clinical setting consists of a relatively crude and subjective determination of gross functional impairment. The clinician typically uses pin prick to assess pain, a tuning fork at one frequency (typically 256 Hz) for vibration, cotton wool or a camel brush for light touch, and two blunt needles for two-point discrimination (often calipers or a compass) (see Chapter 22, this volume). These are held against the skin in various places, and the patient is requested simply to state whether the stimulus was felt. Clinicians may also assess light touch by use of von Frey hairs, which are stiff wires of various diameters, pushed against the skin. Ability to detect temperature may be tested using a cool instrument or a test tube filled with hot or cold water. Stereognosis may be assessed by asking the patient to identify small objects placed in the hands, while graphesthesia is determined by asking the patient to identify numbers or letters written in the hand.[12] These methods suffer from uncontrolled characteristics of stimulus presentation, as well as such unmeasurable and variable psychological parameters as suggestibility, cooperativeness, and responsiveness. The clinician may also question the patient concerning pain, tingling, and numbness; the answers may depend on these same unmeasurable variables.

Clinicians have attempted to introduce more rigor into assessment of cutaneous function with a number of simple instruments. Two-point discrimination may be assessed by use of a pair of tracks or a groove in a V configuration; the point at which the subject can detect the separation of the two sides of the V is considered the threshold.[13,14] Depth discrimination has been assessed in a similar manner: one side of a groove slopes downward relative to the other side, and the subject reports when s/he detects a difference.[13,14] However, such devices still rely on subjective reports by the observer, and are influenced by the same uncontrollable variables mentioned above.

11.18.2.3 Psychophysics

The testing of sensory system function by behavioral means is termed psychophysics. The methodologies available provide powerful tools for assessment of the functional capabilities of sensory systems, including the somatosensory system. Psychophysical methodology enjoys a long history, beginning with the study of sensory physiology in the nineteenth century.[15] Psychophysics greatly benefitted from the field of signal detection theory developed in the 1950s, which provided a theoretical framework with which to discriminate the system's intrinsic discriminative capabilities (sensory function) from the nonsensory factors.

11.18.2.3.1 Psychophysical methods

Psychophysical methods describe the rules by which stimulus intensities are presented. There are a number of sets of rules for the presentation of stimuli. The most commonly used are briefly discussed.

In the method of limits (MOL), stimuli are presented sequentially in ascending and/or descending series. In a descending series, the relevant stimulus dimension (usually intensity) begins above the threshold and is decreased by the experimenter until the first report of "no detection" by the observer. In an ascending series, the stimulus starts below threshold and is increased sequentially until the first "yes" response is given. Several ascending and descending series should be presented, beginning at different stimulus levels to minimize anticipatory responses. The average of the midpoints between "yes" and "no" is determined to estimate threshold. However, this is often not done, at least in clinical research (see below). The method of limits has the advantage of being fast and easy, requiring less technology than other methods. It has the disadvantage of generating errors of anticipation, as mentioned above, and errors of habituation, in which the observer tends to repeat the preceding response. This procedure is used often in clinical research, but is generally unsuitable for experimental investigation with animals.

In the method of constant stimuli, a set of prechosen stimulus levels is presented in a random order; each set is presented a number of times. The threshold is calculated by determining the percent correct at each stimulus level, and the stimulus level corresponding to a predetermined correct response rate is determined by interpolation. This method is not prone to the same types of biases as is the method of limits. The disadvantage of this method is that many of the stimulus presentations do not contribute to the threshold calculation; therefore many presentations of each stimulus must be included to determine threshold accurately. This method is rarely utilized in humans, although it has been used quite extensively in animal models.

In tracking procedures, the order of stimulus presentation depends directly on the observer's responses. The most straightforward procedure is the simple up–down rule (UDR). In this procedure, if the stimulus is detected, the level of the next stimulus is decreased; if it is not detected, the level is increased on the next trial. The stimulus therefore varies around a level approximately equal to a 50% detection level. Threshold may be determined by simply averaging the stimulus levels presented, or averaging the midpoints between each change of direction of stimulus level. The great advantage of this procedure is that almost all stimulus presentations are close to the threshold, and therefore threshold can be determined reasonably quickly. A possible disadvantage is that the software required to control the stimulus level is necessarily more sophisticated than for the previous methods, although inexpensive, user-friendly computers make this less of a consideration than previously.

There are various possible modifications of the simple UDR. In these so-called up–down transformed rule(s) (UDTR), the rules for presentation of stimuli are varied from the UDR in a systematic manner. For example, two consecutive detections may be required before the stimulus decreases, with only one "miss" required to increase the stimulus. Under this rule, the stimulus will be correctly detected 70.7% of the time. There are many other possible rules for stimulus presentation; the relevant issues have been extensively discussed.[16–18] It is important to understand the impact of the chosen method on where threshold will be estimated (percent correct), and how many trials will be required for threshold estimation (efficiency). This method is very suitable for both animal and human testing.

The Parameter Estimation by Sequential Testing (PEST) procedure is also a variant of the UDR. It automatically determines each stimulus level according to a predetermined rule, in order to approach threshold as efficiently as possible. The step size is variable, and the testing period ends when a selected criteria has been met.[19,20] This is usually a time-efficient method, which has enjoyed some use in human research (but see Ref. 21).

11.18.2.3.2   *Behavioral response paradigm*

Psychophysical methods specify the rule governing stimulus presentation within the relevant domain (i.e., intensity, frequency).

The behavioral response paradigm specifies the set of rules governing the response requested of the observer. There are two major paradigms: the forced-choice procedure and the yes–no procedure. In the forced-choice paradigm, the subject is presented with two or more stimuli. In the spatial version (e.g., two-alternative forced-choice) the observer indicates in which position the stimulus is presented. In the temporal version (e.g., two-interval forced-choice), the observer indicates during which time period the stimulus occurred. Abstaining from making a choice is not allowed. The point of no detection in this paradigm is 50%; the threshold must therefore be defined as some point lying between 50% and 100%, usually 75% correct.

In the yes–no procedure, the observer must decide whether or not a stimulus was presented within a specified time period. It has been recommended that a high proportion of negative or "catch" trials be included to assess the frequency of false positive reports.[15,18] Many of the original discussions of this issue were written before modern computer technology was available, and may assume that the stimulus will be presented at a fixed time following a warning signal. Under those circumstances, of course, a large number of negative trials must be included. However, the form of this procedure that is often currently used includes a variable period between the onset of the possibility of stimulus presentation and the actual presentation of the stimulus (foreperiod). The proportion of false "yeses" during this period also provides a measure of false positives,[22] particularly if the range of foreperiods is at least several seconds. False negatives are often not assessed in this paradigm. Inclusion of trials with superthreshold stimuli provides a measure of this variable.[22] In the yes–no paradigm, the threshold is often set at the 50% detection point, although another value may be chosen. (For example, if a UDTR is used, the threshold may be estimated at a point other than 50% (see above)).

In both of these response paradigms, it is recommended that feedback be given to the subject regarding whether the response was correct.

11.18.2.4   Methodologies for Assessing Specific Modalities

11.18.2.4.1   *Vibration*

The modality measured most often by use of psychophysical techniques is vibration sense, probably in large part because the parameters of stimulus presentation are more easily controlled than for other modalities. The apparatus consists of an electromagnetic vibrator which is attached to a blunt probe or rod. The probe is pressed with a specified force against the area to be tested, generally a finger or toe (Figure 1). A piezoelectric accelerometer attached to the shaft generates an electrical signal proportional to the displacement of the probe, thereby permitting accurate measurement of vibration amplitude. The vibration frequency and amplitude are controlled electronically, usually by computer. Some investigators have designed their own systems, and provide a source of detailed description of the hardware system.[23–25] Typically one frequency is tested at a time, and the vibration amplitude is varied between trials. The vibration is typically sinusoidal, and the onset and offset should be gated (i.e., have a non-instantaneous rise and decay time). It is important to ensure that the vibratory stimulus is not detected aurally. This is usually accomplished in animal (monkey) research by covering the ears with headphones through which white noise is delivered; this potential confound is apparently ignored in human testing. Stimulus intensity is presented according to the chosen psychophysical method and behavioral response paradigm (see section on psychophysical techniques). Stimulus intensity may be expressed validly as displacement of the probe in micrometers,

specifying whether this is expressed as peak-to-peak, peak, or root mean square (RMS). Data also may be validly expressed as change in decibels (dB) relative to a reference value (usually 1 μm). It is necessary to calibrate the system by a physical means such as a stroboscope or other device, in order to ensure that theoretical calculations of displacement based on accelerometer calibration are correct. It is also essential to calibrate system components on a regular basis during the course of the study, to verify that performance of the several electronic and mechanical components has not changed over time. Failure to calibrate the system on a regular basis means that there is no guarantee that the data collected at the beginning and end of the study are comparable. Failure to calibrate combined with expressing the data in idiosyncratic (nonsense) units precludes the opportunity to compare data across studies, as well as depriving the informed reader of the ability to assess whether the data seem reasonable (an indication of whether the study was properly performed). It may seem that these points are obvious, but are often not followed by researchers using commercially available systems (see below).

There are a number of systems commercially available for assessing vibration sensitivity. The Vibrameter (currently Vibrameter III, Somedic AB, Stockholm) is a hand-held 13 mm plastic probe that vibrates at 100 Hz. The data output provided by the supplier is "volts." Data are typically expressed in these units, with no indication that the investigators calibrated the instrument.[26,27] A similar device, the Biothesiometer (Biomedical Instrument Co., Newbury, Ohio), has also been used.[28–30] Data are expressed in "units." With both of these instruments, the amplitude of vibration is highly dependent on the pressure with which the probe is applied, which is very difficult to control. An instrument currently in popular use is the Vibratron (Sensortek or Physiotemp, Clifton, NJ).[21,30–37] It has two ports that vibrate at 100 Hz; the two stimuli provide the opportunity to use the two-alternative forced-choice method. The data output is expressed as "units" from 1 to 20. A third instrument in common use is the Optacon (Telesensory Systems, Palo Alto, CA). This consists of a 24 × 6 matrix of rods vibrating at 230 Hz; the amplitude is varied by a potentiometer. The unit of measure provided by the company is "volts"; the actual output is usually not determined.[14,38,39]

Goldberg and Lindblom[28] performed a systematic analysis of variables affecting vibration amplitude. They performed complete calibrations of two Biothesiometer units with the instrument placed against different tissues

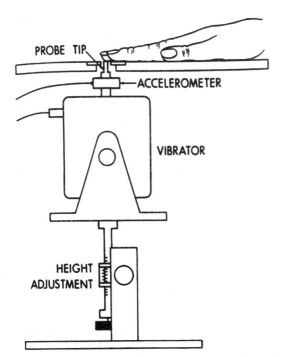

Figure 1 Schematic drawing of an assembly used for assessment of vibration sensitivity (reproduced by permission from *Neurobehav. Toxicol.*, 1979, **1**, 23–31).

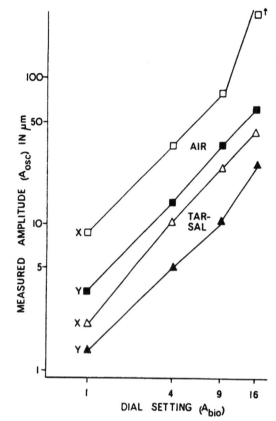

Figure 2 Comparison of two different Biothesiometer units (X and Y): relationship between preset dial setting (1, 4, 9 and 16 μm) and actual (measured) amplitudes, in air and on metatarsal stimulation of the same person. (reproduced by permission of British Medical Journal from *J. Neurol. Neurosurg. Psychiatry*, 1979, **42**, 793–803).

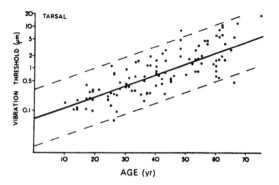

Figure 3 Relationship between age and vibration threshold for metatarsal stimulation in 110 healthy subjects. Vibration threshold and age were highly correlated for both upper and lower extremities (reproduced by permission of British Medical Journal from *J. Neurol. Neurosurg. Psychiatry*, 1979, **42**, 793–803).

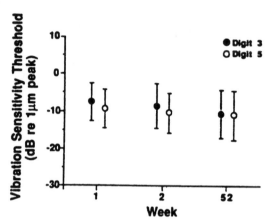

Figure 4 Mean (± 1 SD) of the absolute threshold of vibration sensitivity, in normal subjects, tested in two fingers at 1, 2 and 52 weeks (*n* = 29) (reproduced by permission of Brtitish Medical Journal from *J. Neurol. Neurosurg. Psychiatry*, 1986, **49**, 1191–1194).

and in air. They found that for a series of dial settings on the instrument, the actual amplitude varied several fold as a function of the tissue being tested. Further, the actual amplitudes of the two instruments varied by a factor of two, and were considerably higher than the values supplied by the manufacturer (Figure 2). The observed differences should not be unexpected. Such an analysis underscores the necessity of complete and routine calibration of the system being used.

Several investigators have explored the biological variables affecting vibration thresholds. There is a direct correlation between age and increased vibratactile thresholds[28,32,33,37,40] (Figure 3). There is also a relationship between increased vibration threshold and greater height.[32,40] It appears that neither gender nor skin temperature over the range usually encountered has consistent effects on vibration thresholds. In a study designed to assess the stability of vibration thresholds over time, 29 healthy females were tested using a two-interval

forced-choice procedure and a routinely calibrated investigator-designed system.[23] The third and fifth digits were each tested at 150 Hz, one week and one year apart. There was no change in threshold over the course of a year (Figure 4).

11.18.2.4.2 Temperature

Probably the second most common modality assessed in human clinical and toxicological investigation is temperature sense. Temperature and vibration are often both assessed within a single study, a good strategy since both large (vibration) and small (temperature) diameter fiber damage is thereby assessed.

Systems have been developed by individual investigators.[26,41–44] These systems usually take advantage of a Peltier element controlling the temperature of a metal plate in contact with the skin. When current is passed through the bimetallic junction of dissimilar elements, one side will cool and the other heat, depending on the polarity of the current. Reversal of the temperature of the plate in contact with the skin is accomplished by reversal of the applied current. The rate of change is typically $0.5–1.0° s^{-1}$. Thresholds for both warming and cooling are sensitive to age.[43] There is at least one commercially available unit, the Temperature Sensitivity Tester (TST) (Sensortek Inc., Clifton, NJ). The TST consists of two 4.8×4.6 cm metal plates that are cooled or heated by water circulating through them. The temperature of each plate is controlled separately. The temperature of the plate is displayed on a digital readout; the accuracy of this may not be checked by the investigator.[30,36,45]

11.18.2.4.3 Other modalities

In an attempt to improve on the standard clinical procedure of pushing a hand-held von Frey hair against the skin to assess light touch, an automated method was developed to present modified von Frey hairs with specified degrees of force.[46] Several procedures were examined for stimulus presentation and data analysis, derived from standard psychophysical techniques and signal detection theory. The authors concluded that the forced-choice method was best and the standard clinical procedure the worst of the tested procedures in detecting differences in sensitivity between body parts in normal subjects. A system designed by Dyck et al.[41] assessed touch-pressure sensitivity by applying a controlled wave-form that feels like a light tap or touch. A forced-choice paradigm using a UDTR rule converging on 75% correct detections was employed. Several methods have been developed to assess pain, such as a "pinchometer" in which skin folds are pinched against flat plates with a known force.[29,47] Such methods represent an attempt on the part of clinical investigators to overcome the subjectivity inherent in the standard clinical neurological examination.

11.18.2.4.4 Behavioral methodological errors

In addition to a lack of understanding concerning adequate characterization and control of stimulus presentation, investigators also may confuse psychophysical methods and be-

havioral response paradigms. Gerr and Letz[21] used the Vibraton II to assess vibration sensitivity in 22 healthy volunteers, comparing a yes–no procedure using the MOL with a two-alternative forced-choice procedure using a PEST method developed by the manufacturer. The authors concluded that the "method of limits was found to be more reliable and . . . less time consuming than the recommended forced-choice procedure," a statement which suggests confusion between the choice of the psychophysical method (MOL vs. PEST) and the behavioral paradigm (forced-choice vs. yes–no). In fact, the forced-choice procedure used was a variation of a suboptional UDTR in which two of three correct choices were required before the stimulus intensity was lowered, with two-out-of-three incorrect required for the stimulus intensity to increase. Such a procedure is extremely inefficient, as discussed by Maurissen.[18] Threshold was determined in a nonstandard way based on only eight values. The fault was not with the use of the forced-choice procedure, but with the use of an inefficient method of stimulus presentation coupled with determination of threshold based on too few data points. However, the following year a study in painters was published with one author from the just-described paper in common, discussing the inadequacies of the PEST procedure used previously.[30] A new PEST was developed that calculated the maximum likelihood estimate of the stimulus level likely to elicit a correct response from the subject 75% of the time, based on responses on all previous trials, following each trial. The authors argued that this method provides the maximum amount of information on the location of the subject's threshold (presumably in the least amount of time). The same confusion concerning psychophysical methods and behavioral response paradigm has also been made by other investigators.[48,49] Another frequent error is to include only a descending sequence of method of limits and/or using only two or three determinations for threshold determination.[29,44] Failure to understand these issues of appropriate experimental design may result in time wasted by the use of inefficient methods of stimulus presentation, and/or generation of unreliable data due to systematic bias or thresholds determined on the basis of an insufficient number of data points.

11.18.2.4.5 Correlation between symptoms and assessment techniques

Somatosensory dysfunction is usually heralded by numbness, paresthesia, and/or dysesthesia, as evidenced by the lists of effects in

Tables 2–5. There would not necessarily be expected to be a high correlation between such subjective symptoms and assessments of somatosensory function. Nor would functions subserved by large vs. small fibers be expected to be correlated, depending on the neurotoxic agent. The correlation between subjective clinical examination and more objective means of assessing sensitivity in various modalities is of particular interest. Cherniak *et al.*[35] compared a symptoms score, clinical assessment, functional tests, vibration thresholds, and electrophysiological measures in shipyard workers using pneumatic grinding tools, and reported significant correlations between two-point discrimination, tuning fork, objective assessment of vibration threshold, and reported symptoms. Nerve conduction velocities were not correlated with other measures. A study performed in patients receiving a standard hospital electrodiagnostic evaluation found a relationship between vibration thresholds, clinical neurological examination results, and nerve conduction velocities.[50] However, correlations were not particularly high. Neuropathy symptoms score, vibratory and cooling detection thresholds, nerve conduction velocity, and clinical neurological examination were compared in a study in diabetic patients.[42] The authors concluded that while most measures were associated as determined by regression analysis, the association was not high enough to be predictive. In a study in painters, there was no correlation between the results of qualitative assessment, neurological examination, or reported symptoms in individuals found to have higher vibration and temperature thresholds on objective assessment.[30] The authors speculate that quantitative assessment may have detected early, presymptomatic effects, or that lack of correlation that may have resulted from individual differences in labeling intensity of reported events. Similar results have been found in other populations.[36,51]

11.18.2.5 Techniques for Animal Studies

There are several simple or screening techniques appropriate for assessment of somatosensory function in animal models. Screening tests for neurotoxicity often include assessment of "sensorimotor function." These tests, typically performed with rodents, include such measures as fore- and hindlimb grip strength, foot splay when dropped from a height, and ability to stay on a rotating rod (rotarod). Impairment on these measures may be due to impairment of somatosensory as well as motor function, including feedback from joints and muscles as

well as skin. Such tests may be particularly sensitive to agents that produce both sensory and motor peripheral neuropathy, such as acrylamide.[52] However, it is not possible to differentiate motor from sensory dysfunction using these procedures alone.

Tests for assessment of pain and/or temperature sensation have been used in the pharmaceutical industry since at least the 1950s. The jumping or flinching of a rodent when shock is applied to the feet is considered an indication of the intactness of cutaneous sensation. Pain is often evaluated by pinching the tail or foot to elicit a withdrawal response. Pain and/or temperature sensitivity are assessed by immersing the tail in hot water or placing the animal on a hot plate, and determining the latency of withdrawal. The frequency of head turns toward the flank in guinea pigs with flanks coated with synthetic pyrethroids has been used to quantify the tingling and burning sensations produced by these agents.[53] The results of all of the tests will obviously be confounded by motor deficits. For example, Tilson and Burne[54] found that triethyl tin (TET) produced impairment on the tail flick and hot plate tests, but produced no effect on the ability of the rat to detect a shock applied to the tail when a small motor movement (a nose poke) was required to terminate the shock. They concluded the effects of TET were motor rather than sensory, and cautioned regarding interpretation of presumably sensory assessments that require substantial motor responses.

Operant methodology has been used extensively to assess sensory function in animals by psychophysical means.[55,56] The most commonly assessed somatosensory function in animals is undoubtedly vibration sensitivity. For example, a modified tracking, yes–no procedure was used to assess vibration sensitivity in aging monkeys exposed developmentally to methylmercury or over the lifetime to lead.[25] Vibration was delivered to a fingertip using a blunt needle attached to a computer-controlled system. Threshold was determined over a number of frequencies. Deficits in vibration sensitivity at frequencies above 60–100 Hz were observed in both methylmercury-exposed and lead-exposed monkeys. The effects of acrylamide exposure also have been studied in monkeys using vibration sensitivity.[7,8] A yes–no tracking procedure was used to follow the course of acrylamide intoxication and recovery, using two vibration frequencies corresponding to the two end-organ receptor types. Deficits were observed during dosing at both frequencies, and recovered over the course of months after dosing ceased, outlasting effects such as weight loss and impairment on a visuomotor task. There was no deficit in the ability to detect an electrical stimulus, indicating sparing of small fibers. Nerve biopsy revealed mostly normal-appearing nerve fibers, with a small proportion of fibers exhibiting changes typical of acrylamide toxicity. In a similar experiment,[57] the effects of administration of misonidazole, a radiosensitizing agent used in cancer therapy, were determined over a range of vibration frequencies. Vibration sensitivity was impaired for at least 6 months after cessation of dosing, at which point testing was discontinued. These changes were observed in the absence of changes in sensory or motor nerve conduction velocity.

11.18.3 SELECTED AGENTS

11.18.3.1 Methylmercury

World attention focused on the potential for agents released into the environment to cause severe neurological impairment with the outbreaks of methylmercury poisoning in Minamata and later in Niigata, Japan in the 1950s and 1960s (see Chapter 29, this volume). Methylmercury was discharged from industrial plants into rivers that flowed into ocean areas from which local inhabitants fished. The methylmercury was taken up by marine biota and bioconcentrated by fish, which were ingested by humans. A total of over 1000 people were eventually diagnosed as suffering from methylmercury poisoning. In two out-

breaks of poisoning in Iraq, including a large episode in 1971, over 6000 people suffered severe neurological disease as the result of ingesting organomercury-treated seed grain.[58] In both Japan and Iraq, adult poisoning was characterized by motor and sensory impairment, including somatosensory dysfunction (Table 2). Based on the Iraqi data, paresthesias were determined to be the most sensitive indicator of poisoning in adults, occurring before other symptoms and signs such as ataxia or hearing deficits.[59] Somatosensory impairment is also a cardinal symptom of Minamata disease.[60] Symptoms include decreased sensitivity in a glove-and-stocking pattern, although tingling and hyperesthesia have also been reported. Perioral dysesthesia is present in some cases. Hypoesthesia or dysesthesia of mouth and limbs was still present as long as 30 years after poisoning, contributing to difficulty in daily activities such as eating, bathing, and dressing.[61] These impairments grew relatively worse compared to unexposed individuals as a function of advancing age.

11.18.3.2 Triorthocresyl Phosphate

Triorthocresyl phosphate is an organophosphorus substance that has been responsible for several incidents of mass poisoning. During Prohibition, contamination of an alcoholic beverage containing ginger extract with triorthocresyl phosphate (TOCP) led to an epidemic

Table 2 Environmental agents.

Agents	Effects	Ref.
Chemicals		
Lead	Paresthesia, decreased vibration sense	25,62
Methylmercury	Paresthesia, tingling, dysesthesia, impaired vibration sense, perioral dysesthesia	25,59–61
PCBs	Paresthesia, hypoesthesia	63
TCDD (2,3,7,8-tetachlorobenzo-*p*-dioxin)	Paresthesia, hypoesthesia	64,65
Natural Agents		
β,β′-Imidodiproponitrile (IDPN)	Paresthesia	66
Podophyllin	Paresthesia	67
Pyridoxine	Paresthesia including face, impaired temperature sense	45

that became known as the Ginger-Jake syndrome. In 1930, a bootlegger added a material used in varnish and lacquers which contained TOCP to an alcoholic beverage containing Jamaica ginger, called "Jake." The poisoning episode left 20 000–100 000 people permanently affected.[68] Early symptoms included numbness and tingling in the extremities, with partial permanent paralysis at higher doses (Table 3). Subsequent outbreaks occurred in South Africa in 1937 and Morocco in 1959 as a result of contaminated cooking oil; over 10 000 individuals were affected in Morocco.

11.18.3.3 Hexacarbons (*n*-hexane and methyl *n*-butyl ketone)

n-Hexane is widely used as an inexpensive solvent in many manufacturing processes. It is also a minor component of gasoline and its combustion products. Methyl *n*-butyl ketone was increasingly used as a solvent until an outbreak of peripheral neuropathy in 1973 in a fabric plant in Ohio; it is now recognized to be more toxic than *n*-hexane. Both agents are metabolized by humans and animals to 2,5-hexanedione, the active neurotoxicant. Earliest symptoms of hexacarbon toxicity include sensory dysfunction in the hands and feet, including loss of touch, pinprick, vibration, and thermal sensation (Table 4). In the least affected individuals only sensory effects are present; at higher doses, toxicity progresses to distal weakness. Partial or full recovery usually takes place after removal from the source of exposure, although signs and symptoms may at first intensify over the course of weeks or months. Hundreds of workers have been affected in numerous outbreaks, including 93 people involved in sandal making in 1968–1970 in Japan.[87] Hexacarbon toxicity may also result from inhalant abuse.

11.18.3.4 Acrylamide

Acrylamide polymer is used as a strengthener of paper and cardboard, a grouting agent, and a flocculator.[11] The acrylamide monomer produces a dying-back peripheral neuropathy affecting both sensory and motor nerve fibers, particularly distally. Sensory symptoms include complaints of numbness and paresthesias in the feet and fingers (Table 4); vibration and proprioception are the most commonly affected

Table 3 Pesticides and related agents.

Chemical	Effects	Ref.
Methyl bromide	Paresthesia, tingling, impaired vibration, 2-point discrimination, depth discrimination	14
Organophosphates	Dysesthesia, paresthesia, impaired touch, pain, temperature, vibration, and position sense	3
Tri-*o*-cresyl-phosphate (TOCP)		69–75
Leptophos		76,77
Malathion Chlorpyriphos		78
Pyridostigmine ("pretreatment" for nerve agent poisoning in the Persian Gulf War)	Tingling of extremities	79
Synthetic pyrethroids[a] Flucythrinate Cypermethrin Fenvalerate Permethrin	Paresthesia, dysesthesia, abnormal facial sensation	80–83
Thallium (rodenticide)	Paresthesia, hyperesthesia, impaired light touch, pin prick, and vibration sense	84–86

[a]In decreasing order of toxicity.

Table 4 Chemicals used in manufacturing.

Chemical	Effects	Ref.
Acrylamide	Numbness, paresthesia, impaired vibration, pain, touch, position sense	33,34,38,64, 89–91
Allyl chloride (3-chloropropane)	Paresthesia, impaired pain, touch, vibration, position sense	92,93
Arsenic	Paresthesia, impaired touch, pinprick, temperature sense	94,95
Carbon disulfide	Paresthesia, hypoesthesia, dysesthesia, impaired pain, touch, temperature, vibration sense	96–100
Mercury vapor	Paresthesia	101
Methyl chloroform (1,1,1-trichloroethane)	Paresthesia, impaired pain sense, perioral tingling	102–104
Methyl *n*-butyl ketone[a]	Paresthesia, impaired pain, temperature, touch sense	105,106
n-Heptane	Paresthesia	107
n-Hexane[a]	Paresthesia, impaired pain, touch, vibration, position sense, also affects face	9,10,108–111
Solvent (paint manufacture, painters)	Paresthesia, impaired vibration sense	27,30,112–114
Trichlorethylene	Numbness in face, mouth, hands, feet, impaired pinprick sense	104,115

[a]2,5,Hexanedione active agent.

modalities.[88] Impaired vibration sensitivity has been replicated in monkeys.[7,8] Since it is the monomer rather than the polymer that is neurotoxic, poisoning is usually the result of industrial exposure. Partial or full recovery usually occurs following termination of exposure.

11.18.3.5 Therapeutic Agents

The classes of therapeutic agents that most often have been reported to produce impairment of somatosensory function include anticancer, antibacterial and, more recently, antiviral agents (Table 5). It has been recognized since the mid-1970s that the vinca alkaloids such as vincristine and vinblastine produce paresthesia (see Chapter 5, this volume). More recently, taxoids, a new class of cytotoxic agents, have been recognized to produce serious sensory neuropathy without motor involvement, including numbness, paresthesias, painful sensation, impaired vibration, temperature, position, pain, and temperature sense. It has been reported that impairment of vibration and proprioception occurs before pain and temperature sense,[116–118] suggesting that large fibers

are damaged first. *Cis*-platinum also produces severe damage to all cutaneous and position senses, although large fibers may be affected before small ones.[6,119] In contrast, the vinca alkaloids may affect both large and small fibers equally. For both *cis*-platinum and the taxoids, peripheral neuropathy is the major dose-limiting factor.[120–122] The new antiviral agents 2′,3′-dideoxyinosine (ddI) and 2′,3′-dideoxycitidine (ddC), used in the treatment of AIDS, produce a painful neuropathy, particularly in the lower extremities, including burning and shooting pain in the feet and legs. These drugs have been demonstrated to produce impairment of vibration, pinprick, and touch sense. The antibacterial agents clioquinol and isoniazid have long been known to produce somatosensory abnormalities.[123,124] Clioquinol was marketed as an oral intestinal amoebicide in 1934, and was used widely thereafter. In the mid-1950s a seriously disabling gastrointestinal and neurological syndrome as a result of clioquinol toxicity was identified in Japan; eventually 10 000 were diagnosed with the disease. It included myelopathy and optic nerve damage. Isoniazid is a cheap and effective antituberculosis drug, which exerts its antibacterial influence by interference

Table 5 Therapeutic agents.

Agent	Effects	Ref.
Anticancer		
Cisplatin (*cis*-platinum)	Dysesthesia, paresthesia, impaired vibration, position, and temperature sense	6,26,116,120, 125–130
Docetaxel	Paresthesia	131
Fludarabine and cytosine arabinoside	Paresthesia	132
Misonidazole	Paresthesia, impaired vibration sense	57,133
Paclitaxel (Taxol)	Dysesthesia, paresthesia, perioral numbness, impaired vibration, position, pain, and temperature sense	116–118,122,131, 134–137
Procarbazine	Paresthesia	138,139
Vinblastine	Paresthesia	139,140
Vincristine	Paresthesia, impaired position sense	120,139–142
Antiviral		
Acyclovir (acycloguanosine)	Hyperesthesia, paresthesia	143
ddC (2′,3′-dideoxycytidine)	Dysesthesia, impaired pinprick, vibration, and touch sense	144–147
ddI (2′,3′-dideoxyinosine)	Dysesthesia, paresthesia, impaired pinprick, and touch sense	144,148–153
Antibacterial and antifungal		
Chloramphenicol	Numbness, tingling, impaired touch, vibration sense	154,155
Clioquinol	Paresthesia, dysesthesia, impaired pain, touch, temperature sense, facial numbness	124
Colistimethate	Paresthesia	156
Dapsone	Numbness, decreased light touch, pain, vibration sense, two-point discrimination	157,158
Demeclocycline	Paresthesia, dysesthesia	159
Doxycycline	Paresthesia, dysesthesia	160
Iodochlorhydroxyquin	Numbness, dysesthesia, impaired touch, temperature, pain, and vibration sense	161–163
Isoniazid	Paresthesia, tingling, numbness	123
Metronidazole	Paresthesia, hyperesthesia, decreased touch, pin prick, and temperature sense	164–167
Nitrofurantoin	Numbness, paresthesia, tingling, dysesthesia, impaired pin prick, vibration, temperature, touch sense	168–173
Nitrofurazone	Numbness, paresthesia, impaired touch, pain, and vibration sense	174
Thiophenicol	Tingling, burning, hypoesthesia, dysesthesia, decreased vibration sense	175

with the B vitamins. It is metabolized in humans by addition of an acetyl group (acetylation); therefore peripheral neuropathy is observed more frequently in individuals who acetylate slowly as a result of genetic factors.

11.18.4 CONCLUSIONS AND RECOMMENDATIONS

Somatosensory dysfunction is a frequent consequence of environmental or industrial exposure to neurotoxic agents, as well as a side-effect of therapeutic agents. In recognition of the potential for debilitation as a result of exposure to environmental neurotoxicants, ATSDR has recommended that objective assessment of vibration sensitivity be included in a battery for community testing. Clinicians also recognize somatosensory dysfunction as the dose-limiting factor in therapeutic agents such as antiviral drugs used against AIDS. While the use of subjective examination methodology enjoys a long history, such assessments can only reveal gross neurological impairment, and are subject to a number of uncontrolled and unquantifiable confounding factors. Perhaps more importantly, such techniques are not suitable for experimental studies, where consistency in methodology across time, subjects, and laboratories is essential. Both clinical[42,50] and experimental[22,24] researchers have argued for the use of objective functional assessment of the somatosensory system. Such measures provide the possibility of precisely controlling the parameters of stimulus presentation and response requirement, thereby increasing the reliability, accuracy, and validity of the data. It is encouraging that such techniques have been increasingly applied in human research. Unfortunately, investigators involved in clinical research are often apparently unfamiliar with the appropriate experimental design and procedures consistent with collection of interpretable data. It is important that regulatory and advisory agencies, as well as individual investigators, understand and utilize valid methodology in the study of somatosensory function.

11.18.5 REFERENCES

1. W. K. Anger, 'Worksite behavioral research. Results, sensitive methods, test batteries and the transition from laboratory data to human health.' *Neurotoxicology*, 1990, **11**, 627–717.
2. W. K. Anger, 'Human neurobehavioral toxicology testing: current perspectives.' *Toxicol. Ind. Health*, 1989, **5**, 165–180.
3. W. K. Anger, R. Letz, D. W. Chrislip *et al.*, 'Neurobehavioral test methods for environmental health studies of adults.' *Neurotoxicol. Teratol.*, 1994, **16**, 489–497.
4. R. F. White, F. Gerr, R. F. Cohen *et al.*, 'Criteria for progressive modification of neurobehavioral agents.' *Neurotoxicol. Teratol.*, 1994, **16**, 511–524.
5. R. W. Amler, J. A. Lybarger, W. K. Anger *et al.*, 'Adoption of an adult environmental neurobehavioral test battery.' *Neurotoxicol. Teratol.*, 1994, **16**, 525–530.
6. J. E. Mollman, 'Cisplatin neurotoxicity.' *New Engl. J. Med.*, 1990, **322**, 126–127.
7. J. P. Maurissen, B. Weiss and H. T. Davis, 'Somatosensory thresholds in monkeys exposed to acrylamide.' *Toxicol. Appl. Pharmacol.*, 1983, **71**, 266–279.
8. J. P. Maurissen, B. Weiss and C. Cox, 'Vibration sensitivity recovery after a second course of acrylamide intoxication.' *Fundam. Appl. Toxicol.*, 1990, **15**, 93–98.
9. World Health Organization, 'Environmental Health Criteria 122: n-Hexane,' Geneva, 1991.
10. Y. Takeuchi, 'n-hexane polyneuropathy in Japan: a review of n-hexane poisoning and its preventative measures.' *Environ. Res.*, 1993, **62**, 76–80.
11. P. M. Le Quesne, in 'Experimental and Clinical Neurotoxicology,' eds. P. S. Spencer and H. H. Schaumburg, Williams and Wilkins, Baltimore, MD, 1980, pp. 309–325.
12. A. B. Sterman and H. H Schaumburg, in 'Experimental and Clinical Neurotoxicology,' eds. P. S. Spencer and H. H. Schaumburg, Williams and Wilkins, Baltimore, MD, 1980, pp. 675–680.
13. W. S. Carlson, S. Samueloff, W. Taylor *et al.*, 'Instrumentation for measurement of sensory loss in the fingertips.' *J. Occup. Med.*, 1979, **21**, 4, 260–264.
14. W. K. Anger, L. Moody, J. Burg *et al.*, 'Neurobehavioral evaluation of soil and structural fumigators using methyl bromide and sulfuryl fluoride.' *Neurotoxicology*, 1986, **7**, 137–156.
15. J. P. Maurissen, in 'Neurotoxicology: Approaches and Methods,' eds. L. Chang and W. Slikker, Jr., Academic Press, San Diego, 1995, pp. 239–264.
16. C. D. Kershaw, 'Statistical properties of staircase estimates from two interval forced choice experiments.' *Br. J. Math. Stat. Psychol.*, 1985, **38**, 35–43.
17. G. B. Wetherill and H. Levitt, 'Sequential estimation of points on a psychometric function.' *Br. J. Math. Stat. Psychol.*, part 1, 1965, **18**, 1–10.
18. J. P. Maurissen, 'Quantitative sensory assessment in toxicology and occupational medicine: applications, theory, and critical appraisal.' *Toxicol. Lett.*, 1988, **43**, 321–343.
19. J. M. Findley, 'Estimates on probability functions: a more virulent PEST.' *Percept. Psychophys.*, 1978, **23**, 181–185.
20. A. Pentland, 'Maximum likelihood estimation: the best PEST.' *Percept. Psychophys.*, 1980, **28**, 377–379.
21. F. E. Gerr and R. Letz, 'Reliability of a widely used test of peripheral cutaneous vibration sensitivity and a comparison of two testing protocols.' *Br. J. Ind. Med.*, 1988, **45**, 635–639.
22. D. C. Rice, in 'Neurobehavioral Toxicology,' eds. B. Weiss and J. L. O'Donoghue, Raven Press, New York, 1994, pp. 299–318.
23. J. P. Maurissen and G. J. Chrzan, 'One-year reliability of vibration sensitivity thresholds in human beings.' *J. Neurol. Sci.*, 1989, **90**, 325–334.
24. J. P. Maurissen and B. Weiss, in 'Experimental and Clinical Neurotoxicology,' eds. P. S. Spencer and H. H. Schaumburg, Williams and Wilkins, Baltimore, MD, 1980, pp. 767–774.
25. D. C. Rice and S. G. Gilbert, 'Effects of developmental methylmercury exposure or lifetime lead exposure on vibration sensitivity function in monkeys.' *Toxicol. Appl. Pharmacol.*, 1995, **134**, 161–169.

26. A. Elderson, R. Gerritsou van der Hoop *et al.*, 'Vibration perception and thermoperception as quantitative measurements in the monitoring of cisplatin induced neurotoxicity.' *J. Neurol. Sci.*, 1989, **93**, 167–174.

27. P. Halonen, J. P. Halonen, H. A. Lang *et al.*, 'Vibratory perception thresholds in shipyard workers exposed to solvents.' *Acta Neurol. Scand.*, 1986, **73**, 561–565.

28. J. M. Goldberg and U. Lindblom, 'Standardised method of determining vibratory perception thresholds for diagnosis and screening in neurological investigation.' *J. Neurol. Neurosurg. Psychiatry*, 1979, **42**, 793–803.

29. P. M. Le Quesne and C. J. Fowler, 'A study of pain threshold in diabetics with neuropathic foot lesions.' *J. Neurol. Neurosurg. Psychiatry*, 1986, **49**, 1191–1194.

30. F. J. Bove, R. Letz and E. L. Baker, Jr., 'Sensory thresholds among construction trade painters: a cross-sectional study using new methods for measuring temperature and vibrational sensitivity.' *J. Occup. Med.*, 1989, **31**, 320–325.

31. F. Gerr, D. Hershman and R. Letz, 'Vibrotactile threshold measurement for detecting neurotoxicity: reliability and determination of age-and-height standardized normative values.' *Arch. Environ. Health*, 1990, **45**, 148–154.

32. F. Gerr and R. Letz, 'Vibrotactile threshold testing in occupational health: a review of current issues and limitations.' *Environ. Res.*, 1993, **60**, 145–159.

33. H. Deng, F. He, S. Zhang *et al.*, 'Quantitative measurements of vibration threshold in healthy adults and acrylamide workers.' *Int. Arch. Occup. Environ. Health*, 1993, **65**, 53–56.

34. C. J. Calleman, Y. Wu, F. He *et al.*, 'Relationships between biomarkers of exposure and neurological effects in a group of workers exposed to acrylamide.' *Toxicol. Appl. Pharmacol.*, 1994, **126**, 361–371.

35. M. G. Cherniack, R. Letz, F. Gerr *et al.*, 'Detailed clinical assessment of neurological function in symptomatic shipyard workers.' *Br. J. Ind. Med.*, 1990, **47**, 566–572.

36. J. M. Sosenko, M. T. Gadia, N. Natori *et al.*, 'Neurofunctional testing for the detection of diabetic peripheral neuropathy.' *Arch. Intern. Med.*, 1987, **147**, 1741–1744.

37. J. C. Arezzo and H. H. Schaumburg, 'The vibratron: a simple device for quantitative evaluation of tactile/ vibratory sense.' *Neurology*, 1985, **35**, Suppl. 1, 169.

38. J. C. Arezzo, H. H. Schaumburg and C. A. Petersen, 'Rapid screening for peripheral neuropathy: a field study with the Optacon.' *Neurology*, 1983, **33**, 626–629.

39. M. L. Bleecker, 'Vibration perception thresholds in entrapment and toxic neuropathies.' *J. Occup. Med.*, 1986, **28**, 991–994.

40. J. M. Sosenko, M. Kato, R. Soto *et al.*, 'Determinants of quantitative sensory testing in non-neuropathic individuals.' *Electromyogr. Clin. Neurophysiol.*, 1989, **29**, 459–463.

41. P. J. Dyck, I. R. Zimmerman, P. C. O'Brien *et al.*, 'Introduction to automated systems to evaluate touch-pressure, vibration, and thermal cutaneous sensation in man.' *Ann. Neurol.*, 1978, **4**, 502–510.

42. P. J. Dyck, W. Bushek, E. M. Spring *et al.*, 'Vibratory and cooling detection thresholds compared with other tests in diagnosing and staging diabetic neuropathy.' *Diabetes Care*, 1987, **10**, 432–440.

43. C. J. Fowler, M. B. Carroll, D. Burns *et al.*, 'A portable system for measuring cutaneous thresholds for warming and cooling.' *J. Neurol. Neurosurg. Psychiatry*, 1987, **50**, 1211–1215.

44. L. Ekenvall, B. Y. Nilsson and P. Gustavsson, 'Temperature and vibration thresholds in vibration syndrome.' *Br. J. Ind. Med.*, 1986, **43**, 825–829.

45. J. C. Arezzo, H. H. Schaumburg and C. Laudadio, 'Thermal sensitivity tester. Device for the quantitative assessment of thermal sense in diabetic neuropathy.' *Diabetes*, 1986, **35**, 590–592.

46. R. Sekuler, D. Nash and R. Armstrong, 'Sensitive, objective procedure for evaluating response to light touch.' *Neurology*, 1973, **23**, 1282–1291.

47. B. Lynn and E. R. Perl, 'A comparison of four tests for assessing the pain sensitivity of different subjects and test areas.' *Pain*, 1977, **3**, 353–365.

48. P. J. Dyck, J. L. Karnes, D. A. Gillen *et al.*, 'Comparison of algorithms of testing for use in automated evaluation of sensation.' *Neurology*, 1990, **40**, 1607–1613.

49. H. Muijser, J. Hooisma, E. M. Hoogendijk *et al.*, 'Vibration sensitivity as a parameter for detecting peripheral neuropathy. I. Results in healthy workers.' *Int. Arch. Occup. Environ. Health*, 1986, **58**, 287–299.

50. F. Gerr, R. Letz, D. Hershman *et al.*, 'Comparison of vibrotactile thresholds with physical examination and electrophysiological assessment.' *Muscle Nerve*, 1991, **14**, 1059–1066.

51. R. Tegner and B. Lindholm, 'Vibratory perception threshold compared with nerve conduction velocity in the evaluation of uremic neuropathy.' *Acta Neurol. Scand.*, 1985, **71**, 284–289.

52. S. G. Gilbert and J. P. Maurissen, 'Assessment of the effects of acrylamide, methylmercury, and 2,5 hexanedione on motor functions in mice.' *J. Toxicol. Environ. Health*, 1982, **10**, 31–41.

53. C. M. McKillop, J. A. Brock, G. J. Oliver *et al.*, 'A quantitative assessment of pyrethroid-induced paresthesia in the guinea pig flank model.' *Toxicol. Lett.*, 1987, **36**, 1–7.

54. H. A. Tilson and T. A. Burne, 'Effects of triethyl tin on pain reactivity and neuromotor function of rats.' *J. Toxicol. Environ. Health*, 1981, **8**, 317–324.

55. P. A. Cabe, 'Psychophysical methods for the measurement of somatosensory dysfunction in laboratory animals.' *Environ. Health Perspect.*, 1982, **44**, 93–100.

56. W. C. Stebbins (ed.), 'Animal Psychophysics: the Design and Conduct of Sensory Experiments,' Appleton-Century-Crofts, New York, 1970.

57. J. P. Maurissen, P. J. Conroy, W. Passalacqua *et al.*, 'Somatosensory deficits in monkeys treated with misonidazole.' *Toxicol. Appl. Pharmacol.*, 1981, **57**, 119–126.

58. F. Bakir, S. F. Damluji, L. Amin-Zaki *et al.*, 'Methylmercury poisoning in Iraq.' *Science*, 1973, **181**, 230–241.

59. World Health Organization, 'Environmental Health Criteria 101: Methylmercury,' Geneva, 1990.

60. A. Igata, 'Epidemiological and clinical features of Minimata disease.' *Environ. Res.*, 1993, **63**, 157–169.

61. Y. Kinjo, H. Higashi, A. Nakano *et al.*, 'Profile of subjective complaints and activities of daily living among current patients with Minimata disease after 3 decades.' *Environ. Res.*, 1993, **63**, 241–251.

62. World Health Organization Study Group, 'Recommended Health-Based Limits in Occupational Exposure to Heavy Metals,' Geneva, 1980, pp. 36–80.

63. L. G. Chia and F. L. Chu, 'A clinical and electrophysiological study of patients with polyclorinated biphenyl poisoning.' *J. Neurol. Neurosurg. Psychiatry*, 1985, **48**, 894–901.

64. J. Pazderova-Vejlupková, E. Lukáš, M. Němcova *et al.*, 'The development and prognosis of chronic intoxication by tetrachlorodibenzo-*p*-dioxin in mean.' *Arch. Environ. Health*, 1981, **36**, 5–11.

65. G. Filippini, B. Bordo, P. Crenna *et al.*, 'Relationship between clinical and electrophysiological findings and indicators of heavy exposure to 2,3,7,8-tetrachlorodibenzo-dioxin.' *Scand. J. Work Environ. Health*, 1981, **7**, 257–262.

66. J. W. Griffin and D. L. Price, in 'Experimental and Clinical Neurotoxicology,' eds. P. S. Spencer and H. H. Schaumburg, Williams and Wilkins, Baltimore, 1980, pp. 161–178.

67. M. H. Chang, K. P. Lin, Z. A. Wu *et al.*, 'Acute ataxic sensory neuropathy resulting from podophyllin intoxication.' *Muscle Nerve*, 1992, **15**, 513–514.

68. OTA (Office of Technology Assessment), 'Neurotoxicity: Identifying and Controlling Poisons of the Nervous System,' US Congress, 1990, US Government Printing Office, Washington DC, pp. 43–59.

69. C. R. Bennett, 'A group of patients suffering from paralysis due to drinking Jamaica ginger.' *South Med. J. Birmingham*, 1930, **23**, 371–375.

70. B. T. Burley, 'Polyneuritis from tricresyl phosphate.' *J. Am. Med. Assoc.*, 1932, **98**, 298–304.

71. H. Geoffroy, A. Slomic, M. Benenadji *et al.*, 'Myelopolyneuritis due to tricresyl phosphate—the Moroccan intoxication of 1959: review.' *Presse Med.*, 1960, **68**, 1474–1504.

72. H. H. Merritt and M. Moore, 'Peripheral neuritis associated with ginger extract ingestion.' *N. Engl. J. Med.*, 1930, **203**, 4–12.

73. H. V. Smith and J. M. Spalding, 'Outbreak of paralysis in Morocco due to ortho-cresyl phosphate poisoning.' *Lancet*, 1959, **2**, 1019–1021.

74. E. Svennilson, 'Studies of triorthocresyl phosphate neuropathy, Morocco 1960.' *Acta. Psychiat. Scand.*, 1960, **150**, Suppl., 334–336.

75. D. D. Vora, D. K. Dastur, B. M. Braganca *et al.*, 'Toxic polyneuritis in Bombay due to ortho-cresyl-phosphate poisoning.' *J. Neurol. Neurosurg. Psychiatry*, 1962, **25**, 234–242.

76. H. S. Levin, in 'Behavioral Toxicology—Early Detection of Occupational Hazards,' eds. C. Xintaras, B. L. Johnson and I. de Groot, Department of Health, Education and Welfare, Washington, DC, 1974, pp. 155–164.

77. D. J. Svendsgaard, S. A. Soliman, A. Soffar *et al.*, 'Screening pesticide plant workers for organophosphorus induced delayed neuropathy (OPIDN).' *Toxicology*, 1987, **7**, 538.

78. S. D. Davis and R. J. Richardson, in 'Experimental and Clinical Neurotoxicology,' eds. P. S. Spencer and H. H. Schaumburg, Williams and Wilkins, Baltimore, 1980, pp. 527–544.

79. NIH (National Institutes of Health), 'The Persian Gulf Experience and Health. NIH Technology Assessment Workshop Statement,' 1994, Apr 27–29, 28.

80. S. A. Flannigan, S. B. Tucker, M. M. Key *et al.*, 'Synthetic pyrethroid insecticides: a dermatogical evaluation.' *Br. J. Ind. Med.*, 1985, **42**, 363–372.

81. S. A. Flannigan and S. B. Tucker, 'Variation in cutaneous sensation between synthetic pyrethroid insecticides.' *Contact Dermatitis*, 1985, **13**, 140–147.

82. S. B. Tucker, S. A. Flannigan and C. E. Ross, 'Inhibition of cutaneous paresthesia resulting from synthetic pyrethroid exposure.' *Int. J. Dermatol.*, 1984, **23**, 686–689.

83. P. M. Le Quesne, I. C. Maxwell and S. T. Butterworth, 'Transient facial sensory symptoms following exposure to synthetic pyrethroids—a clinical and electro-physiological assessment.' *Neurotoxicology*, 1981, **2**, 1–11.

84. R. B. Innis and H. Moses, 'Clinical conferences at the Johns Hopkins Hospital. Thallium poisoning.' *Johns Hopkins Med. J.*, 1978, **142**, 27.

85. J. B. Cavanagh, N. H. Fuller, H. R. Johnson *et al.*, 'The effects of thallium salts, with particular reference to the nervous system changes. A report of three cases.' *Q. J. Med.*, 1974, **43**, 293–319.

86. W. J. Bank, D. E. Pleasure, K. Suzuki *et al.*, 'Thallium poisoning.' *Arch. Neurol.*, 1972, **26**, 456–464.

87. T. Inoue, Y. Takeuchi, S. Takeuchi *et al.*, 'A health survey on vinyl sandal manufacturers with high incidence of "n-hexane" intoxication occurred.' *Jap. J. Indust. Health*, 1970, **12**, 73–84.

88. H. A. Tilson, 'The neurotoxicity of acrylamide: An overview.' *Neurobehav. Toxicol. Teratol.*, 1981, **3**, 445–461.

89. E. A. Smith and F. W. Oehme, 'Acrylamide and polyacrylamide: a review of production, use, environmental fate and neurotoxicity.' *Rev. Environ. Health*, 1991, **9**, 215–228.

90. F. S. He, S. L. Zhang, H. L. Wang *et al.*, 'Neurological and electroneuromyographic assessment of the adverse effects of acrylamide on occupationally exposed workers.' *Scand. J. Work Environ. Health*, 1989, **15**, 125–129.

91. World Health Organization, 'Environmental Health Criteria 49: Acrylamide,' Geneva, 1985.

92. F. S. He, D. G. Shen, Y. P. Guo *et al.*, 'Toxic polyneuropathy due to chronic allyl chloride intoxication: a clinical and experimental study.' *Chin. Med. J. (Beijing, Engl. Ed.)*, 1980, **93**, 177–182.

93. F. S. He and S. L. Zhang, 'Effects of allyl chloride on occupationally exposed subjects.' *Scand. J. Work. Environ. Health*, 1985, **11**, Suppl. 4, 43–45.

94. J. P. Conomy and K. L. Barnes, 'Quantitative assessment of cutaneous sensory function in subjects with neurological disease.' *J. Neurol. Sci.*, 1976, **30**, 221–235.

95. R. G. Feldman, C. A. Niles, M. Kelly-Hayes *et al.*, 'Peripheral neuropathy in arsenic smelter workers.' *Neurology*, 1979, **29**, 939–944.

96. O. Aaserud, L. Gjerstad, P. Nakstad *et al.*, 'Neurological examination, computerized tomography, cerebral blood flow and neuropsychological examination in workers with long-term exposure to carbon disulfide.' *Toxicology*, 1988, **49**, 277–282.

97. B. Knave, B. Kolmodin-Hedman, H. E. Persson *et al.*, 'Chronic exposure to carbon disulfide: effects on occupationally exposed workers with special reference to the nervous system.' *Work Environ. Health*, 1974, **11**, 49–58.

98. R. Lilis, in 'Behavioral Toxicology—Early Detection of Occupational Hazards,' eds. C. Xintaras, B. L. Johnson and I. de Groot, Department of Health, Education and Welfare, Washington, DC, 1974, pp. 51–59.

99. A. M. Seppäläinen, in 'Behavioral Toxicology—Early Detection of Occupational Hazards,' eds. C. Xintaras, B. L. Johnson and I. de Groot, Department of Health, Education and Welfare, Washington, DC, 1974, pp. 64–72.

100. R. O. Beauchamp, Jr., J. S. Bus, J. A. Popp *et al.*, in 'Critical Reviews in Toxicology,' ed. L. Golberg, CRC Press, Boca Raton, FL, 1983, pp. 169–278.

101. D. G. Ellingsen, T. Morland, A. Andersen *et al.*, 'Relationship between exposure related indices and neurological and neurophysiological effects in workers previously exposed to mercury vapour.' *Br. J. Ind. Med.*, 1993, **50**, 736–744.

102. R. A. House, G. M. Liss and M. C. Wills, 'Peripheral neuropathy associated with 1,1,1-trichloroethane.' *Arch. Environ. Health*, 1994, **49**, 196–199.

103. D. C. Howse, G. L. Shanks and S. Nag, 'Peripheral neuropathy following prolonged exposure to methyl chloroform.' *Neurology*, 1989, **39**, 242.

104. G. M. Liss, 'Peripheral neuropathy in two workers exposed to 1,1,1-trichloroethane.' *JAMA*, 1988, **260**, 2217.

105. W. J. Krasavage, J. L. O'Donoghue, G. D. DiVincenzo *et al.*, 'The relative neurotoxicity of methyl-n-butyl-ketone, n-hexane and their metabolites.' *Toxicol. Appl. Pharmacol.*, 1980, **52**, 433–441.

106. N. Allen, J. R. Mendell, D. J. Billmaier *et al.*, 'Toxic polyneuropathy due to methyl n-butyl ketone. An industrial outbreak.' *Arch. Neurol.*, 1975, **32**, 209–218.

107. F. Valentini, R. Agnesi, L. Dal Vecchio *et al.*, 'Does n-heptane cause peripheral neurotoxicity? A case report in a shoemaker.' *Occup. Med. (Oxf)*, 1994, **44**, 102–104.

108. C. C. Huang, N. S. Chu, S. Y. Cheng *et al.*, 'Biphasic recovery in n-hexane polyneuropathy. A clinical and electrophysiological study.' *Acta Neurol. Scand.*, 1989, **80**, 610–615.

109. C. Cianchetti, G. Abbritti, G. Perticoni *et al.*, 'Toxic polyneuropathy of shoe-industry workers. A study of 122 cases.' *J. Neurol. Neurosurg. Psychiatry*, 1976, **39**, 1151–1161.

110. A. Herskowitz, N. Ishii and H. Schaumburg, '*N*-hexane neuropathy. A syndrome occurring as a result of industrial exposure.' *N. Engl. J. Med.*, 1971, **285**, 82–85.

111. C. Pastore, D. Marhuenda, J. Marti *et al.*, 'Early diagnosis of n-hexane-caused neuropathy.' *Muscle Nerve*, 1994, **17**, 981–986.

112. M. L. Bleecker, K. I. Bolla, J. Agnew *et al.*, 'Dose-related subclinical neurobehavioral effects of chronic exposure to low levels of organic solvents.' *Am. J. Ind. Med.*, 1991, **19**, 715–728.

113. R. Y. Demers, B. L. Markell and R. Wabeke, 'Peripheral vibratory sense deficits in solvent-exposed painters.' *J. Occup. Med.*, 1991, **33**, 1051–1054.

114. DHHS (Department of Health and Human Services), 'Organic Solvent Neurotoxicity,' Depart. Health Human Services, Washington DC, 1987.

115. R. G. Feldman, R. F. White, J. N. Currie *et al.*, 'Long-term follow-up after single toxic exposure to trichloroethylene.' *Am. J. Ind. Med.*, 1985, **8**, 119–126.

116. E. K. Rowinsky, M. R. Gilbert, W. P. McGuire *et al.*, 'Sequences of taxol and cisplatin: a phase I and pharmacologic study.' *J. Clin. Oncol.*, 1991, **9**, 1692–1703.

117. E. K. Rowinsky, V. Chaudhry, D. R. Cornblath *et al.*, 'Neurotoxicity of Taxol.' *Monogr. Natl. Cancer. Inst.*, 1993, **15**, 107–115.

118. P. H. Wiernik, E. L. Schwartz, A. Einzig *et al.*, 'Phase I trial of taxol given as a 24-hour infusion every 21 days: responses observed in metastatic melanoma.' *J. Clin. Oncol.*, 1987, **5**, 1232–1239.

119. T. J. Walsh, A. W. Clark, I. M. Parhad *et al.*, 'Neurotoxic effects of cisplatin therapy.' *Arch. Neurol.*, 1982, **39**, 719–720.

120. F. P. Harmers, W. H. Gispen and J. P. Neijt, 'Neurotoxic side-effects of cisplatin.' *Eur. J. Cancer*, 1991, **27**, 372–376.

121. V. Chaudhry, E. K. Rowinsky, S. E. Sartorius *et al.*, 'Peripheral neuropathy from taxol and cisplatin combination chemotherapy: clinical and electrophysiological studies.' *Ann. Neurol.*, 1994, **35**, 304–311.

122. R. S. Finley and E. K. Rowinsky, 'Patient care issues: the management of paclitaxel-related toxicities.' *Ann. Pharmacother.*, 1994, **28**, S27–S30.

123. W. F. Blakemore, in 'Experimental and Clinical Neurotoxicology,' eds. P. S. Spencer and H. H. Schaumburg, Williams and Wilkins, Baltimore, 1980, pp. 476–489.

124. H. H. Schaumburg and P. S. Spencer, in 'Experimental and Clinical Neurotoxicology,' eds. P. S. Spencer and H. H. Schaumburg, Williams and Wilkins, Baltimore, 1980, pp. 395–406.

125. S. W. Thompson, L. E. Davis, M. Kornfeld *et al.*, 'Cisplatin neuropathy. Clinical, electrophysiologic, morphologic, and toxicologic studies.' *Cancer*, 1984, **54**, 1269–1275.

126. S. S. Legha and I. W. Dimery, 'High-dose cisplatin administration without hypertonic saline: observation of disabling neurotoxicity.' *J. Clin. Oncol.*, 1985, **3**, 1373–1378.

127. L. Reinstein, S. S. Ostrow and P. H. Wiernik, 'Peripheral neuropathy after cis-platinum (II) (DDP) therapy.' *Arch. Phys. Med. Rehabil.*, 1980, **61**, 280–282.

128. D. D. Von Hoff, R. Schilsky, C. M. Reichert *et al.*, 'Toxic effects of cis-dichlorodiammineplatinum (II) in man.' *Cancer Treat. Rep.*, 1979, **63**, 1527–1531.

129. M. Hemphill, A. Pestronk, T. Walsh *et al.*, 'Sensory neuropathy in cis-platinum chemotherapy.' *Neurology*, 1980, **30**, 429.

130. S. W. Hansen, S. Helweg-Larsen and W. Trojaborg, 'Long-term neurotoxicity in patients treated with cisplatin, vinblastine, and bleomycin for metastatic germ cell cancer.' *J. Clin. Oncol.*, 1989, **7**, 1457–1461.

131. M. Marty, J. M. Extra, S. Giacchetti *et al.*, 'Taxoids: a new class of cytotoxic agents.' *Nouv. Rev. Fr. Hematol.*, 1994, **36**, Suppl. 1, S25–S28.

132. S. M. Kornblau, J. Cortes-Franco and E. Estey, 'Neurotoxicity associated with fludarabine and cytosine arabinoside chemotherapy for acute leukemia and myelodysplasia.' *Leukemia*, 1993, **7**, 378–383.

133. S. Disch, M. I. Saunders, P. Anderson *et al.*, 'The neurotoxicity of misonidazole: pooling of data from five centres.' *Br. J. Radiol.*, 1978, **51**, 1023–1024.

134. D. R. Kohler and B. R. Goldspiel, 'Paclitaxel (taxol).' *Pharmacotherapy*, 1994, **14**, 3–34.

135. H. J. Guchelaar, C. H. ten Napel, E. G. de Vries *et al.*, 'Clinical, toxicological and pharmaceutical aspects of the antineoplastic drug taxol: a review.' *Clin. Oncol. (R. Coll. Radiol.)*, 1994, **6**, 40–48.

136. R. B. Lipton, S. C. Apfel, J. P. Dutcher *et al.*, 'Taxol produces a predominantly sensory neuropathy.' *Neurology*, 1989, **39**, 368–373.

137. K. Swenerton, E. Eisenhauer, W. ten Bokkel Huinink *et al.*, 'Taxol in relapsed ovarian cancer: high vs low dose and short vs long term infusion.' *Proc. Annu. Meet. Am. Soc. Clin. Oncol.*, 1993, A810.

138. H. D. Weiss, M. D. Walker and P. H. Wiernik, 'Neurotoxicity of commonly used antineoplastic agents (second of two parts).' *N. Engl. J. Med.*, 1974, **291**, 127–133.

139. H. D. Weiss, M. D. Walker and P. H. Wiernik, 'Neurotoxicity of commonly used antineoplastic agents (first of two parts).' *N. Engl. J. Med.*, 1974, **291**, 75–81.

140. S. Rosenthal and S. Kaufman, 'Vincristine neurotoxicity.' *Ann. Intern. Med.*, 1974, **80**, 733–737.

141. C. M. Hogan-Dann, W. G. Fellmeth, S. A. McGuire *et al.*, 'Polyneuropathy following vincristine therapy in two patients with Charcot-Marie-Tooth syndrome.' *JAMA*, 1984, **252**, 2862–2863.

142. S. S. Legha, 'Vincristine neurotoxicity. Pathophysiology and management.' *Med. Toxicol.*, 1986, **1**, 421–427.

143. S. Rashiq, L. Briewa, M. Mooney *et al.*, 'Distinguishing acyclovir neurotoxicity from encephalomyelitis.' *J. Intern. Med.*, 1993, **234**, 507–511.

144. S. F. LeLacheur and G. L. Simon, 'Exacerbation of dideoxycytidine-induced neuropathy with dideoxyinosine.' *J. Acquir. Immune Defic. Syndr.*, 1991, **4**, 538–539.

145. A. R. Berger, J. C. Arezzo, H. H. Schaumburg *et al.*, '2′,3′-dideoxycytidine (ddC) toxic neuropathy: a study of 52 patients.' *Neurology*, 1993, **43**, 358–362.

146. O. P. Martinez and M. A. French, 'Acoustic neuropathy associated with zalcitabine-induced peripheral neuropathy.' *AIDS*, 1993, **7**, 901–902.

147. T. C. Merigan, G. Skowron, S. A. Bozzette *et al.*, 'Circulating p24 antigen levels and responses to dideoxycytidine in human immunodeficiency virus (HIV) infections. A phase I and II study.' *Ann. Intern. Med.*, 1989, **110**, 189–194.

148. R. G. Miller, 'Neuromuscular complications of human immunodeficiency virus infection and anti-retroviral therapy.' *West. J. Med.*, 1994, **160**, 447–452.

149. G. J. Moyle, M. R. Nelson, D. Hawkins *et al.*, 'The use and toxicity of didanosine (ddI) in HIV-positive individuals intolerant to zidovudine (AZT).' *Q. J. Med.*, 1993, **86**, 155–163.

150. J. S. Lambert, M. Seidlin, R. C. Reichman *et al.*, '2′,3′-dideoxyinosine (ddI) in patients with the acquired immunodeficiency syndrome of AIDS-related complex. A phase I trial.' *N. Engl. J. Med.*, 1990, **322**, 1333–1340.

151. T. P. Cooley, L. M. Kunches, C. A. Saunders *et al.*, 'Once-daily administration of 2′,3′-dideoxyinosine (ddI) in patients with the acquired immunodeficiency syndrome or AIDS-related complex. Results of a phase I trial.' *N. Engl. J. Med.*, 1990, **322**, 1340–1345.

152. I. M. Pike and C. Nicaise, 'The didanosine Expanded Access Program: safety analysis.' *Clin. Infect. Dis.*, 1993, **16**, S63–S68.

153. R. M. Franssen, P. L. Meenhorst, C. H. Koks *et al.*, 'Didanosine, a new antiretroviral drug. A review.' *Pharm. Weekbl. Sci. Ed.*, 1992, **14**, 297–304.

154. S. Charache, D. Finkelstein, P. S. Lietman *et al.*, 'Peripheral optic neuritis in a patient with hemoglobin SC disease during treatment of salmonella osteomyelitis with chloramphenicol.' *Johns Hopkins Med. J.*, 1977, **140**, 121–124.

155. R. J. Joy, R. Scalettar and D. B. Sodee, 'Optic and peripheral neuritis—probable effect of prolonged chloramphenicol therapy.' *J. Am. Med. Ass.*, 1960, **173**, 1731–1734.

156. J. Koch-Weser, V. W. Sidel, E. B. Fereman *et al.*, 'Adverse effects of sodium colistimethate. Manifestations and specific reaction rates during 317 courses of therapy.' *Ann. Intern. Med.*, 1970, **72**, 857–868.

157. F. W. Epstein and M. Bohm, 'Dapsone-induced peripheral neuritis.' *Arch. Dermatol.*, 1976, **112**, 1761–1762.

158. W. C. Koller, L. K. Gehlmann, F. D. Malkinson *et al.*, 'Dapsone-induced peripheral neuropathy.' *Arch. Neurol.*, 1977, **34**, 644–646.

159. P. Frost, G. D. Weinstein and E. C. Gomez, 'Methacycline and demeclocycline in relation to sunlight.' *JAMA*, 1971, **216**, 326–329.

160. P. Frost, G. D. Weinstein and E. C. Gomez, 'Phototoxic potential of minocycline and doxycycline.' *Arch. Dermatol.*, 1972, **105**, 681–683.

161. I. Sobue, K. Ando, M. Iida *et al.*, 'Myeloneuropathy with abdominal disorders in Japan. A clinical study of 752 cases.' *Neurology*, 1971, **21**, 168–173.

162. J. D. Spillane, 'The geography of neurology.' *Br. Med. J.*, 1972, **2**, 506–612.

163. T. Tsubaki, Y. Honma and M. Hoshi, 'Neurological syndrome associated with clioquinol.' *Lancet*, 1971, **1**, 696–697.

164. A. Coxon and C. A. Pallis, 'Metronidaxole neuropathy.' *J. Neurol. Neurosurg. Psychiatry*, 1976, **39**, 403–405.

165. I. D. Ramsay, 'Endocrine opthalopathy.' *Br. Med. J.*, 1968, **4**, 706.

166. G. Said, J. Goasguen and C. Laverdant, 'Polyneurites au cours des traitments prolonges par le metronidazole. [Neuropathy in long term treatment with metronidazole.]' *Rev. Neurol. (Paris)*, 1978, **134**, 515–521.

167. B. Ursing and C. Kanme, 'Metronidazole for Crohn's disease.' *Lancet*, 1975, **1**, 775–777.

168. F. G. Ellis, 'Acute polyneuritis after nitrofurantoin therapy.' *Lancet*, 1962, **2**, 1136–1138.

169. A. I. Jacknowitz, J. L. Le Frock and R. A. Prince, 'Nitrofurantoin polyneuropathy: report of two cases.' *Am. J. Hosp. Pharm.*, 1977, **34**, 759–762.

170. L. Loughridge, 'Peripheral neuropathy due to nitrofurantoin.' *Lancet*, 1962, **2**, 1133–1135.

171. W. J. Martin, K. B. Corbin and D. C. Utz, 'Parasthesias during treatment with nitrofurantoin—report of case.' *Proc. Mayo Clin.*, 1962, **37**, 288–292.

172. C. J. Rubenstein, 'Peripheral neuropathy caused by nitrofurantoin.' *JAMA*, 1964, **187**, 647–649.

173. J. F. Toole and M. L. Parrish, 'Nitrofurantoin polyneuropathy.' *Neurology*, 1973, **23**, 554–559.

174. H. Collings, 'Polyneuropathy associated with nitrofuran therapy.' *Arch. Neurol.*, 1960, **3**, 656–660.

175. Y. Shinohara, F. Yamaguchi and F. Gotoh, 'Toxic neuropathy as a complication of thiopenicol therapy.' *Eur. Neurol.*, 1977, **16**, 161–164.

11.19
Behavioral Screening
for Neurotoxicity

VIRGINIA C. MOSER
US Environmental Protection Agency, Research Triangle Park, NC, USA

11.19.1 INTRODUCTION

Behavioral evaluations comprise a critical segment of neurotoxicological assessments. As with all methods in toxicity testing, an understanding and appreciation of the elements of the test method is necessary for full and appropriate utilization. Behavior is an organism's actions and responses to its environment: it is required for the organism to perform crucial life functions, that is, finding food and reproduction. Behavior represents the integration of motor, sensory, and associative neuronal functions which cannot be assessed using only neurochemical, histological, or physiological techniques.[1] The study of behavior is the most accessible method for evaluating the integrity of the nervous system in the whole organism; furthermore, it is generally regarded as a suitable indicator of nervous system function.[2,3]

Altered behavior reflects direct and/or indirect effects on the nervous system. The complexity of the nervous system implies that it would be a sensitive indicator of nervous system perturbations produced by chemicals. A wide array of behavioral methods have therefore been adopted for use in neurotoxicology research.[4] Behavioral changes may be the first measurable effect of chemical exposure, being evident at lower doses, or have an earlier onset, than overt clinical signs or structural lesions.[5,6] For example, behavioral deficits preceded structural changes in the neurotoxicity produced by compounds such as acrylamide.[7,8] Typically, measures that are the result of neuronal integration are affected first, for example, gait, activity.[8] The advantage of behavioral endpoints, however, is more that they reflect the subject's overall, integral function rather than being specifically more sensitive or an inherently superior test.[9,10]

In some circumstances, behavioral changes may be difficult to reveal due to wide variations in normal function. Furthermore, the nervous system shows plasticity, residual capacity, and compensatory mechanisms which may occur in the face of permanent pathological damage.[11] Neuronal tissue in the central nervous system (CNS) cannot regenerate, but undamaged neurons may change their function and assume new roles in order to maintain survival.[12] Adaptation, or drawing on functional reserves, may occur but there may also be major consequences of these adaptations; these can be uncovered using stress or drug challenges.[9] On the other hand, there may be clear behavioral changes without evidence of pathological damage. There is no *a priori* indication of which endpoint will be the most appropriate for the neurotoxic assessment of a given chemical, therefore the most conservative screening strategy is to assess both structure and function. In this chapter, the methods used to evaluate behavior and function at the first-tier level of testing are discussed.

11.19.2 HISTORY OF BEHAVIORAL SCREENING

Simply stated, the purpose of screening chemicals in laboratory animals is to predict effects in humans.[13] Chemicals shown to have significant potential activity can then be subjected to further studies. Over the decades, the process of screening chemicals for potential nervous system activity has taken place in a variety of settings and different types of test batteries have been designed. The extent to which behavior has been studied varies, and tests have taken on different forms, with more or less specificity and sensitivity for neurologi-

cal effects. Behavioral assessments in toxicology studies are the result of three converging disciplines: toxicology, pharmacology, and psychology.

The development of methods for the evaluation of functional changes in laboratory animals has taken different paths. The emergence of behavioral testing in general toxicology has been reviewed.[5,14] Behavior was included in the conduct of standard general toxicity studies only as a means of monitoring the subjects in long-term studies to prevent early lethality, resulting in loss of tissues for investigation.[15,16] Thus, these early clinical, or in-life, observations were proposed more as a method of reviewing the subject's health status in order to obtain the desired end product of the study, that is, to keep the subject living long enough to show chronic effects of treatment. In marked contrast, behavioral tests in the field of psychopharmacology were designed specifically to detect potential CNS effects of drugs. There were few early reports recommending actual behavioral tests in toxicity testing.[17–19]

The study of behavior is of course the crux of experimental psychology and behavior analysis, and tests have been developed to study the effects of chemicals as well as using chemicals to study behavior. Two approaches that have been used stress different endpoints. Some felt that innate behaviors are too variable, and therefore not acceptably reliable. Behavior was consequently put under some form of control and shaped, for example, operant or conditioned responding. The ability to manipulate the experimental variables and parameters would lead to a lower variability of the response, which increased the sensitivity of the test. A major drawback of this approach is that these studies often require considerable training of the subjects. Others have studied innate behaviors which are considered stable since they relate to life-continuing functions, for example, food-seeking behavior, detection of danger. For many years, the open field has been used as a method of eliciting specific, reproducible, and reliable behaviors.[20] Tests of emotionality, excitability, and arousal level have typically been evaluated in the open field.[21,22] Some of these measures have also been used for neuroscience research[23] and incorporated into neurobehavioral screening. The open field has now become a standard arena in which to closely monitor the subject while it is exploring.[5] Ethologists have also recommended the inclusion of social behavior and other biological functions for screening.[24]

Since the mid-1970s, there has been a growing concern regarding the vulnerability of the nervous system to toxicants, and the importance of neurotoxicity testing emerged as a factor in human health assessment. A report from the National Academy of Sciences (NAS) in 1975[25] first suggested the concept of tiered testing with preliminary screening for neurobehavioral effects, including tests typically used in the pharmaceutical industry, changes in motor activity, observations, reflexes, and so on. Some investigators also recommended that neurobehavioral evaluations be integrated into ongoing toxicity studies.[26,27] Since then, a variety of expert panels have recommended assessment of new chemicals for their neurotoxic potential.[11,28–30] Tests that were included in these recommendations were observational batteries, along with measures of motor activity and neuropathological assessments. To assess neurobehavioral function, standardized batteries of tests have been developed and these assessments are often employed in routine toxicity studies.

11.19.3 EVOLUTION OF SCREENING BATTERIES

11.19.3.1 Context of Development

The development of specific test batteries for screening chemicals has been influenced by many factors, indicating the evolving interdisciplinary nature of the field of neurobehavioral toxicology. Differences in various batteries indicate the influence of the originating discipline, with emphasis on different functions: veterinary and neurologically based detailed examinations, specific behavioral measures with known neural localization and function resulting from general neuroscience research, and psychologically based, observed behaviors. An evaluation of the usefulness of any battery should include the context in which they were developed and used. This includes an understanding of the following.

(i) The express purpose of the tests. Different batteries have been developed for various purposes, such as evaluating possible side effects of a drug, the neurotoxic potential of a new chemical, or detailed mechanistic studies of specific classes of compounds.

(ii) The setting in which it is used. There may be differences in the attitudes of the testing environment, or the willingness to use these tests, whether it be the chemical industry, pharmaceutical laboratories, or academic research. Likewise, the test species chosen may influence the data, although these tests are applicable for many species.

(iii) Experience and background of the investigator(s) developing the tests. There is a

general hesitancy to use a standardized test,[31] and investigators often make modifications to existing batteries. These are naturally influenced by the training of the investigator (i.e., neurology, pharmacology, etc.). Since neurotoxicology is an interdisciplinary field, it is expected that there would be representatives from many areas of science which may use these types of tests.

(iv) The types of chemicals being examined, and the exposure conditions. For drugs, a typical concern for pharmacologists is the side effects of acute doses (e.g., drowsiness), or of chronic consumption (e.g., dyskinesias). While both acute and chronic effects may be important with environmentally-relevant chemicals, the preference is to study chronic effects using relevant routes of administration (e.g., drinking water or feed); these experimental paradigms make it less likely to obtain acute effects. For some chemicals for which no acute effects are expected, only lifetime studies may be undertaken. Thus, tests which tend to detect more acute effects (e.g., excitability, reactivity, autonomic changes) are utilized more in acute studies, whereas long-term studies may be more directed towards other chronic effects (e.g., neuromuscular, sensory, cognitive).

The various types of test batteries can be arranged according to the context in which they were developed, and the purpose of use; these are presented in Table 1. Details and examples of behavioral test batteries for each situation follow.

11.19.3.2 Categories of Screening Batteries

11.19.3.2.1 Psychopharmacological and pharmaceutical uses

These tests originated in and are still widely used by the pharmaceutical industry. There were generally two different objectives of these tests. One was to screen for nervous system side effects of potential drugs. A classic example of screening for central effects of drugs in mice is the Irwin screen.[32] Irwin described a series of observations and manipulations which were ranked using a common rating scale. Parameters that were evaluated included behavioral (spontaneous activity, motor-affective and sensorimotor responses), neurological (posture, equilibrium, gait, muscle tone, CNS excitation), autonomic (eyes, secretions, and excretions), toxicological, and miscellaneous assessments. In a similar manner, Campbell and Richter[33] described an observational battery using a fewer number of signs, mostly simple autonomic, activity and neuromuscular measures.

Table 1   Context in which various behavioral screening methods were developed and were/are used.

Framework of development	Purpose(s)	Characteristics/examples
Psychopharmacology	Potential therapeutic use Side effects	Drug challenges, models of disease states CNS activity
Mechanistically based	Specific mechanism of action	Stereotypy scales, serotonin syndrome
Neurologically based	Specific functions Specific brain areas/nerves	Neuronal reflex and response
Focused screening	Specific types of neurotoxicity	Delayed neuropathy, ataxia scales
General screening	Neurotoxicity first tier	FOB, neuromuscular screen

They reported that these signs were capable of discriminating drug actions into major pharmacologic classes (e.g., sympathomimetics, sympatholytics).

The other purpose of some pharmacological tests was to screen for efficacy, that is, potential nervous system activity which may be therapeutically beneficial. Sometimes these tests were animal procedures intended to model specific human disease states, and a treatment-related change in the tests could predict a beneficial action in humans.[13] For example, the ability of drugs to reverse the effects of reserpine in rodents is predictive of antidepressant activity in humans. Bastian[34] used a series of tests based on activity against metrazol-induced seizures as a model of CNS depression. This approach is most useful when the animal model used is based on known mechanisms of action for existing drugs.

11.19.3.2.2   Mechanistically based batteries

There are numerous batteries based on specific mechanisms of action of well-studied chemicals. Similarity in responses implies similar mechanisms of action. For example, stereotypy rating scales have been used to quantify the result of dopaminergic activity.[35–38] While this approach has been used extensively in psychopharmacology, it was also used to elucidate the mechanism of a neurotoxicant, triadimefon.[39]

A series of particular behaviors has been described as the serotonin syndrome, which has been widely used to study pharmacological agents acting on this neurotransmitter system.[40,41] Neurobehavioral studies with the neurotoxicant 3,3'-iminodipropionitrile (IDPN) revealed signs of toxicity that were similar to the serotonin syndrome, and these effects were subsequently established to be mediated through the serotonergic system.[42,43] Thus, although these types of tests have considerable utility in psychopharmacological studies, they can and have been applied in research involving neurotoxicants as well.

11.19.3.2.3   Neurologically based batteries

Some batteries are specific neurological tests which are based on human and/or veterinary examinations (see Chapter 22, this volume). These focus on functions of specific nerves, responses, and reflexes, often to the exclusion of reactivity, excitability, and activity assessments. Sterman[44] described the neurological basis of the human clinical examination: close observation of gait, motor, and sensory testing, tendon reflexes, and cerebellar and cranial nerve functions. A battery of tests to be used in rats, analogous to the clinical experience, was then proposed. Detailed descriptions of postures, movements, and fine analysis of goal-directed behaviors has also been proposed as a method of studying rodent behaviors.[45] Observable abnormal movements could be related to the effects produced by brain lesions, which could theoretically be extended to studying the site of action of chemicals.

Tupper and Wallace[46] described a neurological examination consisting of various reflexes, reactions, equilibrium tests, and the auditory startle. The measures were designated according to function of specific brain areas, for example, the flexion reflex which tests spinal cord function. Subsequent lesions in the specific brain areas verified their hypotheses of what each test was presumed to measure (test validity), for example, equilibrium deficits were evident primarily in cerebellar lesioned rats.

They also demonstrated that some reactions were not as reliable in control rats as others.

11.19.3.2.4 Focused screening

Some screening tests include only a few endpoints that are valid and predictive of specific types of neurotoxicity. These tests are used mostly when evaluating the propensity for a specific effect, or when a series of structurally similar chemicals are being tested. For example, the use of a rating scale for ataxia in hens exposed to organophosphate compounds has become an accepted measure of the potential of some chemicals to produce delayed neuropathy.[47]

The battery of tests used by Jolicoeur et al.[48] described various models of ataxia: several neurobehavioral tests were compared to evaluate the ataxia produced by different neuromotor toxicants. Of the tests used—motor activity, catalepsy, rigidity, landing foot splay, gait analysis, and reflexive responses—all were capable of providing quantitative measures of ataxia, but the profile within these measures were different according to the toxicant under study.

11.19.3.2.5 General screening

Most recommendations for general screening include testing as many functions as practical in the first evaluation to increase the probability of detecting changes when there are no expectations as to the outcome. The battery of tests used for neurotoxicity screening has generally become known as a functional observational battery (FOB), a term used when the US Environmental Protection Agency (US EPA) promulgated guidelines for these tests in 1985.[49] Indeed, the term FOB does not refer to any one specific protocol, but rather to an approach used for neurobehavioral screening and including certain key elements. This approach has now been developed for several species of laboratory animals, including the rat,[50–54] mouse,[55] rabbit,[56] dog,[57] and nonhuman primate.[58] These protocols range from expanded clinical observations with a few quantal measures to FOBs with more than 50 tests.[59] Generally, the tests are intended to detect actions of a chemical on both the central and peripheral nervous system.

Fowler et al.[27] described the "rat toxicity screen," which included observations in open field, after provocation, and while handling. Not designed for screening large numbers of rats, it consisted of 49 parameters taken at very close time intervals in order to obtain the most data possible with the least amount of chemical. This screen also included clinical chemistry and hematology studies.

The observational screen described by Gad[50] was designed more to detect neuromuscular changes than to discover all possible neurobehavioral effects. While some clinical signs were recorded (e.g., pupil response, salivation, lacrimation), most of the tests were motor-oriented (e.g., activity, righting reflex, hind-leg splay, extensor thrust, and limb rotation). Gad recommended a second phase of the screen to be a series of in vitro assays to complement the in vivo findings.

A "neurobehavioral checklist" proposed by Alder and Zbinden[5] included observations in the home cage and while handling: standard clinical signs, reflex and coordination tests, rotating rod, swimming test, and catalepsy time.[5,60,61] Subsequent validation studies showed the checklist to be sensitive to psychotherapeutics and drugs with behavioral side effects, but not as successful at detecting neurotoxic agents such as acrylamide,[60] demonstrating that the battery had many tests of excitability and activity but not of neuromuscular function. The authors also proposed an analysis of rodent's social behavior, for example, dominance or aggression of male rats. These tests have not been validated for toxicity studies, however, and few investigators since have included measures enumerated in that study.

Shortly after the US EPA promulgated guidelines for the FOB,[49] Moser et al. constructed a FOB protocol based on the EPA guidelines.[53] Since then they have tested many chemicals in the process of validating the battery and refining the protocol. The measures assessed several functional domains, including neuromuscular/motor/vestibular function, activity levels, sensorimotor reactivity, autonomic function, excitability states, and general health indices.[52,53,62] In addition to observations and manipulations, both grip strength and landing foot splay were routinely quantified.[63,64]

The observational battery used by O'Donoghue[54] contains more specific test measures than many of the aforementioned protocols. However, it was not intended for all subjects in a study, but only those showing impairment for which further evaluation was warranted. The study includes cage-side observations, physical examination, and a neurological examination consisting of evaluations of gait, posture, and reactions to various stimuli, as well as specific reflexes which are traced to relevant neuronal pathways.

Some test batteries are geared more towards long-term effects, which may involve more specifically neuromotor changes. The tests proposed by Kulig[65] included many of the motor and activity measures of other batteries (e.g., grip strength), but in addition incorporated an automated test of hindlimb placement and peripheral nerve conduction velocity. Evaluation of performance of a learned task, two-choice visual discrimination, was also incorporated to increase the sensitivity of the battery. Kulig and Lammers[66] reviewed the available test measures which specifically evaluate neuromuscular and motor function.

A neurobehavioral test battery proposed for use in studies for the National Toxicology Program (NTP) has fewer measures but is intended to be used in studies where there is already a suspicion of neurotoxicity, that is, FOB data are available, structure–activity analysis is suggestive, or human data exist. These measures are somewhat more specific but include neuromuscular, motor, acoustic and nociceptive reactivity, and clinical and health indications.[2]

Motor activity is typically included in neurobehavioral screening, and is sometimes treated as a separate test altogether. For example, the US EPA test guidelines[49] first considered motor activity as a stand-alone test, whereas the revisions to these guidelines[67] strongly suggested that activity measures be conducted in the same rats that undergo the observational and functional procedures. The utility of motor activity has been established over decades of psychopharmacology and neuroscience research (see Chapter 17, this volume).[68] Its addition at the screening level of evaluation points to motor activity as an apical test of neuronal function which can be influenced by a variety of factors including changes in the central or peripheral nervous system as well as general health conditions.[3] Motor activity is most often measured in an automated system, which provides objective, quantitative data. Some investigators, however, have expressed concern that by using these automated measures, subtle but important differences in behavior may be missed.[69]

The numerous neurobehavioral screening batteries in use today incorporate features from many to all of these specific batteries. The functions measured with these behavioral procedures can be separated into domains of neurological effects, although there is significant overlap in many of these tests. These domains can be specified as follows: (i) autonomic—evidenced by observable changes in the function of the sympathetic and parasympathetic nervous systems; (ii) neuromuscular—detects motor dysfunction of central and/or peripheral origin; (iii) reactivity or excitability—can be considered the subject's general responsiveness

to nonspecific stimuli; (iv) sensory—evaluation of the motor response to a sensory stimulus; and (v) others, including convulsions, tremors, and physiological parameters. Tables 2–5 list some of the most common measures included in various screening protocols, grouped according to the functions which they measure. While this list is not exhaustive, it provides specific

Table 2 Test measures which are included in many neurobehavioral screening batteries to assess neuromuscular function and coordination, along with the human symptomatology associated with these forms of neurotoxicity.

Neuromuscular/coordination	
Test measures in animals	*Human symptomatology*
Catalepsy, landing foot splay, gait abnormalities, rotarod, forelimb and hindlimb grip strength, swimming performance, muscle tone, balance beam, extensor thrust, limb rotation, hindlimb traction	Weakness, incoordination, dizziness, fatigue, postural abnormalities

Source: Refs. 5, 8, 27, 32, 46, 48, 50–54, 63, 64, 66, 71, 72, 83, and 84.

Table 3 Test measures which are included in many neurobehavioral screening batteries to assess sensory and motor reactivity, along with the human symptomatology associated with these forms of neurotoxicity.

Reactivity/sensory and motor	
Test measures in animals	*Human symptomatology*
Simple reflexes Pupil reflex, flexor reflex, corneal reflex, grasping reflex, pinna reflex	Paresthesias, footdrop, perceptual dysfunctions, anosmia, auditory deficits, tinnitus, visual deficits, nystagmus, proprioceptive deficits
Complex responses/reactions Righting reaction, hopping response, placing response, nociceptive responses, auditory responses, visual responses, olfactory responses, somatosensory responses, extensor postural thrust response, negative geotaxis	

Source: Refs. 5, 8, 27, 32, 46, 48, 50–54, 66, 71, 72, 83, and 84.

Table 4 Test measures which are included in many neurobehavioral screening batteries to assess autonomic function and general excitability levels, along with the human symptomatology associated with these forms of neurotoxicity.

Test measures in animals	Human symptomatology
Autonomic	
Salivation, lacrimation, urination, defecation, palpebral closure, pupil size, respiration, oculocardiac reflex	Cholinergic crisis, pupil reflex, and size
Activity/reactivity	
Motor activity, arousal, stereotypy, rearing, emergence latency, grooming, handling, and removal reactivity	Hyperactivity, nervousness, irritability, agitation, apathy, lethargy, depression

Source: Refs. 5, 8, 27, 32, 46, 48, 50–54, 66, 71, 72, 83, and 84.

Table 5 Test measures which are included in many neurobehavioral screening batteries, although not specifically behavioral tests, to assess physiological and clinical parameters, along with the associated human symptomatology. Also included are human effects which are not reflected in current screening batteries.

Test measures in animals	Human symptomatology
Physiological/clinical	
Tremor, myoclonus, fasciculations, convulsions, body weight, body temperature, skin and hair coat, tail elevation, inclined plane, eye prominence	Tremor, myoclonus, fasciculations, convulsions, weight changes, hypo- or hyperthermia
Other	
	Headache, nausea and vomiting, sweating, impaired short- or long-term memory, confusion

Source: Refs. 5, 8, 27, 32, 46, 48, 50–54, 66, 71, 72, 83, and 84.

information regarding the types of tests used. Note that many tests actually assess functions from more than one domain, for example, rearing is not only a measure of activity and exploration but the action also requires intact equilibrium and neuromuscular components.

11.19.4 PRINCIPLES OF NEUROBEHAVIORAL SCREENING

11.19.4.1 Strategies for Development

Zbinden et al.[70] described steps or strategies that are common to the development of all toxicity testing. These procedures are: definition of the target, selection and validation of the method, and setting of criteria for interpretation. The target is an evaluation of the integrity of several different neurological functions. Tilson and Cabe[71] recommended testing chemicals which have known effects in humans and that would produce different neurotoxicological profiles. The data would then provide test validity (i.e., that the measure is evaluating what it is expected to), and predictive validity (i.e., that the results predict known effects in humans).

An approach that has been suggested for the selection of methods in neurobehavioral studies[71,72] is to design a test battery that would encompass the range of signs and symptoms of neurotoxic effects in humans. These include changes in sensory (e.g., paresthesias, visual and auditory deficits), motor (e.g., weakness, fatigue, incoordination, tremor), affective (e.g., nervousness, lethargy), and associative (cognitive, e.g., disorientation, confusion, impaired short- and long-term memory) function, as well as physiological responses (e.g., disrupted sleep, loss of appetite). In Tables 2–5, the test procedures in animals are listed along with the human symptoms that they are presumed to detect.

11.19.4.2 Operating Characteristics

In general, screening batteries are based on noninvasive observations and manipulations. These tests are often compared to neurological examinations used in humans to diagnose or distinguish neurological syndromes. All such tests rely on careful evaluation of the subject, whether it be human, dog, or rat. Some tests can include appropriate test devices to provide quantitative, objective data. On the other hand, many measures are often subjective. While some investigators feel that objective data are always preferable, there is the possibility of missing important but unexpected changes when the test becomes too focused.[32] As Irwin stated, "...systematic observation offers the possibility for obtaining a broad range of information from a single system of measurement, for observing unanticipated treatment effects, and for obtaining information difficult or impossible to derive otherwise."[32] Subjective evaluations that are specified as distinct rating scales introduce a semi-quantitative aspect to the data. Graded scales provide more information on the effects than do quantal, or all-or-nothing, measures.[31] Quantal measures usually require a clear description of what constitutes "other than normal." Whereas with lethality, for example, "yes" and "no" are easily distinguished, the case for evaluations such as activity is not so obvious, where "increased" or "decreased" must have associated with them operational definitions or else the measure is meaningless.

Observational procedures are not difficult, but do require a significant level of comprehension, technique, and enthusiasm. It is crucial that the observer be unaware ("blind") of the subject's treatment, in order to prevent deliberate or subconscious bias from entering the data.[31] Adequate training of the observer is equally important. Documented procedures for training, certification, renewal, and update of standards are important elements of testing taking place under the good laboratory practices (GLP); these are indispensable for the non-GLP laboratory settings as well.

The same subject can be tested repeatedly, which is valuable in determining onset or recovery of effects and aids in the interpretation of group comparisons. The response of the whole animal is measured, therefore, there is a multitude of possible influences from various sources. Extraneous factors such as noise, light, and so on, must be rigidly controlled. Since no training of the subject is involved, there is no explicit control over the behavior. There may be more variability, leading to the need for larger sample size and tighter control over experimental variables. The behaviors examined tend to be more robust, however, if the observations include the subject's innate responses or reactions of some function that is important to its survival. Strain and gender of the subjects may also influence the results.[73,74] While many testing guidelines require the use of males and females, almost no studies employ more than one strain to evaluate differences in responsiveness among the groups.

Regardless of the protocol adopted, appropriate validation and standardization is required.[3,72] Screening procedures in neurotoxicity are designed to detect potential activity on the nervous system, not necessarily the lowest effective dose or degree of toxicity.[2] The data from screening batteries should be adequate to provide direction for further testing. These tests are usually used with large numbers of animals, and therefore should not require extensive training of the subjects or be

time-consuming to perform. In many respects, these techniques are very different from secondary tests, which may require special equipment, training, and conditioning. Advantages and disadvantages of both screening studies and second-tier (more complex) measures are listed in Tilson and Mitchell.[3]

11.19.4.3 Statistical Analysis

The use of rating scales or descriptions make statistical analyses of screening data more complex than that used for traditional interval data. Special concerns, for example, distribution of control data and sample size, arise when attempting to derive probability values. Furthermore, there are numerous measures collected on each individual subject. This repeated testing, or within-subject design, provides powerful data to assess onset and recovery of toxicity, but greatly complicates the statistical techniques. While some have proposed analytical methods using commonly available packages,[75] others have been critical of that approach. Currently, there is little agreement on the appropriate statistical methods for these data.

Since screening batteries are typically the first line of testing, sensitivity and specificity of the test method is crucial.[76,77] An effective screen should not allow any false-negatives, that is, the type II error rate should be low. A toxic compound that passes this level of evaluation without concern could be put into use, possibly with dire consequences. On the other hand, the number of false-positives, the type I error rate, should also be low. A test battery that shows almost every compound to be toxic would be quite costly, since in a tiered-testing scheme these compounds would be submitted to secondary evaluations before it became known that they were, in fact, nontoxic. Therefore the toxicologist must always balance the two error rates to achieve a situation that is acceptable and yet conservative.

The question constantly arises regarding correction of probability values to account for the numbers of endpoints taken for each subject. Traditional approaches, for example, Bonferroni corrections, are probably too conservative at this level of detection. One method is to compress the test data into composite scores to decrease the number of overall analyses and for simplicity in interpretation.[78] Severity scores on a one-to-four scale have been proposed to convert the various data from an FOB to a common score,[78] where a score of one reflects data most likely to occur in controls, and a score of four indicates data which are greatly divergent from controls. These scores are based on the mean or median, and the range of data from the concurrent control group. Scores for the measures within each functional domain can be combined to provide overall domain scores, which are compared back to control values. This approach was used to show that the small but significant changes in a few neuromuscular measures produced by one chemical (2-hydroxyethyl acrylate)[78] was not of sufficient magnitude to produce an overall neuromuscular domain effect, whereas the positive control (acrylamide) in that study showed strong neuromuscular domain changes.

A similar method has been used by Tamborini *et al.*,[61] who described a "ToxScore" derived from the magnitude of difference from control groups. For each measure, a "ToxRatio" was calculated (essentially the treated data expressed as a proportion of the control group). The sum of ToxRatios was converted to a relative point system to provide the "total ToxScore." The relationship between "ToxScores" and dose predicted toxicity as well as or better than traditional LD_{50} values. This method was used for studies assessing general indices of toxicity as well as neurobehavioral changes. Unfortunately, this approach was developed for continuous variable data, and is not applicable to endpoints that are ranked or descriptive data.

A somewhat different approach was taken by Koek *et al.*[79] Cluster analysis (similar to factor analysis) was used to differentiate behavioral characteristics of pharmacological agents. The drugs were classified based on how alike or different their effects were on a battery of tests. The underlying hypothesis was that differing functions would be distinguished as different clusters, whereas similar actions would fall into the same or related cluster. This approach has not yet been attempted for neurotoxicological data.

Alternative methods of analyzing the data have been brought forward, which rely more on reviewing the data than on traditional statistical techniques.[70] For example, using historical control values to define "normal values" alleviates the need for concurrent control groups. Considerable drift in these control values can occur, however, due to factors ranging from changes in supplier (i.e., genotype of rats obtained), husbandry (i.e., different feed, bedding), animal care (i.e., new personnel), and so on, and the range of "normal" over a period of years could be quite large; these factors can make interpretation of the data even more difficult. Pre-established determination of test criteria is another approach to assessing toxicity. Zbinden *et al.*[70] proposed a scheme of classifying data based on whether the test criterion was met, and the magnitude of difference from the control group. Results could then be reported on a scale of increasing confidence in the data, ranging from "level of suspicion I–III" to "level of virtual certainty." The test criterion is a critical determination, then, and there may not be consensus among investigators on this issue.

11.19.5 INTERPRETATION OF DATA

Interpretation of results from screening batteries may be difficult, given the large number of endpoints recorded for each subject. However, having many measures provides a "multidimensional approach" which can be useful for the interpretation of effects.[31] Some of the measures are inter-related; it is this redundancy that is most useful for interpreting the effects in individual tests. The investigator should understand the basis for, and characteristics of, each and every observation conducted. Even if there is not an explicit function evaluated by a particular test, its value may be that the data aid in the interpretation of another, similar alteration. The data should be judged in the same manner as is done with other batteries such as clinical chemistries: that is, looking for patterns, or profiles, of change.

Thus, a key point for interpreting these data is to include "information on many functions in order to evaluate the specificity of an effect on one function."[9] Often, however, it may be difficult when using behavioral measures to distinguish the relative contributions of sensory, motor, arousal, or associative factors:[3] this is the reason that several schemes of quantifying the functional domains or relative toxicity have been proposed (see above). Regardless of the statistical approaches used, the neurotoxicologist should review the data and use professional judgment regarding biological significance. The dose at which changes occur, and what other forms of toxicity are produced at the same dose levels, are critical pieces of information for a valid conclusion. As O'Donoghue pointed out,[80] under appropriate conditions almost any compound could appear to be neurotoxic. Sickness can also produce behavioral changes. Responses due to infection have been described.[81] Nonspecific signs including fever are thought to be due to cytokines—but this is probably not the same effect as seen with malaise associated with toxicity. Human health is not well served if the data collected in these neurobehavioral tests cannot be distinguished from other forms of toxicity. Thus, factors to consider for evaluating behavioral data include magnitude of the effect, dose- and time-response characteristics, inherent variability of the test, and generality of the findings; in addition, all data should be considered in the context of other neurotoxicological measures (e.g., neuropathology) and general toxicity results.

11.19.6 EXTRAPOLATION TO HUMANS

There are many instances where extrapolations from animal studies to effects anticipated in humans are valid and appropriate. Some functions, reflexes, and reactions are similar across all species; these functions can be easily extrapolated to humans, such as, autonomic parameters, vestibular changes, and ataxia. For example, many of the effects of acrylamide (primarily neuromotor signs) and parathion (cholinergic and neuromuscular effects) that have been reported in humans could be replicated in particular measures of the FOB in rats.[82] Some extrapolations may not display construct validity but yet be predictive. Such measures may not be completely understood, or not have a known physiological basis, yet they have been shown to be valuable behavioral measures, for example, locomotor activity.[68]

Tilson and Mitchell[3] reviewed human and experimental neurotoxicology and concluded that there is a reasonable correspondence between human effects and what can be detected in laboratory animals. Major points of disparity included sensory and associative (cognitive) deficits in animals, particularly rodents. A reasonable overlap of effects between the experimental and human condition, however, should be sufficient to allow extrapolation, and therefore to test for neurotoxic potential using these methods.

11.19.7 FUTURE DIRECTIONS AND NEEDS

There are a number of important functions in humans that are not reliably measured in laboratory animals. These include tests of subtle sensory dysfunction and cognitive function. Further improvements of tests are needed to improve our ability to adequately screen new chemicals, without an unmanageable burden of cost and labor demands.

Furthermore, there is continued need for studies of the sensitivity, reliability, reproducibility, and cost effectiveness of behavioral screening batteries. Ongoing research on possible improvements should incorporate new knowledge, instrumentation, and techniques as they are developed. Screening assessments in neuro-

toxicity should not be allowed to become so standardized and inflexible that new ideas and innovations are stifled. As long as continual refinements are made, behavioral test methods will continue to be important tools with which to protect human health in the future.

11.19.8 REFERENCES

1. N. K. Mello, 'Behavioral toxicology: a developing discipline.' *Fed. Proc.*, 1975, **34**, 1832–1834.
2. H. A. Tilson, 'Behavioral indices of neurotoxicity.' *Toxicol. Pathol.*, 1990, **18**, 96–104.
3. H. A. Tilson and C. L. Mitchell, 'Neurobehavioral techniques to assess the effects of chemicals on the nervous system.' *Annu. Rev. Pharmacol. Toxicol.*, 1984, **24**, 425–450.
4. D. A. Cory-Slechta, 'Behavioral measures of neurotoxicity.' *Neurotoxicology*, 1989, **10**, 271–295.
5. S. Alder and G. Zbinden, 'Neurobehavioral tests in single- and repeated-dose toxicity studies in small rodents.' *Arch. Toxicol.*, 1983, **54**, 1–23.
6. B. Broxup, K. Robinson, G. Losos *et al.*, 'Correlation between behavioural and pathological changes in the evaluation of neurotoxicity.' *Toxicol. Appl. Pharmacol.*, 1989, **101**, 510–520.
7. V. C. Moser, D. C. Anthony, W. F. Sette *et al.*, 'Comparison of subchronic neurotoxicity of 2-hydroxyethyl acrylate and acrylamide in rats.' *Fundam. Appl. Toxicol.*, 1992, **18**, 343–352.
8. A. B. Sterman and R. C. Sheppard, 'A correlative neurobehavioral-morphological model of acrylamide neuropathy.' *Neurobehav. Toxicol. Teratol.*, 1983, **5**, 151–159.
9. C. L. Mitchell and H. A. Tilson, 'Behavioral toxicology in risk assessment: problems and research needs.' *Crit. Rev. Toxicol.*, 1982, **10**, 265–274.
10. H. A. Tilson, 'Behavioral indices of neurotoxicity: what can be measured?' *Neurotoxicol. Teratol.*, 1987, **9**, 427–443.
11. World Health Organization (WHO), 'Environmental Health Criteria Document 60: Principles and Methods for the Assessment of Neurotoxicity Associated with Exposure to Chemicals,' World Health Organization, Geneva, 1986.
12. S. Norton, 'Is behavior or morphology a more sensitive indicator of central nervous system toxicity?' *Environ. Health Perspect.*, 1978, **26**, 21–27.
13. S. D. Glick, in 'Behavioral Pharmacology,' eds. S. D. Glick and J. Goldfarb, Mosby, St. Louis, MO, 1976, pp. 339–361.
14. R. C. MacPhail and H. A. Tilson, in 'Neurotoxicology: Approaches and Methods,' eds. L. Chang and W. Slikker, Academic Press, New York, 1995, pp. 231–238.
15. D. L. Arnold, S. M. Charbonneau, Z. Z. Zawidzka *et al.*, 'Monitoring animal health during chronic toxicity studies.' *J. Environ. Pathol. Toxicol.*, 1977, **1**, 227–239.
16. J. G. Fox, 'Clinical assessment of laboratory rodents on long term bioassay studies.' *J. Environ. Pathol. Toxicol.*, 1977, **1**, 199–226.
17. J. M. Barnes and F. A. Denz, 'Experimental methods in determining chronic toxicity: a critical review.' *Pharmacol. Rev.*, 1954, **6**, 191–242.
18. E. M. Boyd, 'The acute oral toxicity of acetylsalicylic acid.' *Toxicol. Appl. Pharmacol.*, 1959, **1**, 229–239.
19. J. B. Ruffin, 'Functional testing for behavioral toxicity: a missing dimension in experimental environmental toxicology.' *J. Occup. Med.*, 1963, **5**, 117–121.

20. R. N. Walsh and R. A. Cummins, 'The open-field test: a critical review.' *Psychol. Bull.*, 1976, **83**, 482–504.

21. J. Archer, 'Tests for emotionality in rats and mice: a review.' *Anim. Behav.*, 1973, **21**, 205–235.

22. A. Ivinskis, 'A study of validity of open-field measures.' *Aust. J. Psychol.*, 1970, **22**, 175–183.

23. J. D. Porter, S. M. Pellis and M. E. Meyer. 'An open-field activity analysis of labyrinthectomized rats.' *Physiol. Behav.*, 1990, **48**, 27–30.

24. A. P. Silverman, 'An ethologist's approach to behavioural toxicology.' *Neurotoxicol. Teratol.*, 1988, **10**, 85–92.

25. National Academy of Sciences (NAS), 'Principles for Evaluating Chemicals in the Environment,' National Academy Press, Washington, DC, 1975.

26. R. W. Brimblecombe, 'Behavioral tests in acute and chronic toxicity studies.' *Pharmacol. Ther. B*, 1979, **5**, 413–415.

27. J. S. L. Fowler, J. S. Brown and H. A. Bell, 'The rat toxicity screen.' *Pharmacol. Ther. B*, 1979, **5**, 461–466.

28. R. W. Leukroth (ed.), 'Predicting Neurotoxicity and Behavioral Dysfunction from Preclinical Toxicologic Data,' Life Sciences Research Office, Federation of American Societies for Experimental Biology, Bethesda, MD, 1986.

29. National Academy of Sciences (NAS), 'Toxicity Testing: Strategies to Determine Needs and Priorities,' National Academy Press, Washington, DC, 1984.

30. US Congress, Office of Technology Assessment (OTA), 'Neurotoxicity: Identifying and Controlling Poisons of the Nervous System,' OTA-BA-436, US Government Printing Office, Washington, DC, 1990, pp. 105–144.

31. S. Irwin, 'Drug screening and evaluative procedures.' *Science*, 1962, **136**, 123–128.

32. S. Irwin, 'Comprehensive observational assessment: Ia. A systematic, quantitative procedure for assessing the behavioral and physiologic state of the mouse.' *Psychopharmacology (Berl.)*, 1968, **13**, 222–257.

33. E. S. Campbell and W. Richter, 'An observational method estimating toxicity and drug actions in mice applied to 68 reference drugs.' *Acta Pharmacol. Toxicol. (Copenh.)*, 1967, **25**, 345–363.

34. J. W. Bastian, 'Classification of CNS drugs by a mouse screening battery.' *Arch. Int. Pharmacodyn.*, 1961, **133**, 347–364.

35. P. C. Contreras, K. C. Rice, A. E. Jacobson et al. 'Stereotyped behavior correlates better than ataxia with phencyclidine–receptor interactions.' *Eur. J. Pharmacol.*, 1986, **121**, 9–18.

36. E. H. Ellinwood, Jr. and R. L. Balster, 'Rating the behavioral effects of amphetamine.' *Eur. J. Pharmacol.*, 1974, **28**, 35–41.

37. P. J. Fray, B. J. Sahakian, T. W. Robbins et al., 'An observational method for quantifying the behavioural effects of dopamine agonists: contrasting effects of *d*-amphetamine and apomorphine.' *Psychopharmacology (Berl.)*, 1980, **69**, 253–259.

38. M. H. Lewis, A. A. Baumeister, D. L. McCorkle et al., 'A computer-supported method for analyzing behavioral observations: studies with stereotypy.' *Psychopharmacology (Berl.)*, 1985, **85**, 204–209.

39. Q. D. Walker, M. H. Lewis, K. M. Crofton et al., 'Triadimefon, a triazole fungicide, induces stereotyped behavior and alters monoamine metabolism in rats.' *Toxicol. Appl. Pharmacol.*, 1990, **102**, 474–485.

40. B. L. Jacobs, 'An animal behavior model for studying central serotonergic synapses.' *Life Sci.*, 1976, **19**, 777–785.

41. M. R. Pranzatelli and S. R. Snodgrass, 'Serotonin-lesion myoclonic syndromes. II. Analysis of individual syndrome elements, locomotor activity and behavioral correlations.' *Brain Res.*, 1986, **364**, 67–76.

42. J. L. Cadet, 'The iminodipropionitrile (IDPN)-induced dyskinetic syndrome: behavioral and biochemical pharmacology.' *Neurosci. Biobehav. Rev.*, 1989, **13**, 39–45.

43. S. Przedborski, M. Wright, S. Fahn et al., 'Quantitative autoradiographic changes in 5-[^{3}H]HT-labeled 5-HT$_1$ serotonin receptors in discrete regions of brain in the rat model of persistent dyskinesias induced by iminodipropionitrile (IDPN).' *Neurosci. Lett.*, 1990, **116**, 51–57.

44. A. B. Sterman, 'The pathology of toxic axonal neuropathy: a clinical-experimental link.' *Neurobehav. Toxicol. Teratol.*, 1984, **6**, 463–466.

45. P. Teitelbaum, in 'Changing Concepts of the Nervous System,' eds. A. R. Morrison and P. L. Strick, Academic Press, New York, 1982, pp. 467–487.

46. D. E. Tupper and R. B. Wallace, 'Utility of the neurological examination in rats.' *Acta Neurobiol. Exp. (Warsz.)*, 1980, **40**, 999–1003.

47. M. G. Cherniack, 'Toxicological screening for organophosphorus-induced delayed neurotoxicity: complications in toxicity testing.' *Neurotoxicology*, 1988, **9**, 249–271.

48. F. B. Jolicoeur, D. B. Rondeau, A. Barbeau et al., 'Comparison of neurobehavioral effects induced by various experimental models of ataxia in the rat.' *Neurobehav. Toxicol.*, 1979, **1**, Suppl. 1, 175–178.

49. US Environmental Protection Agency, 'Toxic Substances Control Act Testing Guidelines.' 40 CFR, Part 798, Subpart G Section 798.6050, *Fed. Regist.*, 1985, **50**, 39458–39460.

50. S. C. Gad, 'A neuromuscular screen for use in industrial toxicology.' *J. Toxicol. Environ. Health*, 1982, **9**, 691–704.

51. G. C. Haggerty, 'Development of tier I neurobehavioral testing capabilities for incorporation into pivotal rodent safety assessment studies.' *J. Am. Coll. Toxicol.*, 1989, **8**, 53–69.

52. K. L. McDaniel and V. C. Moser, 'Utility of a neurobehavioral screening battery for differentiating the effects of two pyrethroids, permethrin and cypermethrin.' *Neurotoxicol. Teratol.*, 1993, **15**, 71–83.

53. V. C. Moser, J. P. McCormick, J. P. Creason et al., 'Comparison of chlordimeform and carbaryl using a functional observational battery.' *Fundam. Appl. Toxicol.*, 1988, **11**, 189–206.

54. J. L. O'Donoghue, 'Screening for neurotoxicity using a neurologically based examination and neuropathology.' *J. Am. Coll. Toxicol.*, 1989, **8**, 97–115.

55. J. S. Tegeris and R. L. Balster, 'A comparison of the acute behavioral effects of alkylbenzenes using a functional observational battery in mice.' *Fundam. Appl. Toxicol.*, 1994, **22**, 240–250.

56. J. M. Hurley, P. E. Losco, S. J. Hermansky et al., 'Functional observational battery (FOB) and neuropathology in New Zealand white rabbits.' *Toxicologist*, 1995, **15**, 247.

57. U. Schaeppi and R. E. Fitzgerald, 'Practical procedure of testing for neurotoxicity.' *J. Am. Coll. Toxicol.*, 1989, **8**, 29–34.

58. R. T. O'Keeffe and K. Lifshitz, 'Nonhuman primates in neurotoxicity screening and neurobehavioral toxicity studies.' *J. Am. Coll. Toxicol.*, 1989, **8**, 127–140.

59. H. A. Tilson and V. C. Moser, 'Comparison of screening approaches.' *Neurotoxicology*, 1992, **13**, 1–13.

60. S. Alder, R. Candrian, J. Elsner et al., 'Neurobehavioral screening in rats: validation study.' *Methods Find. Expt. Clin. Pharmacol.*, 1986, **8**, 279–289.

61. P. Tamborini, H. Sigg and G. Zbinden, 'Acute toxicity testing in the nonlethal dose range: a new approach.' *Regul. Toxicol. Pharmacol.*, 1990, **12**, 69–87.

62. V. C. Moser, 'Screening approaches to neurotoxicity: a

functional observational battery.' *J. Am. Coll. Toxicol.*, 1989, **8**, 85–93.

63. P. M. Edwards and V. H. Parker, 'A simple, sensitive, and objective method for early assessment of acrylamide neuropathy in rats.' *Toxicol. Appl. Pharmacol.*, 1977, **40**, 589–591.

64. O. A. Meyer, H. A. Tilson, W. C. Byrd *et al.*, 'A method for the routine assessment of fore- and hindlimb grip strength of rats and mice.' *Neurobehav. Toxicol.*, 1979, **1**, 233–236.

65. B. M. Kulig, 'A neurofunctional test battery for evaluating the effects of long-term exposure to chemicals.' *J. Am. Coll. Toxicol.*, 1989, **8**, 71–83.

66. B. M. Kulig and J. H. C. M. Lammers, in 'Neurotoxicology,' eds. H. Tilson and C. Mitchell, Raven Press, New York, 1992, pp. 147–179.

67. US Environmental Protection Agency, 'Neurotoxicity Test Guidelines Addendum 10, Pesticide Assessment Guidelines Subdivision F,' Publication number PB 91-154617, National Technical Information Services, Springfield, VA, 1991.

68. L. W. Reiter and R. C. MacPhail, in 'Nervous System Toxicology,' ed. C. L. Mitchell, Raven Press, New York, 1982, pp. 45–65.

69. B. Kolb and I. Q. Whishaw, 'An observer's view of locomotor asymmetry in the rat.' *Neurobehav. Toxicol. Teratol.*, 1985, **7**, 71–78.

70. G. Zbinden, J. Elsner and U. A. Boelsterli, 'Toxicological screening.' *Regul. Toxicol. Pharmacol.*, 1984, **4**, 275–286.

71. H. A. Tilson and P. A. Cabe. 'Strategy for the assessment of neurobehavioral consequences of environmental factors.' *Environ. Health Perspect.*, 1978, **26**, 287–299.

72. H. A. Tilson, C. L. Mitchell and P. A. Cabe, 'Screening for neurobehavioral toxicity: the need for and examples of validation of testing procedures.' *Neurobehav. Toxicol.*, 1979, **1**, Suppl. 1, 137–148.

73. V. C. Moser, K. L. McDaniel and P. M. Phillips, 'Rat strain and stock comparisons using a functional observational battery: baseline values and effects of amitraz.' *Toxicol. Appl. Pharmacol.*, 1991, **108**, 267–283.

74. V. C. Moser, 'Rat strain- and gender-related differences in neurobehavioral screening: acute trimethyl tin neurotoxicity.' *J. Toxicol. Environ. Health*, 1996, **47**, 567–586.

75. J. P. Creason, 'Data evaluation and statistical analysis of functional observational battery data using a linear models approach.' *J. Am. Coll. Toxicol.*, 1989, **8**, 157–169.

76. S. C. Gad, 'Principles of screening in toxicology with special emphasis on applications to neurotoxicology.' *J. Am. Coll. Toxicol.*, 1989, **8**, 21–27.

77. S. C. Gad, 'Screens in neurotoxicity: objectives, design, and analysis, with the functional observational battery as a case example.' *J. Am. Coll. Toxicol.*, 1989, **8**, 287–301.

78. V. C. Moser, 'Applications of a neurobehavioral screening battery.' *J. Am. Coll. Toxicol.*, 1991, **10**, 661–669.

79. W. Koek, J. H. Woods and P. Ornstein, 'A simple and rapid method for assessing similarities among directly observable behavioral effects of drugs: PCP-like effects of 2-amino-5-phosphonovalerate in rats.' *Psychopharmacology (Berl.)*, 1987, **91**, 297–304.

80. J. L. O'Donoghue, in 'Neurobehavioral Toxicity: Analysis and Interpretation,' eds. B. Weiss and J. O'Donoghue, Raven Press, New York, 1994, pp. 19–33.

81. S. Kent, R.-M. Bluthé, K. W. Kelley *et al.*, 'Sickness behavior as a new target for drug development.' *Trends Pharmacol. Sci.*, 1992, **13**, 24–28.

82. V. C. Moser, 'Approaches for assessing the validity of a functional observational battery.' *Neurotoxicol. Teratol.*, 1990, **12**, 483–488.

83. L. Shell, M. Rozum, B. S. Jortner *et al.*, 'Neurotoxicity of acrylamide and 2,5-hexanedione in rats evaluated using a functional observational battery and pathological examination.' *Neurotoxicol. Teratol.*, 1992, **14**, 273–283.

84. H. A. Tilson, 'Neurobehavioral methods used in neurotoxicological research.' *Toxicol. Lett.*, 1993, **68**, 231–240.

11.20
Intermittent Schedules of Reinforcement as Toxicological Endpoints

DEBORAH A. CORY-SLECHTA

University of Rochester School of Medicine and Dentistry, NY, USA

11.20.1   INTRODUCTION

11.20.1.1   Schedule of Reinforcement

Intermittent schedules of reinforcement have been used in two contexts. One is characterizing the nature of a toxicant's adverse effects and to begin to determine the behavioral and neuro-biological mechanisms by which they are produced. Behavior under the control of inter-mittent schedules of reinforcement has also been included in screening protocols for the testing of new chemicals prior to their intro-duction into use. The specific advantage of such approaches is that they can offer a broad spectrum assessment of learned behavior. Intermittent schedules generate highly charac-teristic patterns of performance that exhibit considerable generality across species as well as consistency of pattern across time. For these reasons, they offer the possibility of examining the ability of a subject to learn that character-istic pattern of behavior as well as an assess-ment of toxicants effects on the ability to perform in this capacity across time.

Learned (operant) behavior is followed by consequences which determine the probability of that response occurring in the future. Those consequences may be reinforcing events, in which case the future probability of the response increases. If punishing events follow the response, or if a reinforcing event is withheld, the response will decline in frequency. The term "schedule of reinforcement" refers to the specific nature of the rules governing the relationship between a response and the con-sequences maintaining it.[1] Although infre-quently the case, the consequent event may follow every occurrence of the response, in which case the schedule is deemed to be a continuous reinforcement schedule. More typi-cally however, the consequent event occurs on some type of nonregular basis, that is, on an intermittent schedule of reinforcement. For example, seldom is every visit to the mailbox rewarded by the arrival of the letter which we await. In the wild, predation or foraging behavior is not rewarded with the capture of prey or location of a foodsource on its every occurrence. Besides the economy achieved by intermittent schedules of reinforcement, beha-vior maintained under such conditions is, in fact, more robust than that occurring under continuous reinforcement schedule conditions. As one example of this greater robustness, when reinforcement is withheld, behavior that had been maintained under conditions of continuous reinforcement typically will decline more rapidly than will behavior that had been maintained under conditions of intermittent reinforcement.

11.20.1.2   Schedule-controlled Behavior

As might be expected, different "rules" for the scheduling of reinforcement availability can result in marked differences in both the rate and pattern of responding that are controlled by the schedule, that is, in the schedule-controlled behavior that occurs. These rates and patterns of responding are typically highly characteristic of the particular schedule imposed and also exhibit comparatively broad species generality.

11.20.1.3   Significance of Schedule-controlled Behavior

Despite their substantial role in the realm of human behavior, the significance of schedule-controlled behaviors as endpoints in neurotoxi-cology has been overlooked and to some extent misunderstood.[2,3] This no doubt arises in part from the lack of obvious validity of schedule-controlled performance when considered rela-tive to various other behavioral processes such as measures of learning and memory (see Chapter 21, this volume). However, the rate and patterning and allocation of human learned behavior is a function of the operative schedules of reinforcement. As such, schedule-controlled behavior plays a critical role in all behavior. Consider the process of learning. The rate at which a particular response is learned may very well be influenced by the operative reinforcement schedule. Furthermore, whether the response will be acquired at all may depend upon the strength of the competing responses and the concurrent availability of other rein-forcement options. Thus, an understanding of schedule-controlled performance is critical to a full understanding of behavior.

11.20.2   CLASSIFICATION OF SCHEDULES OF REINFORCEMENT

Reinforcement availability is generally scheduled either on the basis of time, as in interval schedules, or on the basis of the number of responses, as in ratio schedules, or on some combination of these two. Further, schedules of reinforcement may be simple, that is, involving a single reinforcement contin-gency, or may involve multiple, even concur-rently operative contingencies of reinforcement. The human behavioral environment includes the potential for some of the most complex scheduling conditions to occur, in that multiple schedule options may be operative simultaneously. Further, the nature of these concurrent schedules may vary widely, and possibly even change with some frequency. In a working environment, for example, one may have multiple goals to accomplish with different priorities and consequences for failure to meet deadlines.

Obviously the laboratory offers the experimenter direct control over reinforcement contingencies and scheduling conditions, and permits a more precise analysis of the ways in which different reinforcement schedules come to control various aspects of responding, including its patterning, temporal distribution, force, rate, resistance to extinction, etc. From an understanding of the mechanisms controlling behavior on simple schedules of reinforcement then comes a basis for understanding schedule-controlled behavior under more complex conditions. Of direct relevance to toxicology, moreover, is the fact that an extensive data base describing the effects of a wide variety of classes of drugs on schedule-controlled behavior has accumulated. This database allows comparisons of patterns of changes in schedule-controlled operant behavior resulting from toxicant treatment to those produced by chemicals or drugs whose mechanisms of action may be far better understood, resulting in potentially useful insights into mechanisms of toxicant action.

11.20.2.1 Simple Schedules of Reinforcement

Four simple schedules of reinforcement have been defined. On fixed interval (FI) and variable interval (VI) schedules, reinforcement availability is based on time. On fixed ratio (FR) and variable ratio (VR) schedules, reinforcement availability is based on completion of a specified number of responses. Characteristic patterns of performance and relative rates of responding emerge on each of these schedules once performance stabilizes. Further, the temporal patterns of responding, as well as the drug effects upon behavior, are often highly comparable across species.

11.20.2.1.1 Fixed interval schedules

The FI schedule stipulates that the first response occurring after a specified interval of time has elapsed results in reinforcement; responses during the interval itself have no programmed consequence. Thus, on an FI 60 s schedule, for example, the first response occurring after 60 s elapses produces reinforcement; occurrences of the designated response during

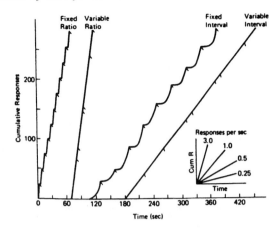

Figure 1 Schematic cumulative records of performance on the fixed ratio, variable ratio, fixed interval, and variable interval schedules. Responses are cumulated vertically over time. Each downward deflection represents reinforcement delivery (reproduced by permission of Van Nostrand Reinhold from 'Psychopharmacology: A Biochemical and Behavioural Approach,' 1977).

the 60 s interval itself are without any specified consequences. The next interval starts after reinforcement. The FI schedule typically engenders a characteristic "scalloped" pattern of responding, as shown in Figure 1,[4] which includes a pause (period of no-responding) after reinforcement delivery, known as the postreinforcement pause (PRP), which is followed by a gradually increasing rate of responding as the next opportunity for reinforcement availability approaches. Studying for an examination might be considered an example of an FI-like schedule in the human environment; little or no studying may occur early in the semester, with the rate of studying gradually escalating as the time of the final examination approaches.

The scalloped pattern of responding on FI schedules exhibits notable similarities across species, and across different types of responses and reinforcers, as illustrated in Figure 2.[5] This figure illustrates that the FI scallop is exhibited in species ranging from rats to monkeys, emitting different response for various reinforcers. Human FI performance also shows a scalloped pattern under similar circumstances.

11.20.2.1.2 Variable interval schedules

Reinforcement on a VI schedule is also available after a specified interval of time has elapsed, but in the case of the VI schedule, the time interval varies from one interval to the next. Thus on a VI 60 s schedule, the average

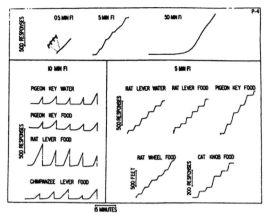

Figure 2 Generality of characteristic fixed interval responding. Responses cumulate vertically, time is represented horizontally. The top panel shows the behavior of an individual pigeon pecking a key for food at three different FI values and the scalloped pattern persists despite the 100-fold change in the time interval. Performance on a FI 5 min schedule is shown in the right panel with reinforcement delivery indicated by downward deflections of the pen. Performance on a FI 10 min is shown in the left panel with the pen resetting to baseline after each reinforcer delivery. For each FI value, performances are very similar across different species, response devices and reinforcers (reproduced by permission of Raven Press from 'Importance of Fundamental Principles of Drug Evaluation,' 1968).

time interval between reinforcement opportunities is 60 s, but each time interval varies in an unpredictable way. The unpredictable nature of reinforcement availability on the VI schedule results in a pattern of relatively continuous responding over time, with little pausing observed after reinforcement delivery, as shown in Figure 1. One example of a VI-type schedule in the human environment is the repeated phone calls (the response) when a busy signal is encountered in attempting to place a telephone call because of the variable length of telephone conversations.

11.20.2.1.3   *Fixed ratio schedules*

The FR schedule requires the completion of a specified number of responses for reinforcement. On an FR 10 schedule, then, the designated response must occur 10 times before reinforcement ensues. Fixed ratio schedules are probably best exemplified by the piecework systems that operated in the early factory production systems where workers were paid per piece. Salespeople who work exclusively on a commission basis represent another example. FR responding is characterized by a pause

immediately after reinforcement delivery, called either the postreinforcement pause or the pre-run pause, and is followed by a very rapid rate of responding until the ratio requirement is completed (Figure 1). The high constant rate of responding generated results from the relationship between rate of reinforcement and rate of responding, that is, the more rapidly the ratio is completed, the sooner reinforcement occurs.

11.20.2.1.4   *Variable ratio schedules*

On a VR schedule, the number of responses required for reinforcement varies from one reinforcement opportunity to the next, with the mean of those numbers designated by the schedule parameter value. Thus, the average number of responses required for reinforcement is 100 on a VR 100 schedule, but the specific number varies from one reinforcement delivery to the next. Thus, as is the case with the VI schedule, the availability of reinforcement is unpredictable, since a reinforced response may be followed immediately by another reinforced response. This results in a high and consistent rate of responding with little or no pausing following reinforcement delivery (Figure 1). Perhaps the best example of a VR schedule in the human environment is that of gambling, where the average number of plays per payoff varies unpredictably (as does the size of the payoff in some cases).

11.20.2.2   Multiple Schedules of Reinforcement

The differences in performance on the four schedules of reinforcement reflect differences in the contingencies of reinforcement. Since ratio schedules require completion of a number of responses for reinforcement to occur, they generate higher rates of responding than do interval schedules, in which technically, a single response occurring after the interval ends can produce reinforcement. Moreover, variable schedules generate higher rates of responding than fixed-interval and ratio schedules because of the unpredictability of reinforcement availability in the former. Comparing the effects of a chemical agent under various schedule conditions, then, can provide a better understanding of the mechanisms by which the toxicant acts. Suppression of response rate across all schedules might suggest a nonspecific effect on factors such as motivation.

The use of a multiple schedule of reinforcement allows the experimenter to measure different types of schedule performances within the confines of a single schedule format in a behavioral session. In a multiple schedule, designated component schedules alternate during the course of an experimental session. FI and FR schedules are typically used as component schedules in a multiple schedule because of the marked differences in the contingencies of reinforcement and consequent patterns of responding that they generate. On a multiple FI–FR schedule, the FI and FR schedule components alternate throughout the session, either on the basis of time (e.g., every 15 min) or after a certain number of reinforcer deliveries. Moreover, different external stimuli, called discriminative stimuli, would be associated with each component. For example, illumination of a red light would signal that the operative schedule was the FI, whereas a blue light would signal that the FR component was in effect. After some period of training, the FI and FR components come to control the same type of behavioral performances engendered when they are used in the context of simple schedules (Figure 3).[6] Under such conditions, experimenters can evaluate whether toxicants induce general effects across schedules or whether certain types of schedules demonstrate enhanced sensitivity, or even selective vulnerability. Figure 4[7] shows the preferential vulnerability in pigeons of FI as compared to FR schedule performance in a multiple schedule of reinforcement; carbon disulfide decreased FI response rates at lower exposure levels than those required to suppress FR response rates.

11.20.2.3 Concurrent Schedules of Reinforcement

In the complexity of the natural environment, organisms may be routinely and simultaneously faced with many contingencies of reinforcement, requiring choices of how the amount of time and responding will be allocated and shifted among them. Concurrent schedules represent an experimental simulation of such circumstances and permit a study of choice among concurrently operative reinforcement schedules. The component schedules used in a concurrent schedule depend upon the experimental question of interest. In addition to assessing preferences among various types of schedules and reinforcement options, these approaches permit analyses of the behavior of switching between options, a category of responding also under the control of the schedules. To date, such schedules have not been systematically utilized in the study of toxicant effects.

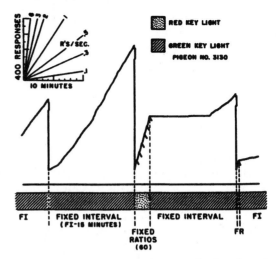

Figure 3 Sample cumulative record of a pigeon's performance on a multiple FI 15 min FR 60 schedule of reinforcement. Responses cumulate vertically, with the pen reset to the baseline at the end of each FI component. Diagonal pips indicate reinforcement delivery (reproduced by permission of the American Industrial Hygiene Association from *Am. Ind. Hyg. Assoc. J.*, 1967, **28**, 482–484).

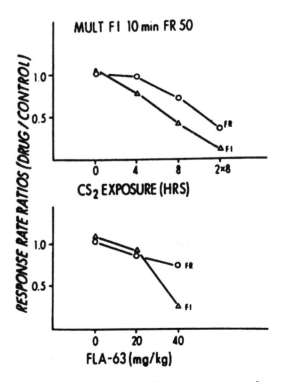

Figure 4 (Top) Ratio of response rates after carbon disulfide exposures to preexposure rates for FI and FR components of a multiple schedule. Each point represents the mean of three pigeons. (Bottom) Same measure showing response rate changes after FLA-63 administration (reproduced by permission of Williams & Wilkins from *J. Pharmacol. Exp. Ther.*, 1976, **199**, 669–678).

11.20.2.4   Measures of Schedule-controlled Behavior

Overall or absolute rate of responding, measured as total number of responses divided by total session time, is considered a global measure of schedule-controlled behavior. Other measures are used to describe the patterns of responding in time. The characteristic pause after reinforcement (postreinforcement pause, PRP) on FI and FR schedules is measured as the time from reinforcement delivery to the first response in the next interval or ratio, respectively. A more localized measure of response rate is provided by the running rate, which represents the rate of responding calculated from the end of the PRP to the time of reinforcement delivery. The advent of computer technology allows a precise measure of the time between successive responses, known as inter-response times (IRTs), which are typically presented in the form of a frequency distribution. Assessments of toxicant-induced changes in IRT distributions have proven to be among the most sensitive and reliable indices of lead effects on FI performance, with lead treatment increasing the frequency of short (<0.5 s) IRTs.[8]

Two measures are used to determine the extent to which characteristic scalloping occurs on the FI schedule. One of these is known as the index of curvature (IOC)[9] and essentially calculates the deviation of the observed slope from a straight line. For the IOC, values of zero indicate a constant rate of responding, while values approaching 1.0 indicate increasing acceleration over the interval. The second measure, quarter life,[10] is defined as the percentage of the interval required for the first 25% of responses to occur.

11.20.3   USE OF SCHEDULE-CONTROLLED BEHAVIOR FOR ASSESSING CHANGES IN VARIOUS BEHAVIORAL PROCESSES

Schedule-controlled operant behavior offers a standardized yet flexible paradigm for evaluating changes in various behavioral processes and for probing behavioral mechanisms of action of toxicants, as described by Cory-Slechta and MacPhail[11] (Table 1). It depicts the controlling relationship in a schedule of reinforcement, $S^D \cdot R \rightarrow S^R$, in which a response is reinforced in the presence of some discriminative stimulus. Using the example of the multiple FI–FR schedule described above, two such controlling relationships exist: in the presence of a red light (discriminative stimulus) a response is reinforced according to an FI schedule, and in the presence of a different discriminative stimulus (blue light) a response is reinforced under FR schedule contingencies. Table 1 also shows the important controlling variables that have been defined,[12] as well as the neurobiological functions that can be evaluated using schedule controlled operant behavior, and manipulating these controlling variables. If, for example, a toxicant is suspected of altering sensory function, then manipulations of the parameters of the external discriminative stimuli such as its intensity or modality, can be made. In this way, changes in visual capacity could be signalled by the fact that greater intensities of illumination of the discriminative stimuli were required to control appropriate schedule responding in a multiple schedule such as described above. The use of tones of differing frequencies as discriminative stimuli would, similarly, allow an assessment of changes in auditory function. Deficits in motor function may be evaluated in the context of schedule-controlled behavior by manipulating parameters of the designated response, including its form, force requirement, required duration or even the required rate. In fact, some experimenters have determined "thresholds" for facets of motor behavior, such as its force, using schedules of reinforcement. In one such example,[13] the force requirement for a lever press was increased each time the force requirement was met, and decreased when the force requirement was not met in a titration procedure. Each instance of the response which exceeded the force requirement resulted in water reinforcement, and by titrating the force, a threshold at which the rat was able to earn 50% of the available water deliveries could be defined. Motivational changes as a behavioral mechanism of toxicant-induced changes in schedule-controlled behavior can be evaluated by modifying aspects of the reinforcer, whether qualitatively (e.g., change from a positive to a negative reinforcer) or quantitatively (i.e., increasing or decreasing the magnitude of reinforcement).

Cognitive functions (see Chapters 21 and 22, this volume) can also be addressed using schedule-controlled responding. In the case of learning, for example, one indication of impaired learning might be an increase in the number of sessions required for the organism to exhibit behavior characteristic of the particular schedule under study. Further, changes in either the particular schedule used or in the parametric value of the schedule can be imposed, and the speed with which the organism attains the performance characteristic of the new schedule type or parameter value can be directly measured. Likewise, schedule-controlled responding can be used as well to evalute memory. Organisms can demonstrate remembering in the form of a more rapid reacquisition of characteristic behavior when a schedule of reinforcement with which the organism has previous experience is once again reimposed. Consider, for example, a case in which behavior is first reinforced under a FR schedule of reinforcement, after which an FI schedule is instead operative. In such a case, remembering would be indicated by a more rapid re-establishment of characteristic FR responding when the schedule is reimposed. The speed (i.e., number of sessions) required of this process can be compared in subjects with and without toxicant treatment.

An example of such an approach is the study by Cory-Slechta[14] which further revealed that toxicant effects which have seemingly disappeared can, under certain conditions, be resurrected during behavioral transitions. In this study, rats exposed to low levels of lead in drinking water initially showed decreased rates of responding on a FI 60 s schedule of food reinforcement which disappeared over the course of 40 experimental sessions. Subsequently, the FI 60 s schedule was changed to a FI 5 min schedule, which induced increases in response rates during the first five sessions following the transition in control rats, but marked decreases in lead-treated rats, as shown in the top panel of Figure 5.[14] These rate differences also subsequently disappeared, at which point the FI 5 min schedule was changed to an FI 5 min clock schedule, a schedule in which a different stimulus signaled each minute of the interval. Again differential changes in control and lead-treated rats emerged (Figure 5(B)). After these rate differences disappeared, the original FI 60 s schedule was reinstated and no differential effects on response rates of control and lead-treated rats were observed (Figure 5(C)). Collectively, these data suggest that the primary effects of lead exposure were on learning (acquisition) and that remembering processes were not affected, since only transitions to new schedule parameters evoked treatment-related effects.

The ability to assess these various behavioral processes using schedule-controlled behavior is made possible by the considerable stability of such performances across extended periods of time. Once characteristic patterns of responding emerge and stabilize, performance of any given subject is typically quite constant. This occurs even though the overall or absolute rates of responding achieved by different subjects will clearly differ.

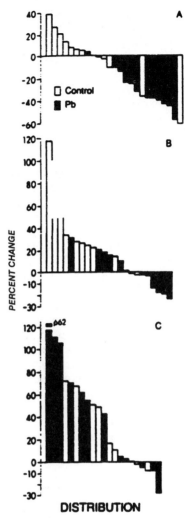

Figure 5 Changes in overall response rate during the first five sessions following the transition from FI 1 min to FI 5 min (A), FI 5 min to FI 5 min clock (B) and FI 5 min clock to FI 1 min (C) plotted as a percentage of the median response rate over the final five sessions in the preceding condition. Each bar shows the percent change for an individual animal; open bars show control rats, solid bars show lead-treated rats. Bars are presented in order of decreasing percentage (reproduced by permission of Intox Press from *Neurotoxicology*, 1990, **11**, 427–441).

11.20.4 FACTORS THAT CAN MODULATE TOXICANT EFFECTS ON SCHEDULE-CONTROLLED BEHAVIOR

11.20.4.1 Nature of the Reinforcement Schedule

Numerous studies examining the effects of both toxicants and drugs on schedule-controlled behavior confirm the fact that the baseline schedule of reinforcement is an important determinant of effect. FI performance, for example, has been shown to be more sensitive to disruption than FR performance across a wide range of drug classes and chemical treatments.[2,15] Levine[7] found that both single and repeated administrations of carbon disulfide to pigeons produced a greater disruption of FI performance than FR behavior; carbon disulfide decreased rates of responding on both schedules but the magnitude of the effect was greater on FI schedules and occurred at lower doses. Similarly, Leander and MacPhail[16] found that the pesticide chlordimeform produced greater disruption of FI than FR performance. It is well documented that FI performance is far more sensitive to the effects of lead than is FR performance, a finding that has been shown both in rodents (see Figure 6)[8,17] and nonhuman primates.[18]

While the basis for the greater sensitivity of FI than FR performance to disruption by drugs or toxicants is yet to be finally established, this differential susceptibility has been related to the nature of the contingencies of reinforcement operative under these two schedules. Specifically, on the FI schedule, response rates can vary widely without affecting the obtained reinforcement density (since, technically, only a single response is required for reinforcement). On the other hand, changes, especially decreases, in response rate on interval schedules

may dramatically decrease the reinforcement density (number of reinforcers per response) obtained, since reinforcement depends upon the completion of a specified number of responses.

Certain toxicant or drug effects on different schedules of reinforcement may actually be manifest as contrasting changes in performance. Such a pattern of effects was reported by Colotla *et al.*[19] who observed that toluene decreased rates of responding on an FR schedule while concurrently increasing response rates on a DRL (differential reinforcement of low rate) schedule of reinforcement, a schedule in which reinforcement depends upon a fixed amount of time elapsing between successive occurrences of the designated response.

11.20.4.2 Baseline Rate of Responding

An additional factor which may modulate the behavioral effects of toxicants on schedule-controlled behavior is the baseline rate of responding controlled by a schedule, a phenomenon known as rate dependency. The most prevalent examples of this involve FI performance. On the FI schedule, as illustrated in Figure 1, response rates tend to be very low in the early part of the interval when no reinforcement is available, and to subsequently increase as the interval progresses.

Drugs with widely differing mechanisms of actions produce similar rate-dependent effects on FI schedules. Amphetamine, for example, increases the low rates of responding early in the interval, while decreasing the high rates exhibited in the later part of the interval. Some neurotoxicants apparently have similar properties. The fungicide triadimefon, via an entirely different mechanism, produces rate-dependent effects[20] that correspond to those produced by amphetamine. Still, other neurotoxicants do not seem to produce rate dependency. Lead increases response rates on FI schedules, but does so by even further increasing the already high response rates late in the interval but having little impact on rates early in the interval.[15,21] Peele and Crofton[22] reported rate-dependent effects of permethrin but not cypermethrin on rates of responding maintained by VI schedules.

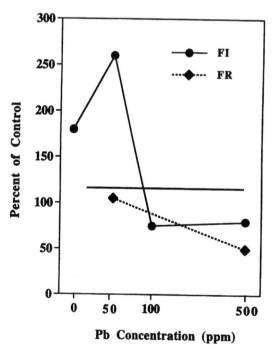

Figure 6 Comparative effects of postweaning lead exposure of rats on FI performance vs. FR performance plotted as a percentage of the control group median response rate (after Cory-Slechta *et al.*[8] and Cory-Slechta[17]).

11.20.4.3 Type of Reinforcer

Few studies have yet to evaluate systematically the extent to which the type of reinforcer used modulates the effects of toxicants on schedule-controlled behavior. However, there are examples suggesting the importance of this variable. Mele *et al.*[23] found that FR performance maintained by milk delivery was more affected by the effects of ionizing radiation than was FR performance maintained by the termination of electric shock delivery. There are reasons to expect that a toxicant may interact with the type of reinforcer, particularly if it induces any type of general malaise that impacts the reinforcing efficacy of food rewards.

11.20.4.4 Type of Response

Similarly having received little systematic attention, the type of response may be another variable modulating the nature of toxicant-induced changes in schedule-controlled behavior. Glowa[24] compared the effects of toluene and *d*-amphetamine on performance under a FI 60 s schedule in which either a nose-poke or a running response was the designated response. The observed effects of both toluene and *d*-amphetamine depended upon the response. At relatively low doses, both toluene and *d*-amphetamine increased rates of nose-poke responding, while higher exposures decreased rates. In contrast, both exerted a dose-related decrease when the response was running. Again, it is not difficult to envisage how a toxicant might interact with the type of response. For example, those requiring greater effort, such as a running or swimming response, may be more readily disrupted by toxicants that have concomitant effects on motor systems.

11.20.5 ADVANTAGES OF SCHEDULE-CONTROLLED BEHAVIOR AS A BASELINE FOR DETECTING BEHAVIORAL TOXICITY

In addition to the obvious advantages offered by the ability to use schedule-controlled behavior as a baseline to determine toxicant effects on various behavioral functions (Table 1), there are several additional benefits offered by schedule-controlled behavior as it relates to defining behavioral toxicity.

11.20.5.1 Cross-species Generality

Characteristic patterns of schedule-controlled behavior exhibit widespread cross-species generality, as depicted for FI performance in Figure 2. In fact, such patterns are not restricted to typical experimental animal species such as rats, pigeons, or monkeys, but are noted

Table 1 Schedule-controlled operant behavior $(S^D \cdot R \to S^R)$.

Controlling variables	Neurobiological function
1. S^D: Discriminative stimuli	
Qualitative	Sensory
Quantitative	
2. R: Response	
Topography	Motor
Force	
Ongoing rate	
3. S^R: Reinforcer	
Qualitative	Motivation
Quantitative	
4. $R \to S^R$: Reinforcement schedule	
Differing schedules	Learning and attention
Schedule parameter	
5. Historical variables	
Long-term	Memory
Short-term (context)	

Source: Cory-Slechta and MacPhail,[11] and MacPhail.[12]
Reprinted with permission from 'Behavioral Measures of Neurotoxicity. Copyright © 1990 by the National Academy of Sciences. Courtesy of the National Academy Press, Washington, DC.

across a wider and even exotic range. Moreover, characteristic FI performance is also seen in humans, particularly under conditions where no instructions are provided to the subjects. This is a feature of schedule-controlled behavior with direct relevance to risk assessment. While the demonstration of cross-species similarity of the behavior itself does not confirm the generality of behavioral toxicity across species, it does, nevertheless, provide assurance that similar behavioral processes are indeed being evaluated across species. This eliminates the uncertainties associated with the use of very different measures across species as is often the case with operant or respondent conditioning approaches used in experimental animals and standardized psychometric tests used in humans.

11.20.5.2 Extensive Database on Drug Effects

An additional advantage of the use of schedule-controlled operant behavior for the evaluation of behavioral toxicity derives from the extensive database currently available describing the effects of a wide variety of pharmacological agents on these baselines. Many of these pharmacological compounds have well-defined neurochemical mechanisms of action. The similarity of toxicant effects on various schedules of reinforcement can then be compared to those produced by various classes of drugs permitting a determination of possible underlying mechanisms of a toxicant's effect.

11.20.5.3   Assessment of Time Course and Reversibility of Toxicant Effects

Once characteristic performance emerges on various schedules of reinforcement, the behavior tends to be relatively stable across long periods of time, providing an enduring behavioral baseline with particular benefits for assessing characteristics of behavioral toxicity.[2] It permits, for example, the determination of the time course of a toxicant's effects. In cases where behavioral impairments exhibit a delayed onset, the time to onset of such an effect can be tracked. Moreover, the fact that behavior is quite stable over prolonged periods of time means that there is a stable baseline from which to determine whether any observed effects are in fact reversible, the time course of that reversibility, and the extent to which reversibility is achieved. Further, by using multiple rather than simple schedules of reinforcements in such experiments it is possible to determine the specificity of observed toxicant effects to particular types of reinforcement contingencies in addition to time course data, thereby gaining substantially greater information. It is conceivable for example that both the time course of effects and the potential for reversibility of effect of a toxicant may differ by reinforcement schedule.

11.20.5.4   Behavioral and Chemical Therapeutics

The opportunity to examine potential behavioral and chemical therapeutic strategies to overcome any toxicant-induced impairment is an additional benefit that can be derived from of the stability of schedule-controlled behavior over extended periods of time. Cory-Slechta and Weiss[25] examined the ability of the chelating agent calcium ethylenediaminetetraacetic acid (CaEDTA) to reverse the changes in FI response rate produced by low-level lead exposure. Contrary to what might have been anticipated, CaEDTA administration actually further exacerbated the decline in FI performance produced by lead. This observation was consistent with findings from additional studies showing that a single injection of CaEDTA increased brain lead concentrations and that a 5 day course of chelation produced no net loss of

lead from the brain.[26] A further example of chemical therapeutics comes from a study by Chambers and Chambers[27] who found that high levels of the centrally acting muscarinic cholinergic receptor antagonist atropine could partially attenuate the rate-decreasing effects of paraoxon on fixed ratio schedule-controlled performance. Such studies help, moreover, to confirm at least a partial muscarinic cholinergic role of the effects of paraoxon on FR responding.

Few studies have yet examined behavioral therapeutic approaches to reversing toxicant-induced behavioral alterations in schedule-controlled responding, but the potential for this approach is suggested by the findings of Newland *et al.*[28] In this study, lead-exposed squirrel monkeys exhibited substantial difficulty in tracking changes in reinforcement densities on a concurrent VI–VI schedule of reinforcement in which reinforcement availability (VI value) differed on the two levers. Specifically, when the relative reinforcement availability on the two levers was switched, that is, the lever with the higher reinforcement density became the one with the lower density, control monkeys, but not lead-treated monkeys exhibited the expected change in relative rates of responding on the two levers. It thus appeared that lead-treated monkeys were not able to make the appropriate behavioral transition, perhaps failing to detect the changes in reinforcement magnitude on the two levers. Only when the investigators imposed additional behavioral training of lead-treated monkeys were they able to more sucessfully follow subsequent changes in reinforcement densities.

11.20.5.5   Assessing Behavioral Mechanisms of Toxicity

Just as there are neurobiological mechanisms of toxicant effects, so too are there behavioral mechanisms of toxicant-induced changes in performance, including schedule-controlled behavior. Drugs or toxicants may alter schedule-controlled behavior by modifying any of the factors which are important in controlling the behavior. These include antecedent factors, current stimulus conditions and the consequences of maintaining behavior. One might ask whether the toxicant has affected the level of motivation of the subject, had an effect on antecedent factors, or whether it caused difficulties in discriminating stimulus conditions associated with the schedule, that is, current stimulus conditions. Alternatively, has it modified response topography, another aspect of current conditions? Or has it changed, for example, the perceived magnitude of reinforcement, that is, the consequences maintaining behavior? Questions about the potential behavioral mechanisms by which toxicants adversely impact schedule-controlled behavior can be addressed using various types of behavioral probes. One approach that has frequently been used in assessing perceived changes in reinforcement magnitude or quality, that is, the consequences maintaining behavior, is to prefeed animals to determine whether inducing a state of satiation alters schedule-controlled responding in the same manner as does a drug or toxicant.

11.20.6 IMPLICATIONS OF CHANGES IN SCHEDULE-CONTROLLED OPERANT BEHAVIOR

11.20.6.1 Interpretation of Response Rate Changes

One question that inevitably arises when considering changes in schedule-controlled behavior as they relate to risk assessment is whether response rate changes are necessarily to be considered as adverse effects indicative of neurotoxicity.[29] To address this issue, several key points must be considered. First, the lever press or other specified response used in the experimental laboratory is an arbitarily chosen response from which generalizations to other classes of operant responses, including those in the human behavioral environment, can be made. In other words, there is little reason to believe that changes in schedule-controlled behavior utilizing a lever press response are restricted to this particular response and to an experimental setting. Further, given that the effects of a toxicant are not for some reason species-specific, we should not assume changes in schedule-controlled performance, as a behavioral endpoint, would be limited to experimental laboratory species. As noted previously, Figure 2 clearly depicts the species generality of the characteristic patterns of behavior seen with schedules of reinforcement.

The risk assessment process requires that changes in schedule-controlled behavior be considered in the context of the human behavioral environment. Behavior in the human environment is also maintained by schedules of reinforcement, albeit schedules far more complex than those typically evaluated in experimental studies. In this regard it is important to note that changes in schedule-controlled behavior of an individual have an impact not only on the functionality of that individual, but also on members of the beha-

vioral community with which that individual interacts. The following section describes some of the potential consequences that could arise from toxicant-induced changes in rates of responding on various types of schedule-controlled behavior; it is important to note that similar consequences could ensue from changes in patterns or alterations in the distribution of responses in time. In addition, while it is not difficult to imagine examples in the human environment in which changes in schedule-controlled performance might actually improve behavior, we must also be cognizant that such a singular type of change is unlikely to be the only consequence of the toxicant-induced change in schedule-controlled performance.

Because ratio schedules typically generate such high rates of responding, toxicants rarely produce any further increases in rates of responding. Should they occur, however, they should not necessarily be construed as improvements in performance. For example, if one is taking a mathematics examination with 100 problems to be completed within 30 min, responding too quickly may be associated with decreased accuracy, that is, an increased probability of making mistakes. Although some facilitation of ratio-based response rates could conceivably be engendered by a toxicant, there is no guarantee that the response increased in frequency is a desirable behavior, or that only desirable responses would be increased in frequency.

Typically, decrements in rate of responding on ratio-based schedules are produced by toxicant exposures. Decreased response rates on ratio schedules will of course lead to a decrease in the frequency of reinforcement, a situation which can clearly be adverse. Consider a learning paradigm in which the designated response or sequence of behavior to be learned is reinforced according to a ratio schedule. Decreased response rates will, in such situations, retard learning, that is, delay acquisition. Examination taking in academic situations frequently includes a ratio-based type of schedule in which the student may be required to complete a certain number of problems within a certain time frame. The less total material compeleted, no matter how accurately answered the remaining questions, the lower the score will be, a scenario unlikely to be associated with success in school.

Many complex behavioral repertoires are achieved only after a very large number of repetitions of responses. The human behavioral environment frequently encourages individuals to "keep trying" or to "persevere," a type of ratio contingency. Decreased perseverance can mean less fullfillment and failure to achieve full potential. The skills of a concert pianist, a prima ballerina, and an Olympic athlete require years of such repetitive responding, skills which could not be achieved in the face of rate decreases. Even relatively simpler skills, such as learning the alphabet or learning to read, skills critical to future success, require an extensive amount of repetitive responding.

Toxicant exposures have been reported to produce both increases and decreases in rates of responding on interval-based schedules of reinforcement. The potential adversity of decreased response rates is not difficult to discern. Even though rates of responding on interval schedules are typically well in excess of those required to obtain reinforcement, decreases in response rates may be sufficient to lengthen the time to reinforcement and thus decrease the overall frequency of reinforcement. In studying for an examination, or preparing for a presentation, two examples of interval-like schedules in the human behavioral environment, decreases in rate at the end of the interval, that is, right before the examination or presentation, may leave the student unprepared for the examination, or leave the speaker with an unpolished presentation.

The detrimental consequences of increased response rates on interval-based schedules are sometimes more difficult to comprehend since increased rates will certainly not decrease reinforcement availability. But it must also be noted that increased rates on interval schedules, where only a single response is required, are highly inefficient and represent an unnecessary expenditure of energy resources. Consider the foraging behavior of a predator which occurs at intervals of time since a previous meal. Successfully snaring prey often requires repeated attempts on the part of the predator, and, in some niches, must be carefully balanced against metabolic factors and resource availability. If a predator begins responding too soon, or at too high of a rate, it could seriously deplete carefully balanced energy resources needed for future predation. This can be a particular problem for a female of the species, threatening not only her viability but that of offspring who may be dependent upon her as their sole food source.

11.20.6.2 Temporary Changes in Response Rates

There are cases in which toxicants induce changes in response rate that appear to be temporary or transient, leading some to state that such changes are not really adverse effects, since they are not sustained. There are at least two arguments against such an interpretation. The first is the fact that under some conditions, changes in schedule-controlled behavior have seemingly disappeared only to reemerge in the face of altered environmental conditions. Figure 4 demonstrated just such a finding in the case of low-level lead-induced changes in FI performance. Such data clearly raise the possibility that effects that have apparently disappeared may instead remain dormant until provoked by environmental conditions. An additional argument against the notion that temporary effects are not necessarily adverse is derived from studies examining the impact of behavioral history on subsequent behavior and on drug effects. The term behavioral history refers to the sum of the past behavioral conditioning of an organism. The demonstration that behavioral history is important to subsequent behavioral performance is probably most dramatically exemplified by numerous studies which have shown that behavioral responding can be established and maintained by response-produced shock in animals with prior experience under schedules of shock postponement or avoidance. Normally, one might expect shock to serve as a punishing stimulus or as a negative reinforcer, the removal of which increases response probability. Barret[30] demonstrated, however, that monkeys would work under a concurrent schedule in which one type of response postponed shock delivery under a Sidman avoidance schedule while another response actually produced shock presentation on an FI schedule. The concurrent schedule was implemented following experience of the monkey working solely under the shock avoidance schedule. A possible form in the human realm may be post-traumatic stress disorder, a syndrome which came to the public attention prominently after the Vietnam War and again after the Gulf War. In this syndrome, the experiences of some individuals during their military service results in a behavioral history with long-term and sometimes dramatic residual consequences, impacting future behavior that can even include violence.

In addition to modifying subsequent behavior, behavioral history can also alter the effects of drugs on behavior, even on behavioral performances quite different from those which comprise the behavioral history. As shown by Barrett,[30] *d*-amphetamine dose–effect curves on a baseline of responding that results in punishment are substantially different in animals with and without prior behavioral experience on a shock avoidance paradigm. Specifically, in monkeys with no avoidance history, *d*-amphetamine administration resulted in dose-related suppression of punished responding, whereas in monkies with an avoidance history, low doses of *d*-amphetamine produced substantial increases in response rates and high doses suppressed responding.

While there are currently few studies which have examined the impact of behavioral history as it relates to neurotoxicant effects upon behavior, there is little biological reason to doubt that similar phenomena would occur. Thus, even temporary behavioral changes produced by toxicant exposure could produce a behavioral history that would potentially modify subsequent behavioral performance, as well as alter behavioral reponsiveness to future central nervous system medications.

11.20.6.3 Magnitude of Response Rate Changes and Adversity

An additional question concerning the implementation of schedule-controlled behavior in the context of risk assessment relates to the magnitude of change in behavior that should be required to be designated as an adverse effect. In other words, is a 10% change in response rate an adverse effect, or should adversity require that such changes exceed, say, 25%? There are different answers to such questions. One answer comes from a behavioral perspective and it argues that the characteristic patterns of behavior seen under schedules of reinforcement constitute normal behavior, and that any deviation from this performance would be defined as an adverse effect. Another answer includes statistical considerations. Certainly if a statistically significant change in schedule-controlled performance can be detected, then it would be considered adverse. In experiments which offer the possibility of using each subject as its own control, the ability to detect toxicant effects may be greater than when between-group comparisons are involved, since smaller changes in performance may be identified, and inter-animal variability is typically substantially less than is intra-animal variability.

In many cases, the ability to use each subject as its own control in studies of toxicant-induced changes in schedule-controlled behavior is precluded because the toxicant of interest induces effects only when administered during the subject's development. For such studies, sample size will be an important consideration with respect to the ability to identify toxicant effects. While schedule-controlled performance of individual animals tend to be fairly stable over time as discussed above, there can sometimes be marked differences between animals in the level of performance.

Figure 7 shows the extent of both within- and between-animal variability that has been typical for FI performance of rats.[8] The rates of responding of most rats exhibit considerable stability across the course of experimental sessions. Also obvious is the more prominent between-group variability: the majority of control animals exhibited response rates ranging from 5 to 20 responses per min, while 2 of 12 (about 17%) showed considerably higher response rates. As this figure should also make clear, a rather considerable toxicant-induced change in behavior would be required to detect differences between-groups as opposed to within-group based on such variability and standard types of statistical approaches.

11.20.6.4 Neurotoxicity Systemic Toxicity

As with other behaviors, it is certainly possible for schedule-controlled behavior to be altered by factors that may be nonspecific, or more accurately, not directly related to changes in central nervous system function. As an example, having a stomach flu or other gastrointestinal disorder would be expected to adversely influence behavior, likely decreasing motivation through alterations in the efficacy of food reward.

Such nonspecific factors as a cause of changes in schedule-controlled behavior can be gauged in several ways. Among the simpler approaches is the measurment of food and fluid consumption and/or body weight following toxicant exposure, since systemic toxicity is likely to be accompanied by decreases in these measures. In line with such assessments, some investigators have used a measure of the speed of ingestion of a highly preferred foodstuff. One must be careful, though, not to assume that decreases in food consumption and/or body weight are necessarily causative of changes in schedule-controlled behavior and that neurotoxicity *per se* is therefore not indicated. Rather, this must be answered empirically through the use of procedures such as prefeeding.

Progressive ratio schedules represent another approach frequently used to evaluate motivational changes. On this schedule, the size of the ratio that must be completed for reinforcement increases after each reinforcer delivery according to a formula. The highest ratio value obtained before some specified period of pausing (nonresponding) occurs is called the "break point" and is considered to be indicative of the efficacy of the reinforcer. Moreover, the multiple schedule format is particularly useful for addressing issues such as specificity (*vide supra*). In such cases, nonspecific effects may be

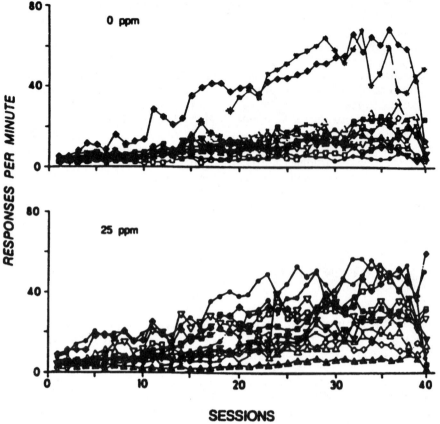

Figure 7 Individual overall response rates (responses per minute) of rats ($n = 12$ per group) exposed from weaning to 25 ppm sodium acetate (top panel) or 25 ppm lead acetate (bottom panel) over 40 sessions on a FI 1 min schedule of food delivery (reproduced by permission of Academic Press from *Toxicol. Appl. Pharmacol.*, 1985, **78**, 291–299).

manifest as rate decreases in all schedule components. Although multiple FI–FR schedules were described earlier, such comparisons need not be restricted to FI–FR and others may be used as the experimental questions warrant.

An additional behavioral approach that may be of assistance in separating specific from non-specific toxicant-induced changes in schedule-controlled performance is embodied in resistance to extinction. In this paradigm, reinforcement for responding is withheld and the time and/or number of responses emitted before behavior is eliminated is assessed. A more rapid extinction, that is, decreased resistance to extinction in toxicant-treated subjects might be considered to be indicative of systemic toxicity, particularly if such an effect was noted concurrently with decreases in body weight and/or food consumption.

In summary, assessment of schedule-controlled operant behaviors particularly under varying experimental contexts, offers an opportunity to examine toxicant-induced changes in a variety of behavioral processes, including alterations in learning, attention, and memory processes, sensory function, motor behavior, and motivational states. In providing information about the nature of the specific behavioral deficit, studies of schedule-controlled behavior yield important information not only about the risks accruing from toxicant exposure, but also about the potential mechanistic bases of such effects. The stability of schedule-controlled behavior across extended periods of time offers the opportunities for assessment of time-course of toxicant effect and the determination of both behavioral and chemical therapeutics. Cross-species generality of the characteristic patterns of performance provides confidence in the continuity of such behavioral processes across species. Such studies have provided significant information about the risks of toxicity and their bases associated with metals such as lead, about solvents, and pesticides.

11.20.7 REFERENCES

1. C. B. Ferster and B. F. Skinner, 'Schedules of Reinforcement,' Appleton-Century-Crofts, 1957.

2. D. A. Cory-Slechta, in 'Neurotoxicology, Target Organ Toxicology Series,' eds. H. L. Tilson and C. L. Mitchell, Raven Press, New York, 1992, pp. 271–294.

3. B. Weiss and D. A. Cory-Slechta, in 'Principles and Methods of Toxicology,' ed. A. W. Hayes, Raven Press, New York, 1994, p. 1091.

4. L. S. Seiden and L. A. Dykstra, 'Psychopharmacology: A Biochemical and Behavioral Approach,' Van Nostrand Reinhold, New York, 1977.

5. R. W. Kelleher and W. H. Morse, in 'Importance of Fundamental Principles of Drug Evaluation,' eds. D. H. Tedeschi and R. E. Tedeschi, Raven Press, New York, 1968, p. 383.

6. R. P. Beliles, R. S. Clark, P. R. Bellusco *et al.*, 'Behavioral effects in pigeons exposed to mercury vapor at a concentration of $0.1 \, mg/m^3$.' *Am. Ind. Hyg. Assoc. J.*, 1967, **28**, 482–484.

7. T. E. Levine, 'Effects of carbon disulfide and FLA-63 on operant behavior in pigeons.' *J. Pharmacol. Exp. Ther.*, 1976, **199**, 669–678.

8. D. A. Cory-Slechta, B. Weiss and C. Cox, 'Performance and exposure indices of rats exposed to low concentrations of lead.' *Toxicol. Appl. Pharmacol.*, 1985, **78**, 291–299.

9. W. Fry, R. W. Kelleher and L. Cook, 'A mathematical index of performance on fixed-interval schedules of reinforcement.' *J. Exp. Anal. Behav.*, 1960, **3**, 193.

10. L. R. Gollub, 'The relations among measures of performance on fixed-interval schedules.' *J. Exp. Anal. Behav.*, 1964, **7**, 337.

11. D. A. Cory-Slechta and R. C. MacPhail, in 'Neurotoxicology: Approaches and Methods,' eds. L. Chang and W. Slikker, Jr., Academic Press, San Diego, CA, 1995, pp. 225.

12. R. C. MacPhail, in 'Behavioral Measures of Neurotoxicity. Report of a Symposium,' eds. R. W. Russel, P. E. Flattau and A. M. Pope, National Academy Press, Washington, DC, 1990, pp. 347.

13. J. C. Elsner, C. Fellmann and G. Zbinden, 'Response force titration for the assessment of the neuromuscular toxicity of 2,5-hexanedione in rats.' *Neurotoxicol. Teratol.*, 1988, **10**, 3–13.

14. D. A. Cory-Slechta, 'Exposure duration modifies the effects of low level lead on fixed-interval performance.' *Neurotoxicology*, 1990, **11**, 427–441.

15. D. A. Cory-Slechta, in 'Basic Principles of Neurotoxicology,' ed. L. Chang, Marcel Dekker, New York, 1994, p. 313.

16. J. D. Leander and R. C. MacPhail, 'Effect of chlordimeform (a formamidine pesticide) on schedule-controlled responding.' *Neurobehav. Toxicol. Teratol.*, 1980, **2**, 315.

17. D. A. Cory-Slechta, 'Prolonged lead exposure and fixed ratio performance.' *Neurobehav. Toxicol. Teratol.*, 1986, **8**, 237–244.

18. D. C. Rice, 'Schedule-controlled behavior in infant and juvenile monkeys exposed to lead from birth.' *Neurotoxicology*, 1988, **9**, 75–87.

19. V. A. Colotla, S. Bautista, M. Lorenzana-Jimenez *et al.*, 'Effects of solvents on schedule-controlled behavior.' *Neurobehav. Toxicol.*, 1979, **1**, Suppl. 1, 113–118.

20. V. C. Moser and R. C. MacPhail, 'Neurobehavioral effect of triadimefon, a triazole fungicide, in male and female rats.' *Neurotoxicol. Teratol.*, 1989, **11**, 285–293.

21. D. A. Cory-Slechta, in 'Advances in Behavioral Pharmacology,' eds. T. Thompson, P. B. Dews and J. E. Barrett, Academic Press, New York, 1984, vol. 4, p. 211.

22. D. B. Peele and K. M. Crofton, 'Pyrethroid effects on schedule-controlled behavior: time and dosage relationships.' *Neurotoxicol. Teratol.*, 1987, **9**, 387–394.

23. P. C. Mele, C. G. Franz and J. R. Harrison, 'Effects of ionizing radiation on fixed-ratio escape performance in rats.' *Neurotoxicol. Teratol.*, 1990, **12**, 367–373.

24. J. R. Glowa, 'Comparisons of some behavioral effects of *d*-amphetamine and toluene.' *Neurotoxicology*, 1987, **8**, 237–247.

25. D. A. Cory-Slechta and B. Weiss, 'Efficacy of the chelating agent CaEDTA in reversing lead-induced changes in behavior.' *Neurotoxicology*, 1989, **10**, 685–697.

26. D. A. Cory-Slechta, B. Weiss and C. Cox, 'Mobilization and redistribution of lead over the course of calcium disodium ethylenediamine tetraacetate chelation therapy.' *J. Pharmacol. Exp. Ther.*, 1987, **243**, 804–813.

27. J. E. Chambers and H. W. Chambers, 'Short-term effects of paraoxon and atropine on schedule-controlled behavior in rats.' *Neurotoxicol. Teratol.*, 1989, **11**, 427–432.

28. M. C. Newland, S. Yezhou, B. Logdberg *et al.*, 'Prolonged behavioral effects of *in utero* exposure to lead or methyl mercury: reduced sensitivity to changes in reinforcement contingencies during behavioral transitions and in steady state.' *Toxicol. Appl. Pharmacol.*, 1994, **126**, 6–16.

29. D. A. Cory-Slechta, in 'Neurobehavioral Toxicity: Analysis and Interpretation,' eds. B. Weiss and J. O'Donoghue, Raven Press, New York, 1994, p. 195.

30. J. E. Barrett, in 'Developmental Behavioral Pharmacology,' eds. N. K. Krasnegor, D. B. Gray and T. Thompson, Lawrence Erlbaum Associates, NJ, 1986, p. 99.

11.21
Learning and Memory

HUGH A. TILSON

US Environmental Protection Agency, Research Triangle Park,
NC, USA

11.21.1 INTRODUCTION

The fact that exposure to chemicals can result in adverse effects on the structure and function of the central and peripheral nervous system of humans has been documented in numerous review papers and book chapters.[1-3] Anger and Johnson[2] reviewed 750 chemicals affecting the nervous system and found some 120 nervous system-related chemical effects associated with exposure to those chemicals. Manifestation of neurotoxicity in humans may be categorized as chemical-induced alterations in motor, sensory, affect, personality, and cognitive function. Of the neurotoxic effects reported, alterations in cognitive function appear to be the most diverse and include changes in learning, mathematical abilities, categorization, coding, concept shifting, distractibility, memory, pattern recognition, reading, spatial relations, vigilance, vocabulary, and intelligence. A review of epidemiological workplace data and behavioral effects of extended-duration chemical exposures of humans indicated that alterations in cognitive function were frequently reported following exposure to many environmentally relevant chemicals, including lead, mercury, carbon disulfide, styrene, and solvent mixtures or combinations.[1]

Due to the potential for chemicals to affect adversely cognitive function in humans, considerable effort has been made to characterize the effects of chemicals in animal models of cognitive function. Information from the study of such models might be useful in developing strategies for identification of chemicals that could affect cognitive function in humans, determine the site of centrally acting neurotoxicants, and improve our understanding of the neurobiological underpinnings of cognitive function. The purpose of this chapter is threefold. First, the chapter attempts to provide working definitions of cognitive functions, such as learning and memory, in terms frequently used by behavioral toxicologists. It is important to have a common vocabulary to assess methods used in this area of research. Second, this chapter provides an overview of some of the procedures commonly used in neurotoxicology to assess the effects of chemicals on cognitive function in animals. It should be noted that this overview is not intended to be comprehensive or complete, but is intended to illustrate specific points by discussing examples. Finally, this chapter discusses some of the critical experimental and conceptual variables that are important for studies on chemical-induced cognitive dysfunction.

11.21.2 DEFINITIONS

11.21.2.1 Principles of Behavioral Analysis

Cognition is a psychological term that includes learning and memory, as well as perception, language, intelligence, and reasoning.[4] Cognitive phenomena are essentially internal psychological processes and, from the experimental point-of-view, must be inferred from overt changes in an organism's behavior. Behavior can be defined as what an organism does or as the integrative product of the sensory, motor, and integrative processes in the nervous system.[5,6] Experimental psychologists have classified behavior as being of two types, that is, respondent or operant (Table 1). In addition, responses can also be unconditioned (unlearned) or conditioned (learned). Respondents are elicited by a specific identifiable stimulus. There is usually a defined temporal relationship between the stimulus and the elicited response. Examples of respon-

Table 1   Classification of behavior.

Unconditioned behavior	Respondent-elicited by known stimulus Operant-emitted with no known eliciting stimulus
Learned behavior	Classical conditioning—response is elicited by stimulus after pairing of unconditioned and conditioned stimuli Operant or instrumental response probability changes as a function of response consequence (reinforcement)

dents include kineses, taxes, reflexes, and species-specific behaviors. Operant behaviors are not elicited by a single, identifiable stimulus, but are emitted or occur voluntarily. Operant behaviors are controlled by their consequences. Many operant behaviors are learned and include discriminated conditional responses and schedule-controlled behaviors.

Respondent and operant behaviors can be modified by the process of conditioning or learning. Respondent learning or conditioning refers to the repeated pairing of an initially neutral stimulus with a stimulus that elicits a specific response, that is, an unconditioned stimulus that elicits an unconditioned response. Eventually, the previously neutral stimulus becomes a conditioned stimulus that is capable of eliciting a conditioned response. Operant responses become conditioned by the presentation or withdrawal of stimuli known as reinforcers. If the probability of a response increases after the presentation of a stimulus, then positive reinforcement has occurred. If the probability of a response increases after a stimulus is removed or terminated, then negative reinforcement has occurred. Learned changes in responding contingent upon reinforcement is referred to as operant or instrumental conditioning.

11.21.2.2 Learning and Memory

Learning can be defined as "an enduring change in the mechanisms of behavior that results from experience with environmental events,"[7] or a "relatively permanent change in an organism's potential for responding that results from prior experience or practice."[8] Inherent in this definition is the fact that learning is something that occurs internally and is suggested or inferred by changes in behavior or potential to behave, and that learning is relatively persistent. That is, behavioral changes associated with learning can be distinguished from behavioral changes due to fatigue, adaptation, or disease.

Traditional definitions of memory include many of the same elements associated with learning, that is, internal phenomenon, behavioral manifestation, change in potentiality, and persistence.[7,8] However, memory differs from learning in that memory concerns persistence of a behavioral change after the change has been acquired.[9] Furthermore, memory studies indicate that there are clear cases in which knowledge of preceding events can have a transient effect on subsequent behavior and that memory can be modified by conditions that do not covary directly with learning, that is, retroactive and proactive interference, stimuli that evoke reminders, and contextual cues.[10] Studies of memory have indicated the existence of a capacity-limited or short-term memory and a long-lasting memory, and that the neurobiological substrate for these forms differ.[11] Long-term memory may further be divided into memory associated with information based on skills or procedures and information based on specific facts or data. Declarative memory consists of explicit facts, episodes, lists and routine information. Procedural memory is implicit and accessible only through performance. Declarative memory can be episodic or working, that is, relative to a specific time and place, or semantic or reference, that is, facts or data collected during the course of specific experiences. Procedural memory includes skill, priming effects, respondent conditioning, habituation, and sensitization.

In summary, the differences between learning and memory are sometimes difficult to discern. Many effects of chemicals on "learning" may, under some conditions, be as easily described as effects on "memory." Nonetheless, it is clear that while learning and memory are closely related, they represent the operation of different cognitive processes.

11.21.3 EXAMPLES OF CHEMICAL-INDUCED COGNITIVE DYSFUNCTION

11.21.3.1 Nonassociative Learning

Nonassociative learning consists of changes in fundamental behaviors (i.e., startle reflex, limb withdrawal) that occur with repeated presentation of environmental stimuli, such as a light or tone. Examples of nonassociative learning include habituation and sensitization

(Table 2). An example of a study involving the effects of a neurotoxicant on habituation is that of Crofton et al.[12] who exposed rats to 3,3′-iminodipropionitrile (IDPN) on postnatal days 5–7 and tested them at various ages up to adulthood. Developmental exposure to IDPN altered the ontogenic profile of motor activity as measured in a figure-eight maze. Control animals showed little habituation of activity at 13 d and 15 d of age. However, starting at 17 d of age, the controls showed a typical pattern of habituation consisting of a high-frequency of motor activity counts in the first 5–10 min of the session followed by a rapid decrease for the rest of the 30 min session. Animals receiving 300 mg kg^{-1} IDPN postnatally did not habituate during the course of the 30 min session on postnatal days 17, 19, 21, and 30. IDPN-treated rats were generally hyperactive on day 60 of age, but showed some indication of habituation during the session at 60 d of age. IDPN-treated animals were also deficient in other measures of learning and memory, including olfactory conditioning, passive avoidance and spatial learning in a T-maze.

Table 2 Some examples of procedures used in neurotoxicological research to study learning and memory in animals.

Cognitive process	Experimental procedure
Nonassociative learning	
Habituation	Motor activity, startle reflex
Sensitization	Kindled behavioral responses
Associative learning	
Classical conditioned	Nictitating membrane reflex, eye-blink, flavor aversion
Instrumental conditioned	
Negatively reinforced	Passive avoidance Active avoidance Y-maze discrimination Morris water maze, Biel water maze
Positively reinforced	Radial arm maze, alternation, T-maze, operant, delayed-matching-to-sample, discrimination reversal, repeated acquisition, Hebb–Williams maze, Wisconsin general test apparatus

Source: Miller and Eckerman,[13] Eckerman and Bushnell,[9] and Spear et al.[14]

Gilbert[15] studied the effects of lindane and found that intermittent exposure produced behavioral sensitization. Rats were dosed with lindane for 30 d or 3 times/wk for 10 wk. Enhanced behavioral responsiveness, as indicated by increased incidence of myoclonic jerks and clonic seizures, emerged over the course of dosing and persisted for 2–4 wk after the last dose. The pattern of the increased behavioral sensitivity indicated that repeated exposure to subconvulsant doses of lindane results in a persistent alteration in the sensitivity of the central nervous system. Similar effects have also been reported for other pesticides, such as dieldrin and endosulfan.[16,17]

11.21.3.2 Associative Conditioning

Associative learning or conditioning is usually categorized as being classical (respondent) or instrumental (operant) and is formed following the association between a stimulus and a behavioral outcome.

11.21.3.2.1 *Classical conditioning*

In classical conditioning, an association develops between a previously neutral stimulus and a response following repeated pairing of the neutral stimulus with a unconditioned stimulus. An example of a classical conditioning paradigm used in neurotoxicology to assess memory is the flavor aversion procedure, which is based on the finding that rats will avoid consuming solutions with flavors previously paired with the consequences of a negative stimulus. Peele *et al.*[18] exposed rats to trimethyltin and assessed them 30 d later for conditioned flavor aversions to lithium chloride paired with sodium saccharin. Two days after being conditioned, the rats received concurrent access to water and saccharin, and the preference for the saccharin solution was measured. Rats dosed with trimethyltin 0.5 h or 3 h after pairing with saccharin did not differ from controls. If a 6 h delay was used, however, there was a significant impairment in the trimethyltin-exposed rats relative to controls. These results are consistent with other studies showing that trimethyltin significantly affects cognitive function, using other behavioral paradigms, such as retention of passive avoidance.[19]

An example of a classically conditioned response that has been used to evaluate learning is the eye-blink response,[20] which uses contingent, temporal pairings of a conditioned stimulus, such as a tone, and an unconditioned stimulus, such as a brief airpuff to the eye. The airpuff elicits a reflexive eye-blink and following repeated conditioning trials, the tone comes to elicit the eye-blink response. In their article on the eye-blink procedure, Stanton and Freeman[21] report that developmental exposure to the antimitotic agent methylazoxymethanol (MAM) interferes with the acquisition of the eye-blink conditioning reflex in infant rats. MAM is known to affect the development of the brain, and the development of the cerebellar granule cells is particularly affected. This is particularly relevant since it is known that the essential neural circuitry mediating the eye-blink response includes elements of the cerebellar and brain stem systems. The eye-blink conditioning paradigm has several advantages for studies concerning the effects of chemicals on learning, that is, the neural substrate has been extensively studied; it can be measured in a number of species, including humans; it lends itself to comparisons across ages; it does not require language competence; and it requires a relatively simple sensorimotor response. A similar response, the conditioned nictitating membrane reflex, has been used extensively to study the effects of aluminum on cognitive function in rabbits.[22] Another classically conditioned paradigm that has been used in neurotoxicology is the olfactory discrimination learning task in preweanling rats.[23]

11.21.3.2.2 *Instrumental conditioning*

(i) *Procedures using negative reinforcement*

(a) *Passive avoidance.* One frequently used passive avoidance method involves placing a rat into a chamber divided into two compartments separated by a door. The two compartments are often of different size and one of them may be lighted, while the other is darkened. Mundy *et al.*[24] infused colchicine into the nucleus basalis magnocellularis of rats to destroy ascending cholinergic fibers in the cortex. Rats were tested for effects on cognitive function 14 d after treatment. The rats were placed into a smaller lighted compartment for 10 s, after which a guillotine door was opened. The latency to enter a connected larger, darkened chamber was recorded. Upon entering the larger chamber, the rats received a negative reinforcer (i.e., mild electric shock to the gridfloor). Forty-eight hours later, memory was assessed by repeating the entire procedure, omitting the negative reinforcer. Prior treatment with colchicine did not affect the initial latency to cross-over during training. On the memory test, however, colchicine-treated animals had significantly lower latencies to enter the larger compartment, suggesting a deficit in memory. In a subsequent paper, Shaughnessy et al.[25] replicated these findings on passive avoidance and confirmed that these effects could be seen in animals that were not hyperactive; changes in shock sensitivity have not been reported in animals with experimental lesions in the nucleus basalis, further supporting the conclusion that colchicine has a specific effect on cognitive function. As discussed by Mundy et al.,[24] one of the most consistent behavioral effects observed after lesions in the nucleus basalis is an impairment in passive avoidance responding, suggesting that cholinergic functioning in the cortex is an important mediator of this response. The passive avoidance task is used frequently in neurotoxicological studies to study memory.[9]

(b)   Active avoidance. In contrast to passive avoidance tasks in which withholding a response is reinforced, active avoidance tasks require that the animal perform a specific response to avoid negative reinforcement. Frequently, the onset of the negative stimulus is preceded by a warning stimulus which is terminated if a conditioned response is made. If the correct response is not made, the negative stimulus is presented for a set period of time or until the correct response is emitted. One-way active avoidance requires the animal to move in one direction to escape or avoid negative reinforcement. In the two-way avoidance task, the animal is allowed to shuttle beween two compartments to escape or avoid negative reinforcement.

For example, the effects of the pesticide lindane on learning were assessed in a shuttle box comprised of two different sized compartments separated by an automated guillotine door.[26] Sessions were initiated by placing the rat into the smaller compartment and the subsequent presentation a warning stimulus; if the rat did not move to the opposite chamber within 10 s (avoidance response), electric shock was applied to the grids of the chamber. Moving to the opposite chamber following shock was defined as an escape response. All stimuli were terminated during a 10–20 s variable intertrial interval (ITI). Shuttle responses during the ITI were recorded, as were frequency and latency of avoidance, and escape responses. Lindane significantly reduced the number of correct avoidance trials in a 60-trial session without increasing the number of shuttle responses during the ITI, or the latency to respond to the shock during trials when an escape response was made. These data suggest that lindane specifically interferes with learning.

Lindane was also found to interfere with the retention of a step-through passive avoidance task.[26]

A more complicated procedure than the one-or two-way shuttle avoidance task is the symmetrical Y-maze. One version of this task involves the presentation of a warning stimulus (e.g., light and/or tone) in one of the two unoccupied arms of the maze. If the animal does not enter the arm with the stimulus within a specified period of time, negative reinforcement is applied. McLamb et al.[27] reported that rats having received intrahippocampal infusions of colchicine were impaired in the acquisition of this task. Disruption of hippocampal function by infusion of neurotoxicants, such as colchicine, has also been reported to affect acquisition of a two-way avoidance response[28] and retention of passive avoidance.[29] Whether or not colchicine directly affects cognitive function in the Y-maze, however, is confounded by the likelihood that colchicine alters the reactivity or responsiveness of rats performing shock-reinforced tasks in such a way as to indirectly facilitate learning.

(c)   Water mazes. The Morris water maze is being used with increasing frequency in neurotoxicology to assess spatial memory. This procedure requires that a rat swim to a submerged platform in a large tank of water. Escape from the water is the (negative) reinforcement and acquisition of this task is thought to be dependent upon the animal's ability to utilize extra maze cues in determining the position of the submerged platform in the maze.[30] An example of the use of the Morris water maze is a study by Lee and Rabe,[31] who administered the antimitotic agent methylazoxymethanol (MAM) to pregnant rats on gestational day 15 and tested the offspring in the water maze at various ages. Gestational exposure to antimitotic agents, such as MAM, results in behavioral deficits and microencephaly. Developmental exposure to MAM impaired the rate of acquisition in the water maze, as measured by increased latencies to find the submerged platform. Control rats acquired the task within 2 d of training, while some MAM-exposed rats had not acquired the task within 5 d of training. When the platform was removed from the pool after training, the MAM-treated rats showed a lesser degree of preference for the quadrant where the platform had been placed, further indicating a deficiency in learning. Subsequent tests for retention indicated that the MAM-treated animals had a retention deficit as well. The effects of MAM on performance in the water maze did not appear to be related to a

deficit in motor function, since measures of swim times indicated that the MAM-treated animals tended to swim faster than controls. The Morris water maze has been used to study the effects of disulfoton,[32] IDPN,[33] and tris(2-chloroethyl)phosphate[34] on learning and memory in rats.

Another type of water maze that has been used in neurotoxicology is the Biel maze. This maze is a multiple-T maze that requires that the animal make a series of position discriminations before finding the arm leading to removal from the water in the maze. The Biel maze has been used to study the developmental neurotoxicity of methylmercury[35] and MAM.[36]

(ii)   Procedures using positive reinforcement

(a)   *Radial arm maze (RAM).* The RAM is a spatial task in which animals are required to recall the location of previously entered and nonentered feeding sites during a free-choice test session.[37] In this procedure, the most effective response strategy is to not enter those arms of the maze from which the food has been removed during a previous entry or in which food has never been present. The typical RAM consists of an open, central area from which several (usually 8) arms radiate like spokes of a wheel. Rats deprived of food are placed in the central arena and permitted to enter the arms to find food or water located at the end of the arms. Usually, test sessions are terminated after all of the reinforcers have been found or after some time-limit has been exceeded. RAMs can be used with a number of species, accommodate a number of experimental manipuluations, for example, delays, conditioned cues, and be used to assess two forms of memory, that is, reference and working memory. Wirsching *et al.*,[38] for example, always put food in the same arms for each daily trial and food-deprived rats learned over time to avoid the arms which never had food, that is, a test for reference memory. This procedure was compared to that in which all arms have food each day and the animal must remember which arms have been visited on that day, that is, working memory. Wirsching *et al.* found that the anticholinergic muscarinic receptor antagonist scopolamine selectively disrupted the working memory component of the task without affecting reference memory. The RAM has been used to study the effects of trimethyltin[39] and IDPN[31] on cognitive functioning in rats.

(b)   *Delayed alternation.* The use of delayed alternation has been proposed as a model of

cognitive development related to the integrity of the septohippocampal cholinergic pathway. Stanton *et al.*[40] used a T-maze to assess the cognitive effects of developmental exposure to chlorpyrifos, a long-acting inhibitor of acetylcholinesterase. In this experiment, rats were trained using a discrete-trial, delayed alternation task. This involves requiring the food-deprived rat to make a forced choice to enter one arm of the T-maze for food reinforcement. At a later time, the rats are given a choice train in which both arms are available for entry, but reinforcement is available only in the arm alternate to that entered on the preceding forced trial. Rats were also tested on a position discrimination task, in which the rats were reinforced for consistently selecting one of the two arms. Acquisition of the position discrimination task is not specifically affected by disruption of hippocampal function. Stanton *et al.* reported that a single exposure to chlorpyrifos produces a dose- and time-dependent deficit in the T-maze delayed alternation, but not position discrimination task. These behavioral effects were associated with dose-related inhibition of cholinesterase activity in the frontal cortex and hippocampus.

A procedure similar to the T-maze alternation task was reported by Rice and Karpinski,[41] who trained monkeys to alternate lever pressing for reinforcement between two stimuli, with correct alternations being reinforced. In this case, the stimuli are the same and there are no cues to indicate the correct position on any given trial. Delays can be inserted between opportunities to respond, as a means of evaluating short-term memory. Monkeys exposed to lead during development and tested at 7–8 years of age were impaired during the initial learning of the task as well as the longer, but not shorter, delays.

(c)   *Delayed matching-to-sample (DMTS).* This procedure measures short-term retention and consists of presenting a sample stimulus (e.g., tone, color, pattern) that is terminated after a specified time or a response by the animal. This is followed by an interval, which is ended by the presentation of the sample and alternative comparison stimuli or test stimuli. Choosing the comparison stimulus that matches the sample results in reinforcement, while selecting the wrong stimuli usually terminates the trial and initiates an intertrial interval. The introduction of a delay between the time when the sample is presented and when it must be selected from among the set of stimuli makes the delayed matching to sample task a test of short-term memory.[14] Bushnell *et al.*[42] used a DMTS procedure in rats to assess the effect of chlorpyrifos. The procedure used by Bushnell *et al.* records several measures including correct responses, latency to respond, accuracy, responding during the intertrial interval, and assesses both working (matching accuracy) and reference (discrimination trials in which the cue light always indicated the correct response at the end of the delay) memory in the same animal during the test session. Rats received a single injection of chlorpyrifos and were tested in the DMTS paradigm for several days afterward. Chlorpyrifos caused motor slowing as measured by increased response latencies and decreased responses during the intertrial interval. Matching accuracy, a measure of working memory, was reduced for 2–3 weeks after chorpyrifos, whereas discrimination accuracy, an index of reference memory, was not affected. The DMTS procedure has been used to study the cognitive effects of diisopropylfluorophosphate in rats[43] and lead in monkeys.[44]

(d) Discrimination reversal. One experimental manipulation that appears to be sensitive to chemical-induced cognitive deficits is the discrimination reversal. In one version of this task, the experimental animal is presented with two or more stimuli which vary in one or more parameters, for example, color, position, or pattern, and the animal must respond to one of the stimuli for reinforcement. Once the animal has learned to respond to the stimulus according to some criterion, the contingency is changed so that responses to the previously incorrect stimulus are now reinforced. One useful feature of the reversal procedure is that a separate learning curve can be generated each time that a reversal is made. Rice and Willes[45] reported that monkeys exposed to lead continuously since birth and tested at 2–3 years of age were not impaired on the learning of a discrimination task, but did not improve over successive reversals as rapidly as control monkeys. Subsequent experiments found that developmental exposure to lead impairs performance of both spatial and nonspatial discrimination tasks.[46]

The reversal procedure appears to support the contention that an increase in task complexity, or a shift in task requirement, may challenge the organism and be more likely to detect chemical-induced alterations in cognitive function.[13] The reversal procedure has been reported to detect cognitive dysfunction produced by a number of chemicals from different classes, including trimethyltin,[47] lead,[48] diisopropylfluorophosphate,[49] styrene,[50] and polychlorinated biphenyls (PCBs).[51]

(e) Repeated acquisition. This procedure is similar to the reversal paradigm in that the animal is required to adapt its performance on the basis of newly introduced requirements for reinforcement.[13] Repeated acquisition procedures require learning a new sequence or chain of responses within a given testing session. For example, Cohn *et al.*[52] used a multiple schedule of repeated acquisition and performance to assess cognitive dysfunction in rats following developmental exposure to lead. In this task, sequences of three responses were reinforced and the correct sequence during the repeated acquisition component changed with each experimental session. The correct sequence during the performance component remained constant across sessions. Changes in accuracy observed only during the repeated acquisition component are thought to reflect relatively specific alterations in learning, while changes in both components reflect nonspecific behavioral alterations. Cohn *et al.* found that developmental exposure to lead significantly decreased accuracy on the repeated acquisition component, but not on the performance component. Further analysis of the data indicated that the effects may be associated with an increase in perseverative responding in the lead-exposed rats.

Bushnell and Angell[53] used a repeated acquisition procedure in the radial arm maze to study the effects of trimethyltin on cognitive function of rats. In this experiment, rats were trained to obtain food pellets at the end of 4 or 8 baited arms of a daily 12-trial session. The set of baited arms was changed each day requiring the rats to learn a new set of baited arms each session. Rats having received trimethyltin were found to have a significantly slower decline in within-session error (i.e., entering nonbaited arms) rates. A repeated acquisition procedure has also been described for the Morris water maze.[34]

(f) Hebb–Williams maze. This maze consists of a series of alternative choices that can be configured to present different learning problems. Swartzwelder *et al.*[54] assessed the effects of trimethyltin on the ability of rats to learn the Hebb–Williams maze. Rats were food deprived and given 10 daily trials for each of 12 different maze configurations. The trimethyltin-treated animals made consistently more errors on each of the 12 problems than did controls. The treated animals, however, also made significantly more perseverative responses than

controls. The behavioral tendency to respond in the same manner on every problem represents a response strategy that is not optimal for solving problems in the Hebb–Williams maze. The effects on performance in the maze was attributed to the trimethyltin-induced disruption of hippocampal function.

(g) Wisconsin general test apparatus (WGTA). This procedure has been used extensively to test cognitive function in non-human primates. In this procedure, monkeys are placed in a cage facing a table upon which a test tray is rolled to within the animal's reach and out of it again. One version of this test has two recessed food wells that can be covered with lids having various visual clues. The experimenter baits the food well in the presence of the monkey and then the test tray is removed from view. After a period of time, the monkey is permitted to view the food wells again and displace one of the lids to obtain any food that may be inside. There are a number of variations of this task, including spatial reversals, color reversal, partial reinforcement, probability learning, progressive probability shifts, object alternation learning, and learning sets. An example of the use of the WGTA is a series of experiments performed by Bowman et al.[55] who attempted to correlate peak PCB body burdens with acquisition and performance of tasks in the WGTA. Bowman et al. reported significant correlations with PCB levels and errors in 5 of 9 learning tasks.

11.21.4 ISSUES IN THE INTERPRETATION OF RESULTS

11.21.4.1 Nonassociative vs. Associative Factors

In animal studies, chemical-induced alterations in learning and memory are usually inferred from changes in behavior. Therefore, it is important to understand the factors that control behavior which could affect the interpretation of results from studies on learning and memory. Both forms of conditioning, respondent and operant, are dependent on the relationship of a response with stimuli in the environment and the ability of the animal to make the appropriate motor response. If a chemical alters the perception of those stimuli by the animal or the ability to perform the required response, then measures in learning and memory may be altered due to non-associative factors. Respondent behaviors are particularly susceptible to the magnitude and

duration of the eliciting stimulus and chemicals that diminish the perception of this stimuli could indirectly influence learning or memory. Operant behavior is particularly influenced by antecedent conditions, such as the past history of reinforcement and the current motivational state of the animal. In addition, there are consequence variables, such as the type of reinforcers, the magnitude and duration of reinforcement, and programmed contingency between stimulus and response, that affect the probability of operant responding. Chemically induced alterations in these nonassociative variables could also indirectly affect measures of learning and memory. Other examples of nonassociative factors include state-dependent effects, interaction with stress, and alterations in exploration or response bias. Finally, effects of chemicals on cognitive function are sometimes task-specific, that is, a deficit in learning may be observed using one type of learning task, but may not be detected using a different task. Table 3 lists a number of nonassociative variables important for the interpretation of results from studies on learning and memory.

11.21.4.2 Examples of Interpretational Problems

11.21.4.2.1 Enhanced reactivity

Mactutus et al.[56] studied the effects of neonatal exposure to the pesticide chlordecone on the acquisition, retention and extinction of a passive avoidance response. Rat pups were given a single dose of the insecticide chlordecone on day 4 postnatally and trained on a passive avoidance task on day 18. Memory was assessed either at 24 h or 72 h later; extinction was assessed 24 h after the 72 h retention test. A

Table 3 Representative nonassociative factors that can influence the interpretation of studies on chemical-induced dysfunction.

Sensory effects	Ability to perceive stimuli used in conditioning or reinforcement
Motor effects	Capability to respond to rest contingencies
Motivational effects	Alters reinforcing capacity of stimuli Changes reactivity to test conditions
Response strategy	Response bias Perseveration Task specificity

second experiment measured retention at 24 h and/or 144 h after training. There was a significant increase in the number of chlordecone-exposed rats required to attain a 60 s response withholding criterion and memory deficits over long, but not short or intermediate retention intervals. The memory deficits were detected by alterations in time spent in the "safe" chamber and decreased response latencies. However, alterations of a nonassociative nature were also evident in chlordecone-exposed animals, including increased training latencies and increased resistance to extinction. The latter was associated with behavioral changes consistent with "freezing", that is, increased response latencies, increased time spent in the "unsafe" chamber and decreased frequency of responding. That the chlordecone-exposed rats may have responded differentially to the behavioral procedure was further reinforced by the observation that serum corticosterone levels were higher in chlordecone-exposed animals when measured immediately after the last retention trial of each experiment.

Changes in "reactivity" of animals to chemicals affecting the nervous system are common effects following exposure to chemicals that act on the central nervous system. It is, therefore, important to consider the possibility that these nonassociative changes in behavior could interact with the contingencies of the learning and or memory test and affect the behavioral measure used to measure cognitive function.

11.21.4.2.2 *Alterations in sensory function*

Environmental stimuli, such as lights or tones, are frequently used in learning and memory studies to control or manipulate behavior. Extra-maze cues are essential for learning and performing spatial tasks, such as the Morris water maze. In addition, aversive stimuli, such as electric shock, applied to the grids of the test chamber are used in studies involving negative reinforcement. Chemicals that affect the ability of the animal to perceive such stimuli could conceivably affect the outcome of studies on learning and memory without affecting cognitive function directly. In the study by Lee and Rabe[31] cited earlier, it was observed that rats exposed to MAM during gestation showed a marked deficit in learning to locate a hidden platform in a Morris water maze. One interpretation of these results is that exposure to MAM during development reduced the rat's capacity to acquire and use spatial information, that is, a form of cognitive dysfunction. Lee and Rabe, however, point out that MAM-induced forebrain dysplasia may

not be directly responsible for the observed deficits in the water maze since it has also been reported that the visual system of MAM-exposed rats is damaged[57] and their ability to discriminate visual patterns is impaired.[58] Therefore, since learning and performance in the Morris water maze is dependent on utilization of visual extramaze cues, it is likely that the inability to use appropriate visual cues to solve the maze contributed to the observed behavioral deficits.

An example of a study that attempted to control for the possibility that the chemical treatment may have affected the perception of a stimulus essential for the assessment of cognitive function is that of Peele *et al.*[59] These investigators gave adult rats injections of IDPN on three successive days and tested them behaviorally 4 weeks after the last injection. Significant deficits in a shock-motivated passive avoidance task were observed in the rats receiving IDPN. Although the treatment may have altered cognitive function, that is, retention, another explanation is that IDPN may have altered the sensitivity of the rats to the electric shock. To test this possibility, Peele *et al.* placed the rats into a two-compartment shuttle box and applied shocks of various intensities to the grids of the chamber where the rat was located at the start of each trial. A rat could escape the shock by moving to the other chamber. This procedure permitted the determination of side preferences and tested the hypothesis that if the treated animals were less sensitive to the shock, then they should spend more time in the side with the shock than controls. Their results indicated that there was no difference between exposed and control rats, indicating that sensitivity to the shock was not changed in the IDPN-treated rats.

11.21.4.2.3 *Response capability*

In order to assess cognitive function using behavioral tests, the ability of the animal to perform the designated response must be taken into consideration. Llorens *et al.*,[33] for example, tested the effects of IDPN on acquisition and performance of rats in several behavioral tests. The retention of passive avoidance and effects on the acquisition of a food-reinforced response in the RAM, steady-state performance in the RAM and RAM repeated acquisition were significantly affected by IDPN. It was difficult to assess learning capabilities in the Morris water maze, however, because the there was a dose-related impairment in swimming ability. It was hypothesized that vestibular toxicity[60] produced by IDPN affected their ability to swim, precluding the assessment of cognitive function using this test.

11.21.4.2.4 *Response bias*

The problem of response bias has already been alluded to in the study by Swartzwelder *et al.*,[54] who reported that trimethyltin-induced increases in response preseveration were not optimal for solving the Hebb–Williams maze. Exposure to some chemicals can also alter the pattern of responding in such a way as to facilitate learning. Tilson *et al.*[61] assessed the ability of rats treated with triethyllead to acquire a two-way shuttle box response in a procedure in which the two compartments were of unequal size. Each training trial began by placing the rat into the smaller compartment; it was then given 60 training trials to avoid the onset of electric footshock, which was signaled by the presentation of a combined light and tone conditioned stimulus. The triethyllead-treated rats made significantly more avoidance responses during the 60 trials, that is, acquisition appeared to be facilitated in the lead-exposed animals. Treated rats did not appear to be hyperactive in the task, since the number of shuttle responses during the intertrial interval was not significantly affected, nor did the lead-treated animals differ in their sensitivity to the electric footshock in a separate study. An analysis of the pattern of responding in the shuttle box, however, indicated that the lead-exposed animals made significantly more responses going from the small to the large compartment; there was no difference between groups in going from the large to the small compartment. One interpretation of these results is that the lead-exposed animals tended to perform the task as a one-way avoidance procedure, that is, they tended to stay in or return to the small chamber after each trial. Control rats exhibited no such response bias.

11.21.4.2.5 *Task specificity*

As discussed previously, there are several types of memory and if each has its own neurobiological substrate, then neurotoxicants might affect some measures of memory, but not others. An example of this can be found in a study by Llorens *et al.*[32] who studied the effects of the repeated exposure to the cholinesterase inhibitor disulfoton. In these experiments, rats were injected for 30 d with various doses of disulfoton and tested for cognitive dysfunction in the Morris water maze and passive avoidance procedures. Disulfoton affected the acquisition of the spatial memory task, but had no effect on acquisition or retention of passive avoidance.

It is evident that some learning and memory tasks may be more complex than others or depend the integrity of different regions of the central nervous system. Therefore, the use of a single-learning test may be inappropriate as an index of chemical-induced cognitive dysfunction. In their studies on the developmental neurotoxicity of hyperphenylalaninemia, Strupp *et al.*[62] suggested that the ability to transfer learning across a series of related tasks may be a more sensitive assessment of cognitive dysfunction. These investigators tested animals with experimentally induced phenylketonuria in a series of behavioral tests, each involving some type of discrimination, and examined a number of variables over the course of the entire study, including learning rate of individual problems, between-task transfer, combined within- and between-task transfer, learning rate for tasks with multiple versions, and cumulative errors. Their results indicated that an animal's performance on a series of related tasks provides a more sensitive assessment of cognitive dysfunction than performance on any single task.

11.21.4.2.6 *Attention*

Eckerman and Bushnell[9] have suggested that the concept of learning cannot be easily separated from the issue of attention. In order for learning to occur, stimuli involved in the learning process have to be observed or noticed. That is, in order for a stimulus to change the probability or occurrence of behavior, it must first be registered by the nervous system. These authors suggest that many of the effects that may initially seem to be effects on acquisition, may be effects on attention instead. Studies on chemically induced changes in learning might attempt to assess this possibility by utilizing differential stimuli in the behavioral paradigm, that is, a positive discriminative stimulus (S^+) that predicts when reinforcement will reliably occur and a negative discriminative stimulus (S^-) associated with withholding reinforcement. Well-controlled performance is dependent on noticing or observing the relevant stimuli, as well as the ability to maintain differential responding. If performance is not well-controlled, it may be difficult to differentiate whether the chemically induced disruption is due to a loss of attention to the relevant cues or cognitive function. Specific tests may have to be performed to determine the role of attention in the interpretation of results from studies on learning, including assessment of latent learning and vigilance.

11.21.5 SUMMARY AND CONCLUSIONS

Humans exposed to chemicals report alterations in cognitive function, including deficits in learning and memory. Conceptually, it is sometimes difficult to separate learning and memory, although these processes can be operationally defined. A number of procedures have been used by neurotoxicologists to detect and quantify the effects of chemicals on learning and memory in animal models, and can be classified generally as procedures that assess nonassociative or associative learning. Examples of nonassociative learning include habituation and sensitization, while examples of associative learning include tests that use positive or negative reinforcement. Examples of tests that use negative reinforcement include shock avoidance or escape from a water maze. Procedures commonly used in neurotoxicology that employ positive reinforcement include tests of radial arm maze, delayed-matching-to-sample, alternation and reversal. Since there are different forms of learning and memory, no single task may be adequate to satisfactorily assess chemically induced cognitive dysfunction. A battery of tests or some assessment of cumulative cognitive dysfunction may be necessary to conclude conclusively that a chemical does or does not affect learning and memory. Finally, the interpretation of results from studies on learning and memory must be considered in the context of other potential changes in nervous system function, that is, sensory, motor, motivation, response strategy, and attention. Neurotoxicological research in this area is closely linked to a better understanding of the neurobiology of learning and memory. Future investigations should attempt to use the available knowledge about the putative sites of neurotoxic effects and underlying neurobiological bases of the various forms of learning and memory.

11.21.6 REFERENCES

1. W. K. Anger, 'Worksite behavioral research: results, sensitive methods, test batteries and the transition from laboratory data to human health.' *Neurotoxicology*, 1990, **11**, 627–717.
2. W. K. Anger and B. L. Johnson, 'Chemicals Affecting Behavior,' ed. J. O'Donoghue, CRC Press, Boca Raton, FL, pp. 51–148.
3. US Congress, Office of Technology Assessment, 'Neurotoxicity: Identifying and Controlling Poisons of the Nervous System,' OTA-BA-436, US Government Printing Office, Washington, DC, 1990, pp. 43–59.
4. G. A. Miller and M. S. Gazzaniga, in 'Handbook of Cognitive Neuroscience,' ed. M. S. Gazzaniga, Plenum, New York, 1984, pp. 284–295.
5. H. A. Tilson and G. J. Harry, in 'Nervous System Toxicology,' ed. C. L. Mitchell, Raven Press, New York, 1982, pp. 1–27.
6. D. A. Cory-Slechta, 'Behavioral measures of neurotoxicity.' *Neurotoxicology*, 1989, **10**, 271–295.
7. M. Domjan and B. Burkhard (eds.), 'The Principles of Learning and Behavior,' Brooks/Cole, Pacific Grove, CA, 1986.
8. W. C. Gordon (ed.), 'Learning and Memory,' Brooks/Cole, Pacific Grove, CA, 1989.
9. D. A. Eckerman and P. J. Bushnell, in 'Neurotoxicology,' ed. H. A. Tilson and C. L. Mitchell, Raven Press, New York, 1992, pp. 213–270.
10. C. F. Mactutus, in ''Principles of Neurotoxicology,' ed. L. Chang, Dekker, New York, 1994, pp. 397–441.
11. L. R. Squire, 'Mechanisms of memory.' *Science*, 1986, **232**, 1612–1619.
12. K. M. Crofton, D. B. Peele and M. E. Stanton, 'Developmental neurotoxicity following neonatal exposure to 3,3'-iminodipropionitrile in the rat.' *Neurotoxicol. Teratol.*, 1993, **15**, 117–129.
13. D. B. Miller and D. A. Eckerman, in 'Neurobehavioral Toxicology,' ed. Z. Annau, Johns Hopkins University Press, Baltimore, MD, 1986, pp. 94–149.
14. N. E. Spear, J. S. Miller and J. A. Jagielo, 'Animal memory and learning.' *Annu. Rev. Psychol.*, 1990, **41**, 169–211.
15. M. E. Gilbert, 'Repeated exposure to lindane leads to behavioral sensitization and facilitates electrical kindling.' *Neurotoxicol. Teratol.*, 1995, **17**, 131–141.
16. M. E. Gilbert, 'A characterization of chemical kindling with the pesticide endosulfan.' *Neurotoxicol. Teratol.*, 1992, **14**, 151–158.
17. R. M. Joy, L. G. Stark, S. L. Peterson *et al.*, 'The kindled seizure: production of and modification by dieldrin in rats.' *Neurobehav. Teratol.*, 1980, **2**, 117–124.
18. D. B. Peele, J. D. Farmer and J. E. Coleman, 'Time-dependent deficits in delay conditioning produced by trimethyltin.' *Psychopharmacology (Berl.)*, 1989, **97**, 521–528.
19. T. J. Walsh, M. Gallagher, E. Bostock *et al.*, 'Trimethyltin impairs retention of a passive avoidance task.' *Neurobehav. Toxicol. Teratol.*, 1982, **4**, 163–167.
20. I. Gormezano, W. F. Prokasy and R. F. Thompson (eds.), 'Classical Conditioning: Behavioral, Physiological and Neurochemical Studies in the Rabbit,' 3rd edn., Erlbaum Press, Hillsdale, NJ, 1987.
21. M. E. Stanton and J. H. Freeman, Jr., 'Eye-blink conditioning in the infant rat: an animal model of learning in developmental neurotoxicology.' *Environ. Health Perspect.*, 1994, **102**, Suppl. 2, 131–139.
22. R. A. Yokel, 'Repeated systemic aluminum exposure effects on classical conditioning of the rabbit.' *Neurobehav. Toxicol. Teratol.*, 1983, **5**, 41–46.
23. M. E. Stanton, 'Neonatal exposure to triethyltin disrupts olfactory discrimination learning in preweaning rats.' *Neurotoxicol. Teratol.*, 1991, **13**, 515–524.
24. W. R. Mundy, S. Barone, Jr. and H. A. Tilson, 'Neurotoxic lesions of the nucleus basalis induced by colchicine: effects on spatial navigation in the water maze.' *Brain Res.*, 1990, **512**, 221–228.
25. L. W. Shaughnessy, S. Barone, Jr., W. R. Mundy *et al.*, 'Comparison of intracranial infusions of colchicine and ibotenic acid as models of neurodegeneration in the basal forebrain.' *Brain Res.*, 1994, **637**, 15–26.
26. H. A. Tilson, S. Shaw and R. L. McLamb, 'The effects of lindane, DDT, and chlordecone on avoidance responding and seizure activity.' *Toxicol. Appl. Pharmacol.*, 1987, **88**, 57–65.
27. R. L. McLamb, W. R. Mundy and H. A. Tilson, 'Intradentate colchicine impairs acquisition of a two-way active avoidance response in a Y-maze.' *Neurosci. Lett.*, 1988, **94**, 338–342.

28. T. J. Walsh, D. W. Schulz, H. A. Tilson *et al.*, 'Colchicine-induced granule cell loss in rat hippocampus: selective behavioral and histological alterations.' *Brain Res.*, 1986, **398**, 23–36.

29. H. A. Tilson, B. C. Rogers, L. Grimes *et al.*, 'Time-dependent neurobiological effects of colchicine administered directly into the hippocampus of rats.' *Brain Res.*, 1987, **408**, 163–172.

30. R. G. M. Morris, P. Garrud, J. N. P. Rawlins *et al.*, 'Place navigation impaired in rats with hippocampal lesions.' *Nature*, 1982, **297**, 681–683.

31. M. H. Lee and A. Rabe, 'Premature decline in Morris water maze performance of aging micrencephalic rats.' *Neurotoxicol. Teratol.*, 1992, **14**, 383–392.

32. J. Llorens, K. M. Crofton, H. A. Tilson *et al.*, 'Characterization of disulfoton-induced behavioral and neurochemical effects following repeated exposure.' *Fundam. Appl. Toxicol.*, 1993, **20**, 163–169.

33. J. Llorens, K. M. Crofton and D. B. Peele, 'Effects of 3,3'iminodipropionitrile on acquisition and performance of spatial tasks in rats.' *Neurotoxicol. Teratol.*, 1994, **16**, 583–591.

34. H. A. Tilson, B. Veronesi, R. L. McLamb *et al.*, 'Acute exposure to tris(2-chloroethyl)phosphate produces hippocampal neuronal loss and impairs learning in rats.' *Toxicol. Appl. Pharmacol.*, 1990, **106**, 254–269.

35. C. V. Vorhees, 'Behavioral effects of prenatal methylmercury in rats: a parallel trial to the Collaborative Behavioral Teratology Study.' *Neurobehav. Toxicol. Teratol.*, 1985, **7**, 717–725.

36. C. V. Vorhees, K. Fernandez, R. M. Dumas *et al.*, 'Pervasive hyperactivity and long-term learning impairments in rats with induced microencephaly from prenatal exposure of methylazoxymethanol.' *Brain Res.*, 1984, **317**, 1–10.

37. D. S. Olton, J. T. Becker and G. E. Handelmann, 'Hippocampus, space and memory.' *Behav. Brain Sci.*, 1979, **2**, 313–365.

38. B. A. Wirsching, R. J. Beninger, K. Jhamandas *et al.*, 'Differential effects of scopolamine on working and reference memory of rats in the radial arm maze.' *Pharmacol. Biochem. Behav.*, 1984, **20**, 659–662.

39. T. J. Walsh, D. B. Miller and R. S. Dyer, 'Trimethyltin, a selective limbic system neurotoxicant, impairs radial-arm maze performance.' *Neurobehav. Toxicol. Teratol.*, 1982, **4**, 177–183.

40. M. E. Stanton, W. R. Mundy, T. Ward *et al.*, 'Time-dependent effects of acute chlorpyrifos administration on spatial delayed alternation and cholinergic neurochemistry in weanling rats.' *Neurotoxicology*, 1994, **15**, 201–208.

41. D. C. Rice and K. F. Karpinski, 'Lifetime low-level lead exposure produces deficits in delayed alternation in adult monkeys.' *Neurotoxicol. Teratol.*, 1988, **10**, 207–214.

42. P. J. Bushnell, C. N. Pope and S. Padilla, 'Behavioral and neurochemical effects of acute chlorpyrifos in rats: tolerance to prolonged inhibition of cholinesterase.' *J. Pharmacol. Exp. Ther.*, 1993, **266**, 1007–1017.

43. P. J. Bushnell, S. S. Padilla, T. Ward *et al.*, 'Behavioral and neurochemical changes in rats dosed repeatedly with diisopropylfluorophosphate.' *J. Pharmacol. Exp. Ther.*, 1991, **256**, 741–750.

44. D. C. Rice, 'Behavioral deficit (delayed matching to sample) in monkeys exposed from birth to low levels of lead.' *Toxicol. Appl. Pharmacol.*, 1984, **75**, 337–345.

45. D. C. Rice and R. F. Willes, 'Neonatal low-level lead exposure in monkeys (*Macaca fascicularis*): effect on two-choice nonspatial form discrimination.' *J. Environ. Pathol. Toxicol.*, 1979, **2**, 1195–1203.

46. D. C. Rice, 'Lead-induced changes in learning: evidence for behavioral mechanisms from experimental animal studies.' *Neurotoxicology*, 1993, **14**, 167–178.

47. P. J. Bushnell, 'Delay-dependent impairment of reversal learning in rats treated with trimethyltin.' *Behav. Neural Biol.*, 1990, **54**, 75–89.

48. P. J. Bushnell and R. E. Bowman, 'Reversal learning deficits in young monkeys exposed to lead.' *Pharmacol. Biochem. Behav.*, 1979, **10**, 733–789.

49. K. Raffaele, D. Olton and Z. Annau, 'Repeated exposure to diisopropylfluorophosphate (DFP) produces increased sensitivity to cholinergic antagonists in discrimination retention and reversal.' *Psychopharmacology (Berl.)*, 1990, **100**, 267–274.

50. P. J. Bushnell, 'Styrene impairs serial spatial reversal learning in rats.' *J. Am. Coll. Toxicol.*, 1994, **13**, 279.

51. S. L. Schantz, E. D. Levin, R. E. Bowman *et al.*, 'Effects of perinatal PCB exposure on discrimination-reversal learning in monkeys.' *Neurotoxicol. Teratol.*, 1989, **11**, 243–250.

52. J. Cohn, C. Cox and D. A. Cory-Slechta, 'The effects of lead exposure on learning in a multiple repeated acquisition and performance schedule.' *Neurotoxicology*, 1993, **14**, 329–346.

53. P. J. Bushnell and K. E. Angell, 'Effects of trimethyltin on repeated acquisition (learning) in the radial-arm maze.' *Neurotoxicology*, 1992, **13**, 429–441.

54. H. S. Swartzwelder, J. Hepler, W. Holahan *et al.*, 'Impaired maze performance in the rat caused by trimethyltin treatment: problem-solving deficits and perseveration.' *Neurobehav. Toxicol. Teratol.*, 1982, **4**, 169–176.

55. R. E. Bowman, M. P. Heironimus and J. R. Allen, 'Correlation of PCB body burden with behavioral toxicology in monkeys.' *Pharmacol. Biochem. Behav.*, 1978, **9**, 49–56.

56. C. F. Mactutus, K. L. Unger and H. A. Tilson, 'Neonatal chlordecone exposure impairs early learning and memory in the rat on a multiple measure passive avoidance task.' *Neurotoxicology*, 1982, **3**, 27–44.

57. K. Ashwell, 'Direct and indirect effects on the lateral geniculate nucleus neurons of prenatal exposure to methylazoxymethanol acetate.' *Brain Res.*, 1987, **432**, 199–214.

58. S. C. Pereira, C. R. Legg, S. Russell *et al.*, in 'Brain Plasticity, Learning and Memory,' eds. B. E. Will, P. Schmitt and J. C. Dalrymple-Alford, Plenum, New York, 1985, pp. 321–330.

59. D. B. Peele, S. D. Allison and K. M. Crofton, 'Learning and memory deficits in rats following exposure to 3,3'iminodipropionitrile.' *Toxicol. Appl. Pharmacol.*, 1990, **105**, 321–332.

60. J. Llorens, D. Dememes and A. Sans, 'The behavioral syndrome caused by 3,3'-iminodipropionitrile and related nitriles in the rat is associated with degeneration of the vestibular sensory hair cells.' *Toxicol. Appl. Pharmacol.*, 1993, **123**, 199–210.

61. H. A. Tilson, C. F. Mactutus, R. L. McLamb *et al.*, 'Characterization of triethyl lead chloride neurotoxicity in adult rats.' *Neurobehav. Toxicol. Teratol.*, 1982, **4**, 671–638.

62. B. J. Strupp, M. Bunsey, D. A. Levitsky *et al.*, 'Deficient cumulative learning: an animal model of retarded cognitive development.' *Neurotoxicol. Teratol.*, 1994, **16**, 71–79.

11.22
Human Nervous System and Behavioral Toxicology

NANCY FIEDLER

Environmental and Occupational Health Sciences Institute, Piscataway, NJ, USA

11.22.1 INTRODUCTION TO TEST METHODOLOGY

Approximately 850 neurotoxic chemicals can be found in workplaces throughout the United States.[1] Job categories in which neurotoxicants are used include agricultural workers, drycleaners, electronics workers, painters, ironworkers, printers, and health care workers.[2] In addition, many home products contain neurotoxicants.

Since the nervous system is exquisitely sensitive to many of these chemicals, systematic methods are needed to quantify their impact on human functions. The purpose of this chapter is twofold: (i) to describe the rationale and methods for assessment of neurotoxicity in adult humans, and (ii) to review representative studies that have evaluated cognitive function among individuals exposed, either acutely or chronically, to neurotoxicants.

11.22.2 NEUROPSYCHOLOGICAL METHODS AND BRAIN INJURY

Since the mid-1930s, clinical neuropsychology has developed as a method to identify deficits due to traumatic brain injury or dementing disorders. Two batteries of tests, the Halstead–Reitan and the Luria Nebraska, represent collections of tests proven useful in identifying the behavioral manifestations of cerebral insults.[3] These batteries incorporate standard tests of intelligence, attention/concentration, sensory motor function, spatial ability, receptive and expressive language, executive function, and memory. However, for the individual with mild head injury, exposure to neurotoxicants, or less apparent pathology, these batteries have not been as effective in detecting behavioral deficits. Tests from cognitive experimental psychology that focus on learning and information processing have been more sensitive to detect subtle deficits (see also Chapter 21, this volume).

11.22.2.1 Defining Abnormality of Cognition

Deficits in an individual's cognitive behaviors are identified by two methods: (i) comparison of performance to a normative standard from a population of demographically similar individuals, and (ii) comparison of the individual's current performance to a preinjury level of ability. Unlike laboratory tests which are classified into dichotomous categories of normal or abnormal (e.g., cell changes), a continuous distribution of scores on tests of cognitive ability are considered normal. Normal is a relative term defined within a context of such important demographics as age, gender, and educational status. For example, age-related changes occur in cognitive functions such as reaction time, attention, and memory.[3] Therefore, abnormality is defined statistically as performance that is more than two standard deviations below the mean for an appropriate comparison group (i.e., comparable age and education). Among tests of intellectual skills, composite scores that place an individual's performance two or more standard deviations below the population mean are considered "abnormal" or functionally retarded.[4] While controversial, these classifications are predictive of the ability to function in school and work settings.

To detect a loss of function due to a toxic exposure, performance is evaluated in the context of the individual's pre-exposure level of function. Given the range of acceptable cognitive function, even relatively large changes in cognitive function resulting from exposure to neurotoxicants may not appear "abnormal." Also, within the normal, unexposed population, an individual's performance may vary by one standard deviation from one test of cognitive ability to another.[5] Therefore, to interpret performance as impaired requires a pattern of reduced performance from several interrelated tests of a function such as memory. If test scores for a given function are two standard deviations below that individual's estimated pre-exposure ability, these scores are considered both a statistically and clinically significant result. When exposed and unexposed workers are matched on the relevant demographic variables, detection of a significant difference may suffice to suggest reduction in cognitive performance due to exposure. Even then, one may question whether such reductions are clinically meaningful.

11.22.3 REVIEW OF TESTS

Approximately 250 different neuropsychological tests have been used to evaluate the effects of neurotoxicants on behavior.[6–10] Several batteries, incorporating traditional neuropsychological tests, are used for clinical and research purposes (e.g., London School Hygiene Battery;[11] Neurobehavioral Core Test Battery of the World Health Organization;[12] Pittsburgh Occupational Exposures Test (POET)[13]). One development is the use of computerized batteries for cognitive testing (e.g., Neuro-behavioral Evaluation System (NES-2)). Unlike many of the traditional, neuropsychological tests, the NES-2 does not require a highly trained examiner, therefore, it can be administered and scored with relative ease. However, it does not have normative data to allow clinical interpretation of an individual's scores. Thus, these tests are only useful for controlled research comparisons.

Table 1 contains a list of the functions assessed and a sample of the tests frequently used. Tests have been arbitrarily categorized into functional categories, but often overlap these categories. The following is a brief overview and description of representative tests for the functions listed.

11.22.3.1 Overall Verbal Ability

To determine cognitive impairment, an estimate of pre-exposure intellectual function is

Table 1 Neuropsychological tests.

Function	Representative tests	Ref.
Overall cognitive ability—verbal	Wide range achievement test—revised	14
	National adult reading test—revised	15
Visuospatial ability	Block design (WAIS-R)	4
	Raven's progressive matrices	16
Concentration/attention	Simple reaction time (NES-2)	17
	Stroop color–word task	17
	Continuous performance test (NES-2)	18
Motor skills and strength	Grooved pegboard	19
	Finger tapping	19
	Dynamometer	19
Visuomotor coordination	Hand-eye coordination test (NES-2)	18
	Digit symbol (WAIS-R)	4
Memory		
Verbal	Logical memory (WMS-R)	20
	Paired associates (WMS-R)	20
	California verbal learning test	21
	Digit span (WAIS-R)	4
Visual	Visual reproduction (WMS-R)	20
	Complex figure test	22
Sensory tests		
Audition	Audiometer	
	Seashore rhythm	19
Vision	Color vision	
Tactile	Finger agnosia	19
	Vibratron	23
Olfaction	University of Pennsylvania smell identification test, olfactory threshold tests	24
Affect/personality	Profile of mood states	25
	MMPI-2	26

needed. Since few, if any, studies have obtained baseline ability data, education level or tests of current verbal ability are used as surrogates for baseline ability. Education is generally a poor surrogate for baseline ability.[11] Most studies cite verbal ability tests as methods that are insensitive to neurotoxicants.[2,3] However, chronic organic solvent exposure may result in a general dementia affecting all aspects of cognitive function, including word knowledge and general information.[27–29] Prospective studies, in which baseline intellectual ability is known rather than estimated, are needed.

11.22.3.2 Visuospatial Ability

Tests of visuospatial ability, such as Block Design (WAIS-R),[4] requires that the individual manipulate blocks to copy a two-dimensional design. Block Design has been used to evaluate the effects of lead and solvents, with mixed results.[30,31] The Raven's Progressive Matrices, also sensitive to the effects of neurotoxicants,[32] assesses intellectual ability by presenting visuospatial, conceptual problems rather than verbal conceptual problems (e.g., proverbs).

11.22.3.3 Concentration/Attention

Tests of concentration and attention assess the ability to orient and sustain attention to visual or auditory stimuli. This ability is the precursor to learning and memory. Poor concentration and memory are among the most frequent complaints reported by individuals exposed to neurotoxicants and partially define the encephalopathies of solvent exposure.[27] Tasks of concentration/attention include simple reaction time in response to auditory (e.g., tone) or visual stimuli (designated letter) and more complex reaction time tasks with distractors, such as the Continuous Performance

Test.[18] The length of tasks can be varied to monitor vigilance. Since the task is relatively simple, lengthening the task or changing signal probability increases the demand to sustain attention to a boring task. Such a task is often valid as a test of the impact that neurotoxicants may have on behaviors required in many work settings (e.g., air traffic control, anesthesia monitors). For example, some studies of low-dose alcohol administration suggested that performance (stimulus sensitivity and reaction times) was compromised under low but not high signal probability (e.g., Jansen *et al.*[33]). Tests of attention/concentration and vigilance are a part of test batteries applied to worksite testing (e.g., Tuff Battery;[34] WHO Battery;[12] POET[35]). Gamberale[7] cites simple reaction time as the most sensitive test for detecting behavioral performance effects due to solvent exposure.

11.22.3.4 Motor Skills

Tests of motor skills assess speed, dexterity, and strength by asking the subject to place pegs in holes while being timed (e.g., Grooved Pegboard;[19] Santa Ana Pegboard[36]) or measure grip strength by pressure against a spring-loaded device (dynamometer).[19] Finger tapping[19] is another simple test of coordination and speed that has been sensitive to the effects of neurotoxicants.[2] While visuomotor coordination is required, abstract perceptual processing is minimal for these tests.

11.22.3.5 Visuomotor Coordination

Tests of visuomotor skills typically involve complex levels of motor coordination in response to visual stimuli. For example, the hand-eye coordination test from the NES battery[18] assesses the ability to move a computer cursor with a joystick along a sine wave pattern on a screen at a constant rate of speed. Another verbally mediated test of eye–hand coordination is a coding task which requires the subject to code symbols with letters while being timed (Digit Symbol[4]). It has proven to be sensitive to the effects of neurotoxicants (e.g., lead, solvents, mercury).[37–40]

11.22.3.6 Memory

Tests of memory for abstract visual designs (e.g., Benton) for verbal materials such as words or numbers (California Verbal Learning Test;[21] Paired Associates[20]) involves presentation of the stimulus to be encoded. The subject is asked to recall the stimulus immediately and after a relatively short delay (e.g., 30 min). A more recent development in memory testing incorporates the tradition of cognitive psychology through the assessment of memory processes such as learning efficiency over several trials (e.g., Spurgeon *et al.*[41]) which may be more sensitive to subtle effects.

11.22.3.7 Sensory Function

The Agency for Toxic Substances and Disease Registry (ATSDR) has recognized the need to evaluate sensory as well as cognitive function in the battery recommended for use in environmental health field studies.[42] Tests of visual, auditory, and tactile sensory function were included in a battery to evaluate individuals in communities potentially exposed to hazardous waste sites. Tests of audition range from simple evaluation of hearing with an audiometer to more complex tests assessing the ability to discern speech or rhythmic patterns (e.g., SCAN-A, competing words test[43]). Tests of tactile perception and vibration sense include simple tactile perception (finger agnosia)[19] and sense of vibration using a device[44] to measure perception of fine vibrations in the finger or toe. These tests have successfully detected losses of peripheral sensory perception due to mercury or solvents. Of interest is the finding of color vision loss among solvent exposed workers.[23,45] Finally, altered sense of smell may occur with exposure to neurotoxicants. Schwartz *et al.*[46] reported dose-related decrements in olfactory function as measured with a test of olfactory discrimination (University of Pennsylvania Smell Identification Test, UPSIT).[24] These deficits were observed among workers exposed chronically to low levels of organic solvents, suggesting that such sensory tests may be sensitive indicators of the effects of long-term, low-level exposure (see also Chapter 15, this volume).

11.22.3.8 Affect and Personality

Mood is reported to be one of the earliest signs of compromised function due to neurotoxicant exposure. The World Health Organization/Nordic Council and the International Solvent Workshop classified the mildest disorder attributable to organic solvents as an organic affective syndrome (i.e., Type 1). Fatigue, irritability, depression, and lability are mood changes reported to occur in the first stages of neurotoxicity.[27] Representative scales that have been used include the Profile of Mood States,[25] Minnesota Multiphasic Personality Inventory—2,[26] and Beck Depression

Inventory.[47] Unfortunately, these measures rely on self-report of symptoms rather than any objective indicators of mood. Therefore, they are subject to reporting biases which may be influenced by the circumstances in which the individual is being evaluated (e.g., litigation).

11.22.4 SUMMARY AND CRITIQUE OF TEST METHODS

It is clear from the neuropsychological literature that many tests are available and sensitive to naturally occurring as well as toxicant-induced variation in human cognitive behavior. However, neuropsychological testing is also time-consuming. First, standardized administration and scoring requires examiner training to achieve reliability. Second, a battery of tests is required to fully assess cognitive function. It remains difficult to predict exactly which cognitive function is likely to be affected by a specific neurotoxicant. Therefore, investigators typically choose to represent each function in a battery of tests. Both of these factors add to the time and expense of administering neuropsychological tests.

Although neuropsychological tests provide a detailed description of cognitive function, attribution of performance deficits to specific causes may be confounded by a number of extraneous factors including: motivation, education, psychiatric disability, other disease processes (e.g., diabetes, head injury, alcoholism), and language and culture. Individuals who are depressed or are motivated by monetary compensation to perform poorly may not achieve valid test results. Methods based on probability models to detect motivational level can determine whether the individual is putting forth a reasonable effort.[48] Most of the tests reviewed also require a minimum level of education/reading ability and may be influenced by language and cultural factors. Therefore, they may not be applicable for poorly educated, non-English speaking populations who are likely to be in most danger of exposures to neurotoxicants. Finally, examiners need to document carefully other diseases and injuries that could potentially influence performance. These include relatively minor hand injuries that could slow visuomotor speed.

Since investigation of neurotoxic effects is time-consuming, the ability to compare results across studies is desirable. In response to this need, the Neurobehavioral Core Test Battery (NCTB) was developed from a collaboration of the World Health Organization and the US National Institute for Occupational Safety and Health.[12] For this battery, a consistent group of tests was recommended for international use. To test the applicability of this battery for cross-cultural studies, the NCTB was administered to healthy adults not exposed to chemicals at work. The countries included were 10 nations from Europe, North and Central America, and Asia. Performance on some tests such as simple reaction time and memory for abstract figures (Benton Visual Retention) was similar across countries. Performance on other tests, however, varied from one country to another suggesting differential understanding of, or response to, the test demands.[47] Unless cultural factors exert a systematic influence that can be identified, comparison across studies could be problematic.

Finally, it may not be readily apparent how minor reductions in speed or learning processes could be meaningful. For example, for workers who must perform fine motor manipulations under time pressure, reduced visuomotor speed due to toxic exposure is expected to be relevant to productivity at work. Information about predictive validity could provide the most compelling evidence for controlling or reducing exposure.

11.22.5 COGNITIVE EFFECTS OF SELECTED NEUROTOXICANTS

11.22.5.1 Pesticides

Pesticides include known neurotoxicants used in the world agricultural community (see also Chapters 26 and 27, this volume). The acute, neurotoxic effects of pesticides are fatigue, headache, nausea, blurry vision, tremor, confusion, and given sufficient exposure, coma and death.[48] Most acute symptoms resolve following cessation of exposure, although they have been documented to persist from days to months.[49,50] For example, certain organophosphates have been associated with an uncommon delayed-onset peripheral neuropathy which can occur one to several weeks after an acute exposure.[48] More controversial, however, is the question of chronic central nervous system effects resulting from repeated exposure to pesticides.

Studies of the long-term consequences of pesticide use have focused largely on investigation of chronic neuropsychological effects following a documented acute overexposure or poisoning.[51,52] Savage *et al.*,[52] Rosenstock *et al.*,[53] and Steenland *et al.*[54] selected workers for study who were identified by a physician or were hospitalized for acute pesticide poisoning. These studies reported that the pesticide poisoned groups performed significantly worse than matched controls on several neuropsychological tests. While the criteria for inclusion in these studies was documentation of an acute poisoning episode, many of the subjects in these studies worked in occupations where organophosphate pesticides were consistently used (e.g., pesticide applicators, farm workers in Nicaragua). Thus, the effects of long-term exposure through occupational use of pesticides could not be separated from the effects of a clinical episode of acute poisoning.

Additional studies have been conducted in which long-term, chronic exposure without documentation of acute poisoning has been investigated.[55] Rodnitzky[51] reported no significant neurobehavioral effects among a group of 23 farmers and commercial pesticide applicators. Daniell et al.[56] also found no significant differences in neuropsychological performance between pesticide applicators and beef slaughter-house workers. In contrast, Stephens et al.[57] reported significant decrements in sustained attention and information processing and greater psychiatric disorder for farmers chronically exposed to organophosphates through the process of dipping sheep for pests. Some methodological limitations in the Rodnitzky[51] and Daniell et al.[56] studies may have compromised their ability to detect effects. These limitations included small sample size, low level of education, non-English speaking subgroups, and potential pesticide exposure among the controls. The Stephens et al.[57] study had a larger sample size and a distinctly unexposed control group, quarry workers.

To date, the literature on chronic pesticide exposure suggests that the most significant cognitive deficits occur when individuals have a high-dose pesticide poisoning which causes them to seek treatment. Long-term exposure without acute illness may not produce the same significant alterations in cognitive function.

11.22.5.2 Human Neurobehavioral Studies of Lead

While the mechanisms of lead toxicity continue to be debated, several studies have evaluated the neurobehavioral effects of lead (see Chapters 28 and 29, this volume).[13,58–62] Among lead-exposed workers, some studies documented significant reductions in memory, learning, verbal concept formation, psychomotor function, and/or mood.[63,64] Others, however, found only marginal dose–response relationships or few significant differences between groups, particularly when the lead exposure remained in the low to moderate range (i.e., blood lead below 70 μg dL^{-1}).[13,59] In both the studies by Ryan et al.[13] and Braun and Daigneault,[62] the only significant differences were in measures of motor as opposed to cognitive function (see

Table 2 for a sample of cross-sectional studies). None of the studies listed in Table 2 were able to match control and exposed samples on relevant demographic variables. Thus, these variables had to be controlled statistically. This probably compromised the investigators' ability to detect differences between the exposed and unexposed groups.

11.22.5.3 Behavioral Studies of Organic Solvents

Solvents have general anesthetic properties, leading to generalized depression or narcosis of the central nervous system (CNS), ranging from lightheadedness at the lowest doses to coma, convulsions, and death at the highest doses. Chronic or recurrent exposure to all of the organic solvents produces a nonspecific array of symptoms including headache, nausea, dizziness, and fatigue.[66] These reflect a generalized CNS depression believed to involve nonspecific changes in membrane function as well as neurotransmitter levels.[68] Long-term, high-level exposure to solvents causes persistent (and to some extent irreversible) impairment in cognitive function associated with an objective loss of CNS tissue.[27] The magnitude of exposure required to produce such changes, particularly the dementia and encephalopathy known in Scandinavia as "painters' syndrome," is highly controversial.[69]

Beginning in the 1970s, numerous studies have evaluated the effects of chronic solvent exposure.[61,70–75] Reductions in memory and concentration, psychological symptoms, and reduced psychomotor speed have been documented in the Scandinavian literature. Based largely on this literature, two international workshops categorized solvent-induced CNS disorders according to the categories listed in Table 3.

Comprehensive reviews of the literature suggest that some form of permanent cognitive impairment results from chronic solvent exposure.[10,72,78–80] Some studies questioned the existence of the "chronic painters' syndrome" described in the Scandinavian literature.[81,82] These investigators reviewed the early cases and found that after controlling for confounders such as age, intelligence, and alcohol use, the attribution of dementia to solvent exposure could be questioned. However, work from other countries supports the chronicity of neuropsychological deficits even after exposure has terminated.[29,72]

While many studies support the contention that workers chronically exposed to solvents have reductions in cognitive performance,

Table 2   Cross-sectional studies of chronic lead exposure.

Author	Exposure	Exposed subjects	Control	Results	Comments
Braun and Daigneault[62]	Time-weighted blood lead $(53 \pm 7.5)\,\text{mg}\,\text{dL}^{-1}$	41 Lead smelter workers (all male)	Age	1. Motor function 2. Mood	1. Education not matched 2. Exposure to solvents in controls
Ryan et al.[13] Parkinson et al.[65]	Time-weighted blood lead $(49 \pm 12)\,\text{mg}\,\text{dL}^{-1}$	288 Lead-exposed factory workers	Age, gender (all male)	1. Psychomotor speed and manual dexterity 2. Interpersonal conflict	1. Education not matched
Stollery et al.[66]	Current blood lead low = <20 mg med = 21–40 mg high = 41–80 mg	91 Male battery and printing factory workers	No unexposed controls	1. Attention 2. Psychomotor function	1. Higher recent alcohol use in most exposed
Williamson and Teo[67]	Current blood lead $(2.36\,\text{mg}\,\text{mL}^{-1} = 0.64)$	59 Male battery and lead smelters	Age, education, alcohol use	1. Visual function 2. Sensory attention 3. Psychomotor performance 4. Memory	1. Gender not matched (17% women in controls)
Baker et al.[64] Baker et al.[61]	Time-weighted blood lead (12 months) current blood lead = 32.8 $\text{mg}\,\text{mL}^{-1}$ (10–80)	99 Lead-exposed foundry workers	Not matched; statistical control	1. Psychomotor speed 2. Memory 3. Verbal ability 4. Mood	1. Some controls were lead exposed 2. Significant findings for workers with blood lead $>60\,\text{mg}\,\text{dL}^{-1}$

Table 3   Categories of solvent-induced CNS disorders.[a]

Severity of condition	Category of CNS disorder	
	Identified by WHO/Nordic Council of Ministries Working Group, Copenhagen, June 1985[b]	*International Solvent Workshop, Raleigh, NC October, 1985*[c]
Minimal	Organic affective syndrome	Type 1
Moderate	Mild chronic toxic encephalopathy	Types 2A or 2B
Pronounced	Severe chronic toxic encephalopathy	Type 3

[a]In view of the difficulty of categorizing these disorders, correspondence between the two systems of nomenclature is not exact.   [b] Source: WHO.[76]   [c] Source: Ref. 77.

conclusions from these studies often have to be qualified due to flaws in design. Most of these investigations were cross-sectional (see Table 4) and rely on appropriate control groups to draw conclusions about the cognitive status of the exposed subjects. Selecting appropriate controls is difficult given the number of variables that must be controlled. At a minimum, subjects should be matched for age, gender, ethnicity, and pre-exposure ability. In addition,

Table 4 Cross-sectional studies of chronic solvent exposure.

Author(s)	Exposure	Subjects	Control matching	Result	Comment
Escalona et al.[84]	Exposure history; mean = 7 years	56 Male 29 Female adhesive factory	Age, education	1. Mood 2. Attention 3. Visuomotor coordination 4. Fine motor coordination	1. Exposure ranged from <5 years to over 10 years 2. Gender not matched
Broadwell et al.[85]	Exposure history only ($x = 9.2$ years; 3–18 years)	25 Micro-electronics	Age, gender, ethnicity, education	1. Mood 2. Motor speed 3. Attention 4. Visuomotor coordination 5. Vibration sensation 6. Visual contrast sensitivity 7. Grip strength	1. Spanish as first language in exposed patients
Daniell et al.[56]	Passive dosimetry	High (39), medium (32), low (29), autobody painters	Education, gender	1. Visual perception 2. Memory	1. Significant age differences 2. Heavy alcohol use 3. Acute and chronic exposure
Spurgeon et al.[41]	Exposure history 1–30 years	90 Brush painters, 144 painters, printers, boat builders	Age, gender	1. Visuomotor coordination 2. Verbal memory	1. Education not matched
Bleeker et al.[73]	Industrial hygiene measures— 13–15 years; duration of exposure: 5–36 years	176 Male paint manufacture workers	No unexposed controls; confounders controlled in regression	1. Visuomotor coordination 2. Attention 3. Vibratory threshold	1. Intensity not lifetime dose associated with effect
Bowler et al.[29]	Exposure history: mean = 6.7 years (± 3.9 years)	67 Micro-electronics workers (93% female)	Age, education gender, ethnicity	1. Visual memory 2. Verbal memory 3. Visuomotor coordination 4. Attention 5. Motor speed and coordination 6. Mood	1. Workers disabled at testing

solvent-exposed workers also use alcohol extensively, which further confuses the picture of cognitive impairment that can be attributed to workplace exposure of solvent mixtures. Among studies evaluating cognitive effects of chronic solvent exposure, the most significant effects were seen among workers who were disabled at the time[29] and/or self-referred to an occupational health clinic for evaluation of cognitive complaints.[83]

Although alcohol use has been regarded as a confound, it could also help answer the question of the chronicity of effect seen with long-term exposure to solvents, albeit through a different route of exposure, ingestion. Numerous studies document the cognitive effects of chronic alcohol use (for reviews of this literature (see Grant,[86] Parsons,[87] and Ryan and Butters[88]). Ingestion of alcohol over years is associated with cognitive deficits such as poor verbal memory and reduced psychomotor speed. These deficits are similar to those observed in solvent-exposed workers.[89] While cognitive performance shows improvement with abstinence, it is not yet known whether individuals recover completely.[86] The parameters of exposure that produce permanent cognitive impairment are not known. Some investigators suggest that duration or frequency of exposure to alcohol, or solvent mixtures, may not be as important as the frequency of high-dose exposure such as that seen with binge drinking or solvent intoxication.[73,90] Workers who have both exposures during their lifetime, that is, organic solvents and alcohol, appear to be at increased risk of organic brain damage.[91]

11.22.6 EXPOSURE ASSESSMENT IN THE NEUROBEHAVIORAL LITERATURE

Accurate assessment of exposure, particularly for the studies of chronic neurotoxic effects, is crucial and may account for the discrepant findings within the neurobehavioral literature. Studies vary in the method by which exposed subjects were selected. Among lead studies, exposure was determined by blood lead levels, either at the time of recruitment or over varied time periods during employment (e.g., time-weighted average), as well as by a history of employment within a lead industry.[64–66] In an epidemiological study by Parkinson *et al.*[65] subjects had to be employed a minimum of one year. While this study and a follow-up published by Ryan *et al.*[13] were largely negative, the only group for which positive neurobehavioral findings were observed was the older group who had worked more years with lead. Workers varied both in body burden and cumulative exposure history (years of exposure) which contributed to variation in the effects between studies. With the advent of K-wave x-ray fluorescence, however, better estimates of the body burden of lead can be obtained. For example, the half-life of lead in bone is approximately 20 years.[82] Therefore, estimates of bone lead provide a better indicator of cumulative long-term exposure.

The occurrence of chronic solvent exposure has been determined by work history and records of industrial hygiene measurements, when available.[41,74,92] As with studies of lead-exposed subjects, exposure histories varied within and between studies. Ryan *et al.*[31]

reported on a group of workers whose exposure to solvents ranged from 1 to 18 years (see Table 4). Industrial hygiene measurements are rarely available. Some of these shortcomings remain confounds since there are no good biomarkers of chronic solvent exposure. However, studies have benefitted from the use of a standardized solvent exposure questionnaire[93] and/or history of solvent monitoring at a worksite.[73]

As for solvent studies, chronic pesticide exposure has been documented primarily by work history.[56,57] Those studies evaluating the neurobehavioral effects of an acute overexposure relied on a physician or hospital diagnosis with little to no quantification of chronic exposure even though many subjects had worked with pesticides over years.[52,53]

Finally, few studies have investigated the interactive effects of exposure to multiple neurotoxicants, a situation often encountered in working environments. While some studies have acknowledged multiple exposures, few investigations have directly assessed the neuropsychological consequences of multiple, simultaneous exposures.[30]

11.22.7 SUMMARY AND NEW DIRECTIONS

Neuropsychological tests have proven their sensitivity for documenting cognitive deficits due to neurotoxicant exposure. However, results across studies have been inconsistent due to variability in the subjects studied, exposure history, and test batteries. Therefore, it has been difficult to identify a pattern of cognitive deficits associated with a given neurotoxicant. The challenge is: (i) to develop better methods to quantify exposure including multiple exposures; (ii) to refine neuropsychological methods to capture the subtle deficits that may arise from exposure at levels significantly lower than those seen in the early studies of solvents and lead; and (iii) to understand the impact of individual differences in susceptibility to exposures.

In the past, batteries of tests were chosen to represent the basic cognitive functions (Table 1). While this method offers an overview of all cognitive behaviors, it is generally atheoretical. Positive advance could be made in the field if specific functions were selected for more in-depth examination based on what is known about the toxicology of the chemical(s). This strategy was used by Mergler and co-workers[75] in her investigation of color vision among solvent-exposed workers. It may be time to employ such a strategy among lead-exposed workers. For example, results from Ryan *et al.*[13]

and Braun *et al.*[45] (see Table 2) suggest focusing on motor rather than cognitive function.

Prospective studies need to be developed in which workers are followed over a period of time to assess changes from baseline. This would obviate the need to estimate pre-exposure ability, a frequent problem in cross-sectional designs. Such studies will require neuropsychological tests suitable for repeated measures conditions. For example, Otto *et al.*[9] found significant practice effects for several tests from the NES battery. To avoid ceiling effects with repeated administration, he suggested that test parameters be altered to make the tasks more difficult and better suited to repeated measures design.[9]

Finally, individual differences in susceptibility[89] may be determined by a host of factors including but not limited to pre-exposure intellectual ability, personality, gender, and other exposures such as alcohol or drug use. It has been suggested that individuals of lower pre-exposure intellectual ability may be at relatively more risk of neurologic injury due to a lack of compensatory abilities. With regard to gender, most studies, up to 1996, have been conducted with males, since men are frequently found in work settings where exposures occur. It is interesting to note that one of the studies reporting the most significant effect on cognitive performance was a study in which women from the microelectronics industry comprised 93% of the sample.[29] As women move into industries where exposures occur, differences in neurologic health effects will be important to investigate. These and other individual differences will enhance our understanding of the variability in effects seen from exposure to neurotoxicants.

11.22.8 REFERENCES

1. Chemical Regulation Reporter (14 March 1986). *Health Hazards*, p. 1598.
2. D. E. Hartman, 'Neuropsychological Toxicology: Identification and Assessment of Human Neurotoxic Syndromes,' 1st edn., Pergamon Press, New York, 1988.
3. M. D. Lezak, 'Neuropsychological Assessment,' 2nd edn., Oxford University Press, New York, 1983.
4. D. Wechsler, 'Wechsler Adult Intelligence Scale—Revised,' Psychological Corporation, San Antonio, TX, 1981.
5. W. Mittenberg, G. B. Thompson and J. A. Schwartz, 'Abnormal and reliable differences among wechsler memory scale—revised subtests.' *Psychol. Assessment*, 1991, **3**, 492–495.
6. W. K. Anger, 'Worksite behavioral research. Results, sensitive methods, test batteries and the transition from laboratory data to human health.' *Neurotoxicology*, 1990, **11**, 627–717.
7. F. Gamberale, 'Use of behavioral performance tests in the assessment of solvent toxicity.' *Scand. J. Work Environ. Health*, 1985, **11**, Suppl. 1, 65–74.
8. R. F. Russell, in 'Behavioral Measures of Neurotoxicity,' eds. R. W. Russell, P. E. Flattau and A. M. Pope, National Academy Press, Washington, DC, 1990, pp. 206–225.
9. D. A. Otto, H. K. Hudnell, D. E. House *et al.*, 'Exposure of humans to a volatile organic mixture. Behavioral Assessment.' *Arch. Environ. Health*, 1992, **47**, 23–30.
10. W. K. Anger and B. L. Johnson, in 'Environmental and Occupational Medicine,' ed. W. N. Rom, Little, Brown & Co., Boston, MA, 1992, pp. 573–593.
11. N. Cherry, H. Venables and H. A. Waldron, 'Description of the tests in the London School of Hygiene test battery.' *Scand. J. Work Environ. Health*, 1984, **10**, Suppl. 1, 18–19.
12. B. L. Johnson, E. L. Baker, M. El Batawi *et al.*, (eds.) 'Prevention of Neurotoxic Illness in Working Populations,' Wiley, New York, 1987.
13. C. M. Ryan, L. A. Morrow, E. J. Bromet *et al.*, 'Assessment of neuropsychological dysfunction in the workplace: normative data from the Pittsburgh Occupational Exposure Test Battery.' *J. Clin. Exp. Neuropsychol.*, 1987, **9**, 665–679.
14. S. Jastak and G. S. Wilkinson, 'Wide Range Achievement Test Administration Manual,' Jastak Associates, Inc., Wilmington, DE, 1984.
15. H. E. Nelson, 'National Adult Reading Test (NART): Test Manual,' NFER, Nelson, UK, 1982.
16. J. D. Raven, 'Guide to Standard Progressive Matrices,' Lewis, London, 1960.
17. M. R. Trenerry, B. Crosson, J. DeBoe *et al.*, 'Stroop Neuropsychological Screening Test Manual.' Psychological Assessment Resources, Odessa, FL, 1989.
18. R. Letz and E. L. Baker, 'NES2 Neurobehavioral Evaluation Systems,' Neurobehavioral Systems, Winchester, MA, 1988.
19. R. K. Heaton, I. Grant and C. G. Matthews, 'Comprehensive norms for an expanded Halstead–Reitan battery: demographic corrections, research findings, and clinical applications,' Psychological Assessment Resources, Inc., Odessa, FL, 1991.
20. D. Wechsler, 'Wechsler Memory Scale—Revised Manual,' The Psychological Corporation, San Antonio, TX, 1987.
21. D. C. Delis, J. H. Framer, E. Kaplan *et al.*, 'California Verbal Learning Test,' The Psychological Corporation, San Antonio, TX, 1987.
22. R. S. H. Visser, 'Manual of the Complex Figure Test CFT: a Test Involving Restructuring Behaviour Developed for the Assessment of Brain Damage,' Sweta & Zeitlinger, Amsterdam, 1973.
23. D. Mergler and L. Blain, 'Assessing color vision loss among solvent-exposed workers.' *Am. J. Ind. Med.*, 1987, **12**, 195–203.
24. R. L. Doty, 'The Smell Identification Test,' Sensonics, Haddon Heights, NJ, 1983.
25. D. M. McNair, M. Lorr and L. F. Droppleman, 'Profile of Mood States,' Educational and Industrial Testing Service, San Diego, CA, 1981.
26. S. F. Hathaway and J. C. McKinley, 'Minnesota Multiphasic Personality Inventory 2,' University of Minnesota Press, MN, 1989.
27. NIOSH US Department of Health and Human Services, 'Organic solvent neurotoxicity,' Bulletin 48, National Institute for Occupational Safety and Health, Cincinnati, OH, 1987.
28. R. G. E. Oberg, H. Udesen, A. M. Thomsen *et al.* (eds.), 'Neurobehavioural Methods in Occupational and Environmental Health,' World Health Organization (Environmental Health Documents 3), World Health Organization, Copenhagen, 1985, pp. 130–135.
29. R. M. Bowler, D. Mergler, G. Huel *et al.*, 'Neuropsychological impairment among former microelectronics workers.' *Neurotoxicology*, 1991, **12**, 87–103.

30. M. L. Bleecker, J. Agnew, J. P. Keogh *et al.*, in 'Advances in the Biosciences, Neurobehavioral Methods in Occupational Health,' eds. R. Gilioli, M. G. Cassito and V. Foa, Pergamon Press, New York, 1983, vol. 46, pp. 255–262.

31. C. M. Ryan, L. A. Morrow and M. Hodgson, 'Cacosmia and neurobehavioral dysfunction associated with occupational exposure to mixtures of organic solvents.' *Am. J. Psychiatry*, 1988, **145**, 1442–1445.

32. M. Istoc-Bobis and S. Gabor, 'Psychological disfunctions in lead- and mercury-occupational exposure.' *Revue Roumaine des Sciences Sociales—Serie de Psychologie*, 1987, **31**, 183–191.

33. A. A. Jansen, J. J. de Gier and J. L. Slangen, 'Alcohol effects on signal detection performance.' *Neuropsychobiology*, 1985, **14**, 83–87.

34. H. Hanninen and K. Lindstrom, 'Neurobehavioral Test Battery of the Institute of Occupational Health,' Institute of Occupational Health, Helsinki, 1989.

35. C. Hogstedt, M. Hane and O. Axelson, in 'Developments in Occupational Medicine,' ed. C. Zenz, Year Book Medical Publishers, Chicago, IL, 1980, pp. 249–258.

36. E. A. Fleishman, 'Dimensional analysis of psychomotor abilities.' *J. Exp. Psychol.*, 1954, **48**, 437–454.

37. P. Grandjean, J. Beckmann and G. Ditlev, in 'Advances in the Biosciences, Neurobehavioral Methods in Occupational Health,' eds. R. Giliol, M. G. Cassitto and V. Foa, Pergamon Press, New York, 1983, pp. 301–308.

38. J. A. Valciukas, R. Lilis, J. Eisinger *et al.*, 'Behavioral indicators of lead neurotoxicity: results of a clinical field survey.' *Int. Arch. Occup. Environ. Health*, 1978, **41**, 217–236.

39. G. Angotzi, D. Camerino, F. Carboncini *et al.*, in 'Advances in the Biosciences, Neurobehavioral Methods in Occupational Health,' eds. R. Giliol, M. G. Cassitto and V. Foa, Pergamon Press, New York, 1983, pp. 247–253.

40. P. Orbaek, M. Lindgren, H. Olivecrona *et al.*, 'Computed tomography and psychometric test performances in patients with solvent induced chronic toxic encephalopathy and healthy controls.' *Br. J. Ind. Med.*, 1987, **44**, 175–179.

41. A. Spurgeon, C. N. Gray, J. Sims *et al.*, 'Neurobehavioral effects of long-term occupational exposure to organic solvents: comparable studies.' *Am. J. Ind. Med.*, 1992, **22**, 325–335.

42. L. J. Hutchinson, R. W. Amler, J. A. Lybarger *et al.*, 'Neurobehavioral Test Batteries for use in Environmental Health Field Studies,' US Department of Health and Human Services, Atlanta, GA, 1992.

43. R. W. Keith, 'SCAN-A—A Test for Auditory Processing Disorders in Adolescents and Adults,' The Psychological Corporation, San Antonio, TX, 1994.

44. 'Vibratron II Operating Manual,' Physitemp Instruments, Clifton, NJ, 1991.

45. C. M. Braun, S. Daigneault and B. Gilbert, 'Color discrimination testing reveals early printshop solvent neurotoxicity better than a neuropsychological test battery.' *Arch. Clin. Neuropsychol.*, 1989, **4**, 1–13.

46. B. S. Schwartz, D. P. Ford, K. I. Bolla *et al.*, 'Solvent-associated decrements in olfactory function in paint manufacturing workers.' *Am. J. Ind Med.*, 1990, **18**, 697–706.

47. A. T. Beck, 'Beck Depression Inventory,' The Psychological Corporation, San Antonio, TX, 1978.

48. M. F. Greiffenstein, W. J. Baker and T. Gola, 'Validation of malingered amnesia measures with a large clinical sample.' *Psychol. Assessment*, 1994, **6**, 218–224.

49. S. D. Murphy, in 'Casarett and Doull's Toxicology: The Basic Science of Poisons,' 3rd edn., eds. C. D. Klaassen, M. O. Amdur and J. Doull, 3rd edn., Macmillan, New York, 1986, pp. 519–581.

50. J. E. Midtling, P. G. Barnett, M. J. Coye *et al.*, 'Clinical management of field worker organophosphate poisoning.' *West. J. Med.*, 1985, **142**, 514–518.

51. R. L. Rodnitzky, 'Occupational exposure to organophosphate pesticides: a neurobehavioral study.' *Arch. Environ. Health*, 1975, **30**, 98–103.

52. E. P. Savage, T. J. Keefe, L. M. Mounce *et al.*, 'Chronic neurological sequelae of acute organophosphate pesticide poisoning.' *Arch. Environ. Health*, 1988, **43**, 38–45.

53. L. Rosenstock, M. Keifer, W. E. Daniell *et al.*, 'Chronic central nervous system effects of acute organophosphate pesticide intoxication.' The Pesticide Health Effects Study Group, *Lancet*, 1991, **338**, 223–227.

54. K. Steenland, B. Jenkins, R. G. Ames *et al.*, 'Chronic neurological sequelae to organophosphate pesticide poisoning.' *Am. J. Public Health*, 1994, **84**, 731–736.

55. B. Eskanazi and N. Maizlish, in 'Medical Neuropsychology. The Impact of Disease on Behavior,' eds. R. E. Tarter, D. H. Van Thiel and K. L. Edward, Plenum, New York, 1988, pp. 223–263.

56. W. Daniell, A. Stebbins, J. O'Donnell *et al.*, 'Neuropsychological performance and solvent exposure among car body repair shop workers.' *Br. J. Ind. Med.*, 1993, **50**, 368–377.

57. R. Stephens, A. Spurgeon, I. A. Calvert *et al.*, 'Neuropsychological effects of long-term exposure to organophosphates in sheep dip.' *Lancet*, 1995, **345**, 1135–1139.

58. J. D. Repko and C. R. Corum, 'Critical review and evaluation of the neurological and behavioral sequelae of inorganic lead absorption.' *CRC Crit. Rev. Toxicol.*, 1979, **6**, 135–187.

59. H. Haenninen, S. Hernberg, P. Mantere *et al.*, 'Psychological performance of subjects with low exposure to lead.' *J. Occup. Med.*, 1978, **20**, 683–689.

60. J. A. Valciukas and R. Lilis, 'A composite index of lead effects. Comparative study of four occupational groups with different levels of lead absorption.' *Int. Arch. Occup. Environ. Health*, 1982, **51**, 1–14.

61. E. L. Baker, R. F. White, L. J. Pothier *et al.*, 'Occupational lead neurotoxicity: improvement in behavioural effects after reduction of exposure.' *Br. J. Ind. Med.*, 1985, **42**, 507–516.

62. C. M. J. Braun and S. Daigneault, 'Sparing of cognitive executive functions and impairment of motor functions after industrial exposure to lead: a field study with control group.' *Neuropsychology*, 1991, **5**, 179–193.

63. P. Grandjean, E. Arnig and J. Beckmann, 'Psychological dysfunctions in lead-exposed workers. Relation to biological parameters of exposure.' *Scand. J. Work Environ. Health*, 1978, **4**, 295–303.

64. E. L. Baker, R. G. Feldman, R. A. White *et al.*, 'Occupational lead neurotoxicity: a behavioural and electrophysiological evaluation. Study design and year one results.' *Br. J. Ind. Med.*, 1984, **41**, 352–361.

65. D. K. Parkinson, C. Ryan, E. J. Bromet *et al.*, 'A psychiatric epidemiologic study of occupational lead exposure.' *Am. J. Epidemiol.*, 1986, **123**, 261–269.

66. B. T. Stollery, H. A. Banks, D. E. Broadbent *et al.*, 'Cognitive functioning in lead workers.' *Br. J. Ind. Med.*, 1989, **46**, 698–707.

67. A. M. Williamson and R. K. C. Teo, 'Neurobehavioural effects of occupational exposure to lead.' *Br. J. Ind. Med.*, 1986, **43**, 374–380.

68. ATSDR, 'Toxicological Profile for Total Xylenes,' Agency for Toxic Substances and Disease Registry, Atlanta, GA, 1990.

69. F. Gerr and R. Letz, in 'Environmental and Occupational Medicine,' ed. W. Rom, Little Brown & Co., Boston, MA, 1992, pp. 843–865.

70. H. Hanninen, L. Eskelinen, K. Husman *et al.*, 'Beha-

vioral effects of long-term exposure to a mixture of organic solvents.' *Scand. J. Work Environ. Health*, 1976, **2**, 240–255.

71. J. A. Valciukas, R. Lilis, R. M. Singer *et al.*, 'Neurobehavioral changes among shipyard painters exposed to solvents.' *Arch. Environ. Health*, 1985, **40**, 47–52.

72. E. L. Baker, R. E. Letz, E. A. Eisen *et al.*, 'Neurobehavioral effects of solvents in construction painters.' *J. Occup. Med.*, 1988, **30**, 116–123.

73. M. L. Bleecker, K. I. Bolla, J. Agnew *et al.*, 'Dose-related subclinical neurobehavioral effects of chronic exposure to low levels of organic solvents.' *Am. J. Ind. Med.*, 1991, **19**, 715–728.

74. H. Hanninen, M. Antti-Poika, J. Juntunen *et al.*, 'Exposure to organic solvents and neuropsychological dysfunction: a study on monozygotic twins.' *Br. J. Ind. Med.*, 1991, **48**, 18–25.

75. R. M. Bowler, D. Mergler, S. S. Rauch *et al.*, 'Affective and personality disturbances among female former microelectronics workers.' *J. Clin. Psychol.*, 1991, **47**, 41–52.

76. WHO Nordic Council of Ministries, 'Organic solvents and the central nervous system, EH5,' World Health Organization and Nordic Council of Ministries, Copenhagen, Denmark, 1985, 1–39.

77. Anonymous, 'Human aspects of solvent neurobehavioral effects.' *Neurotoxicology*, 1986, **7**, 43–56.

78. P. Gregersen, 'Neurotoxic effects of organic solvents in exposed workers: two controlled follow-up studies after 5.5 and 10.6 years.' *Am. J. Ind. Med.*, 1988, **14**, 681–701.

79. C. Edling, K. Ekberg, G. Ahlborg, Jr. *et al.*, 'Long term follow up of workers exposed to solvents.' *Br. J. Ind. Med.*, 1990, **47**, 75–82.

80. F. Gamberale, 'Use of behavioral performance tests in the assessment of solvent toxicity.' *Scand. J. Work Environ. Health*, 1985, **11**, Suppl. 1, 65–74.

81. E. O. Errebo-Knudsen and F. Olsen, 'Organic solvents and presenile dementia (the painters' syndrome). A critical review of the Danish literature.' *Sci. Total Environ.*, 1986, **48**, 45–67.

82. A. Gade, E. L. Mortensen and P. Bruhn, 'Chronic painter's syndrome. A reanalysis of psychological test data in a group of diagnosed cases, based on comparisons with matched controls.' *Acta Neurol. Scand.*, 1988, **77**, 293–306.

83. L. A. Morrow, C. M. Ryan, M. J. Hodgson *et al.*, 'Alterations in cognitive and psychological functioning after organic solvent exposure.' *J. Occup. Med.*, 1990, **32**, 444–450.

84. E. Escalona, L. Yanes, O. Feo *et al.*, 'Neurobehavioral evaluation of Venezuelan workers exposed to organic solvent mixtures.' *Am. J. Ind. Med.*, 1995, **27**, 15–27.

85. D. K. Broadwell, D. J. Darcey, H. K. Hudnell *et al.*, 'Work-site clinical and neurobehavioral assessment of solvent-exposed microelectronics workers.' *Am. J. Ind. Med.*, 1995, **27**, 677–698.

86. I. Grant, 'Alcohol and the brain: neuropsychological correlates.' *J. Consult. Clin. Psychol.*, 1987, **55**, 310–324.

87. O. A. Parsons, 'Cognitive functioning in sober social drinkers: a review and critique.' *J. Stud. Alcohol*, 1986, **47**, 101–114.

88. C. Ryan and N. Butters, 'Alcohol consumption and premature aging: a critical review,' ed. M. Galanter, Plenum, New York, 1984, vol. 2, pp. 223–250.

89. H. Hanninen, 'The psychological performance profile in occupational intoxications.' *Neurotoxicol. Teratol.*, 1988, **10**, 485–488.

90. W. Fals-Stewart, J. Schafer, S. Lucente *et al.*, 'Neurobehavioral consequences of prolonged alcohol and substance abuse: a review of findings and treatment implications.' *Clin. Psychol. Rev.*, 1994, **14**, 755–778.

91. N. M. Cherry, F. P. Labreche and J. C. McDonald, 'Organic brain damage and occupational solvent exposure.' *Br. J. Ind. Med.*, 1992, **49**, 776–781.

92. K. I. Bolla, B. S. Schwartz, J. Agnew *et al.*, 'Subclinical neuropsychiatric effects of chronic low-level solvent exposure in US paint manufacturers.' *J. Occup. Med.*, 1990, **32**, 671–677.

93. A. T. Fidler, E. L. Baker and R. E. Letz, 'Neurobehavioral effects of occupational exposure to organic solvents among construction painters.' *Br. J. Ind. Med.*, 1987, **44**, 292–308.

11.23
The Role of Behavior in the Assessment of Toxicity

BERNARD WEISS

University of Rochester School of Medicine, NY, USA

11.23.1 WHERE IS BEHAVIOR A PRIMARY FOCUS?

Behavior is sometimes viewed as merely another index of nervous system function, equivalent in rank or primacy to indices such as neurochemical measures, morphological classifications, or even *in vitro* assays. Such a position demeans the ultimate purpose of research in neurotoxicity, which is to understand and predict its consequences for the whole organism—especially the human organism. Disorders of function, translated into behavior, are the primary form by which these consequences assert themselves. The other forms may guide us to underlying mechanisms, or alert us to the possible emergence of latent damage[1] or warn us about exposure scenarios whose impact we may have discounted in the past. However, they remain ancillary issues when toxicology has a rendezvous with the public.

11.23.1.1 Determining Readiness to Perform

Performance assays based on neuropsychological test procedures are finding increased popularity in several different settings.[2] Typically, one or more tests are administered to determine readiness-to-perform in critical jobs such as those in transportation, where errors of judgment or slowness in responding jeopardize lives or equipment. They were seen originally, in many situations, as replacements for drug testing because they offered freedom from legal implications. Now they are recognized for additional virtues. Incipient flu, for example, may make a truck driver just as incapacitated as marijuana, but would not be detected by an assay of urine.

The same kind of philosophy is appealing for determining the presence of neurotoxicity as a criterion for deciding whether to remove a worker from a particular work position. Exposure to mercury vapor can produce shifts in the frequency spectrum of tremor.[3] Such information was used to relate the magnitude of shift to urine mercury levels in chlor-alkali workers so that standards could be defined for when it might be appropriate to shift them temporarily to another job.[4] The aim was to intervene before overt toxicity occurred. For a similar purpose, Rahill *et al.*[5] used continuous performance testing to gauge the response of subjects to toluene exposure. They sought to determine the potential magnitude of interference with performance stemming from toluene and other neurotoxic solvents off-gassing from structures in space vehicles. A special problem in such studies is the multiple stressors, such as microgravity, confronting space crews. Small performance decrements contributed by an aggregation of sources can cumulate to produce major threats to mission integrity. These are not the kinds of effects, appearing at low exposure levels, that any secondary index of neurotoxicity is capable of discerning.

11.23.1.2 The Identification of Neurotoxicity at Environmentally Relevant Doses

The situations and findings described above challenge the assumptions of much toxicological research because what are called mechanistic experiments almost always occur in the context of high doses. Few such experiments then take the next step: demonstrating that the phenomena they report apply to the lower dose levels that characterize most exposures in the human environment or that even produce behavioral consequences in animals. Without such a connection, there is no cogent basis for accepting such findings as relevant for risk assessment or as explanations for the results of behavioral assays.

Metal neurotoxicity offers a plethora of examples. Despite the recognition, buttressed by a vast literature from both epidemiological and laboratory studies, that lead exposure at low levels can interfere with cognitive function, mechanistic research still tends to lag such recognition. The primary impediment is reliance of high doses in acute *in vitro* experiments. Notable exceptions are beginning to appear, however. Regan[6] described a variety of studies relating low-level lead exposure to interfere with the contribution of neural cell adhesion molecules (NCAMs) to brain developmental processes such as synaptogenesis (see Chapter 7, this volume). Another laboratory is actively engaged in relating environmentally congruous lead exposure, complex behavioral performance, and neurochemical processes.[7] Unfortunately, papers continue to appear that rely jointly on crude behavioral endpoints and high doses.

Insecticides present corresponding issues. Proposals for exposure standards for organophosphorus (OP) compounds tend to be prescribed on the basis of cholinesterase levels in blood; for example, a statistically significant decrease might be assumed as undesirable. But such a standard is not biologically plausible, not only because pseudocholinesterase is what typically is measured, but also because minor peripheral decreases in cholinesterase may not reflect acetylcholinesterase levels in brain. Furthermore, chronic exposure to OPs (at least those that do not cause delayed neurotoxicity) also leads to compensatory nervous system changes in acetylcholine release[8] (see Chapter 26, this volume). Finally, because acetylcholine is present in such abundance, some behavioral changes become detectable only when its level has been substantially lowered. Because of these factors, the criteria of excessive exposure should be based primarily on behavioral indices, supplemented by biochemical measures. One instance in which a strict reliance on peripheral biochemical measures would have been misleading is the observation that farm workers exposed to OP insecticides scored higher on an anxiety scale than controls.[9]

Explorations of behavior have also uncovered a troublesome aftermath of acute poisoning. Savage *et al.*[10] compared a population of farm workers who had suffered an episode of OP poisoning one year earlier with a matched control group. Their findings, based on a collection of neuropsychological tests, are summarized in Figure 1. They indicate lingering deficits following such an episode despite relatively

Chronic Neurological Sequelae of Acute Organophosphate Pesticide Poisoning

	100 Cases	100 Controls	p-value
WAIS Verbal IQ	105.40	111.86	0.001
WAIS Performance	108.41	110.13	0.242
WAIS Full Scale	107.50	111.77	0.001
Impairment Rating	1.07	0.91	0.001
Halstead Index	0.30	0.23	0.020
Pegboard	148.34	137.96	0.002
Card Sorting	17.07	12.91	0.001

From Savage et al, 1988

Figure 1 Chart depicting consequences of an acute episode of organophosphate intoxication after one year.[10] WAIS = Wechsler Adult Intelligence Scale.

rapid clinical recovery. Studies of other populations have confirmed these results. One seminal paper[11] offered a neurophysiological correlate. It reported persistent changes in brain electrical activity in workers and monkeys previously exposed to the nerve gas, Sarin. One group[12] uncovered enduring deficits in vibration sensitivity in farm workers who, many months prior to testing, had been in contact with the OP insecticide methamidophos. No cogent mechanistic explanation for all of these findings has been forthcoming, which is the typical situation; namely, the intrusion of a behavioral result that seems inconsistent with prevailing mechanistic knowledge.

11.23.1.3 Detection of Incipient Neurodegenerative Disease

Degenerative diseases of the nervous system begin long before they crystallize into clinical entities. Their emergence is typically insidious; family members, for example, in retrospect, can recall evidence of memory difficulties in relatives who develop Alzheimer's disease. In President Reagan's case, because presidents are reported on so assiduously, some observers were able to note indications of the disease even during the time he was in office.

Huntington's disease also takes an insidious course. In one South American community where the disease is endemic,[13] families can identify which members will later be struck with the disease. The cues are subtle, and outsiders are not easily schooled in their discrimination. It is apparently no single cue, but an assortment of them, that provides the diagnosis. Moreover, at this early stage, even neuropathological indices are questionable. Behavioral clues, such as impaired cognitive ability and flexibility, are relied on to predict who will be a victim.

Neurobehavioral measures could surely provide early indices of disease, but prospective surveys are rarely pursued. Their potential is evident, but mostly such techniques have been employed by investigators primarily concerned with comparing populations retrospectively. Rifat *et al.*,[14] on the basis of scores from dementia rating scales, contrasted retired miners advised to inhale aluminum powder decades earlier with controls. The powder was believed to confer protection against silicosis. Figure 2 shows, however, that the formerly exposed miners displayed evidence of the kind of impairment associated with Alzheimer's disease. None had been institutionalized; in fact, none had manifested sufficient memory or other failings that would have led family members or observers to comment. Sample populations can surely be found, as in retirement communities, who might volunteer for longitudinal studies of performance to determine how precisely clinical eruptions of neurodegenerative disease can be predicted. Test batteries comprised of memory components, motor components (such as tremor spectra for Parkinson's disease), and other functions might help establish a basic collection of tests that could then be used, not only for prediction, but to ascertain differences in the rate of disease progression. Rate should be a significant variable in evaluating the efficacy of drug therapy, but the appropriate neurobehavioral methods are unfamiliar to many clinical investigators.

11.23.1.4 Longitudinal Monitoring

Functional monitoring is, in effect, the topic discussed above. But monitoring finds much wider uses. One arena in which periodic behavioral assays has achieved notable success is in developmental neurotoxicity. Consider the lead literature. The full impact of elevated environmental lead was not comprehended until investigators began tracing the outcome of prenatal exposure. Only then was the biomedical community convinced that low lead exposure levels posed a health threat. In one landmark study,[15] over 250 children were divided into three groups on the basis of lead concentrations in umbilical cord blood: low (mean of $1.8 \, \mu g \, dL^{-1}$), medium (mean of $6.5 \, \mu g \, dL^{-1}$), and high (mean of $14.5 \, \mu g \, dL^{-1}$). Scores on an inventory of infant development disclosed a widening gap between the low and high groups from 6 to 24 months of age. This and other studies have also shown another crucial feature of developmental sensitivity to lead: changes in the profile of effects and

correlations as the children grow older. It is an outcome to be expected, of course, but only longitudinal investigations could have revealed it so sharply.

Longitudinal behavioral assessments could also be coordinated with readiness-to-perform assays. In many work situations, the principal concern might be chronic exposure rather than fitness on any particular day, or, both kinds of indices might be sought. In those settings, the same assays, administered weekly or monthly, could provide the requisite information.

11.23.1.5 Validating Clinical Observations

Neurotoxic hazards had been detected and described by astute clinicians many centuries before toxicology emerged as a unique discipline. When dose–response relationships, precise descriptions, and quantitative insights are needed, however, the tools of experimental science need to be invoked. One example confronting clinical medicine is a new set of entities that might be labeled as chemical sensitivity syndromes. I include multiple chemical sensitivity (MCS), sick building syndrome (SBS), chronic fatigue syndrome (CFS), and the Gulf War syndrome (GWS). What these all have in common is a pattern of complaints, like those outlined for MCS in Figure 3, for which no unequivocal organic measures are discernible. CFS comes closest to presenting a distinctive etiology because some of the evidence links it to viral infections, but even assiduous searches for immune system disorders have failed to uncover an unambiguous connection.

**Effect of Exposure of Miners to Aluminum Powder
(Rifat et al, 1990)**

	Mean Adjusted Score*	Proportion Impaired
Exposed (n=261)	78.6000	0.130
Unexposed (n=346)	82.8000	0.050
Difference	6.3000	0.080
p-value	0.0001	0.002

*Mini-Mental State, Raven Matrices, Symbol-Digit

Figure 2 Differences between control miners and miners treated with aluminum powder to prevent silicosis. All subjects had been retired for many years before testing.[14]

Determining the validity of such syndromes seems to have eluded the usual medical techniques. What are left, then, are neurobehavioral probes. How to use these effectively is the enigma. Challenge experiments seem superficially to be the answer, but their design presents issues that seem almost unresolvable. If adverse effects are induced by exposure levels much lower than those prescribed by workplace standards such as threshold limit values (TLVs), can ethically acceptable protocols be designed? The answer lies in adopting appropriate experimental approaches; for example, intermittent challenges in which the outcome of each exposure episode is traced through time, measured by performance, and analyzed by time series techniques.

An equivalent dilemma confronted a group of us when we sought to test the hypothesis that some proportion of hyperactive (ADD) children might actually be responding to certain dietary constituents such as synthetic food dyes.[16] An experiment was designed in which children were repeatedly challenged with a blend of food colors. The selected children's parents had noted improvements in behavior when a diet free of synthetic colors, flavors, and other constituents, claimed by Feingold to trigger reactions, was adopted in the home.

During an 11 week period, the 22 children in the study consumed a soft drink, containing a blend of seven food dyes, on eight occasions. On all other days, they consumed a control drink of equivalent appearance and taste. Outcome measures consisted of parental ratings and systematic recordings of the childrens' behavior. In this double-blind clinical trial, two of the 22 children, neither of whom were clinically hyperactive, showed clear evidence of responsiveness to the dye blend.[17] Figure 4 shows the response of one of these children. Data such as these, confirmed by other studies,[18,19] indicate that challenge studies are feasible when the experimental question is whether substances at prevailing levels of exposure may be toxic. This is another instance in which behavioral measures revealed adverse effects that would have remained obscured had conventional guides to toxicity prevailed.

11.23.2 FUNCTIONS OF BEHAVIORAL ASSESSMENT IN TOXICOLOGY

Behavioral assessments are undertake of neurotoxicants for two purposes. Prediction is the aim of screening procedures in animal studies and aimed at the stage of risk assessment termed hazard identification. But why behavior, with its attendant complexities, at such an early stage? Because we lack a substitute. Chemical structure and properties remain an ambiguous guide to neurotoxicity. *In vitro* assays, unless we already know what to

Key Features of the

Multiple Chemical Sensitivity Syndrome

- **Predominant symptoms refer to central nervous system**
  - **difficulty concentrating**
  - **poor memory**
  - **clumsiness**
  - **paresthesias**
  - **affective disturbances**
- **Prevalence higher in women than in men**
- **Peak age group is 30-50 years old**
- **Symptoms elicited by many different chemical agents**
  - **environmental**
  - **drugs**
  - **foods**
- **Hyperreactivity to sound, light, touch**

Figure 3 Symptoms characterizing patients with multiple chemical sensitivity.

Subject 73: Response to Color Challenge

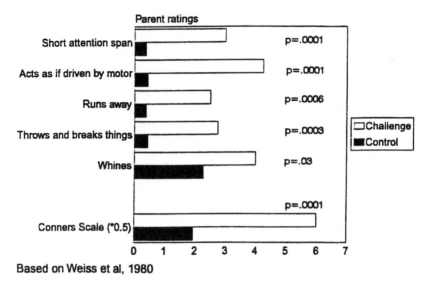

Based on Weiss et al, 1980

Figure 4 Responses to a blend of synthetic food dyes in a 34 month old girl maintained between challenges on an elimination diet prescribing food additives (after Weiss *et al.*[17]).

look for, are not well enough established at present to permit a rational selection among all the possible models. Morphology is a solid criterion, but requires a daunting investment of time. Simple behavioral assays, such as a functional observation battery (FOB), are preferred by many researchers as a fairly reliable, if crude, source of information about potential nervous system activity (see Chapter 13, this volume). In fact, the kind of responses that FOBs include actually dominate the observations that pharmacologists and toxicologists make when they first expose an animal to a new compound.

Verification and evaluation is the second purpose. It might be undertaken, for example, when suggestions arise that a workplace chemical seems to be producing unusual complaints or responses among exposed workers, or when the adequacy of current exposure standards is questioned. The discovery of chlordecone (Kepone) poisoning in a Virginia chemical plant[20] triggered numerous experiments that attempted to find responsible mechanisms and to enlarge the pool of available data about the compound.

Verification of workplace standards, more than any other factor, drove the growth of research on organic solvent neurotoxicity. Because they were known to induce reactions such as narcosis, solvents quickly became the object of neuropsychological research once investigators became aware, from publications by Hanninen *et al.*[21] that chronic workplace

exposure, in the absence of clinical signs, might evoke subtle adverse effects. Hanninen's early work was based on carbon disulfide, indisputably a potent neurotoxicant, and provided a convincing demonstration of the power of neuropsychological testing to reveal impaired function that ordinarily would have been unrecognized.

Such demonstrations encouraged the use of psychological test procedures to determine more adequately the ability of prevailing workplace exposure standards to protect against major or irreversible effects. Before the introduction of such quantitative criteria, the standards for many agents had been based largely on impressionistic observations.

11.23.3   BEHAVIORAL ASPECTS OF RISK ASSESSMENT

The widely-cited National Academy of Sciences report[22] divided the risk assessment process into four stages that by now have become virtual clichés: hazard identification, dose–response assessment or modeling, exposure assessment, and risk characterization. Its basic premises, however, derive from cancer, and, in part, from the regulatory status it enjoys. Labeling a chemical as potentially carcinogenic activates a sequence of procedures and decisions that are not applied to agents not deemed carcinogenic. The Environmental Protection Agency (EPA), for example, proceeds down two scientifically distinct paths for these two categories. Carcinogen risks are formulated as statistical models contrived to protect public health to a level of 10^{-5} to 10^{-6}. Systemic toxicants are currently regulated on the basis of no-effect levels buffered by safety or uncertainty factors. The hazard identification stage effectively divides the science of toxicology into two streams.

11.23.3.1 Hazard Identification and Confirmation

Hazard identification is a simpler process for neurotoxicants, but it possesses far less significance in that context. Cancer is a single outcome. Neurotoxicity can take many forms, as shown in Figure 5. Unlike the assumptions made about the induction of cancer, for which only total, cumulative exposure is modeled, we have to acknowledge that different patterns of exposure can elicit different patterns of toxicity, and, moreover, that reversibility, especially at low levels, is common. Again in contrast to cancer, no single outcome, but a plethora of outcomes, is possible. As noted in Figure 5, behavioral outcomes might include disturbances of learning (of which many kinds exist), interference with constitutionally determined reactions, or diminution of the capacity to perform complex tasks.

Given such an array of possibilities, elementary hazard identification fuses with another stage I called hazard confirmation. This is the stage at which we probe the fundamental properties of the agent and explore its toxic potential for a multiplicity of endpoints.

Hazard Confirmation, Characterization

- **Pattern of toxicity**
  - ▶ **Acute, immediate**
  - ▶ **Chronic, cumulative**
  - ▶ **Delayed**
- **Reversibility**
- **Significant behavioral endpoints**
  - ▶ **Acquisition (operant and classical)**
  - ▶ **Reflexive, naturalistic**
  - ▶ **Complex performance, discriminations**

Figure 5 Additional information required for neurobehavioral toxicity evaluation to extend the hazard identification phase of conventional risk assessment. Such information is not included in cancer risk assessment protocols.

11.23.3.2 Dose–Consequence Modeling

The diversity of possible outcomes turns the dose–response phase of neurobehavioral risk assessment into a rather different creature than the equivalent phase for a presumed carcinogen. A single statistical model is entirely insufficient to describe many different kinds of effects. The modeling, even in as primitive a form as no-observed-adverse-effect-levels (NOAELs), will take different forms with different endpoints. In addition, it should also take account of correlations among measures and individual differences. Such considerations do not arise with cancer, but are essential for the comprehension of all the factors contributing to risk. Figure 6 also shows two other ingredients in the dose–response equation because dose also determines their values. Rate of progression is important because high rates may provide no opportunity for remodeling or countervailing process to occur; conversely, slow rates of functional impairment may obscure the development of adverse effects until they have attained an irreversible status. Similarly, short and prolonged latencies to detectable impairment may, in different ways, either promote or counteract irreversibility.

11.23.3.3 Exposure Assessment

Exposure assessment is also more complicated with neurotoxicants than with carcinogens. For ease of modeling, the pattern of exposure is ignored in cancer risk assessment; only total dose enters the model. High doses administered over a short period of time are considered equivalent to low doses administered over a prolonged period provided the cumulative totals are equal.

For agents such as volatile organic solvents, brief exposures at high levels act as anesthetics. At least by crude clinical criteria, those who have suffered narcosis in such situations seem to recover without permanent damage. However, studies such as those conducted in workers who experienced an episode of insecticide poisoning have not been carried out for solvents.

Dose-Consequence Relationships, Modeling

- **Spectrum and sequence of effects**
  - ▶ **Correlations among endpoints**
  - ▶ **Individual differences**
- **Rate of progression**
- **Latency to specified effects**

Figure 6 Variables to be included in dose modeling for neurobehavioral toxicity.

Studies of the consequences of repeated episodes of high, but not narcotic, doses have not been conducted. Some of the effects described by the organic solvent syndrome, which is attributed to chronic exposure, may have arisen from repeated, acute elevations in exposure levels in work settings not as well regulated as contemporary workplaces (see Chapter 21, this volume). Workers evaluated in the 1970s may have begun employment in the 1940s or even earlier. Truly chronic, or stable workplace levels, for many situations, may be a rather recent development.

Figure 7 lists five situations, based jointly on exposure pattern and reversibility, and associated examples. As noted above, solvents can produce anesthesia at high levels, but the effects are clinically reversible. 1-Methyl-4-phenyl-1,2,3,6-tetrahydropyridine (MPTP), the notorious contaminant of designer drugs that triggered a small epidemic of Parkinson's disease among drug abusers in California, is a prime example of an agent that, after a single exposure, leads to irreversible consequences. A high dose is necessary in this instance because lower doses, administered subchronically to monkeys, fail to evoke such a catastrophic response. Acrylamide is an intriguing example because, with the same dosage schedule, it can produce reversible blunting of vibration sensitivity but irreversible damage to certain cells in the retina. The explanation is probably the difference between the sensitivity and recoverability of axons in the peripheral nervous system and nerve cells in the central nervous system. Prenatal exposures fall into another classification, because, as with methylmercury, the effects may not become detectable until long after birth.

11.23.3.4 Risk Description, Outcome, Characterization

Evaluating the health hazards of neurotoxicant exposure poses many more problems of interpretation than those posed by a relatively uniform entity such as cancer. The sources of interpretive ambiguity are listed in Figure 8. Changes in the prevalence or incidence of a particular manifestation are the most common criteria in epidemiological investigations, but offer some different challenges with neurotoxicants. In the chlordecone episode mentioned earlier, the first indication of toxicity consisted of nervousness. It is hardly a marker that would be considered seriously in a conventional epidemiological study, or even connected to chemical exposures by an occupational physician.

Shifts in population distributions of some indices are also an uncommon measure in conventional studies, but proved especially fruitful in estimating the adverse impact of lead on neurobehavioral development. Figure 9 diagrams the wide impact of even a quite modest shift in the distribution of IQ scores. The distribution at the top depicts a distribution with a mean of 100 and a standard deviation of 15. Because IQ on tests such as the Standford–Binet and Wechsler Intelligence Scale for Children is calculated as the quotient of test age divided by chronological age, the average child will score 100. With a standard deviation of 15, seen on these tests, 2.3 million individuals in a population of 100 million will score above 130. A shift of five IQ points, or 5%, yields a mean IQ of 95 and reduces the number of individuals scoring above 130 to 990 thousand. It also inflates the number of individuals scoring below 70 and requiring special classrooms and remedial interventions. This comparison illustrates the principle that small shifts in the mean of a distribution are greatly amplified at the tails. In the case of lead, it demonstrates that what superficially might seem an almost trivial neurotoxic effect can exert a profound societal influence. Because IQ is a significant determinant of earning power, even effects of this magnitude are multiplied greatly in a working lifetime.

Impaired development is not confined to IQ scores. It could also be expressed in the form of

EXPOSURE/RISK CLASSIFICATIONS

Category	Example
• Acute Reversible	• Solvents, Anesthesia
• Acute Irreversible	• MPTP, Parkinsonism
• Chronic Reversible	• Acrylamide, Vibration
• Chronic Irreversible	• Acrylamide, Vision
• Latent Irreversible	• Prenatal Methylmercury

Figure 7 Types of exposure patterns linked to neurobehavioral consequences.

Risk Description, Outcome, Characterization

- • Elevated incidence, prevalence
- • Shift in population distribution
- • Impaired development
- • Accelerated aging
- • Reduced compensatory capacity
  - ▶ Silent toxicity
- • Rate of recovery from reversible effect

Figure 8 Dimensions of risk relevant to neurobehavioral toxicity.

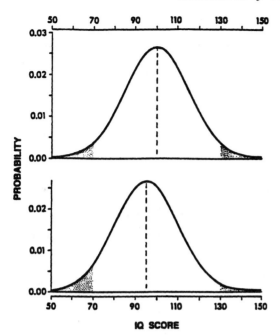

Figure 9 Two distributions of IQ scores. (Top) The standardized distribution with a mean of 100 and standard deviation of 15. In a population of 100 million, 2.3 million individuals will score above 130, and an equal number below 70. (Bottom) IQ distribution after a 5% shift to a mean of 95. In this distribution, the number of individuals scoring above 130 falls to 990 thousand.

conduct disorders. In a seminal paper comparing young primary school students with high and low dentine lead concentrations in deciduous teeth, Needleman et al.[23] obtained ratings from teachers that included items surveying behaviors such as inattention, distractibility, hyperactivity, and others. They obtained a compelling dose–response relationship between the ratings and tooth lead values as well as significant IQ differences between the highest and lowest lead groups. Motor function is another aspect of development responsive to early toxic exposure. Cox et al.,[24] for example, noted a relationship between prenatal methylmercury exposure, as determined by maternal hair levels, and delayed walking in the offspring.

Except for variants that can be traced to genetic etiologies, most neurodegenerative disease offers no clear cause. For Parkinson's disease, because of the MPTP episode, the evidence for an environmental trigger now seems much more cogent than before, but specific agents have so far eluded the search. Neurodegenerative disease is a disease of old age, and likely to emerge when intrinsic compensatory processes, because of reduced reserves, can no longer cope with natural erosion. Any agent that hastens natural neuronal loss will have the effect of speeding up the loss of reserve capacity and reduce the latency to clinical eruption. Weiss and Simon[25] calculated that an acceleration of only 0.1% annually in neuronal loss after the age of 25 would reduce the reserves of a 55 year old brain to that of a 70 year old brain. Such an effect could not be seen in the individual, but it certainly would affect a population.

The final entry is derived, in part, from data indicating that the aging nervous system is much slower to recover from damage than young systems. For example, young rats recovered from an acute challenge with an anticholinesterase compound much more quickly than rats over 2 years of age.[26] Old mice are also much more sensitive to the neurochemical actions of MPTP than young mice;[27] young mice recover from acute dosing and old mice suffer irreversible damage. Any process or event that reduces the compensatory reserve of the brain will impair its ability to recover from additional challenges. Such reductions might be due to a continuing process, as in the model described above,[25] or the legacy of an injury suffered in the past, as in early development.

11.23.4 LIMITATIONS OF CURRENT APPROACHES

Risk assessment is an exercise designed for prediction. Advances in the way we undertake neurotoxic risk assessment have made it more secure and acceptable to other toxicologists and to regulators, but its predictive power is still limited. What kind of information should a neurotoxic risk assessment provide? Most observers would like to see data bearing on the ability to engage in complex human activities and performance such as learning and discriminative processes. Most of the information derived from the risk assessment process is irrelevant for this purpose and mostly suitable for simple hazard identification.

11.23.4.1 Animal Testing

Figure 10 lists the factors that confine the ability of this process to extrapolate to the situations and endpoints of greatest concern in public health. One sturdy impediment is the somewhat crude methods still favored by screening tests. Potential neurobehavioral toxicity is most commonly assayed by a functional observation battery. It typically consists of a series of standardized manipulations and scoring guidelines applied to a rodent. Although it

Limitations of Current Approaches

- **Animal Test Procedures**
  - **Reliance on simple measures**
  - **Neglect of microanalyses**
- **Human Test Procedures**
  - **Diagnostic tests rather than performance**
  - **Lack of indices of judgment, complex decision**
  - **Lack of advanced motor and sensory testing**
- **Experimental Design**
  - **Lack of longitudinal studies; aging**
  - **Emphasis on group statistics**

Figure 10 Limitations imposed on neurobehavioral risk assessment by current approaches.

is an efficient way of obtaining clues about nervous system activity, it is hardly the secure footing required to gauge the possible effects in humans exposed to low doses of a test substance.

A sensible response would be to adopt criteria closer to the endpoints about which humans are concerned; that is, endpoints that represent the complexities of human behavior. Schedule-controlled operant behavior is a popular choice because it is capable of almost infinite variation in parameter choice at the same time that it relies on fairly standard experimental situations. It was the choice of the EPA when it sought additional information about the neurotoxic potential of 10 high-volume solvents.[28] EPA explained its choice as a decision stemming from the need to obtain data on learning, memory, and other key components of human function.

Such a requirement makes even more sense, however, when it is undertaken early rather than late in the assessment process. With the strategy called tiered testing, the process typically moves from crude, simple assays to more complex, sensitive ones. If crude endpoints are used in the first stage, however, it is almost inevitable that more sensitive endpoints will have to be introduced later with any agent for which human exposure decisions have to be made. Even a superficial cost–benefit analysis will confirm that it makes sense to begin with the more sensitive endpoints. Schedule-controlled operant behavior is considered such an endpoint. Most investigations of schedule-controlled operant behavior, however, are themselves too simplistic. They adopt schedules, parameters, and modes of data analysis that, for convenience, lend themselves to cursory examination. Often, however, the most informative details are found in the deep and searching analyses I call microanalysis that most experimenters are reluctant to pursue. It embodies the application of digital computer technology to disclose the underlying features of behavior,[29] and it has proven useful in several key situations. Cohn et al.,[30] for example,

unraveled the details of response sequences emitted by rats that could choose any of three available levers at any time; only one specified four-component sequence was deemed correct and reinforced with a food pellet. The time and position of each lever press was recorded by the computer that controlled the experiment. Post-weaning lead treatment produced patterns that could be described as perseveration and that would have gone unrecognized in conventional analyses that ignore the fine details of behavior.

11.23.4.2 Human Testing

In the area of human neuropsychological testing, as well, investigators seem reluctant to apply detailed, rigorous analyses. One source of their reluctance is understandable; standardized tests are attractive because they might facilitate comparisons across studies, but if they were to be analyzed by unconventional procedures the advantage of standardization would be lost. Another source of reluctance is the common use of batteries, or collections, of tests combined to assess a broad spectrum of functions such as verbal performance, motor function, memory, and other categories of behavior. Detailed analyses of each component would produce enormous amounts of data. The adoption of computer-based test procedures overcomes some of these problems. Some Swedish investigators[31] argue that, despite their shorter history and lack of standardization in clinic populations, computer-based tests offer so many advantages that they eventually will, so to speak, drive the others from the marketplace.

One inherent flaw of many of the venerable standardized tests is their history. They were designed primarily as instruments to be used by clinicians for patient diagnosis, so they emphasize individual scores and persistent traits. In contrast, neurobehavioral toxicology needs instruments designed for comparing the performance of groups subjected to varying exposure conditions. Performance tests, then, examine function from a different vantage point, which is why they do not require the same kind of standardization as diagnostic tests, but place more emphasis on reliability. One example of a diagnostic test might require a patient to try to reproduce a series of line drawings, which are then scored for accuracy and for selected features that seem to arise in certain patient groups. One example of a performance test would be discriminative reaction time, in which the subject is required to select which response is appropriate to a specific stimulus and to emit it as quickly as possible.

Current test batteries, whatever their history, still lack certain features that fail to provide a complete assessment of functional capacity. The ability to make decisions based upon an appraisal of many complex variables, or to judge the suitability of a certain course of action with many ramifications, is considered to be the mark of high-level function. Those kinds of challenges, exemplified by responses to emergencies in the control rooms of nuclear power plants, and calling upon advanced integrative functions, are missing in the current array of batteries. Perhaps they would be difficult to score, and would require more time to learn and complete than most batteries allot to their constituent tasks. They are also the object of many investigations in experimental psychology, and provide the attraction of computer games such as those in which the player takes on the task of air traffic control at LAX. Reasonable surrogates should be practicable for investigations in behavioral toxicology. The series of pioneering studies conducted at the Air Force School of Aviation Medicine by Payne and Hauty in the 1950s demonstrates the power of a single demanding task to reveal the impact of low doses of drugs on performance.[32]

Current batteries are also weak in their ability to determine subtle functional deficits in motor performance and sensory acuity. Motor ability is usually tested by simple tests such as the Purdue Pegboard, a test originally devised for screening assembly line workers. It is hardly comparable to the demands of settings in which workers must make fine adjustments to expensive tools. Nor do such tests gauge aspects of motor function such as tremor, an important facet of some neurodegenerative diseases and neurotoxic exposure. The technology for doing so, although far from inconsequential, is also relatively direct and requires no exotic equipment or novel computer programming.[33]

Sensory testing is an even more neglected area, although a few investigators are attempting to redress this situation. The consequences of chronic exposure to volatile organic solvents have overwhelmingly been examined by cognitive tests with a few simple motor tasks included. Sensory function is typically a secondary concern. When sensory assessment has been included, it usually has been in the form of simple, almost screening tests, rather than in the form of advanced measures that can point to mechanisms. Studies are beginning to show awareness of these deficits. Some have examined the consequences for color vision of chronic exposure to organic solvents,[34] while others have also sought to explore how odor detection and identification are modified by such exposures.[35]

11.23.4.3 Experimental Design

Neurotoxicity is rarely a problem of acute exposure and onset. In most contexts it appears as an emergent phenomenon. Contrast this feature with the literature, which emphasizes high doses and short-term observations. This discrepancy is often defended with the explanation that such an approach yields mechanistic information. It is not a cogent defense and is especially ingenuous in the framework of developmental neurotoxicity. Although the primary public health issues posed by gestational and neonatal neurotoxic exposures arise from their consequences later in life, an unfortunate proportion of animal experiments confine their observations to the period of early development. Immature organisms, however, can be assayed only for relatively gross, immature responses such as reflex development. The implications of developmental toxicity can only be pursued by exploring the kind of functional capacities exemplified, for example, by complex schedule-controlled operant behavior. In adopting such endpoints, however, it is equally important to adopt different standards of analysis and in addition to the kind of microanalysis discussed earlier, it is also essential to examine individual subjects. Distributions of performance measures often reveal differences between treatments that fail to emerge when analyzed by conventional group statistics. Even in allegedly homogeneous groups of animals, wide individual differences typically appear. Clues to treatment effects often appear in what seems to be a subpopulation of outliers. Without an examination of individual response tendencies, the food additive experiment alluded to previously[17] would have revealed little.

Furthermore, as with cancer, the full manifestations of toxic exposure may dawn only as the organism ages. If neurodegenerative disease, as some suspect, originates in some event or process triggered years earlier, even during early development, the typical experimental study is unlikely to provide much predictive power. It will have ended before the impact of whatever neurotoxic process commenced earlier has become detectable. No one disputes the difficulty of longitudinal studies, but no one has yet offered a convincing substitute.

11.23.5 PREDICTION PROTOCOL PHASES

The preceding material is not just a series of admonitions, but is intended to provide a rationale for a more complete approach to neurobehavioral risk assessment than the

Prediction Protocol

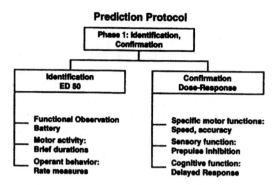

Figure 11 Model protocol for neurobehavioral risk assessment in the hazard identification and hazard confirmation phases.

current mode. Phase 1 of this protocol is outlined in Figure 11. It aims at the two stages described earlier as identification and confirmation. The identification process, unlike the conventional scheme designed for cancer, strives for dose information on the basis of fairly direct behavioral assessments. A functional observation battery and assay of motor activity are included in most test batteries. Simple schedule-controlled operant behavior is usually considered a second-tier assay, but some simple schedules, such as fixed ratio, are easy enough to train that they can easily be incorporated into the early phases of assessment.

From the three endpoints listed, the median effective dose (ED_{50}) should be ascertainable. It can then be used to select doses for the confirmation phase. Here, as for motor function, more specific information is obtained. Instead of simple, spontaneous motor function, speed and accuracy of reaching, for example, can be determined. Rats are easily trained to reach their paws into a food cup to retrieve small pellets of food. Another aspect of motor function is assayed by examining willingness to engage in exercise. Ozone is a deep lung irritant. Rats allowed to run in a running wheel, a highly preferred activity, reduce their running in the presence of ozone at concentrations as low as 0.12 ppm, the value prescribed as an environmental limit by the EPA.[36]

Sensory function also receives more than the cursory examination typical of the functional observation battery. Instead of scoring the response to a finger snap, say, the auditory startle response is used to determine auditory acuity. Prepulse inhibition refers to the ability of less intense sounds preceding the startle stimulus to reduce startle amplitude. Sensitivity to these prepulse stimuli yields measures of auditory function resembling an audiogram. Cognitive assessment expands from simple

operant schedules to assays of learning and memory. The delayed-response model is a venerable technique in psychological science. A subject is first presented with a stimulus (or generates one), then, after a delay whose length is prescribed by the experimenter, the subject is presented with a choice. The longer the delay, the lower the proportion of correct choices. In one study,[37] monkeys were trained to alternate presses between two levers, but a superimposed delay of variable length separated the responses. Monkeys exposed during early development to lead performed less successfully than controls, and tended to persevere on one lever instead of alternating.

For many purposes, phase 1 will be sufficient to provide data from which to calculate benchmark doses and to provide wide margins between anticipated human exposure levels and those producing effects in laboratory animals. Phase 2 (Figure 12) would be undertaken when additional questions about neurotoxic potential arise during phase 1. Advanced testing would explore the endpoints noted under specific functions. Instead of the relatively simple, straightforward prepulse inhibition paradigm of phase 1, phase 2 advances into mature psychophysics such as that required for assessments of vibration sensitivity.[38] Complex motor performance tasks, such as those used to study the chronic effects of neuroleptic drugs,[39] replace the simpler demands and less analytical procedures of phase 1, and venturing beyond even the repeated acquisition model of phase 1, an additional complexity is imposed by studies of choice. Here, a subject is faced with two levers that deliver reinforcement with two different probabilities. The paradigm is called the "matching law"[40] because subjects tend to distribute responses in accord with the density of reinforcement. As used in phase 2, the probabilities associated with each lever shift

Prediction Protocol

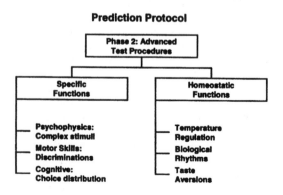

Figure 12 Model protocol for extending and amplifying neurobehavioral endpoints in the risk assessment process.

from time to time, and the subject's relative choices between them provide an index of how responsive it is to these shifts. It is an especially appealing situation because it mimics the natural behavior of foraging.

Phase 2 also includes three homeostatic functions that tend to be overlooked in discussions of neurotoxic risk assessment. Behavioral thermoregulation has a long history in psychophysiology and behavioral pharmacology. A typical situation is one in which a rat presses a lever in the cold to activate a heat lamp.[41] This situation is sensitive to drugs,[42] thyroid function,[43] nutritional variables,[44,45] microwave exposure,[46] and toxic metals.[47]

Many biological functions are governed by pacemakers whose influence can be described in terms of periodicities. Diurnal rhythms are the most visible and their disruption leads to numerous clinical complaints. Television advertisements for medications claiming to counteract insomnia are common. Longer variations appear in the form of the menstrual cycle. In rodents, diurnal cycles appear in the form of motor activity, and, because they are nocturnal animals, activity is vastly greater during the dark than during the light periods. A small body of studies[48] indicates neurotoxic disruption of circadian activity and ingestive behaviors. The rodent estrous cycle is a useful endpoint for questions pertaining to neuroendocrine status. Females greatly increase spontaneous motor activity during estrous, so that disruptions of estrogenic mechanisms are clearly revealed by examining activity periodicities. Because environmental endocrine disrupters such as 2,3,7,8-tetrachlorodibenzo-*p*-dioxin (TCDD) represent a lively research area in toxicology, and because such properties are shared with agents such as organochlorine pesticides, these indices may come to occupy a larger role in future risk assessments. Sexual function is not yet a common component of neurotoxic risk assessment, but the influence of endocrine disruptors on the current scene in environmental toxicology may change that lack of emphasis. One reason for such recognition stems from the finding that impaired male rat copulatory behavior and genital structure is perhaps the most sensitive outcome of developmental exposure to TCDD.[49]

One of the objections lodged against behavior as an index of neurotoxicity is lack of specificity. Especially in animals, changes in behavior, it is claimed, may arise from a generalized malaise arising from effects on a system or organ only indirectly linked to nervous system function. For this reason, independent assessment of such illness may at times prove useful. Conditioned taste aversion is one

technique for doing so. The prototype is a situation in which rats are allowed to consume a saccharin solution, which is highly preferred. Several hours later, the rats are exposed to a substance that brings on some adverse effect. After recovery from these effects, the rats' preference for saccharin is greatly reduced, an outcome described by conditioned or learned taste aversion. If a substance is suspected, at some dose level, of altering a behavioral endpoint such as a stimulus discrimination by producing a sick animal, the hypothesis can be tested with the taste aversion technique.

11.23.6 BEHAVIOR AS A COORDINATE FOCUS

Behavior is rarely considered a useful endpoint for assaying the function of systems other than the central nervous system. It is an unfortunate oversight, because behavioral endpoints can often illuminate mechanisms or provide indications of early toxic processes that would otherwise be missed (Figure 13). One example is hepatic encephalopathy. In this condition, neurons typically shows no consistent evidence of pathology during the early stages, even at the level of ultrastructure. At the same time, psychiatric manifestations and impaired intellectual function may be apparent. Research based on animal models of this disease should profit by tracing its course by parallel biochemical, morphological, and behavioral measures.

Neurodegenerative disease is another area in which behavioral measures would prove useful. Especially now, when the genetic loci of Huntington's disease and the familial form of Alzheimer's disease have been identified, neuropsychological test batteries can play an important role in tracing the course of the disease

Behavior as a Coordinate Focus?

- **Incipient Neurodegenerative Disease**
  - ▶ **Huntington's disease**
- **Diseases of Other Organ Systems**
  - ▶ **Hepatic encephalopathy**
- **Non-CNS Drugs**
- **Food Additives**
- **Air Pollutants**
  - ▶ **Ozone and performance**
- **Immune System Function**
  - ▶ **Sickness behavior**
  - ▶ **Conditioned responses**

Figure 13 Situations in which the application of behavioral techniques is appropriate or necessary for a full assessment.

Drug Classes Inducing Behavioral Changes In Offspring

- Analgesic
- Antiallergen
- Antibiotic
- Cardiovascular
- Anticoagulant
- Anticonvulsant
- Vitamin analogs
- Cytostatic
- Gastrointestinal
- Hyperlipemic
- Bronchiolytic
- Spasmolytic

(Ulbrich and Palmer, 1995)

Figure 14 Drug classes other than those designed for direct action on the nervous system that produce adverse effects on neurobehavioral development.[50]

because, as noted before, the earliest indications consist of behavioral disturbances.

Behavioral measures are important for determining the safety of drugs other than those designed to alter central nervous system function. Figure 14 lists drug classes known to induce behavioral aberrations in the offspring of rodents treated during gestation. It explains why several countries require behavioral teratology assessments for all new agents.[50] But, even for adults, behavioral assays would prove useful. Many agents targeted for other purposes influence nervous system function. Their behavioral actions are termed side effects, but, from the standpoint of safety, might be the dominant effect. Certain antiallergens are prescribed with a warning that they induce somnolence (a property also used to market them as mild hypnotics) and could interfere with the ability to drive. A number of antihypertensive drugs also possess this property. Other drugs induce dizziness. These secondary effects usually emerge during clinical trials. Because their effects on complex performance are rarely quantified, it is up to the patient to judge his or her capacity to carry out demanding tasks.

About 3000 agents are permitted as food additives. None have undergone premarket testing for behavioral toxicity, in fact, relatively few have undergone any toxicity testing at all. Yet, as described earlier, the evidence is overwhelming that at least food dyes are capable of inducing behavioral disturbances in children. Adults are not exempt from such effects. Case reports appear with some consistency in the medical literature about untoward responses, expressed in behavior, in adults given medications containing color additives or other presumably inert ingredients.[51]

The ozone data referred to previously is only one demonstration of how behavioral measures might illuminate the effects of air pollutants on health. Ozone induces adverse

responses not just in lung function, but also leads to complaints implicating the central nervous system. Headaches, dizziness, and confusion are among these. Such subjective complaints are difficult to quantify, but are nonetheless significant. Animal data illuminate some of the more subtle responses. In one experiment, rats were trained to press a lever to release a brake on a running wheel.[52] Every 20 presses released the brake for 15 s. Exposure to 0.08 ppm ozone, a value regularly exceeded in many parts of the United States, reduced both lever pressing and running. These results were interpreted as evidence that the rats reduced their motivation to exercise so as to avoid the unpleasant effects of breathing ozone. A similar interpretation was offered to explain the tendency of cross-country runners in Los Angeles to show performance lags on high oxidant days.[53]

The science of psychoneuroimmunology has developed around the theme that the immune and central nervous systems are intimately linked. The evidence is overwhelming that each influences the other and that they respond in tandem to situations such as psychological stress.[54] Behavioral techniques serve two functions in this new discipline. First, they are able to quantify variables such as the perceived stress provoked during situations such as final examinations. Second, they offer a way to modulate immune system function. Demonstrations of conditioned immunomodulation in animals effectively launched the science[55] and the phenomenon has also been shown in humans.[56] Sickness behavior, evoked in humans by cytokines, is also apparent and measurable in animals as behavioral products. This coupling of the two systems also means that immunotoxicology and neurobehavioral toxicology are inextricably joined even though the connection is recognized only vaguely at present.

Endocrine function is also closely coupled to behavior. Sexual development was alluded to earlier. Prenatal influences on thyroid function influence postnatal behavioral function; hypothyroidism, which can be induced by gestational exposure to certain of the PCBs (polychlorinated biphenyls), leads to impaired brain development. Excessive stimulation of animals during gestation and infancy induce both measurable neuroendocrine changes and alterations in behavioral responses later in life. We can be certain that toxic exposures during early development that modify neuroendocrine function will be expressed later in life by behavioral measures. The wide-ranging review by Ader *et al.*[54] is a blunt reminder that too often, in toxicology, we tend to be captured by narrow perspectives.

11.23.7 SUMMARY: TOXICOLOGY FROM THE STANDPOINT OF BEHAVIOR

This chapter has aimed to examine how behavioral science and toxicology have activated and enriched one another. For toxicology, the influence has been clearer: the incorporation of a new set of toxic criteria and a set of measures that have impelled the science into new areas. Enlarging the purview of toxicology is an unalloyed benefit for the science. The influence of toxicology on behavioral science provides a more abstruse benefit. As with toxicology, it has meant a dilation of boundaries. Unlike the contribution of behavioral science to toxicology, it has not meant new criteria, but new challenges. How equipped is behavioral science to defend claims that chronic workplace exposure to organic solvents, lead, or insecticides has led to enduring neuropsychological deficits? What kinds of behavioral variables in animals parallel measures, such as intelligence test scores, that have provided grounds for asserting developmental neurotoxicity? How are changes in rat motor activity or human reaction time measures related to how people actually function in their accustomed settings? With progress toward the answers to such questions, toxicology itself will benefit.

ACKNOWLEDGMENTS

Preparation supported in part by Grants ES-01247 and ES-05433 from NIEHS, and Grant DA-07737 from NIDA.

11.23.8 REFERENCES

1. B. Weiss and K. R. Reuhl, in 'Handbook of Neurotoxicology,' ed. L. Chang, Dekker, New York, 1994, pp. 765–784.
2. A. Wetherell, 'Performance tests.' *Environ Health Perspect.*, 1996, **104**, Suppl. 2, 247–273.
3. R. W. Wood, A. B. Weiss and B. Weiss, 'Hand tremor induced by industrial exposure to inorganic mercury.' *Arch. Environ. Health*, 1973, **26**, 249–252.
4. G. D. Langolf, D. B. Chaffin, R. Henderson *et al.*, 'Evaluation of workers exposed to elemental mercury using quantitative tests of tremor and neuromuscular functions.' *Am. Ind. Hyg. Assoc. J.*, 1978, **39**, 976–984.
5. A. A. Rahill, B. Weiss, P. E. Morrow *et al.*, 'Human performance during exposure to toluene.' *Aviat. Space Environ. Med.*, 1996, **67**, 640–647.
6. C. M. Regan, 'Lead-impaired neurodevelopment. Mechanisms and threshold values in the rodent.' *Neuro-toxicol. Teratol.*, 1989, **11**, 533–537.
7. J. Cohn and D. A. Cory-Slechta, 'Assessment of the role of dopaminergic systems in lead-induced learning impairments using a repeated acquisition and performance baseline.' *Neurotoxicology*, 1994, **15**, 913–926.
8. H. Michalek, S. Fortuna and A. Pintor, 'Age-related differences in brain choline acetyltransferase, cholines-

terases and muscarinic receptor sites in two strains of rats.' *Neurobiol. Aging, 1989*, **10**, 143–148.
9. H. S. Levin, R. L. Rodnitzky and D. L. Mick, 'Anxiety associated with exposure to organophosphorous compounds.' *Arch. Gen. Psychiatry*, 1976, **33**, 225–228.
10. E. P. Savage, T. J. Keefe, L. M. Mounce *et al.*, 'Chronic neurological sequelae of acute organophosphate pesticide poisoning.' *Arch. Environ. Health*, 1988, **43**, 38–45.
11. F. H. Duffy and J. L. Burchfiel, 'Long term effects of the organophosphate sarin on EEGs in monkeys and humans.' *Neurotoxicology*, 1980, **1**, 667–689.
12. R. McConnell, M. Keifer and L. Rosenstock, 'Elevated quantitative vibrotactile threshold among workers previously poisoned with methamidophos and other organophosphate pesticides.' *Am. J. Ind. Med.*, 1994, **25**, 325–334.
13. J. B. Penney, Jr., A. B. Young, I. Shoulson *et al.*, 'Huntington's disease in Venezuela: 7 years of follow-up on symptomatic and asymptomatic individuals.' *Mov. Dis.*, 1990, **5**, 93–99.
14. S. L. Rifat, M. R. Eastwood, D. R. McLachlan *et al.*, 'Effect of exposure of miners to aluminum powder.' *Lancet*, 1990, **336**, 1162–1165.
15. D. Bellinger, A. Leviton, C. Waternaux *et al.*, 'Longitudinal analyses of prenatal and postnatal lead exposure and early cognitive development.' *N. Engl. J. Med.*, 1987, **316**, 1037–1043.
16. B. F. Feingold, 'Why Your Child is Hyperactive,' Random House, New York, 1975.
17. B. Weiss, J. H. Williams, S. Margen *et al.*, 'Behavioral responses to artificial food colors.' *Science*, 1980, **207**, 1487–1489.
18. B. Weiss, 'Food additives and environmental chemicals as sources of childhood behavior disorders.' *J. Am. Acad. Child. Psychiatry*, 1982, **21**, 144–152.
19. B. J. Kaplan, J. R. McNicol, R. A. Conte *et al.*, 'Dietary replacement in preschool-aged hyperactive boys.' *Pediatrics*, 1989, **83**, 7–17.
20. S. B. Cannon, J. M. Veazey, Jr., R. S. Jackson *et al.*, 'Epidemic Kepone poisoning in chemical workers.' *Am. J. Epidemiol.*, 1978, **107**, 529–537.
21. H. Hanninen, L. Eskelinen, K. Husman *et al.*, 'Behavioral effects of long-term exposure to a mixture of organic solvents.' *Scand. J. Work. Environ. Health*, 1976, **2**, 240–255.
22. National Research Council, 'Risk Assessment in the Federal Government: Managing the Process,' National Academy Press, Washington, DC, 1983.
23. H. L. Needleman, C. Gunnoe, A. Leviton *et al.*, 'Deficits in psychologic and classroom performance of children with elevated dentine lead levels.' *N. Engl. J. Med.*, 1979, **300**, 689–695.
24. C. Cox, T. W. Clarkson, D. O. Marsh *et al.*, 'Dose-response analysis of infants prenatally exposed to methyl mercury: an application of a single compartment model to single-strand hair analysis.' *Environ. Res.*, 1989, **48**, 318–332.
25. B. Weiss and W. Simon, in 'Behavioral Toxicology,' eds. B. Weiss and V. G. Laties, Plenum, New York, 1975, pp. 429–435.
26. H. Michalek, S. Fortuna, M. T. Volpe *et al.*, 'Age-related differences in the recovery rate of brain cholinesterases, choline acetyltransferase and muscarinic acetylcholine receptor sites after a sucbacute intoxication of rats with the anticholinesterase agent, isofluorophate.' *Acta Neurobiol. Exp. (Warsz.)*, 1990, **50**, 237–249.
27. I. Date, M. F. D. Notter, S. Y. Felten *et al.*, 'MPTP-treated young mice but not aging mice show partial recovery of the nigrostriatal dopaminergic system by stereotaxic injection of acidic fibroblast growth factor (aFGF).' *Brain Res.*, 1990, **526**, 156–160.

28. Environmental Protection Agency, 'Multi-substance rule for the testing of neurotoxicity.' *Fed. Regist.*, 1991, **56**, 9105–9119.

29. B. Weiss, in 'Theory of Reinforcement Schedules,' eds. W. N. Schoenfeld and J. Farmer, Appleton-Century-Crofts, New York, 1970, pp. 277–311.

30. J. Cohn, C. Cox and D. A. Cory-Slechta, 'The effects of lead exposure on learning in a multiple repeated acquisition and performance schedule.' *Neurotoxicology*, 1993, **14**, 329–346.

31. A. Iregren, 'Behavioral methods and organic solvents: questions and consequences.' *Environ. Health Perspect.*, 1996, **104**, Suppl. 2, 361–366.

32. B. Weiss and V. G. Laties, 'Enhancement of human performance by caffeine and the amphetamines.' *Pharmacol. Rev.*, 1962, **14**, 1–36.

33. M. C. Newland, 'Quantification of motor function in toxicology.' *Toxicol. Lett.*, 1988, **43**, 295–319.

34. D. Mergler and L. Blain, 'Assessing color vision loss among solvent-exposed workers.' *Am. J. Ind. Med.*, 1987, **12**, 195–203.

35. B. S. Schwartz, D. P. Ford, K. I. Bolla *et al.*, 'Solvent-associated decrements in olfactory function in paint manufacturing workers.' *Am. J. Ind. Med.*, 1990, **18**, 697–706.

36. J. L. Tepper, B. Weiss and C. Cox, 'Microanalysis of ozone depression of motor activity.' *Toxicol. Appl. Pharmacol.*, 1982, **64**, 317–326.

37. D. C. Rice and K. F. Karpinski, 'Lifetime low-level lead exposure produces deficits in delayed alternation in adult monkeys.' *Neurotoxicol. Teratol.*, 1988, **10**, 207–214.

38. J. P. J. Maurissen, B. Weiss and H. T. Davis, 'Somatosensory thresholds in monkeys exposed to acrylamide.' *Toxicol. Appl. Pharmacol.*, 1983, **71**, 266–279.

39. S. C. Fowler, M. M. LaCerra and A. Ettenberg, 'Effects of haloperidol on the biophysical characteristics of operant responding: implications for motor and reinforcement processes.' *Pharmacol. Biochem. Behav.*, 1986, **25**, 791–796.

40. M. Davison and D. McCarthy, 'The Matching Law,' Erlbaum, Hillsdale, NJ, 1988.

41. B. Weiss and V. G. Laties, 'Behavioral thermoregulation.' *Science*, 1961, **133**, 1338–1344.

42. B. Weiss and V. G. Laties, 'Effects of amphetamine, chlorpromazine and pentobarbital on behavioral thermoregulation.' *J. Pharmacol. Exp. Ther.*, 1963, **140**, 1–7.

43. V. G. Laties and B. Weiss, 'Thyroid state and working for heat in the cold.' *Am. J. Physiol.*, 1959, **197**, 1028–1034.

44. B. Weiss, 'Thermal behavior of the subnourished and pantothenic acid deprived rat.' *J. Comp. Physiol. Psychol.*, 1957, **50**, 481–485.

45. S. C. Yeh and B. Weiss, 'Behavioral thermoregulation during vitamin B_6 deficiency.' *Am. J. Physiol.*, 1963, **205**, 857–862.

46. S. Stern, L. Margolin, B. Weiss *et al.*, 'Microwaves: effect on thermoregulatory behavior in rats.' *Science*, 1979, **206**, 1198–1201.

47. C. Watanabe, B. Weiss, C. Cox *et al.*, 'Modification by nickel of instrumental thermoregulatory in rats.' *Fundam. Appl. Toxicol.*, 1990, **14**, 578–588.

48. H. L. Evans, 'Quantitation of naturalistic behaviors.' *Toxicol. Lett.*, 1988, **43**, 345–359.

49. T. A. Mably, R. W. Moore, R. W. Goy *et al.*, '*In utero* and lactational exposure of male rats to 2,3,7,8-tetrachlorodibenzo-*p*-dioxin. 2. Effects on sexual behavior and the regulation of luteinizing hormone secretion in adulthood.' *Toxicol. Appl. Pharmacol.*, 1992, **114**, 108–117.

50. B. Ulbrich and A. K. Palmer, 'Neurobehavioral aspects of developmental toxicity testing.' *Environ. Health Perspect.*, 1996, **104**, Suppl. 2, 407–412.

51. R. I. Shader and D. J. Greenberg, 'User-unfriendly drugs—fillers, additives, excipients.' *J. Clin. Psychopharmacol.*, 1985, **5**, A15–A16.

52. J. S. Tepper and B. Weiss, 'Determinants of behavioral response with ozone exposure.' *J. Appl. Physiol.*, 1986, **60**, 868–875.

53. W. S. Wayne, P. F. Wehrle and R. E. Carrol, 'Oxidant air pollution and athletic performance.' *JAMA*, 1967, **199**, 901–904.

54. R. Ader, D. Felten and N. Cohen, 'Interactions between the brain and the immune system.' *Annu. Rev. Pharmacol. Toxicol.*, 1990, **30**, 561–602.

55. R. Ader and N. Cohen, 'Behaviorally conditioned immunosuppression.' *Psychosom. Med.*, 1975, **37**, 333–340.

56. S. F. Maier, L. R. Watkins and M. Fleshner, 'Psychoneuroimmunology. The interface between behavior, brain, and immunity.' *Am. Psychol.*, 1994, **49**, 1004–1017.

SECTION V

11.24

In Vivo Systems: Animal Models of Neurological Diseases

THOMAS J. WALSH

Rutgers University, New Brunswick, NJ, USA

11.24.1 INTRODUCTION

Neurotoxicology has evolved an integrative perspective on (i) what damages the nervous system, (ii) how the damage is manifested at the molecular, cellular, anatomical, physiological, and behavioral level, and (iii) the developmental profile of vulnerability to neurotoxic injury.[1] Borrowing conceptual and methodological approaches from a number of parent disciplines neurotoxicology has begun to address issues that are germane to all of the clinical neurosciences. Examining the specific toxicity of a given toxic agent is necessary to evaluate the potential impact of the compound on populations at risk. However, the knowledge derived from such mechanistic studies is applicable, in a broader sense, to understanding issues related to neurodegeneration and neurodegenerative diseases, such as selective vulnerability and compensatory responses to neural injury.[2] In addition, neurotoxicology has provided the scientific community with experimental tools and *in vivo* research strategies to evaluate the detrimental effects of compounds on the nervous system.[3]

419

Neurotoxicology focuses on the how, what, where, when, and why of central nervous system (CNS) injury. Understanding the bases of neurotoxic phenomenon allows for the development and use of theoretically important tools. There is a dynamic relationship between unraveling the properties of neurotoxicants and using them, or related compounds, in the exploration of brain diseases. An incident reported in 1990 illustrates this natural sequence. Ingestion of cultivated blue mussels contaminated with domoic acid resulted in an outbreak of food poisoning affecting over 100 people in Canada[4] (see Chapter 31, this volume). The acute phase of toxicity included gastrointestinal, cardiovascular, and neurological disturbances (agitation, disorientation, convulsions). In addition, a persisting dementia was observed in a significant percentage of those who survived the acute toxicity. The dementia was age-related and most evident in men over the age of 50. When some of these people came to autopsy, their brains exhibited damage to the hippocampus, other limbic structures, and neocortex. Domoic acid, the causative agent in this episode, is a potent excitotoxin that interacts with kainic acid receptors. Domoic acid produced a profile of cognitive deficits and neuropathology similar to that observed in Alzheimer's disease. After establishing the role of domoic acid in this episode investigators are now using this agent as a tool to explore the pathophysiology and functional manifestations of Alzheimer's disease and to produce useful models of the disease.[5] This reflects the symbiotic relationship between neurotoxicology and the clinical disciplines that comprise neuroscience.

The main focus of this chapter will be to examine how selective neurotoxins can be used to model neurological disorders (see Table 1). While the primary emphasis will be on Alzheimer's disease, other examples will be used to illustrate the factors that govern the development and use of animal models. What are the factors that guide the development of these models? What criteria can be applied to the evaluation of existing models? Finally, how can models be used to examine the biology of brain disease and to identify new therapeutic regimens?

11.24.2 THE DEFINITION OF ANIMAL MODELS

Animal models simulate some of the defining characteristics of a specific human disease state to aid in the understanding of the pathophysiology of the disease and in the development of efficacious therapies. Every human disease exhibits a profile of unique characteristics: (i) predisposing factors, (ii) time of onset, (iii) rate of progression, (iv) genetic susceptibility, and (v) cellular, molecular, and functional pathologies. Animal models provide a simplified approach to address the complex factors related to human pathology and disease. Kornetsky defines animal models as "an

Table 1   Neurotoxins used to model neurological diseases.

Toxin	Mechanism of action	Disease
Models of the cholinergic pathology of Alzheimer's disease		
AF64A	High affinity choline uptake inhibitor	
Excitotoxins	Agonists at excitatory amino acid receptors	
Colchicine	Microtubule inhibitor	
192-Saporin	Immunotoxin that destroys p75-expressing target neurons	
Models of the striatal pathology of Huntington's disease		
Excitotoxins	Agonists at excitatory amino acid receptors	
Quinolinic acid	Endogenous excitotoxin	
3-Nitropropionic acid	Mitochondrial toxin	
Models of the dopaminergic pathology of Parkinson's disease		
6-Hydroxydopamine	Free radical generation	
MPTP	Mitochondrial toxin	
Methamphetamine	Dopamine releasing agent	
Models of other neurodegenerative and developmental diseases		
Kainic acid	Excitotoxin	Temporal lobe epilepsy
OX7-Saporin	Immunotoxin	Purkinje cell degeneration
Methylazoxymethanol	Antimitotic	Microcephaly
α-Methylphenylalanine	Inhibits phenylalanine hydroxylase	Phenylketonuria

experimental compromise in which a simple experimental system is used to represent a more complex and less available system."[6] While they are unable to reproduce the multifactorial causes and symptoms of most human neurological diseases, they do provide a viable way to explore the causes, manifestations, and treatment of these conditions.

11.24.3 THE NEED FOR ANIMAL MODELS

Most neurodegenerative disorders have an unknown etiology and present a complex picture of pathology. It is difficult to unravel the essential features of a disease due to this inherent complexity. The use of animal models offers several distinct advantages in this effort. First, they simplify the system being studied. Alzheimer's disease is associated with changes in brain systems, neurotransmitters, metabolic processes, cytoskeletal integrity, protein processing, and gene expression.[7] Which of these changes, alone or in combination, should serve as a focus for the development of a model of the disease? Which pathologic process might be targeted for therapeutic intervention? Animal models have a known cause and allow an examination of a specific aspect of the disease. Second, they reduce the number of uncontrollable variables. Animal models limit the influence of genotype, nutritional status, exposure to viruses, and other potentially important environmental influences. In addition, they minimize the interpretive impact of concurrent medical conditions, length and profile of agonal state, and the amount of time between death and autopsy. They generally offer a high degree of experimental control. Last, they allow unlimited access to tissue. Animal models provide ready access to fresh tissue, and also allow for experimental manipulations that would be impossible in humans.

11.24.3.1 Medical Advances Based on Animal Models

In addition to the theoretical and practical issues that influence the development of animal models, there are also ethical considerations that must be addressed. Does the cost-benefit analysis of the beneficial outcome of research outweigh the potential "cost" to the animal being studied? The history of modern medicine and its advances have been dependent upon the humane use of animal models that aided in the

development of antibiotics and vaccines for infectious diseases, and pharmacological and surgical therapies for the conditions that plague mankind: cardiovascular disease, cancer, diabetes, and so on. The increases in life expectancy from 50 yr at the turn of this century to 73–78 yr in the 1990s is in large part the result of knowledge gained from animal models.[8] Animal research has a pervasive beneficial effect on society ranging from reduced infant mortality to promoting longer and healthier lives. Animal research mitigates needless human and animal suffering. The interested reader is referred to a thoughtful chapter on the ethics of animal models of neurological disorders by Olfert.[9]

11.24.4 ALZHEIMER'S DISEASE

This section will provide a brief overview of the defining characteristics of Alzheimer's disease. This disorder will serve as a conceptual focus for a critical discussion of the issues that have guided the development of animal models. The efforts of Walsh and co-workers to develop a model of Alzheimer's disease using the cholinotoxin AF64A and the immunotoxin 192-saporin are described in detail elsewhere.[10–12]

Alzheimer's disease is the most prevalent of the age-related cognitive disorders.[7] It typically presents a slow and progressive neurological decline which necessitates institutional care that is estimated to cost 3 billion dollars per year in the United States alone. Incidence increases with age with approximately 3%, 14%, and 34% of people in the age brackets 60–69, 70–79, and 80+, affected.[13] A variety of factors including advanced age, genetic predispositions, exposure to environmental toxins, neuroactive viruses, and prior brain injury may contribute to the expression of the disease.

The scope of cellular and structural changes that occur in the Alzheimer's disease brain has made it difficult to focus on biological substrates that might be targets for therapeutic intervention. However, the brain is differentially affected by Alzheimer's disease with pathology concentrated in the hippocampus, association cortex, and limbic system.[14,15]

A major focus for the study of Alzheimer's disease has been the cholinergic system. All biochemical markers of presynaptic cholinergic function including acetylcholine synthesis and release, and high-affinity choline transport are decreased in the hippocampus and neocortex in Alzheimer's disease.[16,17] There is also a pronounced loss of cholinergic neurons in the basal forebrain which innervate these regions.[18] These changes are a prominent part of the disease process since there is a significant correlation between the severity of dementia, the number of senile plaques, and the loss of cholinergic markers.[19,20] Experimental studies in animals also support a critical role for the cholinergic system in memory and dementia.[11]

11.24.4.1 Memory Impairment in Alzheimer's Disease

The profile of cognitive changes observed in Alzheimer's disease has provided indirect support for the hypothesis that there are at least two memory processes that are responsible for different types of cognitive operations.[21] One process, referred to as procedural memory, seems to support the performance of skills or habits, a class of operations dependent upon rules or procedures (i.e., learning to ride a bicycle or always parking in lot 49). The other, referred to as declarative or working memory, allows the retention of specific information about events or episodes (i.e., what did I have for breakfast or where exactly did I park in lot 49 today).

Early in the course of Alzheimer's disease patients are impaired in the acquisition and retention of recently presented information (working memory), but are not impaired in the acquisition of cognitive and perceptual skills (procedural memory).[22] This dissociation of memory processes is most evident in those situations where a subject can perform a learned perceptual or motor skill in the absence of any recollection of the specific episodes in which it was acquired.

The profile of memory impairments observed in Alzheimer's disease has been characterized using Baddeley's Working Memory Model as a theoretical framework.[23] Working memory directs cognitive effort to the demands of the existing situation. Baddeley postulates that a Central Executive System directs the activity of two supporting working memory systems; (i) the Articulatory Loop System which retains verbal information, and (ii) the Visual–Spatial System which retains nonverbal information. Early in the course of Alzheimer's disease there appears to be a selective impairment in the Central Executive System. This impairment appears to be dependent upon cholinergic dysfunction since it can be reproduced in health young adults by the muscarinic antagonist scopolamine.[11,24] These observations indicate that the earliest neuropsychological manifestation of Alzheimer's disease is a disruption of working memory processes that depend upon the integrity of cholinergic neurons in the brain.

11.24.4.2 Animal Models of Alzheimer's Disease

There is compelling evidence that the memory impairments in Alzheimer's disease are dependent upon the loss of cholinergic function. Therefore, a logical strategy for developing an animal model of Alzheimer's disease might be to focus on the "target" symptoms of working memory deficits and cholinergic hypofunction.[12] However, the development of a reliable model of Alzheimer's disease has proved to be challenging since the etiology of the disease is obscure and is likely to involve a variety of interacting factors. While it is possible to induce some of the behavioral and neurobiological alterations associated with Alzheimer's disease, the proximal cause of these changes is probably not analogous to the multitude of etiological factors that contribute to the human condition.[12] No one model mimics all the major characteristics of the human disease state. However, there is an ongoing effort to develop models that provide insights into the disease process, and foster the development of new and better therapies. There is also a healthy skepticism regarding the rationale and utility for such models.

11.24.5 CRITERIA FOR ESTABLISHING ANIMAL MODELS

The development of a viable animal model of any neurological disease must be predicated upon a set of pre-established criteria. The literature is plagued by the indiscriminant and imprecise use of the term "animal model." Models are anything the author wishes them to be. Models of dementia have been predicated upon cholinergic hypofunction induced by (i) muscarinic antagonists, (ii) lesions of cholinergic nuclei, (iii) severing of the fimbria-fornix, and (iv) cholinotoxins.[11,12] The impetus for these "models" has often been accident, convenience, and history. This posture devalues the merit and applicability of animal models. Models need to be operationally defined and rigorously evaluated.

In one of the first attempts to formulate meaningful criteria for animal models, McKinney and Bunney argued that they should reproduce the disease in four dimensions: etiology, symptomatology, treatment efficacy, and underlying pathophysiology.[25] A consideration of each of these factors contributes to the understanding of the utility of the model and helps to bridge the gap between the model and the human condition. However, this strategy is too exacting for models of neurological disorders where etiology is obscure and probably combinatorial. A variety of interacting influences might lead to a physiological condition that is conducive to the development of brain changes associated with neurodegenerative diseases.[26] For example, a number of factors can contribute to Parkinson's disease, but there is no evidence that "one" specific factor is the proximate cause. Parkinson's-like pathology can be produced by viruses (*Encephalitis lethargica*), environmental toxins like MPTP (1-methyl-4-phenyl-1,2,3,6-tetrahydropyridine), carbon disulfide, and manganese, and possibly by drugs of abuse such as methamphetamine.[27–29] Parkinson's disease reflects a pattern of brain pathology and motor deficits that can result from a number of individual or interactive factors. Do environmental toxins produce this disease? Perhaps. Are they the only cause? Probably not.

Providing a set of criteria for the development and evaluation of animal models identifies areas for future research and provides an archetype to evaluate the advantages and disadvantages of existing models. Willner suggests that animal models should be evaluated in relation to their face validity, predictive validity, and construct validity.[30]

Face validity reflects the similarities between the model and the disease state. Are some of the characteristic behavioral and neurobiological symptoms of the disorder reproduced with acceptable fidelity in the model? As already mentioned, Alzheimer's disease is associated with the degeneration of cholinergic neurons in the basal forebrain that innervate structures involved in the coordination of memory processes, most notably the hippocampus and cortex. The functional consequence is a loss of presynaptic cholinergic indices and the impairment of memory processes dependent upon these brain systems. To have face validity, a model of Alzheimer's disease must produce a persistent cholinergic hypofunction and associated memory impairments. Due to the complexity of the disease it is recognized that other models focusing on changes in the cytoskeleton or the expression of β-amyloid might also exhibit face validity.[7]

A second consideration is predictive validity which is based upon the concordance between therapeutic strategies that are effective in the model and the disease state. The levorotatory form of dihydroxyphenylalanine (L-DOPA) and other dopamine agonists can improve the motor impairments observed in models of Parkinson's disease.[2] There is a correspondence between the treatment modalities that are used to manage the disease and which are efficacious

in the model. Therefore, the predictive validity of the model is in part based on whether it produces symptoms that are responsive to standard therapy. However, predictive validity is problematic for models of neurological diseases which lack established treatment regimens (i.e., positive controls). There is a conspicuous disparity between the effectiveness of drugs in animal models of Alzheimer's disease and in patients with the disease.[31] A variety of drugs (cholinergic agonists, cholinesterase inhibitors, nootropics) are effective in reversing cognitive deficits in available models of Alzheimer's disease, but they lack clinical efficacy. The value of simple replacement strategies in diseases, such as Alzheimer's disease, might need to be reevaluated in light of our better understanding of the physiology of circuits that are affected by the disease.[32]

A final consideration is the construct validity of an animal model. What are the theoretical considerations and the rationale for the development of a specific animal model? Neurodegenerative diseases can manifest a variety of biological and functional changes. Alzheimer's disease is associated with changes in brain systems, neurotransmitters, metabolic processes, cytoskeletal integrity, protein processing, and gene expression. Which of these changes, alone or in combination, should serve as a focus for the development of a model of the disease? Which pathological event most contributes to the behavioral pathology? Since the etiology of Alzheimer's disease remains obscure, and in fact might be combinatorial in nature, an animal model must focus on relevant behavioral and neurobiological "target" symptoms. However, as more is learned about the disease process other models might prove to be more useful for evaluating the pathophysiology of the disease and for developing new therapies. In fact, a myopic focus on specific target processes can be detrimental to understanding the complexity of human disease. In reference to animal models of schizophrenia Mogensen cautions that "the process of developing a more complete understanding of the neural substrates of schizophrenia may be seriously impaired by the rather one-sided focus on the 'dopaminergic' animal models."[33]

11.24.6 GOALS OF ANIMAL MODELS

Animal models can be used to focus on a variety of unique but interactive questions; (i) mechanisms of pathology, (ii) neuroplasticity following injury, (iii) structure–function relationships, (iv) etiological factors, and (v) new therapeutic strategies. Each of these topics represents unique goals and strategies. They will be discussed in relation to the development of animal models of Alzheimer's disease.

11.24.6.1 Mechanisms of Pathophysiology

A fundamental question in Alzheimer's disease is what causes the neuronal degeneration. Is it consequent to excitotoxicity, β-amyloid expression, apoptosis, calcium dyregulation, or an interplay of these factors? These questions address the proximate cause of cell death, regardless of the precipitating event. For example AIDS, ischemia, epileptic brain injury, and maybe other neurodegenerative diseases might induce cell death through common final mechanisms involving excitotoxicity and calcium dysfunction.[34–37] In fact, site-specific injection of excitotoxins can reproduce some of the pathology and functional deficits observed in Alzheimer's and Huntington's diseases.[38] This line of research is important for revealing common and perhaps ubiquitous mechanisms of CNS cell death. It will be critical to identify drugs that prevent or limit cell death regardless of the initiating event. A disease-modifying approach that targets the "phenomenon" of neurodegeneration rather than the pathology of Alzheimer's disease, *per se*, might be a more rational and realistic strategy. One of the attractions of therapies based upon the use of neurotrophic factors or antioxidants is that they might be useful in treating a wide range of degenerative conditions that affect the nervous system.[39,40] In fact, clinical studies are either underway or planned to evaluate the efficacy of neurotrophins in the treatment of peripheral neuropathies (nerve growth factor, NGF), amyotrophic lateral sclerosis (ciliary neurotrophic factor, CNTF), Parkinson's disease (glia-derived neurotrophic factor, GDNF), and Alzheimer's disease (NGF).[40]

A related issue concerns the resistance of cells to the disease process. For example, association cortex is typically deeply infested with the pathological hallmarks of Alzheimer's disease. However, primary motor and sensory areas remain unaffected until very late in the disease.[14] Why are these cortical areas relatively spared? Does it relate to an intrinsic resistancy of the neurons comprising these areas?

Another example of selective vulnerability concerns cholinergic neurons. Brain cholinergic neurons are organized into two groups of nuclei: (i) the basal forebrain complex consists of cholinergic neurons that innervate the cortex, hippocampus, and amygdala, and (ii) the upper brainstem complex that innervates the thalamus, colliculi, lateral septum, and medial prefrontal cortex. The basal forebrain complex degenerates very early in Alzheimer's disease and its compromise is related to the cognitive deficits observed in the disease.[3,11] There is a strong correlation between senile plaques, neurofibrillary tangles, cholinergic decreases, and dementia scores.[20] However, the upper brainstem complex is not affected in Alzheimer's disease.[41] Cholinergic neurons in both systems share similar morphologies and sizes. What renders one population vulnerable and the other resistant? One factor might be the trophic support required by neurons in the basal forebrain complex. These neurons express the low-affinity neurotrophin receptor, p75, while the upper brainstem complex neurons do not.[42] Neurons expressing p75 appear to be uniquely vulnerable to apoptosis and amyloid toxicity.[43,44] Is this a signpost for vulnerability to Alzheimer's disease? Immunotoxins that target the p75 receptor are being used to explore the molecular and cellular correlates of this selective vulnerability (see below).

11.24.6.2 Extent and Function of Neuroplasticity

Abnormal plasticity might contribute to a variety of diseases. It appears that the brains of Alzheimer's patients exhibit alterations which reflect aberrant forms of neuroplasticity and not simply a diminished response of the aged brain.[26] Atypical growth might be one of the mechanisms contributing to the pathology observed in Alzheimer's disease. Butcher and Woolf caution against interpreting "instances of somal, dendritic, and axonal remodeling in Alzheimer's disease as attempts by the nervous system to repair itself following damage rather than as major pathologic episodes leading to functional impairment and eventual cellular degeneration."[45] In contrast, Cotman and co-workers suggest that sprouting and other forms of plasticity might be compensatory early in the course of the disease, but subsequently contribute to the development of neuritic plaques and other forms of pathology.[46] Both hypotheses suggest that changes in neuroplasticity might actually promote cellular events that result in aberrant neuronal organization (neuritic plaques, neurofibrillary tangles) and the death of vulnerable neurons. For example, it has been shown that alterations in neuroplasticity accompany temporal lobe epilepsy and might serve to initiate or enhance the pathogenic process in the hippocampus and allied structures.[47]

Neurotoxins are being used to examine how the brain reorganizes after injury. Compensatory changes in receptor number, sensitivity, and coupling to signal transduction mechanisms, as well as synaptogenesis are probably common events associated with neurodegeneration.[2,26] How do these changes contribute to the ongoing pathology, the functional deficits, and the likely success of pharmacological therapy? For example, AF64A is a cholinotoxin that decreases neurochemical indices of cholinergic function and produces the loss of cholinergic neurons in the basal forebrain.[12] Potter and Thorne have demonstrated lasting changes in cholinergic receptor function together with shifts in the dose–response curve for both muscarinic and nicotinic agonists following AF64A administration.[48] Therefore, the long-term effectiveness of cholinergic therapies for Alzheimer's disease will be limited by the magnitude of compensatory changes induced by the disease process.

11.24.6.3 Structure–Function Relationships

Another goal of animal models is to reproduce the biological and functional changes observed in human diseases. These models help to relate specific neural alterations to specific symptoms. Which symptoms are associated with injury to which areas? They do not focus on specific etiological factors which predispose to neurodegenerative disorders, but rather examine changes in the brain which induce a similar profile of symptoms.[49]

The importance of this kind of structural analysis in neurobiology is highlighted by the case of H.M.[50] H.M. is probably the most studied individual in the history of psychology. In 1953 he underwent a bilateral temporal lobe resection to control symptoms of epilepsy which were unresponsive to available medical treatment. The results of the surgery epitomize the "good news–bad news" predicament. His epileptic symptoms were dramatically reduced in frequency and severity, but he developed a profound anterograde amnesia in which he could not consolidate new information into memory. From day to day he can read the same magazines and he can interact with psychologists who have studied his memory problem for decades, as if they were newly introduced. His personality and other cognitive and perceptual abilities were intact following the surgery. H.M.'s cognitive deficits provide evidence that memory is not a unitary function, but a family of cognitive processes (declarative memory and procedural memory) subserved by different neural substrates.[21] However, extrapolating the memory deficits to specific components of the temporal lobe damaged by his surgery has proven difficult due to the extent of the resection. Most of his hippocampus was removed together with the amygdala and portions of temporal cortex including parahippocampal gyrus, and perirhinal and entorhinal cortices. Which of these structures, alone or in combination, are responsible for his deficit? The matrix of possibilities is daunting: (i) hippocampus alone, (ii) hippocampus + amygdala, (iii) hippocampus + parahippocampal gyrus, and all of the other permutations of the matrix. Animal models using site-specific injection of excitotoxins have been able to systematically address these questions to define the anatomy and function of the temporal lobe memory system.[51]

In a related context, excitotoxins have been widely used to produce models of a variety of neurological diseases even though excitotoxicity might not be a causative factor in the disease process.[34,49] Excitotoxins such as kainic acid, ibotenic acid, quinolinic acid, and *N*-methyl-D-aspartate (NMDA) can be used to produce axon-sparing lesions in specific brain areas. Sanberg and Coyle have used these compounds in rodent models to reproduce the neuropathology and functional changes observed in Huntington's disease.[38] These studies provide a functional anatomy of brain–behavior relationships that exist in given disease states and also provide a way to explore innovative new therapeutic strategies.

11.24.6.3.1 The use of immunotoxins to model neurodegenerative conditions

Immunotoxins were initially developed as a targeted approach to eliminate tumor cells that express unique antigens.[52] Immunotoxins are conjugates of a monoclonal antibody targeting a specific antigen combined with a ribosome-inactivating proteins (RIP). These toxins catalytically inhibit ribosomal activity and irreversibly halt protein synthesis which results in cell death. Since they recognize and destroy only antibody-targeted cells it is possible to create highly selective lesions which can mimic neurodegenerative disorders and/or address fundamental neurobiological questions.[53]

Cholinergic neurons in the basal forebrain contain p75 neurotrophin receptors which mediate the intracellular effects of NGF.[54] Walsh and co-workers are using these compounds to examine the function of the cholinergic basal forebrain and to model diseases of cholinergic hypofunction, in particular Alzheimer's disease.[10,55] 192-Saporin combines the 192 IgG monoclonal antibody to the p75 low-affinity neurotrophin receptor with saporin, a potent RIP derived from *Saponaria officinalis*. The immunotoxin targets the p75 receptor localized on cholinergic nerve terminals in neocortex and hippocampus, and on cholinergic cell bodies in the basal forebrain, but not on those cholinergic cell groups found in the upper brainstem.[52,53] This regional selectivity is important since the upper brainstem complex of cholinergic neurons is spared in Alzheimer's disease.[41] Site-specific injection of 192-saporin produces a selective loss of cholinergic neurons and neurochemical markers together with cognitive impairments that resemble those observed in Alzheimer's disease.[10,52,53,55] 192-Saporin already provides a model of Alzheimer's disease with both face validity and construct validity. It can also be used to ask fundamental questions about cholinergic biology and to evaluate new therapies to prevent or treat the functional consequences of cholinergic degeneration (predictive validity). The available data suggest that 192-saporin may be the "best" model to study the cholinergic pathology of Alzheimer's disease.

The utility of immunotoxins is also being explored in other contexts. Immunotoxins that target Purkinje cells in the cerebellum (OX7-saporin) or noradrenergic neurons in the locus coeruleus (antidopamine-β-hydroxylase-saporin) are being used to examine the functions of these neuronal populations and their contributions to various disease states.[52] In general, immunotoxins offer a new approach to lesion specific neuronal populations based upon the unique molecular and immunologic identity of these populations. These agents should be useful in studies of neurogenesis, synaptic and behavioral plasticity induced by experience, age, and disease, as well as in the modeling of degenerative diseases.

11.24.6.4   Potential Etiological Factors

Animal model are also used to address specific etiological theories of neurodegenerative disorders. For example, if aluminum is a causative factor in Alzheimer's disease it should produce "Alzheimer disease-like" effects in animal models. However, there is little supporting evidence for such an hypothesis, because aluminum fails to produce these effects in animals. In fact, the available evidence suggests that the brain pathology associated with Alzheimer's disease leads to the accumulation of aluminum in the brain.[56] Increased brain aluminum might be an "effect" of the disease rather than a "cause."

11.24.6.4.1   The promise of transgenic (knockout) mice

There is a growing appreciation that genetic factors contribute to a variety of neurodegenerative disorders. While the genetic contribution to inherited disorders, such as Huntington's disease, is well understood it appears that other degenerative disorders, such as Alzheimer's disease, also have a strong genetic predisposition. To understand the contribution of genotype to neuropathology and functional deficits, transgenic mouse models are being developed.[57] Transgenic mice have an exogenous gene incorporated into their genomes as embryos. The gene becomes permanently incorporated into the germ line and is passed through successive generations. The functional consequences of this manipulation can be explored at the molecular, biochemical, cellular, and behavioral level.

Early-onset Alzheimer's disease can be associated with mutations in the β-amyloid precursor protein gene on chromosome 21 and with a gene (STM2) with unknown function on chromosome 1.[58,59] In addition, the E4 allele of the apolipoprotein E gene on chromosome 19 is associated with an enhanced risk for late-onset Alzheimer's disease, while the E2 allele appears to exert a protective influence.[60] How do these genotypes contribute to the pathological cascade of Alzheimer's disease? Are they causative factors or "permissive" factors? Studies are examining transgenic mice that express the human β-amyloid precursor protein gene to determine the role of amyloid expression in the disease process. A cardinal feature of Alzheimer's disease is enhanced expression of amyloid precursor protein, as well as anomalous intracellular (lysosomal) processing of the protein. This aberrant protein processing results in fragments of β-amyloid precursor protein that form fibrillary aggregates. It is hypothesized that this event is the direct precursor to the formation of amyloid plaques in the brain of Alzheimer's disease patients. A report illustrates mice that are transgenic for the mutant β-amyloid precursor protein gene exhibit a number of pathological changes that characterize Alzheimer's disease including: amyloid plaques, regional synaptic loss, and proliferation of both astrocytes and microglia.[58] These transgenic mice will be used to explore the efficacy of new strategies designed to alter the processing of β-amyloid precursor protein or to minimize its biological impact on vulnerable neural processes. Can the aberrant processing of β-amyloid precursor protein be altered or can aggregate formation be prevented? However, it is important to realize that the genetic contribution to Alzheimer's disease is not absolute. Other factors such as exposure to xenobiotics, neuroactive viruses, head injury, or nutritional anomalies might be necessary to induce Alzheimer's disease-like pathology. Transgenic animals provide a way to explore the dynamic relationship between genetic and epigenetic influences. Transgenic mice models can also be used to address genetic factors related to the vulnerability or resistance to disease processes. For example, mice that are transgenic for the free radical scavenging enzyme, copper/zinc superoxide dismutase, are resistant to ischemia and to the neurotoxic effects of such chemicals as MPTP, 3-nitropropionic acid, and methamphetamine.[61-63]

11.24.6.5 New Therapeutic Strategies

The ultimate goal of animal models is to a understand a disease well enough to prevent it or successfully treat it. Can we learn enough about the pathobiology of a disease to devise therapies that either correct, or compensate for, the underlying pathophysiology and its consequences? Models are used to (i) screen for new compounds, (ii) refine existing therapies, and (iii) explore speculative new strategies. A variety of new therapies for neurodegenerative diseases are being studied: antioxidants, excitatory amino acid antagonists, Ca^{2+}-channel blockers and calpain inhibitors, trophic factors, and neurotransplantation.[39] Animal models are necessary to evaluate effectiveness, mechanisms of action, and potential side effects of new therapies.

The need for new therapies is highlighted by the lack of treatments that prevent, treat, or manage such neurodegenerative disorders as Alzheimer's or Huntington's diseases. While dopamine precursors like L-DOPA or agonists like bromocriptine can treat the symptoms of Parkinson's disease for a limited period of time, they are limited by a high incidence of physical (nausea) and cognitive (psychosis) side effects and they do not mitigate the disease process. The discovery that glia-derived neurotrophic factor (GDNF) provides trophic support for midbrain dopamine neurons and can prevent the dopamine cell loss and hypofunction induced by MPTP, suggests new avenues might be available for retarding the rate or extent of dopamine cell loss in neurodegenerative conditions.[64] In general, trophic factors might be one way to modify the progression of a number of degenerative diseases. Regardless of whether a loss of trophic factor activity contributes to the onset or progression of a disease can exogenous trophic factors limit the course or rate of degeneration? Will this strategy stabilize populations of neurons faced with insult in a variety of models of degeneration?

11.24.7 CONCLUSIONS

Neurotoxins are powerful tools that can explore the structure and function of the nervous system in both health and disease. They can be used to address essential questions at the behavioral, structural, cellular, and molecular level. As tools, they need to be employed with care and precision in the pursuit of specific goals. Neurotoxins have been indispensable in the development of animal models of neurological diseases. They provide an approach to understanding the basic pathophysiology of degeneration, explore the functional manifestations of neural injury, and help develop therapies that some day will prevent or treat the disorders that rob us of our ability to move, think, remember, and communicate. Animal models are a good investment for our health, longevity, and quality of life.

ACKNOWLEDGMENTS

The author would like to thank Chet Gandhi for his comments on this manuscript. The original work reported here was supported by NSF grant (IBN 9514557) and a gift in memory of Colonel Norman C. Kalmar to TJW.

11.24.8 REFERENCES

1. L. W. Chang (ed.), 'Principles of Neurotoxicology,' Dekker, New York, 1994.
2. M. J. Zigmond and E. M. Stricker, 'Parkinson's disease: studies with an animal model.' *Life Sci.*, 1984, **35**, 5–18.
3. T. J. Walsh, K. D. Opello, J. D. Adams *et al.*, in 'Handbook of Neurotoxicology,' ed. L. W. Chang, Dekker, New York, 1994, pp. 525–550.
4. T. M. Perl, L. Bedard, T. Kosatsky *et al.*, 'An outbreak of toxic encephalopathy caused by eating mussels contaminated with domoic acid.' *N. Engl. J. Med.*, 1990, **322**, 1775–1780.
5. R. J. Sutherland, J. M. Hoesing and I. Q. Whishaw, 'Domoic acid, an environmental toxin, produces hippocampal damage and severe memory impairment.' *Neurosci. Lett.*, 1990, **120**, 221–223.
6. C. Kornetsky, in 'Animal Models in Psychiatry and Neurology,' eds. I. Hanin and E. Usdin, Pergamon, New York, 1977, p. 1.
7. R. D. Terry and R. Katzman, 'Senile dementia of the Alzheimer type.' *Ann. Neurol.*, 1983, **14**, 497–506.
8. J. Goldman and L. Cote, in 'Principles of Neural Science,' 3rd edn, eds. E. R. Kandel, J. H. Schwartz and T. M. Jessell, Elsevier, New York, 1991, p. 974.
9. E. D. Olfert, in 'Neuromethods, Vol. 21., Animal Models of Neurological Diseases,' eds. A. Boulton, G. Baker and R. Butterworth, Humana Press, Totowa, NJ, 1992, pp. 1–28.

10. T. J. Walsh, R. M. Kelly, K. D. Dougherty *et al.*, 'Behavioral and neurobiological alterations induced by the immunotoxin 192-IgG-saporin-cholinergic and noncholinergic effects following ICV injection.' *Brain Res.*, 1996, **702**, 233–245.

11. T. J. Walsh and J. J. Chrobak, in 'Current Topics in Animal Learning: Brain, Emotion, and Cognition,' eds. L. Dachowski and C. Flaherty, Lawrence Erlbaum, Hillsdale, NJ, 1991, pp. 347–379.

12. T. J. Walsh and K. D. Opello, in 'Toxin-Induced Models of Neurological Disorders,' eds. M. Woodruff and A. Nonneman, Plenum, New York, 1994, pp. 259–279.

13. D. A. Evans, H. H. Funkenstein, M. S. Albert *et al.*, 'Prevalence of Alzheimer's disease in a community population of older persons. Higher than previously reported.' *JAMA*, 1989, **262**, 2551–2556.

14. A. Brun and E. Englund, 'Regional pattern of degeneration in Alzheimer's disease: neuronal loss and histopathological grading.' *Histopathology*, 1981, **5**, 549–564.

15. B. T. Hyman, G. W. Van Horsen, A. R. Damasio *et al.*, 'Alzheimer's disease: cell specific pathology isolates the hippocampal formation.' *Science*, 1984, **225**, 1168–1170.

16. R. T. Bartus, R. L. Dean, III, B. Beer *et al.*, 'The cholinergic hypothesis of geriatric memory dysfunction.' *Science*, 1982, **217**, 408–414.

17. J. T. Coyle, D. L. Price and M. R. DeLong, 'Alzheimer's disease: a disorder of cortical cholinergic innervation.' *Science*, 1983, **219**, 1184–1190.

18. P. J. Whitehouse, R. G. Struble, J. C. Hedreen *et al.*, 'Alzheimer's disease and related dementias: selective involvement of specific neuronal systems.' *CRC Crit. Rev. Clin. Neurobiol.*, 1985, **1**, 319–339.

19. E. K. Perry, B. E. Tomlinson, G. Blessed *et al.*, 'Correlation of cholinergic abnormalities with senile plaques and mental test scores in senile dementia.' *Br. Med. J.*, 1978, **2**, 1457–1159.

20. G. K. Wilcock, M. M. Esiri, D. M. Bowen *et al.*, 'Alzheimer's disease: correlation of cortical choline acetyltransferase activity with the severity of dementia and histological abnormalities.' *J. Neurol. Sci.*, 1982, **57**, 407–417.

21. D. F. Sherry and D. L. Schacter, 'The evolution of multiple memory systems.' *Psychol. Rev.*, 1987, **94**, 439–454.

22. N. J. Cohen and L. R. Squire, 'Preserved learning and retention of pattern-analyzing skill in amnesia: dissociation of knowing how and knowing that.' *Science*, 1980, **210**, 207–210.

23. A. D. Baddeley, 'Working Memory,' Clarendon Press, Oxford, 1986.

24. J. M. Rusted and D. M. Warburton, 'The effects of scopolamine on working memory in healthy young volunteers.' *Psychopharmacology (Berl.)*, 1988, **96**, 145–152.

25. W. T. McKinney, Jr. and W. E. Bunney, Jr., 'Animal model of depression. I. Review of evidence: implications for research.' *Arch. Gen. Psychiatry*, 1969, **21**, 240–248.

26. T. J. Walsh and K. D. Opello, 'Neuroplasticity, the aging brain, and Alzheimer's disease.' *Neurotoxicology*, 1992, **13**, 101–110.

27. J. W. Langston and P. Ballard, 'Parkinsonism induced by 1-methyl-1,4-phenyl-1,2,3,6-terahydropyridine (MPTP): implications for treatment and the pathogenesis of Parkinson's disease.' *Can. J. Neurol. Sci.*, 1984, **11** Suppl. 1, 160–165.

28. R. T. Ravenholt and W. H. Foege, '1918 influenza, *Encephalitis lethargica*, Parkinsonism.' *Lancet*, 1982, **2**, 860–864.

29. C. M. Tanner, 'The role of environmental toxins in the etiology of Parkinson's disease.' *Trends Neurosci.*, 1989, **12**, 49–54.

30. P. Willner, in 'Behavioural Models in Psychopharmacology, Theoretical, Industrial and Clinical Perspectives,' ed. P. Willner, Cambridge University Press, Cambridge, 1991, pp. 3–18.

31. M. Sarter, J. Hagan and P. Dudchenko, 'Behavioral screening for cognition enhancers: from indiscriminate to valid testing: part I.' *Psychopharmacology (Berl.)*, 1992, **107**, 144–159.

32. T. J. Walsh, 'Site-specific pharmacology for the treatment of Alzheimer's disease.' *Exp. Neurol.*, 1993, **124**, 43–46.

33. J. Mogensen, in 'Handbook of Laboratory Animal Science: Volume II Animal Models,' eds. P. Svendsen and J. Hau, CRC Press, Boca Raton, FL, 1994, p. 125.

34. R. N. Auer, 'Excitotoxic mechanisms, and age-related susceptibility to brain damage in ischemia, hypoglycemia and toxic mussel poisoning.' *Neurotoxicology*, 1991, **12**, 541–546.

35. S. C. Bondy, 'Intracellular calcium and neurotoxic events.' *Neurotoxicol. Teratol.*, 1989, **11**, 527–531.

36. S. A. Lipton, 'Models of neuronal injury in AIDS: another role for the NMDA receptor?' *Trends Neurosci.*, 1992, **15**, 75–79.

37. P. Nicotera and S. Orrenius, 'Ca^{2+} and cell death.' *Ann. NY Acad. Sci.*, 1992, **648**, 17–27.

38. P. R. Sanberg and J. T. Coyle, 'Scientific approaches to Huntington's disease.' *CRC Crit. Rev. Clin. Neurobiol.*, 1984, **1**, 1–44.

39. T. J. Walsh, R. M. Kelly and R. W. Stackman, 'Strategies to limit brain injury and promote recovery of function.' *Neurotoxicology*, 1994, **15**, 467–475.

40. J. E. Springer, 'Experimental evidence for growth factor treatment and function in certain neurological disorders.' *Exp. Neurol.*, 1993, **124**, 2–4.

41. N. J. Woolf, R. W. Jacobs and L. L. Butcher, 'The pontomesencephalotegmental cholinergic system does not degenerate in Alzheimer's disease.' *Neurosci. Lett.*, 1989, **96**, 277–282.

42. N. J. Woolf, E. Gould and L. L. Butcher, 'Nerve growth factor receptor is associated with cholinergic neurons of the basal forebrain but not the pontomesencephalon.' *Neuroscience*, 1989, **30**, 143–152.

43. S. Rabizadeh, C. M. Bitler, L. L. Butcher *et al.*, 'Expression of the low-affinity nerve growth factor receptor enhances beta-amyloid peptide toxicity.' *Proc. Natl. Acad. Sci. USA*, 1994, **91**, 10703–10706.

44. S. Rabizedeh, J. Oh, L. T. Zhong *et al.*, 'Induction of apoptosis by the low-affinity NGF receptor.' *Science*, 1993, **261**, 345–348.

45. L. L. Butcher and N. J. Woolf, 'Neurotrophic agents may exacerbate the pathologic cascade of Alzheimer's disease.' *Neurobiol. Aging*, 1989, **10**, 557–570.

46. J. W. Geddes, K. J. Anderson and C. W. Cotman, 'Senile plaques as aberrant sprout-stimulating structures.' *Exp. Neurol.*, 1986, **94**, 767–776.

47. T. Sutula, G. Cascino, J. Cavazos *et al.*, 'Mossy fiber synaptic reorganization in the epileptic human temporal lobe.' *Ann. Neurol.*, 1989, **26**, 321–330.

48. B. Thorne and P. E. Potter, 'Lesions with the neurotoxin AF64A alters hippocampal cholinergic receptor function.' *Brain Res. Bull.*, 1995, **38**, 121–127.

49. J. W. Olney, in 'Handbook of Neurotoxicology,' ed. L. W. Chang, Dekker, New York, 1994, p. 495.

50. B. Milner, S. Corkin, and H.-L. Teuber, 'Further analysis of the hippocampal amnesic syndrome: 14 year follow-up study of H. M.' *Neuropsychologia*, 1968, **6**, 215–234.

51. L. R. Squire and S. Zola-Morgan, 'The medial temporal lobe memory system.' *Science*, 1991, **253**, 1380–1386.

52. R. G. Wiley and D. A. Lappi, 'Suicide Transport and Immunolesioning,' R. G. Landes, Austin, TX, 1994.

53. R. G. Wiley, 'Neural lesioning with ribosome-inactivating proteins: suicide transport and immunolesioning.' *Trends Neurosci.*, 1992, **15**, 285–290.

54. J. E. Springer, 'Nerve growth factor receptors in the central nervous system.' *Exper. Neurol.*, 1988, **102**, 354–365.

55. T. J. Walsh, C. D. Herzog, C. Gandhi *et al.*, 'Injection of IgG 192-saporin into the medial septum produces cholinergic hypofunction and dose-dependent working memory deficits.' *Brain Res.*, 1996, **726**, 69–79.

56. P. Szerdahelyi and P. Kasa, 'Intraventricular administration of the cholinotoxin AF64A increases the accumulation of aluminum in the rat parietal cortex and hippocampus but not in the frontal cortex.' *Brain Res.*, 1988, **444**, 356–360.

57. A. Aguzzi, S. Brandner, U. Sure *et al.*, 'Transgenic and knockout mice: models of neurological disease.' *Brain Pathol.*, 1994, **4**, 3–20.

58. D. Games, D. Adams, R. Alessandrini *et al.*, 'Alzheimer-type neuropathology in transgenic mice overexpressing V717F β-amyloid precursor protein.' *Nature*, 1995, **373**, 523–527.

59. E. Levy-Lahad, W. Wasco, P. Poorkaj *et al.*, 'Candidate gene for the chromosome 1 familial Alzheimer's disease locus.' *Science*, 1995, **269**, 973–977.

60. W. J. Strittmatter, A. M. Saunders, D. Schmechel *et al.*, 'Apolipoprotein E: high-avidity binding to β-amyloid and increased frequency of type 4 allele in late-onset familial Alzheimer disease.' *Proc. Natl. Acad. Sci. USA*, 1993, **90**, 1977–1981.

61. M. F. Beal, R. J. Ferrante, R. Henshaw *et al.*, '3-Nitroproprionic acid neurotoxicity is attenuated in copper/zinc superoxide dismutase transgenic mice.' *J. Neurochem.*, 1995, **65**, 919–922.

62. H. Kinouchi, C. J. Epstein, T. Mizui *et al.*, 'Attenuation of focal cerebral ischemic injury in transgenic mice overexpressing CuZn superoxide dismutase.' *Proc. Natl. Acad. Sci. USA*, 1991, **88**, 11158–11162.

63. S. C. Bondy, 'Reactive oxygen species: Relation to aging and neurotoxic damage.' *Neurotoxicology*, 1992, **13**, 87–100.

64. A. Tomac, E. Lindqvist, L. F. Lin *et al.*, 'Protection and repair of the nigrostriatal dopaminergic system by GDNF *in vivo*.' *Nature*, 1995, **373**, 335–339.

11.25
In Vitro Systems in Neurotoxicological Studies

GERALD J. AUDESIRK

University of Colorado at Denver, CO, USA

11.25.1 INTRODUCTION

The general goals of toxicological research are to identify toxic substances, the exposure levels at which they produce toxic effects, their target species and organs, and their mechanisms of toxicity. With this information in hand, it may be possible to devise strategies to reduce exposures (e.g., elimination of lead in gasoline), ameliorate the effects of exposure (e.g., by chelation therapy for lead), and, in some cases, identify substances that are likely to be toxicants, based on their chemical structure and/or their known cellular or molecular modes of action. Investigation of mechanisms of toxic action and, to a lesser extent, identification of toxicants have been the primary goals of *in vitro* toxicology.

Defining the mechanisms whereby toxicants exert their toxic effects is important for several reasons. First, many studies and government regulations attempt to extrapolate from effects in nonhuman animals to effects in humans. This is obviously much more justifiable if the mechanisms of action are known and if similar mechanisms operate in human and nonhuman species. Second, knowing the mechanisms of action of a toxicant may help in devising treatments for affected individuals or populations. Third, mechanistic information may allow estimation of relative risks of different exposure levels. Fourth, mechanistic information may provide a basis for determining the types of *in vivo* studies, and possibly the concentration ranges needed for previously untested, but related, substances.

Toxic substances have historically been identified largely because of overt health effects in humans or other animals, usually after environmental or workplace exposures. A more proactive approach, commonly taken in industry in the 1990s, is to attempt to quantify the toxicity of chemicals before their introduction into the marketplace. Laboratory animals, usually rodents, are most frequently used for such studies, with a variety of measures, such as mortality and tumor development in selected organs, used as endpoints. These studies are time-consuming and expensive, and only a small fraction of all of the new chemicals developed each year can undergo such testing. Further, subtle effects, such as slightly diminished learning abilities or immune responses, may be difficult to detect. Finally, there are both ethical and financial incentives to reduce the numbers of animals used in toxicity testing, particularly if discomfort is involved. Therefore, government, industry, and academic institutions have sought alternative *in vitro* methods of screening substances for toxicity.

This chapter will begin with an overview of some of the important advantages and disadvantages of *in vitro* systems for neurotoxicological studies, followed by brief descriptions of the most commonly used *in vitro* systems. Some applications of *in vitro* systems to neurotoxicity screening will be discussed briefly. The principal focus of the chapter will be on the use of *in vitro* systems for mechanistic studies and the extension of *in vitro* results to *in vivo* effects. A catalog of the vast number of *in vitro* studies in neurotoxicology is not presented; instead, this chapter will concentrate on fundamental principles and issues of *in vitro* neurotoxicology, with a few selected examples chosen to illustrate major points.

11.25.2 ADVANTAGES, DISADVANTAGES, AND COMPLEMENTARITY BETWEEN *IN VITRO* AND *IN VIVO* APPROACHES

Ideally, *in vitro* studies and *in vivo* studies should complement one another. On the one hand, it is often difficult to determine the mechanism(s) of toxicant action purely from *in vivo* experiments. On the other hand, purely *in vitro* studies may reveal mechanisms of action that do not apply *in vivo*, because of toxicant metabolism, barriers to toxicant distribution, or lower toxicant concentration *in vivo*. Therefore, *in vitro* experiments should always be designed with a view toward ultimate application to *in vivo* toxicity. The challenges of applying *in vitro* data to *in vivo* toxicity will be further discussed in Section 11.25.6.

11.25.2.1 Advantages of *In Vitro* Approaches in Neurotoxicology

Under certain circumstances, *in vitro* studies can reduce animal use. A completely validated *in vitro* system for testing, for example, cosmetics for dermal irritation, would virtually eliminate the need for live animals in some situations. Even partially validated *in vitro* models could be used as initial screening devices, to limit the number of suspected toxicants that would be tested in live animals.[1] Here, however, the focus will be on the practical and scientific advantages of *in vitro* systems for determining both toxicity *per se* and mechanisms of toxicity, apart from effects on animal usage.

By far the greatest advantage of *in vitro* systems for neurotoxicological studies is their simplicity. Ideally, *in vitro* systems can be much more thoroughly controlled than *in vivo* systems. The nervous system in an intact animal is immensely complex, with numerous interactions not only among neurons, but also with other parts of the body such as the immune and endocrine systems. *In vitro*, these interactions can be minimized or at least well defined. The use of single cells, parts of cells (e.g., isolated axons), or cell extracts can allow the precise determination of cellular and molecular mechanisms of action, for example, that a toxicant alters current flow through a specific ion channel or inhibits a specific enzyme. Slightly more complex *in vitro* systems, such as cell cultures, can be used to investigate somewhat more complex phenenomena, such as synaptic transmission or neuronal or glial differentiation, under well-defined conditions. The level of complexity of *in vitro* studies can be increased further, until the *in vitro* research becomes an extension of an *in vivo* study, such as patch clamping brain slices from *in vivo* exposed animals to determine if ion channel function is compromised. At each step, the increased level of complexity is at least partially controlled, and results can be compared back to results obtained previously in more simplified systems.

Another advantage of *in vitro* approaches is the availability of human tissues. Some tissues, such as placenta and foreskin, are readily available. Although less commonly employed, neuronal cultures have been prepared from aborted fetuses.[2] Several immortal cell lines of human neuronal origin, including one derived from cerebral cortex, are also available.[3,4]

11.25.2.2 Disadvantages of *In Vitro* Approaches in Neurotoxicology

The simplicity of *in vitro* systems and the separation of the cells or tissues under study from the influences of the intact animal are also the greatest disadvantages of *in vitro* studies. Cells or tissues removed from an animal encounter a very different nutrient environment, lacking at least some of the growth factors that may assist cell functioning *in vivo*. They usually lack the support of surrounding cells. There is no blood–brain barrier, nor do many of the metabolic transformations that occur in, for example, the liver, take place *in vitro*. The interactions that occur *in vivo* may enhance toxicity, such as the conversion of parathion to paraoxon,[5,6] or decrease toxicity, such as the metabolism of pyrethroid insecticides to harmless products.[7] If the *in vitro* system compromises the health of the cells or tissues, it may yield false indications of toxicity. Conversely, if interactions between multiple cell types are required for toxicity, *in vitro* systems will fail to indicate this unless the appropriate cell types are present and interacting similarly to the *in vivo* situation.

It is possible to overcome some of these disadvantages, at least partially. For example, neurons may be cocultured with astrocytes or other glial cells, which may both enhance the health and differentiation of the neurons and mimic, to some degree, either protective or harmful interactions of the glia with a neurotoxicant. Organic toxicants may be metabolized by incubation with liver microsomes or liver extracts[8–10] to yield some of the products that would probably be formed *in vivo*. Alternatively, if the toxicologically active product is already known, the *in vitro* system could be exposed to this product instead of the parent compound. Increasingly sophisticated neuronal culture techniques help to assure healthier cells and allow better experimental manipulation of culture conditions, including serum-free, low-protein media that more closely mimic cerebrospinal fluid, compared to more traditional high-protein media containing 10% or more serum.

11.25.3 *IN VITRO* SYSTEMS

For the purposes of this chapter, *in vitro* systems will be considered to be those in which both the exposure to a neurotoxicant and the analysis of effect are made *in vitro*. Therefore, studies of brain slices, cells, or cell extracts taken directly from *in vivo* exposed animals will not be classified as *in vitro*. Studies of brain slices, cells, or cell extracts taken from unexposed animals and subsequently acutely or chronically exposed to toxicants will be classified as *in vitro*.

11.25.3.1 Whole Embryo

Entire embryos can be maintained in culture, exposed to toxicants, and analyzed with a variety of methodologies, from morphological to biochemical. The embryo is removed from the metabolic influences of the mother and toxicant exposure of the embryo may therefore be relatively well controlled by the experimenter. However, the exposures to the target organs may still be unknown or uncontrolled, because the entire embryo is usually bathed in the toxicant, and the distribution to various organs and accumulation in those organs is complex and generally unknown. Further, interactions within the nervous system, metabolic transformations of toxicants, and interactions with other systems of the body are scarcely less complex than, but may be different from, those that would occur *in vivo* during gestation.

11.25.3.2 Organotypic Cultures, Explants, and Brain Slices

Small pieces of nervous tissue, including brain, spinal cord, or dorsal root ganglia, can be maintained in culture for extended periods of time.[11] Ideally, all of the cell types in the explant and their interactions *in vitro* are similar to those *in vivo*. Although this is seldom proven, in small explants it is reasonable to assume that toxicant concentrations and durations of exposure to neurons and glia within the explant are essentially identical to those of the surrounding media, so exposure conditions are probably well-controlled.

One type of explant, the brain slice, is becoming increasingly popular both for basic research in neurobiology and in neurotoxicology.[12] Brain slices may be used immediately after preparation[13–15] or may be kept in long-term culture in roller tubes;[16–18] in either case, the slice is suitable for electrophysiological, morphological, and biochemical experiments. In some brain regions, such as the hippocampus, a significant degree of connectivity and neuronal interaction is maintained in relatively thin slices (approximately 400 μm), with the tissue properly oriented during slicing. Thus the effects of toxicants on ion channels in intact neurons with almost normal *in vivo* architecture, on transmission at central synapses, on metabolic changes such as protein phosphorylation, or on complex interactive processes such as long-term potentiation, may be studied in slices. Particularly attractive is the opportunity to study the same phenomenom, for example, long-term potentiation, in slices taken from toxicant-exposed animals and in slices from unexposed animals that are subsequently exposed to toxicants *in vitro*.

11.25.3.3 Dissociated Cell Cultures and Cocultures

One of the most commonly used *in vitro* systems, and the one most associated with *in vitro* neurotoxicology, is the dissociated cell culture, including cultures of neurons, glial cells, immortal cell lines derived from neurons or glia, or cocultures of two or more cell types. Depending on cell types, media and substrate employed, plating density, and duration of culture, dissociated neuronal cultures can provide a variety of preparations, from well-separated individual neurons with processes that closely resemble axons and dendrites, both morphologically and biochemically, to dense aggregates interconnected with a network of neurites.[19] Quite pure cultures of both astrocytes and oligodendrocytes can also be obtained easily.[20] Finally, neurons are often co-cultured with glial cells, usually astrocytes. The neurons may be grown directly on a "feeder layer" of astrocytes[21,22] or a "sandwich" technique may be used in which the neurons and astrocytes grow on separate, but closely apposed, surfaces.[23] Coculturing of

neurons with astrocytes is used both to promote survival and differentiation of the neurons and to study interactions between the two.

11.25.3.3.1 Primary cells

Primary cultures of neurons or glia are those that are cultured directly from an animal, and usually maintained in culture more than one day (acutely isolated cells are therefore not usually considered to be "cultured"). Neurons or glial cells may be cultured from defined brain regions, allowing the investigator to study the effects of neurotoxicants on neurons from, for example, susceptible and resistant areas. In certain circumstances, relatively homogeneous neuronal cultures may be obtained, such as cerebellar granule neurons isolated from early postnatal rat or mice pups[24] (although granule neurons are not physiologically homogeneous, for example, in their responses to stimulation of metabotropic glutamate receptors[25]). Various manipulations can assist in purifying cultures. For example, using culture dishes that have not been treated with attachment factors, accompanied by judicious shaking at appropriate intervals, can virtually eliminate neurons from astrocyte cultures.[20] Certain types of serum-free media[26] or antimitotic treatments, such as cytosine arabinoside or fluorodeoxyuridine,[19] suppress glial cell proliferation and maintain relatively pure neuronal cultures.

Although primary cultures consist of "normal" neurons or glia (Figure 1), it should be kept in mind that their morphology and metabolic state may differ significantly from the same cell types *in vivo*. For example, cultured rat hippocampal neurons show enhanced dendritic growth when plated on polylysine as an attachment factor and enhanced axonal growth when plated on laminin.[27] Such differences may enhance neurotoxicology studies (e.g., by providing greater ease of studying certain phenomena, such as axonal elongation) or may hinder interpretation of results if the *in vitro* mechanisms cannot be related readily to *in vivo* events.

11.25.3.3.2 Immortal cell lines

Immortal cell lines are an alternative to primary cultures. Frequently used cell lines include neuroblastoma cells such as N1E-115 (mouse) or SH-SY5Y (human), pheochromocytoma (PC12) cells, and glioma cells such as C6. Immortal cell lines have certain advantages over primary cultures. First, because they multiply indefinitely in culture under the right conditions, they can simultaneously eliminate animal use and provide larger quantities of cells

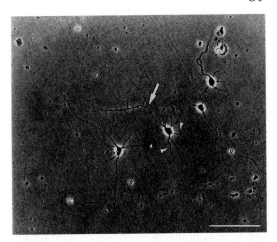

Figure 1 Embryonic rat hippocampal neurons in primary culture for two days show some morphological characteristics that are similar to those seen *in vivo*. Note the single long axon (arrow) and several shorter dendrites (arrowheads). The distinction between axons and dendrites in these neurons also extends to the biochemical level, for example, the segregation of microtubule-associated proteins. Scale bar, 100 μm.

than are usually conveniently available in primary cultures. Second, many can be caused to differentiate in culture "on command," by the addition of appropriate differentiation factors such as nerve growth factor, retinoic acid, or dibutyryl cyclic AMP. Third, they may be relatively homogeneous compared to some types of primary cultures (but see below). Fourth, there is an increasingly large catalog of well-characterized substrains of some cell lines, such as PC12 cells that are deficient in cyclic AMP-dependent protein kinase,[28] providing the equivalent of "knock-out" cells in culture. Finally, many cell lines, such as PC12, are rather easily transfected with foreign genes.[29]

Unfortunately, immortal cell lines have significant drawbacks as well. Immortal cells differ significantly from "normal" neurons, perhaps most strikingly in their ability to multiply indefinitely under the right culture conditions. Further, most neuroblastoma cell lines contain cells with widely differing numbers of chromosomes and therefore differing doses of genes. Immortal cell lines also undergo evolution in culture. The cell lines may mutate, and genetic drift or (usually unknown) selective pressures in culture may promote significant changes in phenotype; for example, PC12 cells from different laboratories may show different reponses compared to the original line.[29] The "same" cell line may therefore differ significantly between laboratories, or in the same laboratory over time. Finally, although individual molecules such as voltage-sensitive calcium channels may

be functionally identical in primary neurons and immortal cell lines, the cell lines may lack some molecules normally found in primary neurons and complex cascades of intracellular signaling may be somewhat different. Depending on the toxicant and cell line under study, these differences may lead to enhanced susceptibility, or enhanced resistance, in the immortal cells.

11.25.3.4 Cell-free Biochemical and Molecular Studies

Finally, biochemical studies of the effects of toxicants on organelles, cell extracts, or purified molecules, although often not considered to be *in vitro* systems, are in fact the ultimate *in vitro* technology. In principle, all of the confounding variables such as hormonal and nutritional status, metabolism of toxicants, and interactions among cells are eliminated, and a "pure" effect of toxicant on organelle or molecule can be determined. Biochemical and molecular studies in defined chemical systems or cell-free extracts will not be considered further in this chapter. The interested reader is referred to Chapters 4–6, this volume.

11.25.4 SCREENING FOR NEUROTOXICITY

Cell cultures, particularly of immortal cell lines, have been proposed as fast, inexpensive methods of screening substances for neurotoxicity. The studies performed have mostly employed known neurotoxicants to determine whether substances that are neurotoxic *in vivo* also cause deleterious effects in culture and whether the dose–response relationships are similar *in vivo* and *in vitro*. Overall, two types of toxicity have been examined: cytotoxic effects that are likely to occur in any cell, and effects that might be specific to neurons or glia.

11.25.4.1 General Cytotoxicity

Measurements that reveal general cytotoxic effects that would probably be applicable to any cell type in culture include trypan blue exclusion[30] (a measure of plasma membrane integrity), release of lactate dehydrogenase[31,32] (LDH; also a measure of plasma membrane integrity), neutral red uptake[33] (a measure of plasma membrane and lysosomal integrity), total protein[30] (as a measure of cell proliferation), MTT conversion[30,34] (a measure of mitochondrial activity), resting membrane potential,[34,35] intracellular free Ca^{2+}

concentrations,[36] ATP content,[37] glucose consumption,[38] synthesis of stress (heat-shock) proteins,[39–42] synthesis of metallothionein,[39] gross cellular morphology (e.g., vacuolization of cytoplasm, degeneration of cells[43]), cell multiplication,[44] uridine uptake,[45] and glutathione levels.[31] Although such measurements are not neuron or glia-specific, they may reveal neuron or glia-specific toxicity if, for example, much higher concentrations of a substance are needed to alter trypan blue exclusion or MTT conversion in glia than in neurons.[30] General cytotoxicity assays usually reveal little of the underlying mechanisms of toxicity.

11.25.4.2 Neuronal and Glial-specific Toxicity

Neurons and glia have several unique characteristics that distinguish them from most or all other cell types, including proteins such as glial fibrillary acidic protein (GFAP) or neuron-specific enolase (NSE). Some *in vitro* screening experiments have employed neuron-specific or glia-specific endpoints as measures of toxicity. It should be noted that an effect on a cell type-specific endpoint does not necessarily imply that the toxicant exerts a cell type-specific effect. For example, in cultured astrocytes, GFAP synthesis per cell might decline either because of a specific effect on GFAP transcription, translation, and degradation or because of generalized effects on protein synthesis which might occur with many proteins in many cell types. With this caveat in mind, a number of neuron-specific or glia-specific neurotoxicity endpoints have been tested, including neurite outgrowth[43,46,47] and concentrations of GFAP,[30,48] beta-S100[38] (glia-specific), and NSE.[38]

In several studies, a variety of substances were evaluated on a battery of two or more tests. In general, the tests correlated well with one another. For example, Xie and Harvey[34] measured the resting membrane potential and MTT conversion rates in NG108–15 neuroblastoma cells exposed to nine substances. The IC_{50}s for the two assays correlated extremely well ($r > 0.9$). However, the absolute IC_{50} values were quite often different, varying by as much as a factor of three; in general, the resting membrane potential was more sensitive than the MTT assay.

11.25.4.3 Validation of *In Vitro* Screening Tests

Validation of *in vitro* screening results is obviously essential if cultures, explants, or other *in vitro* methods are to supplement, let alone replace, testing in live animals. There are various levels of validation, with increasing degrees of confidence that the *in vitro* test is a good predictor of human or animal toxicity. First, only substances that are toxic *in vivo* should be toxic *in vitro*, and the reliability of the *in vitro* tests, particularly the percentages of false positives and false negatives as measured against the human toxicity of a battery of known neurotoxicants, should be no worse than conventional *in vivo* animal tests. Second, when a series of substances are tested in an *in vitro* system, the ranking of relative toxicity should be similar to the ranking of toxicity *in vivo*. For example, Borenfreund and Babich[33] found that the relative cytotoxicity of a series of organotin compounds to N2a neuroblastoma cells corresponded to their toxicity to several types of live organisms. In contrast, a series of snake venom toxins showed very poor correlation between LD_{50}s *in vivo* and depolarization of the resting membrane potential in cultured skeletal muscle cells.[35] Third, the concentration-effect relationship for a series of substances should be the same *in vitro* as *in vivo* (measured, for example, as the slope of IC_{50}s *in vitro* vs. LD_{50}s or other toxic effects *in vivo*[44,49]). Fourth, ideally the absolute concentrations that exert toxic effects *in vitro* should be similar to those that are toxic *in vivo*. This fourth level of validation is quite difficult, particularly for neurotoxicants. Many of the *in vitro* endpoints mentioned above measure either cell death or cell proliferation. Similarly, the *in vivo* endpoint most readily available is often the LD_{50} or human plasma lethal concentration. For some neurotoxicants, such as MPTP[50] or trimethyl tin,[51] neuronal cell death is an important aspect of *in vivo* toxicity and organismal death may occur, so cytotoxicity *in vitro* and LD_{50}s *in vivo* may be well related. For many others, particularly for those that exert chronic, low-level toxicity, such as inorganic lead, significant behavioral effects may occur at concentrations and durations of exposure well below those that produce either neuronal or organismal death. Therefore, many of the most easily measured *in vitro* endpoints, such as LDH release or trypan blue exclusion, may not be relevant to some of the mechanisms of *in vivo* toxicity.

11.25.5 CELLULAR MECHANISMS OF NEUROTOXICITY

Although the use of *in vitro* systems as screening tools to replace live animals in determining toxicity of previously untested substances has not yet gained wide acceptance, *in vitro* systems have secured an important niche in investigations of mechanisms of neurotoxicity.

11.25.5.1 Acute Neurotoxicity *In Vitro*

In vitro preparations, especially brain slices and cultures of both primary neurons and cell lines, are extremely well suited to determining the acute action of neurotoxicants on specific cellular and subcellular processes. In these experiments, the preparation typically is exposed to the toxicant for seconds to hours and the toxicant effect usually is compared to the same phenomenon measured either in parallel control preparations or in the same preparation before toxicant exposure and after toxicant washout.

11.25.5.1.1 Electrophysiological studies

Appropriate *in vitro* preparations have long been used for many electrophysiological studies, such as investigations of the function of voltage-sensitive and ligand-gated ion channels, synaptic transmission, and long-term potentiation. The methods employed include recording with both intracellular and extracellular electrodes, and voltage/patch clamping. Although originally only a few especially favorable preparations were commonly used, such as the frog neuromuscular junction for studies of synaptic transmission or the giant axons and neuron somata of various invertebrates for voltage clamp studies of ion channels, the development of patch clamp techniques[52] has enabled these studies to be applied to a very wide assortment of preparations from mammals.

(i) Ion channels and synaptic transmission

Voltage clamp techniques are used routinely in studies of voltage-sensitive ion channels, including measurements of the voltage dependence of activation and inactivation, kinetics of channel opening and closing, ion permeation, and channel block by both organic and inorganic substances. Voltage and patch clamping are also used to analyze ligand-gated ion channels, such as the ionotropic glutamate, $GABA_A$, and nicotinic acetylcholine receptors, and second messenger-activated channels, such as calcium-activated potassium channels. Voltage and patch clamping have been employed widely in neurotoxicology to determine the acute actions of toxicants, especially pesticides and heavy metals, on voltage-sensitive and ligand-gated ion channels. Only a few examples will be given here, and the interested reader is referred to Chapter 9, this volume and to other reviews.[53-59]

Many heavy metals, including cadmium, cobalt, gadolinium, lanthanum, lead, mercury, nickel, and zinc inhibit current flow through voltage-sensitive calcium channels.[53,60-62] Heavy metals seem to have much less effect on other voltage-sensitive ion channels, for example, most[63-65] but not all[66] studies have found that inorganic lead has little or no effect on voltage-sensitive sodium or potassium channels. With the principal exceptions of inorganic lead[54,63-65] and inorganic and methylmercury,[67-69] these metals have been studied, not from a toxicological viewpoint, but as tools with which to investigate the properties of the calcium channels. The effects of inorganic lead have been studied in primary cultures of hippocampal neurons[70] and dorsal root ganglion cells,[63] in N1E-115 mouse neuroblastoma cells,[64,71] and in SH-SY5Y human neuroblastoma cells.[65] The concentration dependence and degree of inhibition are quite similar in all four cell types, except for hippocampal neurons, which seem to be somewhat more sensitive.[54] There is a multiplicity of calcium channel types, several of which were not recognized at the time of these studies. Therefore, it is possible that there are differences in lead sensitivity among channel subtypes that may account for the increased sensitivity of calcium currents in hippocampal neurons.

In general, heavy metals, including cadmium, cobalt, gadolinium, lanthanum, inorganic lead, nickel, and zinc produce a reversible block of calcium channels, although inorganic lead apparently produces irreversible changes in some cells.[54,71] In the best-documented case, cadmium produces a rapid onset, rapidly reversible reduction in the mean open time of individual L-type calcium channels without altering the single-channel conductance.[72] Inorganic mercury[67,69] and methylmercury,[68] however, cause a slowly developing, irreversible block of calcium channels, suggesting a different mode of action, perhaps a modification of the channel proteins.

Since neurotransmitter release is normally triggered by calcium influx into presynaptic terminals, heavy metals, by reducing calcium influx, would be expected to impair synaptic transmission. In addition, some heavy metals also directly impact postsynaptic receptors and/or their associated ion channels. For example, Pb^{2+} has been found to inhibit current flow through the ion channels associated with the NMDA subclass of glutamate receptors[73-75] (although one study failed to find any inhibition of NMDA receptor-associated channels by quite high Pb^{2+} concentrations[76]) and nicotinic acetylcholine receptors[77,78] in acutely dissociated and cultured hippocampal neurons, and through channels associated with nicotinic acetylcholine receptors and 5-hydroxytryptamine (5-HT$_3$) receptors in cultured N1E-115 neuroblastoma cells.[64] In rat dorsal root ganglion cells, inorganic mercury increases current flow through GABA$_A$ receptor-associated chloride channels, whereas methylmercury decreases current flow through these same channels.[79] Cadmium, zinc, and inorganic mercury also reduce current flow through kainate-sensitive glutamate receptor-associated ion channels, with the effects of zinc and cadmium apparently caused by a different mechanism than the effects of mercury.[80]

Other channels may also be affected by heavy metals. For example, most neurons contain one or two different types of calcium-activated potassium channels. As has been found with some other Ca^{2+}-activated processes, both Cd^{2+} and Pb^{2+} mimic Ca^{2+} and stimulate opening of calcium-activated potassium channels in N1E-115 neuroblastoma cells.[81]

Pyrethroid and certain organochlorine insecticides also alter the function of both voltage-sensitive and ligand-gated ion channels, findings that are based on *in vitro* systems. For example, DDT and pyrethroids (e.g., permethrin, cypermethrin, deltamethrin) all reduce inactivation of voltage-sensitive sodium channels.[82,83] In single-channel experiments performed with neuroblastoma cells, Narahashi and co-workers showed that the pyrethroids deltamethrin[84] and fenvalerate[85] stabilize sodium channel states, thereby increasing the open-channel time. Pyrethroids also inhibit chloride influx through GABA$_A$ receptor-associated ion channels,[86–88] as does lindane.[89–91]

As these selected examples show, *in vitro* systems are ideally suited to discover cellular and subcellular mechanisms of action of neurotoxicants. What is often less obvious, however, is which mechanism or combination of mechanisms is important *in vivo*. For example, there is considerable debate about the *in vivo* importance of pyrethroid effects on GABA$_A$ channels. Inorganic lead, although a potent inhibitor of voltage-sensitive calcium channels and several types of ligand-gated channels *in vitro*, may not exert its toxic effects *in vivo* by any of these mechanisms; in general, the lead concentrations that are effective *in vitro* are considerably greater than the lead concentrations that are ever found *in vivo*. The problem of effective concentrations of lead *in vitro* vs. *in vivo* is briefly discussed by Simons[92] and Audesirk.[70]

(ii) Long-term potentiation

Brain slice preparations, particularly the hippocampal slice, have begun to find use in studies of toxicological effects on more complex phenomena, particularly long-term potentiation (LTP). LTP is a long-lasting enhancement of synaptic transmission triggered by stimulation of specific presynaptic elements in specific patterns or frequencies.[93] LTP is thought to be a possible cellular substrate underlying certain types of learning.[94] Some types of LTP depend on activity of the NMDA subtype of glutamate ionotropic receptors, as well as a cascade of incompletely understood intracellular events, at least some of which are modulated by calcium. Exposure to inorganic lead *in vivo* causes learning deficits. Further, lead has been shown to affect both calcium channels and NMDA receptors *in vitro* and to alter the activity of a variety of intracellular messenger molecules, including calmodulin, which mediates many cellular responses to changes in calcium concentration.[95,96] Therefore, the effects of inorganic lead on LTP has received considerable study. Lead inhibits the development of LTP, both in acutely exposed slices[76,97] and in slices obtained from animals chronically exposed *in vivo*.[98–100] Since the concentrations employed *in vitro* were about two orders of magnitude greater than what would be expected in the cerebrospinal fluid or brain extracellular fluid of lead-exposed animals, it is not clear to what extent the mechanisms that might cause inhibition of LTP *in vitro* contribute to the inhibition of LTP caused by *in vivo* exposure. Indeed, what would seem to be the most likely mechanisms by which lead might inhibit LTP, that is, inhibition of voltage-sensitive calcium channels or NMDA receptors, appear to have been ruled out, at least in the piriform cortex slice.[76]

(iii) Intracellular calcium ion homeostasis

The intracellular concentration of many ions is regulated tightly. The free calcium ion concentration in the cytoplasm is a key modulator (either directly or through activation of calmodulin) of many intracellular molecules, including adenylate cyclase, phosphodiesterase, calcium-calmodulin-dependent protein kinase, calcineurin (a protein phosphatase), and protein kinase C. Changes in intracellular free calcium, brought about by changes in influx, extrusion, or release from intracellular stores, alter the activities of these molecules and consequently alter neuronal physiology and development. High levels of intracellular calcium can trigger cell death.[101] Consequently, perturbation of intracellular calcium homeostasis, either directly or indirectly, is a potential target for many toxicants that alter physiology, development, or survival of neurons.[102]

Calcium homeostasis may be measured in a variety of ways, including measurement of intracellular free calcium ion concentrations with ion-sensitive dyes and measurement of calcium fluxes into and out of cells or organelles. A large number of neurotoxicants, including such diverse classes as organometals,[103–105] inorganic metals,[106–108] organochlorine insecticides,[109–111] and cyanide[112] alter intracellular free calcium concentrations *in vitro*, as measured with calcium-sensitive dyes. Although there are pitfalls to the techniques, including difficulties in absolute calibration and possible effects of other metals on the dyes, including metals normally found within the cell (e.g., zinc[103,113]) and extracellularly applied metals (e.g., lead[114]), estimation of intracellular free calcium ion concentrations, particularly short-term changes in calcium concentration, is a powerful tool for uncovering possible mechanisms of toxic action.

11.25.5.2 Semichronic Neurotoxicity *In Vitro*

Long-term exposures of *in vitro* systems (usually days to weeks) are a somewhat closer approximation to *in vivo* exposure. There are four principal rationales for long-term *in vitro* exposure. First, some phenomena may take days or weeks to develop, such as the growth of axons and dendrites, and therefore of necessity can only be studied with long exposures. Second, the effects of some acute exposures may trigger cellular compensation. For example, several agents that reduce current flow through voltage-sensitive calcium channels (e.g, ethanol[115,116]) stimulate an upregulation of calcium channel density in cultured neurons after several days of exposure.[117,118] Therefore, the net long-term effect of a neurotoxicant may be some combination of acute effects plus cellular compensation. Third, some effects may appear only after relatively long-term exposure. For example, intracellular effects may require a threshold intracellular concentration of a toxicant that builds up slowly over time. Finally, long-term exposure may allow the *in vitro* system to be studied at low concentrations that more closely mimic those encountered *in vivo*, particularly for intracellular effects that may develop slowly at low concentrations and more rapidly at higher concentrations.

These long-term toxicant effects, of course, also occur *in vivo*, and one might question whether long-term *in vitro* experiments are useful compared to *in vivo* exposure followed by subsequent *in vitro* electrophysiological, morphological, or biochemical analysis of brain tissue. The relative simplicity and controllability of *in vitro* systems is once again the key advantage, as is illustrated by studies of the effects of ethanol on calcium channels by Greenfield, Messing, and co-workers.[118,119] As mentioned previously, acute *in vitro* ethanol exposure inhibits current flow through voltage-sensitive calcium channels. Chronic ethanol exposure of PC12 cells for several days upregulates dihydropyridine-sensitive (L-type) calcium channels.[118] Further experiments showed that activation of protein kinase C was required for upregulation to occur.[119] These results suggest experiments that might be performed *in vivo* to see if similar mechanisms are at work in the intact brain.

In addition, some experiments can be carried out *in vitro* that cannot be performed, or at least not easily, in whole animals. For example, enzyme inhibitors that can be applied in culture might not be stable *in vivo*, might not cross the blood–brain barrier, or might cause so many effects in various organs of the body that the health or even survival of the animal might be compromised. Therefore, long-term *in vitro* experiments may complement and extend short-term studies, may suggest mechanisms that could otherwise not readily be discovered, and may serve as trial experiments to assist in the design and execution of *in vivo* studies.

11.25.5.2.1 Electrophysiological studies

Measurements of resting membrane potential are perhaps the simplest electrophysiological technique applied to long-term toxicant exposure of *in vitro* systems. Xie and Harvey[34] exposed NG108–15 neuroblastoma cells to a variety of chemicals and measured the resting potential at 1 and 24 h. In every case, the IC_{50} for a significant reduction of the resting potential was lower in the 24 h exposed cells than in the one hour exposed cells, which supports the use of longer-term cultures to see effects at lower concentrations. Unfortunately, the more powerful techniques of voltage and patch clamping are quite demanding. Further, there is often enormous cell-to-cell variability even within a single culture dish. With acute exposures, voltage-clamp experiments can often use a cell as its own control, which greatly reduces the sample size needed to determine significant effects. With chronic exposures, the large variability among cells demands large sample sizes. Although there are some approaches to reducing variability, such as normalizing currents to cell capacitance, the technical difficulties have discouraged most investigators from pursuing electrophysiological studies of neurons after long-term toxicant exposure in culture.

Despite the difficulties, electrophysiological analysis of long-term cultures offers significant advantages for understanding cellular processes of toxicity. Returning to ethanol as an example, although acute exposure to ethanol inhibits high-threshold voltage-sensitive calcium channels in PC12 cells,[116] chronic exposure increases high-threshold calcium currents in PC12 cells[120] and cultured cerebellar Purkinje neurons,[121] suggesting that upregulation of channel density[118,119] may have overcompensated for the inhibition of function of individual channels.

11.25.5.2.2   *Other physiological, biochemical, and molecular studies*

A large number of physiological, biochemical, and molecular endpoints are amenable to study in long-term *in vitro* cultures. Since many cells (usually thousands to millions) are pooled for each measurement and because some biochemical and physiological measurements are fairly rapid and can be replicated essentially simultaneously in many samples, the problems of variability associated with electrophysiological experiments in long-term cultures are greatly reduced. Most of the screening tests discussed earlier in Section 11.25.4, in fact, consist of a physiological or biochemical measurement (e.g., rate of respiration, concentrations or rate of synthesis of specific proteins) made following toxicant exposure for hours to days, with entire cell populations (usually a dish of cultured cells) serving as the unit of analysis.

A distinction should be drawn between studies designed to test whether a neurotoxicant has a detrimental effect *in vitro* and those designed to investigate the mechanism of action. Screening tests, by definition, utilize endpoints that are relatively easy to measure and provide rapid, cost-effective information about toxicity. In many cases, screening tests use endpoints designed to measure cytotoxicity, such as release of lactate dehydrogenase or exclusion of trypan blue. These endpoints will reveal seriously compromised cellular integrity, regardless of the underlying cause. Studies designed to investigate mechanisms, on the other hand, normally investigate some morphological, biochemical, or physiological parameter for which the researcher hypothesizes a

specific cellular mechanism. For example, one possible mechanism of toxicity of a number of metals is production of reactive oxygen species.[122] Hypothesizing that changes in glutathione concentrations might provide a protective mechanism against inorganic lead toxicity, Legare *et al.*[123] measured glutathione levels in astroglia exposed to lead for up to nine days. Glutathione levels in lead-exposed cultures were initially depressed. By 6–9 days, however, glutathione concentrations had rebounded significantly above the concentrations measured in control cultures, suggesting a possible compensatory mechanism.

11.25.5.2.3   *Developmental and structural studies*

Much of what is currently known about the mechanisms of neuronal differentiation and the production of neuronal cytoarchitecture has been derived from studies of cultured neurons. A large number of intracellular signaling pathways are involved in the initial production of processes from a newly plated neuronal soma and the elaboration of axonal and dendritic trees, including transmembrane calcium fluxes,[124–127] intracellular calcium ion concentrations,[128] protein kinases,[129–133] and protein phosphatases.[134,135] Long-term studies in culture offer the opportunity to study mechanisms whereby neurotoxicants that interact with one or more of these signaling pathways may alter neurite development. Low levels of inorganic lead, for example, inhibit the initial production of neurites from cultured hippocampal neurons.[136,137] Inhibitors of calmodulin, cyclic AMP-dependent protein kinase, and calcium-calmodulin-dependent protein kinase reverse this effect (Figure 2). Stimulators of cyclic AMP-dependent protein kinase and inhibitors of protein phosphatases mimic the effect (Figure 2).[138] These results suggest that lead may alter neurite development through changes in protein phosphorylation.

11.25.6   EXTENDING *IN VITRO* RESULTS TO THE INTACT ANIMAL

As the examples in the above sections have shown, *in vitro* systems are ideally suited to determining (i) whether a suspected toxicant damages cells (screening studies) and (ii) possible cellular effects whereby the toxicant causes damage (mechanistic studies). However, while *in vitro* studies may reveal potential mechanisms of neurotoxicity, the relevance of *in vitro* results

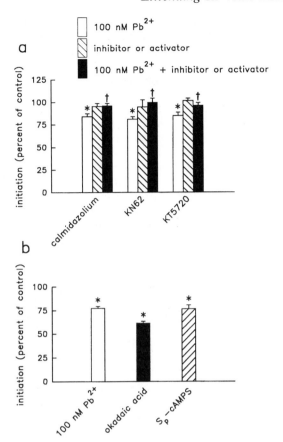

Figure 2 Inorganic lead (100 nM) inhibits neurite initiation in cultured enbryonic rat hippocampal neurons. (a) Reduction of neurite initiation by Pb^{2+} is completely reversed by inhibiting calmodulin (with 250 nM calmidazolium), calcium–calmodulin-dependent protein kinase (with 5 μM KN62), or cyclic AMP-dependent protein kinase (with 250 nM KT5720). Asterisks denote significant differences ($p < 0.05$) between initiation in control cultures and in cultures epxosed to Pb^{2+}; daggers denote significant differences between initiation in cultures exposed to Pb^{2+} alone and in cultures exposed to Pb^{2+} combined with the indicated inhibitor. The inhibitors alone, at these concentrations, had no effect on initiation. (b) Inhibition of neurite initiation by Pb^{2+} is mimicked by treatments that enhance protein phosphorylation, including 0.5 nM okadaic acid (at this concentration, an inhibitor of protein phosphatase 2A) and 10 μM S_p-cAMPS (a membrane-permeable analogue of cyclic AMP that stimulates cyclic AMP-dependent protein kinase). All three treatments significantly reduce neurite initiation compared to initiation rates in control cultures (after Kern and Audesirk[138]).

to *in vivo* neurotoxicity is not always immediately clear. Three important difficulties in relating *in vitro* mechanisms to *in vivo* neurotoxicity are metabolism and distribution of toxicants *in vivo*, the concentration/duration dilemma, and suitable endpoints, often with multiple mechanisms generating a common endpoint.

11.25.6.1 Metabolism and Distribution of Neurotoxicants *In Vivo*

As was pointed out earlier, many organic neurotoxicants are metabolized *in vivo*, often principally by the liver. In many cases, it does not appear that neurons of the central nervous system are important sites of metabolism, so that application of the "parent" substance to, for example, neuronal cultures, will not result in significant metabolism. If metabolism is an important route of detoxification (e.g., pyrethroid insecticides), then data from neuronal cultures may indicate spuriously high toxicity, owing to the presence of relatively high concentrations of the unaltered parent substance. If metabolism generates a toxic substance from a relatively harmless parent substance (e.g., parathion, tetraethyl lead), then data from neuronal cultures may indicate spuriously low toxicity.

The distribution of neurotoxicants *in vivo* must also be considered, particularly as affected by the blood–brain barrier. Many organic neurotoxicants readily cross the blood–brain barrier, while many water-soluble metals cross much less readily. However, immature animals or animals in certain disease states may have much more permeable blood–brain barriers. Further, certain neurotoxicants may accumulate preferentially in certain structures, for example, inorganic lead may accumulate in the hippocampus more than in most other brain regions[139–142] although this has been disputed.[143]

As these examples illustrate, *in vitro* studies, to be roughly comparable to *in vivo* exposures, must utilize carefully chosen substances at appropriate concentrations, using the *in vivo* data that may be available.

11.25.6.2 The Concentration–Duration Dilemma

Unfortunately, even in the rare case in which *in vivo* concentrations and distributions are well known, it is not always clear what the appropriate concentration for *in vitro* studies should be. Even "long term" *in vitro* studies are usually, by *in vivo* standards, quite short term. Cultures and explants are seldom maintained for more than a few days, very rarely for a few weeks. *In vivo* exposures may last for weeks or months in experimental rodent studies, and human exposures may last for many years. Although it is not always true that the biologically relevant exposure is some simple combination of toxicant concentration and duration of exposure, in many instances relatively high concentrations for short durations do appear to give results that are roughly comparable to lower concentrations for longer durations. For example, a 50 day exposure of aggregating embryonic rat brain cultures to 10^{-7} M lead acetate produced a similar reduction in glutamine synthetase activity as a nine day *in vitro* exposure to 10^{-6} M lead acetate.[144]

One reason for using relatively high "nonphysiological" toxicant concentrations *in vitro* is the difficulty of detecting small effects, even though these may be relevant toxicologically. In voltage-clamp studies, it would usually be difficult to achieve statistical significance with a 10% or 20% difference in current magnitude caused by a toxicant, because of large cell-to-cell variability. Thus, if variability is great and/or the experimental paradigm difficult and time-consuming, relatively high toxicant concentrations may be justified.

Whatever the rationale, if the toxicant concentrations employed *in vitro* are much larger than those encountered *in vivo*, the results must be interpreted very conservatively. Although this is becoming less common, occasionally toxicant concentrations are used *in vitro* that are several orders of magnitude greater than the relevant *in vivo* concentrations. Under such circumstances, it is difficult to determine what, if any, significance the *in vitro* findings may have for *in vivo* toxicity.

11.25.6.3 Suitable Endpoints and Multiple Mechanisms with a Common Endpoint

In vitro neurotoxicology studies usually have the goal of discovering underlying mechanisms of toxicant action. In many cases, one might expect a toxicant to have multiple, possibly independent, effects at the subcellular level. For example, pyrethroid insecticides have long been known to delay inactivation of voltage-dependent sodium channels. With considerable justification, this is usually regarded as their primary mode of both insecticidal action and mammalian toxicity. However, Enan and Matsumura[145] discovered that certain pyrethroids are extraordinarily potent inhibitors of the calcium–calmodulin-dependent protein phosphatase, calcineurin, at picomolar concentrations. This is certainly worthy of investigation as a second mechanism of pyrethroid neurotoxicity.

As this example illustrates, one cannot expect any single study, or even a series of studies, to determine the one and only mechanism of action of a toxicant. Indeed, it may often be a desirable strategy to look first at endpoints, such as cell death, resting membrane potential, neuronal differentiation, or intracellular free calcium ion concentration, that are known to be impacted by multiple intracellular pathways. If an effect is found on such a generalized endpoint, then further studies can attempt to elucidate which "upstream" intracellular events and molecules might be the ultimate toxicant targets.

One must also keep in mind that the endpoints measured *in vitro*, although valid indicators of cellular toxicity, and suggestive of possible *in vivo* mechanisms, cannot necessarily be assumed to apply *in vivo*. The energy status of cells, the array of growth factors, hormones, and attachment factors to which cells are exposed, direct interactions with other nearby cells, and even the basic geometry of cells *in vitro* is likely to differ from the situation *in vivo*. Cells *in vitro* may be either more or less supplied with nutrients, growth factors, etc., and may be either more or less susceptible to toxicant action than their counterparts *in vivo*. Therefore, *in vitro* studies are generally most useful if they are designed to investigate cellular mechanisms of action that predict specific toxic effects that can be measured *in vivo*.

11.25.7 REFERENCES

1. B. Veronesi, in 'Neurotoxicology,' eds. H. Tilson and C. Mitchell, Raven Press, New York, 1992, pp. 21–49.
2. F. C. Chiu, R. Rozental, C. Bassallo, *et al.*, 'Human fetal neurons in culture: intercellular communication and voltage- and ligand-gated responses.' *J. Neurosci. Res.*, 1994, **38**, 687–697.
3. G. V. Ronnett, in 'Neurotoxicology: Approaches and Methods,' eds. L. W. Chang and W. Slikker, Jr., Academic Press, San Diego, CA, 1995, pp. 581–593.
4. G. V. Ronnett, L. D. Hester, J. S. Nye *et al.*, 'Human cortical neuronal cell line: establishment from a patient with unilateral megalencephaly.' *Science*, 1990, **248**, 603–605.
5. R. A. Neal, 'Studies on the metabolism of diethyl 4-nitrophenyl phosphorothionate (parathion) *in vitro*.' *Biochem J.*, 1967, **103**, 183–191.
6. T. Kamataki, M. C. M. Lee Lin, D. H. Belcher *et al.*, 'Studies of the metabolism of parathion with an apparently homogeneous preparation of rabbit liver microsomal cytochrome P-450.' *Drug Metab. Dispos.*, 1976, **4**, 180–189.
7. T. Shono, K. Ohsawa and J. E. Casida, 'Metabolism of *trans*- and *cis*-permethrin, *trans*- and *cis*-cypermethrin and deltamethrin by microsomal enzymes.' *J. Agric. Food Chem.*, 1979, **27**, 316–325.
8. B. Ames, J. McCann and E. Yamasaki, 'Methods for detecting carcinogens and mutagens with the *Salmonella*/mammalian-microsome mutagenicity test.' *Mutat. Res.*, 1975, **31**, 347–364.
9. L. M. Segal and S. Fedoroff, 'Cholinesterase inhibition by organophosphorus and carbamate pesticides in aggregate cultures of neural cells from the foetal rat brain: the effects of metabolic activation and pesticide mixtures.' *Toxicol. In Vitro*, 1989, **3**, 123–128.

10. G. L. Sprague and T. R. Castles, 'Estimation of the delayed neurotoxic potential and potency for a series of triaryl phosphates using an *in vitro* test with metabolic activation.' *Neurotoxicology*, 1985, **6**, 79–86.

11. M. B. Bornstein, in 'Neurotoxicology: Approaches and Methods,' eds. L. W. Chang and W. Slikker, Jr., Academic Press, San Diego, CA, 1995, pp. 573–579.

12. S. B. Fountain and T. J. Teyler, in 'Neurotoxicology: Approaches and Methods,' eds. L. W. Chang and W. Slikker, Jr., Academic Press, San Diego, CA, 1995, pp. 517–535.

13. C. Yamamoto and H. McIlwain, 'Electrical activities in thin sections from the mammalian brain maintained in chemically defined media *in vitro*.' *J. Neurochem.*, 1966, **13**, 1333–1343.

14. T. V. Dunwiddie, in 'Electrophysiological Techniques in Pharmacology,' ed. H. M. Geller, Liss, New York, 1986, pp. 65–69.

15. J. S. Kelly, in 'New Techniques in Psychopharmacology,' eds. L. L. Iversen, S. D. Iversen and S. H. Snyder, Plenum, New York, 1982, pp. 95–183.

16. B. H. Gahwiler, 'Organotypic monolayer cultures of nervous tissue.' *J. Neurosci. Methods*, 1981, **4**, 329–342.

17. B. H. Gahwiler, 'Development of the hippocampus *in vitro*: cell types, synapses and receptors.' *Neuroscience*, 1984, **11**, 751–760.

18. B. H. Gahwiler, S. M Thompson, E. Audinat *et al.*, in 'Culturing Nerve Cells,' eds. G. Banker and K. Goslin, MIT Press, Cambridge, MA, 1991, pp. 379–411.

19. G. Banker and K. Goslin, in 'Culturing Nerve Cells,' eds. G. Banker and K. Goslin, MIT Press, Cambridge, MA, 1991, pp. 11–39.

20. S. W. Levison and K. D. McCarthy, in 'Culturing Nerve Cells,' eds. G. Banker and K. Goslin, MIT Press, Cambridge, MA, 1991, pp. 309–336.

21. I. D. Forsythe and G. L. Westbrook, 'Slow excitatory postsynaptic currents mediated by *N*-methyl-D-aspartate receptors on cultured mouse central neurones.' *J. Physiol. (Lond.)*, 1988, **396**, 515–533.

22. K. A. Yamada, J. M. Dubinski and S. M. Rothman, 'Quantitative physiological characterization of a quinoxalinedione non-NMDA receptor antagonist.' *J. Neurosci.*, 1989, **9**, 3230–3236.

23. K. Goslin and G. Banker, in 'Culturing Nerve Cells,' eds. G. Banker and K. Goslin, MIT Press, Cambridge, MA, 1991, pp. 251–281.

24. E. Trenkner, in 'Culturing Nerve Cells,' eds. G. Banker and K. Goslin, MIT Press, Cambridge, MA, 1991, pp. 283–307.

25. D. Milani, P. Candeo, M. Favaron *et al.*, 'A subpopulation of cerebellar granule neurons in culture expresses a functional mGluR1 metabotropic glutamate receptor: effect of depolarizing growing conditions.' *Receptors Channels*, 1993, **1**, 243–250.

26. G. J. Brewer, J. R. Torricelli, E. K. Evege *et al.*, 'Optimized survival of hippocampal neurons in B27-supplemented Neurobasal, a new serum-free medium combination.' *J. Neurosci. Res.*, 1993, **35**, 567–576.

27. B. C. Wheeler and G. J. Brewer, 'Selective hippocampal neuritogenesis: axon growth on laminin or pleiotrophin, dendrite growth on poly-D-lysine.' *Soc. Neurosci. Abstr.*, 1994, **20**, 1292.

28. D. Glowacka and J. A. Wagner, 'Role of the cAMP-dependent protein kinase and protein kinase C in regulating the morphological differentiation of PC12 cells.' *J. Neurosci. Res.*, 1990, **25**, 453–462.

29. L. A. Greene, M. M. Sobeih and K. K. Teng, in 'Culturing Nerve Cells,' eds. G. Banker and K. Goslin, MIT Press, Cambridge, MA, 1991, pp. 208–226.

30. M. R. Cookson, R. McClean, S. P. Williams *et al.*, 'Use of astrocytes for *in vitro* neurotoxicity testing.' *Toxicol. In Vitro*, 1994, **8**, 817–819.

31. C. L. Galli, B. Viviani and M. Marinovich, 'Cell cultures: a tool for the study of mechanisms of toxicity.' *Toxicol. In Vitro*, 1993, **7**, 559–568.

32. G. Repetto, P. Sanz and M. Repetto, 'In vitro effects of mercuric chloride and methylmercury chloride on neuroblastoma cells.' *Toxicol. In Vitro*, 1993, **7**, 353–357.

33. E. Borenfreund and H. Babich, 'In vitro cytotoxicity of heavy metals, acrylamide, and organotin salts to neural cells and fibroblasts.' *Cell Biol. Toxicol.*, 1987, **3**, 63–73.

34. K. Xie and A. L. Harvey, 'Evaluation of nerve cell toxicity *in vitro* by electrophysiological and biochemical methods.' *Toxicol. In Vitro*, 1993, **7**, 275–279.

35. A. L. Harvey, 'Possible developments in neurotoxicity testing *in vitro*.' *Xenobiotica*, 1988, **18**, 625–632.

36. B. Viviani, A. D. Rossi, S. C. Chow *et al.*, 'Organotin compounds induce calcium overload and apoptosis in PC12 cells.' *Neurotoxicology*, 1995, **16**, 19–25.

37. A. Castano and J. V. Tarazona, 'ATP assay on cell monolayers as an index of cytotoxicity.' *Bull Environ. Contam. Toxicol.*, 1994, **53**, 309–316.

38. J. Huang, H. Tanii, K. Kato *et al.*, 'Neuron and glial cell marker proteins as indicators of heavy metal-induced neurotoxicity in neuroblastoma and glioma cell lines.' *Arch. Toxicol.*, 1993, **67**, 491–496.

39. J. W. Bauman, J. Liu and C. D. Klaassen, 'Production of metallothionein and heat-shock proteins in response to metals.' *Fundam. Appl. Toxicol.*, 1993, **21**, 15–22.

40. K. Kato, S. Goto, K. Hasegawa *et al.*, 'Coinduction of two low-molecular-weight stress proteins, αB crystallin and HSP28, by heat or arsenite stress in human glioma cells.' *J. Biochem.*, 1993, **114**, 640–647.

41. W. Levinson, H. Oppermann and J. Jackson, 'Transition series metals and sulfhydryl reagents induce the synthesis of four proteins in eukaryotic cells.' *Biochim. Biophys. Acta*, 1980, **606**, 170–180.

42. H. Yamada and S. Koizumi, 'Induction of a 70-kDa protein in human lymphocytes exposed to inorganic heavy metals and toxic organic compounds.' *Toxicology*, 1993, **79**, 131–138.

43. K. S. Khera and C. Whalen, 'Detection of neuroteratogens with an *in vitro* cytotoxicity assay using primary monolayers cultured from dissociated foetal rat brains.' *Toxicol. In Vitro*, 1988, **2**, 257–273.

44. H. E. Kennah II, D. Albulescu, S. Hignet *et al.*, 'A critical evaluation of predicting ocular irritancy potential from an *in vitro* cytotoxicity assay.' *Fundam. Appl. Toxicol.*, 1989, **12**, 281–290.

45. C. Shopsis and S. Sathe, 'Uridine uptake as a cytotoxicity test: correlations with the Draize test.' *Toxicology*, 1984, **29**, 195–206.

46. E. M. Abdulla and I. C. Campbell, 'L-BMAA and kainate-induced modulation of neurofilament concentrations as a measure of neurite outgrowth: implications for an *in vitro* test of neurotoxicity.' *Toxicol. In Vitro*, 1993, **7**, 341–344.

47. D. J. Brat and S. Brimijoin, 'A paradigm for examining toxicant effects on viability, structure, and axonal transport of neurons in culture.' *Mol. Neurobiol.*, 1992, **6**, 125–135.

48. M. R. Cookson and V. W. Pentreath, 'Alterations in the glial fibrillary acidic protein content of primary astrocyte cultures for evaluation of glial cell toxicity.' *Toxicol. In Vitro*, 1994, **8**, 351–359.

49. F. A. Barile, P. J. Dierickx and U. Kristen, 'In vitro cytotoxicity testing for prediction of acute human toxicity.' *Cell Biol. Toxicol.*, 1994, **10**, 155–162.

50. J. W. Langston, P. Ballard, J. W. Tetrud *et al.*, 'Chronic Parkinsonism in humans due to a product of a meperidine-analog synthesis.' *Science*, 1983, **219**, 979–980.

51. A. W. Brown, W. N. Aldridge, B. W. Street *et al.*, 'The behavioral and neuropathological sequelae of intoxication by trimethyltin compounds in the rat.' *Am. J. Pathol.*, 1979, **97**, 59–82.

52. O. P. Hamill, A. Marty, E. Neher *et al.*, 'Improved patch-clamp techniques for high-resolution current recording from cells and cell-free membrane patches.' *Pflugers Arch.*, 1981, **391**, 85–100.

53. G. Audesirk, in 'Biological Effects of Heavy Metals,' ed. E. C. Foulkes, CRC Press, Boca Raton, FL, 1989, pp. 1–17.

54. G. Audesirk, 'Electrophysiology of lead intoxication: effects on voltage-sensitive ion channels.' *Neurotoxicology*, 1993, **14**, 137.

55. G. Audesirk, in 'Neurotoxicology: Approaches and Methods,' eds. L. W. Chang and W. Slikker, Jr., Academic Press, San Diego, CA, 1995, pp. 137–156.

56. J. van den Bercken, M. Oortgiesen, T. Leinders-Zufall *et al.*, in 'Neurotoxicology: Approaches and Methods,' eds. L. W. Chang and W. Slikker, Jr., Academic Press, San Diego, CA, 1995, pp. 603–612.

57. H. P. M. Vijverberg and J. van den Bercken, 'Neurotoxicological effects and the mode of action of pyrethroid insecticides.' *Crit. Rev. Toxicol.*, 1990, **21**, 105–126.

58. H. P. M. Vijverberg and J. R. de Weille, 'The interaction of pyrethroids with voltage-dependent Na channels.' *Neurotoxicology*, 1985, **6**, 23–34.

59. T. Narahashi, in 'Neuropharmacology and Pesticide Action,' eds. M. G. Ford, G. G. Lunt, R. C. Reay *et al.*, Ellis Horwood, Chichester, 1986, pp. 36–60.

60. S. Hagiwara and L. Byerly, 'Calcium channels.' *Annu. Rev. Neurosci.*, 1981, **4**, 69–125.

61. C. Edwards, 'The selectivity of ion channels in nerve and muscle.' *Neuroscience*, 1982, **7**, 1335–1366.

62. P. R. Stanfield, 'Voltage-dependent calcium channels of excitable membranes.' *Br. Med. Bull.*, 1986, **42**, 359–367.

63. M. L. Evans, D. Busselberg and D. O. Carpenter, 'Pb^{2+} blocks calcium currents in cultured dorsal root ganglion cells.' *Neurosci. Lett.*, 1991, **129**, 103–106.

64. M. Oortgiesen, R. G. van Kleef, R. B. Bajnath *et al.*, 'Nanomolar concentrations of lead selectively block neuronal nicotinic acetylcholine responses in mouse neuroblastoma cells.' *Toxicol. Appl. Pharmacol.*, 1990, **103**, 165–174.

65. E. Reuveny and T. Narahashi, 'Potent blocking action of lead on voltage-activated calcium channels in human neuroblastoma cells SH-SY5Y.' *Brain Res.*, 1991, **545**, 312–314.

66. M. Madeja, N. Binding, R. Musshoff *et al.*, 'Effects of lead on cloned voltage-operated neuronal potassium channels.' *Naunyn Schmiedebergs Arch. Pharmacol.*, 1995, **351**, 320–327.

67. M. Pekel, B. Platt and D. Busselberg, 'Mercury (Hg^{2+}) decreases voltage-gated calcium channel currents in rat DRG and *Aplysia* neurons.' *Brain Res.*, 1993, **632**, 121–126.

68. T. J. Shafer and W. D. Atchison, 'Methylmercury blocks N- and L-type Ca^{2+} channels in nerve growth factor-differentiated pheochromocytoma (PC12) cells.' *J. Pharmacol. Exp. Ther.*, 1991, **258**, 149–157.

69. F. Weinsberg, U. Bickmeyer and H. Wiegand, 'Effects of inorganic mercury (Hg^{2+}) on calcium channel currents and catecholamine release from bovine chromaffin cells.' *Arch. Toxicol.*, 1995, **69**, 191–196.

70. G. Audesirk and T. Audesirk, 'The effects of inorganic lead on voltage-sensitive calcium channels differ among cell types and among channel subtypes.' *Neurotoxicology*, 1993, **14**, 259–265.

71. G. Audesirk and T. Audesirk, 'Effects of inorganic lead on voltage-sensitive calcium channels in N1E-115 neuroblastoma cells.' *Neurotoxicology*, 1991, **12**, 519–528.

72. J. B. Lansman, P. Hess and R. W. Tsien, 'Blockade of current through single calcium channels by Cd^{2+}, Mg^{2+}, and Ca^{2+}. Voltage and concentration dependence of calcium entry into the pore.' *J. Gen. Physiol.*, 1986, **88**, 321–347.

73. M. Alkondon, A. C. S. Costa, V. Radhakrishnan *et al.*, 'Selective blockade of NMDA-activated channel currents may be implicated in learning deficits caused by lead.' *FEBS Lett.*, 1990, **261**, 124–130.

74. H. Ujihara and E. X. Albuquerque, 'Developmental change of the inhibition by lead of NMDA-activated currents in cultured hippocampal neurons.' *J. Pharmacol. Exp. Ther.*, 1992, **263**, 868–875.

75. V. Uteshev, D. Busselberg and H. L. Haas, 'Pb^{2+} modulates the NMDA-receptor-channel complex.' *Naunyn Schmiedebergs Arch. Pharmacol.*, 1993, **347**, 209–213.

76. N. Hori, D. Busselberg, M. R. Matthews *et al.*, 'Lead blocks LTP by an action not at NMDA receptors.' *Exp. Neurol.*, 1993, **119**, 192–197.

77. J. G. Montes, K. Ishihara, M. Alkondon *et al.*, 'Inhibition by lead of nicotinic currents in cultured rat hippocampal neurons.' *Soc. Neurosci. Abstr.*, 1994, **20**, 1138.

78. K. Ishihara, M. Alkondon, J. G. Montes *et al.*, 'Nicotinic currents in neurons acutely dissociated from rat hippocampus and inhibition of the currents by lead.' *Soc. Neurosci. Abstr.*, 1994, **20**, 1138.

79. O. Arakawa, M. Nakahiro and T. Narahashi, 'Mercury modulation of GABA-activated chloride channels and non-specific cation channels in rat dorsal root ganglion neurons.' *Brain Res.*, 1991, **551**, 58–63.

80. J. A. Umback and C. B. Gundersen, 'Mercuric ions are potent noncompetitive antagonists of human brain kainate receptors expressed in *Xenopus* oocytes.' *Mol. Pharmacol.*, 1989, **36**, 582–588.

81. T. Leinders, R. G. D. M. van Kleef and H. P. M. Vijverberg, 'Divalent cations activate small-(Sk) and large-conductance (Bk) channels in mouse neuroblastoma cells: selective activation of Sk channels by cadmium.' *Pflugers Arch.*, 1992, **422**, 217–222.

82. A. E. Lund and T. Narahashi, 'Kinetics of sodium-channel modification as the basis for the variation in the nerve membrane effects of pyrethroids and DDT analogs.' *Pesticide Biochem. Physiol.*, 1983, **20**, 203–216.

83. T. Narahashi, 'Nerve membrane Na$^+$ channels as targets of insecticides.' *Trends Pharmacol. Sci.*, 1992, **13**, 236–241.

84. K. Chinn and T. Narahashi, 'Stabilization of sodium channel states by deltamethrin in mouse neuroblastoma cells.' *J. Physiol. (Lond.)*, 1986, **380**, 191–207.

85. S. F. Holloway, V. L. Salgado, C. H. Wu *et al.*, 'Kinetic properties of single sodium channels modified by fenvalerate in mouse neuroblastoma cells.' *Pflugers Arch.*, 1989, **414**, 613–621.

86. A. T. Eldefrawi and M. E. Eldefrawi, 'Receptors for γ-aminobutyric acid and voltage-dependent chloride channels as targets for drugs and toxicants.' *FASEB J.*, 1987, **1**, 262–271.

87. L. J. Lawrence and J. E. Casida, 'Stereospecific action of pyrethroid insecticides on the γ-aminobutyric acid receptor-ionophore complex.' *Science*, 1983, **221**, 1399–1401.

88. A. A. Ramadan, N. M. Bakry, A. S. M. Marei *et al.*, 'Action of pyrethroids on GABA$_A$ receptor function.' *Pest. Biochem. Physiol.*, 1988, **32**, 97–105.

89. F. Matsumura and S. M. Ghiasuddin, 'Evidence for similarities between cyclodiene type insecticides and picrotoxinin in their action mechanisms.' *J. Environ. Sci. Health B*, 1983, **18**, 1–14.

90. T. Obata, H. I. Yamamura, E. Malatynska *et al.*, 'Modulation of γ-aminobutyric acid-stimulated chloride influx by bicycloorthocarboxylates, bicyclophosphorus esters, polychlorocycloalkanes and other cage convulsants.' *J. Pharmacol. Exp. Ther.*, 1988, **244**, 802–806.

91. N. Tokutomi, Y. Ozoe, N. Katayama *et al.*, 'Effects of lindane (γ-BCH) and related convulsants on GABA_A receptor-operated chloride channels in frog dorsal root ganglion neurons.' *Brain Res.*, 1994, **643**, 66–73.

92. T. J. B. Simons, 'Lead-calcium interactions in cellular lead toxicity.' *Neurotoxicology*, 1993, **14**, 77–85.

93. T. V. P. Bliss and T. Lomo, 'Long-lasting potentiation of synaptic transmission in the dentate area of the anaesthetized rabbit following stimulation of the perforant path.' *J. Physiol. (Lond.)*, 1973, **232**, 331–356.

94. T. J. Teyler and P. DiScenna, 'Long-term potentiation.' *Annu. Rev. Neurosci.*, 1987, **10**, 131–161.

95. G. W. Goldstein and D. Ar, 'Lead activates calmodulin sensitive processes.' *Life Sci.*, 1983, **33**, 1001–1006.

96. E. Habermann, K. Crowell and P. Janicki, 'Lead and other metals can substitute for Ca^{2+} in calmodulin.' *Arch. Toxicol.*, 1983, **54**, 61–70.

97. L. Altmann, K. Sveinsson and H. Weigand, 'Long-term potentiation in rat hippocampal slices is impaired following acute lead perfusion.' *Neurosci. Lett.*, 1991, **128**, 109–112.

98. L. Altmann, M. Gutowski and H. Wiegand, 'Effects of maternal lead exposure on functional plasticity in the visual cortex and hippocampus of immature rats.' *Brain Res. Dev. Brain Res.*, 1994, **81**, 50–56.

99. L. Altmann, F. Weinsberg, K. Sveinsson *et al.*, 'Impairment of long-term potentiation and learning following chronic lead exposure.' *Toxicol. Lett.*, 1993, **66**, 105–112.

100. S. M. Lasley, J. Polan-Curtain and D. L. Armstrong, 'Chronic exposure to environmental levels of lead impairs *in vivo* induction of long-term potentiation in rat hippocampal dentate.' *Brain Res.*, 1993, **614**, 347–351.

101. S. Orrenius, D. J. McConkey, G. Bellomo *et al.*, 'Role of Ca^{2+} in toxic cell killing.' *Trends Pharmacol. Sci.*, 1989, **10**, 281–285.

102. M. A. Verity, 'Ca^{2+}-dependent processes as mediators of neurotoxicity.' *Neurotoxicology*, 1992, **13**, 139–147.

103. M. F. Denny, M. F. Hare and W. D. Atchison, 'Methylmercury alters intrasynaptosomal concentrations of endogenous polyvalent cations.' *Toxicol. Appl. Pharmacol.*, 1993, **122**, 222–232.

104. M. F. Hare and W. D. Atchison, 'Methylmercury mobilizes Ca^{++} from intracellular stores sensitive to inositol 1,4,5-trisphosphate in NG108–15 cells.' *J. Pharmacol. Exp. Ther.*, 1995, **272**, 1016–1023.

105. H. Komulainen and S. C. Bondy, 'Increased free intrasynaptosomal Ca^{2+} by neurotoxic organometals: distinctive mechanisms.' *Toxicol. Appl. Pharmacol.*, 1987, **88**, 77–86.

106. A. D. Rossi, O. Larsson, L. Manzo *et al.*, 'Modifications of Ca^{2+} signaling by inorganic mercury in PC12 cells.' *FASEB J.*, 1993, **7**, 1507–1514.

107. F. A. Schanne, J. R. Moskal and R. K. Gupta, 'Effect of lead on intracellular free calcium ion concentration in a presynaptic neuronal model: ^{19}F-NMR study of NG108–15 cells.' *Brain Res.*, 1989, **503**, 308–311.

108. X. X. Tan, C. Tang, A. F. Castoldi *et al.*, 'Effects of inorganic and organic mercury on intracellular calcium levels in rat T lymphocytes.' *J. Toxicol. Environ. Health*, 1993, **38**, 159–170.

109. C. A. Ferguson and G. Audesirk, 'Non-GABA_A-mediated effects of lindane on neurite development and intracellular free calcium ion concentration in cultured rat hippocampal neurons.' *Toxicol. In Vitro*, 1995, **9**, 95–106.

110. R. M. Joy and V. W. Burns, 'Exposure to lindane and two other hexachlorocyclohexane isomers increases free intracellular calcium levels in neurohybridoma cells.' *Neurotoxicology*, 1988, **9**, 637–643.

111. H. Komulainen and S. C. Bondy, 'Modulation of levels of free calcium within synaptosomes by organochlorine insecticides.' *J. Pharmacol. Exp. Ther.*, 1987, **241**, 575–581.

112. J. D. Johnson, W. G. Conroy and G. E. Isom, 'Alteration of cytosolic calcium levels in PC12 cells by potassium cyanide.' *Toxicol. Appl. Pharmacol.*, 1987, **88**, 217–224.

113. M. F. Denny and W. D. Atchison, 'Methylmercury-induced elevations in intrasynaptosomal zinc concentrations: an ^{19}F-NMR study.' *J. Neurochem.*, 1994, **63**, 383–386.

114. J. L. Tomsig and J. B. Suszkiw, 'Pb^{2+}-induced secretion from bovine chromaffin cells: fura-2 as a probe for Pb^{2+}.' *Am. J. Physiol.*, 1990, **259**, C762–768.

115. D. A. Twombly, M. D. Herman, C. H. Kye *et al.*, 'Ethanol effects on two types of voltage-activated calcium channels.' *J. Pharmacol. Exp. Ther.*, 1990, **254**, 1029–1037.

116. D. Mullikin-Kilpatrick and S. N. Triestman, 'Ethanol inhibition of L-type Ca^{2+} channels in PC12 cells: role of permeant ions.' *Eur. J. Pharmacol.*, 1994, **270**, 17–25.

117. J. C. Harper, C. H. Brennan and J. M. Littleton, 'Genetic up-regulation of calcium channels in a cellular model of ethanol dependence.' *Neuropharmacology*, 1989, **28**, 1299–1302.

118. S. S. Marks, D. L. Watson, C. L. Carpenter *et al.*, 'Comparative effects of chronic exposure to ethanol and calcium channel antagonists on calcium channel antagonist receptors in cultured neural (PC12) cells.' *J. Neurochem.*, 1989, **53**, 168–172.

119. R. O. Messing, A. B. Sneade and B. Savidge, 'Protein kinase C participates in up-regulation of dihydropyridine-sensitive calcium channels by ethanol.' *J. Neurochem.*, 1990, **55**, 1383–1389.

120. A. J. Grant, G. Koski and S. N. Triestman, 'Effect of chronic ethanol on calcium currents and calcium uptake in undifferentiated PC12 cells.' *Brain Res.*, 1993, **600**, 280–284.

121. D. L. Gruol and K. L. Parsons, 'Chronic exposure to alcohol during development alters the calcium currents of cultured cerebellar Purkinje neurons.' *Brain Res.*, 1994, **634**, 283–290.

122. S. J. Stohs and D. Bagchi, 'Oxidative mechanisms in the toxicity of metal ions.' *Free Radic. Biol. Med.*, 1995, **18**, 321–326.

123. M. E. Legare, R. Barhoumi, R. C. Burghardt *et al.*, 'Low-level lead exposure in cultured astroglia: identification of cellular targets with vital fluorescent probes.' *Neurotoxicology*, 1993, **14**, 267–272.

124. G. Audesirk, T. Audesirk, C. Ferguson *et al.*, 'L-type calcium channels may regulate neurite initiation in cultured chick embryo brain neurons and N1E-115 neuroblastoma cells.' *Brain Dev. Dev. Brain Res.*, 1990, **55**, 109–120.

125. M. Rogers and I. Hendry, 'Involvement of dihydropyridine-sensitive calcium channels in nerve growth factor-dependent neurite outgrowth by sympathetic neurons.' *J. Neurosci. Res.*, 1990, **26**, 447–454.

126. R. A. Silver, A. G. Lamb and S. R. Bolsover, 'Calcium hotspots caused by L-channel clustering promote morphological changes in neuronal growth cones.' *Nature*, 1990, **343**, 751–754.

127. B. A. Suarez-Isla, D. J. Pelto, J. M. Thompson *et al.*, 'Blockers of calcium permeability inhibit neurite extension and formation of neuromuscular synapses

in cell culture.' *Dev. Brian Res.*, 1984, **316**, 263–270.

128. S. B. Kater, M. P. Mattson, C. Cohan *et al.*, 'Calcium regulation of the neuronal growth cone.' *Trends Neurosci.*, 1988, **11**, 315–321.

129. J. G. Altin, R. Wetts, K. T. Riabowol *et al.*, 'Testing the *in vivo* role of protein kinase C and c-fos in neurite outgrowth by microinjection of antibodies into PC12 cells.' *Mol. Cell Biol.*, 1992, **3**, 323–333.

130. J. L. Bixby, 'Protein kinase C is involved in laminin stimulation of neurite outgrowth.' *Neuron*, 1989, **3**, 287–297.

131. L. Cabell and G. Audesirk, 'Effects of selective inhibition of protein kinase C, cyclic AMP-dependent protein kinase, and Ca^{2+}-calmodulin-dependent protein kinase on neurite development in cultured rat hippocampal neurons.' *Int. J. Dev. Neurosci.*, 1993, **11**, 357–368.

132. M. A. Cambray-Deakin, J. Adu and R. D. Burgoyne, 'Neuritogenesis in cerebellar granule cells *in vitro*: a role for protein kinase C.' *Brain Res. Dev. Brain Res.*, 1990, **53**, 40–46.

133. V. Felipo, M. D. Minana and S. Grisolia, 'A specific inhibitor of protein kinase C induces differentiation of neuroblastoma cells.' *J. Biol. Chem.*, 1990, **265**, 9599–9601.

134. C. Arias, N. Sharma, P. Davies *et al.*, 'Okadaic acid induces early changes in microtubule-associated protein 2 and phosphorylation prior to neurodegeneration in cultured cortical neurons. *J. Neurochem.*, 1993, **61**, 673–682.

135. J. Y. Chiou and E. W. Westhead, 'Okadaic acid, a protein phosphatase inhibitor, inhibits nerve growth factor-directed neurite outgrowth in PC12 cells.' *J. Neurochem.*, 1992, **59**, 1963–1966.

136. T. Audesirk, G. Audesirk, C. Ferguson *et al.*, 'Effects of inorganic lead on the differentiation and growth of cultured hippocampal and neuroblastoma cells.' *Neurotoxicology*, 1991, **12**, 529–538.

137. M. Kern, T. Audesirk and G. Audesirk, 'Effects of inorganic lead on the differentiation and growth of cortical neurons in culture.' *Neurotoxicology*, 1993, **14**, 319–327.

138. M. Kern and G. Audesirk, 'Inorganic lead may inhibit neurite development of cultured rat hippocampal neurons through hyperphosphorylation.' *Toxicol. Appl. Pharmacol.*, 1995, **133**, 111–123.

139. M. F. Collins, P. D. Hrdina, E. Whittle *et al.*, 'Lead in blood and brain regions of rats chronically exposed to low doses of the metal.' *Toxicol. Appl. Pharmacol.*, 1982, **65**, 314–322.

140. E. J. Fjerdingstad, G. Danscher and E. Fjerdingstad, 'Hippocampus: selective concentration of lead in the normal rat brain.' *Brain Res.*, 1974, **80**, 350–354.

141. P. Grandjean, 'Regional distribution of lead in human brains.' *Toxicol. Lett.*, 1978, **2**, 65.

142. A. M. Scheuhammer and M. G. Cherian, 'The regional distribution of lead in normal rat brain.' *Neurotoxicology*, 1982, **3**, 85–92.

143. D. V. Widzowski and D. A. Cory-Slechta, 'Homogeneity of regional brain lead concentrations.' *Neurotoxicology*, 1994, **15**, 295–307.

144. M. G. Zurich, F. Monnet-Tschudi and P. Honegger, 'Long-term treatment of aggregating brain cell cultures with low concentrations of lead acetate.' *Neurotoxicology*, 1994, **15**, 715–719.

145. E. Enan and F. Matsumura, 'Specific inhibition of calcineurin by type II synthetic pyrethroid insecticides.' *Biochem. Pharmacol.*, 1992, **43**, 1777–1784.

SECTION VIA

11.26
Anticholinesterase Insecticides

DONALD J. ECOBICHON
McGill University, Montreal, PQ, Canada

11.26.1 CHEMICAL CLASSES

The anticholinesterase class of insecticides has evolved from experimental chemicals first synthesized in the mid-to-late 1930s. In the case of the organophosphorus ester subgroup, many of the original chemicals proved to be extremely toxic to all forms of life, some being developed into chemical warfare agents, the "nerve gases." In contrast, the original carbamate esters synthesized had fungicidal properties but no insecticidal activity. The chemicals in use today have evolved as third or fourth generation derivatives of the prototype agents. The anticholinesterase insecticides are represented by an array of structures seeking, by way of structure–activity relationships, to enhance potency and selective insect toxicity while reducing toxicity toward nontarget species. There are over 200 different organophosphorus ester and some 25–30 carbamate ester insecticides available on the market, formulated into thousands of products used worldwide in medicine, agriculture, and forestry for disease and pest control.

With the progress made with these chemicals, have these agents lost any of their inherent toxicity? Qualitatively, the answer is yes; fewer severe acute toxicities are being reported in the literature. Quantitatively, this chapter will highlight not only the immediate or acute toxicity but also, for both chemical classes of anticholinesterase insecticides, the longer-term, transient and/or persistent neurotoxicity observed in individuals exposed to single, high level or repeated low-to-moderate concentrations. These chemicals have not entirely lost their potency/toxicity and, as the techniques of detection improve, this toxicity is being manifested in more subtle ways.

11.26.1.1 Chemical Structures

While the anticholinesterase insecticides have a common mechanism of action, they arise from two distinctly different origins. The basic "backbone" structures of these two chemical classes are shown in Figure 1. The organophosphorus compounds are esters of phosphoric or phosphorothioic acids, suitable substituents at R_1 and R_2 being alkyl or alkoxy groups and, in a few instances, alkylamido groups where nitrogen (N) replaces the oxygen (O), and with substituents at R_3 being alkyl, alkoxy, or aryl in nature. The carbamate compounds are esters of carbamic (the monoamide of carbon dioxide) acid, achieving chemical stability with the introduction of one or two N-alkyl substituents at R_4 and R_5 and with aliphatic or aryl substituents at R_6. Overall shape, steric dimensions, and substituent groups are important factors for the variations in insecticidal potency and species specificity. For detailed discussions of nomenclature, chemistry, and development of these insecticides, the reader is referred to Holmstedt,[1] O'Brien,[2] Fest and Schmidt,[3] Kuhr and Dorough,[4] Matsumura[5] and Ecobichon and Joy.[6]

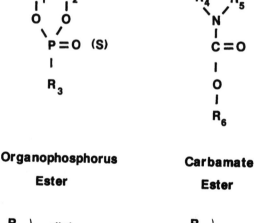

Organophosphorus Ester

R_1	alkyl
R_2	alkoxy
	alkylamido

R_3	alkyl
	alkoxy
	aryl

Carbamate Ester

| R_4 | alkyl |
| R_5 | |

| R_6 | alkyl |
| | aryl |

Figure 1 The "backbone" structures of the two types of anticholinesterase class of insecticides, the organophosphorus and carbamate esters. With organophosphorus compounds, the esters may be of phosphoric (P=O) or of phosphorothioic (P=S) acids. The substituents R_{1-6} denote the variety of moieties directly attached to or through an oxygen to the phosphorus or carbamoyl carbon.

11.26.2 MECHANISM OF ACTION

11.26.2.1 Enzyme Inhibition

Both organophosphorus and carbamate ester insecticides inhibit nervous tissue acetylcholinesterase (AChE), the enzyme responsible for terminating the action of the neurotransmitter, acetylcholine (ACh). However, there are marked contrasts in the nature of enzyme inhibition by the two classes of chemicals (Figure 2).

The interaction between an organophosphorus ester and the active site, a serine hydroxy group, in the AChE macromolecule results in a transient intermediate being formed which subsequently hydrolyzes, with the loss of the R_3 substituent (Figure 1), leaving a relatively stable, phosphorylated, and inactive enzyme. Under usual circumstances, the phosphorylated enzyme is reactivated only at a very slow rate or appears to be irreversibly inhibited. The observed toxicity from the accumulating ACh will be prolonged and persistent until either a sufficient quantity of AChE is reactivated or new AChE is synthesized to efficiently destroy the accumulated ACh. The nature of the substituents at R_1, R_2, and R_3 plays an important role in: (i) the physicochemical properties; (ii) specificity for the active site; (iii) the ease with which the R_3 subgroup can be removed; (iv) tenacity of binding at the active site; and (v) dephosphorylation to yield free enzyme. Many of the earlier insecticides bound tenaciously, the dialkylphosphorylated AChE being exceptionally stable. In certain cases with phosphate, phosphonate, and phosphoroamidate esters, an additional intermediate step may take place, with the dealkylation of one of the R_1 or R_2 substituents, a subsequent rearrangement of the bonding producing an irreversibly inhibited alkylphosphorylated AChE, referred to as an "aged" enzyme.[6,7] Aging is dependent on the size of the R_1 and R_2 substituents, with those esters having alkyl chains longer than methyl or ethyl participating in this phenomenon. Many organophosphorus esters have methyl groups at R_1 and R_2 and are less tightly bound to AChE, permitting the dialkylphosphorylated enzyme to dissociate spontaneously within 1–5 days. In contrast, carbamate esters attach to the active site of AChE to form an unstable intermediate complex followed by the hydrolytic removal of the R_6 substituent, leaving a carbamoylated AChE to undergo a second hydrolysis (decarbamoylation) to yield free enzyme. Essentially, carbamate esters are reversible inhibitors of AChE.

The AChE–insecticide interaction can be presented in another way, examining the rate constants involved in the various steps of inhibition and reaction (Figure 2). The distinct

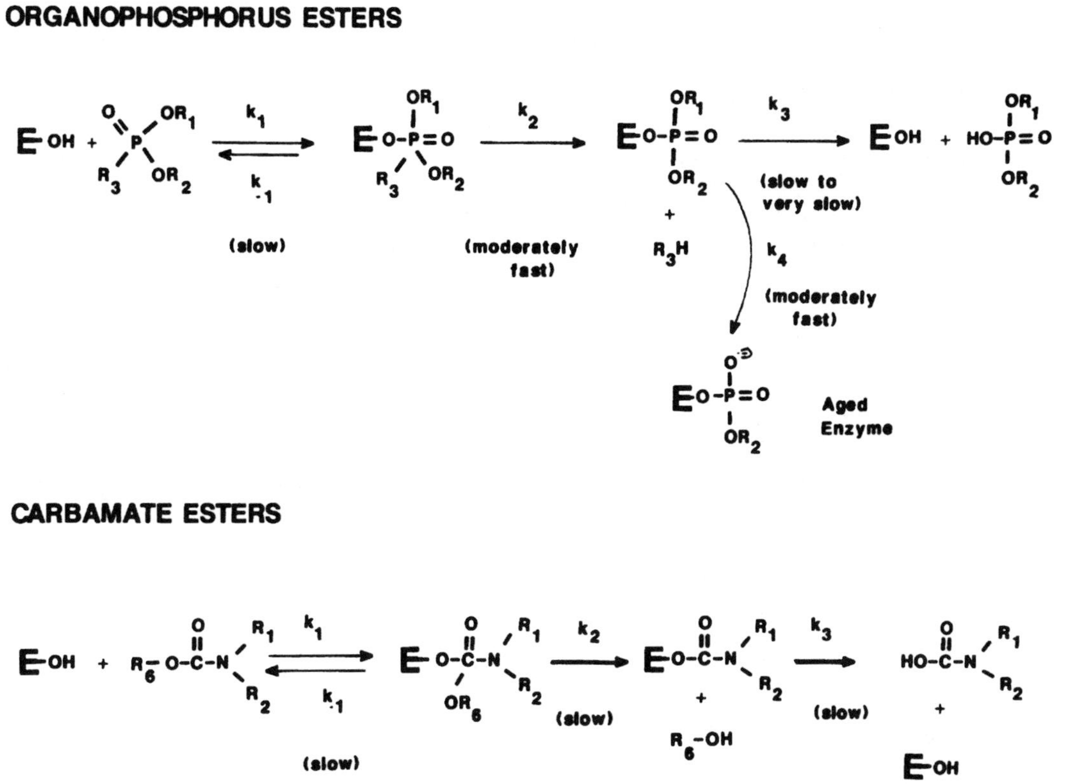

ORGANOPHOSPHORUS ESTERS

CARBAMATE ESTERS

Figure 2 Schematic diagram of the interaction between an organophosphorus or carbamate ester with the serine hydroxy group in the active site of acetylcholinesterase (E—OH), indicating the relative rates of each step in the reaction, the loss of substituent groups (R_3, R_6-OH) from the intermediate complex (second reaction) and the aging, the dephosphorylation or decarbamoylation of the inhibited enzymes.

difference between the two classes of chemicals lies in the rate(s) of dephosphorylation and decarbamoylation, being slow-to-extremely slow for organophosphorus esters and rapid-to-slow for carbamates.[7,8] Carbamate esters are only rather poor substrates for AChE, having a low turnover rate compared to that for ACh. Once the R_6 substituent has been removed, all carbamoylated AChE molecules are essentially identical. With organophosphorus esters, the slow dephosphorylation rate may range over several orders of magnitude because of the different substituents found at R_1 and R_2 (methyl, ethyl, isopropyl, etc.) in different insecticides.

Organophosphorus esters will also interact in much the same manner with the active site serine hydroxy group found in enzymes known collectively as serine proteases, including such enzymes as trypsin, chymotrypsin, nonspecific carboxylesterases, and arylesterases (Figure 2). With the exception of the arylesterases capable of hydrolyzing organophosphate esters, the inhibition of these other enzymes may play little role in the overall toxicity other than as "sinks" to bind the agent. Some nonspecific carboxylesterases hydrolyze certain alkyl sub-

stituents in organophosphorus esters, this being a detoxification reaction. One enzyme, a carboxylesterase known as neuropathy target esterase (NTE), can be inhibited by certain organophosphorus esters and appears to play some role in organophosphate-induced, delayed polyneuropathy (OPIDP, previously known as OPIDN).[9,10] In contrast, carbamate esters frequently undergo hydrolytic degradation catalyzed by nonspecific carboxylesterases, the route and rate of biotransformation being both chemical- and species-specific.[6,8]

11.26.2.2 Interaction with Acetylcholine Receptors

While the OPIDP can be partially explained, other anticholinesterase-induced, neuropathic sequelae are poorly understood. Therefore, the search for other mechanisms of action of organophosphorus and carbamate esters continues. Unfortunately, much of the attention has been focused only on organophosphorus esters.

Considerable evidence points toward a direct interaction between organophosphorus esters and secondary targets such as muscarinic (mAChR) and nicotinic (nAChR) acetylcholine receptors, the results showing a competitive block (antagonism) at mAChR and an induced desensitization of nAChR following a partial agonist effect directly attributed to the organophosphorus ester (see Chapter 8, this volume).[11] In one early example, high concentrations of diisopropylfluorophosphate (DFP) caused a hyperpolarization of the nAChR of eel (Torpedo) electric organ concomitant with an inhibition of acetylcholine binding.[12] Evidence suggests that desensitization of receptors may be related to the binding of the organophosphorus ester at a third site on the nAChR.[13] At micromolar levels in the circulation, certainly attainable in poisonings, organo-

phosphorus esters may directly induce toxicity at mAChR and nAChR whereas, at nanomolar concentrations, toxicity is due to inhibition of nervous tissue AChE.[14] However, in many studies, the agents tested (nerve gases, DFP, echothiophate, etc.) are not representative of the insecticide molecule beyond sharing the same chemical classification. Affinities of nerve gases and insecticides for muscarinic receptors differ by a 1000-fold range.[11]

Since many of the delayed, persistent signs/symptoms of both organophosphorus and carbamate intoxications to be described later are quite similar, can these be attributed to some action(s) of the accumulated ACh on receptors? The phenomenon, observed with other neurotransmitters released at high concentration for

Table 1 Signs and symptoms of organophosphorus insecticide poisoning.

Nervous tissue and receptors affected	*Site affected*	*Manifestations*
Parasympathetic autonomic (Muscarinic receptors) post-ganglionic nerve fibers	Exocrine glands	Increased salivation, lacrimation, perspiration
	Eyes	Miosis (pinpoint and nonreactive), ptosis, blurring of vision, conjunctival injection, "bloody tears"
	Gastrointestinal tract	Nausea, vomition, abdominal tightness, swelling and cramps, diarrhea, tenesmus, fecal incontinence
	Respiratory tract	Excessive bronchial secretions, rhinorrhea, wheezing, edema, tightness in chest, bronchospasms, bronchoconstriction, cough, bradypnea, dyspnea
	Cardiovascular system	Bradycardia, decrease in blood pressure
	Bladder	Urinary frequency and incontinence
Parasympathetic and sympathetic autonomic fibers (nicotinic receptors)	Cardiovascular system	Tachycardia, pallor, increase in blood pressure
Somatic motor nerve fibers (nicotinic receptors)	Skeletal muscles	Muscle fasciculations (eyelids, fine facial muscles), cramps, diminished tendon reflexes, generalized muscle weakness in peripheral and respiratory muscles, paralysis, flaccid or rigid tone
		Restlessness, generalized motor activity, reaction to acoustic stimuli, tremulousness, emotional lability, ataxia
Brain (acetylcholine receptors)	Central nervous system	Drowsiness, lethargy, fatigue, mental confusion, inability to concentrate, headache, pressure in head, generalized weakness
		Coma with absence of reflexes, tremors, Cheyne-Stokes respiration, dyspnea, convulsions, depression of respiratory centers, cyanosis

Source: Ecobichon and Joy.[6]

prolonged periods, is one of desensitization or even down-regulation of the receptor population. Evidence points to a 15–40% decrease in mAChR in rat brain and/or a reduced sensitivity (binding, affinity) for muscarinic agonists, although this was not seen uniformly throughout the brain.[15] While some investigations have shown an enhanced sensitivity of nAChR following chronic treatment with organophosphorus compounds, others have demonstrated a decreased population of nAChR in brain tissue.[16] At least one study has suggested that nAChR were more resistant to change than were mAChR. However, much of the original research was examining tolerance to organophosphorus ester insecticides and was carried out before the isolation and characterization of the subtypes of cholinergic receptors. Studies involving repeated administration of parathion have shown the down-regulation of mAChR in rat and mouse brain without changing agonist affinities, this effect occurring through reduced synthesis of messenger RNA at the transcription level.[17,18] Some of the central and peripheral neurological signs/symptoms associated with insecticide intoxication (Table 1) could be attributed to the down-regulation of receptors, the desensitization or even the enhanced sensitivity of receptors in the presence of accumulating high levels of ACh.

Evidence suggests an interaction between dopaminergic and gamma-aminobutyric acid (GABAergic) neurons and cholinergic neurons following acute or subacute exposure to organophosphorus esters.[16] Following acute exposure, increased cholinergic activity was countered by increased dopaminergic and GABAergic activities whereas, in subacute exposure, this did not occur, possibly due to the down-regulation of receptors with less excitatory activity and a normal inhibitory response of the other systems.[16] Acute treatment of rats with DFP caused increased brain glutamate and GABA levels whereas subacute treatment resulted in decreased GABA turnover. Dopamine turnover was increased with acute treatment but was decreased during subacute treatment. Changes in these two systems appear to be secondary to the increased ACh and may act singularly or in combination to counteract the increased cholinergic activity caused by organophosphorus ester inhibition of AChE.[16]

Hypotheses involving second messengers responding either as a consequence of a direct action of organophosphorus esters on mAChR or indirectly via ACh have been explored to a

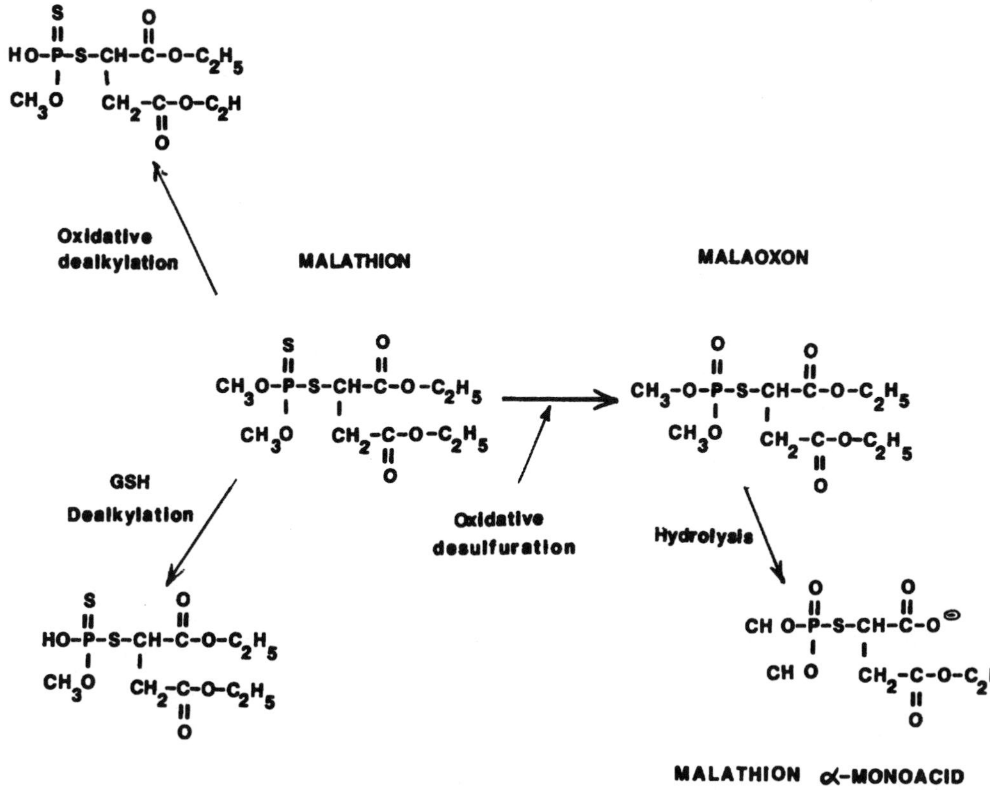

Figure 3 The biotransformation of the organophosphorus ester, malathion, showing key Phase I (oxidative dealkylation and desulfuration, ester hydrolysis) and Phase II (glutathione (GSH) dealkylation) pathways involved either in the detoxification or activation (forming malaoxon) of this ester.

limited extent.[19] The results are confusing because of regional distribution of receptor subtypes in the brain and differential effects on brain structures. The interaction of ACh with mAChR can activate phosphoinositidase C which hydrolyzes phosphatidylinositol-4,5-biphosphate to diacyl-glycerol and inositol-1,4,5-triphosphate, the latter causing a mobilization of cystosolic calcium ions when bound to specific receptor sites in or near the endoplasmic reticulum.[19] Both DFP and malaoxon, the oxygen analogue of the insecticide malathion, caused increased regional brain levels of inositol-1-phosphate, the hydrolysis product of inositol-1,4,5-triphosphate, concomitant with convulsions in the treated animals. A more refined study revealed elevated inositol-1,4,5-triphosphate in rat neocortex and striatum immediately after treatment with soman. Major, induced changes in intracellular concentrations of calcium have an important role in synaptic transmission and cell functions.

There are proposed mechanisms by which organophosphorus esters caused increased brain tissue levels of cyclic AMP and cyclic GMP, mediated through ACh-related actions on receptors and the release of catecholamines.[19] The question has been raised as to whether or not these second messenger effects are due to an ACh-like action or to some other, as yet unidentified, direct mechanism.[19]

11.26.3 BIOTRANSFORMATION

The biotransformation of the anticholinesterase insecticides can proceed by complicated, multiple Phase I (oxidative, reductive, hydrolytic) pathways contributing to both the activation and/or detoxification of the agents, and by Phase II (conjugative, dealkylation) pathways, generally associated with detoxification. A relatively simple example is given in Figure 3 for the organophosphorus ester, malathion. The pathways and the rates of biotransformation are species-specific and highly dependent upon the nature of the substituents attached to the nucleus of the ester. Those interested in specific details should consult Kuhr and Dorough,[4] Ecobichon and Joy,[6] Eto[20] or Chambers and Levi.[21]

Arylesterases cleave substituent phenols from organophosphorus esters while nonspecific carboxylesterases can hydrolyze aliphatic ester side chains (Figure 3). Oxidative enzymes can desulfurate phosphorothioate esters, converting them to the more potent phosphate esters. Thioether substituents can be oxidized to

sulfoxides and sulfones. Reductive pathways can convert nitro- to amino-groups. Phase I oxidative enzymes can directly dealkylate, particularly methyl substituents, a reaction shared with glutathione S-alkyl transferase, a Phase II enzyme, resulting in detoxification of the agent.[6] Phase II enzymes, glucuronyl- and sulfo-transferases, are involved in the conjugation of hydrolyzed phenolic substituents, forming water-soluble, readily excreted products.[6]

Carbamate esters are susceptible to Phase I hydrolysis involving nonspecific carboxylesterases in various tissues in a species-specific manner. The products, methyl or dimethylcarbamic acid and an aliphatic or aromatic alcohol are relatively water soluble.[6] The unstable methylated carbamic acids rapidly decompose into carbon dioxide and methyl- or dimethyl-amine while the alcoholic moiety may be conjugated and excreted. Phase I oxidative reactions can occur at appropriate groups, for example, hydroxylation of the methyl on carbamic acid or of substituent groups on the aromatic ring moiety.[6] Phase II glucuronidation, sulfation, glucosidation, conjugation with glutathione or amino acids, can take place at appropriate, available substituent groups.

11.26.4 NEUROTOXICOLOGY

11.26.4.1 Acute Toxicity

With inhibition of nervous tissue AChE by either organophosphorus or carbamate ester insecticides, accumulation of free ACh at the nerve endings of all cholinergic nerves results in a continual stimulation of electrical activity. The signs of toxicity include those arising from: (i) stimulation of the muscarinic receptors of the parasympathetic autonomic nervous system (ANS); (ii) stimulation and subsequent blockage of nicotinic receptors (ganglia of the sympathetic and parasympathetic divisions of the ANS as well as neuromuscular junctions); and (iii) receptors in the central nervous system (Table 1). Appropriate management of the acute intoxication crisis usually alleviates the signs and symptoms quickly, although with organophosphorus agents some of these may persist for a few days. In contrast, carbamate-induced acute toxicity may be intense but persists for approximately 6–12 h, with reactivation of the inhibited AChE terminating the neurotransmitter action. Close follow-up of acutely poisoned patients has shown that there are residual neurological effects persisting far beyond the acute crisis stage(s) of intoxication.[6]

11.26.4.2 Intermediate Syndrome

Usually related to a high level exposure to an organophosphorus ester, the onset of this "syndrome" occurs between 24 h and 96 h after poisoning, frequently affecting conscious patients showing no characteristic cholinergic signs of intoxication. The syndrome is characterized by marked weakness in musculature innervated by motor cranial nerves II to VII and X, with sudden appearance of weakness of the neck flexor muscles, weakness of muscles in the upper (proximal) limbs (shoulder abduction, hip flexion), absence of fasciculations but occasional spasticity, hyperreflexion and dystonic reactions and a decrease or absence of tendon reflexes.[22] Respiratory insufficiency may develop rapidly over a 6 h period, the patient using accessory muscles for ventilation accompanied by an increase in ventilatory rate, cyanosis, coma, and death if these symptoms are not recognized early enough to initiate supportive treatment. There is no evidence that carbamate esters elicit this syndrome, perhaps because the inhibition of AChE does not persist long enough.

It has been suggested that the "intermediate syndrome" may be associated with postsynaptic neuromuscular junction dysfunction, the excess accumulating ACh causing a depolarizing blockade of neuromuscular function, not unlike the effects seen in myasthenia gravis, a neuromuscular condition in which there is a shortage of ACh receptors.[23] In the presence of this excess ACh, there may be a rapid desensitization, even a down-regulation, of ACh receptors, reducing muscle function.

11.26.4.3 Organophosphate-induced Delayed Polyneuropathy

Historically, OPIDP has been recognized for approximately 100 years, associated originally with the chemical tri-*o*-tolyl phosphate (TOTP). This literature has been reviewed.[6] However, certain phosphate, phosphonate, and phosphoramidate esters can cause the same syndrome. Few such agents have been used as commercial insecticides, this peculiar neurotoxicity usually having been identified during mandatory testing by standard protocols used to assess all insecticides.

Following acute exposure, usually to a high level, of a suitable organophosphorus ester, (diisopropylfluorophosphate, mipafox, leptophos, diazinon, merphos, methamidophos), a flaccidity, muscle weakness in arms and legs begins in approximately 7–14 days, ataxia giving way to a spasticity, hypotonicity, hyper-reflexia, clonus and abnormal reflexes indicative of damage to the pyramidal tracts and permanent upper motor neuron damage.[6] Histological examination of the nervous system reveals Wallerian, "dying back" degeneration of the longest and large diameter axons and their myelinic sheaths in distal regions of the peripheral nerves and in the spinal cord (rostral ends of ascending tracts and distal ends of descending tracts).[6] The domestic chicken and the cat show morphological and physiological changes in the nervous system similar to those observed in exposed humans. While other species, particularly rodents, may not readily show the ataxia and hindlimb paralysis, the morphological changes described above are seen.[24]

Biochemical studies have demonstrated an agent-selective, rapid inhibition of a neuronal, nonspecific carboxylesterase called neuropathic target esterase (NTE) which appears to have some, as yet unidentified, role in neuronal lipid metabolism. If the acute exposure results in >70% inhibition of NTE, the characteristic OPIDP progresses over a period of 7–21 days even though the activity of NTE returns to normal values within 14 days. A second stage, the aging of the phosphorylated NTE to form an irreversibly inhibited enzyme also must occur to cause OPIDP. While any phosphate ester or phosphorothioate ester converted *in vivo* to a phosphate should be capable of causing OPIDP, the highly toxic agents cause rapid death before inhibition of the NTE. Insufficient amounts of the oxon analogues of phosphorothioates are formed at any one time to inhibit the NTE to the required level. There also appear to be subtle structure–activity relationships between organophosphorus esters and the active site of NTE, many esters being poor inhibitors of the enzyme.[24] Other esters, carbamates, phosphates, sulfonates, may inhibit NTE to greater than 70% but are not neuropathic since the NTE does not undergo aging.[24] While the inhibition of NTE remains a useful biomarker for potential OPIDP-producing organophosphorus esters, histopathologic evidence of axonal damage is a requirement of regulatory protocols.

11.26.4.4 Other Persistent Neuropathic Effects

Exposure to high concentrations of organophosphorus ester insecticides, for example, intentional suicide, occupational exposure to concentrates, and accidental ingestion, has resulted in a number of reported incidents involving sensory, neurobehavioral, cognitive, and neuromuscular deficits which have persisted for months or years after exposure. There has been much effort to study these single, often bizarre, anecdotal intoxications, but evidence points to persistent dysfunctions of both peripheral and central nervous systems.[6] A classic study is that of 19 farm workers poisoned by a combination of mevinphos and phosphamidon.[25] Many subjective symptoms (blurred vision, night sweats, headaches, muscular weakness, nausea) were evident and "quantifiable" some 4–5 months after poisoning. Unfortunately, no direct neurological assessment was conducted. However, the observations made were consistent with earlier reports in the literature. Other collections of cases have shown a mixture of sensory, motor and neuropsychological signs/symptoms.[6]

Psychological difficulties (hallucinations, delusions, irritability, nervousness, anxiety, memory loss, depression, dissociation, schizophrenic reactions, behavioral or mood changes) have been reported in a large number of earlier studies related to occupational exposure to agricultural insecticides. Some of these effects have persisted for months and years before disappearing.[26] Better designed studies have confirmed the earlier, anecdotal observations. Some 24 months after poisoning, the patients performed significantly below matched controls in neuropsychological functions (auditory attention, visual memory, visuomotor speed, sequencing, problem solving, motor steadiness, reaction, dexterity).[27] Other studies have revealed persistent alterations in electroencephalograms in monkeys and in humans exposed to organophosphorus esters, the latter showing prolonged psychological impairment (anxiety, psychomotor depression, unusual dreams, intellectual impairment).[6]

There is little evidence of prolonged neurotoxicity following the exposure of humans to carbamate ester insecticides although, in one anecdotal report of dermal intoxication with a formulation of carbaryl, a polyneuropathy resulted with symptoms including a persistent photophobia, mild and persistent paresthesia, memory loss, muscular weakness, lassitude, and fatigue.[6] A second anecdotal case, of a 75 year old male exposed continuously over an 8–10 month period to 10% carbaryl dust applied repeatedly to control fleas, confirmed the persistent signs/symptoms observed previously.[28] Following initial acute intoxication with characteristic signs, the patient's health deteriorated while he continued to live in the home, showing such signs/symptoms as muscle spasms, headache with severe pressure, tinnitus, vertigo, rhinorrhea and excessive lacrimation, weakness in major skeletal muscle groups with fascicula-

tions, somnolence, and mental confusion. When the patient abandoned the home, symptoms which failed to abate included altered sleep patterns, headache, tinnitus, mental confusion, and a stocking-and-glove peripheral neuropathy. This anecdotal case was perhaps confounded by the age of the patient and concomitant treatment of gastric ulcers with cimetidine, an agent known to inhibit the biotransformation of carbaryl.[28]

Severe neuromuscular effects have been observed in swine subchronically treated with carbaryl, the animals showing a progressive myasthenia, incoordination, ataxia, tremor and clonic muscular contractions terminating in prostration and paraplegia.[29] An exaggerated flexion of the rear legs caused considerable difficulty in the animals' backing up or sitting down, and the other symptoms became more prominent when the swine were forced to move. Histological examination revealed lesions confined to the CNS and skeletal muscles. The predominant lesions in the CNS involved edema and fragmentation of the myelin tracts in the brainstem and cerebellar peduncles with swollen or ruptured axons. The muscular lesions consisted of traumatic or ischemic myodegeneration, acute hyaline and vacuolar degeneration and dystrophic calcification. A delayed neuropathy can be elicited by several different carbamates in the chicken, although repeated oral administration is necessary.[30] In contrast to the OPIDP-inducing esters, carbamates are weaker and more ephemeral neurotoxicants, while still causing axonal demyelination with prolonged exposure. Carbamates, as has been indicated, do not inhibit NTE to the level (>70%) required to elicit OPIDP. However, young chicks, treated subacutely (7 days) with either aldicarb or carbaryl, show a persistently abnormal gait, problems balancing and some peripheral paralysis for up to 40 days post-treatment.[31] A variety of animal studies, both acute and subacute, have demonstrated the ability of a number of carbamate insecticides to induce neurobehavioral problems, particularly in cognitive skills. These studies have been reviewed.[6]

11.26.5 TREATMENT

11.26.5.1 Organophosphorus Esters

Atropine is particularly important in acute, life-threatening intoxications, frequent, small doses (1.0 mg s.c. or i.v.) being administered to control the initial muscarinic signs (Table 1). Relatively large doses of atropine, up to 50 mg day^{-1}, may be necessary to control severe muscarinic signs and, since atropine is highly toxic, the status of the patient must be monitored continually by examining for dilatation of the pupils (mydriasis), absence of secretions (dry mouth), facial flushing and/or disappearance of sweating.[7]

Diazepam should be included in the treatment regimen of all but the mildest cases of intoxication since it counteracts some aspects of CNS-derived and neuromuscular signs not antagonized by atropine. Doses of 10 mg s.c. or i.v. are appropriate and may be repeated as required.[7] Other centrally acting drugs are contraindicated since they may depress respiration.

The entire spectrum of signs/symptoms can be controlled by the administration of specific antidotes, the oxime compounds, that reactivate the inhibited nervous tissue AChE, permitting the biological destruction of the accumulated acetylcholine. The most common agent used is pralidoxime (2-PAM, Protopam), administered doses of 1.0 g by slow i.v. infusion over 20 min.[7] Early treatment is more effective than delayed administration since, with time, the organophosphorus ester may become tightly or irreversibly bound to the enzyme, and the reactivating agent will be less effective. A single treatment with pralidoxime will reduce the amount of atropine needed. Pralidoxime treatment can be repeated but care should be taken since this agent binds calcium ions, causing severe muscle cramping, particularly in the legs, that bear some similarity to the signs/symptoms of intoxication but can be alleviated by the oral or i.v. administration of calcium solutions.

11.26.5.2 Carbamate Esters

The clinical treatment of carbamate-induced toxicity is similar to that for organophosphorus ester intoxication with the exclusion of the oxime antidote.[7] The short-lived, carbamate inhibition of nervous tissue AChE precludes the use of the oximes. However, with certain carbamates, notably carbaryl, pralidoxime enhances the toxicity. With other carbamates, pralidoxime has little or no beneficial effect.

11.26.6 REFERENCES

1. B. Holmstedt, 'Pharmacology of organophosphorus cholinesterase inhibitors,' *Pharmacol. Rev.*, 1959, 567.
2. R. D. O'Brien, 'Insecticides: Action and Metabolism,' Academic Press, New York, 1967.
3. C. Fest and K. J. Schmidt, 'The Chemistry of Organophosphorus Pesticides; Reactivity, Synthesis, Mode of Action, Toxicology,' Springer-Verlag, New York, 1973.
4. R. J. Kuhr and H. W. Dorough, 'Carbamate Insecticides: Chemistry, Biochemistry and Toxicology,' CRC Press, Cleveland, OH, 1976.
5. F. Matsumura, 'Toxicology of Insecticides,' 2nd edn., Plenum, New York, 1985.
6. D. J. Ecobichon and R. M. Joy, 'Pesticides and Neurological Diseases,' 2nd edn., CRC Press, Boca Raton, FL, 1994.
7. D. J. Ecobichon, in 'Casarett and Doull's Toxicology. The Basic Science of Poisons,' 4th edn., eds. M. O. Amdur, J. Doull and C. D. Klaassen, Pergamon Press, New York, 1991, pp. 565–622.
8. W. N. Aldridge and E. Reiner, 'Enzyme Inhibitors as Substrates. Interactions of Esterases With Esters of Organophosphorus and Carbamic Acids,' North-Holland/Elsevier, Amsterdam, 1972.
9. M. K. Johnson, 'Organophosphates and delayed neuropathy—is NTE alive and well?' *Toxicol. Appl. Pharmacol.*, 1990, **102**, 385–399.
10. M. Lotti, 'The pathogenesis of organophosphate polyneuropathy.' *Crit. Rev. Toxicol.*, 1992, **21**, 465–487.
11. A. T. Eldefrawi, D. Jett and M. E. Eldefrawi, in 'Organophosphates. Chemistry, Fate and Effects,' eds. J. E. Chambers and P. E. Levi, Academic Press, San Diego, CA, 1992, pp. 257–270.
12. E. Bartels and D. Nachmansohn, 'Organophosphate inhibitors of acetylcholine-receptor and -esterase tested on the electroplax.' *Arch. Biochem. Biophys.*, 1969, **133**, 1–10.
13. M. E. Eldefrawi, G. Schweizer, N. M. Bakry *et al.*, 'Desensitization of the nicotinic acetylcholine receptor by diisopropylfluorophosphate.' *J. Biochem. Toxicol.*, 1988, **3**, 21–32.
14. N. M. Bakry, A. H. el-Rashidy, A. T. Eldefrawi *et al.*, 'Direct actions of organophosphate anticholinesterases on nicotinic and muscarinic acetylcholine receptors.' *J. Biochem. Toxicol.*, 1988, **3**, 235–259.
15. L. Churchill, T. L. Pazdernik, F. Samson *et al.*, 'Topographical distribution of down-regulated muscarinic receptors in rat brains after repeated exposure to diisopro-pylfluorophosphate.' *Neuroscience*, 1984. **11**, 463–472.
16. B. Hoskins and I. K. Ho, in 'Organophosphates. Chemistry, Fate and Effects,' eds. J. E. Chambers and P. E. Levi, Academic Press, San Diego, CA, 1992, pp. 285–297.
17. D. A. Jett, E. F. Hill, J. C. Fernando *et al.*, 'Down-regulation of muscarinic receptors and the m3 subtype in white-footed mice by dietary exposure to parathion.' *J. Toxicol. Environ. Health*, 1993, **39**, 395–415.
18. D. A. Jett, J. C. Fernando, M. E. Eldefrawi *et al.*, 'Differential regulation of muscarinic receptor subtypes in rat brain regions by repeated injections of parathion.' *Toxicol. Lett.*, 1994, **73**, 33–41.
19. L. G. Costa, in 'Organophosphates. Chemistry, Fate and Effects,' eds. J. E. Chambers and P. E. Levi, Academic Press, San Diego, CA, 1992, pp. 271–284.
20. M. Eto, 'Organophosphorus Pesticides: Organic and Biological Chemistry,' CRC Press, Cleveland, OH, 1974.
21. J. E. Chambers and P. E. Levi (eds.), 'Organophosphates. Chemistry, Fate and Effects,' Academic Press, San Diego, CA, 1992.
22. N. Senanayake and L. Karalliedde, 'Neurotoxic effects of organophosphorus insecticides. An intermediate syndrome.' *N. Engl. J. Med.*, 1987, **316**, 761–763.
23. L. Karalliedde and N. Senanayake, 'Organophosphorus insecticide poisoning.' *Br. J. Anaesth.*, 1989, **63**, 736–750.
24. B. Veronesi and S. Padilla, in 'Organophosphates. Chemistry, Fate and Effects,' eds. J. E. Chambers and P. E. Levi, Academic Press, San Diego, CA, 1992, pp. 353–366.

25. M. D. Whorton and D. L. Obrinsky, 'Persistence of symptoms after mild to moderate acute organophosphate poisoning among 19 farm field workers.' *J. Toxicol. Environ. Health*, 1983, **11**, 347–354.

26. L. Rosenstock, W. Daniell, S. Barnhart *et al.*, 'Chronic neuropsychological sequelae of occupational exposure to organophosphate insecticides.' *Am. J. Ind. Med.*, 1990, **18**, 321–325.

27. L. Rosenstock, M. Keifer, W. E. Daniell *et al.*, 'Chronic central nervous system effects of acute organophosphate pesticide intoxication. The Pesticide Health Effects Study Group.' *Lancet*, 1991, **338**, 223–227.

28. R. A. Branch and E. Jacqz, 'Subacute neurotoxicity following long-term exposure to carbaryl.' *Am. J. Med.*, 1986, **80**, 741–745.

29. H. E. Smalley, P. J. O'Hara, C. H. Bridges *et al.*, 'The effects of chronic carbaryl administration on the neuromuscular system of swine.' *Toxicol. Appl. Pharmacol.*, 1969, **14**, 409–419.

30. S. W. Fisher and R. L. Metcalf, 'Production of delayed ataxia by carbamic acid esters.' *Pestic. Biochem. Physiol.*, 1983, **19**, 243–253.

31. M. Farage-Elawar, M. F. Ehrich, B. S. Jortner *et al.*, 'Effects of multiple doses of two carbamate insecticides on esterase levels in young and adult chickens.' *Pestic. Biochem. Physiol.*, 1988, **32**, 262–268.

11.27
Organochlorine and Pyrethroid Insecticides

†ROBERT M. JOY
University of California, Davis, CA, USA

11.27.1 INTRODUCTION

The synthesis of dichlorodiphenyltrichloroethane (DDT) by Zeidler in 1874 marked the beginning of a new era characterized by the large scale utilization of synthetic chemicals to control various pests and pest-borne diseases. The success of DDT quickly led to the intro- duction of other chlorinated hydrocarbons including lindane, aldrin, dieldrin, endrin, and chlordane. These were used in massive amounts during World War II for the control of mosquito-borne diseases, such as malaria, dengue fever, and filariasis, and up to 1970, were the standard agents used in insect control. The hazards inherent in the large scale use of the organochlorine insecticides began to emerge during the same period.[1] The environmental impact of massive organochlorine insecticide application was addressed publicly and dramatically in 1962 by Carson[2] in "Silent Spring." Organochlorines can persist unchanged in the environment for long periods of time, accumulate in soils and translocate from their point of application into rivers, lakes, and oceans. Due to their persistence and high lipid-solubility, they can accumulate along food chains, which has led to widescale loss of fish and birds.[3,4] These undesirable characteristics, coupled with concerns that the continuous presence of organochlorine residues in man and animals might have oncogenic, mutagenic, and teratogenic consequences years later have led to a dramatic reduction in their use, and their replacement by less persistent chemicals.

Pyrethrins are natural products produced by various species of Chrysanthemums. Their insecticidal activity was known by the nineteenth century, and they were used worldwide by 1850.[5] Pyrethrum, prepared from ground flowers, consists of a mixture of pyrethrins. Pyrethrum is essentially nontoxic to mammals but is an effective, rapid acting insecticide. Prior to the development of DDT and other organochlorine insecticides, pyrethrum was extensively used for both agricultural and domestic purposes, even though it was rapidly broken down in light requiring continual reapplication. Now it is only used as a domestic insecticide. However, the high selectivity of pyrethrum for insects and its low mammalian toxicity led to a search for synthetic derivatives, called pyrethroids, possessing more stable chemical properties. The modern pyrethroids are potent, selective insecticides which comprise a large part of the present world foliar market.

This chapter will focus on the neurotoxic effects of organochlorine and pyrethroid insecticides. For more general reviews of organochlorine insecticide toxicology the reader is referred to the works of Hayes,[6,7] Smith,[8] Matsumura,[9] Baker and Wilkinson,[10] and Joy.[11] For more general reviews of pyrethroids the reader is referred to the works of Ray,[12] Ruigt,[13] Soderlund and Bloomquist,[14] and Joy.[15]

11.27.2 COMPARATIVE TOXICITY OF THE ORGANOCHLORINE AND PYRETHROID INSECTICIDES

Table 1 indicates exposure levels of organochlorine and pyrethroid insecticides producing acute lethal effects in animals and man. These data are approximate since actual toxicity is affected by many factors. In humans, acute toxicity results most often from ingestion or skin contact, whereas chronic toxicity occurs via ingestion, skin contact, or inhalation. The cyclodienes and lindane are the most toxic, and cause poisoning from any route of exposure. The DDT analogues, cage compounds, and pyrethroids possess lower acute toxicities and are relatively nontoxic on dermal or inhalational exposure.

Organochlorine insecticides can be separated into three groups. (i) DDT and its analogues produce similar neurotoxic effects and share similar mechanisms. (ii) Lindane and the cyclodienes produce similar effects and also share similar mechanisms. (iii) Mirex and chlordecone form a third group. The pyrethroids also break down into two groups. The type I compounds produce effects similar to DDT and act by a similar mechanism. The type II compounds produce a distinctly different symptom profile.

11.27.3 NEUROTOXICITY OF THE DICHLORODIPHENYLETHANE DERIVATIVES

11.27.3.1 Signs and Symptoms of Poisoning in Insects and Animals

11.27.3.1.1 Acute exposure

The major symptoms of acute exposure to DDT and its analogues are of nervous origin. Insects typically exhibit dose-related effects which include: (i) hyperextension of the legs and uncoordinated movements; (ii) general tremulousness; (iii) ataxia; (iv) hyperreactivity to external stimulation; (v) loss of righting capabilities; (vi) development of two types of leg movements including a high-frequency tremor and a slower flexion and extension movement sequence; (vii) the eventual disappearance of tremors leaving only sporadic, isolated movements of the head, tarsi, palpi, cerci, and antennae; and (viii) immobility followed shortly by death.[19]

In rodents the first perceptible effect is hyperexcitability and increased spontaneous activity. As intoxication becomes more severe, blepharospasm, hyperreflexia, and twitching of the ears and vibrissae occur. Massive myoclonic jerks may appear in response to sensory stimulation. A prominent feature is tremor, which develops as an intention tremor during voluntary movement, and may become so intense that purposeful movements become impossible. Spontaneous and stimulus-induced myoclonus accompany the tremor at high doses.[20,21] Tremoring may continue for hours or days depending upon the severity of exposure. Death may occur following a convulsive

Table 1 Acute toxicity of organochlorine and pyrethroid insecticides.[a]

Insecticide	Rat-oral LD_{50} mg kg^{-1}	Rat-dermal LD_{50} mg kg^{-1}	Human-lowest[b] LD_{50} mg kg^{-1}
Dichlorodiphenylethanes[c]			
DDT	250	500	50
DDD (Rhothane)	3400		500
DMC (Dimite)	500		500
Dicofol (Kelthane)	575	1000[b]	500
Chlorobenzylate	4000		500
Methoxychlor	6000		430
Hexachlorocyclohexanes			
Lindane (gamma-BHC)	88	500	180
Cyclodienes			
Endrin	3	15	5
Telodrin	5	5	5
Isodrin	7	23	5
Endosulfan	18	74	50
Heptachlor	40	195	50
Dieldrin	46	60	5
Toxaphene	90	780[d]	40
Aldrin	67	98	1.25
Strobane	200		50
Chlordane	283	700	40
Cage compounds			
Mirex	365		
Chlordecone (Kepone)	125	2000	
Pyrethroids[e]			
Type I			
Pyrethrin II	>600 (1)		
Tetramethrin	>4000 (2.3)		
Allethrin	200 (3.5)		
Cismethrin	63 (6.5)		
Type II			
Deltamethrin	52 (2.3)		
Cypermethrin	900 (55)		
Fenvalerate	450 (75)		

[a]Toxicity values taken from Fairchild[16] unless otherwise indicated.   [b]The lowest reported dose introduced by any route other than inhalation over any given period of time which resulted in death.   [c]Toxicity values taken from Hayes.[17]   [d]All dichlorodiphenyltrichloroethane derivative oral rat LD_{50} values from Matsumura.[18]   [e]All pyrethrin toxicity data from Ray.[12] Values in parentheses are i.v. LD_{50} values.

seizure, or it may follow a terminal period characterized by progressive exhaustion, flaccid immobility, and unconsciousness. Recovery is possible from any of these stages, and, if it occurs, all symptoms disappear, usually without signs of irreversible damage.[1,9,22]

Cats, dogs, and monkeys show similar symptoms, and convulsive episodes occur early and more frequently. Convulsions develop abruptly and are generalized clonic convulsions. Death may occur suddenly during a convulsive episode or during the period of exhaustion following many convulsive seizures.[23,24]

In man the earliest effects are paresthesias of the mouth and face. Altered motor function leading to ataxia and abnormal stepping reac-

tions also tends to occur early. These symptoms are typically followed by dizziness, confusion, general malaise, headache, and fatigue. Tremor, particularly of the hands, is a common symptom of DDT poisoning. Convulsions have resulted from ingesting large quantities of DDT. Very few deaths have been reported.[1]

11.27.3.1.2 Chronic exposure

In animals chronic exposure to DDT produces many of the same symptoms seen with acute exposure. However, their onset, intensity, and progression vary depending upon the kinetics of exposure. Chronic intoxication may also be associated with loss of weight, anorexia, mild anemia, muscular weakness, and tremors. At sufficiently high exposure levels, convulsive episodes with subsequent muscular weakness, paralysis, coma, and death may occur.[25,26] Pathological changes can be observed in various organs including the central nervous system (CNS).

11.27.3.2 Sites of Action

DDT analogues are poisons which act primarily at the level of the CNS in man and higher animals. Important peripheral nerve effects also occur, particularly in insects. The combination of central and peripheral actions is responsible for the hyperexcitability, generalized tremors, spasticity, and convulsions that characterize their toxicity.

In insects, DDT and its analogues produce an increase in and prolongation of the negative afterpotential in nerve axons leading to repetitive discharge.[27] Afferent fibers and/or their associated sensory receptors are more susceptible than motor fibers, but both can be affected.[28] CNS actions are also involved in poisoning in insects.[19,29]

In vertebrates, effects of DDT and analogues on peripheral nerves is overshadowed by effects developing in the CNS. The spinal cord and brainstem are the primary loci for the development of abnormal motor function after DDT exposure. This is particularly true for hyperreflexia, tremors, and myoclonus. Convulsive phenomena rely more heavily upon higher centers, particularly the cortex. Intact subjects exhibit a lower threshold for seizure development than do decerebrate animals.[24]

In animals DDT also produces changes in behavior.[11] During brain growth and development, animals appear to be much more sensitive to DDT than as adults. Subtle changes, such as alterations in rates of habituation have been reported.[30,31] This is reminiscent of behavioral changes reported in rats fed salmon from Lake Ontario that were "heavily" contaminated with polychlorinated biphenyls (PCBs) and other organochlorine residues.[32,33] Rats given the contaminated diet developed a preference for predictable food rewards more quickly than did the control rats, and were much more affected by the introduction of mild electric shocks into the feeding paradigm.

11.27.3.3 Mechanisms of Action

In arthropod axons DDT prolongs the falling phase of the action potential, increases the amplitude and duration of the negative after-

potential, and produces repetitive discharge.[27,34] The mechanisms responsible were subsequently clarified by Narahashi and Haas[35,36] in lobster giant axons, and by Shanes[37] and Hille[38] in frog sciatic nerves (see Chapter 9, this volume). The primary effect of DDT is to change the kinetics of inactivation of sodium channels. The activation of sodium channels is unaffected by DDT, but channel inactivation is delayed by about 10–30-fold, leading to a persisting depolarized state. Potassium conductance is also depressed, further slowing axonal repolarization. These lead to repetitive discharge. This action is not unique to DDT, but is shared by other DDT derivatives and by pyrethroid insecticides.

In insects, a good correlation exists between exposure levels evoking repetitive discharge in the abdominal nerve cord, those producing convulsive behavior, and those which are insecticidal.[39] In vertebrates delayed sodium channel inactivation appears to be the cause of changes observed in peripheral nerve, skeletal muscle, myocardium, and probably in the CNS. Axons near presynaptic terminals are of small diameter and may be more susceptible to developing repetitive discharge with DDT than larger fibers simply through their geometry.[40] Nerve terminals contain sodium channels, also. Their prolonged depolarization in the presence of DDT could lead to increased calcium entry and increased transmitter release. This would produce major amplifications of synaptic processes, perhaps at exposure levels below those needed to produce repetitive action potentials in larger axons. If true, the effects of DDT would be expected to be diverse and widespread and not be restricted to a single type of neurotransmitter. Type I pyrethroids produce an intoxication that is indistinguishable from that produced by DDT and affect sodium channels in a similar manner.

Although most of the effects of DDT appear to stem directly from its action on sodium channel kinetics, other factors may well be involved. Independent effects of DDT upon synaptic activity, transmitter synthesis, kinetics of release or degradation, neurogenic amines or various enzyme systems participating in ion transport, metabolism, oxidative phosphorylation, and so on may exist, but have not yet been clearly demonstrated.[8,11]

DDT changes the levels or effects of many chemical and neurochemical substances.[22,41] In most studies where catecholamine systems were evaluated, increases in the turnover of norepinephrine, dopamine, and serotonin have been reported. Attempts to correlate these changes with toxic symptoms suggest that increases in the turnover of serotonin correlate best with the temporal development of hyperthermia and myoclonus, whereas increases in turnover of both serotonin and norepinephrine correlate best with the temporal development of tremor.[11]

Changes in cholinergic function have been described after DDT exposure. In the behavioral studies of Eriksson *et al.*,[30,31] exposure of 10 day old mice to a single dose of DDT $(0.5\,mg\,kg^{-1})$ led to a permanent hyperactive condition in the mice as adults. In rats, single oral doses of DDT can decrease acetylcholine levels in cerebellum[42] and striatum,[43] while two month exposure to DDT in food can produce a selective decrease in the number of muscarinic receptor sites in the cerebellum.[44] These various changes in cholinergic function are of interest, but they need to be further evaluated before their relationship to DDT poisoning can be determined.

11.27.3.4 Neurotoxicity of DDT and its Analogues to Humans

Information about the neurotoxicity of DDT to man has risen from various sources including purposeful exposure by volunteers to known concentrations of DDT and involuntary, accidental exposure to estimable quantities of DDT. Many comprehensive reviews[1,6–8,11] on DDT exposure in man have been written. They should be consulted by anyone wishing more information on exposures to DDT. Ingestion of DDT by volunteers or accidental ingestion indicates that amounts below $10\,mg\,kg^{-1}$ usually produce few or no symptoms, from $10\,mg\,kg^{-1}$ to $16\,mg\,kg^{-1}$ produce moderate to severe symptoms, while exposure to greater than $20\,mg\,kg^{-1}$ can induce convulsions and death. Commonly reported symptoms include apprehension and hyperexcitability, a moderately increased respiration and bradycardia. Vomiting is common. Numbness and partial paralysis may develop and are most apparent in the distal portions of the extremities. Proprioception and vibratory sensation can be diminished or lost in the fingers and toes. Convulsions occur with large exposures.[3] Far fewer complications have been reported with either dermal or respiratory exposures, even to massive amounts.[1]

Chronic toxicity to DDT may emerge slowly and show certain additional complications not part of the acute poisoning syndrome. It may be characterized by loss of weight, anorexia, mild anemia, muscular weakness, and tremors. Anxiety, nervous tension, hyperexcitability, and fear are commonly expressed by the patient. Myoclonic jerks may occur,[45] but these are more commonly observed with cyclodienes and lindane.

The persistence of DDT toxicity following acute ingestion is generally only 1–2 d in duration. Exceptions to this have been observed when idiosyncratic responses have developed. These are rare, although a few reports suggest that an idiosyncratic response, usually consisting of variable polyneuropathic phenomena, can occur. These are more frequently seen on chronic exposure and may possess an allergic component.

Chronic exposure to DDT has been implicated as a cause for various allergic and hypersensitivity reactions. Although most of these are dermatologic, some polyneuropathies have been described in which exposure to DDT has been present.[1,11,45] In none of these reports is there an uncomplicated exposure to DDT alone. The patients have invariably been exposed to many different chemicals simultaneously, some of which have also been implicated as producing polyneuropathies. Since the use of DDT and derivatives has been very severely restricted, there have been very few reports of poisoning since the early 1980s.

11.27.4 NEUROTOXICITY OF CYCLODIENES AND LINDANE

Lindane, the gamma isomer of hexachlorocyclohexane, and cyclodienes exhibit a different neurotoxic profile. Whereas DDT has actions on sensory nerve fibers and peripheral receptors which play an important role in the expression of toxicity in most species, cyclodienes have little or no peripheral effects. The CNS is the only important site of action. Lindane exhibits cross-tolerance with the cyclodienes,[46] implying similar mechanisms of action. This has been amply confirmed.

Species differ in their susceptibility to the cyclodienes and lindane. Fish and birds are very susceptible. For mammals, the order from most to least sensitive is roughly dog, man, monkey, cat, guinea pig, rabbit, rat, hamster, and mouse. The larger domesticated animals are relatively insensitive. Young animals are usually more susceptible than older animals, lean animals more susceptible than obese animals, and females are more susceptible than males in some species, notably the rat.

11.27.4.1 Signs and Symptoms of Poisoning in Insects and Animals

11.27.4.1.1 Acute exposure

Insects show five stages of toxicity following exposure.[47] These include (i) hyperexcitability,

(ii) uncoordinated movements, (iii) convulsive periods, (iv) prostration and paralysis, and (v) death. Insects poisoned with lindane or a cyclodiene frequently display fanning movements of the wings insufficient to cause flight. At a late stage of poisoning, the legs are all tightly contracted under the body and the wings are deflected downwards, the insect equivalent of convulsion. This is followed by progressive exhaustion and death. Hyperactivity is usually greater and lasts longer than that seen with DDT.

In most vertebrates the outstanding feature of toxicity is convulsions, of sudden onset and extreme violence. Death may occur during convulsions or during the postictal period, or recovery may occur. After recovery the neurological symptoms appear completely reversible in the vast majority of cases.[48] In man, sudden, precipitous convulsions without prior symptoms can be the first observable sign of lindane or cyclodiene exposure.[49]

The temporal development and the severity of symptoms depend upon the route of administration and the dose. Intravenous administration of convulsive doses produces seizures within 2–10 min preceded by signs of hyperexcitability.[50] Oral administration produces symptoms after 0.5–2 h. In most cases, symptoms progress in a stereotyped manner through phases of hyperexcitability to convulsions, followed by recovery or death. Due to their relative potencies, exposure to the cyclodienes or lindane by the dermal or respiratory routes is more frequently associated with toxicity than is exposure to DDT by those routes.[48]

11.27.4.1.2 *Chronic exposure*

With chronic exposure hyperexcitability and convulsions are frequently observed after a variable delay. Additional effects may develop which also require time for their expression. As one example, Worden[51] found that feeding telodrin or endrin to rats at levels of 25–100 ppm resulted in dose-related effects, which included hyperresponsiveness to sensory stimuli, audiogenic seizures, swelling of the subcutaneous tissues of the head, staring eyes, bloody incrustations over the eyelids, sporadic mild convulsions lasting about 30 s, and in fatal cases, violent convulsions leading to death.

Chronic exposure to high levels of these compounds can cause increased liver weights, weight loss, impaired survival, impaired mating performance, and impaired progeny survival.[8,48] Lindane can induce acute hepatic porphyria in humans and animals.[52]

11.27.4.2 Sites of Action

The primary site of action of cyclodienes and lindane is the CNS. All are capable of increasing spontaneous activity within the insect CNS.[53] The frequency of spontaneous discharge in the central nerve cord of the cockroach is increased and the synaptic afterdischarge is greatly prolonged.[29]

In mammals, the cyclodienes and lindane produce an increase in neuronal excitability similar to that observed for pentylenetetrazol type convulsants.[50] A prominent feature of poisoning is the development of inappropriate motor responses to sensory stimuli. Sudden sensory input results in progressively exaggerated reflexive movements which become frank myoclonic jerks. These may precipitate a convulsive seizure. Even in paralyzed subjects, where the proprioceptive consequences of motor activity are eliminated, tactile, auditory, and stroboscopic light stimulation are effective precipitants of electrical seizure activity.[50,54]

Although all of these compounds will produce convulsions in normal animals at some dose, subconvulsant exposures also increase the probability of eliciting epileptiform behavior due to other causes. Dieldrin precipitates seizures in sheep exposed to stroboscopic light stimulation at exposure levels which produce no overt evidence of clinical toxicity.[55] Subconvulsant exposures to cyclodienes and lindane facilitate the development of convulsive responses to other chemicals, such as pentylenetetrazol[56] and picrotoxin.[57] Most significantly they potentiate the development and expression of kindling, and kindled seizures, an animal model of epilepsy, in subjects exposed *in utero*,[58] prior to weaning[58] and when mature.[59–61] To the extent that kindling represents a good model for epileptogenesis in man, the kindling data on dieldrin and lindane suggest that they might predispose susceptible individuals to develop epilepsy, however, there is no evidence from epidemiological studies in man that this has occurred.

At high levels of exposure some very detrimental effects have been reported upon mating and progeny-caring behavior. Wynn[62] has reported that dieldrin produces nest desertion and other dysfunctional behavior in mallard ducks. In mice, dieldrin at 10 ppm decreased the tendency of mothers to nurse. At exposure levels above 15 ppm, pups were killed or simply neglected.[62] A dose dependent reduction in sexual receptivity has been observed after exposure to lindane.[63] At lower exposure levels, various behavioral deficits have been described in a number of species. These various findings allow three generalizations. The first is that the capacity of cyclodienes and lindane to modify behavior is dose-dependent. The second is that performance of complex tasks is more readily disrupted than performance of simpler tasks, and the third is that transitional behaviors, such as occur during acquisition or extinction of responses, are much more susceptible to disruption than are stable behaviors.[64]

11.27.4.3 Mechanisms of Action

In both insects[65,66] and higher animals[67-69] excitatory synaptic activity is greatly enhanced. Synapses with many converging presynaptic elements whose net activity serves to modulate the excitability of the postsynaptic cell are most sensitive. At these synapses, the response of the postsynaptic cell to excitatory afferent input increases, leading to a lowered threshold of excitation and to an increase in the number and frequency of action potentials generated as a response. This process "avalanches" along polysynaptic pathways, ultimately producing responses 10–100 times more intense than normal. This exaggerated neuronal activity is transmitted by way of motor outflow to the various organs, glands, and muscles. Stimulation results in inappropriately exaggerated responses, first transient, but eventually sustained in a convulsive seizure.

It is now well established that a major target for cyclodienes and lindane is the $GABA_A$-receptor–chloride channel complex that subserves one type of GABA-mediated inhibition (see Chapters 8 and 9, this volume). This hypothesis has been reviewed by Joy[11] from an historical perspective, and that source should be consulted for more complete references. All of these compounds bind reversibly to the picrotoxin binding site which is located on the chloride channel portion of the complex.[70-74] All inhibit GABA-induced chloride uptake into mouse or rat brain vesicles.[70,75,76] Potencies for inhibiting GABA-induced chloride uptake and binding to the picrotoxin site show high correlation and share a similar stereospecificity.[75,77] Cyclodienes and lindane depress or block permeability changes induced in intact cells *in vitro* by GABA,[77,78] and lindane and dieldrin block $GABA_A$-mediated inhibition in the mammalian hippocampus *in vitro*[79] and *in vivo*.[80-82] Single channel analysis has established that lindane does not affect single channel conductances or channel open times, but does reduce the number of active channels.[83] Kinetic considerations suggest this occurs by stabilizing the channel in the closed state.

Additional mechanisms have been proposed based on the fact that cyclodienes and lindane produce changes in the concentrations and turnover of many chemical substances in the brain, block ATPases, and cause alterations in calcium homeostasis. Reviews of these possibilities are available.[8,11]

11.27.4.4 Neurotoxicity of Cyclodienes and Lindane to Humans

Since the cyclodienes and lindane are toxic at lower exposure levels than are the DDT analogues, their toxicity in man is more common. Reviews providing information concerning voluntary, usually with suicidal intent, and involuntary exposure by the oral, inhalational, or dermal routes to these compounds are available.[1,8,11,48]

The most typical response to the ingestion of toxic amounts of all the compounds is the development of convulsive seizures. These are often sudden and abrupt in onset, and they may occur without warning or any prodromal signs. In some subjects, convulsions may be preceded by hyperexcitability, myoclonic jerks, and general malaise. Headache, nausea, vomiting, and dizziness may or may not be experienced.

Two types of chronic exposure syndromes have been described by Jager.[48] In one of these, exposure occurs constantly and results in a slow accumulation of the insecticide coupled with a progressing symptomology. In the second type the intake of insecticide remains below that required for overt symptoms. However, the subject is rendered extremely sensitive to additional acute exposure. Repeated, chronic exposure may induce histopathological changes in liver and kidney. The compounds and their metabolic products are stored in fat, and mobilization of these stores during intensive activity and starvation may cause toxic symptoms to recur.[17] In subjects dying during acute exposure, a number of pathological effects can be attributed to convulsions induced by the toxicant.

The presence of convulsions is a potentially disturbing consequence because convulsions, regardless of origin, can directly cause neuronal cell death and persisting neurologic symptoms. Convulsions may be particularly dangerous to children.[84-86] Isolated cases of latent polyneuropathies have been tentatively related to exposures to cyclodienes and lindane.[45,87]

11.27.5 NEUROTOXICITY OF MIREX AND CHLORDECONE

Mirex was introduced in 1959 as a bait for fire ants. It is toxic when ingested, but has little effect by contact. As its acute LD_{50} indicates (Table 1), it is not particularly toxic. It does accumulate with chronic exposures, however, and in animals displays a high chronicity index.[8] Animals fed near lethal doses exhibit reduced rates of weight gain, abnormal blood chemistry values, increased liver-to-body weight ratios, and decreased spleen-to-body weight ratios. Few neurological signs emerge. There have been no reports of human poisoning by mirex.

Chlordecone is a complex structured chlorinated hydrocarbon insecticide developed in the early 1950s and registered as a pesticide in 1955. Between 1966 and 1973 $5 \times 10^4 - 2 \times 10^5$ kg were produced annually. In 1973 Life Science Products Company of Hopewell, VA became the sole producer of chlordecone in the world, the annual production reaching a level of 4×10^5 kg of greater than 90% purity by 1975.

Little of this chemical was used in the US, the major home use being in a formulation at a concentration of 0.125% in ant and roach traps. Some 99% of the world production was exported to West Germany where it was used in the formulation of a pesticide mixture, Kelevan, which was exported to Central and South America for use in controlling the banana borer weevil.

11.27.5.1 Toxicity to Animals

Chlordecone is readily absorbed from the gastrointestinal tract of laboratory animals, and absorption through the skin has been reported.[88] In rodents tremors and abnormal gait are early features of toxicity.[89] Startle responses are exaggerated. Muscle weakness develops and can continue for a long period after termination of exposure.

Chlordecone has a major impact on female reproductive function and behavior.[88,90] Animals show constant estrus, the development of large follicles, the absence of corpora lutea in the ovaries, and failure to reproduce. Chlordecone administration to female hamsters or rat pups masculinizes them. When chlordecone is given to female rats after impregnation, fertility is substantially reduced.

Behavioral effects of chlordecone in adult and young animals have been reviewed.[11] A dose-related increase in startle response, decrease in forelimb grip strength, and impairment in acquisition, retention, and extinction of conditioned-avoidance responding have been reported.

11.27.5.2 Mechanisms of Action

The mechanism of action of kepone is not well understood. As tremor is a major symptom, it might be suspected that chlordecone affects the nervous system in a manner similar to DDT. This is not the case. The quality of tremor is quite different.[91] Pharmacological antagonism of chlordecone tremor also has a different profile.[91]

Measurement of levels and turnover of brain peptides and biogenic amines have generated variable results.[11] The many and variable effects reported for chlordecone are reminiscent of changes reported after other organochlorines and neurotoxic chemicals. It is difficult to believe that any of these changes are primary effects. It is more likely that they are consequences of functional changes evoked by chlordecone through some more basic mechanism. As is true for other organochlorines, chlordecone inhibits a number of important ATPases. Membrane bound Na^+-K^+-ATPase and the oligomycin sensitive, mitochondrial Mg^{2+}-ATPase are among the most sensitive.[92,93] Chlordecone is a potent inhibitor of Ca^{2+}-ATPase (calcium pump) in heart and brain[94,95] and inhibits the uptake of labeled Ca^{2+} by sarcoplasmic reticulum vesicles. *In vivo*, symptomatic doses of chlordecone inhibit brain Ca^{2+}-ATPase activity by 50% and reduce total-brain and synaptosomal calmodulin levels.[94] While these findings are of interest, it is of concern that a great many compounds, including DDT, cyclodienes, and lindane also have marked effects on these same systems, yet differ so significantly in their symptoms of poisoning.

11.27.5.3 Toxicity to Humans

What has become known as the "Hopewell epidemic" occurred during the summer of 1975. This is the only known human intoxication with chlordecone. It resulted from the industrial exposure of 133 individuals, employed during a 17 month period of chlordecone synthesis under very unsatisfactory hygienic and health conditions. The factory was ultimately closed, but it was subsequently found that massive amounts of chlordecone had also been discharged into the James river. This resulted in the contamination of the river system and the tidal areas of the bottom end of Chesapeake Bay. In consequence, chlordecone was bioaccumulated in the aquatic life and seriously affected the local fishing industry. The clinical aspects of the disaster have been described in detail in several papers.[96–99]

Patients with prolonged exposure to chlordecone developed nervousness, tremors, chest pains, weight loss, arthralgia, skin rash, mental changes, opsoclonus, muscle weakness, ataxic gait, incoordination, and slurred speech.[99] In severe cases, the tremor was present at rest and in all cases was increased by use of the affected limb. The hands were chiefly involved, but fine tremor of the head and trembling of the entire body (called "Kepone shakes" by the workers) were observed. A greatly exaggerated startle response was also noted in severely affected people. Another prominent complaint was visual difficulty, characterized by an inability to fixate and focus. Personality changes were observed, the most common being irritability, difficulty with recent memory, and mild depression which, in some cases, approached frank disorientation. In addition, semen analysis in four patients revealed severe impairment of spermatogenesis. Oligospermia was seen in 13 men, the severity ranging from no motile sperm in 2 workers to an intermediate sperm count in the rest which was lower than normal values.[97]

11.27.6 NEUROTOXICITY OF PYRETHRINS AND PYRETHROID INSECTICIDES

Pyrethrum is the product prepared from ground flowers of Chrysanthemum species by extraction with organic solvents. It consists of 6 related esters derived from 2 acids and 3 alcohols. The proportions of each varies depending upon the strain of flower, conditions of culture and method of extraction and concentration.[5,12] Pyrethrins decompose rapidly in light to inactive compounds. They are liquids at room temperature, are insoluble in water but are soluble in many organic solvents. They undergo rapid hydrolysis in water, particularly in the presence of acid or alkali. They are primarily contact poisons and penetrate chitin rapidly. They have almost no activity when given orally because they are so readily hydrolyzed to inactive products. Pyrethroids are synthetic compounds that resemble pyrethrins in insecticidal activity but are chemically much more stable. Their chemistry is complex, and has been reviewed by Ray.[12]

11.27.6.1 Signs and Symptoms of Poisoning in Animals

Table 1 compares oral and intravenous LD_{50} of various pyrethroids. Although many of these compounds are potent toxicants when injected, their rapid metabolism results in large differences in the oral vs. intravenous toxicity. Their

dermal toxicity is even lower than their oral toxicity due to slow penetration through skin and rapid metabolism.

Taken as a group, the pyrethroids evoke a number of overlapping effects at toxic levels. The evaluation of a great number of compounds led to the recognition that there are two primary syndromes of poisoning.[100-102] Type I pyrethroids produce poisoning symptoms very similar to DDT and cause the development of a progressive fine whole-body tremor, exaggerated startle response, uncoordinated twitching of the dorsal muscles, hyperexcitability, and death. Associated with the tremor is a large increase in metabolic activity which results in hyperthermia.

Type II pyrethroids produce a more diverse response. In rodents there is a progressive development of repetitive chewing, nosing, and washing, excessive salivation, coarse whole-body tremor, increased extensor tone in the hind limbs, choreiform movements of the limbs and tail, convulsions, and death.[101] Body temperature is either not affected or decreased. Most pyrethroids can be placed into one or the other of these two categories, although pyrethroids intermediate in symptomology exist.

11.27.6.2 Mechanisms of Action

The reviews of Ray[12] and Joy[15] should be consulted for more complete source references in the area (see Chapter 9, this volume). All of the pyrethroids that produce mammalian toxicity are known to interact with sodium channels in excitable tissues.[103-104] Like DDT, some pyrethroids, such as allethrin and tetramethrin, cause the development of spontaneous or evoked repetitive discharges in nerve fibers. However, not all pyrethroids affect axons in the same way. Some produce repetitive discharge for long periods, others produce a short period of repetitive discharge with rapidly decrementing action potential amplitudes and others cause only a progressive conduction block. Which occurs depends on the extent to which a given pyrethroid increases the time constant for inactivation of sodium currents. Those that produce the shortest increases (DDT, allethrin) produce an afterdepolarization which is sufficient to evoke repetitive discharge, but the transient persistence of current flow attenuates so rapidly that the resting potential is relatively unaffected. Those that produce intermediate increases in the duration of sodium current (phenothrin, tetramethrin) also induce an afterdepolarization which evokes repetitive discharge. If the interval between action potentials is smaller than the duration of enhanced current flow, then the afterdepolarizations summate slowly, producing a burst of decrementing action potentials followed by a conduction block. For those producing very persistent current flows (cypermethrin, fenvalerate, deltamethrin), each action potential leads to a maintained, summating depolarization which rapidly produces conduction block.

There is a correlation with symptoms of poisoning. Type I pyrethroids and DDT produce short increases in the time-constant for sodium current inactivation and produce nearly identical symptoms during exposure. Type II pyrethroids extend time constants for inactivation hundreds of milliseconds to seconds. This has been proposed by some to be the basis for the symptoms produced by type II agents.[104-105] Only a fraction of the sodium channels in a nerve need be affected to produce repetitive discharge or conduction block. Narahashi[103] has calculated that symptoms of poisoning should be observable when only 0.1% of channels are affected.

While most investigators would agree that type I pyrethroids produce their toxic effects through their effects on sodium channels, there is less consensus regarding type II compounds. Gammon *et al.*[106] proposed that type II compounds might be acting on the $GABA_A$-receptor–chloride channel complex. Subsequently, Lawrence and Casida[74,107] showed that type II, but not type I pyrethroids, inhibited the binding of *t*-butylbicyclophosphorothionate (TBPS) to the picrotoxin binding site on the complex. The relative potency for displacing TBPS showed a good correlation to acute toxicity in mice with intracerebral injection. Unfortunately the potencies of active and inactive pyrethroids for TBPS binding, GABA-evoked chloride flux and mammalian LD_{50} values show poor agreement.[108,109] Moreover, pyrethroids are much less potent inhibitors of GABA-dependent function than they are of sodium channel function.[78,110] Differences in binding potencies between 200–1000 have been reported. Electrophysiological studies in the hippocampus of intact rats also contradict this theory. Instead of reducing $GABA_A$-mediated inhibition at that locus, type II pyrethroids greatly enhanced inhibition as measured by a paired pulse technique.[111-113]

Pyrethroids displace the binding of specific ligands at other sites besides the sodium and GABA-gated chloride channel, and change concentrations of or turnover of many chemicals substances in the brain.[15] However, there is no compelling reason at present to think these play a significant role in toxicity. Most appear to be secondary effects.

11.27.6.3 Toxicity to Humans

Considering their widespread use, the safety of pyrethroids is remarkable. With very few exceptions, occupational or unintentional exposures have not produced serious deleterious effects. Injury to humans from pyrethrum, the natural product, most frequently results from its allergenic properties rather than its direct toxicity. Contact dermatitis is by far the most common.[114] Irritative or allergic reactions have also been reported for synthetic pyrethroids.[115] These consist primarily of dermal symptoms which include itching, burning, and paresthesias, as well as respiratory and ocular signs of irritation. The dermal effects develop after a delay of 0.5–3 h and last from 0.5 to 8 h, depending upon the extent of exposure. While paresthesia has been reported by people using type I or type II pyrethroids, this phenomenon is much more pronounced with type II compounds.[116] Because of their widespread use, fenvalerate and deltamethrin are common offenders. Other neurological effects are rare.

The most comprehensive report of poisoning by pyrethroids is a Chinese study that reviews 573 cases of acute pyrethroid poisoning in China between 1983 and 1988.[15,117] This includes 325 cases of acute deltamethrin poisoning, 196 cases of fenvalerate poisoning, 45 cases of cypermethrin poisoning, and 7 cases of poisoning by other pyrethroids. About half of the poisonings resulted from occupational exposure and half from accidental exposure. Occupational exposures resulted from inappropriate handling, use of highly concentrated solutions, spraying against the wind, clearing stoppage of sprayers by mouth or hands, and by lack of personal protection. The accidental exposures were most often the result of ingestion. The severity of exposure ranged from those producing only paresthesia of the face and mouth to death. Seven deaths were reported in all.

Symptoms were quite similar for all the pyrethroids. Following occupational exposure the first symptoms were burning or itching sensations of the face which usually developed about 6 h after exposure. When ingested, the first symptoms were usually epigastric pain, nausea and vomiting. Systemic symptoms developing after absorption of sufficient pyrethroid included dizziness, headache, weakness, and fatigue. More serious cases developed coarse muscular fasciculations, disturbances in consciousness, convulsions, coma, and death.[117]

11.27.7 REFERENCES

1. W. J. Hayes, Jr., in 'DDT, das Insektizid Dichlordiphenyltrichlorathan und seine Bedeutung' (The Insecticide Dichlorodiphenyltrichloroethane and Its Significance),' ed. P. Muller, Birkhauser Verlag, Basel, 1959, vol. 2, pp. 11–247.
2. R. Carson, 'Silent Spring,' Houghton Mifflin, Boston, MA, 1962.
3. C. F. Wurster, 'Aldrin and dieldrin.' *Environment*, 1971, **13**, 33–45.
4. E. A. Fendick, E. Mather-Mihaich, K. A. Houck *et al.*, 'Ecological toxicology and human health effects of heptachlor.' *Rev. Environ. Contam. Toxicol.*, 1990, **111**, 61–142.
5. R. D. O'Brien, 'Insecticides Action and Metabolism,' Academic Press, New York, 1967.
6. W. J. Hayes, Jr., 'Toxicology of Pesticides,' Williams & Wilkins, Baltimore, MD, 1975.
7. W. J. Hayes, Jr., 'Pesticides Studied in Man,' Williams & Wilkins, Baltimore, MD, 1982.
8. A. G. Smith, in 'Handbook of Pesticide Toxicology,' eds. W. J. Hayes, Jr. and E. R. Laws, Jr., Academic Press, San Diego, CA, 1991, pp. 731–915.
9. F. Matsumura, 'Toxicology of Insecticides,' 2nd edn., Plenum Press, New York, 1985.
10. S. R. Baker and C. F. Wilkinson, 'The effects of pesticides on human health.' *Adv. Modern Environ. Toxicol.*, 1990, **18**, 1–438.
11. R. M. Joy, in 'Pesticides and Neurological Diseases,' eds. D. J. Ecobichon and R. M. Joy, CRC Press, Boca Raton, FL, 1994, pp. 81–170.
12. D. E. Ray, in 'Handbook of Pesticide Toxicology,' eds. W. J. Hayes, Jr. and E. R. Laws, Jr., Academic Press, San Diego, CA, 1991, pp. 585–636.
13. G. S. F. Ruigt, in 'Comprehensive Insect Physiology, Biochemistry and Pharmacology,' eds. G. A. Kerkut and L. I. Gilbert, Pergamon Press, Oxford, 1985, vol. 12, pp. 183–239.
14. D. M. Soderlund and J. R. Bloomquist, 'Neurotoxic actions of pyrethroid insecticides.' *Annu. Rev. Entomol.*, 1989, **34**, 77–96.
15. R. M. Joy, in 'Pesticides and Neurological Diseases,' 2nd edn., eds. D. J. Ecobichon and R. M. Joy, CRC Press, Boca Raton, FL, 1994, pp. 291–312.
16. E. G. Fairchild (ed.), 'Agricultural Chemicals and Pesticides,' A subfile of the NIOSH Registry of toxic effects of chemical substances, National Institute of Occupational Safety and Health, US Department of Health, Education and Welfare, Public Health Service, Cincinatti, OH, July, 1977.
17. W. J. Hayes, Jr., 'Clinical Handbook on Economic Poisons: Emergency Information for Treating Poison,' Public Health Services Publ. No. 476, Toxicology Section, Communicable Disease Center, Atlanta, GA, 1963.
18. F. Matsumura, 'Toxicology of Insecticides,' Plenum Press, New York, 1975.
19. J. M. Tobias and J. J. Kollros, 'Loci of action of DDT in the cockroach (*Periplanata americana*).' *Biol. Bull.*, 1946, **91**, 247–255.
20. E. C. Hwang and M. H. Van Woert, '*p,p'*-DDT-induced neurotoxic syndrome: experimental myoclonus.' *Neurology*, 1978, **28**, 1020–1025.
21. E. C. Hwang, A. Plaitakis, I. Magnussen *et al.*, 'Relationship of inferior olive-climbing fibers to *p, p'*-DDT-induced myoclonus in rats.' *Neurosci. Lett.*, 1981, 24, 103–108.
22. P. D. Hrdina, R. L. Singhal and G. M. Ling, 'DDT and related chlorinated hydrocarbon insecticides: pharmacological basis of their toxicity in mammals.' *Adv. Pharmacol. Chemother.*, 1975, **12**, 31–88.
23. F. S. Phillips and A. Gilman, 'Studies on the pharmacology of DDT (2,2 bis-(*p*-chlorophenyl)-1,1,1-trichloroethane). I. The acute toxicity of DDT following intravenous injection in mammals with observations on the treatment of acute DDT poisoning.' *J. Pharmacol. Exp. Ther.*, 1946, **86**, 213–221.
24. R. B. Bromiley and P. Bard, 'Tremor and changes in reflex status produced by DDT in decerebrate, decerebrate-decerebellate and spinal animals.' *Bull. Johns Hopkins Hosp.*, 1949, **84**, 414–429.
25. R. J. Bing, B. McNamara and F. H. Hopkins, 'Studies on the pharmacology of DDT (2,2-bis-(*p*-chlorophenyl)-1,1,1-trichloroethane). The chronic toxicity of DDT in the dog.' *Bull. Johns Hopkins Hosp.*, 1946, **78**, 308–315.
26. O. G. Fitzhugh and A. A. Nelson, 'The chronic oral toxicity of DDT (2,2-*bis*(*p*-chlorophenyl)-1,1,1-trichloroethane).' *J. Pharmacol. Exp. Ther.*, 1947, **89**, 18–30.
27. T. Narahashi and T. Yamasaki, 'Mechanism of increase in negative after-potential by dicophanum (DDT) in the giant axons of the cockroach.' *J. Physiol. (Lond.)*, 1960, 152, 122–140.
28. K. D. Roeder and E. A. Weiant, 'The effect of DDT on sensory and motor structures in the cockroach leg.' *J. Cell. Comp. Physiol.*, 1948 32, 175–186.
29. T. Narahashi, in 'Advances in Insect Physiology,' eds, J. W. L. Beament, J. E. Treherne and V. B. Wigglesworth, Academic Press, London, pp. 1–93.
30. P. Eriksson, T. Archer and A. Fredriksson, 'Altered behavior in adult mice exposed to a single low-dose of DDT and its fatty acid conjugate as neonates.' *Brain Res.*, 1990, **514**, 141–142.
31. P. Eriksson, L. Nilsson-Hakansson, A. Nordberg *et al.*, 'Neonatal exposure to DDT and its fatty acid conjugate: effects on cholinergic and behavioral variables in the adult mouse.' *Neurotoxicology*, 1990, **11**, 345–354.
32. H. B. Daly, D. R. Hertzler and D. M. Sargent, 'Ingestion of environmentally contaminated Lake Ontario salmon by laboratory rats increases avoidance of unpredictable aversive nonreward and mild electric shock.' *Behav. Neurosci.*, 1989, **103**, 1356–1365.
33. H. B. Daly, 'Reward reductions found more aversive by rats fed environmentally contaminated salmon.' *Neurotoxicol. Teratol.*, 1991, **13**, 449–453.
34. T. Narahashi and T. Yamasaki, 'Behaviors of membrane potential in the cockroach giant axons poisoned by DDT.' *J. Cell. Comp. Physiol.*, 1960, **55**, 131–142.
35. T. Narahashi and H. G. Haas, 'DDT: interaction with nerve membrane conductance changes.' *Science*, 1967, **157**, 1438–1440.
36. T. Narahashi and H. G. Haas, 'Interaction of DDT with the components of lobster nerve membrane conductance.' *J. Gen. Physiol.*, 1968, **51**, 177–198.
37. A. M. Shanes, 'Electrical phenomena in nerve. III. Frog sciatic nerve.' *J. Cell. Comp. Physiol.*, 1951, **38**, 17–40.
38. B. Hille, 'Pharmacological modifications of the sodium channels of frog nerve.' *J. Gen. Physiol.*, 1968, **51**, 199–219.
39. M. Uchida, H. Naka, Y. Irie *et al.*, 'Insecticidal and neuroexciting actions of DDT analogs.' *Pestic. Biochem. Physiol.*, 1974, 4, 451–455.
40. J. H. Welsh and H. T. Gordon, 'The mode of action of certain insecticides on the arthropod nerve axon.' *J. Cell. Comp. Physiol.*, 1947, **30**, 147–172.
41. S. M. Ghiasuddin and D. M. Soderlund, 'Pyrethroid insecticides: potential, stereospecific enhancers of mouse brain sodium channel activation.' *Pestic. Biochem. Physiol.*, 1985, **24**, 200–206.
42. W. N. Aldridge, B. Clothier, P. Forshaw *et al.*, 'The effect of DDT and the pyrethroids cismethrin and decamethrin on the acetyl choline and cyclic nucleo-

tide content of rat brain.' *Biochem. Pharmacol.*, 1978, **27**, 1703–1706.

43. M. A. Matin, F. N. Jaffery and R. A. Siddiqui, 'A possible neurochemical basis of the central stimulatory effects of *p,p'*-DDT.' *J. Neurochem.*, 1981, **36**, 1000–1005.

44. M. I. Fonseca, J. S. Aguilar, C. Lopez *et al.*, 'Regional effect of organochlorine insecticides on cholinergic muscarinic receptors of rat brain.' *Toxicol. Appl. Pharmacol.*, 1986, **84**, 192–195.

45. R. B. Jenkins and R. F. Toole, 'Polyneuropathy following exposure to insecticides.' *Arch. Int. Med.*, 1964, **1**, 691–695.

46. J. R. Busvine, 'Houseflies resistant to a group of chlorinated hydrocarbon insecticides.' *Nature (London)*, 1954, 174, 783–785.

47. C. C. Roan and T. L. Hopkins, 'Mode of action of insecticides.' *Ann. Rev. Entomol.*, 1948, **6**, 333–367.

48. K. W. Jager, 'Aldrin, Dieldrin, Endrin, and Telodrin, an Epidemiological Toxicological Study of Long-Term Occupational Exposure,' Elsevier, New York, 1970.

49. Y. Coble, P. Hildebrandt, J. Davis *et al.*, 'Acute endrin poisoning.' *JAMA*, 1967, **202**, 489–493.

50. R. M. Joy, 'Convulsive properties of chlorinated hydrocarbon insecticides in the cat central nervous system.' *Toxicol. Appl. Pharmacol.*, 1976, **35**, 95–106.

51. A. N. Worden, 'Toxicity of telodrin.' *Toxicol. Appl. Pharmacol.*, 1969, **14**, 556–573.

52. E. J. Gralla, R. W. Fleischman, Y. K. Luthra *et al.*, 'Toxic effects of hexachlorobenzene after daily administration to beagle dogs for one year.' *Toxicol. Appl. Pharmacol.*, 1977, **40**, 227–239.

53. C. M. Wang and F. Matsumura, 'Relationship between the neurotoxicity and *in vivo* toxicity of certain cyclodiene insecticides in the German cockroach.' *J. Econ. Entomol.*, 1970, **63**, 1731–1734.

54. R. M. Joy, 'Electrical correlates of preconvulsive and convulsive doses of chlorinated hydrocarbon insecticides in the CNS.' *Neuropharmacology*, 1973, **12**, 63–76.

55. G. A. Van Gelder, B. E. Sandler, W. B. Buck *et al.*, 'Convulsive seizures in dieldrin exposed sheep during photic stimulation.' *Psychol. Rep.*, 1969, **24**, 502.

56. L. Hulth, L. Hoglund, A. Bergman *et al.*, 'Convulsive properties of lindane, lindane metabolites, and the lindane isomer alpha-hexachlorocyclohexane: effects on the convulsive threshold for pentylenetetrazol, and the brain content of gamma-aminobutytric acid (GABA) in the mouse.' *Toxicol. Appl. Pharmacol.*, 1978, **46**, 101–108.

57. B. E. Fishman and G. Gianutsos, 'CNS biochemical and pharmacological effects of the isomers of hexachlorocyclohexane (lindane) in the mouse.' *Toxicol. Appl. Pharmacol.*, 1988, **93**, 146–153.

58. T. E. Albertson, R. M. Joy and L. G. Stark, 'Facilitation of kindling in adult rats following neonatal exposure to lindane.' *Brain Res.*, 1985, **349**, 263–266.

59. R. M. Joy, L. G. Stark and T. E. Albertson, 'Proconvulsant effects of lindane: enhancement of amygdaloid kindling in the rat.' *Neurobehav. Toxicol. Teratol.*, 1982, **4**, 347–354.

60. R. M. Joy, L. G. Stark and T. E. Albertson, 'Proconvulsant action of lindane compared at two different kindling sites in the rat—amygdala and hippocampus.' *Neurobehav. Toxicol. Teratol.*, 1983, **5**, 465–468.

61. R. M. Joy, L. G. Stark and T. E. Albertson, 'Proconvulsant actions of lindane: effects on afterdischarge thresholds and durations during amygdaloid kindling in rats.' *Neurotoxicology*, 1983, **4**, 211–219.

62. B. B. Virgo and G. D. Bellward, 'Effects of dietary dieldrin on offspring viability, maternal behaviour, and milk production in the mouse.' *Res. Comm. Chem. Pathol. Pharmacol.*, 1977, **17**, 399–409.

63. L. Uphouse, 'Decreased rodent sexual receptivity after lindane.' *Toxicol. Lett.*, 1987, **39**, 7–14.

64. R. M. Joy, 'Mode of action of lindane, dieldrin and related insecticides in the central nervous system.' *Neurobehav. Toxicol. Teratol.*, 1982, **4**, 813–823.

65. C. M. Wang, T. Narahashi and M. Yamada, 'The neurotoxic action of dieldrin and its derivatives in the cockroach.' *Pestic. Biochem. Physiol.*, 1971, **1**, 84–91.

66. D. L. Shankland and M. E. Schroeder, 'Pharmacological evidence for a discrete neurotoxic action of dieldrin (HEOD) in the American cockroach. *Periplaneta americana* (L),' *Pestic. Biochem. Physiol.*, 1973, **3**, 77–86.

67. R. M. Joy, 'Alteration of sensory and motor evoked responses by dieldrin.' *Neuropharmacology*, 1974, **13**, 93–110.

68. R. M. Joy, 'The alteration by dieldrin of cortical excitability conditioned by sensory stimuli.' *Toxicol. Appl. Pharmacol.*, 1976, **38**, 357–368.

69. R. M. Joy, 'Comparative effects of convulsants on the antidromic cortical response to pyramidal tract stimulation.' *Neuropharmacology*, 1975, **14**, 869–881.

70. S. M. Ghiasuddin and F. Matsumura, 'Inhibition of gamma-aminobutyric acid (GABA)-induced chloride uptake by gamma-BHC and heptachlor epoxide.' *Comp. Biochem. Physiol.*, 1982, **C73**, 141–144.

71. F. Matsumura and K. Tanaka, in 'Cellular and Molecular Neurotoxicology,' ed. T. Narahashi, Raven Press, New York, 1984, pp. 225–240.

72. F. Matsumura, 'Involvement of picrotoxinin receptor in the action of cyclodiene insecticides.' *Neurotoxicology*, 1985, **6**, 139–163.

73. L. J. Lawrence and J. E. Casida, Interactions of lindane, toxaphene and cyclodienes with brain-specific *t*-butylbicyclophosphorothionate receptor.' *Life Sci.*, 1984, **35**, 171–178.

74. J. E. Casida and L. J. Lawrence, 'Stucture-activity correlations for interactions of bicyclophosphorus esters and some polychlorocycloalkane and pyrethroid insecticides with the brain-specific *t*-butylbicyclophosphorothionate receptor.' *Environ. Health Perspect.*, 1985, **61**, 123–132.

75. I. M. Abalis, M. E. Eldefrawi and A. T. Eldefrawi, 'Effects of insecticides on GABA-induced chloride influx into rat brain microsacs.' *J. Toxicol. Environ. Health*, 1986, **18**, 13–23.

76. J. R. Bloomquist, P. M. Adams and D. M. Soderlund, in 'Sites of Action for Neurotoxic Pesticides,' eds., R. M. Hollingworth and M. B. Green, American Chemical Society Symposium Series No. 356, Washington, DC, 1987, 97–106.

77. K. A. Wafford, S. C. R. Lummis and D. B. Sattelle, 'Block of an insect central nervous system GABA receptor by cyclodiene and cyclohexane insecticides.' *Proc. R. Soc. Lond. B Biol. Sci.*, 1989, **237**, 53–61.

78. N. Ogata, S. M. Vogel and T. Narahashi, 'Lindane but not deltamethrin blocks a component of GABA-activated chloride channels.' *FASEB J.*, 1988, **2**, 2895–2900.

79. R. M. Joy, W. F. Walby, L. G. Stark *et al.*, 'Lindane blocks $GABA_A$-mediated inhibition and modulates pyramidal cell excitability in the rat hippocampal slice.' *Neurotoxicology*, 1995, **16**, 217–228.

80. R. M. Joy and T. E. Albertson, 'Interactions of lindane with synaptically mediated inhibition and facilitation in the dentate gyrus.' *Neurotoxicology*, 1987, **8**, 529–542.

81. R. M. Joy and T. E. Albertson, 'Convulsant-induced changes in perforant path-dentate gyrus excitability in urethane-anesthetized rats.' *J. Pharmacol. Exp. Ther.*, 1988, **2**, 887–895.

82. T. E. Albertson and R. M. Joy, in 'Insecticide Action from Molecule to Organism,' eds., T. Narahashi and J. E. Chambers, Plenum, New York, 1989, pp. 115–137.

83. F. Zufall, C. Franke and H. Hatt, 'Similarities between the effects of lindane (γ-HCH) and picrotoxin on ligand-gated chloride channels in crayfish muscle membrane.' *Brain Res.*, 1989, **503**, 342–345.

84. P. C. Gupta, 'Neurotoxicity of chronic chlorinated hydrocarbon insecticide poisoning—a clinical and electroencephalographic study in man.' *Indian J. Med. Res.*, 1975, **63**, 601–606.

85. C. R. Angle, M. S. McIntyre and R. L. Meile, 'Neurologic sequelae of poisoning in children.' *J. Pediatr.*, 1968, **73**, 531–539.

86. H. Jacobziner and H. W. Raybin, 'Poisoning by insect-icide (endrin).' *NY State J. Med.*, 1959, **59**, 2017–2022.

87. T. M. Onifer and J. P. Whisnant, 'Cerebellar ataxia and neuronitis after exposure to DDT and lindane.' *Proc. Mayo Clinic*, 1957, **32**, 67–72.

88. J. J. Huber,' Some physiological effects of the insecticide Kepone in the laboratory mouse.' *Toxicol. Appl. Pharmacol.*, 1965, **7**, 516–524.

89. J. L. Egle, Jr., P. S. Guzelian and J. F. Borzelleca, 'Time course of the acute toxic effects of sublethal doses of chlordecone (Kepone).' *Toxicol. Appl. Pharmacol.*, 1979, **48**, 533–536.

90. V. Sierra and L. Uphouse, 'Long-term consequences of neonatal exposure to chlordecone.' *Neurotoxicology*, 1986, **7**, 609–621.

91. H. A. Tilson, J. S. Hong, J. M. Gerhart *et al.*, 'Animal models in neurotoxicology: the neurobehavioral effects of chlordecone (Kepone).' *Adv. Behav. Pharmacol.*, 1986, **8**, 2249–2273.

92. D. Desaiah, 'Biochemical mechanisms of chlordecone neurotoxicity: a review.' *Neurotoxicology*, 1982, **3**, 103–110.

93. D. Desaiah, in 'Cellular and Molecular Neurotoxicology,' ed. T. Narahashi, Raven Press, New York, 1984, pp. 257–265.

94. D. Desaiah, C. S. Chetty and K. S. Prasada Rao, 'Chlordecone inhibition of calmodulin activated calcium ATPase in rat brain synaptosomes.' *J. Toxicol. Environ. Health*, 1985, **16**, 189–195.

95. P. R. Kodavanti, J. A. Cameron, P. R. Yallapragada *et al.*, 'Effect of chlordecone (Kepone) on calcium transport mechanisms in rat heart sarcoplasmic reticulum.' *Pharmacol. Toxicol.*, 1990, **67**, 227–234.

96. S. B. Cannon, J. M. Veazey, Jr., R. S. Jackson *et al.*, 'Epidemic kepone poisoning in chemical workers.' *Am. J. Epidemiol.*, 1978, **1**, 529–537.

97. J. R. Taylor, J. B. Selhorst and V. P. Calabrese, in 'Experimental and Clinical Neurotoxicology', eds. P. S. Spencer and H. H. Schaumberg, Williams & Wilkins, Baltimore, MD, 1980, pp. 407–421.

98. J. R. Taylor, J. B. Selhorst, S. A. Houff *et al.*, 'Chlordecone intoxication in man. I. Clinical observations.' *Neurology*, 1978, **28**, 626–630.

99. A. J. Martinez, J. R. Taylor, S. A. Houff *et al.*, in 'Neurotoxicology,' eds. L. Roizin, H. Shiraki and N. Grcevic, Raven Press, New York, 1977, vol. 1, pp. 443–459.

100. R. D. Verschoyle and W. N. Aldridge, 'Structure-activity relationships of some pyrethroids in rats.' *Arch. Toxicol.*, 1980, **45**, 325–329.

101. D. E. Ray, 'The contrasting actions of two pyrethroids (deltamethrin and cismethrin) in the rat.' *Neurobehav. Toxicol. Teratol.*, 1982, **4**, 801–804.

102. J. G. Scott and F. Matsumura, 'Evidence for two types of toxic actions of pyrethroids on susceptible and DDT-resistant German cockroaches.' *Pestic. Biochem. Physiol.* 1983, **19**, 141–150.

103. T. Narahashi, 'Nerve membrane ionic channels as the primary target of pyrethroids.' *Neurotoxicology*, 1985, **6**, 3–22.

104. H. P. M. Vijverberg, J. R. de Weille, G. S. F. Ruigt *et al.*, in 'Neuropharmacology and Pesticide Action,' eds. M. G. Ford, G. G. Lunt, R. C. Reay *et al.*, Ellis Horwood, Chichester, 1986, pp. 267–285.

105. T. Narahashi, in 'Neuropharmacology and Pesticide Action,' eds. M. G. Ford, G. G. Lunt, R. C. Reay *et al.*, Ellis Horwood, Chichester, 1986, pp. 36–60.

106. D. W. Gammon, M. A. Brown and J. E. Casida, 'Two classes of pyrethroid action in the cockroach.' *Pestic. Biochem. Physiol.*, 1981, **15**, 181–191.

107. L. J. Lawrence and J. E. Casida, 'Stereospecific action of pyrethroid insecticides on the γ-aminobutyric acid receptor-ionophore complex.' *Science*, 1983, **221**, 1399–1401.

108. A. A. Ramadan, N. M. Bakry, A. S. M. Marei *et al.*, 'Action of pyrethroids on GABA$_A$ receptor function.' *Pestic. Biochem. Physiol.*, 1988, **32**, 97–105.

109. J. R. Bloomquist, P. M. Adams and D. M. Soderlund, 'Inhibition of γ-aminobutyric acid-stimulated chloride flux in mouse brain vesicles by polychlorocycloalkane and pyrethroid insecticides.' *Neurotoxicology*, 1986, **7**, 11–20.

110. A. E. Chalmers, T. A. Miller and R. W. Olsen, 'Deltamethrin: a neurophysiological study of the sites of action.' *Pestic. Biochem. Physiol.*, 1987, **27**, 136–141.

111. M. E. Gilbert, C. M. Mack and K. M. Crofton, 'Pyrethroids and enhanced inhibition in the hippocampus of the rat.' *Brain Res.*, 1989, **477**, 314–321.

112. R. M. Joy, T. E. Albertson and D. E. Ray, 'Type I and type II pyrethroids increase inhibition in the hippocampal dentate gyrus of the rat.' *Toxicol. Appl. Pharmacol.*, 1989, **98**, 398–412.

113. R. M. Joy, T. Lister, D. E. Ray *et al.*, 'Characteristics of the prolonged inhibition produced by a range of pyrethroids in the rat hippocampus.' *Toxicol. Appl. Pharmacol.*, 1990, **103**, 528–538.

114. C. P. McCord, C. H. Kilker and D. K. Minster, 'Pyrethrum dermatitis: a record of the occurrence of occupational dermatoses among workers in the pyrethrum industry.' *JAMA*, 1921, **77**, 448–449.

115. S. A. Flannigan, S. B. Tucker, M. M. Key *et al.*, 'Synthetic pyrethroid insecticides: a dermatological evaluation.' *Br. J. Ind. Med*, 1985, **42**, 363–372.

116. S. A. Flannigan and S. B. Tucker, 'Variation in cutaneous perfusion due to synthetic pyrethroid exposure.' *Br. J. Ind. Med.*, 1985, **42**, 773–776.

117. F. He, S. Wang, L. Liu *et al.*, 'Clinical manifestations and diagnosis of acute pyrethroid poisoning.' *Arch. Toxicol.*, 1989, **63**, 54–58.

11.28

Neuronal Targets of Lead in the Hippocampus: Relationship to Low-level Lead Intoxication

KAREN L. SWANSON, MURILO MARCHIORO,
KUMATOSHI ISHIHARA, MANICKAVASAGOM ALKONDON, and
EDNA F. R. PEREIRA
University of Maryland at Baltimore, MD, USA

and

EDSON X. ALBUQUERQUE
University of Maryland at Baltimore, MD, USA
and Federal University of Rio de Janeiro, Brazil

11.28.1 INTRODUCTION

The most common form of lead toxicity is a subtle and often undetected syndrome of delayed mental maturation, lowered intelligence, and poor performance in school. By the time clinical symptoms of low-level lead intoxication become apparent, there may already have been a long history of lead exposure. Intermediate lead exposure impairs hematopoiesis and damages the kidney, and at the extreme, severe lead exposure can cause convulsion, coma, and death. Unfortunately, much of the damage to the brain may be irreversible. Although many neurophysiological processes including transmitter release, neurogenesis, and plasticity are altered by lead, the mechanisms underlying these changes have been equivocal. Evidence has come to light about the molecular targets and mechanisms of lead neurotoxicity. Lead has specific affinity for many of the Ca^{2+}-binding sites in the brain that are important for regulation of growth and maturation, as well as plastic processes such as learning and memory. Also, the function of certain transmitter-gated ion channels are inhibited by Pb^{2+}. By highlighting the molecular targets of Pb^{2+} in neurons, this report endeavors to show some of the ways by which lead may alter neurophysiological processes to exert its neurotoxic effects.

11.28.1.1 Relationship Between Lead Exposure and Neuropathological Syndromes

Prior to learning of the neurological consequences of exposure to lead, mankind carelessly contaminated the urban environment with this toxic heavy metal.[1] Two major forms of lead are the natural salts of Pb^{2+} and the synthetic organic lead compounds. ("Pb^{2+}" and "Zn^{2+}" refer specifically to the inorganic salts of lead and zinc; these are used in contexts where the chemical forms were not critical or were not identified.) Historically, lead was included in pottery and glazes, and lead acetate has been used as a sweetener. Human exposure now results from several modern sources of lead, chiefly: lead-based paint on the walls of old buildings, where the Pb^{2+} content of dry paint applied before 1960 can be as high as 50%; pipe-joining solders can leach lead into water supplies; atmospheric lead contamination caused by the use of organolead antiknock fuel additives before 1978, which not only caused direct lead exposure by inhalation but was high enough to contaminate rain, water supplies, and soil; food plants grown in lead contaminated soils; and the production of storage batteries.[2] Although lead-based paint was banned in the USA in 1977, old paint remains the most concentrated nonindustrial source of lead. Millions of children are still at high risk of lead intoxication.

The patterns and severity of lead-induced clinical neurotoxicity comprise a spectrum of acute and chronic intoxications that are greatly affected by the age of the individual, a fact that has been attributed to several age-dependent steps in the process of intoxication. Initially, exposure of individuals to lead is largely determined by age-related behaviors. For example, infants and young children frequently sample non-food materials including lead-based paint, whereas some adults are employed by industries utilizing lead. The absorption of Pb^{2+} by immature animals is greater than by adult animals, according to physiological processes that facilitate Ca^{2+} absorption and regulate blood Ca^{2+} levels and Ca^{2+} deposition in growing bones. At the level of the nervous system, the brain has limited abilities to recover from damage and, therefore, symptoms may persist for a long period after a short, substantial exposure, or damage may become evident when the effects accumulate during a prolonged period of low-level exposure. Neurotoxicity in the infant has been attributed to these fundamental factors, as well asthe greater vulnerability of the developing brain to neurological insults that can retard or arrest maturation. Although different patterns of experimental neurotoxicity may be elicited by organolead compounds and Pb^{2+} salts, environmental degradation and internal metabolism during chronic low-level exposure appear to enable both forms of lead to contribute to the same syndrome of impaired mental performance.

The accumulation of lead upon chronic exposure contributes to the prevalence of lead poisoning. Whereas the average American's intake is about $100-350\,\mu g\,d^{-1}$, an individual's blood lead concentrations can be elevated greatly by ingestion of lead-based paint or contaminated water. Adults absorb only up to 10% of the ingested lead, but infants can absorb up to 50% of dietary lead. Therefore, ingestion of a nearly imperceptible amount of paint dust or chips can dramatically increase a child's risk of lead intoxication. For example, daily intake is doubled when an infant ingests 1 mg of 1960's lead-based paint dust that contains 50% lead and absorbs 50% ($250\,\mu g$ Pb^{2+}). Thus, paint remains a major source of lead poisoning of infants and small children. With a half-life of 30 days in plasma, and a half-life of 30 years in bone, lead accumulates readily. The largest portion (90%) of the body burden of lead is stored in bone.

There is no evidence of a physiological requirement for Pb^{2+}, and precivilized man is estimated to have had a blood lead level of about $0.01\,\mu M = 0.2\,\mu g\,dL^{-1}$, which contrasts markedly with the 1990s population average blood lead levels from $0.04-1.3\,\mu M = 0.8-27\,\mu g\,dL^{-1}$, with the higher average levels being found in urban and industrial areas.[3] Legal actions have mandated changes that have reduced exposure to lead. Blood lead screening and medical treatment programs and lead-paint abatement programs were established in the 1970s. For an example of the conditions in the 1980s, an estimated 18% of the urban poor under six years of age had blood lead levels between 25 and $50\,\mu g\,dL^{-1}$ in the state of Maryland, and the cost for detection, medical care, abatement of lead-paint hazard, and special education was $4 million per year. Through eliminating leaded fuels, the urban atmospheric lead levels and the human blood levels declined in parallel in the 1980s.[4] However, because the blood lead level recognized by the Center for Disease Control (CDC) as a risk for intoxication has also been reduced progressively from $60\,\mu g\,dL^{-1}$ in 1970 to $10\,\mu g\,dL^{-1}$ in 1991, the number of children considered to be at risk for lead intoxication is still large. Currently, the CDC recognizes the following classes of lead intoxication in children: Class I, not poisoned; Class II, $\geq 10\,\mu g\,dL^{-1}$, frequent monitoring of the child and advice to parents about environmental lead and nutrition are recommended; Class IIB, $\geq 15\,\mu g\,dL^{-1}$, the child is at risk for a lower IQ and its environment should be investigated; Class III, $\geq 20\,\mu g\,dL^{-1}$, full medical evaluation and environmental remediation are necessary; Class IV, $\geq 45\,\mu g\,dL^{-1}$, chelation therapy becomes important; and Class V, $\geq 70\,\mu g\,dL^{-1}$ represent a medical emergency.[4] The fact that over 50% of African–American black children in poverty and 17% of all children, independent of race or economic status, have blood lead levels exceeding $10\,\mu g\,dL^{-1}$ is cause for concern.[5,6] Because large numbers of children have blood lead levels above $10\,\mu g\,dL^{-1}$, controversy has arisen among practicing clinicians, many of whom consider this level excessively low.[7] However, the new standard has the economic advantage of providing access for people at risk of lead intoxication to social programs; for example, "Women, Infants, and Children" programs in all states can provide for education and improved nutrition that may minimize lead intoxication.[8] The CDC calculated that an economic benefit to society as a whole, in excess of costs, would arise from the elimination of lead poisoning.[5] The calculation incorporated the findings of numerous studies suggesting that intelligence decreases by 0.25 IQ units for each μg lead per dL blood.[9–11]

11.28.1.2 Relationships Between Solute and Tissue Lead Concentrations *In Vivo* and *In Vitro*

Experimental treatments of animals and isolated neuronal tissues have shown that Pb^{2+} can produce many toxic effects, a fact that is not surprising considering that Pb^{2+} and micronutrients such as Zn^{2+}, Co^{2+}, and Cr^{2+} have high affinity for many proteins, owing to their ionic and covalent interactions with anionic side groups with the preference sulfhydryl > amino > hydroxyl. In contrast, the macronutrients Ca^{2+} and Mg^{2+} have weaker ionic interactions with the preference hydroxyl > amino > sulfhydryl.[12] Thus, the chemical properties of Pb^{2+} allow many interactions at carboxyl sites normally occupied by Ca^{2+} and sulfhydryl sites normally occupied by Zn^{2+}. Yet, to show that an experimentally observed effect of Pb^{2+} is pertinent to clinical toxicity, it would be helpful to demonstrate that the amount of Pb^{2+} bound to the target molecules and/or concentrations of free Pb^{2+} ([free Pb^{2+}]) near this target are similar under the clinical and experimental conditions.

The toxic threshold blood lead level of $10\,\mu g\,dL^{-1}$ corresponds mathematically to $0.48\,\mu M$, but since most of the lead is bound to intracellular hemoglobin (>80%) or phospholipid and proteins in membranes of erythrocytes, the concentrations of lead in plasma and CSF are much lower, about $30\,nM$.[13] Furthermore, much of the extracellular lead is also bound to proteins, so that extracellular [free Pb^{2+}] may be only about $1\,nM$.[14,15] Although low, the level of extracellular [free Pb^{2+}] is maintained by rapid equilibration with a repository of Pb^{2+} absorbed by soluble ligands and adsorbed by tissues.[13] The intraneuronal [free Pb^{2+}] in rat cerebral cortex slices has been shown to be only $\approx 175\,pM$ when the total brain lead levels were $3\,\mu g\,g^{-1}$ ($= 1.44\,\mu M$), using NMR for selective and simultaneous identification of divalent metal cations chelated by 5F-BAPTA ((1,2-*bis*(21-amino-5-fluorophenoxy)ethane-N,N,N',N'-tetra-acetic acid)).[16] Given that extracellular and intracellular [free Pb^{2+}] are much lower than total brain lead, most of the Pb^{2+} in brain tissue is bound to high-affinity sites or sequestered in compartments of subcellular organelles not accessed by 5F-BAPTA. The existence of such repositories of lead is supported by evidence that the lead-buffering capacity of live neuronal tissue can be exceeded. After the start of experimental Pb^{2+} treatment of adult rats via drinking water, the blood lead level rises rapidly for three days and then it rises slowly, at a time when the lead level in brain tissue continues to rise steadily.[17,18] When total blood lead exceeds $50\,\mu g\,dL^{-1}$, the [free Pb^{2+}] in blood increases rapidly.[19] Thus, the ratio of total blood lead:total brain lead depends upon the duration of lead exposure, and blood lead levels may not adequately reflect the brain lead levels.

Of particular relevance to studies demonstrating deficits of memory and learning, several studies measuring the regional distribution of total lead in the rat brain have found elevated levels of lead in the hippocampus when lead treatment was given at a low level for several weeks and brain lead was measured after at least six weeks of age (Table 1).[20–24] Because of the ubiquity of lead, there are no animals that truly are not exposed to lead; lead levels vary considerably among control groups, and some untreated populations can more appropriately be classified as lead intoxicated (see Table 1). Under low-level lead treatment conditions, the hippocampi contained from about 0.08 to $1.6\,\mu g\,lead\,g^{-1}$ tissue ($=0.038$ to $0.77\,\mu M$) and

held from twice to seven-times the lead concentration of other brain regions (elevated levels were also found in the rat cortex [24] and in the murine hypothalamus[25]). This relatively high concentration of lead in the hippocampus disappears when the exposure level is increased and lead is then also concentrated in other brain areas.[21,22,26]

The storage of lead appears to be linked to the storage of zinc. Whereas most brain zinc is a fixed, integral part of many proteins and enzymes and the total brain zinc is about 145 ppm, the level of free Zn^{2+} in brain interstitial fluid is only about $0.15\,\mu M$. Levels of total zinc vary little from brain region to brain region, but there is a pool of histochemically reactive Zn^{2+} that is found in limbic and cerebrocortical regions.[28] This pool of Zn^{2+} is restricted to the presynaptic vesicles of a subset of glutamatergic neurons that innervate these regions. Lead appears to be colocalized with Zn^{2+} in the hilus of the dentate gyrus,[27] where the glutamatergic mossy fibers contain high concentrations of Zn^{2+} in clear round synaptic vesicles.[28] Therefore, Pb^{2+} may also be concentrated at microscopic sites *in vivo*. The presynaptic vesicular concentration of Zn^{2+} is based upon an energy-dependent, high-affinity

Table 1   Chronic lead treatment and blood and brain lead levels in rats.

	Lead treatments		Lead concentrations					
Ref.	Interval	Level and route	Blood ($\mu g\,dL^{-1}$)	Blood[a] (μM)	Brain[b] ($\mu g\,g^{-1}$)	Hipp ($\mu g\,g^{-1}$)	Hipp Brain	CDC class[c]
20	young adults, untreated	(0.7 ppm in diet)			0.2	0.49	2.5	
21	postnatal days 3 to 42	control	7	3.4	0.04	0.45	1.1	I
		$0.1\,mg\,kg^{-1}\,d^{-1}$ *per os*	14	6.8	0.06	0.11	1.8	IIA
		$1\,mg\,kg^{-1}\,d^{-1}$ *per os*	31	15	0.17	0.19	1.1	III
22	postnatal days 21 to 90	control	4					I
		50 ppm diet	15	7.2	0.35	0.7	2.0	IIB
		500 ppm diet	50	24	0.75	1	1.3	IV
23	45 days old "untreated"	(0.2–0.7 ppm in diet)[d]			0.4	1.6	4.0	
17	adults 42 days	control	2.8	1.3	1.32[e]			I
		1% in water	68	32	5.32[e]			IV
16	prenatal day 7 to postnatal day 14	control	2–3	12	0.002			I
		$1\,mg\,kg^{-1}\,d^{-1}$	31	15	0.15			III
		$100\,mg\,kg^{-1}\,d^{-1}$	521	251	2.7			V

[a]The micromolar blood lead concentrations were calculated using the relationship $10\,\mu g\,dL^{-1}=0.48\,\mu M$.   [b]"Brain" tissues were whole brain, whole forebrain, or cortex, except forhippocampal levels.   [c]The CDC classification for childhood lead intoxication based on blood lead concentrations is provided for comparison.   [d]The "untreated" rats in this study were apparently subject to a high-level lead exposure; the source was not identified but did not appear to be dietary.   [e]Brain lead levels were reported in ppm dry weight, causing them to be about threefold higher that other levels reported in ppm wet weight.

uptake of Zn^{2+}; K_M estimates range from 1 to $23\,\mu M$. Zn^{2+} is released from these terminals along with neurotransmitter,[29,30] and, in contrast to generally low concentrations of interstitial Zn^{2+}, it has been estimated that the concentration of Zn^{2+} in these synapses may become as high as $150–300\,\mu M$. Thus, levels of zinc may be 1000 times that found in other brain areas.[28] A similar uptake and release of Pb^{2+} during synaptic transmission could greatly elevate the $[Pb^{2+}]$ within the very small confines of a synaptic cleft. Although the mechanism for concentration of lead in the hippocampus has yet to be shown, all conditions suggest that the $[Pb^{2+}]$, like the $[Zn^{2+}]$, in different subcellular fractions may encompass a wide range of concentrations that change dynamically over small increments of time.

Most experimental reports specify the amount of Pb^{2+} salt added to a solution, even though Pb^{2+} has well-known physical properties that cause it to bind to glassware, to precipitate, and to form soluble complexes in addition to binding to tissue. The effective extracellular concentration of free $[Pb^{2+}]$ in solution will be further reduced by binding of Pb^{2+} to various buffers and amino acids in the extracellular solutions used with *in vitro* preparations and to various phospholipids and proteins of the tissue; thus, the extracellular $[free\ Pb^{2+}]$ could be considerably below the stated concentration of lead salt.[14,31] Thus, comparison among various experimental conditions and with the clinically lead-poisoned condition is hampered by the difficulty of identifying the $[free\ Pb^{2+}]$ at target sites, both *in vivo* and *in vitro*, and one is obliged to consider the molecular targets affected by a wide range of Pb^{2+} concentrations.

11.28.2 MOLECULAR TARGETS OF Pb²⁺ IN NEURONAL TISSUE

Many Ca^{2+}-sensitive physiological processes have been shown to be affected by Pb^{2+}. In neuronal membranes, the currents of various Ca^{2+}-permeable channels, including both voltage-gated Ca^{2+} channels and transmitter-gated channels such as N-methyl-D-aspartate (NMDA) receptors and the α-bungarotoxin-sensitive neuronal nicotinic acetylcholine receptors (nAChR) are inhibited by Pb^{2+} (Figure 1). Alteration by Pb^{2+} of the Ca^{2+} currents in presynaptic nerve terminals may mediate inhibition of evoked transmitter release and influx of Ca^{2+} into synaptosomes. Pb^{2+} inside presynaptic nerve terminal may mediate the stimulation of spontaneous transmitter release. Direct actions of Pb^{2+} on intracellular targets

involved in Ca^{2+} regulation and Ca^{2+}-mediated events have also been reported, and some of these targets may have sensitivities to Pb^{2+} near the level of intracellular $[Pb^{2+}]$. The combination of direct actions of Pb^{2+} with indirect actions on intracellular Ca^{2+} may account for the deficits in complex cognitive and mnemonic functions, which are highly sensitive to Ca^{2+} (see Chapters 8, 9, 13, and 29, this volume).

11.28.2.1 NMDA Receptor Function: Potentiation and Inhibition by Pb²⁺

The effects of Pb^{2+} on several types of ionotropic glutamate receptors were first studied at the molecular level in hippocampal neurons cultured for 7–21 days. The peak amplitude of NMDA-induced currents is preferentially inhibited with an $IC_{50} \approx 10\,\mu M\ Pb^{2+}$, whereas the quisqualate- and kainate-induced currents are only slightly decreased in amplitude, even when the $[Pb^{2+}]$ is as high as $250\,\mu M$.[32] Kinetic analysis of NMDA single-channel currents revealed that the mechanism of inhibition is unrelated to membrane voltage, ion-channel blockade, or the channel closing rate; instead, Pb^{2+} decreases the rate of receptor activation (channel opening) by agonists. Furthermore, the onset of the effect of Pb^{2+} is rapid (seconds), but the offset is slow (>30 min). Pb^{2+} inhibits the binding of MK-801 ((+)-5-methyl-10,11-dihydro-5*H*-dibenzo[a,d]-cyclohepten-5,10-imine)) to the NMDA receptor-ion channel equally in the absence or presence of added glutamate. Thus, Pb^{2+} lowers the rate of NMDA receptor-ion channel activation by acting on the closed-channel conformation at an extracellular site, different from the glutamate-binding site.

A particularly important finding about age-related neurotoxicity is that the NMDA-evoked current in immature hippocampal neurons is particularly sensitive to the inhibitory effects of Pb^{2+} (Figure 2).[33,34] The whole-cell currents evoked by a rapid-pulse application of NMDA comprise two kinetic components: a rapidly decaying ($\tau < 1\,s$) "fast" component and a more sustained "slow" component.[35] These components are thought to arise from different types of NMDA receptors, and the proportions of the components change during development. The fast component contributes more than half the peak whole-cell current in young neurons cultured for less than 10 days. Maturation of the neurons increases mostly the size of the slow component, so that the fast component contributes only 21% of the peak current in neurons cultured for 21–30 days.[35] Similar proportions of the fast and slow components are also

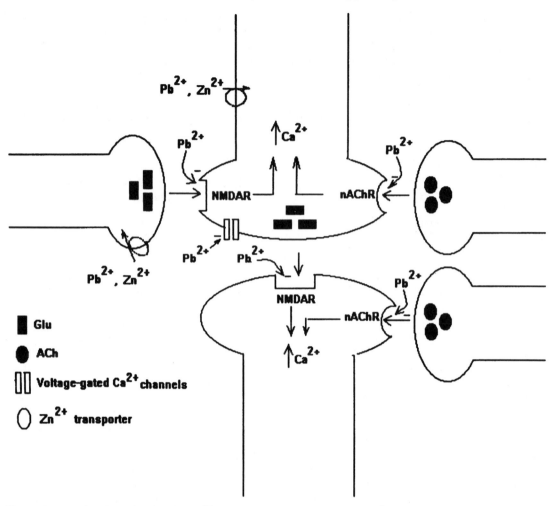

Figure 1   Putative sites of action for Pb^{2+} and a possible mechanism for Pb^{2+} accumulation in glutamatergic synaptic terminals. A model complex synapse is shown with a dendrite that is innervated by a cholinergic terminal and a glutamatergic terminal, which is itself innervated by glutamatergic and cholinergic terminals. For simplicity, the only receptors shown are those that are sensitive to Pb^{2+}. The text describes how actions of Pb^{2+} on NMDA receptors, nAChRs, and voltage-gated Ca^{2+} channels can affect Ca^{2+} levels in ways that might disturb transmitter release, postsynaptic responses, LTP, and the cues for morphological development.

present in acutely dissociated hippocampal neurons from 3–4 day old and 21–30 day old rats, respectively (Figure 3).[34] Although it appeared that the slow component in 21–31 day old neurons was resistant to Pb^{2+} (Figure 3),[34] a study (unpublished) using rapid-pulse application of Pb^{2+} to 11–15 day old neurons during the slow component shows that this current can also be inhibited rapidly by Pb^{2+} and it recovers from inhibition within a fraction of a second. Thus, the inhibitory effect of Pb^{2+} results in a greater reduction of NMDA-evoked currents in neurons under 15 days of age than in neurons over 21 days of age. These observations receive support from a biochemical study showing that the $[Pb^{2+}]$ required to inhibit $[^3H]$-MK-801 binding to NMDA receptors in rat brain cortex doubles from an IC_{50} of $\approx 34\,\mu M$ to $\approx 78\,\mu M$ between 14 and 56 days of age.[36] Thus,

the NMDA receptors in the immature rat are twice as sensitive to inhibition by Pb^{2+} as the NMDA receptors in the mature rat.

To understand the mechanism by which Pb^{2+} preferentially affects the immature form of the NMDA receptor, it is important to consider other characteristics that distinguish the NMDA receptors of immature and mature neurons. First, the fast component of the current is elicited at higher concentrations of NMDA ($EC_{50} = 27\text{–}39\,\mu M$) and glycine ($EC_{50} = 1.8\,\mu M$) than are necessary to elicit the slow component of the current (NMDA $EC_{50} = 13\text{–}14\,\mu M$ and glycine $EC_{50} = 0.3\,\mu M$), consistent with the greater potency and higher affinity of agonists at the receptors underlying the slow component.[35] Second, only the affinity of the slow component receptor for NMDA and glycine was reduced after treatment with

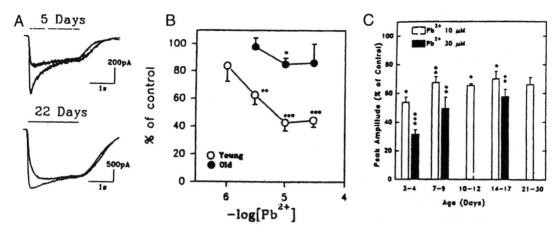

Figure 2 Preferential inhibition by lead of the peak NMDA currents in immature neurons. (A) Each pair of traces shows the transient whole-cell currents of one hippocampal neuron, cultured for the specified number of days, that were elicited by application of NMDA (50 μM) and glycine (10 μM) during the time shown by the bar. The larger current in each pair was obtained under the control condition, and the smaller current in each pair was obtained after Pb²⁺ (10 μM) application via the bath solution. The initial peak current was reduced by Pb²⁺, but the sustained current at 2 s was nearly unaffected. (B) The concentration–response relationship for inhibition of peak current amplitudes in young (1 to 10 days old) and old (21 to 30 days old) cultured hippocampal neurons. (C) Relationship of development *in vivo* to inhibition by Pb²⁺ of the peak NMDA currents in acutely dissociated hippocampal neurons. Each point represents the mean for 3 to 14 neurons, and the bar represents the SE Statistically significant differences from control were *, p < 0.05; **, p < 0.01; ***, p < 0.001, according to the student's t-test (after Ujihara and Albuquerque[33] and Ishihara *et al.*[34]).

trypsin.[34] Third, the slow component of the NMDA-induced current was more susceptible to inhibition by Zn^{2+} ($IC_{50} \approx 9.6$ μM) than was the fast component ($IC_{50} \approx 20$ μM).[35] Guilarte *et al.*[37] also found high and low Zn^{2+}-affinity [³H]-MK-801 binding sites in forebrain membrane preparations, but the two populations cannot readily be equated with the fast and slow components of the NMDA currents because the receptor populations and Zn^{2+}-sensitivity do not follow the same pattern. Ujihara and Albuquerque[35] found a large increase in the amplitude of the Zn^{2+}-sensitive slow component of the NMDA current with aging from 10–30 days in culture, but the populations defined by [³H]-MK-801 binding were of similar sizes in 14 day old rats and adult rats. The two populations of [³H]-MK-801 sites contrast in that the majority (>80%) is associated with high affinity for Zn^{2+} ($IC_{50} = 0.77$ μM in normal 14 day old rats) and the minority (16% in 14 day old rats and 11% in adult rats) had low affinity for Zn^{2+} ($IC_{50} = 57$ μM) that became threefold less sensitive as the rats matured.[37] The potency of Mg^{2+} in causing voltage-dependent blockade of the NMDA channel was unchanged by maturation.[36] One factor contributing to the difference between the electrophysiologically and biochemically defined populations is that MK-801 also binds with high affinity to α-bungarotoxin (α-BGT)-insensitive nicotine acetylcholine

receptors (nAChRs) ($IC_{50} \approx 25$ μM),[38] so that the [³H]-MK-801-binding sites may not be associated exclusively with NMDA receptors. These studies demonstrate marked changes in the population of NMDA receptors during the postnatal development of the hippocampus and the dentate gyrus in the rat, which in humans occurs during the late fetal and postnatal periods.[39]

Pb²⁺ has been shown to be a noncompetitive inhibitor of glycine and of NMDA at the NMDA receptors in hippocampus and cortex.[32,33,40] Other analyses have revealed that Pb²⁺ acts *in vitro* in a concentration-dependent manner to decrease the number of [³H]-MK-801 sites,[37] suggesting a noncompetitive action, and to decrease the affinity of MK-801,[36] suggesting a competitive action with respect to an ion channel site. Zn^{2+} exhibits both voltage-dependent and voltage-independent inhibition of NMDA currents, showing actions within the ion channel and outside the ion channel,[35] and so the possible interaction of Pb²⁺ and Zn^{2+} has been tested. The actions of Pb²⁺ and Zn^{2+} in rat brain membranes are similar in several ways in that both cations decrease the rate of dissociation of MK-801 from the NMDA receptor, independently inhibit the binding of MK-801 to the same total number of sites, and bind to sites that are sensitive to covalent modification of histidine residues.[37] In the 14 day old rat, the low affinity for Zn^{2+} is decreased an additional

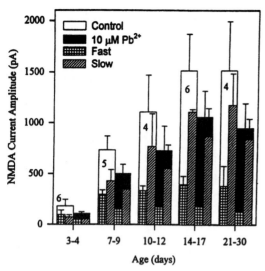

Figure 3 Development *in vivo* of NMDA-elicited currents and inhibition by Pb^{2+}. The peak amplitude of the NMDA-elicited currents in acutely dissociated hippocampal neurons, represented by white columns, increase during postnatal development *in vivo*. In front of the peak current amplitudes are the slow component amplitude, which is the amplitude 2 s after the start of the NMDA pulse, and the fast component amplitude, which is the peak amplitude minus the 2 s amplitude. The increasing amplitude of the slow component accounts for more of the increase in peak amplitude than does the increasing amplitude of the fast component. The effect of age is enhanced in this plot because the hippocampal neurons represented here were treated with trypsin (0.25%, 30 min) before being dissociated, and the fast component is more sensitive to trypsin. The amplitudes of the NMDA currents (white columns) were reduced by superfusion with 10 μM Pb^{2+} (black columns), and the inhibition of the fast component (checked columns) was greater than the inhibition of the slow component (diagonal-hatched column). At each age, the control and Pb^{2+}-treated values are from the same neurons; the number of neurons is given in the figure. The columns and bars represent the mean and SE of the current amplitudes.

one-half by 40 μM Pb^{2+}, without affecting the number of sites, suggesting competition between Zn^{2+} and Pb^{2+}. On the other hand, the high affinity for Zn^{2+} of the receptors in immature and mature neurons is increased by Pb^{2+} four- and twofold,[37] respectively. Some of the changes caused by Pb^{2+} in affinity for Zn^{2+} could partly arise from the decreased proportion of high-affinity Zn^{2+} sites, which changes the discrimination between the two sites whose affinities for Zn^{2+} differ <100-fold in the absence of Pb^{2+}. Although the noncompetitive interactions of Pb^{2+} and Zn^{2+} just described may suggest that these ions act on different sites, one should also consider that Pb^{2+} and Zn^{2+} might act at the

same site but do not appear to be competitive because Pb^{2+} dissociates slowly from or acts irreversibly on a Zn^{2+}-binding site on the NMDA receptor. Supporting this possibility, irreversible inhibition of NMDA currents has been seen after 10–15 min Pb^{2+} treatments of isolated neurons [41] and outside-out patches.[32] The minor population of [^{3}H]-MK-801 sites, which are associated with low affinity for Zn^{2+}, may be associated with two affinities for Pb^{2+}, as follows. Pb^{2+} binds with only a single affinity ($IC_{50} \approx 4$ μM in 14 day old rats and 9 μM in adult rats) at the lowest concentrations (≤ 2.5 μM) of Zn^{2+}, which would bind to the high-affinity Zn^{2+} sites ($IC_{50} < 1$ μM) to cause the significant decrease in the number of [^{3}H]-MK-801 sites in adult rat brain membranes that can be inhibited by Pb^{2+}. However, at 5–10 μM Zn^{2+}, which would largely saturate the MK-801 sites coupled with high-affinity for Zn^{2+}, the remaining MK-801 sites displayed two affinities for Pb^{2+} by IC_{50}'s of ≈ 1.2 and 51–94 μM Pb^{2+}.[37] In summary, [^{3}H]-MK-801 binding sites appear to comprise at least three types that may be located on different states of the receptor or different receptors: (i) an NMDA receptor with high affinity for Zn^{2+} (0.77 μM) and low affinity for Pb^{2+} (≈ 20 μM), which is present in greater numbers in mature neurons; (ii) a receptor with low affinity for Zn^{2+} (≥ 50 μM) and high affinity for Pb^{2+} (1 μM), which may be the immature NMDA receptor and/or the α-BGT-sensitive nAChR; and (iii) a receptor with low affinities for both Zn^{2+} (≥ 50 μM) and Pb^{2+} (≥ 50 μM), which could be the α-BGT-insensitive nAChR. One may be tempted to focus attention on the larger population of sites with high affinity for Zn^{2+} (category 1), but the receptors with lower affinity for Zn^{2+} (categories 2 and 3) will have a greater proportion of sites available to Pb^{2+} under resting conditions. Zn^{2+} released from synapses along with glutamate could displace Pb^{2+} from its low affinity sites (category 3), but the inhibition of the subset of MK-801-binding receptors with low-affinity sites for Zn^{2+} and high-affinity sites for Pb^{2+} (category 2) would be more stable.

Competition between Pb^{2+} and Ca^{2+} has been found to affect the amplitude of NMDA whole-cell currents. The Pb^{2+} concentration-inhibition curves had Hill coefficients near unity, suggesting a single site of action for Pb^{2+}, and these curves were shifted to the right by a step-wise increase in the concentration of Ca^{2+}, from 0.2 mM to 2 mM and 20 mM (Figure 4).[42]

The actions of Zn^{2+} and Ca^{2+} on NMDA receptor activation are mediated by effects on glycine binding, which is essential for sustained

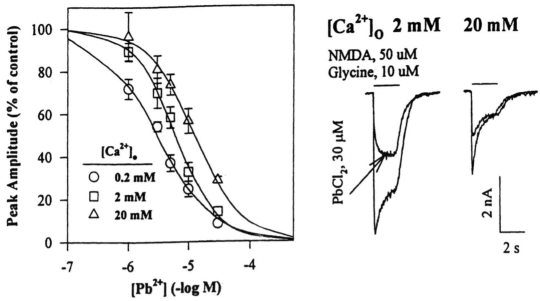

Figure 4 Ca²⁺ reverses the Pb²⁺-induced inhibition of NMDA currents. Whole-cell currents were elicited by 2 s pulse applications of NMDA (50 μM) and glycine (10 μM) onto cultured fetal rat hippocampal neurons. At extracellular Ca²⁺ concentrations of 0.2, 2 and 20 mM, currents were elicited in the control condition and after superfusion with Pb²⁺ (1 to 30 μM). The maximum amplitude in the Pb²⁺-treated condition was compared to the maximum amplitude before Pb²⁺ treatment at the same Ca²⁺ concentration. As suggested by the waveform shown for the 2 mM Ca²⁺ condition, a selective reduction of the fast component of the current was under emphasized by this method of analysis. At the physiological concentration of extracellular Ca²⁺ (2 mM), Pb²⁺ inhibited the NMDA currents with an IC₅₀ of 5.9 μM. Pb²⁺ was a more potent inhibitor (IC₅₀ = 3 μM) at 0.2 mM extracellular Ca²⁺ and Pb²⁺ was a less potent inhibitor (IC₅₀ = 12.7 μM) at 20 mM extracellular Ca²⁺. In each case, the Hill coefficient was close to unity (1.0–1.3).

receptor activity. Under physiological conditions, glycine in interstitial fluid ($\approx 1\,\mu M$) is sufficient to facilitate nearly maximal NMDA currents.[43] Zn²⁺ inhibits the binding of glycine,[43] and inhibits the NMDA current at a saturating concentration of glycine.[35] At low concentrations of glycine ($<0.1\,\mu M$), the small NMDA whole-cell currents decay rapidly due to desensitization,[33,44] and Ca²⁺ reduces the dissociation of glycine and potentiates the response evoked by NMDA.[45,46] The effect of Pb²⁺ on NMDA-evoked currents of neurons *in vitro* is also dependent upon glycine concentration. The inhibitory effect of Pb²⁺ on NMDA-evoked currents increases at higher glycine concentrations ($\geq 0.1\,\mu M$),[33] but the currents elicited at a lower glycine concentration (0.05 μM) are potentiated by Pb²⁺ (Figure 5). The potency of Pb²⁺ for these effects was of the same order of magnitude. The coagonist, potentiating effect glycine (10 μM) can be antagonized by kynurenic acid. Whereas kynurenic acid (50 μM) did not inhibit the current elicited in 0.05 μM glycine, after kynurenic acid did reverse the potentiating effect of Pb²⁺ (Figure 5). An explanation may be that in the presence of glycine (0.5 μM), Pb²⁺ acts via the extracellular Ca²⁺ site of the receptor to enhance the affinity

of glycine, and that kynurenic acid blocks the potentiating mechanism, disclosing that the inhibitory effect of Pb²⁺ is also present at the lower concentration of glycine. The mechanism of these interactions is currently being explored further.

Metal ion–amino acid complexes are weak by comparison to complexes such as Pb²⁺–EDTA and cannot account for the Pb²⁺-induced inhibition of NMDA receptor activation at moderate concentrations ($<100\,\mu M$) of NMDA and glycine. They also cannot account for the differences between NMDA whole-cell currents or receptor binding where the critical variable was age of the preparation. However, when the concentrations of divalent cations and amino acids are in the millimolar range, Pb²⁺ and Zn²⁺ can form stable complexes with glycine and glutamate. This might be a concern for glutamatergic transmission where high concentrations of glutamate (1 mM) and Zn²⁺ (300 μM) are achieved in the synapse under physiological conditions.[29,47]

Combining all of the above observations on NMDA currents, Pb²⁺ appears to discriminate among NMDA receptor subtypes. For the mature hippocampal NMDA receptor subtype with high affinity for Zn²⁺, the reduced

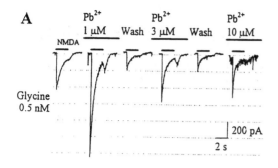

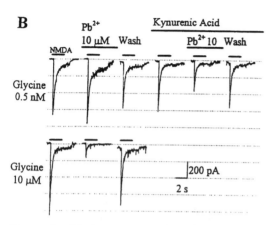

Figure 5 Glycine concentration determines the effect of Pb^{2+} on NMDA currents. A: in contrast to the previously shown inhibitor effect of Pb^{2+}, under low glycine conditions Pb^{2+} causes potentiation of the NMDA current. The responses of a single neuron are shown. The bath solution contained 0.5 nM glycine at all times. Addition of micromolar concentrations of Pb^{2+} clearly potentiated the NMDA-elicited current, and the potentiating effect was reversed by washing, although a residual inhibition is evident in this figure after the first Pb^{2+} treatment. The average effect of 10 μM Pb^{2+} on several neurons (not shown) was a potentiation of about 50%. The neuron represented here responded to concentrations from 1 to 10 μM Pb^{2+} with similar degrees of potentiation. The concentration–response relationship is as yet unclear. B: the mechanism underlying the potentiation of NMDA-elicited currents by 10 μM Pb^{2+} are illustrated by the responses of a single neuron. A small potentiation by Pb^{2+} in low glycine was reversed by washing. When the same neuron was superfused with 10 μM glycine, Pb^{2+} was inhibitory. Although kynurenic acid alone had little effect in the absence of Pb^{2+}, it blocked the potentiating effect of Pb^{2+}, leaving an inhibitory effect that was also reversible.

amplitude of the slow whole-cell current amplitudes reflects a stable reduction of single channel activation within seconds after Pb^{2+} (>10 μM) application, an effect that was not readily reversed by addition of Ca^{2+}. Pb^{2+} is a more potent inhibitor of the NMDA receptors that are abundant in immature neurons and have low affinity for Zn^{2+}, and the currents of

these receptors are seen briefly after a pulse application of NMDA. The inhibition of NMDA currents in immature neurons by Pb^{2+} was expressed best when the neurons were exposed to high concentrations of glycine (>0.1 μM), and the inhibition was antagonized by elevation of external Ca^{2+} concentration. Thus, the inhibition of NMDA currents by Pb^{2+} could signal competition whereby Pb^{2+} prevents a potentiating action of Ca^{2+}, which in normal conditions increases the affinity of glycine.[45,46] On the other hand, the effect on NMDA currents of Pb^{2+} plus dilute glycine shows that there is an action by which Pb^{2+} can potentiate NMDA responses. Thus, the binding of Pb^{2+} to the closed-channel state of the NMDA receptor inhibits the large NMDA-plus-glycine-activated currents, but potentiates the smaller NMDA-plus-low glycine-activated currents. The results thus far emphasize the importance of understanding the fundamental interactions of various cotransmitters and divalent cations including Ca^{2+}, Zn^{2+}, and Pb^{2+}.

These effects of Pb^{2+} on the NMDA receptors are highly relevant to clinical toxicity, because NMDA receptors play key roles in memory and learning processes by modulating synaptic plasticity and long-term potentiation. Furthermore, the presence of certain NMDA receptors during early development suggests that they may have a role in neurotrophic processes. Consequently, functional impairment by Pb^{2+} of the receptor present in immature animals could slow or arrest maturation. Chronic postnatal exposure to Pb^{2+} was apparently sufficient to alter the expression of NMDA receptors, because the severity of NMDA-induced seizures was altered in 15 and 25 day old rat pups fostered to Pb^{2+}-treated dams.[48] In fact, chronic Pb^{2+} exposure of rats during gestation and during the neonatal period altered the cortical populations of NMDA receptors labeled by [3H]MK-801 and of muscarinic receptors labeled by [3H]-N-methylscopolamine, but similar exposure of mature rats had no effect on [3H]MK-801-labeling.[36,49]

11.28.2.2 Nicotinic Acetylcholine Receptors (nAChRs)

Several studies of neuromuscular transmission have demonstrated effects of Pb^{2+} on the release of ACh (see Section 11.28.3.1), but at the same concentrations Pb^{2+} has no direct effect on the muscle type nAChR[50] or the ganglionic nAChR.[51,52] The first observation that neuronal nAChR is sensitive to Pb^{2+} was made in neuroblastoma cells.[53,54] In that study, nicotinic currents were inhibited as $[Pb^{2+}]$ increased from 10 nM to 3 μM. Above 3 μM $[Pb^{2+}]$, there was less blockade of the nicotinic current and the currents decayed more slowly. These actions of lead could be associated with two distinct actions on different populations of neuronal nAChRs that are heterogenous in their structural, kinetic, and pharmacological properties. For example, at least two subtypes of neuronal nAChRs are found in hippocampal neurons. Most hippocampal neurons respond to nicotinic agonists with a rapidly desensitizing cationic whole-cell current, referred to as the type IA current; this current is subserved by α7-bearing receptors with low sensitivity to ACh and high sensitivity to the antagonists methyllycaconitine and α-BGT.[55,56] Less frequently, hippocampal neurons respond with a slowly desensitizing current, referred to as a type II current that can be blocked by dihydro-β-erythroidine and may be subserved by the α4β2 nAChR. By studying the nicotinic currents of isolated hippocampal neurons, it has been possible to assess the effects of Pb^{2+} on the function of different nAChRs from a brain region of great importance to cognitive function.

Low concentrations of Pb^{2+} were found to block the type IA nicotinic current of cultured hippocampal neurons in a dose-dependent manner (IC_{50} 3 μM; Figure 6).[50] The mechanism of Pb^{2+} action on the type IA nicotinic current has not yet been fully elucidated. The possibility was considered that the effect of Pb^{2+} may be related to the high permeability of the α-BGT-sensitive nAChR to Ca^{2+},[57] but the inhibitory effect of Pb^{2+} was found to be independent of voltage, suggesting that Pb^{2+} binds to a site outside of the membrane electric field.[50] The inhibition caused by lead was not competitive with respect to ACh. A study has shown that extracellular Ca^{2+} regulates the cooperativity and efficacy of ACh at the α-BGT-sensitive nAChR.[58] By competing for the external site of this receptor, where the action of Ca^{2+} potentiates nicotinic currents, Pb^{2+} could reduce the efficacy of the agonist. This neuronal nAChR may also be subject to modulation by internal Ca^{2+}-dependent processes that inhibit receptor function. Therefore, entry of Pb^{2+} into neurons and internal actions of Pb^{2+} that modulate second messengers could be important.

By comparison to the type IA nicotinic current, the type II nicotinic currents were

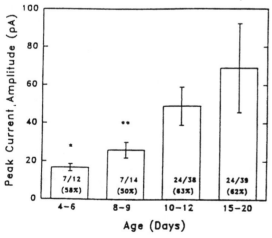
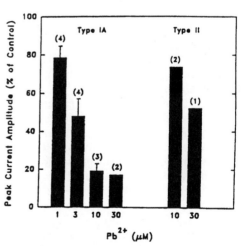

Figure 6 Maturation of the nicotinic currents of hippocampal neurons, and inhibition of nicotinic currents by lead. Left: Neurons were dissociated from postnatal rat hippocampi, immediately prior to testing. Each ratio indicates the number of neurons that responded to a brief pulse (0.2–1 s) of ACh relative to the total number of neurons tested in that age group, and the ratio is also expressed as a percentage. The height of each bar indicates the average peak amplitudes of inward currents induced by ACh (1 mM) when the neuron was held at 56 mV, and the vertical error bar indicates the SEM. The average peak amplitudes increased with the age of the neurons acutely dissociated from the hippocampi of postnatal rats of the specified ages. *Significantly different from the 10 to 12 day old and 15 to 20 day old groups, $p < 0.05$; **significantly different from the 10 to 12 day old group, $p < 0.05$. Right: pulse application of ACh (1 s, 1 mM) elicits rapidly desensitizing type IA currents, which are subserved by an α-BGT-sensitive nicotinic receptor, and nondesensitizing type II currents, which are subserved by an α-BGT-insensitive receptor. Addition of micromolar concentrations of Pb^{2+} to the extracellular solution reduced the amplitude of both types of nicotinic currents. The concentration-response relationship for the effect on type IA currents yielded an IC_{50} of 3 μM and a Hill coefficient of 1.0. Although insufficient data were available to determine these parameters for the inhibition of type II currents, they were clearly less sensitive to Pb^{2+} (after Ishihara *et al*.[50]).

much less sensitive to Pb^{2+} ($IC_{50} \geq 30\,\mu M$). When both types of nAChR are present in a single neuron and give rise to a mixed current, Pb^{2+} still selectively impairs the $\alpha7$-mediated current. This shows that intracellular effects mediated through the α-BGT-sensitive receptor, either permeation of Pb^{2+} or indirect effect of Ca^{2+} currents, do not affect the α-BGT-insensitive nAChR. Therefore, the resistance of the latter receptor to Pb^{2+} is not a function of Ca^{2+} permeation, but could be related in other ways to differences between the receptors in their responses to Ca^{2+}. Whereas the α-BGT-sensitive nAChR current is potentiated by low concentrations of Ca^{2+} and inhibited by high concentrations of Ca^{2+}, higher Ca^{2+} concentrations ($EC_{50} \approx 2$–$3\,mM$) increase the efficacy of the α-BGT-insensitive receptors.[59] Perhaps, in neuroblastoma cells, the inhibition by $Pb^{2+} < 3\,\mu M$, but less so by $Pb^{2+} > 3\,\mu M$, of nicotinic currents in neuroblastoma cells[54] is related to nAChRs with differ sensitivities to Ca^{2+}.

The peak amplitude of the nicotinic currents increases for 3–4 weeks under normal conditions *in vivo* (Figure 6),[50] as well as in tissue culture.[60] It is likely that the developmental pattern is related to the growth of neurites and expression of nAChRs in synaptic regions. This notion is supported by the fact that the peak amplitude of nicotinic currents in cultured neurons is an order of magnitude greater than that in acutely dissociated neurons, suggesting that some of the nAChRs are lost during the acute dissociation procedure, which causes the loss of most of the neurites.[50] Furthermore, evidence using robotic micromanipulators to apply nicotinic agonists to the apical dendrites of pyramidal neurons in hippocampal slices indicates that many of the nAChRs of more mature hippocampal neurons are located outside the soma.[61] The combination of these observations suggests that Pb^{2+} blocks nAChRs, which are likely to lie in or near synaptic regions and may have important roles in synaptic transmission or in regulating other types of neurotransmission.[62]

11.28.2.3 5-HT₃ Receptor

The transient serotonin-induced currents of neuroblastoma cells are carried through 5-HT₃ receptor-ion channels. Pb^{2+} reduces reversibly the amplitude of the 5-HT₃ (5-hydroxytryptamine serotonin) currents: $1\,\mu M\ Pb^{2+}$ reduces the peak current amplitude by 25%, and $100\,\mu M$ decreases it by more than 50%. Thus, the IC_{50} is about $50\,\mu M$, but the slope of the concentration–inhibition curve is quite low.[54]

These facts suggest that Pb^{2+} is a less potent inhibitor of the serotonin-induced currents than of other ligand-gated receptors and voltage-gated ion channels.

11.28.2.4 Metal Ion-activated Cation Channel

Neuroblastoma cells respond to superfusion of 1–$200\,\mu M\ Pb^{2+}$ or Cd^{2+}, as well as to superfusion of higher concentrations of Al^{3+}, with noninactivating whole-cell and single-channel currents that are characterized by a linear current–voltage relationship with a reversal potential near $0\,mV$ and a unit conductance of $24\,pS$.[63] Neither antagonists of nicotinic, muscarinic, or 5-HT₃ receptors (D-tubocurarine, atropine, and ICS 205–930, respectively) nor Na^+- or K^+-channel blockers (tetrodotoxin and tetraethylammonium, respectively) could prevent the current, although the current was reversed by removal of the heavy metal. Also the current is unaffected by the Na^+-K^+ ATPase inhibitor ouabain, but the current was blocked by replacing both extracellular and intracellular Na^+ with K^+. Thus, the current is attributed to a novel Na^+-dependent metal ion-activated cation channel. The site of action of the metal appears to be extracellular, since the possibility of internal action of Pb^{2+} was discarded based on the fact that internal Mg^{2+}-EGTA did not alter the heavy metal-induced currents. This channel also seems to be inhibited by an action of the metal ion at low-affinity sites in the channel. Neither the physiological function nor the distribution of this channel in normal neuronal tissue is known. The pretreatment of hippocampal neurons with concentrations of Pb^{2+} up to $10\,\mu M$ did not stimulate steady-state currents that could be measured using the whole-cell patch-clamp technique (unpublished observation of Albuquerque and co-workers).

11.28.2.5 Ca²⁺-dependent K⁺ Channels (K⁺_Ca Channels)

Various heavy metal cations have different potencies for activation and blockade of the subtypes of K^+_{Ca} channels.[64] Pb^{2+}, applied to the internal cytoplasmic side of inside-out membrane patches, is more potent than Ca^{2+}, Cd^{2+}, Co^{2+}, Fe^{2+}, and Mg^{2+} as an activator of small and large conductance K^+_{Ca} channels of mouse neuroblastoma cells. Compared to the potency in membrane patches (1–$90\,\mu M$), Pb^{2+}

Table 2 Effects of Ca^{2+} and Pb^{2+} on parameters of single K$^+_{Ca}$ channels.

Metal	μM	P$_O$	f$_O$	τ$_O$	n
Small-conductance, voltage-insensitive channels[a]					
Ca^{2+}	1	0.57	0.54	1.00	5
Pb^{2+}	1	0.99	0.69	1.41	4
	90	1.03	0.83	1.44	3
Large-conductance, voltage-sensitive channels[b]					
Pb^{2+}	1	0.11	0.85	0.16	2
	90	0.92	0.94	0.91	2

Source: Vijverberg *et al.*[64]
The parameters measured were: relative open probability (P$_O$), relative opening frequency (f$_O$), and relative mean open time (τ$_O$). Parameters are normalized to their maximum values obtained in the presence of Ca^{2+} at ([a]) 14.4 μM or ([b]) 115 μM in the same patch. Values are the means, and *n* is the number of patches.

had a much more potent effect on the voltage-sensitive, charybdotoxin-sensitive K$^+_{Ca}$ currents in intact adrenal chromaffin cells.[65] The K$^+_{Ca}$ current was increased fivefold by increasing intracellular EGTA-buffered free Pb^{2+} from 0.1 to 10 nM, with half-maximal effect at $K_{0.5}$ of 0.5 nM free Pb^{2+}. The effects on these K$^+_{Ca}$ channels contrast with the absence of activation of K$^+_{Ca}$ channels from rat skeletal muscle by Pb^{2+}.[66]

The small-conductance, voltage-insensitive K$^+_{Ca}$ channel, which is blocked by the bee venom peptide apamin, is extremely sensitive to Ca^{2+} (EC$_{50}$ = 1.1 μM) and is maximally activated (open probability, P$_O$ = 0.99) by 1 μM of Pb^{2+}, the lowest concentration tested (Table 2).[64] The large-conductance, voltage-sensitive K$^+_{Ca}$ channel, which is blocked by tetraethylammonium, is less sensitive to Ca^{2+} (EC$_{50}$ = 18 μM). At a holding potential of 0 mV, this voltage-sensitive K$^+_{Ca}$ channel

opens nearly as often at 1 μM Pb^{2+} (relative opening frequency, f$_O$ = 0.85) as it does at 115 μM Ca^{2+}, although the duration of the channel opening is markedly shorter (16%) than when activated by Ca^{2+}. At 90 μM Pb^{2+}, the channel open time is prolonged, so that the open probability (P$_O$) matches that obtained at high Ca^{2+} concentration. Thus, 1 μM of Pb^{2+} can increase K$^+$ currents through the small-conductance K$^+_{Ca}$ channels, and 90 μM can produce large currents through both small and large conductance K$^+_{Ca}$ channels (in membranes held at 0 mV).

11.28.2.6 Ca^{2+} Channels

Voltage-gated Ca^{2+} channels have different sensitivities to activation by different voltages and different sensitivities to inhibition by various heavy metal ions (Table 3). Therefore,

Table 3 Inhibition of voltage-activated Ca^{2+} channels by Pb^{2+}.

Channel type	Activation (mV)	Inactivation rate (ms)	Antagonists		IC$_{50}$Pb^{2+} (μM)
			Organic	Metal	
T	−100 to −30	rapid		Ni^{2+}	6[a]
L	−40 to 0	>100	dihydropyridine, nifedipine, ω-conotoxin	Cd^{2+}	1[a], 0.03[b]
N	large	20–40	ω-conotoxin	Cd^{2+}	0.64[a], 0.8[b]
P	−40 to 20	moderate	ω-agatoxin, funnel-web spider toxin	Cd^{2+}	24[c], 3[d]

The IC$_{50}$ values for Pb^{2+} inhibition of Ca^{2+}-channel currents in the last column are approximate. For T-, L-, and N-type channels, the currents were carried by Ba^{2+} (10 mM) and were measured ([a]) in dorsal root ganglion cells[69,92] and ([b]) in neuroblastoma cells.[67] For the P-type channels, the observations were of ([c]) Ca^{2+} currents of *Aplysia californica* measured in 10 mM Ca^{2+} artificial seawater and ([d]) agatoxin-sensitive channels of hippocampal neurons measured in 5 mM Ba^{2+}.[70] The difference in the latter two is expected because the higher concentrations of Ca^{2+} increases the IC$_{50}$ of Pb^{2+}.

it is not surprising that many studies have shown inhibition by Pb^{2+} of the L-, N-, T-, and P-types of vertebrate Ca^{2+} channels. In embryonic rat hippocampal neurons, the L-type (nondecaying) current appeared to be selectively inhibited by 20 nM Pb^{2+}, whereas the N-type (moderately rapid-decaying) current was also inhibited at 200 nM Pb^{2+} (Figure 7).[67] However, different results from experiments performed on other neurons suggest that the sensitivity to blockade by Pb^{2+} may be more a function of cell type than of Ca^{2+} channel type.

Several voltage-clamp studies have shown that Pb^{2+} has very little effect on the rates of activation and inactivation of whole-cell Ca^{2+} currents or on the rates of opening and closing of single Ca^{2+} channels. The inhibitory effect of Pb^{2+} on Ca^{2+} currents is antagonized by

elevating the concentration of Ca^{2+} or Ba^{2+}, suggesting instead that lead competitively inhibits the passage of currents through open Ca^{2+} channels. Whereas Ca^{2+} may have affinity for two binding sites along the ion channel pathway, Pb^{2+} may have high affinity for one or more of these Ca^{2+} sites.[14,68] In contrast to the findings of potent Pb^{2+} inhibition of T-, L-, and N-type channels in other types of neurons (Table 3),[69] the inhibition of Ca^{2+} currents in hippocampal neurons is mediated by a selective action of Pb^{2+} (3 μM) on the P-type channels.[70] In a study of the *Aplysia californica* abdominal ganglion Ca^{2+} channels (similar to the P-type channel) at 10 mM Ca^{2+}, a concentration of 24 μM Pb^{2+} reduced the Ca^{2+} current by 50%, and the value of the Hill coefficient for the concentration–response curve was one.[68] Although there was no evidence in this study that Pb^{2+} actually permeated the P-type channel, the inhibition of Ca^{2+} current was voltage dependent and was rapidly reversible, suggesting a single site of action within the membrane electric field.

Pb^{2+} itself is highly permeable through some Ca^{2+} channels. In fact, the permeability of Pb^{2+} through the Ca^{2+} channels of chromaffin cells is approximately 10 times that of Ca^{2+}.[14,31] The Pb^{2+}-stimulated quantal release of transmitter can be blocked by Cd^{2+};[71] this interaction may indicate that Cd^{2+} blockade of some Ca^{2+} channels prevents the permeation of Pb^{2+} into neuronal terminals, thereby precluding an intracellular action of Pb^{2+} that could account for the stimulation of neurotransmitter (see Section 11.28.3.1).

11.28.2.7 Second Messengers and Regulatory Proteins

A large number of intracellular pathways are regulated by Ca^{2+} and many of these have been shown to be affected by Pb^{2+}. Due to its affinity for Ca^{2+} sites, Pb^{2+} in some cases mimics the effect of Ca^{2+}, but in others Pb^{2+} inhibits Ca^{2+}-dependent mechanisms. In yet other cases, Pb^{2+} may alter Ca^{2+}-dependent mechanisms via allosteric sites.

The Ca^{2+}-binding protein calmodulin is affected by the same concentrations of free Pb^{2+} or free Ca^{2+}, each buffered by EGTA.[72] Interaction of Pb^{2+} with this protein has resulted in many changes in neurons: activation of calmodulin-dependent phosphodiesterase, phosphorylation of similar membrane-associated proteins such as Ca^{2+}/ calmodulin-dependent kinase (at 0.5 mM Ca^{2+} plus 0.2 mM EGTA); binding of calmodulin to membrane proteins, and stimulation of

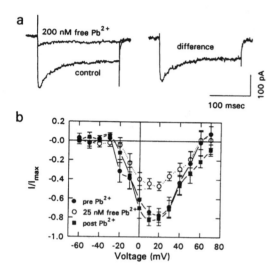

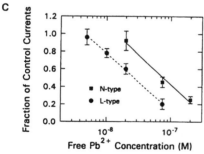

Figure 7 Inhibition by Pb^{2+} of voltage-gated Ca^{2+} channels in cultured fetal rat hippocampal neurons. A: a voltage step from −80 to 0 mV activated both N-type (slowly inactivating) and L-type (rapidly inactivating) barium currents, and both were inhibited by 200 nM free Pb^{2+}. B: the current-voltage relationship shows the reversible inhibition by 25 nM free Pb^{2+}. The data were from 10 neurons. C: the concentration-response relationship shows that both types of channels are sensitive to Pb^{2+}, although the N-type channels (circles) are inhibited at lower concentrations of Pb^{2+} than are the L-type channels (squares) (after Audesirk[14]).

calmodulin-dependent K^+ channels.[72,73] It is not known whether the free-Pb^{2+} level during intoxication is sufficient to activate calmodulin in neurons, however, it seems unlikely that free Pb^{2+} is maintained at the same level as free Ca^{2+}.

Among a dozen heavy metal ions tested at 100 pM in one report, Pb^{2+} appeared to be unique in its ability to stimulate partially purified rat brain protein kinase C (PKC), and 100 pM of Pb^{2+} was as effective as 10 μM Ca^{2+}.[74] Because this low concentration of Pb^{2+} does not stimulate the calmodulin-dependent protein kinase, PKC might contribute to the phosphorylation of different membrane proteins in the presence of a low concentration of Pb^{2+} (0.4 μM Pb^{2+} plus 0.2 mM EGTA).[72] However, another study using three types of purified PKC found no stimulation, but instead Pb^{2+} inhibited the phosphatidylserine- and cis-fatty acid-stimulated activities of these types of PKC with IC_{50} values of 2–10 μM Pb^{2+}.[75] The action of Pb^{2+} was found to be not via the Ca^{2+}-binding site on the regulatory domain of PKC, but via a reversible inhibition of the catalytic activity, which was apparent from Pb^{2+}'s inhibition of constitutive activity of the proteolytically generated catalytic fragment. Thus, the target(s) of Pb^{2+} that is responsible for the phosphorylation induced by low intracellular concentrations of Pb^{2+} (submicromolar Pb^{2+};[72] picomolar Pb^{2+}[74]) remains to be identified.

The adenyl cyclase activity of rat cerebellar homogenates is inhibited by exposure to Pb^{2+} for 3 min, yielding an apparent IC_{50} of 3 μM.[76] Furthermore, the action of 100 μM Pb^{2+} was irreversible, suggesting that a more prolonged exposure to lower concentrations, even below the reported IC_{50}, may provide a greater level of inhibition. As a consequence of adenyl cyclase inhibition, one may expect that cAMP-dependent kinase activity would be reduced. Perhaps opposing this action, Pb^{2+} directly inhibits cGMP-phosphodiesterase of rat retinal rods.[77]

Pb^{2+} alters the level of nitric oxide (NO) of cultured brain endothelial cells by decreasing the constitutive production by Ca^{2+}- and calmodulin-dependent nitric oxide synthase (NOS) ($\approx$30% reduction of NO at 10 nM of Pb^{2+}), and increasing Ca^{2+} concentration reverses the Pb^{2+}-inhibition of Ca^{2+}- and calmodulin-dependent NOS.[78] The production of NO by cytokine-inducible, Ca^{2+}-independent NOS increased 38% in brain endothelial cells at 1 μM Pb^{2+},[78] but decreased in macrophages ($IC_{50} = 0.35$–0.95 μM).[79] Although it does not appear to be known what effect Pb^{2+} has on NOS of neurons such as granule cells of the dentate gyrus,[80] the inhibition of endothelial NOS could impair neuronal plasticity.

Besides the acute effects of Pb^{2+} on specific intracellular Ca^{2+}-sensitive proteins, chronic Pb^{2+} exposure (1 mg lead acetate/kg diet 10 weeks) of neonatal rats reduces the number of IP_3 receptors on endoplasmic reticulum of cortical neurons to 40% of normal.[81] The number of IP_3 receptors of adults rats was not changed by the same diet, despite the adult brain Pb^{2+} levels achieving two-thirds of the neonatal brain Pb^{2+} levels. In the neonate, the maximum intracellular Ca^{2+} concentration achieved by IP_3-induced release of Ca^{2+} from intracellular storage sites was reduced to only 54% of control. Also, the ability of IP_3 to enhance the GTP-evoked increase of intracellular Ca^{2+} concentration was blocked in the neonatal rat.[81] These results suggest that chronic Pb^{2+} exposure may reduce the function of phospholipase C-coupled receptors, including α_1-adrenergic, muscarinic cholinergic, and serotonergic receptors.[81] In contrast, the IP_4-regulated distribution of intracellular Ca^{2+} was unaffected by chronic Pb^{2+} treatment.

11.28.3 NEUROPATHOPHYSIOLOGY OF LEAD

11.28.3.1 Dual Effects of Pb^{2+} on Transmitter Release: Stimulation of Spontaneous Release and Inhibition of Evoked Release

It has been known for some time that extracellular Pb^{2+} at low micromolar concentrations enhances spontaneous or basal release of neurotransmitter from presynaptic nerve endings. In frog neuromuscular preparations, Pb^{2+} at concentrations between 5 μM and 50 μM increased the frequency of miniature endplate potentials.[82,83] A similar enhancement of transmitter release was observed in rat phrenic nerve-diaphragm preparations.[84] Also, micromolar concentrations of Pb^{2+} were shown to increase the basal rate of release of radio-labeled transmitters from preloaded synaptosomal preparations.[85–87] It has been observed that Pb^{2+} also enhances the frequency of miniature synaptic events in hippocampal neurons (Figure 8). Since Pb^{2+} does not activate presynaptic Ca^{2+} channels or receptors (with the possible exception of the metal-ion activated receptor, see above) that could mediate Ca^{2+} influx and initiate transmitter release, it is likely that Pb^{2+} has an intraterminal target. In fact, Pb^{2+} acts at nanomolar concentrations on digitonin-permeablized rat brain synaptosomes to specifically enhance the release of vesicular transmitter.[88] Facilitation of transmitter release

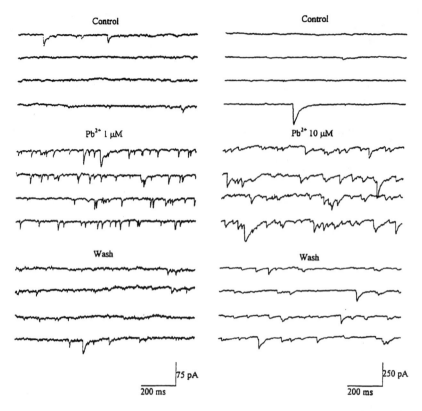

Figure 8 Increase in frequency of miniature postsynaptic currents induced by Pb^{2+}. Hippocampal neurons were cultured in standard media[55] and were treated with tetrodotoxin and atropine prior to the recording of spontaneous currents by the whole-cell patch-clamp technique. Studies of similar preparations have found that the spontaneous miniature postsynaptic currents are mediated by the AMPA type of glutamate receptors and/or GABA receptors. Each column shows the spontaneous synaptic currents of a single neuron, and the currents shown for each condition are from a continuous 4 s sample. Under the control condition the frequency of spontaneous events is on the order of 1 s. Addition of 1 µM or 10 µM $PbCl_2$ to the extracellular milieu for 2–3 min elicited a large increase in the frequency of miniature synaptic currents. Following the treatment with Pb^{2+} for a total of 5 min, the stimulatory effects were partly reversed by washing the preparation for 3–4 min. Residual activity is more noticeable after treatment with 10 µM Pb^{2+}.

could arise from a disturbance of intracellular Ca^{2+} homeostasis,[31] mobilization of intracellular Ca^{2+}, or an action in which Pb^{2+} substitutes for Ca^{2+}.[89] However, at the time of onset of enhanced spontaneous transmitter release from hippocampal synaptosomes (15–30 s after Pb^{2+} treatment), no mobilization of internal ^{45}Ca was observed.[87] Several observations support actions of Pb^{2+} at Ca^{2+} targets. The most sensitive response to Pb^{2+} that has been found is activation at subnanomolar Pb^{2+} concentration of protein kinase C (PKC),[74] which may account for the facilitated transmitter release via a number of phosphorylated enzymes, although another study found no direct activation and only inhibition of PKC.[75]

Although there is no evidence that the presynaptic effect of Pb^{2+} is specific to the type of transmitter contained within the terminal for release, only the evoked release of the fast-acting and intermediate-acting transmitters

ACh, GABA, DA,[87] and glutamate (unpublished) have been demonstrated. There is a strong suggestion that the regulation of type of transmitter and cotransmitter release is dependent upon the pattern of neuronal stimulation and the site of Ca^{2+} action within the terminal.[90] For example, the close proximity of the Ca^{2+} channel to the synaptic vesicle docking complexes in the active zone may account for the rapid release of classical excitatory transmitters such as glutamate and ACh, whereas saturation of the Ca^{2+} buffering capacity of the neuron, which is normally caused by repetitive stimulation, will yield a greater release of neuropeptide transmitters from the same terminal. Therefore, the type of transmitter released may indicate the target for Pb^{2+}. Also, the termination of, for example, glutamate release is due to low Ca^{2+}-affinity of the target linked to the release of docked vesicles, but a higher affinity action of Pb^{2+} at the same target could result in more persistent stimulation of glutamate release.

The question arises as to how Pb^{2+} enters the nerve terminal. Although Pb^{2+} competitively inhibits each type of voltage-sensitive Ca^{2+} channel, Pb^{2+} is itself highly permeable through these channels. The permeability of Pb^{2+} through the Ca^{2+} channels of the chromaffin cells was approximately 10 times that of Ca^{2+} (see Section 11.28.2.6). On the other hand, since the increased spontaneous release of transmitter from motoneurons and cultured brain neurons occurred under resting conditions when the Ca^{2+} channels (particularly the N- and L-types with high voltage thresholds) should be inactive, another possibility is that Pb^{2+} enters the terminals through weakly activated presynaptic receptor-ion channels. Nicotinic receptors, especially of the α-BGT-insensitive α4β2 subtype, are known to be located on presynaptic terminals in several brain regions, where their role is to facilitate transmitter release. In the hippocampus, [³H]-nicotine-binding receptors have been located in the molecular layer of the dentate gyrus and the pyramidal cell layer of the hippocampal areas CA1-CA4.[91] In contrast to the NMDA receptors, these nicotinic receptors are not blocked by Mg^{2+} at resting and hyperpolarized potentials. Furthermore, the nicotinic receptors are highly permeable to Ca^{2+}: the permeability to Ca^{2+} relative to the permeability to Cs^+ of the α-BGT-sensitive nAChR (α7 subtype) is about 6.[57] Although Pb^{2+} at micromolar concentrations is a noncompetitive antagonist of α7-nicotinic, NMDA, and 5-HT_3 receptors, the voltage independence of Pb^{2+}-induced inhibition of these receptors shows that blockade does not occur via an ion channel site. Therefore, where transmitter-gated receptors are present on nerve terminals, they may contribute to a small influx of Pb^{2+}, at low concentrations of Pb^{2+}. On the other hand, the mature NMDA receptor, the kainate receptor, the quisqualate receptor, the α4β2 nicotinic receptor, and even the metal ion-activated receptor are all relatively uninhibited by Pb^{2+} and so may also pass this cation at higher concentrations. Pb^{2+} may also be taken into the terminal by the active Zn^{2+} transporter.

The stimulatory effect of micromolar Pb^{2+} on spontaneous transmitter release is accompanied by a reduction of the quantal content and the endplate potential amplitude during neuromuscular transmission[84] and a reduction of the release of dopamine during K^+-depolarization of synaptosomes.[85] It has been proposed that the depressant effect of Pb^{2+} is mediated by extracellular targets located presynaptically, where Pb^{2+} (1 μM) blocks high-threshold voltage-gated Ca^{2+} channels[68,92] to inhibit competitively Ca^{2+}-dependent neurotransmitter release.[52] Since the depression of evoked transmitter release and the facilitation of spontaneous transmitter release are linked by their occurrence at the same concentration of Pb^{2+}, consideration should be given to whether these reciprocal effects of Pb^{2+} could be intimately linked, both depending on a single mechanism of action. First, the inhibition by Pb^{2+} of Ca^{2+} currents through voltage-gated channels and consequent blockade of evoked transmitter release potentially could increase the intracellular transmitter stores and possibly enhance transmitter release indirectly. However, it is helpful to notice that Cd^{2+}, which also blocks a Ca^{2+} channel but has no specific action on PKC, blocks evoked transmitter release but reverses the stimulatory effect of Pb^{2+} on spontaneous transmitter release.[71] Therefore, the inhibition of evoked release is unlikely to stimulate indirectly spontaneous release. Second and conversely, the enhanced spontaneous release via an intracellular action of Pb^{2+} (see Section 11.28.2.7) could deplete the available stores of transmitter, so that the quantal size and/or the quantal content could be reduced more than in the case of Ca^{2+} channel blockade alone. To the authors' knowledge, the effect of preloaded intraterminal Pb^{2+} on evoked transmitter release when Pb^{2+} is absent from the extracellular milieu has not been evaluated.

11.28.3.2 Pb^{2+} Inhibits Long-term Potentiation

The basis for learning and memory is thought to be the plastic process of long-term potentiation (LTP) whereby the frequent use of a selected pathway initiates within it a prolonged potentiation of synaptic responses. Many studies have examined the mechanisms underlying LTP, as well as the effect of Pb^{2+} on LTP, in the hippocampus because: this region of the brain is important for cognitive functions and memory; discrete complex synapses on large numbers of dendritic spines provide the morphological basis for focal interactions; and the laminar organization and the development of LTP in the hippocampus provide an *in vitro* preparation in which neurotransmission can be studied electrophysiologically and can be subjected to pharmacological manipulations. Several studies are in agreement that LTP in mature brain tissue is reduced by Pb^{2+} *in vitro*[93,94] and *in vivo*.[95] A set of stimulus conditions and a sequence of biochemical steps, which vary somewhat with brain regions under study, are necessary for LTP to develop and persist. Therefore, any of several mechanisms could account for the inhibitory effect of Pb^{2+}. Post-tetanic potentiation (PTP), which is an increase in transmitter release caused by Ca^{2+} in the presynaptic terminal, provides a basis for the first step of potentiation that appears with all preparations. The LTP of glutamatergic trans-mission in the CA1 area of the hippocampus requires the stimulation of AMPA receptors that provides the depolarization necessary to remove the Mg^{2+} blockade of NMDA recep-tors and the stimulation of NMDA receptors that provides the Ca^{2+} influx that triggers the LTP process. A persistent increase in the pre-synaptic transmitter release may be evoked by a retrograde factor released from the postsynap-tic neuron (NO or arachidonic acid). The targets of Pb^{2+} that are accountable for its inhibition of LTP remain to be clarified.

Superfusion of hippocampal slices with Pb^{2+} ($15\,\mu M$, effective in 5 min) consistently decreases the amplitude of electrically evoked CA1 responses (field excitatory postsynaptic potentials and population spikes) by changing the input–output relationship in a manner that suggests an elevated threshold for transmitter release. However, in the continued presence of Pb^{2+}, the response in the CA1 area immediately following tetanic stimulation is variably affected; sometimes (50% of preparations) PTP is decreased by superfusion of neurons with Pb^{2+}, and at other times PTP appears normal.[93] Surprisingly, after the removal of Pb^{2+}, LTP was induced in preparations where PTP had been inhibited, but LTP was inhibited in preparations where PTP had been normal. These data suggest that presynaptic facilitation (requiring Ca^{2+} channel function) is not re-quired for induction of LTP, and furthermore, PTP occurring in the presence of Pb^{2+} reduces the level of LTP following the removal of Pb^{2+}. The differences between preparations could signify different actions of Pb^{2+} within pre-synaptic terminals. Considering the evidence from studies of neuromuscular transmission showing that the facilitation of transmitter release by 60 Hz stimulation is increased by Pb^{2+}, it may be that under conditions of repeti-tive stimulation intracellular Pb^{2+} mimics the action of Ca^{2+}.[96] Thus, when no PTP is seen in the hippocampal slice preparation, little Ca^{2+} or Pb^{2+} has entered the presynaptic terminal, and LTP develops as an independent process; but when PTP is seen, both Ca^{2+} and Pb^{2+} have entered the terminals to cause PTP, but the release of transmitter by single test stimuli could be inhibited by the intraterminal Pb^{2+}, counteracting the appearance of LTP. In the

piriform cortex slice from mature rats, LTP of the lateral olfactory tract-piriform cortex exci-tatory postsynaptic potential is inhibited by Pb^{2+} ($5\,\mu M$), even though PTP in this region is not affected by Pb^{2+} (up to $20\,\mu M$) and the NMDA-elicited response is insensitive to Pb^{2+} ($10\,\mu M$).[94] These results suggest that the inhibi-tion of LTP in the piriform cortex by Pb^{2+} can arise from an intracellular change in the post-synaptic neuron, a blockade of the feedback to the presynaptic neuron, or an action preventing the long-term presynaptic facilitation.

LTP of the excitatory postsynaptic potential generated *in vivo* in the dentate gyrus by stimulation of the perforant path is completely absent at 85–105 days of age in postnatally Pb^{2+}-exposed rats.[95] In addition to any acute effect of the Pb^{2+} present in the dentate gyrus at the time the response was measured, alteration of hippocampal synaptic morphology by Pb^{2+} no doubt contributes to the absence of LTP.

11.28.3.3 Effects of Pb^{2+} Exposure During Development on Neurogenesis, Postnatal Growth, and Maturation

Exposure to highly toxic concentrations of Pb^{2+} has been shown to induce encephalopathy and to alter the integrity of the capillaries in the nervous system.[97] Lead administration to lactating rats (1% in the maternal diet) produced in pups a blood lead level of $100\,\mu g\,dL^{-1}$ that did not cause hemorrhage or necrosis but decreased food intake, body weight, and brain weight.[98] Under the lead-treated and pair-fed conditions, the number of cortical synapses was reduced and a large percentage of the synapses were incomplete (i.e., few or no synaptic vesicles were present) at 10 and 21 days.[98] After weaning to normal diet, the difference between the numbers of synapses in lead-treated pups and pair-fed pups increased up to 60 days (the end of the observation period). Undernourishment was a larger contributor than lead treatment to inhibition of cortical synaptogenesis, which normally increases rapidly between 14 and 21 days of age. Several studies on low-level lead intoxication have focused on the susceptibility of the hippocampus, a brain region strongly implicated in cognition and memory.

The effects of chronic Pb^{2+} treatment on the development of two regions of the hippocam-pus have received particular attention: the dentate granular cell/mossy fiber and the CA3 pyramidal neuron/Schaffer collateral. The den-tate gyrus does not mature morphologically in the rat until 25 days of age. Lead dosing

$(7.5\,\mathrm{mg\,kg^{-1}})$ for 10 days after parturition reduced the dimensions of the dentate gyrus, hippocampal formation, and the cerebral cortex, as well as the acetylcholinesterase activity of the hippocampus, cerebrum, and midbrain.[99] Morphometic analysis of the developing hippocampus showed that Pb^{2+} exposure impaired differentiation and growth of specific subfields.[100] The infrapyramidal field of the 15 day old rat dentate gyrus is affected more by low-level lead poisoning (postnatal 0.2% Pb^{2+} in drinking water) than is the suprapyramidal field. The cross-sectional areas of the infrapyramidal stratum molecular and the dentate hilus were reduced in size, and the neurons in the hilar portion of CA3 stratum pyramidal were increased in number despite a small decrease in the area of the region.[100] Each of these changes indicates underdevelopment of the mossy fiber axons arising from granular neurons of the infrapyramidal leaf of the dentate gyrus, which are the last neurons in the dentate gyrus to develop. The effect of lead on terminal boutons was a reduction in density concomitant with a tendency to increase in size. After weaning to a normal diet, compensatory changes were indicated from the histological examination of 90 day old rats.[1,101] In another study of rats exposed postnatally to a higher level of Pb^{2+} (4% lead salt in drinking water), the heavy metal treatment did not alter brain size but significantly reduced the width of the hippocampus, the length of the dentate gyrus, and the overall length of the mossy fiber pathway (identified by Timm's stain) at day 25, and these signs of dentate gyrus hypoplasia persisted, albeit to a lesser degree, without further Pb^{2+} treatment at day 60.[102] (Timm's stain identifies tissues with high levels of heavy metals, particularly Zn^{2+}. The concentration of Pb^{2+} is probably not high enough to affect the detection by Timm's stain of Zn^{2+} in glutamatergic neurons or terminals. Therefore, the effects of chronic Pb^{2+} exposure on Timm's staining reflect changes in the function and structure of the affected neurons.) Such changes in neuronal structure and synapse formation may be related to the impairment of the conversion of neural cell adhesion molecules, which occurs at blood lead levels above $20\,\mu\mathrm{g\,dL^{-1}}$.[103] In apparent contrast, lower levels of postnatal treatment (0.02–0.4% lead acetate in drinking water for 21 days) increased the volume of the proximal dendritic zone of the dentate molecular layer, the granule cell layer, and the mossy fiber zone; whereas intermediate levels (0.08 and 0.16%) produced no effects, suggesting that the effects of early postnatal lead exposure are bimodal.[104] In the latter study, the exposure producing hyper-

trophy (0.02%) yielded blood lead level about $21\,\mu\mathrm{g\,dL^{-1}}$, i.e., twice the current action level in humans established by the CDC. When adult rats were exposed for three months to a similarly low concentration of lead in water (0.01%), the forebrain synaptosomes contained about half the number of synaptic vesicles found in synaptosomes from control rats.[105]

Pb^{2+} also produced concentration-related effects on morphology of cultured embryonic hippocampal neurons at 24 hours. The lower concentrations of Pb^{2+} (0.02–1.0 μM) *in vitro* were associated with a decreased (about 25–50%) initiation of neurites.[15,106] Higher concentrations of Pb^{2+} (500 μM and 1000 μM), which promoted survival, also caused decreased initiation of neurites, but stimulated neurite elongation and dendritic branching, suggesting relatively specific effects of Pb^{2+} against the initial differentiation of the neuron.[106] All of the above effects were diminished by increasing from 2% to 10% the amount of fetal calf serum used in the culture medium. Neurite initiation could be related to mechanisms triggered by or mediated by intracellular Ca^{2+} and protein kinases that elicit changes in the cytoskeletal structure. Comparison of the effects of various calmodulin and specific protein kinase inhibitors alone or in combination with Pb^{2+} on neurite initiation in hippocampal cultures showed that Pb^{2+} acts similarly to a PKC inhibitor, the effects of Pb^{2+} were reversed by calmodulin and PKA inibitors, but there was no significant effect on sustained whole-cell intracellular Ca^{2+} concentration.[15] Thus, according to these studies, an intracellular target of Pb^{2+}, perhaps calmodulin or a Ca^{2+}/calmodulin-dependent enzyme, may produce protein hyperphosphorylation to decrease neurite initiation.[15]

It seems appropriate to consider briefly the fact that abnormal functional activity can alter the synaptic organization, even in the adult rat limbic system. Lasting modifications of activity were found to induce Timm granules in the neuronal terminals of the supragranular layer of the dentate gyrus, demonstrating that axonal growth, terminal sprouting, and synaptic reorganization are induced by propagation of the activity to the dentate gyrus. Low-level stimulation of the perforant path produced a few Timm granules after five to six days, and epileptiform activity induced by kindling stimulation of the perforant path, the olfactory bulb, or the amygdala produced greater changes that lasted from 18 hours to as long as five months.[107] These effects could be seen without neuronal degeneration in CA3/CA4. On the other hand, when intense stimulation evoked afterdischarges of the perforant path for 24 hours, there was a loss of Timm stain in the mossy fiber pathway and a loss of CA3/CA4 neurons that receive afferent input from granule cells.[107]

11.28.4 SUMMARY AND CONCLUSIONS

Low-level lead poisoning is an insidious process that impairs neuronal development of the fetus and the neonate and can result in a long-lasting impairment of mental performance. The molecular sites of lead action are sites for physiological Ca^{2+} and Zn^{2+} and other novel sites of many proteins. Most lead is bound, and very low concentrations of free Pb^{2+} are present extracellularly or intracellularly, but Pb^{2+} can be released from protein sites and perhaps also vesicular pools, so that Pb^{2+} concentrations may be heterogeneous and dynamic. The inhibition of NMDA receptors and nAChRs may be a first step in the impairment of neuronal growth in the hippocampus. Even though the NMDA receptor and nAChR populations are known to change during normal development, more fundamental information is needed about how these Ca^{2+}-permeable receptors feedback the signals regulating neuronal growth. The vulnerability of the NMDA receptors and α7 nAChRs present on hippocampal neurons to Pb^{2+} is noteworthy. Inhibition of Ca^{2+} currents through these receptors and through Ca^{2+} channels will alter many Ca^{2+}-dependent processes. Glutamate receptors, nAChRs, and Ca^{2+} channels may provide pathways for entry of Pb^{2+} into neurons, where lower levels of Pb^{2+} target directly calmodulin cyclic nucleotide synthesis and degradation, protein kinase C, and K^+_{Ca} channels. By these means, changes in Ca^{2+} signals and internal Pb^{2+} can stimulate the spontaneous release of transmitter, inhibit the evoked release of transmitter, and inhibit LTP.

Investigation of Pb^{2+} intoxication may provide important clues to the physiological processes of neuronal development. Pharmacological stimulation of neuronal development and synaptogenesis may be an approach useful for chronic low level lead intoxication and other cognitive disorders such as Alzheimer's disease. Nevertheless, prevention of lead exposure remains the best clinical approach, since it seems unlikely that a drug can be discovered to reverse the many acute actions of lead.

ACKNOWLEDGMENTS

The authors thank Mrs. Mabel Zelle for helpful comments on the manuscript. This study was supported by NIEHS grant ES05730.

11.28.5 REFERENCES

1. J. S. Lin-Fu, in 'Low Level Lead Exposure: The Clinical Implication of Current Research,' ed. H. L. Needleman, Raven Press, New York, 1980, pp. 5–16.
2. R. P. Beliles, in 'Toxicology: The Basic Science of Poisons,' eds. L H. Casarett and J. Doull, 1975, pp. 454–502.
3. J. M. Christensen and J. Kristiansen, in 'Metals in Clinical and Analytical Chemistry,' eds. H. G. Seiler, A. Sigel and H. Sigel, Dekker, New York, 1994, pp. 425–550.
4. Centers for Disease Control 1991, 'Preventing Lead Poisoning in Young Children: A Statement by the Centers for Disease Control—October 1991,' US Department of Health and Human Services, Public Health Service, Atlanta, GA, 1991, pp. 1–105.
5. H. L. Needleman, 'Childhood lead poisoning: a disease for the history texts.' *Am. J. Public Health*, 1991, **81**, 685–687.
6. Maryland Lead Paint Poisoning Commission, 'Report of the Lead Paint Poisoning Commission: Prepared for Governor William Donald Schaefer and the Maryland General Assembly,' 1994, pp. 1–45 and App. A-1.
7. S. J. Schaffer and J. R. Campbell, 'The new CDC and AAP lead poisoning prevention recommendations: consensus versus controversy.' *Pediatr. Ann.*, 1994, **23**, 592–599.
8. J. D. Sargent, 'The role of nutrition in the prevention of lead poisoning in children.' *Pediatr. Ann.*, 1994, **23**, 636–642.
9. D. Faust and J. Brown, 'Moderately elevated blood lead levels: effects on neuropsychologic functioning in children.' *Pediatrics*, 1987, **80**, 623–629.
10. H. L. Needleman, 'The current status of childhood low-level lead toxicity.' *Neurotoxicology*, 1993, **14**, 161–166.
11. D. Bellinger and K. N. Dietrich, 'Low-level lead exposure and cognitive function in children.' *Pediatr. Ann.*, 1994, **23**, 600–605.
12. P. L. Goering, 'Lead-protein interactions as a basis for lead toxicity.' *Neurotoxicology*, **14**, 1993, 45–60.
13. C. N. Ong, W. O. Phoon, B. L. Lee *et al.*, 'Lead in plasma and its relationships to other biological indicators.' *Ann. Occup. Hyg.*, 1986, **30**, 219–228.
14. G. Audesirk, 'Electrophysiology of lead intoxication: effects on voltage-sensitive ion channels.' *Neurotoxicology*, 1993, **14**, 137–147.
15. M. Kern and G. Audesirk, 'Inorganic lead may inhibit neurite development in cultured rat hippocampal neurons through hyperphosphorylation.' *Toxicol. Appl. Pharmacol.*, 1995, **134**, 111–123.
16. A. K. Singh, 'Neurotoxicity in rats chronically exposed to lead ingestion: Measurement of intracellular concentrations of free calcium and lead ions in resting or depolarized brain slices.' *Neurotoxicology*, 1995, **16**, 133–138.
17. H. Björklund, B. Lind, M. Piscator, *et al.*, 'Lead, zinc, and copper levels in intraocular brain tissue grafts, brain, and blood of lead-exposed rats.' *Toxicol. Appl. Pharmacol.*, 1981, **60**, 424–430.
18. A. Y. P'an and C. Kennedy, 'Lead distribution in rats repeatedly treated with low doses of lead acetate.' *Environ. Res.*, 1989, **48**, 238–247.
19. A. J. Al-Modhefer, M. W. Bradbury and T. J. Simons, 'Observations on the chemical nature of lead in human blood serum.' *Clin. Sci. (Colch.)*, 1991, **81**, 823–829.
20. A. M. Scheuhammer and M. G. Cherian, 'The regional distribution of lead in normal rat brain.' *Neurotoxicology*, 1982, **3**, 85–92.
21. M. F. Collins, P. D. Hrdina, E. Whittle *et al.*, 'Lead in

blood and brain regions of rats chronically exposed to low doses of the metal.' *Toxicol. Appl. Pharmacol.*, 1982, **65**, 314–322.

22. S. V. Kala and A. L. Jadhav, 'Region-specific alterations in dopamine and serotonin metabolism in brains of rats exposed to low levels of lead.' *Neurotoxicology*, 1995, **16**, 297–308.

23. E. J. Fjerdingstad, G. Danscher and E. Fjerdingstad, 'Hippocampus: selective concentration of lead in the normal rat brain.' *Brain Res.*, 1974, **80**, 350–354.

24. J. M. Lefauconnier, G. Bernard, F. Mellerio *et al.*, 'Lead distribution in the nervous system of 8-month-old rats intoxicated since birth by lead.' *Experientia*, 1983, **39**, 1030–1031.

25. M. Cholewa, A. L. Hanson, K. W. Jones *et al.*, 'Regional distribution of lead in the brains of lead-intoxicated adult mice.' *Neurotoxicology*, 1986, **7**, 9–18.

26. D. V. Widzowski and D. A. Cory-Slechta, 'Homogeneity of regional brain lead concentrations.' *Neurotoxicology*, 1994, **15**, 295–307.

27. G. Danscher, E. J. Fjerdingstad, E. Fjerdingstad *et al.* 'Heavy metal content in subdivisions of the rat hippocampus (zinc, lead and copper).' *Brain Res.*, 1976, **112**, 442–446.

28. C. J. Frederickson, 'Neurobiology of zinc and zinc-containing neurons.' *Int. Rev. Neurobiol.*, 1989, **31**, 145–238.

29. S. Y. Assaf and S. H. Chung, 'Release of endogenous Zn^{2+} from brain tissue during activity.' *Nature*, 1984, **308**, 734–736.

30. G. A. Howell, M. G. Welch and C. J. Frederickson, 'Stimulation-induced uptake and release of zinc in hippocampal slices.' *Nature*, 1984, **308**, 736–738.

31. T. J. Simons, 'Lead-calcium interactions in cellular lead toxicity.' *Neurotoxicology*, 1993, **14**, 77–85.

32. M. Alkondon, A. C. Costa, V. Radhakrishnan *et al.*, 'Selective blockade of NMDA-activated channel currents may be implicated in learning deficits caused by lead.' *FEBS Lett.*, 1990, **261**, 124–130.

33. H. Ujihara and E. X. Albuquerque, 'Developmental change of the inhibition by lead of NMDA-activated currents in cultured hippocampal neurons.' *J. Pharmacol. Exp. Ther.*, 1992, **263**, 868–875.

34. K. Ishihara, M. Alkondon, J. G. Montes *et al.*, 'Nicotinic responses in acutely dissociated rat hippocampal neurons and the selective blockade of fast-desensitizing nicotinic currents by lead.' *J. Pharmacol. Exp. Ther.*, 1995, **273**, 1471–1482.

35. H. Ujihara and E. X. Albuquerque, 'Ontogeny of N-methyl-D-aspartate-induced current in cultured hippocampal neurons.' *J. Pharmacol. Exp. Ther.*, 1992, **263**, 859–867.

36. T. R. Guilarte and R. C. Miceli, 'Age-dependent effects of lead on [^{3}H]MK-801 binding to the NMDA receptor-gated channel ionophore: *in vitro* and *in vivo* studies.' *Neurosci. Lett.*, 1992, **148**, 27–30.

37. T. R. Guilarte, R. C. Miceli and D. A. Jett, 'Biochemical evidence of an interaction of lead at the zinc allosteric sites of the NMDA receptor complex: effects of neuronal development.' *Neurotoxicology*, 1995, **16**, 63–71.

38. A. S. Ramoa, M. Alkondon, Y. Aracava *et al.*, 'The anticonvulsant MK-801 interacts with peripheral and central nicotinic acetylcholine receptor ion channels.' *J. Pharmacol. Exp. Ther.*, 1990, **254**, 71–82.

39. S. A. Bayer, J. Altman, R. J. Russo *et al.*, 'Timetables of neurogenesis in the human brain based on experimentally determined patterns in the rat.' *Neurotoxicology*, 1993, **14**, 83–144.

40. T. R. Guilarte, R. C. Miceli and D. A. Jett, 'Neurochemical aspects of hippocampal and cortical Pb^{2+} neurotoxicity.' *Neurotoxicology*, 1994, **15**, 459–466.

41. D. Büsselberg, D. Michael and B. Platt, 'Pb^{2+} reduces voltage- and N-methyl-D-aspartate (NMDA)-activated calcium channel currents.' *Cell. Mol. Neurobiol.*, 1994, **14**, 711–722.

42. M. Marchioro, A. T. Eldefrawi and E. X. Albuquerque, 'Lead (Pb^{2+}) interactions with the N-methyl-D-aspartate (NMDA) receptor: modulation by Ca^{2+} and glycine.' *Soc. Neurosci. Abs.*, 1995, **21**, 351.

43. G. C. Yeh, D. W. Bonhaus and J. O. McNamara, 'Evidence that zinc inhibits N-methyl-D-aspartate receptor-gated ion channel activation by noncompetititve antagonism of glycine binding.' *Mol. Pharmacol.*, 1990, **38**, 14–19.

44. L. Vyklicky Jr., M. Benveniste and M. L. Mayer, 'Modulation of N-methyl-D-aspartic acid receptor desensitization by glycine in mouse cultured hippocampal neurons.' *J. Physiol. (Lond.)*, 1990, **428**, 313–331.

45. Y. Gu and L. Y. M. Huang, 'Modulation of glycine affinity for NMDA receptors by extracellular Ca^{2+} in trigeminal neurons.' *J. Neurosci.*, 1994, **14**, 4561–4570.

46. L. Y. Wang and J. F. MacDonald, 'Modulation by magnesium of the affinity of NMDA receptors for glycine in murine hippocampal neurones.' *J. Physiol. (Lond.)*, 1995, **486**, 83–95.

47. G. Tong and C. E. Jahr, 'Multivesicular release from excitatory synapses of cultured hippocampal neurons.' *Neuron*, 1994, **12**, 51–59.

48. T. L. Petit, L. C. LeBoutillier and W. J. Brooks, 'Altered sensitivity to NMDA following developmental lead exposure in rats.' *Physiol. Behav.*, 1992, **52**, 687–693.

49. D. A. Jett and T. R. Guilarte, 'Developmental lead exposure alters N-methyl-D-aspartate and muscarinic cholinergic receptors in the rat hippocampus: an autoradiographic study.' *Neurotoxicology*, 1995, **16**, 7–18.

50. K. Ishihara, M. Alkondon, J. G. Montes *et al.*, 'Ontogenically related properties of N-methyl-D-aspartate receptors in rat hippocampal neurons and the age-specific sensitivity of developing neurons to lead.' *J. Pharmacol. Exp. Ther.*, 1995, **273**, 1459–1470.

51. K. Kostial and V. B. Vouk, 'Lead ions and synaptic transmission in the superior cervical ganglion of the cat.' *Brit. J. Pharmacol.*, 1957, **12**, 219–222.

52. T. E. Kober and G. P. Cooper, 'Lead competitively inhibits calcium-dependent synaptic transmission in the bullfrog sympathetic ganglion.' *Nature*, 1976, **262**, 704–705.

53. M. Oortgiesen, R. G. van Kleef, R. B. Bajnath *et al.*, 'Nanomolar concentrations of lead selectively block neuronal nicotinic acetylcholine responses in mouse neuroblastoma cells.' *Toxicol. Appl. Pharmacol.*, 1990, **103**, 165–174.

54. M. Oortgiesen, T. Leinders, R. G. van Kleef *et al.*, 'Differential neurotoxicological effects of lead on voltage-dependent and receptor-operated ion channels.' *Neurotoxicology*, 1993, **14**, 87–96.

55. M. Alkondon and E. X. Albuquerque, 'Diversity of nicotinic acetylcholine receptors in rat hippocampal neurons: I. Pharmacological and functional evidence for distinct structural subtypes.' *J. Pharmacol. Exp. Ther.*, 1993, **265**, 1455–1473.

56. M. Alkondon, S. Reinhardt, C. Lobron *et al.*, 'Diversity of nicotinic acetylcholine receptors in rat hippocampal neurons. II. The rundown and inward rectification of agonist-elicited whole-cell currents and identification of receptor subunits by *in situ* hybridization.' *J. Pharmacol. Exp. Ther.*, 1994, **271**, 494–506.

57. N. G. Castro and E. X. Albuquerque. 'α-Bungarotoxin-sensitive hippocampal nicotinic receptor channel has a high calcium permeability.' *Biophys. J.*, 1995, **68**, 516–524.

58. R. Bonfante-Cabarcas, K. L. Swanson, M. Alkondon

et al., 'Diversity of nicotinic acetylcholine receptors in rat hippocampal neurons. IV. Regulation by external Ca^{2+} of α-bungarotoxin-sensitive receptor function and rectification induced by internal Mg^{2+}.' *J. Pharmacol. Exp. Ther.*, 1996, **277**, 432–444.

59. M. Amador and J. A. Dani, 'Mechanism for modulation of nicotinic acetylcholine receptors that can influence synaptic transmission.' *J. Neurosci.*, 1995, **15**, 4525–4532.

60. M. Alkondon and E. X. Albuquerque, 'Initial characterization of the nicotinic acetylcholine receptors in rat hippocampal neurons.' *J. Recept. Res.*, 1991, **11**, 1001–1021.

61. E. X. Albuquerque, E. F. R. Pereira, R. Bonfante-Cabarcas *et al.*, 'Nicotinic acetylcholine receptors on hippocampal neurons: cell-compartment specific expression and modulatory control of channel activity.' *Prog. Brain Res.*, 1996, in press.

62. E. X. Albuquerque, E. F. R. Pereira, N. G. Castro *et al.*, 'Neuronal nicotinic receptors: Function, modulation and structure.' *Semin. Neurosci.*, 1995, **7**, 91–101.

63. M. Oortgiesen, R. G. van Kleef and H. P. Vijverberg, 'Novel type of ion channel activated by Pb^{2+}, Cd^{2+} and Al^{3+} in cultured mouse neuroblastoma cells.' *J. Membr. Biol.*, 1990, **113**, 261–268.

64. H. P. Vijverberg, T. Leinders-Zufall and R. G. van Kleef, 'Differential effects of heavy metal ions on Ca^{2+}-dependent K^+ channels.' *Cell. Mol. Neurobiol.*, 1994, **14**, 841–857.

65. L. R. Sun and J. B. Suszkiw, 'Pb^{2+} activates potassium currents in bovine adrenal chromaffin cells.' *Neurosci. Lett.*, 1994, **182**, 41–43.

66. A. Oberhauser, O. Albarex and R. Latorre, 'Activation by divalent cations of a Ca^{2+}-activated K^+ channel from skeletal muscle membrane.' *J. Gen. Physiol.*, 1988, **92**, 67–86.

67. G. Audesirk and T. Audesirk, 'The effects of inorganic lead on voltage-sensitive calcium channels differ among cell types and among channel subtypes.' *Neurotoxicology*, 1993, **14**, 259–265.

68. D. Büsselberg, M. L. Evans, H. Rahmann *et al.*, 'Lead and zinc block a voltage-activated calcium channel of *Aplysia* neurons.' *J. Neurophysiol.*, 1991, **65**, 786–795.

69. D. Büsselberg, M. L. Evans, H. L. Hass *et al.*, 'Blockade of mammalian and invertebrate calcium channels by lead.' *Neurotoxicology*, 1993, **14**, 249–258.

70. H. Ujihara, M. Sasa and T. Ban, 'Selective blockade of P-type calcium channels by lead in cultured hippocampal neurons.' *Jpn. J. Pharmacol.*, 1995, **67**, 267–269.

71. G. P. Cooper and R. S. Manalis, 'Interactions of lead and cadmium on acetylcholine release at the frog neuromuscular junction.' *Toxicol. Appl. Pharmacol.*, 1984, **74**, 411–416.

72. E. Habermann, K. Crowell and P. Janicki, 'Lead and other metals can substitute for Ca^{2+} in calmodulin.' *Arch. Toxicol.*, 1983, **54**, 61–70.

73. G. W. Goldstein, 'Evidence that lead acts as a calcium substitute in second messenger metabolism.' *Neurotoxicology*, 1993, **14**, 97–101.

74. J. Markovac and G. W. Goldstein, 'Picomolar concentrations of lead stimulate brain protein kinase C.' *Nature*, 1988, **334**, 71–73.

75. K. Murakami, G. Fend and S. G. Chen, 'Inhibition of brain protein kinase C subtypes by lead.' *J. Pharmacol. Exp. Ther.*, 1993, **264**, 757–761.

76. J. A. Nathanson and F. E. Bloom, 'Lead-induced inhibition of brain adenyl cyclase.' *Nature*, 1975, **255**, 419–420.

77. D. A. Fox, D. Srivastave and R. L. Hurwitz, 'Lead-induced alterations in rod-mediated visual functions and cGMP metabolism: new insights.' *Neurotoxicology*, 1994, **15**, 503–512.

78. M. E. Blazka, G. J. Harry and M. I. Luster, 'Effect of lead acetate on nitrite production by murine brain endothelial cell cultures.' *Toxicol. Appl. Pharmacol.*, 1994, **126**, 191–194.

79. L. Tian and D. A. Lawrence, 'Lead inhibits nitric oxide production *in vitro* by murine splenic macrophages.' *Toxicol. Appl. Pharmacol.*, 1995, **132**, 156–163.

80. D. S. Bredt, P. M. Hwang and S. H. Snyder, 'Localization of nitric oxide synthase indicating a neural role for nitric oxide.' *Nature*, 1990, **347**, 768–770.

81. A. K. Singh, 'Age-dependent neurotoxicity in rats chronically exposed to low levels of lead: calcium homeostasis in central neurons.' *Neurotoxicology*, 1993, **14**, 417–427.

82. R. S. Manalis and G. P. Cooper, 'Letter: presynaptic and postsynaptic effects of lead at the frog neuromuscular junction.' *Nature*, 1973, **243**, 354–356.

83. G. P. Cooper, J. B. Suszkiw and R. S. Manalis, 'Heavy metals: effects on synaptic transmission.' *Neurotoxicology*, 1984, **5**, 247–266.

84. W. D. Atchison and T. Narahashi, 'Mechanism of action of lead on neuromuscular junctions.' *Neurotoxicology*, 1984, **5**, 267–282.

85. D. J. Minnema, R. D. Greenland and I. A. Michaelson, 'Effect of *in vitro* inorganic lead on dopamine release from superfused rat striatal synaptosomes.' *Toxicol. Appl. Pharmacol.*, 1986, **84**, 400–411.

86. D. J. Minnema and I. A. Michaelson, 'Differential effects of inorganic lead and δ-aminolevulinic acid *in vitro* on synaptosomal γ-aminobutyric acid release.' *Toxicol. Appl. Pharmacol.*, 1986, **86**, 437–447.

87. D. J. Minnema, I. A. Michaelson and G. P. Cooper, 'Calcium efflux and neurotransmitter release from rat hippocampal synaptosomes exposed to lead.' *Toxicol. Appl. Pharmacol.*, 1988, **92**, 351–357.

88. Z. Shao and J. B. Suszkiw, 'Ca^{2+}-Surrogate action of Pb^{2+} on acetylcholine release from rat brain synaptosomes.' *J. Neurochem.*, 1991, **56**, 568–574.

89. J. P. Bressler and G. W. Goldstein, 'Mechanisms of lead neurotoxicity.' *Biochem. Pharmacol.*, 1991, **41**, 479–484.

90. M. Verhage, W. E. Ghijsen and F. H. Lopes da Silva, 'Presynaptic plasticity: the regulation of Ca^{2+}-dependent transmitter release.' *Prog. Neurobiol.*, 1994, **42**, 539–574.

91. P. Seguela, J. Wadiche, K. Dineley-Miller *et al.*, 'Molecular cloning, functional properties and distribution of rat brain α7: a nicotinic cationic channel highly permeable to calcium.' *J. Neurosci.*, 1993, **13**, 596–604.

92. M. L. Evans, D. Busselberg and D. O. Carpenter, 'Pb^{2+} blocks calcium current of cultured dorsal root ganglion cells.' *Neurosci. Lett.*, 1991, **129**, 103–106.

93. L. Altman, K. Sveinsson and H. Wiegand, 'Long-term potentiation in rat hippocampal slices is impaired following acute lead perfusion.' *Neurosci. Lett.*, 1991, **128**, 109–112.

94. N. Hori, D. Bsselberg, M. R. Matthews *et al.*, 'Lead blocks LTP by an action not at NMDA receptors.' *Exp. Neurol.*, 1993, **119**, 192–197.

95. S. M. Lasley, J. Polan-Curtain and D. L. Armstrong, 'Chronic exposure to environmental levels of lead impairs *in vivo* induction of long-term potentiation in rat hippocampal dentate.' *Brain Res.*, 1993, **614**, 347–351.

96. J. B. Picket and J. C. Bornstein, 'Some effects of lead at mammalian neuromuscular junction.' *Am. J. Physiol.*, 1984, **246**, C271-C276.

97. A. Pentschew and F. Garro, 'Lead encephalo-myelopathy of the suckling rat and its implications on the porphyrinopathic nervous diseases. With special

reference to the permeability disorders of the nervous system's capillaries.' *Acta Neuropathol. (Berl.)*, 1966, **6**, 266–278.

98. D. R. Averill Jr. and H. L. Needleman, in 'Low Level Lead Exposure: The Clinical Implications of Current Research,' ed. H. L. Needleman, Raven Press, New York, 1980, pp. 201–210.

99. R. T. Louis-Ferdinand, D. R. Brown, S. F. Fiddler *et al.*, 'Morphometric and enzymatic effects of neonatal lead exposure in the rat brain.' *Toxicol. Appl. Pharmacol.*, 1978, **43**, 351–360.

100. J. B. Campbell, D. E. Woolley, V. K. Vijayan *et al.*, 'Morphometric effects of postnatal lead exposure on hippocampal development of the 15-day-old rat.' *Brain Res.*, 1982, **255**, 595–612.

101. J. B. Kawamoto, S. R. Overmann, D. E. Woolley *et al.*, 'Morphometric effects of preweaning lead exposure on the hippocampal formation of adult rats.' *Neurotoxicology*, 1984, **255**, 125–148.

102. D. P. Alfano, J. C. LeBoutillier and T. L. Petit, 'Hippocampal mossy fiber pathway development in normal and postnatally lead-exposed rats.' *Exp. Neurol.*, 1982, **75**, 308–319.

103. C. M. Regan, 'Neural cell adhesion molecules, neuronal development, and lead toxicity.' *Neurotoxicology*, 1993, **14**, 69–74.

104. L. Slomianka, J. Rungby, J. J. West *et al.*, 'Dose-dependent bimodal effect of low-level lead exposure on the developing hippocampal region of the rat: a volumetric study.' *Neurotoxicology*, 1989, **10**, 177–190.

105. L. Jablonska, M. Walski and U. Rafalowska, 'Lead as an inductor of some morphological and functional changes in synaptosomes from rat brain.' *Cell. Mol. Neurobiol.*, 1994, **14**, 701–709.

106. M. Kern, T. Audesirk and G. Audesirk, 'Effects of inorganic lead on the differentiation and growth of cortical neurons in culture.' *Neurotoxicology*, 1993, **14**, 319–327.

107. T. Sutula, X. X. He, J. Cavazos *et al.*, 'Synaptic reorganization in the hippocampus induced by abnormal functional activity.' *Science*, 1988, **239**, 1147–1150.

11.29
Neurotoxicology of Metals

LOUIS W. CHANG and GRACE LIEJUN GUO
University of Arkansas for Medical Sciences, Little Rock, AR, USA

11.29.1 INTRODUCTION

A large number of elements in nature can be classified as metals. Many of these, such as copper, iron, and zinc, are important to life while others, for example mercury, lead, and cadmium, have adverse effects on the biological system and may be referred to as toxic metals.

The most noticeable neurotoxic metals are mercury, lead, arsenic, cadmium, aluminum, and manganese. Some metals exist as organo-metals such as methylmercury, alkyl leads, triethyltin, and trimethyltin which are also potent neurotoxicants.

The neurotoxicity of mercury, lead, cadmium, aluminum, manganese, organoleads, and organotins have been reviewed extensively by Chang and Verity,[1] Cory-Slechta and Pounds,[2] Hastings,[3] Lukiw and McLachlan,[4] Chu et al.,[5] and Chang,[6,7] respectively. Readers are encouraged to refer to these reviews if more detailed information is desired. Other publications[8,9] also represent sources for readers who are interested in metal toxicology (see also Chapters 2, 4, 13, and 28, this volume).

11.29.2 MERCURY

Mercury may exist as elemental (metallic) mercury, mercury vapor, inorganic mercury salts (mercurous or mercuric), and organic mercury compounds (aryl or alkyl mercury). Mercury vapor and alkyl mercuric compounds are found to be most neurotoxic.

11.29.2.1 Mercury Vapor

Elemental mercury ("quicksilver"), when ingested, is poorly absorbed from the gastrointestinal tract and poses few toxic consequences. Metallic mercury, however, is very volatile and vaporizes readily. Mercury vapor, when inhaled, is absorbed efficiently[10] and has a high affinity for the central nervous system (CNS). Most mercury is distributed to the gray matter of the cerebral and cerebellar cortices and to various brainstem nuclei.[10-13]

Mercury vapor poisoning is characterized by nonspecific symptoms such as fatigue, gingivitis, gastrointestinal disturbance, insomnia, shyness, increased excitability, loss of memory, personality changes, and depression. These syndromes are sometimes referred to as erethism or micromercuralism.[14,15] In more severe conditions, a fine intentional tremor involving fingers, tongue, eyelids, and lips usually follows. The tremors may develop into a generalized body tremor with spasms of the extremities.[16] Amyotrophic lateral sclerosis (ALS)-like symptoms have also been reported.[17]

11.29.2.2 Inorganic Mercury Salts

11.29.2.2.1 Mercurous salt

Mercurous mercury poisoning has been reported in children who used teething powder containing calomel.[18,19] Patients exhibited redness of the hands and feet ("Pink disease") with painful extremities (acrodynia).[20] Some patients also displayed photophobia, anorexia, profuse sweating, and insomnia.

11.29.2.2.2 Mercuric salt

Micromercuralism, erethism, tremor, and incoordination ("mad hatter" syndrome) similar to that observed in mercury vapor poisoning can also be observed in chronic exposures to inorganic mercuric salts such as mercuric oxide and mercuric nitrate.[16,21] Neuronal changes in the cerebellum and in the dorsal root ganglia have been described in rats exposed to mercuric chloride.[22]

11.29.2.3 Organomercury Compounds

11.29.2.3.1 Arylmercury and alkoxyalkylmercury

Arylmercury (e.g., phenylmercury) and alkoxyalkylmercury (e.g., methoxyethylmercury) are biologically unstable and are degraded rapidly, mainly in the liver, into inorganic mercury (Hg^{2+}).[23-25] Therefore, they exert toxic actions similar to those of inorganic mercuric salts.

Induction of ALS-like or motor neuron disease-like syndromes by inorganic mercury and by phenylmercuric compounds has been reported.[17,26-28] Observation by Arvidson[29] showing an accumulation of mercury (Hg^{2+}) in spinal and brainstem motor neurons following intramuscular mercury injection further supports this hypothesis. However, controversial findings and observations also exist. A distributional study by Gage and Swain[30] failed to demonstrate significant mercury in the brain and cord following systemic exposure to arylmercury. Other investigators also failed to find any pathological lesions in the CNS in animals or humans exposed to phenylmercury.[16,21,31-37] In view of such controversial observations, further investigation is needed to clarify this phenomenon.

11.29.2.3.2 Alkylmercury

The best known neurotoxic examples of alkylmercury are methylmercury and ethylmercury; both of these are short-chain organomercuric compounds. Methylmercury was found to be associated with the massive outbreak of poisonings in Japan ("Minamata disease")[38,39] and in Iraq in the 1970s.[40-43]

The most consistent pathological lesions induced by methylmercury were found in the calcarine cortices (visual cortices), dorsal root ganglia, and cerebellum.[39,44,45] This topographical distribution of lesions correlates well with the primary neurological signs and symptoms (constriction of visual field, sensory disturbance, and cerebellar ataxia) observed in the patients with Minamata disease.

In addition to the CNS lesions, Chang and Hartmann first demonstrated the extensive damage of dorsal root ganglion neurons and fibers in rats after exposure to methylmercury.[22] These observations were later confirmed by other investigators.[46,47]

The pattern of cerebellar lesion may serve as a characteristic diagnostic criterion for methylmercury poisoning. Cerebellar granule cell lesion acquires a characteristic cell loss at the depth of the sulci with proliferation of Bergmann's glial fibers. Widespread destruction of the granule cells eventually occurs in prolonged intoxication. Most of the Purkinje neurons, however, are spared.

11.29.2.4 Mechanism of Actions for Mercury Neurotoxicity

The mechanistic basis for mercury neurotoxicity is complex and multifactorial. The major pathways which may involve in the pathogenesis of methylmercury neurotoxicity include: (i) inhibition of protein and macromolecular synthesis, (ii) mitochondrial dysfunction, (iii) defective calcium and ion flux, (iv) abnormal neurotransmitter homeostasis, (v) oxidative injury, (vi) cytoskeletal (microtubule) disaggregation, and (vii) modified post-translation phosphorylation. Detailed discussion of these aspects have been well summarized and presented in reviews by Chang and Verity[1] and Verity.[48] Readers are encouraged to seek more complete information from these reviews.

11.29.3 LEAD

Lead, like mercury, is one of the most insidious neurotoxicants known to man. Lead-related toxicity is most prominent in occupational exposure. Health effects on children, however, constitute the greatest concern in environmental exposure. Since the elimination of the use of lead (organolead) in gasoline, most of the health concerns related to lead are inorganic leads. Because of space limitations, only inorganic lead neurotoxicity will be reviewed in this chapter.

11.29.3.1 Inorganic Lead

In the past, inorganic lead poisoning was a serious concern in industrial and in occupational situations (e.g., in smelting processes or in paint strippings). Because of preventive legislation in industrial and occupational health, lead poisoning is now mainly a nonindustrial problem. The greatest concerns are (i) exhaust fumes from automobiles still using gasoline containing lead, (ii) water supplies via pipes containing lead, (iii) paint containing lead in old buildings or on toys that children may chew on, and (iv) industrial release exhausts containing lead to air.

11.29.3.1.1 Acute lead encephalopathy

Lead acts directly on the cerebral vasculatures inducing blood–brain barrier dysfunction and cerebral edema.[49,50] The clinical symptoms are consistent with those of acute cerebral edema, namely headaches, dizziness, papilledema, convulsions, obnubilation, and coma.[51–53]

Autopsy examination of the brain reveals cerebral swelling with flattening of the convolutions and compression of the sulci.[53] Congestion of the vessels and petechial hemorrhages in various areas of the brain are also evident.

Microscopically, the pathological features are capillary congestion, thickening of the walls of the veins, and edema of the endothelium. Hyalinization of the arteriole walls, endothelial proliferation, and adventitial fibrosis are occasional findings.[49]

The edema is more prominent in the white matter than in the gray matter. Foci of softening and edematous necrosis may also be found.[51] Glial proliferation in the cortex and Ammon's horn are prominent.[54]

11.29.3.1.2 Chronic lead intoxication

In chronic lead encephalopathy, the gross alterations of the brain are more subtle than those observed in acute situations. Chronic state exhibits a predilection for mesodermal reactions, including proliferation of vascular walls, endothelial hyperplasia, adventitial fibrosis, and secondary deposit of calcium. Glial reaction is even more extensive than that observed in acute conditions.

11.29.3.1.3 Developmental lead neurotoxicity

The developing nervous system is found to be very vulnerable to the toxicity of lead and constitutes the greatest concern in lead exposure. Significant neurobehavioral changes and learning disability have been reported in animals and children following lead exposure during development. Reviews in these aspects have been published.[2,55–57] The neuropathology in developmental lead neurotoxicity is primarily in three brain areas: cerebral cortex, cerebellum, and hippocampus.

(i) Cerebral cortex

Cerebral cortical edema is the most prominent finding. This edematous condition is found mainly in the white matter of the corpus striatum. Such neuropathological changes were only observed in very young animals.[58] A delayed synaptogenesis and cerebral cortical development have also been reported by McCauley *et al.*[59] An impaired desialylation of the D2-CAM and N-CAM proteins was suggested as the mechanism on lead-induced inhibition of neural cell acquisition and reduced synaptic elaboration.[60–64]

(ii) Cerebellum

Edema and focal hemorrhages in the gray matter were observed. Purkinje cell loss and gitter cell (macrophages) proliferation are prominent.[2,55] Hypoplasia, regional atrophy, and degeneration of the white matter also occur.

(iii) Hippocampus

Lead exposure during the developmental period has significant effects in the hippocampal weight and size.[65,66] Lead also decreases the size of the mossy fiber zone, density of the mossy fiber synapses, and granule cell layer. Loss of dendritic spine and reduction of dendritic branching are also observed in the pyramidal and granule cells.[65–68]

The changes on the mossy fiber zone, granule cell layer, and dendritic developments, together with changes in neurotransmitters,[2,55] may be the bases of the various neurobehavioral changes in animals and children exposed to lead.

11.29.3.1.4 *Other neuropathological involvements*

(i) Peripheral nerve changes

Peripheral neuropathy is one of the common neuropathological involvements exhibited by lead poisoned individuals. The effect is predominantly on large myelinated nerve fibers, and motor nerve fibers are especially affected.

The most characteristic change is segmental demyelination where degeneration is usually limited to a single internodal segment, with normal myelin above and below.[69,70] Axonal degeneration, especially the distal parts of the nerves, also occurs. These changes were more frequent in large diameter axons and can be traced to the motor endplates.

(ii) Spinal cord changes

Changes in the spinal cord in chronic lead poisoning have also been reported.[71,72] The involvement of the cord can mimic that in motor neuron disease or in ALS.

11.29.3.2 Mechanistic Considerations for Lead Neurotoxicity

The mechanism of actions for lead neurotoxicity are multifaceted and complex. The influences of lead on neurotransmitters and on ion channels have been reviewed in great detail by Mailman *et al.*[73] and by Narahashi,[74] respectively. Bimolecular studies also revealed the impact of lead on gene expression, signal transduction, and the calcium messenger system.[2] A summary on the neurotoxicity and on mechanistic actions of lead was presented in excellent reviews by Cory-Slechta and Pounds[2] and by Chang.[75] Interested readers are encouraged to seek further information from these reviews (see also Chapter 28, this volume).

11.29.4 ARSENIC

Arsenical compounds were used as war gases (chlorophenol-dichloroarsine or lewisite). In acute poisoning, involvement of the nervous system is indicated by headache, vomiting, confusion, epileptic convulsions, and coma. In the CNS, petechial hemorrhages are the main pathological features.[76] Focal perivascular demyelination and necrosis in cerebral cortex and white matter also occurred.[77,78]

In chronic exposures, peripheral polyneuropathy is the predominant neurotoxic consequence. A syndrome known as Ronnskar disease has been associated with chronic arsenic poisoning.[79] Three phases of the disease may be identified: an initial phase of weaknesses, anorexia, and vomiting, followed by a phase of cutaneous pigmentation and hyperkeratosis. The final stage is represented by peripheral neuropathy characterized by paraesthesias of lower and upper extremities, extensor palsy, distal weakness, and wasting of forearms and legs, especially smaller muscle of hands and feet. Myelin and axonal degeneration resemble those in distal axonopathy. Sensory fibers appear to be more affected than motor fibers. Involvement of cranial nerves, posterior root ganglia, and anterior horn neurons have also been reported.[78]

The toxicity of arsenic may be the result of its action on the tricarboxylic acid cycle at the level of pyruvate metabolism by inactivating the SH-rich cofactor thionic acid. Succinate is sometimes found in urine of patients with arsenic poisoning. Therefore, arsenic may also react with the SH-groups in the enzyme succinate dehydrogenase.

11.29.5 CADMIUM

Headache, vertigo, sleep disturbance, tremor, sweating, and sensory disturbance are associated with cadmium poisoning. Inhalation and oral ingestion are the most common routes of entry of cadmium (Cd) into the body. Once absorbed, Cd binds to plasma proteins, primarily albumin and β_2-microglobulin.[80–83] The biological half-life of Cd is about 200 days in rat and 16–33 years in humans.[83]

Newborn rats are more susceptible to neurotoxicity from Cd than adult rats. The less well-developed blood–brain carrier in the newborn may contribute to this difference.[84–86] Increased levels of Cd in the CNS of young animals may also result from the decreased CSF bulk flow, differences in plasma protein binding, or lesser amounts of glial cells and metallothionein at the circumventricular area of the younger animals.[87,88]

Cd competes with certain essential metals such as zinc (Zn) and calcium (Ca) at metalloenzymes, proteins, and ion channels.[81–83] Cd can produce symptoms mimicking those of Zn-deficiency and Zn is known to modulate acute Cd toxicity.[81–84] Cd inhibits Ca-ATPases, binds calmodulin, and inhibits mitochondrial Ca transport.[81,83,89–91] Brain Zn–Ca superoxide dismutase is reduced in rats upon exposure to Cd and lipid peroxidation in brain is increased by Cd.[92–94]

In 1967, Gabbiani *et al.*[95] first described hemorrhagic lesions in the dorsal root and Gasserian ganglia of rats exposed to Cd. Neuronal necrosis in some of the ganglia was observed. Similar pathological lesions were seen in guinea pigs, hamsters, and mice. Vascular changes involving arterioles, capillaries, and venules were also reported.[84,96] Cd does not cross the blood–brain barrier readily. The high ganglionic concentration of Cd may be related to the highly fenestrated and permeable ganglionic vessels. This condition may contribute to the hemorrhagic and neuronal changes in peripheral ganglia while the CNS of the adult animals remains relatively intact.

Some of the neuronal changes in the ganglia may occur as secondary changes to hemorrhagic and vascular lesions.[96] Neuronal changes are also demonstrated in the cerebellum during early developmental life of the animals exposed to Cd.[97] While granule cells of the cerebellum appear to be most vulnerable, Purkinje cell damage was also evident.

Cd exposure generally increases the levels of dopamine and norepinephrine.[83,92,98] Acetylcholine and serotonin, on the other hand, can be increased or decreased by Cd pending on the duration of exposure, dose, and other factors.[98–100] Cd also alters synaptic transmission.[89] Evolved transmitter release is reduced by Cd. This reduction probably results from blockage of presynaptic Ca channels. This effect can be reversed by increasing extracellular Ca. Cd may also alter synaptic transmission by inhibition of mitochondrial metabolism, Na-K ATPases, acetylcholinesterase, and monoamine oxidase, or by increasing membrane rigidity.[81,83,90,101]

The activities of calmodulin-dependent enzymes, such as cyclic nucleotide phosphodiesterase and synaptic Ca-ATPase in the cerebral cortex, are reduced by Cd.[91] These inhibitions probably result from interaction of Cd with calmodulin.

One of the prominent pathological effects of Cd is alterations in the vasculature. Acute Cd exposure induces vacuolation, endothelial gaps, and increased permeability of the cerebral blood vessels.[84,102] Widespread endothelial vacuolation and hemorrhage have been observed in newborns after Cd exposure.[84] This phenomenon reduces with age of the animals. In adults, however, endothelia in the sensory ganglia are vulnerable to Cd toxicity.[84] Choroid plexus accumulates high levels of toxic metals.[85,86,103,104] Damages of the choroid plexus by Cd have also been reported.[86]

Astrocytes have a high uptake of metals entering the CNS.[105] These cells are rich in metallothionein (MT)[87,105] which probably provides Zn to Zn-dependent enzymes in the glial cells. It has been postulated that damaged glial cells will release glutamate or induce lipid peroxidation which have an adverse effect on the neuronal function.[105]

Behavioral alterations are the most sensitive indicators of Cd neurotoxicity in animals.[3] The behavioral alterations may vary with age at exposure. Decreases in performance and activity occurred after *in vitro* exposure while increased locomotor activity occurred after neonatal exposure. Detailed behavioral neurotoxicology of Cd has been reviewed by Hastings[3] and is not further discussed here.

11.29.6 ALUMINUM

It is now well accepted that aluminum is a neurotoxic metal inducing significant biochemical and cytoskeletal neuropathology. The topography and extent of neuropathological degeneration is influenced by the aluminum compound involved, the route of administration, the duration of exposure, and the animal species involved.

Under acute exposure conditions, New Zealand white rabbits develop a fulminant encephalomyelopathy marked by extensive intraneuronal and dendritic accumulations of neurofilaments (NF) (see Chapter 5, this volume).[106] Motor neurons, anterior thalamic nuclei, and neurons of the parasubiculum appear to be most sensitive to such pathological changes. Oculomotor nuclei and Purkinje cells are usually not involved.[106] The neuropathology is reflected clinically by generalized motor slowing with a loss of toxic mobility, followed by quadriparesis, spasticity, convulsions, and coma.

In *in vitro* situations, neuroblastoma cells develop intraneuronal argentophilic NF upon exposure to aluminum.[107–109] However, a significant increase in cAMP and choline acetyltransferase was observed in the absence of alterations in adenylate cyclase and in NF phosphorylation.[110] This observation suggests that the increase in cAMP activity in *in vitro* model of aluminum neurotoxicity may not be mediated via activation of G-protein for adenylate cyclase. Cortical and dorsal root ganglion (DRG) neurons are found to be relatively insensitive to aluminum toxicity.[111–113] In contrast, monolayer motor neuron cultures are extremely sensitive to aluminum toxicity.[114]

At molecular level, aluminum alters DNA transcription by inducing chromatin condensation and reduces the amount of transcribable DNA.[115–117] Binding to chromatin, aluminum inhibits nuclear binding of corticosterone receptor complexes, chromosomal puffing, and chromatin-dependent RNA synthesis.[118,119] Based on various experiments, Strong and co-workers[120–123] concluded that aluminum induces the formation of a pool of phosphatase-resistant, highly phosphorylated NF. It is further proposed that NF aggregation induced by aluminum in New Zealand white rabbits reflects perturbation in the post-translational processing of NF. Through both phosphatase resistance and phosphatase inhibition,

large numbers of highly phosphorylated NF, which is resistant to proteolysis, is produced.

In chronic exposures, the New Zealand white rabbits develop NF inclusions in spinal motor neurons and selected brainstem nuclei.[124] In monkeys, the pathological lesion is seen primarily as anterior horn cell chromatolysis with argentophilic bodies and axonal and dendritic swellings with accumulation of NF.[125,126]

It must be pointed out that aluminum neurotoxicity may occur without the presence of NF aggregation. This is observed especially in rats, a species that does not develop significant cytoskeletal pathology, but rather a concomitant fulminant neurobehavioral deficit. In this situation, the impact on Ca-mediated pathways and phosphorylation is probably the mechanistic path.[127–134]

11.29.7 MANGANESE

Unlike many toxic metals such as lead, mercury, and cadmium, manganese (Mn) is a normal biological constituent and an essential trace element. Under physiological conditions, Mn is fairly uniformly distributed in the brain.[135] In toxic overexposure conditions, high concentration of Mn was found to accumulate in the basal ganglia.[136]

In humans, acute Mn intoxication induces irritability, speech disturbance, compulsive actions, and hallucinations. In chronic situations, the neurological signs of manganism resemble Parkinson's syndrome and dystonia.[5] Neuropathological changes are found mainly in the basal ganglia with neuronal degeneration in the putamen, caudate, and globus pallidus.[137] Decreased dopamine concentrations in the striatum are found in both humans and experimental animals.[137–140] Chronic occupational exposure to Mn produces a slowly progressive and irreversible extrapyramidal movement disorder.[141,142] Typical pathology is basal ganglia atrophy characterized by neuronal loss and gliosis with involvement of the striatum.

In monkeys intoxicated by Mn, neurochemical analysis showed that the Mn levels were highest in the pallidum and putamen with a reduction of dopamine and DOPAC, but not homovanillic acid, in the lentiform nuclei.[136,143,144] It is found that Mn had the greatest toxic effect on dopaminergic neurons in these brain regions.

Dopaminergic and GABAergic neurons were found to uptake and transport Mn ions in an anterograde, but not retrograde, direction.[145] It has been shown that Mn is predominantly distributed to area of the brain with high levels of nonheme iron,[146] including the caudate-putamen, globus pallidus, substantia nigra, and subthalamic nucleus.

The toxicity of Mn may be related to the biochemical transformation of Mn^{+2} to Mn^{+3} with autoxidation of dopamine to semiquinone and the production of free radicals.[147–149] Mn neurotoxicity may also result from an impairment of cellular antioxidant system with a decreased GSH level and GSH peroxidase activity.[148–151]

Mn also accumulates in mitochondria.[152] Such accumulation of Mn in mitochondria will also interfere with mitochondrial functions such as increased activity of mitochondrial cytochrome P450[153] and disruption of mitochondrial calcium regulation with inhibition of oxidative phosphorylation[154] leading to excessive amounts of extracellular glutamate, increased NMDA receptor stimulation, and eventually excitotoxic damage.

11.29.8 ALKYLTINS

While inorganic tin is not known to be neurotoxic, organic tin compounds, especially triethyl- and trimethyltin, are potent neurotoxicants capable of inducing characteristic neural lesions in the mammalian CNS.

11.29.8.1 Triethyltin

The most important human episode of human triethyltin (TET) toxicity occurred in France in the 1950s. A medication, Stalinon, was inadvertently contaminated with 10% TET. Patients showed various neurological problems, including persistent headache, vertigo, visual disturbances, abdominal pain, psychiatric disturbances, muscular weakness, EEG changes, increased cerebral spinal fluid (CSF) pressure, and convulsion.[155] Some patients progressed to develop a flaccid type of paraplegia, sensory loss, absence of reflexes, coma, and death. Autopsies revealed severe edema in the white matter of the brain and spinal cord.[156–158]

Animal studies confirmed that massive cerebral edema, confined to the white matter of the CNS, is the primary and characteristic lesion induced by TET.[159–165] Electron microscopic examination revealed that the edema is intramyelinic, with splitting the myelin sheath at the interperiod line to form fluid-filled vacuoles.[165–168] The edematous effect apparently is quite specific in the CNS with little or no effect on the peripheral nerves.[167]

The precise mechanism of actions of TET on the central myelin is still not clear. Various investigations suggest that these actions would include: (i) lipid peroxidation on the myelin membranes as a result of free radical formation;[169,170] (ii) an inhibition on mitochondrial phosphorylation and ATP-production;[171–173] and (iii) a reduction of ATPase, 5-nucleotidase, and phosphodiesterase in various brain regions[174] which would contribute to meylinic edema. A more detailed discussion on these aspects has been presented in separate reviews.[6,175]

11.29.8.2 Trimethyltin

Fortemps and co-workers[176–178] described several cases of accidental human exposures to trimethyltin (TMT). The patients suffered mental confusion, headaches, seizures, and psychic disturbances. Some of these patients also displayed a wide range of psychomotor symptoms including personality changes, irritability, memory deficits, insomnia, aggressiveness, headaches, tremors, convulsion, and changes of libido. However, no histopathological information on the CNS in these patients was available.

TMT also shows prominent and characteristic neurotoxic effects on animal models. The behavioral changes in rats exposed to TMT include aggression, hyperirritability, tremor, spontaneous seizures, hyperreactivity, and changes in schedule-controlled behavior.[179–183] Rats showed selective sensitivity to TMT in areas of the limbic system, including the entorhinal cortex and the hippocampus.[184]

Mice were found to be more sensitive to TMT toxicity than rats.[184] Aside from lesions in the hippocampus, pathological changes were also found in the brainstem and spinal cord.[185,186] Within the limbic system, mice showed lesion involvement primarily in the fascia dentate (granule cells) with little involvement of the hippocampal Ammon's horn (pyramidal neurons) and the entorhinal cortex. Rats, on the other hand, showed more prominent involvement in the pyramidal neurons of the Ammon's horn with less involvement in the fascia dentate as well as considerable pathology at the entorhinal cortex.[187]

Subsequent investigations with neonatal rats,[188,189] further demonstrated that the vulnerability of the Ammon's horn to TMT was associated closely with and heavily dependent upon the functional maturity and integrity of the neurons and the neuronal circuitry in the hippocampal formation.

Chang[190] proposed that the unique pattern of neuronal damage in the limbic system induced by TMT is related to hyperexcitation of neuron groups along the neural circuitry of the limbic system. This hypothesis was well supported by various biochemical studies demonstrating changes in the GABA and glutamate systems by TMT.[191,192] Biochemical studies revealed a reduction in glutamate and GABA uptake and synthesis[193–198] with an increased synaptic release of glutamate in the hippocampus. A study by Andersson *et al.*[198] demonstrated that TMT exposure leads to a regionally selective loss of NMDA and KA receptors in the limbic system. These changes suggest that there may be an increased extracellular level and over-stimulation of excitatory amino acid receptors. The release of glutamate, together with a depletion of hippocampal zinc[199] and a decrease in inhibitory synaptic functions[200] with damage of dentate basket cells by TMT[201] and an altered GABA compartmentalization in the hippocampus,[198] will promote neuronal hyperexcitation and neuronal damage.

The toxicopathological mechanism of TET and TMT neurotoxicity are complex and have been reviewed by Chang.[6]

11.29.9 REFERENCES

1. L. W. Chang and M. A. Verity, in 'Handbook of Neurotoxicology,' eds. L. W. Chang and R. S. Dyer, Dekker, New York, 1995, pp. 31–60.
2. D. A. Cory-Slechta and J. Pounds, in 'Handbook of Neurotoxicology,' eds. L. W. Chang and R. S. Dyer, Dekker, New York, 1995, pp. 61–90.
3. L. Hastings, in 'Handbook of Neurotoxicology,' eds. L. W. Chang and R. S. Dyer, Dekker, New York, 1995, pp. 171–212.
4. W. J. Lukiw and D. R. McLachlan, in 'Handbook of Neurotoxicology,' eds. L. W. Chang and R. S. Dyer, Dekker, New York, 1995, pp. 105–143.
5. N. S. Chu, F. H. Hochberg, D. B. Calne *et al.*, in 'Handbook of Neurotoxicology,' eds. L. W. Chang and R. S. Dyer, Dekker, New York, 1995, pp. 91–104.
6. L. W. Chang, in 'Handbook of Neurotoxicology,' eds. L. W. Chang and R. S. Dyer, Dekker, New York, 1995, pp. 143–170.
7. L. W. Chang and R. S. Dyer (eds.), 'Handbook of Neurotoxicology,' Dekker, New York, 1995.
8. L. W. Chang and L. Magos (eds.), 'Toxicology of Metals,' CRC Press, Boca Raton, FL, 1996.
9. M. Yasui, M. J. Strong, K. Ota *et al.* (eds.), 'Mineral and Metal Neurotoxicology,' CRC Press, Boca Raton, FL, 1996.
10. M. H. Berlin, G. F. Nordberg and F. Serenius, 'On the site and mechanism of mercury vapor resorption in the lung. A study in the guinea pig using mercuric nitrate Hg 203.' *Arch. Environ. Health*, 1969, **18**, 42–50.
11. G. F. Nordberg and F. Serenius, 'Distribution of inorganic mercury in the guinea pig brain.' *Acta Pharmacol. Toxicol.*, 1969, **27**, 269–283.
12. N. Takahata, H. Hayashi, B. Watanabe *et al.*, 'Accumulation of mercury in the brains of two autopsy cases with chronic inorganic mercury poisoning.' *Folia Psychiatr. Neurol. Jpn.*, 1970, **24**, 59–69.
13. M. Berlin, J. Carlsen and T. Norseth, 'Dose-dependence of methylmercury metabolism. A study of distribution: biotransformation and excretion in the squirrel monkey.' *Arch. Environ. Health*, 1975, **30**, 307–313.
14. I. M. Trachtenberg, *Zdorv'ja. Kiev*, 1969, pp. 292–301 (in Russian: translation available through EPA).
15. L. Friberg and J. Vostal (eds.), 'Mercury in the Environment,' CRC Press, Boca Raton, FL, 1972.
16. W. Stopford, in 'The Biogeochemistry of Mercury in the Environment,' ed. J. O. Nriagu, Elsevier/North Holland, Amsterdam, 1979, pp. 367–397.
17. F. Q. Vroom and M. Greer, 'Mercury vapour intoxication.' *Brain*, 1972, **95**, 305–318.
18. H. Swift, *Australian Med. Cong.*, Trans. 10th Session, Auckland, New Zealand, 1914.
19. J. Warkany and D. M. Hubbard, *J. Pediatr.*, 1953, **42**, 365–369.
20. D. Cheek, in 'Brennemann's Practice of Pediatrics,' Harper & Row, Hagerstown, New York, 1980, pp. 110–124.
21. WHO, 'Environmental Health Criteria I. Mercury,' World Health Organization, Geneva, 1976.
22. L. W. Chang and H. A. Hartmann, 'Ultrastructural studies of the nervous system after mercury intoxication. II. Pathological changes in the nerve fibers.' *Acta Neuropathol. (Berl.)*, 1972, **20**, 316–334.
23. J. W. Daniel, J. C. Gage and P. A. Lefevre, 'The metabolism of methoxyethylmercury salts.' *Biochem. J.*, 1971, **121**, 411–415.
24. J. C. Gage, 'Mechanisms for the biodegradation of organic mercury compounds: the action of ascorbate and of soluble proteins.' *Toxicol. Appl. Pharmacol.*, 1975, **32**, 225–238.
25. R. P. Beliles, in 'Toxicology—The Basic Sciences of Poisons,' eds. L. T. Casarett and Doull, Macmillan, New York, 1975, pp. 454–502.
26. I. Brown, *A.M.A. Arch. Neurol. Psych.*, 1954, pp. 674–681.
27. A. Kantarjian, *Neurology*, 1964, **15**, 639–644.
28. C. R. Adams, D. K. Ziegler and J. T. Lin, 'Mercury intoxication simulating amyotrophic lateral sclerosis.' *JAMA*, 1983, **250**, 642–643.
29. B. Arvidson, 'Inorganic mercury is transported from muscular nerve terminals to spinal and brainstem motoneurons.' *Muscle Nerve*, 1992, **15**, 1089–1094.
30. J. C. Gage and A. A. B. Swain, *Biochem. Pharmacol.*, 1961, **8**, 77 (Abst. No. 250).
31. L. F. Goldwater, in 'Mercury. Mercurials, and Mercaptans,' ed. C. C. Thomas, Springfield, IL, 1963, pp. 56–67.
32. A. C. Ladd, L. J. Goldwater and M. B. Jacobs, *Arch. Environ. Health*, 1964, **9**, 43–52.
33. R. D. Currier and A. F. Haerer, 'Amyotrophic lateral sclerosis and metallic toxins.' *Arch. Environ. Health*, 1968, **17**, 712–719.
34. M. C. Roberts, A. A. Seawright and J. C. Ng, 'Chronic phenylmercuric acetate toxicity in a horse.' *Vet. Hum. Toxicol.*, 1979, **21**, 321–327.
35. S. Conradi, D. Ronnevi and F. Norris, in 'Human Motor Neuron Diseases,' ed. L. P. Rowland, Raven Press, New York, 1982, pp. 201–231.
36. P. S. Spencer and H. H. Schaumberg, in 'Human Motor Neuron Diseases,' ed. L. P. Rowland, Raven Press, New York, 1982, pp. 249–266.
37. R. Yanagihara, in 'Human Motor Neuron Diseases,' ed. L. P. Rowland, Raven Press, New York, 1982, pp. 233–247.
38. T. Takeuchi, in 'Minamata Disease,' ed. M. Kutsuma, Study Group of Minamata Disease, Kumamoto University, Japan, 1968, pp. 141–228.
39. T. Takeuchi, in 'Neurotoxicology,' eds. L. Roizin, H. Shiraki and N. Grcevic, Raven Press, New York, 1977, vol. 1, pp. 235–246.

40. F. Bakir, S. F. Damluji, L. Amin-Zaki *et al.*, 'Methyl-mercury poisoning in Iraq.' *Science*, 1973, **181**, 230–241.

41. L. Amin-Zaki, S. Elhassani, M. A. Majeed *et al.*, 'Intra-uterine methylmercury poisoning in Iraq.' *Pediatrics*, 1974, **54**, 587–595.

42. L. Amin-Zaki, S. Elhassani, M. A. Majeed *et al.*, 'Perinatal methylmercury poisoning in Iraq.' *Am. J. Dis. Child.*, 1976, **130**, 1070–1076.

43. L. Amin-Zaki, M. A. Majeed, T. W. Clarkson *et al.*, 'Methylmercury poisoning in Iraqi children: clinical observations over two years.' *Br. Med. J.*, 1978, **1**, 613–616.

44. L. W. Chang, in 'Biogeochemistry of Mercury,' ed. J. O. Nriagu, Elsevier, New York, 1979, 519–580.

45. L. W. Chang, in 'Experimental and Clinical Neuro-toxicology,' eds. P. S. Spencer and H. H. Schaumberg, Williams & Wilkins, Baltimore, MD, 1980, pp. 508–526.

46. S. P. Herman, R. Klein, F. A. Talley *et al.*, 'An ultrastructural study of methylmercury-induced primary sensory neuropathy in the rat.' *Lab Invest.*, 1973, **28**, 104–118.

47. J. M. Jacobs, N. Carmichael and J. B. Cavanagh, 'Ultrastructural changes in the nervous system of rabbits poisoned with methyl mercury.' *Toxicol. Appl. Pharmacol.*, 1977, **39**, 249–261.

48. M. A. Verity, in 'Mineral and Metal Neurotoxicol-ogy,' eds. M. Yasui, M. J. Strong, K. Ota *et al.*, CRC Press, Boca Raton, FL, 1996, pp. 159–168.

49. M. Meyer, in 'Neuropathology,' ed. J. G. Greenfield, Edward Arnold, London, 1958, pp. 1401–1427.

50. A. Pentschow, in, 'Handbuch der Speziellen Patholo-gischen Anatomic und Histologie,' Springer-Verlag, Berlin, 1958, vol. 13, Pt. 2B, pp. 175–184.

51. H. B. Marsden and V. K. Wilson, *Br. Med. J.*, 1955, **1**, 324–336.

52. N. Popoff, S. Winberg and S. Feigin, *Neurology*, 1963, **13**, 101–107.

53. J. F. Smith, R. L. McLaurin, J. B. Nichols *et al.*, *Brain*, 1960, **83**, 411–420.

54. K. A. Koryyey, *St. Dtsch. Ztschr. Nervenheilk*, 1931, **122**, 18–25.

55. D. Cory-Slechta, in 'Toxicology of Metals,' ed. L. W. Chang, CRC Press, Boca Raton, FL, 1996, pp. 537–560.

56. D. Rice and E. Silbergeld, in 'Toxicology of Metals,' ed. L. W. Chang, CRC Press, Boca Raton, FL, 1996, pp. 659–676.

57. H. L. Needleman, in 'Toxicology of Metals,' ed. L. W. Chang, CRC Press, Boca Raton, FL, 1996, pp. 405–4l4.

58. G. W. Goldstein, A. K. Asbury and I. Diamond, 'Pathogenesis of lead encephalopathy. Uptake of lead and reaction of brain capillaries.' *Arch. Neurol.*, 1974, **13**, 382–389.

59. P. T. McCauley, R. J. Bull, A. P. Tonti *et al.*, 'The effect of prenatal and postnatal lead exposure on neonatal synaptogenesis in rat cerebral cortex.' *J. Toxicol. Environ. Health*, 1982, **10**, 639–651.

60. F. Hasan, G. R. Cookman, G. J. Keane *et al.*, 'The effect of low level lead exposure on the postnatal structuring of the rat cerebellum.' *Neurotoxicol. Teratol.*, 1989, **11**, 433–440.

61. C. M. Regan, 'Lead-impaired neurodevelopment. Mechanisms and threshold values in the rodent.' *Neurotoxicol. Teratol.*, 1989, **11**, 533–537.

62. C. M. Regan, G. R. Cookman, G. J. Keane *et al.*, in 'Lead Exposure and Child Development,' eds. M. A. Smith, L. D. Grant and A. I. Sors, Kluwer Academic Publishers, Dordrecht, 1989, pp. 440–452.

63. R. A. Clasen, J. F. Hartmann, A. J. Starr *et al.*, 'Electron microscopic and chemical studies of the

64. vascular changes and edema of lead encephalopathy. A comparative study of the human and experimental disease.' *Am. J. Pathol.*, 1974, **74**, 215–240.

64. A. Hirano and J. A. Kochen, in 'Progress in Neuropathology,' ed. H. M. Zimmerman, Grune & Stratton, New York, 1979, vol. 3, ch. 2.

65. T. L. Petit, D. P. Alfano and J. C. LeBoutillier, 'Early lead exposure and the hippocampus: a review and recent advances.' *Neurotoxicology*, 1983, **41**, 79-84.

66. J. B. Campbell, D. E. Woolley, V. K. Vijayan *et al.*, 'Morphometric effects of postnatal lead exposure on hippocampal development of the 15-day-old rat.' *Dev. Brain Res.*, 1982, **255**, 595–612.

67. D. P. Aliano, J. C. Le Boutillier and T. L. Petit, 'Hippocampal mossy fiber pathway development in normal and postnatally lead-exposed rats.' *Exp. Neurol.*, 1982, **75**, 308–319.

68. L. Slomianka, J. Rungby, M. J.West *et al.*, 'Dose-dependent bimodal effect of low-level lead exposure on the developing hippocampal region of the rat: a volu-metric study.' *Neurotoxicology*, 1989, **10**, 177–190.

69. P. J. Dyck, A. J. Windebank, P. A. Low *et al.*, 'Blood nerve barrier in rat and cellular mechanisms of lead-induced segmental demyelination.' *J. Neuropathol. Exp. Neurol.*, 1980, **39**, 700–712.

70. C. Winder and I. Kitchen, 'Lead neurotoxicity: a review of the biochemical, neurochemical and drug induced behavioural evidence.' *Prog. Neurobiol.*, 1984, **22**, 59–87.

71. J. A. Simpson, D. A. Seaton and J. F. Adams, *Neurol. Neurosurg. Psychiat.*, 1964, **27**, 536–541.

72. B. Livesley and C. E. Scissons, 'Chronic lead intoxi-cation mimicking motor neurone disease.' *Br. Med. J.*, 1968, **4**, 387–388.

73. R. B. Mailman, M. Maylebon and C. P. Lawler, in 'Toxicology of Metals,' ed. L. W. Chang, CRC Press, Boca Raton, FL, 1996, pp. 627–638.

74. T. Narahashi, in 'Toxicology of Metals,' ed. L. W. Chang, CRC Press, Boca Raton, FL, 1996, pp. 677–698.

75. L. W. Chang, in 'Toxicology of Metals,' ed. L. W. Chang, CRC Press, Boca Raton, FL, 1996, pp. 511–536.

76. L. Osetowska, in 'Pathology of the Nervous System,' ed. J. Minckler, McGraw-Hill, New York, 1971, vol. 2, pp. 1644–1651.

77. A. Meyer, in 'Neuropahtology,' eds. W. Blackwood, W. H. McMenemey, A. Meyer *et al.*, Year Book Medical Publishers, Chicago, IL, 1963, pp. 261–262.

78. T. Smith, in 'Greenfield's Neuropathology,' eds. W. Blackwood and J. A. Corsellis, Year Book Medical Publishers, Chicago, IL, 1976, pp. 152–153.

79. M. J. Politis, H. H. Schaumburg, and P. S. Spencer, in 'Experimental and Clinical Neurotoxicology,' eds. P. S. Spencer and H. H. Schaumburg, Wllliams & Wilkins, Baltimore, MD, 1980, pp. 613–630.

80. V. A. Murphy, in 'Mineral and Metal Neuro-toxicology,' CRC Press, Boca Raton, FL, 1996, pp. 229–242.

81. J. A. Babitch, in 'Metal Neurotoxicity,' ed. J. C. Bondy and K. N. Prasad, CRC Press, Boca Raton, FL, 1988, pp. 141–166.

82. J. Taylor and F. K. Ennever, Toxicological Profile for Cadmium, Atlanta, Agency for Toxic Sub-stances and Disease Registry, US Public Health Service, 1993.

83. R. Nath, R. Prasad, V. K. Palinal *et al.*, 'Molecular basis of cadmium toxicity.' [Review] *Prog. Food Nutr. Sci.*, 1984, **8**, 109–163.

84. B. Arvidson, in 'Neurobiology of the Trace Elements,' I. E. Dreosti and R. M. Smith, Humana Press, Clifton Heights, NJ, 1983, vol. 2, pp. 51–78.

85. B. Arvidson, 'Autoradiographic localization of

cadmium in the rat brain.' *Neurotoxicology* 1986, **7**, 89–96.

86. A. A. Valois and W. S. Webster, 'The choroid plexus and cerebral vasculature as target sites for cadmium following acute exposure in neonatal and adult mice: an autoradiographic and gamma counting study.' *Toxicology*, 1987, **46**, 43–55.

87. N. Nishimura, H. Nishimura, A. Ghaffar *et al.*, 'Localization of metallothionein in the brain of rat and mouse.' *J. Histochem. Cytochem.*, 1992, **40**, 309–315.

88. W. M. Pardridge, in 'Peptide Drug Delivery to the Brain,' Raven Press, New York, 1991, pp. 52–98.

89. W. D. Atchison, 'Effects of neurotoxicants on synaptic transmission: lessons learned from electrophysiological studies.' *Neurotoxicol. Teratol.*, 1988, **10**, 393–416.

90. K. I. Ahammadsahib, R. R. Jinna and D. Desaiah, 'Protection against cadmium toxicity and enzyme inhibition by dithiothreitol.' *Cell Biochem. Func.*, 1989, **7**, 185–192.

91. P. J. S. Vig and R. Nath, '*In vivo* effects of cadmium on calmodulin and calmodulin regulated enzymes in rat brain.' *Biochem. Int.*, 1991, **23**, 927–934.

92. R. Pal, R. Nath and K. D. Gill, 'Lipid peroxidation and antioxidant defense enzymes in various regions of adult rat brain after co-exposure to cadmium and ethanol.' *Pharmacol. Toxicol.*, **73**, 209–214.

93. R. Pal, R. Nath and K. D. Gill, 'Influence of ethanol on cadmium accumulation and its impact on lipid peroxidation and membrane bound functional enzymes (Na$^+$, K($+$)-ATPase and acetylcholinesterase) in various regions of adult rat brain.' *Neurochem. Int.*, 1993, **23**, 451–458.

94. J. R. Prohaska, 'Functions of trace elements in brain metabolism.' *Physiol. Rev.*, 1987, **67**, 858–901.

95. G. Gabbiani, A. Gregory and D. Baic, 'Cadmium-induced selective lesions of sensory ganglia.' *J. Neuropathol. Exp. Neurol.*, 1967, **26**, 498–506.

96. C. V. Nolan and Z. A. Shaikh, 'The vascular endothelium as a target tissue in acute cadmium toxicity.' *Life Sci.*, 1986, **39**, 1403–1409.

97. G. Gabbiani, D. Baic and C. Deziel, 'Toxicity of cadmium for the central nervous system.' *Exp. Neurol.*, 1967, **18**, 154–160.

98. G. S. Shukla and R. L. Singhal, 'The present status of biological effects of toxic metals in the environment: lead, cadmium, and manganese.' [Review] *Can. J. Physiol. Pharmacol.*, 1984, **62**, 1015–1031.

99. K. P. Das, P. C. Das, S. Dasgupta *et al.*, 'Serotonergic-cholinergic neurotransmitters' function in brain during cadmium exposure in protein restricted rat.' *Biol. Trace Elem. Res.*, 1993, **36**, 119–127.

100. A. Gupta and R. C. Murthy, 'Comparative neurotoxicity of cadmium in growing and adult rats after repeated administration.' *Biochem. Int.*, 1990, **21**, 97–105.

101. A. Gupta and S. Chandra, 'Gestational cadmium exposure and brain development: a biochemical study.' *Ind. Health*, 1991, **29**, 65–71.

102. K. L. Wong and C. D. Klaassen, 'Neurotoxic effects of cadmium in young rats.' *Toxicol. Appl. Pharmacol.*, 1982, **63**, 330–337.

103. W. Zheng, D. F. Perry, D. L. Nelson *et al.*, 'Choroid plexus protects cerebrospinal fluid against toxic metals.' *FASEB J.*, 1991, **5**, 2188–2193.

104. A. A. Valois and W. S. Webster, 'The choroid plexus as a target site for cadmium toxicity following chronic exposure in the adult mouse: an ultrastructural study.' *Toxicology*, 1989, **55**, 193–205.

105. J. K. Young, 'Glial metallothionein.' *Biol. Signals*, 1994, **3**, 169–175.

106. M. J. Strong, R. Yanagihara and A. V. Wolff, in 'Amyotrophic Lateral Sclerosis: New Advances in Toxicology and Epidemiology,' eds. F. C. Rose and F. H. Norris, Smith-Gordon, London, 1990, pp. 157–173.

107. T. B. Shea, J. F. Clarke and T. R. Wheelock, 'Aluminum salts induce the accumulation of neurofilaments in perikarya of NB2a/d1 neuroblastoma.' *Brain Res.*, 1989, **492**, 53–64.

108. C. A. Miller and E. M. Levine, 'Effects of aluminum salts on cultured neuroblastoma cells.' *J. Neurochem.*, 1974, **22**, 751–758.

109. G. M. Cole, K. Wu and P. S. Timiras, *Int. J. Dev. Neurosci.*, 1985, **3**, 23–32.

110. H. S. Singer, C. D.Searles and I. H. Hahn *et al.*, 'The effect of aluminum on markers for synaptic neurotransmission, cyclic AMP, and neurofilaments in a neuroblastoma × glioma hybridoma (NG108-15).' *Brain Res.*, 1990, **528**, 73–79.

111. M. R. Gilbert, B. L. Harding and D. L. Price, *J. Neuropathol. Exp. Neurol.*, 1989, **48**, 362.

112. D. Langui, B. H. Anderson, J. P. Brion *et al.*, 'Effects of aluminium chloride on cultured cells from rat brain hemispheres.' *Brain Res.*, 1988, **438**, 67–76.

113. D. Langui, A. Probst and B. Anderton *et al.*, 'Aluminium-induced tangles in cultured rat neurones. Enhanced effect of aluminium by addition of maltol.' *Acta Neuropathol. (Berl.)*, 1990, **80**, 649–655.

114. M. J. Strong and R. M. Garruto, 'Neuron-specific thresholds of aluminum toxicity *in vitro*. A comparative analysis of dissociated fetal rabbit hippocampal and motor neuron-enriched cultures.' *Lab. Invest.*, 1991, **65**, 243–249.

115. W. J. Lukiw, T. P. A. Kruck and D. R. McLachlan, 'Alterations in human linker histone-DNA binding in the presence of aluminum salts *in vitro* and in Alzheimer's disease.' *Neurotoxicology*, 1987, **8**, 291–302.

116. W. J. Lukiw, T. P. A. Kruck and D. R. C. McLachlan, 'Linker histone–DNA complexes: enhanced stability in the presence of aluminum lactate and implications for Alzheimer's disease.' *FEBS Lett.*, 1989, **253**, 59–62.

117. W. J. Lukiw, P. St. George-Hyslop and D. R. Crapper McLachlan, in 'Regulation of Gene Expression and Brain Function,' ed. P. J. Harrison, Springer-Verlag, New York, 1994, pp. 31–45.

118. C. Sanderson, D. R. C. McLachlan and U. De Boni, 'Inhibition of corticosterone binding *in vitro*, in rabbit hippocampus, by chromatin bound aluminum.' *Acta Neuropathol. (Berl.)*, 1982, **57**, 249–254.

119. C. Sanderson, D. R. C. McLachlan and U. De Boni, *Can. J. Genet. Cytol.*, 1982, **24**, 27–36.

120. M. J. Strong, *Rev. Biochem. Toxicol.*, 1994, **11**, 75–115.

121. M. J. Strong and D. M. Jakowec,. '200 kDa and 160 kDa neurofilament protein phosphatase resistance following *in vivo* aluminum chloride exposure.' *Neurotoxicology*, 1994, **15**, 799–808.

122. J. Savory, M. M. Herman and J. C. Hundley *et al.*, 'Quantitative studies on aluminum deposition and its effects on neurofilament protein expression and phosphorylation, following the intraventricular administration of aluminum maltolate to adult rabbits.' *Neurotoxicology*, 1993, **14**, 9–12.

123. H. Yamamoto, Y. Saitoh, S. Yasugawa *et al.*, 'Dephosphorylation of tau factor by protein phosphatase 2A in synaptosomal cytosol fractions, and inhibition by aluminum.' *J. Neurochem.*, 1990, **55**, 683–690.

124. M. J. Strong, A. V. Wolff, I. Wakayama *et al.*, 'Aluminum-induced chronic myelopathy in rabbits.' *Neurotoxicology*, 1991, **12**, 9–22.

125. L. Yano, S. Yoshida and Y. Uebayashi, *Biomed. Res.*, 1989, **10**, 33–41.

126. R. M. Garruto, S. K. Shankar and R. Yanagihara *et al.*, 'Low-calcium, high-aluminum diet-induced motor

neuron pathology in cynomolgus monkeys.' *Acta Neuropathol. (Berl.)*, 1989, **78**, 210–219.

127. G. V. W. Johnson, K. W. Cogdill and R. S. Jope, 'Oral aluminum alters *in vitro* protein phosphorylation and kinase activities in rat brain.' *Neurobiol. Aging*, 1990, **11**, 209–216.

128. M. Cochran, D. C. Elliott, P. Brennan *et al.*, 'Inhibition of protein kinase C activation by low concentrations of aluminium.' *Clin. Chim. Acta*, 1990, **194**, 167–172.

129. M. L. Koenig and R. S. Jope, 'Aluminum inhibits the fast phase of voltage-dependent calcium influx into synaptosomes.' *J. Neurochem.*, 1987, **49**, 316–320.

130. N. Siegel and A. Haug, 'Aluminum interaction with calmodulin. Evidence for altered structure and function from optical and enzymatic studies.' *Biochim. Biophys. Acta*, 1983, **744**, 36–45.

131. C. G. Suhayada and A. Haug, 'Organic acids prevent aluminum-induced conformational changes in calmodulin.' *Biochem. Biophys. Res. Commun.*, 1984, **119**, 376–381.

132. B. J. Farnell, D. R. Crapper McLachlan and K. Baimbridge *et al.*, 'Calcium metabolism in aluminum encephalopathy.' *Exp. Neurol.*, 1985, **88**, 68–83.

133. M. Deleers, J. P. Servais and E. Wulfert, 'Micromolar concentrations of A13+ induce phase separation, aggregation and dye release in phosphatidylserine-containing lipid vesicles.' *Biochim. Biophys. Acta*, 1985, **813**, 195–200.

134. S. D. Provan and R. A. Yokel, 'Aluminum inhibits glutamate release from transverse rat hippocampal slices: role of G proteins, Ca channels and protein kinase C.' *Neurotoxicology*, 1992, **13**, 413–420.

135. F. C. Wedler and R. B. Denman, 'Glutamine synthetase: the major Mn(II) enzyme in mammalian brain.' *Curr. Top. Cell. Regul.*, 1984, **24**, 153–169.

136. H. Eriksson, K. Magiste and L. O. Plantin *et al.*, 'Effects of manganese oxide on monkeys as revealed by a combined neurochemical, histological and neurophysiological evaluation.' *Arch. Toxicol.*, 1987, **61**, 46–52.

137. M. Yamada, S. Ohno and I. Okayasu *et al.*, 'Chronic manganese poisoning: a neuropathological study with determination of manganese distribution in the brain.' *Acta Neuropathol. (Berlin)*, 1986, **70**, 273–278.

138. G. Gianutsos and M. T. Murray, 'Alterations in brain dopamine and GABA following inorganic or organic manganese administration.' *Neurotoxicology*, 1982, **3**, 75–81.

139. E. D. Bird, A. H. Anton and B. Bullock, 'The effect of manganese inhalation on basal ganglia dopamine concentrations in Rhesus monkey.' *Neurotoxicology*, 1984, **5**, 59–65.

140. N. Autissier, L. Rochette and P. Dumas *et al.*, 'Dopamine and norepinephrine turnover in various regions of the rat brain after chronic manganese chloride administration.' *Toxicology*, 1982, **24**, 175–182.

141. C. C. Huang, C. S. Lu and N. S. Chu, 'Progression after chronic manganese exposure.' *Neurology*, 1993, **43**, 1479–1483.

142. N. S. Chu, F. H. Hochberg, D. B. Calne *et al.*, in 'Handbook of Neurotoxicology,' eds. L. W. Chang and R. S. Dyer, Dekker, New York, 1995, pp. 91–103.

143. H. Eriksson, P. G. Gillberg and S. M. Aquilonius *et al.*, 'Receptor alterations in manganese intoxicated monkeys.' *Arch. Toxicol.*, 1992, **66**, 359–364.

144. H. Eriksson, J. Tedroff and K. A. Thuomas, 'Manganese induced brain lesions in *Macaca fascicularis* as revealed by positron emission tomography and magnetic resonance imaging.' *Arch. Toxicol.*, 1992, **66**, 403–407.

145. W. N. Sloot and J. B. P. Gramsbergen, 'Axonal transport of manganese and its relevance to selective neurotoxicity in the rat basal ganglia.' *Brain Res.*, 1994, **657**, 124–132.

146. J. M. Hill and R. C. Switzer, III, 'The regional distribution and cellular localization of iron in the rat brain.' *Neuroscience*, 1984, **11**, 595–603.

147. J. Donaldson, D. McGregor and F. LaBella, 'Manganese neurotoxicity: a model for free radical mediated neurodegeneration?' *Can. J. Physiol. Pharmacol.*, 1982, **60**, 1398–1405.

148. D. G. Graham, 'Catecholamine toxicity: a proposal for the molecular pathogenesis of manganese neurotoxicity and Parkinson's disease.' *Neurotoxicology*, 1984, **5**, 83–95.

149. F. S. Archibald and C. Tyree, 'Manganese poisoning and the attack of trivalent manganese upon catecholamines.' *Arch. Biochem. Biophys.*, 1987, **256**, 638–650.

150. J. J. Liccione and M. D. Maines, 'Selective vulnerability of glutathione metabolism and cellular defense mechanisms in rat striatum to manganese.' *J. Pharmacol. Exp. Ther.*, 1988, **247**, 156–161.

151. J. Donaldson, 'The physiopathologic significance of manganese in brain: its relation to schizophrenia and neurodegenerative disorders.' *Neurotoxicology*, 1987, **8**, 451–462.

152. C. E. Gavin and T. E. Gunter, in 'Mineral and Metal Neurotoxicology,' eds. M. Yasui, M. Strong, K. Ota *et al.*, CRC Press, Boca Raton, FL, 1996, pp. 305–310.

153. J. J. Liccione and M. D. Maines, 'Manganese-mediated increase in the rat brain mitochondrial cytochrome P-450 and drug metabolism activity: susceptibility of the striatum.' *J. Pharmacol. Exp. Ther.*, 1989, **248**, 222–228.

154. C. E. Gavin, K. K. Gunter and T. E. Gunter, 'Mn^{2+} sequestration by mitochondria and inhibition of oxidative phosphorylation .'*Toxicol. Appl. Pharmacol.*, 1992, **115**, 1–5.

155. T. Alajouanine, L. Derobert and S. Thieffry, *Rev. Neurol.*, 1958, **98**, 85–96.

156. J. M. Barnes and H. B. Stoner, *Pharmacol. Rev.*, 1959, **11**, 211–231.

157. P. Cossa, Duplay, Fischgold *et al.*, *Rev. Neurologique*, 1958, **98**, 97–108.

158. H. B. Stoner, J. M. Barnes and J. I. Duff, *Br. J. Pharmacol.*, 1955, **10**, 16–24.

159. P. N. Magee, H. B. Stoner and J. M. Barnes, *J. Pathol. Bac.*, 1957, **73**, 102–124.

160. R. M. Torack, R. D. Terry and H. M. Zimmerman, *Am. J. Pathol.*, 1960, **36**, 273–288.

161. R. M. Torack, J. Gordon and J. Prokop, 'Pathobiology of acute triethyltin intoxication.' *Int. Rev. Neurobiol.*, 1970, **12**, 45–86.

162. G. R. Wenger, D. E. McMillan and L. W. Chang, 'Effects of triethyltin on responding of mice under a multiple schedule of reinforcement.' *Neurobehav. Toxicol. Teratol.*, 1986, **8**, 659–665.

163. D. E. McMillan, L. W. Chang, S. O. Ideumdia *et al.*, 'Effects of trimethyltin and triethyltin on lever pressing, water drinking and running in an activity wheel: associated neuropathology.' *Neurobehav. Toxicol. Teratol.*, 1986, **8**, 499–507.

164. L. W. Chang, in 'Structural and Functional Effects of Neurotoxicants: Organometals,' eds. H. A. Tilson and S. B. Sparber, Wiley, New York, 1987, pp. 82–116.

165. F. P. Aleu, R. Katzman and R. D. Terry, 'Fine structure and electrolyte analyses of cerebral edema induced by alkyl tin intoxication.' *J. Neuropathol. Exp. Neurol.*, 1963, **22**, 403–413.

166. A. Hirano, H. M. Zimmerman and S. Levine, 'Intramyelinic and extracellular spaces in triethyltin intoxication.' *J. Neuropathol. Exp. Neurol.*, 1968, **27**, 571–580.

167. D. I. Graham and N. K. Gonatas,. 'Triethyltin sulfate-induced splitting of peripheral myelin in rats.' *Lab. Invest.*, 1973, **29**, 628–632.

168. J. M. Jacobs, J. E. Cremer and I. B. Cavanagh, *Neuropathol. Appl. Neurobiol.*, 1977, **3**, 169–181.

169. R. A. Prough, M. A. Stalmach, P. Wiebkin *et al.*, 'The microsomal metabolism of the organometallic derivatives of the group-IV elements, germanium, tin and lead.' *Biochem. J.*, 1981, **196**, 763–770.

170. P. Wiebkin, R. A. Prough and J. W. Bridges, 'The metabolism and toxicity of some organotin compounds in isolated rat hepatocytes.' *Toxicol. Appl. Pharmacol.*, 1982, **62**, 409–420.

171. M. Stockdale, A. P. Dawson and M. J. Selwyn, 'Effects of trialkyltin and triphenyltin compounds on mitochondrial respiration.' *Eur. J. Biochem.*, 1970, **15**, 342–351.

172. M. S. Rose and W. N. Aldridge, 'Oxidative phosphorylation. The effect of anions on the inhibition by triethyltin of various mitochondrial functions, and the relationship between this inhibition and binding of triethyltin.' *Biochem. J.*, 1972, **127**, 51–59.

173. D. A. Kirschner and V. S. Sapirstein, 'Triethyl tin-induced myelin oedema: an intermediate swelling state detected by X-ray diffraction.' *J. Neurocytol.*, 1982, **11**, 559–569.

174. J. S. Wassenaar and A. M. Kroon, 'Effects of triethyltin on different ATPases, 5'-nucleotidase and phosphodiesterases in grey and white matter of rabbit brain and their relation with brain edema.' *Eur. Neurol.*, 1973, **10**, 349–370.

175. L. W. Chang, in 'Handbook of Neurotoxicology,' eds. L. W. Chang and R. S. Dyer, Dekker, New York, 1995, pp. 143–170.

176. E. Fortemps, G. Amand, A. Bombois *et al.*, 'Trimethyltin poisoning. Report of two cases.' *Int. Arch. Occup. Environ. Health*, 1978, **41**, 1–6.

177. W. D. Ross, E. A. Emmett, J. Steiner *et al.*, 'Neurotoxic effects of occupational exposure to organotins.' *Am. J. Psych.*, 1981, **138**, 1092–1095.

178. C. H. Rey, H. J. Reinecke and R. Besser,'Methyltin intoxication in six men; toxicologic and clinical aspects.' *Vet. Hum. Toxicol.*, 1984, **26**, 121–122.

179. A. W. Brown, W. N. Aldridge, B. W. Street *et al.*, 'The behavioral and neuropathologic sequelae of intoxication by trimethyltin compounds in the rat.' *Am. J. Pathol.*, 1979, **97**, 59–82.

180. G. R. Wenger, D. E. McMillan and L. W. Chang, 'Behavioral toxicology of acute trimethyltin exposure in the mouse.' *Neurobehav. Toxicol. Teratol.*, 1982, **4**, 157–161.

181. G. R. Wenger, D. E. McMillan and L. W. Chang, 'Behavioral effects of trimethyltin in two strains of mice. I. Spontaneous motor activity.' *Toxicol. Appl. Pharmacol.*, 1984, **73**, 78–88.

182. G. R. Wenger, D. E. McMillan and L. W. Chang, 'Behavioral effects of trimethyltin in two strains of mice. II. Multiple fixed ratio, fixed interval.' *Toxicol. Appl. Pharmacol.*, 1984, **73**, 89–96.

183. R. S. Dyer, W. F. Wonderlin and T. L. Deshields, 'Trimethyltin-induced changes in gross morphology of the hippocampus.' *Neurobehav. Toxicol. Teratol.*, 1982, **4**, 141–147.

184. L. W. Chang, G. R. Wenger, D. E. McMillan *et al.*, 'Species and strain comparison of acute neurotoxic effects of trimethyltin in mice and rats.' *Neurobehav. Toxicol. Teratol.*, 1983, **5**, 337–350.

185. L. W. Chang, T. M. Tiemeyer, G. R. Wenger *et al.*, 'Neuropathology of trimethyltin intoxication. III. Changes in the brain stem neurons.' *Environ. Res.*, 1983, **30**, 399–411.

186. L. W. Chang, G. R. Wenger and D. E. McMillan, 'Neuropathology of trimethyltin intoxication. IV. Changes in the spinal cord.' *Environ. Res.*, 1984, **34**, 123–134.

187. L. W. Chang and R. S. Dyer, 'A time-course study of trimethyltin induced neuropathology in rats.' *Neurobehav. Toxicol. Teratol.*, 1983, **5**, 443–459.

188. L. W. Chang, 'Hippocampal lesions induced by trimethyltin in the neonatal rat brain.' *Neurotoxicology*, 1984, **5**(2), 205–215.

189. L. W. Chang, 'Trimethyltin induced hippocampal lesions at various neonatal ages.' *Bull. Environ. Contam. Toxicol.*, 1984, **33**, 295–301.

190. L. W. Chang, 'Neuropathology of trimethyltin: a proposed pathogenetic mechanism.' *Fundam. Appl. Toxicol.*, 1986, **6**, 217–232.

191. S. V. Doctor, L. G. Costa, D. A. Kendall *et al.*, *Toxicologist*, 1982, **2**, 86 (Abstract).

192. S. V. Doctor, L. G. Costa, D. A. Kendall *et al.*, 'Trimethyltin inhibits uptake of neurotransmitters into mouse forebrain synaptosomes.' *Toxicology*, 1982, **25**, 213–221.

193. S. V. Doctor, L. G. Costa and D. A. Kendall *et al.*, 'Effect of trimethyltin on chemically-induced seizures.' *Toxicol. Lett.*, 1982, **13**, 217–227.

194. D. L. DeHaven, T. J. Walsh and R. B. Mailman, 'Effects of trimethyltin on dopaminergic and serotonergic function in the central nervous system.' *Toxicol. Appl. Pharmacol.*, 1985, **75**, 182–189.

195. R. B. Mailman, M. R. Krigman, G. D. Frye *et al.*, 'Effects of postnatal trimethyltin or triethyltin treatment on CNS catecholamine, GABA, and acetylcholine systems in the rat.' *J. Neurochem.*, 1983, **40**, 1423–1429.

196. L. U. Naalsund, C. N. Allen and F. Fonnum, 'Changes in neurobiological parameters in the hippocampus after exposure to trimethyltin.' *Neurotoxicology*, 1985, **6**, 145–158.

197. M. Patel, B. K. Ardelt, G. K. W. Yim *et al.*, 'Interaction of trimethyltin with hippocampal glutamate.' *Neuro-toxicology*, 1990, **11**, 601–608.

198. H. Andersson, A. C. Radlsater and Luthman, *J. Amino Acids*, 1995, **8**, 23–35.

199. L. W. Chang and R. S. Dyer, in 'Neurobiology of Zinc,' eds. C. Frederickson and G. Howell, Alan R. Liss, New York, 1984, pp. 275–290.

200. R. S. Dyer, W. F. Wonderlin, T. J. Walsh *et al.*, *Soc. Neurosci. Abstr.*, 1982, **8** (No. 23.7), 82.

201. L. W. Chang and R. S. Dyer, 'Early effects of trimethyltin on the dentate gyrus basket cells: a morphological study.' *J. Toxicol. Environ. Health*, 1985, **16**, 641–653.

11.30
Botanical Neurotoxins

MICHAEL A. PASS
University of Queensland, St. Lucia, Qld, Australia

11.30.1 INTRODUCTION

A wide range of compounds in plants are toxic to the nervous system of mammals. Their mechanisms of action vary from functional disturbances to necrosis of parts of the nervous system. In some instances the central nervous system is affected and in others the toxin exerts its effect on the peripheral nerves. Some other plant toxins, which do not have their primary effect on the nervous system, produce clinical syndromes which may be confused with neurotoxicoses. For instance, cardiac glycosides induce acute heart failure. The clinical signs include muscular weakness, trembling, and distress, signs that also occur with some acute neurointoxications. Hepatic encephalopathy is a neurological dysfunction secondary to liver disease which can be caused by plant toxins.

Botanical neurotoxins may be synthesized by the plant or may be a product of bacteria or fungi that infest the plant. Annual ryegrass

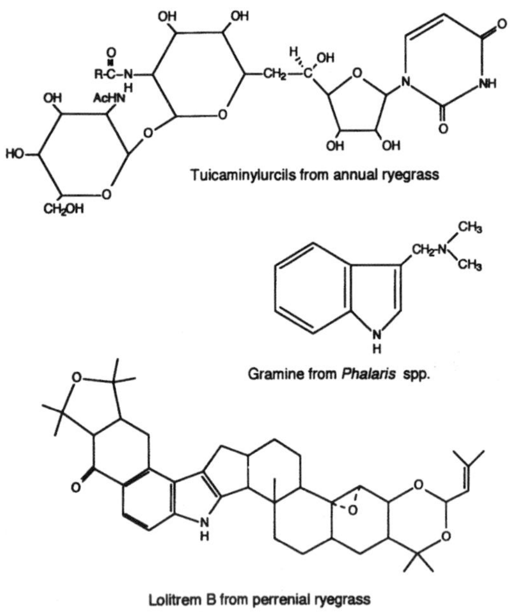

Tuicaminylurcils from annual ryegrass

Gramine from *Phalaris* spp.

Lolitrem B from perrenial ryegrass

Figure 1   Toxins causing staggers.

γ-coniceine   coniine

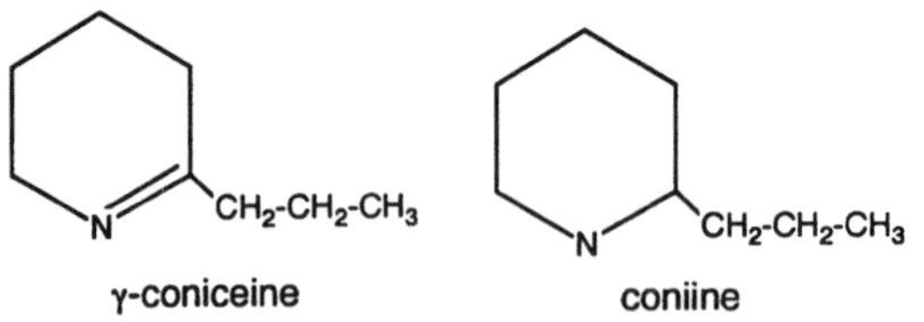

Figure 2   Conium alkaloids.

11.30.2   TOXINS ACTING AT SYNAPSES

11.30.2.1   Conium Alkaloids

Conium alkaloids including γ-coniceine and coniine (Figure 2) are the toxins in poison hemlock (*Conium maculatum*). Humans and grazing animals, especially cattle, have been poisoned by this plant. Many other species are also susceptible to poisoning.[13]

These alkaloids interfere with synaptic transmission at the neuromuscular junction of skeletal muscle and autonomic ganglia. Initially there is stimulation of ganglia and muscular excitation. This is followed by blockade of ganglionic transmission.[14] Clinically, the initial stimulatory action of these alkaloids is manifest as frequent urination and defecation, and by twitching.[13] Affected animals then become staggery and eventually recumbent due to flaccid paralysis. The pupils are dilated, respiration weak, and cardiovascular activity is depressed. Death occurs from respiratory paralysis. Convulsions are not usually a feature of poison hemlock intoxication.[12,13] The pharmacological activity of conium alkaloids[13] bears similarities to the action of nondepolarizing muscle relaxants such as succinylcholine.[15] Affected animals can be treated by removing as much toxin as possible from the stomach and administering tannins to inactivate any residual alkaloid in the gastrointestinal tract.[12,16]

toxicity is an example of neurotoxicity associated with a bacterium growing on a plant.

Many plant toxins are secondary compounds produced by metabolism in the plant.[1] A large number of secondary compounds are waste products of metabolism but their toxicity can be beneficial to plants as a means of protecting them from attack by herbivores. In particular, they play an important role in controlling herbivorous insects.[1] Their toxicity to mammals is probably more an incidental effect rather than an evolved protective mechanism.[1] For instance, the toxic protoalkaloids in Graminaceae which are responsible for phalaris staggers in sheep, act as feeding deterents for insects.[2]

There is no systematic relationship between the chemical structure of most plant toxins and the toxic syndromes that they induce. For example, diseases described as "staggers" occur in sheep grazing *Phalaris* spp., perennial ryegrass, and annual ryegrass but the toxins are structurally dissimilar (Figure 1). Thus, it is difficult to develop a systematic classification of toxic plant diseases based on the chemical nature of the toxins.

This chapter focuses on plant intoxications of economically important animals for which the toxin(s) has been identified and some information is known on the mechanism of action of the toxin. Other neurotoxic plants have been identified but little is known about the offending toxin. The reader is referred to several publications that identify many of these plants.[3–12]

11.30.2.2   Diterpene Alkaloids from Larkspur

Delphinium spp. contain a range of diterpene alkaloids[17,18] which act at the neuromuscular junction to cause paralysis of skeletal muscles. The most potent alkaloid appears to be methyllycaconitine (Figure 3).[17]

Many of these alkaloids cause flaccid paralysis by inhibiting synaptic transmission. Whether this is a competitive or noncompetitive blockade is unclear[17] although anticholinesterase inhibitors are recommended for treating poisoned animals.[12,19]

Cattle are most commonly poisoned by larkspurs and the syndrome is acute in onset. Clinical signs include hypersensitivity, muscle

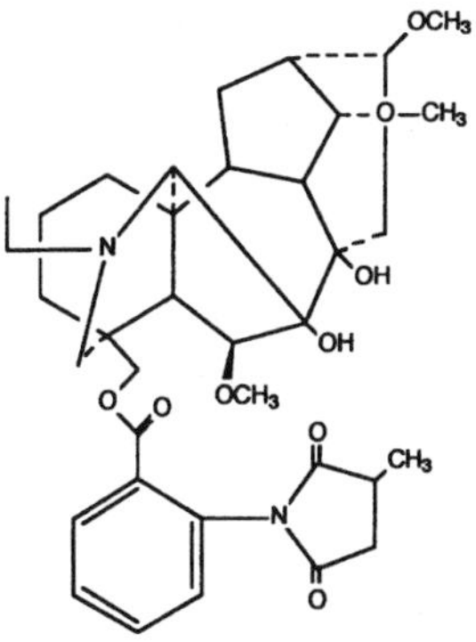

Figure 3 Methyllcaconitine.

tremors, abdominal pain, collapse, convulsions and death from respiratory paralysis.[20]

11.30.2.3 Nicotine Alkaloids

Nicotiana spp. including commercial tobacco (*Nicotiana tabacum*) and wild tobacco, contain a number of alkaloids with the most important being nicotine, anatabine, anabasine, and nornicotine.[21] Of these, nicotine (Figure 4) is the major neurotoxin while compounds such as anabasine are teratogenic.[22] In Australia, *Duboisia hopwoodii* also contains nicotine alkaloids and can poison livestock.[5]

Grazing animals are the most frequently affected group of animals and acute toxicity is most common in hungry animals which consume the relatively unpalatable plants.

Nicotine is an agonist of nicotinic cholinergic receptors. These receptors are located on ganglionic postsynaptic membranes of the sympathetic and parasympathetic nervous systems. Low concentrations of nicotine stimulate these receptors and high concentrations inhibit them.[16] Nicotinic receptors are also present on the postsynaptic membrane of neuromuscular junctions of skeletal muscle. Stimulation of nicotinic receptors in the central nervous system (CNS) causes the release of other neurotransmitters including acetylcholine, dopamine, noradrenaline, and serotonin.

The clinical signs of nicotine poisoning will vary depending on which nicotinic receptors are affected. The receptors on skeletal muscle cells are most readily stimulated causing muscle fasciculations, incoordinated movement and eventually paralysis. Effects on the autonomic nervous system cause dilated pupils, increased cardiovascular activity, and increased gastrointestinal motility and secretion. Animals often recover if left undisturbed.[5] Convulsions can occur in the most severely affected animals[16] presumably due to the effects on the CNS.

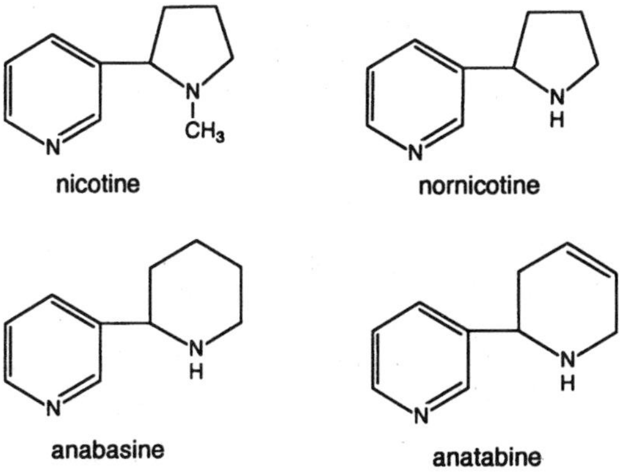

11.30.2.4 Tropane Alkaloids

This group of alkaloids includes atropine, scopolamine (hyoscine), and hyoscyamine (Figure 5). These compounds occur in plants such as *Atropa belladonna* (deadly nightshade), *Datura* spp. (thorn apples and jimson weed), *Duboisia* spp. (corkwoods), and *Brugmansia* spp. (Angel's trumpet). All mammals are susceptible to toxicity.

The toxins are competitive antagonists of acetylcholine at muscarinic receptors and the clinical signs of toxicity reflect this action. The signs include dilated pupils, reduced salivation and consequent dryness of the mouth, tachycardia, and reduced gastrointestinal motility. Signs of central neurotoxicity are evident in the most severely intoxicated animals due to inhibition of acetylcholine-mediated synaptic transmission in the CNS. Delerium, convulsions, and coma result from the effects on the CNS.[5]

Tropane alkaloid poisoning can be treated with acetylcholinesterase inhibitors such as physostigmine.[5] This enzyme is responsible for degrading acetylcholine within synapses. Inhibition of the enzyme raises the concentration of acetylcholine in the synaptic cleft leading to

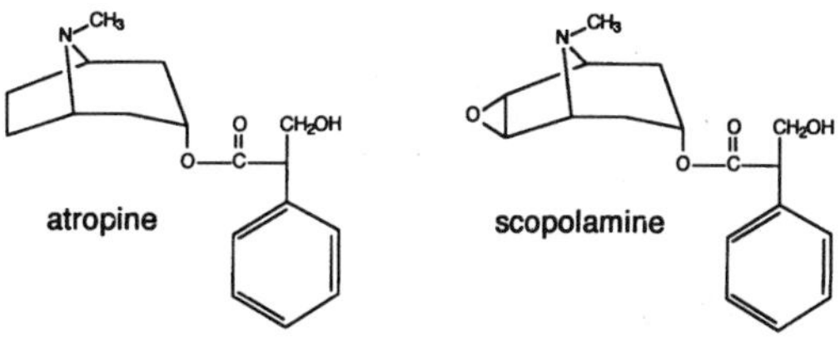

Figue 4 Nicotine alkaloids.

Figure 5 Tropane alkaloids.

competitive displacement of the tropane alkaloid from muscarinic receptors. This restores synaptic transmission.

Grazing animals are most likely to be intoxicated by tropane alkaloid-containing plants but other animals occasionally are poisoned. People who use these plants recreationally for their hallucinogenic properties can be poisoned, sometimes fatally.

11.30.3 TOXINS CAUSING DIRECT NEURONAL INJURY

11.30.3.1 Aliphatic Nitrotoxins

3-Nitropropionic acid (NPA) and its glucose esters, and miserotoxin (Figure 6) are important aliphatic nitrotoxins present in a variety of plants including *Astragalus* spp. and *Indigofera* spp.[23,24]

Miserotoxin is hydrolyzed by ruminal bacteria to 3-nitropropanol, the latter is then converted to NPA by hepatic alcohol dehydrogenase.[23,25] Thus, miserotoxin is toxic only to ruminants. Glucose esters of NPA can be hydrolyzed by mammalian esterases and are, therefore, potentially toxic to all mammals.[23]

Toxicity to ruminants can be acute or chronic.[26] The acute form is characterized by rapid onset of incoordination, distress, labored breathing, cyanosis, muscular weakness, and collapse with few distinctive pathological findings. The clinical signs of chronic intoxication are weight loss, poor hair coat, hindlimb paresis, knuckling of the fetlocks, poor exercise tolerance, and terminally, respiratory distress. In cattle, the main pathological findings are congestion of the liver, abomasal ulceration, emphysema, Wallerian degeneration of the spinal cord and peripheral nerves, and focal hemorrhages in the brain.[27] Chronic toxicity in rats and mice is associated with focal necrosis in the brain, particularly in the caudate putamen.[28–30]

The most important toxic action of NPA is to irreversibly inhibit the enzyme succinate dehydrogenase, a key enzyme in the Kreb's cycle, causing a reduction in cellular energy produc-

tion.[31,32] Two theories have been proposed to explain the focal brain lesions seen in laboratory animals. One suggests that excitotoxins acting in concert with depressed neuronal energy production is responsible.[28] The other suggests that the lesions may be a result of differences between brain regions in their ability to replenish succinate dehydrogenase damaged by NPA.[32,33]

11.30.3.2 Cyanide

Cyanogenic compounds occur in over 100 species of plants.[34] The HCN occurs as glycosides with about 55 known compounds.[34,35] Examples of cyanogenic glycosides are shown in Figure 7.

Acute cyanide poisoning causes cardiotoxicity. However, chronic consumption of cyanide containing compounds can result in a degenerative neuropathy in people and animals.[5,35] Cassava is the major cause of poisoning in people and occurs when cassava products are not properly prepared before being eaten. Cooking or soaking in water will detoxify or remove the cyanide.

Several plants have been implicated as poisoning animals (Table 1).[3,5,34] Ruminants are most susceptible to poisoning because enzymatic hydrolysis of the glycosides to release HCN occurs in the rumen.

For cyanogenic glycosides to induce toxicity, the HCN must be released and this occurs enzymatically. The initial step is hydrolysis of the glycosidic bond by β-glucosidase (Figure 8).

The HCN is then released from the aglycone by hydroxynitrile lyase (Figure 9).

These enzymes are present in some cyanogenic glycoside containing plants and the reactions can also be performed by ruminal microorganisms.[34] Release of HCN occurs when the green plant is crushed exposing the glycoside to the enzymes or when the glycoside is metabolized by ruminal microbes. It has been suggested that intracellular enzymes may also be able to release HCN and that this may be important for the development of the chronic

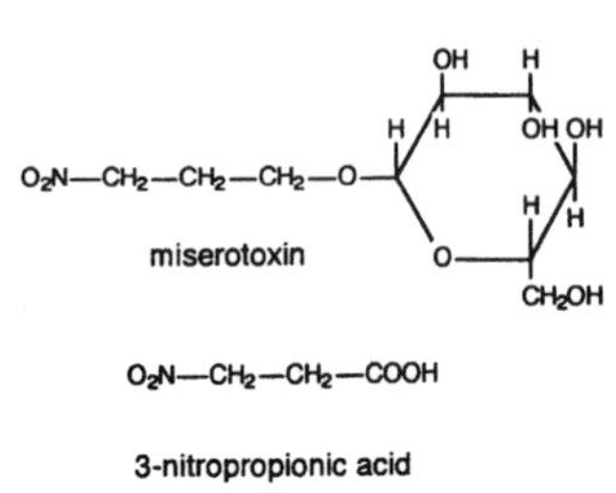

Figure 6 Aliphatic nitrotoxins.

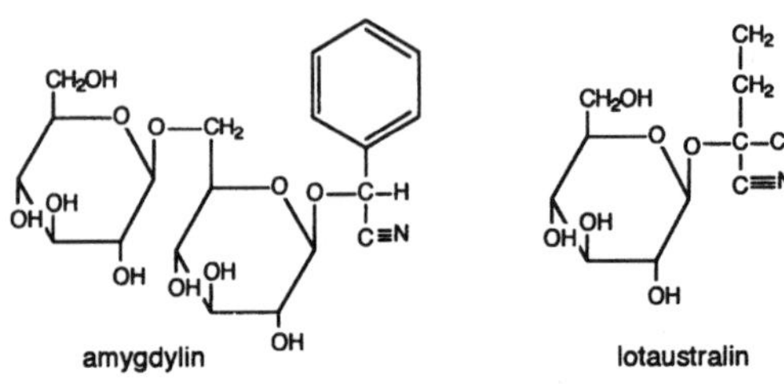

Figure 7 Cyanogenic glycosides.

Table 1 Cyanogenic glycoside containing plants poisonous to animals.

Plant	Glycoside
Rosaceous sp. (wild cherry, choke cherry)	Amygdalin
Mountain mahogany, apricot seeds, bitter almonds	Prunasin
Sambucus nigra (black elderberry), *Acacia cunninghamii*	Dhurrin
Taxus sp. (Canadian and Japanese yew), *Bambusa* sp. (bamboo)[a]	Taxiphyllin
Trifolium repens (white clover), *Linum* sp. (flax), *Lotus* sp., *Manihot* sp. (cassava), *Phaseolus lunatus* (lima bean)	Linamarin Lotaustralin
Triglochin sp. (arrow grass)	Triglochinin

Source: Conn.[34]
[a]Toxicity attributed to an alkaloid.

neurological disease.[34] Cyanide can be detoxified by mammalian cells by the following reaction:

$$CN^- + S_2O_3^{2-} \rightarrow SO_3^{2-} + SCN^-$$

cyanide   thiosulfate   sulfite   thiocyanate

HCN exerts its toxic action by interfering with cellular respiration through the inhibition of cytochrome oxidase.[35]

Conn summarized the factors which influence the toxicity of cyanogenic compounds.[34] "These include: the concentration of the compound(s) in the plant; the amount of the toxic plant ingested; the size and kind of animal involved; the type of food ingested simultaneously; the possibility of the plant enzymes remaining active in the digestive tract of the animal; and the ability of the animal to detoxify the HCN which it encounters."

The neurological syndrome caused by chronic exposure to HCN has been reported

in people, horses, cattle, and pigs.[5,36–38] Ataxia is associated with an axonal neuropathy.

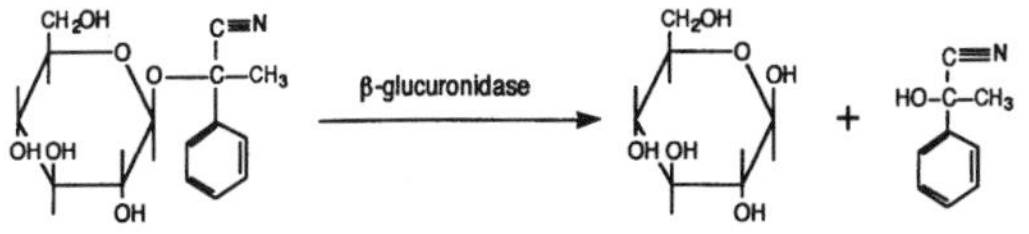

11.30.3.3 Cycad Toxins

Cycads of the genera *Cycas, Macrozamia, Lepidozamia,* and *Bowenia* have been recorded as causing toxicity in mammals. Neurotoxicity occurs most often in cattle and humans. Toxicity in cattle is manifest as hepatic and gastrointestinal disease, or as a neurological disease.[39,40] Humans develop a neurological syndrome consisting of amyotrophic lateral sclerosis, Parkinsonism, and dementia. It occurs most commonly in people of the Western Pacific region.[41,42] The hepatic and gastrointestinal syndrome in cattle is caused by the methyl-azoxymethanol glycosides cycasins and macrozamins[39,43] and will not be discussed further here.

Poisoning of humans occurs if they consume flour made from cycad seeds. There is a long latency of up to 10 years between consumption of the plant and onset of clinical signs. The disease is characterized by muscular weakness and atrophy progressing to spastic paralysis.[44] The Parkinsonism develops later and is evident as slow voluntary movements, tremors, rigid muscles, and dementia.[44]

The disease in cattle is characterized by hindleg ataxia.[39] The hindquarters sway laterally when the animal walks and the feet tend to knuckle over. Paralysis of the hindlimbs can occur in the most severely affected animals and complete recovery is unusual. Demyelination and axonal degeneration of the spinal cord and some peripheral nerves have been reported.[39,40]

The cause of the neurological syndrome has not been confirmed but α-amino-β-methyl-aminopropionic acid (BMAA) (Figure 10) has been suggested as being the toxin responsible for the human syndrome.[45] This is based on studies of the toxicity of BMAA in primates.[45] Although BMAA is neurotoxic, the quantities of it in plants that have intoxicated cattle are considered to be too low to induce toxicity.[43,46] The role of BMAA in the human disease has also been disputed.[47]

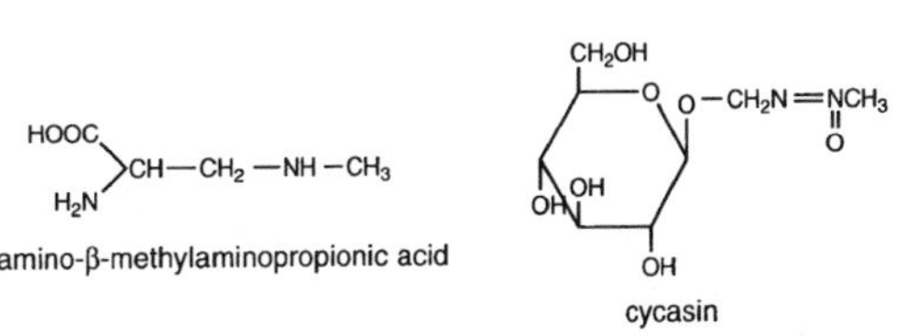

Figure 8 Hydrolysis of cyanogenic glycosides.

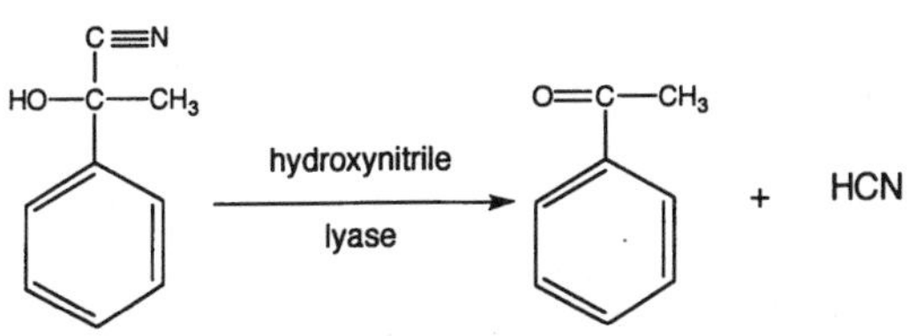

Figure 9 Metabolic release of hydrogen cyanide.

α-amino-β-methylaminopropionic acid

cycasin

Figure 10 Cycad toxins.

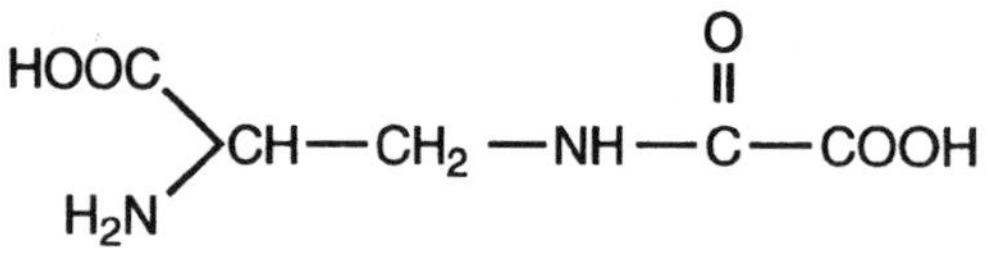

Figure 11 β-Oxalylamino-L-alanine.

Figure 13 Swainsonine.

Despite the doubts as to whether BMAA causes clinical disease, it is certainly neurotoxic under experimental conditions. Administration of BMAA to experimental animals causes damage to GABAergic neurones in the cerebellum.[43] It has been proposed that the amino acid acts as an excitotoxin on the NMDA receptor.[43]

An alternative hypothesis to explain cycad-associated neurotoxicity is that cycasin (Figure 10) forms DNA adducts which inhibit genomic expression and result in neuronal degeneration.[41] According to Spencer *et al.*,[41] this could explain the long latency of the disease.

11.30.3.4 Lathyrogenic Amino Acids

Neurolathyrism has been reported in humans, cattle, and horses that have consumed *Lathyrus* spp.[48] Chick pea (*Lathyrus. sativus*) is the plant most commonly implicated in the human syndrome. β-Oxalylamino-L-alanine (BOAA) (Figure 11) is considered to be the most likely cause of neurolathyrism.[41,49]

Lathyrism is manifest as a spastic paraplegia as a result of symmetrical degeneration of corticospinal tracts from the thoracic to the sacral regions of the spinal cord.[44] Clinical signs of intoxication have been described by Spencer *et al.*[49] and include muscular weakness, tremor, paresthesia, aching, and abnormalities of urination.

The toxicity of BOAA is related to its ability to stimulate the quisqualate receptor on neurones.[50] The natural agonist for this receptor is glutamate, an excitatory amino acid. Excessive stimulation of these receptors results in neuronal degeneration.

11.30.3.5 Stypandrol

Stypandrol (Figure 12) is the toxic agent in blindgrass (*Stypandra imbricata*), a poisonous

plant of Western Australia.[51] It is poisonous to grazing livestock and chickens.[52]

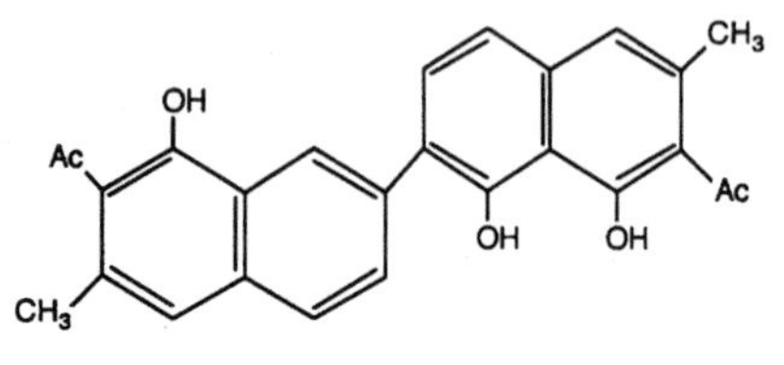

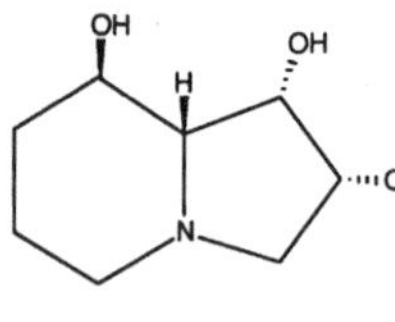

The clinical syndrome caused by this compound exhibits two phases. The acute phase is characterized by paresis of the hind limbs and is associated with disruption of the structure of myelin in the central nervous system.[53] The animal may die at this stage or progress to the second phase. The acute syndrome subsides and over the next 6–8 weeks the animal becomes blind. Blindness is due to Wallerian degeneration of the optic nerves and retinal deterioration.[52–54] It has been suggested that the neurological lesions are due to free radical damage to myelin proteins.[55]

11.30.3.6 Swainsonine

Swainsonine is an indolizidine alkaloid (Figure 13).[56] It is the cause of poisoning of livestock by *Swainsonia* spp. in Australia and the cause of "locoism" in North America. The latter is a syndrome in livestock resulting from the ingestion of certain *Astragalus* and *Oxytropis* spp.[57]

Affected animals lose weight or grow poorly, become depressed although demonstrate hyperexitability if stressed, and become ataxic.[58] Toxicity follows consumption of the plant for several weeks. It has long been held that animals become addicted to locoweeds but behavioural studies suggest that physiological and psychological addiction does not occur.[59]

Swainsonine poisoning is essentially a lysosomal mannoside storage disease.[60] The toxin inhibits lysosomal α-mannosidase[61] the enzyme responsible for hydrolyzing mannose-containing oligosaccharides. This results in accumulation of oligosaccharides in lysosomes including those of neurons. The toxin also inhibits Golgi mannosidase II[62] resulting in the inhibition of complex oligosaccharide synthesis and the synthesis of abnormal glycoproteins. Neuronal abnormalities occur as a consequence of the lysosomal storage disease and are probably the basis for the neurological syndrome.[63]

11.30.3.7 Tryptamine Alkaloids

Indole protoalkaloids such as gramine (Figure 14) are considered to be the cause of

Figure 12 Stypandrol.

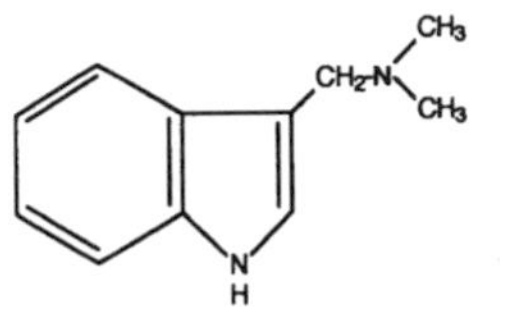

Figure 14 Gramine.

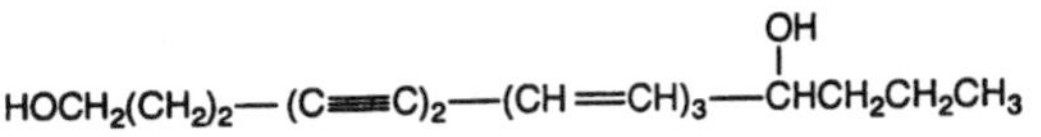

Figure 15 Cicutoxin.

the neurological syndrome known as "phalaris staggers" caused by *Phalaris* spp.[64-68]

Phalaris toxicity is manifest as sudden death or as a chronic neurological disease, phalaris staggers. The sudden death syndrome has two causes, cardiotoxicity and polioencephalomalacia.[69] Cardiotoxicity has been attributed to *N*-methyltyramine.[70] Polioencephalomalacia from phalaris intoxication is associated with edematous vacuolation in the gray matter of the cerebral cortex.[71]

Sheep and cattle are susceptible to developing the neurological syndrome of phalaris staggers although the two species differ in the clinical manifestations of the disease.[71] Sheep develop twitching of muscles, head shaking, and paresis leading to staggering. Cattle develop a mild ataxia but can become hyperexcited. Cattle often have difficulty chewing and swallowing with consequent weight loss. The neurological syndrome may develop several weeks after exposure to the plant.[68,72]

The most prominent pathological finding is accumulation of indole-like pigments in neurons of the brain and spinal cord.[65,67,73] It has been suggested that phalaris staggers results from a biochemical rather than an anatomical disturbance to neurons. The functional disturbance has been attributed to the action of the toxins on serotonergic receptors.[73]

Phalaris staggers can be prevented by dietary supplementation with cobalt.[71,74] This promotes ruminal detoxification of the phalaris toxins.

11.30.4 OTHER MECHANISMS

11.30.4.1 Cicutoxin

Cicutoxin (Figure 15) occurs in the water hemlocks which include *Cicuta maculata*, *C. douglasii* and *C. virosa*. Water hemlocks are considered to be the most poisonous plants in the USA. The roots are the most toxic part of the plant.[19] Hulbert and Oehme quote a report by Gress[75] that poisoning can occur by drinking water into which the plant has been crushed.[12]

Cicutoxin is described as being a "convulsive poison."[16] Humans have been intoxicated by

these plants.[7,76] Cattle and sheep are most frequently poisoned but other species such as pigs and ponies are also susceptible.[12,77,78] The plants induce vomiting which tends to protect monogastric animals such as pigs from intoxication.[12,79] Death can occur within minutes of consumption of the plant.[80] The clinical signs of poisoning include salivation, muscle twitching, and violent muscle spasms and convulsions.[81] Coma occurs and death appears to be a result of respiratory paralysis.[12]

11.30.4.2 Lolitrems

Perennial ryegrass staggers of sheep, cattle, horses, and deer[82,83] are caused by lolitrems, particularly lolitrem B (Figure 16). These compounds are synthesized by the endophytic fungus *Acremonium loliae* which grows on perrenial ryegrass (*Lolium perenne*).[84]

The clinical syndrome has been described as "a tremoring syndrome upon which may be superimposed head shaking, degrees of incoordination, abnormal staggering gait, eventual stumbling, and collapse to the ground sometimes to be followed by severe muscular spasms after which the sheep will regain its feet after a variable time and walk away."[82] Clinical signs are most evident in animals which are stimulated to move about. Death is uncommon and animals recover if removed from the toxic pasture.

No specific neuropathology has been identified although degeneration of cerebellar Purkinje cells has been observed.[85,86] These are thought to be secondary changes rather than a direct effect of lolitrems.[83]

Interestingly, the endophyte protects the ryegrass plants from infestation by Argentine stem weevils and eradication of the endophyte reduces the ryegrass pasture production.[83]

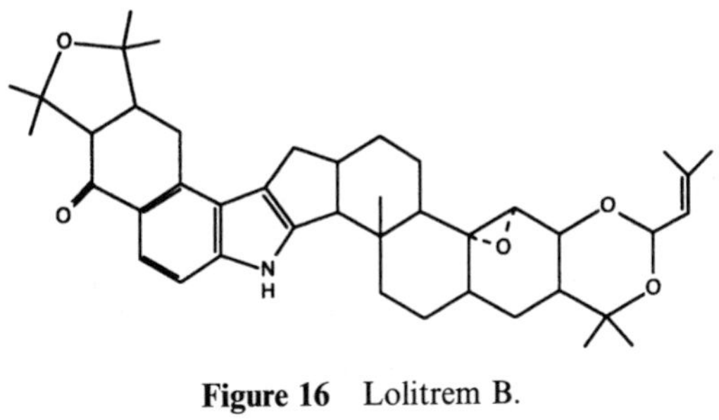

Figure 16 Lolitrem B.

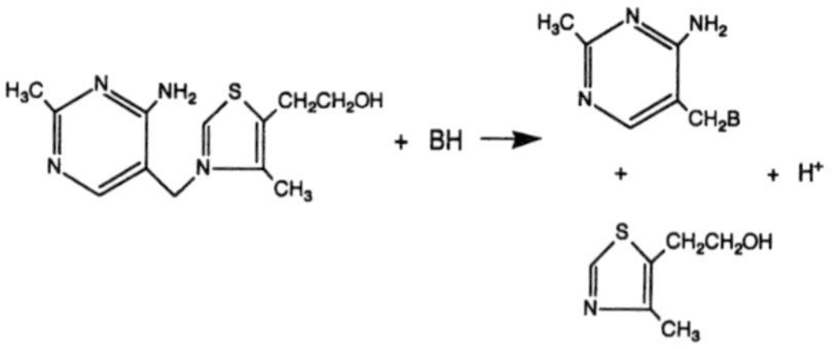

Figure 17 Metabolism of thiamine by thiaminase.

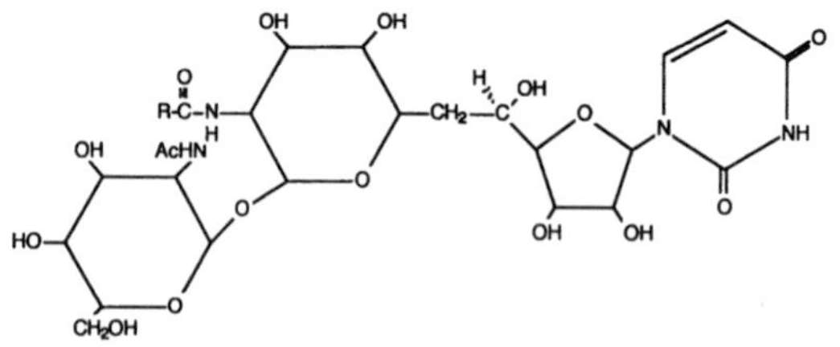

Figure 18 Tunicaminyluracils.

11.30.4.3 Thiaminase

Some ferns including Bracken fern (*Pteridium* spp.), Mulga fern (*Cheilanthes sieberi*), and Nardoo fern (*Marsilia drummondii*) contain thiaminase I which degrades thiamine.[5,87] Thiaminase toxicity has also been associated with horsetail (*Equisetum* spp.).[16,88]

Monogastric species such as the horse and pig are more susceptible to poisoning than ruminants because the latter synthesize thiamine in relatively large amounts in the rumen. Nevertheless, sheep and cattle can develop thiamine deficiency if a sufficient quantity of thiaminase is ingested. More often though, thiamine deficiency in ruminants is a result of ruminal synthesis of thiaminase than as a result of consumption of plants which contain the enzyme.[5]

Affected animals develop muscle tremors, ataxia and eventually cannot stand. Clonic spasms and opisthotonus develop as the disease progresses. Apparent blindness has been reported in poisoned sheep and horses.[87,89] The disease is often fatal if animals are not treated with thiamine.

Toxic ferns also contain agents that depress bone marrow function and carcinogens. These syndromes occur in ruminants. The former may involve thiaminase as a causal agent.[88] Depression of bone marrow causes enzootic hematuria a disease characterized by bleeding.[16,87] Bracken fern poisoning in cattle is more commonly manifest as enzootic hematuria than neurological thiamine deficiency. Carcinogenicity is due to ptaquiloside which promotes tumor development in the intestines and urinary bladder in animals and people.[90]

The pathology of thiamine deficiency in ruminants is dominated by polioencephalomalacia which is a softening of the cerebrocortical gray matter.[91] Neuronal necrosis occurs in the affected tissue. Neurological lesions are less characteristic in horses and pigs and the pathology is suggestive of heart failure.[16]

Thiamine is essential for the activity of the micochondrial enzymes pyruvate dehydrogenase, α-ketoglutarate dehydrogenase, and cytoplasmic transketolase.[88] Thiaminase acts in the gastrointestinal tract to split the thiamine molecule (Figure 17).

The co-factor BH is an amine or sulfhydryl compound in the diet.[88] The resulting thiamine deficiency impairs cellular energy synthesis by reducing the conversion of pyruvate to acetyl-CoA and inhibiting the Kreb's cycle through the effect on mitochondrial enzymes. The effect of thiamine deficiency on transketolase is to reduce NADPH synthesis from the pentose pathway which impairs anabolic reactions and nucleic acid synthesis.

11.30.4.4 Tunicaminyluracils

A neurological syndrome associated with several plant species is caused by tunicaminyluracil compounds (Figure 18) which are synthesized by bacteria associated with the plants.[92,93]

Three plants have been identified in Australia as causing this syndrome. These are: annual ryegrass (*Lolium rigidum*) which causes annual ryegrass toxicity (ARGT), blown grass (*Agrostis avenacea*) the cause of flood plain staggers, and annual beard grass (*Polypogon monspeliensis*) which causes Stewarts Range syndrome.[94] The toxins are synthesized by the bacterium *Clavibacter toxicus*[95] which infects a nematode living on the plant. Originally the bacterium *Corynebacterium rathayi* was thought to be the causative organism[96] and hence, the toxic compounds are often referred to as corynetoxins.[97] ARGT has also been reported from South Africa[98] and the disease also occurs in the USA where the offending plant is *Festuca rubra commutata*.[99] *Streptomyces* spp. also synthesize tunicaminyluracils and it has been suggested that this may have been the cause of toxicity in pigs consuming moldy grain.[92,100]

Clinically, the disease is characterized by muscle tremors and incoordination. Convulsions and death are common in affected animals.[93,101] The clinical signs become more severe if animals are stressed.

Pathological changes are observed in the brain and the liver. Perivascular edema is a common finding particularly in the cerebral meninges. The Purkinje and granular cells of the cerebellum are the most commonly affected cells in the brain[102] and the injury to these cells is a result of ischemia. The pathogenesis of injury has been well described[93] and essentially the disease is a result of injury to capillary endothelial cells. Injury to these cells leads to capillary obstruction, localized ischemia and hypoxic neuronal damage. It appears that the endothelial cell damage is due to impaired synthesis of *N*-glycosylated glycoproteins as a result of the toxic action of the tunicaminyluracils.[103] Hepatotoxicity is sometimes seen and is characterized by individual hepatocyte necrosis and mild bile duct hyperplasia.[104]

11.30.5 HEPATIC ENCEPHALOPATHY

The neurological syndrome of hepatic encephalopathy is caused by inadequate hepatic metabolism of ammonia as a consequence of liver disease. Ammonia, from bacterial protein metabolism in the intestinal tract and from tissue metabolism of amino acids, is normally converted to urea in the liver before it is excreted in the urine. If this metabolism is inadequate, the ammonia concentration increases in the blood and the body tissues leading to neurotoxicity.[105] Other compounds, such as mercaptans and short chain fatty acids, may also be involved.[105]

There are numerous poisonous plants that cause liver disease. If liver dysfunction is substantial, the major presenting signs are essentially neurological as a result of hepatic encephalopathy. Examples are many and include pyrrolizidine alkaloidosis in horses,[5] poison peach (*Trema aspera*) intoxication,[106] and noogoora burr (*Xanthium pungens*) poisoning of livestock.[107,108]

11.30.6 CONCLUSIONS

Poisonous plants are responsible for major economic losses to the grazing industries throughout the world. While it is difficult to make accurate estimates of these losses, it is generally accepted that they amount to about 2% of livestock production in the USA and in excess of $80 million dollars annually in Australia.[109,110] No estimates of the proportion of these losses attributable to neurotoxicants have been published. However, the relatively large literature on neurotoxic poisonous plants suggests that it is substantial.

The occurrence of human poisonings by plant neurotoxins in the western world is relatively uncommon, but not so in the underdeveloped countries. For instance, cycad toxicity occurs in almost epidemic proportions and with devastating results in areas of the Western Pacific. Human poisonings appear to be more common in underdeveloped countries where a single food is relied upon for a major proportion of the diet.[41]

Treatment is available for some neurotoxic plant poisonings, particularly when the toxin acts on synaptic receptors. However, when structural damage results from the action of the toxin, residual neurological incapacity is common and treatment is unlikely to be effective.

11.30.7 REFERENCES

1. J. M. Kingsbury, in 'Effects of Poisonous Plants on Livestock,' eds. R. F. Keeler, K. R. Van Kampen and L. F. James, Academic Press, New York, 1978, pp. 81–91.
2. L. J. Concuera, 'Effects of indole alkaloids from gramineae on aphids.' *Phytochemistry*, 1984, **23**, 539–541.
3. S. L. Everist, 'Poisonous Plants of Australia,' Angus and Robertson, London, 1974.
4. J. M. Kingsbury, 'Poisonous Plants of the United States and Canada,' Prentice-Hall, Englewood Cliffs, NJ, 1964.
5. A. A. Seawright, 'Animal Health in Australia, Vol. 2, Chemical and Plant Poisons,' 2nd edn., Australian Government Publishing Service, Canberra, 1989.
6. T. S. Kellerman, J. A. W. Coetzer and T. W. Naude, 'Plant Poisonings and Mycotoxicoses of Livestock in Southern Africa,' Oxford University Press, Cape Town, 1988.
7. L. F. James, R. F. Keeler, A. E. Johnson *et al.*, 'Plants Poisonous to Livestock in the Western States,' US Department of Agriculture, Agricultural Information Bulletin 415, 1980.
8. R. F. Keeler, K. R. Van Kampen and L. F. James (eds.), 'Effects of Poisonous Plants on Livestock,' Academic Press, New York, 1978.
9. A. A. Seawright, M. P. Hegarty, L. F. James *et al.* (eds.), 'Plant Toxicology,' Queensland Poisonous Plants Committee, Yeerongpilly, 1985.
10. L. F. James, R. F. Keeler, E. M. Bailey, Jr., *et al.* (eds.), 'Poisonous Plants,' Iowa State University Press, Ames, IA, 1992.
11. S. M. Colgate and P. R. Dorling, 'Plant-Associated Toxins Agricultural, Phytochemical and Ecological Aspects,' CAB International, Wallingford, 1994.
12. L. C. Hulbert and F. W. Oehme, 'Plants Poisonous to Livestock,' Kansas State University, Manhattan, 1988.
13. K. E. Panter and R. F. Keeler, 'in 'Toxicants of Plant Origin, I. Alkaloids,' ed. P. R. Cheeke, CRC Press, Boca Raton, FL, 1989, pp. 109–132.
14. W. C. Bowman and I. S. Snaghvi, 'Pharmacological actions of hemlock (*Conium maculatum*) alkaloids.' *J. Pharm. Pharmacol.*, 1963, **15**, 1.
15. H. R. Adams, in 'Veterinary Pharmacology and Therapeutics,' 6th edn., eds. N. H. Booth and L. E. McDonald, Iowa State University Press, Ames, IA, 1988, pp. 137–151.
16. D. J. Humphreys, 'Veterinary Toxicology,' 3rd edn., Baillière Tindall, London, 1988.

17. J. D. Olsen and G. D. Manners, in 'Toxicants of Plant Origin, I, Alkaloids,' ed. P. R. Cheeke, CRC Press, Boca Raton, FL, 1989, pp. 291–326.
18. Y. Bai, M. N. Benn and W. Majak, in 'Poisonous Plants,' eds. L. F. James, R. F. Keeler, E. M. Bailey, Jr. *et al.*, Iowa State University Press, Ames, IA, 1992, pp. 304–313.
19. J. W. Hardin, 'Stock Poisoning Plants of North Carolina,' Agricultural Experiment Station, North Carolina State University at Raleigh, NC, 1973, Bulletin 414.
20. J. D. Olsen, in 'Effects of Poisonous Plants on Livestock,' eds. R. F. Keeler, K. R. Van Kampen and L. F. James, Academic Press, New York, 1978, pp. 535–543.
21. L. P. Bush and M. W. Crowe, in 'Toxicants of Plant Origin, Vol. I. Alkaloids,' ed. P. R. Cheeke, CRC Press, Boca Raton, FL, 1989, pp. 87–107.
22. R. F. Keeler and M. W. Crowe, in 'Plant Toxicology,' eds. A. A. Seawright, M. P. Hegarty, L. F. James *et al.*, Queensland Poisonous Plants Committee, Yeerongpilly, 1985, pp. 324–333.
23. W. Majak and M. A. Pass, in 'Toxicants of Plant Origin, Vol II. Glycosides,' ed. P. R. Cheeke. CRC Press, Boca Raton, FL, 1989, pp. 143–159.
24. F. R. Stermitz and G. S. Yost, in 'Effects of Poisonaus Plants on Livestock,' eds. R. F. Keeler, K. R. Van Kampen and L. F. James, Academic Press, New York, 1978, pp. 371–378.
25. M. A. Pass, A. D. Muir, W. Majak *et al.*, 'Effect of alcohol and aldehyde dehydrogenase inhibitors on the toxicity of 3-nitropropanol in rats.' *Toxicol. Appl. Pharmacol.*, 1985, **78**, 310–315.
26. L. F. James, in 'Handbook of Natural Toxins,' eds. R. F. Keeler and A. T. Tu, Marcel Dekker, New York, 1983, vol. 1, pp. 445–462.
27. L. F. James, W. J. Hartley, M. C. Williams *et al.*, 'Field and experimental studies in cattle and sheep poisoned by nitro-bearing *Astragalus* or their toxins.' *Am. J. Vet. Res.*, 1980, **41**, 377–382.
28. D. H. Gould and D. L. Gustine, 'Basal ganglia degeneration, myelin alterations, and enzyme inhibition induced in mice by the plant toxin 3-nitropropionic acid.' *Neuropathol. Appl. Neurobiol.*, 1982, **8**, 377–393.
29. D. H. Gould, M. P. Wilson and D. W. Hamer, 'Brain enzyme and clinical alterations in rats and mice by nitroaliphatic toxicants.' *Toxicol. Lett.*, 1985, **27**, 83–89.
30. B. F. Hamilton, D. H. Gould, M. P. Wilson *et al.*, 'Enzyme and structural alterations in brains of rats intoxicated with 3-nitropropionic acid.' *Fed. Proc.*, 1984, **43**, 380.
31. T. A. Alston, L. Mela and H. J. Bright, '3-Nitropropionate, the toxic substance of *Indigofera*, is a suicidal inactivator of succinate dehydrogenase.' *Proc. Natl. Acad. Sci. USA*, 1977, **74**, 3767–3771.
32. M. A. Pass, C. H. Carlisle and K. R. Reuhl, '3-Nitropropionic acid toxicity in cultured murine embryonal carcinoma cells.' *Nat. Toxins*, 1994, **2**, 386–394.
33. M. A. Pass, in 'Plant-Associated Toxins,' eds. S. M. Colgate and P. R. Dorling, CAB International, Wallingford, UK, 1994, pp. 541–545.
34. E. E. Conn, in 'Effects of Poisonous Plants on Livestock,' eds. R. F. Keeler, K. R. Van Kampen and L. F. James, Academic Press, New York, 1978, pp. 301–310.
35. O. O. Tewe and E. A. Iyayi, 'Toxicants of Plant Origin, Vol. II. Glycosides,' ed. P. R. Cheeke, CRC Press, Boca Raton, FL, 1989, pp. 43–60.
36. B. O. Osuntokun, 'An ataxic neuropathy in Nigeria. A clinical, biochemical and electrophysiological study.' *Brain*, 1968, **91**, 215–248.
37. P. R. Knight, 'Equine cystitis and ataxia associated with grazing of pastures dominated by *Sorghum* species.' *Aust. Vet. J.*, 1968, **44**, 257.
38. R. A. McKenzie and L. I. McMicking, 'Ataxia and urinary incontinence in cattle grazing sorghum.' *Aust. Vet. J.*, 1977, **53**, 496–497.
39. P. T. Hooper, in 'Effects of Poisonous Plants on Livestock,' eds. R. F. Keeler, K. R. Van Kampen and L. F. James, Academic Press, New York, 1978, pp. 337–347.
40. P. T. Hooper, in 'Handbook of Natural Toxins, Vol. 1, Plant and Fungal Toxins,' eds. R. F. Keeler and A. T. Tu, Dekker, New York, 1983, pp. 463–471.
41. P. S. Spencer, A. C. Ludolph and G. E. Kisby, 'Neurologic diseases associated with use of plant components with toxic potential.' *Environ. Res.*, 1993, **62**, 106–113.
42. Z.-X. Zhang, D. W. Anderson and N. Mantel, 'Geographic patterns of Parkinsonism-Dementia complex in Guam 1956 through 1985.' *Arch. Neurol.*, 1990, **47**, 1069–1074.
43. A. A. Seawright, A. W. Brown, C. C. Nolan *et al.*, in 'Toxins and Targets, Effects of Natural and Synthetic Poisons on Living Cells and Fragile Ecosystems,' eds. D. Watters, M. Lavin, D. Maguire *et al.*, Harwood Academic Publishers, Chur, 1992, pp. 81–86.
44. J. M. Jacobs and P. M. Le Quesne, in 'Greenfield's Neuropathology,' 5th edn. eds. J. H. Adams and L. W. Duchen, Oxford University Press, London, 1992, pp. 881–987.
45. P. S. Spencer, P. B. Nunn, J. Hugon *et al.*, 'Guam amyotrophic lateral sclerosis-Parkinsonism-Dementia linked to a plant excitant neurotoxin.' *Science*, 1987, **237**, 517–522.
46. A. A. Seawright, A. W. Brown, C. C. Nolan *et al.*, 'Selective degeneration of cerebellar cortical neurones caused by cyad neurotoxin, L-BMAA in rats.' *Neuropathol. Appl. Neurobiol.*, 1990, **16**, 153–169.
47. M. W. Duncan, J. C. Steele, I. J. Kopin *et al.*, '2-Amino-3-(methylamino)-propanoic acid (BMAA) in cycas flour: an unlikely cause of amyotrophic lateral sclerosis and parkinsonism-dementia of Guam.' *Neurology*, 1990, **40**, 767–772.
48. M. P. Hegarty, in 'Effect of Poisonous Plants on Livestock,' eds. R. F. Keeler, K. R. Van Kampen and L. F. James, Academic Press, New York, 1978, pp. 575–585.
49. P. S. Spencer, D. N. Roy, A. Ludolph *et al.*, 'Lathyrism: evidence for role of the neuroexcitatory amino acid BOAA.' *Lancet*, 1986, **2**(8515), 1066–1067.
50. S. M. Ross, D. N. Roy and P. S. Spencer, 'β-N-Oxalylamino-L-alanine action on glutamate receptors,' *J. Neurochem.*, 1989, **53**, 710–715.
51. C. R. Huxtable, S. M. Colegate and P. R. Dorling, in 'Toxicants of Plant Origin, Vol. IV. Phenolics,' ed. P. R. Cheeke, CRC Press, Boca Raton, FL, 1989, pp. 83–84.
52. C. R. Huxtable, P. R. Dorling and S. M. Colegate, in 'Plant Toxicology,' eds. A. A. Seawright, M. P. Hegarty, L. F. James *et al.*, Queensland Poisonous Plants Committee, Yeerongpilly, 1985, pp. 381–385.
53. C. R. Huxtable, P. R. Dorling and S. M. Colegate, in 'Poisonous Plants,' eds. L. F. James, R. F. Keeler, E. M. Bailey, Jr. *et al.*, Iowa State University Press, Ames, IA, 1992, pp. 464–468.
54. D. C. Main, D. H. Slatter, C. R. Huxtable *et al.*, '*Stypandra imbricata* ("Blindgrass") toxicosis in goats and sheep—clinical and pathologic findings in 4 field cases.' *Aust. Vet. J.*, 1981, **57**, 132–135.
55. P. R. Dorling, S. M. Colegate and C. R. Huxtable, in 'Poisonous Plants,' eds. L. F. James, R. F. Keeler, E. M. Bailey, Jr. *et al.*, Iowa State University Press, Ames, IA, 1992, pp. 469–473.

56. S. M. Colegate, P. R. Dorling and C. R. Huxtable, 'A spectroscopic investigation of swainsonine: an α-mannosidase inhibitor isolated from *Swainsona canescens*.' *Aust. J. Chem.*, 1979, **32**, 2257–2264.

57. L. F. James, in 'Handbook of Natural Toxins,' eds. R. F. Keeler and A. T. Tu, Marcel Dekker, New York, 1983, vol. 1, pp. 445–462.

58. P. R. Dorling, S. M. Colegate and C. R. Huxtable, in 'Toxicants of Plant Origin, Vol. I. Alkaloids,' ed. P. R. Cheeke, CRC Press, Boca Raton, FL, 1989, pp. 237–256.

59. M. H. Ralphs and J. A. Pfister, in 'Plant-Associated Toxins,' eds. S. M. Colgate and P. R. Dorling, CAB International, Wallingford, UK, 1994, pp. 478–483

60. P. R. Dorling, C. R. Huxtable, S. M. Colgate *et al.*, in 'Plant Toxicology,' eds. A. A. Seawright, M. P. Hegarty, L. F. James *et al.*, Queensland Poisonous Plants Committee, Yeerongpilly, 1985, pp. 255–265.

61. P. R. Dorling, C. R. Huxtable and S. M. Colegate, 'Inhibition of lysosomal α-mannosidase by swainsonine, an indolizidine alkaloid from *Swainsona canescens*.' *Biochem. J.*, 1987, **191**, 649–651.

62. A. D. Elbein, R. Solf, P. R. Dorling *et al.*, 'Swainsonine: an inhibitor of glycoprotein processing.' *Proc. Natl. Acad. Sci. USA*, 1981, **78**, 7393–7397.

63. C. R. Huxtable, in 'Veterinary Clinical Toxicology,' The Postgraduate Committee in Veterinary Science, 1987, Proceedings 103, pp. 85–90.

64. L. J. Concuera, in 'Toxicants of Plant Origin, Vol. I. Alkaloids,' ed. P. R. Cheeke, CRC Press, Boca Raton, FL, 1989, pp. 169–177.

65. E. Odriozola, C. Campero, T. Lopez *et al.*, 'Neuropathological effects and deaths in cattle and sheep in Argentina from *Phalaris angusta*.' *Vet. Human Toxicol.*, 1991, **33**, 465–467.

66. C. A. Bourke, M. J. Carrigan and R. J. Dixon, 'Experimental evidence that tryptamine alkaloids do not cause *Phalaris aquatica* sudden death syndrome in sheep.' *Aust. Vet. J.*, 1988, **65**, 218–220.

67. A. Van-Halderan, J. R. Green and D. J. Schneider, 'An outbreak of suspected Phalaris staggers in sheep in the western Cape Province.' *J. S. Afr. Vet. Assoc.*, 1990, **61**, 39–40.

68. S. S. Nicholson, B. M. Olcott, E. A. Usenik *et al.*, 'Delayed phalaris grass toxicosis in sheep and cattle.' *J. Am. Vet. Med. Assoc.*, 1989, **195**, 345–346.

69. C. A. Bourke and M. J. Carrigan, 'Mechanisms underlying *Phalaris aquatica* "sudden death syndrome" in sheep.' *Aust. Vet. J.*, 1992, **69**, 165–167.

70. N. Anderton, P. A. Cockrum, D. W. Walker *et al.*, in 'Plant-Associated Toxins Agricultural, Phyochemical and Ecological Aspects,' eds. S. M. Colegate and P. R. Dorling, CAB International, Wallingford, UK, 1994, pp. 269–274.

71. C. A. Bourke, in 'Plant-Associated Toxins Agricultural, Phyochemical and Ecological Aspects,' eds. S. M. Colegate and P. R. Dorling, CAB International, Wallingford, UK, 1994, pp. 523–528.

72. C. A. Bourke, M. J. Carrigan, J. T. Seaman *et al.*, 'Delayed development of clinical signs in sheep affected by *Phalaris aquatica* staggers.' *Aust. Vet. J.*, 1987, **64**, 31–32.

73. C. A. Bourke, M. J. Carrigan and R. J. Dixon, 'The pathogenesis of the nervous syndrome of *Phalaris aquatica* toxicity in sheep.' *Aust. Vet. J.*, 1990, **67**, 356–358.

74. D. W. Dewey, H. J. Lee and H. R. Marston, 'Provision of cobalt to ruminants by means of heavy pellets.' *Nature (London)*, 1958, **181**, 1367.

75. E. M. Gress, 'Poisonous Plants of Pennsylvania,' Pennsylvania Department of Agriculture, 1935, Bulletin 18, General Bulletin 531.

76. M. J. Ball, M. L. Flather and J. C. Forfar. 'Hemlock water dropwort poisoning.' *Postgrad. Med. J.*, 1987, **63**, 363–365.

77. R. A. Smith and D. Lewis, 'Cicuta toxicosis in cattle: case history and simplified analytical method.' *Vet. Hum. Toxicol.*, 1987, **29**, 240–241.

78. R. G. Dükstra and R. Falkena, 'Een geval van cicutoxine-intoxicatie bij pony's.' [A case of cicutoxine poisoning in ponies] *Tijdschr. Diergeneeskd*, 1981, **106**, 1037–1039.

79. F. W. Oehme, in 'Effects of Poisonous Plants on Livestock,' eds. R. F. Keeler, K. R. Van Kampen and L. F. James, Academic Press, New York, 1978, pp. 67–80.

80. R. A. Smith, in 'Poisonous Plants,' eds. L. F. James, R. F. Keeler, E. M. Bailey, Jr. *et al.*, Iowa University Press, Ames, IA, 1992, pp. 293–297.

81. K. E. Panter, R. F. Keeler and D. C. Baker, 'Toxicosis in livestock from the hemlocks (*Conium* and *Cicuta* spp.).' *J. Anim. Sci.*, 1988, **66**, 2407–2413.

82. P. H. Mortimer, in 'Effects of Poisonous Plants on Livestock,' eds. R. F. Keeler, K. R. Van Kampen and L. F. James, Academic Press, New York, 1978, pp. 353–361.

83. P. H. Mortimer and M. E. di Menna, in 'Plant Toxicology,' eds. A. A. Seawright, M. P. Hegarty, L. F. James *et al.*, Queensland Poisonous Plants Committee, Yeerongpilly, 1985, pp. 604–611.

84. R. T. Gallagher, E. P. White and P. H. Mortimer, 'Ryegrass staggers: isolation of potent neurotoxins lolitrem A and lolitrem B from staggers-producing pastures.' *N.Z. Vet. J.*, 1981, **29**, 189–190.

85. B. L. Munday and R. W. Mason, 'Lesions in ryegrass staggers in sheep.' *Aust. Vet. J.*, 1967, **43**, 598–599.

86. R. W. Mason, 'Axis cylinder degeneration associated with ryegrass staggers in sheep and cattle.' *Aust. Vet. J.*, 1968, **44**, 428.

87. B. F. Chick, C. Quin and B. V. McCleary, in 'Plant Tocicology,' eds. A. A. Seawright, M. P. Hegarty, L. F. James *et al.*, Queenland Poisonous Plants Committee, Yeerongpilly, 1985, pp. 453–464.

88. B. F. Chick, B. V. McCleary and R. J. Beckett, in 'Toxicants of Plant Origin, Vol. III. Proteins and Amino Acids,' ed. P. R. Cheeke, CRC Press, Boca Raton, FL, 1989, pp. 73–91.

89. K. C. Barnett and W. A. Watson, 'Bright blindness in sheep. A primary retinopathy due to feeding bracken (*Pteris aquilina*).' *Res. Vet. Sci.*, 1970, **11**, 289–290.

90. B. L. Smith, A. A. Seawright, J. Ng *et al.*, in 'Plant-Associated Toxins,' eds. S. M. Colgate and P. R. Dorling, CAB International, Wallingford, UK, 1994, pp. 45–50.

91. K. V. F. Jubb and C. R. Huxtable, in 'Pathology of Domestic Animals,' eds. K. V. F. Jubb, P. C. Kennedy and N. Palmer, Academic Press, San Diego, CA, 1993, pp. 267–439.

92. C. A. Bourke, in 'Plant-Associated Toxins Agricultural, Phyochemical and Ecological Aspects,' eds. S. M. Colegate and P. R. Dorling, CAB International, Wallingford, UK, 1994, pp. 399–404.

93. J. W. Finnie, in 'Plant-Associated Toxins Agricultural, Phyochemical and Ecological Aspects,' eds. S. M. Colegate and P. R. Dorling, CAB International, Wallingford, UK, 1994, pp. 405–409.

94. W. L. Bryden, C. L. Trengove, E. O Davis *et al.*, in 'Plant-Associated Toxins Agricultural, Phyochemical and Ecological Aspects,' eds. S. M. Colegate and P. R. Dorling, CAB International, Wallingford, UK, 1994, pp. 410–415.

95. I. T. Riley and K. M. Ophel, '*Clavibacter toxicus* sp. nov., the bacterium responsible for annual ryegrass toxicity in Australia.' *Int. Syst. Bact.*, 1992, **42**, 64–68.

96. P. Vogel, D. S. Petterson, P. H. Berry *et al.*, 'Isolation of a group of glycolipid toxins from seedheads of

annual ryegrass (*Lolium rigidum* Gaud.) infected by *Corynebacterium rathayi*.' *Aust. J. Exp. Biol. Med. Sci.*, 1981, **59**, 455–467.

97. J. A. Edgar, P. A. Cockrum, P. L. Stewart *et al.*, in 'Plant-Associated Toxins Agricultural, Phyochemical and Ecological Aspects,' eds. S. M. Colegate and P. R. Dorling, CAB International, Wallingford, UK, 1994, pp. 393–398.

98. D. J. Schneider, 'First report of annual ryegrass toxicity in the Republic of South Africa.' *Onderst. J. Vet. Res.*, 1981, **48**, 251–255.

99. J. H. Galloway, 'Grass seed nematode poisoning in livestock.' *J. Am. Vet. Med. Ass.*, 1961, **139**, 1212.

100. C. A. Bourke, 'A naturally occurring tunicamycin-like intoxication in pigs eating water damaged wheat.' *Aust. Vet. J.*, 1987, **64**, 127–128.

101. E. O. Davis, G. E. Curran, W. T. Hetherington *et al.*, 'Clinical, pathological and epidemiological aspects of flood plain staggers, a corynetoxicosis of livestock grazing *Agrostis avenacea*.' *Aust. Vet. J.*, 1995, **72**, 187–190.

102. P. H. Berry, J. M. Howell and R. D. Cook, 'Morphological changes in the central nervous system of sheep affected with experimental annual ryegrass (*Lolium rigidum*) toxicity.' *J. Comp. Pathol.*, 1980, **90**, 603–617.

103. C. C. J. Culvenor and M. V. Jago, in 'Tricothecenes and Other Mycotoxins,' ed. J. Lacey, Wiley, New York, 1985, pp. 159–168.

104. P. H. Berry, R. B. Richards, J. M. Howell *et al.*, 'Hepatic damage in sheep fed annual ryegrass, *Lolium rigidum* parasitised by *Anguina agrestis* and *Corynebacterium rathayi*.' *Res. Vet. Sci.*, 1982, **32**, 148–156.

105. R. M. Hardy, 'Hepatic coma.' *Semin. Vet. Med. Surg. (Small Anim.)*, 1988, **3**, 311–320.

106. C. R. Mulhearn, 'Poison Peach (*Trema aspera*): a plant poisonous to stock.' *Aust. Vet. J.*, 1942, **18**, 68–72.

107. G. C. Kenny, S. L. Everist and A. K. Sutherland, 'Noogoora burr poisoning of cattle.' *Queensland Agric. J.*, 1950, **70**, 172–177.

108. A. A. Seawright, J. Hrdlicka, J. S. Lee *et al.*, 'Toxic substances in the food of animals. Some recent findings of Australian poisonous plant investigations.' *J. Appl. Toxicol.*, 1982, **2**, 75–82.

109. L. F. James, in 'Plant Associated Toxins Agricultural, Phytochemical and Ecological Aspects,' eds. S. M. Colegate and P. R. Dorling, CAB International, Wallingford, 1994, pp. 1–6.

110. C. C. J. Culvernor, in 'Plant Toxicology,' eds. A. A. Seawright, M. P. Hegarty, L. F. James *et al.*, Queensland Poisonous Plants Committee, Yeerongpilly, 1985, pp. 3–13.

11.31
Excitotoxicity

M. FLINT BEAL

Massachusetts General Hospital, Boston, MA, USA

11.31.1 INTRODUCTION

Excitotoxicity refers to neuronal death caused by activation of excitatory amino acid receptors. Several lines of evidence have linked excitotoxicity to the pathogenesis of both acute and chronic neurologic diseases. The initial observation that glutamate was neurotoxic was that of Lucas and Newhouse, who found that administration of glutamate to mice resulted in retinal degeneration.[1] Subsequent studies of Olney and colleagues linked neurotoxicity to the activation of excitatory amino acid receptors, and the term "excitotoxin" was coined.[2] Further advances were those of Rothman linking release of excitatory amino acids to anoxic cell death in hippocampal cultures,[3] and of Choi linking calcium influx to delayed cell death caused by excitatory amino acids.[4] More work has linked activation of excitatory amino acid receptors to free radical generation and nitric oxide, both of which may lead to oxidative stress.[5,6] The role of excitatory amino acids in acute neurologic diseases such as stroke, trauma, and hypoglycemia has received strong support. A possible role of excitotoxicity in chronic neurologic diseases has been supported by both studies in animal models and work with a glutamate release inhibitor in amyotrophic lateral sclerosis.

11.31.2 MOLECULAR BIOLOGY OF EXCITATORY AMINO ACID RECEPTORS

Excitatory amino acid receptors were initially classified pharmacologically into four distinct classes of binding sites in mammalian brain, named according to the agonists: α-amino-3-hydroxy-5-methyl-4-isoxazolepropionic acid (AMPA), kainate, *N*-methyl-D-aspartate (NMDA), and the quisqualate-sensitive metabotropic site.[7,8] AMPA and kainate sites mediate conventional fast synaptic transmission through an ionotropic channel. The NMDA sites regulate Ca^{2+} and Na^+ influx, are gated by Mg^{2+}, and have been implicated in both learning and synaptic plasticity. The metabotropic site acts through G-proteins to either activate phospholipase C or to decrease cyclic AMP (see Chapter 8, this volume).

Receptor proteins representing each of the major excitatory amino acid receptors have been cloned and sequenced. The AMPA, kainate and NMDA receptors are members of the superfamily of ion-gated ligand channels. All known members are hetero-oligomers composed of several subunits. Four subunits of approximately 900 amino acids in length with four membrane-spanning regions were initially identified as AMPA receptors (GluR1–4).[9–12] Each subunit has a unique distribution in brain. There are alternatively spliced subunits termed "flip" and "flop" isoforms.[13] Subsequent studies identified kainate subunits (GluR5–7, KA1, KA2),[14–16] which participate in the formation of high-affinity kainate receptors. KA2 forms active heteromeric channels when expressed with GluR5. These subunits are localized to neurons in the hippocampus where they frequently colocalize with NMDA subunits.[17]

A subunit for the NMDA receptor termed NMDAR1 was initially cloned and sequenced in 1991.[18] This 938 amino acid protein has about 25% homology to the AMPA and kainate subunits. This receptor when expressed shows appropriate agonist and antagonist pharmacology, a glycine coagonist site, calcium permeability, zinc inhibition, and voltage-dependent block by Mg^{2+}. Subsequently four further NMDA subunits (NR2A–D) were cloned.[19–21] Expression of these subunits with the initial subunit to form heteromeric com-

plexes results in NMDA receptor channels of high activity. These complexes show unique distributions in the brain, and differences in gating behavior, affinities for agonists, and sensitivity to antagonists. NR2A is widely distributed in the brain while NR2B is expressed only in the forebrain and NR2C is found predominantly in the cerebellum.

The metabotropic excitatory amino acid receptor family encodes subunits termed mGluR1–7.[22,23] Two subunits (mGluR1 and mGluR5) are linked by a G protein to phospholipase C, which generates inositol phosphate which then releases calcium from intracellular stores. The proteins have 1199 amino acids and appear to have seven transmembrane domains. The remaining metabotropic glutamate receptors are linked to inhibition of cAMP formation.

The cloned AMPA receptors vary in their calcium permeability. GluR1, GluR3, and GluR4 all display strong inwardly rectifying current–voltage and calcium permeability.[24] In contrast GluR2 has a linear current–voltage relation, and when coexpressed with other subunits it suppresses their strong inward rectification, and abolishes their calcium permeability. An arginine to glutamine mutation changes the rectification and confers calcium permeability to GluR2.[25,26] The gene sequence codes for glutamine but this is changed to arginine as a result of RNA editing.[27]

A similar amino acid sequence results in an asparagine at amino acid 598 in NMDAR1. A mutation of the asparagine to glutamine results in decreased calcium permeability.[28,29] The consequences of this mutation for excitotoxicity have been examined in non-neuronal kidney cells which express NMDA receptors following transfection.[30,31] Following expression of NMDA receptors in these cells they undergo cell death which is blocked by NMDA antagonists. Transfected cells expressing the mutated NMDA receptor which conducts less Ca^{2+} are less vulnerable to cell death. In these cells heterodimers expressing NMDAR1 with NR2A show more cell death than those with NR2B, and those with NR2C are resistant to cell death.[30]

The NMDA receptor is also regulated by extracellular pH, which may be important during seizures and ischemia. Protons inhibit the receptor by interacting with the NMDAR1 subunit, and polyamines potentiate receptor function by relief of the tonic proton inhibition. A single amino acid (lysine 211) was identified in exon 5, which mediates the pH sensitivity and polyamine relief of tonic inhibition, indicating that it serves as a pH-sensitive constitutive modulator of NMDA receptor function.[32]

These findings provide experimental evidence that alterations in the amino acid sequences of excitatory amino acid subunits could alter calcium permeability or other properties which could then lead to excitotoxicity. Improved knowledge of the pharmacology and distribution of excitatory amino acids may also lead to the development of improved receptor antagonists for the treatment of neurologic disease.

11.31.3 CELLULAR MECHANISMS OF EXCITOTOXICITY

11.31.3.1 Calcium

An important advance in clarifying mechanisms of excitotoxicity was that of Choi that delayed glutamate neurotoxicity was calcium dependent.[4] Subsequent studies showed that calcium load in cultured cortical neurons correlates with subsequent neuronal degeneration,[33,34] whereas intracellular calcium concentrations do not.[35] This supports the idea that much of the glutamate-induced Ca^{2+} load is sequestered into mitochondria rather than free in the cytoplasm. Evidence has shown that mitochondria and Na^+/Ca^{2+} exchange, which is the major means of efflux of mitochondrial Ca^{2+}, buffer glutamate-induced calcium loads in cultured cortical neurons.[36,37] The increases in mitochondrial calcium also lead to metabolic dysfunction as shown by a lowering in intracellular pH.[37]

Randall and Thayer showed that there are three phases of change in intracellular calcium preceding cell death in cultured hippocampal neurons.[38] There is an initial phase of increased intracellular calcium lasting 5–10 min followed by a latent phase of approximately 2 h in which calcium returns to normal. The third phase consists of a gradual sustained rise in intracellular sodium that reaches a plateau associated with cell death. Tymianski *et al.* showed that cell-permeant Ca^{2+} chelators reduced excitotoxic cell injury.[39] These authors also demonstrated "source specificity" of the Ca^{2+} load, showing the Ca^{2+} entering through the NMDA channel was more toxic than that entering through other sources.[40] This observation is consistent with the earlier work of Choi and colleagues demonstrating that the NMDA receptor mediates most of the excitotoxic effects of glutamate.[4] It has been suggested that Ca^{2+} influx by the NMDA receptor may have access to proteins or compartments which make this Ca^{2+} more efficacious in producing cell death. One possible compartment is the mitochondria.

The release of calcium from intracellular stores may also contribute to excitotoxicity.[41,42]

The means by which increased intracellular calcium leads to cell death may involve several mechanisms. These include activation of protein kinases, phospholipases, nitric oxide synthase, proteases, endonucleases, inhibition of protein synthesis, mitochondrial damage, and free radical generation.[43] Evidence to support or refute these various possibilities is limited, particularly *in vivo*. Kainic acid neurotoxicity is associated with activation of calpain *in vivo*.[44] Furthermore calpain inhibitors reduce AMPA-induced neurotoxicity *in vitro*, although they are not effective against glutamate toxicity.[45,46] Calpain inhibitors show efficacy in a gerbil model of global ischemia and in models of focal ischemia in rats.[47–49] Following excitotoxic activation of the NMDA receptor there is inhibition of calcium/calmodulin kinase II activity in cultured hippocampal neurons.[50] On the other hand a specific inhibitor of calcium/calmodulin kinase II protects against NMDA and hypoxia/hypoglycemia induced cell death in cultured cortical neurons.[51]

11.31.3.2 Free Radicals

The initial report linking free radicals to excitotoxicity was that of Dykens, who showed that kainate-induced damage to cerebellar neurons could be attenuated by superoxide dismutase, allopurinol, and hydroxy radical scavengers such as mannitol.[52] Other studies showed that glutathione depletion exacerbates excitotoxicity while the free radical scavengers α-tocopherol, ascorbic acid, and ubiquinone show neuroprotective effects.[53–55] The 21-aminosteroids and α-phenyl-N-*tert*-butylnitrone which scavenge free radicals also protect against excitotoxicity *in vitro*. The vitamin E analogue trolox protects cultured neurons from AMPA toxicity.[56] Cultured cortical neurons which overexpress superoxide dismutase are resistant to both glutamate and ischemia induced neurotoxicity.[57]

Direct evidence linking excitotoxicity to free radical generation comes from studies using electron paramagnetic resonance which show that NMDA dose-dependently increases superoxide formation in cultured cerebellar neurons.[6] The effects are blocked by NMDA antagonists or removing extracellular Ca^{2+}. This is consistent with the findings of Dykens that exposure of isolated cortical mitochondria to $2.5\,\mu M$ Ca^{2+}, which is similar to concentrations which occur in the setting of excitotoxicity, leads to free radical generation.[58] In synaptosomes NMDA, kainic acid and AMPA all stimulate free radical generation.[59] Electron paramagnetic resonance also showed generation of free radicals *in vivo* following systemic administration of kainic acid.[60]

Direct evidence linking increases in intracellular calcium to mitochondrial production of reactive oxygen species has been obtained *in vitro*.[61,62] Dugan and colleagues used the oxidation-sensitive dye dihydrorhodamine 123 with confocal microscopy to demonstrate that exposure to NMDA, but not kainate, ionomycin or elevated potassium led to oxygen radical production in cultured neurons.[61] This was confirmed by studies using electron paramagnetic resonance. The increase in oxygen radical production was blocked by inhibitors of mitochondrial electron transport and mimicked by an uncoupler of electron transport. In contrast, inhibitors of nitric oxide synthase and arachidonic acid metabolism had no effect. The study of Reynolds and Hastings used the oxidation-sensitive dye dichlorodihydrofluorescein to study the effects of glutamate in neuronal cultures.[62] Glutamate at excitotoxic concentrations caused localized areas of increased fluorescence at the margins of the cell body which were dependent on NMDA receptor activation and calcium entry, and which were blocked by an uncoupler of mitochondrial electron transport. These two studies therefore suggest a critical role of Ca^{2+}-dependent uncoupling of neuronal mitochondrial electron transport in the production of reactive oxygen species following glutamate exposure.

Beal and co-workers examined the relationship of excitotoxicity to free radical production *in vivo*.[63] They showed that malonate, which produces excitotoxic lesions, leads to increased hydroxy radical generation as assessed by the salicylate trapping method. The free radical spin trap *N-tert*-butyl-α (2-sulfophenyl)-nitrone (S-PBN) attenuated both hydroxy radical generation and neurotoxicity. It also attenuated striatal lesions produced by NMDA, AMPA, and kainate. These findings provide direct *in vivo* evidence for a role of free radicals in excitotoxicity.

11.31.3.3 Nitric Oxide

The role of nitric oxide in excitotoxicity is under intense investigation. Dawson and colleagues originally demonstrated that nitric oxide synthase inhibitors and hemoglobin, which scavenges nitric oxide, block glutamate neurotoxicity *in vitro*.[5] They subsequently showed that pretreatment of cultures with quisqualate, which preferentially kills nitric oxide synthase neurons, blocks glutamate neurotoxicity in the cultures.[64] Subsequent studies however have been controversial with several groups reporting that inhibition of nitric oxide synthase had no effect on excitotoxicity *in vitro*.[65–69] Excitotoxicity can occur in the absence of nitric oxide synthase since cultured kidney neurons which lack the enzyme show excitotoxicity when transfected with NMDA receptors.[30]

Initial studies of the effects of inhibition of nitric oxide synthase on NMDA-induced excitotoxicity *in vivo* were also conflicting.[70,71] Similar problems were encountered in studies of focal ischemia. This appears to be due to the nonspecificity of the nitric oxide synthase inhibitors utilized, which have effects on both the neuronal and endothelial isoforms, leading to vascular effects.[72] Evidence strongly favoring a role of neuronal nitric oxide synthase in focal ischemic lesions has come from studies showing that lesions are attenuated in mice with a knockout of the enzyme.[72]

Several studies showed that 7-nitroindazole is a relatively specific inhibitor of the neuronal isoform of nitric oxide synthase *in vivo*. It has no effects on blood pressure or on acetylcholine-induced vasorelaxation.[9,73,74] 7-Nitroindazole reduces focal ischemic lesions.[74] Beal and co-workers found that it significantly attenuated excitotoxicity produced by NMDA but not by AMPA or kainate.[75] This is consistent with the observations of Dawson and colleagues *in vitro*,[5,64] and suggests that Ca^{2+} influx via the NMDA receptor leads to activation of neuronal nitric oxide synthase.

11.31.4 EXCITOTOXIC MECHANISMS IN ACUTE NEUROLOGIC DISEASE

Direct evidence for the role of excitatory amino acids in human neurologic disease was made after observations concerning domoic acid neurotoxicity. In 1987 on Prince Edward Island in Canada a number of individuals consumed mussels contaminated with domoic acid, a potent agonist of the kainic acid subtype of excitatory amino acid receptors.[76] Many of the afflicted patients developed an encephalopathy with complex partial seizures and memory disturbance. Older individuals were particularly susceptible, consistent with experimental evidence showing age-dependent susceptibility to kainate toxicity. Several deaths occurred and necropsy findings showed a loss of neurons in the CA3 field of the hippocampus, which has large numbers of kainic acid receptors.[76,77]

A compelling case for a role of excitotoxicity in stroke has been made based on observations in experimental animals. These studies show that experimental models of stroke are associated with increased glutamate in the extracellular fluid, and that it is attenuated by glutamatergic denervation or by glutamate antagonists.[78] Levels of extracellular glutamate measured by microdialysis are in the range known to be neurotoxic *in vitro*. The distribution of ischemic cell changes corresponds roughly with that of NMDA receptors, however cerebellar Purkinje cells which are vulnerable are devoid of NMDA receptors. The Purkinje cells have AMPA receptors and may be particularly vulnerable because the receptors undergo less complete desensitization to AMPA than those of other cerebellar neurons.[79] Both competitive and noncompetitive NMDA antagonists are effective in focal models of ischemia but show little effect in global models of ischemia. In global models of ischemia non-NMDA antagonists are effective.[80] The experimental studies are sufficiently promising that a number of excitatory amino acid receptor antagonists are currently being tested in human stroke trials.[81,82]

The role of excitotoxicity in head trauma has also received considerable experimental support. Concussive brain injury results in marked increases in extracellular glutamate concentrations.[83] MK-801 attenuates focal brain edema following fluid-percussion brain injury in rats and results in improved brain metabolic status.[84] Dextromethorphan attenuates declines in magnesium and brain bioenergetics following trauma.[85] Similarly kynurenate and indole-2-carboxylic acid reduce cerebral edema and improve cognitive and motor dysfunction induced by trauma.[86] A study showed that kynurenate protects against hippocampal cell loss induced by fluid-percussion injury.[87] Clinical studies of excitatory amino acid antagonists in head trauma in man are underway. Initial results have shown that they reduce increased intracranial pressure. Substantial experimental evidence has also demonstrated that hypoglycemia-induced brain damage is attenuated by NMDA antagonists.[88]

11.31.5 SLOW EXCITOTOXICITY

The role of excitotoxicity in neurodegenerative diseases is speculative. In these diseases there is no evidence for an increase in glutamate concentrations, with the exception of glutamate in the CSF in ALS patients. Furthermore increases in glutamate concentrations by themselves may not be sufficient to cause excitotoxicity.[89] A search for increases in concentrations of other endogenous excitotoxins such as quinolinic acid has been unsuccessful.[90] The concept of slow or weak excitotoxicity has therefore been proposed.[43,91] One possibility to account for this would be a receptor abnormality which could lead to increased calcium influx.

Another possibility is that slow excitotoxicity could occur as a consequence of an impairment in energy metabolism. This could occur by a variety of mechanisms including genetic mutations in mitochondrial electron transport or Krebs cycle enzymes. Another possibility would be oxidative damage to components of the electron transport chain or mitochondrial membranes. The possibility that impaired energy metabolism could result in excitotoxicity was originally demonstrated by the work of Novelli and co-workers.[92] They showed that inhibitors of oxidative phosphorylation or Na^+-K^+ ATPase allowed glutamate to become neurotoxic at concentrations which ordinarily exhibited no neurotoxicity. This was felt to be due to a reduction in ATP leading to partial neuronal depolarization. This may then lead to relief of the voltage-dependent Mg^{2+} block of the NMDA receptor leading to persistent receptor activation by ambient levels of glutamate.

Consistent with this possibility Zeevalk and Nicklas showed that partial energy impairment in cultured chick retina with either iodoacetate (a glycolysis inhibitor) or with cyanide (an inhibitor of oxidative phosphorylation) leads to NMDA receptor activation and excitotoxicity in the absence of any increase in extracellular concentrations of glutamate.[93] Furthermore graded titration of membrane potential with potassium mimicked the toxicity produced by graded metabolic inhibition.[94] Potassium channel activators, which hyperpolarize the cell membrane, can block excitotoxicity *in vitro*.[95] Inhibitors of the Na^+-K^+ ATPase produce lesions in rat substantia nigra and striatum.[96] Other studies showed that metabolic inhibition in hippocampal slices not only increased depolarization in response to glutamate, but also inhibited neuronal repolarization.[97] This could be a consequence of mitochondrial uptake of calcium, which then further impairs mitochondrial function and ATP production. A study of cultured fibroblasts from patients with the mitochondrial disorder MELAS showed impaired calcium buffering.[98] There is therefore a complex interrelationship between excitotoxicity and mitochondrial function, in that mitochondrial dysfunction can lead to excitotoxicity which then further impairs mitochondrial function. Beal and co-workers have carried out a series of studies of the effects of mitochondrial toxins in animals which have further validated these concepts *in vivo*.

11.31.6 EXCITOTOXICITY AND NEURODEGENERATIVE DISEASES

The role of excitotoxicity in neurodegenerative diseases is based on circumstantial evidence. Some of the best evidence is for a role of excitotoxicity in Huntington's disease (HD). Initial observations showed that kainic acid striatal lesions could mimic many of the neuropathologic features of HD.[99,100] They however do not produce sparing of striatal interneurons containing the histochemical marker NADPH-diaphorase which are spared in HD.[101,102] Beal and co-workers and others subsequently showed that quinolinic acid and other NMDA agonists produce an improved animal model, since they result in relative sparing of NADPH-diaphorase neurons.[103,104] The relative sparing is much more dramatic with chronic striatal lesions in which there is striatal shrinkage.[105,106] Parvalbumin neurons which are spared in HD are also relatively preserved by NMDA agonists, but preferentially vulnerable to kainate.[106–108] In primates quinolinic acid produces striking sparing of NADPH-diaphorase neurons as well as an apomorphine inducible movement disorder.[109]

Further support for an NMDA excitotoxic process comes from studies of NMDA receptors in HD postmortem tissue. If the neurons containing these receptors are preferentially vulnerable one would expect a depletion of NMDA receptors. This was shown to be the case in HD striatum[110,111] as well as in the striatum of an asymptomatic at-risk patient, who showed a 50% depletion of NMDA receptors, suggesting that this occurs early in the disease process.[112]

The case for a role of excitotoxicity in Parkinson's disease (PD) is based on two observations. There is a loss of NMDA receptors in PD substantia nigra with a lesser reduction of AMPA sites and no change in metabotropic sites.[113] Furthermore several studies have suggested that the toxicity of MPTP, which models PD, is reduced by excitatory amino acid antagonists. The initial report was that of Turski and colleagues, who showed that MPP$^+$ neurotoxicity in the substantia nigra was attenuated by excitatory amino acid antagonists, although this was later disputed by Sonsalla and colleagues.[114,115] Subsequent studies of attenuation of MPTP neurotoxicity in mice by excitatory amino acid antagonists have

been conflicting, however two studies in primates showed neuroprotective effects.[116,117]

The case for excitotoxicity in ALS is based on three lines of evidence. First, activity of the glutamate-metabolizing enzyme glutamate dehydrogenase was reported to be reduced in leukocytes of ALS patients.[118,119] This finding however appears to be found in a variety of neurologic disorders and is not specific to ALS. Second, glutamate concentrations may be increased in both plasma and CSF of ALS patients.[119,120] Rothstein and colleagues found reduced synaptosomal glutamate uptake in ALS postmortem tissue, and showed reductions in the astrocytic glutamate transporter.[121] Lastly, ingestion of beta-*N*-oxalyl-amino-L-alanine may play a role in lathyrism,[122] and alpha-amino-beta-methylamino-propionic acid (BMAA), which is found in cycads, was linked to the Western Pacific form of ALS.[123] Further evidence supporting a role of excitotoxicity in ALS are findings of reduced numbers of NMDA receptors and increases in kainic acid receptors.[124–126] Studies in organotypic cultures of spinal cord showed that selective inhibition of glutamate transport results in slow degeneration of motor neurons over several weeks, which is blocked by antagonists of non-NMDA glutamate receptors.[127]

A possible role of excitotoxicity in Alzheimer's disease has been proposed. The toxicity of β-amyloid in cell culture is enhanced by both glutamate and by glucose deprivation.[128–130] A loss of NMDA receptors was found in AD cerebral cortex and hippocampus.[131,132] Glutamatergic neurons are prone to neurofibrillary tangles,[133] and tangles are localized to neurons involved in cortico-cortical pathways, which are known to utilize glutamate as a transporter.[134,135] Considerable evidence also exists in AD for a defect in energy metabolism which could render cells more vulnerable to secondary excitotoxicity.[136]

11.31.7 SECONDARY EXCITOTOXICITY DUE TO MITOCHONDRIAL TOXINS

A number of mitochondrial toxins are available which have been used to model neurodegenerative diseases. Some of these compounds can produce selective neuronal degeneration which can model neurodegenerative diseases. Furthermore, studies of these toxins have supported the notion that impairment of energy metabolism can lead to secondary excitotoxic neuronal injury.

11.31.7.1 Amino-oxyacetic Acid

Amino-oxyacetic acid is a nonselective inhibitor of transaminases, including kynurenine transaminase, GABA transaminase, and aspartate transaminase.[137,138] Amino-oxyacetic acid injections lead to a depletion of ATP, which most likely is due to inhibition of aspartate transaminase, an essential component of the malate–aspartate shunt across mitochondrial membranes. This shuttle is the predominant means of moving NADH from the cytoplasm to mitochondria for oxidation. Inhibition of aspartate transaminase in brain slices and synaptosomes results in decreased oxygen consumption and ATP generation.[139–141] Consistent with an effect on energy metabolism we found that pretreatment with either coenzyme Q_{10} or 1,3-butanediol, which can improve ATP generation, could attenuate striatal lesions produced by amino-oxyacetic acid.[142] The lesions are blocked by NMDA antagonists.

The lesions are not caused by direct excitatory amino acid receptor activation since amino-oxyacetic acid has no direct depolarizing effects in either hippocampal slices or cultured striatal neurons.[137,143] The lesions produce sparing of NADPH-diaphorase neurons, consistent with an NMDA receptor-mediated excitotoxic process.[137] Lesions in the hippocampus and in neonatal rats are also blocked by NMDA receptor antagonists.[143,144]

11.31.7.2 Malonate

Malonate is a reversible inhibitor of succinate dehdyrogenase.[145] Succinate dehydrogenase plays a central role in both the tricarboxylic acid cycle and as part of complex II of the electron transport chain. Beal and co-workers and others have examined the effects of intrastriatal injections of malonate in rats.[146,147] Intrastriatal injections of malonate produced dose-dependent excitotoxic lesions which were attenuated by both competitive and noncompetitive NMDA antagonists. Coinjection with succinate blocks the lesions, consistent with an effect on succinate dehydrogenase.[148] More work showed that coinjection of subtoxic malonate with nontoxic concentrations of NMDA, AMPA, and L-glutamate produced large lesions, showing that metabolic inhibition can exacerbate both NMDA and non-NMDA receptor mediated excitotoxicity *in vivo*.[147] An NMDA antagonist reduced the glutamate toxicity by 40% but a non-NMDA antagonist had no effect, suggesting that the NMDA receptor

may play a major role in situations of metabolic compromise *in vivo*.

Beal and co-workers showed that malonate lesions were accompanied by a significant reduction in ATP levels and a significant increase in lactate *in vivo* as shown by chemical shift magnetic resonance imaging.[149,150] Furthermore they showed that pretreatment with coenzyme Q_{10} or nicotinamide, which block malonate-induced ATP depletions, blocked the lesions.[150] Histologic studies showed that the lesions spare both NADPH–diaphorase neurons and somatostatin concentrations, consistent with observations in HD.[146,149] The lesions are strikingly age-dependent and *in vivo* magnetic resonance imaging shows a significant correlation between increasing lesion size and lactate production.[149]

11.31.7.3 3-Nitropropionic Acid

3-Nitropropionic acid (3-NP) is a naturally occurring, abundant plant mycotoxin that is associated with neurological illnesses in grazing animals and humans.[151] Ingestion in livestock results in hindlimb weakness, knocking together of the hindlimbs while walking, and goose stepping. Outbreaks of illness in man occurred in China after ingestion of mildewed sugar cane that contains the fungus Arthrinium. The illness is characterized by initial gastrointestinal disturbance followed by encephalopathy with stupor and coma. Patients who recover have delayed onset of nonprogressive dystonia 7 d to 40 d after regaining consciousness. The patients also show facial grimacing, torticollis, dystonia, and jerk-like movements. Computed tomography scans show bilateral hypodensities in the putamen and, to a lesser extent, in the globus pallidus.

3-Nitropropionic acid is an irreversible inhibitor of succinate dehydrogenase, which is part of both the tricarboxylic acid cycle and the electron transport chain.[152,153] *In vitro* studies showed that 3-NP has no direct depolarizing effects on neurons in hippocampal slices.[154] Rather it leads to hyperpolarization by activation of ATP-sensitive potassium channels. Studies in cortical explants showed that 3-NP produces cellular ATP depletion and neuronal damage by an excitotoxic mechanism.[155] Pathologic changes were significantly attenuated by pretreatment with excitatory amino acid antagonists. Similarly, in primary mesencephalic cultures MK-801 attenuated 3-NP toxicity to dopaminergic neurons.[156] In cultured cerebellar neurons 3-NP produces neurotoxicity which is delayed but not prevented by the competitive NMDA antagonist 2-amino-5-phosphonovaleric acid.[157] In cultured striatal and cortical neurons 3-nitropropionic acid neurotoxicity was unaffected by glutamate antagonists, but was reduced by the macromolecular synthesis inhibitors cycloheximide, emetine or actinomycin D, consistent with apoptotic cell death.[158] Exposure to 3-nitropropionic acid also produced cell body shrinkage and DNA fragmentation on agarose gels, suggesting apoptotic cell death.

Studies in mice of 3-NP toxicity showed damage in the striatum and CA1 field of the hippocampus.[159] Ultrastructural studies showed dendrosomatic swelling, chromatin clumping, and mitochondrial swelling considered to be consistent with an excitotoxic injury. There were no effects on blood pressure or arterial oxygen levels.[160]

Beal and co-workers found that intrastriatal injections of 3-NP in rats produced dose-dependent lesions with neuronal loss and gliosis.[161] There were depletions of ATP and focal increases in lactate in the basal ganglia. The lesions were strikingly age-dependent, with much larger lesions in young adult and adult animals than in juvenile animals, which has been confirmed by others. The age-dependence correlated directly with the degree of increase in lactate concentrations following administration of a uniform dose of 3-NP to animals of various ages.

Subacute systemic administration resulted in the development of motor slowing and dystonic posturing, which has been confirmed by other authors.[162] There were large symmetric lesions in the basal ganglia which were nonselective in that Nissl and NADPH–diaphorase neurons were equally vulnerable. The lesions were accompanied by focal accumulations of lactate in the basal ganglia as shown by magnetic resonance spectroscopy. The lactate accumulations preceded lesions detectable with T_2-weighted imaging.

Consistent with an excitotoxic mechanism the lesions are attenuated by prior decortication or treatment with lamotrigine, which blocks glutamate release. Turski and Ikonomidou showed that the AMPA antagonist NBQX could block the lesions.[163] Microdialysis studies showed that there was no significant increase in extracellular glutamate concentrations after administration of neurotoxic doses of 3-NP, which resulted in twofold increases in lactate.[162] These results are therefore consistent with 3-NP inducing a secondary excitotoxicity by making neurons more vulnerable to endogenous levels of glutamate. Furthermore, *in vivo* ^{3}H-MK-801 receptor autoradiography showed that systemic administration of 3-NP was associated with activation of NMDA-receptors.[164]

Chronic low-dose administration of 3-NP over 1 month by subcutaneous osmotic pumps produced subtle lesions in the dorsolateral striatum in which there was neuronal loss, gliosis and sparing of NADPH–diaphorase neurons, similar to findings in HD.[162] This finding was confirmed using *in situ* hybridization for somatostatin mRNA, which was preserved as compared with both substance P and enkephalin mRNA.[164] Dopamine terminals were preserved with lower doses of 3-NP as shown by ^{3}H-mazindol autoradiography.

To investigate the mechanism of the lesions we administered 3-NP to transgenic mice over-expressing the enzyme superoxide dismutase as compared with littermate controls.[165] In the animals overexpressing superoxide dismutase 3-NP neurotoxicity was significantly attenuated. Furthermore, 3-NP-induced increases in hydroxy radical generation and 3-nitrotyrosine, a marker for peroxynitrite induced damage, were significantly attenuated in the mice overexpressing superoxide dismutase. These findings suggest that oxidative damage, perhaps mediated by peroxynitrite, plays a role in 3-NP toxicity.

Beal and co-workers extended our studies to nonhuman primates to attempt to produce chorea, the cardinal clinical feature of HD.[161] Following 3–6 weeks of 3-NP administration, apomorphine induced a movement disorder closely resembling that seen in HD. The animals showed orofacial dyskinesia, dystonia, dyskinesia of the extremities, and choreiform movements. Both a clinical rating scale and quantitative analysis of individual movement velocities confirmed that 3-NP treated animals had a significant increase in choreiform and dystonic movements. More prolonged 3-NP treatment in two additional primates resulted in spontaneous dystonia and dyskinesia accompanied by lesions in the caudate and putamen seen on magnetic resonance imaging.

Histologic evaluation showed changes reminiscent of HD with a depletion of Nissl-stained and calbindin-stained neurons, yet sparing of NADPH-diaphorase and large neurons. There were proliferative changes in the dendrites of spiny neurons on Golgi studies, and preservation of the striosomal organization of the striatum, similar to changes in HD. Lastly, there was sparing of the nucleus accumbens. These findings therefore show that 3-NP neurotoxicty in primates can replicate many of the characteristic features of HD, strengthening the possibility that slow excitotoxicity might be involved in its pathogenesis.

11.31.8 THERAPEUTIC DIRECTIONS

11.31.8.1 Excitatory Amino Acid Antagonists

If excitotoxicity plays a role in neurologic diseases then it should be possible to prevent neurologic damage with glutamate release blockers or with excitatory amino acid antagonists. Several compounds have been developed which are sodium channel blockers and which block glutamate release. One such compound is lamotrigine, which has been approved for use as an antiepileptic in man.[130] In rat brain cortical slices it inhibits glutamate release. It is neuroprotective against kainate striatal lesions and against MPTP dopaminergic neurotoxicity *in vivo*.[166–168] Beal and co-workers found that it attenuates malonate, MPP$^+$, and 3-NP neurotoxicity.[149]

Several analogues of lamotrigine show efficacy in models of cerebral ischemia. The related compounds BW1003C87 and BW619C89 decrease ischemia-induced glutamate release and infarct volumes following middle cerebral artery occlusion.[166,169,170] BW1003C87 is also effective in global ischemia.[171]

Riluzole is another compound which blocks glutamate release in hippocampal slices *in vitro* and in the cat caudate nucleus *in vivo*.[172,173] It has significant neuroprotective effects in both focal and global models of cerebral ischemia.[174,175] It protects against MPTP-induced decreases in dopamine levels in mice.[176] Of great interest, it was shown to slow the progression of ALS.[177]

The approach of using glutamate-release blockers therefore appears to be very promising. In our hands *in vivo* and in a model of glutamate-induced cortical lesions *in vitro* these compounds show equal or better efficacy than glutamate-receptor antagonists.[178] These compounds also appear to be well tolerated in man.

A large number of compounds have been developed which modulate NMDA receptors. These compounds include competitive antagonists for the glutamate binding site, noncompetitive antagonists which bind to the ion channel, glycine site antagonists and polyamine site antagonists.[179,180] Compounds from all of these classes are highly effective in experimental models of stroke, epilepsy, and traumatic brain injury. Among the most potent competitive NMDA antagonists are *cis*-4-phosphonomethyl-2-piperidine-carboxylic acid (CGS 19755) and D-3 (2-carboxypiperazin-4-yl) propenyl-1-phosphonic acid (D-CPPene). Agents such as MK-801, CNS 1102, and phencyclidine interact with a site within the ion channel to produce a noncompetitive blockade of glutamate. Agents such as 7-chloro-kynurenic, and 3-amino-1-hydroxy-2-pyrrolidone (HA 966),

and ACEA 1021 attenuate NMDA receptor function by blocking the site through which glycine allosterically enhances NMDA receptor function.[181] Another allosteric regulatory site is the polyamine site at which compounds such as ifenprodil appear to act. In addition, work has led to the development of several non-NMDA receptor antagonists such as NBQX. Non-NMDA antagonists are particularly effective in preventing hippocampal damage in global models of ischemia.[80,182] One concern however is that they result in widespread reductions in cerebral glucose utilization in experimental animals.[183]

MK-801 has shown efficacy in both cat and rat models of permanent middle cerebral artery occlusion. The extent of reduction in infarct size ranges between 40–65%.[184–186] In most studies the caudate nucleus is not protected. Gill *et al.* showed that maximal infarct reduction of 60% was achieved with plasma MK-801 concentrations of 19 ng mL^{-1}. Higher levels led to less protection, which may have been due to hypotension.[187] Protection is less effective in strains of rats such as spontaneous hypertensive rats which have less collateral.[188] CNS 1102 administered 1 h after ischemia reduced infarct volume by over 50% and improved neurological outcome.[189] Dextrorphan showed efficacy in rabbits after 1 h of multiple vessel occlusions and dextromethorphan was effective when administered 1 h after the onset of focal ischemia.[190,191]

In focal ischemia rats with permanent middle cerebral artery occlusion showed a 70–80% reduction in infarct area when given CGS 19755 5 min before or 5 min after vascular occlusion.[192] The competitive antagonist D-CPPene reduced cortical infarct volume by more than 75% when therapy was begun prior to permanent MCA occlusion in cats, but had no effect if therapy was delayed for 1 h.[193] Ifenprodil reduced infarct volume by 40% in a cat middle cerebral artery occlusion model.[194] A glycine site antagonist (L-687,414) produced a 41% decrease in cortical infarct volume in a rat model of focal ischemia.[195] Another glycine antagonist, ACEA 1021, produced a 35% reduction in infarct volume in focal ischemia but was ineffective in global ischemia.[196]

As noted above, NBQX is effective in models of global ischemia in which NMDA antagonists are typically ineffective. NBQX has efficacy in focal ischemia models. In a rat permanent middle cerebral artery occlusion model, NBQX reduced infarct volume by 25%.[197] Similarly, delayed treatment at 90 min produced a 25% protection in a middle cerebral artery occlusion model in spontaneously hypertensive rats.[198]

The major concern regarding the use of NMDA antagonists in man is that they may exert adverse behavioral side-effects. NMDA receptors play a critical role in learning and memory, and some noncompetitive NMDA antagonists such as phencyclidine can produce psychotomimetic effects in man.[199,200] Initial phase II studies of both a competitive NMDA antagonist (CGS 19755) and a noncompetitive NMDA antagonist (dextrorphan) showed significant psychoactive effects at therapeutic doses.[81,82] As many as 50% of the patients experienced agitation, somnolence, or hallucinations. It is as yet uncertain whether similar effects are associated with all NMDA antagonists. Some NMDA antagonists such as memantine appear to be well tolerated in man.[179] It is possible that lower-affinity NMDA antagonists which block and unblock the receptor more rapidly may exhibit less toxicity.[179,201] A potential advantage of noncompetitive NMDA antagonists is that they are use-dependent. They increase their block of NMDA receptors proportionately with glutamate concentrations, which may be advantageous under conditions of excessive glutamate release. Both glycine site antagonists and polyamine site antagonists may also exhibit less behavioral toxicity.

Both competitive and noncompetitive NMDA antagonists have been reported to attenuate MPP+ and MPTP neurotoxicity in mice and rats, however this has been controversial.[114,115,202,203] Two studies in primates using MK-801 and the competitive NMDA antagonist CPP showed protection against MPTP neurotoxicity in primates.[116,117] In our studies both competitive and noncompetitive antagonists were neuroprotective against aminooxyacetic acid, malonate, and 3-acetylpyridine neurotoxicity.[137,149,204] The non-NMDA antagonist NBQX protects against 3-NP neurotoxicity.[163]

11.31.8.2 Free Radical Scavengers

As discussed above, there is now substantial evidence linking excitotoxicity to free radical generation. A number of studies showed that a variety of free radical scavengers can attenuate excitotoxicity *in vitro*.[205,206] There however is very limited evidence *in vivo*. One promising approach is to use free radical spin traps such as α-phenyl-N-*tert*-butyl-nitrone (PBN), N-*tert*-butyl-α-(2-sulfophenyl)-nitrone (S-PBN), and 5,5-dimethyl-1-pyroline N-oxide (DMPO). These compounds react with free radicals to form more stable adducts. They are widely used to detect the generation of free radicals using electron paramagnetic resonance. Following systemic administration, PBN is widely distributed in all tissues with a plasma half-life of about 12 h.[207] It appears to be concentrated in mitochondria. PBN is effective in reducing ischemia-reperfusion injury in gerbils,[208,209] focal ischemia in rats,[210] and oxidative damage to proteins in aged gerbils.[211]

Beal and co-workers found that pretreatment with S-PBN significantly attenuates striatal excitotoxic lesions in rats produced by N-methyl-D-aspartate, kainic acid, and AMPA.[63] In a similar manner striatal lesions produced by malonate were dose-dependently blocked by S-PBN. Lesions produced by MPP+ and 3-acetylpyridine were also protected. DMPO administration protects against systemic administration of 3-NP. The neuroprotective effects of S-PBN were additive with MK-801 against both malonate and 3-acetylpyridine neurotoxicity. This suggests that a combination of compounds acting at sequential steps in the excitotoxic cascade might have improved efficacy. An advantage of this approach is that one might be able to use lower doses of compounds to avoid behavioral toxicity. S-PBN has no effects on malonate-induced ATP depletions or on spontaneous striatal electrophysiologic activity, showing that it does not act at excitatory amino acid receptors. It does however attenuate malonate-induced increases in hydroxy radical generation.

Another group of compounds which show promise as free radical scavengers are the 21-aminosteroids or lazeroids, which can attenuate glutamate toxicity *in vitro*.[212] They are effective in both ischemic neuronal damage and experimental head injury in mice.[213-215] They produce dose-dependent neuroprotection in a rat model of focal ischemia.[216] Dihydrolipoate is an antioxidant which is well tolerated in man and which is effective in experimental models of stroke, and against malonate and NMDA-induced striatal lesions.[217,218]

A potential therapeutic advantage of free radical spin traps such as PBN and S-PBN is that they appear to have a much improved therapeutic window. Most studies of either focal ischemia or excitotoxic lesions showed a maximal therapeutic window (the time in which one can administer therapy after the insult and still achieve efficacy) of 1–2 h. PBN however was reported to exert neuroprotective effects when administered as long as 12 h after focal ischemic insult.[210] In another study PBN showed efficacy when administered at 3 h after a middle cerebral artery occlusion.[219] Beal and co-workers found that S-PBN showed efficacy when administered up to 6 h after malonate injections in rat striatum.[220]

11.31.8.3 Nitric Oxide Synthase Inhibitors

As discussed above there is evidence linking excitotoxicity to nitric oxide generation. Results using nonselective nitric oxide synthase (NOS) inhibitors however have been controversial. The absence of consensus may be due to the prior lack of inhibitors with specificity for the various isoforms of the enzyme. The three isoenzymes are the constitutive neuronal and endothelial isoforms and the inducible form localized to macrophages. An initial report showed that L-nitroarginine at $1\,mg\,kg^{-1}$ reduced infarct volume in a mouse model of focal cerebral ischemia.[221] The same dosing regimen was effective in reducing size following middle cerebral artery occlusion in rats.[222] Other studies showed that NOS inhibition reduced caudate injury in cats following focal ischemia, although there was no effect on cortical infarct volume.[223] In contrast several studies showed that inhibition of NOS increased infarct volume following focal ischemia in rats.[224–226] These studies generally used higher doses of NOS inhibitors which will inhibit endothelial NOS, leading to vasoconstriction and reduced cerebral perfusion. Consistent with this effect NO donors increase blood flow and reduce brain damage in focal ischemia.[227] These findings have been clarified by the observation that low doses of NOS inhibitors are neuroprotective, whereas higher doses are ineffective in the mouse focal ischemic model, consistent with adverse vascular effects at higher dose levels.[228] Similar controversy exists concerning neuroprotective effects of NOS inhibition on survival of hippocampal CA1 neurons following global ischemia,[229–231] with more consistent protection with lower dosage of NOS inhibitors.[232] Strong evidence for a role of the neuronal isoform of NOS in focal ischemia came from studies in mice with a knockout of the neuronal isoform of NOS. These mice show a significant attenuation in the size of focal ischemic lesions.[72]

Improved inhibitors of NOS have been described. One of these is 7-nitroindazole (7-NI), which is a relatively specific inhibitor of the neuronal isoform of NOS *in vivo*.[9] *In vivo* studies showed no effects on blood pressure and on endothelium-dependent and acetylcholine-induced blood vessel relaxation. 7-NI is effective against focal ischemic lesions *in vivo*.[74]

Beal and co-workers found that 7-NI significantly attenuated NMDA striatal excitotoxic lesions, but not those induced by kainic acid or AMPA.[75] 7-NI dose-dependently reduced striatal malonate lesions and the protection was reversed by L-arginine but not by D-arginine. 7-NI produced nearly complete protection lesions produced by systemic administration of 3-NP. 7-NI protected against malonate-induced decreases in ATP and increases in lactate. Its effects were not mediated by excitatory amino acid receptors since it had no effect on spontaneous electrophysiologic activity in the striatum *in vivo*. One mechanism by which NO· is thought to mediate its toxicity is by interacting with superoxide to form peroxynitrite which then may nitrate tyrosine residues. 7-NI attenuated increases in hydroxy radical and 3-nitrotyrosine generation *in vivo*, which may be a consequence of peroxynitrite formation. Beal and co-workers also found that 7-NI dose-dependently protected against MPTP-induced dopaminergic toxicity in mice.[63] 7-NI also attenuated increases in striatal 3-nitrotyrosine induced by MPTP. Another study showed that inhibition of NOS with the nonselective inhibitor nitro-L-arginine-reduced MPP^+-induced increases in hydroxy radical generation and showed mild protection against MPTP-induced depletions of dopamine.

11.31.9 CONCLUSIONS

A role of excitotoxicity in acute neurologic diseases is strongly supported by experimental studies. The role of excitotoxicity in neurodegenerative diseases is more speculative but evidence favoring this possibility continues to accrue. There is now strong evidence linking excitotoxicity to both free radical and nitric oxide generation. Therapeutic studies in animals have firmly established the efficacy of excitatory amino acid antagonists in models of focal cerebral ischemia and head trauma. Studies have also shown efficacy of free radical scavengers and neuronal NOS inhibitors. These studies are now being extended to therapeutic trials in man and should definitively establish the role of excitotoxicity in neurologic illnesses.

11.31.10 REFERENCES

1. D. R. Lucas and J. P. Newhouse, 'The toxic effect of sodium L-glutamate on the inner layers of the retina.' *Arch. Ophthalmol.*, 1957, **58**, 193–201.
2. J. W. Olney, 'Brain lesions, obesity and other disturbances in mice treated with monosodium glutamate.' *Science*, 1969, **164**, 719–721.
3. S. Rothman, 'Synaptic release of excitatory amino acid neurotransmitter mediates anoxic neuronal death.' *J. Neurosci.*, 1984, **4**, 1884–1891.
4. D. W. Choi, 'Ionic dependence of glutamate neurotoxicity.' *J. Neurosci.*, 1987, **7**, 369–379.
5. V. L. Dawson, T. M. Dawson, E. D. London *et al.*, 'Nitric oxide mediates glutamate neurotoxicity in primary cortical cultures.' *Proc. Natl. Acad. Sci. USA*, 1991, **88**, 6368–6371.

6. M. Lafon-Cazal, S. Pietri, M. Culcasi *et al.*, 'NMDA-dependent superoxide production and neurotoxicity.' *Nature*, 1993, **364**, 535–537.

7. D. T. Monaghan, R. J. Bridges and C. W. Cotman, 'The excitatory amino acid receptors: their classes, pharmacology, and distinct properties in the function of the central nervous system.' *Annu. Rev. Pharmacol. Toxicol.*, 1989, **29**, 365–402.

8. A. B. Young and G. E. Fagg, 'Excitatory amino acid receptors in the brain: membrane binding and receptor autoradiographic approaches.' *Trends Pharmacol. Sci.*, 1990, **11**, 126–133.

9. R. C. Babbedge, P. A. Bland-Ward, S. L. Hart *et al.*, 'Inhibition of rat cerebellar nitric oxide synthase by 7-nitro indazole and related substituted indazoles.' *Br. J. Pharmacol.*, 1993, **110**, 225–228.

10. J. Boulter, M. Hollmann, A. O'Shea-Greenfield *et al.*, 'Molecular cloning and functional expression of glutamate receptor subunit genes.' *Science*, 1990, **249**, 1033–1037.

11. M. Hollmann, A. O'Shea-Greenfield, S. W. Rogers *et al.*, 'Cloning by functional expression of a member of the glutamate receptor family.' *Nature*, 1989, **342**, 643–648.

12. K. Kleinanen, W. Wisden, B. Sommer *et al.*, 'A family of AMPA-selective glutamate receptors.' *Science*, 1990, **249**, 556–560.

13. B. Sommer, K. Keinänen, T. A. Verdoorn *et al.*, 'Flip and flop: a cell-specific functional switch in glutamate-operated channels of the CNS.' *Science*, 1990, **249**, 1580–1585.

14. J. Egebjerg, B. Bettler, I. Hermans-Borgmeyer *et al.*, 'Cloning of a cDNA for a glutamate receptor subunit activated by kainate but not AMPA.' *Nature*, 1991, **351**, 745–748.

15. A. Herb, N. Burnashev, P. Werner *et al.*, 'The KA-2 subunit of excitatory amino acid receptors shows widespread expression in brain and forms ion channels with distantly related subunits.' *Neuron*, 1992, **8**, 775–785.

16. P. Werner, M. Voight, K. Keinänen *et al.*, 'Cloning of a putative high-affinity kainate receptor expressed predominantly in hippocampal CA3 cells.' *Nature*, 1991, **351**, 742–744.

17. S. J. Siegel, W. G. Janssen, J. W. Tullai *et al.*, 'Distribution of the excitatory amino acid receptor subunits GluR2(4) in monkey hippocampus and colocalization with subunits GluR5–7 and NMDAR1.' *J. Neurosci.*, 1995, **15**, 2707–2719.

18. K. Moriyoshi, M. Masu, T. Ishii *et al.*, 'Molecular cloning and characterization of the rat NMDA receptor.' *Nature*, 1991, **354**, 31–37.

19. H. Meguro, H. Mori, K. Araki *et al.*, 'Functional characterization of a heteromeric NMDA receptor channel expressed from cloned cDNAs.' *Nature*, 1992, **357**, 70–74.

20. H. R. S. Monyer, R. Sprengel, R. Schoepfer *et al.*, 'Heteromeric NMDA receptors: molecular and functional distinction of subtypes.' *Science*, 1992, **256**, 1217–1221.

21. T. Kutsuwada, N. Kashiwabuchi, H. Mori *et al.*, 'Molecular diversity of the NMDA receptor channel.' *Nature*, 1992, **358**, 36–41.

22. K. M. Houamed, J. L. Kuijper, T. L. Gilbert *et al.*, 'Cloning, expression and gene structure of a G protein-coupled glutamate receptor from rat brain.' *Science*, 1991, **252**, 1318–1321.

23. Y. Tanabe, M. Masu, T. Ishii *et al.*, 'A family of metabotropic glutamate receptors.' *Neuron*, 1992, **8**, 169–179.

24. M. Hollmann, M. Hartley and S. Heinemann, 'Ca^{2+} permeability of Ca-AMPA-gated glutamate receptor channels depends on subunit composition.' *Science*, 1991, **252**, 851–853.

25. R. I. Hume, R. Dingledine and S. F. Heinemann, 'Identification of a site in glutamate receptor subunits that controls calcium permeability.' *Science*, 1991, **253**, 1028–1031.

26. T. A. Verdoorn, N. Burnashev, H. Monyer *et al.*, 'Structural determinants of ion flow; through recombinant glutamate receptor channels.' *Science*, 1991, **252**, 1715–1718.

27. B. Sommer, M. Kohler, R. Sprengel *et al.*, 'RNA editing in brain controls a determinant of ion flow in glutamate-gated channels.' *Cell*, 1991, **67**, 11–19.

28. N. Burnashev, R. Schoepfer, H. Monyer *et al.*, 'Control by asparagine residues of calcium permeability and magnesium blockade in the NMDA receptor.' *Science*, 1992, **257**, 1415–1419.

29. K. Sakurada, M. Masu and S. Nakanishi, 'Alteration of Ca^{2+} permeability and sensitivity to Mg^{2+} and channel blockers by a single amino acid substitution in the *N*-methyl-D-aspartate receptor.' *J. Biol. Chem.*, 1993, **268**, 410–415.

30. N. J. Anegawa, D. R. Lynch, T. A. Verdoorn *et al.*, 'Transfection of *N*-methyl-D-aspartate receptors in a nonneuronal cell line leads to cell death.' *J. Neurochem.*, 1995, **64**, 2004–2012.

31. M. Cik, P. L. Chazot and F. A. Stephenson, 'Expression of NMDAR1-1a (N598Q)/NMDAR2A receptors results in decreased cell mortality.' *Eur. J. Pharmacol.*, 1994, **266**, R1–R3.

32. S. F. Traynelis, M. Hartley and S. F. Heinemann, 'Control of proton sensitivity of the NMDA receptor by RNA splicing and polyamines.' *Science*, 1995, **268**, 873–876.

33. D. M. Hartley, M. C. Kurth, L. Bjerkness *et al.*, 'Glutamate receptor-induced ^{45}Ca^{2+} accumulation in cortical cell culture correlates with subsequent neuronal degeneration.' *J. Neurosci.*, 1993, **13**, 1993–2000.

34. S. Eimerl and M. Schramm, 'The quantity of calcium that appears to induce neuronal death.' *J. Neurochem.*, 1994, **62**, 1223–1226.

35. M. R. Witt, K. Dekermendjian, A. Frandsen *et al.*, 'Complex correlation between excitatory amino acid-induced increase in the intracellular Ca^{2+} concentration and subsequent loss of neuronal function in individual neocortical neurons in culture.' *Proc. Natl. Acad. Sci. USA*, 1994, **91**, 12303–12307.

36. R. J. White and I. J. Reynolds, 'Mitochondria and Na$^+$/Ca^{2+} exchange buffer glutamate-induced calcium loads in cultured cortical neurons.' *J. Neurosci.*, 1995, **15**, 1318–1328.

37. G. J. Wang, R. D. Randall and S. A. Thaymer, 'Glutamate-induced intracellular acidification of cultured hippocampal neurons demonstrates altered energy metabolism resulting from Ca^{2+} loads.' *J. Neurophysiol.*, 1994, **72**, 2563–2569.

38. R. D. Randall and S. A. Thayer, 'Glutamate-induced calcium transient triggers delayed calcium overload and neurotoxicity in rat hippocampal neurons.' *J. Neurosci.*, 1992, **12**, 1882–1895.

39. M. Tymianski, M. C. Wallace, I. Spigelman *et al.*, 'Cell-permeant Ca^{2+} chelators reduce early excitotoxic and ischemic neuronal injury *in vitro* and *in vivo*.' *Neuron*, 1993, **11**, 221–235.

40. M. Tymianski, M. P. Charlton, P. L. Carlen *et al.*, 'Source specificity of early calcium neurotoxicity in cultured embryonic spinal neurons.' *J. Neurosci.*, 1993, **13**, 2085–2104.

41. A. Frandsen and A. Schousboe 'Dantrolene prevents glutamate cytotoxicity and Ca^{2+} release from intracellular stores in cultured cerebral cortical neurons.' *J. Neurochem.*, 1991, **56**, 1075–1078.

42. S. Z. Lei, D. Zhang, A. E. Abele *et al.*, 'Blockade of NMDA receptor-mediated mobilization of intracel-

lular Ca^{2+} prevents neurotoxicity.' *Brain Res.*, 1992, **598**, 196–202.

43. M. F. Beal, 'Does impairment of energy metabolism result in excitotoxic neuronal death in neurodegenerative illnesses?' *Ann. Neurol.*, 1992, **31**, 119–130.

44. R. Siman and J. C. Noszek, 'Excitatory amino acids activate calpain I and induce structural protein breakdown *in vivo*.' *Neuron.*, 1988, **1**, 279–287.

45. H. Caner, J. L. Collins, S. M. Harris *et al.*, 'Attenuation of AMPA-induced neurotoxicity by a calpain inhibitor.' *Brain Res.*, 1993, **607**, 354–356.

46. H. Manev, M. Favaron, R. Siman *et al.*, 'Glutamate neurotoxicity is independent of calpain 1 inhibition in primary cultures of cerebellar granule cells.' *J. Neurochem.*, 1991, **57**, 1288–1295.

47. R. T. Bartus, K. L. Baker, A. D. Heiser *et al.*, 'Postischemic administration of AK275, a calpain inhibitor, provides substantial protection against focal ischemic brain damage.' *J. Cereb. Blood Flow Metab.*, 1994, **14**, 537–544.

48. S. C. Hong, Y. Goto, G. Lanzino *et al.*, 'Neuroprotection with a calpain inhibitor in a model of focal cerebral ischemia.' *Stroke*, 1994, **25**, 663–669.

49. K. S. Lee, S. Frank, P. Vanderklish *et al.*, 'Inhibition of proteolysis protects hippocampal neurons from ischemia.' *Proc. Natl. Acad. Sci. USA*, 1991, **88**, 7233–7237.

50. S. B. Churn, D. Limbrick, S. Sombati *et al.*, 'Excitotoxic activation of the NMDA receptor results in inhibition of calcium/calmodulin kinase II activity in cultured hippocampal neurons.' *J. Neurosci.*, 1995, **15**, 3200–3214.

51. I. Hajimohammadreza, A. W. Probert, L. L. Coughenour *et al.*, 'A specific inhibitor of calcium/calmodulin-dependent protein kinase-II provides neuroprotection against NMDA- and hypoxia/hypoglycemia-induced cell death.' *J. Neurosci.*, 1995, **15**, 4093–4101.

52. J. A. Dykens, A. Stern and E. Trenkner, 'Mechanisms of kainate toxicity to cerebellar neurons *in vitro* is analogous to reperfusion tissue injury.' *J. Neurochem.*, 1987, **49**, 1222–1228.

53. R. J. Bridges, J. Y. Koh, C. G. Hatalski *et al.*, 'Increased excitotoxic vulnerability of cortical cultures with reduced levels of glutathione.' *Eur. J. Pharmacol.*, 1991, **192**, 199–200.

54. A. Favit, F. Nicoletti, U. Scapagnini *et al.*, 'Ubiquinone protects cultured neurons against spontaneous and excitotoxin-induced degeneration.' *J. Cereb. Blood Flow Metab.*, 1992, **12**, 638–645.

55. M. D. Majewska and J. A. Bell, 'Ascorbic acid protects neurons from injury induced by glutamate and NMDA.' *NeuroReport.*, 1990, **1**, 194–196.

56. H. S. Chow, J. J. Lynch III, K. Rose *et al.*, 'Trolox attenuates cortical neuronal injury induced by iron, ultraviolet light, glucose deprivation, or AMPA.' *Brain Res.*, 1994, **639**, 102–108.

57. P. H. Chan, L. Chu, S. F. Chen *et al.*, 'Reduced neurotoxicity in transgenic mice overexpressing human copper–zinc-superoxide dismutase.' *Stroke*, 1990, **21**, III80.

58. J. A. Dykens, 'Isolated cerebral and cerebellar mitochondria produce free radicals when exposed to elevated Ca^{2+} and Na$^+$: implications for neurodegeneration.' *J. Neurochem.*, 1994, **63**, 584–591.

59. S. C. Bondy and D. K. Lee, 'Oxidative stress induced by glutamate receptor agonists.' *Brain Res.*, 1993, **610**, 229–233.

60. A. Y. Sun, Y. Cheng, Q. Bu *et al.*, 'The biochemical mechanisms of the excitotoxicity of kainic acid. Free rad-ical formation.' *Mol. Chem. Neuropathol.*, 1992, **17**, 51–63.

61. L. L. Dugan, S. L. Sensi, L. M. Canzoniero *et al.*, 'Mitochondrial production of reactive oxy-gen species in cortical neurons following exposure to *N*-methyl-D-aspartate.' *J. Neurosci.*, 1995, **15**, 6377–6388.

62. I. J. Reynolds and T. G. Hastings, 'Glutamate induces the production of reactive oxygen species in cultured forebrain neurons following NMDA receptor activation.' *J. Neurosci.*, 1995, **15**, 3318–3327.

63. J. B. Schulz, D. R. Henshaw, D. Siwek *et al.*, 'Involvement of free radicals in excitotoxicity *in vivo*.' *J. Neurochem.*, 1995, **64**, 2239–2247.

64. V. L. Dawson, T. M. Dawson, D. A. Bartley *et al.*, 'Mechanisms of nitric oxide mediated neurotoxicity in primary brain cultures.' *J. Neurosci.*, 1993, **13**, 2651–2661.

65. C. Demerle-Pallardy, M. O. Lonchampt, P. E. Chabrier *et al.*, 'Absence of implication of L-arginine/nitric oxide pathway in neuronal cell injury induced by L-glutamate or hypoxia.' *Biochem. Biophys. Res. Commun.*, 1991, **181**, 456–464.

66. S. J. Hewett, J. A. Corbett, M. L. McDaniel *et al.*, 'Inhibition of nitric oxide formation does not protect murine cortical cell cultures from *N*-methyl-D-aspartate neurotoxicity.' *Brain Res.*, 1993, **625**, 337–341.

67. P. J. Pauwels and J. E. Leysen, 'Blockade of nitric oxide formation does not prevent glutamate-induced neurotoxicity in neuronal cultures from rat hippocampus.' *Neurosci. Lett.*, 1992, **143**, 27–30.

68. P. S. Puttfarcken, W. E. Lyons and J. T. Coyle, 'Dissociation of nitric oxide generation and kainate-mediated neuronal degeneration in primary cultures of rat cerebellar granule cells.' *Neuropharmacology*, 1992, **31**, 565–575.

69. R. F. Regan, K. E. Renn and S. S. Panter, 'NMDA neurotoxicity in murine cortical cell cultures is not attenuated by hemoglobin or inhibition of nitric oxide synthesis.' *Neurosci. Lett.*, 1993, **153**, 53–56.

70. C. Moncada, B. Lekieffre, B. Arvin *et al.*, 'Effect of NO synthase inhibition on NMDA- and ischaemia-induced hippocampal lesions.' *Neuroreport*, 1992, **3**, 530–532.

71. M. Lerner-Natoli, G. Rondouin, F. de Block *et al.*, 'Chronic NO synthase inhibition fails to protect hippocampal neurones against NMDA toxicity.' *Neuroreport*, 1992, **3**, 1109–1112.

72. Z. Huang, P. L. Huang, N. Panahian *et al.*, 'Effects of cerebral ischemia in mice deficient in neuronal nitric oxide synthase.' *Science*, 1994, **265**, 1883–1885.

73. P. K. Moore, P. Wallace, Z. Gaffen *et al.*, 'Characterization of the novel nitric oxide synthase inhibitor 7-nitroindazole and related indazoles, 'Antinociceptive and cardiovacular effects.' *Br. J. Pharmacol.*, 1993, **110**, 219–224.

74. T. Yoshida, V. Limmroth, K. Irikura *et al.*, 'The NOS inhibitor, 7-nitroindazole, decreases focal infarct volume but not the response to topical acetylcholine in pial vessels.' *J. Cereb. Blood Flow Metab.*, 1994, **14**, 924–929.

75. J. B. Schulz, R. T. Matthews, D. R. Henshaw *et al.*, 'Inhibition of neuronal nitric oxide synthase (NOS) protects against neurotoxicity produced by 3-nitropropionic acid, malonate, and MPTP.' *Soc. Neurosci Abst.*, 1994, **20**, 1661.

76. J. S. Teitelbaum, R. J. Zatorre, S. Carpenter *et al.*, 'Neurologic sequelae of domoic acid intoxication due to the ingestion of contaminated mussels.' *N. Engl. J. Med.*, 1990, **322**, 1781–1787.

77. F. Cendes, F. Andermann, S. Carpenter *et al.*, 'Temporal lobe epilepsy caused by domoic acid intoxication: evidence for glutamate receptor-mediated excitotoxicity in humans.' *Ann. Neurol.*, 1995, **37**, 123–126.

78. J. McCulloch, 'Glutamate receptor antagonists in

cerebral ischemia.' *J. Neural. Transm. Suppl.*, 1994, **43**, 71–79.

79. J. R. Brorson, P. A. Manzolillo, S. J. Gibbons *et al.*, 'AMPA receptor desensitization predicts the selective vulnerability of cerebellar Purkinje cells to excitotoxicity.' *J. Neurosci.*, 1995, **15**, 4515–4524.

80. M. J. Sheardown, E. O. Nielson, A. J. Hansen *et al.*, '2,3-Dihydroxy-6-nitro-7-sulfamoyl-benzo (F) quinoxaline: a neuroprotectant for cerebral ischemia.' *Science*, 1990, **247**, 571–574.

81. G. W. Albers, R. P. Atkinson, R. E. Kelley *et al.*, 'Safety, tolerability, and pharmacokinetics of the *N*-methyl-D-aspartate antagonist dextrorphan in patients with acute stroke. Dextrophon Study Group.' *Stroke*, 1995, **26**, 254–258.

82. J. Grotta, W. Clark, B. Coull *et al.*, 'Safety and tolerability of the glutamate antagonist CGS 19755 (Selfotel) in patients with acute ischemic stroke. Results of a Phase IIa randomized trial.' *Stroke*, 1995, **26**, 602–605.

83. Y. Katayama, D. P. Becker, T. Tamura *et al.*, 'Massive increases in extracellular potassium and the indiscriminate release of glutamate following concussive brain injury.' *J. Neurosurg.*, 1990, **73**, 889–900.

84. T. K. McIntosh, R. Vink, H. Soares *et al.*, 'Effect of noncompetitive blockade of *N*-methyl-D-aspartate receptors on the neurochemical sequelae of experimental brain injury.' *J. Neurochem.*, 1990, **55**, 1170–1179.

85. E. M. Golding and R. Vink, 'Efficacy of competitive vs. noncompetitive blockade of the NMDA channel following traumatic brain injury.' *Mol. Chem. Neuropathol.*, 1995, **24**, 137–150.

86. D. H. Smith, K. Okiyama, M. J. Thomas *et al.*, 'Effects of the excitatory amino acid receptor antagonists kynurenate and indole-2-carboxylic acid on behavioral and neurochemical outcome following experimental brain injury.' *J. Neurosci.*, 1993, **13**, 5383–5392.

87. R. R. Hicks, D. H. Smith, T. A. Gennarelli *et al.*, 'Kynurenate is neuroprotective following experimental brain injury in the rat.' *Brain Res.*, 1994, **655**, 91–96.

88. T. Wieloch, 'Hypoglycemia-induced neuronal damage prevented by an *N*-methyl-D-aspartate antagonist.' *Science*, 1985, **230**, 681–683.

89. L. Massieu, A. Morales-Villagran and R. Tapia, 'Accumulation of extracellular glutamate by inhibition of its uptake is not sufficient for inducing neuronal damage: an *in vivo* microdialysis study.' *J. Neurochem.*, 1995, **64**, 2262–2272.

90. M. P. Heyes, K. J. Swartz, S. P. Markey *et al.*, 'Regional brain and cerebrospinal fluid quinolinic acid concentrations in Huntington's disease.' *Neurosci. Lett.*, 1991, **122**, 265–269.

91. R. L. Albin and J. T. Greenamyre, 'Alternative excitotoxic hypotheses.' *Neurology*, 1992, **42**, 733–738.

92. A. Novelli, J. A. Reilly, P. G. Lysko *et al.*, Glutamate becomes neurotoxic via the *N*-methyl-D-aspartate receptor when intracellular energy levels are reduced.' *Brain Res.*, 1988, **451**, 205–212.

93. G. D. Zeevalk and W. J. Nicklas, 'Chemically induced hypoglycemia and anoxia: relationship to glutamate receptor-mediated toxicity in retina.' *J. Pharmacol. Exp. Ther.*, 1990, **253**, 1285–1292.

94. G. D. Zeevalk and W. J. Nicklas, 'Mechanisms underlying initiation of excitotoxicity associated with metabolic inhibition.' *J. Pharmacol. Exp. Ther.*, 1991, **257**, 870–878.

95. A. E. Abele and R. J. Miller, 'Potassium channel activators abolish excitotoxicity in cultured hippo-

96. G. J. Lees and W. Leong, 'The sodium–potassium ATPase inhibitor ouabain is neurotoxic in the rat substantia nigra and striatum.' *Neurosci. Lett.*, 1995, **188**, 113–116.

97. M. W. Riepe, N. Hori, A. C. Ludolph *et al.*, 'Failure of neuronal ion exchange, not potentiated excitation, causes excitotoxicity after inhibition of oxidative phosphorylation.' *Neuroscience*, 1995, **64**, 91–97.

98. A. M. Moudy, S. D. Handran, M. P. Goldberg *et al.*, 'Abnormal calcium homeostasis and mitochondrial polarization in a human encephalomyopathy.' *Proc. Natl. Acad. Sci. USA*, 1995, **92**, 729–733.

99. J. T. Coyle and R. Schwarcz, 'Lesions of striatal neurons with kainic acid provides a model for Huntington's chorea.' *Nature*, 1976, **263**, 244–246.

100. E. G. McGeer and P. L. McGeer, 'Duplication of biochemical changes of Huntington's chorea by intrastriatal injections of glutamic and kainic acids.' *Nature*, 1976, **263**, 517–519.

101. D. Dawbarn, M. E. De Quidt and P. C. Emson, 'Survival of basal ganglia neuropeptide Y-somatostatin neurons in Huntington's disease.' *Brain Res.*, 1985, **340**, 251–260.

102. R. J. Ferrante, N. W. Kowall, M. F. Beal *et al.*, 'Selective sparing of a class of striatal neurons in Huntington's disease.' *Science*, 1985, **230**, 561–563.

103. M. F. Beal, N. W. Kowall, K. J. Swartz *et al.*, 'Differential sparing of somatostatin-neuropeptide Y and cholinergic neurons following striatal excitotoxic lesions.' *Synapse*, 19893, **3**, 38–47.

104. M. F. Beal, N. W. Kowall, D. W. Ellison *et al.*, 'Replication of the neurochemical characteristics of Huntington's disease with quinolinic acid.' *Nature*, 1986, **321**, 168–171.

105. M. F. Beal, R. J. Ferrante, K. J. Swartz *et al.*, 'Chronic quinolinic acid lesions in rats closely resemble Huntington's disease.' *J. Neurosci.*, 1991, **11**, 1649–1659.

106. T. J. Bazzett, J. B. Becker, R. C. Falik *et al.*, 'Chronic intrastriatal quinolinic acid produces reversible changes in perikaryal calbindin and parvalbumin immunoreactivity.' *Neuroscience*, 1994, **60**, 837–841.

107. K. M. Harrington and N. W. Kowall, 'Parvalbumin immunoreactive neurons resist degeneration in Huntington's disease striatum.' *J. Neuropathol. Exp. Neurol.*, 1991, **50**, 309.

108. H. J. Waldvogel, R. L. M. Faull, M. N. Williams *et al.*, 'Differential sensitivity of calbindin and parvalbumin immunoreactive cells in the striatum to excitotoxins.' *Brain Res.*, 1991, **546**, 329–335.

109. R. J. Ferrante, N. W. Kowall, P. B. Cipolloni *et al.*, 'Excitotoxin lesions in primates as a model for Huntington's disease: histopathologic and neurochemical characterization.' *Exp. Neurol.*, 1993, **119**, 46–71.

110. L. S. Dure IV, A. B. Young and J. B. Penney, 'Excitatory amino acid binding sites in the caudate nucleus and frontal cortex of Huntington's disease.' *Ann. Neurol.*, 1991, **30**, 785–793.

111. A. B. Young, J. T. Greenamyre, Z. Hollingsworth *et al.*, NMDA receptor losses in putamen from patients with Huntington's disease.' *Science*, 1988, **241**, 981–983.

112. R. L. Albin, A. B. Young, J. B. Penney *et al.*, 'Abnormalities of striatal projection neurons and *N*-methyl-D-aspartate receptors in presymptomatic Huntington's disease.' *N. Engl. J. Med.*, 1990, **322**, 1293–1298.

113. M. C. Difazio, Z. Hollingsworth, A. B. Young *et al.*, 'Glutamate receptors in the substantia nigra of

Parkinson's disease brains.' *Neurology*, 1992, **42**, 402–406.

114. P. K. Sonsalla, G. D. Zeevalk, L. Manzino *et al.*, 'MK-801 fails to protect against the dopaminergic neuropathology produced by systemic 1-methyl-4-phenyl-1,2,3,6-tetrahydropyridine in mice or intranigral 1-methyl-4-phenylpyridinium in rats.' *J. Neurochem.*, 1992, **58**, 1979–1982.

115. L. Turski, K. Bressler, K. J. Rettig *et al.*, 'Protection of substantia nigra from MPP$^+$ neurotoxicity by *N*-methyl-D-aspartate antagonists.' *Nature*, 1991, **349**, 414–418.

116. K. W. Lange, P. A. Loschmann, E. Sofic *et al.*, 'The competitive NMDA antagonist CPP protects substantia nigra neurons from MPTP-induced degeneration in primates.' *Naunyn–Schmiedebergs Arch. Pharmacol.*, 1993, **348**, 586–592.

117. A. Zuddas, G. Oberto, F. Vaglini *et al.*, 'MK-801 prevents 1-methyl-4-phenyl-1,2,3,6-tetrahydropyridinine-induced parkinsonism in primates.' *J. Neurochem.*, 1992, **59**, 733–739.

118. S. Malessa, P. N. Leigh, O. Bertel *et al.*, 'Amyotrophic lateral sclerosis: glutamate dehydrogenase and transmitter amino acids in the spinal cord.' *J. Neuro. Neurosurg. Psych.*, 1991, **54**, 984–988.

119. A. Plaitakis, E. Constantakakis and J. Smith 'The neuroexcitotoxic amino acids glutamate and aspartate are altered in the spinal cord and brain in amyotrophic lateral sclerosis.' *Ann. Neurol.*, 1988, **24**, 446–449.

120. J. D. Rothstein, G. Tsai, R. W. Kuncl *et al.*, 'Abnormal excitatory amino acid metabolism in amyotrophic lateral sclerosis.' *Ann. Neurol.*, 1990, **28**, 18–25.

121. J. D. Rothstein, L. J. Martin and R. W. Kuncl, 'Decreased glutamate transport by the brain and spinal cord in amyotrophic lateral sclerosis.' *N. Engl. J. Med.*, 1992, **362**, 1464–1468.

122. P. S. Spencer, D. N. Roy, A. Ludolph *et al.*, 'Lathyrism: evidence for role of the neuroexcitatory aminoacid BOAA.' *Lancet*, 1986, **2**, 1066–1067.

123. P. S. Spencer, P. B. Nunn, J. Hugon *et al.*, 'Guam amyotrophic lateral sclerosis–Parkinsonism–dementia linked to a plant excitant neurotoxin.' *Science*, 1987, **237**, 517–522.

124. P. J. Shaw, R. M. Chinnery and P. G. Ince, 'Non-NMDA receptors in motor neuron disease (MND): a quantitative autoradiographic study in spinal cord and motor cortex using [^{3}H]CNQX and [^{3}H]kainate.' *Brain Res.*, 1994, **655**, 186–194.

125. C. Krieger, R. Wagey and C. Shaw, 'Amyotrophic lateral sclerosis: quantitative autoradiography of [^{3}H]MK-801/NMDA binding sites in spinal cord.' *Neurosci. Lett.*, 1993, **159**, 191–194.

126. H. Allaoua, I. Chaudieu, C. Krieger *et al.*, 'Alterations in spinal cord excitatory amino acid receptors in amyotrophic lateral sclerosis patients.' *Brain Res.*, 1992, **579**, 169–172.

127. J. D. Rothstein, L. Jin, M. Dykes-Hoberg *et al.*, 'Chronic inhibition of glutamate uptake produces a model of slow neurotoxicity.' *Proc. Natl. Acad. Sci. USA*, 1993, **90**, 6591–6595.

128. A. Copani, J. Y. Koh and C. W. Cotman, 'β-amyloid increases neuronal susceptibility to injury by glucose deprivation.' *Neuroreport*, 1991, **2**, 763–765.

129. J. Y. Koh, L. L. Yang and C. W. Cotman, 'β-amyloid protein increases the vulnerability of cultured cortical neurons to excitotoxic damage.' *Brain Res.*, 1990, **533**, 315–320.

130. M. P. Mattson, B. Cheng, D. Davis *et al.*, 'β-amyloid peptides destabilize calcium homeostasis and render human cortical neurons vulnerable to excitotoxicity.' *J. Neurosci.*, 1992, **12**, 376–389.

131. J. T. Greenamyre, J. B. Penney, A. B. Young *et al.*, 'Alterations in L-glutamate binding in Alzheimer's and Huntington's disease.' *Science*, 1985, **227**, 1496–1499.

132. J. T. Greenamyre, J. B. Penney, C. J. D'Amato *et al.*, 'Dementia of the Alzheimer type: changes in hippocampal L-[^{3}H] glutamate binding.' *J. Neurochem.*, 1987, **48**, 543–551.

133. N. W. Kowall and M. F. Beal, 'Glutamate-, glutaminase-, and taurine-immunoreactive neurons develop neurofibrillary tangles in Alzheimer's disease.' *Ann. Neurol.*, 1991, **29**, 162–167.

134. J. Rogers and J. H. Morrison, 'Quantitative morphology and regional laminar distributions of senile plaques in Alzheimer's disease.' *J. Neurosci.*, 1985, **5**, 2801–2808.

135. R. C. Pearson, M. M. Esiri, R. W. Hiorns *et al.*, 'Anatomical correlates of the distribution of the pathological changes in the neocortex in Alzheimer disease.' *Proc. Natl. Acad. Sci. USA*, 1985, **82**, 4531–4534.

136. M. F. Beal, 'Mechanisms of excitotoxicity in neurologic diseases.' *FASEB J.*, 1992, **6**, 3338–3344.

137. M. F. Beal, K. J. Swartz, B. T. Hyman *et al.*, 'Aminooxyacetic acid results in excitoxin lesions by a novel indirect mechanism.' *J. Neurochem.*, 1991, **57**, 1068–1073.

138. E. Urbanska, C. Ikonomidou, M. Sieklucka *et al.*, 'Aminooxyacetic acid produces excitotoxic lesions in the rat striatum.' *Synapse*, 1991, **9**, 129–135.

139. A. J. Cheeseman and J. B. Clark, 'Influence of the malate–aspartate shuttle on oxidative metabolism in synaptosomes.' *J. Neurochem.*, 1988, **50**, 1559–1565.

140. R. A. Kauppinen, T. S. Sihra and D. G. Nicholls, 'Aminooxyacetic acid inhibits the malate–aspartate shuttle in isolated nerve terminals and prevents the mitochondria from utilizing glycolytic substrates.' *Biochim Biophys. Acta*, 1987, **930**, 173–178.

141. S. M. Fitzpatrick, A. J. Cooper and T. E. Duffy, 'Use of β-methylene-D,L-aspartate to assess the role of aspartate amino-transferase in cerebral oxidative metabolism.' *J. Neurochem.*, 1983, **41**, 1370–1383.

142. E. Brouillet, D. R. Henshaw, J. B. Schulz *et al.*, 'Aminooxyacetic acid striatal lesions attenuated by 1,3-butanediol and coenzyme Q10.' *Neurosci. Lett.*, 1994, **177**, 58–62.

143. O. G. McMaster, F. Du, E. D. French *et al.*, 'Focal injection of aminooxyacetic acid produces seizures and lesions in rat hippocampus: evidence for mediation by NMDA receptors.' *Exp. Neurol.*, 1991, **113**, 378–385.

144. J. W. McDonald and D. D. Schoepp, 'Aminooxyacetic acid produces excitotoxic brain injury in neonatal rats.' *Brain Res.*, 1991, **624**, 239–244.

145. J. L. Webb, 'Enzyme and Metabolic Inhibitors,' Academic Press, New York, 1963.

146. M. F. Beal, E. Brouillet, B. Jenkins *et al.*, 'Age-dependent striatal excitotoxic lesions produced by the endogenous mitochondrial inhibitor malonate.' *J. Neurochem.*, 1993, **61**, 1147–1150.

147. J. G. Greene, R. H. Porter, R. V. Eller *et al.*, Inhibition of succinate dehydrogenase by malonic acid produces an "excitotoxic" lesion in rat striatum.' *J. Neurochem.*, 1993, **61**, 1151–1154.

148. J. G. Greene and J. T. Greenamyre, 'Exacerbation of NMDA, AMPA, and L-glutamate excitotoxicity by the succinate dehydrogenase inhibitor malonate.' *J. Neurochem.*, 1995, **64**, 2332–2338.

149. R. Henshaw, B. G. Jenkins, J. B. Schulz *et al.*, 'Malonate produces striatal lesions by indirect NMDA receptor activation.' *Brain Res.*, 1994, **647**, 161–166.

150. M. F. Beal, D. R. Henshaw, B. G. Jenkins *et al.*, 'Coenzyme Q_{10} and nicotinamide block striatal lesions produced by mitochondrial toxin malonate.' *Ann. Neurol.*, 1994, **36**, 882–888.

151. A. C. Ludolph, F. He, P. S. Spencer *et al.*, '3-Nitropropioinic acid—exogenous animal neurotoxin and possible human striatal toxin.' *Can. J. Neurol. Sci.*, 1991, **18**, 492–498.

152. T. A. Alston, L. Mela and H. J. Bright, '3-Nitropropionate, the toxic substance of *Indigofera*, is a suicide inactivator of succinate dehydrogenase.' *Proc. Natl. Acad. Sci. USA*, 1977, **74**, 3767–3771.

153. C. J. Coles, D. E. Edmondson and T. P. Singer, 'Inactivation of succinate dehydrogenase by 3-nitropropionate.' *J. Biol. Chem.*, 1979, **254**, 5161–5167.

154. M. Riepe, N. Hori, A. C. Ludolph *et al.*, 'Inhibition of energy metabolism by 3-nitropropionic acid activates ATP-sensitive potassium channels.' *Brain Res.*, 1992, **586**, 61–66.

155. A. C. Ludolph, M. Seelig, A. Ludolph *et al.*, '3-Nitropropionic acid decreases cellular energy levels and causes neuronal degeneration in cortical explants.' *Neurodegeneration*, 1992, **1**, 155–161.

156. G. D. Zeevalk, E. Derr-Yellin and W. J. Nicklas, 'NMDA receptor involvement in toxicity to dopamine neurons *in vitro* caused by the succinate dehydrogenase inhibitor 3-nitropropionic acid.' *J. Neurochem.*, 1995, **64**, 455–458.

157. M. Weller and S. M. Paul, '3-Nitropropionic acid is an indirect excitotoxin to cultured cerebellar granule neurons.' *Eur J. Pharmacol.*, 1993, **248**, 223–228.

158. M. I. Behrens, J. Koh, L. M. Canzoniero *et al.*, '3-Nitropropionic acid induces apoptosis in cultured striatal and cortical neurons.' *Neuroreport*, 1995, **6**, 545–548.

159. D. H. Gould and D. L. Gustine, 'Basal ganglia degeneration, myelin alterations, and enzyme inhibition in mice by the plant toxin 3-nitropropionic acid.' *Neuropathol. Appl. Neurobiol.*, 1982, **8**, 377–393.

160. B. F. Hamilton and D. H. Gould, 'Correlation of morphologic brain lesions with physiologic alterations and blood–brain barrier impairment in 3-nitropropionic acid toxicity in rats.' *Acta Neuropathol. (Berl.)*, 1987, **74**, 67–74.

161. E. Brouillet, P. Hantraye, R. Ferante *et al.*, 'Chronic administration of 3-nitropropionic acid induced selective striatal degeneration and abnormal choreiform movements in monkeys.' *Proc. Natl. Acad. Sci.*, 1995, **92**, 7105–7109.

162. M. F. Beal, E. Brouillet, B. G. Jenkins, 'Neurochemical and histologic characterization of excitotoxic lesions produced by the mitochondrial toxin 3-nitropropionic acid.' *J. Neurosci.*, 1993, **13**, 4181–4192.

163. L. Turski and H. Ikonomidou, 'Striatal toxicity of 3-nitropropionic acid prevented by the AMPA antagonist NBQX.' *Soc. Neurosci. Abst.*, 1994, **20**, 1677.

164. U. Wullner, A. B. Young, J. B. Penney *et al.*, '3-Nitropropionic acid toxicity in the striatum.' *J. Neurochem.*, 1994, **63**, 1772–1781.

165. M. F. Beal, R. J. Ferrante, R. Henshaw *et al.*, '3-Nitropropionic acid neurotoxicity is attenuated in copper/zinc superoxide dismutase transgenic mice.' *J. Neurochem.*, 1995, **65**, 919–922.

166. M. J. Leach, J. H. Swan. D. Eisenthal *et al.*, 'BW619C89, a glutamate release inhibitor, protects against focal cerebral ischemic damage.' *Stroke*, 1993, **24**, 1063–1067.

167. E. G. McGeer and S. G. Zhu, 'Lamotrigine protects against kainate but not ibotenate lesions in rat striatum.' *Neurosci. Lett.*, 1990, **112**, 348–351.

168. S. A. Jones-Humble, P. F. Morgan and B. R. Cooper, 'The novel anticonvulsant lamotrigine prevents dopa-

169. S. H. Graham, J. Chen, F. R. Sharp *et al.*, 'Limiting ischemic injury by inhibition of excitatory amino acid release.' *J. Cereb. Blood Flow Metab.*, 1993, **13**, 88–97.

170. S. H. Graham, J. Chen, J. Lan *et al.*, 'Neuroprotective effects of a use-dependent blocker of voltage-dependent sodium channels, BW619C89, in rat middle cerebral artery occlusion.' *J. Pharmacol. Exp. Ther.*, 1994, **269**, 854–859.

171. D. Lekieffre and B. S. Meldrum, 'The pyrimidine-derivative, BW1003C87, protects CA1 and striatal neurons following transient severe forebrain ischaemia in rats. A microdialysis and histological study.' *Neuroscience*, 1993, **56**, 93–99.

172. A. Cheramy, L. Barbeito, G. Godeheu *et al.*, 'Riluzole inhibits the release of glutamate in the caudate nucleus of the cat *in vivo*.' *Neurosci. Lett.*, 1992, **147**, 209–212.

173. D. Martin, M. A. Thompson and J. V. Nadler, 'The neuroprotective agent riluzole inhibits release of glutamate and aspartate from slices of hippocampal area CA1.' *Eur. J. Pharmacol.*, 1993, **250**, 473–476.

174. C. Malgouris, F. Bardot, M. Daniel *et al.*, 'Riluzole, a novel antiglutamate, prevents memory loss and hippocampal neuronal damage in ischemic gerbils.' *J. Neurosci.*, 1989, **9**, 3720–3727.

175. J. Pratt, J. Rataud, F. Bardot *et al.*, 'Neuroprotective actions of riluzole in rodent models of global and focal cerebral ischemia.' *Neurosci. Lett.*, 1992, **140**, 225–230.

176. A. Boireau, P. Dubedat, F. Bordier *et al.*, 'Riluzole and experimental parkinsonism: antagonism of MPTP-induced decrease in central dopamine levels in mice.' *Neuroreport*, 1994, **5**, 2657–2660.

177. G. Bensimon, L. Lacomblez, V. Meininger *et al.*, 'A controlled trial of riluzole in amyotrophic lateral sclerosis. ALS/Riluzole Study Group.' *N. Engl. J. Med.*, 1994, **330**, 585–591.

178. H. Fujisawa, D. Dawson, S. E. Browne *et al.*, 'Pharmacological modification of glutamate neurotoxicity *in vivo*.' *Brain Res.*, 1993, **629**, 73–78.

179. S. A. Lipton and P. A. Rosenberg, 'Excitatory amino acids as a final common pathway for neurologic disorders.' *N. Engl. J. Med.*, 1994, **330**, 613–622.

180. J. McCulloch, 'Excitatory amino acid antagonists and their potential for the treatment of ischaemic brain damage in man.' *Br. J. Clin Pharmacol.*, 1992, **34**, 106–114.

181. D. W. Newell, A. Barth and B. A. T. Malouf, 'Glycine site NMDA receptor antagonists provide protection against ischemia-induced neuronal damage in hippocampal slice cultures.' *Brain Res.*, 1995, **675**, 38–44.

182. R. Bullock, D. I. Graham, S. Swanson *et al.*, 'Neuroprotective effect of the AMPA receptor antagonist LY-293558 in focal cerebral ischemia in the cat.' *J. Cereb. Blood Flow Metab.*, 1994, **14**, 466–471.

183. S. E. Browne and J. McCulloch, 'AMPA receptor antagonists and local cerebral glucose utilization in the rat.' *Brain Res.*, 1994, **641**, 10–20.

184. E. Ozyurt, D. I. Graham, G. N. Woodruff *et al.*, 'Protective effect of the glutamate antagonist, MK-801 in the focal cerebral ischemia in the cat.' *J. Cereb. Blood Flow Metab.*, 1988, **8**, 138–143.

185. C. K. Park, D. G. Nehls, D. I. Graham *et al.*, 'Focal cerebral ischaemia in the cat: treatment with the glutamate antagonist MK-801 after induction of ischaemia.' *J. Cereb. Blood Flow Metab.*, 1988, **8**, 757–762.

186. C. K. Park, D. G. Nehls, D. I. Graham *et al.*, 'The glutamate antagonist MK-810 reduces focal ischemic

---

mine depletion in C57 black mice in the MPTP animal model of Parkinson's disease.' *Life Sci.*, 1994, **54**, 245–252.

brain damage in the rat.' *Ann. Neurol.*, 1988, **24**, 543–551.

187. R. Gill, C. Brazell, G. N. Woodruff *et al.*, 'The neuroprotective action of dizocilpine (MK-801) in the rat middle cerebral artery occlusion model of focal ischaemia.' *Br. J. Pharmacol.*, 1991, **103**, 2030–2036.

188. S. Roussel, E. Pinard and J. Seylaz, 'Effect of MK-801 on focal brain infarction in normotensive and hypersensitive rats.' *Hypertension*, 1992, **19**, 40–46.

189. M.-E. Meadows, M. Fisher and K. Minematsu, 'Delayed treatment with a noncompetitive NMDA antagonist CNS-1102, reduces infarct size in rats.' *Cerebrovasc Dis.*, 1994, **4**, 26.

190. G. K. Steinberg, D. Kunis, R. DeLaPaz *et al.*, 'Neuroprotection following focal cerebral ischaemia with the NMDA antagonist dextromethorphan, has a favorable dose response profile.' *Neurol. Res.*, 1993, **15**, 174–180.

191. G. K. Steinberg, C. P. George, R. DeLaPaz *et al.*, 'Dextromethorphan protects against cerebral injury following transient focal ischemia in rabbits.' *Stroke*, 1988, **19**, 1112–1118.

192. R. Simon and K. Shiraishi, '*N*-methyl-D-aspartate antagonist reduces stroke size and regional glucose metabolism.' *Ann. Neurol.*, 1990, **27**, 606–611.

193. M. Chen, R. Bullock, D. I. Graham *et al.*, 'Evaluation of a competitive NMDA antagonist (D-CPPene) in feline focal cerebral ischaemia.' *Ann. Neurol.*, 1991, **30**, 62–70.

194. B. Gotti, D. Duverger, J. Bertin *et al.*, 'Ifenprodil and SL-82.0715 as cerebral anti-ischemic agents. I. Evidence for efficacy in models of focal cerebral ischemia.' *J. Pharmacol. Exp. Ther.*, 1988, **247**, 1211–1221.

195. R. Gill, R. J. Hargreaves and J. A. Kemp, 'The neuroprotective effect of the glycine site antagonist 3*R*-(+)-*cis*-4-methyl-HA966 (L-687,414) in a rat model of focal ischemia.' *J. Cereb. Blood Flow Metab.*, 1995, **15**, 197–204.

196. D. S. Warner, H. Martin, P. Ludwig *et al.*, '*In vivo* models of cerebral ischemia: effects of parenterally administered NMDA receptor glycine site antagonists.' *J. Cereb. Blood Flow Metab.*, 1995, **15**, 188–196.

197. R. Gill, L. Nordhom and D. Lodge, 'The neuroprotective actions of 2,3-dihydroxy-6-nitro-7-sulfamoyl-benzo(F)quinoxaline (NBQX) in a rat focal ischaemia model.' *Brain Res.*, 1992, **580**, 35–43.

198. D. Xue, Z. G. Huang, K. Barnes *et al.*, 'Delayed treatment with AMPA, but not NMDA, antagonists reduces neocortical infarction.' *J. Cereb. Blood Flow Metab.*, 1994, **14**, 251–261.

199. W. Koek and F. C. Colpaert, 'Selective blockade of *N*-methyl-D-aspartate (NMDA)-induced convulsions by NMDA antagonists and putative glycine antagonists: relationship with phencyclidine-like behavioral effects.' *J. Pharmacol. Exp. Ther.*, 1990, **252**, 349–357.

200. R. G. Morris, E. Anderson, G. S. Lynch *et al.*, 'Selective impairment of learning and blockade of long-term potentiation by an *N*-methyl-D-aspartate receptor antagonist, AP5.' *Nature*, 1986, **319**, 774–776.

201. M. A. Rogawski, 'Therapeutic potential of excitatory amino acid antagonists: channel blockers and 2,3-benzodiazepines.' *Trends Pharmacol. Sci.*, 1993, **14**, 325–331.

202. A. Kupsch, P. A. Loschmann, H. Sauer *et al.*, 'Do NMDA receptor antagonists protect against MPTP-toxicity? Biochemical and immunocytochemical analyses in black mice.' *Brain Res.*, 1992, **592**, 74–83.

203. E. Brouillet and M. F. Beal, 'NMDA antagonists partially protect against MPTP induced neurotoxicity in mice.' *Neuroreport*, 1993, **4**, 387–390.

204. J. B. Schulz, D. R. Henshaw, B. G. Jenkins *et al.*, '3-Acetylpyridine produces age-dependent excitotoxic lesions in rat striatum.' *J. Cereb. Blood Flow Metab.*, 1994, **14**, 1024–1029.

205. M. Lafon-Cazal, M. Culcasi, F. Gaven *et al.*, 'Nitric oxide, superoxide and peroxynitrite: putative mediators of NMDA-induced cell death in cerebellar granule cells.' *Neuropharmacology*, 1993, **32**, 1259–1266.

206. T. L. Yue, J. L. Gu, P. G. Lysko *et al.*, 'Neuroprotective effects of phenyl-*t*-butyl-nitrone in gerbil global brain ischemia and in cultured rat cerebellar neurons.' *Brain Res.*, 1992, **574**, 193–197.

207. G. Chen, T. M. Bray, E. G. Janzen *et al.*, 'Excretion, metabolism and tissue distribution of a spin trapping agent, α-phenyl-*N*-*tert*-butyl-nitrone (PBN) in rats.' *Free Radic Res. Commun.*, 1990, **9**, 317–323.

208. C. N. Oliver, P. E. Starke-Reed, E. R. Stadtman *et al.*, 'Oxidative damage to brain proteins, loss of glutamine synthetase activity, and production of free radicals dur-ing ischemia/reperfusion-induced injury to gerbil brain.' *Proc. Natl. Acad. Sci. USA*, 1990, **87**, 5144–5147.

209. J. W. Phillis and C. Clough-Helfman, 'Protection from cerebral ischemia injury in gerbils with spin trap agent *N*-*tert*-butyl-α-phenylnitrone.' *Neurosci. Lett.*, 1990, **116**, 315–319.

210. X. Cao and J. W. Phillis, 'α-Phenyl-*tert*-butyl-nitrone reduces cortical infarct and edema in rats subjected to focal ischemia.' *Brain Res.*, 1994, **644**, 267–272.

211. J. M. Carney, P. E. Starke-Reed, C. N. Oliver *et al.*, 'Reversal of age-related increase in brain protein oxidation, decrease in enzyme activity, and loss in temporal and spatial memory by chronic administration of the spin-trapping compound *N*-*tert*-butyl-α-phenylnitrone.' *Proc. Natl. Acad. Sci. USA*, 1991, **88**, 3633–3636.

212. H. Monyer, D. M. Hartley and D. W. Choi, '21-Aminosteroids attenuate excitotoxic neuronal injury in cortical cell cultures.' *Neuron*, 1990, **5**, 121–126.

213. E. D. Hall, 'Cerebral ischemia, free radicals and antioxidant protection.' *Biochem. Soc. Trans.*, 1988, **21**, 334–339.

214. E. D. Hall, P. A. Yonkes, J. M. McCall *et al.*, 'Effects of the 21-aminosteroid U74006F on experimental head injury in mice.' *J. Neurosurg.*, 1988, **68**, 456–461.

215. E. D. Hall and P. A. Yonkes, 'Attenuation of post-ischemic cerebral hypoperfusion by the 21-amino steroid U74006.' *Stroke*, 1988, **19**, 340–344.

216. C. K. Park and E. D. Hall, 'Dose-response analysis of the effect of 21-aminosteroid tirilazad mesylate (U-74006F) upon neurological outcome and ischemic brain damage in permanent focal cerebral ischemia.' *Brain Res.*, 1994, **645**, 157–163.

217. J. H. Prehn, C. Karkoutly, J. Nuglisch *et al.*, 'Dihydrolipoate reduces neuronal injury after cerebral ischemia.' *J. Cereb. Blood Flow Metab.*, 1992, **12**, 78–87.

218. J. T. Greenamyre, M. Garcia-Osuna and J. G. Greene, 'The endogenous cofactors, thioctic acid and dihydrolipoic acid, are neuroprotective against NMDA and malonic acid lesions of striatum.' *Neurosci. Lett.*, 1994, **171**, 17–20.

219. Q. Zhao, K. Pahlmark, M. L. Smith *et al.*, 'Delayed treatment with the spin trap alpha-phenyl-*n*-*tert*-butyl nitrone (PBN) reduces infarct size following transient middle cerebral artery occlusion in rats.' *Acta Physiol. Scand.*, 1994, **152**, 349–350.

220. J. B. Schulz, R. T. Matthews, B. G. Jenkins *et al.*, 'Improved therapeutic window for treatment of histotoxic hypoxia with a free radical spin trap.' *J. Cereb. Blood Flow Metab.*, 1995, **15**, 948–952.

221. J. P. Nowicki, D. Duval, H. Poignet *et al.*, 'Nitric oxide mediates neuronal death after focal ischemia in the mouse.' *Eur. J. Pharmacol.*, 1991, **204**, 339–340.

222. T. Nagafuji, T. Matsuii, T. Koide *et al.*, 'Blockade of nitric oxide formation by N^{ω}-nitro-L-arginine mitigates ischemic brain edema and subsequent cerebral infarction in rats.' *Neurosci. Lett.*, 1992, **147**, 159–162.

223. T. Nishikawa, J. R. Kirsch, R. C. Koehler *et al.*, 'Nitric oxide synthase inhibition reduces caudate injury following transient focal ischemia in cats.' *Stroke*, 1994, **25**, 877–885.

224. D. A. Dawson, K. Kusumoto, D. I. Graham *et al.*, 'Inhibition of nitric oxide synthesis does not reduce infarct volume in a rat model of focal cerebral ischaemia.' *Neurosci. Lett.*, 1992, **142**, 151–154.

225. J. W. Kuluz, R. J. Prado, W. D. Dietrich *et al.*, 'The effect of nitric oxide synthase inhibition on infarct volume after reversible focal cerebral ischemia in conscious rats.' *Stroke*, 1993, **24**, 2023–2039.

226. S. Yamamoto, E. V. Golanov, S. B. Berger *et al.*, 'Inhibition of nitric oxide synthesis increases focal ischemic infarction in rat.' *J. Cereb. Blood Flow Metab.*, 1992, **12**, 717–726.

227. F. Zhang, J. G. White and C. Iadecola, 'Nitric oxide donors increase blood flow and reduce brain damage in focal ischemia: evidence that nitric oxide is beneficial in the early stages of cerebral ischemia.' *J. Cereb. Blood Flow Metab.*, 1994, **14**, 217–226.

228. A. Carreau, D. Duval, H. Poignet *et al.*, 'Neuroprotective efficacy of N^{ω}-nitro-L-arginine after focal cerebral ischemia in the mouse and inhibition of cortical nitric oxide synthase.' *Eur. J. Pharmacol.*, 1994, **256**, 241–249.

229. A. M. Buchan, S. Z. Gertler, Z. G. Huang *et al.*, 'Failure to prevent selective CA1 neuronal death and reduce cortical infarction following cerebral ischemia with inhibition of nitric oxide synthase.' *Neuroscience*, 1994, **61**, 1–11.

230. M. Caldwell, M. O'Neill, B. Earley *et al.*, 'N^{G}-Nitro-L-arginine protects against ischaemic-induced increases in nitric oxide and hippocampal neurodegeneration in the gerbil.' *Eur. J. Pharmacol.*, 1994, **260**, 191–200.

231. G. Sancesario, M. Iannone, M. Morello *et al.*, 'Nitric oxide inhibition aggravates ischemic damage of hippocampal but not of NADPH neurons in gerbils.' *Stroke*, 1994, **25**, 436–444.

232. S. Shapira, T. Kadar and B. A. Weissman, 'Dose-dependent effect of nitric oxide synthase inhibition following transient forebrain ischemia in gerbils.' *Brain Res.*, 1994, **668**, 80–84.

Subject Index

Every effort has been made to index as comprehensively as possible, and to standardize the terms used in the index in line with the following standards:

EMTREE Thesaurus as a general guide to the selection of preferred terms.

IUPAC Recommendations for the nomenclature of chemical terms, with trivial names being employed where normal usage dictates.

In view of the diverse nature of the terminology employed by the different authors, the reader is advised to search for related entries under the appropriate headings.

The index entries are presented in letter-by-letter alphabetical sequence. Chemical terms are filed under substituent prefixes, where appropriate, rather than under the parent compound name; this is in line with the presentation given in the EMTREE Thesaurus.

The index is arranged in set-out style, with a maximum of three levels of heading. Location references refer to page number; major coverage of a subject is indicated by bold, italicized, elided page numbers; for example,

Risk assessment (RA), pesticides *1234–55*
  toxicological data 345

See cross-references direct the user to the preferred term; for example,

Vitamin A *See* Retinol

See also cross-references provide the user with guideposts to terms of related interest, from the broader term to the narrower term, and appear at the end of the main heading to which they refer; for example

Smoking
  See also Cigarette smoking; Pipe smoking